D1441035

The Geology of North America
Volume D-2

Sedimentary Cover—
North American Craton: U.S.

Edited by

L. L. Sloss
Department of Geological Sciences
Northwestern University
Evanston, Illinois 60201

1988

Acknowledgment

Publication of this volume, one of the Centennial Field Guide Volumes of *The Decade of North American Geology Project* series, has been made possible by members and friends of the Geological Society of America, corporations, and government agencies through contributions to the Decade of North American Geology fund of the Geological Society of America Foundation.

Following is a list of individuals, corporations, and government agencies giving and/or pledging more than $50,000 in support of the DNAG Project:

Amoco Production Company
ARCO Exploration Company
Chevron Corporation
Exxon Production Research Company
Getty Oil Company
Gulf Oil Exploration and Production Company
Paul V. Hoovler
Kennecott Minerals Company
Kerr McGee Corporation
Marathon Oil Company
Maxus Energy Corporation
McMoRan Oil and Gas Company
Mobil Oil Corporation
Pennzoil Exploration and Production Company
Phillips Petroleum Company
Shell Oil Company
Caswell Silver
Standard Oil Company of Indiana
Sun Exploration and Production Company
Superior Oil Company
Tenneco Oil Company
Texaco, Inc.
Union Oil Company of California
Union Pacific Corporation and its operating companies:
Union Pacific Resources Company
Union Pacific Railroad Company
Upland Industries Corporation
U.S. Department of Energy

Published by the Geological Society of America, Inc.
3300 Penrose Place, P.O. Box 9140, Boulder, Colorado 80301

Printed in U.S.A.

Library of Congress Cataloging-in-Publication Data

Sedimentary cover, North American Craton.

(The Geology of North America ; v. D-2)
Bibliography: p.
Includes index.
1. Geology—United States. 2. Cratons—United States. 3. Geology, Stratigraphic. I. Sloss, L. L. (Laurence Louis), 1913– . II. Series.
QE71.G48 1986 vol. D-2 557.3 s [557.3] 88-10957
[QE77]
ISBN 0-8137-5205-1

Front Cover: Grand Canyon, Arizona, view north from south rim, showing Proterozoic rocks of the inner gorge overlain by cratonic cover rocks of the Sauk, Kaskaskia, and Absaroka sequences. Photo by Lee Gladish.

Contents

THE WESTERN MID-CONTINENT REGION

THE EASTERN MID-CONTINENT REGION

CONCLUSIONS

Plates
(in accompanying slipcase)

Preface

The Geology of North America series has been prepared to mark the Centennial of The Geological Society of America. It represents the cooperative efforts of more than 1,000 individuals from academia, state and federal agencies of many countries, and industry to prepare syntheses that are as current and authoritative as possible about the geology of the North American continent and adjacent oceanic regions.

This series is part of the Decade of North American Geology (DNAG) Project, which also includes eight wall maps at a scale of 1:5,000,000 that summarize the geology, tectonics, magnetic and gravity anomaly patterns, regional stress fields, thermal aspects, seismicity, and neotectonics of North America and its surroundings. Together, the synthesis volumes and maps are the first coordinated effort to integrate all available knowledge about the geology and geophysics of a crustal plate on a regional scale.

The products of the DNAG Project present the state of knowledge of the geology and geophysics of North America in the 1980s, and they point the way toward work to be done in the decades ahead.

In addition to the contributions from organizations and individuals acknowledged at the front of this book, major support has been provided to the editors of this volume by the Department of Geological Sciences, Northwestern University.

A. R. Palmer
General Editor for the volumes
published by the Geological
Society of America

J. O. Wheeler
General Editor for the volumes
published by the Geological
Survey of Canada

Foreword

This volume is part of the harvest from seeds planted in late-1970s meetings of the Society's Centennial Committee under the chairmanship of Richard H. Jahns. Actual germination waited until a workshop was held in Norman, Oklahoma, bringing together interested parties in mid-1982; guidance was provided by Charles J. Mankin, Peter R. Vail, and this reporter. At this stage the planned volume was tentatively titled "The Midcontinent" or "The Continental Interior"; both are useful in identifying the subject areas to be covered in the U.S., but neither is particularly apt for application to a planned Canadian companion volume. In consultation with James D. Aitken both books adopted *Sedimentary Cover—North American Craton,* to be followed by subtitles designating the U.S. or Canada.

Meanwhile, other working groups were simultaneously preparing volumes on the Atlantic Continental Margin; the Appalachian, Ouachita, and Cordilleran Orogenic Belts; and on the Gulf of Mexico Basin, but no lines were established to define the outboard limits of the areas of responsibility of the cadre assembled to describe and discuss the craton. Averting a classic turf war, the authors of this volume abdicated responsibility for the Atlantic coastal province and for Paleozoic orogens (Appalachian and Ouachita) to the south and southeast, but the mountainous blocks that extended from the Ouachita foreland to the "Ancestral Rockies" in late Paleozoic time are clearly cratonic elements. By the same token, the elevation of blocks on the early Cenozoic Cordilleran foreland does not exclude the region from the pre-Cenozoic craton. Much of the Gulf of Mexico Basin and its sedimentary fill are demonstrably extracratonic; responsibility for coverage is relinquished without serious reservation even though it is difficult to rationalize a separation between Gulf Coast elements such as the Sabine Uplift or the Mississippi Embayment and the craton proper.

By the time the preliminaries of planning and execution were resolved, events conspired to reduce the governing troika of this volume to a single individual—the one who had signed on originally in a kind of emeritus/advisory capacity. Readers will note that the loss of adequate oversight and direction is not without cost. Perhaps the most serious deficiency is the brevity of treatment in description and analysis of regions between basins. Teams of investigators who volunteered (or were coerced) to prepare chapters on this and that area of the craton were largely drawn from the community of basin stratigraphers. Basins are where cover rocks accumulate in fascinating variety and where they are preserved most completely. Basins are also the places where fossil fuels, saline mineral resources, and ground water are most widely available, providing a drive for subsurface exploration and a motivation for concern by industry and state and federal agencies. Thus, it is not remarkable that contributors to this volume, who were identified and organized in terms of specific cratonic basins, have emphasized basin stratigraphy and tectonics, producing a corresponding shortfall in the attention paid to such elements as the Ozark Uplift, the Transcontinental Arch, or the Nashville Dome.

A further management deficiency is identifiable in the degree to which the authors of individual chapters fail to take into account the variations in interpretation and concept expressed in other chapters. This circumstance is the natural result of having all chapters in preparation simultaneously without adequate communication among the authors of the parts. Given a higher level of structure, the diversity of opinion would be balanced by inclusion of alternative views where controversy exists. Perhaps, when the Society celebrates its second Centennial, the successor to this volume will profit from this experience.

The authors and editors of chapters in this book were urged to keep in mind a target readership distinct from that normally addressed. With varying degrees of success we have tried to write for graduate students and for non-North Americans who have not matured with an osmotically acquired knowledge of the geography of this continent and who lack famliarity with our multitudinous stratigraphic names. At the same time, we hope that our words and plates and figures will be of value to geologists here and abroad who can profit from an introduction to a major part of this continent. To these ends we have attempted to keep the weight of stratigraphic terminology reduced to its elements and to include on tables and diagrams the position of each stratigraphic unit mentioned. Similarly, we have tried to hew to the principle that local place names employed in the text be identifiable on accompanying maps. Readers with a limited acquaintance with North American political geography will find it useful to have access to an atlas displaying the "lower forty-eight" states.

Not enough can be said in gratitude to the many men and women who contributed to this effort. A special appreciation is due to those authors whose manuscripts arrived on time and who have waited patiently for their laggard colleagues to complete their tasks. It must be said that this latter group was noteworthy for the absence of overt malice with which they accepted the abuse leveled against them. Thanks are gladly bestowed on the many reviewers who labored in anonymity to improve and refine the manuscripts represented here. And, of course, we must recognize the fact that many of the chapters of this volume were made possible by the cooperation of the U.S. Geological Survey and the several state surveys and bureaus who made available their people's data and time for this joint effort.

Finally, let it be acknowledged that none of this would have come to pass without the unremitting travail of A. R. Palmer, the science coordinator of the Society's Centennial Program. Pete has been an unfailing source of optimism where pessimism was the only rational mood; no detail has been too trivial for his attention, no wayward contributor too shirking to deserve a pep talk, and no aging editor too old to benefit from words of encouragement.

L. L. Sloss
Department of Geological Sciences
Northwestern University
Evanston, Illinois 60208

The Geology of North America
Vol. D-2, Sedimentary Cover—North American Craton: U.S.
The Geological Society of America, 1988

Chapter 1

Introduction

L. L. Sloss
Department of Geological Sciences, Northwestern University, Evanston, Illinois 60208

This volume is devoted to the Phanerozoic sedimentary strata covering the largely, but by no means exclusively, crystalline rocks of that part of the North American craton in the United States (exclusive of Alaska). Readers will note that this opening sentence implies constraints of space, time, and lithology on the subject matter dealt with here. The geographic qualifications derive from the organization structure of The Decade of North American Geology, and it only remains to point out that cratonic rocks and their histories are not delimited by national borders—nor by continental margins where these represent the lines of fission of former megacontinents. The qualifications of age and rock type are discussed on later pages, but first it is necessary to raise the question: "What is a craton?"

DEFINITION OF A CRATON

The term *craton* stems from the same Greek root as autocrat and bureaucrat and was applied geologically by Hans Stille (1936, 1941) in the sense of a strong, unyielding buckler or *shield.* In Stille's time of pre–plate tectonic innocence, he envisaged oceanic as well as continental cratons, the two terranes separated by *orthogeosynclines,* these being the lineal descendants of the 19th century "geosynclinals" of Dana and Hall. Further, Stille divided orthogeosynclines into *eugeosynclines* ("true geosynclines"), volcanic-dominated trends outboard with reference to continents, and *miogeosynclines* ("lesser geosynclines"), inboard nonvolcanic belts. Thus, continental cratons, by original definition, were bounded at the inboard margins of miogeosynclines.

Stille advanced these ideas late in his career, and continued development was inhibited by overriding political and military circumstances. With the cessation of hostilities, the cause of orthogeosynclines and their bedfellows, continental cratons, was taken up by Marshall Kay (1947, 1951), a son of the stable interior, Iowa, transported by Fate to the site of an ancient geosyncline, the Atlantic coast. (Kay, perhaps in deference to Stille or to classical Greek, invariably pronounced "craton" with a broad "a", as in "granite," but usage and custom in North America has settled on a long "a," as in "aged.") In addition to erecting a considerable lexicon of varied types of geosynclines (autogeosyncline, zeugogeosyncline, etc.), Kay recognized bounding flexures (Wasatch line, Adirondack line) that marked the western and eastern inboard limits of North American miogeosynclines; the broad stable region between the flexures was defined as the craton (distinguished, in the North American case, as the "hedreocraton," meaning especially long-lasting or steadfast, in testimony to its long history of stability).

The orthogeosyncline/craton paradigm was a popular forward step toward the understanding of continental margins and continental interiors, but many problems remained for stratigraphers, sedimentologists, and tectonists. In theory, eugeosynclines and miogeosynclines were separated by "geanticlines," and numerous hypothetical cross sections were constructed to illustrate the relationship. In reality, suites of rocks identified as "eugeosynclinal" were found to be almost everywhere in tectonic contact with "miogeosynclinal" successions; thus the duplex nature of orthogeosynclines was placed in doubt. Early on (*e.g.,* Pettijohn, 1943) it was recognized that many volcanic successions are accompanied by great thicknesses of immature detrital sediments, called "graywackes," in rhythmically repeated graded cycles, commonly including bedded cherts. Over the succeeding years, flume experiments combined with observations in outcrop (*e.g.,* Kuenen and Migliorini, 1950), in the subsurface (*e.g.,* Natland and Kuenen, 1951), and by deep-sea sampling demonstrated that many "eugeosynclinal" associations were matched by deep-sea analogs—a model presented with clarity by Drake and others (1959) who showed the kinship between ancient "eugeosynclinal" successions and the present-day accumulations of the Atlantic slope and rise.

Meanwhile, tectonists experienced continuing difficulties in locating the positions of the Wasatch and Adirondack lines, the syn-depositional flexures that were presumed to mark the inboard limits of miogeosynclines and, thus, the margins of the craton. Accelerations of the rates of change of thickness at "flexures" separating orthogeosynclines from cratons were shown to be products of crustal shortening, or where stacked thrust sheets are not present, the increase in thickness of this or that part of the stratigraphic succession is not significantly greater than can be noted deep within the cratonic interior. Further, miogeosynclines could not be shown to be geographically continuous trends of

Sloss, L. L., 1988, Introduction, *in* Sloss, L. L., ed., Sedimentary Cover—North American Craton; U.S.: Boulder, Colorado, Geological Society of America, The Geology of North America, v. D-2.

uniformly rapid subsidence and thick sedimentary accumulation; rather, more detailed study revealed the presence of basins and arches different in no important characteristic from equivalent elements of the cratonic interior.

Application of the concept of a craton circumscribed by surrounding miogeosynclines was particularly vexing to stratigraphers whose study areas extended across the line theoretically delineating distinct tectonic terranes. Here it was common to find that the stratigraphy on either side of the line of demarcation, although often obfuscated by regional nomenclature, was essentially the same for distances deep into regions declared to be "geosynclinal" and "extracratonic."

Many of these problems moved toward resolution in the 1960s. Dietz and Holden (1965), building on the earlier findings of Drake and others (1959), declared the greater part of the sedimentation referred to as miogeosynclinal in the Stille/Kay model to be the product of deposition on continental shelves, the prime exemplar being the largely submerged continental shelf off the mid-Atlantic states. This sedimentary-tectonic setting was termed a *miogeocline* by Dietz and Holden and was to be distinguished from the oceanic *eugeocline.* With the enhanced clarity of vision afforded by plate tectonics, we now identify miogeoclines (also known as *shelf prisms* or *shelf wedges*) as subsident passive (or divergent) continental margins.

Thus, the craton, a child of geosynclinal theory, is left orphaned by the demise of its defining limits, the "orthogeosynclines." Yet, "craton" as an identifying term and as a concept remains useful for application to the more or less stable interiors of continents—regions like Kansas or Saskatchewan. All that is needed is a redefinition of cratonic margins, but there is no great measure of agreement on the choice of criteria to establish the boundaries of cratons in a post-geosynclinal era. At one extreme are the tectonists who look toward Kansas or Saskatchewan from an active oceanic margin (as from the U.S. West Coast, for example); in such a group the tendency is to confine the craton to terranes beyond the topographic, structural, and magmatic influences of the adjacent active margin. In this view, all of North America west of the Rocky Mountain front is extracratonic, including the Colorado Plateau and the Laramide uplifts and basins of Colorado and Wyoming (all the way to the Black Hills?). At the other extreme are stratigraphers who track significant packages of strata from the continental interior deep into peripheral regions of structural deformation before familiar stratigraphic entities are lost to facies change or tectonic truncation.

Neither of the above positions is totally tenable, although as a stratigrapher, this reporter adheres more closely to the school influenced by the view from the craton. Through all but the last 70 to 80 million years of Phanerozoic time the Colorado Plateau and central Wyoming enjoyed the same degree of tectonic stability as Kansas, and the area of the overthrust belt in western Montana was scarcely less stable than Saskatchewan. By Late Cretaceous time, however, these regions were being encroached upon and involved with the clearly noncratonic deformational and magmatic behavior of the Sevier orogenic episodes. In other words, after hundreds of millions of years of existence in tectonic states like those of Kansas or Saskatchewan, a significant area of North America was de-cratonized. By the same token, it can be argued that the Appalachian margin of North America, a region identifiable as obviously extracratonic from Late Ordovician through much of the remainder of the Paleozoic, became restabilized (cratonized) in the Mesozoic and remains so today.

The message here is that the boundaries of cratons shift with the passage of time, not only by tectonic erosion and by the accretion of exotic lithospheric fragments but also by alteration of the tectonic behavior of autochthonous crust. Therefore, the definition of the margins of a particular craton is incomplete without identification of a specific span of geologic time—a principle long accepted in the interpretation of Precambrian history and no less applicable to the Phanerozoic.

What are the major identifying hallmarks of cratons as these are exemplified by regions such as Kansas and Saskatchewan? One such attribute is the maintenance over millions to tens of millions of years of the solid surfaces of cratonic lithosphere within tens of meters (or, at most, one or two hundred meters) of sea level—a measure of remarkable stability on a dynamic Earth with a radius of some 6,400 km. This is not to suggest that cratonic lithosphere is not subject to vertical motions; the implication, rather, is that for extraordinary spans of geologic time the uplift of cratonic terranes rarely exceeds the rate of erosion, and for even longer periods the subsidence of cratonic negative elements rarely surpasses rates of sedimentation. Under these conditions, erosion and deposition on cratons are extremely sensitive to eustatic changes of sea level at amplitudes that are of trivial influence in ocean basins and their active margins. Similarly, geoidal deformations or shifts of continents with respect to the high- versus low-latitude geometry of the geoid are potentially significant in the distribution of erosional and depositional environments on cratons.

The hypsographic elevation of existing cratons above ocean floors is, of course, directly attributable to the relative buoyancy of continental crust. It follows that the long-term continuity of shallowly submerged or slightly emergent ancient cratons is ascribable to the presence of continental crust in thicknesses of tens of kilometers. Thus, cratons are synonymous with extensive regions of thick continental crust, suggesting that cratons extend oceanward of continental interiors as far (and as long) as the presence of continuous thick continental crust is indicated for the time span in question by near-sea-level hypsography as described above. Such a definition would place the limits of the North American craton at the opening of Phanerozoic time at, or just inboard of, the shelf breaks marking the then passive margins of the ancient continent. Here, a belt of transitional crust, produced by stretching and thinning or by injection of mantle-derived magma or by both processes, formed the outboard edge of buoyant crust and its tectonic consequences.

The position of the Paleozoic western margin of cratonic North America, for example, can be approximated by investigation of the degree to which Mesozoic intrusives have been affected by partial assimilation of ancient continental crust (as indicated by $^{87}Sr/^{86}Sr$ ratios, *e.g.,* Kistler and Peterman, 1973). Exploration of the preserved Mesozoic eastern margin of this continent (*e.g.,* Swift and others, 1987) shows that thick (35 km) crust easily accepted as cratonic gives way to thin (7 km), obviously oceanic crust across a zone of transitional crust that may vary in width from a few tens of km to more than 200 km. Typically, the topographic shelf edge moved back and forth narrowly across the zone of transitional crust in Jurassic and later time, such that "cratonist" students of the lithologic facies and thickness of strata would be unable to draw a fixed line defining the edge of the craton by these criteria, although the band of shifting margins is narrow in terms of cratonic scales.

In summation, the concept of a continental heartland, the craton, is philosophically sound and is pragmatically applicable over huge areas. Further, it is urged that the principle of terming any terrane as cratonic that behaves cratonically (following the model of Kansas and Saskatchewan) during any significant span of time be accepted, difficulties of precise definition of outboard limits not withstanding.

CHRONOLOGIC SUBDIVISIONS FOR ANALYSIS OF CRATONIC HISTORY

Readers will note a dichotomy in the identification and selection of chronologic subdivisions for description and analysis of successive intervals of Phanerozoic time among authors of individual chapters of this volume. One group of workers treats the stratigraphy of the areas of their concern, and the geologic history represented by the stratigraphic successions involved, in segments delimited by the systems and periods (Ordovician, Cretaceous, etc.) of formal chronostratigraphy and geochronology as these are defined by biostratigraphic criteria. Other chapters apply a subdivision of rocks and time based on assemblages (*sequences*) of strata bounded by major interregional unconformities.

The utility of unconformity-bounded sequences in the analysis of cratonic history is discussed at greater length in a following chapter on tectonic evolution. Here, it should be pointed out that the great cratonic unconformities, those which define "natural" packages of cratonic strata and which permit definition of cratonically significant episodes of geologic history, tend to decay in identity and significance as cratonic margins are approached.

Where passive margins are involved, the time values of the lacuna formed by a cratonic unconformity and the physical evidence of such an unconformity decline, often monotonically, toward the paleo-position of the shelf break, outboard of which continuous sub-base-level deposition prevailed. Under these conditions, biostratigraphically defined systems, periods, and their subdivisions form reasonable units for analysis.

The practicality of cratonic sequences as the basis of historical reconstruction at active margins is further complicated by the common development there of local or regional unconformities, often presenting dramatic petrologic and structural discontinuities imposed by orogenesis, accretion, and accompanying events. Here, again, chronostratigraphy based on biostratigraphy is the reasonable alternative.

The forces noted above drive craton-margin stratigraphers toward reliance on classical units of rocks and time, and away from a hierarchy of units based on interregional unconformities. In this volume, as a consequence, chapters covering terrains adjacent to cratonic margins adhere to the classical system of classification and subdivision while chapters on the cratonic interior hew to the sequence-stratigraphy line wherever federal and state surveys supporting the work presented have policies condoning the recognition of a nonclassical segmentation of the stratigraphic column.

REFERENCES CITED

Dietz, R. S., and Holden, J. C., 1965, Miogeoclines (miogeosynclines) in space and time: Journal of Geology, v. 74, p. 566–583.

Drake, C. L., Ewing, M., and Sutton, G. H., 1959, Continental margins and geosynclines, the east coast of North America north of Cape Hatteras, *in* Ahrens, L. H., and others, eds., Physics and chemistry of the Earth: New York, Pergamon Press, v. 3, p. 110–198.

Kay, M., 1947, Geosynclinal nomenclature and the craton: American Association of Petroleum Geologist Bulletin, v. 31, p. 1283–1289.

——, 1951, North American Geosynclines: Geological Society of America Memoir 48, 143 p.

Kistler, R. W., and Peterman, Z. E., 1973, Variations in Sr, Rb, K, Na, and initial Sr^{87}/Sr^{86} in Mesozoic granitic rocks and intruded wall rocks in central California: Geological Society of America Bulletin, v. 84, p. 3489–3512.

Kuenen, P. H., and Migliorini, C. I., 1950, Turbidity currents as a cause of graded bedding: Journal of Geology, v. 58, p. 91–126.

Natland, M. L., and Kuenen, P. H., 1951, Sedimentary history of Ventura Basin, California, and the action of turbidity currents: Society of Economic Paleontologists and Mineralogists Special Publication 2, p. 78–107.

Pettijohn, F. J., 1943, Archean sedimentation: Geological Society of America Bulletin, v. 54, p. 925–972.

Stille, H., 1936, Wege und Ergebnisse der geologisch-tektonischen Forschung: Festschrift Kaiser-Wilhelm Gesellschaft Fordhandlung Wissenschaften, Band 2, 617 p.

——, 1941, Einfuhrung in den Bau Amerikas: Berlin, Borntraeger, 717 p.

Swift, B. A., Sawyer, D. S., Grow, J. A., and Klitgord, K. D., 1987, Subsiding crustal structure and thermal evolution of Georges Bank Basin: American Association of Petroleum Geologists Bulletin, v. 71, p. 702–718.

MANUSCRIPT ACCEPTED BY THE SOCIETY JANUARY 6, 1988

Printed in U.S.A.

The Geology of North America
Vol. D-2, Sedimentary Cover—North American Craton: U.S.
The Geological Society of America, 1988

Chapter 2

Geophysical aspects of the craton: U.S.

William J. Hinze and Lawrence W. Braile
Department of Earth and Atmospheric Sciences, Purdue University, West Lafayette, Indiana 47907

INTRODUCTION

The midcontinent province between the Appalachian-Ouachita orogen and the Rocky Mountains is that portion of the United States craton which has been relatively stable tectonically for roughly the past billion years. Diastrophism has been restricted largely to broad, slow, vertical movements and highly attenuated, passive response to orogenic activity, which has occurred at plate margins around the North American craton. Thus, except for very localized structures associated with rare anorogenic intrusions and meteorite impacts, the Phanerozoic sedimentary rocks are only disturbed by minor faulting and gentle warping. As a result, regional geophysical investigations of the midcontinent have primarily focused on the study of the more laterally variable physical properties of the crystalline rocks of the basement and deeper portions of the crust. However, a great number of detailed geophysical investigations have been conducted over cratonic basins in the search for structural and stratigraphic variations within the Phanerozoic sedimentary rocks. With a few notable exceptions, these surveys, largely seismic reflection and gravity, have been conducted to locate hydrocarbon accumulations and consequently are represented by proprietary data.

Thus, although geophysical studies, particularly those employing modern, common-depth-point, seismic-reflection techniques, have great utility in mapping the minor structural and stratigraphic variations that occur within the Phanerozoic sedimentary rocks of the midcontinent, our principal source of information regarding the subsurface nature of these rocks is drillhole data. In contrast, our knowledge of the basement igneous and metamorphic rocks of the craton is based on regional geophysical studies working in concert with the irregularly distributed basement drillholes and, by extrapolation, from the crystalline rock outcrops of the Canadian Shield and the few other areas where such rocks crop out through the sedimentary rock cover.

The basement rocks record the evolution of a large section of the North American continental nucleus through Precambrian time. As such, they provide our principal clue to the origin of subsequent, broad-scale, Phanerozoic, tectonic features of the crust. Furthermore, the geophysical signatures of the basement rocks can, in a general way, be translated into their relative strength characteristics and thus can be used to map zones of weakness, which have controlled crustal deformation and localized resurgent tectonics throughout Phanerozoic time. It is these strength characteristics that have localized the release of widespread stress fields and focused stresses, which are manifested in the Phanerozoic sedimentary rock cover by structural deformation and related stratigraphic effects as well as in observed contemporary tectonic activity.

The quantity and quality of regional geophysical data are highly variable over the midcontinent, but, in general, the data are limited to reconnaissance or, at best, regional coverage. Gravity and magnetic data provide the best overall coverage, while seismic and neotectonic (e.g., heat flow and crustal movement) data are sparse and poorly distributed.

MIDCONTINENT CRUSTAL CHARACTERISTICS

Introduction

The structural and compositional characteristics of the midcontinent crust reflect the long and complex geologic history of the region. Archean rocks of the midcontinent date back to over 3 billion years (Denison and others, 1984), and interpretation of basement geology suggests that during Precambrian time the crust accreted by collision and welding of continental margin volcano/sedimentary belts. The crust was modified by superimposed anorogenic thermal-tectonic events and, later, orogenic processes. The principal characteristics of the crust as we observe them today were established by the end of Precambrian time. Subsequent events in the Phanerozoic led to the deposition of an extensive blanket of primarily shelf-type sedimentary rocks with accumulations within basins reaching thicknesses of up to 5 to 10 km.

The sedimentary rock column records a complex vertical-movement history throughout the Phanerozoic, during which the crust probably behaved in a largely passive manner. Only very locally was the crust disturbed by igneous intrusive events. As a result, we do not envision that the crust has changed in a significant way since Precambrian time; however, there are important

Hinze, W. J., and Braile, L. W., 1988, Geophysical aspects of the craton: U.S., *in* Sloss, L. L., ed., Sedimentary Cover—North American Craton; U.S.: Boulder, Colorado, Geological Society of America, The Geology of North America, v. D-2.

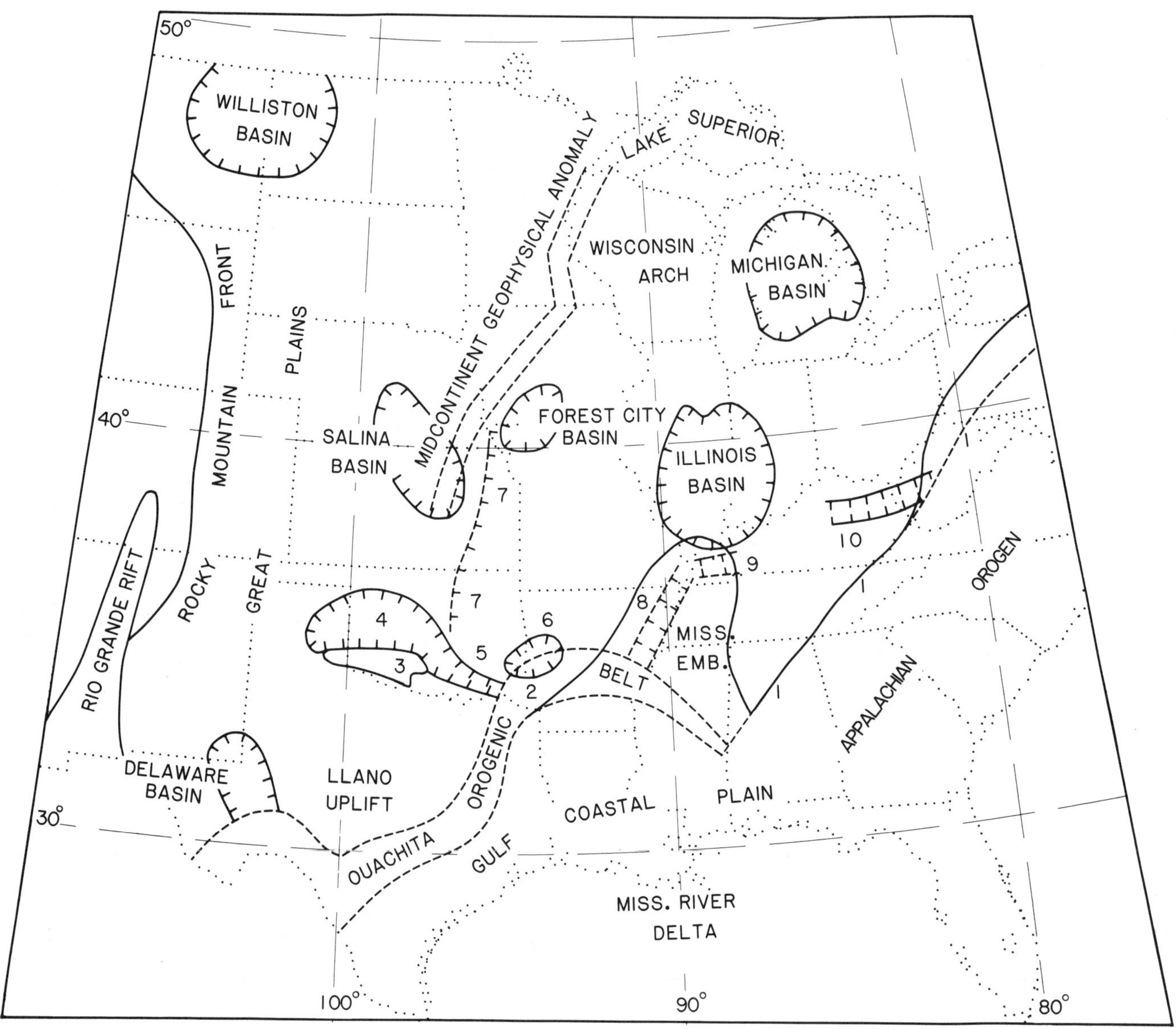

Figure 1. Location of major geologic features of the midcontinent (1—New York-Alabama Lineament; 2—Ouachita Foldbelt; 3—Wichita Uplift; 4—Anadarko Basin; 5—Ardmore Basin; 6—Arkoma Basin; 7—Nemaha Uplift; 8—Reelfoot Rift; 9—Rough Creek Graben; 10—Rome Trough).

exceptions to this generalization. The crust of the craton was altered in Late Proterozoic time with the breakup of the continent to form the Proto-Atlantic Ocean and the resulting development of a series of failed-arm rifts extending into the craton from the then-continental margin (Keller and others, 1983). The Reelfoot Rift underlying the Mesozoic Mississippi Embayment (Ervin and McGinnis, 1975) and the Southern Oklahoma Aulacogen (Hoffman and others, 1974) (Fig. 1) with Early Cambrian-age, bimodal, igneous rocks of the Wichita Mountains, are the most widely known examples of this event. More recent changes in the crust and the upper mantle may be reflected in the considerable uplift of regions such as the Great Plains (Trimble, 1980), Colorado Plateau, and the Adirondack Mountains, which have been uplifted roughly 1500 m or more over the past 10 m.y. (Gable and Hatton, 1983).

The crustal characteristics of the midcontinent are known only through geophysical investigations because of the paucity of deep drillholes, particularly into the crystalline crust. Geophysical investigations are only useful in mapping the horizontal and vertical characteristics of the crust in an indirect way. Physical properties of the crust such as velocity, acoustic impedance, density variations, and magnetic polarization contrasts determined by geophysical measurements must be translated into structure and composition. There is considerable ambiguity in this translation,

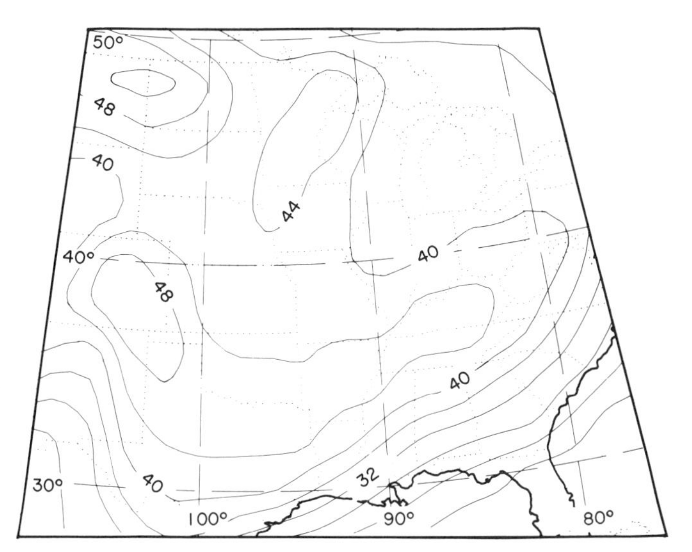

Figure 2. Crustal thickness map (after Braile and others, 1986a). Contour interval is 4 km.

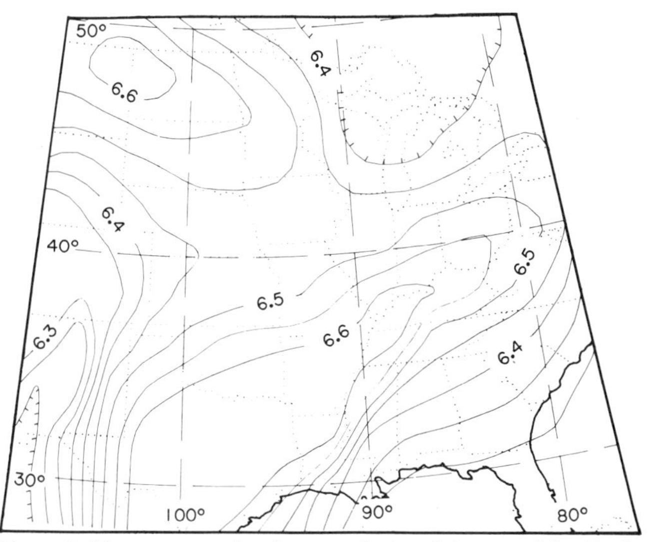

Figure 3. Mean compressional wave velocity of the crust (after Braile and others 1986). Contour interval is 0.05 km/s.

so great care must be used to minimize this problem by tying the interpretation to known geology wherever possible and interrelating the results from different geophysical observations.

Seismic Data

Introduction. Information on the overall nature, composition, and configuration of the midcontinent crust is derived primarily from seismic data. Although numerous seismic techniques have been used for this purpose, including seismic reflection, travel-time residuals, and analysis of local earthquake-generated seismic waves, the bulk of the regional crustal information has been obtained from refraction and surface-wave studies. The results of the latter studies have been evaluated and analyzed by Braile and others (1986). The midcontinent portion of the contoured maps of the crustal thickness (Moho depth), the mean compressional wave crustal velocity, excluding the Phanerozoic sedimentary rock column, and the upper mantle (P_n) compressional wave velocity are presented in Figures 2, 3, and 4 respectively. Contouring is a subjective process, which is particularly ambiguous where data points are sparse, as in the midcontinent. These maps focus on regional variations by giving little weight in the contouring process to significant variations based on only one observation and to discrepant values.

Marked variations are noted by Braile and others (1986) between the crustal thickness, average crustal velocity, and P_n velocity of the midcontinent and the western United States. Mean values for the midcontinent are approximately: crustal thickness, 42 km; P_n velocity, 8.1 km/s; and average crustal velocity, 6.5 km/s. These are compared to average United States and adjacent Canada values of roughly 36 km, 8.0 km/s, and 6.4 km/s respectively. The differences in these means primarily reflect variations in temperature and composition of the relatively stable midcontinent and eastern United States in contrast to the tectonically active Rocky Mountains. The craton within the Cordillera, as defined by Sloss (1982), is indistinguishable from the rest of the western United States crust on the basis of the seismic-derived characteristics.

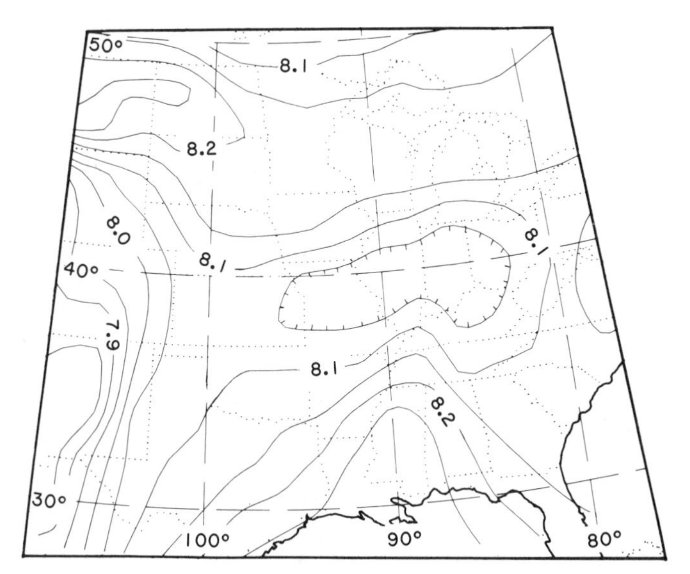

Figure 4. Compressional wave velocity of the upper mantle (after Braile and others, 1986). Contour interval is 0.05 km/s.

Crustal Thickness. Within the midcontinent, the thinnest crust is found in the northeast adjacent to Canada. This relationship is unanticipated because this region, in the vicinity of the western Great Lakes, is undergoing regional uplift associated with glacial rebound (Clark and Persoage, 1970). Apparently, the isostatic adjustment to the mass of the Pleistocene glaciers is not reflected in an observable way by a thickening of the crust as would be suggested by Airy-type compensation. It is also interesting to note that the region of thinner crust in the Great Lakes area is also the region which has the lowest average crustal velocity (Fig. 3). Observed seismic velocities are directly related to densities, thus it is surprising to find the thinnest crust associated with the lowest density crust in obvious contradiction to simple Pratt-type isostatic relationships. These observations suggest that isostasy is achieved by a combination of Airy- and Pratt-type compensation with the level of compensation within the upper mantle.

The thickest crust in the midcontinent is found immediately east of the eastern margin of the Rocky Mountains in the vicinity of the Great Plains, where mean surface elevations (Fig. 5) gradually rise from about 400 m in the east to approximately 1600 m along the eastern edge of the Cordillera. On the crustal-thickness contour map (Fig. 2), two maxima are recognized—one centered over eastern Colorado and the other over the boundary between Montana and North Dakota—but the minimum between them may simply reflect the lack of data because no seismic-refraction profiles are available from this region. Another area of regionally thickened crust is observed in Minnesota, Wisconsin, and Iowa approximately along the axis of the northern portion of the Midcontinent Geophysical Anomaly. In a more local sense, Smith and others (1966) find crustal thicknesses of up to 55 km along seismic-refraction profiles in the Lake Superior Basin. More recently, Halls (1982) found a similar thickness of the crust in the Lake Superior Basin and its extensions to the southeast and southwest. A zone of thickened crust is also observed extending easterly from the southern Great Plains across northern Texas, Oklahoma, and Arkansas into the northern Mississippi Embayment, and northeasterly along the eastern margin of the Appalachian Mountains. This trend crosses several major crustal structures and, therefore, probably represents the combined effect of several crustal processes, including thickened crust associated with the Mississippi Embayment (Austin and Keller, 1982; Mooney and others, 1983), Southern Oklahoma Aulacogen, Ouachita Foldbelt, and the subsurface extension of the Grenville basement province, which occurs in the subsurface west of the Appalachian Mountains in the eastern United States and west of the subsurface Ouachita orogenic belt in Texas (King, 1975).

Average Crustal Velocity. The average crustal velocity is a particularly significant parameter because, as discussed by Pakiser and Robinson (1966) and Smithson and others, (1981), it is a function of mean crustal composition, which in turn provides important evidence for the interpretation of crustal genesis and evolution. The mean crustal velocity of the midcontinent, 6.5 km/s, suggests an average composition between granite (≃6.1 km/s) and diorite (≃6.7 km/s), but observed increasing velocities with depth within the crust indicate an enhanced mafic content of the crust at depth.

The variations in the average, compressional, seismic-wave velocity of the crust (Fig. 3) show a rough correspondence to the crustal thickness (Fig. 2). As noted above, the thinner crust over

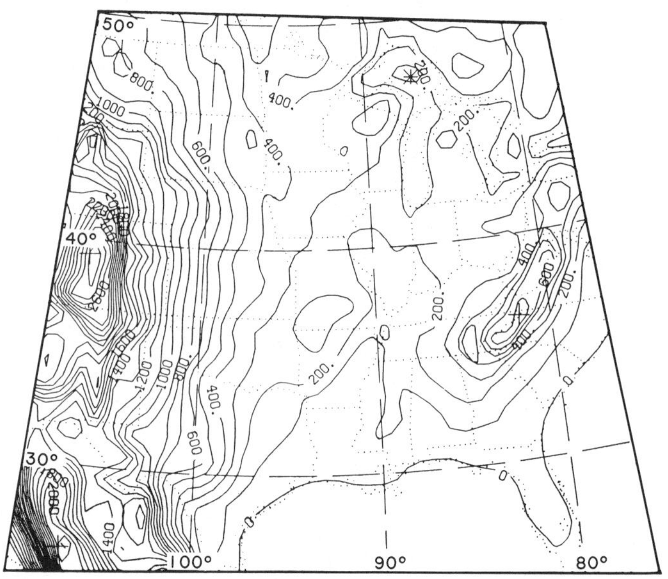

Figure 5. Mean surface elevation of 1° (latitude/longitude) areas. Contour interval is 100 m.

the western Great Lakes region corresponds to lower crustal velocities in the 6.4 km/s range. In contrast, the thickened crust beneath the Great Plains correlates with higher mean crustal velocities with values generally greater than 6.5 km/s. The depression in the mean crustal velocity over the central Great Plains (Kansas, Nebraska, and South Dakota) may be more apparent than real because of the lack of seismic refraction data in this region. The higher crustal velocity, ranging up to 6.6 km/s, also corresponds to the zone of thickened crust extending across Texas to the Mississippi Embayment and then northeastward along the western margin of the Appalachian orogen. Relatively higher average crustal velocities are noted in the Mississippi Embayment (Mooney and others, 1983) and for the Grenville province in Canada, in contrast to the adjacent older basement province (Mereu and Jobidon, 1971).

Sub-Moho Characteristics. One of the most contrasting geophysical properties between the midcontinent and the western United States is the velocity of the compressional seismic waves within the uppermost mantle (P_n). Beneath the Cordillera, these velocities generally are in the 7.8–7.9 km/s range (Fig. 4), while in the midcontinent and eastern United States the values are roughly 8.1 km/s. Higher P_n velocities (>8.2 km/s) are observed in the northern Great Plains and the southern portion of the Mississippi Embayment and the Mississippi River Delta. Black and Braile (1982) suggest that the variation in continental P_n velocity may be primarily a reflection of temperature at the Moho and, thus, the relatively constant P_n velocity in the midcontinent region may be related to the average heat flow of the craton.

Compressional velocities increase in a variable manner with depth within the mantle of the midcontinent and lateral inhomogeneities are indicated for depths down to 200 km (Kurita,

1976). No low-velocity zones for compressional waves are observed in the upper mantle (McMechan, 1979), but marked velocity gradients occur at roughly 400 and 600 km. McMechan (1979) also notes a general progression toward higher velocities within the upper 800 km of the mantle from the Cordillera to the east.

Gravity and Magnetic Anomalies

Introduction. Gravity and magnetic surveying have proven to be efficient procedures for characterizing the crust of the midcontinent. As a result, considerable data are available from detailed surveys of limited areas as well as from regional investigations of the entire midcontinent.

The midcontinent anomaly patterns of both the gravity and magnetic maps are complex and highly diversified. This is somewhat unanticipated because the mass and magnetic polarization variations in the relatively undisturbed, surficial, Phanerozoic, sedimentary rock column are limited in magnitude and areal dimensions. Gravity anomalies derived from the Phanerozoic sedimentary rocks are generally less than 5 mGals in amplitude and magnetic anomalies are less than a few tens of gammas (nT). Anomalies originating from basement topographic variations generally are of similar orders of magnitude in the midcontinent (Hinze and Merritt, 1969). Thus, the gravity (Plate 1-A) and magnetic (Plate 1-B) anomalies measured in tens of milligals and hundreds of gammas are derived from the underlying, igneous and metamorphic, crustal, basement rocks. These rocks, where observed in outcrop, have marked density and magnetic polarization changes, which are directly related to intense anomalies, and the anomaly patterns generally mimic the structural trend of crystalline rocks.

Gravity-Magnetic-Geologic Relationships. Although the magnetic-anomaly pattern (Plate 1-B) consists of a complex array of irregular anomalies, anomalies commonly are located along linear trends. One of the most striking of these magnetic lineaments is the New York-Alabama Lineament (Fig. 1; King and Zietz, 1978) which strikes northeasterly for more than 1600 km from southern Alabama into New England. The lineament approximately coincides with the western limit of the geological and geophysical trends of the Appalachian Mountains and, therefore, marks an important structural break within the craton. It occurs along the western margin of the regional Appalachian gravity minimum and the Appalachian Foldbelt. Southeast of the lineament the magnetic anomalies are broadened and subdued because the basement crystalline rocks are overlain by the thick sedimentary rock column of the Appalachian Foldbelt. The regional gravity minimum, which extends over the Appalachian Plateau and the Blue Ridge of the Appalachian orogen, is commonly related to a thickened crust (James and others, 1968), while the buried edge of the Precambrian craton is placed along the prominent, regional, gravity gradient that separates the minimum from the eastern positive-gravity anomaly (Thomas, 1983). The steep gravity gradient and the positive anomaly of the southern Appalachian Mountains can be modeled in the North Carolina region by a suture zone, a thinned crust, or a shallow mafic body (Hutchinson and others, 1983). The southern margin of the craton, associated prior to subduction with the Appalachian and Ouachita orogenies at the end of the Paleozoic Era, is observed by the truncation of the Appalachian magnetic trends in southern Alabama and by a termination of anomaly patterns along a postulated transform fault (Cebull and others, 1976) connecting the Appalachian and Ouachita orogens. The anomaly associated with the subsurface Ouachita orogenic belt continues southwest from central Arkansas to southeast Oklahoma and then south-southeasterly to south Texas. South of this belt, gravity anomalies increase, reflecting a decrease in the crustal thickness beneath the Gulf Coastal Plain (Keller and Shurbet, 1975).

Within the craton, the magnetic- and gravity-anomaly patterns are quite diverse, reflecting the long and complex history of the Precambrian basement rocks. The anomalies can generally be grouped into units or terrains based on characteristic patterns of anomalies with similar dimensions, amplitudes, and strike direction. In addition, there are linear anomalies that transect the more pervasive anomaly patterns. The Bouguer gravity-anomaly patterns are generally less obvious because of the distorting effect of the long-wavelength anomalies (measured in hundreds of kilometers) associated with isostatic effects. These effects are particularly prominent west of 100°W longitude, due to the subsurface sources that approximately balance the increasing elevation isostatically to the west of the Great Plains region. These effects are largely eliminated in the short-wavelength (λ <250 km) gravity-anomaly map (Plate 1-C) prepared by Hildenbrand and others (1982a).

The most prominent gravity anomaly within the craton is the Midcontinent Geophysical Anomaly which extends north-northeasterly from eastern Kansas to the west end of Lake Superior. Its great length, steep gradients, and large amplitude make it particularly conspicuous on the gravity map and only slightly less so on the magnetic-anomaly map. The correlative gravity- and magnetic-positive anomalies, which are traced to outcropping Keweenawan basalts and late Precambrian clastic basins in the Lake Superior region, are the geophysical expression of a 1100-Ma paleo-rift zone, the Midcontinent Rift System (Halls, 1978). The anomalies have been traced in an arcuate pattern through Lake Superior to more subdued anomalies in southeastern Michigan (Hinze and others, 1975). Geophysical modeling, outcrop, and drillhole data indicate that the anomalies of the Midcontinent Rift System are due to the complex superposition of effects from troughs filled with sedimentary rocks and mafic volcanic rocks, occasional overlying basins filled with clastics, and pervasive intrusions from the mantle into the lower crust. The eastern arm of the Midcontinent Rift System has been continued south from southern Michigan into Tennessee on the basis of discontinuous, north-south trending, correlative, positive gravity and magnetic anomalies (Halls, 1978; Keller and others, 1982 and 1983; Lidiak and others, 1985).

Another prominent rift-related anomaly occurs over the

Southern Oklahoma Aulacogen, which strikes west-northwestward from the Ouachita Foldbelt in southern Oklahoma along the Oklahoma-Texas border into the panhandle of Texas. Although a major crustal-scale feature is required to explain the observed 100-mGal anomaly, the positive anomalies correlate with late Precambrian-Cambrian basalt, gabbro, anorthosite, and diorite of the Amarillo-Wichita Uplift.

A contrasting gravity- and magnetic-anomaly pattern is observed over the Mississippi Embayment and its axial Reelfoot Rift (Ervin and McGinnis, 1975). The Mississippi Embayment is observed on the gravity-anomaly map as a broad positive anomaly, largely caused by a thickened, high-density, lower crust (Mooney and others, 1983). A sedimentary rock-filled graben associated with the rift has been mapped by Hildenbrand and others (1977) on the basis of two parallel trends of localized gravity- and magnetic-positive anomalies that occur along the margin of the rift. The wavelength of anomalies is greater over the graben than on the margins. Recent studies utilizing detailed gravity- and magnetic-anomaly maps indicate that this feature breaks up into a complex of similar features extending into Indiana, Kentucky, Illinois, and Missouri. Braile and others (1982b) suggest that the gravity and magnetic anomalies of these segments, which make up the New Madrid Rift Complex plus the Rome Trough in eastern Kentucky and West Virginia, coalesce to form a series of east-west anomalous features providing a possible explanation for the so-called 38th-Parallel Lineament (Heyl, 1972).

Roughly from western Ohio eastward, the gravity anomalies and the belts of magnetic anomalies, made up of an intricate "birds-eye" pattern, trend north to north-northeast, on strike with similar anomaly patterns observed over the outcropping Grenville province in Canada. The eastern limit of this anomaly pattern is the New York-Alabama Lineament, while the western limit—the Grenville Front—extends north-south through western Ohio into Kentucky and Tennessee. The Grenville trend occurs as far west as 87°W in what is identified by Lidiak and others (1985) as a foreland zone where Grenville orogenesis has affected the older, 1500-Ma, felsic terrane to the west. The actual Grenville Front is identified as far west as 85°W at latitude 36°N with the western termination of the "birds-eye" magnetic-anomaly pattern. This pattern is also observed over the Grenville-age rocks of the Llano Uplift and its extensions into the subsurface to the north and southwest. These anomalies in Texas are truncated to the south and southeast along the trend of the buried Ouachita Foldbelt, which is marked by a curvilinear positive-gravity anomaly. The anomalies extend to the northwest into western Texas to a northeast-striking magnetic minimum, which extends from southwestern Oklahoma into southwest Texas. The position of the Grenville Front in Texas, as suggested by King (1975), may coincide with this northeast-trending magnetic minimum.

In the western Great Lakes region, the regional gravity and magnetic anomalies occur in alternating bands of maxima and minima that strike east to northeast. Broad positive-magnetic anomalies generally correlate with granitic and gneissic belts; negative-magnetic anomalies are identified with meta-volcanic/sedimentary rock terrains, while the gravity anomalies generally have an inverse relationship to the magnetic anomalies. However, direct correlation of gravity and magnetic anomalies is observed over the Keweenawan basalts of the Midcontinent Rift System, mafic intrusives, and intrusive granites. These anomaly trends are interrupted by the Midcontinent Rift System, and central Wisconsin is dominated by negative-gravity anomalies probably related to relatively low-density, Proterozoic, granitic intrusives. These east to northeast trends broaden and attenuate under the Phanerozoic sedimentary rocks of the Michigan Basin, and their strike shifts to the southeast in the vicinity of Lake Michigan. They extend across Michigan and are terminated at the Grenville Front. The anomalies extend from the Canadian border south to roughly the 43°N parallel and westerly across the eastern one-third of the Dakotas; there they are truncated at the boundary with the extension of the Churchill basement province from Canada into North Dakota (Green and others, 1979; Lidiak, 1982). The buried Churchill Province anomalies change from a south-southeasterly strike in eastern Montana and western North Dakota to a southeasterly strike and extend across Nebraska, Iowa, and Missouri, possibly as far as the New York-Alabama Lineament in eastern Tennessee and Kentucky (Hinze and Zietz, 1985). The character of these anomalies changes over this extensive length; they are disrupted by the anomalies of the Midcontinent Rift System and the New Madrid Rift Complex. North of these anomalies, in Illinois, Indiana, and adjacent states, the anomaly trends are not well defined, but there is a tendency for the regional anomalies to strike northwest. The boundary between this trend of anomalies and the easterly to northeasterly striking anomalies is poorly defined, but occurs in the vicinity of the Illinois-Wisconsin state boundary.

A broad, ill-defined, positive-magnetic anomaly with a few local anomalies extends from Tennessee to eastern Colorado. This anomaly lies to the south of the zone of southeasterly striking anomalies associated with the buried extension of the Churchill Province discussed above. The southern limit of this anomaly is associated with the interpreted transform fault connecting the southern Appalachian and the Ouachita Mountains (Thomas, 1977). The northern margin is marked by a discontinuous, bordering, negative anomaly, which is most intense in south-central Kansas. This band of anomalies in Kansas has been correlated by Yarger (1981) with a boundary between an older, mesozonal granite terrain to the north and a younger, epizonal granite terrain to the south. The broad, positive-magnetic anomaly is a dominant long-wavelength feature of satellite-elevation (400-km) magnetic observations. It is observed to extend with diminishing intensity northeastward from the southwestern United States to the vicinity of Lake Michigan. Starich and others (1984) correlate the anomaly with enhanced lower-crustal magnetizations associated with the thermal event that caused the widespread felsic intrusions and extrusions in the midcontinent approximately 1500 Ma. There is no coherent gravity-anomaly pattern in this region, but marked

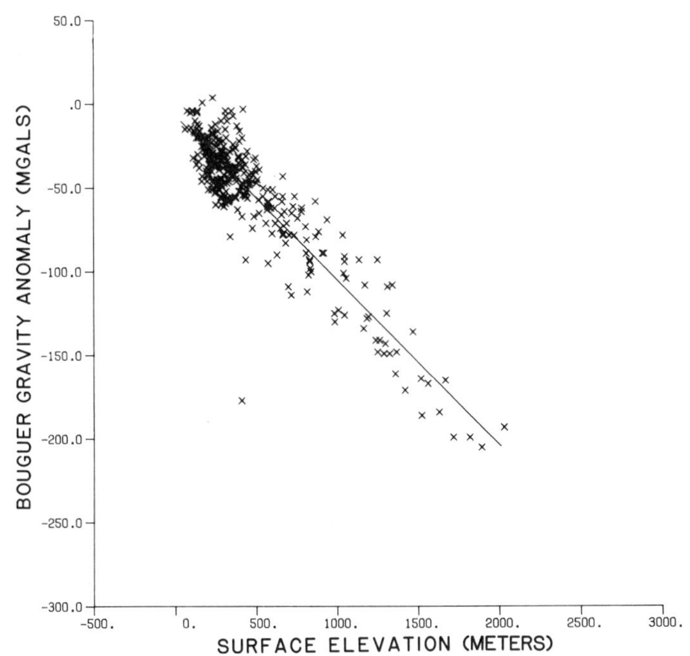

Figure 6. Scatter diagram of relationship between mean 1° (latitude/longitude) Bouguer gravity anomaly and surface elevation of the midcontinent.

anomalies are related to the Ouachita Mountains and the Southern Oklahoma Aulacogen. An intense gravity minimum in southeastern Oklahoma and west-central Arkansas overlies the Ouachita Mountains and the adjacent Arkoma Basin. The gravity minimum is probably a result of the thick sequence of upper Paleozoic sedimentary rocks as well as a structurally related thickening of the crust. Similar, but less-intense minima are related to the Anadarko and Ardmore Basins. These basins had their origin as grabens associated with the Southern Oklahoma Aulacogen and were accentuated when late Paleozoic compression deformed the region. The Arbuckle Uplift gravity high, which lies west of the Ouachita Mountains, east of the Anadarko Basin, and north of the Ardmore Basin, is probably a result of mafic intrusions related to the development of the Southern Oklahoma Aulacogen in late Precambrian-Cambrian time (Kruger, 1983).

Gravity-Elevation Relationships. Visual inspection of the surface elevation map (Fig. 5) and the Bouguer gravity-anomaly map (Plate 1-A) shows that there is a general inverse relationship between these two parameters. This is anticipated in regions that are in or are approaching isostatic equilibrium such that regional changes in elevation are offset by deficiencies in mass at depth so that, at some level within the earth, the weight of overlying rock columns is constant. In the computation of the Bouguer gravity anomaly, the mass of the earth from sea level to the surface is accounted for. Thus, if the earth approaches isostatic equilibrium, the mean Bouguer gravity anomaly decreases with elevation. This inverse relationship is clearly observed in a plot of mean one-degree (latitude/longitude) surface elevation and Bouguer gravity anomaly (Fig. 6). Regression analysis indicates that the mean Bouguer gravity values across the midcontinent, from the Rocky Mountains to the Appalachian Foldbelt, decrease in an essentially linear manner (correlation coefficient is –0.91) at a rate of roughly –98 mGals/km. This rate is less than anticipated for a region in isostatic equilibrium (–112 mGals/km, assuming a surface rock density of 2.67 g/cm^3) and indicates undercompensation for the surface relief. An explanation for this observation is suggested by a consideration of the free-air gravity anomaly (Fig. 7), which also provides useful geologic information and constraints upon the isostatic balance of an area.

The free-air anomaly, unlike the Bouguer gravity anomaly, does not take into account the mass of the earth between sea level and the surface of the earth and, therefore, measures the mass of the subjacent earth. If the earth is in perfect isostatic equilibrium, free-air gravity anomalies will be uncorrelated with surface topography. To eliminate the effect of local topographic changes, we have calculated the mean free-air anomaly in one-degree (latitude/longitude) areas and compared the results with the corresponding average elevation. The results, which are presented in Figure 8, indicate that mean anomalies of most one-degree areas are within 20 mGals of the isostatically ideal situation, and the correlation between mean free-air anomalies and surface elevation is poor (correlation coefficient is –0.15). However, there is a tendency for the higher-elevation areas (>1000 m) to have greater average free-air anomalies. This can also account for the Bouguer gravity/surface elevation gradient, which is lower than anticipated for a region in isostatic equilibrium. The source of this variation can be seen by comparing Figure 5, the elevation map, with Figure 7, a free-air gravity-anomaly map based on passing all anomaly wavelengths greater than 2 degrees. The northern Great Plains region, from eastern Montana to the Mississippi River, is generally a positive free-air–anomaly area, suggesting that the enhanced elevation is isostatically undercompensated. A factor in accentuating this effect is the strong positive-gravity anomaly associated with the Midcontinent Geophysical Anomaly, but, even excluding the area of this feature from consideration, the northern Great Plains from Montana to Minnesota is isostically undercompensated. A major factor contributing to the undercompensation of this region is the north-south linear-positive anomaly, which extends south from the Canadian border to western Nebraska. The source of this anomaly is unknown, but it does correspond to the position of the Churchill basement province boundary as it extends into the United States from Canada. This is also the location of an abnormally high electrical conductor within the crust (Charpayungpun and Landisman, 1977).

Several additional observations can be made from Figure 7. The central part of the midcontinent, including the southern Great Plains, tends toward negative free-air anomalies and thus slight isostatic overcompensation. In addition to the positive anomalies discussed above, local, positive, free-air gravity anomalies occur over the Michigan Basin, the Mississippi Embayment and the deepest part of the Illinois Basin, the Southern Oklahoma

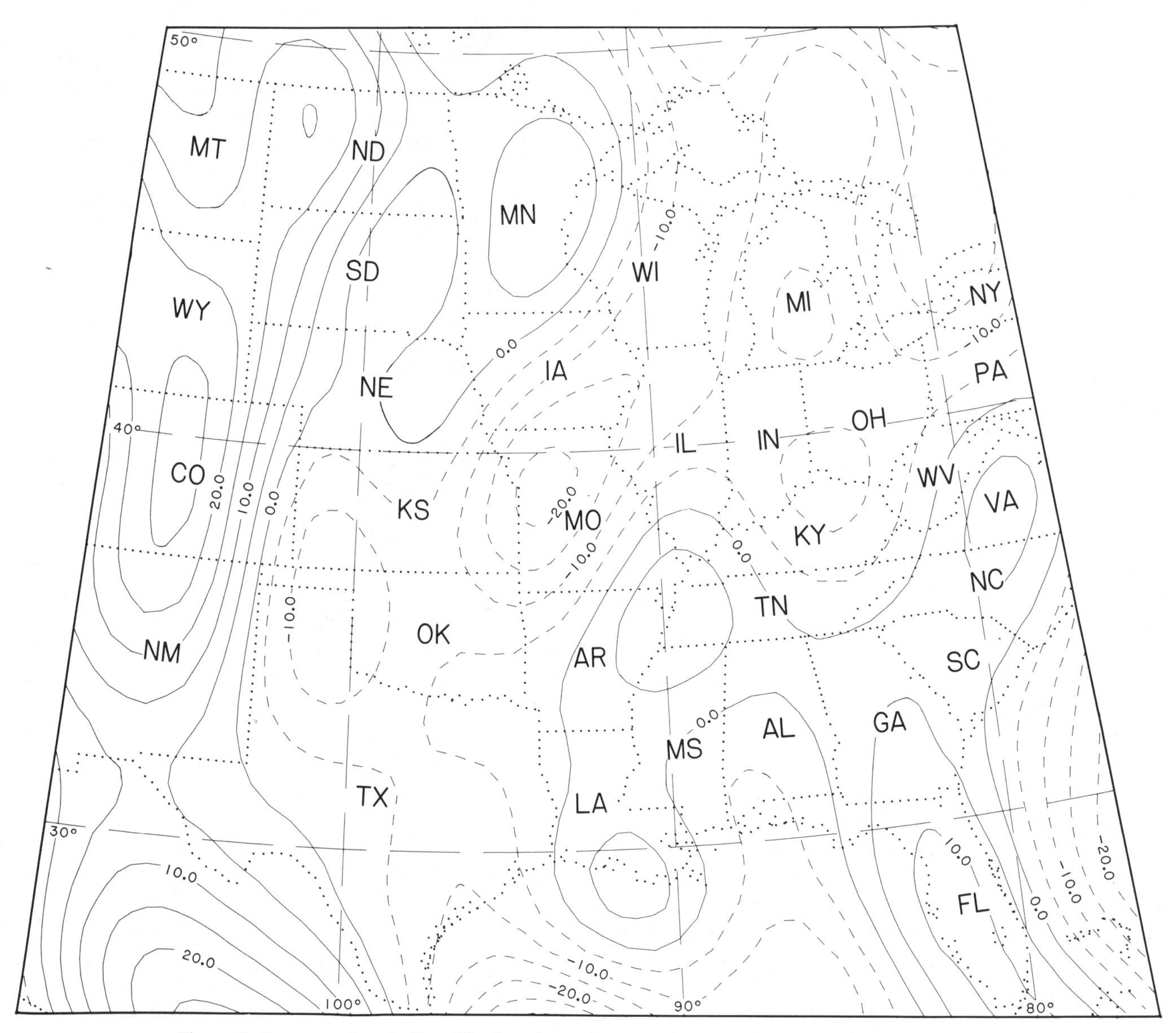

Figure 7. Long-wavelength (λ >2°), free-air gravity-anomaly map of the midcontinent. Contour interval is 5 mGals.

Aulacogen area, and the Mississippi River Delta. The bordering regions of the Appalachian Mountains and especially the Cordillera are also positive free-air anomalies. There are no consistent relationships observed between the free-air anomalies of Figure 7 and maps of the crustal thickness (Fig. 2) and the average crustal P-wave velocity of the crust (Fig. 3), suggesting that isostatic compensation takes place beneath the crust, within the upper mantle, and/or by a complex interaction of crustal thickness and crustal composition.

Geophysical Characteristics Along 90° W and 37°N

To obtain an integrated view of the several geophysical characteristics of the midcontinent crust discussed above, we have prepared profiles of these parameters along 90°W longitude and 37°N latitude, together with a simplified upper-crustal section which illustrates the mean surface elevation and the mapped basement elevation (Figs. 9 and 10). The surface Bouguer gravity anomaly and near-surface magnetic-anomaly profiles are shown together with a profile of their long-wavelength components derived from upward continuation of the profiles to an elevation of 40 km. The profiles permit the visual correlation of parameters, but are limited by the effect of geophysical observations that extend into the profile from sources adjacent to, but not crossed by, the profile.

Several generalizations regarding the profiles are worthy of note, but the lack of obvious general correlation between parameters is particularly striking. The profile along 90°W longitude (Fig. 9) extends from the Canadian border to the Gulf of Mexico, where the crust thins to about one-half (20 km) of the normal thickness. A thickened crust south of the Central Lowlands corre-

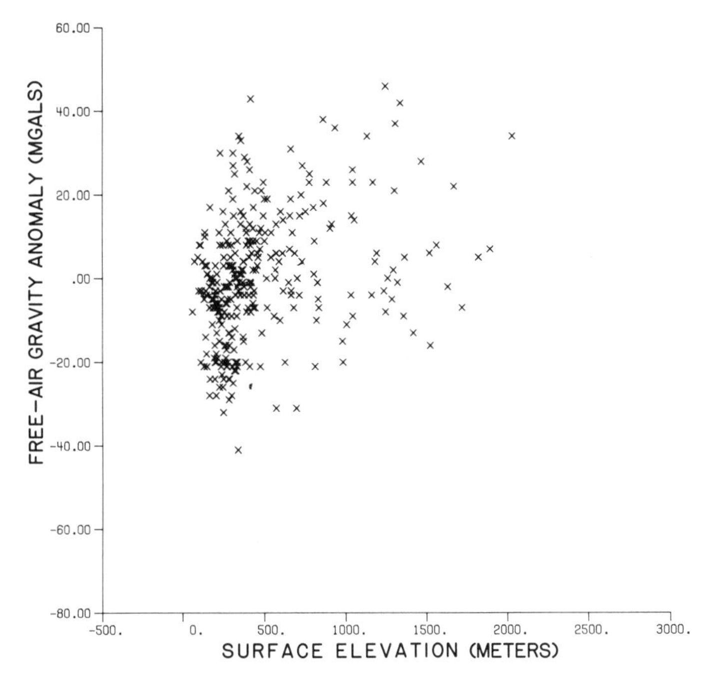

Figure 8. Scatter diagram of relationship between mean 1° (latitude/longitude), free-air gravity anomaly and surface elevation of the midcontinent.

lates with an increase in the average crustal velocity in the Mississippi Embayment. There is no obvious relationship between the P_n velocity variations and other displayed parameters. The free-air gravity anomalies oscillate around zero with maximum amplitudes observed over the exposed Precambrian Shield in Wisconsin at the northern end of the profile and near the Gulf of Mexico coast. The free-air and Bouguer gravity anomalies are both intensely negative over the Wisconsin Arch. The Bouguer gravity anomaly increases to the south from Wisconsin, reaching a maximum in the Gulf of Mexico. The gradients of the Bouguer gravity anomalies are generally inversely proportional to the depth to the basement. This relationship is even more evident on the near-surface magnetic-anomaly profile where the amplitudes of the anomalies vary inversely to the basement depth. A decrease in the gradients of the near-surface magnetic anomalies is especially evident over the Coastal Plain and the nearly featureless anomaly field over the Gulf.

The 37°N profile (Fig. 10) extends from the Atlantic Continental Shelf across the midcontinent to the southern Rocky Mountains. The crust thickens along the Atlantic Coast region to approximately 44 km and maintains this level to the thickened zone beneath the Great Plains in western Kansas and eastern Colorado. The average seismic velocity increases from the east to a maximum beneath the Mississippi Embayment in the vicinity of 90°W and then decreases to the west, mainly under the influence of the crustal effect associated with the Rio Grande Rift where the P_n velocity also decreases. The free-air gravity anomaly oscillates near zero, except in the western Great Plains and southern Rocky Mountains where it reaches a maximum along the profile. The long-wavelength components of the Bouguer gravity anomaly show a general inverse correlation with surface elevation, with the western minimum correlating with decreased crustal velocity and lowered upper-mantle (P_n) velocity. Intense local gravity and magnetic anomalies are observed over the Piedmont and Blue Ridge Provinces of the Appalachian orogen and also over the relatively undeformed, Paleozoic, sedimentary rocks in Kentucky. The latter anomalies are associated with dense, highly magnetic, basement rocks.

GEOPHYSICAL CHARACTERISTICS OF MAJOR GEOLOGICAL FEATURES

The previous review of the general characteristics of the United States craton, as interpreted from available geophysical data, provides us with an overview that is useful for setting the geologic and tectonic framework of the area. Numerous published studies of a more intensified nature back up these discussions with detailed interpretations. In an attempt to provide a more comprehensive review of the geophysical aspects of the craton, we have selected for more detailed discussion a few geological features that are of particular interest because of their importance to the geological community or their controversial nature.

Basins and Arches

During the Phanerozoic, the craton from the Front Range of the Rockies to the Appalachian Foldbelt has been dominated by slow, broad, vertical movements with related periods of either sedimentation or erosion. Sedimentation was concentrated in the midcontinent during the Paleozoic Era and in more or less restricted regions—intracratonic basins—separated by arches. The basin-arch relationship developed as a result of differential subsidence; the arches were rarely above sea level in Paleozoic time. The basins had a long history of development, but subsidence was not necessarily at a constant rate. Meanwhile, extensive sedimentation was taking place in foreland basins around the margins of the craton and in aulacogens. Subsequently, these foreland basins and some aulacogens were deformed by Appalachian, Ouachita, Sevier, and Laramide orogenic activity. Their origin and the history of these basins is considerably different from intracratonic basins such as the Illinois, Michigan, and Williston Basins.

The origin of the slow, long-duration, vertical movements that produced the basin-arch system of the midcontinent is one of the more perplexing problems of the geology of the craton. Several mechanisms have been proposed that are based on localized thermal events (e.g., Sleep and Snell, 1976; Haxby and others, 1976; McKenzie, 1978; and Jarvis and McKenzie, 1980). As pointed out by Lidiak (1982), although these theories account for several observed features of basins, the rocks of the basins show no evidence of requisite thermal events. For example, this has been pointed out by Sleep and Sloss (1980) in the case of the Michigan Basin. In an attempt to focus on a viable hypothesis for

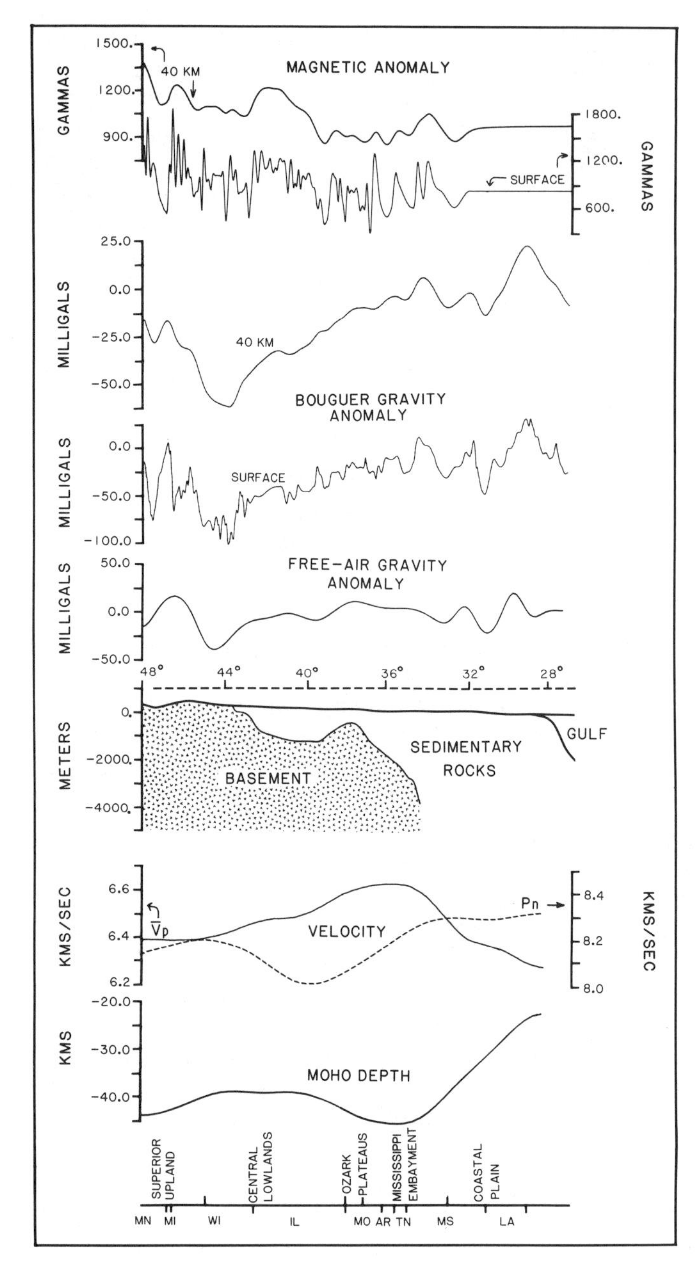

Figure 9. Geophysical profiles along 90°W longitude from the Canadian border to the Gulf of Mexico.

the origin of basins, increasing emphasis has been placed on characterizing the underlying crystalline basement rocks. These rocks are inadequately sampled by drilling; therefore, geophysical studies have played a particularly prominent role in their study. Lidiak (1982) has recently summarized this information.

A review of the gravity-anomaly data over several regions of the midcontinent has shown that positive anomalies predominate over basins. This has led Lyons (1959) to suggest that Paleozoic basins developed by isostatic sinking. Hinze (1963) postulated a similar relationship between the Michigan Basin and the linear gravity high that transects it, and McGinnis (1966, 1970) developed a theory for the midcontinent in which relatively dense material from the mantle intruded into the crust, causing isostatic sinking and basin development. The striking correlation between the depth of the basement and the gravity anomalies is evident in Figures 11A, B, and C, which show the plots of the mean basement elevation of one-degree (latitude/longitude) areas derived from the Basement Rock Map of the United States (Bayley and Muehlberger, 1968) against the mean free-air gravity anomalies of the Michigan, Williston, and Illinois Basins. Free-air gravity anomalies are used rather than Bouguer gravity anomalies to eliminate the elevation (isostatic effect) from the plots. There is no precise relationship, but it is clear that in these basinal areas the gravity anomalies increase as the basement depths in the basins increase. The rate is roughly 15 mGals/km. Correction for the sedimentary/basement rock contrast would, if anything, increase this rate, but the effect is unlikely to significantly alter the results because of the large volumes of high-density carbonate rocks that occur in these basins. In fact, Hinze and others (1978) have shown that the mean density of the Phanerozoic sedimentary rocks in the Michigan Basin is essentially equivalent to the mean density of the crust.

The striking inverse relationship observed over the Williston, Michigan, and Illinois Basins between gravity anomaly and basement elevation is far from universal. For example, the Salina Basin and the Forest City Basin, which lie west and east of the Midcontinent Geophysical Anomaly at roughly the 40°N Parallel, show no obvious correlation between basement elevation and gravity anomaly. Furthermore, linear uplifts occur in intracratonic basins. These structural features formed during the later stages of basin development, coincident or parallel to major gravity and magnetic anomalies whose sources occur within the basement rocks. The uplifts tend to form in basin centers and are not necessarily continuous through it. In the Michigan Basin, the Howell Anticline, which is correlated with the most intense region of the transecting geophysical anomaly, has over 300 m of relief. The LaSalle Anticlinal Belt of the Illinois Basin, which has a similar order of magnitude of relief, also extends along a linear positive-gravity anomaly. A similar north-trending feature, the Nesson Anticline, correlates with a positive geophysical anomaly in the Williston Basin. The occurrence of these anticlines or basement ridges, coincident with Precambrian structural features, suggests that they originate late in basin development due to compressional forces reactivating zones of weakness causing horst-type features to develop, associated primarily with large-mass imbalances in the crust.

Additional evidence of the lack of generality in the basement elevation versus gravity-anomaly relationship is presented in Figure 11D which plots the mean anomalies and basement elevation of one-degree (latitude/longitude) areas from the Front Range and the Canadian border to the foreland basins. When considering this entire region, no correlation between these parameters is observed.

A further complication arises because the previously cited discussions, which have related the basin-arch relation in the

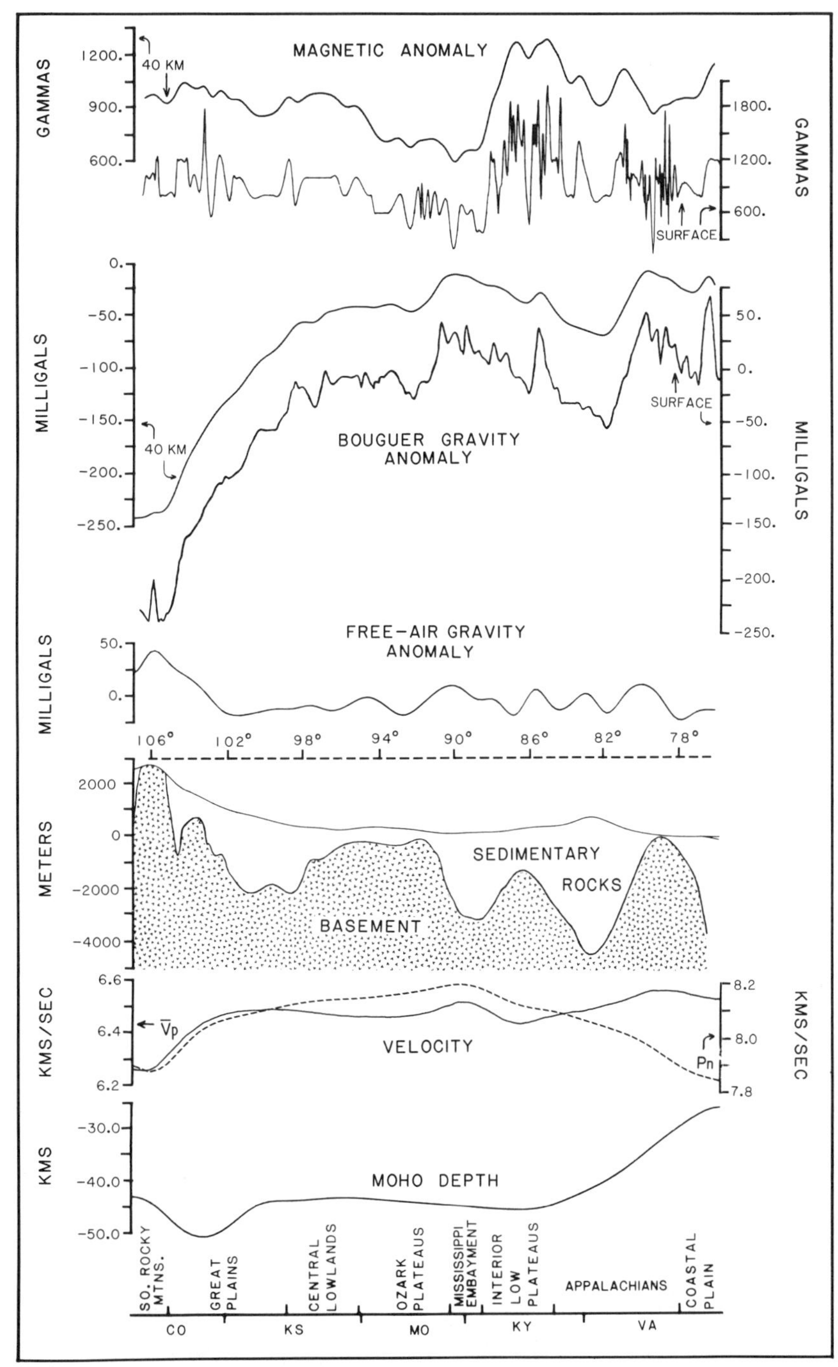

Figure 10. Geophysical profiles along 37°N latitude from the Rocky Mountains to the Atlantic Ocean.

midcontinent to isostatic sinking, have largely ignored the vast time span from the origin of the positive mass features in the Middle Proterozoic and the development of the basins over a period of several hundred m.y. in the Paleozoic Era (Hinze and others, 1972). The relaxation time for isostatic adjustment is short compared to the interval between the emplacement of the positive mass features and basin development. The relaxation time is the time required for isostatic deviation to be reduced to 1/e of its initial value. A reasonable relaxation time of the order of 10^5 years, based on McConnell's (1968) studies, would cause isostatic response to be essentially achieved in a few m.y. Furthermore, the cycles of sedimentation observed in basins, which cannot be ascribed to any combination of eustatic or sediment loading (Sloss, 1982), suggest that there has been periodic application of an alien stress field to the basins.

A solution to this dilemma has been proposed by DeRito

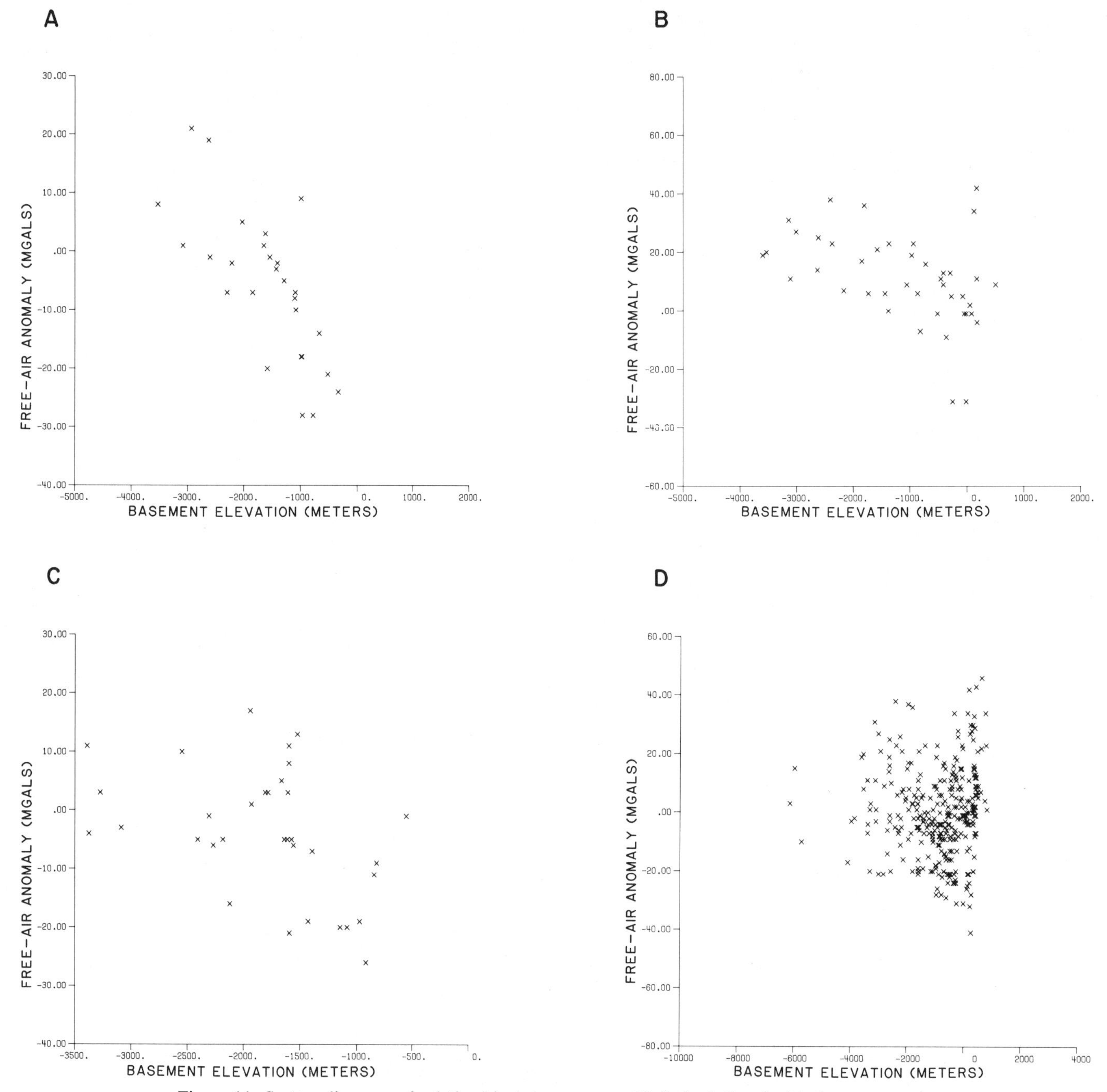

Figure 11. Scatter diagrams of relationship between mean 1° (latitude/longitude), basement rock elevation and free-air gravity anomaly. A—Michigan Basin, B—Williston Basin, C—Illinois Basin, D—midcontinent.

and others (1983). Their mechanical modeling of the lithosphere shows that regional, compressive, stress fields, too low in magnitude to cause lithospheric buckling, may cause renewed subsidence by their interaction with a pre-existing basin and by causing lithospheric thinning by stress-induced viscosity changes in the upper mantle. In this manner, areas of strong positive mass imbalance within the crust are periodically and synchronously downwarped during periods of tectonic reactivation associated with the application of regional stress fields derived from plate motions (e.g., ridge push, slab pull, collision, transform motion, etc.). The variations in the direction and magnitude of the regional stress field and perturbations in the strength of the lithospheric materials provide ample room to accommodate the vagaries between basins and in the movement of depocenters with time within a basin. This solution also does not require that the process start with an arrested continental rift—although many are related to rifts—but only a positive mass imbalance. It is important to observe in the 2°-long-wavelength, free-air gravity-anomaly map (Fig. 7) that major positive anomalies are located over the Michigan, Williston, and Illinois Basins, considered in Figure 11A, B, and C. Other positive anomalies are located over the Mississippi Embayment, the Southern Oklahoma Aulacogen, and the Delaware Basin. Subsidence in these regions may have also been caused by similar imbalances and stress history mechanisms. No specific, long-wavelength, free-air anomaly is observed directly over the Midcontinent Geophysical Anomaly; however, a broad positive anomaly occurs to the west and northwest of the axis of the anomaly, which undoubtedly incorporates the effect of this feature. The lack of a regional mass imbalance is believed to be due to compensation for the rift-related, high-density, volcanic and intrusive rocks by the thickening of the crust and the late-rift-stage, clastic, sedimentary rock basins. It is interesting to speculate that the lack of well-defined Phanerozoic basins over the Midcontinent Geophysical Anomaly from western Lake Superior to Kansas is related to the absence of a localized, specific mass imbalance over the feature. In contrast to the positive, free-air gravity anomalies over the Phanerozoic basins, the basement arches, such as the Wisconsin and Cincinnati Arches and the Nashville Dome in Ohio, Kentucky, and Tennessee, are located within negative, free-air gravity anomalies (Fig. 7).

The Mississippi Embayment

The Mississippi Embayment is a broad, spoon-shaped reentrant of Mesozoic and Cenozoic sedimentary rocks pointing into the Paleozoic terrain of the North American craton from the Gulf Coastal Plain with its axis roughly coincident with the Mississippi River (Fig. 1). Geological evidence shows that the embayment was initiated in late Cretaceous time as a broad downwarp that continued to receive sediments into the Tertiary. The thickness of the sediments along the axis of the embayment near its junction with the Ouachita tectonic belt is of the order of several kilometers. The sediments thin toward the edges and to the northeast along the axis to the margin of the embayment in southern Illinois. Presumably this feature is associated with the opening of the Gulf of Mexico in the Mesozoic Era and thus was described by Burke and Dewey (1973) as a Mesozoic failed rift. In itself, the embayment is an interesting feature, but it becomes a particularly critical feature because geophysical evidence based on numerous recent studies indicates a complex tectonic history from the Precambrian to the present.

Ervin and McGinnis (1975), utilizing crustal seismic studies, gravity data, and geologic information, proposed that the embayment is the site of an earlier rift, the late Precambrian Reelfoot Rift, which was reactivated in the Mesozoic to create the Mississippi Embayment. Subsequently, Hildenbrand and others (1977, 1982a,b) and Kane and others (1981) utilized gravity and magnetic data to better define the Reelfoot Rift and its associated central graben, which contains a thick section of Late Precambrian or Early Cambrian sedimentary rocks. In addition, they noted that the New Madrid Seismic Zone (Nuttli, 1979), the zone of most-intense, present-day, seismic activity west of the Cordillera, lies along the center of the buried Reelfoot Rift.

The sedimentary rocks of the buried graben have been mapped with seismic reflection studies (Zoback and others, 1980) and seismic surface waves (Austin and Keller, 1982); crustal seismic studies (Mooney and others, 1983) confirm the concept originally proposed by Ervin and McGinnis (1975) that the axis of the embayment is underlain with anomalously high-velocity/density lower crust. Braile and others (1982a,b) have utilized primarily gravity and magnetic data to extend the Precambrian rift northeast into southwestern Indiana (Fig. 12), where recent seismic reflection studies confirm the presence of thick, faulted, pre–Late Cambrian sedimentary rocks (Sexton and others, 1986). Furthermore, Braile and others (1982b) suggest that the rift breaks into a multi-arm feature at the head of the embayment, which they term the New Madrid Rift Complex. Segments of the complex extend not only northeasterly, but northwesterly along the Mississippi River to St. Louis, and easterly into the Rough Creek graben (Soderberg and Keller, 1981) (Fig. 12). Braile and others (1988) provide evidence that the New Madrid Rift Complex has had a profound influence on the depositional history of the Illinois Basin and the Mississippi Embayment and on the tectonism of this intraplate region since the Precambrian. They suggest that the cause of this intraplate activity is related to plate interaction along the eastern margin of the North American craton and plate-motion-related stresses in the North American Plate.

The identification of Reelfoot Rift as a potential crustal zone of weakness along which there is considerable seismic activity, plus focal mechanisms of earthquakes (Herrmann and Canas, 1978) and mapping of the regional stress pattern (Zoback and Zoback, 1981), has led to the "zone of weakness model" (Sbar and Sykes, 1973; Sykes, 1978) for the contemporary earthquake activity. According to this model (Braile and others, 1988), the seismicity of the New Madrid Seismic Zone is due to reactivation of ancient rift-originated faults within the crystalline crust by an appropriately oriented stress field. The seismicity of the area may be enhanced by focusing of the regional stress field by regional (crustal) and local basement inhomogeneities.

The Midcontinent Rift System

The interest of the geoscience community in ancient continental rifts continues to grow because these features provide significant information on the relative importance and mechanisms of plate-tectonic processes, particularly in Precambrian time. In addition, continental rifts are commonly related to ore and hydrocarbon deposits and are noted for their propensity to be reactivated by alien stress fields—even after hundreds of millions of years—and deep burial beneath sedimentary rocks. Increasing evidence indicates that a major continental rift, the Midcontinent Rift System, developed in the Middle Proterozoic (Keweenawan) approximately 1100 Ma over a vast area of central North America (Van Schmus and Hinze, 1985). Although the gravity anomaly associated with this feature, the Midcontinent Geophysical

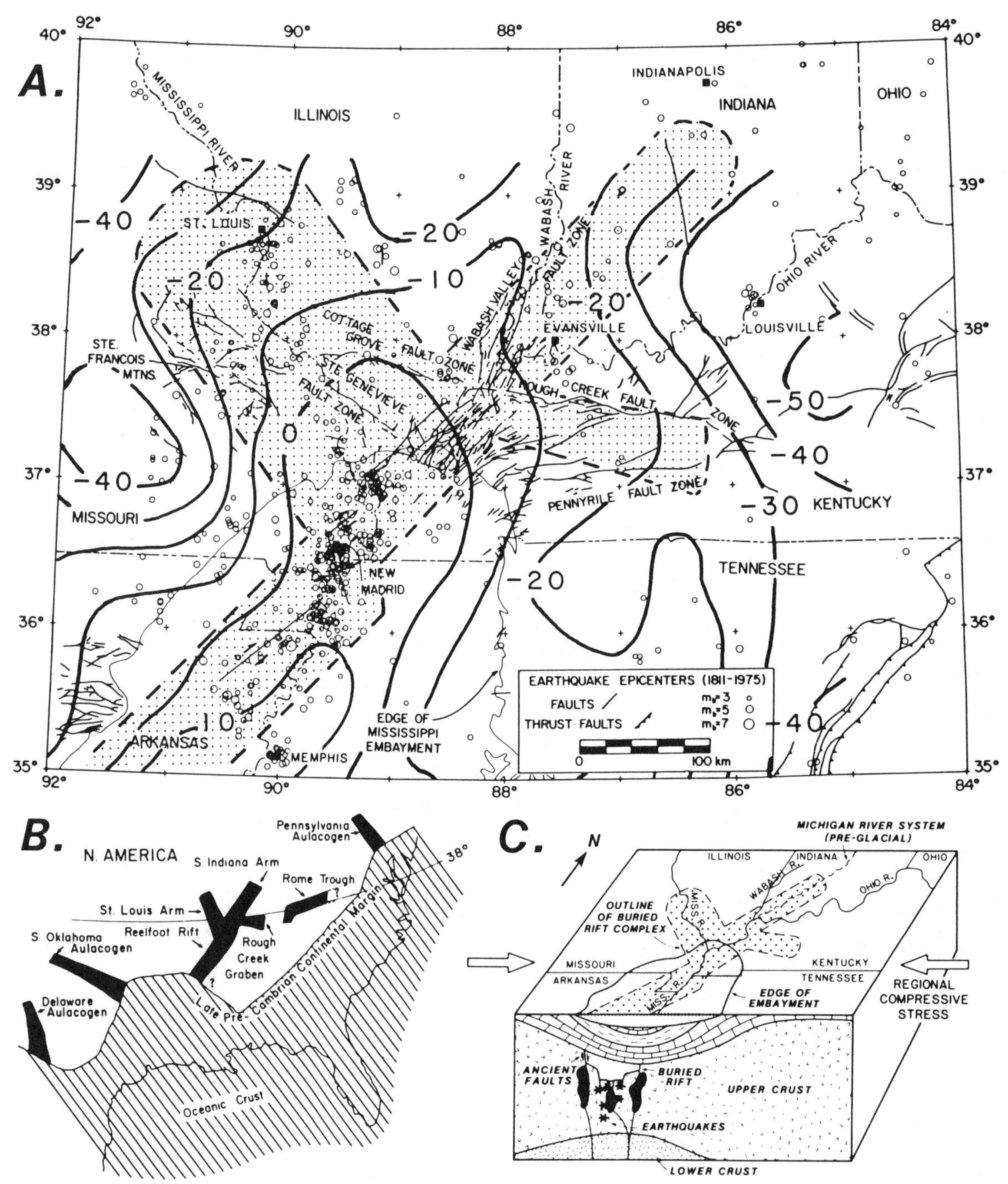

Figure 12. New Madrid Rift Complex (after Braile and others, 1988). A—Location of New Madrid Rift Complex (stippled area) with respect to regional Bouguer gravity-anomaly contours, earthquake epicenters (circles), and surface geologic structure. B—New Madrid Rift Complex and associated rifts and their relationship to the 38th Parallel Lineament. C—Schematic block diagram of New Madrid Rift Complex region.

Anomaly, is probably the most striking anomaly on the Bouguer gravity-anomaly map (Plates I-A and C), the source of the anomaly is observed only around the shores of Lake Superior and the area immediately to the southwest of the lake. Elsewhere, it is hidden from direct view by Phanerozoic sedimentary rocks of the craton. Thus, its location and geological characteristics are primarily known from interpretation of geophysical data and extrapolation from the outcrop area and a few basement drillholes (Halls, 1978). In the Lake Superior region where it crops out, the accumulated geological and geophysical data (Wold and Hinze,

1982) indicate that the rift originated with extensional forces. The event was initiated with extrusion of largely basaltic rocks into a series of lava basins along the rift axis. Subsequently, the subsiding basins were the loci of extensive clastic sedimentation and major longitudinal and transverse faulting. The Lake Superior Basin, which is made up of several independent lava basins (Green, 1982), follows the axis of the lake and in its deepest portion contains up to 15 km of tholeiitic basalts and olivine tholeiites overlain by largely red, clastic sedimentary rocks. Normal faults, which mark the margins of grabens in classic rift models, are not observed. The vast quantities of intruded and extruded mafic rock along a long segmented belt, which transects the regional pre-existing geologic pattern over most of its length, as observed in the anomaly maps (Plates I-A, B and C), is the principal evidence for the rift interpretation of the Lake Superior Basin and the Midcontinent Geophysical Anomaly.

The Great Plains

The Great Plains, an interesting tectonic feature of the North American craton, has received only modest attention. The elevation of the regional surface of the Great Plains increases from roughly 400 m near the eastern edge of the Dakotas, Nebraska, and Kansas at approximately 97°W longitude to the order of 1600 m at the eastern margin of the Rocky Mountains. Trimble (1980), utilizing geomorphic evidence, suggests that this uplift occurred after the end of the deposition of the Miocene Ogallala Formation, 5 to 6 m.y. ago. Recently, Gable and Hatton (1983) considered all available lines of evidence and concluded that the western Great Plains from Montana southward was uplifted from 1000 to 1500 m in the last 10 m.y. This rise in surface elevation is associated with a general uplift of the Cordillera during that period, which reaches up to 3000 m or more in the Basin and Range Province during the last 10 m.y.

One of the interesting aspects of the observed uplift is the broad region of the craton that is involved in the process—a distance of up to 1000 km from the Cordillera. Much of this region appears to be nearly in isostatic equilibrium because free-air gravity anomalies (Fig. 7) average roughly zero milligals except for the western portion of the Dakotas. Our previous discussion indicates that in general the Great Plains has a thickened, higher-than-average-velocity crust. Heat-flow data from the region are sparse, but Swanberg and Morgan (1981) based on their heat-flow map of the United States from silica geothermometry, point out that a major midcontinent heat-flow high (the "Ogallala High") extends north along the Great Plains from the panhandle of Texas to the Canadian border.

These relationships bring to mind several unanswered questions. Are the elevated heat flow and the abnormal crust related to the source of the increased surface elevation? If so, what is the process by which they originated? Can they be explained by a mantle thermal perturbation or crustal underplating? What is the next stage in the process and will it lead to greater surface inflation or deflation? These and many related questions will have to be answered before we can start to understand the geophysical character of the cratonic crust and the processes at work upon it.

The Colorado Plateau

The "craton", as defined by Sloss (1982), ". . . identifies the nuclear region of North America underlain by unmodified and relatively undeformed continental crust." Although the lithosphere has been affected, the crust of the Colorado Plateau has remained sufficiently unmodified during the Phanerozoic that we can consider it part of the craton. Thus, despite the fact that it is virtually surrounded by Cenozoic tectonism, we shall briefly discuss the geophysical characteristics of the Colorado Plateau.

The Colorado Plateau is one of the major tectonic and physiographic provinces of western North America. It is centered near the common corners of New Mexico, Colorado, Utah, and Arizona, is surrounded on three sides by the extensional provinces of the Basin and Range and Rio Grande Rift, and is bordered to the north and northeast by the middle and southern Rocky Mountains. It has remained relatively stable during the major, late Phanerozoic tectonic activity that has shaped western North America. Deformation of the plateau during Laramide and subsequent tectonism has been limited to gentle warping and marginal faulting. However, it has been epeirogenically uplifted roughly 2 km during latest Mesozoic and Cenozoic time.

Several recent studies have summarized the geophysical studies of the Colorado Plateau and adjacent regions (e.g., Keller and others, 1979; Thompson and Zoback, 1979; and McGetchin and others, 1980). Regional, free-air gravity anomalies indicate that the Colorado Plateau, unlike the adjacent Rocky Mountains, is in near isostatic equilibrium. The pattern of gravity and magnetic anomalies over the plateau has a prominent northeasterly trend, particularly in Arizona and southern Utah, which apparently mimics the structural trends of the basement crystalline rocks as well as Tertiary to recent volcanic rocks (Mabey and others, 1978). The area is a regional magnetic high (Hinze and Zietz, 1985), perhaps reflecting the relatively low heat flow, deep Curie point isotherm (Shuey and others, 1973), and the resulting thick magnetic crust.

Numerous crustal seismic investigations show anomalously thick crust, roughly 45 km, beneath the Colorado Plateau, thinning to roughly 30 km in the Basin and Range Province and 30 to 35 km beneath the Rio Grande Rift. Moho depths as shallow as 25 km are indicated under the generally anomalous western margin of the plateau. Seismic studies indicate that the crust of the plateau is typical of the craton (Keller and others, 1979), but the upper-mantle velocity is lower than normal for these areas. Direct measurements and silica geothermometry data indicate that the heat flow through the plateau is slightly higher than in the Great Plains, but considerably less than in the Basin and Range Province. Thermal and gravity modeling of the Colorado Plateau suggest that the thickness of the lithosphere is intermediate between the Basin and Range and Great Plains (Keller and others, 1979) at roughly 60 km, which is consistent with both deep seismic and electrical sounding.

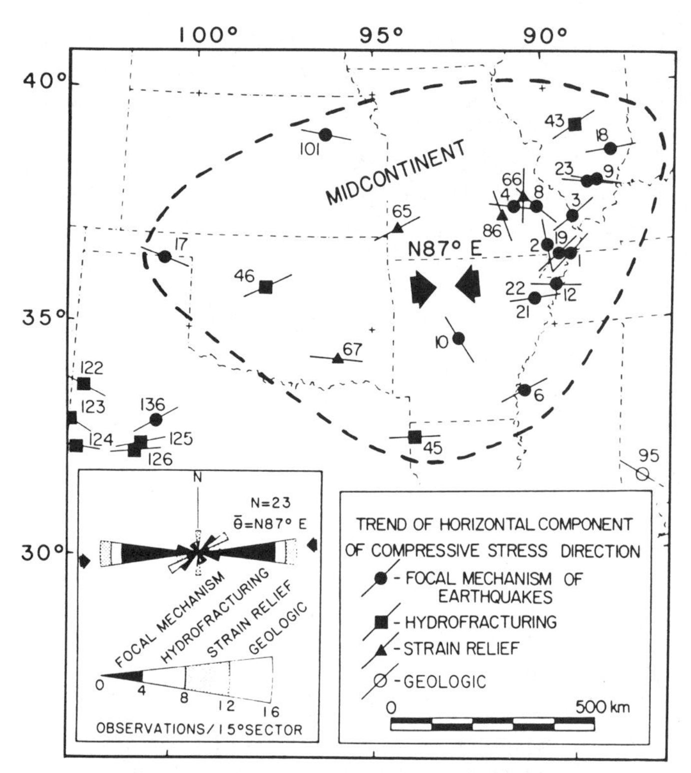

Figure 13. Trends of the horizontal component of compressive stress directions (dashed line encloses region of similar stress direction data for which mean value of compressive stress directions is shown by solid arrow) (after Harrison and others, 1983). Numbers correspond to table in original reference.

The outstanding problem of the Colorado Plateau is its origin. However, as pointed out by McGetchin and others (1980), the plateau is part of the broad regional Cenozoic uplift of the western United States. Thus, the problem is really related to the cause of the regional uplift and why the plateau has remained relatively intact during this process. McGetchin and others (1980) list numerous possible uplift mechanisms, but the available data do not contain the critical information that make it possible to provide a definitive solution to the problem.

GEOPHYSICAL EVIDENCE OF NEOTECTONISM

Introduction

Much of the previous discussion relates to the geologic history of the midcontinent. However, geophysical measurements provide input critical to evaluating the present-day tectonic processes—the neotectonism. The current low level of tectonic activity in the midcontinent has restricted the number and breadth of studies; as a result, our knowledge of the neotectonism of the midcontinent is limited. Only during the past decade has this situation begun to change as the possible profound effect of natural hazards, particularly earthquakes, on the siting of critical man-made structures and population centers has become a matter of concern.

Heat Flow

Heat-flow determinations in the midcontinent from the combination of rock thermal conductivity and temperature gradient are sparse (Combs and Simmons, 1973; Sass and others, 1976); only a few reduced heat-flow measurements (in which a correction has been made for the heat generation of the crystalline rocks) have been reported. However, Swanberg and Morgan (1981) have prepared a heat-flow map of the United States from a detailed distribution of water-geochemical determinations (roughly 200,000). Their method utilizes the silica/heat-flow interpretation technique based on the temperature dependence of quartz solubility and the empirical relation between these temperatures and observed heat flow. Their map is quite consistent with observed-heat-flow maps, but shows more detail in the midcontinent, where direct heat-flow observations are lacking. As discussed previously, their map shows an abnormal zone of high heat flow extending north from the Texas panhandle to the Canadian border. This feature could be related to deep crustal or mantle sources, but it may also be accounted for by hydrothermal circulation of water along appropriately oriented sedimentary rocks or fractures (e.g., Gosnold and German, 1982). Elsewhere in the eastern and central United States, normal heat flow (<60 mWm^{-2}) prevails, except for higher values in the Mississippi Embayment (Swanberg and Morgan, 1981).

The higher heat-flow values of the Mississippi Embayment are reinforced by a local thermal anomaly, which has been defined in the New Madrid Seismic Zone area by Swanberg and others (1982). The possibility of a slightly enhanced thermal gradient, together with the major faults of the area and the thick sedimentary section of the embayment, suggest that the localized higher heat flow may be associated with a hydrothermal system in which groundwater is heated during circulation deep within the crust (Hinze and others, 1984). The current seismic activity in the Mississippi Embayment could possibly be related to reactivation of the original Reelfoot Rift and, thus, the enhanced heat flow could be associated with heat derived from cooling intrusive magmatic bodies. However, this possibility is not compatible with the focal mechanisms observed in the New Madrid Seismic Zone (Herrmann and Canas, 1978), which indicate right-lateral fault displacements rather than extensional mechanism.

Regional Stress Pattern

Evaluation of the regional stress pattern of the craton has taken on special significance, with increasing attention given to earthquake-hazard prediction. Several studies (e.g., Sbar and Sykes, 1973; Sykes, 1978; and Hinze and others, 1980) indicate that ancient zones of weakness within the crust are likely candidates to localize seismicity, and those features that are appropriately oriented to the regional stress pattern (Zoback and Zoback, 1981; Braile and others, 1982b) are most likely to fail. Thus, definition of the regional stress field is an important factor in neotectonic investigations. Although little is known concerning the

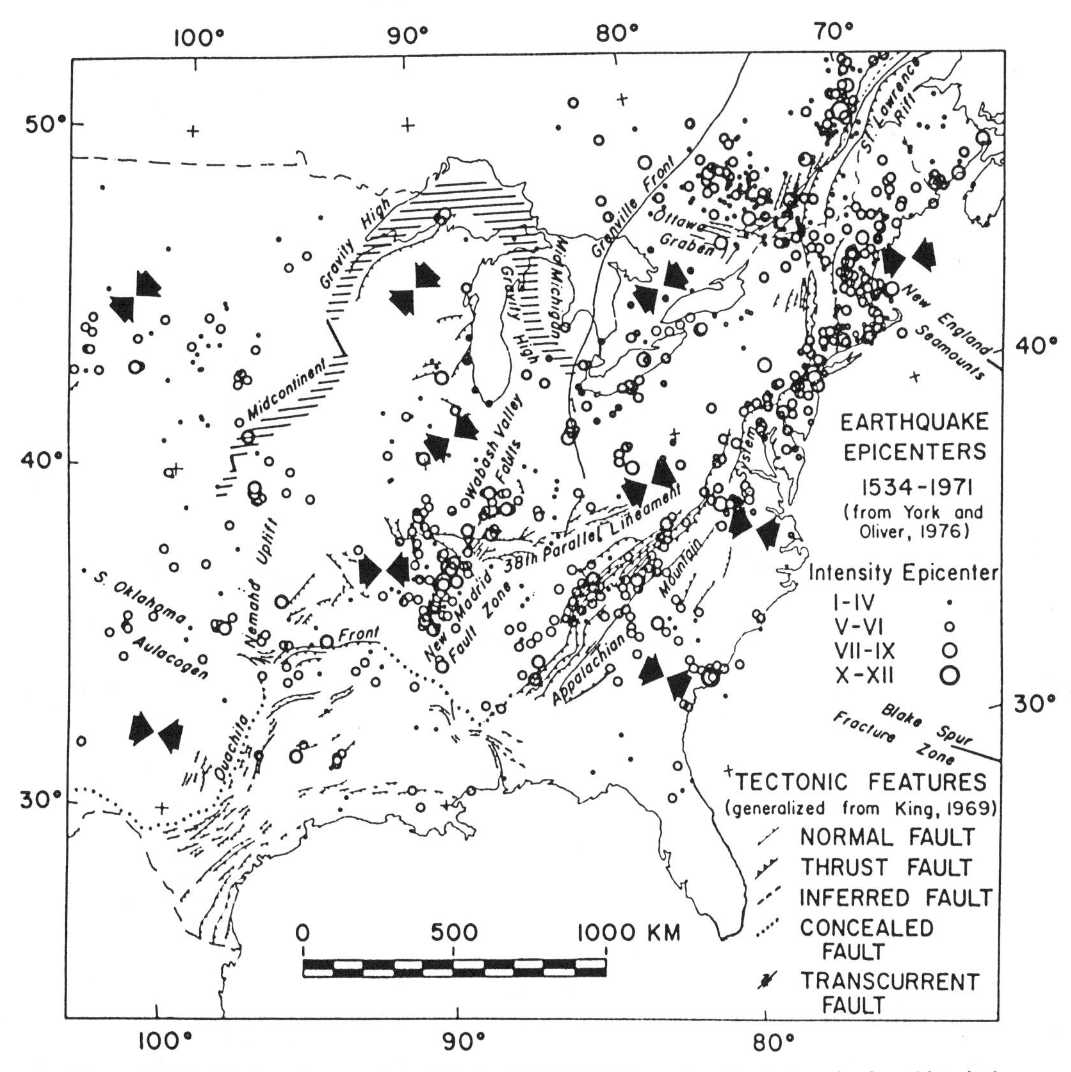

Figure 14. Distribution of reported earthquakes (1534–1971) in eastern North America from historical and instrumental data (from York and Oliver, 1976), major tectonic features (modified from King, 1969), and mean trend of the horizontal-component compressive stress directions (solid arrow).

magnitude of the stress or its relative change with depth, several recent studies (Sbar and Sykes, 1973; Zoback and Zoback, 1980) have compiled the trend of the maximum compressive stress directions. The results of a recent compilation of the trend of the horizontal component of compressive stress direction, as reported in the literature, is presented in Figure 13 (Harrison and others, 1983). The compilation includes results from focal mechanisms of earthquakes, hydrofracturing, strain-relief, and geologic indicators. Results from geologic indicators and surface strain-relief measurements are most susceptible to error. The solid arrows show the mean stress direction within major geographic areas (indicated by the closed, dashed lines). The means of the areas within the midcontinent—the western Great Lakes, the eastern Great Lakes, the southern Appalachians, and west Texas—range from N59°E to N82°W. These directions are roughly parallel to those predicted from plate-tectonic models (Zoback and Zoback, 1980) and reflect the absolute plate velocities as calculated from poles of rotation of plate motion following the model of Minster and Jordan (1978) or Richardson and others (1979).

Seismicity

Earthquake activity provides an important source of information on the recent tectonism of the midcontinent. However, the historical record is short (<200 years, except near major population centers along the eastern seaboard), and the observed recurrence intervals are long. Therefore, the mapped seismicity patterns must be considered as essentially indications of the "instantaneous" earthquake activity. Significant earthquakes, such as the Sharpsburg, Kentucky earthquake (m_b = 5.1) of 1980, continue to occur in regions of the midcontinent without a notable historical earthquake record. One of the recent earthquake-epicenter compilations of eastern North America (York and Oliver, 1976) is presented in Figure 14, together with the major tectonic features of the region and mean trends of the horizontal component of compressive directions. The pattern of seismicity is rather diffuse, with several zones of more intense activity (e.g., New Madrid, southern Appalachians, St. Lawrence River Valley). In the central midcontinent, the most intense earthquake activity is centered in the New Madrid Seismic Zone at the head of the Mississippi Embayment, and the historical record is dominated by the 1811–1812 earthquake sequence and associated aftershocks. The major shocks of the 1811–1812 sequence had magnitudes (m_b) of 7.2, 7.1, and 7.4 (Nuttli, 1973), major earthquakes by any standard of comparison. Magnitude-recurrence relations indicate a given-magnitude earthquake is many times less frequent than an equivalent earthquake in active seismic zones in the western United States, but the low absorption of seismic-wave energy in eastern North America greatly increases the affected area over that of an equivalent earthquake in the western United States (Nuttli, 1981). Earthquake focal depths are poorly known in the eastern United States. Herrmann (1979) has computed a range of 2 to 20 km in the central United States.

Perhaps the most striking aspect of Figure 14 is the poor correlation between the location of individual epicenters and mapped geologic features. There are many areas in which earthquake epicenters do not appear to be correlated with mapped geologic features and, on the contrary, mapped structures are present that do not appear to be seismically active. This observation is confirmed by review of more-detailed larger-scale maps. Although there is a crude correlation between zones of concentrated seismicity and mapped tectonic features (Braile and others, 1982c), the weak correlation is not surprising because the geologic structures are primarily pre-Cenozoic in age and, as such, are derived from tectonic processes and stresses completely different from those of the present. However, several studies have investigated the correlation of earthquake epicenters in eastern North America to pre-Cenozoic structures (e.g., Woollard, 1958). Kanasewich (1965) and Wesnousky and Scholz (1980) indicate that seismicity is concentrated at the edge of the central North American craton and at major basement province boundaries. Sbar and Sykes (1973) suggest that intraplate seismicity is associated with ancient zones of weakness that are reactivated within the present-day stress field caused by on-going plate motions. The zone-of-weakness hypothesis has been extended to account for eastern United States seismicity in numerous investigations (e.g., Sykes, 1978; Hamilton, 1981; and Braile and others, 1982c). It has been particularly useful as a seismo-tectonic model for the New Madrid Seismic Zone (Braile and others, 1982b and c), where considerable geologic, geophysical, stress, and seismicity data are available for an integrated analysis. These studies offer hope that geologic and geophysical investigations of ancient and thus often hidden zones of weakness, combined with analysis of contemporary tectonism obtained from stress and seismicity observations, will provide the clue to understanding the major seismicity of the North American craton.

REFERENCES

Austin, C. B., and Keller, G. R., 1982, A crustal structure study of the northern Mississippi Embayment, *in* McKeown, F. A., and Pakiser, L. C., eds., Investigations of the New Madrid, Missouri, earthquake region: U.S. Geological Survey Professional Paper 1236, p. 39–54.

Bayley, R. W., and Muehlberger, W. R., compilers, 1968, Basement rock map of the United States (exclusive of Alaska and Hawaii): Washington, D.C., U.S. Geological Survey, 2 sheets, scale 1:2,500,000.

Black, P. R., and Braile, L. W., 1982, P_n velocity and cooling of the continental lithosphere: Journal of Geophysical Research, v. 87, p. 10, 557–10, 568.

Braile, L. W., Hinze, W. J., Keller, G. R., and Lidiak, E. G., 1982a, The northeastern extension of the New Madrid seismic zone, *in* McKeown, F. A., and Pakiser, L. C., eds., Investigations of the New Madrid, Missouri, earthquake region: U.S. Geological Survey Professional Paper 1236, p. 175–184.

Braile, L. W., Keller, G. R., Hinze, W. J. and Lidiak, E. G., 1982b, An ancient rift complex and its relation to contemporary seismicity in the New Madrid seismic zone: Tectonics, v. 1, p. 225–237.

—— 1982c, Seismicity and tectonics of the midcontinent United States, Proceedings Third International Earthquake Microzonation Conference: Washington, D.C., National Science Foundation, p. 25–38.

Braile, L. W., Hinze, W. J., Sexton, J. L., Keller, G. R., and Lidiak, E. G., 1986, Tectonic development of the New Madrid seismic zone: Tectonophysics, v. 131, p. 1–21.

Braile, L. W., Hinze, W. J., von Frese, R.R.B., and Keller, G. R., 1988, Seismic properties of the crust and upper-most mantle of North America, *in* Pakiser, L. C., and Mooney, W., eds., Geophysical framework of the continental United States: Geological Society of America Memoir 172 (in press).

Burke, K., and Dewey, J. F., 1973, Plume-generated triple junctions; Key indicators in applying plate tectonics to old rocks: Journal of Geology, v. 81, p. 406–433.

Cebull, S. E., Shurbet, D. H., and Keller, G. R., 1976, Possible role of transform faults in the development of apparent offsets in the Ouichita–southern Appalachian tectonic belt: Journal of Geology, v. 84, p. 107–114.

Charpayungpun, W., and Landisman, M., 1977, Crust and upper mantle near the Great Plains: American Geophysical Union Geophysics Monograph 20, p. 553–575.

Clark, R. H., and Persoage, N. P., 1970, Some implications of crustal movement in engineering planning: Canadian Journal of Earth Sciences, v. 7, p. 628–633.

Combs, J., and Simmons, G., 1973, Terrestrial heat flow determinations in the north central United States: Journal of Geophysical Research, v. 78, p. 441–460.

Denison, R. E., Lidiak, E. G., Bickford, M. E., and Kisvarsanyi, E. B., 1984, Geology and geochronology of Precambrian rocks in the central interior region of the United States, *in* Harrison, J. E., and Z. E. Peterman, Z. E., eds., Correlation of Precambrian rocks of the United States and Mexico: U.S. Geological Survey Professional Paper 1241-C, p. C-1–C-20.

DeRito, R. F., Cozzarelli, F. A., and Hodge, D. S., 1983, Mechanism of subsidence of ancient cratonic rift basins: Tectonophysics, v. 94, p. 141–168.

Ervin, C. P., and McGinnis, L. D., 1975, Reelfoot Rift; Reactivated precursor to the Mississippi Embayment: Geological Society of America Bulletin, v. 86, p. 1287–1295.

Gable, D. J., and Hatton, T., 1983, Maps of vertical crustal movements in the United States over the last 10 million years: U.S. Geological Survey Map I-1315, 2 sheets, scale 1:5,000,000 and 1:10,000,000, 25 p.

Gosnold, W. D., and German, K. E., 1982, Geothermal investigations in Nebraska, *in* Hot Dry Rock Geothermal Exploration Workshop, June

21-23, 1982, Los Alamos National Lab, Los Alamos, New Mexico.

Green, A. G., Cumming, G. L., and Cedarwell, D., 1979, Extension of the Superior–Churchill boundary zone into southern Canada: Canadian Journal of Earth Sciences, v. 16, p. 1691–1701.

Green, J. C., 1982, Geology of Keweenawan extrusive rocks, *in* Wold, R. J., and Hinze, W. J., eds., Geology and tectonics of the Lake Superior Basin: Geological Society of America Memoir 156, p. 47–56.

Halls, H. C., 1978, The late Precambrian central North American rift system; A survey of recent geological and geophysical investigations, *in* Newmann, E. R., and Ramberg, I., eds., Tectonics and geophysics of continental rifts, NATO Advanced Study Institute, Series C, 37D: Boston, Reidel, p. 111–123.

Halls, H. C., 1982, Crustal thickness in the Lake Superior region, *in* Wold, R. J., and Hinze, W. J., eds., Geology and tectonics of the Lake Superior Basin: Geological Society of America Memoir 156, p. 239–244.

Hamilton, R. M., 1981, Geologic origin of eastern U.S. seismicity, *in* Beavers, J. E., ed., Earthquakes and earthquake engineering; Eastern United States, v. 1: Ann Arbor, Michigan, Ann Arbor Science, p. 3–24.

Harrison, W., and 9 others, 1983, Geology, hydrology, and mineral resources of crystalline rock areas of the Lake Superior region, United States: Argonne National Laboratory Report ANLIES-134, p. 1, 384 p.

Haxby, W. F., Turcotte, D. L., and Bird, J. M., 1976, Thermal and mechanical evolution of the Michigan basin: Tectonophysics, v. 36, p. 57–75.

Herrmann, R. B., 1979, Surface wave focal mechanisms for eastern North American earthquakes with tectonic implications: Journal of Geophysical Research, v. 84, p. 3543–3552.

Herrmann, R. B., and Canas, J. A., 1978, Focal mechanism studies in the New Madrid seismic zone: Seismological Society of America Bulletin, v. 68, p. 1095–1102.

Heyl, A. V., 1972, The 38th Parallel Lineament and its relationship to ore deposits: Economic Geology, v. 67, p. 879–894.

Hildenbrand, T. G., Kane, M. F., and Stauder, W., 1977, Magnetic and gravity anomalies in the northern Mississippi Embayment and their spatial relation to seismicity: U.S. Geological Survey Map MF-914.

Hildenbrand, T. G., Simpson, R. W., Godson, R. H., and Kane, M. F., 1982a, Digital colored residual and regional Bouguer gravity maps of the conterminous United States; Comments on significant features: U.S. Geological Survey Open-File Report 82–284.

Hildenbrand, T. G., Kane, M. F., and Hendricks, J. D., 1982b, Magnetic basement in the upper Mississippi Embayment region; A preliminary report, *in* McKeown, F. A., and Pakiser, L. C., eds., Investigations of the New Madrid, Missouri, earthquake region, *U.S. Geological Survey Professional Paper 1236,* 39–54.

Hinze, W. J., 1963, Regional gravity and magnetic anomaly maps of the southern peninsula of Michigan: Michigan Geological Survey Report Investigation 1, 26 p.

Hinze, W. J., and Merritt, D. W., 1969, Basement rocks of the southern peninsula of Michigan, *in* Studies of the Precambrian of the Michigan Basin: Michigan Basin Geological Society, Field Excursion Guidebook, p. 28–59.

Hinze, W. J., and Zietz, I., 1985, The composite magnetic anomaly map of the conterminous United States, *in* Hinze, W. J., ed., The utility of regional gravity and magnetic anomalies: Society of Exploration Geophysicists, p. 1–24.

Hinze, W. J., Roy, R. F., and Davidson, D. M., 1972, The origin of late Precambrian rifts: Geological Society of America Abstracts with Programs, v. 4, no. 7, p. 723.

Hinze, W. J., Kellogg, R. L., and O'Hara, N. W., 1975, Geophysical studies of basement geology of southern peninsula of Michigan: American Association of Petroleum Geologists Bulletin, v. 59, p. 1562–1584.

Hinze, W. J., Bradley, J. W., and Brown, A. R., 1978, Gravimeter survey in the Michigan Basin deep borehole: Journal of Geophysical Research, v. 83, p. 5864–5868.

Hinze, W. J., Braile, L. W., Keller, G. R., and Lidiak, E. G., 1980, Models for midcontinent tectonism, *in* Continental tectonics: Washington, D.C., National Academy of Science, p. 73–83.

Hinze, W. J., and 9 others, 1984, Exploration for hot dry rock geothermal resources in the midcontinent, U.S.A.: Purdue University Final Report of Subcontract No. 9-X60-2133K-1 (Los Alamos National Lab.), v. 2, 79 p.

Hoffman, D., Dewey, J. F., and Burke, K.A.C., 1974, Aulacogens and their genetic relation to geosynclines with a Proterozoic example from Great Slave Lake, Canada, *in* Dott, R. H., Jr., and Shaver, R. H., Modern and ancient geosynclinal sedimentation: Society of Economic Paleontologists and Mineralogists Special Publication 19, p. 38–55.

Hutchinson, D. R., Grow, J. A., and Klitgord, K. D., 1983, Crustal structure beneath the southern Appalachians; Nonuniqueness of gravity modeling: Geology, v. 11, p. 611–615.

James, D. W., Smith, T. J., and Steinhart, J. S., 1968, Crustal structure of the middle Atlantic states: Journal of Geophysical Research, v. 73, p. 1983–2007.

Jarvis, G. T., and McKenzie, D., 1980, Sedimentary basin formation with finite extension rates: Earth and Planetary Science Letters, v. 48, p. 42–52.

Kanasewich, E. R., 1965, Seismicity and other properties of geological provinces: Nature, v. 208, p. 1275–1278.

Keller, G. R., and Shurbet, D. H., 1975, Crustal structure of the Texas Gulf Coast: Geological Society of America Bulletin, v. 86, p. 807–810.

Keller, G. R., Braile, L. W., and Morgan, P., 1979, Crustal structure, geophysical models and contemporary tectonism of the Colorado Plateau: Tectonophysics, v. 61, p. 131–147.

Keller, G. R., Bland, A. E., and Greenberg, J. K., 1982, Evidence for a major late Precambrian tectonic event (rifting?) in the eastern midcontinent region, U.S.A.: Tectonics, v. 1, p. 213–223.

Keller, G. R., Lidiak, E. G., Hinze, W. J., and Braile, L. W., 1983, The role of rifting in the tectonic development of the midcontinent, U.S.A.: Tectonophysics, v. 94, p. 391–412.

King, E. R., and Zietz, I., 1978, The New York–Alabama Lineament; Geophysical evidence for a major crustal break in the basement beneath the Appalachian basin: Geology, v. 6, p. 312–318.

King, P. B., compiler, 1969, Tectonic map of North America: U.S. Geological Survey, 2 sheets, scale 1:5,000,000.

King, P. B., 1975, Ancient southern margin of North America: Geology, v. 3, p. 732–734.

Kruger, J. N., 1983, Regional gravity anomalies in the Ouachita System and adjacent areas, [M.S. thesis]: El Paso, University of Texas, 196 p.

Kurita, T., 1976, Crustal and upper mantle structure in the central United States of America from body-wave spectra, surface wave dispersion, travel-time residuals, and synthetic seismograms: Physics of the Earth and Planetary Interiors, v. 12, p. 65–86.

Lidiak, E. G., 1982, Basement rocks of the main interior basins of the midcontinent: University of Missouri, Rolla Journal, v. 3, p. 5–24.

Lidiak, E. G., Hinze, W. J., Keller, G. R., Reed, J. E., Braile, L. W., and Johnson, R. W., 1985, Geologic significance of regional gravity and magnetic anomalies in the east-central midcontinent, *in* Hinze, W. J., ed., The utility of regional gravity and magnetic anomalies: Society of Exploration Geophysicists, p. 287–307.

Lyons, P. L., 1959, The Greenleaf anomaly; A significant gravity feature: Kansas Geological Survey Bulletin, v. 137, p. 105–120.

Mabey, D. R., Zietz, I., Eaton, G. P., and Kleinkopf, M. D., 1978, Regional magnetic patterns in part of the Cordillera in the western United States, *in* Smith, R. B., and Eaton, G. P., eds., Cenozoic tectonics and regional geophysics of the western Cordillera: Geological Society of America Memoir 152, p. 93–106.

McConnell, R. K., Jr., 1968, Viscosity of the mantle from relaxation time spectra of isostatic adjustment: Journal of Geophysical Research, v. 73, p. 7089–7105.

McGetchin, T. R., Burke, K. C., Thompson, G. A., and Young, R. A., 1980, Mode and mechanisms of plateau uplifts, *in* Bally, A. W., Bender, P. L., McGetchin, T. R., and Walcott, R. I., eds., Dynamics of plate interiors: Washington, D.C., American Geophysical Union, Geodynamics, Ser. 1, p. 99–110.

McGinnis, L. D., 1966, Crustal tectonics and Precambrian basement in northeast-

ern Illinois: Illinois State Geological Survey Report Investigation 219, 29 p.
——1970, Tectonics and gravity field in the continental interior: Journal of Geophysical Research, v. 75, p. 317–331.
McKenzie, D., 1978, Some remarks on the development of sedimentary basins: Earth and Planetary Science Letters, v. 40, p. 25–32.
McMechan, G. A., 1979, An amplitude constrained P-wave velocity profile for the upper mantle beneath the eastern United States: Seismological Society of America Bulletin, v. 69, p. 1733–1744.
Mereu, R. F., and Jobidon, G., 1971, A seismic investigation of the crust and Moho on a line perpendicular to the Grenville front: Canadian Journal of Earth Sciences, v. 8, p. 1553–1583.
Minster, J. B., and Jordan, T. H., 1978, Present-day plate motions: Journal of Geophysical Research, v. 83, p. 5331–5354.
Mooney, W., Andrews, M., Ginzburg, A., Peters, D., and Hamilton, R., 1983, Crustal structure of the northern Mississippi Embayment and a comparison with other continental rift zones: Tectonophysics, v. 94, p. 327–348.
Nuttli, O. W., 1973, The Mississippi Valley earthquakes of 1811 and 1812; Intensities, ground motion, and magnitudes: Seismological Society of America Bulletin, v. 63, p. 227–248.
——1979, Seismicity in the central United States: Geological Society of America Reviews in Engineering Geology, v. 4, p. 67–93.
——1981, Similarities and differences between western and eastern United States earthquakes and their consequences for earthquake engineering, *in* Beavers, J. E., ed., Earthquakes and earthquake engineering; Eastern United States: Ann Arbor, Michigan, Ann Arbor Science, v. 1, p. 25–52.
Pakiser, L. C., and Robinson, R., 1966, Composition of the continental crust as estimated from seismic observations: American Geophysical Union Geophysical Monograph 10, p. 620–626.
Richardson, R. M., Solomon, S. L., and Sleep, N. H., 1979, Tectonic stress in plates: Reviews of Geophysics and Space Physics, v. 17, p. 981–1020.
Sass, J. H., and 6 others, 1976, A new heat-flow contour map of the conterminous United States: U.S. Geological Survey Open-File Report. 76-756, 24 p.
Sbar, M. L., and Sykes, L. R., 1973, Contemporary compressive stress and seismicity in eastern North America; An example of intraplate tectonics: Geological Society of America Bulletin, v. 84, p. 1861–1882.
Sexton, J. L., Braile, L. W., Hinze, W. J., and Campbell, M. J., 1986, Seismic reflection profiling studies of a buried Precambrian rift beneath the Wabash Valley Fault Zone: Geophysics, v. 51, p. 640–660.
Shuey, R. T., Schellinger, D. K., Johnson, E. H., and Alley, L. B., 1973, Aeromagnetics and the transition between the Colorado Plateau and the Basin and Range Provinces: Geology, v. 1, p. 107–110.
Sleep, N. H., and Sloss, L. L., 1980, The Michigan Basin, *in* Bally, A. W., Bender, P. L., McGetchin, T. R., and Walcott, P. I., eds., Dynamics of plate interiors: Washington, D.C., American Geophysical Union, Geodynamics, Ser. 1, p. 93–98.
Sleep, N. H., and Snell, N. S., 1976, Thermal contraction and flexure of midcontinent and Atlantic marginal basins: Geophysical Journal of the Royal Astronomical Society, v. 45, p. 125–154.
Sloss, L. L., 1982, The Midcontinent Province; United States, *in* A. R. Palmer, A. R., ed., Perspectives in regional geological synthesis: Geological Society of America, DNAG Special Publication 1, p. 27–39.
Smith, T. J., Steinhart, J. S., and Aldrich, L. T., 1966, Crustal structure under Lake Superior: American Geophysical Union Geophysical Monograph 10, p. 181–197.
Smithson, S. B., Johnson, R. A., and Wong, Y., 1981, Mean crustal velocity; A critical parameter for interpreting crustal structure and crustal growth: Earth and Planetary Science Letters, v. 53, p. 323–332.
Society of Exploration Geophysicists, 1982, Gravity anomaly map of the United States (exclusive of Alaska and Hawaii): Tulsa, Oklahoma, Society of Exploration Geophysicists, 2 sheets, scale 1:2,500,000.
Soderberg, R. K., and Keller, G. R., 1981, Geophysical evidence for deep basin in western Kentucky: American Association of Petroleum Geologists Bulletin, v. 65, p. 226–334.
Starich, P. J., Hinze, W. J., and Braile, L. W., 1984, The south-central United States magnetic anomaly: EOS American Geophysical Union Transactions, v. 65, p. 202.
Swanberg, C. A., and Morgan, P., 1981, Heat flow map of the United States based on silica geothermometry, *in* Sass, J. H., and 6 others, Heat flow from the crust of the United States, *in* Touloukian, Y. S., Judd, W. R., and Roy, R. F., eds., Physical properties of rocks and minerals; McGraw-Hill/CINDAS data series on material properties, v. II-2, p. 540–544.
Swanberg, C. A., Mitchell, B. J., Lohse, R. L., and Blackwell, D. D., 1982, Heat flow in the upper Mississippi Embayment: U.S. Geological Survey Professional Paper 1236, p. 185–189.
Sykes, L. R., 1978, Intraplate seismicity, reactivation of pre-existing zones of weakness, alkaline magmatism, and other tectonism postdating continental fragmentation: Reviews of Geophysics and Space Physics, v. 16, p. 621–688.
Thomas, M. D., 1983, Tectonic significance of paired gravity anomalies in the southern and central Appalachians, *in* Hatcher, R. D., Jr., Williams, H., and Zietz, I., eds., Contributions to the tectonics and geophysics of mountain chains: Geological Society of America Memoir 158, p. 113–124.
Thomas, W. A., 1977, Evolution of Appalachian–Ouachita salients, and recesses from reentrants and promontories in the continental margin: American Journal of Science, v. 277, p. 1233–1278.
Thompson, G. A., and Zoback, M. L., 1979, Regional geophysics of the Colorado Plateau: Tectonophysics, v. 61, p. 149–181.
Trimble, D. E., 1980, Cenozoic tectonic history of the Great Plains contrasted with that of the southern Rocky Mountains; A synthesis: The Mountain Geologist, v. 17, p. 59–69.
U.S. Geological Survey, 1982, Composite magnetic anomaly map of the United States, Part A; Conterminous United States: U.S. Geological Survey Map GP 954A, 2 sheets, scale 1:2,500,000.
Van Schmus, W. R., and Hinze, W. J., 1985, The Midcontinent Rift System: Annual Review of Earth and Planetary Sciences, v. 13, p. 345–383.
Wesnousky, S. G., and Scholz, C. H., 1980, The craton; Its effect on the distribution of seismicity and stress in North America: Earth and Planetary Science Letters, v. 48, p. 348–355.
Wold, R. J., and Hinze, W. J., eds., 1982, Geology and tectonics of the Lake Superior Basin: Geological Society of America Memoir 156, 280 p.
Woollard, G. P., 1958, Areas of tectonic activity in the United States as indicated by earthquake epicenters: EOS American Geophysical Union Transactions, v. 39, p. 1135–1140.
Yarger, H. L., 1981, Aeromagnetic survey of Kansas: EOS American Geophysical Union Transactions, v. 62, p. 173–178.
York, E., and Oliver, J. E., 1976, Cretaceous and Cenozoic faulting in eastern North America: Geological Society of America Bulletin, v. 87, p. 1105–1114.
Zoback, M. D., Hamilton, R. M., Crone, A. J., Russ, D. P., McKeown, F. A. and Brockman, S. R., 1980, Recurrent intraplate tectonism in the New Madrid seismic zone: Science, v. 209, p. 971–976.
Zoback, M. L., and Zoback, M. D., 1980, State of stress in the conterminous United States: Journal of Geophysical Research, v. 85, p. 6113–6156.
Zoback, M. D., and Zoback, M. L., 1981, State of stress and intraplate earthquakes in the United States: Science, v. 213, p. 96–104.

MANUSCRIPT ACCEPTED BY THE SOCIETY SEPTEMBER 15, 1986

ACKNOWLEDGMENTS

Preparation of this paper was in part supported by NASA Grants NAS5-25030 and NCC5-21, NRC Grant NRC-04-81-195-01, and a grant from the Chevron Oilfield Research Corporation. The authors were assisted in the preparation of the illustrations by Dhananjay Ravat, David Taylor, and Mark Brumbaugh. We acknowledge the cooperation of the Society of Exploration Geophysicists and the U.S. Geological Survey for providing color separates for the preparation of the colored gravity and magnetic anomalies shown in Plates 1-A, B, and C.

Printed in U.S.A.

The Geology of North America
Vol. D-2, Sedimentary Cover—North American Craton: U.S.
The Geological Society of America, 1988

Chapter 3

Tectonic evolution of the craton in Phanerozoic time

L. L. Sloss
Department of Geological Sciences, Northwestern University, Evanston, Illinois 60201

INTRODUCTION

It is the intent of this chapter to summarize the tectonic history of the North American Craton south of the Canadian border and north of the Mojave-Sonora megashear (see Anderson and Schmidt, 1983) of northern Mexico from the latest Proterozoic to the present. The tectonic evolution of cratons is most readily deciphered by analysis of the preserved cover rocks, their distributions and thickness, their petrologic characteristics and fossils, and the regional and interregional unconformities that punctuate the stratigraphic record.

The pattern of preserved distribution of cover strata of a particular age is a measure of the minimum area below depositional base level during the time in question. Such patterns, especially where marine and intertidal deposits are involved, and the extremes of distributions represented by erosional outliers, isolated fault blocks, and xenoliths in diatremes, are important to studies of ancient sea levels. The form and pattern of preserved thicknesses of strata are of greater utility in tectonic analysis than geographic distribution alone. The great majority of preserved cratonic cover rocks were deposited at or near depositional base level and thus within a few tens of meters of sea level. Dewatering and compaction are relatively rapid processes at cratonic time scales as continuing sedimentation imposes loads in excess of hydrostatic pressure; therefore, the thickness of strata representing a significant span of geologic time in the absence of evidence of monotonic deepening or shallowing is a meaningful approximation of the degree of subsidence of the depositional site relative to base level (and, by extension, relative to sea level). Inasmuch as stratigraphic thicknesses per unit of time vary widely from place to place, such differences cannot be explained by changes in sea level alone but must be attributed to differential *tectonic* subsidence. The history and geography of differential subsidence (complemented by differential uplift) are the dominant themes of cratonic evolution and of the interpretations embodied in this chapter.

Interpretations of cratonic evolution are further refined by consideration of depositional environments and provenance identity derived from the petrologic and paleobiologic content of accumulated sediment and from the deduced distributary paths of detrital materials. These essential elements of paleogeographic and tectonic reconstruction demand a level of detail inappropriate to this cratonwide review and are not treated exhaustively here; readers will find supporting particulars in the regional chapters of this volume.

Similarly, regional and subregional unconformities, especially (but not exclusively) those that are the consequences of plate interactions at or adjacent to cratonic margins, are not within the province of this chapter, although such stratal discontinuities are commonly of first-order importance in regional studies. Rather, emphasis is placed here on truly interregional surfaces marking interruptions in the continuity of sedimentation across the entire craton. These are the "sequence-bounding unconformities" identified some decades ago (e.g., Sloss, 1963) to which are added intervening significant interregional unconformities deriving from seismic stratigraphy and supporting studies by P. R. Vail and colleagues (e.g., Vail and others, 1977; Haq and others, 1987).

CHRONOLOGIC SUBDIVISIONS FOR ANALYSIS OF CRATONIC EVOLUTION

The classical periods of Phanerozoic time (Cambrian, etc.) and the chronostratigraphic systems representing them are poorly suited to investigation of the tectonic history of the craton. Too often depositional and tectonic phases pass without change or interruption from one period to the next (as at the Cambrian–Ordovician boundary, for example), or major interregional unconformities and shifts in tectonic mode occur within system/period limits (as in the Devonian). Interregional unconformities, whether created by tectonism or eustasy, provide identifiable discontinuities in the stratal record that make possible the recognition of rational subdivisions of the Phanerozoic history of cratons. The surfaces of unconformity are rarely isochronous in cratonic stratigraphy since such surfaces typically involve truncation and loss of record in the subjacent units and time-consuming progressive onlap of overlying strata. It is important to note, however, that all of the rocks below an interregional cratonic unconformity are everywhere older than the point in time marked by the minimum chronostratigraphic lacuna commonly marked at craton

Sloss, L. L., 1988, Tectonic evolution of the craton in Phanerozoic time, *in* Sloss, L. L., ed., Sedimentary Cover—North American Craton; U.S.: Boulder, Colorado, Geological Society of America, The Geology of North America, v. D-2.

TABLE 1. CORRELATION CHART SHOWING THE RELATION OF SEQUENCES AND SUBSEQUENCES TO THE STANDARD CHRONOSTRATIGRAPHIC SCHEME FOR THE PHANEROZOIC AND LATEST PROTEROZOIC
(Adapted from Palmer, 1983)

System	Series	Stage	End (Ma)	Subsequence (Ma)	Subsequence
Quaternary		Pleistocene	0		
Tertiary	Neogene	Pliocene	2		Tejas III
		Miocene	5.1		
	Paleogene	Oligocene	24.6	29	Tejas II
		Eocene	38	39	Tejas I
		Paleocene	54.9	60	
Cretaceous	Upper	Maastrichtian	65		
		Campanian	73		
		Santonian	83		Zuni III
		Coniacian	87.5		
		Turonian	88.5		
		Cenomanian	91	96	
	Lower	Albian	97.5		
		Aptian	113		Zuni II
		Barremian	119		
		Hauterivian	125		
		Valanginian	131		
		Berriasian	138	134	
Jurassic	Upper	Portlandian	144		
		Kimmeridgian	150		
		Oxfordian	156		Zuni I
	Middle	Callovian	163		
		Bathonian	169		
		Bajocian	175		
	Lower	Aalenian	181		
		Toarcian	188	186	
		Pleinsbachian	194		
		Sinemurian	200		
		Hettangian	206		
Triassic	Upper	Rhaetian	213		Absaroka III
		Norian	219		
		Carnian	225		
	Middle	Ladinian	231		
		Anisian	238		
	Lower	Scythian	243	245	
Permian	Ochoa	Tatarian			
	Guadalupe		248		Absaroka II
		Kazanian	253		
	Leonard	Kungurian	258		
		Artinskian	263		
	Wolfcamp	Sakmarian	268	268	
Pennsylvanian	Virgin	Stephanian	286		
	Missouri				
	Des Moines	Westphalian D			
		Westphalian C			
		Westphalian B	296		Absaroka I
	Atoka	Westphalian A			
	Morrow	Numurian C			
		Numurian B			
Mississippian	Chester	Numurian A		330	
	Valmayer	Visean	333		Kaskaskia II
	Kinderhook	Tournaisian	352		

TABLE 1. CORRELATION CHART SHOWING THE RELATION OF SEQUENCES AND SUBSEQUENCES TO THE STANDARD CHRONOSTRATIGRAPHIC SCHEME FOR THE PHANEROZOIC AND LATEST PROTEROZOIC (continued)
(Adapted from Palmer, 1983)

System	Series	Stage	End (Ma)	Subsequence (Ma)	Subsequence
				362	
Devonian	Upper	Famennian	360		
		Frasnian	367		
	Middle	Givetian	374		
		Eifelian	380		Kaskaskia I
		Emsian	387		
	Lower	Siegenian	394	401	
		Gedinnian	401		
Silurian	Upper	Pridolian	408		
		Ludlovian	414		Tippecanoe II
	Middle	Wenlockian	421		
	Lower	Llandoverian	428		
				438	
Ordovician	Upper	Ashgillian	438		
		Caradocian	448		
	Middle	Llandeilan	458		Tippecanoe I
		Llanvirnian	468		
	Lower	Arenigian	478	488	
		Tremadocian	488		
					Sauk III
Cambrian	Upper	Trempealeauan	505		
		Franconian		515	
		Dresbachian			
					Sauk II
	Middle		523	548	
	Lower		548		Sauk I
Precambrian	Upper Proterozoic	Ediacaran	590	600	

margins by the "turn around" from offlap to onlap; similarly, all of the beds overlying such an unconformity are everywhere younger than the same chronostratigraphic position. Thus, interregional unconformities place unambiguous bounds on time slices for analysis of cratonic evolution.

The packages of strata between interregional unconformities and, by extension, the chronologic equivalents of the packages, have been termed *synthems* (Chang, 1975); recently, the International Subcommission on Stratigraphic Classification (1987) has formalized the adoption of "synthem" and proposed for such bodies of strata the erection of a new category of stratigraphic unit, neither litho nor chronostratigraphic, based on unconformity-bounded successions. This proposition is too new to judge the acceptability of "synthem" and a revised stratigraphic hierarchy. For present purposes, *sequence* (e.g., Sauk sequence) continues in application to the six major unconformity-bounded successions; subdivisions are called *subsequences,* identified by Roman numerals (e.g., Zuni III). Table 1 lists the sequences and subsequences applied here and their chronologic limits trivially modified from the time scale of Harland and others (1982).

Sequence terminology and sequence stratigraphy used in this chapter, as noted in the introduction to this volume, are not uniformly applied by all contributors. In some instances the adherence to time-honored chronologic and chronostratigraphic subdivisions is a consequence of the necessity to conform to stratigraphic classifications approved by governmental agencies; in other cases the first-order importance of regional unconformities, particularly those related to extracratonic orogenic events, reduces the ready applicability of "standard" sequence terminology. By reference to the chronostratigraphic chart (Table 1) accompanying this chapter, readers will be able to place all discussions into the context of cratonic sequence stratigraphy.

SEQUENCE AND SUBSEQUENCE MAPS

Much of the discussion that follows is based on consideration of a series of maps (Figs. 1 through 10) designed to portray the rates of subsidence of the craton as these vary geographically and from time slice to time slice. The maps presented derive from a synthesis of regional isopach studies, many of which are embodied in this volume. Inasmuch as the concern here is for *rates,* rather than absolute thicknesses, isopach values are normalized per unit time (m/m.y.) on the accompanying maps. The great majority of available isopach maps are based on present-day thicknesses, commonly extended across surficial valleys and local uplifts, but without reconstruction of the effects of ancient episodes of regional and interregional erosion. Therefore, the thicknesses mapped are minima, as are the interpreted amounts of subsidence required for deposition. It follows, then, that the rates expressed on Figures 1 through 10 are also minimum values representing syndepositional subsidence, less erosion and compaction. Nevertheless, the displays constitute reasonable characterizations of cratonic subsidence history; further, divorced from the time dependency of isopachs, the maps permit comparison of long- and short-span time slices.

Designed for reproduction at the approximate scale of $1{:}17.5 \times 10^6$, the maps are necessarily generalized, demanding the glossing over of many fascinating details of preeminent importance in local or subregional studies. Readers are urged to seek, in the appropriate companion chapters of this volume, maps of larger scale and supporting discussion. Considerations of drafting and reproduction force yet another departure from the ideal—areas indicating subsidence rates of less than 5 m/m.y. are not shown. This operational decision results in the nonrepresentation of a number of slowly subsiding but nonetheless significant basins, particularly in the central midcontinent regions. Again, attention is directed to the pertinent chapters for coverage of these missing but not forgotten elements.

As noted in the introductory chapter of this volume, the denigration of classical geosynclinal theory, and thus the lessening of faith in the inboard margin of a "miogeosyncline" as the limiting boundary of a craton, leaves tectonists without a broadly accepted definition of craton margins. The accompanying maps of this chapter bear hachured lines that approximate the preserved inboard limits of major emergent "thin-skinned" thrusts of the Sevier, Ouachita-Marathon, and Appalachian Orogens. These mountain systems are the subjects of other volumes of this series, and their anorogenic prehistories are not, in principle, the business of this chapter. The Blue Ridge Thrust and its extensions in the Appalachian Province make a convenient southeast limit for the purposes of this chapter, although reflection-seismic data suggest significant southeasterly continuity of craton-related strata below the thrusts. In any case, that the Appalachian Province outboard of the Blue Ridge has been orogenic and extracratonic (at least with reference to the North American Craton) throughout much of Paleozoic time is not in dispute.

Similarly, the Ouachita-Marathon Thrust Belts form an agreeable southern limit for this study. Here, the pre–late Carboniferous stratigraphy of the allochthonous sheets is markedly different from that immediately to the north and is logically considered (e.g., Thomas, 1976; Shurbet and Cebull, 1987) to represent rifted passive margin with deep-water basins outlined by right-lateral transform faults, which interrupt the northeast trend of the Appalachian-Ouachita-Marathon continental margin.

The Cordilleran cratonic edge is not so easily defined by Late Mesozoic–Early Cenozoic thrusts of the Sevier trend. Instead, the thrust belt performs a kind of random walk across thickness and facies trends, conforming only locally to the position and orientation of high rates of change in the subsidence and depositional environments of preorogenic times. Thus, the responsibility of the student of the pre-Cretaceous craton extends well to the west of the thrust belt and into the realm of the companion volume on the Cordilleran Orogen. By the same token, Cordilleran tectonists cannot escape a concern for structural and magmatic responses to Sevier (largely Cretaceous) and Laramide (mostly Eocene) orogenic pulses that are felt far into the pre-Cretaceous craton. One consequence of the lack of definition of the western margin of the craton is a partial duplication of coverage in this volume and the companion volume on the U.S. Cordilleran Orogen (Burchfiel and others, 1988). So far as tectonic synthesis is concerned, however, the overlap provides opportunities for exposing differing approaches derived from analysis of differing time slices and divergent points of view.

Although this chapter attempts to summarize the tectonosedimentary history of a major part of a craton over the greater part of Phanerozoic time, the competence of any single reporter cannot extent to cover so great a scope of history and geography. Therefore, the pages that follow record the findings of many investigators, not all of whom can be individually identified and acknowledged. The maps and accompanying text are heavily dependent on the materials of the companion chapters of this volume, especially those that were completed before this writing.

Published works that constitute invaluable sources for this synthesis include those by Peterson and Smith (1986) and Stewart (1980), and the compendia assembled by the Rocky Mountain Association of Geologists (Mallory, 1972) and the Shell Oil Company (Cook and Bally, 1975).

In spite of the acknowledged and unacknowledged contributions to the body of facts and interpretations that are included in this chapter, it remains a highly personal document full of statements subject to challenge, if not outright refutation, by superior knowledge.

LATE PROTEROZOIC PROLOGUE

A review of the Proterozoic history of the craton in the conterminous United States is presented by Bickford and others (1986); with consideration of latest pre-Phanerozoic margins by Speed (1983), Speed and others (1988), and by Thomas (1976). Reduced to essentials, the middle North American Craton in latest Proterozoic time had achieved its Phanerozoic form. The aulacogenic* invaginations of the Cordilleran margin, represented by the Belt Supergroup and its correlatives (~1,350 Ma), and the slightly younger extensional manifestations of the Midcontinent Rift system had long since ceased activity, as had the collisional effects of the Grenville episode of the Appalachian region (~1,100 Ma) and the coeval Llano Province of Texas. On the western and eastern borders of the craton, the Windermere and Ocoee successions and their equivalents, occupying the approximate interval 750 to 650 Ma, are dominated by craton-derived

*The term *aulacogen* was introduced by the Soviet geologist N. S. Shatski (e.g., Shatski, 1946) to recognize trenchlike features that originate as complex grabens oriented more or less normal to continental margins and evolve into flexural downwarps that gradually die out 200 km or more into cratonic interiors. Hoffman (e.g., Hoffman and others, 1974) bears responsibility for expanded application of the term in North America.

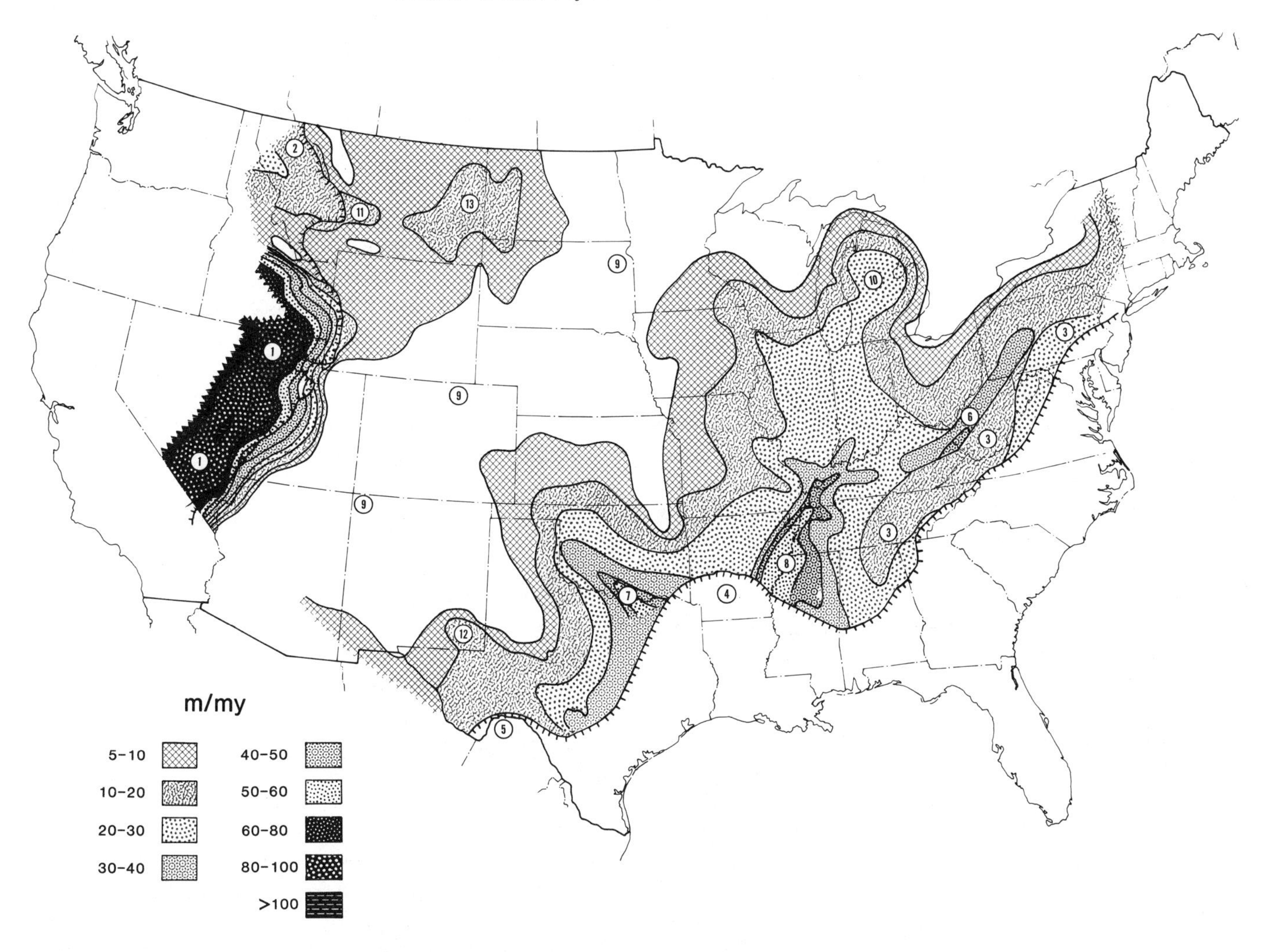

Figure 1. Sauk sequence (latest Proterozoic to Early Ordovician) net subsidence rate. 1, Cordilleran Shelf Basin; 2, area of delayed Sauk subsidence; 3, Appalachian Shelf; 4, Ouachita margin; 5, Marathon margin; 6, Rome Trough; 7, Southern Oklahoma Aulacogen; 8, Mississippi River–Reelfoot Rift system; 9, Transcontinental Arch; 10, Michigan Basin; 11, Central Montana (Belt) Trough; 12, Tobosa Basin; 13, Williston Basin.

detritus apparently deposited along passive continental margins, establishing a pattern that prevailed to and beyond the opening of the Paleozoic Era.

Sauk sequence (Latest Proterozoic, Ediacaran, to Early Ordovician, Llanvirnian)

The Sauk sequence is defined as those strata lying above an unconformity cut on rocks as old or older than the Late Proterozoic Windermere, Ocoee, and equivalent strata noted above and below the sub-Tippecanoe or a younger unconformity.

Although Sauk depositional time extends over more than 150 m.y., this span is represented here by a single map (Fig. 1), a consequence of the lack of resolution of subsequence boundaries available for synthesis at an interregional scale. It should be noted, however, that by far the greater area of the craton carries no preserved record of the first half of the Sauk depositional episode and the blurring of detail by integration over an excessive time interval is confined to the cratonic margins.

Sub-Sauk surface

Over vast areas of the cratonic interior, the Sauk sequence rests with profound unconformity on Proterozoic and older Precambrian crystalline rocks or, to a more limited degree, on the bevelled edges of truncated Proterozoic sediments and volcanics of the Midcontinent Rift system. Body fossils are commonly abundant not far above the base of the Sauk, and it was believed for decades that none existed below. In recognition of this condition, C. D. Walcott, some 80 years ago, identified the pre-Sauk gap in the record as the "Lipalian Interval" and ascribed to it the time of development of hard parts by marine invertebrates. Regardless of evolutionary significance, the pre-Sauk hiatus amounts to tens of millions of years over an enormous region, and to as much as 100 m.y. where latest Cambrian and Early Ordovician strata form the base of the sequence. Cratonwide nondepositional episodes of similar dimensions are rare in younger Phanerozoic time.

Conceptually, given the absence of strongly rooted land plants, and the length of the hiatal episode, it would be reasonable to predict that the sub-Sauk surface would be a featureless plain. Such appears to be the case in many areas of discontinuous outcrop and moderate subsurface observations; commonly, however, where exposure or drilling density permit, abrupt local departures from a planar surface are seen as buried hills or are inescapably deduced from the facies and overlap relationships of basal Sauk units. Paleotopographic relief, supported by differences in the resistance to erosion of petrologies at the unconformity surface, is commonly measured in tens of meters, but it is remarkable that any discernible relief is present. Evidence of an integrated stream drainage pattern developed on the erosion surface is lacking, and thick paleosols are rare (both characteristics stand in contrast to conditions at many younger interregional unconformities). Almost everywhere, basal Sauk units are dominated by medium- to coarse-grained quartz arenites; pebbles and boulders of local derivation are not uncommon. Most observers would agree with Dott and others (1986) that the basal sands represent a combination of dune and beach processes; thus, the sub-Sauk surface can be envisaged as the product of unimpeded wind erosion succeeded ultimately by marine planation of insufficient longevity to erase the inequalities of the eolian landscape.

Sauk I (latest Proterozoic to Early Cambrian)

A preserved record of the latest Proterozoic through Early Cambrian time span of Sauk I is limited to the passive shelf margins of the Cordilleran and Appalachian borders of the craton. A simple continuous miogeocline is not evident along the western margin; instead, a rapidly subsiding embayment (1)* extended from southern Nevada and adjoining southeastern California to northern Utah and southern Idaho and received kilometer-thick deposits of craton-derived sand. To the north, in western Montana and the panhandle of Idaho (2), no Sauk I rocks are known, and transgression of the cratonic margin was delayed until mid-Cambrian (Sauk II) time. On the southeastern margin, latest Proterozoic and Early Cambrian onlap similarly involved thick quartz sands from the cratonic interior, succeeded by shales and carbonates. Students of the Appalachian (3) Ouachita (4) Marathon (5) margin (e.g., Thomas, 1976) view the sinuosities of the trend as reflections of transform displacements of the craton margin related to the Late Proterozoic–Early Cambrian opening of Iapetus Ocean, the proto-Atlantic. In addition, the divergent motion was effective inboard of the Appalachian margin to form the extensional Rome trough (6), a site of thick Early Cambrian accumulation.

Further evidence of extensional tectonics related to cratonic margins is found in elongate graben systems that penetrate the craton from its southern edge. The best documented of these, called the Southern Oklahoma Aulacogen (7) by Hoffman and others (1974) is ancestral to the Anadarko Basin and is marked by high-angle faulting and the emplacement of rhyolite flows and comagmatic plutons in Early and Middle Cambrian time. An equally impressive feature, the Mississippi River Graben–Reelfoot Rift system (8), is known from geophysical observations and sparse deep drilling. No unequivocal evidence of activity during Sauk I deposition is available, but it seems untuitively reasonable that Early and Middle Cambrian tectonic and igneous processes, as in southern Oklahoma, may be represented. Regardless of time of initiation, the graben/rift system is clearly ancestral to the Illinois Basin and the Mississippi Embayment. The east-directed arm of the system, the Rough Creek Graben (see Collinson and others, this volume, Figs. 5 and 7), is believed by some workers to be continuous with the southwestern extension of the Rome Trough.

Sauk II (Middle and earliest Late Cambrian)

Middle and earliest Late Cambrian (Sauk II) time was characterized by renewed onlap of the craton with basal quartz sands of interior derivation reaching the flanks of the Transcontinental Arch (9) and much of the currently exposed edges of the Canadian Shield by the close of the episode. West and northwest of the Transcontinental Arch the basal sands are typically succeeded by silt and shale units, whereas carbonates of the onlap succession follow directly on sandstone inboard of the Appalachian shelf. The general lack of fine-grained detritals on the eastern and southeastern cratonic interior has been attributed to scour and transport by west-flowing (in terms of present geography) winds but may also represent lower rates of submergence or higher rates of sedimentation leading to the bypassing of fines to the eastern or southern cratonic margin.

Sauk II sediments, dominated by arkosic sandstones, extend in kilometer and greater thicknesses from the Mississippi Valley/Reelfoot Grabens north across the length of Illinois and western Indiana, but it is not clear whether the high rate of subsidence was related to faulting at significant distances north of the Ohio River. Rift-related subsidence continued in the Rome Trough, but there is a regional lack of Sauk II sedimentation in the area of the Southern Oklahoma Aulacogen, perhaps as a consequence of continued igneous activity early in the episode and relict thermal uplift.

The Michigan Basin (10), the archetype of craton-interior basins, appears for the first time during the Sauk II cycle, separated from the Illinois-Indiana area of downwarp by a low-relief isopach divide at a position well north of the younger Kankakee Arch. Similarly, the west flank of the basin appears to have been defined by an axis tens of kilometers east of the mid-Paleozoic

*Refers to tectonic elements identified by number on Figures 1 to 10.

Wisconsin Arch. It has been suggested (e.g., Adler, personal communication, 1987) that the position of the Sauk Michigan Basin and, perhaps, the isopach anomaly shown by Collinson and others (this volume, Fig. 7) in adjoining northeastern Illinois, may represent a left-lateral offset of the Mississippi Valley–Reelfoot trend.

At the Cordilleran margin the shelf basin of Late Proterozoic and Sauk I times continued to subside, while downwarp was initiated in the previously emergent region to the north, extending eastward along the axis of the much older (1,350 Ma) Belt aulacogenic trough (11).

Sauk II deposition closed with craton-wide offlap and marine regression, accompanied by stripping and locally severe truncation as, for example, in southwestern Montana (Sloss and Moritz, 1951).

Sauk III (Mid–Late Cambrian to Early Ordovician)

Sauk transgression was renewed in mid–Late Cambrian time, the Franconian of North American classification, and continued, with pauses and minor regressions, into the Early Ordovician, onlapping Sauk II along the flanks of the Transcontinental Arch and the edges of the Canadian Shield. The dominant features of Sauk III stratigraphy are carbonate masses that are museums of varied carbonate petrology, calcitic, and commonly greater than a kilometer thick at cratonic margins and the axes of the Southern Oklahoma and Mississippi Graben–Reelfoot Rift Aulacogens, thinner and typically dolomitized on less-rapidly subsiding elements (e.g., the Tobosa Basin, 12) and cratonic-interior platforms. Approach to the Wisconsin Arch (or, more accurately, its more easterly predecessor) is signalled by regressive tongues of quartz arenites, which represent retreats of the strandline of 400 km or more. During Sauk III accumulation, net subsidence with respect to sea level brought about the burial of high-standing topographic remnants of the pre-Sauk surface, such as the quartzite ridges of southern Wisconsin and the "porphyry" knobs of the Ozark region.

Among other tectonic elements of the craton, note the appearance of the Williston Basin (13) as a definable basin bearing the same relationship to the Belt trough as that expressed by the Michigan Basin with reference to the Mississippi Valley–Reelfoot system.

Sauk deposition terminated with apparently rapid emergence of the craton above base level before the end of Early Ordovician time.

TIPPECANOE SEQUENCE (MIDDLE ORDOVICIAN TO EARLY DEVONIAN)

The time span of the Tippecanoe sequence extends from Middle Ordovician (Llandeilan) through earliest Devonian (Gedinnian). Although the sequence is readily divisible (by the effects of latest Ordovician glacioeustatic regression) into Ordovician (Tippecanoe I) and Silurian–earliest Devonian (Tippecanoe II) segments, there is slight distinction between the subsequences in terms of tectonic evolution; therefore, the sequence is treated here by a single map (Fig. 2).

Sub-Tippecanoe surface

The sub-Tippecanoe unconformity is pervasive throughout the cratonic interior and is marked by a significant stratigraphic lacuna near cratonic margins, as in the Appalachian Shelf Basin. Only locally does the unconformity involve angular discordance or truncation observable at outcrop scale. Detailed subcrop mapping (e.g., Mai and Dott, 1985) reveals as much as 50 m of local incision, including integrated stream channels as well as buried hills. On a regional scale, however, the sub-Tippecanoe subcrop pattern shows that basal Tippecanoe units overlie upper Sauk (Lower Ordovician) strata over very large areas except where Sauk is broadly overstepped, as on the edges of the Canadian Shield. In general, the few elements of positive tectonic tendency, marked by anomalous depths of erosion, are those active during Sauk deposition, and these seldom coincide with positive elements of later Paleozoic time.

Tippecanoe I (Middle and Late Ordovician)

Tippecanoe I, the Middle and Late Ordovician segment of Tippecanoe accumulation, is preserved as the most widely distributed subsequence in the cratonic record in spite of massive pre-Kaskaskia (Early Devonian) erosion, which was particularly effective west of the Transcontinental Arch. Where preservation is relatively complete, subsidence rates higher than 5 m/m.y. were not widely achieved, as along the flanks of the Transcontinental Arch, and very broad areas of thin accumulation are not represented on the accompanying map (Fig. 2). Readers' attention is directed to the regional chapters (e.g., Bunker and others, this volume) for illuminating details of Middle and Upper Ordovician stratigraphy and tectonics not treated here.

Almost everywhere in the U.S. portion of the cratonic interior, the sub-Tippecanoe erosion surface is covered by a sheet of very pure quartz arenite, exhibiting uniform thicknesses except where affected by pre-Tippecanoe topography, suggesting a low level of tectonic differentiation in the early stages of the episode. Rapid subsidence at this time was confined to the cratonic margins and the now flexurally subsident Anadarko (1), Tobosa (2), and Mississippi Valley (3)–Illinois Basin (4) trends. Basal Tippecanoe deposition is well documented in the Anadarko Basin area where, following a brief post-Sauk hiatus, the earliest Tippecanoe strata are carbonates quickly supplanted by repeated tongues of clean quartz sands intercalated with marine shales and minor carbonates, a pattern indicative of pulsatory transgression and regression. Later, in mid-Ordovician time, continued subsidence led to the progressive onlap of dune, strandline, and marine depositional environments across the cratonic interior. Grossly similar circumstances appear to have governed the beginnings of Tippecanoe deposition in the shelf basins of the Cordilleran area

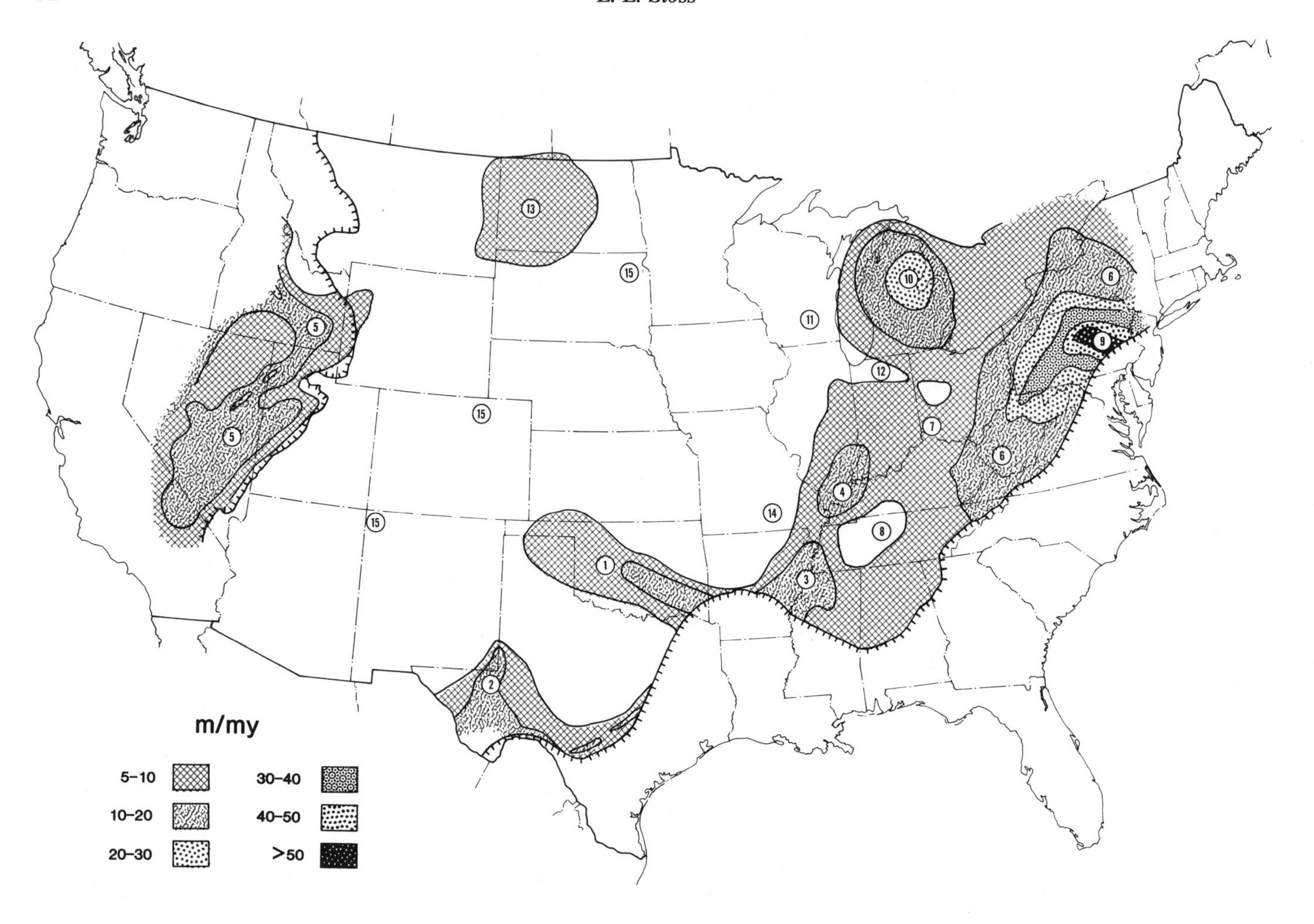

Figure 2. Tippecanoe sequence (Middle Ordovician to Early Devonian) net subsidence rate. 1, Anadarko Basin; 2, Tobosa Basin; 3, Mississippi Valley Graben; 4, Illinois Basin; 5, Cordilleran Shelf basin; 6, Appalachian Shelf; 7, Cincinnati Arch; 8, Nashville Dome; 9, Queenston clastic wedge; 10, Michigan Basin; 11, Wisconsin Arch; 12, Kankakee Arch; 13, Williston Basin; 14, Ozark Dome; 15, Transcontinental Arch.

(5), but by the time the Rocky Mountain region passed below base level the supply of interior-source sand appears to have been depleted, and basal sandstones of significant thickness present a patchy distribution.

Initial Tippecanoe deposition in the area of the Appalachian Shelf (6) is notably deficient in sand and, typically, Sauk carbonates, predominantly dolomites, are succeeded with demonstrable unconformity by Tippecanoe limestones. Indeed, basal sands are generally lacking southeast of the Cincinnati Arch (7)–Nashville Dome (8) axis, although this "positive" axis does not appear to have been effective during Tippecanoe I deposition (the marginally lower rates of subsidence indicated on Fig. 2 are artifacts of post-Tippecanoe erosion). Basal Tippecanoe sands, although markedly time transgressive from early Middle Ordovician to Late Ordovician, mimic the westerly directed distribution pathways and dune/strand environments of Sauk detritals and presumably derive from similar areas of provenance on the Canadian Shield. It has been popular to ascribe the extreme maturity of Tippecanoe sands to recycling of Sauk quartz grains. However, consideration of the volume of clean quartz produced and of the very large areas of crystalline rocks exposed on the shield during Tippecanoe onlap makes it difficult to cling to the hypothesis; further negative evidence is provided by the fact that the basal sandstones are of undiminished maturity where Tippecanoe oversteps Sauk for great distances, as in North Dakota and Manitoba. Again, recourse must be made to the lack of vegetational cover, onlap punctuated by multiple episodes of offlap, and a high degree of tectonic stability.

Basal Tippecanoe strata are succeeded southeast of the Transcontinental Arch by sheetlike carbonates intercalated with shales; the latter largely, but not exclusively, represent west-thinning distal deposits of the Queenston clastic wedge (9) of the now-convergent and emergent northern Appalachian margin. On the Cordilleran Shelf and adjoining western cratonic interior, carbonate deposition prevailed. With advancing Ordovician time the cratonal tectonic framework gradually evolved from the state inherited from the Sauk to a mid-Paleozoic pattern. The Michigan Basin (10) assumed its modern form, separated from regions to the west and south by the Wisconsin (11) and Kankakee (12) Arches; basins appeared in the Iowa area, although subsidence rates were not sufficiently high to appear on the sequence map (Fig. 2; see Bunker and others, this volume), and the Williston Basin (13) resumed flexural subsidence, albeit at so low a rate as to have little effect on depositional environments. Concomitantly, minor uplift was initiated on the Ozark Dome (14). As the tectonic framework became differentiated into a complex geography of large flexural arches and basins, the entire craton subsided with reference to sea level, such that marine conditions transgressed much if not all of the Transcontinental Arch (15) and huge areas of the Canadian Shield. However, the *preserved* distribution of Tippecanoe strata is anomalously circumscribed on the Transcontinental Arch and inboard of the Cordilleran margin where broad areas of western Montana and the Colorado Plateau lack Ordovician and Silurian rocks (except as rare diatreme xenoliths). The truncated edges of Tippecanoe strata overstepped by Kaskaskia rocks exhibit no hint of approach to shorelines or depositional limits, and it must be assumed that prior to Devonian erosion the Tippecanoe sequence was present over much of the western craton.

Farther west, the Cordilleran cratonic margin remained passive; although complicated by thrusts of later date, the record indicates a gradual westward transition from the quartz arenites and relatively pure carbonates of interior facies to an increasingly shaly succession, presumably reprsenting continental-slope deposition. Shelf-slope transition is less obvious at the Ouachita-Marathon margin, perhaps because of thrust-related facies telescoping, or alternatively, there may have been shelf-edge carbonate prominences bordering steep slopes. As noted earlier, the Appalachian margin became the site of active convergence, transforming the passive shelf-prism margin into a foreland Appalachian Basin (6) loaded in part by the molassic Queenston clastic wedge (9). Details of the Caledonide Taconian Orogen are treated in the accompanying volume by Hatcher and others (1988); here, interest centers on the far-reaching spread of Taconian-derived mud and the nearly equivalent reach of ash falls over the eastern areas of the craton.

The top of the Tippecanoe I subsequence is marked by a commonly obscure disconformity, probably related to the episode of Gondwana continental glaciation of latest Ordovician–earliest Silurian time.

Tippecanoe II (Early Silurian to Early Devonian)

The upper Tippecanoe subsequence ranges in age from Early Silurian through earliest Devonian (Gedinnian). The tectonic geography is virtually unchanged from the prevailing during the later phases of Tippecanoe I deposition so that the subsidence map (Fig. 2) of the entire sequence is a satisfactory representation. The Tippecanoe II subsequence, however, suffered the brunt of pre-Kaskaskia erosion, and its preserved distribution is significantly dwarfed when compared to Tippecanoe I.

As noted, the sub–Tippecanoe II unconformity is characteristically undramatic, betraying a lack of cratonal involvement in the development of the discontinuity except in the area of the Hudson Bay Basin where a significant unconformity separates Upper Ordovician and Middle Silurian strata (Dimian and others, 1983). Progressive onlap from Early into Middle Silurian time is evident, however, covering a greater time span than is attributable to recovery from a glacial episode and suggesting renewed reduction of cratonic freeboard by other processes. In view of the occurrence of Middle and Late Silurian rocks as erosional remnants on the shield and as diatreme xenoliths far from the zero isopach lines of regional maps, it is probable that the original distribution of Tippecanoe II deposition was at least as widespread as that of Tippecanoe I.

The major distinction between the two segments of the Tippecanoe sequence lies in the degree to which lithologic facies conform to the geometry of tectonic elements, clearly indicating control of depositional and diagenetic environments by subsidence rates. The Michigan basin (10) is a classic example (see Fisher and others, this volume). Here, although the precise chronology of the facies remains in dispute, evaporites, including halite and potash salts, are fringed by and partially surround a closely spaced halo of "pinnacle" reefs; the basin margin and the surrounding stable region is covered by a bank-and-reef–dominated platforn now largely eroded from the Wisconsin Arch (11) but well preserved on an expanded Kankakee Arch (12). Accelerated subsidence leading to restriction and evaporite deposition is also reflected in sub-basins at the western margin of the Appalachian Basin (6), in the Iowa–northern Kansas area (see Bunker and others, this volume), and to a lesser degree in the Williston Basin (13). The proto-Illinois Basin (4) remained in open communication with southern seaways, as did the Tobosa Basin (2) and the Anadarko region (1); in all three aulacogenic terranes, subsidence rates in excess of sedimentation led to deeper waters and the accumulation of evaporite-free silty and marly carbonates.

The end of Tippecanoe deposition was heralded by sporadic episodes of regression, reaching a climax in earliest Devonian time with supra–base level emergence of the entire cratonic interior, leaving relict seas confined to the most subsident areas of the Cordilleran Shelf, the Appalachian foreland, and the axes of the formerly aulacogenic reentrants.

KASKASKIA SEQUENCE (MID-EARLY DEVONIAN TO LATEST MISSISSIPPIAN)

In North American chronologic terms, the Kaskaskia sequence opened with initiation of transgression of an interregional unconformity in mid–Early Devonian time and closed with the beginning of a major regression in latest Mississippian. Like the Tippecanoe, the Kaskaskia is divisible into two subsequences by a

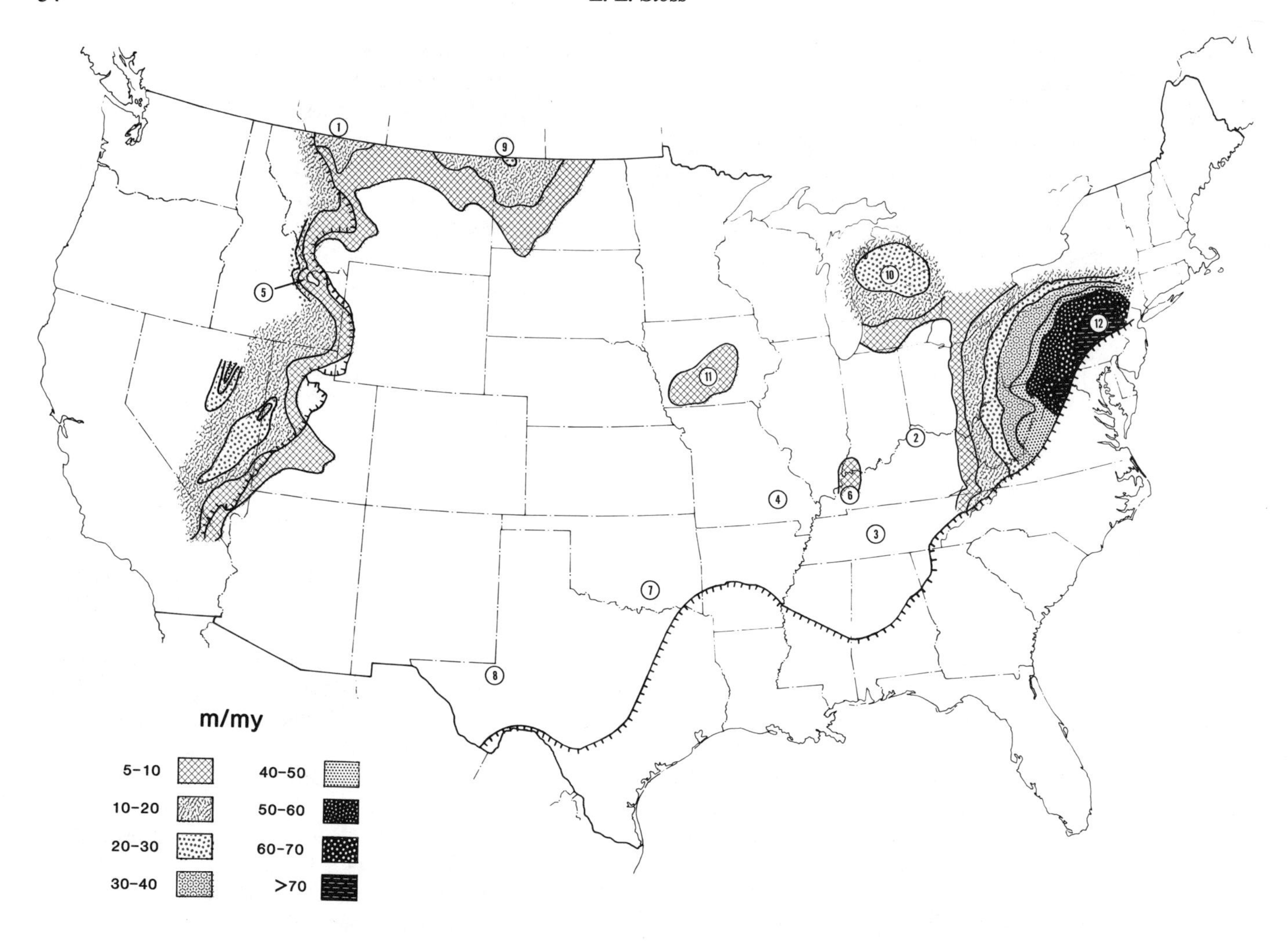

Figure 3. Kaskaskia I subsequence (mid–Early Devonian to Late Devonian) net subsidence rate. 1, Sweetgrass Arch; 2, Cincinnati Arch; 3, Nashville Dome; 4, Ozark Dome; 5, Lemhi Arch; 6, Illinois Basin; 7, Anadarko Basin; 8, Tobosa Basin; 9, Elk Point Basin; 10, Michigan Basin; 11, Iowa Basin; 12, Catskill clastic wedge.

eustatic event relatively uncomplicated by tectonic factors. The subsequences, however, were deposited under distinct tectonic regimes, requiring two maps (Figs. 3 and 4) for adequate treatment.

Sub-Kaskaskia surface

That the hiatus represented by the sub-Kaskaskia unconformity is not a simple consequence of sea-level lowering is made clear by consideration of the complexity of the subcrop geology at the unconformity surface (see Bunker and others, this volume, for a regional example; Bally, 1980, presents a craton-wide view). New positive elements, on which older sequences were stripped or thinned, emerged during the hiatal interval. The ancestral Sweetgrass Arch (or West Alberta Ridge) (1) of northern Montana is an example; here, Tippecanoe strata (and youngest Sauk) were eliminated, although there is no evidence of pre-Kaskaskia existence of the arch. East of the arch the subcrop pattern exhibits a concentric array of the truncated edges of Williston Basin–Tippecanoe units with a remnant patch of Upper Silurian preserved at the basin center.

Similar evidences of differential vertical tectonic motions, regionally broad and flexural, locally abrupt and fault controlled, are found throughout the craton (see, e.g., the pre-Kaskaskia paleogeologic map by Bunker and others, plate 4E). The Transcontinental Arch (Fig. 1, #9, Fig. 2, #15) as well as the Cincinnati Arch (2)–Nashville Dome (3) axis and the Ozark Uplift (4) among others, became significant elements of the tectonic framework during the pre-Kaskaskia hiatus. In addition, as noted in the discussion of the preserved distribution of the Tippecanoe sequence, significant areas of the present-day Colorado Plateau and Central and Southern Rocky Mountains suffered erosion, apparently as the result of uplift of blocks that may be considered precursors of the "Ancestral Rockies" of later Paleozoic time.

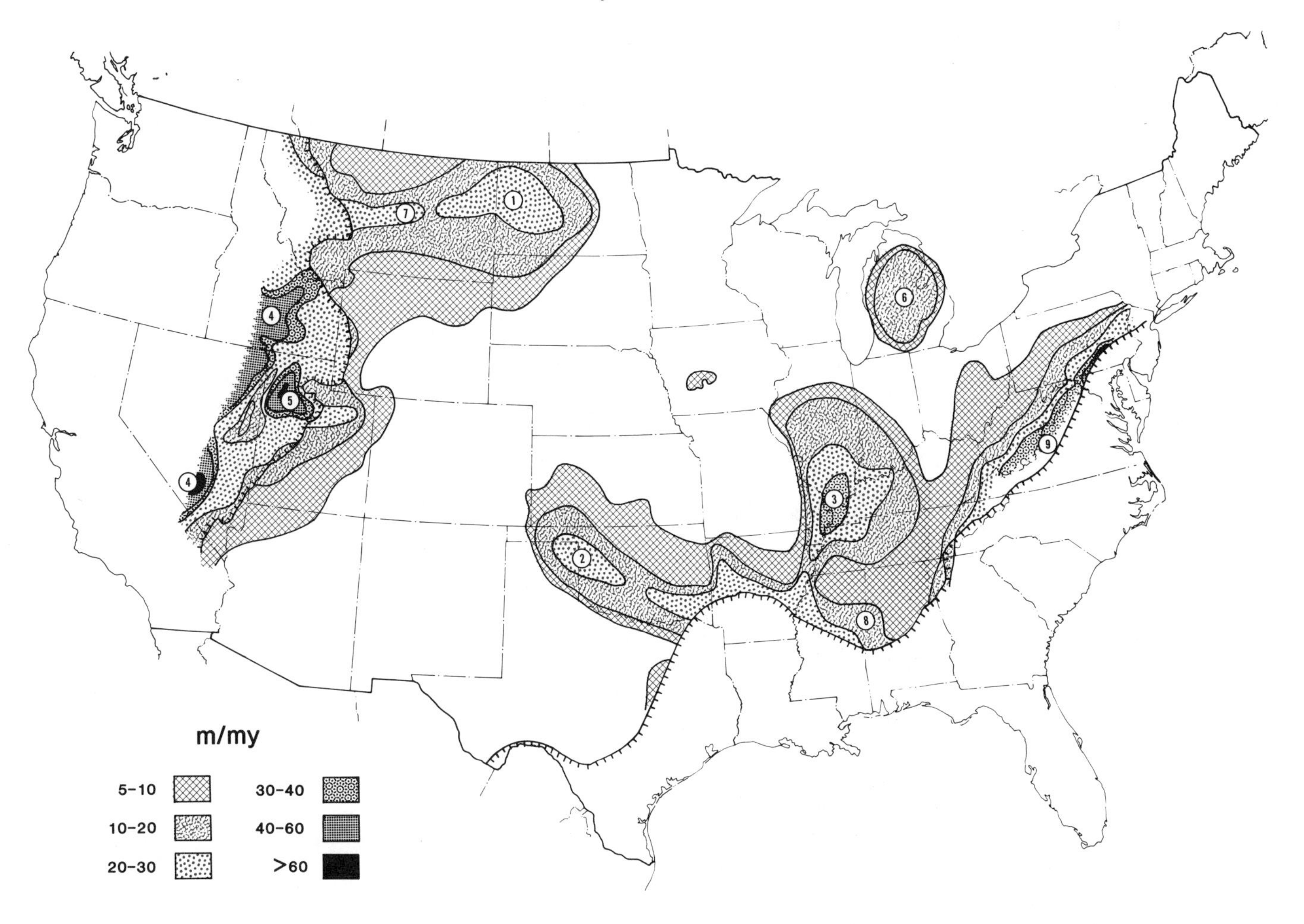

Figure 4. Kaskaska II subsequence (latest Devonian to Late Mississippian) net subsidence rate. 1, Williston Basin; 2, Anadarko Basin; 3, Reelfoot-Illinois Basin; 4, Cordilleran Foreland Basin; 5, Oquirrh Basin; 6, Michigan Basin; 7, Central Montana Trough; 8, Black Warrior Basin; 9, Appalachian Foreland Basin.

Even the Cordilleran continental margin was not spared the effects of differential uplift, as exemplified by such features as the Lemhi Arch (5).

The sub-Kaskaskia surface appears to have been relatively free of abrupt topographic relief, and preserved or slightly reworked paleosols are more common than at older interregional unconformities, perhaps as a consequence of the emergence of rooted land plants. Channels cut into pre-Kaskaskia units and filled with nonmarine Early Devonian sediments are identifiable but are not as prominent as similar features at the sub-Tippecanoe surface.

Kaskaskia I (Mid-Early Devonian to latest Devonian)

The depositional record of Kaskaskia I extends from mid-Early Devonian to near the close of Devonian time as this is defined biostratigraphically. Slow subsidence rates characterized the cratonic interior within the United States, and great areas failed to reach the arbitrary 5 m/m.y. threshold for display on the subsequence map (Fig. 3). The absence of thick accumulation on the southern reaches of the craton, the site of the aulacogenic proto-Illinois (6), Anadarko (7), and Tobosa (8) Basins, is particularly striking in contrast to the marked subsident tendencies of these elements in earlier (and later) depositional episodes. Also worthy of attention is the tectonic pattern of the Williston Basin region (9); in contrast to previous and subsequent behavior (see Gerhard and Anderson, this volume), in Devonian time the region was the distal prolongation of a largely flexural trough, the Elk Point Basin, which extended southeastward across the Prairie Provinces of Canada.

Basal strata of Kaskaskia I are markedly diachronous, reflecting widely differing times of submergence below depositional base level. In general, late Early and early Middle Devonian (Siegenian and Emsian) beds are confined to basins at the cratonic margins and the old re-entrants to the cratonic interior (such as the Reelfoot–Illinois–Michigan Basin trend). With advancing mid-Devonian time, deposition spread widely over the craton southeast of the Transcontinental Arch and from northwest to southeast down the axis of the Elk Point Basin; elsewhere northwest of the arch and east of the Cordilleran margin, subsidence below base level was delayed until early Late Devonian (Frasnian) or remained emergent until latest Devonian–Early Mississippian transgression.

Basal clean quartz arenites are common at the unconformity surface on the eastern craton, but are not as pervasive as older transgressive sands. Some are clearly attributable to reworking of Sauk and Tippecanoe quartz grains, but other sand bodies require a provenance from crystalline rocks of the shield. Basal sands are less widely distributed in the west; here and in eastern areas lacking basal sandstones, it is characteristic to find Kaskaskia I carbonates on carbonates of older sequences, sometimes separated by reworked soil or terra rossa, sometimes without readily discernible contacts.

Differential subsidence of the interior basins was sharply renewed in mid-Devonian (Eifelian) time, leading to isolation of the tectonically closed elements and the deposition of basin-center chlorides and basin-margin reefs in the southern reaches of the Elk Point Basin (9), and in the Michigan (10) Basin and to lesser restriction of the more slowly subsiding Iowa (11) Basin (east-central Iowa Basin of Bunker and others, this volume). Outside of the cratonic basins and west of the reach of the clastic influx to the Appalachian Basin, Middle and early Late Devonian deposition was dominated by carbonates under conditions of general tectonic stability interrupted by episodes of eustatic(?) regression, the latter marked by widespread anhydrites (or their solution-breccia equivalents) of presumed sabkha origin. Finally, particularly in basins southeast of the Transcontinental Arch and beyond the influence of the Catskill clastic wedge, Kaskaskia I deposition closed with the accumulation of black and brown organic-rich shales as noted in greater detail in the several regional chapters of this volume.

Meanwhile, back in the Appalachian Basin and for much of the time span of the subsequence, the dominant sedimentational phenomenon was the Catskill "delta" (12), an enormous accumulation of piedmont sediments derived from the Acadian Orogen. It has been pointed out (e.g., Faill, 1985) that the piedmont complex lies a considerable distance southwest of apparent centers of diastrophic and igneous activity accompanying the Acadian orogeny; thus, the Catskill complex is not a classic foreland accumulation in the sense of the Queenston clastic wedge. Nevertheless, a prodigious volume of detrital sediment was produced, and this affected deposition and depositional environments on cratonic regions far to the west, interdigitating shale and silt units with the prevailing interior carbonates.

As has been noted, Kaskaskia I deposition was subject to a number of interruptions; some of these were accompanied by modest regional rearrangements of the tectonic framework. The close of deposition of the subsequence in latest Devonian (Late Famennian) time was marked by tectonic events differing only in degree from those that preceded it. It is clear, however, that previously established arches and domes tended to emerge above base level and suffer erosion during the pre–Kaskaskia II hiatus, although the pause in sedimentation appears to have been slight in basinal areas.

Kaskaskia II (Latest Devonian to Late Mississippian)

The maximum time of accumulation of the Kaskaskia II subsequence very nearly coincides with the Mississippian Period (Lower Carboniferous), although in terms of detailed chronostratigraphy, there are latest Devonian units included, and latest Mississippian is not represented. The subsidence map of the subsequence (Fig. 4) illustrates the changes in tectonic geography between Kaskaskia I and II. Note the reversion of the Williston Basin (1) to a geometry similar to that controlling Tippecanoe deposition and the disappearance of a Prairie Province connection via an Elk Point Basin. Note also the re-establishment of the Anadarko (2) and Reelfoot-Illinois Basin (3) trends as loci of significant subsidence. Other major differences are seen in the transformation of the Cordilleran Shelf Basin (4) to a foreland basin as a concomitant to the Antler Orogeny and the first appearance of the Oquirrh Basin (5) as a closed tectonic entity.

Kaskaskia II deposition began with renewed and expanded accumulation under anoxic conditions, which overspread interior areas lacking Kaskaskia I representation as a result of nondeposition or erosion. With advancing Early Mississippian time, environments changed to open-circulation carbonate-dominated seas as progressive transgression covered ever-larger regions, especially on the northwest flanks of the Transcontinental Arch, which had remained above depositional base level in earlier Kaskaskia episodes. Southeast of the Transcontinental Arch, in Illinois, Indiana, and Ohio, relatively short-lived pulses of prodelta and deltaic mud and sand invaded the cratonic interior from source areas to the northeast, whereas elsewhere on the craton and its margins, broad limestone seas reached a climax in mid-Mississipian time. Concurrently, revived subsidence created restriction of the Williston (1) and Michigan (6) Basins, resulting in evaporites, including thick salt.

In the Late Mississippian (Late Visean–Early Namurian) the carbonate-generating seaways of the cratonic interior, particularly in the area of the Illinois (3) and Michigan (6) Basins were displaced once and forever by recurrent advances of detrital materials from the northeast. Repeated delta progradations, followed by brief marine transgressions, created a succession of mini-cyclothems, some including coals. Concurrently, the Williston Basin (1) and its western conduit to the cratonic margin, the Central Montana Trough (7), also suffered the replacement of long-continuing carbonate-evaporite environments by sand and shale. Coeval deltaic clastics also invaded the Black Warrior Basin (8) of Mississippi and Alabama (see Thomas, this volume), but these are of extracratonic derivation and herald the appearance of orogenic highlands at the southern margin of the craton.

Foreland clastics continued to occupy the southeast flank of the Appalachian Basin (9) but did not penetrate far to the northwest, while a thick clastic wedge inboard of the Antler Orogen developed at the Cordilleran margin (4).

Kaskaskia II closed with emergence of the craton before the end of Mississippian time. Truncation and thinning of exposed units was severe in the interior, less penetrative but demonstrable at cratonic margins.

ABSAROKA SEQUENCE (LATEST MISSISSIPPIAN TO EARLY JURASSIC)

Beginning in latest Mississippian (Namurian A) time and continuing into Early Jurassic, the history of the Absaroka sequence records the longest and most complex of the Phanerozoic tectonostratigraphic subdivisions treated here. Indeed, although the sequence is illustrated and discussed in three parts, the tectonic and depositional patterns shifted so radically in both space and time that the broad-brush approach used for earlier episodes is quite inadequate, at least for the first two subsequences. Readers requiring a better insight into the complex events of this exciting part of cratonic history will wish to make recourse to all of the regional chapters of this volume.

Sub-Absaroka surface

Profound changes in the tectonic framework of the North American Craton, presaged in the later phases of Kaskaskia deposition by renewed emergence of the Canadian Shield as a source region and by the appearance of orogenic highlands at the Ouachita margin, accelerated markedly during the brief pre-Absaroka hiatus. The results are documented by the extreme complexity of subcrop geology at the sub-Absaroka erosion surface (regional sub-Absaroka patterns are shown on the map by Bunker and others, Plate 4H; and by Sloss, 1984), revealing the stripping of hundreds of meters to kilometers of older sedimentary cover from pre-existent and new positive elements before again passing below base level to receive Absaroka sediments. As might be anticipated, the complexity and variability of pre-Absaroka vertical movements precluded uniform, synchronous, and ubiquitous uplift of the entire craton and its margins. Indeed, certain regions adjacent to margins and marked by obvious unconformities at the bases of preceding sequences are characterized by minimal nondepositional hiatuses prior to Absaroka deposition; the same is true in parts of the Central Montana Trough and at depth in the Anadarko and Ardmore Basins. (For further discussion of the contrasts between the pre-Absaroka and other cratonic unconformities see Sloss, 1984.)

Much, but not all, of the extraordinary cratonic tectonism represented by the sub-Absaroka unconformity is attributable to stresses generated by collisional convergence at the Ouachita (1)–Marathon (2) margin, principally during Absaroka I deposition.

Absaroka I (Latest Mississippian to Early Permian)

The lowest subsequence of the Absaroka extends from very Late Mississippian through Early Permian. As would be expected in the cover of an unconformity of high tectonic and physiographic relief, the ages of lowermost-preserved Absaroka I units are highly variable, reflecting the time required for overlap; furthermore, the subsequence is riddled by a multitude of local and regional unconformities, many of which penetrate to pre-Absaroka rocks. Commonly, the base of Absaroka I is marked by reworked paleosols, deeply incised stream channels, or by karst developed on underlying carbonate units.

From Utah and Wyoming to southern Oklahoma the tectonic map of Absaroka I is dominated by uplifts of sufficient amplitude to expose crystalline basement rocks; the most prominent of these are shown on the accompanying map (Fig. 5): Uncompahgre (3), Front Range (4)–Apishapa (5)–Sierra Grande (6), and Amarillo (7)-Wichita (8)-Arbuckle (9). All are fault bounded, commonly by high-angle reverse faults; the degree of transcurrent motion is significant but difficult to measure. The uplifts are yoked to adjacent down-dropped blocks or wrench-related pull-apart basins such as the Paradox (10), Eagle, Maroon, or Central Colorado (11), and Denver (12) in the Rocky Mountain region. The flanks of these basins received very large volumes of detrital materials eroded from the adjacent uplifts. Inasmuch as the pre-Absaroka cover of Paleozoic sediments was relatively thin, the basement crystallines were quickly exposed; rapid uplift and rapid subsidence maintained steep slopes down which largely unweathered feldspar-rich debris cascaded to form alluvial cones and fan deltas to mingle with the sediments accumulating in the basin interiors (dominated by evaporites in the Paradox Basin and, to a lesser degree, in the Eagle Basin).

The Amarillo-Wichita-Arbuckle trend is a complex of isolated en echelon blocks, including many developed during and buried by Absaroka I deposition; these uplifts define the present configuration of the Anadarko (13) and Fort Worth (14) Basins and a number of smaller southern Oklahoma basins (e.g., Ardmore, Marietta) that cannot be shown at the scale of Figure 5. The uplifts of the Amarillo-Wichita-Arbuckle trend rose from the depths of the old aulacogenic proto-Anadarko Basin, a region of very thick older Paleozoic strata. Therefore, 2 km or more of sedimentary rocks had to be stripped before crystalline rocks were exposed on the uplifts. As a result, much of the detritus mantling the flanks of the uplifts is chert/carbonate conglomerate; arkose, although common, is not as dominating as in the fill of the western basins.

Other fault-bounded blocks appeared to disrupt previous basinal or stable regions. The Central Basin Platform (15) arose from the axis of the Tobosa Basin, separating the Delaware (16) and Midland (17) Basins, with sufficient vertical movement to expose basement rocks; similarly, the Nemaha Uplift (18), defining the Salina (19) and Forest City (20) Basins, emerged high enough to be stripped to crystalline rocks. All of the positive elements identified above attained mountainous relief

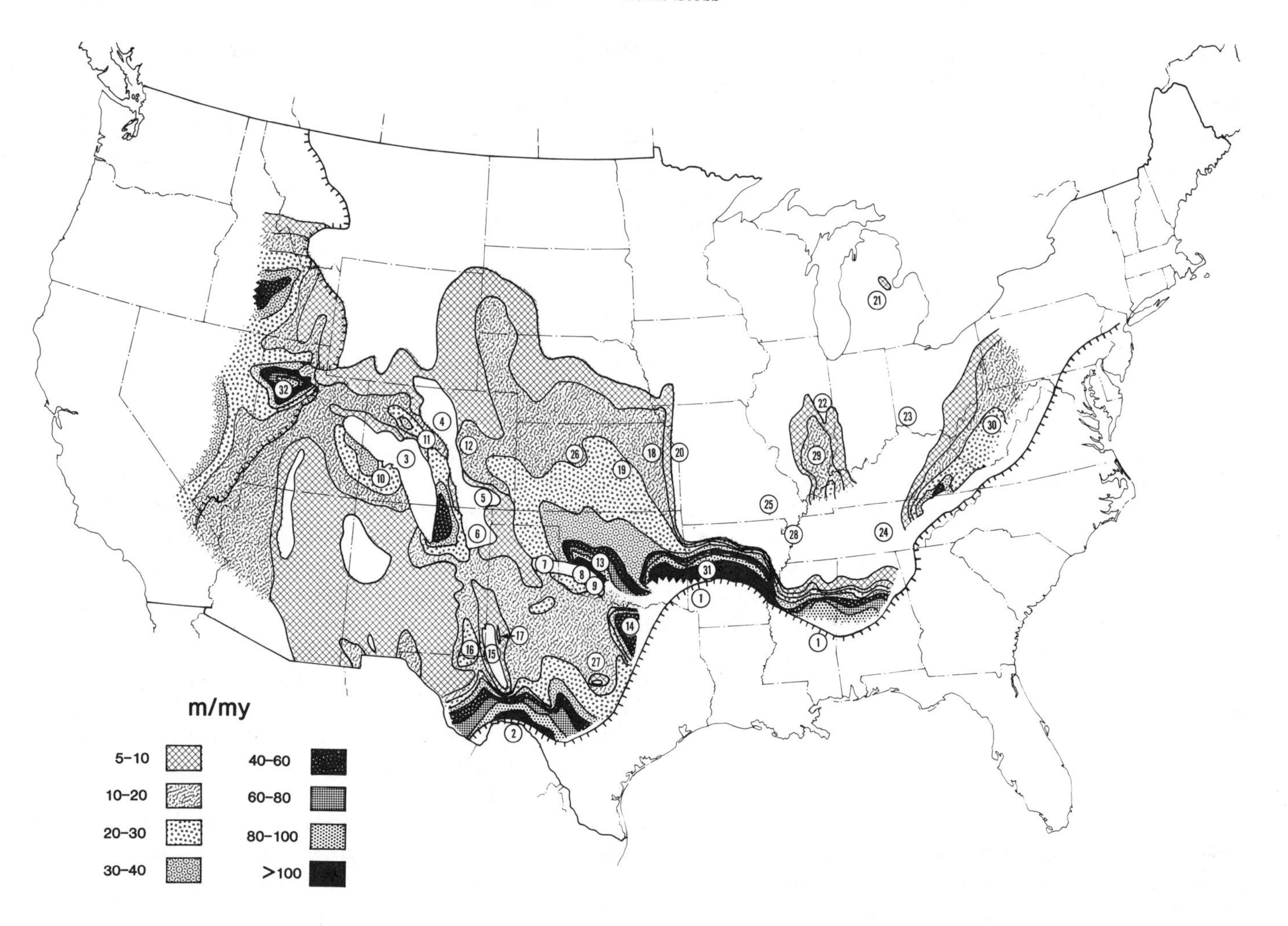

Figure 5. Absaroka I subsequence (latest Mississippian to Early Permian) net subsidence rate. 1, Ouachita margin; 2, Marathon margin; 3, Uncompahgre Uplift; 4, Front Range Uplift; 5, Apishapa Uplift; 6, Sierra Grande Uplift; 7, Amarillo Uplift; 8, Wichita Uplift. 9, Arbuckle Uplift; 10, Paradox Basin; 11, Eagle Basin; 12, Denver Basin; 13, Anadarko Basin; 14, Fort Worth Basin; 15, Central Basin Platform; 16, Delaware Basin; 17, Midland Basin; 18, Nemaha Uplift; 19, Salina Basin; 20, Forest City Basin; 21, Michigan Basin; 22, La Salle Anticline; 23, Cincinnati Arch; 24, Nashville Dome; 25, Ozark Dome; 26, Central Kansas Uplift; 27, Llano Uplift; 28, Pascola Arch; 29, Illinois Basin; 30, Appalachian Basin; 31, Arkoma Basin; 32, Oquirrh Basin.

(the complex of uplifts in Utah, Colorado, and New Mexico are known collectively as the "Ancestral Rockies"), although the Nemaha Uplift existed only briefly as a significant topographic feature, and contributed only a small volume of arkosic debris to the adjacent Forest City Basin. Elsewhere, a host of faulted and tilted blocks developed with lesser amplitudes than those reached by the mountain-forming elements but of great regional and local importance in controlling Absaroka I sedimentary facies and the structure of older strata. The faulted anticlines of the Michigan Basin (21) and the La Salle Anticline (22) are examples.

Many ancient arches and domes, such as Cincinnati (23), Nashville (24), Ozark (25), and Central Kansas (26), were reactivated. The ancient Texas Arch, which can be seen plunging southeast across the state on the Tippecanoe map (Fig. 2), became the site of the much more sharply emergent Llano Uplift (27) prior to and during Absaroka deposition. The Pascola Arch (28), bridging the gap between the Ozark and Nashville Domes, may have had an earlier history among the complex events of the Reelfoot Rift of northeastern Arkansas and western Tennessee, but its major uplift is post-Absaroka I, presumably mid-Permian. Notably, the late Paleozoic rise of the Pascola Arch finally provided the southern closure of the Illinois Basin.

On the eastern half of the U.S. cratonic interior during the time span of Absaroka I, the scene was dominated by multiple southwestward progradations of deltas alternating with eastward and northeastward marine transgressions. In the Illinois Basin (29) the ratio of marine to nonmarine beds approximates 1:1, producing classic coal cyclothems; to the east, in the Appalachian

Basin (30), the proportion of marine strata drops to no more than 10 percent; to the west, as in the Salina (19) and Forest City (20) Basins, marine environments were strongly predominant. Clearly, the major area of detrital provenance of the delta systems lay in eastern Canada where, as suggested by the lithic and feldspathic content of sandstones, large high-relief areas of crystalline rocks were exposed, presumably as a result of the same kind of (fault-involved?) vertical tectonics active elsewhere on the craton. Decades of coal exploration demonstrate that individual subaerial deltas were very large, some covering several states in a single episode of progradation; areas covered by individual marine-transgressive pulses were even more extensive.

The origin of the cyclical marine-nonmarine record of Absaroka I has been debated for generations. A strong case can be made for eustatic, perhaps glacio-eustatic, control, especially in view of the similarity of the cyclical behavior of other cratons in Carboniferous time. Advocacy of glacio-eustatic control, put forward by Wanless and Shepard (1936), has received recent strong support from the work of Veevers and Powell (1987). Consideration must be given, however, to the extreme tectonic activity of the craton during Absaroka I accumulation and to the evidence of erosion at intrasequence unconformities to depths unattainable by any suggested eustatic mechanism. Moreover, successive deltaic progradations demand successive rejuvenations of detrital-source regions, and there is no indication of systematic secular decay of topographic relief of source areas or of stream gradients from cycle to cycle as might be anticipated if eustasy were the controlling process.

Pre-Zuni (pre–Middle Jurassic and especially pre-Cretaceous) erosion has left a depleted record of Absaroka I deposition such that little is known about the sedimentational history of the Williston Basin or the northern reaches of the Appalachian Basin. Fission-track, vitrinite-reflection, and conodont-color measures of paleotemperatures indicate that 1 km or more of cover, presumably Absaroka I and largely Pennsylvanian, was stripped from the Michigan Basin; several kilometers may have been eroded from the north end of the Appalachian Basin (Johnsson, 1986), although there is dispute over the relative influence of elevated heat flow created by Mesozoic igneous activity (Karig, 1987).

The preserved record on continental-interior areas west of the distal deposits of the great delta systems and north of the "Ancestral Rockies," as in southern Montana, Wyoming, South Dakota, western Nebraska and Kansas, and eastern Colorado, reveals little evidence of continued influence of the Transcontinental Arch, a dominating feature of the tectonic geography in earlier Paleozoic time. The Wyoming area was occupied by a broad and stable shelf on which a blanket of quartz arenite was deposited. These sands persist to the sub-Zuni subcrop in southern Montana but intertongue to the east with finer grained detritals and marine carbonates; the suggestion is made that the blanket sands represent the distal deposits of debris shed from the mountainous uplifts to the south and "cleaned up" by reworking under conditions of very slow subsidence.

Tectonic activity at the Appalachian margin appears to have been subdued; clastic wedges from orogenic highlands southeast of the Appalachian Basin (30) were major contributors to the basin fill, but a significant proportion seems to have been transported parallel to the basin axis and along the eastern flank of the Cincinnati-Nashville trend from the same region that supplied the Illinois Basin and the midcontinent. Conditions were markedly different at the Ouachita-Marathon margin; here, collisional convergence created and maintained highlands, the source of massive clastic-wedge aprons that invaded rapidly subsiding foreland basins (e.g., Arkoma Basin, 31). The cratonic border of the Cordilleran trend was largely unaffected by extracratonic events. Subsiding elements, such as the Oquirrh Basin (32), differed from adjacent troughs of the "Ancestral Rockies" in the lack of complementary mountainous uplifts but accumulated extraordinary thicknesses of unexceptional fine sand, silt, and carbonate, nevertheless.

The end of Absaroka I deposition is most easily described in negatives. No younger Absaroka strata are preserved east of Nebraska and Kansas (roughly, eastern half of the craton); the spread of clastic wedges from the Ouachita-Marathon margin ended; rapid differential vertical movements of uplifts and basins of the "Ancestral Rockies" and southern Oklahoma trends ceased. In many places a significant disconformity separates Absaroka I and II, but the physical discontinuity is no more prominent than many that punctuate Absaroka I stratigraphy.

Absaroka II (Middle Permian to Upper Permian)

The middle subsequence of the Absaroka is essentially Middle and Upper Permian in age. Although this is the time of the ultimate closing of the proto-Atlantic (Iapetus) Ocean and the climax of the Alleghanian Orogeny, there is no record of a deeply subsident foreland basin bowed down by an excess mass of thrust sheets and sediment load. As is displayed on Figure 6, basins characterized by rapid subsidence are confined to the west Texas–southeastern New Mexico area and to basins on the [passive?] Cordilleran shelf margin.

The Texas–New Mexico complex of basins, known collectively as the Permian Basin, is the descendant of tectonic states prevailing during Absaroka I deposition but marked by greatly accelerated rates of subsidence, particularly in the Delaware Basin (1), and to a lesser degree in the Midland Basin (2). It is noteworthy that the Permian Basin, although on the foreland of the Marathon Orogen during Absaroka I deposition, underwent the highest subsidence rates after the cessation of Ouachita-Marathon activity. This region, one of the most important oil provinces of the North American Craton, has been documented by thousands of drillholes and described and interpreted in a library of publications. Interested readers will find the cursory treatment here to be unsatisfying and should seek further enlightenment in the Permian Basin chapter by Frenzel and others (this volume). Note that the Delaware Basin subsided at a higher *average* rate (>120 m/m.y.) than any other strictly cratonic negative element in this or any other time span covered in this chapter

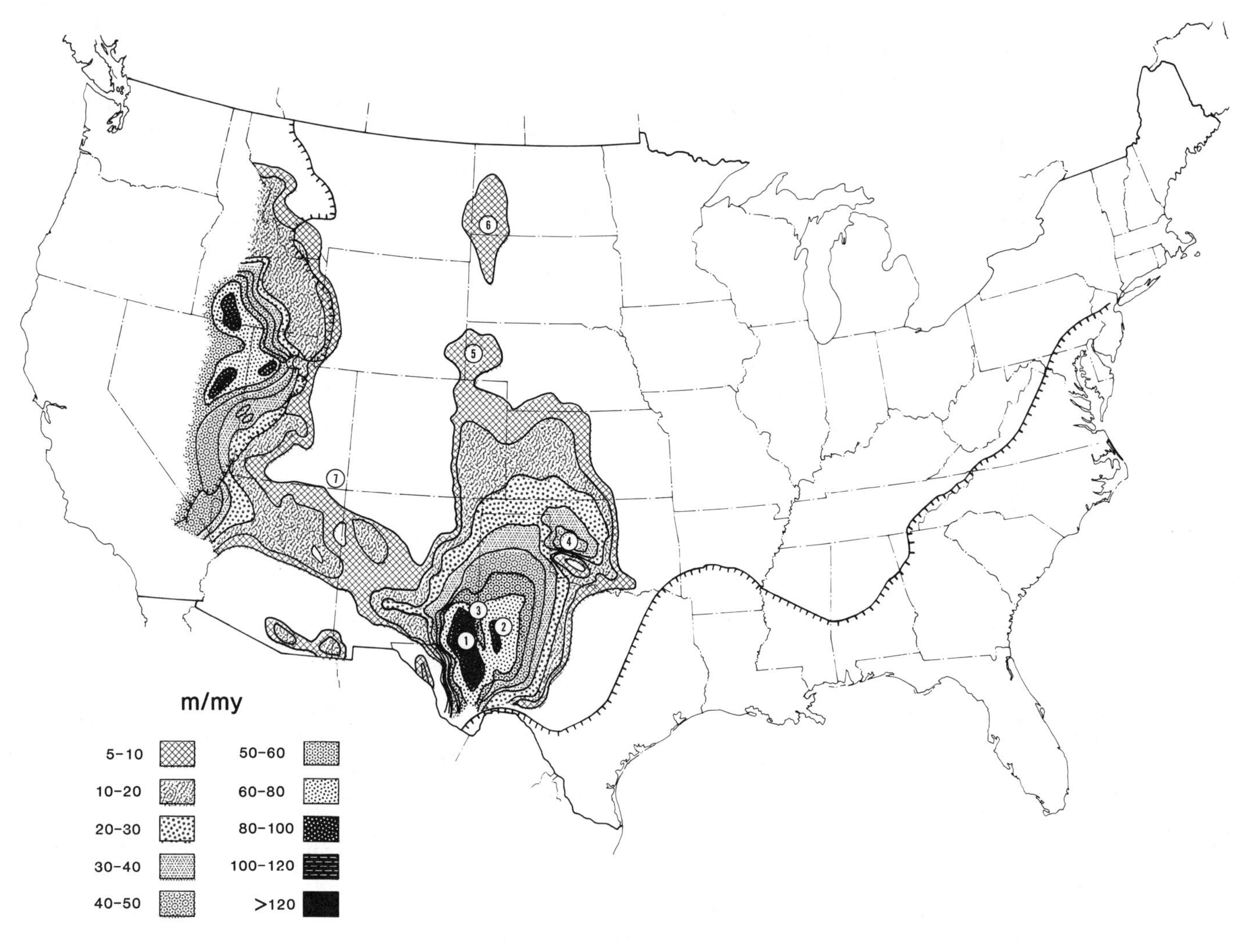

Figure 6. Absaroka II subsequence (Middle and Late Permian) net subsidence rate. 1, Delaware Basin; 2, Midland Basin; 3, Central Basin Platform; 4, Anadarko Basin; 5, Alliance Basin; 6, Williston Basin; 7, Colorado Plateau area.

(possibly excepting the Laramide Uinta Basin of Eocene time). Meanwhile, the surrounding shelves and the Central Basin Platform subsided more slowly. As a result of this differential and the absence of adjacent supra–base level uplifts, the basin, for much of mid-Permian time, was sediment-starved, and water depths far below wave base were persistent. Concurrently, carbonate-prone environments were established on the surrounding shelf areas and the Central Basin Platform, leading to the growth of massive shelf-edge limestone prominences, including algal-bound reefs, behind which sabkhas and evaporitic lagoons were formed. Ultimately, reef growth restricted communication with the open sea to the south, and the entire basin complex became the site of thick evaporite deposition, including halite and bittern salts, filling the void created by the previous excess of subsidence over sedimentation.

The Anadarko Basin (4) continued to subside, but more slowly than before, and receive Absaroka II sediments derived from the Amarillo-Wichita trend, but in lesser volume and finer grain size. To the north, as in the Nebraska-Kansas region of the midcontinent, the previous high level of tectonic differentiation into positive and negative elements virtually disappeared, resulting in a broad stable area affected by repeated alternations of marine and continental deposition. Commonly, episodes of marine regression led to restriction and evaporitic states, while the advances of nonmarine detrital material from the east and northeast are represented by red beds. Thus, circumstances not unlike those that prevailed during Absaroka I accumulation continued in force, but climatic change prevented the development of coal cyclothems. Similar conditions extended north to the feebly subsident Alliance (5) and Williston (6) Basins and their enclosing shelf areas.

The area of the present-day Colorado Plateau (7) and of the "Ancestral Rockies" uplifts was also the site of mixed continental and marine environments. The old uplifted blocks were now reduced to low hills capable of yielding detrital sediments that intertongue with carbonate-evaporite products of seas transgressing from the west to form a complex of lithic facies characterized by difficult correlations and a plethora of stratigraphic terminology. Here, as in the western midcontinent, red beds are important components of the nonmarine deposits, but more impressive units are formed by quartz arenites, representative of dune and strandline environments. Some of the sand tongues and sheets derive from reworking of Absaroka I sediments and from erosion of the basement-rock cores exposed on the stumps of the "Ancestral Rockies" blocks; other units present evidence of transport from uncertainly identified regions of provenance to the south.

Absaroka II deposition on the Cordilleran cratonic margin was characterized by carbonate-dominated marine environments and the presence of rapidly subsiding continental-shelf basins. The eastern Idaho–western Wyoming–southwestern Montana region is of special interest because of the occurrence of phosphatic black shale, pisolitic phosphorite, and bedded chert as a related subset of facies interrupting the normal marine succession. It is widely accepted (e.g., Peterson, 1980) that the phosphate/chert facies is the product of upwelling of enriched waters from a deep shelf basin to the west. East of this region, and elsewhere inboard of the Cordilleran cratonic margin, marine and littoral sediments pass into continental red beds.

East of the continental shelf margin and in areas of the craton proper not affected by post-Absaroka erosion, the top of the Absaroka II subsequence is marked by a minor disconformity that only locally displays significant stripping. The discontinuity is commonly difficult to identify, especially in a nonmarine succession complicated by many channeled interruptions. Clearly, there was no craton-wide Early Triassic episode of cratonic emergence above base level.

Absaroka III (Triassic to Early Jurassic)

The terminal Absaroka subsequence is the record of deposition from Early Triassic through most of Early Jurassic. In terms of cratonic tectonism, and with the exception of the Atlantic/Gulf of Mexico margin, the history of the subsequence mimics that of Absaroka II, only more so. The mountainous blocks of the "Ancestral Rockies" of Late Carboniferous–Early Permian time were very nearly stabilized and all but mantled by deposition while the somewhat younger episode of deep subsidence in the Permian Basin complex was largely completed before the initiation of Absaroka III deposition. Indeed, inboard of the shelf basins of the Cordilleran margin, there is little evidence of tectonic differentiation of the craton during this early Mesozoic time. Furthermore, as the highest package of strata below the profound sub-Zuni unconformity, the subsequence as preserved in many areas is but a shadow of what it may once have been. As a result, only a few words of description are required at the scale employed here.

Figure 7 displays those areas where Absaroka III, as preserved, accumulated at a rate greater than 5 m/m.y. As a nod to completion of the record, some of the 15 or more extensional basins (1) of the now-cratonized Appalachian margin are shown. These appear to anticipate the Jurassic opening of the North Atlantic and, along with similar troughs in the subsurface of the Gulf margin, are treated more exhaustively and more appropriately in the Gulf of Mexico Basin (Salvador, 1988), Atlantic Continental Margin (Sheridan and Grow, 1988), and Appalachian and Ouachita Orogen (Hatcher and others, 1988) volumes of this series.

Nowhere on the U.S. part of the craton, even where Absaroka III is relatively untouched by erosion, is there a record of net subsidence at rates greater than 20 m/m.y. except along the Cordilleran shelf margin. This lack of significant differential subsidence of the cratonic interior persists through the following depositional episodes of Zuni I and II (i.e., from Early Triassic through Early Cretaceous, an interval of 150 m.y. devoid of craton-interior uplifts and basins south of the James Bay Basin of the Hudson Bay region of Canada). Transient and trivial episodes of Absaroka III interior downwarp are marked by the preservation of red beds in the Permian Basin (2) and by salt in the Williston Basin (3), although the assignment of the latter to a Triassic age is subject to debate.

Early Triassic sea occupied the Cordilleran shelf margin of the craton, lapping eastward from southeastern Idaho and northern Utah, where thick carbonates accumulated, to intertongue with red continental sediments that dominate regions to the east. Farther south, as in the area of the Colorado Plateau and the adjoining shelf margin, Early Triassic environments were characterized by extensive tidal flats, punctuated by repeated marine transgressions from the west and continental progradation from the east. As in the more northerly areas, fluviatile conditions became increasingly prevalent to the east. Here, minor contributions of detritals were provided by topographic remnants of the "Ancestral Rockies."

Marine Absaroka III strata younger than Early Triassic are unknown on the Cordilleran shelf or on the craton proper to the east. Overlying Triassic rocks, generally assigned to Late Triassic, but with minimal biostratigraphic control, commonly rest on a surface of regional unconformity. Global patterns of coastal onlap indicate a major sea-level–lowering event at the close of the Middle Triassic (beginning of Supercycle Upper Absaroka A-3 of Haq and others, 1987); this, plus other eustatic events of mid-Triassic time, may explain the apparent absence of Middle Triassic sediments. In any case, much of the Late Triassic record is characterized by the spread of fluviatile deposits, commonly red, with irregularly distributed gypsum beds. Near to the close of the period, and continuing through Early Jurassic time, fluviatile and ephemeral lacustrine environments gave way to shifting dune fields only occasionally interrupted by the reappearance of streams and ponds. Such was the degree of tectonic stability that substantial blankets of quartz arenite cover virtually the entire area of preservation of upper Absaroka III strata and extend westward in disjunct patches to west-central Nevada (4), far beyond the cratonic margin.

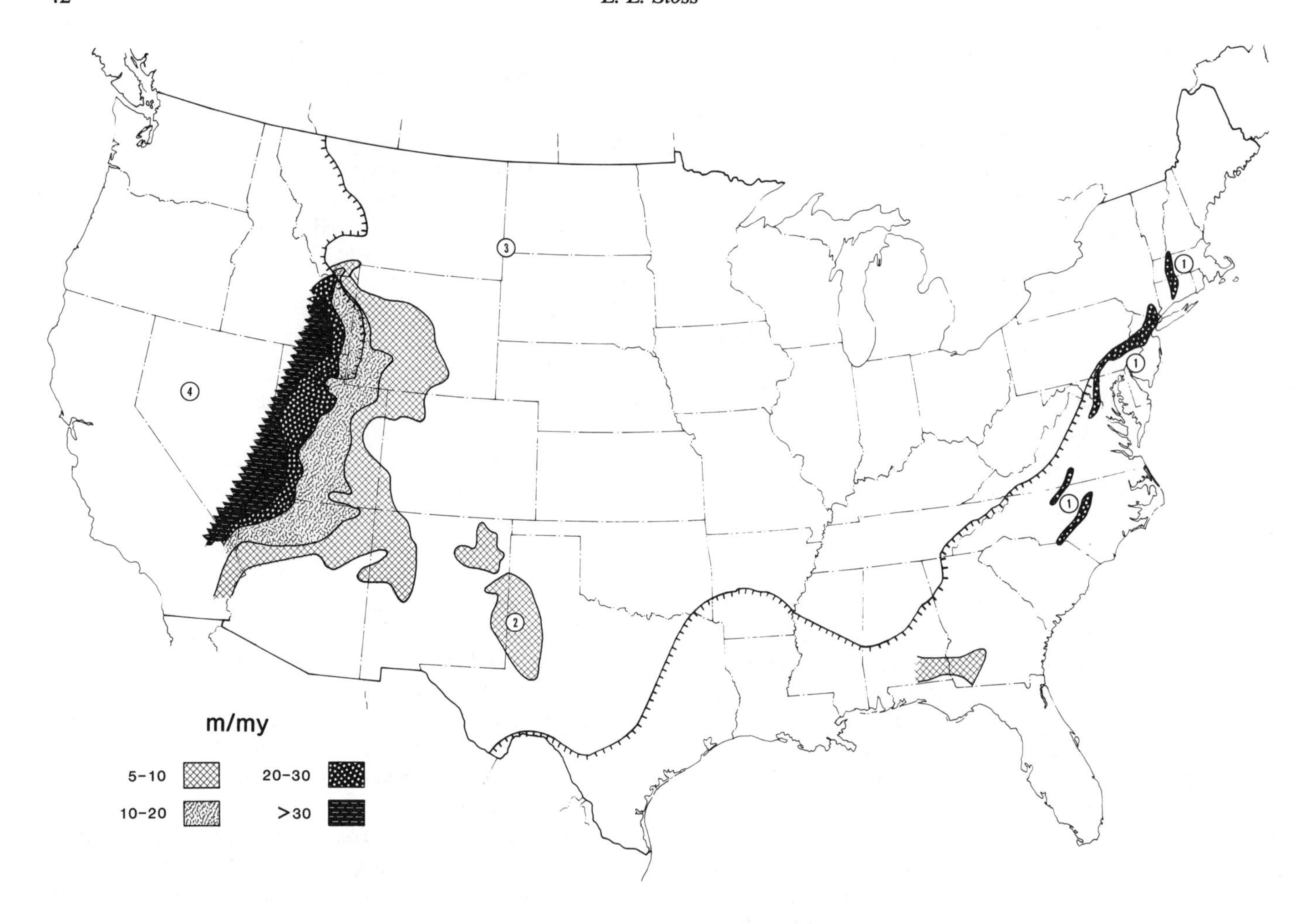

Figure 7. Absaroka III subsequence (Triassic to Early Jurassic) net subsidence rate. 1, Extensional basins; 2, Permian Basin; 3, Williston Basin. 4, Western limit eolian sands.

Absaroka deposition was terminated near the close of Early Jurassic time by cratonic emergence above base level. Concomitant erosion was relatively slight on southwestern regions of the craton, as in the area of the Colorado Plateau, and increased markedly to the north and east. Because of the imposition of repeated intra-Absaroka and post-Zuni erosional episodes, it is difficult to quantify the degree of pre-Zuni stripping, but it must have been substantial. Stearns and Reesman (1986), for example, employ paleotemperature data to estimate that approximately 1.5 km of sedimentary rocks were stripped from the Nashville Dome between Permian and Cretaceous time. Erosion of the same order of magnitude is suggested at the sub-Zuni unconformity in south-central Saskatchewan where Devonian rocks are overlain by Middle Jurassic.

ZUNI SEQUENCE (MIDDLE JURASSIC TO EARLY PALEOCENE)

The Zuni sequence extends in time from latest Early or earliest Middle Jurassic (Aalenian) through Early Paleocene (Danian). This is the time of establishment and evolution of passive continental shelves and slopes at the eastern and southeastern margins of the craton, regions covered by the Atlantic Continental Margin (Sheridan and Grow, 1988) and Gulf of Mexico Basin (Salvador, 1988) volumes of this series. Therefore, much of the discussion of the Zuni sequence is here confined to western areas of the craton and the Cordilleran cratonic margin, as are the accompanying maps (Figs. 8, 9, and 10).

Sub-Zuni surface

As noted above, erosion of older rocks during the pre-Zuni hiatus was severe throughout the craton except for the area of the Colorado PLateau. Indeed, much of the complexity of the geologic map of the United States east of the Laramide uplifts and beyond the cover of Cretaceous strata is no more than trivially modified from patterns that must have existed beneath a Zuni cover, which encompassed very nearly the entire area of the craton in the United States. These exhumed areal-geologic complexities, and their continuation as the sub-Zuni subcrop patterns of the western midcontinent and Great Plains regions (see the pre-Cretaceous paleogeologic map by Bunker and others, Plate 4J), are more the inheritance of syn-Absaroka tectonics (particularly Pennsylvanian–Early Permian events) than of differential vertical movements immediately prior to Zuni deposition.

Regional subcrop patterns at the sub-Zuni unconformity in central and western Montana (e.g., Perry and Sloss, 1943, Fig. 2) are clearly attributable to pre-Zuni erosion, which strongly overprints the effects of intra-Absaroka erosional episodes. In this region, emergence of the Sweetgrass Arch, a broad, south-plunging positive flexural element, is responsible for progressive mid-Jurassic overstep on older and older units from Triassic in northern Wyoming to Early Mississippian in north-central Montana and, as mentioned earlier, to Devonian in south-central Saskatchewan. The degree of pre-Zuni erosion declines southward to the Colorado Plateau region and the physical and chronostratigraphic evidence of unconformity becomes more obscure, particularly where continental red beds or eolianites are succeeded by deposits of the same environments.

The overstep and burial of the sub-Zuni surface has strong parallels with the covering of the sub-Sauk surface. Some 80 or more m.y. elapsed between the initiation of Zuni onlap near the end of the Early Jurassic and ultimate submergence of areas of the Canadian Shield in Late Cretaceous time, a span comparable to that involved in the expansion of Sauk deposition. In both cases the preceeding hiatus was a time of broad cratonic uplift without marked differentiation of craton-interior positive and negative elements. The early history of both sequences is marked by limitation of submergence at a distance inboard of the margins to areas adjacent to the Cordilleran border. Both sequences record pulsatory and progressive expansion of areas subsiding below depositional base level, broken by episodes of offlap and minor erosion. A final point of comparison is the imposition of the Mississippi Embayment over the rejuvenated Sauk-age aulacogenic Mississippi Graben–Reelfoot Rift trend.

Zuni I (latest Early Jurassic, Aalenian, to Early Cretaceous Berriasian)

In conventional terms the Zuni I subsequence is best described as comprising deposits of Middle and Late Jurassic age. In the classification of Haq and others (1987), Zuni I is the equivalent of lower Zuni A plus the first supercycle of lower Zuni B and thus extends from Aalenian to the major eustatic withdrawal of earliest Cretaceous time (Late Berriasian). In actuality, the oldest datable Zuni strata on the craton are Bajocian, and the detail of timing of the close of Zuni I deposition is obscured by the imprecision of nonmarine biostratigraphy near the Jurassic-Cretaceous boundary.

As shown by the accompanying map (Fig. 8) of the western cratonic area considered here, the subsidence pattern is remarkably similar to that of Absaroka III; that is, areas of subsidence at rates in excess of 20 m/m.y. are confined to the Cordilleran margin. Note, however, the shrinkage of this area of moderate

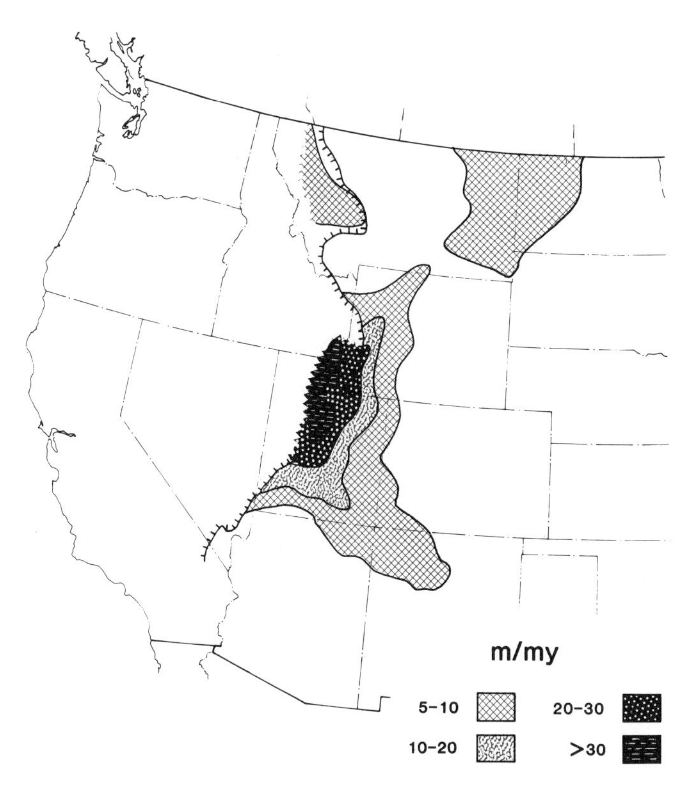

Figure 8. Zuni I subsequence (Middle Jurassic to earliest Cretaceous) net subsidence rate.

subsidence; note also the eastward shift in Wyoming and Utah of the line along which there is an acceleration of the rate of change of subsidence rates—a phenomenon discussed in a later paragraph. A further modification of the Absaroka III subsidence geography is the reappearance of the Williston Basin as a negative element surpassing the 5 m/m.y. threshold for inclusion on the map.

The Middle and early Late Jurassic history of the western craton is a variation on the theme made familiar by preceding subsequences; that is, persistent seaways occupied the more rapidly subsiding areas of the cratonic border, spreading inland to form carbonate and evaporitic tongues intercalated with continental deposits. Marine limestones form an important facies of the overlap succession, overstepping the truncated limits of older strata on the Sweetgrass Arch and commonly succeeding basal evaporites elsewhere. In places, as in central Utah and the Williston Basin, salt is a significant component of the evaporites, indicating isolation by tectonic subsidence. Pre-Oxfordian marine and strandline sandstones and their eolian equivalents are largely clean quartz arenites derived from the interior, from exhumed "Ancestral Rockies" sources, and from putative areas of provenance in southern Arizona. These sands intertongue with fluviatile red beds, which dominate the easternmost areas of preservation of the subsequence.

The Oxfordian was a time of renewed marine transgression of the craton from the north accompanied by the invasion from the west of sands bearing abundant chert grains, suggesting derivation from Permian chert units of the cratonic margin and from extracratonic sources beyond the continental shelf and slope. This marks the first occasion of the cratonward spread of detritus (however distal with respect to source) from Cordilleran orogenic highlands since clasts eroded from craton-margin thrust sheets were carried to foreland-basin sites in the Mississippian. The date of initiation of the Sevier Orogeny is clouded by a debate that is more semantic than geologic; nevertheless, from Oxfordian to Danian, sedimentation on the western reaches of the craton was repeatedly influenced, and eventually dominated, by floods of detritals from orogenic sources in linear trends more or less parallel to the cratonic margin.

Mention was made above of the eastward creep of the belt of rapid subsidence near the cratonic margin in Wyoming and Utah. Similar marginal downwarps are shown on the maps of the Tippecanoe sequence (Fig. 2) and of the Kaskaskia I (Fig. 3) at the Appalachian margin and Absaroka I (Fig. 5) at the Ouachita-Marathon margin; these are concomitants of episodes of plate convergence and, at least in the Tippecanoe and Absaroka examples, can be related to foreland loading by stacked thrusts as well as by the mass imposed by externally derived detritals. It would be easy to attribute the accelerated subsidence of Jurassic time to an equivalent combination of sedimentary and tectonically emplaced loading; however, the major subsidence was pre-Oxfordian, no thrusts of appropriate age and position are known, and the sedimentary fill is dominated by carbonates and evaporites.

The importance of the western provenance region was temporarily overshadowed in latest Jurassic time by the rise to prominence of an as-yet-unidentified source to the south and southwest of the map area of this study. Major stream systems carried coarse clastic debris north- and northeastward across the present site of the Colorado Plateau, across Wyoming, and, with decreasing grain size and increasing paludal contributions, toward the Canadian border in Montana. Popular paleogeographic reconstructions suggest that the southern source for these latest Jurassic detritals was southeasterly extension of Nevadan (or early Sevier) orogenic highlands across the Mogollon area of south-central Arizona; however, the igneous petrologies of clasts in these deposits are incompatible with a Mogollon source, and a provenance region must have existed south of the Mexican border in Sonora (L. T. Silver, personal communication, 1986). The same cryptic region may be considered responsible for tuffaceous material intermingled with fluviatile and lacustrine deposits near the top of the subsequence in the southwestern area of the craton.

The globally effective episodes of lowered sea level in earliest Cretaceous time are commonly difficult to identify by physical criteria amid the many evidences of channeling and scouring inherent in the ubiquitous continental deposits of the closing phases of Zuni I and the initial sedimentation of Zuni II. It is generally agreed, however, that a lacuna involving 10 m.y. or more separates subsequences Zuni I and II over much of the Western Interior Basin, the cratonic region considered here, increasing in amplitude to the south. An equivalent stratigraphic discontinuity is manifest at the Atlantic and Gulf of Mexico margins.

Zuni II (Early Cretaceous, Valanginian, to Late Cretaceous, early Cenomanian)

The second Zuni subsequence is essentially the Lower Cretaceous succession. In the detail provided by Haq and others (1987), Zuni and others (1987), Zuni II encompasses chronostratigraphic divisions from early Neocomian through the greater part of Albian, but in terms of the Western Interior record of concern here, pre-Aptian strata are of little consequence inboard of the cratonic margin. It would appear that much of Neocomian time was occupied by nondeposition and erosion accompanying broad cratonic uplift and the initiation of differential movement of blocks on the Sevier foreland. Regional isopach, paleogeographic, and subcrop maps (e.g., McGookey and others, 1972) indicate continued supra–base level conditions in the Colorado Plateau and adjacent southern Rocky Mountain areas during much of the time of Zuni II accumulation to the north.

The accompanying subsidence map (Fig. 9) illustrates the emergent state of the area noted above, which was bounded near the Mexican border in southeastern Arizona and southwestern New Mexico by relatively rapid downwarp of the Pedregosa Basin. Also worthy of attention is the evidence of deceleration of subsidence in the adjoining southern stretch of the cratonic margin and the coeval spread of increased subsidence northward from southeastern Idaho into western Montana. Fluviatile deposits cratonward of the belt of rapid subsidence include chert-pebble conglomerates that thicken and coarsen to the west, suggesting provenance in tectonic highlands along the east-migrating front of the Sevier belt. Complex facies patterns inferred to represent fluviatile deposition in the same area are interpreted (DeCelles, 1986) as foreland responses to Sevier tectonics.

Concurrently with the distribution of detrital material eroded from Sevier highlands, other, westward-flowing, stream systems carried smaller but noteworthy volumes of sediment derived from the emergent Canadian Shield. These sediments from an eastern source mingle distally with the chert-rich detritus from the west.

The time span of the Zuni II subsequence coincides with one of the great episodes of marine transgression on continents. In western North America this episode is signalled by the southward invasion of seaways from the Arctic, confined between the Canadian Shield and the Sevier Orogen, and the northward spread of marine waters from the Gulf of Mexico. In the relevant area of Figure 9, however, a combination of emergent stability of the Colorado Plateau–Southern Rockies region and the flood of

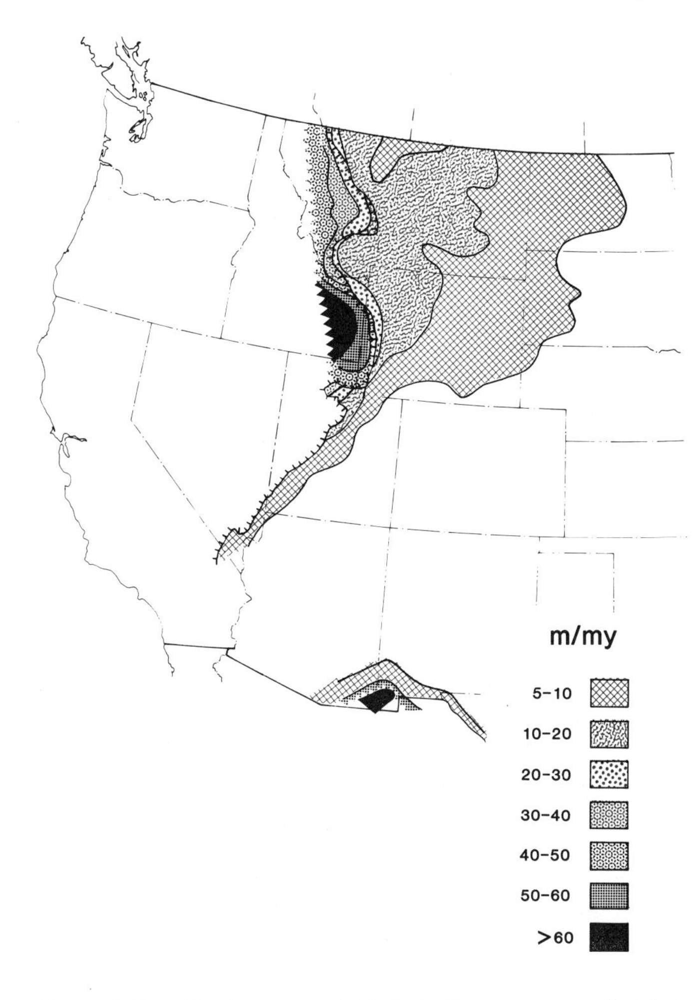

Figure 9. Zuni II subsequence (mid–Early Cretaceous to earliest Late Cretaceous) net subsidence rate.

clastic-wedge detritals from the Sevier trend operated to delay marine transgression south of the central latitudes of the Prairie Provinces of Canada and north of southeastern Colorado until near the end of Zuni II deposition. Subsidence of northern areas of the craton in the western United States was substantial, however, and extensive lacustrine basins filled with carbonates were developed. With approach to the close of deposition here allocated to Zuni II, continued subsidence and rising sea levels brought about the juncture of marine waters from the Arctic and the Gulf. This, the first major transgression of the Western Interior region since early in Late Jurassic time, was short lived. Regional and localized uplift in combination with eustatic sea-level fall late in the Albian was accompanied by regression and the development of an erosion surface marked by incised stream channels.

Zuni III (Late Cretaceous, late Cenomanian, to early Paleocene)

The Zuni III subsequence is represented by sedimentary and volcanic rocks that range in age from latest Albian (latest Early Cretaceous) to Danian (Early Paleocene). Many factors conspire to make any effort to produce a capsule review of the subsequence ineffectual, if not misleading. The time involved is characterized by an intricate history of eustatic events documented chronologically in exquisite detail by a highly developed biostratigraphy, supported and amplified by the wealth of event stratigraphy provided by numerous traceable and identifiable bentonites. It is also the time of maximum activity of the Sevier Orogen whose concomitants include sedimentational, volcanic, and tectonic influences on the adjacent craton to a degree that tends strongly to override or at least modulate the effects of global eustasy. Further, Zuni III strata of the Rocky Mountain and Great Plains states include some of the world's greatest coal reserves and are both the sources and reservoirs of important oil and gas resources. As a result, these rocks have been the subjects of innumerable local and regional outcrop and subsurface studies, which are integrated and synthesized in a library of reports. The paragraphs that follow cannot begin to expose the complexity of what is known about this fascinating segment of cratonic history.

A glance at the Zuni III subsidence map (Fig. 10) reveals the intensity of downwarp of the western areas of the craton; the rates exceed anything recorded inboard of the margins since the Absaroka II subsidence of the Permian Basin some 160 m.y. earlier. Figure 10 is itself an integration of the effects of vertical tectonic events that succeeded one another over a span of more than 30 m.y., such that only the high-amplitude features are visible. The Wind River (1), Front Range (2), and Kaibab (3) Uplifts are among the positive elements that are resolved, as are the Red Desert–Hanna (4), Piceance-Washakie (5), Denver (6), and Pedregosa (7) Basins. (Note that these include elements that are further subdivided and delimited by Laramide events of Cenozoic time.)

Pertinent details of Late Cretaceous tectonic evolution and its sedimentational consequences in the Western Interior region are presented in the studies of individual basins and in separate chapters (Molenaar and Rice; J. Peterson, this volume) that are important parts of this volume. Many workers have published papers of regional scope attempting to relate eustatic and tectonic events to form a cohesive picture; it cannot be stated, however, that a clear consensus has emerged. The spectrum of opinion is exemplified by recent reports from highly qualified experts. Kauffman (e.g., 1986 and references therein) places great emphasis on sea levels and tectonics controlled by mechanisms external to the region (i.e., mid-ocean ridge volume, Pacific-margin convergence and subduction). Weimer (e.g., 1986) stresses the overriding but not exclusive influences of sedimentation rates and dispersal paths and of localized movements of discrete basement blocks, which are not necessarily precursors of Laramide elements. Most workers agree that there were repeated episodes of basinal subsi-

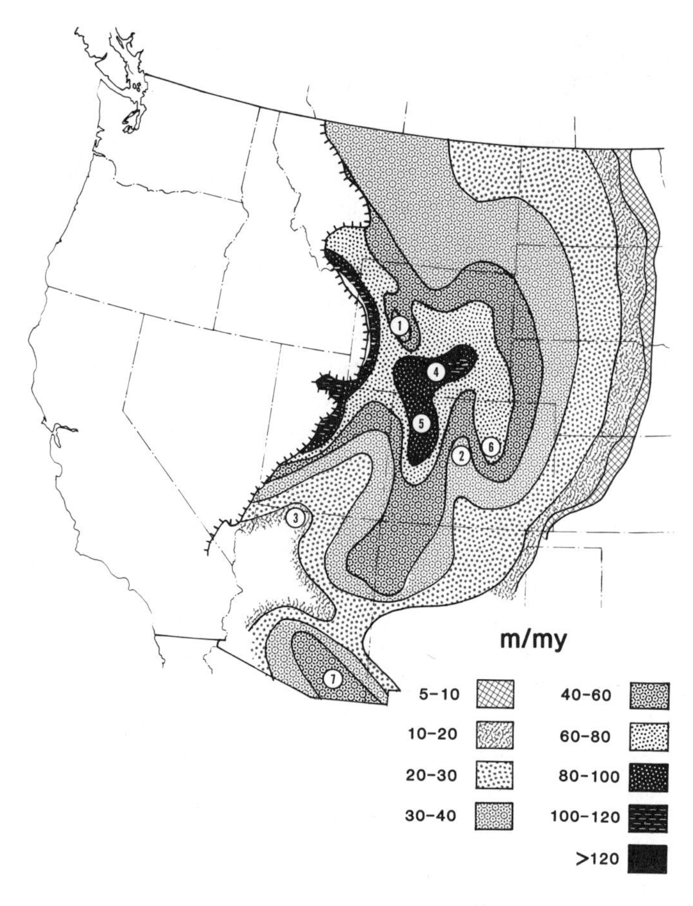

Figure 10. Zuni III subsequence (Late Cretaceous to early Paleocene) net subsidence rate. 1, Wind River Uplift; 2, Front Range Uplift; 3, Kaibab Uplift; 4, Red Desert–Hanna Basin; 5, Piceance-Washakie Basin; 6, Denver Basin; 7, Pedregosa Basin.

dence and that these coincided with eustatic highstands; the tectonic downwarp and high sea levels combined to create transient anoxic pockets of deep water. The kerogen-rich products of deep-water sedimentation, in turn, are responsible for the oil and gas resources that make the Zuni III strata of the Western Interior a target for continued study.

Returning to Figure 10, attention is directed to the strip of very rapid subsidence along the trace of the Sevier thrusts in central Utah, western Wyoming, and southwestern Montana. Conventional wisdom dictates that this most rapidly subsiding keel of a markedly asymmetric foreland (or retro-arc) basin represents isostatic adjustment to the excess mass imposed by stacked thrust sheets and great thicknesses of piedmont deposits. It is beyond denial that these sources of crustal loading must have found a response in subsidence; there are difficulties, however, in easy acceptance of the excess-mass hypothesis as the unique explanation of recorded subsidence rates. Note, for example, that the preserved evidence of maximum downwarp is confined to the arcuate salient centered on the southwest corner of Wyoming. There is no indication in geophysical data or thrust-belt stratigraphy to suggest that equal thicknesses representing equal subsidence rates continue along the orogenic front southward into southern Nevada. (Interpretation of a similar lack in western Montana is complicated by the synorogenic Idaho batholith and its satellites, but similar relative deficiencies in subsidence continue north of the batholith.) Moreover, different segments of the orogenic trend, as delineated by shifting centers of sediment dispersal, were loci of uplift at different times and indicate no particular concentration of activity adjacent to the position of maximum subsidence. Finally, the region of rapid downbuckling is in remarkable conformity with patterns that existed in earlier episodes of Phanerozoic time and which bore no relation to sedimentational and tectonically emplaced loads.

Regardless of the influences on subsidence from Sevier-belt activity, it is clear that repeated pulses of uplift and thrust imbrication at the cratonic margin resulted in the spread of clastic-wedge aprons whose deltaic distal fronts contested with eustatic events to establish the shifting western strand of the interior sea. At the same time, stream systems arising in the continental interior carried detrital sediment to the eastern shores of the same seaway, forming lesser, but significant, deltaic deposits at positions more directly subject to changes in sea levels.

Certain clastic wedges, particularly in southwestern Montana and western Wyoming are preserved close to their proximal limits. Here, they are coarse piedmont fanglomerates that grade eastward to fluviatile deposits of choked and braided streams, passing to meander-belt and point-bar accumulations interspersed with coal-swamp deposits. Distally, and commonly 300 km or more from the Sevier highlands, the clastic wedge complex ends in a typical delta-front assemblage of beach and barrier-island sands intertongued with marine shales.

The majority of clastic wedges are dominated by regressive deposits, which are covered by very thin transgressive deposits, indicating rapid marine onlap following slow progradation. Some of the marine transgressions that abruptly close episodes of clastic-wedge progradation can be identified as times of global coastal onlap and presumed eustatic highstands. Others, however, show no such correlation and appear to represent the cessation of uplift of segments of the Sevier trend, cutting off the detrital supply and permitting continued subsidence to promote transgression.

Magmatic-arc volcanism influenced sedimentation on western areas of the craton from at least Late Jurassic time. Marine strata of Zuni III are punctuated by many bentonites marking ash falls carried eastward from the Sevier orogen by prevailing westerly winds. Eventually, major andesitic extrusions and volcanogenic sediments displaced normal fluviatile deposition on accumulating clastic wedges, as in western Montana, and volcanic contributions to sedimentation spread far to the east by Paleocene time.

On their global cycle chart, Haq and others (1987) place the top of the Zuni sequence at a eustatic-fall event late in Maastrichtian (latest Cretaceous) time, some 3 to 8 m.y. earlier (and significantly lower in the stratigraphic succession) than the sequence boundary as previously defined. Motivated by more than pride of authorship, I find little to recommend this revisionist position other than the questionable recognition of a degree of parity and regular periodicity of chronostratigraphic segments defined by sea-level events. In any case, the biologically catastrophic K/T extinction event finds no counterpart in the gross physical stratigraphy of the Western Interior region.

Driven by deposition in excess of subsidence, the ultimate marine regression and clastic-wedge progradation of the region began in the Late Cretaceous and carried on into the Paleocene (briefly punctuated by a transient and areally circumscribed Danian resurgence of the inland sea). Sediment-distribution paths, although locally influenced by rising foreland elements, continued to be fed by Sevier-trend highlands and, to a lesser degree, by craton-interior sources as before. Thus, the continuity of tectonic and depositional modes into Paleocene time mandates the inclusion of Danian strata in the Zuni sequence.

The major contrast between the sedimentary tectonics of the Zuni sequence and those prevailing during Tejas accumulation is the dominance of detrital sources in the Sevier trend in the former and from Laramide foreland uplifts in the latter. Western provenance areas in Sevier highlands continued to supply the greater volume of sediment to the Western Interior Basin throughout the time of Zuni III. Through Late Cretaceous and early Paleocene time, intrabasin uplifts became increasingly important, and precursors of Laramide elements became identifiable. Ultimately, before the close of the Paleocene, uplift within the Sevier belt and concomitant sediment supply were greatly reduced, to be replaced by rapid emergence of foreland blocks subject to erosion to form the sedimentary fills of intermontane Laramide basins. As a result, a major unconformity, commonly with marked angular discordance, characterizes the surface separating Late Paleocene/Eocene strata of the Tejas sequence from underlying pre-Tejas rocks.

TEJAS SEQUENCE (LATE PALEOCENE TO PRESENT)

Deposition of the uppermost and youngest of the major unconformity-bounded successions of cratonic stratigraphy, the Tejas sequence, began at or near the close of Danian (early Paleocene) time; at the present time, almost all of the North American Craton is above depositional base level, and sub-base level cratonic accumulation of sediment is limited to the passive margins of Atlantic and Gulf coastal regions. Some future depression of the craton below depositional base level will form the supra-Tejas unconformity. The chronostratigraphic boundaries of the post-Tejas nondepositional hiatus are indeterminate, but it is predictable that most of the younger Tejas now well above sea level, including much of the glaciogenic Pleistocene sediments, will be stripped by erosion and expunged from the permanent geologic record before burial and subsidence below a younger sequence.

As in the case of the Zuni sequence, separate volumes of the Decade of North American Geology treat the Atlantic Shelf and the Gulf of Mexico Basin, leaving the areas of the Rocky Mountains/Colorado Plateau and Great Plains for inclusion here. The greater part of Tejas stratigraphy of this region, and the tectonic evolution it records, is preserved in the well-documented Laramide basins; these are described in companion chapters by acknowledged experts (Baars and others, this volume), thereby reducing the responsibility for exhaustive coverage here. Thus, by limiting the area of concern and by reliance on other chapters, the doubtful competence of this writer in matters of Cenozoic history may be mercifully veiled.

Tectonic evolution during Tejas accumulation is here treated in two segments appropriate for discussion of the Rocky Mountain, Colorado Plateau, and Great Plains region: Tejas I includes late Paleocene and early and middle Eocene (approximately the time span 60 to 39 Ma); Tejas II and III are combined for coverage in this essay. In effect, the treatment recognizes Laramide and post-Laramide phases. In terms of the marine sequence stratigraphy and derived chronology of eustatic events delineated by Haq and others (1987), the base of Tejas I in the totally nonmarine succession considered here is at or near the base of Tejas supercycle A2, and the contact between Tejas I and II is close to the separation between Tejas supercycles A3 and A4 as these are defined by Haq and colleagues.

Sub-Tejas surface

Both the Absaroka I and Tejas I subsequences accumulated during times of elevated tectonic activity of the craton, including the uplift of mountainous blocks adjacent to rapidly subsiding basins. The time spans of both subsequences were preceded by premonitory movements deep within the craton, as well as at active margins, while passive margins were relatively unaffected. In the Tejas case, broad emergence of the cratonic interior, marine regression, and passage of former depositional surfaces above base level began in Late Cretaceous time; concomitantly, elements of the Sevier foreland (as shown on Fig. 10, the Zuni III map) began to affect the distribution and thickness of sedimentary accumulation, and some blocks proximal to the belt of advancing overthrusts were carried above base level and into an erosional regime. As a result, the time of initiation of the pre-Tejas hiatal (nondepositional) interval ranges from late Late Cretaceous to late early Paleocene, and the severity of pre-Tejas erosion varies from stripping of all Phanerozoic strata from many uplifted blocks to trivial degradation at the Atlantic and Gulf margins and at the axes of deeply subsident Laramide basins. Sub-Tejas subcrop geology, reflecting these factors, is thus very complex in the regions of active cratonal tectonics (the Laramide uplifts) where basal Tejas beds rest with angular discordance on any and all older rocks, whereas over large passive areas mild disconformity separates Zuni and Tejas strata. In addition, equally large cratonic areas record the overstep of Tejas I strata across truncated Zuni rocks of the cratonic interior from the Great Plains to the Mississippi Embayment.

Tejas I (late Paleocene to middle Eocene)

The tectonic significance of the Laramide uplifts and intervening basins that dominate the history of Tejas I accumulation in the area of concern here has been reviewed by Rodgers (1987). He points out that similar discontinuous mountain ranges, commonly lacking in systematic orientation, form a clan of basement-involved uplifts on many cratons at several times in geologic history, characteristically adjacent to segments, but not the entirety, of certain curvilinear orogenic belts. Rodgers notes that among North American examples the relationship of the Ouachita-Marathon trend with the late Paleozoic mountainous blocks of the southern Midcontinent and Rocky Mountain areas of the craton bears a striking parallelism with the apparent couple represented by the Sevier trend and the adjacent Laramide movements. There are significant differences, however. The timing of crustal shortening along the Ouachita-Marathon Orogen (late Carboniferous and Early Permian) is roughly coincident with the span of repeated uplift episodes from the Arbuckle-Wichita ranges to the "Ancestral Rockies," encompassing perhaps 30 m.y. In contrast, while the climax of Sevier activity (Late Cretaceous and early Paleocene) was matched by mild deformation of the adjacent foreland, Sevier-related thrusting and magmatic activity had largely subsided before the time of late Paleocene and early Eocene maximum vertical movement of the Laramide ranges—an episode scarcely longer than about 10 m.y. Another apparent difference is observed in the seeming dominance of strike-slip displacement of blocks in the late Paleozoic cratonic mountain building versus thrusting and compression evident in at least some of the Laramide ranges. Finally, the late Paleozoic events of the southern Midcontinent and Rocky Mountain regions were notably unaccompanied by and unrelated to igneous activity, whereas Laramide deformation was coincident with or quickly followed by volcanic and plutonic manifestations (including ore-forming episodes) of some magnitude.

It may be reasoned that the differences between late Paleozoic and early Cenozoic mountain building on the North American Craton derive from the distinctions to be anticipated between continent-continent collision (the older event) and arc-continent collision (the Sevier-Laramide scenario). Even if this is accepted as a rational explanation, a number of problems remain in terms of the expression of Laramide movements in separate areas of the western margin of the North American Craton. The Colorado Plateau, although bearing paleomagnetic evidence of modest rotation, is a relatively undeformed region of Kansas-like craton immediately adjacent to the Sevier Orogen. Further, the plateau is at least partially surrounded by Laramide uplifts and is the site of extensive Cenozoic volcanism and less-widespread plutonism. Why was this terrane spared the violent differential vertical displacements that affected neighboring provinces?

A further question is raised by the difference in the mode of Laramide deformation encountered in Wyoming, Utah, and regions to the south, as compared to Montana north of its southern border and to the Black Hills of South Dakota. The southern region is dominated by uplifted blocks and deeply subsident basins bounded by reverse faults. Here the uplifts became the prolific sources of debris that filled the intermontane basins, commonly under conditions of closed drainage systems leading to the development of large lakes and attendant peculiar sedimentation, as recounted in companion chapters of this volume. The mountains of central Montana, superimposed on the trend of the Proterozoic Belt Aulacogen and the Paleozoic Central Montana Trough, are domal uplifts relatively uncomplicated by faulting and not clearly related to yoked subsiding basins in source/sink couples. Laramide activity peters out north of the central Montana array of uplifts, to be represented by isolated small domal masses commonly cored by peralkalic plutons; north of the international border the Prairie Provinces of Canada are devoid of significant Laramide influences.

Hamilton (1981) has suggested that a few degrees of clockwise rotation of the Colorado Plateau about a Euler pole to the east of the plateau, as supported by analysis of paleomagnetic data (Bryan and Gordon, 1986), could well be the cause of the stresses responsible for the Laramide ranges and basins, their predominant orientation, and the northward and northeastward decay in intensity observed. This attractive explanation does not, however, account for the stability of the plateau block while transmitting stress to adjacent areas, nor does it explain why an equivalent level of structural integrity was not maintained in the late Paleozoic.

There are, of course, a number of alternative explanations for Laramide cratonic deformation. Among these it has been suggested (e.g., Dickinson and Snyder, 1978; Henderson and others, 1984) that the basement-involved foreland blocks of the Wyoming area are responses to a very shallowly dipping subduction zone. Analogy has been drawn (e.g., Jordan and Allmendinger, 1986) with "thick-skinned" blocks of the Andean foreland that coincide with flattened segments of the Benioff zone in Peru and Argentina. This scenario is brought into question, however, by consideration of the Tien Shan Mountains of central Asia, a region of active cratonic deformation separated from Himalayan subduction zones by at least 1,500 km. Tien Shan earthquake data, as analyzed by Nelson and others (1987), show basement-involved thrust faults much like those identified among the Laramide uplifts of Wyoming but lacking any indication of a flat or shallowly dipping Benioff zone.

Tejas II/III (late Eocene to present)

Evolution of the western craton in the United States following the late Eocene decline and cessation of compressional forces involves a complex and fast-moving history of erosional degradation of mountainous uplifts, the establishment of integrated drainage and sediment-transport systems and the neotectonic renewal of uplift—all modulated by pervasive volcanism and climaxed by the erosional and depositional effects of repeated glacial episodes and their isostatic concomitants. In the broadest possible terms, the most recent 30-odd m.y. of cratonic development can be divided into two major phases. The first, beginning in late Eocene, is dominated by near planation of the Laramide mountains, in part by erosional bevelling, in part through burial by accumulating basin fill. As planation progressed, masses of coarse- and medium-grained fluviatile sediments were spread as great aprons across the High Plains region (to become vitally important regional aquifers) while volumes of silt and clay were bypassed to prograding deltas of the Gulf of Mexico. Thus, by late in Paleogene time, the Laramide highlands had been reduced to low hills, much in the manner of their predecessors, the "Ancestral Rockies," in Triassic time, and generally low elevations prevailed, such that subtropical regimes were established over broad areas of the Rocky Mountain–Great Plains states.

The cessation of Laramide uplift, accompanied by basin filling, reduction of topographic relief, and regional approach to base level is roughly coincident with deceleration of the convergence of the Farallon and North American plates, as this relative motion has been analyzed by Engebretson and others (1985). Conceivably, the relaxation of compressional stresses transmitted to the craton from the continental margin is responsible for the tectonic interregnum of the late Paleogene.

Regardless of theories to explain the rise, decline, and decay of cratonic mountains on the western craton, this phase of tectonic evolution was replaced in the Neogene by regional uplift that has carried the Great Plains to elevations of 1.5 km, the Colorado Plateau to at least another kilometer higher, and reactivated Rocky Mountain blocks to heights in excess of 4 km. the results are evident in the striking rejuvenation of streams that had meandered in consequent patterns across the nearly featureless landscape of the previous phase. Tributaries to the Colorado and Missouri systems, as well as other rivers flowing from the Southern Rocky Mountains to join the Mississippi or to empty into the Gulf of Mexico, became deeply incised. Commonly, streams superposed their ancestral courses across ancient Laramide blocks as these were stripped of covering and flanking deposits by progressive erosion. The results are seen in the incised meanders of the San Juan and in spectacular gorges cut through great mountain ranges—a familiar example is the Royal Gorge incised by the Arkansas River to cross from the basin of South Park to the Rocky Mountain front in Colorado.

Much of the impressive topographic relief displayed today by the Central and Southern Rocky Mountains is the product of regional uplift and concomitant differential erosion of poorly consolidated basin fill versus the crystalline cores of exhumed Laramide blocks. An important component of relief, however, is provided by the relative vertical movement of mountain blocks and intermontane basins. Such movement is largely accomplished by normal (or at least very high-angle) faulting ascribed to regional extension. The timing of the inception of extension in the Rocky Mountain/Plateau region is roughly synchronous with the beginning of extension in the Great Basin—the region immediately adjoining on the west where crustal extension is measured as 65 percent in the south of the province (Wernicke and others, 1982) to 77 percent in the northeast (Gans, 1987).

It is convenient to think of the western reaches of the craton during Laramide events as being held in the vise formed by a west-moving North American Plate and an east-directed Farallon Plate. Slowing of these relative motions may be thought of as leading to the late Paleogene cessation of Laramide mountain building as discussed above. Finally, Neogene subduction of the Pacific ridge crest and reorientation of relative plate motions to be satisfied by transcurrent slip along the San Andreas system would foster the establishment and continuing maintenance of an extensional regime in the region here considered. Again, the reader is presented with a potentially attractive, if simplistic, scenario that resolves part, but by no means all, of the questions posed by the post-Laramide evolution of cratonal tectonism.

Prominent among unsolved neotectonic problems is the matter of the prevailing extraordinary elevation of much of the North American Craton above sea level. The anomalous hypsography of Holocene time is not confined to active continental margins and adjacent forelands but is pervasive on shields, platforms, and revivified ancient mountain belts on several cratons (excepting, perhaps, Australia) to an extent unattainable by glacially imposed eustacy. There exists no supportable body of theory to rationalize this phenomenon. Possibly, value may yet be found in a concept tentatively put forth some years ago (Sloss and Speed, 1974), which suggested that episodes of broad cratonic elevation represent times of storage of magma created by radial heat flow and trapped below an insulating cover of thick continental crust. Conversely, in this scenario, cratonic submergence results when magma is released to oceanic and continental-margin volcanism.

SUMMARY

What is the message transmitted by a review of 500-odd m.y. of cratonic evolution? For one thing, it becomes clear that "evolution" is an appropriate word to apply to its continuing history long after the North American craton was created by the Proterozoic fusion of Archean minicratons and the accretion of cratonized orogens. For the first 120 m.y. of Phanerozoic time, the span of the Sauk sequence, the tectonic mode of the craton was Late Proterozoic in character. That is, the cratonic margins were largely passive and the interior was riven by extensional forces; flexural cratonic downwarps were rare except as sequelae to rifting, and great thicknesses of relatively uniform sandstones and carbonates were deposited near cratonic margins and in the postrift phase of aulacogenic reentrants to the craton.

In the succeeding 150 m.y., the times of the Tippecanoe and Kaskaskia sequences, the Cordilleran and Appalachian margins became sites of active plate convergence, foreland basins appeared to receive sediments derived from orogenic sources, the craton submerged to a degree not previously attained in cratonic history, and flexure-dominated syndepositional arches and basins characterized the tectonic differentiation of the cratonic interior. Among these mid-Paleozoic elements, the most prominent influence on isopach and sediment-distribution patterns of the cratonic interior is the Transcontinental Arch, which became an important feature during the pre-Kaskaskia reorganization of the tectonic framework.

The next 150 m.y. were occupied by the accumulation of the Absaroka sequence under conditions dramatically different from those prevailing in earlier cratonic episodes. First the Ouachita-Marathon and then the Appalachian margins were assaulted by violent collision events attending the closure of the Iapetus Ocean. Collision at the southern margin of the craton was accompanied by large-scale vertical and lateral displacements hundreds of kilometers inboard of the Ouachita-Marathon margin at a level comparable to the effects of the Cenozoic impact of India on the Asian craton. Concurrently, but without necessary genetic connection, the cratonic interior was subject to repeated oscillations with respect to base level.

The record of the Zuni sequence, encompassing another 120 m.y., is the record of a craton bounded on the east and southeast by the divergent Atlantic and Gulf of Mexico margins and on the west by the convergent Cordilleran margin. The early stages, while Cordilleran convergence remained remote, are reminiscent of Sauk conditions—an elevated but tectonically uncomplicated cratonic interior slowly and erratically submerged below base level. Later Zuni history is more like that of mid-Paleozoic (Tippecanoe and Kaskaskia) times. That is, foreland sedimentation became dominated by detritus from an adjacent orogenic trend while the passive margins were the sites of subsiding shelf basins and the interior was progressively transgressed by marine waters until almost all of the shield was covered.

The 60 m.y. of tectonic evolution responsible, thus far, for the Tejas sequence, may be considered a kind of reversion to conditions prevailing during accumulation of the Absaroka. Foreland uplifts and basins as sediment source and sink, major uplift and erosion of the cratonic interior, repeated marine transgressions, episodic progradation of continental environments toward cratonic margins, all these mimic Absaroka patterns, but very substantial differences remain. Such distinctions are noted in the discussion of the complexities of tectonic modes and tempos represented by Tejas rocks, structures, Cenozoic physiographic evolution, and the emergence of ecosystems within which hominids developed and rose to dominance.

REFERENCES

Anderson, T. H., and Schmidt, V. A., 1983, The evolution of Middle America and the Gulf of Mexico-Caribbean Sea region during Mesozoic time: Geological Society of America Bulletin, v. 94, p. 941–966.

Bally, A. W., 1980, Basins and subsidence; a summary, *in* Bally, A. W., Bender, P. L., McGetchin, T. R., and Wolcott, R. I., eds., Dynamics of plate interiors: American Geophysical Union and Geological Society of America Geodynamics Series, v. 1, p. 5–20.

Bickford, M. E., Van Schmus, W. R., and Zietz, I., 1986, Proterozoic history of the midcontinent region of North America: Geology, v. 14, p. 492–496.

Bryan, P., and Gordon, R. G., 1986, Rotation of the Colorado Plateau; An analysis of paleomagnetic data: Tectonics, v. 5, p. 661–667.

Burchfiel, B. C., Lipman, P. W., and Zoback, M. L., eds., 1988, The Cordilleran orogen; Conterminous United States: Boulder, Colorado, Geological Society of America, The Geology of North America, v. G-3 (in press).

Chang, K. H., 1975, Unconformity-bounded stratigraphic units: Geological Society of America Bulletin, v. 86, p. 1544–1552.

Cook, T. D., and Bally, A. W., eds., 1975, Stratigraphic atlas of North and Central America; Princeton, New Jersey, Princeton University Press, 272 p.

DeCelles, P. G., 1986, Sedimentation in a tectonically partioned, non-marine foreland basin; The lower Cretaceous Kootenai Formation, southwestern Montana: Geological Society of America Bulletin, v. 97, p. 911–931.

Dickinson, W. R., and Snyder, W. S., 1978, Plate tectonics of the Laramide orogeny, *in* Mathews, V., III., ed., Laramide folding associated with basement block faulting in the western United States: Geological Society of America Memoir 151, p. 355–366.

Dimian, M. V., Gray, R., Stout, J., and Wood, B., 1983, Hudson Bay basin, *in* Bally, A. W., ed., Seismic expression of structural styles: American Association of Petroleum Geologists Studies in Geology Series no. 15, v. 2, p. 2.2.4-1.

Dott, R. H., Jr., Byers, C. W., Fielder, G. W., Stenzel, S. R., and Winfree, K. E., 1986, Aeolian to marine transition in Cambro-Ordovician cratonic sheet sandstones of the northern Mississippi Valley, U.S.A.: Sedimentology, v. 33, p. 345–367.

Engebretson, D. C., Cox, A., and Gordon, R. G., 1985, Relative motions between oceanic and continental plates in the Pacific Basin: Geological Society of America Special Paper 206, 59 p.

Faill, R. T., 1985, The Acadian orogeny and the Catskill delta, *in* Woodrow, D. L., and Sevon, W. D., eds., The Catskill delta: Geological Society of America Special Paper 201, p. 15–38.

Gans, P. B., 1987, An open-system two-layer stretching model for the eastern Great Basin: Tectonics, v. 6, p. 1–12.

Hamilton, W., 1981, Plate-tectonic mechanism of Laramide deformation: Laramie, University of Wyoming Contribution to Geology, v. 19, p. 87–92.

Haq, B. U., Hardenbol, J., and Vail, P. R., 1987, Chronology of fluctuating sea levels since the Triassic (250 m.y. to present): Science, v. 235, p. 1156–1167.

Harland, W. B., Cox, A. V., Llewellyn, P. G., Pickton, C.A.G., Smith, A. G., and Walters, R., 1982, A Geologic Time Scale: Cambridge Earth Science Series, Cambridge University Press, 131 p.

Hatcher, R. D., Jr., Thomas, W. A., and Viele, G., eds., 1988, The Appalachian-Ouachita Orogen in the United States: Boulder, Colorado, Geological Society of America, The Geology of North America, v. F-2 (in press).

Henderson, L. J., Gordon, R. G., and Engebretson, D. C., 1984, Mesozoic aseismic ridges of the Farallon plate and southward migration of shallow subduction during the Laramide orogeny: Tectonics, v. 3, p. 121–132.

Hoffman, P. F., Dewey, J. F., and Burke, K., 1974, Aulacogens and their genetic relation to geosynclines; with a Proeterozoic example from Great Slave Lake, Canada, *in* Dott, R. H., Jr., and Shaver, R. H., eds., Modern and ancient geosynclinal sedimentation: Society of Economic Paleontologists and Mineralogists Special Publication 19, p. 38–55.

International Subcommission on Stratigraphic Classification (Amos Salvador, Chairman), 1987, Unconformity-bounded stratigraphic units: Geological Society of America Bulletin, v. 98, p. 232–237.

Johnsson, M. J., 1986, Distribution of maximum burial temperatures across northern Appalachian Basin and implications for Carboniferous sedimentation patterns: Geology, v. 14, p. 384–387.

Jordan, T. E., and Allmendinger, R. W., 1986, The Sierras Pampeanas of Argentina; a modern analogue of Rocky Mountain foreland deformation: American Journal of Science, v. 286, p. 737–764.

Karig, D. E., 1987, Comment *on* 'Distribution of maximum burial temperatures across northern Appalachian Basin and implications for Carboniferous sedimentation patterns': Geology, v. 15, p. 278–279.

Kauffman, E. G., 1986, High-resolution event stratigraphy; Regional and global Cretaceous bio-events, *in* Walliser, O. H., ed., Lecture notes in earth sciences, no. 8, Global bio-events: Berlin, Springer-Verlag, p. 279–335.

Mai, H., and Dott, R. H., Jr., 1985, A subsurface study of the St. Peter Sandstone: Wisconsin Geological and Natural History Survey Information Circular 47, 26 p.

Mallory, W. W., ed., 1972, Geological Atlas of the Rocky Mountain region: Denver, Colorado, Rocky Mountain Association of Geologists, 331 p.

McGookey, D. P., and 6 others, 1972, The Cretaceous System, *in* Mallory, W. W., ed., Geologic atlas of the Rocky Mountain region: Denver, Colorado, Rocky Mountain Association of Geologists, p. 190–228.

Nelson, M. R., McCaffrey, R., and Molnar, P., 1987, Source parameters for 11 earthquakes in the Tien Shan, Central Asia, determined by P and Sh waveform inversion: Journal of Geophysical Research, v. 92, p. 12629–12648.

Palmer, A. R., compiler, 1983, The Decade of North American Geology 1983 Geologic Time Scale: Geology, v. 11, p. 503–504.

Perry, E. S., and Sloss, L. L., 1943, Big Snowy Group; Lithology and correlation in north-Great Plains: American Association of Petroleum Geologists Bulletin, v. 27, p. 1287–1304.

Peterson, J. A., 1980, Permian paleogeography and sedimentary provinces, west-central United States, *in* Symposium on Paleozoic paleogeography of the west-central United States: Denver, Colorado, Rocky Mountain Section, Society of Economic Paleontologists and Mineralogists, p. 271–292.

Peterson, J. A., and Smith, D. L., 1986, Rocky Mountain paleogeography through geologic time, *in* Peterson, J. A., ed., Paleotectonics and sedimentation in the Rocky Mountain region, United States: American Association of Petroleum Geologists Memoir 41, p. 3–20.

Rodgers, J., 1987, Chains of basement uplifts within cratons marginal to orogenic belts: American Journal of Science, v. 287, p. 661–692.

Salvador, A., 1988, The Gulf of Mexico Basin: Boulder, Colorado, Geological Society of America, The Geology of North America, v. J (in press).

Shatski, N. S., 1946, Basic features of the structures and development of the East European Platform, *in* Comparative tectonics of ancient platforms; S.S.S.R.: Akademia Nauk Isvestiya, Geologicheskaya Seriya no. 1, p. 5–62 (in Russian).

Sheridan, R. E., and Grow, J. A., eds., 1988, The Atlantic Continental Margin: U.S.: Boulder, Colorado, Geological Society of America, The Geology of North America, v. I-2, 611 p.

Shurbet, D. H., and Cebull, S. E., 1987, Tectonic interpretation of the westernmost part of the Ouachita-Marathon (Hercynian) orogenic belt, west Texas-Mexico: Geology, v. 15, p. 458–461.

Sloss, L. L., 1963, Sequences in the cratonic interior of North America: Geological Society of America Bulletin, v. 74, p. 93–114.

—— , 1984, Comparative anatomy of cratonic unconformities, *in* Schlee, J. S., ed., Interregional unconformities and hydrocarbon accumulation: American Association of Petroleum Geologists Memoir 36, p. 1–6.

Sloss, L. L., and Moritz, C. A., 1951, Paleozoic stratigraphy of southwestern Montana: American Association of Petroleum Geologists Bulletin, v. 35, p. 2135–2169.

Sloss, L. L., and Speed, R. C., 1974, Relationship of cratonic and continental-margin tectonic episodes, *in* Dickinson, W. R., ed., Tectonics and sedimentation: Society of Economic Paleontologists and Mineralogists Special Publication 22, p. 38–55.

Speed, R. C., 1983, Evolution of the sialic margin in the central western United States, *in* Watkins, J. S., and Drake, C. L., eds., Studies in continental margin geology: American Association of Petroleum Geologists Memoir 34, p. 457–468.

Speed, R. C., Elison, M. W., and Heck, F. R., 1988, Phanerozoic tectonic evolution of the Great Basin, *in* Ernst, W. G., ed., Rubey Volume VII, Metamorphism and crustal evolution, Western conterminous United States: New York, Prentice-Hall (in press).

Stearns, R. G., and Reesman, A. L., 1986, Cambrian to Holocene structural and burial history of Nashville dome: American Association of Petroleum Geologists Bulletin, v. 70, p. 143–154.

Stewart, J. H., 1980, Geology of Nevada: Nevada Bureau of Mines and Geology Special Publication 4, 136 p.

Thomas, W. A., 1976, Evolution of the Ouachita-Appalachian margin: Journal of Geology, v. 84, p. 323–342.

Vail, P. R., Mitchum, R. M., and Thomson, S., III, 1977, Seismic stratigraphy and global changes of sea level; Part 4, Global cycles of relative changes of sea level, *in* Payton, C., ed., Seismic stratigraphy; Applications to hydrocarbon exploration: American Association of Petroleum Geologists Memoir 26, p. 83–97.

Veevers, J. J., and Powell, C. M., 1987, Late Paleozoic glacial episodes in Gondwanaland reflected in transgressive-regressive depositional sequences in Euroamerica: Geological Society of America Bulletin, v. 98, p. 475–487.

Walcott, C. D., 1910, Abrupt appearance of the Cambrian fauna on the North American continent: Smithsonian Miscellaneous Collections, v. 57, no. 1, p. 1–15.

Wanless, H. R., and Shepard, F. P., 1936, Sea level and climatic changes related to Late Paleozoic cycles: Geological Society of America Bulletin, v. 47, p. 1177–1206.

Weimer, R. G., 1986, Relationship of unconformities, tectonics, and sea-level changes in the Cretaceous of western interior United States, *in* Peterson, J. A., ed., Paleotectonics and sedimentation in the Rocky Mountain region, United States: American Association of Petroleum Geologists Memoir 41, p. 397–422.

Wernicke, B., Spencer, J. E., Burchfiel, B. C., and Guth, P. L., 1982, Magnitude of crustal extension in the southern Great Basin: Geology, v. 10, p. 499–502.

Manuscript Accepted by the Society December 29, 1987

Printed in U.S.A.

The Geology of North America
Vol. D-2, Sedimentary Cover—North American Craton: U.S.
The Geological Society of America, 1988

Chapter 4

Triassic and older stratigraphy; Southern Rocky Mountains and Colorado Plateau

D. L. Baars
29056 Histead Drive, Evergreen, Colorado 80439

PRECAMBRIAN

Paleozoic strata of the Southern Rocky Mountains and Colorado Plateau overlie a structurally complex basement of metamorphic, igneous, and metasedimentary rocks. The basement complex was completed in Proterozoic time by major continental-scale wrench faulting that produced a northwesterly trending swarm of faults (the Olympic-Wichita Lineament of Baars, 1976) and a northeasterly set (the Colorado Lineament of Warner, 1978). The apparently conjugate sets of fractures intersect in the heart of the Paradox Basin of the eastern Colorado Plateau. Precambrian sense of displacement was dextral along the Olympic-Wichita trend, at least on the Colorado Plateau (Baars and Stevenson, 1982), and sinistral along the Colorado Lineament (Warner, 1978). Thus, sigma one was directed in a north-south orientation (Fig. 1). Age of the northwesterly faults is bracketed to 1.72 Ga to 1.46 Ga (Baars and Ellingson, 1984); younger granitic intrusives (~1.46 Ga) invade older metamorphics, metasediments, and faults without discretion in the San Juan Mountains. Warner (1978) dated the northeast set at about 1.7 Ga.

Metasedimentary rocks (quartzites and metapelites) of the Uncompahgre Formation are juxtaposed with the metamorphic basement (~1.78 Ga) along the fault blocks in the San Juan Mountains of southwest Colorado. The shape and location of the sedimentary basin for these Proterozoic rocks is enigmatic, but stratigraphic similarities and speculative dating would relate the metasediments with the Mazatzal Group of central Arizona and the Ortega Group of north-central New Mexico (C. W. Harris, personal communication, 1987). Ritzma (1987) argued that the age of the Uinta Mountain Group of northern Utah is greater than 1.65 Ga, suggesting a possible equivalency to the Uncompahgre metasediments. Also, the age of the lower strata of the Belt Supergroup is not well defined. All of these rocks may have been deposited in one very large basin. To complicate the issue, the Uncompahgre metasediments are obviously allochthonous. Thickness of the Uncompahgre Formation has been estimated by various workers to be at least 2,400 to 3,000 m, but accurate measurements are impossible to obtain.

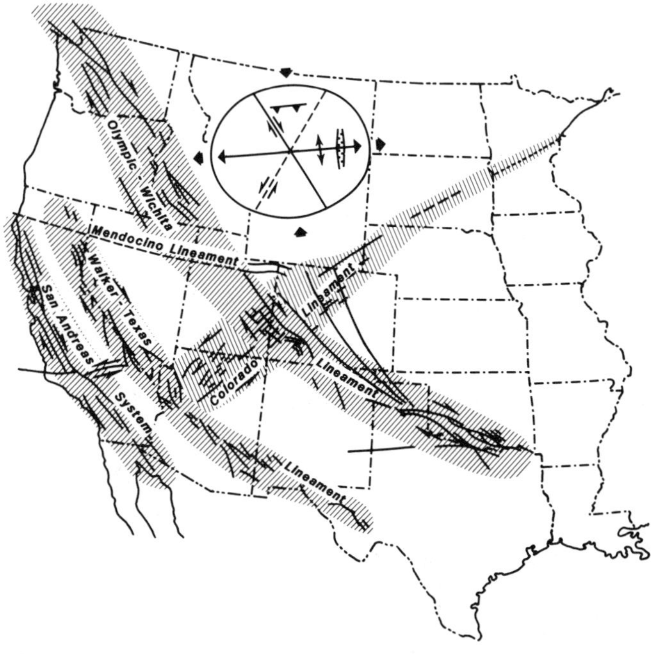

Figure 1. Regional relationships of the Olympic-Wichita and Colorado Lineaments (after Baars, 1976). Note the diagrammatic representation of the main bounding faults of the Ancestral Rockies uplifts in central Colorado as "horsetail splays" off the Olympic-Wichita fault system.

About 4,000 m of Late Precambrian sedimentary rocks occur in the Grand Canyon in northeast-trending fault blocks of the Colorado Lineament. Although largely red beds, the lower Unkar Group contains a thick quartzite sequence (Shinumo Formation), and thin carbonate beds of tidal flat origin occur in both the Dox Formation of the Unkar Group (Stevenson and Beus, 1982) and the Chuar Group (Ford and others, 1969). Various dikes and sills have been intruded into the Unkar Group, and several lava flows are interbedded with the red beds. These have been dated by numerous workers at about 1.0 Ga, making the Supergroup considerably younger than the metasedimentary

Baars, D. L., 1988, Triassic and older stratigraphy; Southern Rocky Mountains and Colorado Plateau, *in* Sloss, L. L., ed., Sedimentary Cover—North American Craton; U.S.: Boulder, Colorado, Geological Society of America, The Geology of North America, v. D-2.

rocks to the east. The faulting that preserved the large tilted fault blocks of the Grand Canyon Supergroup is of pre-Tapeats (Cambrian) age and is presumed to be Late Precambrian. These faults can be traced northeastward across the entire Colorado Plateau and into the Colorado Mineral Belt (Warner, 1978).

Basement faults in the central Colorado Mountains that would later bound the Ancestral Rockies uplifts also have Precambrian ancestry (Tweto, 1980). They are generally north–northwest-trending major faults that may have strike-slip displacement, and appear to be "horsetail splays" off the master Olympic-Wichita megashear (Fig. 1). Orthogonal crustal blocks of minor proportions bounded by conjugate fractures related to the megashears are ubiquitous in the Precambrian basement. Both master shears and minor fractures were continuously, or episodically, reactivated throughout Phanerozoic time, controlling sedimentation on both major and minor scales. This structural fabric is reflected in the two major Phanerozoic tectonic episodes, the Ancestral Rockies orogeny of Pennsylvanian age and the Laramide orogeny of Late Cretaceous–Early Tertiary age.

CAMBRIAN

There was considerable topographic relief on the basement as the Early to Middle Cambrian sea began to encroach upon the western shelf of the craton. Numerous fault-bounded ridges and islands punctuated the shoreline in the vicinity of the Grand Canyon, some rising more than a hundred meters above the Cambrian sea level (McKee and Resser, 1945). The regional slope was such that it took the shoreline from Early to Middle Cambrian to transgress the length of the Grand Canyon, and it did not reach southwestern Colorado until latest Cambrian time (Baars and See, 1968). There, in the San Juan Mountains of today, the Late Cambrian shoreline encountered two large, quartzite-dominated, high fault blocks—the Grenadier and Sneffels horsts—that shed boulder conglomerates into the fringing sea. Paleotopographic relief may have been considerable, for the quartzite boulders are angular and up to a meter in diameter along the bounding faults.

The basal formation (the Tapeats Sandstone of the Grand Canyon, the Ignacio Quartzite of the Paradox Basin and San Juan Mountains, the Tintic of the Wasatch Mountains, the lower Lodore Formation of the Uinta Mountains, and the Sawatch Formation in the central Colorado Basin) is a transgressive littoral sandstone that becomes regionally younger toward the east. It is typically a light-colored, fine- to medium-grained, well- to poorly sorted sandstone, with locally abundant glauconite. Surface exposures are limited to the Grand Canyon, the San Juan Mountains to the south, the Wasatch-Uinta Mountains on the north, and the central Colorado Basin on the east. Subsurface data is meager at best between the outcrops, and no subsurface faunal control is known, making the intricate history of complex transgressions and regressions proposed by Lochman-Balk (1972) impossible to verify.

The basal sandstone is usually overlain by drab-colored mudstones with interfingering and gradational contacts as a result of the regional eastward transgression. The finer-grained nature of the formation, variously called the Bright Angel Shale in the Grand Canyon or the Ophir Formation of central Utah, reflects an offshore, deeper-water environment that developed westward from the littoral sand environment. Both the basal sand and the Bright Angel–Ophir facies may be locally thin or missing by nondeposition across the higher paleostructures.

Transgressive interfingering relationships between the Bright Angel Shale and the overlying Muav carbonate facies were clearly demonstrated in the magnificent exposures in the Grand Canyon by McKee and Resser (1945). Elsewhere, details of the general eastward transgression of the Muav carbonate shelf, and its Utah equivalent Maxfield Limestone, is speculative. Muav-like limestones generally overlie the Bright Angel facies in the western Paradox Basin (Baars, 1958), but the exact relationships with the true Muav and Maxfield Limestone of central Utah are clouded. Although subtidal shelf limestone dominates the sequence, occasional intertidal dolomites document brief periods of sediment buildup to or above mean sea level in the Grand Canyon (McKee, 1969). The Muav-Maxfield Limestones give way upward along the entire Cordilleran hingeline to dolomites that extend eastward into the central Paradox Basin (Fig. 2). This interval has been called "undifferentiated Cambrian" by McKee and Resser (1945), "Supra Muav" by Lochman-Balk (1972), and Lynch Dolomite in the subsurface of the Paradox Basin (Loleit, 1963). Controls on the timing details of carbonate deposition are almost totally lacking. (A. R. Palmer, personal communication, 1987).

ORDOVICIAN AND SILURIAN

Ordovician strata are not known to exist on the Colorado Plateau, but are present in the central Colorado Basin. The section is subdivided into the Lower Ordovician Dotsero Formation (above the Clinetop Algal Member) and the Manitou Limestone, the Middle Ordovician Harding Quartzite, and the Upper Ordovician Fremont Formation. The Manitou Limestone is an intertidal to subtidal carbonate rock that is usually dense and nonporous (Gerhard, 1972). The Harding is typically a thoroughly silica-cemented siltstone or very fine-grained sandstone that is hard and tight. The Fremont is a dolomicrite with common open marine fossils (Baars, 1975).

There are no rocks of Silurian age in the region, with the exception of local occurrences within scattered diatremes near the Colorado-Wyoming border (Chronic, 1972).

DEVONIAN

The Devonian System of the Colorado Plateau and southern Rockies is represented by a thin sequence of stable-shelf sandstones, shales, and carbonates of Late Devonian age. If they were ever present, the Early and Middle Devonian series were re-

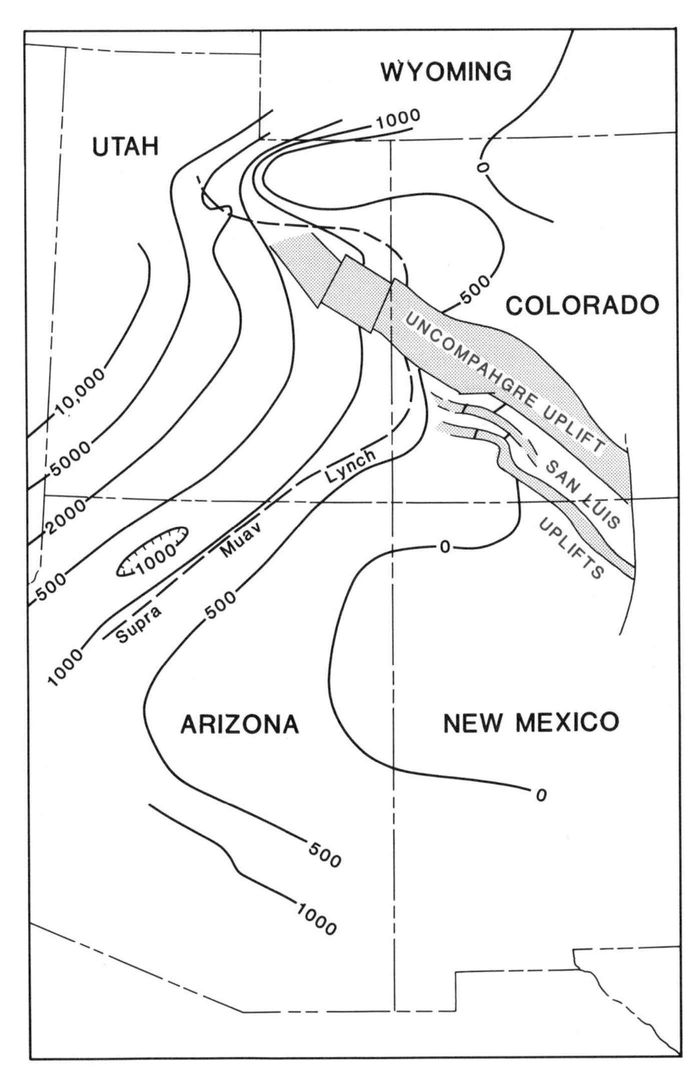

Figure 2. Regional isopach map of the Cambrian System of the southern Rockies and Colorado Plateau provinces. The dashed line represents the southeastern limit of the Supra Muav–Lynch Dolomite facies of Late Cambrian age. Thickness in feet.

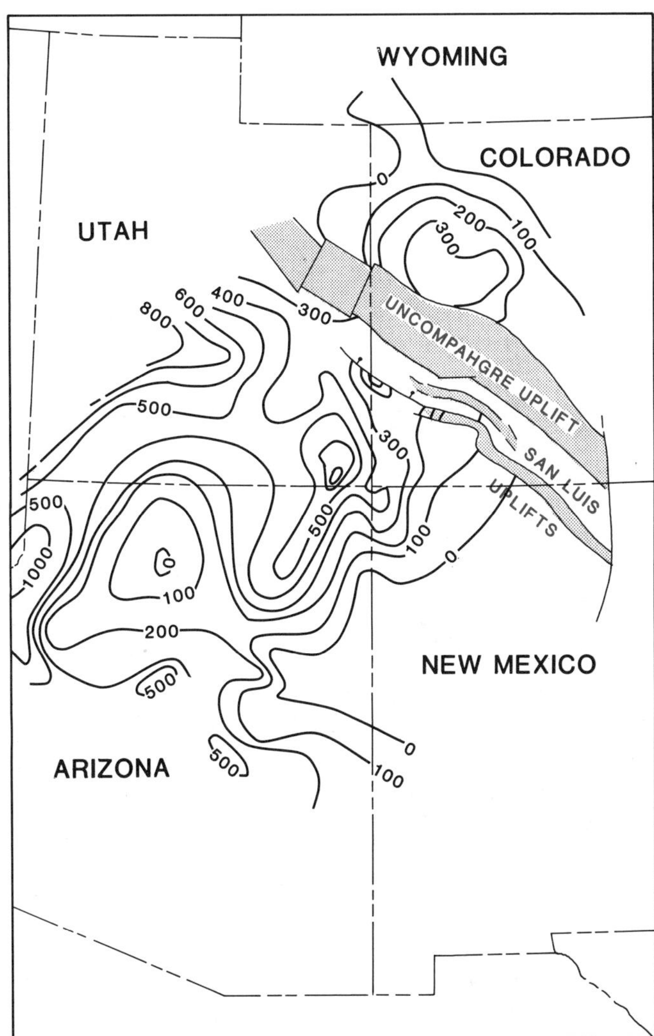

Figure 3. Regional isopach map of the Devonian System in the southern Rockies and Colorado Plateau province. Note the local thin to zero area in north-central Arizona on the East Kaibab Monocline and the thick area of southeastern Utah where the Aneth Formation is locally present. Thickness in feet.

moved, along with Ordovician and Silurian rocks, by pre-Frasnian erosion. However, the major disconformity is obscure. On the Colorado Plateau, the oldest Devonian rocks are in the Aneth Formation, which is composed of dark-colored argillaceous dolomite and interbedded dark shale of early Frasnian age. The Aneth was apparently deposited in a local depression in southeastern Utah (Fig. 3) where temporary euxinic conditions prevailed (Baars and Campbell, 1968).

The Elbert Formation overlies the Aneth where it is present, but elsewhere overlies Late Cambrian strata. The Elbert consists of a basal sandstone, the McCracken Member, and an upper unnamed dolomite and shale member. Both are probably Frasnian in age. The McCracken is best developed in a local basin in southeastern Utah and on the high flanks of rejuvenated basement fault blocks in the San Juan Mountains and eastern Paradox Basin. Dolomite and shaley dolomite of the upper Elbert commonly contain salt casts and stromatolites suggesting a tidal flat depositional environment (Baars and Campbell, 1968; Baars and See, 1968).

The overlying Ouray Limestone is a dense, dark, carbonate sequence that is gradational with the Elbert Formation throughout the Colorado Plateau. It typically is a pelletal micrite containing a distinctive *Paurorhyncha endlichii* brachiopod fauna. Deposition of the Ouray occurred in a low-energy marine environment during Famennian to Early Mississippian (Tournaisian) time (Baars and Campbell, 1968).

In central Colorado, all Devonian rocks are included in the Chaffee Group. The lower formation is the Parting, consisting of quartzose sandstone and lesser amounts of arenaceous shale and some thin, dense, dolomite stringers. The Parting is correlated with the Elbert Formation and is believed to be Frasnian in age based on coexistent fragments of the Late Devonian fish *Bothryolepis coloradensis*. It was deposited in marginal marine, transgressive environments within the central Colorado paleobasin (Baars and Campbell, 1968).

Conformably overlying the Parting Formation is the dolomitic limestone and dolomite of the Dyer Formation. The lower Dyer consists of lime mud containing the *Cyrtiopsis-Paurorhyncha* fauna of the Ouray Limestone. The upper Dyer consists of stromatolitic penecontemporaneous dolomite, which formed in a supratidal (tidal flat above mean high tide) environment during a regressive marine phase. It is correlated with the Ouray and is Famennian to Early Mississippian (Tournaisian) in age (Baars and Campbell, 1968).

Vestigial basins of Late Devonian sedimentation approximately coincide with the present-day Paradox and central Colorado Basins (Fig. 3). The two depocenters were probably only partially segregated by the paleotectonically active Grenadier, Sneffels, and Uncompahgre uplifts (Baars and See, 1968). Devonian strata thin dramatically toward the southwest and the Kaibab uplift, where the system is missing or preserved only within local erosional channels in lower Marble Gorge and eastern Grand Canyon. The Devonian (probably Frasnian) Temple Butte Formation thickens westward in the Grand Canyon toward the Cordilleran miogeocline to over 400 m (Beus, 1969). Thus the Kaibab uplift was a significant paleotectonic high in Late Devonian time.

MISSISSIPPIAN

Carbonate rocks of Early to Middle Mississippian age (Kinderhookian to Osagian) are widespread throughout the Colorado Plateau and southern Rockies. The formation is called the Redwall Limestone in the Grand Canyon and the Leadville Formation in the Paradox and central Colorado Basins. Either term is applicable throughout the region, as the formations are regionally similar in lithology and age. Both are directly correlatable with the Madison Limestone of northern Utah and Wyoming.

A subtle erosional disconformity separates the Redwall from the underlying Devonian in the western Grand Canyon, and a marked disconformity separates the Mississippian from the Cambrian Muav Limestone in the eastern Grand Canyon (McKee and Gutschick, 1969). If the Temple Butte is indeed Frasnian, as interpreted by Beus (1969), then the latest Devonian (Famennian) is missing at the disconformity. Where the Famennian series is represented by the Ouray Limestone in the central Colorado Plateau and Paradox Basin, no Devonian-Mississippian disconformity is discernible (Baars, 1966). However, a disconformity again marks the temporal boundary in the central Colorado Basin, and the basal member, the Gilman, contains quartzose sandstone and limestone pebble conglomerate derived from the ancestral Sawatch uplift (Nadeau, 1972).

McKee and Gutschick (1969) subdivided the Redwall into four formal members in the Grand Canyon based on lithologic differences and disconformities. The Leadville of the Paradox and central Colorado Basins could be similarly subdivided, but for regional and subsurface correlations, Baars (1966) divided the Leadville into informal upper and lower members. The intervening boundary is a subtle disconformity that apparently corresponds to the disconformity between the Thunder Springs and the Mooney Falls Members of McKee and Gutschick (1969). The disconformity in the Grand Canyon was recognized as a physical erosional surface; in the Paradox Basin it is represented by a thin (15 to 30 cm) limestone pebble conglomerate that is widespread in cored wells. The disconformity is seen as a thin paleosol in the San Juan Mountains (Rockwood Quarry). The widespread disconformity is of approximately middle Osagian age throughout the Colorado Plateau and southern Rockies.

The carbonates beneath the mid-Mississippian (Osagian) disconformity are generally fine- to medium-crystalline dolomite (Whitmore Wash and Thunder Springs Members of the Redwall Limestone) to central Colorado (lower member of the Leadville Formation). McKee and Gutschick (1969) interpreted the sequence to represent a transgressive-regressive cycle in the Grand Canyon, terminated by an intraformational disconformity. The transgressive-regressive cycle appears to be a plausible interpretation for the lower member of the Leadville as well, and the intraformational disconformity in the Paradox Basin seems to be coextensive with that of the Grand Canyon. The transgressive Whitmore Wash Member is believed to be latest Kinderhookian to the west and lower to middle Osagian in the eastern Grand Canyon on the basis of endothyrid foraminifera. However, the lower member of the Leadville contains a typical Kinderhookian endothyrid fauna (*Chernyshinella granulosa*) in the Paradox Basin (Baars, 1966). Perhaps the true ranges of the species are uncertain and/or the species are more environmentally distributed.

The upper member of the Leadville of Baars (1966) and the Mooney Falls–Horseshoe Mesa Members of McKee and Gutschick (1969) occur above the mid-Osagian disconformity, marking the return of normal marine conditions (transgression) to the entire Colorado Plateau and southern Rockies regions. The lower, or main mass of the unit, is a massive cliff-forming limestone where exposed, and is regionally composed of fossiliferous lime mudstone to wackestone. Fossils include abundant crinoidal debris, bryozoa, brachiopods, solitary and colonial corals, and abundant foraminifers.

In the subsurface of the Paradox Basin, large crinoidal bioherms (Waulsortian banks) occur on paleotectonic high features, mainly the upthrown sides of basement faults. When dolomitized, the mud component of the biohermal sediments recrystallizes to a saccharoidal texture, and the crinoid columnals dissolve, forming

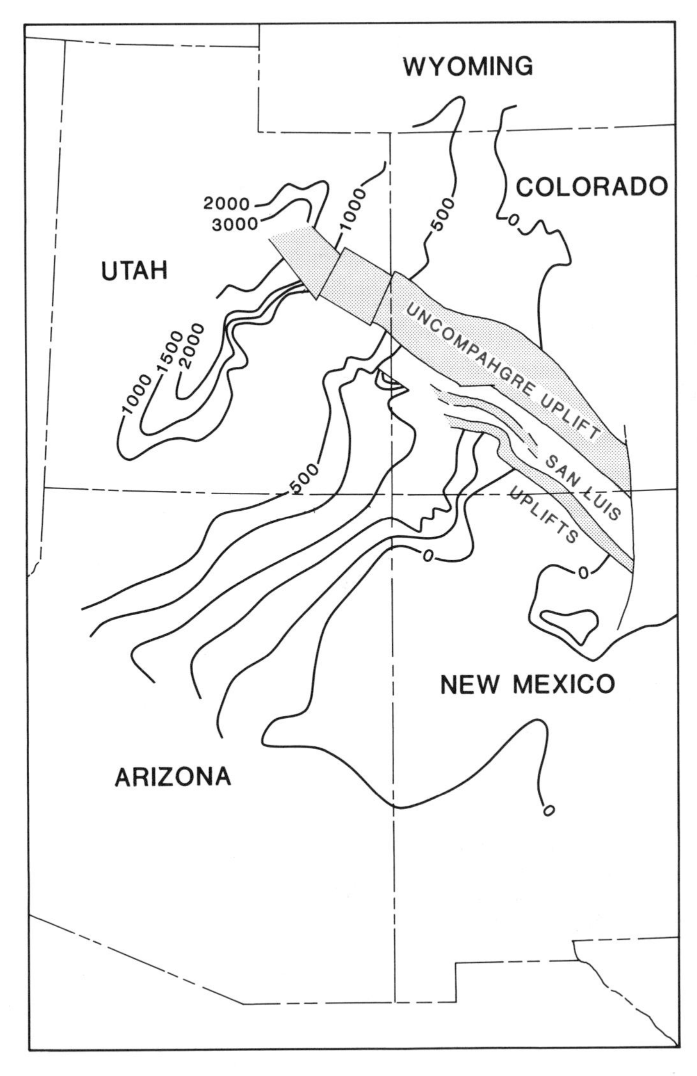

Figure 4. Regional isopach map of the Mississippian System in the southern Rockies and Colorado Plateau provinces. Thickness in feet.

a very effective reservoir rock for petroleum. The Lisbon Valley, McIntyre Canyon, Hook and Ladder, Big Flat, Bartlet Flat, and Salt Wash fields all produce, or have produced, oil from Mississippian reservoirs of this type. Except at Lisbon Valley and McIntyre Canyon, the gas has been of disappointingly poor quality (low BTU), containing a high nitrogen content. In southwestern Colorado, namely the McElmo dome and Dolores anticline areas, vast amounts of carbon dioxide are being produced from Leadville reservoirs to be utilized for enhanced recovery programs for oil production in carbonate reservoirs.

The uppermost member of the Leadville Formation of Baars (1966) is stratigraphically equivalent to the Horseshoe Mesa Member of the Redwall of McKee and Gutschick (1969). Endothyrid foraminifers indicate that it is of Meramecian age. While relatively uniformly distributed throughout the Grand Canyon, the interval is only locally preserved to the east.

At some time following Meramecian time and prior to Middle Pennsylvanian time, the Mississippian shelf seas withdrew from the Colorado Plateau–southern Rockies region, westward toward the Cordilleran miogeocline. Part of the record of the regression is preserved in the western Grand Canyon in the Surprise Canyon Formation of Billingsly and Beus (1985, 1986), where deep erosional channels filled with estuarine sediments of Chesterian age have recently been discovered. The resulting subaerially exposed flatlands of the eastern Colorado Plateau–southern Rockies regions were weathered in a warm, humid climate to form a thick lateritic regolith. The resulting paleosol is known as the Molas Formation, which is recognized throughout the province. Paleosol development was complete by mid-Atokan (early Middle Pennsylvanian) time. On the higher paleostructures, the entire Mississippian System has been removed by erosion (Baars, 1966) (Fig. 4).

PENNSYLVANIAN

The relative paleotectonic quiescence of the earlier Paleozoic gave way to widespread rejuvenation of basement structures during Pennsylvanian time. Mountain ranges, the Ancestral Rockies, rose to considerable elevations and became the source regions for vast quantities of arkosic clastic sediments. Intervening basins sagged along the pre-existing basement fabric to become repositories for the clastics. The Paradox Basin lay to the west of the Uncompahgre uplift, and the Eagle Basin sagged into existence between the Uncompahgre and Front Range uplifts, approximately occupying the position of the central Colorado Basin. Both became restricted marine evaporite basins as the climate became more arid in Middle Pennsylvanian (Desmoinesian) time. Smaller positive structural features, not generally considered to be parts of the Ancestral Rockies, were emergent, but shed little in the way of clastic sediments. Such structures as the Nacimiento, Defiance, Zuni, Kaibab, Circle Cliffs, and Emery uplifts (Fig. 5) were low islands that served to restrict normal marine circulation in the Paradox Basin.

Sedimentary rocks of Early Pennsylvanian (Morrowan) age are known only in the distal reaches of the provinces. Basal beds of the Sandia Formation and its stratigraphic equivalents of central New Mexico are Morrowan, but their distribution to the north was limited. The lower Watahomigi Formation of the Supai Group in the Grand Canyon is also Morrowan in age (McKee, 1979). Lower strata of the Morgan Formation in the Uinta Mountains of northern Utah contain Morrowan faunas, and extend southeastward into the Eagle Basin, where they are called the lower Belden Shale. Elsewhere, the Early Pennsylvanian is represented only by the Molas lateritic paleosol, indicating that humid lowlands existed over most of the Colorado Plateau and southern Rocky Mountain provinces.

By the close of Atokan (early Middle Pennsylvanian) time, the basins were under shallow-marine environments and receiv-

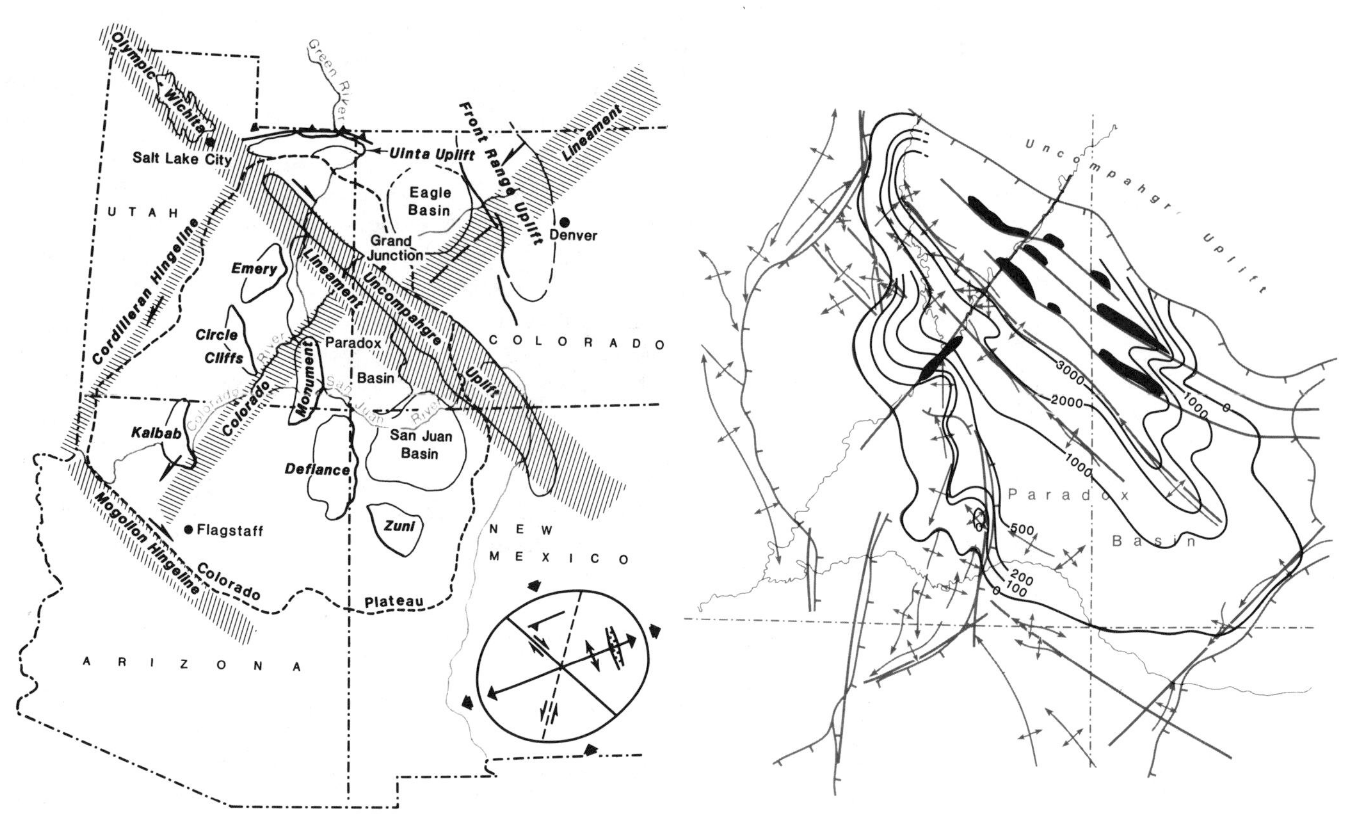

Figure 5. Regional relationships of the Paradox and Eagle Basins to major uplifts of the southern Rockies and Colorado Plateau provinces. Hachured zones are the principal basement-controlled lineaments of the region; arrows indicate relative sense of strike-slip displacement in Precambrian time. (After Baars and Stevenson, 1982).

Figure 6. Isolith map of salt deposits of the Paradox Formation of Middle Pennsylvanian age and their relationships to surface structural features of the Colorado Plateau Province. Isoliths in feet; black areas are salt diapirs. (After Baars and Stevenson, 1982).

ing small amounts of sediment. The first documented uplift of the Ancestral Rockies occurred in the San Luis uplift of northern New Mexico and southern Colorado (Baars and Stevenson, 1984), as extensional east-to-west tectonics commenced along the Olympic-Wichita basement wrench fault zone. The initial pulse exposed fault blocks of predominantly Proterozoic quartzites that supplied relatively nonfeldspathic quartzose sands to the Sandia shallow-marine and deltaic environments. Elsewhere, thin marine limestones and shales are characteristic of the epoch.

By Desmoinesian (Middle Pennsylvanian) time, open marine conditions prevailed throughout the basins. All of the main ranges of the Ancestral Rockies were high and shedding coarse clastics to the adjacent basins; structural relief was perhaps as much as 6,000 m in places. The Paradox and Eagle Basins were almost completely cut off from the surrounding open marine shelves, and evaporites accumulated under "deep" water stagnation in a very arid climate, perhaps imposed by rain shadows of the Front Range and Uncompahgre uplifts. Basin restriction resulted entirely from tectonic isolation that was directly related to the east-to-west pull-apart of the distal margins of the craton (Stevenson and Baars, 1986). These evaporite basins are discussed in more detail elsewhere in this volume.

Although the two basins are similar in many ways, they were not directly connected. The larger of the basins, the Paradox, contained perhaps 2,500 m of evaporites (Fig. 6), much of which is halite, in 29 clearly defined cycles (Hite, 1960). By the end of Desmoinesian time, salt diapirism was altering the configuration of the deeper part of the Paradox Basin. The Eagle Basin underwent a similar history, but on a much smaller scale.

All of the major present-day uplifts of the Colorado Plateau Province were emergent during Middle Pennsylvanian time with the exception of the Monument Upwarp, and it's lesser folds formed paleotectonic shoals. Although it is tempting to connect the uplifts to form extensive positive features (e.g., the Paiute and Zuni-Defiance platforms of Mallory, 1972), sparse subsurface data indicate that open marine sedimentation occurred around and between the relatively small paleo-uplifts, and normal marine communication was prevalent with the seaways of the Cordilleran miogeocline to the west and the southern Arizona miogeocline to the south.

Late Pennsylvanian sedimentation was dominated by coarse, arkosic clastics from the Ancestral Rockies uplifts throughout the southern Rockies province and the eastern Paradox Basin. Although marine conditions prevailed in the western and southern Colorado Plateau Province through Missourian time, clastic sedimentation finally pushed the shoreline to the western and southern provincial borders by Virgilian time. The only known marine strata of Virgilian age occur in the deepest eastern trough of the Paradox Basin, and in central New Mexico.

Uninterrupted clastic sedimentation continued across the Pennsylvanian-Permian boundary immediately adjacent to the Ancestral Rockies uplifts. However, the systemic boundary is marked by an unconformity throughout most of the Colorado Plateau Province. Lower Permian (Wolfcampian) sedimentary rocks rest on Missourian or older strata except in the western Grand Canyon, central New Mexico, and the easternmost Paradox Basin. On the Emery uplift (western San Rafael Swell), the entire Pennsylvanian section has been removed by late Virgilian erosion, and elsewhere Missourian rocks mark the top of the Pennsylvanian System.

PERMIAN

Continental environments dominated the Early Permian landscapes of the southern Rocky Mountains and Colorado Plateau provinces. Arkosic clastic debris was still being shed in colossal quantities from the Ancestral Rockies uplifts, and finer-grained derivatives were spread across the emergent lowlands by largely fluvial processes. At the onset of Permian time, seaways were restricted to the regions west and south of the Cordilleran and Mogollon hingelines of the miogeoclines. However, a marine reentrant extended southeastward from the Oquirrh Basin into the northwestern Colorado Plateau, where the Wolfcampian Elephant Canyon Formation (interbedded marine limestones and fine clastics) was deposited. A stratigraphically similar interval, the Pakoon Formation, occurs in the westernmost Grand Canyon (Baars, 1962).

The intermontane basins of the Ancestral Rockies were filled with reddish brown arkose in Wolfcampian to Leonardian time. The various accumulations of coarse clastics all look alike. Except for their individual geographic occurrences, the Cutler Formation of the Colorado Plateau, the Maroon and Sangre de Cristo Formations of the central Colorado Basin, and the Fountain Formation of the Colorado Front Range would be virtually impossible to distinguish. All were derived from a gneissic and granitic basement; all were deposited by fluvial processes immediately adjacent to prominent high basement blocks; all were stained red by in situ weathering of ferromagnesian minerals in an arid climate; all are Middle Pennsylvanian to Middle Permian in age.

A vast coastal lowland lay to the southwest of the Uncompahgre uplift in Early Permian time. At first a fluvial-tidal flat complex dominated the region (Halgaito Shale) and interfingered with the marine Elephant Canyon Formation to the northwest (Fig. 7) (Baars, 1962). These relatively fine-grained red beds extend as far westward as the Grand Canyon where they are a basal part of the Esplanade Sandstone of the Supai Group. The intimate relationships with the Elephant Canyon Formation and scattered vertebrate remains in the Halgaito date the sequence as lower to middle Wolfcampian (Vaughn, 1962).

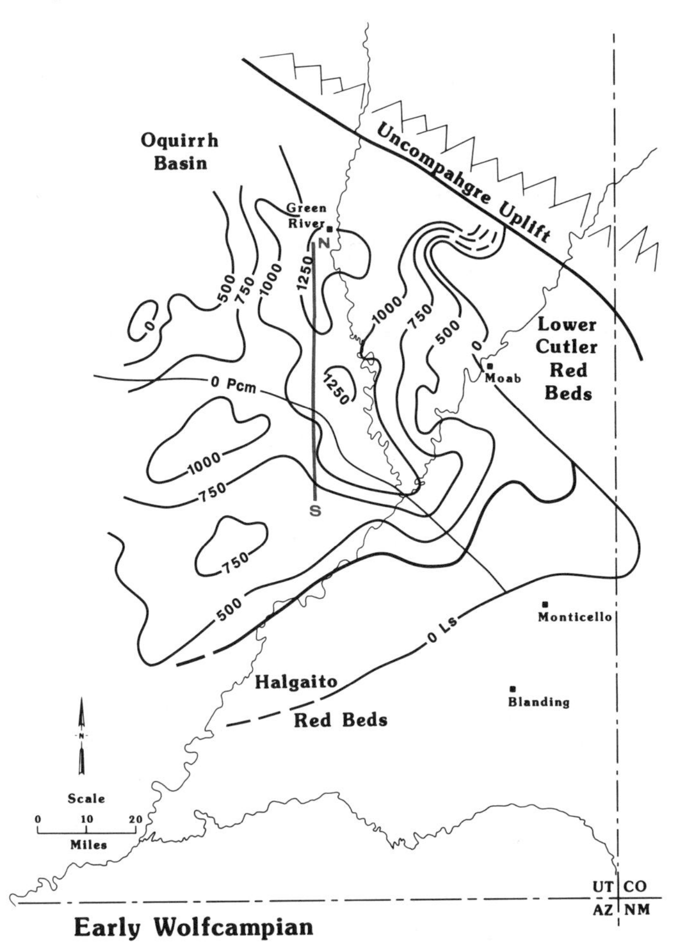

Figure 7. Isopach map of the early Wolfcampian Elephant Canyon Formation, not including the Cedar Mesa Sandstone facies, in east-central Utah. The marine strata interfinger eastward with arkosic red beds of the lower Cutler Group and southeastward with Halgaito red beds. The northern limit of distinguishable Cedar Mesa Sandstone is indicated by the "0 Pcm" line. (From Baars, 1987.)

Gradually, the red-bed lowlands gave way to fine-grained quartzose sands in the coastal environments adjacent to the upper Elephant Canyon sea. The vast "sand sea" of the Cedar Mesa Sandstone extends far to the southwest into the Grand Canyon, where it is known as the main body of the Esplanade Sandstone of the Supai Group. The provenance and depositional environments of this enormous accumulation of sand have been in question for many years. Baars (1962) thought the sand was derived

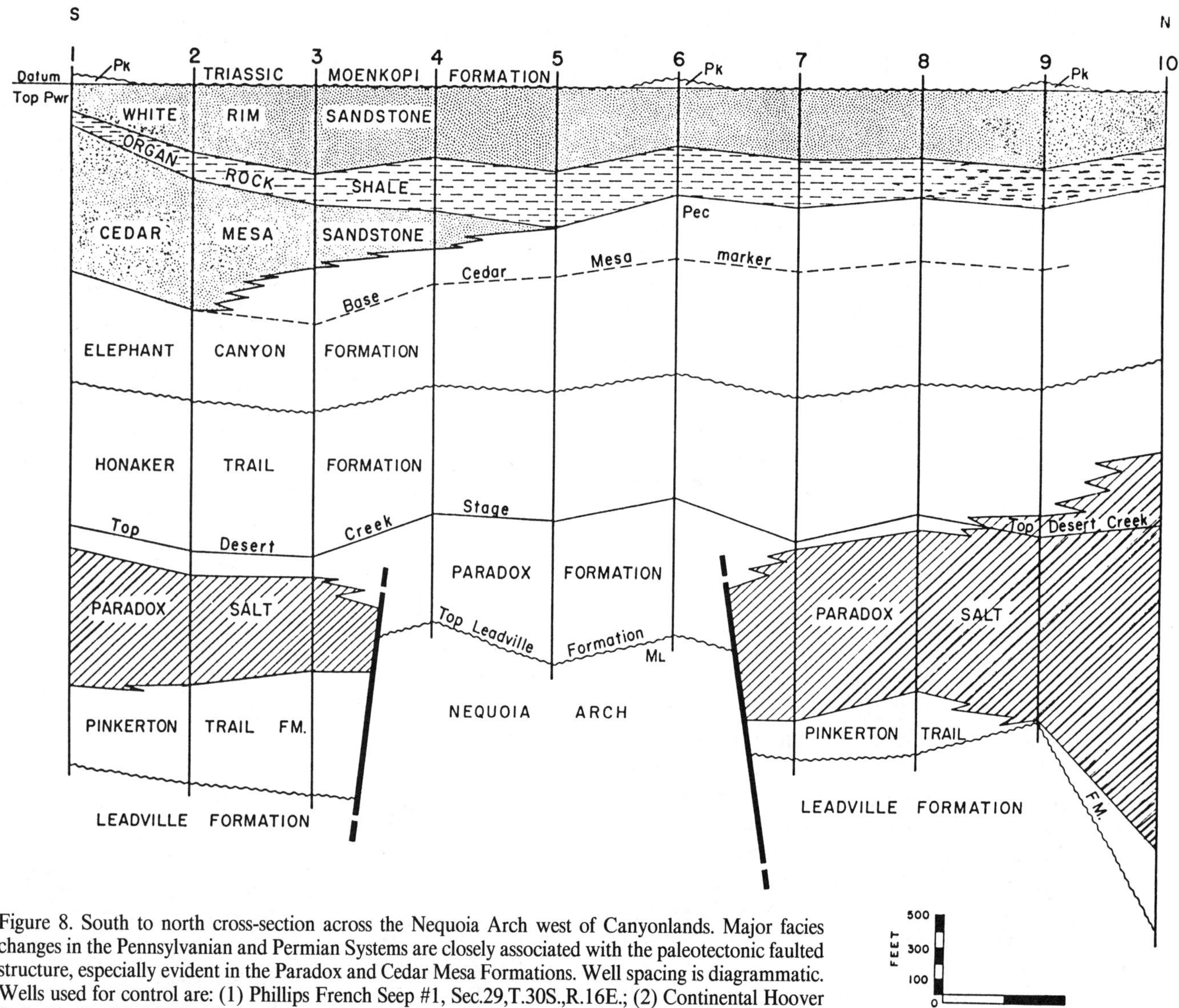

Figure 8. South to north cross-section across the Nequoia Arch west of Canyonlands. Major facies changes in the Pennsylvanian and Permian Systems are closely associated with the paleotectonic faulted structure, especially evident in the Paradox and Cedar Mesa Formations. Well spacing is diagrammatic. Wells used for control are: (1) Phillips French Seep #1, Sec.29,T.30S.,R.16E.; (2) Continental Hoover Fed. #1, Sec.20,T.29S,R.15E.; (3) Murphy Nequoia Arch #4, Sec.14,T.28S.,R. 14E.; (4) Texaco Nequoia Arch #6, Sec.32,T.27S.,R.15E.; (6) Continental Moonshine Wash #2, Sec.22,T.25S.,R.15E.; (7) General Petroleum #45–5–G, Sec.5,T.24S.,R.15E.; (8) Shell Chaffin #1, Sec.21,T.23S.,R.15E.; (9) Amax Petroleum Green River Desert #9–7, Sec.9,T.22S.,R.15E.; (10) Superior Grand Fault #14–24, Sec.24,T.21S.,R.15E. (From Baars, 1987.)

from the northwest, rather than from the Uncompahgre uplift, because of strongly oriented southeast-dipping cross-stratification structures, and the lack of typical Cutler petrology. It was further postulated that the nature of the cross-beds, the subaqueous ripple marks, the highly contorted slump structures, and abundant sand-size skeletal debris indicated a largely marine environment of deposition. However, those who see only the large-scale cross-stratification in the Cedar Mesa continue to insist on an eolian origin for the formation (Loope, 1984). The answer is probably somewhere between the two extreme viewpoints. It is now realized that the Cedar Mesa Sandstone is a coastal facies of the marine upper Elephant Canyon Formation (Baars, 1987), a new correlation that dates the sands of the Cedar Mesa as middle to late Wolfcampian (Fig. 8). The light-colored Cedar Mesa sands also interfinger with red, arkosic, fluvial sands from the Uncompahgre uplift (Cutler Formation) in the Needles District of Canyonlands National Park (Fig. 9), forming prominent red-and-white-banded spires and knobs.

The Lyons Sandstone of the Colorado Front Range is a similar unit, resting on the Fountain arkoses and underlying marginal marine red beds of the Lykins Formation. Like the Cedar Mesa, the Lyons Sandstone has undergone a series of environmental reinterpretations, from eolian to littoral marine and back to eolian.

Continental processes again became dominant, and the lowland red beds of the Organ Rock Shale, and its equivalent Hermit Shale of Grand Canyon, were spread across much of the Colorado Plateau (Baars, 1962). Fossil plants (White, 1929) and some vertebrates (Vaughn, 1964) suggest an early Leonardian age for the red mudstones and siltstones. The widespread slope-forming shaley formation is a prominent marker in the base of the buttes in Monument Valley, through the length of the Grand Canyon, and in western Canyonlands country.

There were more "sand seas." A general lack of diagnostic fossils anywhere on the Colorado Plateau make dating and correlating this part of the stratigraphic sequence difficult, to say the least, and many controversies regarding stratigraphic relationships have arisen. The Coconino Sandstone of northern Arizona overlies the Hermit Shale in Grand Canyon, where the Coconino is obviously an eolian deposit (McKee, 1933). It is generally conceded that the Coconino is in large part, if not wholly, correlative with the water-laid Glorieta Sandstone of northern New Mexico (Baars, 1962). The source of the quartzose sands was in central Arizona, and probably the Mazatzal highland.

But what are the relationships between the Coconino-Glorieta and the DeChelly Sandstone of the Four Corners region? It has been the contention of Baars (1962; 1979) that the DeChelly, a red, fine-grained, largely eolian sandstone derived from the reworked Cutler red beds to the northeast, underlies the Coconino-Glorieta on the Defiance uplift along the Arizona–New Mexico border, and along the Mogollon Rim in north-central Arizona. But the DeChelly is not present in the Grand Canyon because of onlap onto tectonically-induced paleotopography of the East Kaibab monocline by the older sands. Blakey (1979) largely concurs for the Mogollon Rim–eastern Grand Canyon regions, and has named the approximate stratigraphic equivalent of the DeChelly the Schnebley Hill Formation along the Mogollon Rim. The Schnebley Hill is a red sandstone-siltstone sequence formerly considered to be the "Supai Formation."

A further complication is the relationship between the DeChelly Sandstone and the White Rim Sandstone of Canyonlands country. This prominent bench-forming unit of western Canyonlands National Park and adjacent regions was interpreted to be a littoral marine deposit by Baars and Seager (1970), another controversial interpretation, and hosts several billion barrels of tarry petroleum along its updip pinchout just west of the Colorado River. Irwin (1971) made a convincing case that the White Rim is younger than the DeChelly and grades westward into the Toroweap Formation (Fig. 10), a largely marine formation in the Grand Canyon. According to this interpretation, the White Rim Sandstone represents a nearshore bar-sand and/or coastal eolian dune complex that formed along the eastern Toroweap shoreline.

It now seems safe to consider the stratigraphic sequence in question to be, in ascending order: (1) the Organ Rock-Hermit Shale, (2) the DeChelly Sandstone of the Four Corners region, (3) the Coconino-Glorieta Sandstone of northern Arizona and New Mexico (The "Coconino Sandstone" of the San Rafael Swell and Circle Cliffs uplift is the White Rim Sandstone), and (4) the White Rim–Toroweap sedimentary rocks of the western Colorado Plateau (Baars, 1979).

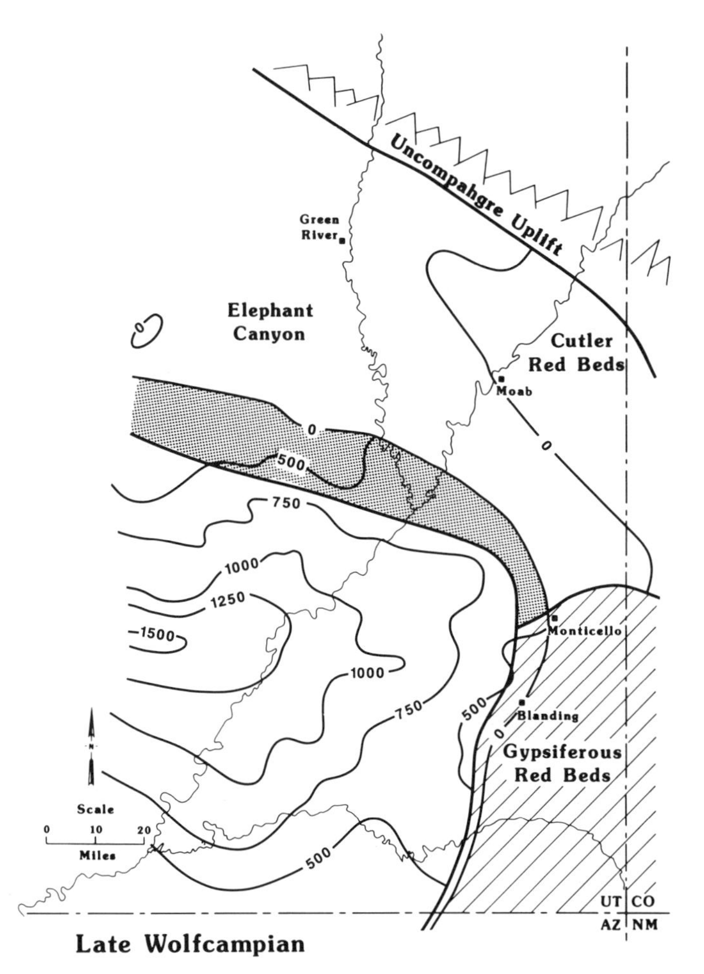

Figure 9. Isopach map of the Cedar Mesa Sandstone in southeastern Utah. The stippled area represents the trend of facies change northward to marine rocks of the upper Elephant Canyon Formation; the ruled area is the lagoonal gypsiferous red-bed facies of the Cedar Mesa in the Blanding Basin. Zero lines north of the Cedar Mesa facies represent the limits of the partly equivalent Elephant Canyon Formation. (From Baars, 1987.)

The youngest Permian rocks of the Colorado Plateau–southern Rockies provinces are the marine formations that border the Plateau country on the south and west. These are the San Andres Limestone of central New Mexico and the Kaibab Limestone of northern Arizona. The stratigraphically equivalent formations have been dated as late Leonardian, based largely on their brachiopod faunas (McKee, 1938; Baars, 1962; 1979).

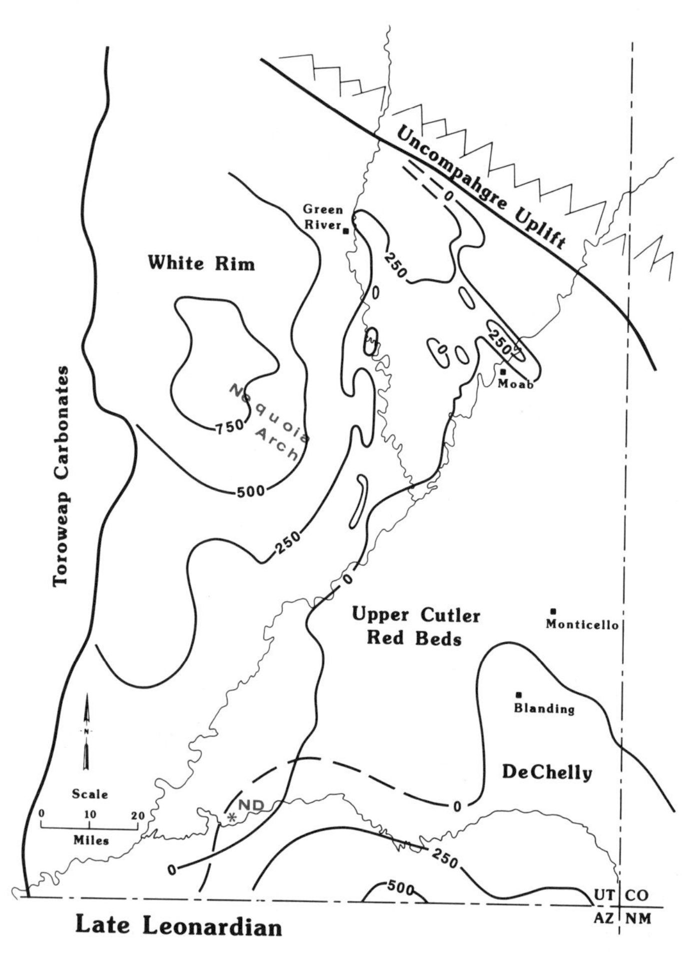

Figure 10. Isopach map of the known areal extent of the White Rim Sandstone of Leonardian age. The local isopach thicks west of, and along, the Green River are believed to be offshore and barrier bars in the Toroweap-equivalent sandstone. Note that the White Rim thickens regionally on the Emery uplift (San Rafael Swell) and locally overlies the eolian DeChelly Sandstone at Nokai dome near the San Juan River to the south. (From Baars, 1987.) ND = Nokai Dome.

No strata of younger Permian (Guadalupian-Ochoan) age have been recognized with any certainty in the province. The youngest Permian carbonate rocks of the San Rafael Swell of east-central Utah still pose a problem. The "Kaibab Limestone" of that region has faunal affinities to the Park City Formation of central northern Utah, and may be somewhat younger than the type Kaibab (McKee, 1938; Baars, 1962). Consequently, the "Kaibab" of the San Rafael Swell may be lower Guadalupian in age (Rascoe and Baars, 1972).

PERMO-TRIASSIC BOUNDARY

Although the Guadalupian and Ochoan epochs are well represented in west Texas, rocks of this age are notably missing over large regions of the western United States. With the possible exception of the "Kaibab" (Park City Formation?) of the San Rafael Swell region, there are no known occurrences of rocks of Middle to Late Permian age in the southern Rockies and Colorado Plateau provinces. Thus a hiatus exists between the Middle Permian and Lower Triassic series. In the eastern Paradox Basin where salt diapirs were actively growing during this time interval, local angular unconformities mark the temporal boundary. Elsewhere, the regional unconformity is less than obvious.

A maverick red bed unit named the Hoskinnini Member is the culprit. By definition, the Hoskinnini overlies the DeChelly Sandstone west of Monument Valley, and was considered to be a member of the Permian Cutler Group (Baker and Reeside, 1929). However, Stewart (1959) changed its status to that of a member of the Moenkopi Formation of Early Triassic age. Later, Thaden and others (1964) reported that the "Hoskinnini is known to grade laterally into the DeChelly . . ." north of the type section, but mapped the Hoskinnini as a lateral equivalent of the White Rim Sandstone in the White Canyon Mining District near Hite, Utah. Red beds considered to be Hoskinnini by Stewart (1959) west of the confluence of the Green and Colorado Rivers overlie the White Rim Sandstone. Baars and Seager (1970) presented evidence to indicate that these strata are more closely related to the Permian section than to the Moenkopi. It has become clear that there is more than one "Hoskinnini Member" in Canyonlands country, and that the stratigraphic position of the Permo-Triassic boundary is still in question on the eastern Colorado Plateau as elsewhere in the western United States.

TRIASSIC

Regardless of the stratigraphic position and age of the "Hoskinnini Member(s)," the Early to Middle(?) Triassic series consists of the Moenkopi Formation. It is a widespread, reddish brown, mudstone-siltstone unit that thickens westward from the eastern Utah and eastern Arizona borders toward the Cordilleran miogeocline. The Moenkopi is fluvial near its eastern limits, but represents intertidal environments over most of the Colorado Plateau; beautifully preserved oscillation ripples, mud cracks, raindrop impressions, and burrows are ubiquitous. It becomes more marine toward the west and includes marine limestone and gypsum members along the western margin of the province, where the Moenkopi attains a thickness of more than 300 m (1,000 ft) (McKee, 1954; MacLachlan, 1972). Age of the Moenkopi was established in large part by the occurrences of ammonites of the *Tirolites* and *Meekoceras* zones in south-central Utah, verifying a correlation with the marine Thaynes Formation of the Cordilleran miogeocline of the northwest Utah–southeast Idaho region (McKee, 1954).

An erosional disconformity with pronounced channels cut into the top of the Moenkopi indicates the contrast with the overlying Chinle Formation of Late Triassic age. Fluvial channel-fill sandstones and conglomerates form basal members of the Chinle Formation on a regional basis. South and west of

White Canyon and North Wash, near Hite, Utah, the channel-fill deposits are known as the Shinarump Member. To the north and east of that area, a younger but similar fluvial deposit, the Moss Back Member, occurs at the base of the Chinle. Unnamed basal fluvial deposits are found in the Chinle and its equivalent Dolores Formation in Colorado, but their exact correlations with Utah and Arizona units is unknown. The Shinarump and Moss Back Members host secondary uranium deposits throughout large regions of the Colorado Plateau. Mineralization occurs where fossil plant debris creates reducing conditions, and the uranium oxides precipitate from the circulating ground waters.

Several named members overlie the basal channel-fill deposits in different geographic areas. They represent color variations, mineralogic content, grain-size variations, and/or depositional environments. Colors vary from light gray to pale purples, greens, and reds. Rock types vary from mudstone to siltstone to sandstone, with widely varying amounts of bentonitic clays and volcanic ashes. Thin red to gray limestone beds occur locally in the middle of the formation. Depositional environments include fluvial channel, overbank, and flood plain deposits intermixed with lacustrine sediments. The Chinle is typically a variegated slope-forming formation that underlies the massive vertical brown cliffs of the Wingate Sandstone. Slopes are usually covered with fallen masses of Wingate debris, and the unstable slopes are often badly slumped. The Chinle Formation varies in thickness from about 150 to 300 m (500 to 1,000 ft), and averages about 250 m (800 ft) (Maclachlan, 1972).

The Lukachukai Member of the Wingate Sandstone forms magnificent vertical cliffs that cap the multicolored slopes of the Chinle Formation over much of the Colorado Plateau. It rests disconformably on the Chinle where present in eastern Utah and western Colorado, and disconformably on the Rock Point Member of the Wingate in northeastern Arizona and northwestern New Mexico. This, the J-O unconformity of Peterson and Pipiringos (1979), is now believed to mark the base of the Jurassic system on the basis of recently discovered fossil palynomorphs. Thus the Chinle and the Rock Point Member of the Wingate are believed to be Triassic in age, and the Lukachukai Member of the Wingate Sandstone is considered to be Lower Jurassic (Peterson and Pipiringos, 1979).

REFERENCES CITED

Baars, D. L., 1958, Cambrian stratigraphy of the Paradox Basin region: Intermountain Association of Petroleum Geologists 9th Annual Field Conference Guidebook, p. 93–101.

——, 1962, Permian System of Colorado Plateau: American Association of Petroleum Geologists Bulletin, v. 46, p. 149–218.

——, 1966, Pre-Pennsylvanian paleotectonics; Key to basin evolution and petroleum occurrences in Paradox Basin, Utah and Colorado: American Association of Petroleum Geologists Bulletin, v. 50, p. 2082–2111.

——, 1975, Pre-Pennsylvanian reservoir rocks of the eastern Colorado Plateau and southern Rocky Mountains, *in* Bolyard, D. W., ed., Deep drilling frontiers of the central Rocky Mountains: Rocky Mountain Association of Geologists Symposium, p. 71–74.

——, 1976, The Colorado Plateau aulocogen; Key to continental-scale basement rifting: Proceedings Second International Conference on Basement Tectonics, p. 157–164.

——, 1979, The Permian System, *in* Baars, D. L., ed., Permianland: Four Corners Geological Society 9th Field Conference Guidebook, p. 1–6.

——, 1987, The Elephant Canyon Formation revisited, *in* Campbell, J. A., ed. Cataract Canyon: Four Corners Geological Society 10th Field Conference Guidebook, p. 81–90.

Baars, D. L., and Campbell, J. A., 1968, Devonian System of Colorado, northern New Mexico, and the Colorado Plateau: The Mountain Geologist, v. 5, no. 1, p. 31–40.

Baars, D. L., and Ellingson, J. A., 1984, Geology of the western San Juan Mountains: Rocky Mountain Section, Geological Society of America 37th Annual Meeting Field Trip Guidebook, p. 1–45.

Baars, D. L., and Seager, W. R., 1970, Stratigraphic control of petroleum in White Rim Sandstone (Permian) in and near Canyonlands National Park: American Association of Petroleum Geologists Bulletin, v. 54, p. 709–718.

Baars, D. L., and See, P. D., 1968, Pre-Pennsylvanian stratigraphy and paleotectonics of the San Juan Mountains, southwestern Colorado: Geological Society of America Bulletin, v. 79, p. 333–350.

Baars, D. L., and Stevenson, G. M., 1982, Subtle stratigraphic traps in Paleozoic rocks of Paradox Basin, *in* Halbouty, M., ed., Deliberate search for the subtle trap: American Association of Petroleum Geologists Memoir 32, p. 131–158.

——, 1984, The San Luis uplift, Colorado and New Mexico; An enigma of the ancestral Rockies: The Mountain Geologist, v. 21, no. 2, p. 57–67.

Baker, A. A., and Reeside, J. B., Jr., 1929, Correlation of the Permian of southern Utah, northern Arizona, and southwestern Colorado: American Association of Petroleum Geologists Bulletin, v. 13, p. 1413–1448.

Billingsley, G. H., 1986, Erosional patterns prior to deposition of the Surprise Canyon Formation in Grand Canyon, Arizona: Geological Society of America Abstracts with Programs, v. 18, no. 5, p. 341.

Billingsley, George H. and Stanley S. Beus, 1985, The Surprise Canyon Formation - an Upper Mississippian and Lower Pennsylvanian(?) rock unit in the Grand Canyon, Arizona, *in* Stratigraphic Notes, 1984: U.S. Geological Survey Bulletin 1605-A, p. A27–A33.

Beus, S. S., 1969, Devonian stratigraphy in northwestern Arizona: Four Corners Geological Society 5th Field Conference Guidebook, p. 127–133.

——, 1986, Biostratigraphy of the Surprise Canyon Formation (Latest Mississippian and earliest Pennsylvanian) in western Grand Canyon, Arizona: Geological Society of America Abstracts with Programs, v. 18, no. 5, p. 341.

Blakey, R. C., 1979, Lower Permian stratigraphy of the southern Colorado Plateau, *in* Baars, D. L., ed., Permianland: Four Corners Geological Society 9th Field Conference Guidebook, p. 115–130.

Chronic, J., 1972, Middle Paleozoic kimberlite diatremes in Colorado and Wyoming: Colorado School of Mines Quarterly, v. 67, no. 4, p. 63–76.

Ford, T. D., Breed, W. J., and Downey, C., 1969, Preliminary geologic report of the Chuar Group, Grand Canyon, Arizona: Four Corners Geological Society 5th Field Conference Guidebook, p. 114–122.

Gerhard, L. C., 1972, Canadian depositional environments and paleotectonics, central Colorado: Colorado School of Mines Quarterly, v. 67, no. 4, p. 1–36.

Hite, R. J., 1960, Stratigraphy of the saline facies of the Paradox Member of the Hermosa Formation of southeastern Utah and southwestern Colorado, *in* Smith, K. G., ed., Geology of the Paradox Basin fold and fault belt: Four Corners Geological Society 3rd Field Conference Guidebook, p. 86–89.

Irwin, C. D., 1971, Stratigraphic analysis of Upper Permian and Lower Triassic strata in southern Utah: American Association of Petroleum Geologists Bulletin, v. 55, p. 1976–2007.

Lochman-Balk, C., 1972, Cambrian System, *in* Mallory, W. W., ed., Geologic atlas of the Rocky Mountain region: Rocky Mountain Association of Geolo-

gists, p. 60–75.
Loleit, A. J., 1963, Cambrian stratigraphic problems of the Four Corners area, *in* Bass, R. O., and Sharps, S. L., eds., Shelf carbonates of the Paradox Basin: Four Corners Geological Society Symposium, p. 21–30.
Loope, D. B., 1984, Eolian origin of upper Paleozoic sandstones, southeastern Utah: Journal of Sedimentary Petrology, v. 54, no. 2, p. 563–580.
MacLachlan, M. E., 1972, Triassic System, *in* Mallory, W. W., ed., Geologic atlas of the Rocky Mountain region: Rocky Mountain Association of Geologists, p. 167–176.
Mallory, W. W., 1972, Pennsylvanian arkose and the ancestral Rocky Mountains, *in* Mallory, W. W., ed., Geologic atlas of the Rocky Mountain region: Rocky Mountain Association of Geologists, p. 131–132.
McKee, E. D., 1933, The Coconiono Sandstone; Its history and origin: Carnegie Institute of Washington Publication 440, p. 77–115.
—— , 1938, The environment and history of the Toroweap and Kaibab Formations of northern Arizona and southern Utah: Carnegie Institute of Washington Year Book, no. 36, 1936–37, Publication 492, 268 p.
—— , 1954, Stratigraphy and history of the Moenkopi Formation of Triassic age: Geological Society of America Memoir 61, 133 p.
—— , 1969, Paleozoic rocks of Grand Canyon: Four Corners Geological Society 5th Field Conference Guidebook, p. 78–90.
—— , 1979, The Esplanade Sandstone of Grand Canyon, *in* Baars, D. L., ed., Permianland: Four Corners Geological Society 9th Field Conference Guidebook, p. 67–80.
McKee, E. D., and Gutschick, R. C., 1969, History of the Redwall Limestone of northern Arizona: Geological Society of America Memoir 144, 612 p.
McKee, E. D., and Resser, C. E., 1945, Cambrian history of the Grand Canyon region: Carnegie Institute of Washington Publication 563, 232 p.
Nadeau, J. E., 1972, Mississippian stratigraphy of central Colorado: Colorado School of Mines Quarterly, v. 67, no. 4, p. 77–102.
Peterson, F., and Pipiringos, G. N., 1979, Stratigraphic relations of the Navajo Sandstone to Middle Jurassic formations, southern Utah and northern Arizona, U.S. Geological Survey Professional Paper 1035-B, 43 p.
Rascoe, B., Jr., and Baars, D. L., 1972, Permian System, *in* Mallory, W. W., eds., Geologic atlas of Rocky Mountain region: Rocky Mountain Association of Geologists, p. 143–165.
Ritzma, H. R., 1987, Basement in the Uintas; An enigma? [abs.]: American Association of Petroleum Geologists Bulletin, v. 71, p. 1014.
Stevenson, G. M., and Baars, D. L., 1986, The Paradox; A pull-apart basin of Pennsylvanian age, *in* Peterson, J. A., ed., Paleotectonics and sedimentation: American Association of Petroleum Geologists Memoir 41, p. 513–539.
Stevenson, G. M., and Beus, S. S., 1982, Stratigraphy and depositional setting of the upper Precambrian Dox Formation in Grand Canyon: Geological Society of America Bulletin, v. 93, p. 163–173.
Stewart, J. H., 1959, Stratigraphic relations of Hoskinnini Member (Triassic?) of Moenkopi Formation on Colorado Plateau: American Association of Petroleum Geologists Bulletin, v. 43, p. 1852–1868.
Thaden, R. E., Trites, A. F., Jr., and Finnell, T. L., 1964, Geology and ore deposits of the White Canyon area, San Juan and Garfield Counties, Utah: U.S. Geological Survey Bulletin 1125, 166 p.
Tweto, O., 1980, Precambrian geology of Colorado, *in* Kent, H. C., and Porter, K. W., eds., Colorado geology: Rocky Mountain Association of Geologists Symposium, p. 37–46.
Vaughn, P. P., 1962, Vertebrates from the Halgaito Tongue of the Cutler Formation, Permian of San Juan County, Utah: Journal of Paleontology, v. 36, no. 3, p. 529–539.
—— , 1964, Vertebrates from the Organ Rock Shale of the Cutler Group, Permian of Monument Valley and vicinity, Utah and Arizona: Journal of Paleontology, v. 38, p. 567–583.
Warner, L. A., 1978, The Colorado Lineament: A middle Precambrian wrench fault system: Geological Society of America Bulletin, v. 89, p. 161–171.
White, D., 1929, Flora of the Hermit Shale, Grand Canyon, Arizona, Carnegie Institute of Washington Publication 405, 221 p.

Manuscript Accepted by the Society September 21, 1987

ACKNOWLEDGMENTS

Special thanks are due to Gene M. Stevenson, whose innumerable contributions of basic ideas, basic interpretations, and critical discussions have made this summary chapter possible. Allison R. "Pete" Palmer and Larry Sloss critically reviewed various forms of the manuscript and have kept me honest as well as greatly improving the final product, and Fred "Pete" Peterson has been of considerable assistance in the summary of the Triassic-Jurassic boundary problems.

Printed in U.S.A.

The Geology of North America
Vol. D-2, Sedimentary Cover—North American Craton: U.S.
The Geological Society of America, 1988

Chapter 5

A synthesis of the Jurassic system in the southern Rocky Mountain region

Fred Peterson
U.S. Geological Survey, MS 939, Box 25046, Denver Federal Center, Denver, Colorado 80225

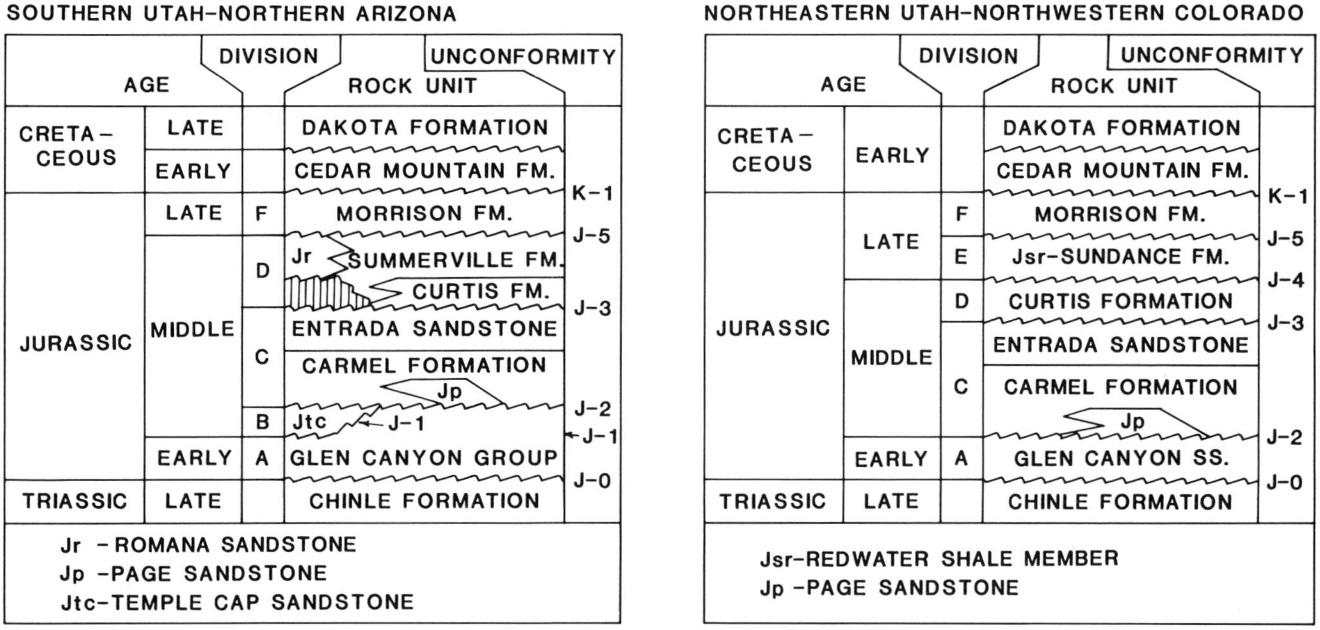

SOUTHERN UTAH–NORTHERN ARIZONA

AGE		DIVISION	ROCK UNIT	UNCONFORMITY
CRETACEOUS	LATE		DAKOTA FORMATION	
	EARLY		CEDAR MOUNTAIN FM.	K-1
JURASSIC	LATE	F	MORRISON FM.	J-5
	MIDDLE	D	Jr SUMMERVILLE FM.	
			CURTIS FM.	J-3
		C	ENTRADA SANDSTONE	
			CARMEL FORMATION	
			Jp	J-2
		B	Jtc J-1	J-1
	EARLY	A	GLEN CANYON GROUP	J-0
TRIASSIC	LATE		CHINLE FORMATION	

Jr - ROMANA SANDSTONE
Jp - PAGE SANDSTONE
Jtc - TEMPLE CAP SANDSTONE

NORTHEASTERN UTAH–NORTHWESTERN COLORADO

AGE		DIVISION	ROCK UNIT	UNCONFORMITY
CRETACEOUS	EARLY		DAKOTA FORMATION	
			CEDAR MOUNTAIN FM.	K-1
JURASSIC	LATE	F	MORRISON FM.	J-5
		E	Jsr-SUNDANCE FM.	J-4
	MIDDLE	D	CURTIS FORMATION	J-3
		C	ENTRADA SANDSTONE	
			CARMEL FORMATION	
			Jp	J-2
	EARLY	A	GLEN CANYON SS.	J-0
TRIASSIC	LATE		CHINLE FORMATION	

Jsr - REDWATER SHALE MEMBER
Jp - PAGE SANDSTONE

Figure 1. Schematic columns showing the vertical sequence of unconformities and stratigraphic divisions with representative formations in the northern and southern parts of the Colorado Plateau region. Not to scale.

INTRODUCTION

Jurassic strata in the southern part of the Rocky Mountain region include such colorful and well-known sedimentary rocks as the Glen Canyon and San Rafael Groups and the Morrison Formation—rock units that contribute greatly to the scenic beauty of the many popular recreation areas in the region. Most of these rocks were deposited in continental environments bordering an epeiric seaway that, at times, migrated into the region from the north or northwest.

For purposes of synthesis and discussion, the rocks are here separated into six divisions, labeled A through F from oldest to youngest, that are bounded by the J-0 to J-5 unconformities of Pipiringos and O'Sullivan (1978) and by sub-Cretaceous unconformities at the top (Fig. 1). Where not cut out by younger erosion surfaces, most of the Jurassic unconformities extend throughout much of the region, although the J-3 unconformity fades out eastward in southeasternmost Utah (O'Sullivan, 1980a). Additionally, most of these unconformities correspond fairly well to the short-term eustatic falls in sea level postulated by Vail and others (1984). An exception is the J-4 unconformity that apparently was not related to any of the major sea-level fluctuations and, instead, may have been produced by tectonic processes entirely within the Western Interior.

This chapter includes important stratigraphic revisions and nomenclatural changes made especially in the Colorado Plateau

Peterson, F., 1988, A synthesis of the Jurassic system in the southern Rocky Mountain region, *in* Sloss, L. L., ed., Sedimentary Cover—North American Craton; U.S.: Boulder, Colorado, Geological Society of America, The Geology of North America, v. D-2.

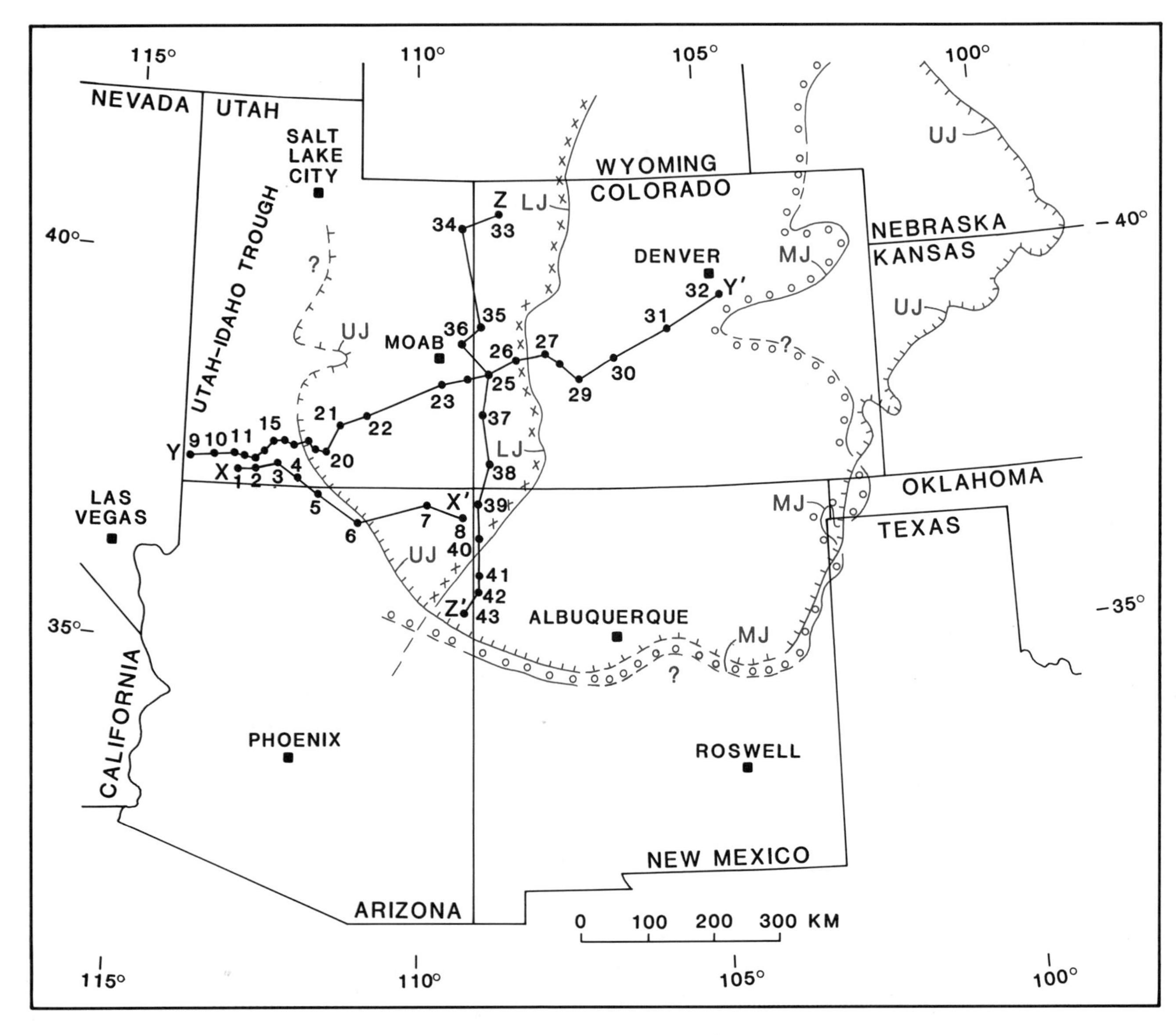

Figure 2. Distribution of the Lower (LJ), Middle (MJ), and Upper (UJ) Jurassic Series in the Southern Rocky Mountains region; each series is present on the side of the line with the symbol; after Peterson (1972), Pipiringos and O'Sullivan (1978), and F. Peterson (unpublished data). Also shown are lines of stratigraphic sections in Figures 3, 4, and 5.

region in recent years (Green, 1974; O'Sullivan, 1980a, b; Pipiringos and O'Sullivan, 1978; Peterson and Pipiringos, 1979; O'Sullivan and Pierce, 1983; Condon and Peterson, 1986; Condon and Huffman, 1988; Peterson, 1988a). Scarcity of age-diagnostic fossils precludes accurate dating of most of these rocks, so the age designations are based on rather long-distance correlations to better dated strata elsewhere in the United States. Thus, Division A is considered Early Jurassic in age based on comparison of faunas and florules with strata in the eastern United States (Olsen and Galton, 1977; Galton, 1978; Peterson and Pipiringos, 1979); Divisions B, C, and D are Middle Jurassic, and Divisions E and F are Late Jurassic based largely on correlation with more fossiliferous strata in the northern part of the Western Interior (Imlay, 1980). Recent fission track dating of zircons in bentonite beds of the Morrison Formation (Division F) suggests that part of that division may extend into the earliest Cretaceous (Kowallis and Heaton, 1987).

Rocks of the Lower, Middle, and Upper Jurassic Series can be viewed broadly as three eastward-thinning sedimentary wedges, each of which tends to be beveled out beneath overlying strata (Figs. 2, 3, 4). The Lower and Middle Jurassic Series thicken westward into the easternmost Basin and Range Prov-

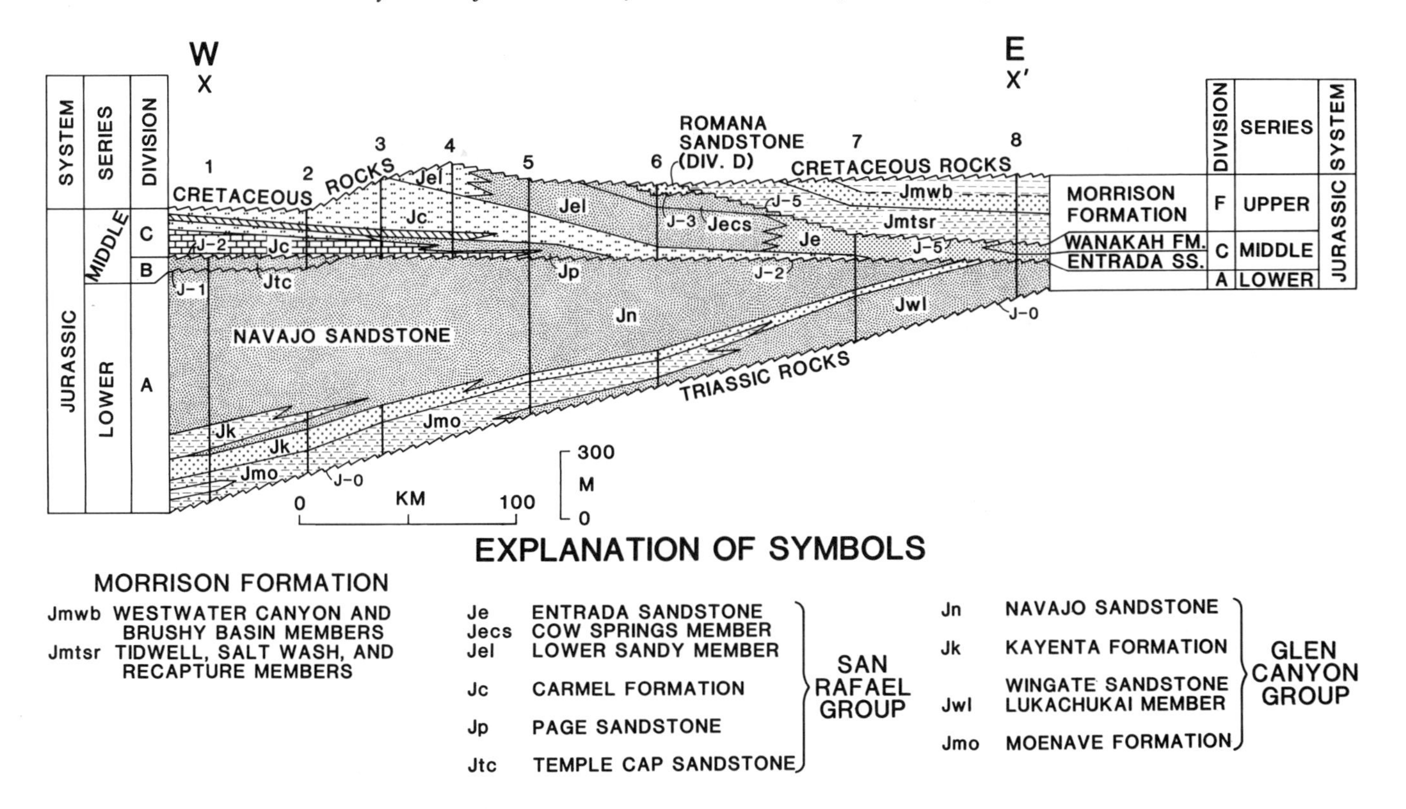

Figure 3. Stratigraphic section of the Jurassic System along the Arizona-Utah state line (after Peterson and Pipiringos, 1979). For location see Figure 2; J-0 to J-5 are unconformities; Glen Canyon Group includes Division A; San Rafael Group includes Divisions B, C, and D. See Figure 4 for explanation of patterns.

ince. There, Lower Jurassic strata are known only from scattered outcrops, but their incomplete and yet considerable thickness (roughly 600 to 800 m or more; Marzolf, 1983; Blakey and Middleton, 1983) suggests deposition in a broad structural depression that probably was an early manifestation of the Utah-Idaho trough. Although Lower and Middle Jurassic formations tend to thicken westward, the overall thickness of the System remains surprisingly close to constant throughout much of the region owing to westward beveling beneath the Cretaceous System. In a broader aspect, however, the Jurassic was largely or entirely beveled out to the west, southwest, and south during the earlier part of the Cretaceous owing to broad uplifts in those directions, especially during the time immediately preceding deposition of the Dakota Formation (Figs. 4, 5).

The Utah-Idaho trough extended south-southwestward from north-central Utah to the southwestern corner of the state and into southeastern Nevada (Figs. 6, 7). The position of the western edge of the trough is poorly understood because Lower and Middle Jurassic rocks were eroded from most of that area. However, two features suggest that the western side lay close to the Nevada-Utah border: (1) the westward increase in coarse clastics in some of the Middle Jurassic formations (Imlay, 1967; Peterson and Pipiringos, 1979; Peterson, 1988a), and (2) the distribution of faunas in the Middle Jurassic Carmel Formation, which has yielded a fairly normal marine fauna in a central belt lying in the middle of the trough that is flanked to the east and west by belts containing depauperate faunas suggestive of proximity to shorelines (Imlay, 1964).

In late Middle to Late Jurassic time, uplift in eastern Nevada, southeastern California, westernmost Utah, and west-central Arizona produced highland regions (the Sevier and Mogollon highlands) that, by the Late Jurassic, were the source of widespread fluvial deposits shed eastward and northeastward onto the southwestern part of the present-day Colorado Plateau. By the end of the Jurassic, the Sevier uplift in southeastern Nevada and westernmost Utah had completely destroyed the Utah-Idaho trough. Although long thought to have been produced by thrusting, recent studies suggest that some other process, as yet unclear, was responsible for the Sevier uplift (Dickinson and others, 1986; Heller and others, 1986).

Significant differences in sediment distribution by fluvial processes distinguish each of the three Jurassic epochs on or near the present-day Colorado Plateau. During Early Jurassic time, streams rose in the ancestral Rocky Mountain highlands east of

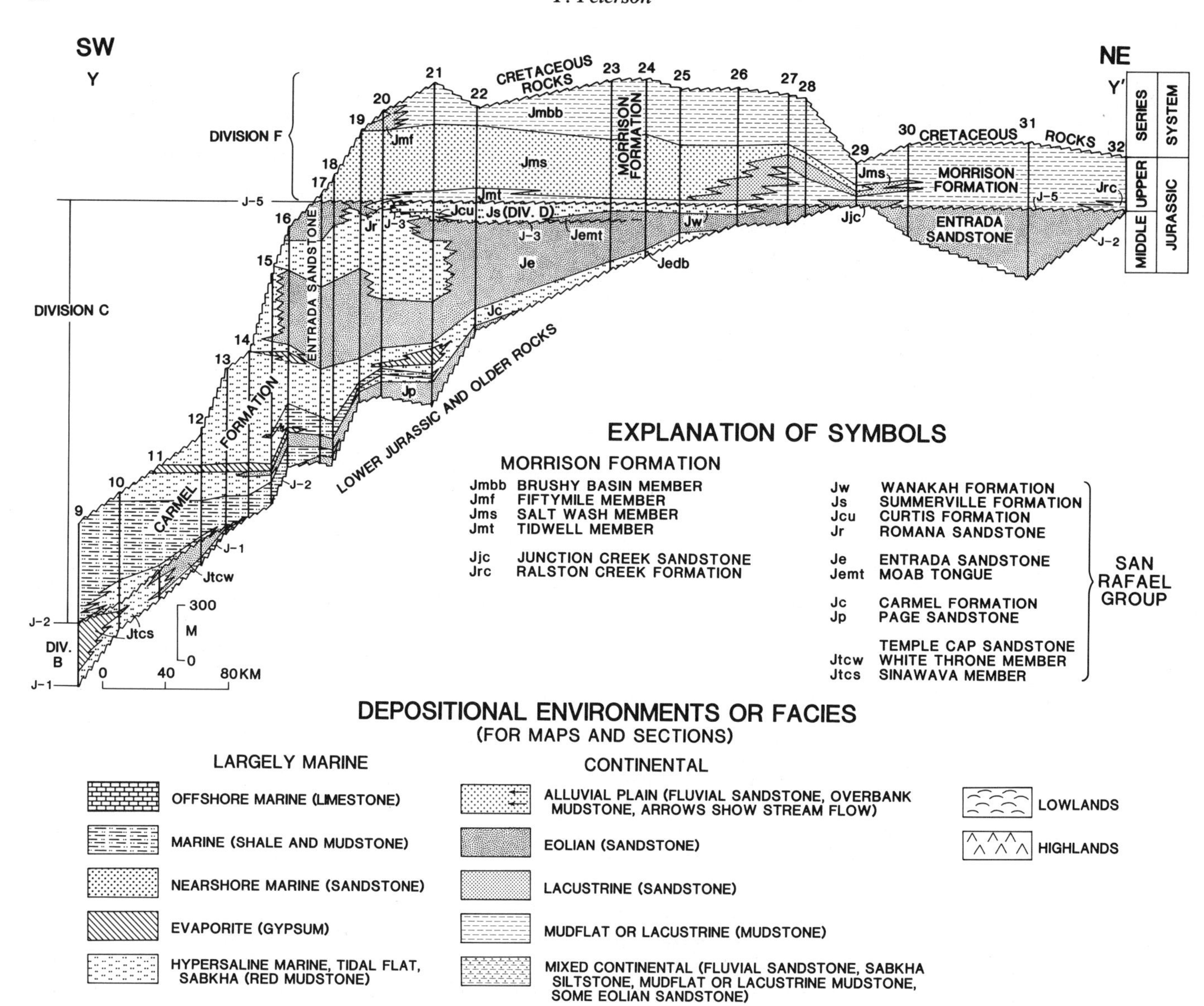

Figure 4. Stratigraphic section of Middle and Upper Jurassic rocks from southwestern Utah to Denver, Colorado, showing westward thickening of Divisions B and C into the Utah-Idaho trough. Note abrupt westward truncation beneath Cretaceous rocks. See Figure 2 for location; J-0 to J-5 are unconformities; Divisions B, C, and D are in the San Rafael Group. Compiled from Gregory and Moore (1931), Stark and others (1949), Hunt and others (1950), Langenheim (1957), Craig and others (1959), Scott (1963), Carter and Gaultieri (1965), Cashion (1967), Hansen (1968, 1971), Cater (1970), Peterson (1980c, 1988a) and unpublished data), Bryant and others (1981), and O'Sullivan (1981).

the Plateau in central Colorado as well as in the eastern Mogollon highlands farther south and flowed westward or northwestward toward the Utah-Idaho trough (Fig. 6; Poole, 1961; Luttrell, 1987). In contrast, scarce Middle Jurassic fluvial conglomerates in the Carmel Formation of south-central Utah that are composed largely of clasts of ash-flow tuff came from the south, presumably off the Mogollon highlands or nearby areas (Fig. 7; Chapman, 1987). Scarce chert and quartzite pebble conglomerate beds of possible fluvial origin and scattered chert and quartzite pebbles in southernmost exposures of Middle Jurassic rocks in northern New Mexico (Dobrovolny and Summerson, 1946; Maxwell, 1982) have a source to the south and, again, probably came from the Mogollon highlands. Although certainly south of the present-day Colorado Plateau, the location of the Mogollon highlands is equivocal; Harshbarger and others (1957) suggest that they lay in central Arizona, whereas Bilodeau (1986) suggests that they were farther south in southernmost Arizona. There are reasonable arguments to support either interpretation.

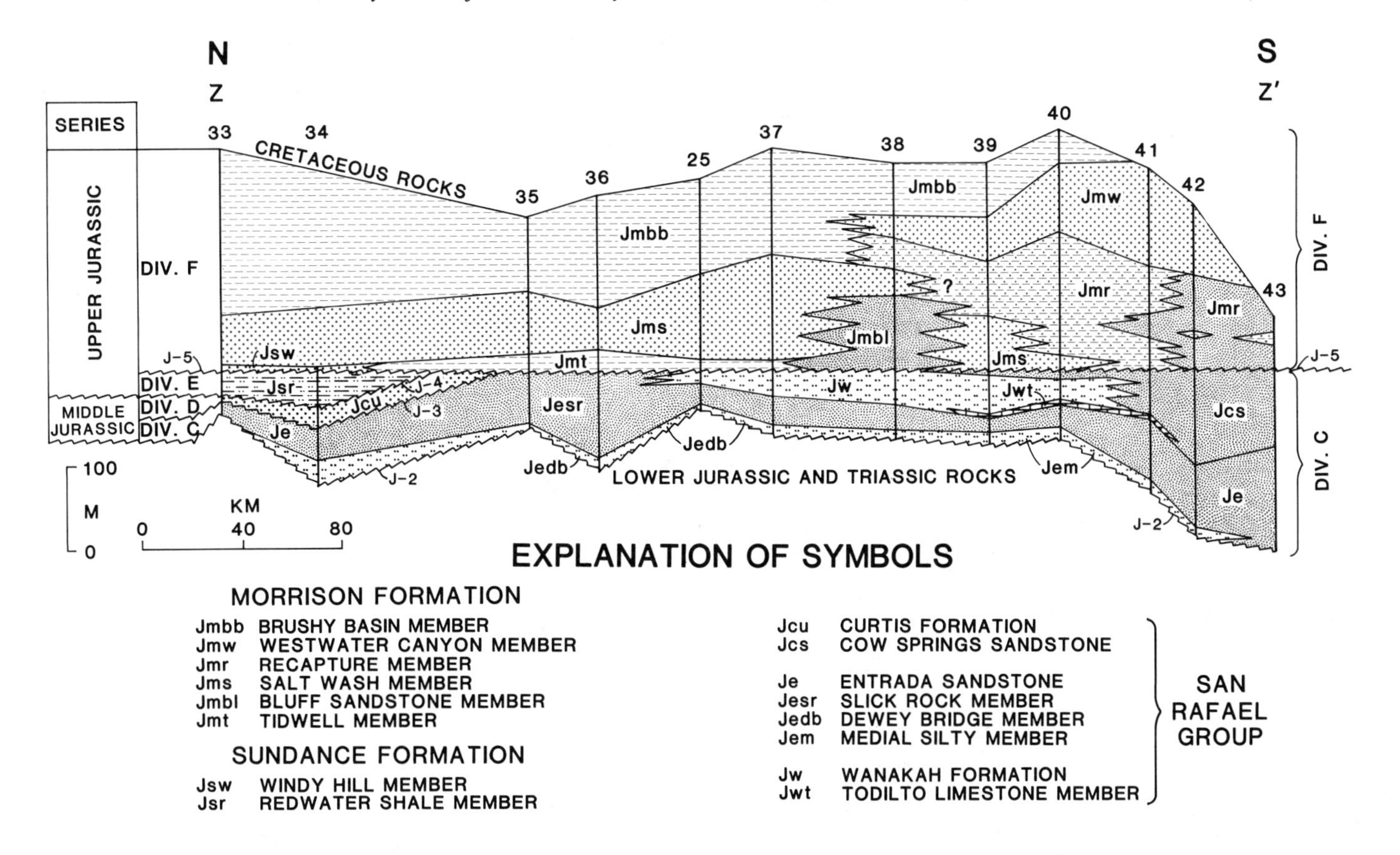

Figure 5. Stratigraphic section of Middle and Upper Jurassic rocks along the Utah-Colorado and Arizona-New Mexico state lines. Note abrupt southward truncation of rocks beneath Cretaceous rocks. For location see Figure 2; J-2 to J-5 are unconformities; San Rafael Group includes Divisions C and D. Compiled from Craig and others (1959), Ekren and Houser (1959), O'Sullivan (1981, 1980b, 1984), and Condon and Peterson (1986).

A remnant of the ancestral Rocky Mountains, which by this time was a series of low hills in central Colorado, also contributed small and perhaps negligible quantities of detritus (Kocurek, 1981). Appreciable uplift began in the Sevier highlands to the west of the Plateau in late Middle Jurassic time, as demonstrated by the influx of coarse detritus in the Curtis and Summerville Formations (Gilluly and Reeside, 1928; Peterson, 1988a). Late Jurassic streams flowed eastward and northeastward off the rising Sevier highlands and Mogollon highlands, and the resulting alluvial deposits were thicker and more widespread than in earlier times (Fig. 8).

In summary, the timing of uplifts that contributed significant quantities of detritus to the area of the present-day Colorado Plateau during the Jurassic was: (1) east and southeast of the Plateau during the Early Jurassic, (2) south of the Plateau during most of the Middle Jurassic, and (3) west and southwest of the Plateau during the late Middle and Late Jurassic.

The climate during Jurassic time probably was warm and dry, judging from some of the lithologies. Beds of gypsum, anhydrite, halite, or eolian sandstone are present in most of the divisions, suggesting a warm and dry environment. Thin beds of authigenic zeolites and feldspar formed as alteration products of volcanic ash that fell in a large playa lake and have been found in Upper Jurassic strata (Turner-Peterson, 1985, 1987; Bell, 1986). The playa lake is also highly suggestive of warm and semiarid to arid conditions. The warm and dry climate hypothesized for the Southern Rocky Mountains region during the Jurassic is compatible with the global warm and dry climate of the Jurassic envisioned by Hallam (1982) and Parrish and others (1982).

STRATIGRAPHY AND DEPOSITIONAL ENVIRONMENTS

Division A lies between the J-0 and J-1 unconformities (Figs. 1, 3) and includes the Aztec, Glen Canyon, and Nugget Sandstones and the part of the Glen Canyon Group that includes the Navajo Sandstone, Kayenta, and Moenave Formations, and the Lukachukai Member of the Wingate Sandstone. The Aztec,

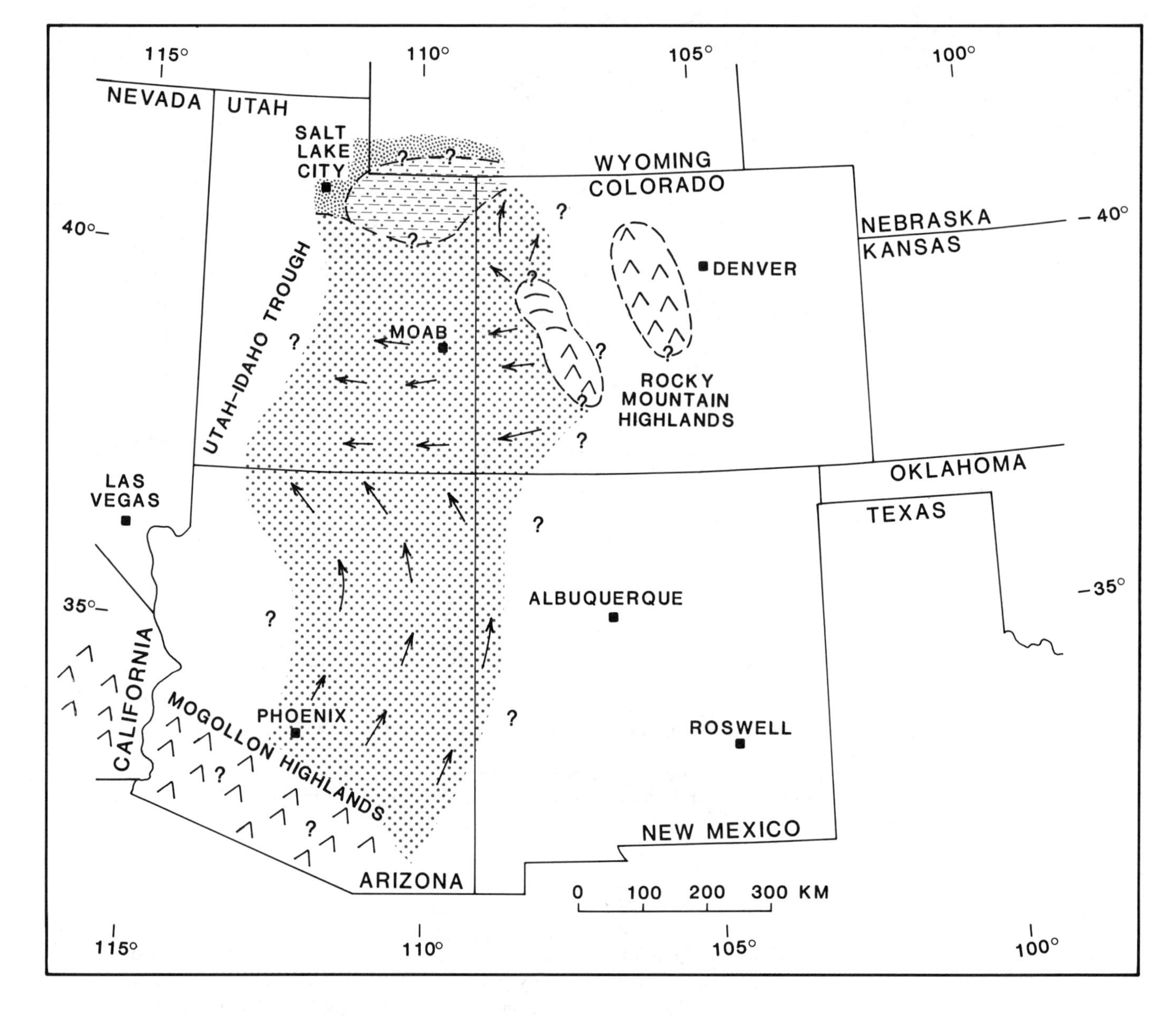

Figure 6. Paleogeographic map of Division A showing the distribution of environments during deposition of the Kayenta Formation. For explanation of patterns see Figure 4.

Glen Canyon, Navajo, Nugget, and Wingate Sandstones consist largely of eolian sandstone deposited by winds that blew from the northwest or northeast (Poole, 1962; Peterson, 1988b). By latest Early Jurassic time, windblown sand covered a large part of the Western Interior of the United States, extending at least from southeastern Idaho to southern Arizona and from the ancestral Rockies of central Colorado to eastern Nevada and possibly into western Nevada (Blakey and others, 1988; Peterson, 1988b). During the earliest part of Early Jurassic time, fluvial deposits of the Moenave Formation, derived from the Mogollon highlands (Edwards, 1985), were deposited in northwestern Arizona and southwestern Utah while eolian sands of the Wingate Sandstone were deposited farther east and north. By about the middle of the Early Jurassic, when the fluvial and partly lacustrine Kayenta Formation was deposited, the ancestral Rockies became a more impressive source area (Fig. 6), whereas relatively minor quantities of detritus came from the Mogollon highlands (Luttrell, 1987). Although predominately fluvial, overbank, and lacustrine in origin, the Kayenta also contains minor quantities of eolian sandstone or intertongues with other eolian formations, attesting to a continuation of eolian processes throughout most of the Early Jurassic wherever they were not overwhelmed by fluvial and related processes.

Division B, lying between the J-1 and J-2 unconformities, is represented by the Temple Cap Sandstone, the oldest formation in the San Rafael Group and restricted to southwestern Utah

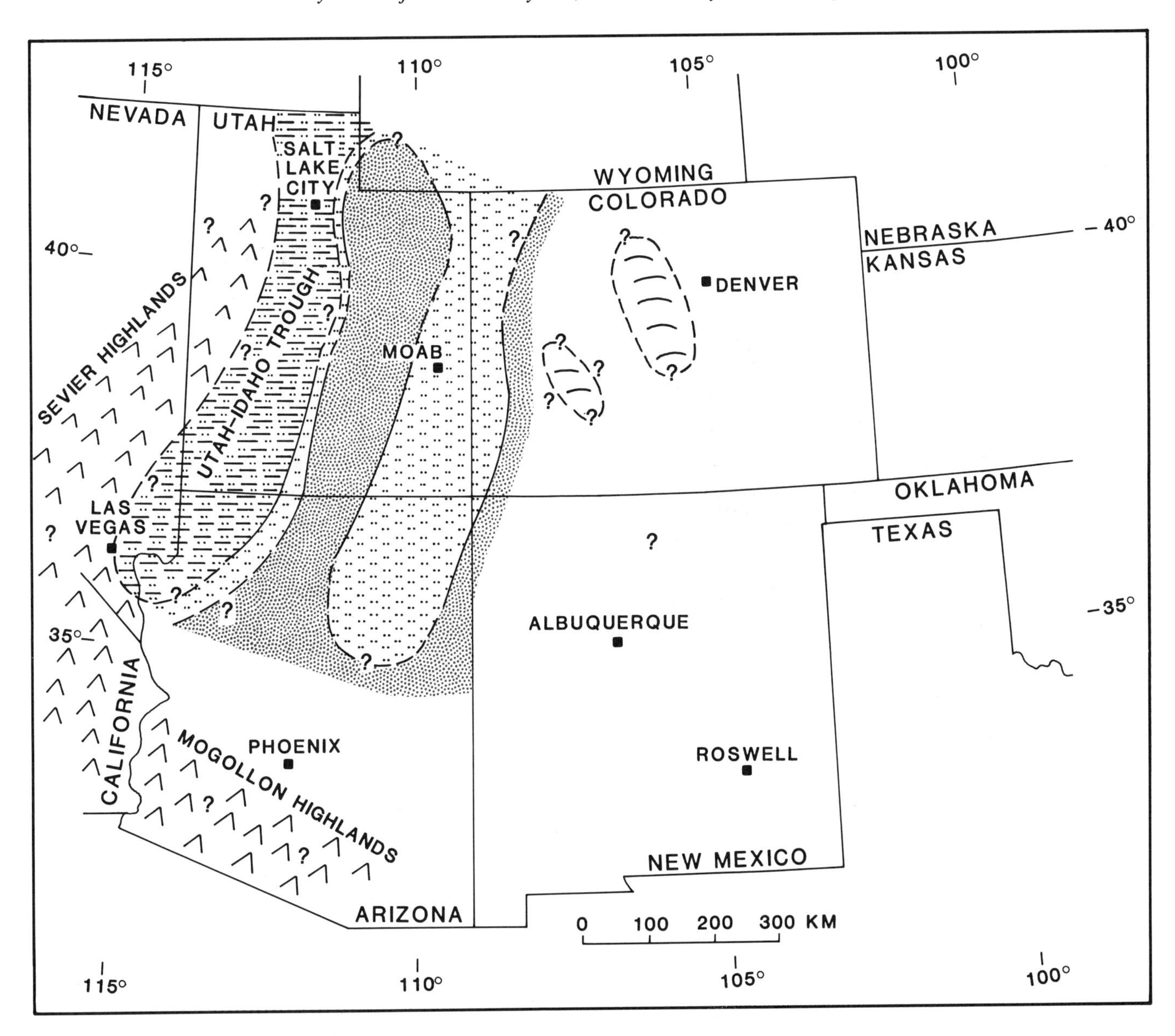

Figure 7. Paleogeographic map of Division C during deposition of the lower parts of the Page and Carmel formations. For explanation of patterns see Figure 4.

(Figs. 1, 3, 4), and the Gypsum Springs Member of the Twin Creek Limestone in northern Utah. Marine limestone was deposited in the northern part of the Utah-Idaho trough at this time, and mudstone, sandstone, and some gypsum or anhydrite were deposited farther south in southwesternmost Utah in hypersaline marine and sabkha environments. Except for the small area in southwestern Utah, these rocks were largely eroded from, or never deposited on, the platform east of the trough during the brief period of emergence that preceded deposition of Division C. The eolian White Throne Member of the Temple Cap is preserved on the platform in a small part of southwestern Utah between the Hurricane Cliffs and Johnson Canyon near Kanab, Utah (Fig. 4; Peterson and Pipiringos, 1979). The original extent of the White Throne dune field is unknown; it may have extended considerably farther east onto the Colorado Plateau.

Division C is bounded by the J-2 and J-3 unconformities. It includes the Exeter, Page, and Carmel Formations, most of the Endrada and Wanakah Formations (including the Pony Express and Todilto Limestone Members of the Wanakah), the Twin Creek Limestone above the Gypsum Spring Member, the Lak and Canyon Springs sandstone members of the Sundance Formation, and the Twelvemile Canyon Member of the Arapien Shale. Also included in this division is the Cow Springs Member of the Entrada Sandstone once thought to be an eolian facies of the Entrada, Summerville, and Morrison Formations in Divisions C, D, and F. It is now known to be a bleached eolian sandstone unit

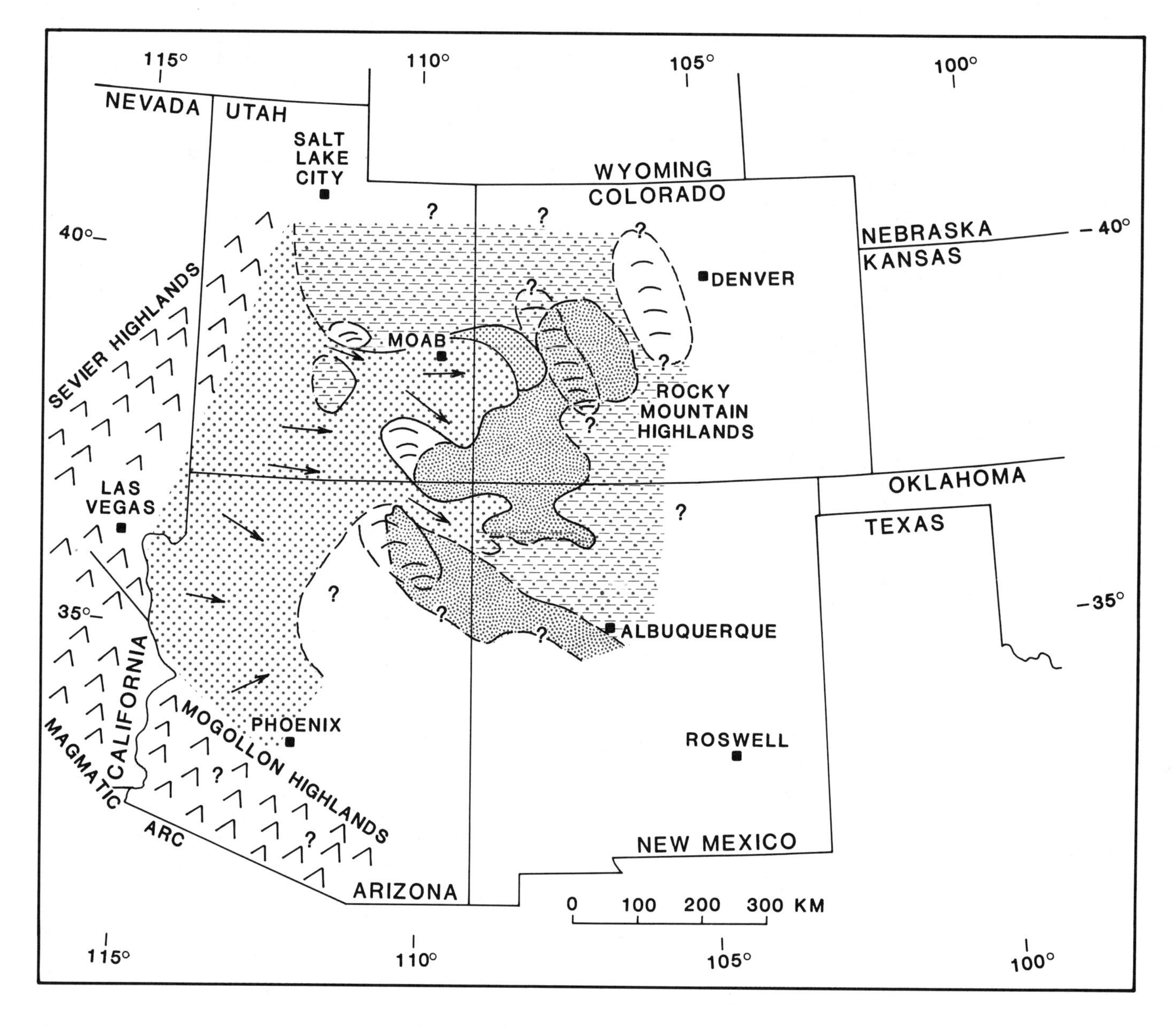

Figure 8. Paleogeographic map of Division F during deposition of the lower part of the Salt Wash and Bluff sandstone members of the Morrison Formation. For explanation of patterns see Figure 4.

in the upper part of the Entrada Sandstone and entirely within Division C (Peterson, 1988a). The Bell Ranch Formation and the lower part of the Ralston Creek Formation are here included in Division C, although part or all of these poorly dated Great Plains units may belong to a younger division.

During deposition of the Carmel Formation in the lower part of Division C (Fig. 4), marine waters occupied the Utah-Idaho trough and extended as far south as southern Nevada (Fig. 7). Older Jurassic rocks, chiefly the Glen Canyon Group and the Temple Cap Sandstone, were tilted gently to the west and beveled out eastward during formation of the J-2 erosion surface. Strata of the Page, Entrada, and Carmel Formations were then deposited in onlapping stages onto this erosion surface. The considerable thickness of Carmel and equivalent beds in west-central Utah (as much as 1,210 m, according to Standlee, 1982) that were deposited in shallow marine, sabkha, and intertidal environments indicates that considerable subsidence occurred in the Utah-Idaho trough at this time. By the end of deposition of the Carmel, sabkha deposits extended eastward from the trough to western Colorado (Blakey and others, 1983). At about this time, the marine waters retreated to the northern part of the Utah-Idaho trough, and the central part became an evaporating basin where thick beds of halite precipitated (Moulton, 1975; Standlee, 1982; Witkind, 1983).

The upper part of Division C, represented chiefly by the Entrada Sandstone below the Moab Tongue, represents a time in which the trough was occupied by sabkhas and possibly tidal flats. The platform area farther east was largely covered by a vast eolian sand sea (Blakey and others, 1988; Peterson, 1988b), although this was interrupted briefly by deposition of limestone, gypsum, or anhydrite (Pony Express or Todilto Limestone Members of the Wanakah Formation) in parts of northern New Mexico and southwestern Colorado. The origin of the Todilto or Pony Express Members is controversial. They are interpreted as having been deposited in either a large lake (Baker and others, 1936; Anderson and Kirkland, 1960; Tanner, 1965; Rawson, 1980) or a large marine embayment connected to the Sundance seaway in Wyoming through a narrow inlet in western Colorado (Baker and others, 1947; Harshbarger and others, 1957; Ridgley and Goldhaber, 1983; J. L. Ridgley, oral communication, 1984). It may well be that the body of water was connected with the Sundance seaway but that the freshwater influence increased away from the seaway. Following deposition of the Todilto and Pony Express Members, sabkha and possible shallow marine deposits in about the middle of the Wanakah Formation (Condon and Peterson, 1986) and perhaps in the Bell Ranch and lower Ralston Creek Formations were deposited in northern New Mexico, southwestern Colorado, and southeastern Utah. The sabkha deposits, as well as the Todilto Limestone, grade laterally into eolian beds of the Entrada Sandstone in northeastern Arizona and northwestern New Mexico (Figs, 3, 4, 5).

Although small quantities of coarse clastics were contributed from highlands flanking parts of the Southern Rocky Mountain region and, at times from lowlands within the region, cross-bedding studies of eolian strata in Division C indicate that most of the windblown sand came from the north (Peterson, 1988b), apparently from the southern margin of the Sundance sea, and only minor quantities came from internal source areas. The bulk of the sand apparently originated in or near south-central Idaho (Hileman, 1973) and was transported to southern Wyoming by shallow, nearshore marine currents along the southern edge of the Sundance seaway. Northerly winds then picked up the sand from the shoreline and carried it inland, depositing it as a large and nearly continuous eolian sand sheet that covered a large part of the Southern Rocky Mountain region.

Division D is defined by the J-3 unconformity below and the J-4 unconformity above, or by the J-5 unconformity where Division E is missing (Figs. 4, 5). The Summerville and Curtis Formations are the best-known representatives of Division D, although it also includes the Moab Tongue of the Entrada Sandstone, the Pine Butte Member of the Sundance Formation, the upper part of the Wanakah Formation in southeastern Utah and nearby areas, the upper part of the Twist Gulch Member of the Arapien Shale, and the newly recognized Romana Sandstone in south-central Utah and north-central Arizona (Peterson, 1988a).

Division D was deposited during a transgressive-regressive cycle in which an embayment of the Sundance sea extended southward into the northern and central parts of Utah-Idaho trough. The marine waters spilled over onto the platform area to the east where glauconitic sandstone and scarce thin limestone or gypsum beds of the Curtis Formation were deposited. In the southern part of the Kaiparowits basin of south-central Utah, eolian strata lie on marginal marine sandstone or siltstone beds at the base of the Romana Sandstone, indicating that regression was accompanied, at least in this region, by windblown sand that prograded into the embayment. Eolian processes were also responsible for deposition of the Moab Tongue of the Entrada Sandstone in east-central Utah during the early stages of deposition of the division. Farther east and southeast, beds in the uppermost part of the Wanakah Formation may be equivalent to the lower part of this division and, if so, would indicate deposition in sabkha and possibly shallow marine environments.

Paleowind directions changed markedly by the beginning of deposition of Division D. Cross-bedding studies indicate that winds generally blew toward the southeast, south, or southwest during deposition of Divisions A, B, and C; whereas winds generally blew toward the east or northeast during deposition of Divisions D and F (Peterson, 1988b; eolian strata are not present in Division E). The change is interpreted to have been caused by northward migration of the continent across wind belts possessing different atmospheric circulation patterns (Parrish and Peterson, 1988).

Division E, lying between the J-4 and J-5 unconformities, consists solely of the Redwater Member of the Sundance or Stump Formations (Figs. 1, 5). The rocks are largely marine mudstone and siltstone deposited in the southern part of the Sundance seaway, which advanced southward into the northernmost part of the Southern Rocky Mountain region during the earliest Late Jurassic. Correlative units elsewhere in the region are unknown.

Division F, containing the youngest Jurassic rocks in the region, is bounded by the J-5 unconformity at the base and by unconformities at the base of the Cretaceous System. Division F consists of the Morrison Formation, Junction Creek Sandstone, upper part of the Ralston Creek Formation, and the Windy Hill Member of the Sundance Formation (Figs. 3, 4, 5). Coarse Morrison deposits of conglomerate and pebbly to conglomeratic sandstone were deposited on the area of the present-day Colorado Plateau by streams with headwaters in the Sevier highlands farther west (Fig. 8). Fluvial deposits that make up the Salt Wash, Recapture, and Westwater Canyon Members of the Morrison Formation (Fig. 5) were once thought to represent alluvial "fan" deposits (Craig and others, 1955), but they are now considered simply as alluvial complexes because they lack the shape, lithologies, and coarseness that are characteristic of alluvial fans (Peterson and Turner-Peterson, 1987). The alluvial complexes prograded northeastward and eastward across more distal mudflats (represented by red mudstone), a local evaporative basin or marine embayment in south-central Utah (gypsum), lacustrine (gray mudstone and limestone), and eolian (cross-bedded clean sandstone) deposits. Jenkins (1957) and Tweto (1957) showed that local sand bodies in central Colorado were derived from what probably were low hills at the site of the former ancestral Rocky Mountains. Detailed sedimentological studies indicate that a small but notable amount of structural deformation occurred in the Colorado Plateau region at this time, which not only influenced the distribution of lithologies but also played a role in governing the distribution of many of the uranium deposits in the Morrison Formation (Kirk and Condon, 1986; Peterson, 1980b, 1984; Santos and Turner-Peterson, 1986).

Sedimentation in Division F ended with deposition of voluminous quantities of volcanic ash, most of which was carried into the region by winds from a magmatic arc that lay along the southwestern edge of the continent. The ash altered to smectite and various zeolites and feldspars in a playa lake that extended northward from northwestern New Mexico into southwestern Colorado and southeastern Utah (Turner-Peterson, 1985; Bell, 1986). Mapping the distribution of the alteration products indicates that the lake was approximately 500 km long (north-south) and 300 km wide (east-west), making it the largest known playa lake ever to have existed (Turner-Peterson, 1987).

ECONOMIC RESOURCES

The most noteworthy economic resource in Jurassic rocks of the Southern Rocky Mountain region is the large quantity of uranium in the Morrison Formation, which amounts to slightly over 50 percent of the uranium resources of the United States. Most of the uranium is in the Westwater Canyon Member of the Morrison Formation (Fig. 5) in the southern San Juan Basin, New Mexico (U.S. Department of Energy, 1983; McCammon and others, 1986; Turner-Peterson, 1985, 1986), but other significant deposits are present in the Salt Wash and Jackpile Sandstone Members of the Morrison. All of the significant ore deposits are in porous and permeable fluvial sandstone beds that allowed the passage of ore-forming fluids. An essential feature for the formation of these deposits appears to have been nearby lake beds that provided some of the organic constituents necessary for ore precipitation (Peterson, 1980a, 1980b); Turner-Peterson, 1985; Spirakis and Hansley, 1986, 1987).

Small but significant quantities of hydrocarbons have been discovered in Jurassic rocks of the Southern Rockies. Several oil fields have been discovered in north-central Utah that produce from the Nugget Sandstone and Twin Creek Limestone (Maher, 1976; Lelek, 1982). Small quantities of oil and gas also have been produced from the Entrada Sandstone in northeastern Utah, north-central Colorado, and northwestern New Mexico (Henkes, 1957; Otto and Picard, 1976; Vincelette and Chittum, 1981). In most cases, the hydrocarbons appear to have migrated into these formations from source rocks in another geologic system. A notable exception is in northwestern New Mexico where oil migrated into the Entrada Sandstone from overlying organic-rich limestone beds of the Todilto Limestone Member of the Wanakah Formation (Vincelette and Chittum, 1981).

Other significant commercial resources recovered from these rocks include gypsum and gold ores. Gypsum is mined from the Arapien Shale in central Utah and the Todilto Limestone Member of the Wanakah Formation in northwestern New Mexico. In addition, the Pony Express Limestone Member of the Wanakah is the host for gold telluride ore deposits once mined in southwestern Colorado (Eckel, 1949).

REFERENCES CITED

Anderson, R. Y., and Kirkland, D. W., 1960, Origin, varves, and cycles of Jurassic Todilto Formation, New Mexico: American Association of Petroleum Geologists Bulletin, v. 44, p. 37–52.

Baker, A. A., Dane, C. H., and Reeside, J. B., Jr., 1936, Correlation of the Jurassic formations of parts of Utah, Arizona, New Mexico, and Colorado: U.S. Geological Survey Professional Paper 183, 66 p.

——, 1947, Revised correlations of Jurassic formations in parts of Utah, Arizona, New Mexico, and Colorado: American Association of Petroleum Geologists Bulletin, v. 31, no. 9, p. 1664–1677.

Bell, T. E., 1986, Deposition and diagenesis in the Brushy Basin Member and upper part of the Westwater Canyon Member of the Morrison Formation, San Juan Basin, New Mexico, *in* Turner-Peterson, C. E., Santos, E. S., and Fishman, N. S., eds., A basin analysis case study; The Morrison Formation, Grants uranium region, New Mexico: American Association of Petroleum Geologists Studies in Geology 22, p. 77–91.

Billodeau, W. L., 1986, The Mesozoic Mogollon highlands, Arizona: An Early Cretaceous rift shoulder: Journal of Geology, v. 94, no. 5, p. 724–735.

Blakey, R. C., and Middleton, L. T., 1983, Lower Mesozoic stratigraphy and depositional systems, southwest Colorado Plateau, *in* Gurgel, K. D., ed., Geologic excursions in stratigraphy and tectonics: Utah Geological and Mineral Survey Special Studies 60, p. 33–39.

Blakey, R. C., Peterson, F., Caputo, M. V., Geesaman, R. C., and Voorhees, B. J., 1983, Paleogeography of Middle Jurassic continental, shoreline, and shallow marine sedimentation, southern Utah, *in* Reynolds, M. W., and Dolly, E. D., eds., Mesozoic paleogeography of the west-central United States: Rocky Mountain Section of the Society of Economic Paleontologists and Mineralogists Symposium 2, p. 77–100.

Blakey, R. C., Peterson, F., and Kocurek, G., 1988, Late Paleozoic and Mesozoic eolian deposits of the western United States: Sedimentary Geology (in press).

Bryant, B., McGrew, L. W., and Wobus, R. A., 1981, Geologic map of the Denver 1° × 2° Quadrangle, north-central Colorado: U.S. Geological Survey Miscellaneous Investigations Series Map I-1163, scale 1:250,000.

Carter, W. D., and Gaultieri, J. L., 1965, Geology and uranium-vanadium deposits of the La Sal Quadrangle, San Juan County, Utah, and Montrose County, Colorado: U.S. Geological Survey Professional Paper 508, 82 p.

Cashion, W. R., 1967, Carmel Formation of the Zion Park region, southwestern Utah; A review: U.S. Geological Survey Bulletin 1244-J, p. J1–J9.

Cater, F. W., 1970, Geology of the Salt Anticline region in southwestern Colorado: U.S. Geological Survey Professional Paper 637, 80 p.

Chapman, M. G., 1987, Depositional and compositional aspects of volcanogenic clasts in the upper member of the Carmel Formation, southern Utah [M.S. thesis]: Flagstaff, Northern Arizona University, 93 p.

Condon, S. M., and Huffman, A. C., Jr., 1988, Revision of Middle Jurassic stratigraphic nomenclature, northwestern New Mexico and northeastern Arizona: U.S. Geological Survey Bulletin 1633-A (in press).

Condon, S. M., and Peterson, F., 1986, Stratigraphy of Middle and Upper Jurassic rocks in the San Juan Basin; Historical perspective, current ideas, and remaining problems, *in*, Turner-Peterson, C. E., Santos, E. S., and Fishman, N. S., eds., A basin analysis case study; The Morrison Formation, Grants uranium region, New Mexico: American Association of Petroleum Geolo-

gists Studies in Geology 22, p. 7–26.
Craig, L. C., Holmes, C. N., Cadigan, R. A., Freeman, V. L., Mullens, T. E., and Weir, G. W., 1955, Stratigraphy of the Morrison and related formations, Colorado Plateau region; A preliminary report: U.S. Geological Survey Bulletin 1009-E, p. 125–168.
Craig, L. C., Holmes, C. N., Freeman, V. L., Mullens, T. E., 1959, Measured sections of the Morrison and adjacent formations: U.S. Geological Survey Open-File Report 485.
Dickinson, W. R., Klute, M. A., and Swift, P. N., 1986, The Bisbee basin and its bearing on Late Mesozoic paleogeographic and paleotectonic relations between the Cordilleran and Caribbean regions, *in* Abbot, P. L., ed., Cretaceous stratigraphy, western North America: Pacific Section of the Society of Economic Paleontologists and Mineralogists, p. 51–62.
Dobrovolny, E., and Summerson, C. H., 1946, Geology of northwestern Quay County, New Mexico: U.S. Geological Survey Oil and Gas Investigations Preliminary Map 62, scale 1:63,360.
Eckel, E. B., 1949, Geology and ore deposits of the La Plata district, Colorado: U.S. Geological Survey Professional Paper 219, 179 p.
Edwards, D. P., 1985, Controls on deposition of an ancient fluvial/eolian depositional system; The Early Jurassic Moenave Formation of north-central Arizona [M.S. thesis]: Flagstaff, Northern Arizona University, 243 p.
Ekren, E. B., and Houser, F. N., 1959, Relations of Lower Cretaceous and Upper Jurassic rocks in the Four Corners area, Colorado: American Association of Petroleum Geologists Bulletin, v. 43, no. 1, p. 190–201.
Galton, P. M., 1978, Fabrosauridae, the basal family of ornithischian dinosaurs (Reptilia: Ornithopoda): Paleontologische Zeitschrift, v. 52, p.138–159.
Gilluly, J., and Reeside, J. B., Jr., 1928, Sedimentary rocks of the San Rafael Swell and some adjacent areas in eastern Utah: U.S. Geological Survey Professional Paper 150-D, p. 61–110.
Green, M. W., 1974, The Iyanbito Member (a new stratigraphic unit) of the Jurassic Entrada Sandstone, Gallup-Grants area, New Mexico: U.S. Geological Survey Bulletin 1395-D, p. D1–D12.
Gregory, H. E., and Moore, R. C., 1931, The Kaiparowits region, a geographic and geologic reconnaissance of parts of Utah and Arizona: U.S. Geological Survey Professional Paper 164, 161 p.
Hallam, A., 1982, Studies in geophysics; Climate in earth history: Washington, D.C., National Academy Press, p. 159–162.
Hansen, W. R., 1968, Geologic map of the Black Ridge Quadrangle, Delta and Montrose Counties, Colorado: U.S. Geological Survey Geologic Quadrangle Map GQ-747, scale 1:62,500.
——, 1971, Geologic map of the Black Canyon of the Gunnison River and vicinity, western Colorado: U.S. Geological Survey Miscellaneous Geologic Investigations Map I-584, scale 1:62,500.
Harshbarger, J. W., Repenning, C. A., and Irwin, J. H., 1957, Stratigraphy of the uppermost Triassic and Jurassic rocks of the Navajo Country: U.S. Geological Survey Professional Paper 291, 74 p.
Heller, P. L., and 7 others, 1986, Time of initial thrusting in the Sevier orogenic belt, Idaho, Wyoming, and Utah: Geology, v. 14, no. 5, p. 388–391.
Henkes, W. C., 1957, Coalmont area, Jackson County, Colorado, *in* Finch, W. C., ed., Guidebook to the geology of North and Middle parks basins, Colorado: Denver, Colorado, Rocky Mountain Association of Geologists, p. 51–54.
Hileman, M. E., 1973, Stratigraphy and paleoenvironmental analysis of the Upper Jurassic Preuss and Stump formations, western Wyoming and southeastern Idaho [Ph.D. thesis]: Ann Arbor, University of Michigan, 203 p.
Hunt, C. B., Averitt, P., and Miller, R. L., 1950, Geology and geography of the Henry Mountains region, Utah: U.S. Geological Survey Professional Paper 228, 234 p.
Imlay, R. W., 1964, Marine Jurassic pelecypods from central and southern Utah: U.S. Geological Survey Professional Paper 483-C, p. C1–C42.
——, 1967, Twin Creek Limestone (Jurassic) in the Western Interior of the United States: U.S. Geological Survey Professional Paper 540, 105 p.
——, 1980, Jurassic paleobiogeography of the conterminous United States in its continental setting: U.S. Geological Survey Professional Paper 1062, 134 p.
Jenkins, M. A., Jr., 1957, Stratigraphy of the Red Dirt Creek area, Grand County, Colorado, *in* Finch, W. C., ed., Guidebook to the geology of North and Middle parks basins, Colorado: Denver, Colorado, Rocky Mountain Association of Geologists, p. 51–54.
Kirk, A. R., and Condon, S. M., 1986, Structural control of sedimentation patterns and the distribution of uranium deposits in the Westwater Canyon Member of the Morrison Formation, northwestern New Mexico; A subsurface study, *in* Turner-Peterson, C. E., Santos, E. S., and Fishman, N. S., eds., A basin analysis case study; The Morrison Formation, Grants Uranium region, New Mexico: American Association of Petroleum Geologists Studies in Geology 22, p. 105–143.
Kocurek, G., 1981, Erg reconstruction; The Entrada Sandstone (Jurassic) of northern Utah and Colorado: Palaeogeography, Palaeoclimatology, Palaeoecology, v. 36, no. 1/2, p. 125–153.
Kowallis, B. J., and Heaton, J. S., 1987, Fission-track dating of bentonites and bentonitic mudstones from the Morrison Formation in central Utah: Geology, v. 15, no. 12, p. 1138–1142.
Langenheim, R. L., 1957, Jurassic stratigraphy in Elk Mountains, west-central Colorado: American Association of Petroleum Geologists Bulletin, v. 41, no. 11, p. 2576–2583.
Lelek, J. J., 1982, Anschutz Ranch East Field, northeast Utah and southwest Wyoming, *in* Powers, R. B., ed., Geologic studies of the Cordilleran thrust belt: Denver, Colorado, Rocky Mountain Association of Geologists, v. 2, p. 619–631.
Luttrell, P. R., 1987, Basin analysis of the Kayenta Formation (Lower Jurassic), central portion Colorado Plateau [M.S. thesis]: Flagstaff, Northern Arizona University, 217 p.
Maher, P. D., 1976, The geology of the Pineview Field area, Summit County, Utah, *in* Hill, J. G., ed., Symposium on geology of the Cordilleran hingeline: Denver, Colorado, Rocky Mountain Association of Geologists, p. 345–350.
Marzolf, J. E., 1983, Early Mesozoic eolian transition from crational margin to orogenic-volcanic arc, *in* Gurgel, K. D., ed., Geologic excursions in stratigraphy and tectonics: Utah Geological and Mineral Survey Special Studies 60, p. 39–46.
Maxwell, C. H., 1982, Mesozoic stratigraphy of the Laguna-Grants region, *in* Grambling, J. A., and Wells, S. G., eds., Albuquerque Country II: 33rd Annual Field Conference, New Mexico Geological Society Guidebook, p. 261–266.
McCammon, R. B., Finch, W. I., Kork, J. O., and Bridges, N. J., 1986, Estimation of uranium endowment in the Westwater Canyon Member, Morrison Formation, San Juan Basin, New Mexico, using a data-directed numerical method, *in* Turner-Peterson, C. E., Santos, E. S., and Fishman, N. S., eds., A basin analysis case study; The Morrison Formation, Grants uranium region, New Mexico: American Association of Petroleum Geologists Studies in Geology 22, p. 331–355.
Moulton, F. S., 1975, Lower Mesozoic and upper Paleozoic petroleum potential of the hingeline area, central Utah, *in* Bolyard, D. W., eds., Deep drilling frontiers of the central Rocky Mountains: Denver, Colorado, Rocky Mountain Association of Geologists, p. 87–97.
Olsen, P. E., and Galton, P. M., 1977, Triassic-Jurassic tetrapod extinctions; Are they real?: Science, v. 197, no. 4307, p. 983–985.
O'Sullivan, R. B., 1980a, Stratigraphic sections of Middle Jurassic San Rafael Group and related rocks from the Green River to the Moab area in east-central Utah: U.S. Geological Survey Miscellaneous Field Studies Map MF—1247.
——, 1980b, Stratigraphic sections of Middle Jurassic San Rafael Group from Wilson Arch to Bluff in southeastern Utah: U.S. Geological Survey Oil and Gas Investigations Chart OC-102.
——, 1981, Stratigraphic sections of some Jurassic rocks from near Moab, Utah, to Slick Rock, Colorado: U.S. Geological Survey Oil and Gas Investigations Chart OC-107.
——, 1984, Stratigraphic sections of Middle Jurassic San Rafael Group and related rocks from Dewey Bridge, Utah, to Uravan, Colorado: U.S. Geological Survey Oil and Gas Investigations Chart OC-124.

O'Sullivan, R. B., and Pierce, F. W., 1983, Stratigraphic diagram of Middle Jurassic San Rafael Group and associated formations from the San Rafael Swell to Bluff in southeastern Utah: U.S. Geological Survey Oil and Gas Investigations Chart OC-119.

Otto, E. P., and Picard, M. D., 1976, Petrology of Entrada Sandstone (Jurassic), northeastern Utah, *in* Hill, J. G., ed., Symposium on geology of the Cordilleran hingeline: Denver, Colorado, Rocky Mountain Association of Geologists, p. 231–245.

Parrish, J. T., and Peterson, F., 1988, Wind directions predicted from global circulation models and wind directions determined from eolian sandstones of the western United States; A comparison: Sedimentary Geology (in press).

Parrish, J. T., Ziegler, A. M., and Scotese, C. R., 1982, Rainfall patterns and the distribution of coals and evaporites in the Mesozoic and Cenozoic: Palaeogeography, Palaeoclimatology, Palaeoecology, v. 40, p. 67–101.

Peterson, F., 1980a, Sedimentology of the uranium-bearing Salt Wash Member and Tidwell unit of the Morrison Formation in the Henry and Kaiparowits basins, Utah, *in* Picard, M. D., ed., Henry Mountains Symposium: Salt Lake City, Utah Geological Association Publication 8, p. 305–322.

—— , 1980b, Sedimentology as a strategy for uranium exploration; Concepts gained from analysis of a uranium-bearing depositional sequence in the Morrison Formation of south-central Utah, *in* Turner-Peterson, C. E., ed., Uranium in sedimentary rocks; Application of the facies concept to exploration: Denver, Colorado, Rocky Mountain Section of the Society of Economic Paleontologists and Mineralogists, p. 65–126.

—— , 1980c, Geologic map and coal deposits of the Big Hollow Wash Quadrangle, Kane County, Utah: U.S. Geological Survey Coal Investigations Map C-84, scale 1:24,000.

—— , 1984, Fluvial sedimentation on a quivering craton; Influence of slight crustal movements on fluvial processes, Upper Jurassic Morrison Formation, western Colorado Plateau: Sedimentary Geology, v. 38, no. ¼, p. 21–49.

—— , 1988a, Stratigraphy and nomenclature of Middle and Upper Jurassic rocks in the western Colorado Plateau, Utah and Arizona: U.S. Geological Survey Bulletin 1633-B (in press).

—— , 1988b, Pennsylvanian to Jurassic eolian transportation systems in the western United States: Sedimentary Geology (in press).

Peterson, F., and Pipiringos, G. N., 1979, Stratigraphic relations of the Navajo Sandstone to Middle Jurassic formations, southern Utah and northern Arizona: U.S. Geological Survey Professional Paper 1035-B, p. B1–B43.

Peterson, F., and Turner-Peterson, C. E., 1987, The Morrison Formation of the Colorado Plateau; Recent advances in sedimentology, stratigraphy, and paleotectonics: Hunteria, v. 2, no. 1, 18 p.

Peterson, J. A., 1972, Jurassic System, *in* Mallory, W. W., ed., Geologic atlas of the Rocky Mountain region: Denver, Colorado, Rocky Mountain Association of Geologists, p. 177–189.

Pipiringos, G. N., and O'Sullivan, R. B., 1978, Principal unconformities in Triassic and Jurassic rocks, Western Interior United States; A preliminary report: U.S. Geological Survey Professional Paper 1035-A, p. A1–A29.

Poole, F. G., 1961, Stream directions in Triassic rocks of the Colorado Plateau: U.S. Geological Survey Professional Paper 424-C, p. C139–C141.

—— , 1962, Wind directions in late Paleozoic to middle Mesozoic time on the Colorado Plateau: U.S. Geological Survey Professional paper 450-D, D147–D151.

Rawson, R. R., 1980, Uranium in the Jurassic Todilto Limestone of New Mexico; An example of a sabkha-like deposit, *in* Turner-Peterson, C. E., ed., Uranium in sedimentary rocks; Application of the facies concept to exploration: Denver, Colorado, Rocky Mountain Section of the Society of Economic Paleontologists and Mineralogists, p. 127–147.

Ridgley, J. L., and Goldhaber, M., 1983, Isotopic evidence for a marine origin of the Todilto Limestone, north-central New Mexico: Geological Society of America Abstracts with Programs, v. 15, p. 414.

Santos, E. S., and Turner-Peterson, C. E., 1986, Tectonic setting of the San Juan Basin in the Jurassic, *in* Turner-Peterson, C. E., Santos, E. S., and Fishman, N. S., eds., A basin analysis case study; The Morrison Formation, Grants uranium region, New Mexico: American Association of Petroleum Geologists Studies in Geology 22, p. 27–33.

Scott, G. R., 1963, Bedrock geology of the Kassler Quadrangle, Colorado: U.S. Geological Survey Professional Paper 421-B, p. 71–125.

Spirakis, C. S., and Hansley, P. L., 1986, The formation and alteration of tabular-type uranium-vanadium deposits as a variant of normal diagenetic processes in organic-rich sediments, *in* Carter, L.M.H., ed., U.S.G.S. research on energy resources, 1986, program and abstracts: U.S. Geological Survey Circular 974, p. 64–65.

—— , 1987, Diagenesis of amorphous organic matter as an essential aspect of genesis and alteration of tabular-type uranium-vanadium deposits, Colorado Plateau: American Association of Petroleum Geologists Bulletin, v. 71, no. 5, p. 616.

Standlee, L. A., 1982, Structure and stratigraphy of Jurassic rocks in central Utah; Their influence on tectonic development of the Cordilleran foreland thrust belt, *in* Powers, R. B., ed., Geologic studies of the Cordilleran thrust belt: Denver, Colorado, Rocky Mountain Association of Geologists, v. 1, p. 357–382.

Stark, J. T., and 5 others, 1949, Geology and origin of South Park, Colorado: Geological Society of America Memoir 33, 188 p.

Tanner, W. F., 1965, Upper Jurassic paleogeography of the Four Corners region: Journal of Sedimentary Petrology, v. 35, no. 3, p. 564–574.

Turner-Peterson, C. E., 1985, Lacustrine-humate model for primary uranium ore deposits, Grants uranium region, New Mexico: American Association of Petroleum Geologists Bulletin, v. 69, p. 1999–2020.

—— , 1986, Fluvial sedimentology of a major uranium-bearing sandstone; A study of the Westwater Canyon Member of the Morrison Formation, San Juan Basin, New Mexico, *in* Turner-Peterson, C. E., Santos, E. S., and Fishman, N. S., eds., A basin analysis case study; The Morrison Formation, Grants uranium region, New Mexico: American Association of Petroleum Geologists Studies in Geology 22, p. 47–75.

—— , 1987, Sedimentology of the Westwater Canyon and Brushy Basin Members, Upper Jurassic Morrison Formation, Colorado Plateau, and relationship to uranium mineralization [Ph.D. thesis]: Boulder, University of Colorado, 169 p.

Tweto, O., 1957, Geologic sketch of southern Middle Park, Colorado, *in* Finch, W. C., ed., Guidebook to the geology of North and Middle Parks basins, Colorado: Denver, Colorado, Rocky Mountain Association of Geologists, p. 18–31.

U.S. Department of Energy, 1983, Statistical data of the uranium industry: U.S. Department of Energy Report GJO-100(83), 77 p.

Vail, P. R., Hardenbol, J., and Todd, R. G., 1984, Jurassic unconformities, chronostratigraphy, and sea-level changes from seismic stratigraphy and biostratigraphy, *in* Schlee, J. S., ed., Interregional unconformities and hydrocarbon accumulation: American Association of Petroleum Geologists Memoir 36, p. 129–144.

Vincelette, R. R., and Chittum, W. E., 1981, Exploration for oil accumulations in Entrada Sandstone, San Juan Basin, New Mexico: American Association of Petroleum Geologists Bulletin, v. 65, no. 12, p. 2546–2570.

Witkind, I. J., 1983, Overthrusts and salt diapirs, central Utah, *in* Miller, D. M., Todd, V. R., and Howard, K. A., eds., Tectonic and stratigraphic studies in the eastern Great Basin: Geological Society of America Memoir 157, p. 45–59.

MANUSCRIPT ACCEPTED BY THE SOCIETY FEBRUARY 2, 1988

ACKNOWLEDGMENTS

My thoughts on the Jurassic of the southern Rockies were sharpened and improved considerably over the past two and a half decades by discussions and the exchange of ideas with R. C. Blakey, S. M. Condon, L. C. Craig, J. E. Marzolf, R. B. O'Sullivan, G. N. Pipiringos, C. E. Turner-Peterson, H. D. Zeller, and many others. The manuscript was improved considerably through helpful reviews by A. R. Palmer, J. A. Peterson, and L. L. Sloss.

Printed in U.S.A.

The Geology of North America
Vol. D-2, Sedimentary Cover—North American Craton: U.S.
The Geological Society of America, 1988

Chapter 6

Cretaceous rocks of the Western Interior Basin

C. M. Molenaar and D. D. Rice
U.S. Geological Survey, M.S. 940, Box 25046, Denver Federal Center, Denver, Colorado 80225

INTRODUCTION

During Cretaceous time, the Western Interior of the United States was the site of part of an elongated epicontinental seaway that extended from the Arctic Ocean to the Gulf of Mexico (Fig. 1). On the west, the seaway was separated from the Pacific Ocean by the tectonically active Cordilleran highland, which was the main provenance for terrigenous sediment shed into the seaway. In contrast, the east side of the seaway was flanked by the low-lying, stable platform of the central and eastern United States and Canadian Shield, which, except during late Early to early Late Cretaceous time, provided only minor amounts of detritus to the seaway.

The sedimentary fill of the Western Interior Basin is dominantly Cretaceous in age, ranging from Aptian through Maastrichtian. The basin fill was asymmetric, with thicknesses of as much as 5,000 m along the west flank. The asymmetry of the basin resulted from varying rates of subsidence relative to sedimentation, sediment loading adjacent to the major bounding orogenic belts, and tectonic loading by thrust plates in the orogenic belt (Jordan, 1981).

Much of the Cretaceous section is preserved in the many early Tertiary or Laramide structural basins within the Western Interior of the U.S. Because of vast amounts of gas, oil, and coal contained in them, Cretaceous rocks of the Western Interior have been widely studied throughout much of the area.

Although many of the areas of preserved Cretaceous rocks are now separated, fairly accurate biostratigraphic correlations are made on the basis of good to excellent faunal zonations that are based primarily on ammonites and bivalves (Cobban and Reeside, 1952). Foraminiferal and palynomorph zonations are also of great value for correlations, especially from drill-hole samples. More recently, radiometric age dating of bentonite beds associated with faunal zones has permitted determination of absolute ages and age spans of the faunal zones (Obradovich and Cobban, 1975).

Most of the following discussion is generalized and is a condensation of the work of many authors over many years; to cite all the authors is impractical. Although the Western Interior Basin extends from the Arctic to the Gulf of Mexico, only the general characteristics of the Cretaceous rocks of the conterminous United States are discussed in this report.

DEPOSITIONAL FACIES AND LITHOLOGY

Cretaceous rocks of the Western Interior Basin consist primarily of shale, siltstone, and sandstone; significant amounts of conglomerate are present only along the western margin, and limestone is generally restricted to the eastern shelf. These rocks were deposited in marine and nonmarine environments as the shoreline advanced and retreated across the western part of the basin during several major depositional cycles. Most of the clastic deposition occurred during regressions, or when the seaway was being filled and the shoreline was migrating seaward. Limestone deposition, which was inhibited by clastic influx from the west, occurred in the central and eastern part of the basin during times of maximum transgression and continued on the eastern shelf during times of regressions (Rice and Shurr, 1983).

Two regional stratigraphic cross sections (Figs. 2 and 3) show the regional facies relations across the northern and central part of the Western Interior Basin, respectively. A third cross section across the San Juan Basin in the southern part of the Western Interior is shown in Molenaar (this volume).

The following is a generalized description of the facies and their depositional setting in order of their occurrence from west to east across the basin. Walther's principal of the vertical depositional sequence reflecting the lateral sequence of strata is especially applicable to the regressive cycles of Cretaceous strata.

Nonmarine deposits

Nonmarine rocks grade from locally thick fanglomerates and conglomeratic sandstones that were deposited adjacent to bounding highlands along the western margin of the basin to alluvial-plain and coastal–delta-plain deposits marginal to the seaway. The lower delta- or coastal-plain facies consists of paludal carbonaceous shale, coal, lacustrine shales and sandstones, and fluvial or distributary channel sandstones with associated levee

Molenaar, C. M., and Rice, D. D., 1988, Cretaceous rocks of the Western Interior Basin, *in* Sloss, L. L., ed., Sedimentary Cover—North American Craton; U.S.: Boulder, Colorado, Geological Society of America, The Geology of North America, v. D-2.

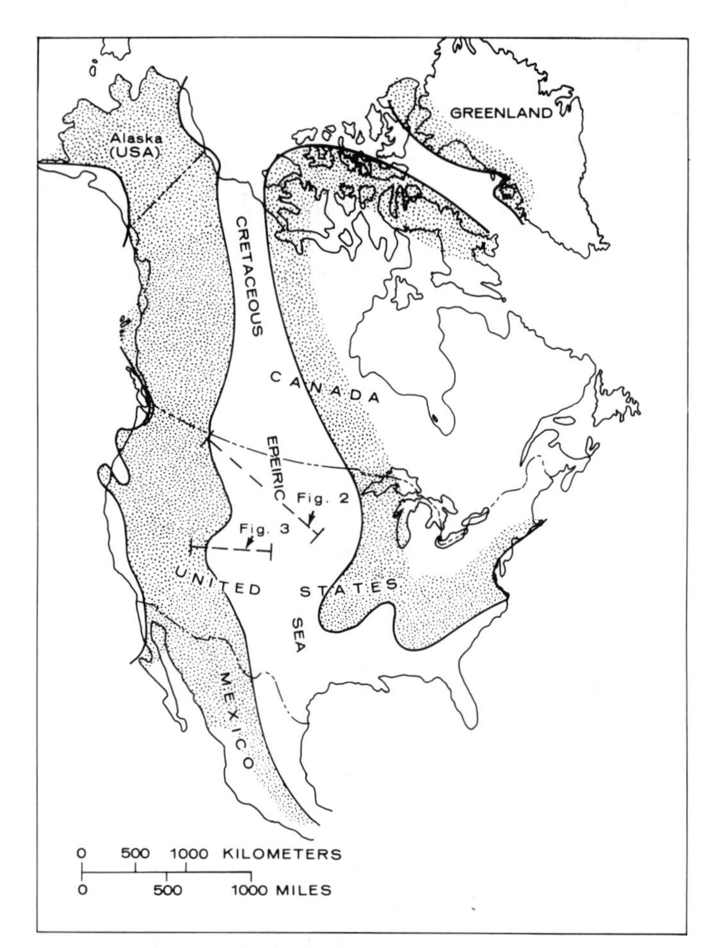

Figure 1. Map showing extent of the Western Interior epicontinental seaway during Late Cretaceous (Campanian) time. (Modified from Gill and Cobban, 1973.) Shading indicates land areas. Lines of sections of Figures 2 and 3 are indicated.

and splay deposits. Farther landward on the upper delta or alluvial plain, carbonaceous shales and coal are much less common, and flood-plain or interchannel shales and fluvial-channel sandstones predominate. Discontinuity of facies, both vertically and laterally, characterizes the nonmarine sequence.

Coastal-barrier sandstones

Coastal-barrier sandstones commonly form conspicuous outcrops throughout the Western Interior in that, because of their weathering resistance and continuity, they form prominent cliffs and hogbacks. In vertical and lateral sequences, they separate the soft-weathering, slope-and-valley–forming, open-marine shales below from the slope-and-ledge–forming nonmarine sequence above, and, as such, make convenient formational units for mapping. Coastal-barrier sandstones have been studied by many workers and, as used here, the term includes all types of shoreline and shoreface sandstone units, including those deposited in barrier-island, strandplain, distributary-mouth bar, delta-front, and tidal-delta and tidal-channel settings. Except for the last two types, which are less common, most coastal-barrier sandstones are generally coarsening-upward sequences indicating shoaling and increasingly higher energy conditions. Textures range from very fine grained at the base to fine grained with some lower medium grained in the upper part. The coarsest part is commonly in the upper shoreface and not the overlying foreshore part of the sequence. Deposition of coastal-barrier sandstones, ranging from 10 to 40 m thick, was dominated by marine processes, mainly longshore currents, which commonly produced fairly straight shorelines. Because of the high energy of depositional processes, these sandstones are generally well sorted and contain little clay when deposited.

Offshore siltstones and shales

Thick deposits of offshore siltstone and shale are a major part of the Cretaceous sequence. They form most of the large valleys and flat areas, the sites of the few cities and towns of the intermountain area. Because the bentonite content of the shale is generally high, expansion and contraction occurs with wetting and drying in the weathered zone. Consequently, special measures must be taken in constructing buildings and roads on these surfaces.

Mixtures of silt and clay were deposited primarily on outer shelf and possibly basinal environments. The source of these fine-grained sediments was the river-borne suspended load that bypassed the coastal zone. Because the direction of transport initially was perpendicular to the shoreline in an offshore direction, there is a regular decrease in grain size to the east in this facies. Once the fine-grained sediments bypassed the coastal zone, they probably were transported parallel to the coast by storm-generated currents. The silt-size material occurs either in lenses and laminae, which probably represent storm bed-load deposits, or is dispersed throughout the shale. The lenses are discontinuous, may be ripple bedded or bioturbated, and have sharp bases. Beds with carbonate concretions, which are commonly fossiliferous, are scattered throughout the offshore siltstone and shale sequence.

Shelf sandstones

Sandstone makes up a minor part of the shelf sequence, which is dominated by siltstone and shale. The sand-size fraction probably was transported by intense geostrophic currents in response to strong wind forcing during winter storms. These sandstone units are often concentrated as lag deposits on or adjacent to mildly positive tectonic elements that were experiencing recurrent movement.

Two types of sandstone are differentiated in the shelf sequence. The first type occurs as thin beds or laminae interbedded with shale and discontinuous at a small scale. These sandstone units are either parallel- or ripple-laminated with sharp bases, or are bioturbated. Although individual beds or laminae are not

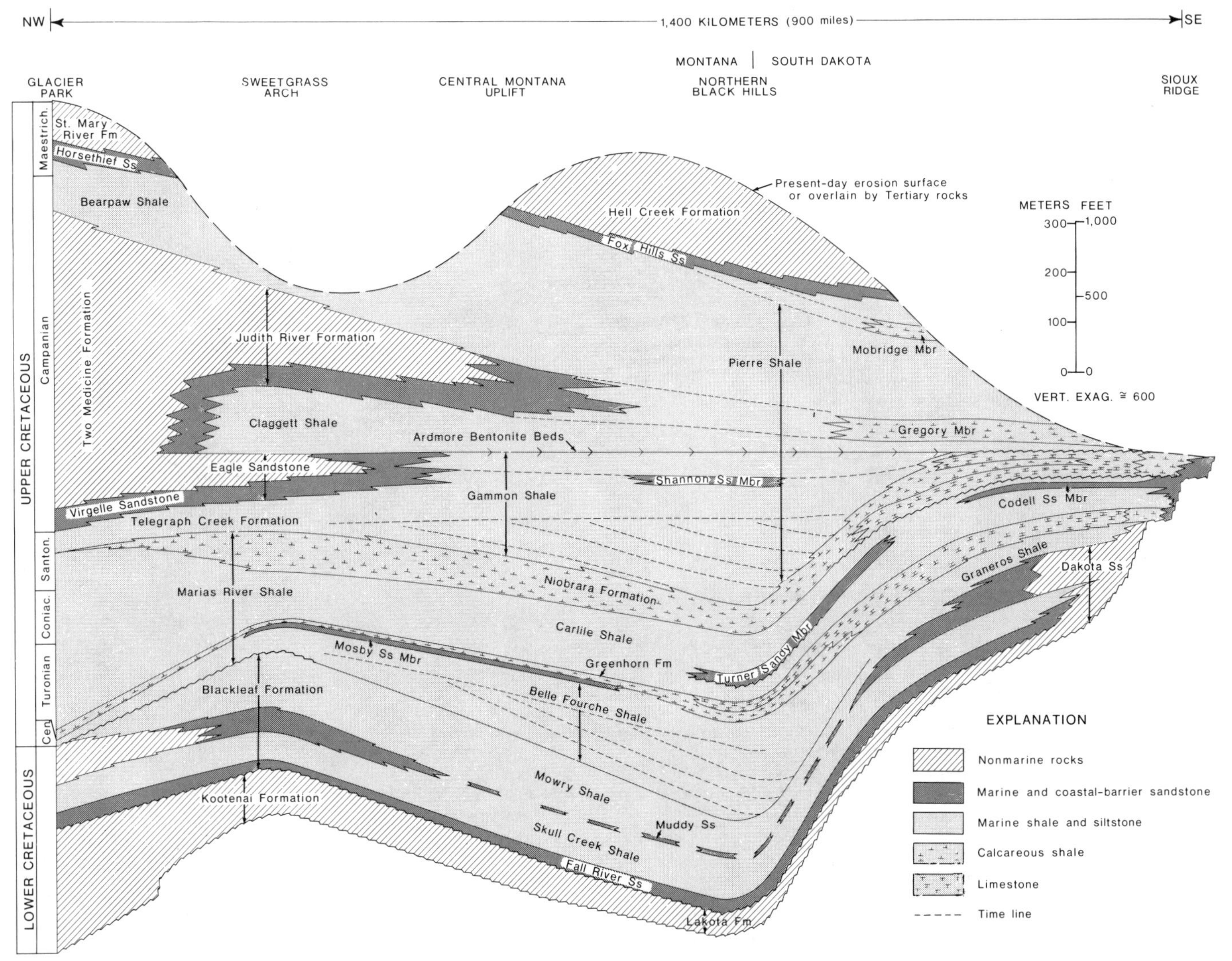

Figure 2. Stratigraphic cross section of Cretaceous rocks from northwestern Montana to southeastern South Dakota. Thicknesses are based on well and outcrop control. Line of section is shown in Figure 1.

broadly traceable, they generally occur in a shaly interval that maintains a fairly constant character over a large area.

The second type of shelf sandstone occurs as thick (average 15 m), elongate bodies deposited as sand ridges, such as the Shannon Sandstone Member of the Gammon Shale (Fig. 2). They are comparable to coastal-barrier sandstones in thickness, vertical sequence, and energy level at time of deposition. Sand ridges can be distinguished from coastal-barrier sandstones by the following criteria: (1) sand ridges occur at considerable distances (as much as 320 km) seaward of the shoreline, (2) they form elongate discontinuous bodies that trend subparallel to the shoreline, (3) they are enclosed in all directions by offshore marine siltstone and shale, and (4) there is no evidence of subaerial exposure.

Turbidites

Sandstone beds deposited by turbidity currents constitute only a minor part of the Cretaceous sequence, and their recognition has only recently been cited in the literature (Winn and others, 1987). These sandstones are generally thin-bedded and of limited overall thickness. Some units exhibit Bouma sequences, mostly Tb-c, and others are bioturbated. Some of the thin-bedded sandstones described as shelf sandstones above may be turbidites. Most of the turbidites probably represent reworking and displacement into deeper water (below storm-wave base) of shoreface sediments by storm-generated turbidity currents. Basinal, canyon-fed, turbidite fans are not represented or have not been reported.

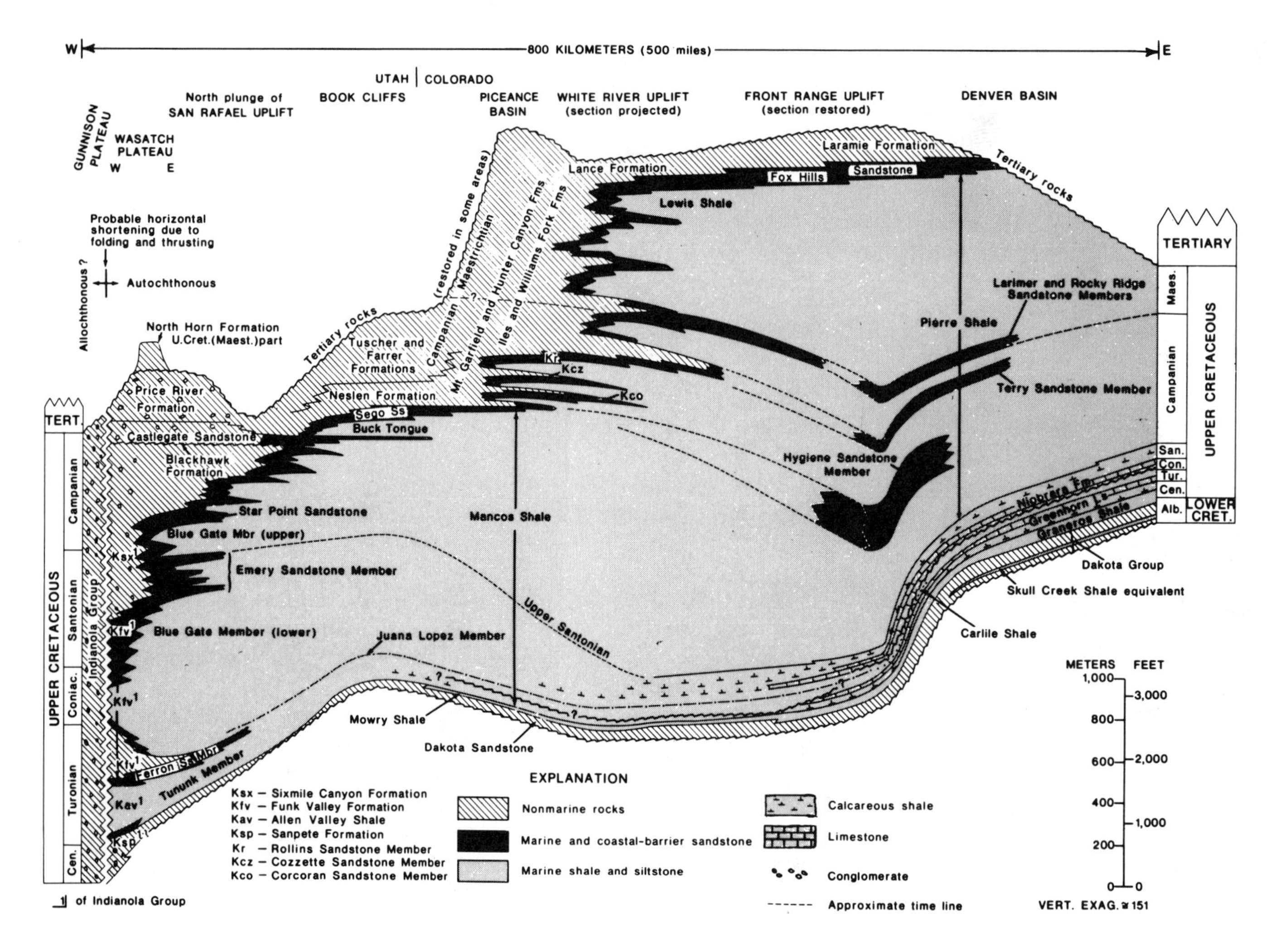

Figure 3. Stratigraphic cross section of Cretaceous rocks from central Utah to northeastern Colorado. Thicknesses are based on well and outcrop control. Line of section is shown in Figure 1.

Limestones

Limestone deposits are common on the eastern shelf of the Western Interior. The westward-extending tongues correspond to major transgressions, for example, the Greenhorn Limestone and the Fort Hays Limestone Member of the Niobrara Formation. Much of the limestone is a chalk, a type of fine-grained carbonate composed of calcareous microfossils and nannofossils (Scholle, 1977). Chalks form only where calcareous nannofossils can live in the overlying water column. Normally, chalks form where terrigenous input is low enough that it does not overwhelm the biologic production rate of calcium carbonate. Although chalks have often been considered to be of deep-water origin, they can form in water depths ranging from tens to thousands of meters if not diluted by constituents such as clastic debris, siliceous microfossils, or coarser carbonate grains (Scholle, 1977). The limestones or chalks grade to argillaceous limestone and calcareous shale vertically and also laterally in the direction of clastic influx (generally from the west).

MAJOR DEPOSITIONAL CYCLES OR CYCLOTHEMS

Several major and numerous minor transgressive–regressive depositional cycles are recognized in the Cretaceous depositional sequence. Five major cycles are recognized and described in the central and northern Western Interior sequence (Weimer, 1960; Kauffman, 1969, 1977; McGookey and others, 1972; Gill and Cobban, 1973). In ascending order, these cycles have been designated the Kiowa–Skull Creek, Greenhorn, Niobrara, Claggett, and Bearpaw cyclothems. In addition, a locally significant cycle (the D-Cross/Pescado transgression and Gallup regression) occurs between the Greenhorn and Niobrara cycles in northwest New Mexico (Molenaar, 1983; Molenaar, this volume). The cyclothems range in age from Albian to Maastrichtian and consist of westward-thinning tongues of marine shale or limestone separated by thicker deltaic or coastal wedges that generally prograded to the east in a subsiding basin. In general, after the Cenomanian-early Turonian (Greenhorn) transgressive cycle, each succeeding cycle transgressed less far landward than the preceding cycle, thus indicating an overall filling of the Western Interior seaway (Figs. 2 and 3).

Attempts are being made to correlate many of the cyclothems of the Western Interior Basin to global sea-level curves presented by Vail and others (1977) and refined by Haq and others (1987). The global curves are based on studies of seismic sequences along several continental margins, augmented with biostratigraphic control from drill holes. Weimer (1984) related three mid-Cretaceous unconformities (at 97, 90, and 80 to 81 Ma) in the east-central part of the basin to low stands of sea level, although he also showed evidence that local tectonics were partly responsible. Undoubtedly, the time of maximum transgression of the Greenhorn cyclothem, a closely correlatable event throughout the Western Interior Basin, is related to a global sea-level high stand (Hancock and Kauffmann, 1979). Some of the later cyclothems, however, are more prominent in some areas of the Western Interior than others, and indeed, the transgressive and regressive maxima of the cycles are often not synchronous. In fact, transgression was sometimes occurring in one area while regression was occurring in another (McGookey and others, 1972; Molenaar, 1983). This could be explained by differences in timing or magnitude of the tectonic events in the major source areas. Differences of timing are especially apparent between the cycles in the southern Rockies and those to the north. The provenance for much of the Upper Cretaceous sediments in the southern Rockies was to the southwest—probably in the area of southwestern Arizona, southern Nevada, southeastern California, and parts of adjoining Mexico. This area probably had a different tectonic history from that of the orogenic belt of western parts of Utah, Wyoming, and Montana, which was the provenance for Cretaceous sediments of the central and northern Rockies.

Clearly, basinal subsidence and differential depositional rates obscures the effects of eustatic sea-level changes in some areas, especially along the western side of the basin where subsidence rates were greater.

Certainly, relative sea-level rise is necessary to accommodate the thick regressive deposits that rise stratigraphically as they prograde, and a greater rate of sea-level rise or basin subsidence (or possibly a decrease in sediment supply) is necessary to cause the major transgressions. Much work on a worldwide basis is necessary to separate the effects of basinal tectonics from eustatic sea-level changes.

ECONOMIC RESOURCES

Oil and gas

Approximately 40 percent of all oil and gas production in the Rocky Mountain province has come from Cretaceous reservoirs, which usually are sandstones. Most of the oil occurs in reservoirs of Early and mid-Cretaceous age, whereas most of the gas occurs in younger Cretaceous sandstones. This relation is due to oil-prone source rocks being more common in the transgressive shales and limestones concentrated in the lower part of the Cretaceous section, such as those in the Mowry Shale, Greenhorn Limestone, and Niobrara Formation. Rates of sedimentation were generally much lower in these parts of the section than in those deposited during regressions or during later Cretaceous time. The marine shales deposited during regressions contain humic (nonmarine) organic matter and tend to generate mostly gas. In addition, nonmarine carbonaceous shales and coals are source rocks for gas.

Some of the major oil fields (100 million barrels or greater of recoverable oil) that produce from Cretaceous reservoirs are Salt Creek field (mainly inner shelf sandstone of the Upper Cretaceous Frontier Formation), Highlight field (marine bar of the Lower Cretaceous Muddy Sandstone), Patrick Draw field (transgressive offlap sandstone of the Upper Cretaceous Almond Formation), and Hartzog Draw field (outer shelf sandstone ridges of the Upper Cretaceous Shannon Sandstone Member of Steele Shale), all in Wyoming, and Bell Creek field (barrier island of the Lower Cretaceous Muddy Sandstone) in Montana. The second largest gas field in the lower 48 states is the Blanco gas field of the San Juan Basin in northwestern New Mexico–southwestern Colorado. This field, with an estimated ultimate production of 23 trillion cubic feet of gas, produces principally from fluvial or distributary-channel sandstones of the Upper Cretaceous Dakota Sandstone, and coastal-barrier sandstones of the Upper Cretaceous Mesaverde Group and Pictured Cliffs Sandstone.

Coal

Upper Cretaceous rocks of the Western Interior contain large reserves of coal. The larger coal mines are in Utah, western and southern Colorado, New Mexico, and Arizona—the Colorado Plateau part of the Western Interior. This area was uplifted in late Tertiary time, and much of the coal-bearing section is exposed in low-dipping areas amenable to strip mining. In general, the Cretaceous coals of Utah and Colorado are of medium- and high-volatile bituminous rank, and the coals of northwestern New Mexico and northeastern Arizona are of subbituminous rank (Averitt, 1972). Locally, coal of anthracite rank occurs in areas that have been heated by intrusive igneous rocks, such as in the Crested Butte coal field of west-central Colorado. A few areas, such as the Somerset coal field west of Crested Butte, the western Book Cliffs, and the Raton Basin, have produced coal of coking quality for use in steel production.

Other mineral deposits

In addition to containing oil, gas, and coal, Cretaceous rocks are the host rocks for minor amounts of uranium. Uranium production has come from the Upper Cretaceous Dakota Sandstone on the south side of the San Juan Basin, and from the Lower Cretaceous Inyan Kara Group in the Black Hills area of South Dakota.

In addition, by virtue of their proximity to igneous intrusions, Cretaceous rocks are host rocks for some base metals in the mining districts of Colorado, such as gold from the famous Breckenridge district and the molybdenum near Crested Butte.

REFERENCES CITED

Averitt, P., 1972, Coal, *in* Geologic atlas of the Rocky Mountains: Rocky Mountain Association of Geologists, p. 297–299.

Cobban, W. A., and Reeside, J. B., Jr., 1952, Correlation of the Cretaceous formations of the Western Interior of the United States: Geological Society of America Bulletin, v. 63, no. 10, p. 1011–1044.

Gill, J. R., and Cobban, W. A., 1973, Stratigraphy and geologic history of the Montana Group and equivalent rocks, Montana, Wyoming, and North and South Dakota: U.S. Geological Survey Professional Paper 766, 37 p.

Hancock, J. M., and Kauffman, E. G., 1979, The great transgressions of the Late Cretaceous: Journal of the Geological Society of London, v. 136, p. 175–186.

Haq, B. U., Hardenbol, J., and Vail, P. R., 1987, Chronology of fluctuating sea levels since the Triassic: Science, v. 235, p. 1156–1166.

Jordan, T. E., 1981, Thrust loads and foreland basin evolution, Cretaceous, western United States: American Association of Petroleum Geologists Bulletin, v. 65, no. 12, p. 2506–2520.

Kauffman, E. G., 1969, Cretaceous marine cycles of the Western Interior: Mountain Geologist, v. 6, no. 4, p. 227–245.

—— , 1977, Geological and biological overview; Western Interior Cretaceous basin: Mountain Geologist, v. 14, nos. 3 and 4, p. 75–99.

McGookey, D. P., and others, 1972, Cretaceous system, *in* Geologic atlas of the Rocky Mountains: Rocky Mountain Association of Geologists, p. 190–228.

Molenaar, C. M., 1983, Major depositional cycles and regional correlations of Upper Cretaceous rocks, southern Colorado Plateau and adjacent areas, *in* Reynolds, M. W., and Dolly, E. D., eds., Mesozoic paleogeography of the west-central United States: Rocky Mountain Section, Society of Economic Paleontologists and Mineralogists Rocky Mountain Paleogeography, Symposium 2, p. 201–224.

Obradovich, J. D., and Cobban, W. A., 1975, A time scale for the Late Cretaceous of the Western Interior of North America, *in* Caldwell, W.G.E., ed., The Cretaceous system in the Western Interior of North America: Geological Association of Canada Special Paper 13, p. 31–54.

Rice, D. D., and Shurr, G. W., 1983, Patterns of sedimentation and paleogeography across the Western Interior seaway during time of deposition of Upper Cretaceous Eagle Sandstone and equivalent rocks, northern Great Plains, *in* Reynolds, M. W., and Dolly, E. D., eds., Mesozoic paleogeography of the west-central United States: Rocky Mountain Section, Society of Economic Paleontologists and Mineralogists Rocky Mountain Paleogeography, Symposium 2, p. 337–358.

Scholle, P. A., 1977, Chalk diagenesis and its relation to petroleum exploration; Oil from chalks, a modern miracle?: American Association of Petroleum Geologists Bulletin, v. 61, no. 7, p. 982–1009.

Vail, P. R., Mitchum, R. M., Jr., and Thompson, S., III, 1977, Global cycles of relative changes of sea level, (Pt. 4), *in* Payton, C. E., ed., Seismic stratigraphy; Application to hydrocarbon exploration: American Association of Petroleum Geologists Memoir 26, p. 83–97.

Weimer, R. J., 1960, Upper Cretaceous stratigraphy, Rocky Mountain area: American Association of Petroleum Geologists Bulletin, v. 44, no. 1, p. 1–20.

—— , 1984, Relation of unconformities, tectonics, and sea-level changes, Cretaceous of Western Interior, U.S.A., *in* Schlee, J. S., ed., Interregional unconformities and hydrocarbon accumulation: American Association of Petroleum Geologists Memoir 36, p. 7–35.

Winn, R. D., Jr., Bishop, M. G., and Gardner, P. S., 1987, Shallow-water and sub-storm-base deposition of Lewis Shale in Cretaceous Western Interior seaway, south-central Wyoming: American Association of Petroleum Geologists Bulletin, v. 71, no. 7, p. 859–880.

MANUSCRIPT ACCEPTED BY THE SOCIETY SEPTEMBER 9, 1987

Printed in U.S.A.

The Geology of North America
Vol. D-2, Sedimentary Cover—North American Craton: U.S.
The Geological Society of America, 1988

Chapter 7

Phanerozoic stratigraphy of the northern Rocky Mountain region

J. A. Peterson
U.S. Geological Survey, Department of Geology, University of Montana, Missoula, Montana 59812

INTRODUCTION

The northern Rocky Mountain region (Figs. 1, 2), contains a relatively complete Phanerozoic stratigraphic section ranging in age from Cambrian to Holocene (Figs. 3 to 5). Thicknesses of Paleozoic rocks are as much as 9,000 m in southeast Idaho, 3,000 m in southwest Montana, 2,500 m in the central part of the Williston Basin, and generally less than 1,500 m in most of Wyoming, South Dakota, and the remainder of Montana. Thicknesses of Mesozoic rocks are as much as 11,000 m in southeast Idaho and western Wyoming, 6,000 m in southwest Montana, and generally less than 3,000 m in northern and eastern Wyoming, South Dakota, and central and eastern Montana. Thicknesses of Tertiary rocks are as much as 300 m in the western part of the Williston Basin, more than 900 m in the deeper parts of the Powder River Basin, and as much as 3,000 m or more in the Tertiary basins of western Montana and southwestern Wyoming.

CAMBRIAN

Cambrian strata unconformably overlie Proterozoic sedimentary rocks in western Montana, east-central and southeastern Idaho, and overlie older Precambrian metamorphic and igneous rocks in the remainder of the region. Thicknesses are more than 1,500 m in south-central Idaho and more than 900 m in western Montana, thinning relatively uniformly eastward and pinching out along the flank of the Transcontinental Arch (Figs. 3 to 6). Cambrian rocks comprise a sequence of marine sandstone, shale, and limestone, which represent the shelf facies of a broad eastward transgression of the Cambrian Cordilleran sea across the early Paleozoic Rocky Mountain shelf.

Cambrian rocks span all of Cambrian time in the miogeocline to the west, where they are in uninterrupted succession with Proterozoic rocks. They become progressively younger eastward until only Upper Cambrian rocks, primarily sandstones, are present along the eastern flank of the shelf province in North Dakota, South Dakota, eastern Wyoming, and Nebraska.

A basal transgressive sandstone (€f, Flathead Sandstone), is of Middle Cambrian age in the west, and equivalent rocks become Late Cambrian to the east (Lochman-Balk, 1972). In the western areas, these beds grade upward to marine green and gray shale followed by a series of carbonate formations, of Middle and Late Cambrian age. These grade eastward across western Montana and western Wyoming from primarily dolomite into limestones alternating with increasingly prominent shale formations. Farther east, in the Williston Basin and most of the Powder River Basin, equivalent beds are assigned to the Deadwood Formation of Late Cambrian and Early Ordovician age (Ross, 1957; Carlson, 1960; Lochman-Balk and Wilson, 1967; Gerhard and Anderson, this volume). The Deadwood Formation and equivalents are more than 300 m thick in western Wyoming, the Central Montana Trough, and the central part of the Williston Basin. These rocks thin eastward, reaching a zero edge in eastern North Dakota and central South Dakota (Fig. 6). Limestone content in the Deadwood and equivalents decreases eastward.

Prominent paleostructures influencing sediment thickness and facies of Cambrian rocks include the Lemhi, Sweetgrass, and Transcontinental Arches, the Beartooth high, the Central Montana Trough, and the Sublett and Williston Basins (Figs. 3 to 6). The relatively uniform eastward thinning is interrupted in several areas where Cambrian beds are absent or thin; for example, on the Sweetgrass arch, in south-central Montana, in the southern part of the Cedar Creek Anticline, and in the vicinity of the Nesson Anticline in North Dakota. Because of inadequate subsurface control, it is uncertain whether these areas of thinning are related to erosion over Cambrian structural highs or to draping over buried hills on the Precambrian surface. Lochman-Balk (1972) estimated that relief on the irregular Precambrian surface may have been as much as 100 to 125 m.

Regional studies of the Cambrian include those by Hanson (1952), Theodosis (1955), and Lochman-Balk (1972).

ORDOVICIAN

Ordovician rocks are more than 300 m thick in east-central and southeast Idaho and in the Williston Basin (Fig. 7). However, they are absent, except for a few remnants, in southeastern Wyoming, southwestern South Dakota, and western Montana,

Peterson, J. A., 1988, Phanerozoic stratigraphy of the northern Rocky Mountain region, *in* Sloss, L. L., ed., Sedimentary Cover—North American Craton; U.S.: Boulder, Colorado, Geological Society of America, The Geology of North America, v. D-2.

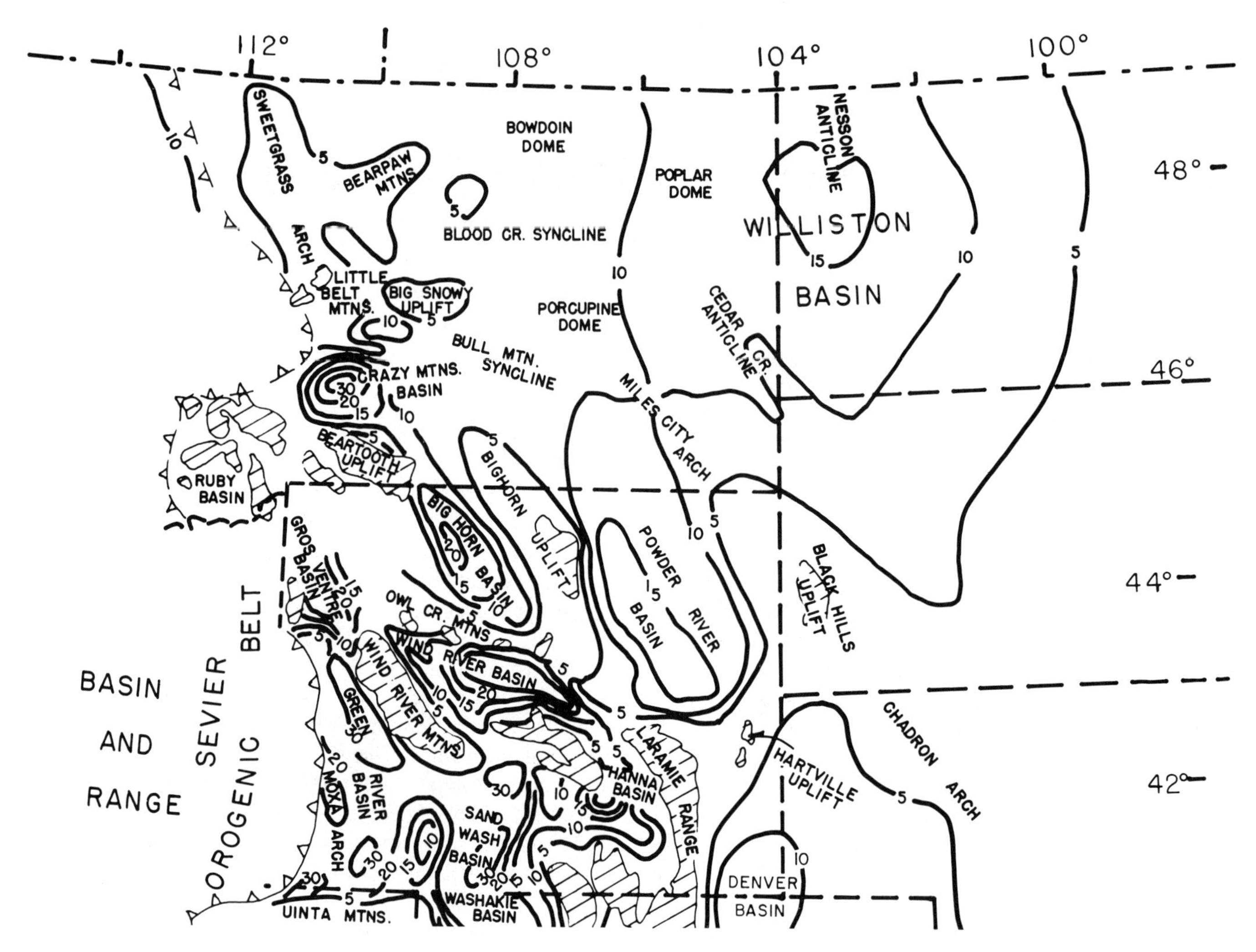

Figure 1. Basins and uplifts of the northern Rocky Mountains showing approximate thickness of Phanerozoic sedimentary cover in thousands of feet. Areas of exposed Archean rocks east of the thrust belt are shown by cross-hatching; eastern edge of thrust belt is shown by barbed line. (From Peterson and Smith, 1986.)

because of nondeposition or erosion during Silurian, Devonian, and Early Mississippian time. Carbonate facies dominated across the entire shelf region during the Ordovician and continued until late Paleozoic time when clastic influx from the interior of the craton returned.

The depocenter of the Williston Basin in northwestern North Dakota (Gerhard and Anderson, this volume) became clearly defined in Ordovician time, and the basin shape remained essentially the same throughout the remainder of Paleozoic and Mesozoic time. Thickness of Ordovician rocks increases eastward and northward into the Williston Basin from the zero line in central Montana and northeastern Wyoming (Fig. 7). The thicknesses also increase to the southwest into the miogeocline.

Lower Ordovician rocks are present in the Williston Basin as marine shale and some limestone in the upper part of the Deadwood Formation, and in southeastern Idaho as the Garden City Limestone. These rocks are disconformably overlain by Middle Ordovician sandstone and quartzite of the Winnepeg Formation (Williston Basin), Swan Peak Quartzite (southeastern Idaho), and Kinnikinic Quartzite (east-central Idaho). This unconformity is related to widespread regression and erosion that affected much of interior North America (Sloss, 1963; Branson, 1964; Ostrom, 1970; Ross, 1975; Witzke, 1980) between Early and Middle Ordovician time.

The regression was followed by marine transgression, which deposited the basal Tippecanoe sequence of Sloss (1963). These Middle Ordovician rocks are more widespread than the earlier Ordovician rocks. The initial deposits are primarily sandstone or

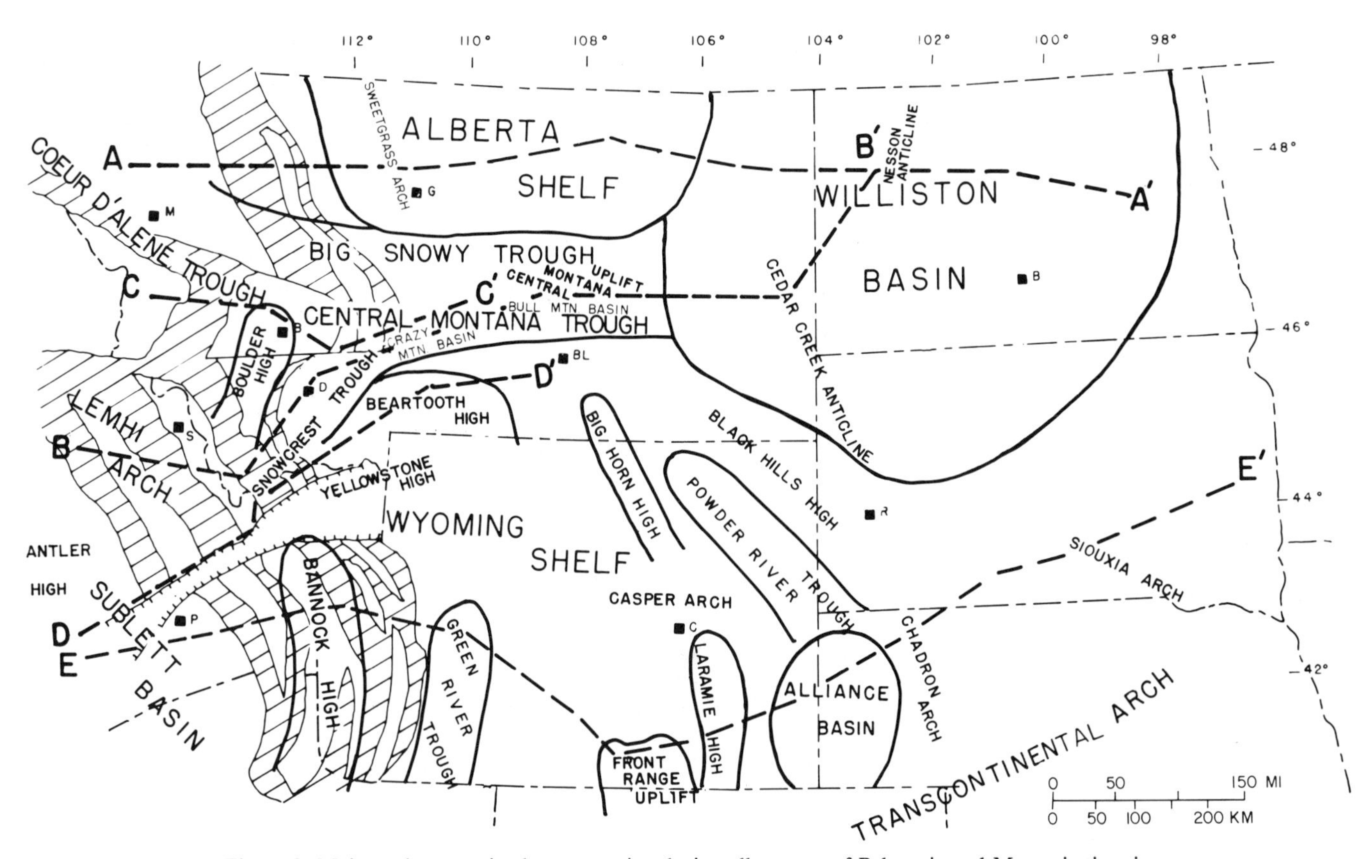

Figure 2. Major paleotectonic elements active during all or part of Paleozoic and Mesozoic time in northern Rocky Mountain region. Lines of cross sections of Figures 3, 4, 5, 15, and 16 are shown. Cities are as follows: Montana: B, Butte; M, Missoula; G, Great Falls; D, Dillon, BL, Billings. Idaho: S, Salmon; P, Pocatello. Wyoming: C, Casper, Wyoming. South Dakota: R, Rapid City, South Dakota. Map is palinspastically reconstructed in thrust belt. (From Peterson, 1985, 1987.)

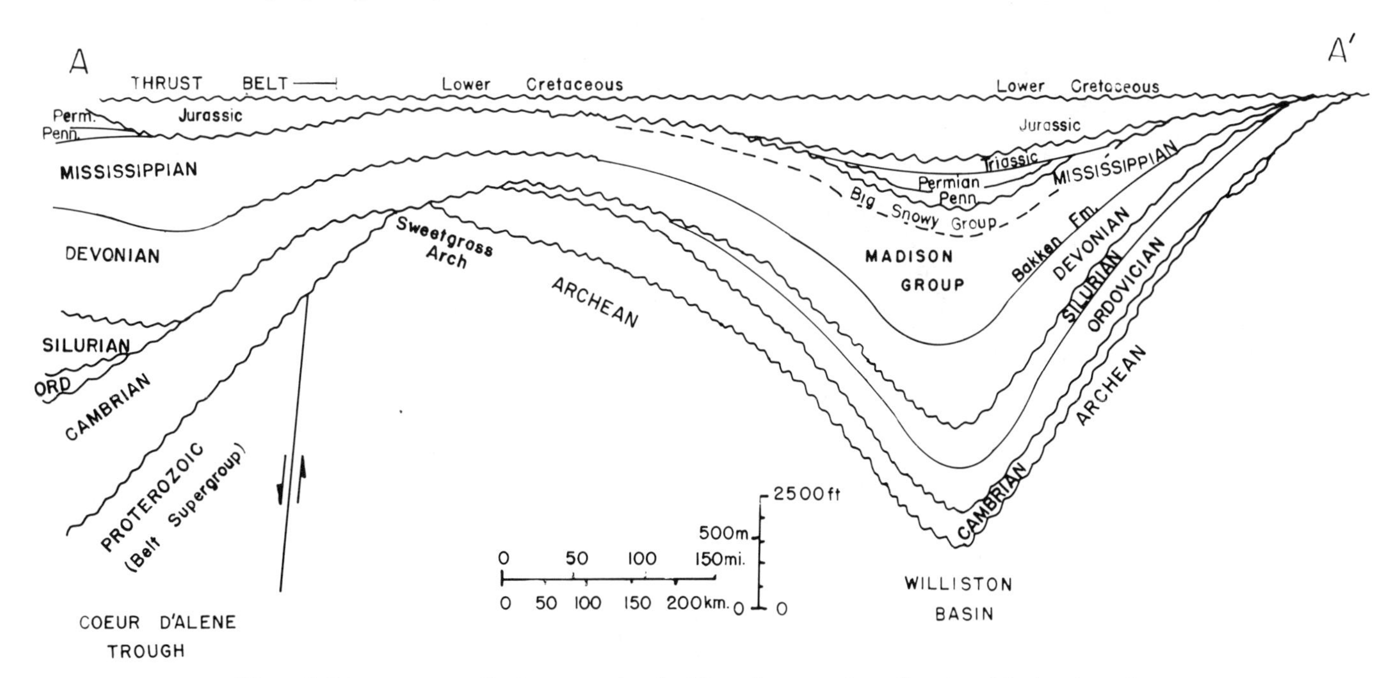

Figure 3. West-east generalized cross section A–A″, partly restored, northwestern Montana to eastern North Dakota. Datum, top of Jurassic. Location of cross section shown on Figure 2. (From Peterson, 1985.)

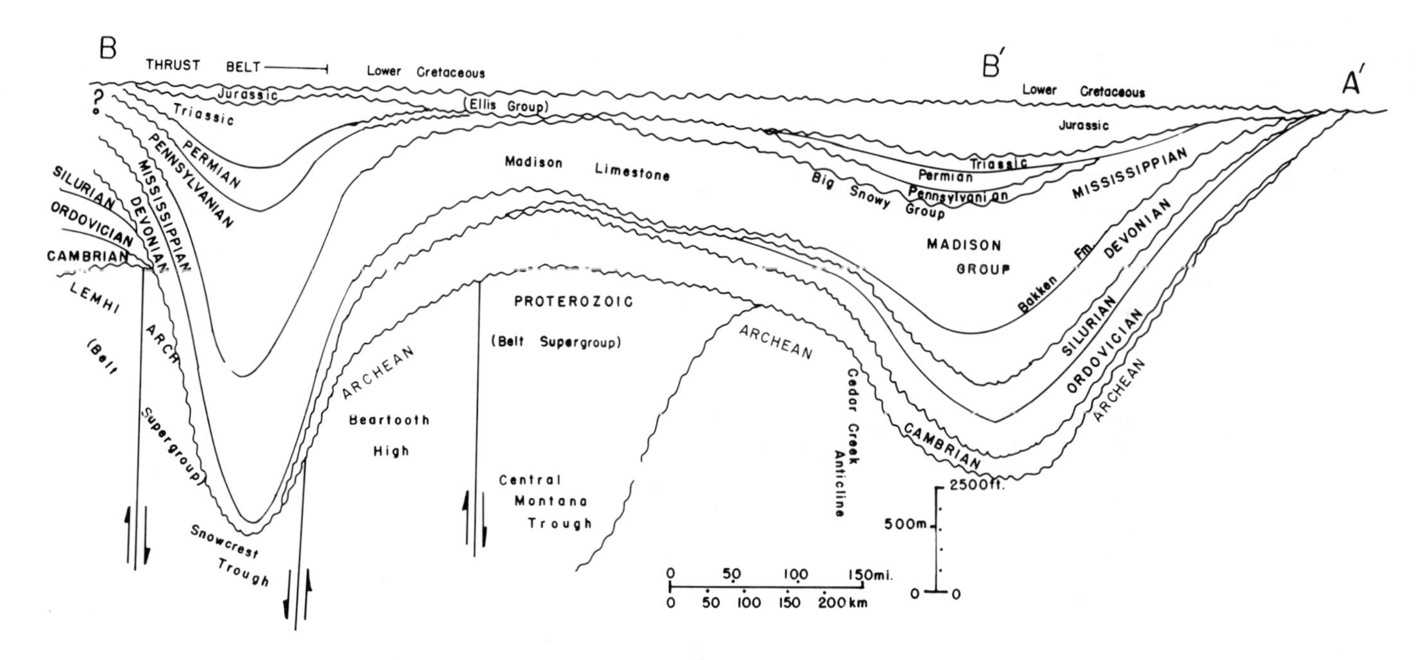

Figure 4. West-east generalized cross section B–B′–A″, partly restored, east-central Idaho to eastern North Dakota. Datum, top of Jurassic. Location of cross section shown on Figure 2. (From Peterson, 1985.)

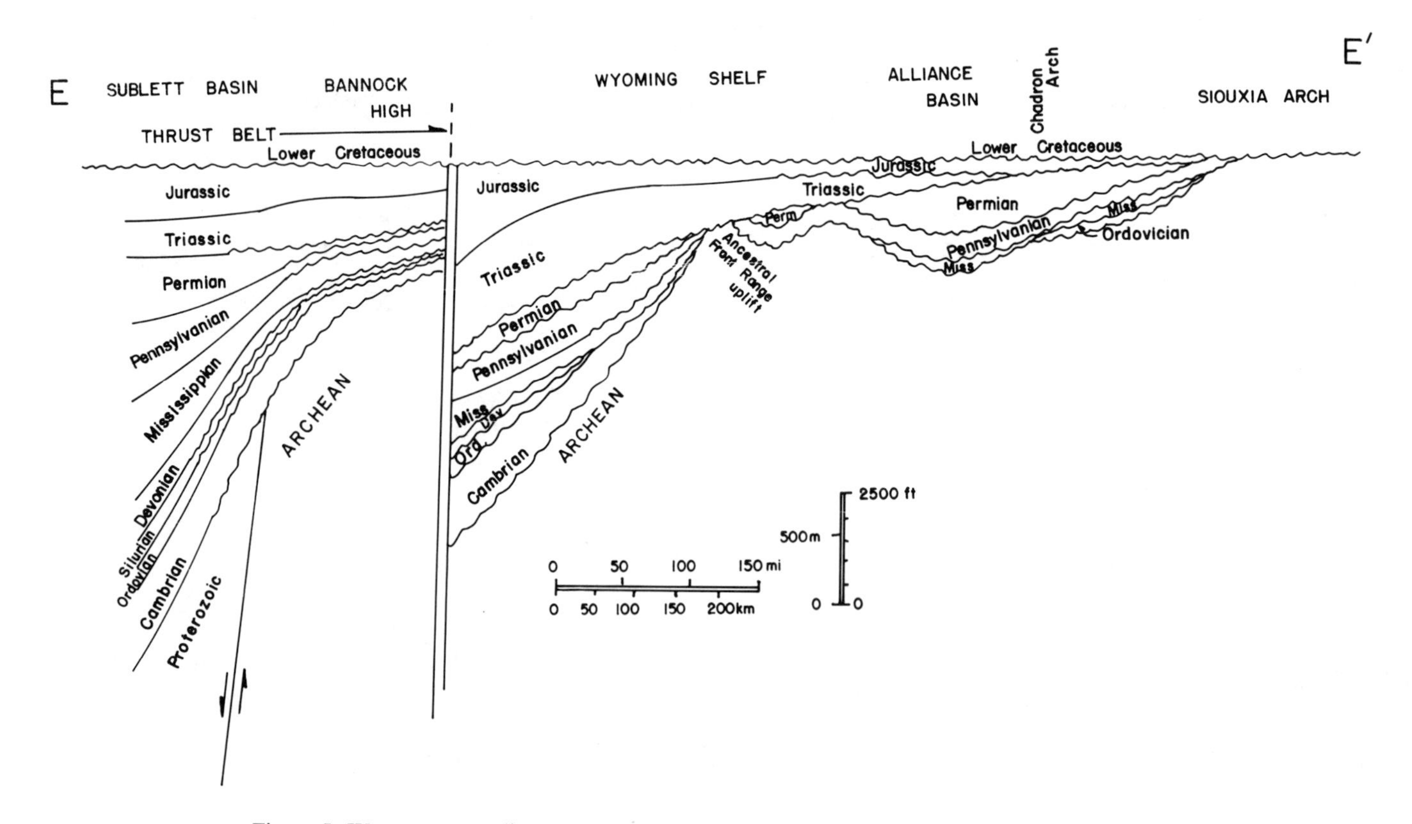

Figure 5. West-east generalized cross section E–E′, from Sublett basin, southeastern Idaho, across southern Wyoming to northeastern Nebraska. Datum, top of Jurassic. Location of cross section shown on Figure 2.

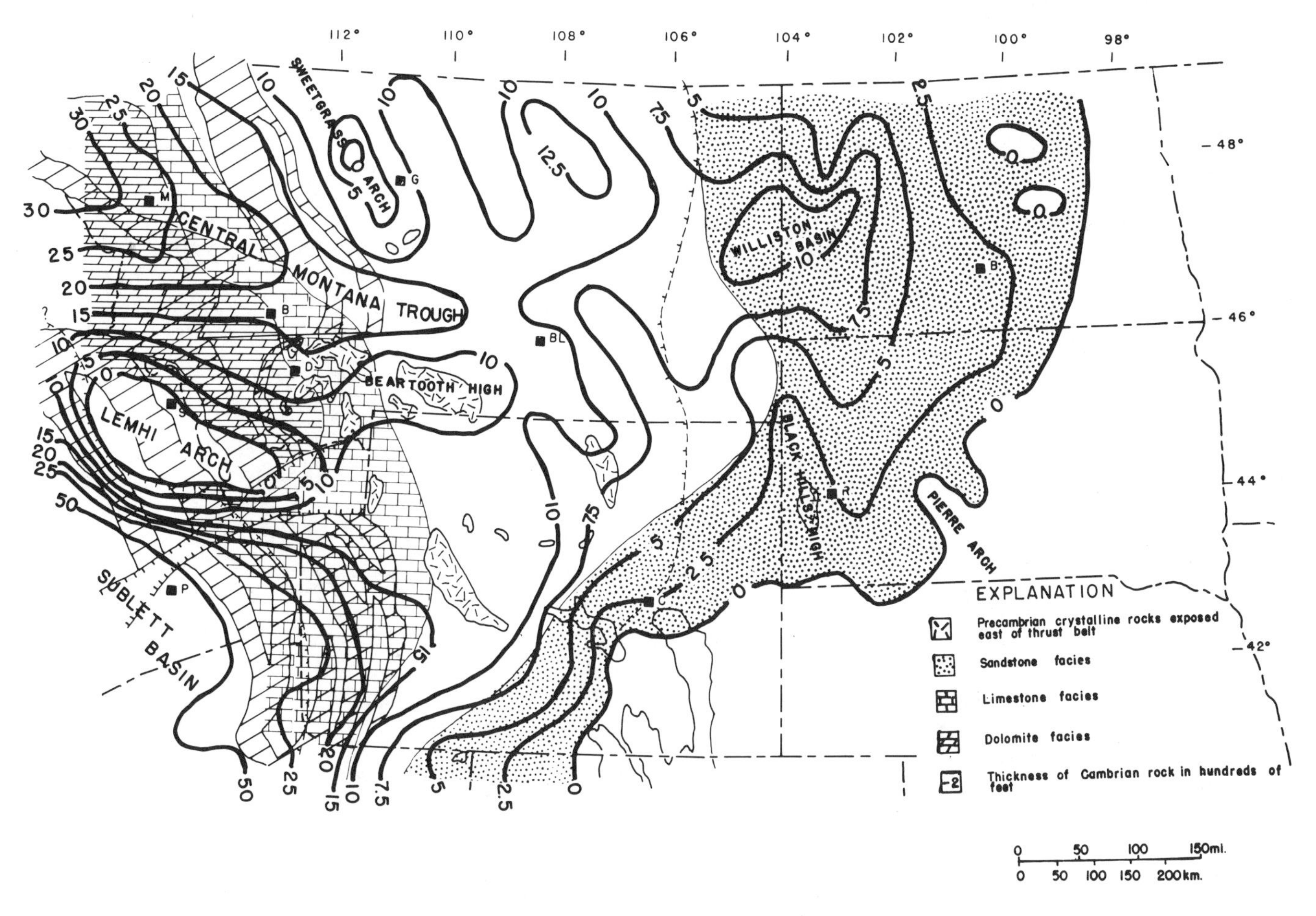

Figure 6. Cambrian System. Thickness (in hundreds of feet) and general facies. Partly restored; palinspastically reconstructed in thrust belt. Eastern limit of Middle Cambrian rocks shown by dashed barbed line. Unpatterned areas are mixed green and gray shale, carbonate and sandstone. (From Peterson, 1985.)

quartzite and shale with minor siltstone and limestone mostly of marine origin. These beds grade upward to widespread marine limestone or dolomite units of mostly Middle and Late Ordovician age that include the Saturday Mountain and Fish Haven dolomites in Idaho, the Bighorn Dolomite in north-central and western Wyoming and south-central Montana, and the Late Ordovician Red River Formation in the Williston Basin. Cyclically deposited anhydrite, carbonate, and shale beds are present at the top of the Red River Formation.

Highest Upper Ordovician rocks are absent primarily because of pre–Late Devonian erosion in most of the northern Rocky Mountain region, except for the Williston Basin where they are represented by carbonate and shale of the Stony Mountain Formation (Gerhard and Anderson, this volume).

Regional studies of Ordovician rocks include those by Ross (1957), Lochman-Balk and Wilson (1967), Foster (1972), Fuller (1961), and Oaks and others (1977).

Active tectonic features during this time include the Lemhi Arch, the Transcontinental Arch, and the Williston Basin.

SILURIAN

Except in the Williston Basin where as much as 300 m of carbonate rocks of Silurian age (Interlake Formation, latest Ordovician and Silurian) conformably overlie Upper Ordovician rocks, Silurian rocks are absent on most of the craton in the northern Rocky Mountain region, primarily because of pre–Late Devonian erosion. A comparably thick section of Silurian dolomite (Laketown Dolomite) in the miogeocline of eastern Idaho pinches out eastward very close to the southwestern Montana–east-central Idaho border, a short distance west of the truncated edge of Ordovician rocks, in the vicinity of the Lemhi Arch (Figs. 4, 8). This erosional offlap relationship between the Bighorn-Interlake and Kinnikinic-Laketown Formations and the

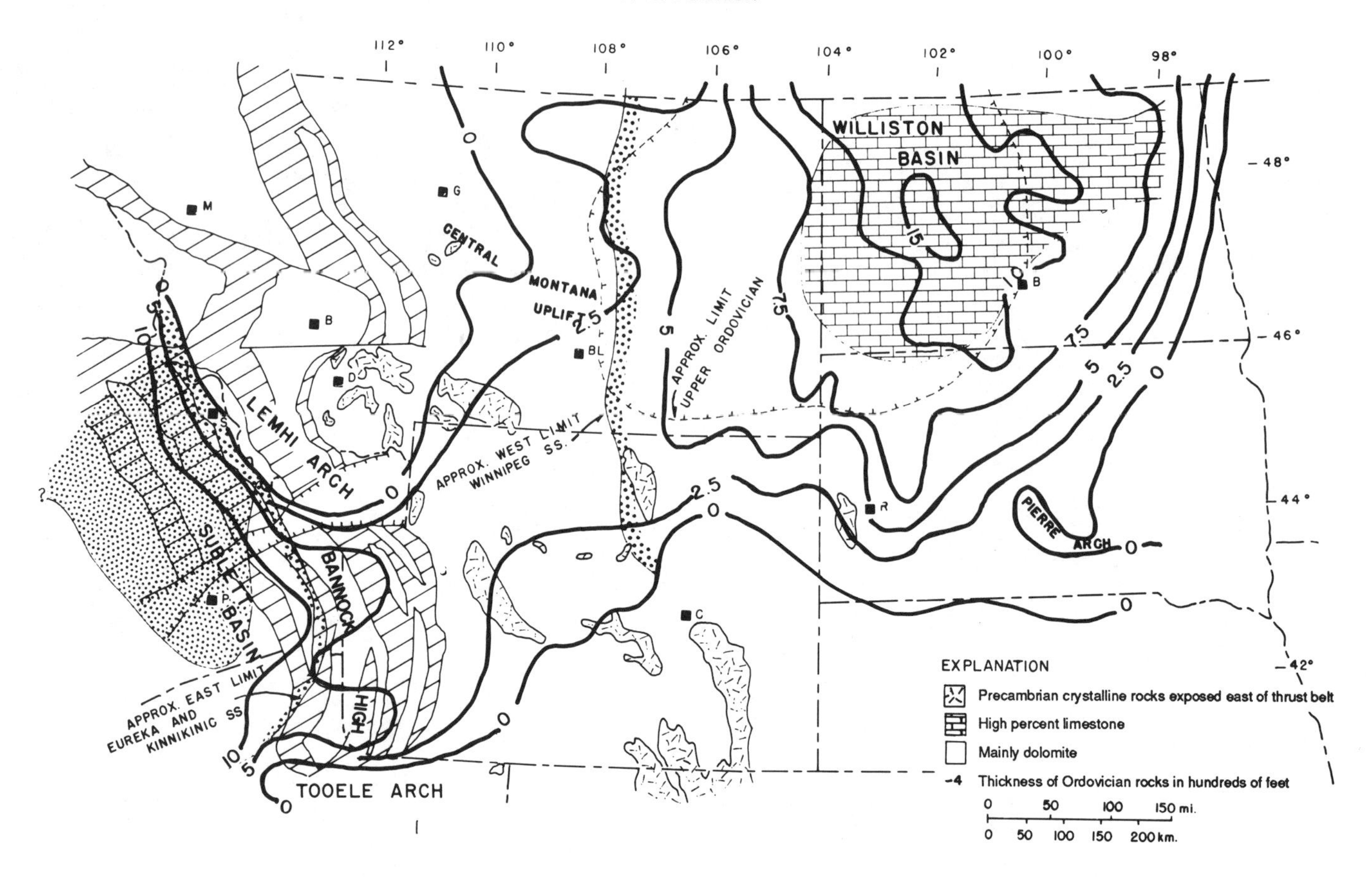

Figure 7. Ordovician System. Thickness (in hundreds of feet) and general facies. Partly restored; palinspastically reconstructed in thrust belt. Unpatterned areas are primarily dolomite. (From Peterson, 1985.)

overlying Upper Devonian beds, plus the total absence of sand or silt in the Silurian and Ordovician carbonate beds, suggests that the erosional edges of both units are some distance from the Ordovician or Silurian shorelines and that a broader area was originally covered by carbonate rocks. Many of these were subsequently removed by pre–Middle or pre–Late Devonian erosion.

DEVONIAN

During most of Late Silurian and Early Devonian time, the northern Rocky Mountain region was emergent. Regional transgression of the continent began in the Middle Devonian (lower part of Kaskaskia sequence of Sloss, 1963), but primarily Upper Devonian rocks disconformably overlie Upper Cambrian and Ordovician rocks in most of the northern Rocky Mountain region. Exceptions are the Williston Basin where Middle Devonian rocks overlie the Silurian Interlake Formation, and in southeast Idaho, along the margin of the miogeocline, where Lower Devonian rocks (Water Canyon Formation) overlie Silurian rocks (Skipp and Sandberg, 1975; Johnson and Sandberg, 1977).

Clastic beds, including conglomerate and breccia of probable Early Devonian age, occur in scattered remnants overlying earlier Paleozoic rocks in parts of southern Montana, northern Wyoming, and eastern Idaho (Sandberg and Mapel, 1967).

The worldwide Late Devonian transgression spread completely across the northern Rocky Mountain region. In western Montana, these rocks comprise a transgressive sequence of fine-to medium-grained red to green clastic marine beds (Maywood or lower Souris River Formations) overlain by a prominent sequence (Jefferson Formation) of generally porous stromatoporoid and coral-bearing dolomite, fossiliferous limestone, evaporite beds, and shaly limestone, which make up the bulk of the Devonian section. The strongly cyclic nature of this sequence has been described by several authors (Wilson, 1955; Sandberg and Hammond, 1958; Rose, 1976). The Jefferson is overlain by regressive green and red shale, siltstone, evaporite beds, and shaly limestone beds of the Three Forks Formation, which grades upward into Upper Devonian-Mississippian dark organic shale and siltstone of the Bakken and equivalent beds. Lithologic aspects and facies distribution of these beds have been described in detail

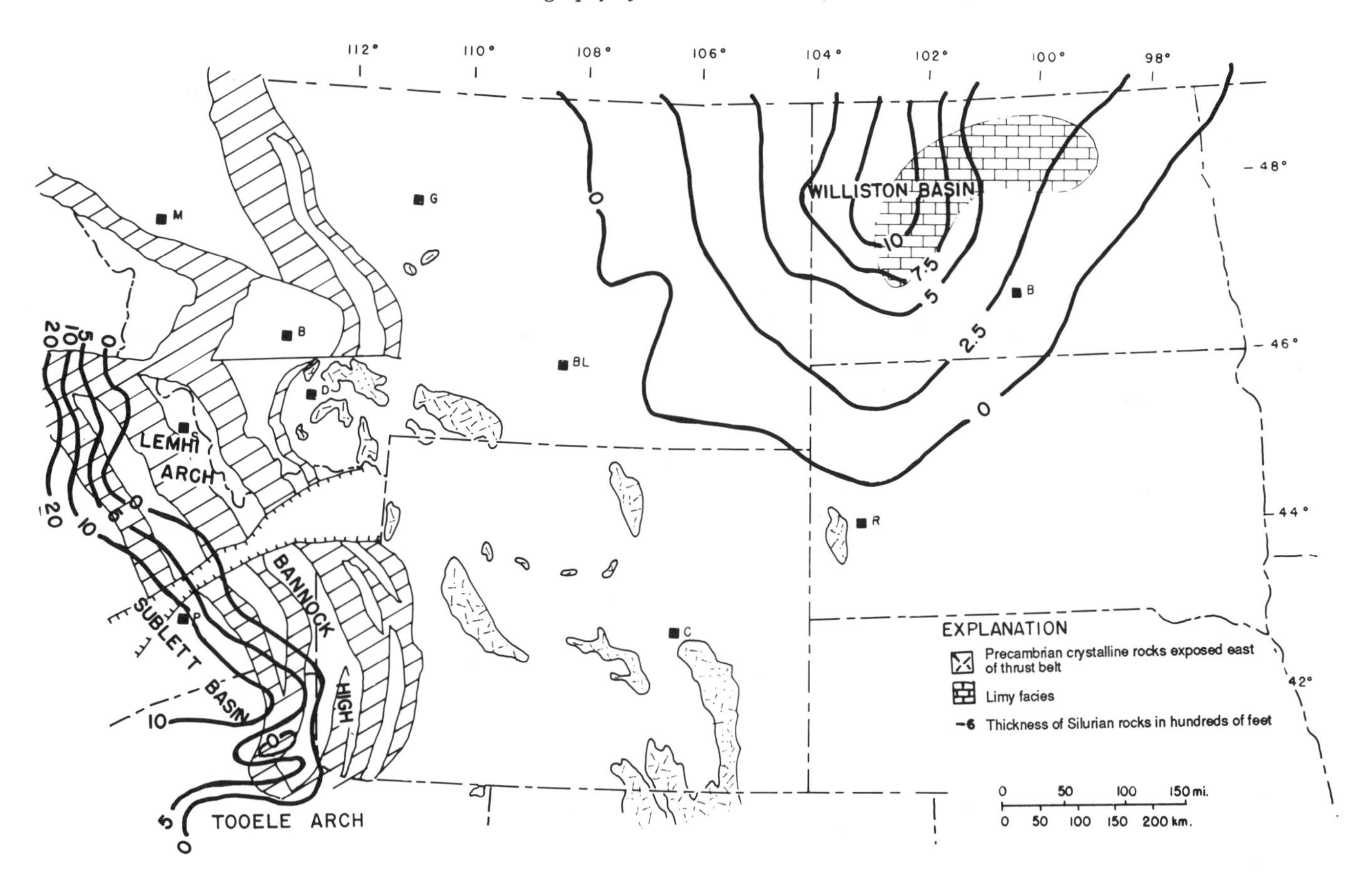

Figure 8. Silurian System. Thickness (in hundreds of feet) and general facies. Partly restored; palinspastically reconstructed in thrust belt. Unpatterned areas are primarily dolomite. (From Peterson, 1985.)

by Rau (1962), McMannis (1962), and Gutschick and others (1962).

The Upper Devonian rock sequence is more than 300 m thick in most of western Montana, and more than 600 m thick in much of the western thrust belt area; it thickens rapidly westward into the Sublett Basin (Muldoon trough of Rose, 1976) at the edge of the miogeocline along the southwestern Montana–Idaho border (Fig. 9). The western facies of the Upper Devonian is dominated by dolomitized carbonate bank deposits characterized by stromatoporoids and tabulate corals and commonly containing thick beds of highly porous sucrosic dolomite. Upper Devonian rocks (Darby Formation) overlie Upper Ordovician rocks (Bighorn Dolomite) in northwestern and west-central Wyoming. The Darby is primarily dolomite or dolomitic limestone less than 100 m thick, and wedges out eastward beneath Mississippian rocks approximately across central Wyoming (Fig. 9).

The continued influence of the Central Montana Trough is evident from the west-east thick trend (Fig. 9) in about the same position as that of the Cambrian (Fig. 6). A relatively prominent thin area is present in the vicinity of the Lemhi Arch in southwestern Montana ("Tendoy high") and an accompanying southwest-northeast thick trend is present just to the east in the Snowcrest trough (Figs. 2, 9). The relatively prominent northwest-southeast thin area that coincides closely with the northwestern overthrust belt in Montana (Fig. 9) may be related to outcrop leaching of gypsum or anhydrite beds of the "Potlatch" facies and similar beds in the underlying Jefferson Formation. Marked thinning or absence of Upper Devonian beds occurs along the Central Montana uplift and extends westward along the Little Belt Mountains area south of Great Falls. The western part of this belt may also be related to evaporite solution of outcrop sections, although the absence of the "Potlatch" breccias in the Little Belt Mountains suggests otherwise. The accompanying absence of the Devonian-Mississippian Bakken Formation beds in this area suggests that broad, gentle uplift occurred here at this time. The distribution of the Bakken and equivalent facies in south-central Montana also demonstrates the continued influence of the Central Montana trough and the Beartooth shelf (Figs. 2, 9).

The regional distribution of evaporite beds and related solution breccias in the Jefferson and Three Forks Formations

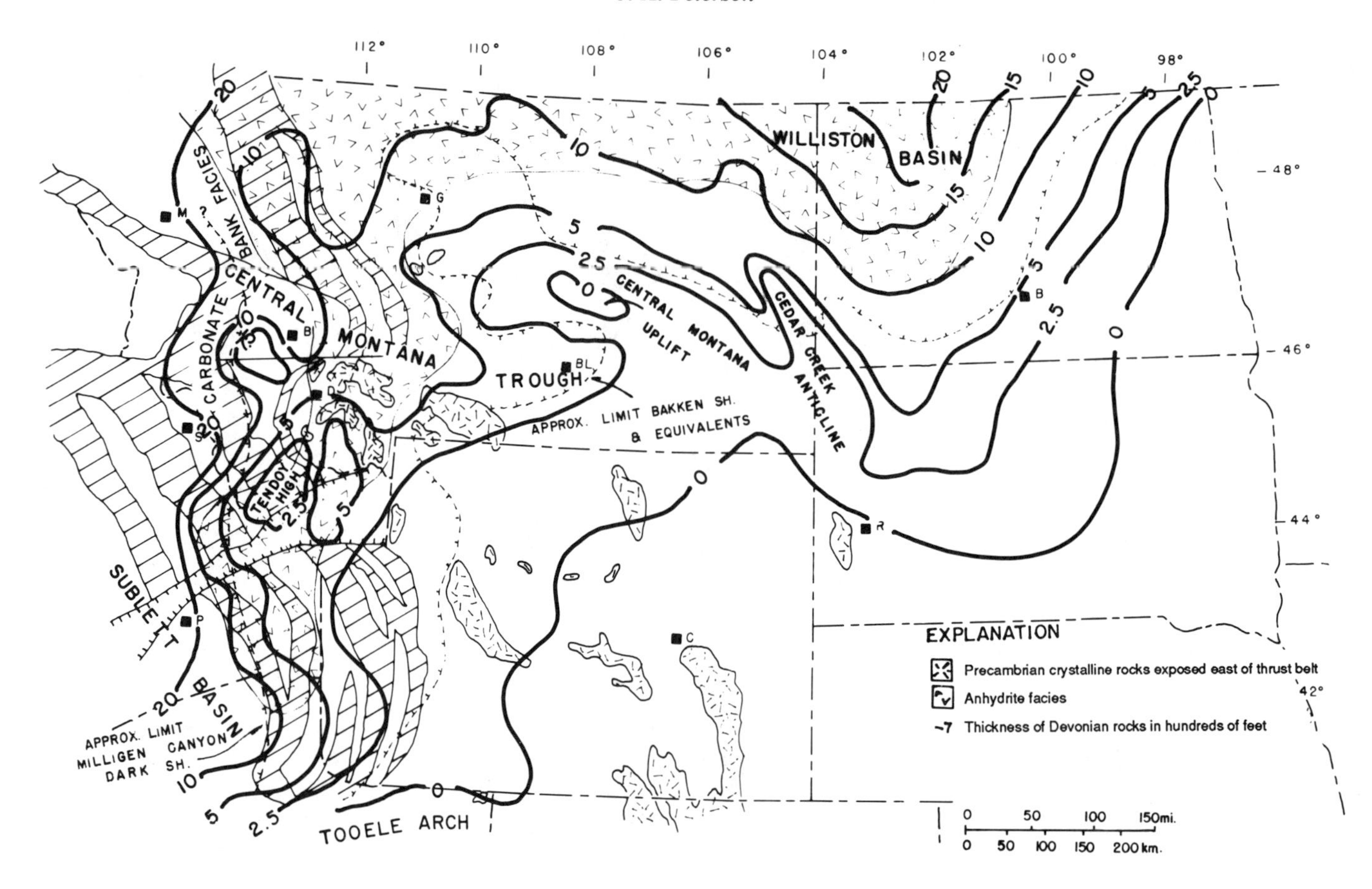

Figure 9. Devonian System. Thickness (in hundreds of feet) and general facies. Partly restored; palinspastically reconstructed in thrust belt. Unpatterned areas are mainly carbonate rocks. (From Peterson, 1985.)

suggests that evaporitic conditions occurred in a back-bank environment immediately east of the north-south–trending belt of carbonate bank facies buildup in westernmost Montana (Figs. 3, 4, 9). This paleogeographic relationship may have resulted in the establishment of a westward-directed hydrodynamic gradient that enabled high-magnesium waters of the shelf environment to move westward during low sea-level evaporitic stages of the cycles, resulting in early dolomitization of the thick carbonate facies by seepage reflux processes. Such a mechanism may have been in effect during much of the early to middle Paleozoic, as suggested by the similarity in gross dolomitic facies distribution of the Cambrian sequence; however, evidence for shelf back-bank evaporitic environments is lacking in the earlier Paleozoic sequences.

Active Devonian tectonic elements include the "Tendoy high" (perhaps a remnant of the Lemhi arch), the central Montana uplift, and the Cedar Creek Anticline (Fig. 9).

Regional studies of Devonian rocks include those by Wilson (1955, 1967), Sandberg and Hammond (1958), Benson (1966), Sandberg and Mapel (1967), and Rascoe and Baars (1972). Detailed lithologic and stratigraphic aspects of Devonian rocks are described by Sandberg and McMannis (1964), Sandberg and Mapel (1967), Rau (1962), Sandberg (1962, 1965), Wilson (1955), and Benson (1966).

DEVONIAN AND MISSISSIPPIAN

The Upper Devonian Three Forks Formation is overlain by black organic shale and siltstone beds of the Bakken Formation in most of the Williston Basin and northern Montana (Fig. 9). These beds are thickest (more than 30 m) in the central part of the basin (Meissner, 1978; Gerhard and Anderson, this volume). The Devonian and Mississippian systemic boundary is generally placed within the Bakken Formation, which is believed to rest disconformably on the Three Forks Formation; although the evidence for a break in sedimentation between the Bakken Formation and Three Forks Formation is not always evident, especially in the subsurface. Brief emergence of most of the shelf area probably took place at the close of Three Forks deposition, and dark organic shale and siltstone beds of the overlying Bakken Forma-

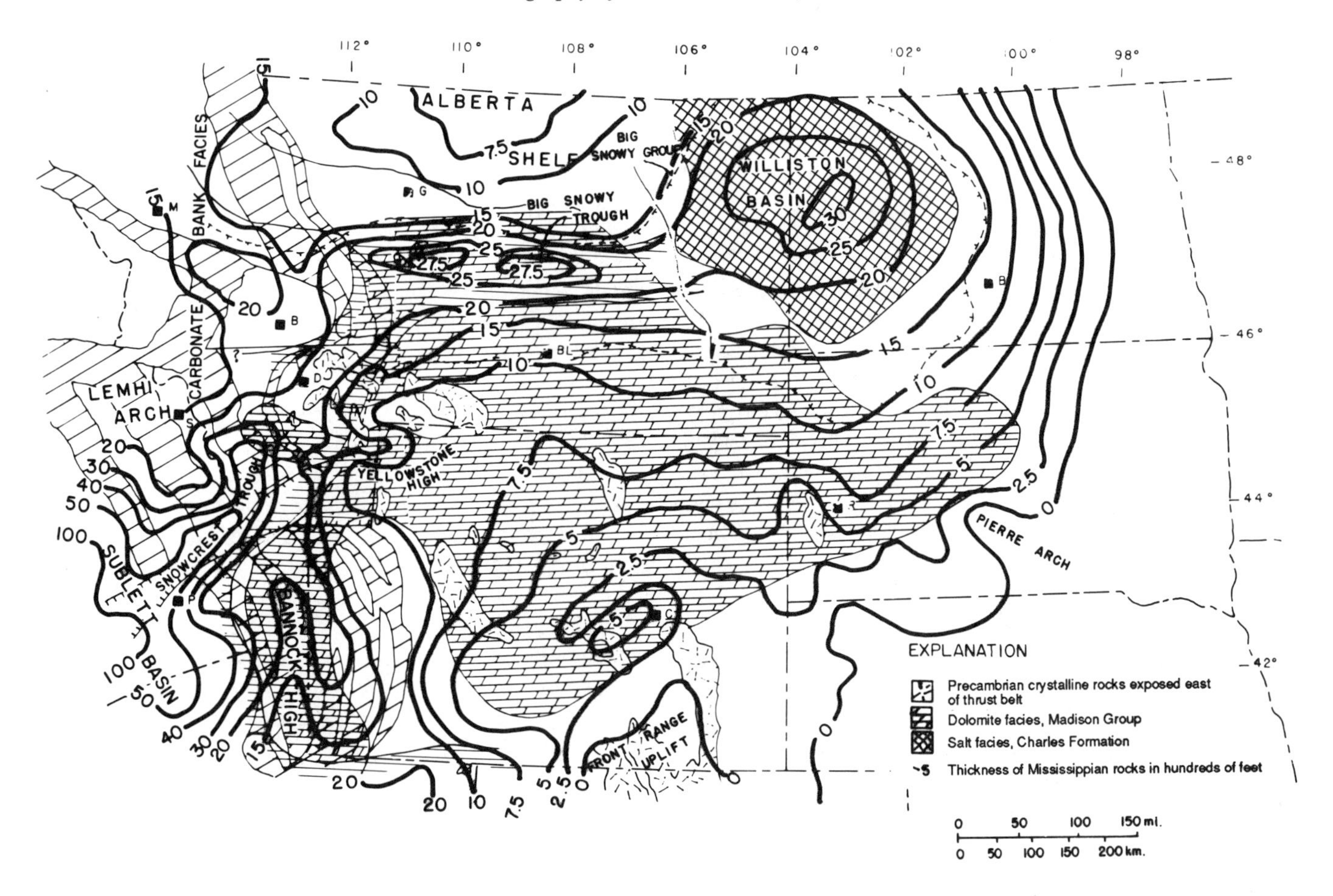

Figure 10. Mississippian System. Thickness (in hundreds of feet) and general facies. Partly restored; palinspastically reconstructed in thrust belt. Limit of Upper Mississippian Big Snowy Group shown by barbed line. Unpatterned areas are mainly carbonate rocks in the Lower Mississippian, and fine siliciclastics, carbonates, and evaporites in the Upper Mississippian. (From Peterson, 1985.)

tion represent initial early transgressive deposits of the advancing Late Devonian and Early Mississippian seaway.

The widespread organic shale facies of the Bakken Formation and its equivalents (Chattanooga, New Albany, and Ohio Shales of central and eastern North America, Exshaw Shale of Canada and northwestern Montana, Pilot Shale of the Great Basin, etc.) is characteristic of Upper Devonian and Lower Mississippian beds in many parts of the world. The Bakken Formation is overlain conformably by argillaceous carbonate beds of the basal Lodgepole Limestone in most of the Williston Basin; in north-central Montana and the northeastern part of the Williston Basin, crinoid mounds overlie the Bakken Formation in several places.

MISSISSIPPIAN

After deposition of the dark, organic-rich shale of the Bakken Formation and equivalents, cyclic shelf carbonate deposition again dominated the northern Rocky Mountain region, but the carbonate-building organic assemblages underwent major changes with the disappearance of important earlier Paleozoic life forms and the rise in dominance of new ones. Mississippian carbonate rocks are characterized by accumulations of crinoid and bryozoan skeletal remains and oolitic bank deposits, in contrast to the abundant stromatoporoid and tabulate coral carbonate bodies of Devonian age.

Widespread cyclic shelf carbonate deposition dominated the Mississippian rock sequences. Widespread anhydrite marker beds are found at the tops of the cycles and are particularly prominent in the Williston Basin and Central Montana Trough. These cycles become progressively more evaporitic in the upper part of the Madison Limestone, culminating in the deposition of the halite-bearing Charles Formation in the Williston Basin. Cyclicity without clear evidence of evaporites characterizes the Madison section in most of Wyoming, where the rocks are predominantly dolomite (Fig. 10).

Madison deposition was followed by widespread emergence and exposure of the Rocky Mountain shelf and development of a regional karst surface. Following this event, clastic deposition dominated sedimentary processes to a large degree. The Upper Mississippian (Chesterian) beds are mainly clastics of the Big Snowy Group, which is restricted in distribution to the trough and depression areas of the Williston Basin, Central Montana Trough, and southwestern Montana (Fig. 10).

Thicknesses of Mississippian beds range from more than 1,500 m in the Sublett Basin and more than 600 m in western Montana, the Big Snowy Trough, and the Williston Basin, to less than 150 m in southeastern Wyoming, south-central South Dakota, and eastern North Dakota (Fig. 10). Mississippian rocks pinch out beneath Pennsylvanian rocks along the flank of the Transcontinental Arch and in the area of the ancestral Rocky Mountains. The depocenter of the Williston Basin is connected to the Sublett Basin depocenter of the Cordilleran miogeocline through the Big Snowy Trough in central Montana and the Snowcrest Trough in southwestern Montana and east-central Idaho. Substantial thinning of Mississippian rocks is evident on the Alberta shelf (chiefly by pre–Middle Jurassic erosion), the Wyoming shelf (and subsidiary elements, the Beartooth and Yellowstone highs), the Lemhi Arch, and the Bannock high, demonstrating probably continuing paleotectonic effect of these elements during Mississippian time.

Regional studies of the Mississippian include those by Sando (1967, 1976b), Craig (1972), Rose (1976), Skipp and others (1979), Smith and Gilmour (1979), Lageson and others (1979), Gutschick and others (1980), and Sandberg and others (1982).

Madison Group

In western Montana, the paleogeographic setting characteristic of the Devonian persisted into the Mississippian, and cyclic carbonate deposition predominated through Early and into Late (Meramecian) Mississippian time. These carbonate beds are not nearly as highly dolomitized as those of the Devonian, and evaporite deposits are not as widespread in the shelf area. Partly for these reasons, cycles in Mississippian rocks are more difficult to recognize. A thick crinoidal-bioclastic bank facies is present in approximately the same position as the thick Devonian stromatoporoid-coral bank facies (Figs. 9, 10). Evaporite beds east of the bank facies, however, are generally restricted to the Central Montana Trough where a prominent anhydrite or gypsum unit ("mid-Madison anhydrite") is widely recognized in the subsurface and is represented by a prominent breccia interval in outcrop sections ("lower solution zone" of Sando, 1972, 1976a). A thick trend of dolomite and anhydrite facies is generally prevalent in the area of the Central Montana Trough, which indicates continuing relatively greater subsidence in this area. Thinning of the Madison section occurs on the Alberta shelf (Figs. 2, 10), which is characterized by crinoidal bank limestone facies. Likewise, thinning occurs in the Beartooth shelf area, south of the evaporitic trough, where the carbonate beds are dominated by oolitic bank limestone facies.

The thick Madison carbonate bank facies in western Montana contains much less dolomite than the bank facies of the Devonian, which is almost entirely dolomite. The reasons for this may be related partly to generally diminished evaporitic conditions on the shelf during Madison deposition, thus decreasing the potential for development of westward-gradient seepage reflux systems. An additional factor is the probability that the Madison crinoid-oolite bank sediments contained considerably less original porosity than did the stromotoporoid-coral reefoid-bank Devonian facies.

The carbonate depositional cycles in the Madison reflect relatively stable tectonic conditions in the continental interior at this time. The absence of medium or coarse clastics and the widespread nature of individual stratigraphic units suggests that eustatic sea-level changes, perhaps coupled with gentle epeirogenic subsidence of the basin region, were the main causes of cyclic deposition.

Widespread emergence that affected much of the continental interior occurred at the close of Madison deposition. Erosion and solution were prevalent, resulting in common solution-breccia beds and karst features at the upper Madison surface (Sando, 1967, 1974, 1976b). In the Williston Basin and Central Montana Trough, the Madison is unconformably overlain by the Big Snowy Group of Chesterian age.

Big Snowy Group

Upper Mississippian and Pennsylvanian strata, which include the Big Snowy Group, and the Amsden, Quadrant, and Tensleep Formations, are dominated by clastic sediments and represent a marked change in regional depositional and tectonic conditions of the western North American continental shelf at this time. These beds contrast sharply with the underlying stable-shelf carbonate sediments that dominated the Rocky Mountain shelf and the adjacent border of the Cordilleran miogeosyncline during Ordovician through middle (Meramecian) Mississippian time. Tectonic activity in the middle and southern Rocky Mountain areas, beginning in Late Mississippian time, resulted in active growth of several prominent local uplifts and accompanying basins (Ancestral Rocky Mountains) and the rise of many new clastic source terranes. At the same time, the supply of clastic material from the Canadian Shield and Transcontinental Arch increased and spread across the northern Rocky Mountain shelf. In Montana, the regional paleogeographic and paleotectonic patterns appear to have remained essentially the same, but carbonate deposition was greatly diminished because of the increased supply of clastic material into the shallow-water marine environment. Most of the organic carbonate bank growth, characteristic of the earlier Paleozoic, was eliminated on the shelf or craton edge at this time. Farther west, however, at greater distances from clastic source areas, a prominent carbonate bank facies is present along the western border of the shelf and the eastern flank of the Sublett Basin (Muldoon trough) in east-central Idaho (Fig. 10).

The Big Snowy Group, consisting in ascending order of the Kibbey, Otter, and Heath Formations, is more than 300 m thick in the Central Montana Trough, and more than 185 m thick in the central part of the Williston Basin in eastern Montana and western North Dakota. In north-central and south-central Montana, northern Wyoming, and around the periphery of the Williston Basin, these rocks were eroded during latest Mississippian, Early Pennsylvanian, and early Mesozoic time.

The Kibbey Formation consists of 60 to 100 m of red shale, siltstone, and sandstone, primarily of near-shore marine origin. Sandstone content is highest in the northern and eastern parts of the Williston Basin, indicating that clastic sources lay in those directions. A widespread, relatively thin carbonate marker unit, "Kibbey lime" (informal subsurface usage), is present in the approximate middle of the formation throughout most of the Williston Basin and Central Montana Trough.

The middle unit of the Big Snowy Group, the Otter Formation, consists of 60 to 100 m of marine, near-shore, and tidal-flat beds of green shale and minor finely crystalline, commonly stromatolitic, dense limestone or dolomite.

The overlying Heath Formation is composed of 60 to 125 m of dark gray to black, highly organic marine shale, limestone, and minor siltstone or fine-grained sandstone and gypsum. These beds were deposited primarily under conditions of restricted marine circulation. The unit is present in the interior of the Central Montana Trough and in the western part of the Williston Basin, but it was removed locally by Early Pennsylvanian erosion in many places, especially in North Dakota.

PENNSYLVANIAN

Pennsylvanian rocks disconformably overlie Mississippian rocks in most of the northern Rocky Mountain region but are probably conformable with Mississippian rocks in the Sublett Basin. The unconformity may be relatively minor in the area of the Central Montana (Big Snowy) Trough and in the central part of the Williston Basin, where the Mississippian and Pennsylvanian boundary is difficult to pick.

In central and northern Montana and in the eastern part of the Williston Basin, all or part of the Pennsylvanian System was removed by pre–Middle Jurassic erosion (Fig. 11). Thickness of the Pennsylvanian System ranges from more than 1,500 m in the Sublett basin to less than 180 m on the Rocky Mountain shelf and in the central Williston Basin. Marked thickening of these beds occurs in the Snowcrest Trough, and evidence of thickening is also present along the axis of the Big Snowy Trough, although much of the section here was eroded before Middle Jurassic deposition. Thinning occurs in the vicinity of several tectonic elements, including the Lemhi Arch, Bannock high, Yellowstone high, Beartooth high, and the Front Range and Laramie Uplifts (Figs 2, 11).

During late Paleozoic (Pennsylvanian and Permian) time, extensive clastic influx from the eastern craton occurred The shelf carbonate depositional environment, which had dominated the entire northern Rocky Mountain shelf from Late Ordovician through Mississippian time, retreated westward to the shelf margin region extending across western Utah, southeastern Idaho, and western Wyoming. On most of the Rocky Mountain shelf in Montana, Wyoming, and central to eastern Utah, Pennsylvanian sedimentary facies are dominated by extensive clean quartzose sandstone deposits in the Tensleep, Quadrant, and Weber Sandstones of nearshore-marine and eolian origin. These beds grade eastward into red-bed, clean sandstone, and evaporite deposits (lower Minnelusa Formation) and westward to mixed clastic and carbonate deposits (Oquirrh, Wells, Morgan, Amsden Formations) in southeast Idaho.

Lower Pennsylvanian (lower Morrowan) beds include marine, deltaic and fluviolacustrine quartzose sandstone, shale, and minor limestone of the Tyler Formation and equivalent rocks, which are primarily restricted in distribution to the Central Montana Trough and the central part of the Williston Basin (Maughan, 1984). The Tyler Formation grades southward into red shale, siltstone, and sandstone of the lower part of the Amsden Formation in south-central Montana and north-central Wyoming and into the lower part of the Minnelusa Formation in southeastern North Dakota and northeastern Wyoming. Evidence of unconformity and erosion is well documented at the base of the Tyler Formation, which cuts deeply into the underlying Upper Mississippian (Chesterian) Otter and Heath Formations in many places. This unconformable relation has been interpreted as representing folding of underlying Mississippian strata at the close of Mississippian time (Fanshawe, 1978), and as stream channeling by others (e.g., Kranzler, 1966). According to Grenda (1978), the faunas and lithofacies of the Tyler Formation and equivalent rocks in western North Dakota suggest a deltaic environment of deposition. Fanshawe (1978) suggests a similar interpretation for the Tyler strata in the Central Montana Trough. The principal source of terrigenous material was the Canadian shield and the Transcontinental Arch (Maughan, 1984).

In the Central Montana Trough, carbonate and dark shale beds of the Lower and Middle Pennsylvanian Alaska Bench Limestone and the Middle Pennsylvanian Devils Pocket Formation of the Amsden Group overlie the Tyler Formation (Maughan, 1975, 1984). To the south, the carbonate beds intermix with red shale beds in the lower to middle part of the Minnelusa Formation and in the Ranchester Limestone Member of the Amsden Formation, and perhaps in the lower part of the Tensleep Sandstone.

The upper part of the Middle Pennsylvanian section is represented mainly by the Tensleep and Quadrant Sandstones in southern Montana and north-central Wyoming, and most of the middle part of the Minnelusa Formation in the Williston Basin, northeastern Wyoming, and the Black Hills. Middle and Upper Pennsylvanian rocks are absent in most of central Montana because pre-Jurassic erosion has increasingly removed them northward, reaching a zero line approximately near the axis of the Central Montana Trough (Fig. 11).

In southwestern Montana, the Central Montana Trough

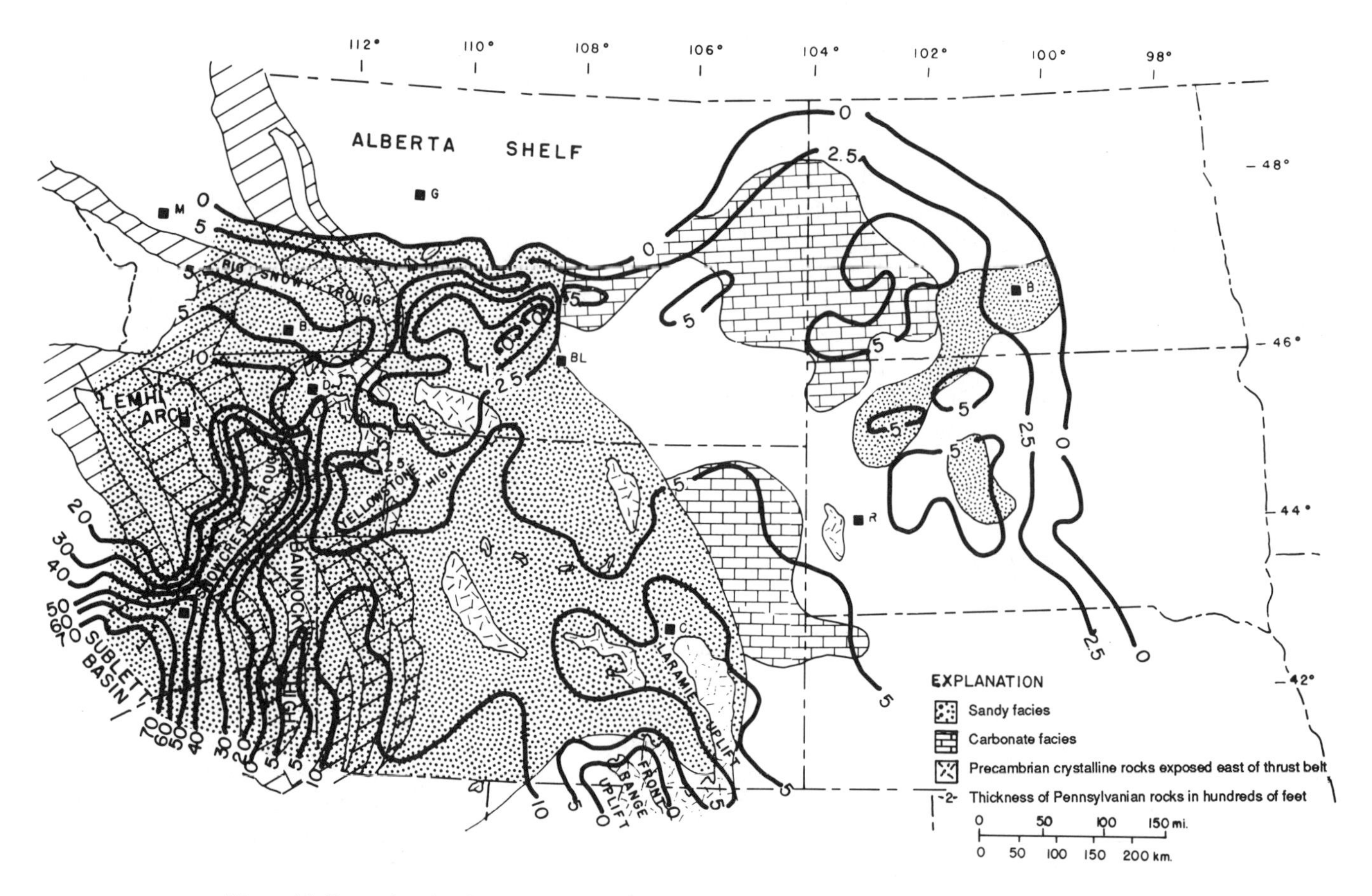

Figure 11. Pennsylvanian System. Thickness (in hundreds of feet) and general facies. Partly restored; palinspastically reconstructed in thrust belt. Unpatterned areas are mainly red beds and some carbonates. (From Peterson, 1985.)

merges westward with the northern extension of the Snowcrest Trough, which contains a substantially thickened section of Upper Mississippian and Pennsylvanian beds and tends to confine the southern extension of the Big Snowy Group (Figs. 2, 10, 11). A markedly thinner sequence is present on the Beartooth shelf, and the Upper Mississippian and Pennsylvanian section is absent on the Alberta shelf. Thinning in both these areas is partly caused by pre–Middle Jurassic erosion, but they probably were also mildly elevated submarine shelf areas at this time. Restored thickness patterns show probable thinning in the vicinity of the Boulder high and the general area of the Lemhi Arch (Figs. 1, 11).

A steadily increasing supply of quartzose sand from the northern source terrane is demonstrated by the almost complete dominance of clean shallow-water marine and eolian sandstone beds in the upper part of the Upper Mississippian–Pennsylvanian siliciclastic sequence (Quadrant Formation). This facies intertongues southwestward with dolomite and limestone beds of the western carbonate bank facies along the east flank of the Sublett Basin (Muldoon Trough of Rose, 1976) and adjacent shelf margin. Much of the thick quartzose sand section is quartzitic or dolomitic, but substantial parts of it are composed of clean, well-sorted, and porous sandstone units in parts of southwestern Montana.

The stratigraphy of Pennsylvanian rocks in the northern Rocky Mountain region and adjacent areas has been studied in detail by several authors, including Agatston (1954), Bates (1955), Foster (1958), Willis (1959), Carlson and Anderson (1965), Ziebarth (1964), Tenney (1966), Maughan and Roberts (1967), Harris (1972), Mallory (1972), Maughan (1975, 1984), Land (1976, 1979), Grenda (1978), Fanshawe (1978), and Smith and Gilmour (1979).

PERMIAN

Sedimentary rocks of Permian age disconformably overlie Middle to Upper Pennsylvanian rocks in most of the northern Rocky Mountain region. These beds comprise a complexly intertonguing sequence, including red beds, evaporites, shelf carbonates, phosphatic shale and phosphorite, bedded chart, and sandstone (Fig. 12). Red beds and evaporites (Permian and Lower Triassic Goose Egg Formation) dominate in the eastern part of

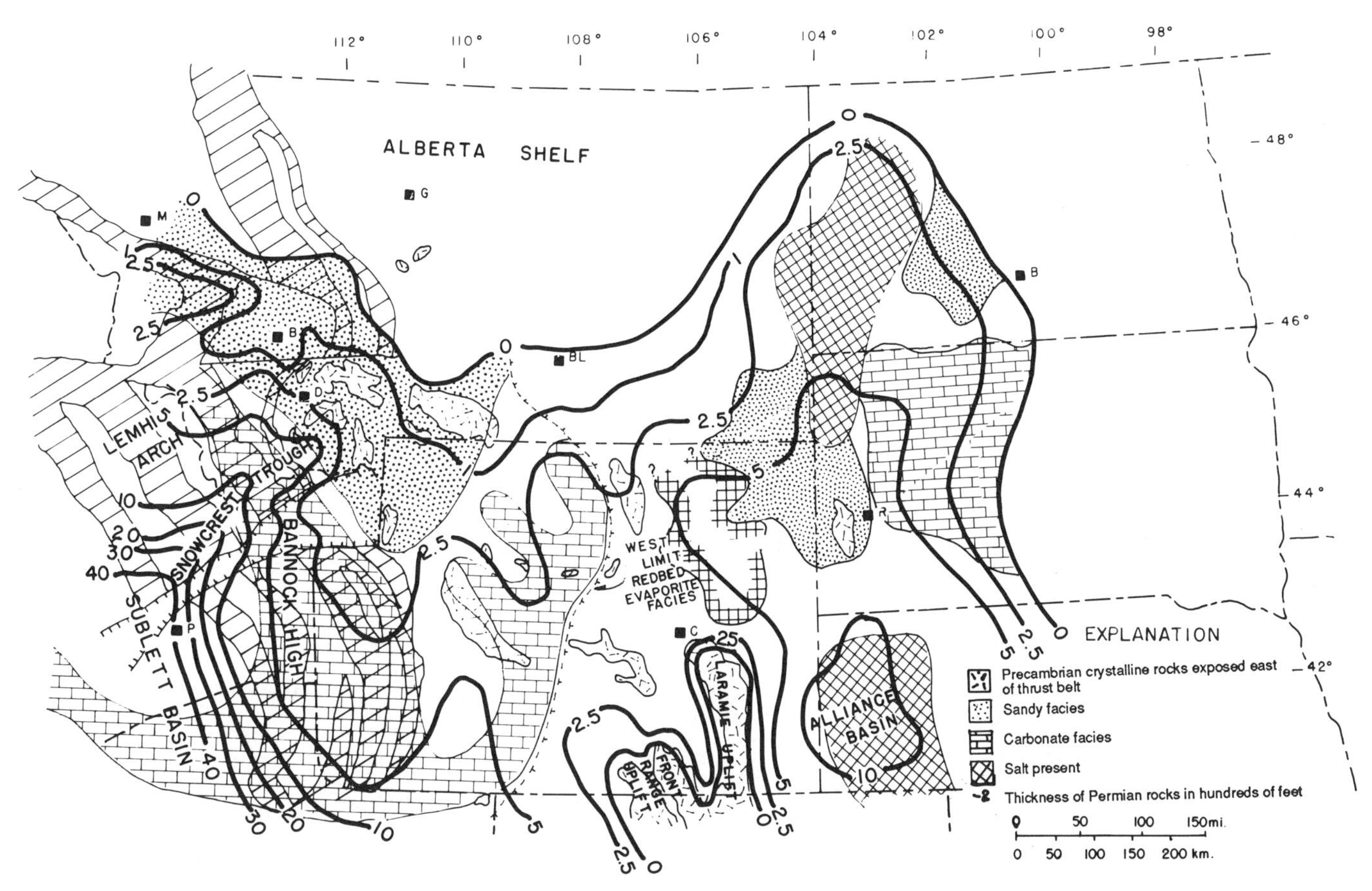

Figure 12. Permian System. Thickness (in hundreds of feet) and general facies. Partly restored; palinspastically reconstructed in thrust belt. Unpatterned areas are mainly red beds, evaporites, and some carbonates to the east; dark shale, chert, and carbonate in western Wyoming; and phosphatic shale and chert in southeastern Idaho. (From Peterson, 1985.)

the Permian sedimentary province, shelf carbonates (Lower Permian Park City Formation or Group) and nearshore marine sandstone (Permian Shedhorn Sandstone) are prevalent in central to western Wyoming, southwestern Montana, and southeastern Idaho. Dark shale, chert, and phosphorite (Lower Permian Phosphoria Formation) is dominant in the Sublett Basin and adjacent west flank of the Bannock high.

Thickness of the Permian System is more than 900 m in the Sublett Basin, thinning markedly to less than 150 m on the Wyoming shelf and in west-central and southern Montana, and is more than 150 m thick in the southwestern Williston Basin and Alliance Basin where evaporites, including salt, are present (Fig. 12). The main elements of the ancestral Rockies (Front Range and Laramie Uplifts), which were emergent clastic source areas, extend northward into southeast Wyoming. Permian rocks are absent in central and northern Montana and in eastern North and South Dakota, largely because of pre–Middle Jurassic erosion. There is some evidence of thinning and a probable clastic sediment source in the vicinity of the Lemhi Arch, and influence of the Bannock high is indicated by the absence of the uppermost Permian beds along this feature. Some evidence is present for early growth of many of the present-day uplifts and basins of the northern Rocky Mountains at this time (Peterson, 1984).

Organic-rich rocks of Permian age (Meade Peak Phosphatic Shale and Retort Phosphatic Shale Members of the Phosphoria Formation) are documented as the major source of oil in Paleozoic reservoirs of the Bighorn and Wind River Basins of central Wyoming (Maughan, 1975; Claypool and others, 1978). The phosphatic shale facies of the Meade Peak Member is thickest in the western Thrust Belt region of southeastern Idaho and northern Utah, but thins markedly across the Thrust Belt region and changes eastward to a red-bed facies approximately along the eastern borders of the Bighorn, Wind River, and Green River Basins (Figs. 2, 12). The same facies of the Retort Member is absent or thins in much of the Thrust Belt region, but thickens somewhat to the east in the Green River Basin and to the north in west-central Wyoming.

Regional analyses of Permian rocks in the northern Great

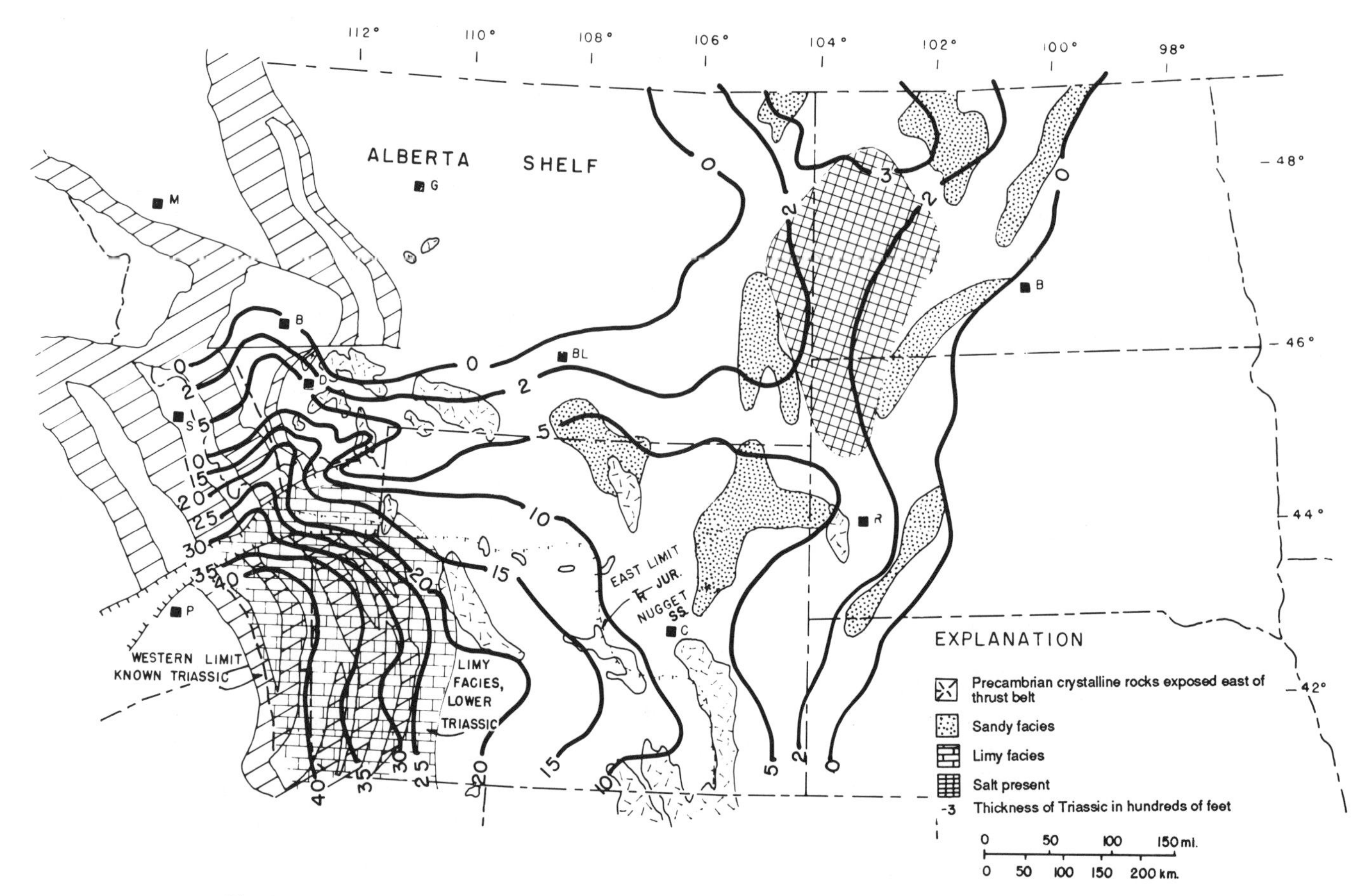

Figure 13. Triassic System. Thickness (in hundreds of feet) and general facies. Partly restored; palinspastically reconstructed in thrust belt. Unpatterned areas are mainly red beds with some evaporites and carbonates. (From Peterson, 1985.)

Plains and northern Rocky Mountains have been published by Foster (1958), McKelvey and others (1959), Sheldon (1963), Cressman and Swanson (1964), Tenney (1966), Maughan (1966, 1967), McKee and others (1967), Yochelson (1968), Rascoe and Baars (1972), and Peterson (1972, 1980a, 1980b, 1984).

TRIASSIC

Rocks of Triassic age are dominated by a red-bed sequence (Chugwater and upper Spearfish Formations and other units) of red shale, siltstone, and sandstone, with lesser amounts of gypsum or anhydrite and carbonate. Halite of probable Triassic age is present at the Permian-Triassic boundary in the Williston Basin (Fig. 13). A Lower Triassic limestone facies (Dinwoody and Thaynes Formations) is present in western Wyoming and southeast Idaho and intertongues eastward with the red-bed facies. In marked contrast with the underlying Permian carbonate beds, marine fossils are not common in the Triassic carbonate facies.

The Triassic System is more than 1,200 m thick in southeast Idaho and thins eastward across the Wyoming shelf to a zero line in northwestern Nebraska, western South Dakota, and central North Dakota (Fig. 13). In central Montana, Triassic beds are truncated beneath the Middle Jurassic unconformity. In southwestern Wyoming, the lower part of the Nugget Sandstone may be of Late Triassic age (Imlay, 1980).

Triassic rocks disconformably overlie Permian rocks over most of the Wyoming shelf, but may be conformable with Permian rocks in the Sublett and Williston Basins. Regional studies of the Triassic include those by Moritz (1951), Kummel (1954), McKee and others (1959), Carlson (1968), and MacLachlan (1972).

JURASSIC

After Triassic time, marine waters withdrew from the Rocky Mountain shelf, resulting in regional erosion of the underlying beds until the Middle Jurassic transgression (Zuni sequence of Sloss, 1963), which covered most of the Rocky Mountains and adjacent areas (Imlay, 1957, 1980). Lower Jurassic strata are absent in most of the northern Rocky Mountain region, except in

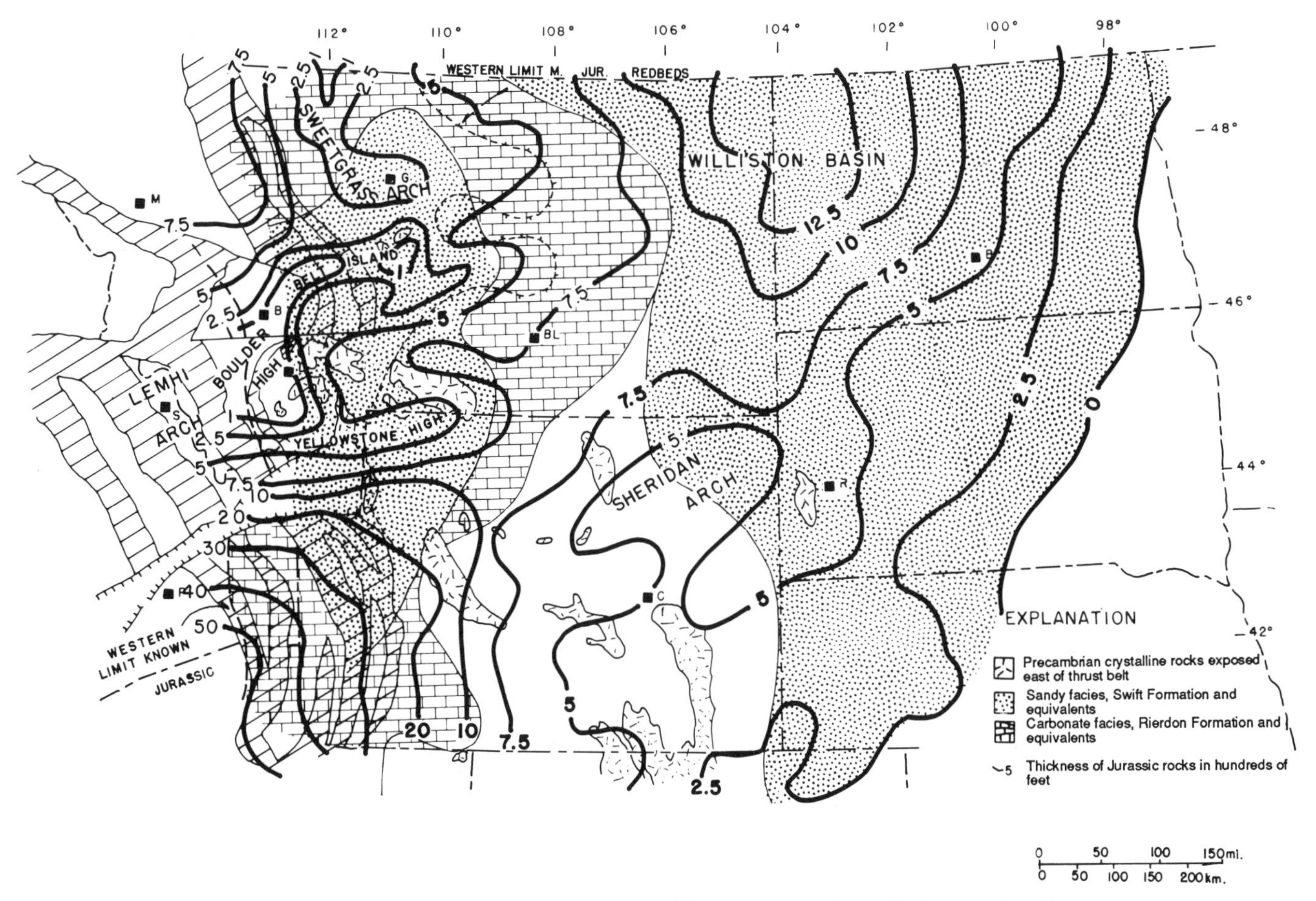

Figure 14. Jurassic System. Thickness (in hundreds of feet) and general facies. Partly restored; palinspastically reconstructed in thrust belt. Unpatterned area is mainly marine shale, siltstone, and sandstone. (From Peterson, 1985.)

southwest Wyoming and southeast Idaho where the upper part of the Nugget Sandstone is of Early Jurassic(?) age, and the northern Williston basin where the upper part of the red-bed and evaporite section commonly assigned to the Triassic may be of Early Jurassic age.

The Jurassic System, sometimes referred to as the "great transgressor" in the Rocky Mountain region, is represented by four main transgressive-regressive marine cycles: (1) Gypsum Spring–Nesson cycle of early Middle Jurassic age, (2) the Sawtooth-lower Sundance–lower Twin Creek cycle of middle Middle Jurassic age, (3) the Rierdon–upper Twin Creek–Preuss–middle Sundance cycle of late Middle Jurassic age, and (4) the Swift-Stump–upper Sundance cycle of early Late Jurassic (Oxfordian) age. The cycles are separated by local to regional unconformities, and each tends to be more widespread than the previous. Latest Jurassic time (Kimmeridgian and Tithonian) is represented by the continental varicolored shale and sandstone beds of the dinosaur-bearing Morrison Formation, which overlies marine Oxfordian rocks in all of the northern Rocky Mountain region.

Jurassic facies are dominated by marine shale, limestone, and sandstone in the west and intertonguing marine sandstone, shale, and red beds to the east. The western carbonate facies (Middle Jurassic Twin Creek Limestone) is more than 800 m thick in southeast Idaho and thins markedly eastward on the Wyoming shelf. In central and northern Montana and eastern North and South Dakota, Middle Jurassic rocks are disconformable on rocks ranging in age from Mississippian to Permian. Elsewhere, the Jurassic rests on Triassic rocks, disconformably in much of the area.

Jurassic paleogeography in central and western Montana (Fig. 14) was dominated by the Belt Island complex (not to be confused with "Belt Island" of late Precambrian age) of gentle uplifts where variable parts of the Ellis Group (Sawtooth, Piper, Rierdon, and Swift Formations) are thinned or absent because of erosion or nondeposition, which occurred during all or parts of the three main depositional cycles. Vestiges of the Paleozoic and early Mesozoic paleostructural features are present during Jurassic time but are much subdued. These include the Big Snowy Trough and possibly the Central Montana and Snowcrest Troughs, all of which show minor thickening of Ellis deposits (Fig. 2, 14). Rejuvenated growth of the Sweetgrass Arch is reflected in thinning and absence of some Ellis units coinciding with the trend of the arch. A major clastic source area appears in southwestern Montana and adjacent Idaho at this time, approximately in the position of the Paleozoic Lemhi Arch (see Fig. 2, 17). This event probably reflects the beginning of tectonic activity that ultimately resulted in development of the western North American thrust and fold belt. The Boulder high, which at this time may represent the initiation of the igneous-tectonic event that culminated with emplacement of the Boulder batholith, retained its expression as part of an eastward projection off the Lemhi Arch and forms a part of the Belt Island trend.

Marine Jurassic rocks in western Montana are dominated by fine to medium clastic sediments, primarily originating from the western source area. Significant carbonate deposits are present in the upper part of the Middle Jurassic sequence (Rierdon Formation), which are the northern extension of the thick carbonate facies of the Twin Creek Limestone of southeastern Idaho. These beds represent the final deposition of important marine carbonate sediments on the Rocky Mountain shelf. Marine clastic deposits of the Swift Formation grade into the basal beds of the overlying nonmarine Morrison Formation without evidence of unconformity.

Detailed analyses of the marine Jurassic of Montana include those by Cobban (1945), Imlay (1945, 1948, 1952, 1957, 1980), Moritz (1951, 1960), Nordquist (1955), McKee and others (1956), and Peterson (1957, 1972).

UPPER JURASSIC AND LOWER CRETACEOUS CONTINENTAL BEDS

The Upper Jurassic Morrison Formation and the Lower Cretaceous Kootenai and Cloverly Formations, and their equivalents are 150 m to more than 450 m thick in western Montana and central to western Wyoming. These units represent an interruption of marine deposition in the western interior that affected the entire Rocky Mountain shelf.

During this time, the shelf was covered with a blanket of varicolored muds, sands, gravels, silts, and lacustrine continental deposits. Much of the fine fraction of this sedimentary blanket may have originated from fallout of fine volcanic debris generated in the western North American volcanic field where important batholith emplacement occurred at this time (Stokes, 1950; Peterson, 1966, 1972; Moberly, 1960; Suttner, 1969; Suttner and others, 1981).

Prominent fluvial channel sand and gravel deposits are present in several parts of the Morrison, Kootenai, and Cloverly sections in all areas of the Rocky Mountains, and a widespread blanket deposit of boulder, gravel, and coarse sand commonly is present at the base of the Kootenai and its equivalents almost everywhere on the Rocky Mountain shelf. A western source for this material is well documented, and the environmental aspects of its origin have been the subject of some discussion (Stokes, 1950; Moberly, 1960; Peterson, 1966, 1972; Suttner, 1969; Suttner and others, 1981; Walker, 1974).

Lacustrine limestone beds are frequently found interbedded with Morrison clastic beds, and evidence has been presented for relatively widespread, probable lacustrine dark shale and coaly deposits in the uppermost Morrison, generally covering the area of the Central Montana Trough (Peterson, 1966, 1972; Walker, 1974). Relatively thick, widespread lacustrine limestone beds are also present in the lower and upper parts of the Kootenai Formation and Cloverly Formation and equivalent beds in western Montana, southeastern Idaho, western Wyoming, and central Utah (Fig. 15; Holm and others, 1977). These deposits probably represent widespread north-south–oriented lake systems that formed in the foredeep area of the western tectonic belt, which was progressively increasing in activity by this time.

At times of lower clastic influx or higher precipitation, the subsiding foredeep filled with nonmarine waters that persisted long enough to allow deposition of thick ostracod, charaphyte, and gastropod-bearing limestone units. These beds grade upward into the continental and nearshore marine sediments of the initial Cretaceous marine transgression.

Detailed studies of Morrison and Kootenai stratigraphy and facies distribution include those by Moberly (1960), Suttner (1969), McGookey and others (1972), Walker (1974), James and Oaks (1977), and Holm and others (1977).

CRETACEOUS

The post-Kootenai Cretaceous rocks were deposited in a sequence of at least five main transgressive-regressive cycles (Figs. 15, 16), which were defined by McGookey and others (1972) in ascending order as follows:

1. Skull Creek cycle (Late Aptian to Early Albian)
2. Greenhorn cycle (Late Albian, Cenomanian and Turonian)
3. Niobrara cycle (Coniacian and Santonian)
4. Claggett cycle (Early Campanian)
5. Bearpaw cycle (Late Campanian and Early Maastrichtian)

The rocks between the Late Albian Mowry Shale (Km) and Telegraph Creek Formation (Ktc) constitute the Colorado Group of the Western Interior reference sequence of Cobban and Reeside (1952). The remaining Cretaceous formations, represent the Montana Group.

The Colorado Group comprises a sequence of well-defined fossiliferous marine transgressive-regressive cyclic clastic deposits. These beds grade westward into a complex intertonguing marine and nonmarine and finally a nonmarine clastic, nonfossiliferous, partly volcanic facies. The origin and distribution of these clastic facies and the chalk deposits have been described by McMannis

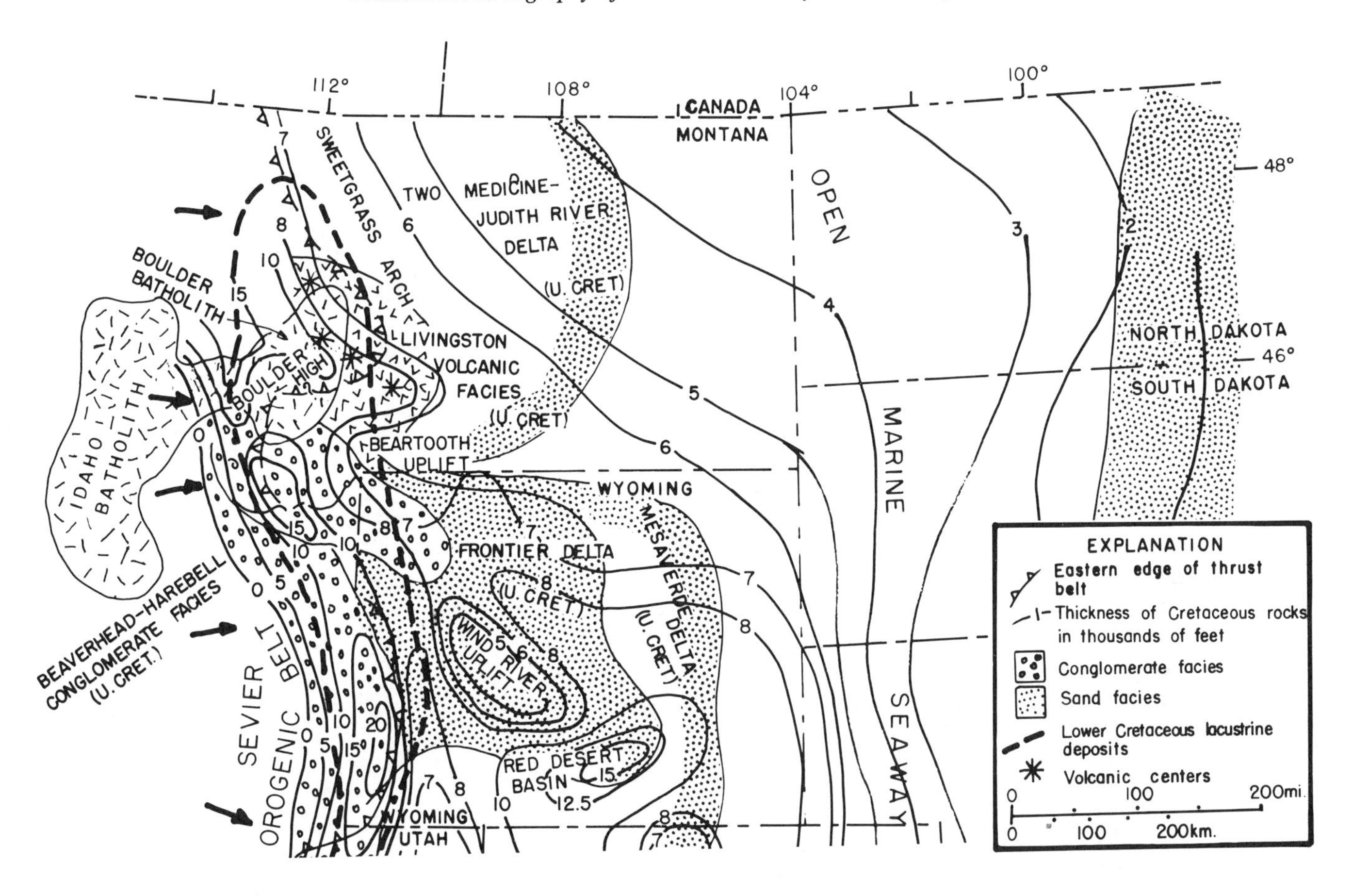

Figure 15. Cretaceous System. Thickness (in thousands of feet) and general facies, partly restored. (Modified from Peterson and Smith, 1986.)

(1965), McGookey and others (1972), Rice and Shurr (1978, 1983), Shurr (1984b), and Shurr and Reiskind (1984).

The main belt of marine-nonmarine transition occurs roughly in the vicinity of the central and northern parts of the western Montana disturbed belt (Figs. 15 to 17) where the eastern open-marine section grades westward into the upper Albian Blackleaf (Kb) and the Cenomanian to Santonian Coberly, Jens, and Carter Creek Formations and their equivalents. Volcanic debris is common in both these units, which as pointed out by McMannis (1965), may be related to the early stages of emplacement of the Idaho batholith and associated volcanic activity. In southwestern Montana, the marine units equivalent to the Colorado Group extend somewhat farther west than in the west-central and northern areas. In Wyoming, the marine-nonmarine transition occurs near the western overthrust belt where units of the marine Cody Shale (Kc) grade westward into the nearshore Frontier and Mesa Verde fluvial-deltaic sandstone facies (Fig. 15).

Major tectonic factors involved in these sedimentary distribution patterns are the progresive growth and eastward spreading of the western thrust and fold belt, and in Montana the emplacement of the Late Cretaceous Idaho and Boulder batholith systems and their satellite intrusives. The influence of lineaments and other paleotectonic elements—such as the Central Montana Uplift, Black Hills, Cedar Creek Anticline, and Transcontinental Arch—on the distribution of coastal and shelf sandstones, offshore shales, and chalks has been demonstrated by Rice and Shurr (1978, 1983) and Shurr (1984a, 1984b).

Thickness of the Colorado Group and equivalents increases westward from less than 500 m east of the overthrust belt in central Montana and Wyoming to more than 2,000 m in western Montana, Wyoming, and southeastern Idaho. The axis of maximum deposition was near the Wyoming-Idaho boundary (Fig. 15). Part of the section in Montana is removed by Cenozoic erosion along a broad belt roughly corresponding with the trend of the Sweetgrass Arch and Little Belt Mountains. This thin trend extends southwestward for some distance across the approximate position of the Boulder high, indicating continued growth of this paleostructural feature. The section is entirely removed by erosion or nondeposition in northwestern Montana and central

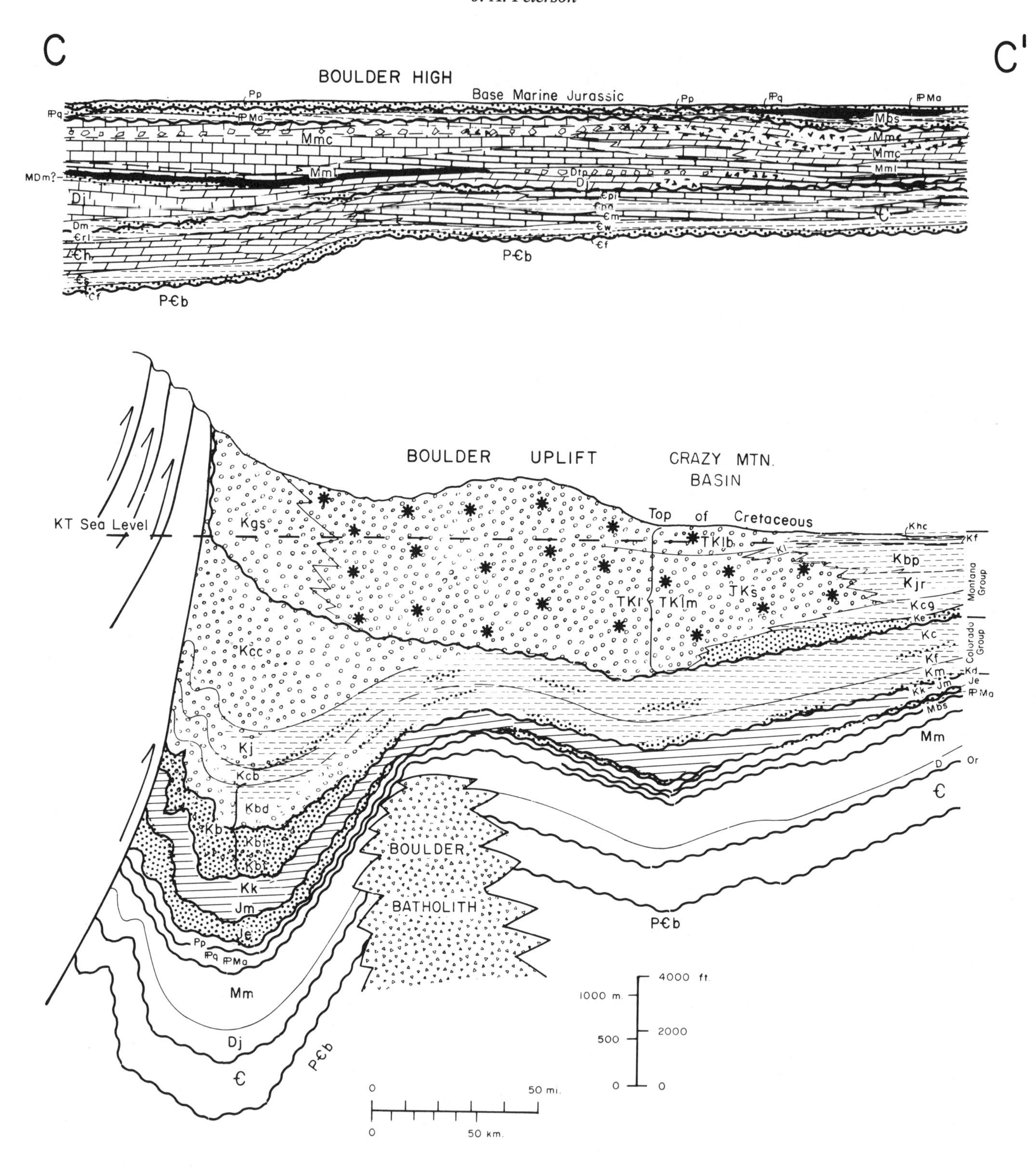

Figure 16. West-east lithofacies cross section C–C′, west-central to east-central Montana. Datum for upper section is base of marine Jurassic; datum for lower section is estimated sea level at close of Cretaceous time. Section is palinspastically restored in thrust belt. Line of cross section shown on Figure 2. Lithologic explanation shown on Figure 17 and formation symbols explained in Table 1. (From Peterson, 1985.)

TABLE 1. FORMATION ABBREVIATIONS USED ON FIGURES 16 AND 17

Abbreviation	Formation
Cretaceous-Tertiary	
TKl	Livingston Group
KTbh	Beaverhead Formation
TKlb	Billman Creek Formation
TKlm	Maudlow Formation
Cretaceous	
Kgs	Golden Spike Formation
Kcc	Carter Creek Formation
Kj	Jens Formation
Kcb	Coberly Formation
Kbd	Dunkelberg Formation
Kfx	Fox Hills Sandstone
Khc	Hell Creek Formation
Kjr	Judith River Formation
Kcg	Claggett Formation
Kt	Thermopolis Shale
Kd	Dakota Formation
Ktc	Telegraph Creek Formation
Kf	Frontier Formation
Kc	Cody Shale
Kmo	Mowry Shale
Kbp	Bearpaw Shale
Ke	Eagle Sandstone
Kb	Blackleaf Formation
Kbt	Taft Hill Member
Kbf	Flood Member
Kk	Kootenai Formation
Jurassic	
Jm	Morrison Formation
Je	Ellis Group
Triassic	
Trt	Thaynes Formation
Trw	Woodside Formation
Trd	Dinwoody Formation
Permian	
Pp	Phosphoria Formation
Pennsylvanian	
IPq	Quadrant Formation
IPMa	Amsden Formation
Mississippian	
Mbs	Big Snowy Group
Mcb	Copper Basin Formation
Mmc	Mission Canyon Formation
Mml	Lodgepole Formation
Msr	Surrett Canyon Formation
Msc	South Creek Formation
Msp	Scott Peak Formation
Mmcy	Middle Canyon Formation
MDm	Milligen Canyon Formation
Devonian	
Dt	Three Forks Formation
Dtp	Potlatch Member
Dj	Jefferson Formation
Dm	Maywood Formation
Silurian	
Sf	Fish Haven Formation
Sl	Laketown Formation
Ordovician	
Ob	Big Horn Formation
Ok	Kinnikinnic Formation
Os	Summerhouse Formation
Cambrian	
Ꞓpa	Park Shale
Ꞓm	Meagher Limestone
Ꞓw	Wolsey Shale
Ꞓf	Flathead Sandstone
PꞒb	Belt Supergroup (Proterozoic)

Idaho, which in large part became a major clastic source area during this time. A general thickening trend extends eastward from the central overthrust belt approximately in the position of the Central Montana Trough.

Continued tectonic activity in westernmost Montana and Idaho expanded the size and elevation of the western source terrane, increasing the influx of coarse clastic material by high gradient streams off the western highland (Fig. 15). Some foredeep lacustrine deposits are also reported in part of the Colorado Group equivalent section in west-central Montana and western Wyoming. Volcanic debris is common in parts of the sequence (Figs. 15 to 17). According to Armstrong and Oriel (1965), approximately two-thirds of the Upper Cretaceous section in the western foredeep was deposited in Colorado time, whereas in the eastern basins most of these rocks were deposited in Montana time.

During deposition of the overlying Montana Group (and equivalents), coarse boulder conglomerate debris, derived primarily from the erosion of Precambrian and Paleozoic rocks, was transported eastward from the rising highland belt associated with the rapidly expanding western thrust and fold belt (Figs. 15 to 17). Volcanic debris and flows are concentrated around the periphery of the Boulder high in southwestern Montana, which at this time was the main site of emplacement of the Boulder Batholith and associated intrusives. The marine Cretaceous sea transgressed westward into this complex of volcanic and conglomerate debris with frequent rapid fluctuations of the western shoreline. At the same time, much volcanic ash and tuff fallout was incorporated with normal shallow-water marine deposits to the east, resulting in the frequent layers of bentonite and bentonitic shale, which are characteristic of both the Colorado and Montana Groups (and their equivalents) in central and eastern Montana and Wyoming.

Five major facies of thick, coarse-grained nonmarine sediments and volcanics dominate Montana equivalent stratigraphy in western Montana (Fig. 16,17):

1. The Beaverhead Group (Ktbh) is a complexly intertonguing facies, more than 4,500 m thick, of quartzite, limestone-boulder conglomerate, and sandstone on the Idaho border in southwestern Montana; here, according to Ryder and Scholten (1973), these beds are probably as old as middle Colorado (Turonian) and as young as Paleocene or Eocene.

2. The Golden Spike Formation (Kgs) (Gwinn and Mutch, 1965), which occurs to the north of the Beaverhead in southwestern Montana, is at least 2,300 m thick and similar in nature to the conglomerate facies of the Beaverhead. A substantial amount of the Golden Spike Formation probably was derived from the Boulder high (Figs. 14, 17).

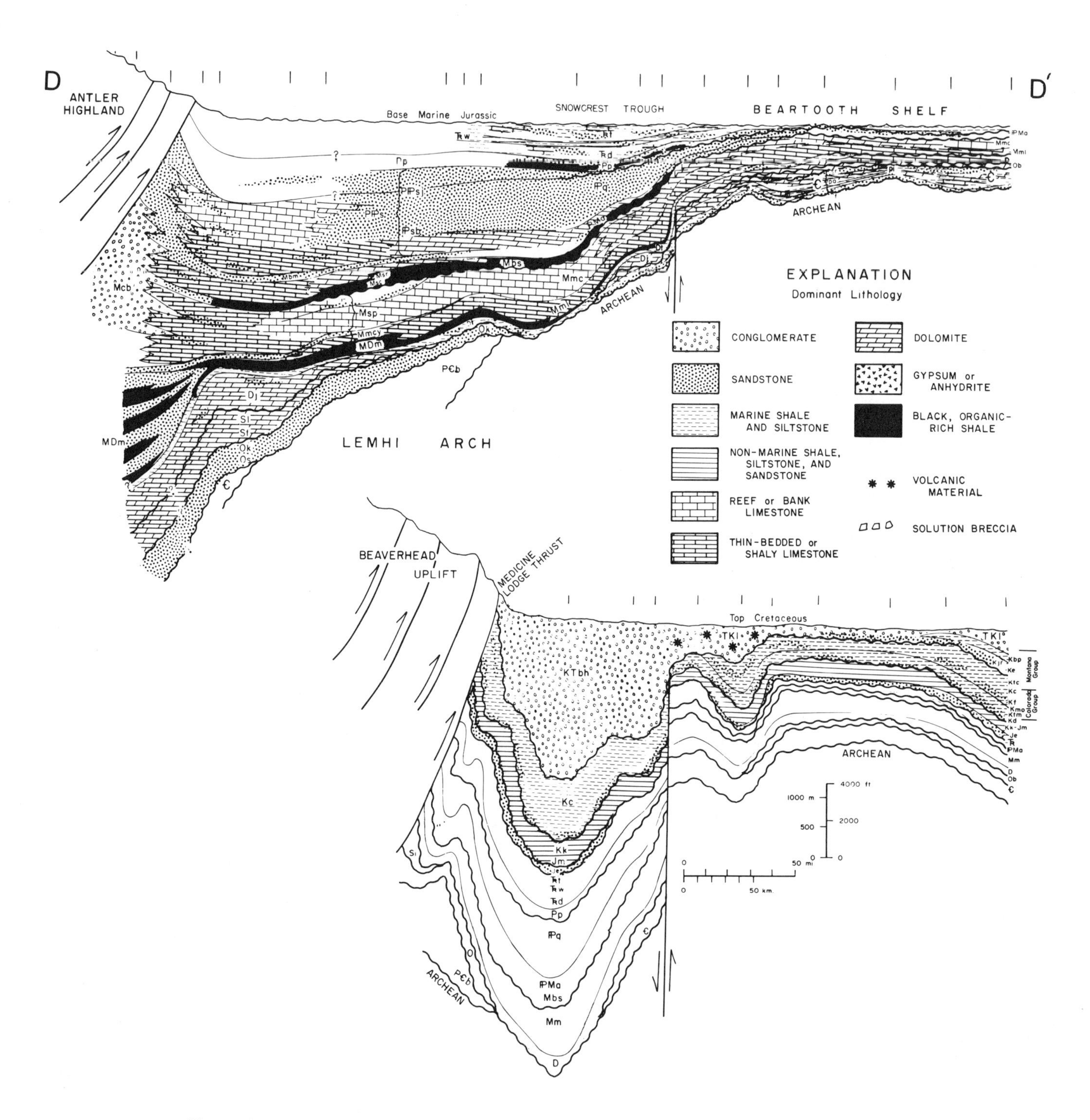

Figure 17. Southwest-northeast lithofacies cross section D–D′, east-central Idaho to central Montana. Datum for upper section is base of marine Jurassic; datum for lower section is estimated sea level at close of Cretaceous time. Section is palinspastically restored in thrust belt. Line of cross section shown on Figure 2. (From Peterson, 1985.)

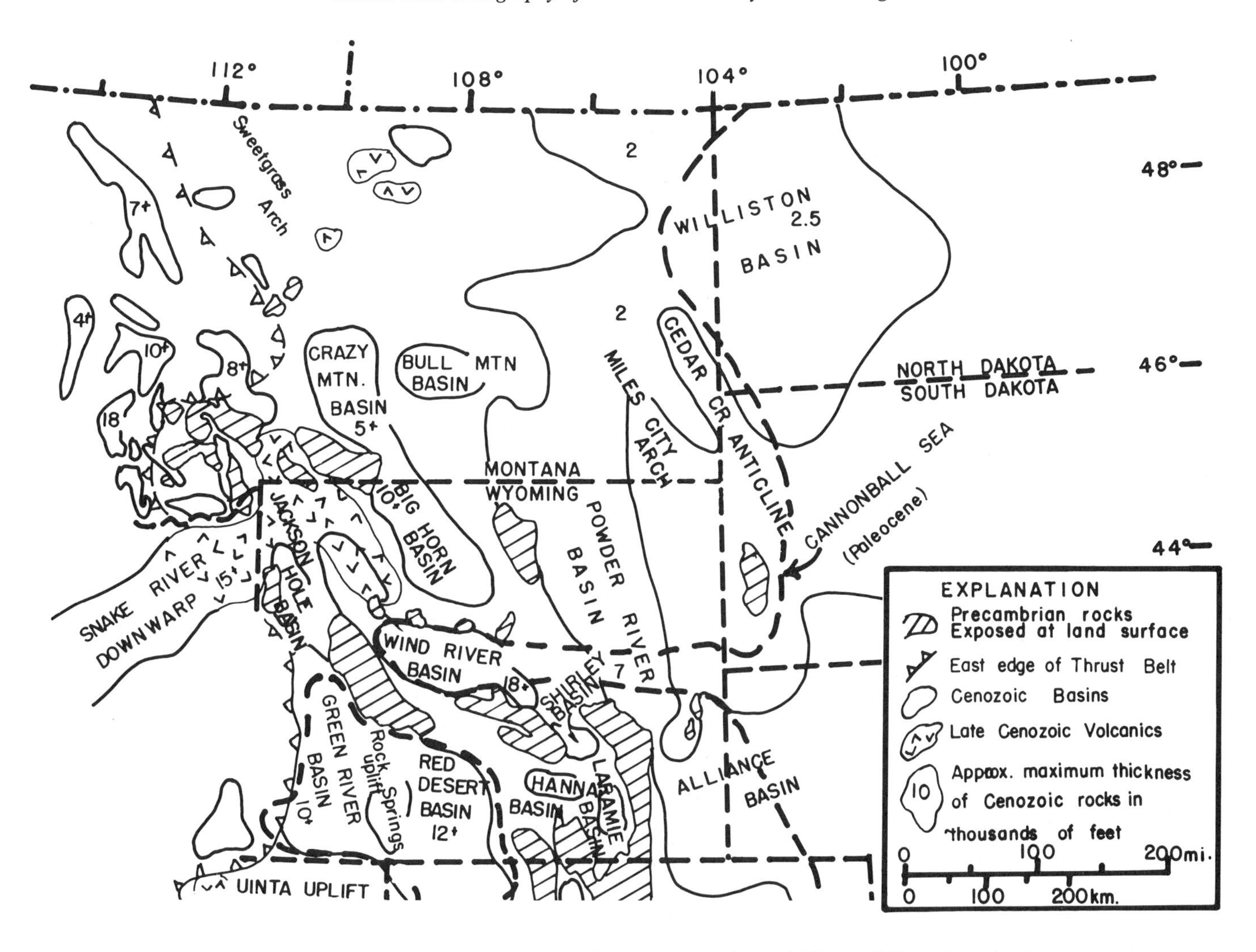

Figure 18. Cenozoic basins. Locations of main basins, approximate thickness of Cenozoic rocks (in thousands of feet), and areas of late Cenozoic volcanism. (Modified from Peterson and Smith, 1986.)

3. The Elkhorn Mountains Volcanics occupy the frontal part of the overthrust belt in the vicinity of Helena and Butte in western Montana and intertongue westward with the "Golden Spike facies". This facies is a complex pile of volcanic flows, tuffs, breccias, and tuffaceous sandstone and mudstone that intertongues and grades eastward into units of the Livingston Group (TKl) (McMannis, 1965; Roberts, 1963; Skipp and McGrew, 1977). Origin of the material is closely related to the emplacement of the Boulder batholith complex (Figs. 15, 17; Smedes, 1966; Robinson and others, 1969).

4. The Livingston Group (TKl), more than 3,000 m thick, occupies the Crazy Mountain Basin and adjacent areas (Fig. 16). The Livingston is a complex of mostly nonmarine volcaniclastic sandstone and mudstone with some flows and tuffs. The lower part intertongues rapidly eastward with marine units of the Montana Group on the east flank of the Crazy Mountain Basin. The upper two-thirds of the Livingston intertongues with nonmarine beds of the Hell Creek (Khc) Formation (Roberts, 1963).

5. The Two Medicine Formation is preserved in outcrop remnants, some of which are at least 600 m thick, in the frontal zone of the northern thrust belt near and south of Glacier National Park. Eastward in the subsurface, the nonmarine Two Medicine thins to 300 m or less and grades into marine units of the Montana Group along the west flank of the Sweetgrass Arch (Schmidt, 1978). To the south it intertongues with the Elkhorn Mountains volcanic facies.

In Wyoming, equivalent rocks are dominated on the east by thick open-marine shale (Pierre Shale; Fig. 15), which grades westward into intertonguing nearshore-marine sandstone, nonmarine sandstone, and coal facies. Along the Wyoming-Idaho-Utah border, the western facies becomes coarsely conglomeratic in the upper part. Much of the Montana Group equivalent section has been removed by Cenozoic erosion in western Montana and southeastern Idaho.

Detailed studies of Montana equivalent facies and correlative rocks in the northern Rocky Mountains include those by Weimer (1960, 1961), Klepper and others (1957), Roberts (1963, 1965), Armstrong and Oriel (1965), Gwinn (1965), Gwinn and Mutch (1965), Viele and Harris (1965), Ryder and Ames (1970), McGookey and others (1972), Skipp and McGrew (1977), Ryder and Scholten (1973), Gill and Cobban (1973), Rice and Shurr (1983), Shurr (1984a, 1984b), Shurr and Reiskind (1984).

TERTIARY

Sedimentation patterns of the early Tertiary reflect the continuation without significant interruption of the Late Cretaceous paleogeographic framework and depositional basins of the northern Rocky Mountains. The major factor in change of depositional environments was the final withdrawal of Late Cretaceous marine waters from the Rocky Mountain shelf. Marine deposition was replaced by widespread lacustrine and fluvial deposition across the emergent, gently sloping Late Cretaceous sea bottom with extensive lacustrine and coastal swamp deposition of coaly beds and clastics in the more active basinal areas of the shelf. These deposits make up the Hell Creek Formation (Upper Cretaceous) and overlying Fort Union Formation (Paleocene) and their equivalents. In western Montana and western Wyoming, remnants of equivalent age rocks are scarce but are probably represented in the upper part (Maastrichtian) of the Beaverhead Group of southwestern Montana and nonmarine Upper Cretaceous and Paleocene coarse clastics and volcanics south of Glacier National Park.

By late Eocene time, the downfaulted continental Tertiary basins had been relatively well established, and the remainder of Tertiary sedimentation was involved with basin-fill deposition in the many downwarped valley areas of the Rocky Mountains (Fig. 18). Middle to upper Tertiary fluvial and lacustrine beds reach thicknesses of more than 1,000 m in several of the basins. Burfiend (1967) reported gravity data that indicated some of the valleys between Bozeman and Butte are filled with at least 1,500 m of unconsolidated material. According to Kuenzi and Fields (1971), the valley fill in the Jefferson Basin east of Butte contains a lower, relatively fine-grained clastic sequence of late Eocene or Oligocene to early Miocene age and an upper Miocene and Pliocene sequence of coarser material, which are separated by a marked unconformity. Similar stratigraphic sequences are present in other Tertiary valleys, and evidence suggests that the unconformity may be a regional feature common to valley-fill history in many parts of the Rocky Mountains (Robinson, 1961, 1963; Kuenzi and Richard, 1969).

REFERENCES

Agatston, R. S., 1954, Pennsylvanian and lower Permian of northern and eastern Wyoming: American Association of Petroleum Geologists Bulletin, v. 38, p. 508–583.

Armstrong, F. C., and Oriel, S. S., 1965, Tectonic development of Idaho-Wyoming thrust belt: American Association of Petroleum Geologists Bulletin, v. 49, p. 1847–1866.

Bates, R. L., 1955, Permo-Pennsylvanian formations between Laramie Mountains, Wyoming, and the Black Hills, South Dakota: American Association of Petroleum Geologists Bulletin, v. 39, p. 1972–2002.

Benson, A. L., 1966, Devonian stratigraphy of western Wyoming and adjacent areas: American Association of Petroleum Geologists Bulletin, v. 50, p. 2566–2603.

Branson, C. C., 1964, Cyclicity in Oklahoma Paleozoic rocks, *in* Merrian, D. F., ed., Symposium on Cyclic Sedimentation: State Geological Survey of Kansas Bulletin 169, v. I, p. 57–62.

Burfiend, W. J., 1967, A gravity investigation of the Tobacco Root Mountains, Jefferson Basin, Boulder Batholith, and adjacent areas [Ph.D. thesis]: Bloomington, Indiana University, 146 p.

Carlson, C. E., 1960, Stratigraphy of the Winnipeg and Deadwood formations in North Dakota: North Dakota Geological Survey Bulletin 35, 149 p.

—— , 1968, Triassic-Jurassic of Alberta, Saskatchewan, Manitoba, Montana, and North Dakota: American Association of Petroleum Geologists Bulletin, v. 52, p. 1969–1983.

Carlson, C. G., and Anderson, S. B., 1965, Sedimentary and tectonic history of North Dakota part of Williston basin: American Association of Petroleum Geologists Bulletin, v. 49, p. 1833–1846.

Claypool, G. E., Love, A. H., and Maughan, E. K., 1978, Organic geochemistry, incipient metamorphism, and oil generation in black shale members of the Phosphoria Formation, Western Interior United States: American Association of Petroleum Geologists Bulletin, v. 62, p. 98–120.

Cobban, W. A., 1945, Marine Jurassic formations of Sweetgrass arch, Montana: American Association of Petroleum Geologists Bulletin, v. 29, p. 1262–1303.

Cobban, W. A., and Reeside, J. B., Jr., 1952, Correlation of the Cretaceous formations of the Western Interior of the United States: Geological Society of America Bulletin, v. 63, p. 1011–1043.

Craig, L. C., 1972, Mississippian System, *in* Mallory, W. W., ed., Geologic atlas of the Rocky Mountain region: Denver, Colorado, Rocky Mountain Association of Geologists, p. 100–110.

Cressman, E. R., and Swanson, R. W., 1964, Stratigraphy and petrology of the Permian rocks of southwestern Montana: U.S. Geological Survey Professional Paper 313-C, p. 275–569.

Fanshawe, J. R., 1978, Central Montana tectonics and the Tyler Formation, *in* The economic geology of the Williston basin: Montana Geological Society 24th Annual Conference, Williston Basin Symposium, September 1978, p. 239–248.

Foster, D. I., 1958, Summary of the stratigraphy of the Minnelusa Formation, Powder River Basin, Wyoming: Wyoming Geological Association 13th Annual Field Conference Guidebook, p. 39–44.

Foster, N. H., 1972, Orodovician system, *in* Geologic Atlas of the Rocky Mountain region: Denver, Colorado, Rocky Mountain Association of Geologists, p. 76–85.

Fuller, J.G.C.M., 1961, Ordovician and contiguous formations in North Dakota, South Dakota, Montana, and adjoining areas in Canada and the United States: American Association of Petroleum Geologists Bulletin, v. 45, p. 1334–1363.

Gill, J. R., and Cobban, W. A., 1973, Stratigraphy and geologic history of the Montana Group and equivalent rocks, Montana, Wyoming, and North and South Dakota: U.S. Geological Survey Professional Paper 776, 37 p.

Grenda, J. C., 1978, Paleozoology of oil-well cores from the Tyler Formation

(Pennsylvanian) in North Dakota, United States of America, *in* The economic geology of the Williston basin: Montana Geological Society 24th Annual Conference, Williston Basin Symposium, September 1978, p. 249–260.

Gutschick, R. C., Suttner, L. J., and Switek, M. J., 1962, Biostratigraphy of transitional Devonian-Mississippian Sappington Formation of southwest Montana, *in* Three Forks–Belt Mountains area, and symposium—The Devonian system of Montana and adjacent areas: Billings Geological Society 13th Annual Field Conference, p. 79–89.

Gutschick, R. C., Sandberg, C. A., and Sando, W. J., 1980, Mississippian shelf margin and carbonate platform from Montana to Nevada, *in* Fouch, T. D., and Magathan, E. R., eds., Paleozoic paleogeography of the west-central United States: Denver, Colorado, Society of Economic Paleontologists and Mineralogists, Rocky Mountain Section, p. 111–128.

Gwinn, V. E., 1965, Cretaceous rocks of the Clark Fork Valley, central-western Montana: Billings Geological Society 16th Annual Field Conference Guidebook, p. 34–37.

Gwinn, V. E., and Mutch, T. A., 1965, Intertongued Upper Cretaceous volcanic and non-volcanic rocks, central-western Montana: Geological Society of America Bulletin, v. 76, p. 1125–1144.

Hanson, A. M., 1952, Cambrian stratigraphy in southwestern Montana: Montana Bureau of Mines and Geology Memoir 33, 46 p.

Harris, W. L., 1972, Upper Mississippian and Pennsylvanian sediments of central Montana [Ph.D. thesis]: Missoula, University of Montana, 251 p.

Holm, M. R., James, W. C., and Suttner, L. J., 1977, Comparison of the Peterson and Draney limestones, Idaho and Wyoming, and the calcareous members of the Kootenai Formation, western Montana, *in* The economic geology of the Williston basin: Montana Geological Society 24th Annual Conference, Williston Basin Symposium, September 1978, p. 259–270.

Imlay, R. W., 1945, Occurrence of Middle Jurassic rocks in western interior of the United States: American Association of Petroleum Geologists Bulletin, v. 29, p. 1019–1027.

—— , 1948, Characteristic marine Jurassic fossils from the western interior of the United States: U.S. Geological Survey Professional Paper 214-B, p. 13–33.

—— , 1952, Summary of Jurassic history in the western interior of the United States: Billings Geological Society 3rd Annual Field Conference Guidebook, p. 79–85.

—— , 1957, Paleocology of Jurassic seas in the western interior of the United States: Geological Society of America Memoir 67, v. 2, p. 469–504.

—— , 1980, Jurassic paleobiogeography of the conterminous United States in its continental setting: U.S. Geological Survey Professional Paper 1062, 134 p.

James, W. C., and Oaks, R. Q., Jr., 1977, Petrology of the Kinnikinnic Quartzite, east-central Idaho: Journal of Sedimentary Petrology, v. 47, p. 1491–1511.

Johnson, J. G., and Sandberg, C. A., 1977, Lower and Middle Devonian continental-shelf rocks of the Western United States, *in* Murphy, M. A., Berry, W.B.N., and Sandberg, C. A., eds., Western North America; Devonian: California University, Riverside Campus Museum Contributions 4, p. 121–143.

Klepper, M. R., Weeks, R. A., and Ruppel, E. T., 1957, Geology of the southern Elkhorn Mountains, Jefferson and Broadwater Counties, Montana: U.S. Geological Survey Professional Paper 292, 82 p.

Kranzler, I., 1966, Origin of oil in lower member of Tyler Formation of central Montana: American Association of Petroleum Geologists Bulletin, v. 50, p. 2245–2259.

Kuenzi, W. D., and Fields, R. W., 1971, Tertiary stratigraphy, structure, and geologic history, Jefferson Basin, Montana: Geological Society of America Bulletin, v. 82, p. 3374–3394.

Kuenzi, W. D., and Richard, B. H., 1969, Middle Tertiary unconformity, North Boulder and Jefferson basins, southwestern Montana: Geological Society of America Bulletin, v. 80, p. 111–120.

Kummel, B., 1954, Triassic stratigraphy of southwestern Montana: Billings Geological Society, 11th Annual Field Conference Guidebook, p. 239–243.

Lageson, D. R., Maughan, E. K., and Sando, W. J., 1979, The Mississippian and Pennsylvanian (Carboniferous) Systems in the United States—Wyoming: U.S. Geological Survey Professional Paper 1110-U, p. U1–U38.

Land, C. B., 1976, Stratigraphy and petroleum accumulation, Tyler sandstones (Pennsylvanian), Dickinson area, North Dakota [abs.]: American Association of Petroleum Geologists Bulletin, v. 60, p. 1401–1402.

—— , 1979, Tyler Sandstones (Pennsylvanian), Dickinson area, North Dakota; A 24 million barrel soil-zone stratigraphic trap [abs.]: American Association of Petroleum Geologists Bulletin, v. 60, p. 485.

Lochman-Balk, C., 1972, Cambrian system, *in* Geologic Atlas of the Rocky Mountain Region: Denver, Colorado, Rocky Mountain Association Geologists, p. 60–75.

Lochman-Balk, C., and Wilson, J. L., 1967, Stratigraphy of Upper Cambrian-Lower Ordovician subsurface sequence in the Williston basin: American Association of Petroleum Geologists Bulletin, v. 52, p. 883–917.

MacLachlan, M. E., 1972, Triassic system, *in* Geological Atlas of the Rocky Mountain Region: Denver, Colorado, Rocky Mountain Association Geologists, p. 166–176.

Mallory, W. W., 1972, Regional synthesis of the Pennsylvanian system, *in* Geologic Atlas of the Rocky Mountain Region: Denver, Colorado, Rocky Mountain Association of Geologists, p. 111–127.

Maughan, E. K., 1966, Environment of deposition of Permian salt in Williston and Alliance basins, *in* 2nd Symposium of Salt Proceedings; V. 1. Geology, geochemistry, mining: Cleveland, Ohio, Northern Ohio Geological Society, p. 35–47.

—— , 1967, Eastern Wyoming, eastern Montana, and the Dakotas, Chapter G, *in* McKee, E. D., and others, Paleotectonic investigations of the Permian system in the United States: U.S. Geological Survey Professional Paper 515-G, p. 125–152.

—— , 1975, Montana, North Dakota, northwestern Wyoming, and northern South Dakota, Chapter O, *in* McKee, E. D., and others, Paleotectonic investigations of the Pennsylvanian system in the United States; Part I, Introduction and regional analyses of the Pennsylvanian System: U.S. Geological Survey Professional Paper 853-O, p. 279–293.

—— , 1984, Paleogeographic setting of Pennsylvanian Tyler Formation and relation to underlying Mississippian rocks in Montana and North Dakota: American Association of Petroleum Geologists Bulletin, v. 68, p. 1778–1795.

Maughan, E. K., and Roberts, A. E., 1967, Big Snowy and Amsden Groups and the Mississippian-Pennsylvanian boundary in Montana: U.S. Geological Survey Professional Paper 554-B, 27 p.

McGookey, D. P., and 8 others, 1972, Cretaceous System, *in* Geologic Atlas of the Rocky Mountain Region: Denver, Colorado, Rocky Mountain Association Geologists, p. 190–228.

McKee, E. D., and others, 1956, Paleotectonic maps, Jurassic System: U.S. Geological Survey Miscellaneous Geological Investigations Map I-300, 33 p.

—— , 1959, Paleotectonic maps of the Triassic System: U.S. Geological Survey Miscellaneous Geological Investigations Map I-300, 33 p.

McKee, E. D., and others, 1967, Paleotectonic maps of the Permian System: U.S. Geological Survey Miscellaneous Geological Investigations Map I-450, 164 p.

McKelvey, V. E., and 5 others, 1959, The Phosphoria, Park City, and Shedhorn Formations in the western phosphate field: U.S. Geological Survey Professional Paper 313-A, p. 1–47.

McMannis, W. J., 1962, Devonian stratigraphy between Three Forks, Montana, and Yellowstone Park: Billings Geological Society 13th Annual Field Conference Guidebook, p. 4–12.

—— , 1965, Resume of depositional and structural history of western Montana: American Association of Petroleum Geologists Bulletin, v. 49, p. 1801–1823.

Meissner, F. F., 1978, Petroleum geology of the Bakken Formation, Williston basin, North Dakota and Montana, *in* The economic geology of the Williston basin: Montana Geological Society, 24th Annual Conference, Williston Basin Symposium, September 1978, p. 17–31.

Moberley, R., Jr., 1960, Morrison, Cloverly, and Sykes Mountain formations, northern Big Horn Basin, Wyoming and Montana: Geological Society of America Bulletin, v. 71, p. 1137–1176.

Moritz, C. A., 1951, Triassic and Jurassic stratigraphy of southwestern Montana: American Association of Petroleum Geologists, v. 35, p. 1781–1814.
——, 1960, Summary of Jurassic stratigraphy of southwestern Montana: Billings Geological Society 11th Annual Field Conference Guidebook, p. 239–243.
Nordquist, J. W., 1955, Pre-Rierdon Jurassic stratigraphy in northern Montana and Williston basin: Billings Geological Society 6th Annual Field Conference Guidebook, p. 96–106.
Oaks, R. Q., Jr., James, W. C., Francis, G. G., and Schulingkamp, W. J., II, 1977, Summary of Middle Ordovician stratigraphy and tectonics, northern Utah, southern and central Idaho, *in* Rocky Mountain Thrust Belt Geology and Resources: Wyoming Geological Association, 29th Annual Field Conference, p. 101–118.
Ostrom, M. E., 1970, Sedimentation cycles in the Lower Paleozoic rocks of western Wisconsin: Wisconsin Geological and Natural History Survey Information Circular, no. 11, p. 10–34.
Peterson, J. A., 1957, Marine Jurassic of northern Rocky Mountains and Williston basin: American Association of Petroleum Geologists Bulletin, v. 41, p. 399–440.
——, 1966, Sedimentary history of the Sweetgrass Arch: Billings Geological Society Proceedings, 17th Annual Field Conference and Symposium, Great Falls, Montana, August 1966, p. 112–133.
——, 1972, Jurassic System, *in* Geologic Atlas of the Rocky Mountain Region: Denver, Colorado, Rocky Mountain Association of Geologists, p. 177–189.
——, 1980a, Depositional history and petroleum geology of the Permian Phosphoria and Park City Formations, Wyoming and southeastern Idaho: U.S. Geological Survey Open-File Report 80–667, 42 p.
——, 1980b, Permian paleogeography and sedimentary provinces, west-central United States, *in* Symposium on Paleozoic paleogeography of west-central United States: Denver, Colorado, Society of Economic Paleontologists and Mineralogists, p. 271–292.
——, 1984, Permian stratigraphy, sedimentary facies, and general petroleum geology, Wyoming and adjacent areas, *in* Symposium on Permian and Pennsylvanian of Wyoming: Casper, Wyoming Geological Association, p. 25–64.
——, 1985, Regional stratigraphy and general petroleum geology of Montana and adjacent areas, *in* Tonnsen, J. J., ed., Montana Oil and Gas Fields Symposium, 1985: Billings, Montana Geological Society, p. 5–45.
——, 1987, Pre-thrusting regional stratigraphy and paleotectonics, Northern Rocky Mountains and adjacent Cordilleran shelf, U.S., *in* Miller, W. R., ed., The Thrust Belt Revisited: Wyoming Geological Association, 38th Field Conference Guidebook, p. 17–26.
Peterson, J. A., and Smith, D. L., 1986, Rocky Mountain paleogeography through geologic time, *in* Peterson, J. A., ed., Paleotectonics and Sedimentation in the Rocky Mountain Region, United States: American Association of Petroleum Geologists Memoir 41, p. 3–19.
Rascoe, B., Jr., and Baars, D. L., 1972, Permian System, *in* Geologic atlas of the Rocky Mountain region: Denver, Colorado, Rocky Mountain Association of Geologists, p. 143–165.
Rau, J. L., 1962, The stratigraphy of the Three Forks Formation: Billings Geological Society, 7th Annual Field Conference Guidebook, p. 35–45.
Rice, D. D., and Shurr, G. W., 1978, Potential for major natural gas resources in shallow, low-permeability reservoirs of the Northern Great Plains, *in* The economic geology of the Williston basin: Montana Geological Society, 24th Annual Conference, Williston Basin Symposium, September 1978, p. 265–281.
——, 1983, Patterns of sedimentation and paleogeography across the Western interior seaway during time of deposition of Upper Cretaceous Eagle Sandstone and equivalent rocks, Northern Great Plains, *in* Reynolds, M. W., and Dolly, E. D., eds., Mesozoic paleogeography of west-central United States: Denver, Colorado, Society of Economic Geologists Rocky Mountain Section, p. 337–358.
Roberts, A. E., 1963, The Livingston Group of south-central Montana: U.S. Geological Survey Professional Paper 475-B, p. B86–B92.
——, 1965, Correlation of Cretaceous and lower Tertiary rocks near Livingston, Montana, *in* Geological Survey Research, 1965: U.S. Geological Survey Professional Paper 525-B, p. B54–B63.
Robinson, G. D., 1961, Origin and development of the Three Forks basin, Montana: Geological Society of America Bulletin, v. 72, p. 1003–1014.
——, 1963, Geology of the Three Forks Quadrangle, Montana: U.S. Geological Survey Professional Paper 370, 143 p.
Robinson, G. D., Klepper, M. R., and Obradovich, J. D., 1969, Overlapping plutonism, volcanism, and tectonism in the Boulder batholith region, western Montana: Geological Society of America Memoir 116, p. 557–576.
Rose, P. R., 1976, Mississippian carbonate shelf margins, western United States: U.S. Geological Survey Journal of Research, v. 4, p. 449–466.
Ross, R. J., Jr., 1957, Ordovician fossils from wells in the Williston basin, eastern Montana: U.S. Geological Survey Bulletin 1021-M, p. 439–510.
——, 1975, Ordovician sedimentation in the western United States, *in* Bassett, M. G., ed., The Ordovician System, Proceedings of a Paleontological Association Symposium, Birmingham: Cardiff, University of Wales Press and National Museum of Wales, p. 73–105.
Ryder, R. T., and Ames, H. T., 1970, Palynology and age of Beaverhead Formation and their paleotectonic implications in Lima region, Montana-Idaho: American Association of Petroleum Geologists Bulletin, v. 54, p. 1155–1171.
Ryder, R. T., and Scholten, R., 1973, Syntectonic conglomerates in southwestern Montana; Their nature, origin, and tectonic significance: Geological Society of America Bulletin, v. 84, p. 773–796.
——, 1986, Syntectonic conglomerates in southwestern Montana; their nature, origin and tectonic significance (with an update), *in* Peterson, J. A., ed., Paleotectonics and sedimentation in the Rocky Mountain region, United States: American Association of Petroleum Geologists Memoir 41, p. 131–149.
Sandberg, C. A., 1962, Geology of the Williston basin, North Dakota, Montana, and South Dakota, with reference to subsurface disposal of radioactive wastes: U.S. Geological Survey Report TEI-809, 148 p.
——, 1965, Nomenclature and correlation of lithologic subdivisions of the Jefferson and Three Forks Formations of southern Montana and northern Wyoming: U.S. Geological Survey Bulletin 1994-N, p. N1–N18.
Sandberg, C. A., and Hammond, C. R., 1958, Devonian system in Williston basin and central Montana: American Association of Petroleum Geologists Bulletin, v. 42, no. 10, p.2293–2334.
Sandberg, C. A., and Mapel, W. J., 1967, Devonian of the northern Rocky Mountain plains, *in* Oswald, D. H., ed., International Symposium on the Devonian System: Calgary, Alberta Society of Petroleum Geologists, v. 1, p. 843–877.
Sandberg, C. A., and McMannis, W. J., 1964, Occurrence and paleogeographic significance of the Maywood Formation of Late Devonian age in the Gallatin Range, southwestern Montana: U.S. Geological Survey Professional Paper 501-C, p. C50–C54.
Sandberg, C. A., Gutschick, R. C., Johnson, J. G., Poole, F. G., and Sando, W. J., 1982, Middle Devonian to Late Mississippian geologic history of the overthrust belt region, western United States, *in* Powers, R. B., ed., Geologic studies of the Cordilleran thrust belt: Denver, Colorado, Rocky Mountain Association of Geologists, v. 2, p. 691–720.
Sando, W. J., 1967, Mississippian depositional provinces in the northern Cordilleran region: U.S. Geological Survey Professional Paper 575-D, p. D29–D38.
——, 1972, Madison Group (Mississippian) and Amsden Formation (Mississippian and Pennsylvanian) in the Beartooth Mountains, northern Wyoming and southern Montana: Montana Geological Society 21st Annual Geological Conference Guidebook, p. 57–63.
——, 1974, Ancient solution phenomena in the Madison Limestone (Mississippian) of north-central Wyoming: U.S. Geological Survey Journal of Research, v. 2, no. 2, p. 133–141.
——, 1976a, Madison Limestone, east flank of Bighorn Mountains, Wyoming: Wyoming Geological Association 28th Annual Field Conference Guidebook, p. 45–52.
——, 1976b, Mississippian history of the northern Rocky Mountain region: U.S. Geological Survey Journal of Research, v. 4, p. 317–338.

Schmidt, R. G., 1978, Rocks and mineral resources of the Wolf Creek area, Lewis and Clark and Cascade Counties, Montana: U.S. Geological Survey Bulletin 1441, 91 p.

Sheldon, R. P., 1963, Physical stratigraphy and mineral resources of Permian rocks in western Wyoming: U.S. Geological Survey Professional Paper 313-B, 273 p.

Shurr, G. W., 1984a, Geometry of shelf-sandstone bodies in the Shannon Sandstone of southeastern Montana, *in* Tillman, R. W., ed., Siliciclastic shelf sediments: Society of Economic Paleontologists and Mineralogists Special Publication 34, p. 63–83.

—— , 1984b, Regional setting of Niobrara Formation in northern Great Plains: American Association of Petroleum Geologists Bulletin, v. 68, p. 598–609.

Shurr, G. W., and Reiskind, J., 1984, Stratigraphic framework of the Niobrara Formation (Upper Cretaceous) in North and South Dakota, *in* Scott, D. F., ed., Mesozoic of middle North America: Canadian Society of Petroleum Geologists Memoir 9, p. 205–219.

Skipp, B., and McGrew, L. W., 1977, The Maudlow and Sedan Formations of the Upper Cretaceous Livingston Group on the west edge of the Crazy Mountain basin, Montana: U.S. Geological Survey Bulletin 1422-B, 68 p.

Skipp, B., and Sandberg, C. A., 1975, Silurian and Devonian miogeosynclinal and transitional rocks of the Fish Creek Reservoir Window, central Idaho: U.S. Geological Survey Journal of Research, v. 3, no. 6, p. 691–706.

Skipp, B. W., Sando, W. J., and Hall, W. E., 1979, The Mississippian and Pennsylvanian (Carboniferous) Systems in the United States; Idaho: U.S. Geological Survey Professional Paper 1110-AA, p. AA1–AA2.

Sloss, L. L., 1963, Sequences in the cratonic interior of North America: Geological Society of America Bulletin, v. 34, p. 423–451.

Smedes, H. W., 1966, Geology and igneous petrology of the northern Elkhorn Mountains, Jefferson and Broadwater Counties, Montana: U.S. Geological Survey Professional Paper 510, 116 p.

Smith, D. L., and Gilmour, E. H., 1979, The Mississippian and Pennsylvanian (Carboniferous) Systems in the United States; Montana: U.S. Geological Survey Professional Paper 1110-X, p. X1–X31.

Stokes, W. L., 1950, Pediment concept applied to Shinarump and similar conglomerates: Geological Society of America Bulletin, v. 61, p. 91–98.

Suttner, L. J., 1969, Stratigraphic and petrographic analysis of Upper Jurassic-Lower Cretaceous Morrison and Kootenai Formations, southwest Montana: American Association of Petroleum Geologists Bulletin, v. 53, p. 1391–1410.

Suttner, L. J., Schwartz, R. K., and James, W. C., 1981, Late Mesozoic to Early Cenozoic foreland sedimentation in southwest Montana, *in* Montana Geological Society Field Conference, southwest Montana: Billings, Montana, Montana Geological Society, p. 93–103.

Tenney, C. S., 1966, Pennsylvanian and Lower Permian depositon in Wyoming and adjacent areas: American Association of Petroleum Geologists Bulletin, v. 50, no. 2, p. 227–250.

Theodosis, S. D., 1955, Belt series of northwestern Montana: Billings Geological Society 6th Annual Field Conference Guidebook, p. 58–63.

Viele, G. W., and Harris, F. G., 1965, Montana Group stratigraphy, Lewis and Clark County, Montana: American Association of Petroleum Geologists Bulletin, v. 49, p. 379–417.

Walker, T. F., 1974, Stratigraphy and depositional environments of the Morrison and Kootenai Formations in the Great Falls area, central Montana [Ph.D. thesis]: Missoula, University of Montana, 195 p.

Weimer, R. J., 1960, Upper Cretaceous stratigraphy, Rocky Mountain area: American Association of Petroleum Geologists Bulletin, v. 44, p. 1–20.

—— , 1961, Spatial dimensions of Upper Cretaceous sandstones, Rocky Mountain area, *in* Geometry of sandstone bodies—a symposium: Tulsa, Oklahoma, American Association of Petroleum Geologists, p. 82–97.0.

Willis, R. P., 1959, Upper Mississippian-lower Pennsylvanian stratigraphy of central Montana and Williston basin: American Association of Petroleum Geologists Bulletin, v. 43, p. 1940–1966.

Wilson, J. L., 1955, Devonian correlations in northwestern Montana: Billings Geological Society 6th Annual Field Conference Guidebook, p. 70–77.

—— , 1967, Carbonate-evaporite cycles in lower Duperow Formation of Williston basin: Canadian Petroleum Geologists Bulletin, v. 15, p. 230–312.

Witzke, B. J., 1980, Middle and Upper Ordovician paleogeography of the region bordering the Transcontinental arch, *in* Fouch, T. D., and Magathan, E. R., eds., Paleozoic Paleogeography of the west-central United States: Denver, Colorado, Society of Economic Paleontologists and Mineralogists Rocky Mountain Section, p. 1–18.

Yochelson, E. L., 1968, Biostratigraphy of the Phosphoria, Park City, and Shedhorn Formations: U.S. Geological Survey Professional Paper 313–D, p. 571–660.

Ziebarth, H. C., 1964, The Tyler Formation of southwestern North Dakota, *in* 3rd International Williston Basin Symposium: Billings, Montana, Billings Geological Society, p. 119–126.

MANUSCRIPT ACCEPTED BY THE SOCIETY NOVEMBER 24, 1987

ACKNOWLEDGMENTS

This chapter is compiled from several previous publications by the author. The original work was critically reviewed by S. B. Anderson, L. C. Bortz, W. H. Curry III, M. W. Longman, E. K. Maughan, T. S. McClellan, W. R. Miller, R. B. Powers, C. A. Sandberg, G. W. Shurr, J. R. Taylor, J. J. Tonnsen, T. E. Tucker, R. M. Weidman, and J. L. Wilson. The assistance of A. R. Palmer and L. L. Sloss in compiling the chapter from previous publications is gratefully acknowledged.

Printed in U.S.A.

The Geology of North America
Vol. D-2, Sedimentary Cover—North American Craton: U.S.
The Geological Society of America, 1988

Chapter 8

Basins of the Rocky Mountain region

D. L. Baars
29056 Histead Dr., Evergreen, Colorado 80439
B. L. Bartleson
Department of Geology, Western State College, Gunnison, Colorado 81230
C. E. Chapin
New Mexico Bureau of Mines and Mineral Resources, Socorro, New Mexico 87801
B. F. Curtis
375 Harvard Lane, Boulder, Colorado 80303
R. H. De Voto
Department of Geology and Geological Engineering, Colorado School of Mines, Golden, Colorado 80401
J. R. Everett
Earth Satellite Corporation, Chevy Chase, Maryland 20815
R. C. Johnson, C. M. Molenaar, F. Peterson, and C. J. Schenk
U.S. Geological Survey, Box 25046, Denver Federal Center, Denver, Colorado 80225
J. D. Love
U.S. Geological Survey, Box 3007, University Station, Laramie, Wyoming 82071
I. S. Merin
Io Geological Consultants, 3041 White Birch Court, Fairfax, Virginia 22031
P. R. Rose
Telegraph Exploration, Telegraph, Texas 76883
R. T. Ryder
U.S. Geological Survey, National Center, Reston, Virginia 22092
N. B. Waechter
Suite 415, 820 16th St., Denver, Colorado 80202
L. A. Woodward
Department of Geology, University of New Mexico, Albuquerque, New Mexico 87131

INTRODUCTION

D. L. Baars

Each of the sedimentary basins of the Rocky Mountain province is geologically unique. All have been structurally controlled but by different episodes and styles of tectonic activity. At least some of the structures have been reactivated one or more times during the Phanerozoic.

Because of the great geologic diversity, no one worker could be authoritative about the geology of the province. Consequently, each basin in this chapter has its own specialist authors. They are identified with their appropriate texts and listed at the beginning of this chapter. Although an attempt has been made to standardize the format of the basin subchapters, some differences in style and geological focus appropriately reflect the differences among the various authors. Little attempt was made to force uniformity in details of approach to the basin analyses.

Organization of the chapter is geographic. Basins west of the Rio Grande Rift system are stratigraphically more closely related to the Cordilleran miogeocline than to the craton and are treated first. These basins are described in order by age, and geographically from south to north. Those basins lying east of the Rio Grande Rift and more closely related to the craton are presented in order from south to north.

Prior to Pennsylvanian time, the site of the modern Rocky Mountains was largely that of a broad, shallow marine shelf at the inner margin of the Cordilleran miogeocline. Late Paleozoic basins developed concurrently with the uplift of the Ancestral Rocky Mountains as the Precambrian basement structural fabric was strongly reactivated, perhaps by an intercontinental plate collision. The fault-bounded depocenters received large amounts

Baars, D. L., and 15 others, 1988, Basins of the Rocky Mountain region, *in* Sloss, L. L., ed., Sedimentary Cover—North American Craton; U.S.: Boulder, Colorado, Geological Society of America, The Geology of North America, v. D-2.

of clastic debris from the juxtaposed uplifts; evaporite sedimentation dominated distal basin-fill processes; and penecontemporaneous salt tectonics modified the basins. Clastic sedimentation gradually filled and obscured these ancient basins from Permian through Jurassic times.

During Cretaceous time, tectonic uplifts west of the province provided copious amounts of sediment to the developing intracratonic seaway that spread across the sites of the present-day Rocky Mountains and western Great Plains. All pre-existing geologic features were deeply buried by thick accumulations of clastic sediments. As tectonic activity spread eastward during the Laramide orogeny, Cretaceous seas gradually waned, and fault-block uplifts and related basins again dominated the region.

During Tertiary time, these numerous deep basins were filled with sediments derived from the bounding uplifts, all in continental environments. Fluvial sedimentation dominated basin-filling processes in most fault-bounded basins, while lacustrine environments prevailed in the Uinta, Piceance, and Greater Green River Basins (Plate 2). In the southern Rocky Mountains, much of the Laramide tectonic fabric resulted from reactivation of older structures, but tectonic rejuvenation is more difficult to document in the Rocky Mountains of Wyoming and Montana. A fundamental flaw in the crust, the Rio Grande Rift, perhaps influenced the geologic history of the region from the beginning, but it emerged as a significant entity only in middle to later Tertiary time. Epeirogenic elevation of the entire Rocky Mountain region is still continuing.

The fascinating details of this complex geologic region are unraveled, basin by basin, in succeeding subchapters. Although highly generalized due to space limitations, the individual discussions provide excellent summaries of the basins of the Rocky Mountain region.

TECTONICS OF THE ROCKY MOUNTAIN REGION

Lee A. Woodward

GEOLOGIC SETTING

The Rocky Mountain region considered here lies east of the Cordilleran thrust and fold belt (Plate 2) and is bounded on the east by the Williston, Denver, and Raton Basins. This tectonic province is commonly called the Rocky Mountain foreland inasmuch as it lies in front of (i.e., in the direction of vergence of) the western thrust and fold belt. On the north, this province appears to end just north of the Little Belt and Big Snowy Uplifts in central Montana; the southern termination is not well defined, but Kelley (1972) presented evidence that the province extends nearly to the Guadalupe Mountains of southern New Mexico.

A structurally unique part of this region, the Colorado Plateau, has been only moderately deformed, compared to the more intensely deformed areas around it, and is described in more detail later.

Most of the features shown on the tectonic map of this region are of Cretaceous and Cenozoic age. However, deformation during Precambrian and late Paleozoic times resulted in crustal anisotropy that locally had strong influence on younger structures.

STRUCTURE OF UPPER CRUST

Principal structures of the Rocky Mountains region are elongate uplifts generally trending north or northwest and large, circular to elongate basins. West-trending uplifts, less common, include the Uinta (northern Utah), Little Belt and Big Snowy (central Montana), and Owl Creek (north-central Wyoming); basins adjacent to these uplifts mostly have similar trends. Structural relief between adjacent uplifts and basins is greatest in Wyoming and parts of adjacent states, where Love (this chapter) notes a maximum of about 11,500 m.

Many basins and uplifts are asymmetrical and have the general appearance of tilted blocks that are bounded on one side by a narrow zone of strongly deformed rocks. Uplifts tend to be fault bounded or broadly anticlinal with draping of sedimentary strata over faults in the basement rocks. Some of the anticlinal uplifts are flat-topped (e.g., Little Belt Uplift) and have monoclines on their flanks. Range-margin faults commonly grade into folds upsection, as seen along plunging uplifts, or along strike as the fault displacement diminishes. As seen at the surface, these faults range from moderately dipping thrusts to high-angle reverse faults. Their geometries at depth are a matter of considerable controversy; recent seismic investigations of the Wind River (Smithson and others, 1978) and the Laramie (Brewer and others, 1982) Uplifts of Wyoming suggest that the faults bounding these uplifts continue at moderate dips into the crystalline basement rocks to depths of 24 and 10–12 km, respectively. In contrast, surface mapping has shown range-marginal thrust faults to steepen with depth along the Beartooth Uplift of southern Montana (Foose and others, 1961).

Range-marginal faults dip toward the uplifts and have yielded toward the adjacent basins. Many uplifts are bounded by faults on only one side, although a few, such as the Uinta and Beartooth Uplifts, are bounded by faults on two or more sides. Thus, there is no uniform direction of yielding; rather, the direction of thrust movement is controlled by the geometric relations of the uplift to the adjacent basins.

A sequence of development includes an early monocline that is later cut by a reverse or thrust fault. During formation of the monocline there is stretching of the limb between the anticlinal and synclinal bends that may lead to development of antithetic and synthetic normal faults that are older than and may be offset by the younger reverse or thrust fault (Lowell, 1970). Also, longitudinal normal faults commonly form in the upthrown block, presumably due to stretching above the thrust as the uplift yields over the adjacent basin. Brown (1983), describing the sequential development of range-margin fold-thrusts, noted the formation of subsidiary folds and faults in the hanging wall because of compression in the synclinal part of the monocline. Thus, the detailed structure of the deformed zone separating an uplift from a basin can be very complex.

In areal extent, the basins are much larger than the uplifts. Also, with respect to sea level, many of the basins are deeper than the adjacent uplifts are high. Thus, much of the present structural relief is due to subsidence of the basins rather than rise of the uplifts with respect to sea level.

Second-order structures within the basins may form in several different ways. Some are the result of faults separating tilted blocks in the basement (e.g., Delaney Butte and Sheep Mountain in the North Park Basin, Colorado); these faults may cut through the overlying strata or may have the strata drape-folded over them. In other cases, disharmonic folding that involves only the sedimentary cover appears to form primarily because of basin subsidence and (or) regional compression in the basin; examples that have been documented using seismic and drill-hole data include Wildhorse Butte Anticline in the Bighorn Basin and Golden Goose oil field in the Wind River Basin (Peterson, 1983).

Radial folds tend to form in response to greater subsidence in the basin center than along the margins; the circumference of strata on the basin margin is forced to occupy a smaller circumference as strata are depressed and pulled toward the center of the basin. The result is local compressional stress fields tangential to the margin of the basin. These folds tend to have small amplitudes and large wave lengths; they occur in most basins but are particularly prominent in the San Juan Basin of northwestern New Mexico and southern Colorado.

Dallmus (1958) demonstrated that when the axis of a large elongate basin—where the earth's curvature is a factor—subsides with respect to the basin margins, the strata will undergo compression perpendicular to the axis. This is because the chord is shorter than the arc it subtends, and as the arc subsides and approaches the chord, there will be folds that are parallel to the axis of the basin. Folds that form by flexure folding may lead to small-scale thrust faults that begin as bedding-plane faults but steepen upward and cut across bedding near anticlinal crests. These listric faults are rootless and are fundamentally different from the major reverse and thrust faults that involve basement rocks and commonly separate basins from uplifts.

Recurrent movement on basement faults has been contemporaneous with deposition of sedimentary units in some of the basins in Colorado (Weimer, 1980) and Wyoming (Slack, 1981). Contemporaneous (or growth) faults that are extensional and tend to flatten at depth, with movement dying out in the sedimentary cover, are well known from the Gulf Coast region (Hardin and Hardin, 1961) but have not been widely documented in the Laramide Rocky Mountain basins, where they have much less stratigraphic separation than those in the Gulf Coast. These listric faults have thicker stratigraphic sections on the downthrown side and may have minor folding in the hanging wall due to rotation of beds along a curved fault. Gravity is the driving mechanism, and generally the sediments are poorly indurated during deformation.

Rise of uplifts commonly leads to second-order structures due to gravitational gliding from the high area and extension of the uplifted block as it is unconstrained and free to expand horizontally over the adjacent basin. Crests of uplifts may have grabens or tilted fault blocks, as in the Owl Creek Uplift (Wise, 1963) and in the Nacimiento Uplift (Woodward, 1976).

Gravitational gliding may result in diverse structures, ranging from small-scale nearly isoclinal recumbent folds (Woodward and others, 1976) to large gravity-slide plates. Two well-known cases of gravity slides occur in the Rocky Mountains region near the Bearpaw Mountains of Montana and at Heart Mountain in Wyoming. During volcanism in the Bearpaw Mountains, doming allowed Cretaceous strata to glide away from the dome along bentonite beds (Reeves, 1946). The faults are listric and follow bedding from the breakaway area to where they cut upsection, becoming thrusts that mostly have less than 1.0 km of displacement. These faults occur up to 40 km from the intrusion that caused the doming.

Pierce (1957) presented a strong case for gravitational sliding of large blocks of Paleozoic carbonates along the Heart Mountain detachment thrust. The breakaway zone for the carbonates in the Absaroka Mountains is about 110 km from Heart Mountain in the Bighorn Basin where the Paleozoic blocks rest on Tertiary beds.

The Colorado Plateau is structurally unique in that it has been only moderately deformed compared to the more intensely deformed regions around it and appears to have behaved as a relatively stable structural unit during Laramide deformation (Late Cretaceous–early Tertiary). Monoclines separating broad, gently tilted blocks characterize the plateau (Kelley, 1955a), with wide structural basins and uplifts being the main tectonic divisions of the plateau. Most uplifts are bounded on one side by a major monocline, and the other side of the uplift is broadly gradational with the adjacent basin through a gentle slope. The basin-uplift boundary on the gentle slope is thus arbitrarily chosen, usually on the basis of the extent of a given stratigraphic unit. There are all gradations from gentle, open monoclines to overturned and thrust structures, with total structural relief as much as 4,200 m (Kelley, 1955a). Basement faults ranging from high-angle faults to low-angle thrusts commonly grade upsection or along strike into monoclines, indicating draping of strata over faulted basement blocks. Most monoclines have sinuous traces, and a few, such as the Defiance, are contorted because of many short, diagonal cross folds. This contortion is generally considered to be evidence of lateral shift. Minor monoclines occur within basins or uplifts.

Late Cenozoic normal faults, superimposed on and commonly controlled by older structures, bound tilted fault-blocks of the Basin and Range province, the Rio Grande Rift, and a few other scattered basins. Most of these faults appear to be listric, flattening at depths less than 15 km (Woodward, 1977; Eaton, 1979). Seismic studies indicate that listric normal faults of the Rio Grande Rift near Socorro, New Mexico, flatten at about 5 km of depth (Cape and others, 1983). The Albuquerque Basin forms the deepest part of the Rio Grande Rift and has approximately 8,500 m of structural relief next to the Sandia Uplift to the east (Black and Hiss, 1974). Wernicke and others (1982) reviewed the evidence concerning the amount of crustal extension in the Basin and Range province and concluded that there is a minimum of 65 percent (140 km) in the southern Great Basin. Woodward (1977) estimated about 8 km of crustal extension across the Rio Grande Rift near Albuquerque, New Mexico. Late Cenozoic sediments fill many of these extensional basins in southern New Mexico, Arizona, western Utah, central Colorado, and southwestern Montana (Plate 2). Those basins of particular interest in this chapter are part of the Rio Grande Rift in Colorado and New Mexico.

CRUSTAL THICKNESS AND STRUCTURE OF LOWER CRUST

Thickness of the continental crust, shown on Figure 1, ranges from about 20 km to around 52 to 54 km, as interpreted from seismic investigations (Smith 1978; Prodehl, 1979; Prodehl and Pakiser, 1980). In general, the crust is thin beneath the Basin and Range province, ranging from about 20 to nearly 40 km. The Colorado Plateau is underlain by a somewhat thicker crust of about 40 to 45 km. Beneath the Southern Rocky Mountains of Colorado and northern New Mexico, the crust is thicker yet, reaching a maximum of 52 to 54 km near the Front Range. In Wyoming the Central Rocky Mountains are underlain by crust about 40 km or slightly less in thickness, and in Montana the crustal thickness ranges from around 35 to nearly 50 km. To the east of the Southern Rocky Mountain front, the Great Plains appear to have a crustal thickness nearly as great as that of the Southern Rocky Mountains. Keller and others (1978) reported that the crust of the Rio Grande Rift has a thickness of 30 to 35 km, considerably thinner than that of the Colorado Plateau (about 45 km) to the west and the Great Plains (about 50 km) to the east.

Eaton (1979) interpreted the crust of the Basin and Range province to be divided into three layers: an upper brittle layer cut by listric normal faults, a thin and ductile middle layer marking the base of the listric faults, and a lower layer that is fairly strong and emplaced with numerous mafic dikes. The ductile layer may be up to 3 km thick and occurs at a depth of 5–15 km, averaging 8 ± 3 km (Eaton, 1979, p. 14).

Based on seismic velocities, there appear to be two distinct layers in the Precambrian crystalline rocks of the Front Range in Colorado, but to the west of the Front Range and in the Great Plains to the east the crystalline crust is not distinctly layered (Prodehl and Pakiser, 1980, p. 154).

North of the Front Range the crust becomes markedly thinner, about 37 km, beneath the Laramie Uplift (Prodehl and Pakiser, 1980). Allmendinger and others (1982) noted that this northward thinning of the crust generally coincides with the Mullen Creek–Nash Fork Shear Zone, a major tectonic boundary in the Medicine Bow Uplift separating Archean rocks on the northwest from Proterozoic rocks on the southeast (Houston, 1971), and also marks the boundary between the north-trending uplifts of the Southern Rocky Mountains and the northwesterly to westerly trends of uplifts in the Central Rocky Mountains of Wyoming. A possible subsurface continuation of the Mullen Creek–Nash Fork Shear Zone may be manifested by the North American Central Plains Conductive Anomaly (Camfield and Gough, 1977), which trends northeasterly from the exposed shear zone and wraps around the east side of the Black Hills Uplift. Hills and others (1975) interpreted the Mullen Creek–Nash Fork Shear Zone as a major Proterozoic plate boundary.

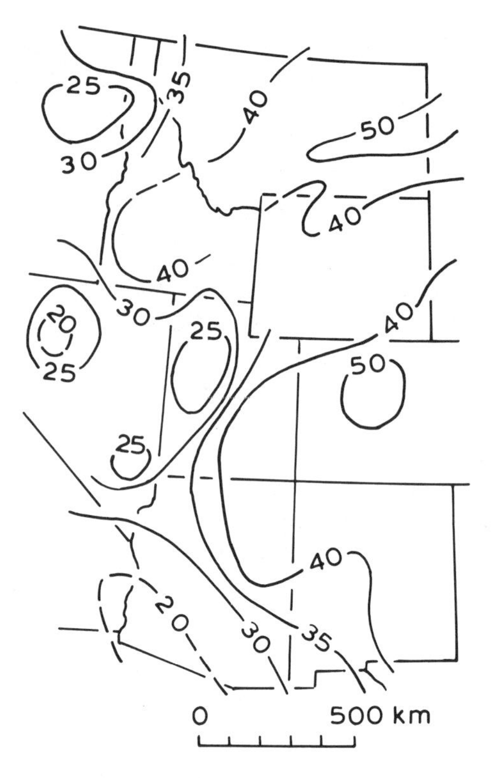

Figure 1. Crustal thickness of the Rocky Mountains region, in kilometers (modified from Smith, 1978).

Prodehl and Pakiser (1980) noted that isostatic compensation in most of the western United States is probably accomplished in part by variations in density in the crust and upper mantle as well as by differences in crustal thicknesses. Gravity lows associated with several areas, such as the San Juan Mountains of southwestern Colorado, have been explained as batholiths in the subsurface beneath numerous calderas rather than as areas with deep crustal roots (Plouff and Pakiser, 1972). Other areas shown on the tectonic map (Plate 2) that probably have concealed batholiths beneath caldera complexes include Yellowstone Park in Wyoming and the Mogollon Plateau of southwestern New Mexico.

PRECAMBRIAN BASEMENT INFLUENCE ON LATER DEFORMATION

Zones of weakness related to anisotropy in Precambrian basement rocks are commonly cited as controlling the orientation of Laramide structures; however, there is little agreement concerning the specific features or their degree of control. Elements of anisotropy that have been noted include schistosity or foliation, bedding or compositional layering, joints and microjoints, faults, shear zones, and Precambrian plutons that contrast markedly in competence with their host rocks.

Major structures of Precambrian age that have been well documented with regard to having strong influence on younger tectonic events include the northeast-trending Colorado Lineament (Warner, 1978), a broad zone marked by the Mullen Creek–Nash Fork Shear Zone in the Medicine Bow Uplift of southern Wyoming and the Colorado Mineral Belt (Tweto and Sims, 1963). Additional northeast-trending belts of major mineralization in the western United States have been described by Landwehr (1967), who suggested a Precambrian ancestry for them. Northwest trends of probably Precambrian age in Colorado, Utah, Arizona, and New Mexico have had a strong influence on Paleozoic sedimentation (Baars and Stevenson, 1982).

Well-documented zones of weakness of Precambrian age appear to have had little influence on the trends of uplifts in the Rocky Mountain region in many cases; notable exceptions include the Uncompahgre, San Luis, and Uinta Uplifts. Hoppin and Palmquist (1965) presented an excellent discussion on the problem of basement influence on later deformation and noted that in the Bighorn Uplift of Wyoming the overall structure developed in response to post-Precambrian regional stresses and that basement anisotropy influenced principally the trends of smaller scale folds and faults.

PALEOTECTONICS

This discussion is concerned only with Phanerozoic events in the Rocky Mountain region. Marine platform or shelf sediments that merge westward into the Cordilleran miogeocline were deposited here during much of the Paleozoic and early Mesozoic, although parts of the region were locally emergent or even strongly uplifted, particularly during the late Paleozoic (see inset map, Plate 2).

During Cambrian, Ordovician, and Silurian time the Transcontinental Arch, extending in a broad and irregular fashion from Wisconsin to Arizona, received less sediment than adjacent areas or was locally emergent. Minor movement on basement structures influenced the sedimentation patterns during the early Paleozoic in Colorado (Tweto, 1980; Baars and See, 1968). A few arches and broad, intervening low areas characterized the tectonic pattern during Devonian time (Baars, 1972). A linear east-trending downwarp in central Montana has a greater thickness of Mississippian strata than to the north or south; this trough connected the Williston Basin with the Cordilleran miogeocline to the west. Elsewhere, the Mississippian appears to have been a time of tectonic stability in the Rocky Mountain region, where a thin carbonate shelf sequence was deposited.

Crowley and others (1985) presented compelling evidence (fission-track analysis of apatite) that the epeirogenic subsidence of the Williston Basin since the Middle Ordovician is due to decay of a thermal anomaly in the lithosphere, accompanied by lithospheric densification and thickening of the elastic lithosphere as proposed by Ahern and Mrkvicka (1984). Episodic epeirogenic subsidence of some Rocky Mountain basins during much of Phanerozoic time suggests that perhaps similar decay of thermal anomalies may have been a factor in their early development also.

In the late Paleozoic, basement-cored uplifts rose mainly in Colorado and New Mexico and shed coarse clastic debris into adjacent marine basins. The uplifts in Colorado were strongly elevated and may have been up to 3,000 m high (Mallory, 1972b). There was also deformation within some of the basins, particularly the Paradox (Baars and Stevenson, 1982), where older faults localized the salt anticlines. Salt flowage may have continued into the Jurassic.

Kluth and Coney (1981) suggested that the late Paleozoic uplifts (Ancestral Rocky Mountains) formed as a result of collision of North America with South America–Africa during the Ouachita-Marathon Orogeny. As noted by Woodward and Ingersoll (1979), the timing of continental collision corresponds with Ancestral Rocky Mountain deformation, and a plate tectonic origin seems very likely, but the precise mechanisms by which stresses and strains are transmitted through hundreds of kilometers of continental crust are poorly understood.

An area of late Paleozoic erosion in southern Colorado continued to be a source area for sediments (continental) during Triassic time (MacLachlan, 1972), although of smaller area and less elevation than previously. In the Jurassic the Rocky Mountain region was relatively stable tectonically.

The previous patterns of sedimentation were changed markedly with the advent of the Cordilleran Thrust Belt on the west, mainly during Late Cretaceous and early Tertiary time. An immense foredeep that formed on the eastern edge of the thrust belt was filled with a great thickness (up to 6,000 m) of clastic sediments (Reeside, 1944). Loading by thrust sheets with resultant downbowing of the adjacent brittle crust of the foreland is commonly accepted as the reason for the foredeep.

TECTONIC DEVELOPMENT

The present tectonic framework of the Rocky Mountain region is a result of deformation during late Mesozoic and Cenozoic time. Late Cretaceous–early Tertiary (Laramide) deformation is responsible for most of the major structural elements in the region, although Neogene extension has resulted in local modification of the older structures and development of fault-block uplifts and associated basins of the Rio Grande Rift, Basin and Range province, and related features.

There is little agreement concerning the geometry of the deep, concealed Laramide structures of the Rocky Mountain region. Two extreme points of view concerning this problem have been presented. One is that low-angle thrusts that extend downward into the crystalline crust dominate the tectonics and were induced by horizontal compression (e.g., Hamilton, 1981). The other is that structural relief is greater than the amount of crustal shortening due to thrusting and folding (Stearns, 1978). Part of the controversy stems from lack of understanding that horizontal compression involves forces or stresses, whereas structural relief is due to movements or kinematics; horizontal compression and large vertical rise of uplifts relative to adjacent basins are not mutually exclusive, and horizontal compression undoubtedly leads to large structural relief. This problem will not be resolved until the deep structures of most basins and uplifts have been studied in more detail, mainly by seismic methods. Meanwhile it seems likely that there may be different geometries for different uplift-basin boundaries; some, such as the Wind River Uplift, are bounded by low-angle thrusts having large amounts of horizontal displacement (Smithson and others, 1978), whereas others, such as the Beartooth Uplift, are bounded at least in part by steeply dipping reverse faults at depth that flatten upward to become low-angle thrusts at high structural and stratigraphic levels (Foose and others, 1961).

Nearly all current workers agree that there is a temporal and spatial relationship between Laramide deformation of the Rocky Mountain foreland and subduction processes along the west coast of North America, with a regional compressional stress field in the foreland leading to reverse, thrust, and subordinate strike-slip faults. Any hypothesis to explain the observed or inferred deformation must also account for the following observations: (1) volumetrically, the geometry of the Rocky Mountain foreland is due largely to basin subsidence, and rise of uplifts, with respect to sea level, is subordinate; (2) many, but not all, basins have subsided episodically during Phanerozoic time; (3) the crust is thicker beneath the Rocky Mountains and Great Plains than it is to the west beneath the Cordilleran foldbelt; and (4) the entire foreland rose epeirogenically after Laramide deformation with minor rejuvenation of some uplifts.

Laramide deformation of the Rocky Mountain foreland has been difficult to explain in terms of the more common kinds of subduction-related deformation inasmuch as amagmatic subsidence of basins and rise of basement-cored uplifts occurred far inland from the subduction zone. Coney and Reynolds (1977) and Dickinson and Snyder (1978) proposed that this style of deformation could be explained by a gently dipping subducted slab analogous to amagmatic parts of the Andes. Hamilton (1981) suggested that compressive deformation north and east of the Colorado Plateau was caused by clockwise rotation of about 2 or 4 degrees of the Colorado Plateau relative to the continental interior, and that the rotation absorbed a small amount of the convergence between the North American Plate and the Farallon Plate being subducted beneath it. Kelley (1955b) demonstrated that the Colorado Plateau was shoved northeast relative to the Rocky Mountains on the east and the north with resultant shift on its northern and eastern margins. However, Hamilton's (1981) hypothesis seems too simplistic to account for the evolution of the Rocky Mountain foreland, particularly in view of basin subsidence and the thick crust. The model involving subduction of a gently dipping slab better accounts for the observations concerning the foreland province, as it explains basin subsidence and thickening of the crust. Gries (1983b) proposed that north-trending foreland structures formed early in the Laramide because of compression due to westward movement of the North American Plate, but as plate motion changed to the southwest and then to the south, the northwest- and east-trending structures developed later in the Laramide.

Neogene extensional deformation was dominated by listric normal faulting in the upper crust, with development of the Rio Grande Rift, the Basin and Range province, and a few other related basins, and was superimposed on and partly controlled by the older structures. Most areas of strong extension are characterized by thin crust and lithosphere, high heat flow, and low seismic wave velocities in the upper mantle. Regional extension, beginning after 40 Ma, has been related to intra-arc extension, back-arc extension, and extension related to lateral slip of the Pacific Plate past the North American Plate. Intra-arc extension may have begun as the rate of plate convergence decreased and the angle of subduction steepened between 30 and 40 Ma. Back-arc extension with diapiric rise and lateral spreading of mantle rocks is favored by many because of the associated high heat flow and volcanic rocks. Eaton (1979), however, suggested that the extensional state of stress evolved from intra-arc extension through back-arc extension to back-transform extension related to the development of the San Andreas right-slip boundary between the Pacific and North American Plates (Atwater, 1970). Ingersoll (1982) favors an unstable triple junction as the cause of extension and notes that the occurrence of the extension in the back-arc setting is not a necessary cause for the extension.

In summary, the present tectonic framework of the Rocky Mountain region is dominated by subsidence of large basins and the rise of areally less-extensive uplifts that formed mainly in Laramide time during regional compression related to movement of the North American plate over a gently dipping subducted slab. Change in direction of plate motion may account for many of the north-south, northwest, and east-west trends of basins and uplifts; some structures, as well as some sedimentation patterns, were strongly influenced by basement anisotropy induced by Precambrian and (or) late Paleozoic deformation. The current controversy concerning the geometry and inferred tectonic style of the Rocky Mountain foreland may stem from the circumstances of various uplifts being founded by faults of different styles; deep seismic lines across critical uplift-basin boundaries are needed to help resolve this problem. Neogene extension that was superimposed on some of the Laramide features and resulted in Basin and Range structures, including the Rio Grande Rift, is related to slowing of plate convergence and oblique motion of the western edge of the North American Plate with possible interplay of an unstable triple junction.

PARADOX BASIN

D. L. Baars

The Paradox Basin is a northwest-trending, subrectangular

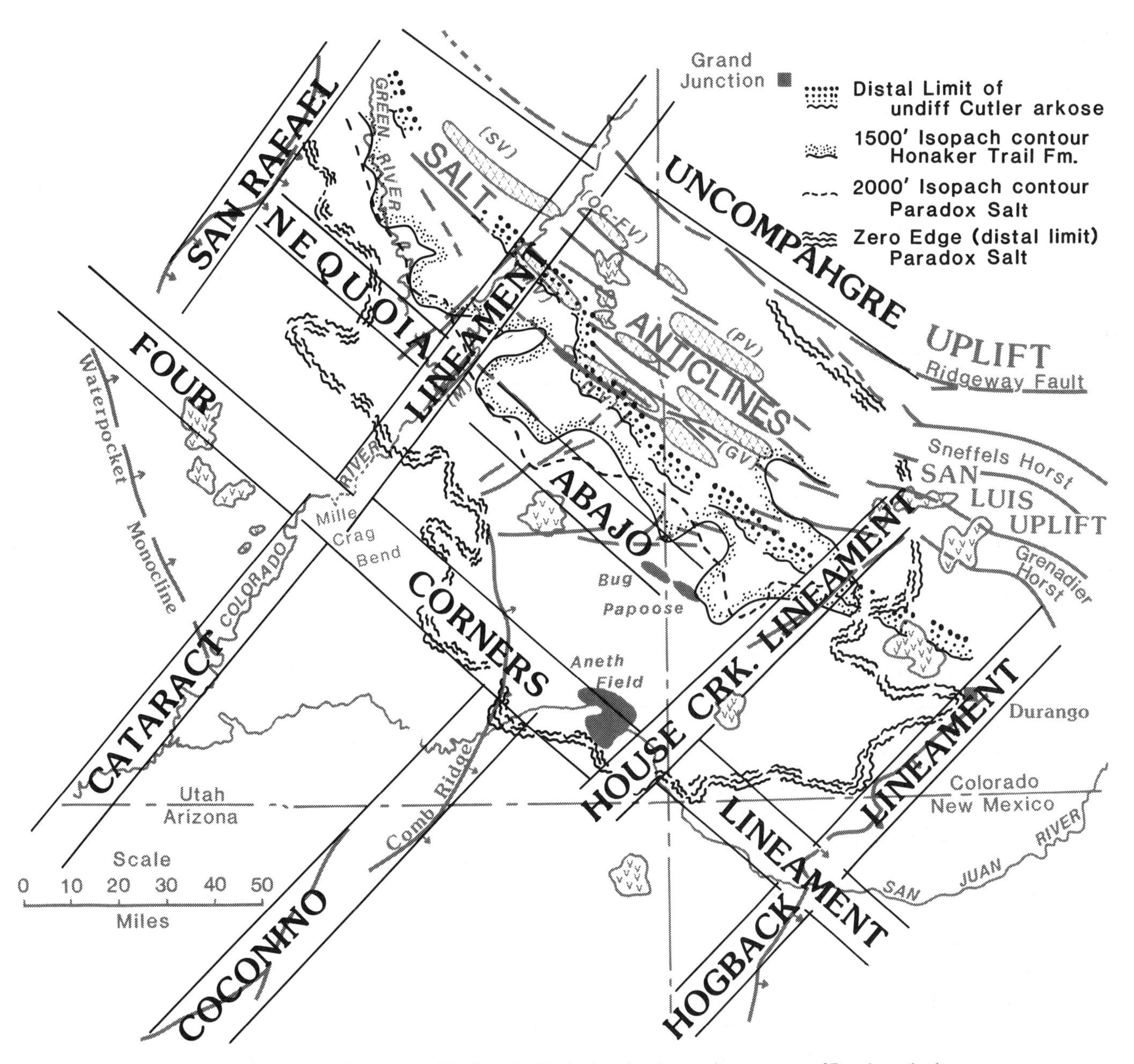

Figure 2. Generalized index map of the Paradox Basin showing the maximum extent of Paradox salt, the larger salt-intruded anticlines, selected surface structures, and paleotectonically activated lineaments that affected sedimentation during Paleozoic time. Note the location of the greater Aneth oil field complex on the Four Corners Lineament and the Bug-Papoose Fields on the Nequoia-Abajo shelf. (SV = Salt Valley; OC-FV = Onion Creek–Fisher Valley; PV = Paradox Valley; GV = Gypsum Valley; LV = Lisbon Valley; MV = Moab Valley salt structures). From Stevenson and Baars, 1986.

evaporite basin of Pennsylvanian age in southeastern Utah and southwestern Colorado (Plate 2). There is no surface expression of the basin except for the occurrence of elongate diapiric anticlines in the deepest central part of the basin. Evaporites attain a thickness of nearly 5,000 m within the diapirs and grade laterally to shallow shelf carbonate rocks along the southern and western shelves of the basin (Fig. 2). Prolific petroleum production from bioherms within these shelf carbonates makes the Paradox an economically important Paleozoic feature in the Southern Rocky Mountains province. The evaporite basin is bordered on the east by the Uncompahgre and San Luis Uplifts, segments of the Ancestral Rocky Mountains, which shed vast quantities of arkosic clastic sediments into the subsiding fault-bounded depression throughout Paleozoic time. Rejuvenation of a complex Precambrian fault system in Pennsylvanian time resulted in the localization and orientation of the basin.

PRE-PENNSYLVANIAN TECTONIC HISTORY

A northwest-oriented basement dextral wrench fault system, which dates circa 1.6 to 1.7 Ga, transects the length of the Paradox Basin and forms the axis of the late Paleozoic structural depression (Baars and Stevenson, 1982; Stevenson and Baars, 1986). This fault system is part of the Olympic-Wichita Lane of Baars (1976) that dominates the paleotectonic pattern in the basin (Fig. 2). A less obvious northeast-trending swarm of basement wrench faults, part of the Colorado Lineament of Warner (1978), offsets the NW-SE set in a sinistral sense in the central part of the basin (Hite, 1975). Warner dated earliest movement along these faults of the Colorado Lane at about 1.7 Ga. The conjugate fracture pattern formed by the NW-SE and the NE-SW faults intersects in the heart of the Paradox Basin. This basic orthogonal tectonic fabric was repeatedly reactivated throughout Paleozoic time and affected sedimentation patterns well into Mesozoic time across the Colorado Plateau (Peterson, this chapter).

Although Cambrian through Mississippian time is noted for its general tectonic quiescence, Baars (1966) and Baars and See (1968) demonstrated that the basement structural features were rejuvenated repeatedly, at least during Late Cambrian, Late Devonian, and Early Mississippian times. Paleotectonically controlled seafloor topography affected depositional patterns in the McCracken Sandstone Member of the Elbert Formation (Late Devonian) and the Leadville Limestone (Mississippian) (Fig. 3), producing porous facies that later were traps for petroleum accumulation. The traps are in the form of offshore sand bars in the McCracken and of dolomitized Walsortian banks (mud-supported crinoidal bioherms) in the Leadville, localized on the high flanks of faulted paleotectonic features (Baars, 1966). No Ordovician or Silurian rocks are preserved in the Paradox Basin.

A red paleosol, the Molas Formation, marks the regional disconformity that developed throughout the Southern Rocky Mountains province in Late Mississippian and Early Pennsylvanian time. This emergent episode was followed by the Ancestral Rocky Mountains orogeny that reactivated basement structure.

BASIN DEVELOPMENT

Strike-slip reactivation along the NW-SE basement wrench faults initiated the extensional opening of the basin in Atokan time (early Middle Pennsylvanian). If plate collision was responsible for the tectonic rejuvenation, as suggested by Kluth and Coney (1981), the resulting stress field was released along the pre-existing basement fractures. Extensional tectonics were initiated at the prominent "kink" in the fault pattern now exposed in the San Juan Mountains of southwestern Colorado (San Luis Uplift of Figs. 2 and 4), where two fault blocks composed predominantly of Precambrian quartzites were uplifted to form the San Luis Uplift (Baars and Stevenson, 1984). Clastic sediments derived from the San Luis Uplift were shed onto adjacent low-

PALEOZOIC CORRELATION CHART
OF
PARADOX BASIN

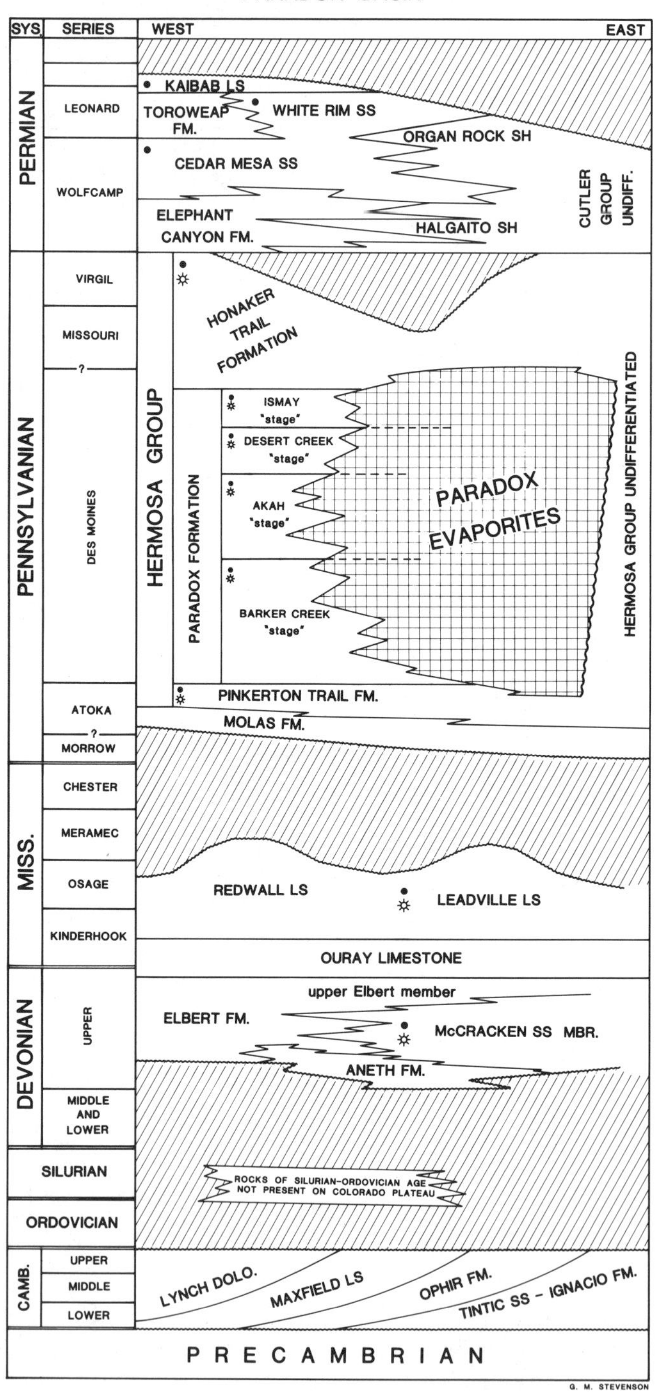

Figure 3. Correlation chart for Paleozoic rocks of the Paradox Basin, showing petroleum-producing horizons. From Baars and Stevenson, 1982.

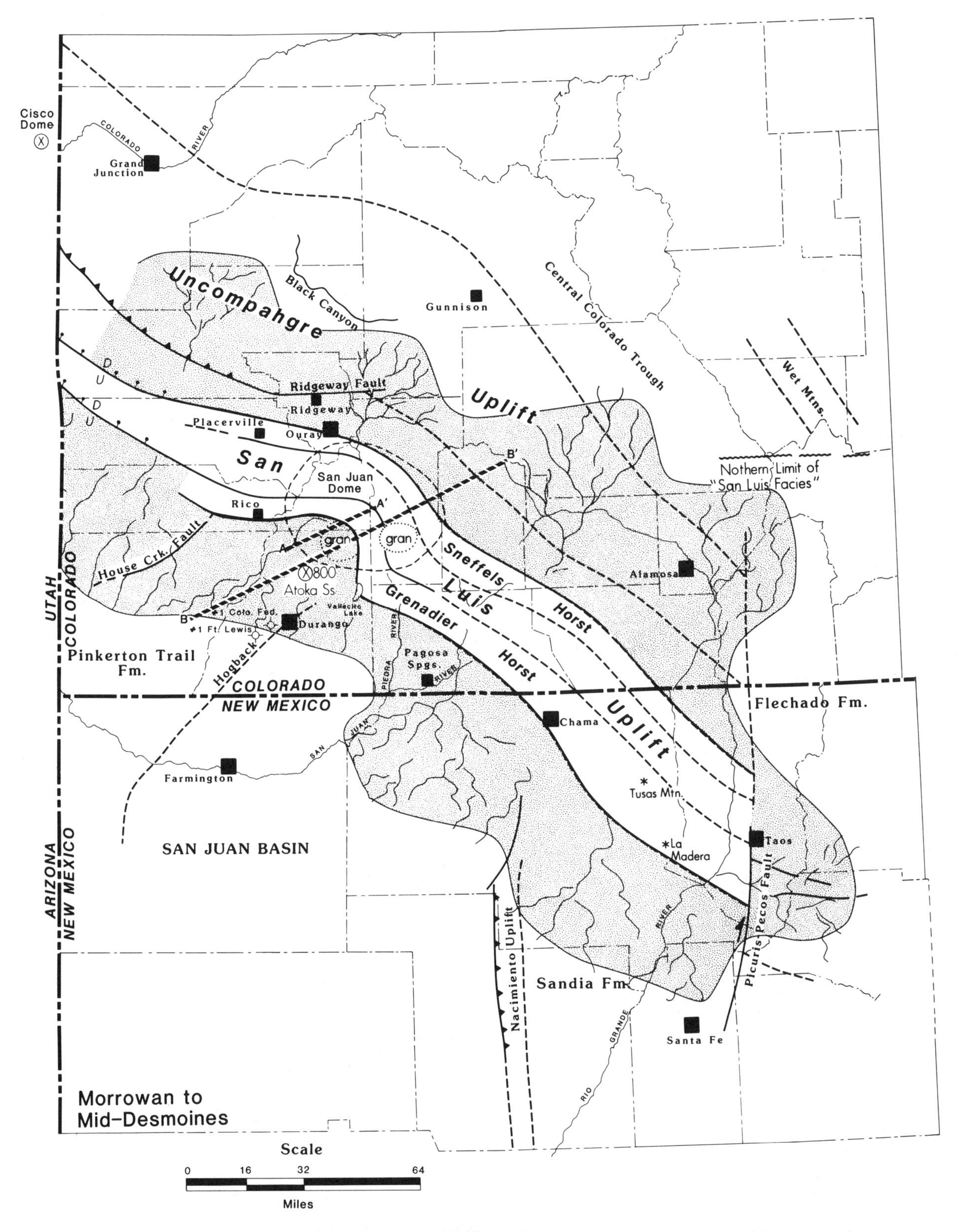

Figure 4. Generalized map of the San Luis Uplift, an Early Pennsylvanian precursor of the Ancestral Rocky Mountains. The prominent "kink" in the basement fault blocks in the San Juan Mountains were the releasing bend for the initial extensional opening of the Paradox Basin. The San Luis Uplift was the source for Morrowan through early Des Moinesian clastic sediments in southwestern Colorado and north-central New Mexico, as shown by the stippled pattern. From Baars and Stevenson, 1984.

lands as sandstones in the lower Hermosa Group and Sandia Formation of southern Colorado and northern New Mexico (Fig. 4).

Although the primary mechanism for basin opening and collapse was extension along the NW-SE wrench faults, extension was facilitated and enhanced by release along the NE-SW trending faults of the Colorado Lane, especially along the House Creek Lineament (Fig. 2) (Stevenson and Baars, 1986). Marine conditions encroached upon the NW-SE trough in Atokan time, and by the early Desmoinesian (late Middle Pennsylvanian) marine waters dominated the incipient Paradox Aulacogen of Gorham (1975). Localized subsidence of the orthogonally fault-bounded basin restricted normal marine circulation, and evaporites began to accumulate.

The halite-dominated evaporites are highly cyclical, and 29 salt cycles have been formally recognized (Hite, 1960). A typical cycle was initiated with a relatively thin black shale (technically a sapropelic carbonate), and as evaporation proceeded, thin dolomite (primary?) and anhydrite beds graded upward into beds of halite 30–100 m thick. Some cycles locally contain commercial beds of potash salts (Hite, 1960). A typical salt cycle includes a cap of thin anhydrite and dolomite. Individual cycles can be readily correlated in the subsurface throughout the basin. Salt sedimentation ceased by the close of Desmoinesian time, as the basin was nearly filled and normal marine conditions prevailed. The net accumulation of halite is shown on Figure 5, and it is notable that ancestors of present-day structures controlled salt thickness along the southwestern paleotectonic shelf.

Generally speaking, the salt cycles grade laterally from restricted marine into normal marine shelf carbonate rocks along the structurally controlled southern and western shelves of the Paradox Basin. Limestone predominates on the shelves, but dolomitization has occurred locally in areas of intertidal deposits. Algal bioherms are commonly distributed along fault-controlled paleo-shoals and produce prolific amounts of petroleum, especially along the Four Corners Lineament (Fig. 2) (Baars and Stevenson, 1982). These facies relationships are shown diagrammatically on Figure 6. Oolitic and skeletal lime sands are locally productive of hydrocarbons, and intertidal carbonate shoals are productive in the Bug Field vicinity along the Nequoia-Abajo Shelf (Fig. 2). The Nequoia-Abajo paleotectonic shelf forms the southwestern hingeline of the deeper Paradox salt basin (Uncompahgre Trough of some writers).

Little is known about the northern Paradox Basin due to a general paucity of well control, but the paleotectonic style changes markedly north of the Cataract Lineament that is generally followed by the course of the Colorado River (Fig. 2). The basin configuration changes from an extensional regime of the southern basin to a generally compressional paleostructural setting. Thrust faulting is recognizable in seismic studies; the Uncompahgre front is distinctly thrusted north of the Colorado River as shown by recent drilling (Frahme and Vaughn, 1983). Relative subsidence was somewhat younger in the northern basin than to the south. The Cataract Lineament appears to have been

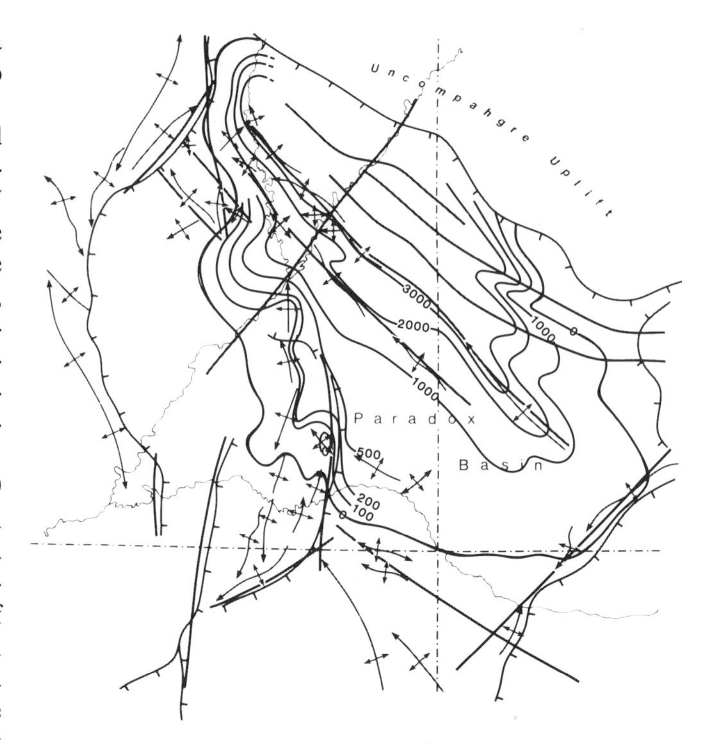

Figure 5. Map showing relationship of salt isoliths of the Paradox Formation to present-day structures of the Four Corners area. Similar structures (shown by standard map symbols) on the Monument Upwarp along the southwestern margin of the basin and the Nequoia-Emery Uplift in the northwestern margin of the basin controlled salt deposition. Depositional thickness of Paradox salt cannot be determined accurately in the region of salt anticlines (shown in solid black) because of massive salt flowage in the area. Most anticlines in the 610 to 914-m belt are pillowed salt structures with relatively minor thickening not indicated at this scale. From Baars and Stevenson, 1982.

the primary pivoting element in the Paradox pull-apart basin, the basin having opened south of the lineament and converged to the north toward the Oquirrh Basin of central Utah.

SALT-INTRUDED STRUCTURES

Salt flowage began in the deep Paradox Basin as marine conditions returned in late Desmoinesian time (Elston and Landis, 1960). The true maximum depositional thickness of evaporites is impossible to determine because of the extensive flowage, but normally bedded salt attains thicknesses greater than 1,000 m outside the area of salt anticlines (Fig. 4). Thickness varies from zero to perhaps 5,000 m in areas of maximum flowage. It is clear, however, that the earliest and thickest depositional salt accumulated within the pre-existing grabens and half-grabens (Hite, 1960) developed by reactivation of Precambrian faults (Figs. 5 and 7). The same faults were responsible for the localization of the NW-SE elongated diapirs.

Diapirism was evidently triggered by differential loading of

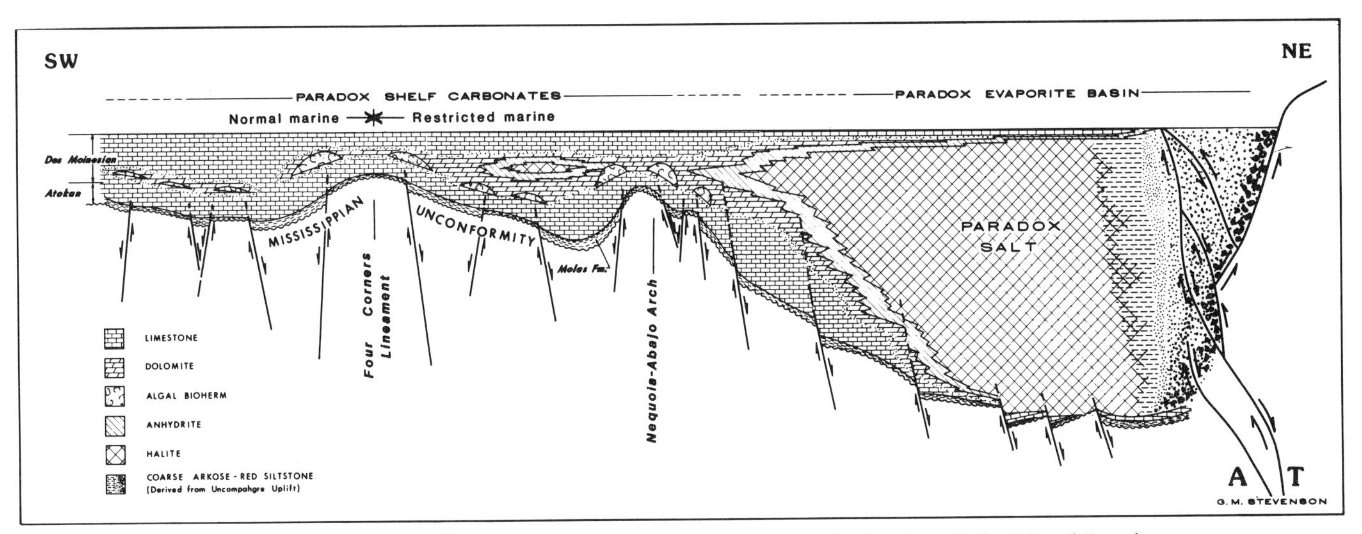

Figure 6. Generalized SW-NE cross section across the Paradox Basin showing relationships of depositional sequences from the Uncompahgre Uplift source area on the right across the salt basin to the southwestern Paradox shelf. Biohermal carbonate rocks associated with paleostructures on the shelf produce large quantities of oil and gas. Modified after Baars and Stevenson, 1982.

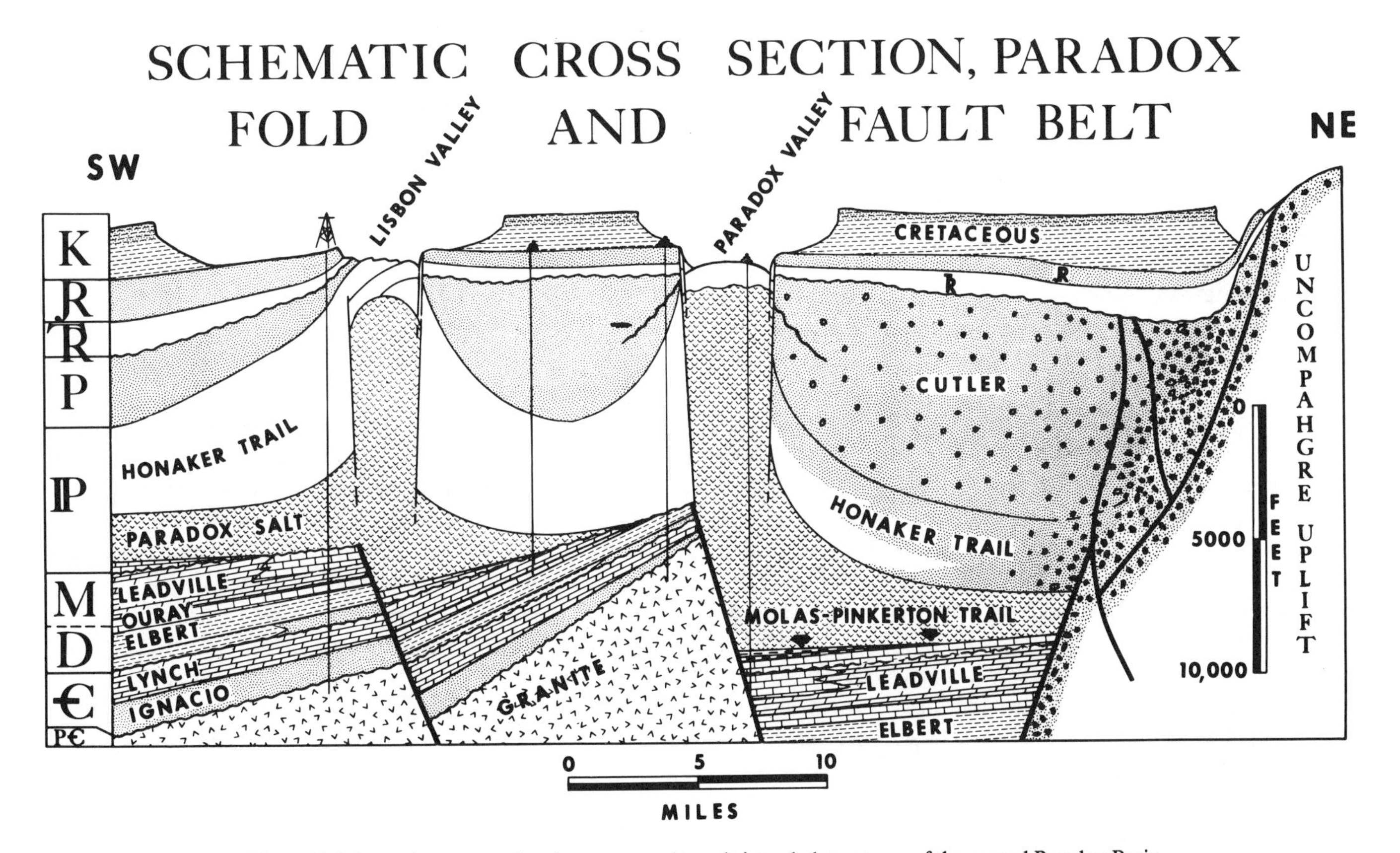

Figure 7. Schematic cross section drawn normal to salt-intruded structures of the central Paradox Basin. Deep-seated faulted structures beneath the Paradox salt were positive during deposition of the Cambrian through Mississippian section as shown by depositional and erosional thinning of the various formations and localized reservoir rocks beneath the Paradox and Lisbon Valley structures. The large fault blocks localized salt flowage from Pennsylvanian through Jurassic times and formed fault-related elongate NW-SE trending salt anticlines. Modified after Baars, 1966.

the salt (Cater and Elston, 1963). By Desmoinesian time the southern Uncompahgre highlands were elevated and shed vast amounts of arkosic debris into the adjacent basin (Baars and Stevenson, 1984). The thick clastic wedge spread from the source area into the basin, interfingering with the evaporites and in effect crowding the salt-depositing environment away from the uplift. The deepest graben adjacent to the Uncompahgre highland contains very thick arkose, but little or no salt (Frahme and Vaughn, 1983). Differential loading by those same clastic wedges apparently initiated salt movement southwestward toward areas of lesser overburden and toward the large basement faults that underlay the salt basin. The mobile salt was deflected upward along faults that now have vertical displacements approaching 2,000 m (Fig. 7). More or less circular salt flowage cells merged to form salt walls, and elongate "salt anticlines" were the result (Hite and Cater, 1972).

Diapirism continued from latest Desmoinesian through Jurassic times, an interval of perhaps 160 m.y. Local angular unconformities and formational thinning along the flanks of the salt structures attest to periodic growth of the diapirs. Nearly every depositional unit from the Late Pennsylvanian Honaker Trail Formation through the Late Jurassic Morrison Formation (Figs. 3 and 7) was affected. Inter- and intraformational unconformities abound, and thickness variations in the strata indicate syndepositional movements. The time of greatest diapirism appears to have been during the Early Permian, however, as the depositional thickness of the Cutler arkosic clastics varies markedly around the diapirs and salt pillows. Distributary streams that carried the arkosic clastics were forced to flow around the salt bulges, restricting the coarse facies of the Cutler to the synclinal troughs between the various salt structures. Some fine-grained clastics bypassed the synsedimentary traps to form the Halgaito and Organ Rock Formations to the west. Thicknesses vary from zero along the diapiric highs to several thousand meters in the immediately adjacent synclines, severely complicating seismic interpretations. Salt-flowage–induced paleotopography drastically affected the depositional patterns of all post-Pennsylvanian formations. Salt may have periodically flowed from the structures as in the present-day Zagros diapirs in Iran. A unique occurrence of salt in the Triassic Moenkopi Formation in the Courthouse Wash Syncline adjacent to the Salt Valley diapir north of Moab suggests that there may have been a local source of extrusive salt in Triassic time.

Near-surface groundwaters dissolved salt from the crests of the diapirs in Pleistocene and Recent times, leaving a residual cap of gypsum, black shale, and thin stringers of dolomite. These cap rocks are thoroughly churned and contorted as seen vividly along Onion Creek east of Moab. Overlying strata, mostly of Cretaceous age, collapsed into the solution voids, creating massive jumbles of highly fractured and collapsed rocks. The resulting surface depressions form elongate NW-SE trending valleys that bear little or no relation to the regional drainage patterns. The early pioneers noticed that the larger rivers in the region cross these collapse valleys at right angles, rather than flowing along the valleys as any respectable river would have done. They considered this situation, and rightly so, to be paradoxical, and so named one of the larger features "Paradox Valley." The evaporite sequence was later named the Paradox Formation for exposed gypsiferous cap rock in the valley (Baker, 1933), and the basin of deposition of the formation naturally came to be known as the Paradox Basin. Superposition of the drainage is only one of many paradoxical features of the basin, so the name is most appropriate.

POST-PARADOX BASIN FILL

The Paradox evaporite basin began to shrink in size after the Desert Creek stage (middle Desmoinesian) (Fig. 3), and by the late Ismay stage salt deposition was restricted to the northern basin, north of the Cataract Lineament. Normal marine, open-water environments were re-established throughout the region by the close of the Desmoinesian. Thick alluvially deposited clastics derived from the Uncompahgre Uplift dominated sedimentation in the eastern Paradox Basin through the Late Pennsylvanian (Missourian-Virgilian). The deltaic clastic wedge interfingered westward with interbedded fossiliferous limestones, shales, and sandstones of the Honaker Trail Formation of Wengerd and Matheny (1958). The marine facies is widespread across the eastern Colorado Plateau, averaging about 300 m in thickness. The basin continued to subside at a moderate rate, but open circulation of the normal marine waters was maintained through the remainder of Pennsylvanian time.

Sedimentation was apparently continuous across the Permian-Pennsylvanian boundary in the eastern Paradox Basin, in the area of the salt-intruded anticlines, but an erosional unconformity marks the systemic boundary elsewhere in the region. Early Permian (Wolfcampian) red beds of the Cutler Group directly overlie Missourian marine strata west and southwest of the salt structures, with Virgilian and locally upper Missourian rocks missing. More than a hundred meters of Missourian Honaker Trail beds have been stripped from paleotectonically elevated structures in the Four Corners region, and Wolfcampian strata of the marine Elephant Canyon Formation (Baars, 1962) rest directly on the Mississippian Leadville Limestone on the Emery Uplift (San Rafael Swell) at the northwestern termination of the Paradox Basin. Stratigraphically transitional strata that cross the Permian-Pennsylvanian boundary or are environmentally transitional from marine to continental (referred to as the "Rico formation" by some writers) are not known to occur outside the salt anticline region.

PERMIAN SYSTEM

Fluvial red beds derived directly from the Uncompahgre Uplift continued to blanket the Paradox Basin through Early Permian time (Fig. 8). These undifferentiated clastics of the Cutler Group (Fig. 3) interfinger westward with marine rocks (Elephant Canyon Formation) and marine and/or eolian light-colored sandstones (Cedar Mesa and White Rim Sandstones) that are

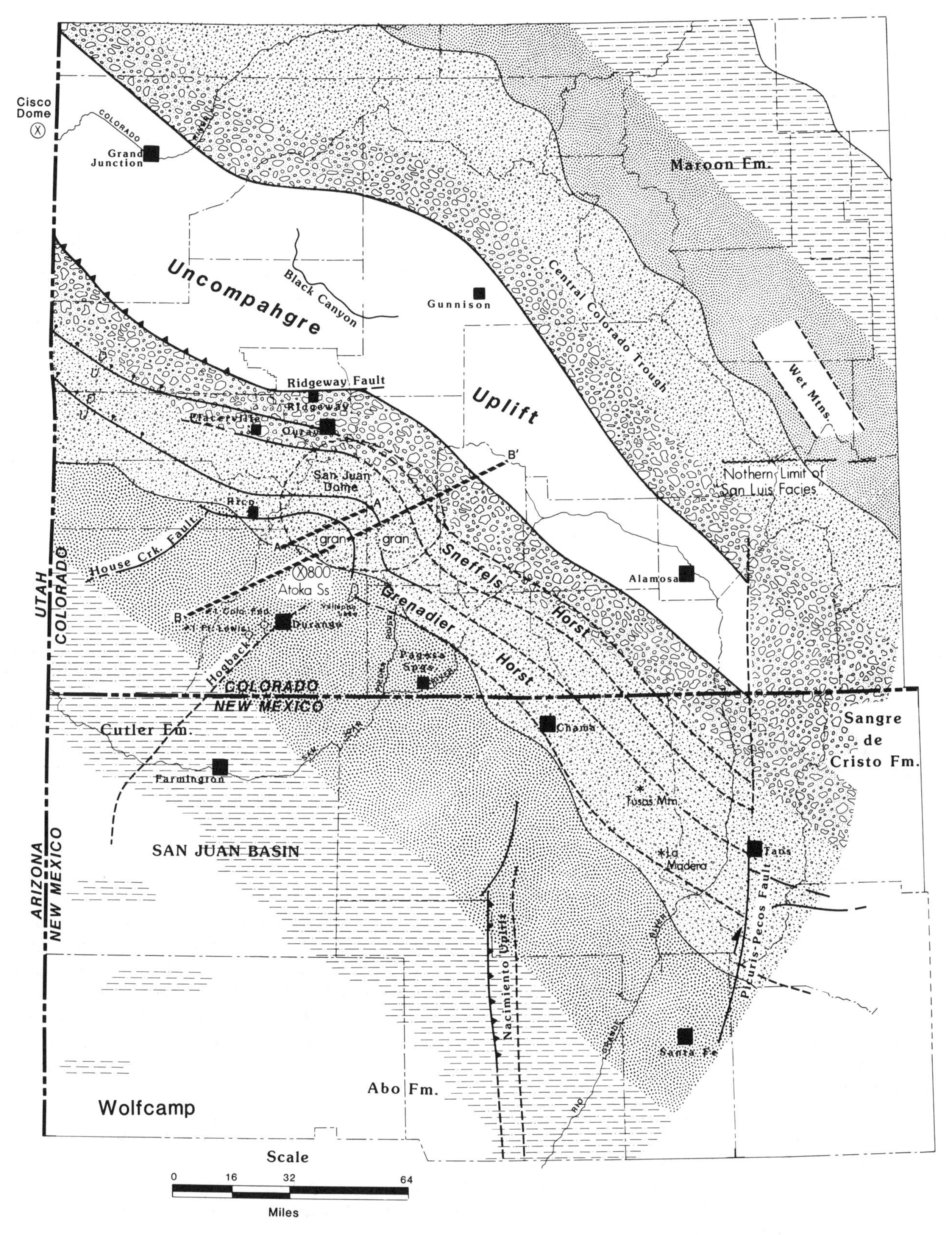

Figure 8. Generalized map of the southeastern Paradox region showing the complete burial in Early Permian (Wolfcampian) time of pre-existing structural elements by continental clastic sediments derived from the Uncompahgre Uplift. From Baars and Stevenson, 1984.

unrelated to events of the Paradox Basin. These formations have closer affinities to regionally distributed sediments in the Oquirrh Basin and Cordilleran miogeocline than to the underlying locally derived sediments in the Paradox Basin. For discussions of the regional Permian stratigraphy, see Baars (1962) modified by Baars (1985).

ECONOMIC GEOLOGY

The Paradox Basin has been a prolific producer of oil and gas from Middle Pennsylvanian rocks since the discovery in 1956 of the giant Aneth Field in southeastern Utah. By far the most important reservoir rocks have been algal bioherms of the Desert Creek and Ismay stages. The primary rock builder is the calcareous alga *Ivanovia* sp. that proliferated on paleotectonically controlled shoals along the southwestern shelf of the salt basin. Total production through 1984 is about 400 million barrels of oil and one trillion cubic feet of gas.

Porosity is usually primary, with grain-supported algal fragments supporting interparticle pore spaces; secondary dolomitization often greatly enhances reservoir quality. However, north of the Aneth area, production in the Bug and Papoose Canyon Fields (Fig. 2) is associated with dolomitized intertidal carbonates along the Nequoia-Abajo Shelf.

The best pre-Pennsylvanian petroleum production is from the Mississippian Leadville Limestone at the Lisbon Field in the salt anticline country. The reservoir is a crinoidal bioherm (Walsortian bank) that developed on a fault-controlled paleotopographic shoal. Porosity is developed where dolomitization has recrystallized crinoidal packstone and wackestone, altering the mud matrix to a saccharoidal texture and dissolving the skeletal particles (Baars, 1966). The Late Devonian McCracken Sandstone Member of the Elbert Formation also produces oil at Lisbon, and elsewhere, from paleotectonically localized offshore sand bars.

For a complete summary of the hundreds of oil and gas fields of the Paradox Basin and related parts of the Four Corners Area, see Fassett (1978 and 1983).

AFTERMATH

The Paradox Basin, like the rest of the Colorado Plateau province, was subjected to the west-east compressional processes of the Laramide Orogeny. Most of the major structures seen at the surface today have generally been attributed to this episode, but in fact the Late Cretaceous–Early Tertiary folding merely modified and enhanced pre-existing structures. The tectonic fabric of the province was well established in Precambrian time, reactivated in late Paleozoic time, and finally exaggerated in Laramide time. The magnificent monoclines of the Colorado Plateau are drape folds across reactivated basement high-angle faults; even minor folds on the monoclinal uplifts were periodically active during Paleozoic time (Fig. 5). The major uplifts such as the Defiance, Zuni, Monument, San Rafael, Kaibab, Uncompahgre, and San Luis were significant features during the Paleozoic, as were the San Juan and Paradox Basins. Certainly Laramide forces generally uplifted and enhanced the structures to form magnificent, textbook quality exhibits. However, the structural configuration of the Colorado Plateau and the excellent examples evident in the Paradox Basin resulted from a long history of reactivations of pre-existing Precambrian basement tectonics.

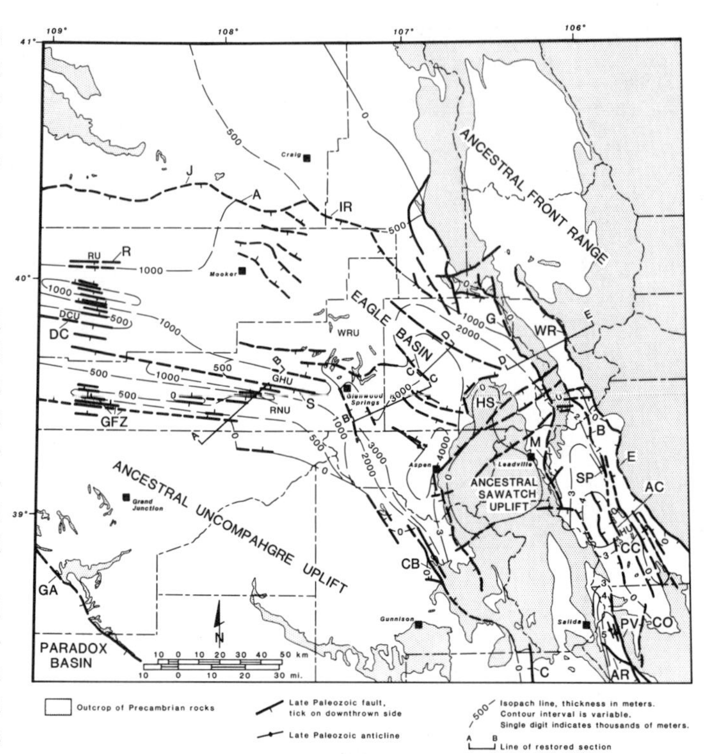

Figure 9. Late Paleozoic tectonic features and thickness of Late Paleozoic rocks in the Eagle Basin, northwestern Colorado. Labeled faults: A, Axial; AC, Agate Creek; AR, Arkansas River; B, Boreas; C, Chester; CB, Crested Butte; CC, Currant Creek; CO, Cotopaxi; DC, Douglas Creek; E, Elkhorn; G, Gore; GA, Gateway; GFZ, Garmesa fault zone; HS, Homestake shear zone; IR, Indian River; J, Juniper; M, Mosquito; PV, Pleasant Valley; R, Rangely; S, Silt; SP, South Park; WR, Williams Range. Labeled uplifts: DCU, Douglas Creek; GHU, Grand Hoback; HU, Hartsel; RU, Rangely; RNU, Rulison. Laramide uplift: WRU, White River.

LATE PALEOZOIC EAGLE BASIN AND ADJACENT AREAS, NORTHWESTERN COLORADO

Richard H. De Voto, Bruce L. Bartleson, Christopher J. Schenk, and Noel B. Waechter

Northwestern Colorado was broken into a mosaic of fault blocks in late Paleozoic time (Fig. 9). Tectonic activity on these faults produced the sedimentary and structural basin of the northern Central Colorado Trough (Plate 2), outlined by the zero isopach of Pennsylvanian and Permian rocks on Figure 9, its many subbasins, the two principal adjacent mountainous erosional areas (the ancestral Uncompahgre and Front Range Uplifts), the basin-center ancestral Sawatch Uplift and many other uplifts within the basin. The Eagle Basin is a part of the northern Central Colorado Trough, generally considered to be the area of occurrence of the Eagle Valley Evaporite (see Fig. 13). The late Paleozoic faults in central Colorado have a dominant north-northwest to northwest trend, and those in northwestern Colorado are east-west trending (Fig. 9).

The Pennsylvanian and Permian strata of northwestern Colorado (Fig. 10) record a history of active basin subsidence along numerous faults and contemporaneous sedimentation of as much as 5,000 m of nonmarine to marine strata (Fig. 11). Abrupt thickness variations and facies changes, onlap and overlap relationships, and angular discordances in a few localities occur adjacent to faults and uplift blocks (De Voto, 1980b). Faults with significant syndepositional offset throughout the Pennsylvanian and Permian include the Garmesa and Juniper–Axial–Indian River Fault Zones and the Gore, Boreas Pass, Elkhorn, South Park, Agate Creek, Pleasant Valley, Arkansas River, and Crested Butte Faults (De Voto and others, 1986; Stone, 1977, 1986). Many other intrabasinal faults were particularly active during the Desmoinesian, producing local horsts, grabens, and half grabens. Movement on these structures significantly influenced accumulation of thick evaporite sequences and abrupt facies and thickness changes (De Voto and others, 1986; Dodge and Bartleson, 1986; Schenk, 1988; Waechter and Johnson, 1986).

MORROWAN-ATOKAN

The Morrowan-Atokan strata (0–500 m thick) generally record a transgression into the Central Colorado Trough from the northwest over a low-relief Late Mississippian humid-climate, karst erosion surface. A residual deposit of red, hematitic silt and clay with chert breccia fragments and rounded clasts of limestone and dolomite occurs discontinuously on the karst surface beneath Pennsylvanian strata. This residual deposit, the Molas Formation, ranges up to 40 m thick where it occurs in Late Mississippian sinkholes (De Voto, 1980a, 1985, 1988).

Deposition of marine shale and siltstone and sandstone turbidites of the Belden Formation occurred in the basin center (Brill, 1944, 1952, 1958; Langenheim, 1952), while marine carbonates and marine and eolian sandstones of the Morgan Formation were deposited in more basin-marginal positions and in the northwestern part of the basin (Figs. 10 an 12) (Driese and Dott, 1984; Irtem, 1977). Uplift and erosion of portions of the ancestral Front Range and Sawatch Uplifts caused coarse-grained fluvial and fan-delta sediments containing clasts of Precambrian rocks to be shed to local basin-edge areas (De Voto, 1972b, 1980b; Kelly, 1984). To the south within the Central Colorado Trough, the Belden strata change facies to dominantly transitional and nonmarine strata of the Kerber (Morrowan) and Sharpsdale (Atokan) Formations (Fig. 12) (De Voto and Peel, 1972; Pierce, 1969). The Kerber Formation comprises carbonaceous shale and siltstone, quartz sandstone, and coal. The Sharpsdale Formation includes red, arkosic sandstone, conglomerate, siltstone, and shale.

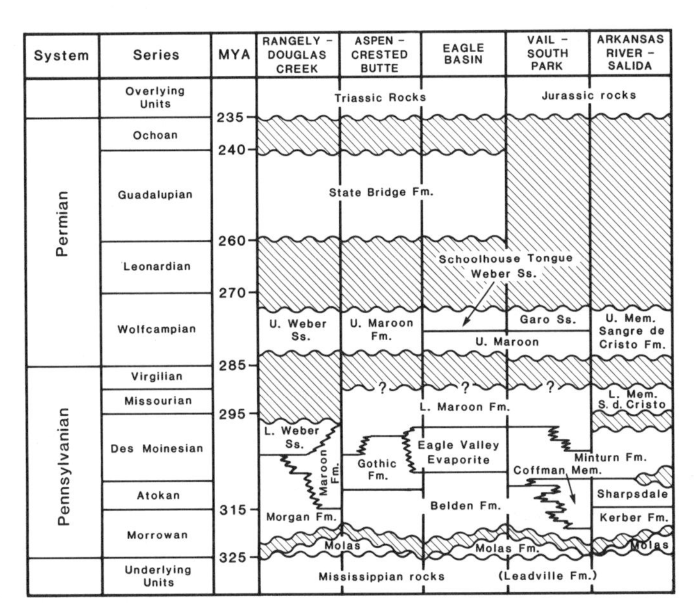

Figure 10. Correlation chart of Pennsylvanian and Permian rocks of northwestern Colorado.

Numerous shoaling-upward, marine to nonmarine cycles of deposition are widespread within the basin, within the Atokan portion of the Morgan Formation and the Coffman Member of the Minturn Formation (Fig. 12) (Driese and Dott, 1984; Kelly, 1984). These probably represent the sedimentary response to repetitive eustatic sea-level changes within the Central Colorado Trough during the Atokan.

DESMOINESIAN

The Desmoinesian strata record a history of vigorous tectonic activity and basinal subsidence that produced complex patterns of alluvial-plain sedimentation at the edge of high-relief, mountainous uplift blocks (particularly the ancestral Uncompahgre and Front Range Uplifts) and basinal, coastal to marine and local salt-basin sedimentation (Fig. 13). The Pennsylvanian climate of the Central Colorado Trough was becoming increasingly arid as a result of the plate movement of Colorado into trade-wind latitudes and the orographic effects of the adjacent mountainous areas.

Glacial-eustatic cycles of marine carbonates to nonmarine eolian sandstones dominate the Desmoinesian strata of the Morgan Formation of the northwestern part of the Central Colorado Trough (Driese and Dott, 1984). Interbedded marine to nonmarine, gray-green sandstone, siltstone, shale, and carbonate rocks of

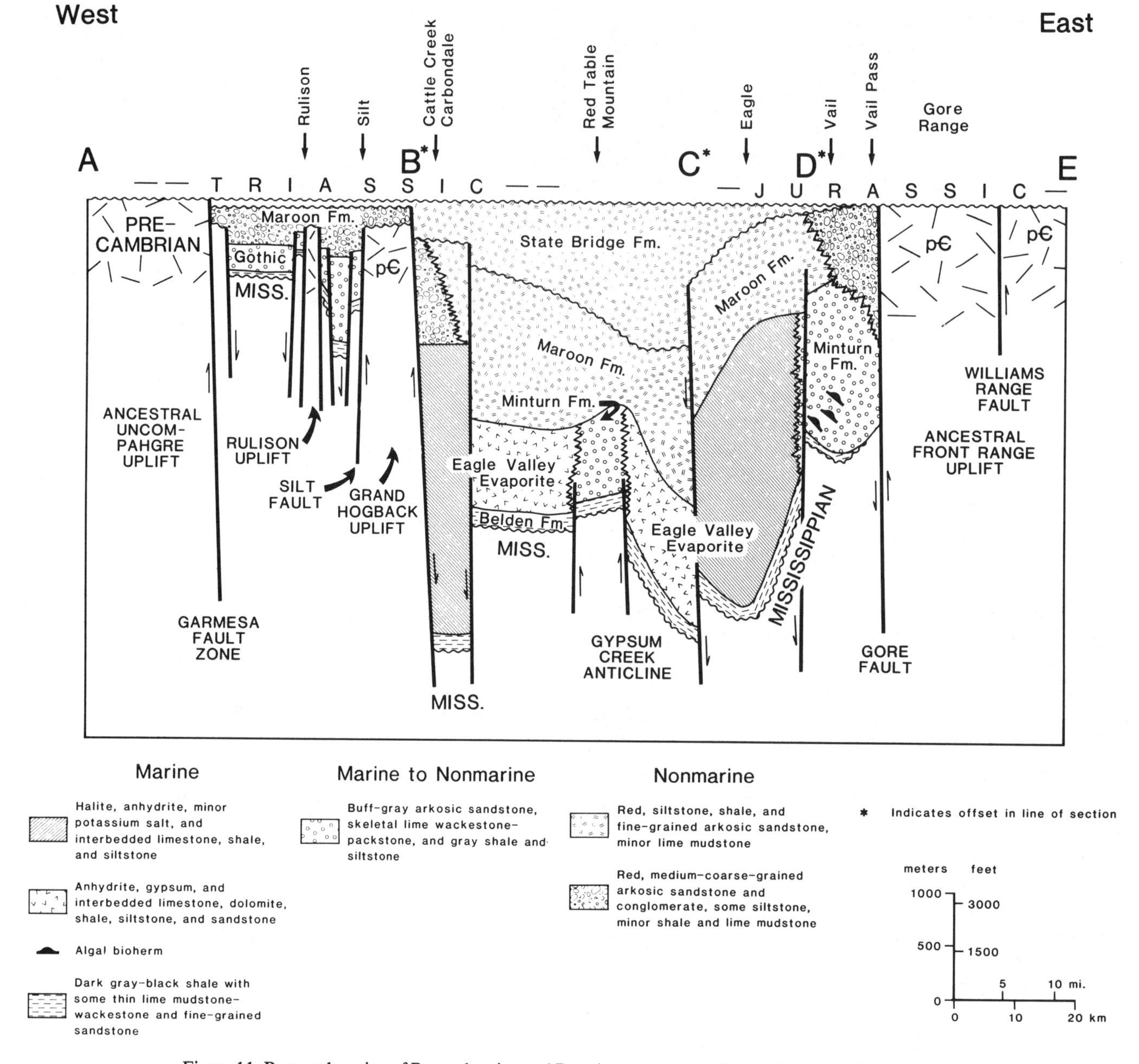

Figure 11. Restored section of Pennsylvanian and Permian rocks, Eagle Basin, Colorado. Line of section shown on Figure 9.

the Gothic and Minturn Formations occur in basin-marginal positions from Meeker southward within the trough (Fig. 13) (Tweto, 1949, 1958; Murray, 1958; Chronic and Stevens, 1958; Langenheim, 1952, 1954; Mallory, 1958, 1960, 1972a; Bartleson, 1968, 1972; De Voto, 1972b, 1980b; Boggs, 1966; Tillman, 1971; Taranik, 1974). The lithofacies generally increase in grain size and siliciclastic content from basin center to basin margin. The Gothic and Minturn strata include at least ten shoaling-upward depositional sequences (cycles); from base to top they are generally composed of marine limestone; marine to shoreline shale, siltstone, and sandstone; and fluvial sandstone and conglomerate. Algal bioherms are conspicuous in several of the depositional sequences, particularly in close proximity to the facies change with and within the Eagle Valley Evaporite (Fig. 13) (Walker, 1972; Irtem, 1977; Wray, 1985; Dodge and Bartleson, 1986).

As much as 2,700 m of interbedded evaporite, gray shale, siltstone, limestone, dolomite, and sandstone of the Eagle Valley Evaporite of Desmoinesian age occupy the center of the Eagle

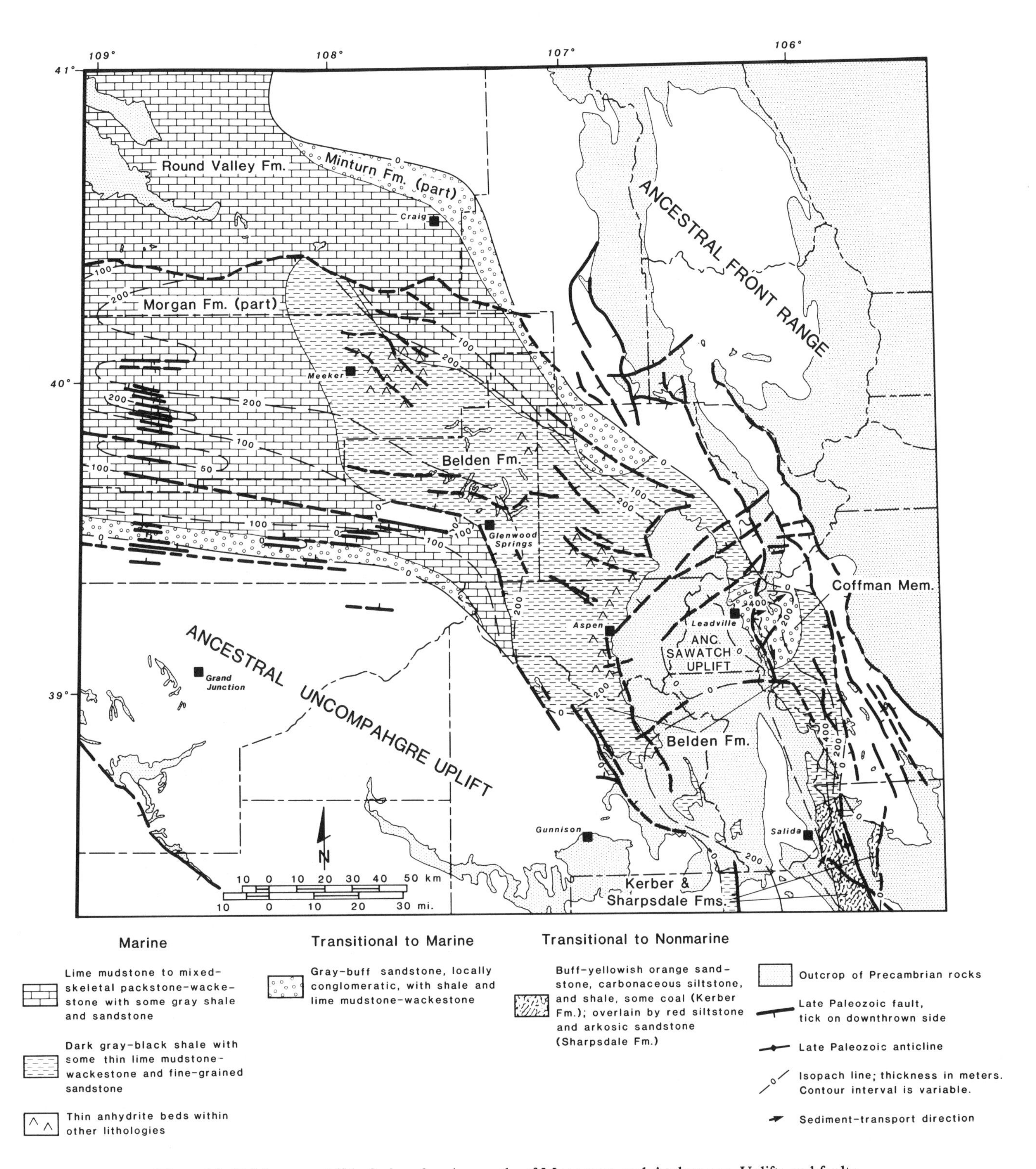

Figure 12. Thickness and lithofacies of rocks mostly of Morrowan and Atokan age. Uplifts and faults labeled on Figure 9.

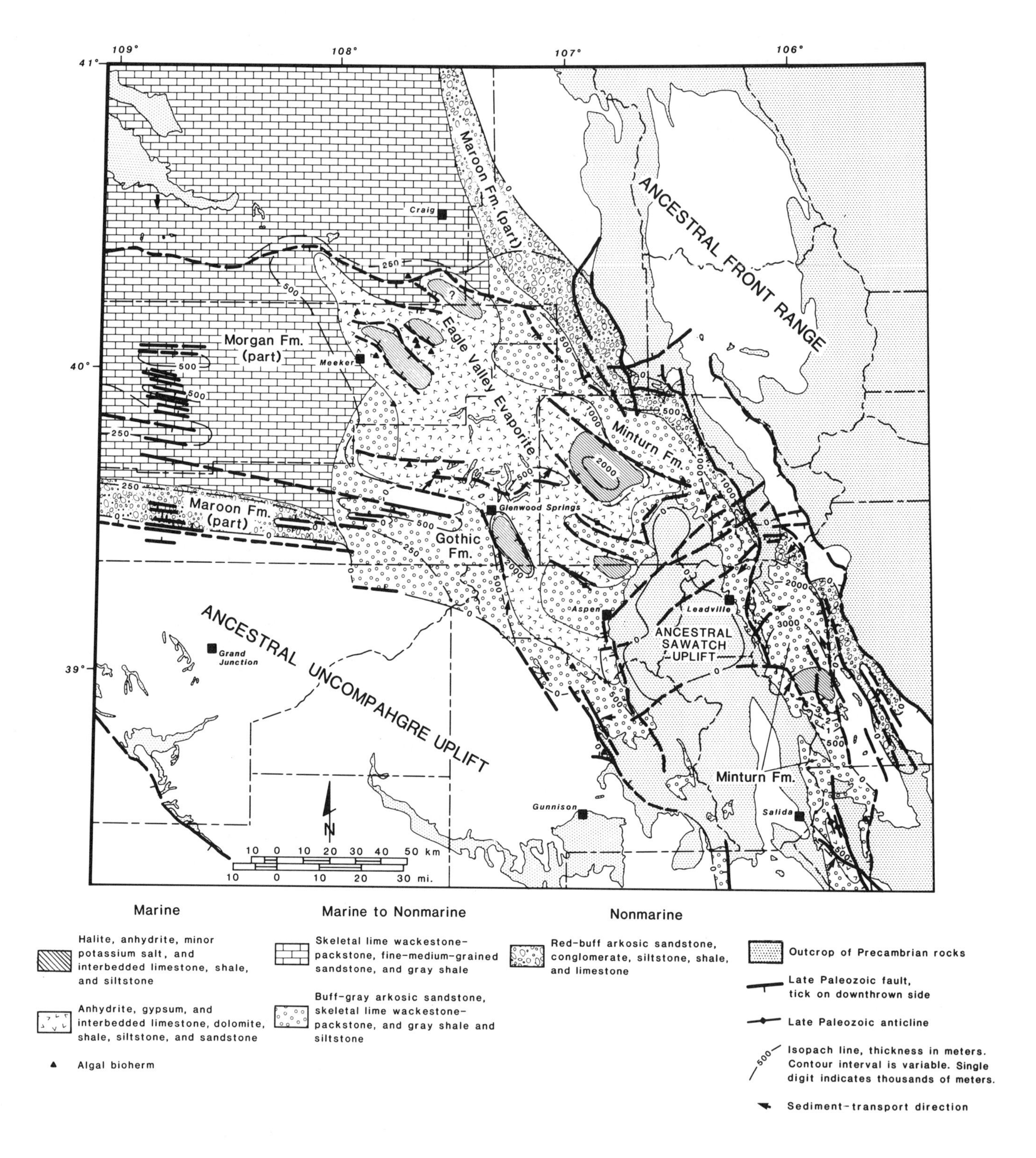

Figure 13. Thickness and lithofacies of rocks mostly of Desmoinesian age. Uplifts and faults labeled on Figure 9.

Basin (Katich, 1958; Mallory, 1971, 1972a, 1977). Lithologies are complexly interbedded and change facies and thickness abruptly in apparent response to local topographic changes caused by syndepositional faulting (Figs. 11 and 13) (De Voto and others, 1986; Dodge and Bartleson, 1986). At least ten shoaling-upward depositional sequences (cycles) occur in the Eagle Valley Evaporite. Several of these cycles generally include basal black shale and siltstone interval, marine carbonate interval, and upper well-laminated anhydrite (Dodge and Bartleson, 1986; De Voto and others, 1986). Some cycles contain shoaling-upward sandstones, ranging from marine turbidites to shoreline, eolian, and fluvial deposits (Schenk, 1986). Locally these rocks are laterally replaced by thick interbedded anhydrite and halite sequences. Some of the cycles pass vertically from well-laminated gypsum to a shoaling-upward carbonate sequence with an abrupt, upper, solution-erosion and brecciated karst surface (Schenk, 1988). The evaporite strata were deposited during particular phases of eustatic cycles, probably from salinity- and density-stratified brines below wave base in actively subsiding, locally tectonically influenced, subbasinal areas within the central part of the Eagle Basin. The presence of glacial-eustatic, marine to nonmarine depositional sequences throughout the basin suggests that cycles of marine flooding and alternate subaerial emergence of much of the basin may have occurred periodically throughout the Desmoinesian (De Voto and others, 1986).

LATE PENNSYLVANIAN AND PERMIAN

The Upper Pennsylvanian and Lower Permian strata of the northern Central Colorado Trough are dominated by a thick (as much as 4,500 m) sequence of nonmarine red beds and eolian sandstones. The eolian fine-grained sandstones of the Weber Sandstone of northwestern Colorado change facies south and southeast within the basin to red, arkosic sandstone, siltstone, and shale and at basin-margin positions to red, coarse-grained sandstone and conglomerate of the Maroon Formation (Figs. 11 and 14) (Bissell and Childs, 1958; Bissell, 1964; Fryberger, 1978, 1979; Whitaker, 1975; De Voto, 1980b; De Voto and others, 1986). Transport directions of the fluvial sandstones indicate that the topographic low of the basin was near its northern margin (Johnson, 1987).

Desmoinesian and Wolfcampian faunas occur in Weber and Maroon strata (Bissell, 1964), whereas Missourian and Virgilian faunas are sparse. These data suggest that a significant Late Pennsylvanian stratigraphic hiatus may exist within the Weber and Maroon strata (Fig. 10) (De Voto and others, 1986). Several unconformities are locally developed within the Desmoinesian to Lower Permian strata adjacent to the Pleasant Valley and Arkansas River Faults east of Salida (Pierce, 1969; De Voto and others 1971; De Voto and Peel, 1972). Adjacent to the Pleasant Valley Fault, a 90-degree angular unconformity exists between the lower and upper members of the Sangre de Cristo Formation. The amount of truncation of underlying strata (0–2,000 m), the angular discordance at the Late Pennsylvanian unconformity, and the thickness of onlapping strata (0–1,525 m) increase toward the Pleasant Valley Fault (Pierce, 1969; De Voto, 1980b).

Significant post-Wolfcampian uplift and folding of the Maroon Formation occurred prior to deposition of the red siltstone, sandstone, shale, and limestone strata of the State Bridge Formation in the Guadalupian (Sharps, 1955, 1962; Freeman, 1971; Freeman and Bryant, 1977).

ECONOMIC GEOLOGY

More than 600 million barrels of oil have been produced from the Weber Sandstone on a doubly plunging Laramide anticlinal structure at the Rangely oil field in northwestern Colorado. Several workers have attributed the oil at Rangely to pre-Laramide stratigraphic entrapment at the facies change from well-sorted sandstones of the Weber Sandstone to poorly sorted arkosic sandstones and siltstones of the Maroon Formation (Fig. 14) (Hoffman, 1957; Whitaker, 1975; Fryberger, 1979; Koelmel, 1986). Most workers believe that the oil was derived from organic-rich shales of the stratigraphically higher Phosphoria Formation that were deposited 100 to 300 kilometers to the west during the Late Permian. However, a few workers have recently suggested that the oil at Rangely may be derived from stratigraphically lower organic-rich shales in the Eagle Valley Evaporite or Belden Formation (Figs. 10 and 13) and that the oil at Rangely may have accumulated in a pre-Laramide, late Paleozoic structural trap (Waechter and Johnson, 1986). Several other smaller oil fields and numerous oil-stained outcrops occur in the Weber Sandstone throughout the northern part of the Central Colorado Trough.

Since 1980, several oil wells have been completed in sandstones of the Eagle Valley Evaporite north of the White River Uplift (Dodge and Bartleson, 1986). Oil staining occurs in algal bioherms of the Eagle Valley Evaporite in outcrops and the subsurface near Meeker (Irtem, 1977, 1983). Oil staining is abundant in sandstones of the Belden, Eagle Valley, and Minturn Formations in outcrops throughout the basin. It is likely that the oil has been derived from organic-rich shales within the Eagle Valley Evaporite.

De Voto (1983) has suggested that many of the lead-zinc-silver-barite deposits in central Colorado may have been formed as a result of late Paleozoic tectonism, basin subsidence, sedimentation, and fluid evolution within the basin (this is not, however, the prevailing opinion). Through the generation of hydrocarbons and overpressured conditions in the basin in the late Paleozoic, metal-bearing brines could have developed and could have moved along basin-margin faults, permeable aquifers, and the underlying Late Mississippian karst-solution features to favorable sites for metal deposition in the Late Mississippian karst features within the Mississippian Leadville Limestone in basin-edge uplift locales. Hundreds of these karst-controlled deposits occur in the Leadville Limestone and other carbonate units below the Late Mississippian karst surface at outcrops throughout central Colorado.

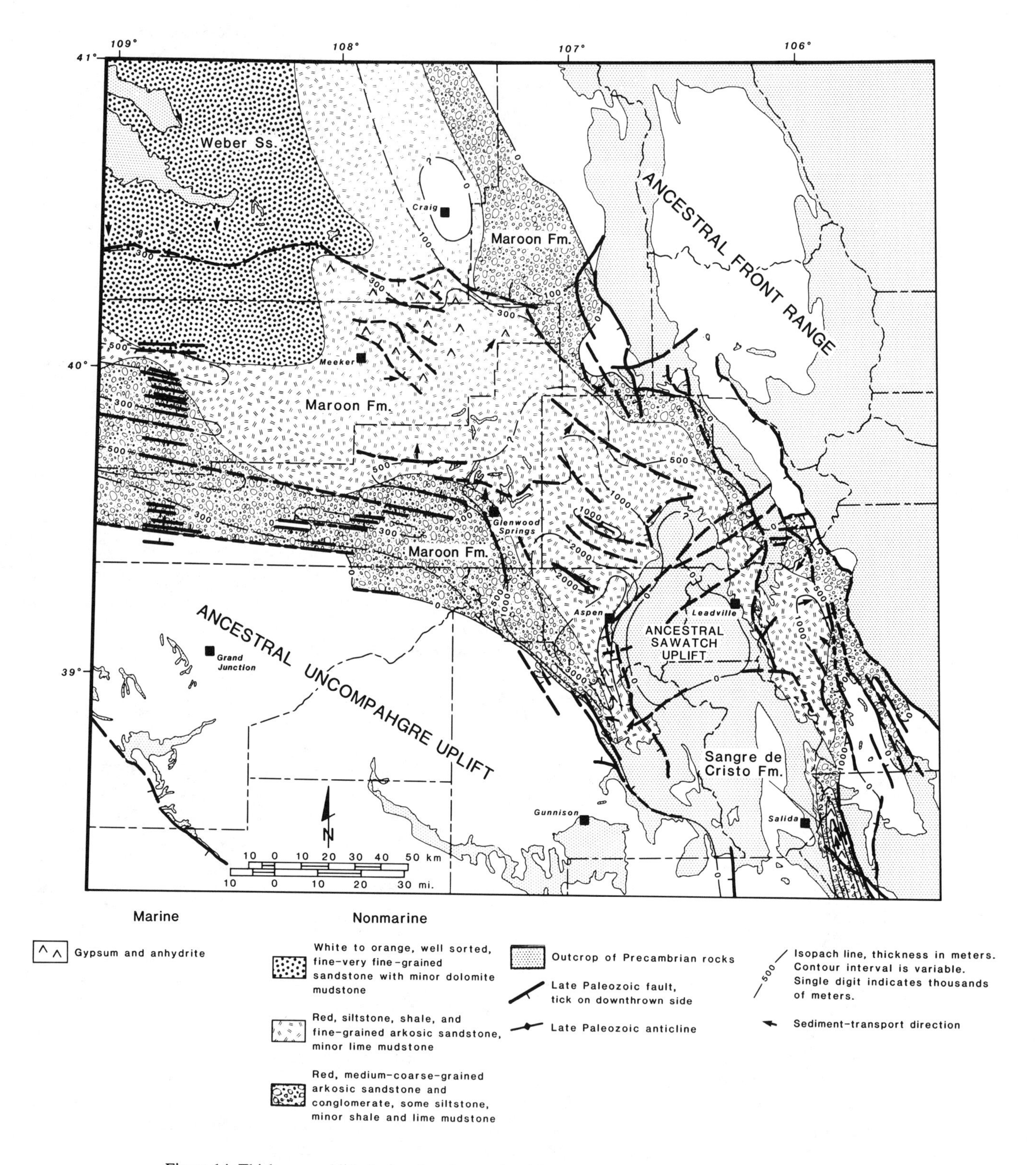

Figure 14. Thickness and lithofacies of rocks mostly of Late Pennsylvanian (Missourian-Virgilian) and Early Permian (Wolfcampian) age. Uplifts and faults labeled on Figure 9.

CRETACEOUS AND TERTIARY ROCKS OF THE SAN JUAN BASIN

C. M. Molenaar

INTRODUCTION

The San Juan Basin of northwestern New Mexico and southwestern Colorado (Plate 2) is an early Tertiary structural depression in which thick sections of Cretaceous and early Tertiary rocks are preserved. The basin is 160 to 240 km across and is elliptical to nearly circular in form. It is bounded by steeply dipping monoclinal segments on the west, north, and east sides and by the low-dipping flank of the Zuni Uplift on the south (Fig. 15).

As much as 4,600 m of sedimentary section overlies Precambrian granitic and metamorphic basement rocks in the deep northern part of the basin near the Colorado–New Mexico border. Pre-Cretaceous sedimentary rocks are as thick as 1,900 m. The extensive outcrops, the large number of well penetrations (more than 13,000 wells), and the simple structure of the basin make this area ideal for scientific as well as economic studies of Cretaceous and Tertiary rocks. Consequently, the stratigraphy and geometry of these rocks—especially the Upper Cretaceous rocks, which contain large reserves of gas, oil, and coal in or flanking the San Juan Basin—have been worked out in great detail by a succession of workers. This chapter is a synthesis of the many fine papers published by these workers.

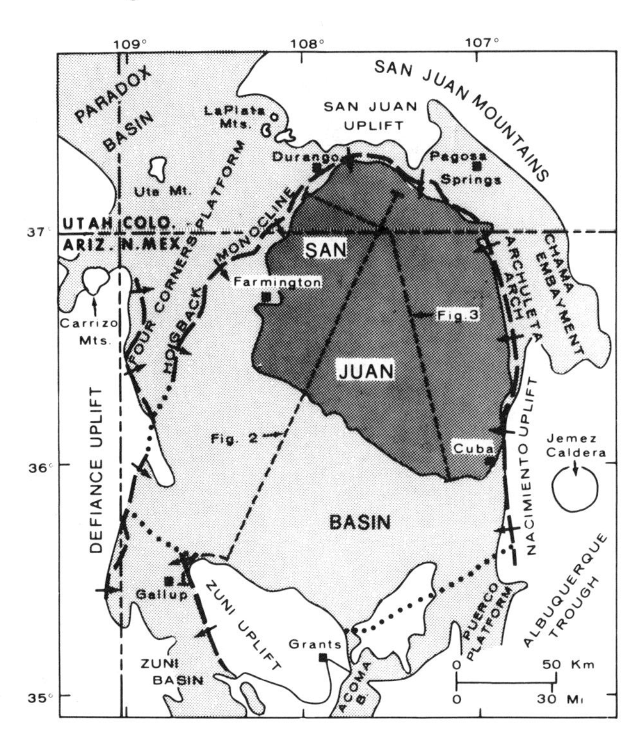

Figure 15. Map of San Juan Basin area showing surrounding structural elements, Tertiary outcrop pattern (dark shading) within San Juan Basin, and Cretaceous outcrop pattern (light shading) within and surrounding the San Juan Basin. Modified from Kelley (1951). Lines of sections of Figures 16 and 17 are indicated.

UPPER CRETACEOUS ROCKS

During Late Cretaceous time, the San Juan Basin area was part of the large Western Interior Basin that transected North America. As much as 2,000 m of Upper Cretaceous strata, consisting of shale, siltstone, and sandstone with lesser amounts of coal and limestone, are preserved in the San Juan Basin. These rocks were deposited in marine and nonmarine environments as a generally northwest-trending shoreline advanced and retreated several times across parts of the area. The depositional environments range from alluvial or upper delta plain landward to the southwest, through coastal or lower delta plain, coastal barrier, and shoreface in the middle, to offshore marine to the northeast.

Four major transgressive-regressive cycles are recognized in the San Juan Basin; a fifth cycle is recognized in the lower part of the marine sequence just south of the San Juan Basin (Molenaar, 1983). Figure 16 is a stratigraphic cross section based on detailed well control and measured surface sections showing the regional relations of Upper Cretaceous strata across the San Juan Basin. (See Molenaar, 1983, fig. 4, for time-stratigraphic relations along this same line of section.) The ages of the various units are based primarily on megafossil control from outcrops around the basin (Peterson and Kirk, 1977). Detailed well-log correlations of bentonite beds, calcareous zones, or silty zones in marine strata have helped in understanding details of the transgressive and regressive cycles of the San Juan Basin. Some of these features are shown in Figure 16. In addition, from the cross section several other depositional features can be inferred, such as (1) shelf breaks and very low-angle slope or clinoform bedding in parts of the thick shale sections; (2) depositional thickening indicating increased subsidence to the south or southwest; (3) minor irregularities in the overall uniformity of subsidence, which may indicate recurrent tectonism; and (4) stratigraphic rises of the different transgressive and regressive units. The term "stratigraphic rise" is used here to denote the vertical difference in stratigraphic position of a time-transgressive lithologic unit between two geographic points. Because shoreline sandstones are deposited in similar water depths, i.e., near sea level, the stratigraphic rise plus the amount of differential compaction is a measure of basin subsidence or relative sea-level rise.

The following are some general comments on the better known or significant Upper Cretaceous rock units in the San Juan Basin. Some of the thickness variations are shown on the cross section (Fig. 16).

Dakota Sandstone (Middle to Upper Cenomanian)

The Dakota Sandstone comprises deposits associated with the first transgression of the Cretaceous seaway into the San Juan

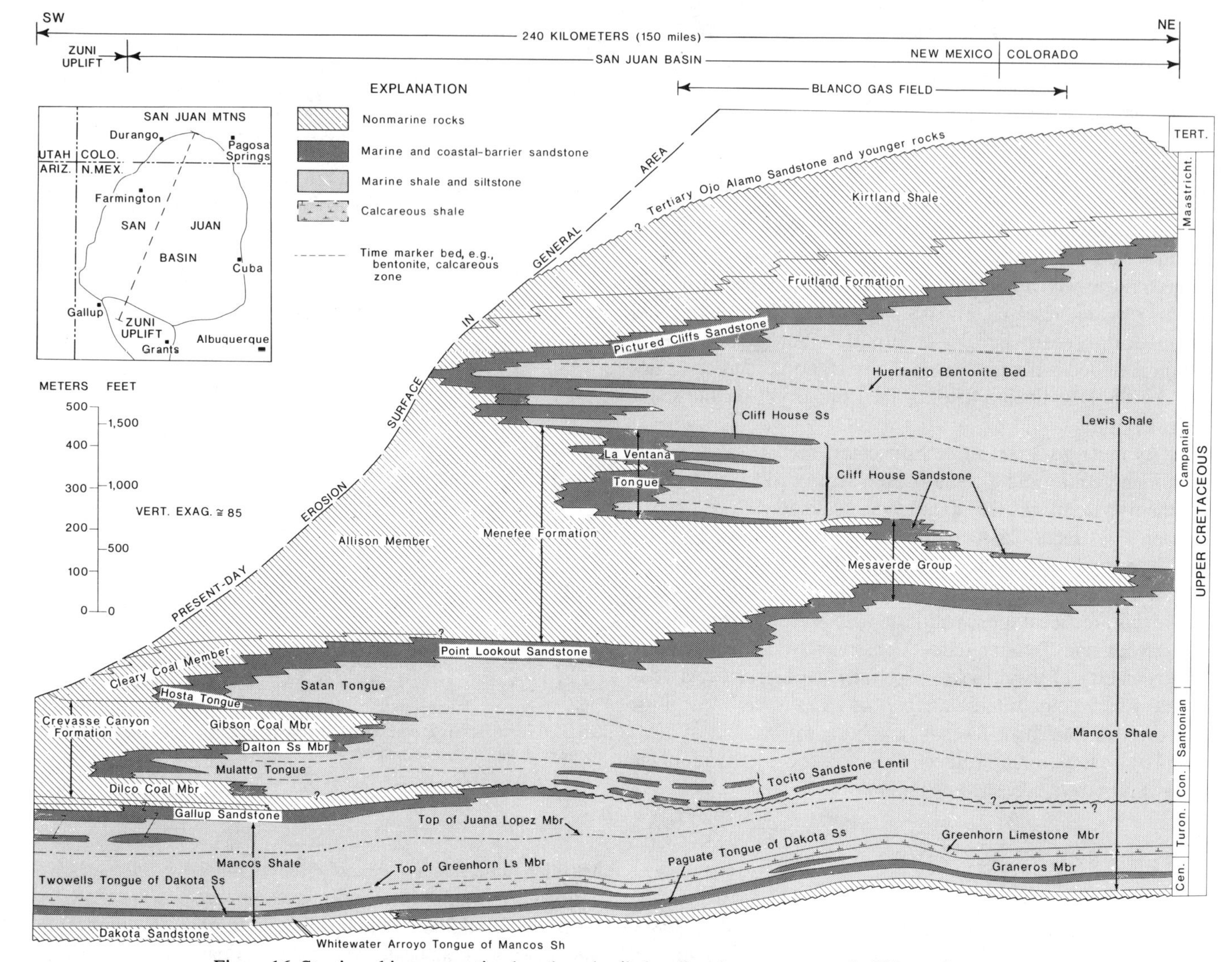

Figure 16. Stratigraphic cross section based on detailed well and outcrop control of Upper Cretaceous rocks across the San Juan Basin, northwestern New Mexico–southwestern Colorado. Cen. = Cenomanian, Turon. = Turonian, Con. = Coniacian, Maastricht. = Maastrichtian. Location of section is shown in Figure 15.

Basin area. The Dakota is a complex unit ranging in thickness from about 20 to 120 m. It consists generally of fluvial sandstone and carbonaceous shale with thin beds of coal in the lower part and a thinner section of marine sandstone in the upper part. The Dakota unconformably overlies the Lower Cretaceous Burro Canyon Formation (a 0- to 40-m-thick fluvial unit) in the northern part of the San Juan Basin and the Jurassic Morrison Formation in the southern part. The transgression advanced generally from east to west across the San Juan Basin. This north-trending shoreline differed from northwest-trending shorelines of later Cretaceous time, probably because of pre-Dakota topographic relief. On the southeast side of the basin, the Dakota is about 120 m thick and has been subdivided into four dominantly sandstone members or tongues, separated by tongues of Mancos Shale (Landis and others, 1973). The sandstone tongues are of shallow-marine shelf origin and extend as much as 200 km eastward from their respective shorelines on the west. These sandstone tongues are discontinuous and are as much as 25 m thick. The lower tongues grade westward into nonmarine deposits within the San Juan Basin. The uppermost and most widespread tongue, the Twowells, merges with the nonmarine Dakota farther west near the Arizona–New Mexico border. The total stratigraphic rise from east to west of the base of the marine deposits, resulting from this first transgression across the San Juan Basin, is about 110 m.

Mancos Shale (middle Cenomanian to lower Campanian)

The Mancos Shale was deposited in deeper water in offshore

areas where energy levels were low, allowing finer clastics to settle out. The maximum water depth in the distal offshore areas is estimated to have been 100–150 m, as suggested by presently compacted low-angle clinoforms (less than 0.5°) (Fig. 16) and with allowances made for compaction.

In addition to shale and siltstone, the Mancos contains zones of calcareous concretions, a few thin limestone beds, many thin bentonite beds, and a few offshore sandstone units. Some of these beds or markers can be correlated for many kilometers on well logs in the subsurface. Some units are prominent enough both in outcrops and in the subsurface to warrant formal member status. Two of the southwesterly extending tongues of Mancos Shale are named the Mulatto Tongue and the Satan Tongue (Fig. 16).

Greenhorn Limestone Member (lower Turonian). The Greenhorn Limestone Member of the Mancos Shale is composed of calcareous shale and thin beds of argillaceous limestone. The limestone beds pinch out in the southwestern part of the basin, but the calcareous shales are still recognizable, especially on electric (resistivity) logs. The Greenhorn ranges in thickness from about 12 to 20 m and probably closely approximates an isochronous unit.

Juana Lopez Member (upper Turonian). The Juana Lopez Member of the Mancos Shale is composed of thinly interbedded calcarenite, very fine grained quartzose sandstone, and shale. The calcarenite is a hash of fragmented fossils. The Juana Lopez is 33 m thick at a reference section on the east side of the San Juan Basin (Dane and others, 1966) and contains several species and genera of cephalopods and pelecypods, some of which are index fossils used in zoning this thin unit.

The calcarenite bed that defines the top of the Juana Lopez forms a small cuesta in outcrops and is a good electric-log marker in the San Juan Basin; it probably represents an isochronous unit over much of the area. Faunal zonation, however, indicates that the top of the Juana Lopez in outcrops in the southern part of the basin is time-equivalent to the middle part of the reference section in the eastern part of the basin. Apparently the lithologic top of the Juana Lopez drops to a lower calcarenite bed to the southwest. Deposition of the lower part of the Juana Lopez was associated with a major transgressive episode that was occurring in areas south of the San Juan Basin (Molenaar, 1983).

Gallup Sandstone (upper Turonian to lower Coniacian)

The Gallup Sandstone consists of coastal-barrier sandstones, paludal carbonaceous shales and coal, and distributary or fluvial channel sandstones. The Gallup comprises deposits associated with the first major regression of the seaway in the San Juan Basin area. This regressive episode is represented only in northwestern New Mexico and northeasternmost Arizona. The Gallup has no known time-equivalent regressive counterparts in other parts of the Western Interior. The stratigraphic rise of the marine part of the Gallup Sandstone from its landward extent, about 75 km southwest of the San Juan Basin, to its seaward pinchout within the San Juan Basin, a distance of about 130 km, is about 100 m.

Tocito Sandstone Lentil of the Mancos Shale (middle to upper Coniacian)

The Tocito Sandstone Lentil of the Mancos Shale consists of a band of elongate northwest-trending offshore sandstone lenses extending across the central part of the San Juan Basin. Abundant medium-scale tabular cross beds indicate that the sandstone was deposited by southeast-flowing currents, which approximately parallel the shoreline some distance to the southwest. These sandstone lenses, which individually are generally less than 15 m thick and are fine to coarse grained, are the major oil producers in the central San Juan Basin—the so-called Gallup oil fields or pools. These sandstone lenses are younger and are much coarser grained than the fine-grained regressive sandstones of the Gallup and are now known as the Tocito Sandstone Lentil of the Mancos Shale. Sandstone lenses of the Tocito are developed within a 25-m-thick interval above an unconformity, which is above the seaward extent of the Gallup Sandstone or within the Mancos Shale farther seaward (Fig. 16). In places, the unconformity cuts as far down section as the Juana Lopez (McCubbin, 1969; Molenaar, 1973).

Crevasse Canyon Formation (Coniacian to middle Santonian)

The Crevasse Canyon Formation consists of the predominantly nonmarine deposits between the Gallup Sandstone and the Point Lookout Sandstone in the southern part of the San Juan Basin. In ascending order, the various members of the Crevasse Canyon are the Dilco Coal Member; the Dalton Sandstone Member, a regressive coastal-barrier sandstone; the Bartlett Barren Member; and the Gibson Coal Member. The Bartlett Barren Member is not shown in Figure 16 because of space limitations and also because it merges with the Gibson through facies change in the vicinity of the section.

Mesaverde Group

The Mesaverde Group of the type area at Mesa Verde National Park in southwestern Colorado is made up of, in ascending order, the Point Lookout Sandstone, Menefee Formation, and Cliff House Sandstone. In this report, the Gallup Sandstone and Crevasse Canyon Formation of the southern part of the San Juan Basin are not included in the Mesaverde Group. The name Mesaverde as used in other parts of the Western Interior refers only generally to the thick marine and nonmarine units overlying the main thick Upper Cretaceous shale.

Point Lookout Sandstone (middle Santonian to lower Campanian). The Point Lookout Sandstone is the most extensive regressive coastal-barrier sandstone in the San Juan Basin. This regressive episode was widespread throughout the Western Interior. In prograding across the San Juan Basin, a distance of about 210 km, the Point Lookout rises stratigraphically about 365 m (Fig. 16). It merges with the overlying Cliff House Sandstone on the north side of the basin. In outcrops west of Pagosa Springs, the combined unit consists of about 30–45 m of thin-bedded marine sandstone. The pinchout probably was not much farther to the northeast.

Menefee Formation (middle Santonian to upper Campanian). The Menefee Formation, the middle unit of the Mesaverde Group, includes nonmarine-paludal to alluvial-plain deposits that accumulated landward from the Point Lookout and Cliff House shorelines to the northeast. The Menefee is composed of paludal carbonaceous shales and coals, fluvial sandstones, and floodplain shales. The thicker coals occur in proximity to the shorelines, where environmental conditions were more conducive to the longevity of coal swamps. In the southern part of the San Juan Basin, the Menefee is divided into two members: the Cleary Coal Member at the base and the generally less coal-bearing Allison Member above. However, the upper part of the Allison is very carbonaceous and coaly, as coastal-swamp conditions were again repeated landward from the Cliff House shoreline. This zone is referred to as the upper coal-bearing unit of the Menefee.

Cliff House Sandstone (lower to upper Campanian). The Cliff House Sandstone is a complex sequence of sandstone tongues of variable thickness that were deposited in a generally transgressive cycle. However, even though the Cliff House is considered an overall transgressive sandstone, genetically the thick sandstone buildups are offlap or regressive deposits. These sandstone units clearly intertongue with the marine Lewis Shale to the northeast and the nonmarine Menefee Formation to the southwest (Fig. 16). The largest buildup, which attains a maximum thickness of about 240 m, is known as the La Ventana Tongue of the Cliff House. Several smaller buildups or steps that are present to the northeast in the deeper part of the basin are gas productive. The total stratigraphic rise of the Cliff House (or basal Lewis) from northeast to southwest is about 400 m over a distance of about 140 km (Fig. 16).

Lewis Shale (lower to upper Campanian)

The Lewis Shale was deposited as a deep-water, offshore shale associated with a significant transgressive episode that has equivalents in other parts of the Western Interior. However, it should not be confused with the "Lewis Shale" of Wyoming and northwest Colorado, which was deposited during a later transgression that is not represented in the San Juan Basin. Like the Mancos Shale, the Lewis contains many thin bentonite beds that serve as reliable time-marker beds for subsurface correlation. One of the thickest and most widespread of these has been named the Huerfanito Bentonite Bed (Fassett and Hinds, 1971). The Huerfanito is a few meters thick and can be correlated throughout the basin on well logs. It is an excellent time marker bed to which stratigraphy of underlying and overlying formations can be related. As shown in Figure 16, these time marker beds do not necessarily represent horizontal surfaces.

Pictured Cliffs Sandstone (upper Campanian)

The Pictured Cliffs Sandstone is a regressive coastal-barrier sandstone associated with the final retreat of the Western Interior sea from the San Juan Basin area. The continued regression of the sea to the northeast of the San Juan Basin resulted in deposition of the younger Trinidad Sandstone in the Raton Basin in southeastern Colorado and northeastern New Mexico. The total stratigraphic rise of the Pictured Cliffs Sandstone across the San Juan Basin from outcrops on the southwest to outcrops on the northeast, a distance of about 130 km, is about 350 m (Fassett and Hinds, 1971).

Fruitland Formation (upper Campanian to Maastrichtian[?])

The Fruitland Formation comprises nonmarine, lower coastal plain deposits that accumulated landward from the Pictured Cliffs shoreline. It ranges in thickness from 0 to 150 m and consists of paludal carbonaceous shale and coal interbedded with siltstone and sandstone (Fassett and Hinds, 1971). The thickest coal beds in the basin, some as much as 13 m thick, occur in the Fruitland.

Kirtland Shale (upper Campanian to Maastrichtian [?])

The Kirtland Shale conformably overlies the Fruitland Formation and consists of upper coastal plain or alluvial-plain deposits that accumulated landward from the Fruitland coal swamps. It has been divided into three units called, in ascending order, the lower shale member, Farmington Sandstone Member, and upper shale member. The lower shale member consists predominantly of shale containing a few thin interbeds of sandstone and siltstone, the Farmington Sandstone consists of a number of fluvial sandstone beds interbedded with shale, and the upper shale member consists of shale and lesser amounts of sandstone. Owing to truncation prior to Tertiary deposition, the Kirtland Shale thins across the San Juan Basin from about 450 m on the northwest to zero on the southeast (Fassett and Hinds, 1971).

McDermott Member of the Animas Formation (Maastrichtian)

The McDermott Member of the Animas Formation is a local unit present in outcrops along the northwest side of the San Juan Basin, primarily in the Colorado part, where it overlies, probably conformably, the Kirtland Shale (Fassett, 1985). It is primarily a volcaniclastic unit of andesitic composition consisting of purple to brown sandstone, conglomerate, shale, and ash beds and is about 90 m thick in the Durango area. On the basis of dinosaur bones reported by Reeside (1924), the McDermott has been considered Late Cretaceous in age. It is now mapped as a basal member of the Animas Formation, which otherwise is Paleocene in age.

The McDermott represents a dramatic change in provenance from the distant southwesterly source for underlying Upper Cretaceous units to a local north or northwestern source as igne-

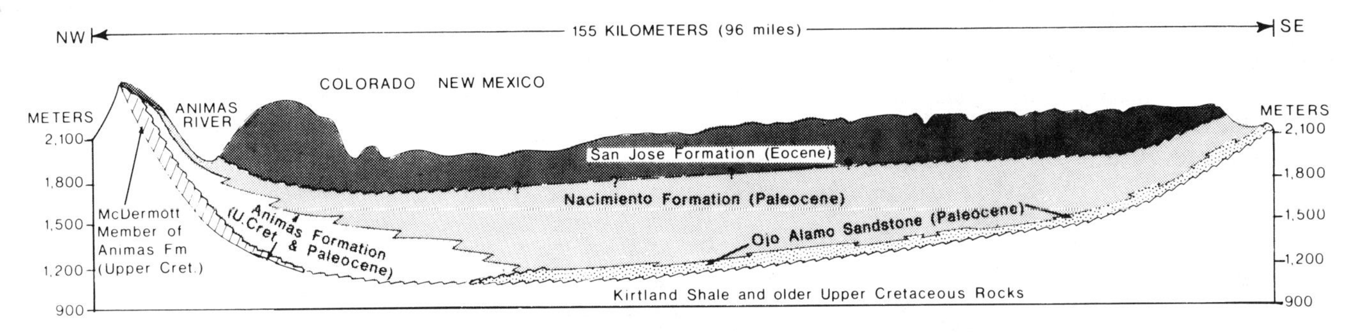

Figure 17. Cross section showing relations of Tertiary rocks across the San Juan Basin (modified from McDonald, 1972). Vertical scale is elevation above sea level. Location of section is shown in Figure 15.

ous activity began in the Rico, La Plata, Ute, and Carrizo mountains area. Late Cretaceous to earliest Tertiary radiometric age dates have been determined for intrusive rocks in those areas (Armstrong, 1969). Most likely, extrusive equivalents of the intrusives in the La Plata Mountains were the source for the andesitic material in the McDermott (Fassett, 1985).

TERTIARY ROCKS

As much as 1,200 m of Paleocene and Eocene continental rocks are preserved in the San Juan Basin. These rocks, which unconformably overlie the Cretaceous section, are composed of debris shed from flanking uplifts, mostly from the north, into the subsiding San Juan Basin. In ascending order, the Tertiary section consists of the Ojo Alamo Sandstone, the upper part of the Animas Formation, and the laterally equivalent Nacimiento Formation, all of Paleocene age, and the San Jose Formation of Eocene age. Figure 17 shows the relations of these rocks across the basin.

Ojo Alamo Sandstone (Paleocene)

The Ojo Alamo Sandstone, which is a major aquifer in the San Juan Basin, is a fluvial unit consisting of massive, fine- to coarse-grained sandstone and conglomeratic sandstone interbedded with shale. It generally ranges from 15 to 45 m in thickness but is as thick as 120 m in some places and is absent in the northern part of the basin. The provenance for the Ojo Alamo was probably the San Juan–La Plata Mountains area to the north and northwest (Powell, 1973; Sikkink, 1986; Fassett, 1985). The exact placement of the lower contact and age of the Ojo Alamo Sandstone have been controversial for many years. The nature of the lower contact, unconformable or not, has also been debated. The various interpretations of many workers are summarized in Fassett and Hinds (1971), O'Sullivan and others (1972), Fassett (1982, 1985), and Fassett and others (1986). Originally, a Late Cretaceous age had been assigned to the Ojo Alamo by some workers on the basis of dinosaur fossils (possibly reworked?); more recently it has been considered by other workers to be Paleocene in age on the basis of pollen. Perhaps the Ojo Alamo may be diachronous in age across the basin. Subsurface data indicate that as much as 640 m of Upper Cretaceous rocks were removed by erosion in the southeast part of the San Juan Basin prior to Ojo Alamo deposition (Fassett and Hinds, 1971).

Upper part of Animas Formation (Paleocene)

In the northern part of the San Juan Basin, where the Ojo Alamo Sandstone is absent as a result either of facies change or truncation, the upper part of the Animas Formation unconformably overlies the McDermott Member, Kirtland Shale, and Fruitland Formation (Fassett and Hinds, 1971; Fassett, 1985). The upper part of the Animas is Paleocene in age, but the lowermost part, the McDermott Member, is considered Late Cretaceous in age on the basis of dinosaur fossils and stratigraphic relations (Reeside, 1924). The upper part of the Animas is as much as 800 m thick and consists mostly of volcaniclastic andesitic detritus of conglomerate and sandstone interbedded with variegated tuffaceous shale. The provenance for the Animas was the San Juan Mountains area north of the San Juan Basin (Fassett, 1985).

Nacimiento Formation (Paleocene)

The Animas Formation grades southward into the Nacimiento Formation, which conformably overlies the Ojo Alamo Sandstone and is Paleocene in age. The Nacimiento Formation is as much as 600 m thick and is composed of variegated to dark-gray floodplain and lacustrine shale and lesser amounts of fluvial sandstone. Minor thin beds of coal also are present.

San Jose Formation (Eocene)

The San Jose Formation unconformably overlies, at least on the basin margins, Paleocene to Upper Cretaceous rocks. The unconformity decreases in magnitude toward the basin center where the San Jose is conformable with underlying rocks. The San Jose is as much as 700 m thick and consists of fluvial conglomeratic and arkosic sandstone and interbedded variegated shale of floodplain origin. On the basis of different proportions of sandstone and shale, Baltz (1967) subdivided the San Jose Formation into four complexly intertonguing members on the east side of the basin. The upper surface of the San Jose is a Holocene erosion surface.

MAJOR ECONOMIC RESOURCES

Gas and Oil

The San Juan Basin contains the second largest (in ultimate production) gas field in the lower 48 states. The Blanco (or San Juan Basin) gas field, which has an estimated ultimate recovery of 23 trillion cubic feet of gas plus significant amounts of condensate, covers about 543,000 ha of the central and northern part of the basin. The field is unusual because the accumulation is thought to be, at least in part, hydrodynamically controlled in the structurally low part of the basin. Isotopic and chemical analyses of the gas indicate that it was generated by thermal processes rather than biogenic processes (Rice, 1983). The major production is from the Dakota, Point Lookout, Cliff House, and Pictured Cliffs Sandstones, with minor production from the Fruitland Formation, Farmington Sandstone Member of the Kirtland Shale, and Nacimiento Formation.

Besides gas and condensate production in the central and northern part of the San Juan Basin, the Dakota Sandstone produces oil in several small, structurally controlled fields on the flanks of the basin. The two largest fields, Hogback and Rattlesnake, are in the 5–7 million barrel range.

Most of the oil in the basin (excluding condensate) is produced from stratigraphic traps in the Tocito Sandstone Lentil of the Mancos Shale, a series of elongate offshore bars (listed as Gallup Sandstone in production records). Over two dozen fields or pools, two as large as 40 million barrels in size, have a total estimated ultimate recovery of about 150 million barrels. In addition, a few fields produce from fractured shale and shaly sandstone reservoirs in the El Vado Sandstone Member of the Mancos Shale (Coniacian-Santonian) in the eastern part of the San Juan Basin. The only production from the true (regressive) Gallup Sandstone is at Hospah field, a structural accumulation in the southern part of the basin.

Channel sandstones in the Menefee Formation produce a small amount of oil in stratigraphic-structural traps at two or three small fields in the southern part of the San Juan Basin. Fassett (1978) provides detailed information on the many oil and gas fields in the San Juan Basin.

Coal

The Fruitland Formation, which contains some 200 billion tons of coal resources (Fassett and Hinds, 1971), has by far the largest minable coal reserves in the San Juan Basin area. The Navajo mine, which supplies coal for the Four Corners power plant southwest of Farmington, is one of the largest coal mines in the country. Along its south and west outcrop belt, the Fruitland contains large resources of strippable coal.

The Menefee Formation contains much smaller, but still significant, reserves of minable coal. These occur in the Cleary Coal Member at the base and in the upper coal member near the top. Because of steep dips in all but the southern outcrop belt, much of the Menefee coal cannot be strip mined.

The Gallup Sandstone and the Dilco and Gibson Members of the Crevasse Canyon Formation have been mined underground for coal since the 1880s in the Gallup area immediately southwest of the San Juan Basin. No underground mines are active today; however, the McKinley mine northwest of Gallup is strip-mining coal from the undivided Cleary and Gibson Members.

Except for some bituminous coal in the Gallup area, most of the coals of the San Juan Basin are low-sulfur, subbituminous coals.

SEDIMENTOLOGIC AND PALEOTECTONIC ANALYSIS OF THE HENRY, KAIPAROWITS, AND BLACK MESA BASINS, UTAH AND ARIZONA

Fred Peterson

INTRODUCTION

The interaction between sedimentation and growing structures at the site of deposition has been noted in Paleozoic and Triassic rocks in many places on the Colorado Plateau (for example, Walcott, 1890; Heylmun, 1958; Fetzner, 1960; Baars, 1966). However, previous studies of tectonic controls on sedimentation during the Jurassic and pre-Laramide Cretaceous generally were not specific and definitive. In this analysis, therefore, emphasis is given to the influence of slight crustal movements on the sedimentology of Jurassic and Cretaceous rocks in and near the three-basin area that includes the Henry, Kaiparowits, and Black Mesa Basins (Fig. 18).

For purposes of sedimentologic and tectonic analysis, Jurassic and Cretaceous formations are grouped into divisions that are bounded by unconformities or depositional surfaces along which little intertonguing is known. Sabkha, coal swamp, or shallow water strata within these divisions, originally deposited on horizontal or nearly horizontal surfaces, permit an interpretation of the style of deformation from variations in thickness of the divisions. The distribution of sedimentary facies, which in many cases was also governed by growing folds, also aids in the tectonic interpretations.

The term uplift is used in a sense relative to adjacent downwarps. Thinning of conformable strata across positive areas indicates subsidence of the positive areas that occurred more slowly than adjacent downwarps. Tectonic squeezing is not a factor because the strata presently dip less than about two degrees throughout most of the region. Similarly, the effects of differential compaction have been shown to be insignificant (Peterson,

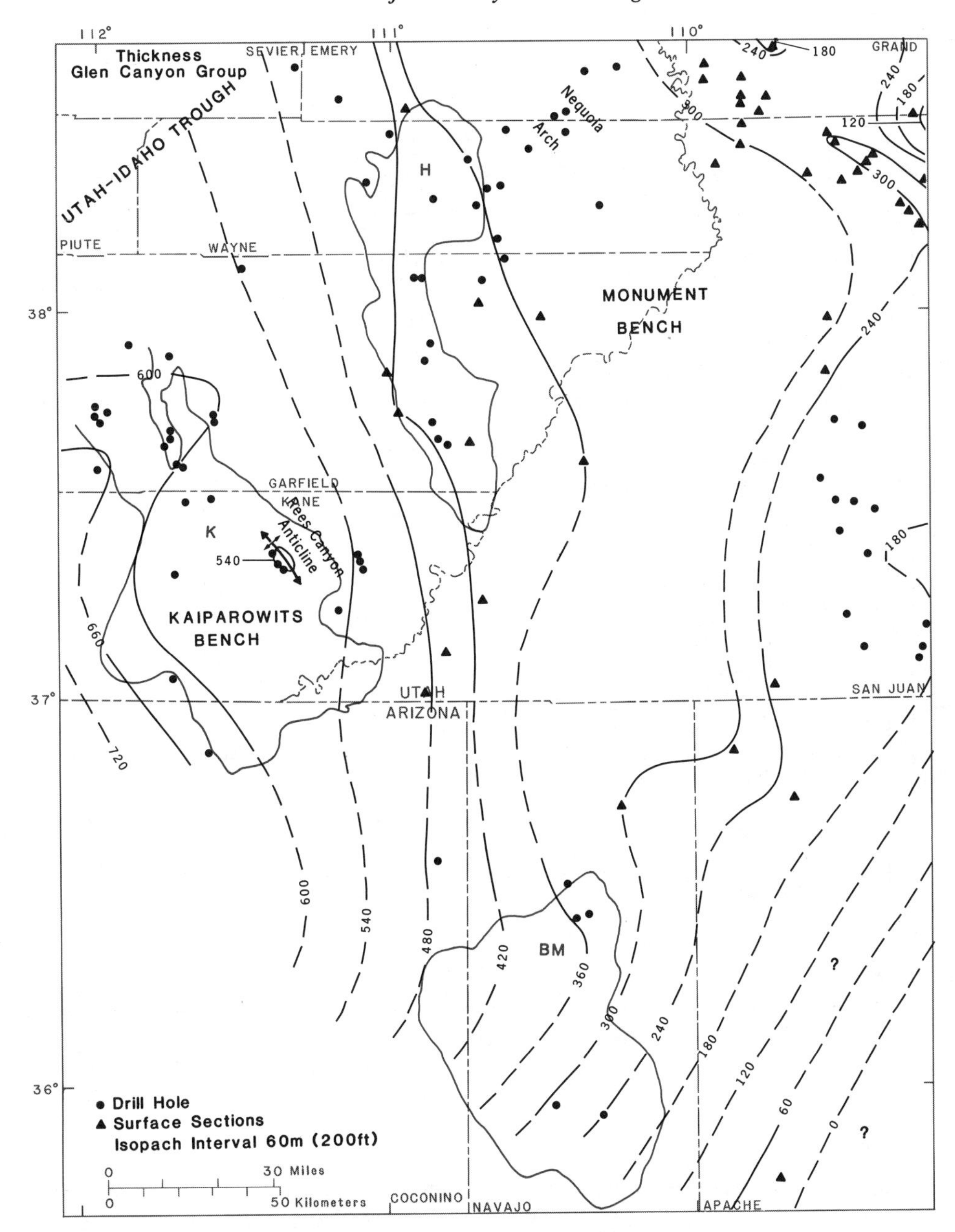

Figure 18. Isopach map of the Glen Canyon Group (excluding the Rock Point Member of the Wingate Sandstone). The present-day Henry Mountains (H), Kaiparowits (K), and Black Mesa (BM) structural basins are outlined in red. The red line is the 1,676-m structure contour line on the base of the Dakota Formation (Kelly, 1955b; O'Sullivan and Beikman, 1963; Williams and Hackman, 1971; Hackman and Wyant, 1973; Hackman and Olson, 1977; Haynes and Hackman, 1978).

1984). The various isopach maps and stratigraphic sections were compiled from data in reports by Gilluly and Reeside (1928), Baker and others (1936), Hunt and others (1953), Harshbarger and others (1957), McFall (1955), Phoenix (1963), Smith and others (1963), O'Sullivan (1965, 1978, 1980a, 1980b, 1981a, 1981b), Davidson (1967), and Peterson (1969b); additional data is from unpublished work by J. C. Wright, D. D. Dickey, and Fred Peterson.

LOWER JURASSIC SERIES

Glen Canyon Group

The oldest rocks of the Jurassic are the Glen Canyon Division, which consists of four formations (Fig. 19) and is bounded by unconformities (Pipiringos and O'Sullivan, 1978; Peterson and Pipiringos, 1979). These rocks are considered Early Jurassic

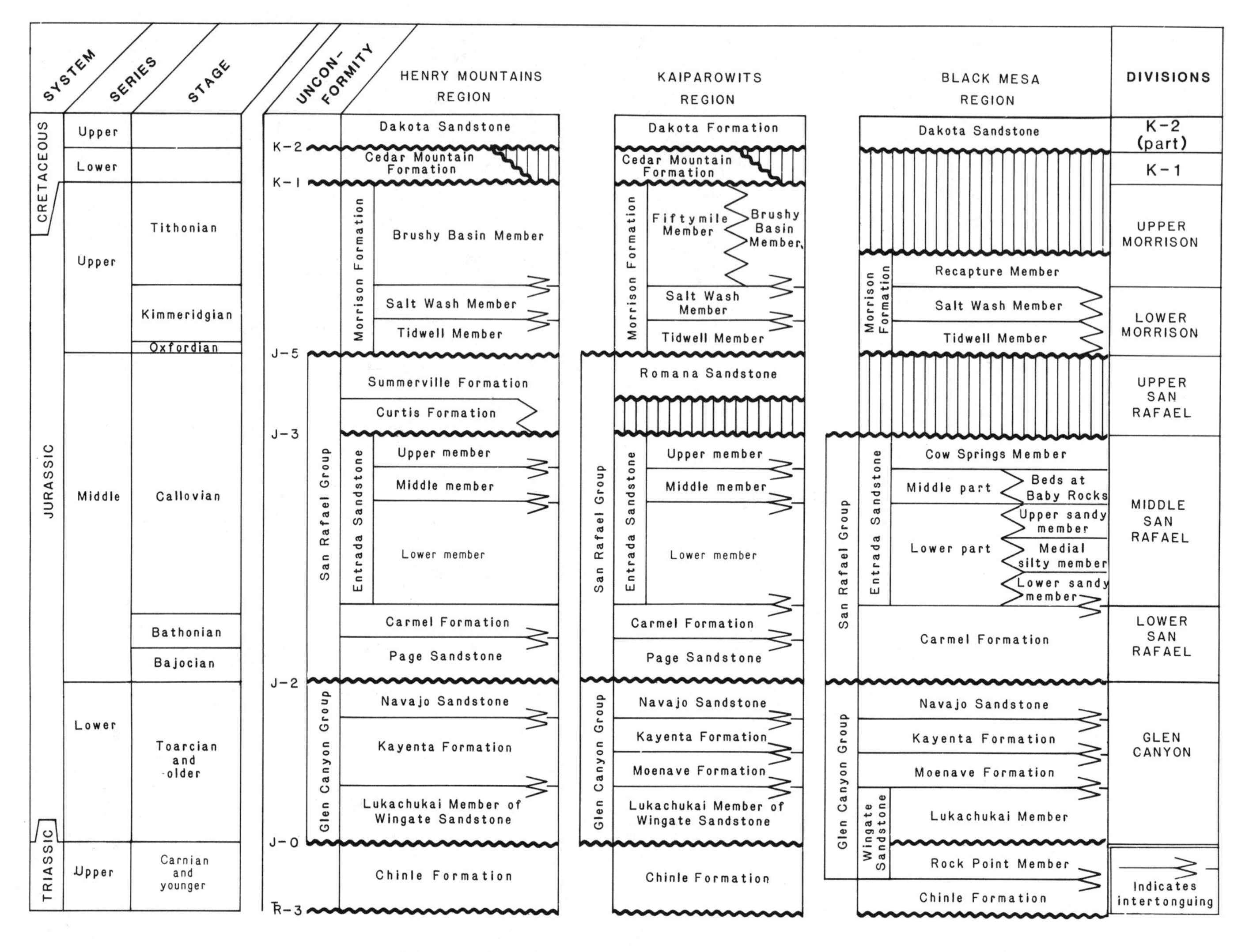

Figure 19. Jurassic formations in the three-basin region.

in age, following recent paleontologic and stratigraphic evaluations by Olsen and Galton (1977), Peterson and others (1977), Galton (1978), Peterson and Pipiringos (1979), and E. C. Colbert (oral communication, 1985).

The Wingate and Navajo Sandstones consist largely of eolian sandstone, although they also contain infrequent limestone lenses deposited in small playa lakes. They are separated by a fluvial, minor overbank floodplain, sabkha, and lacustrine complex that consists of the Moenave and Kayenta Formations, which were shed from highland source regions farther east and southeast.

The Glen Canyon Group thickens westward toward the Utah-Idaho trough (Fig. 18). Westward and northwestward stream flow during deposition of the Kayenta and Moenave Formations (Poole, 1961) indicates regional slopes toward the west and northwest. However, it is not clear if the slope was due to topographic and (or) structural effects. The entire Glen Canyon Group was truncated eastward during the early Middle Jurassic erosion interval, indicating that at least some westward regional tilting occurred shortly after deposition. Wide separation of isopach lines just east of the Henry Basin probably reflects an early manifestation of the Monument Uplift as a structural bench. The Henry Mountains and Black Mesa regions were not downwarped at this time, although a structural bench with a small anticline on it was present in the Kaiparowits region.

MIDDLE JURASSIC SERIES

The Middle Jurassic consists entirely of the San Rafael Group, which is separated into three divisions (Fig. 19).

Lower Division

The lower division of the San Rafael Group consists largely of mudstone, sandstone, limestone, and gypsum deposited during

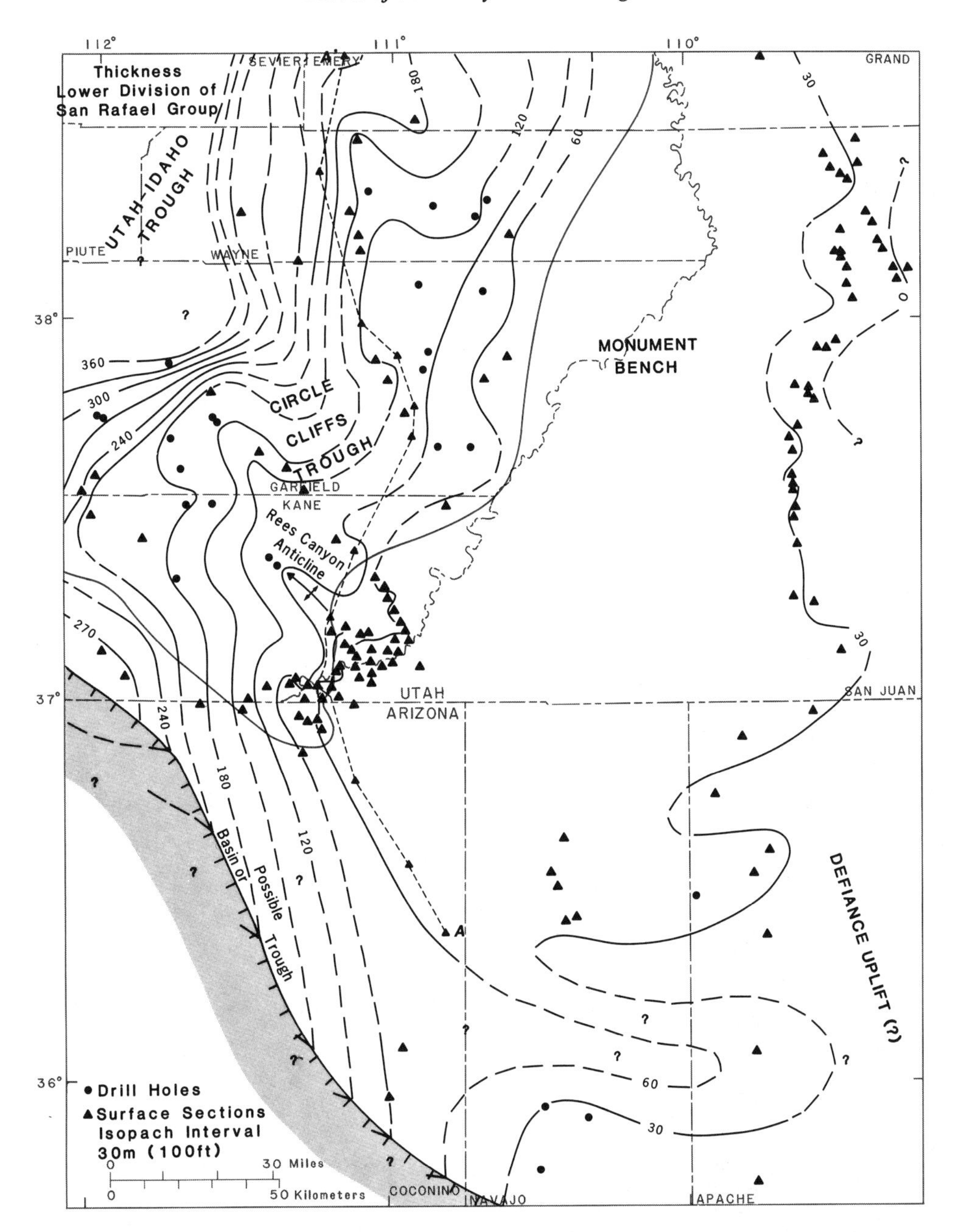

Figure 20. Isopach map of the lower division of the San Rafael Group. The red line is the approximate southeastern limit of limestone in the upper part of the Carmel Formation. The shaded area is where the Dakota Formation bevels the lower division. A–A′, line of section in Figure 21.

two stages of advance and retreat of the sea across part of the region (Figs. 20, 21; Blakey and others, 1983). Marine limestone and mudstone deposits in the westernmost part of the Henry and Kaiparowits regions are flanked farther east and southeast by coastal sabkha deposits of mudstone and lenticular beds of gypsum. During the early stages of deposition of the Page Sandstone, eolian sands accumulated just east of coastal sabkha deposits represented by strata in the lower part of the Carmel Formation. At its greatest extent, the dune field covered most of the Henry Mountains and Kaiparowits regions and probably extended southwestward into the area west of Black Mesa (Peterson, 1986).

Structurally, the region was influenced by the Utah-Idaho Trough and the Monument Uplift. The presence of extensive coastal sabkha and tidal flat strata deposited at or near sea level shows that the westward increase in thickness of this division reflects

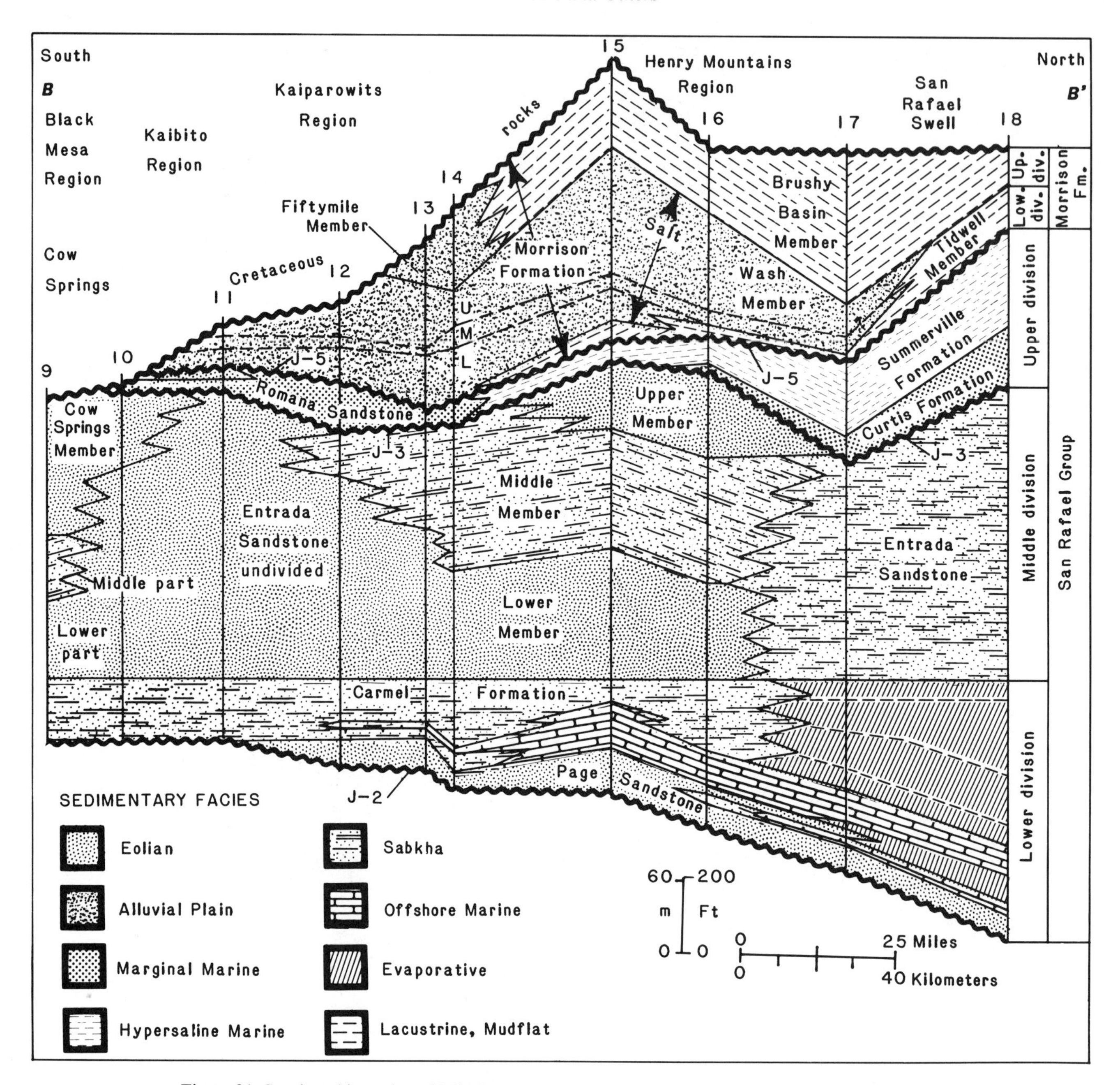

Figure 21. Stratigraphic section of Middle and Upper Jurassic rocks across the Henry Mountains and Kaiparowits regions to Black Mesa. Location shown in Figure 20. J-2, J-3, J-5 are unconformities of Pipiringos and O'Sullivan (1978).

westward regional tilting toward the Utah-Idaho Trough (Fig. 20). Wide separation of isopach lines across the present-day Monument Uplift indicates a structural bench in that area. Eastward pinchout of the Page Sandstone and of limestone in the upper part of the Carmel Formation on the west flank of the bench (Fig. 20) also supports this conclusion.

Several downwarps also are apparent. Greater thicknesses of the lower division across the Circle Cliffs and southern Henry Mountains regions suggest that this area subsided as a northwest-plunging trough (Fig. 20). A trough in the Circle Cliffs area is also suggested by the slightly greater thickness and southeastward extent of limestone in the upper part of the Carmel Formation in the trough as compared to nearby areas. The division also thickens toward a structural low west of Black Mesa. The form and trend of this downwarp are uncertain because so little of the division is preserved there. The southeastward-extending lobe of limestone in the upper part of the Carmel in the Kaiparowits region may reflect deposition that was partly controlled by this structural depression. The lobe does not extend very far south into the structural low, apparently because the Carmel grades into a non-marine (eolian, sabkha, fluvial) facies in this area.

Middle Division

The middle division of the San Rafael Group consists solely of the Entrada Sandstone, which is divided into lower, middle, and upper members in the Kaiparowits and Henry Mountains region (Fig. 19). Stratigraphic reconstructions suggest that the upper member of the Kaiparowits region may be slightly younger than the Cow Springs Member at Black Mesa (Fig. 21).

The lower member of the Entrada consists of a sabkha facies of flat-bedded siltstone and sandstone in the northwestern part of the region and an eolian crossbedded sandstone facies farther east (Fig. 22). The silt and sand were deposited mostly in a broad coastal sabkha shoreward or south of marine environments that lay in the northern part of the Utah-Idaho Trough. Another sabkha facies is also present in the eastern Black Mesa region and appears to represent an isolated inland sabkha. The crossbedded eolian sandstone of the lower member was deposited in a large dune field that stretched eastward from the Henry Mountains region well into Colorado (Kocurek and Dott, 1983).

The middle member of the Entrada formed at a time when the broad coastal sabkha in and adjacent to the Utah-Idaho Trough advanced southeastward across the Henry and Kaiparowits regions and into the eastern part of Black Mesa (Fig. 22). Crossbedded eolian sandstone interfingers with local sabkha deposits in the Kaibito and southwestern Kaiparowits regions, indicating that a dune field lay there and probably farther southwest.

The upper and Cow Springs members consist largely of crossbedded sandstone deposited primarily in eolian environments, but both members include some flat-bedded siltstone and sandstone deposited in sabkha environments. The upper member and the Cow Springs apparently grade into mixed eolian and sabkha strata in the Kaibito region (Fig. 22).

Essentially the same style of structural deformation that began during deposition of the lower division of the San Rafael Group continued during deposition of the middle division. Thus, subsidence continued in the Utah-Idaho Trough, the Monument region continued as a structural bench, a northwest-plunging trough remained in the Circle Cliffs region, and the downwarp west of the Black Mesa region remained, although the axis shifted eastward into the northwestern Black Mesa region (Peterson, 1986).

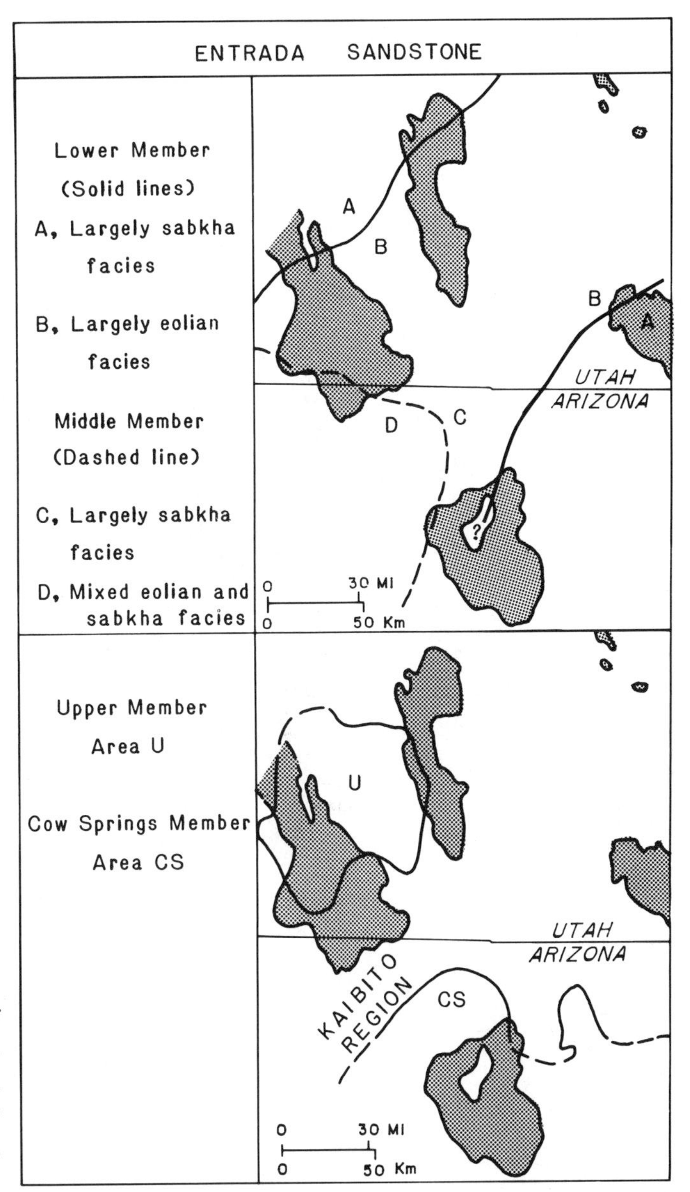

Figure 22. Distribution of facies in the Entrada Sandstone with respect to present-day basins.

Upper Division

The upper division of the San Rafael Group (Fig. 19) was deposited after a brief erosion interval during a general southward advance and northward retreat of a seaway that lay in the Utah-Idaho Trough and northwestern part of the Colorado Plateau region. The Curtis Formation consists largely of glauconitic greenish-gray sandstone and scarce limestone deposited in marine environments. The overlying Summerville consists of red mudstone and siltstone; in the northwest part of the Henry Mountains region it also includes minor thin lenses of gypsum and, near the top of the formation, scarce fluvial sandstone beds. The Summerville was deposited in shallow, restricted marine conditions and (toward the close of deposition in the northwestern Henry Mountains region) in fluvial, overbank, and probably mudflat environments. The Romana Sandstone is a landward facies of the Summerville Formation found only in the Kaiparowits and Kaibito regions (Figs. 19, 23). It consists largely of light-gray sandstone deposited in marginal marine and eolian environments. A

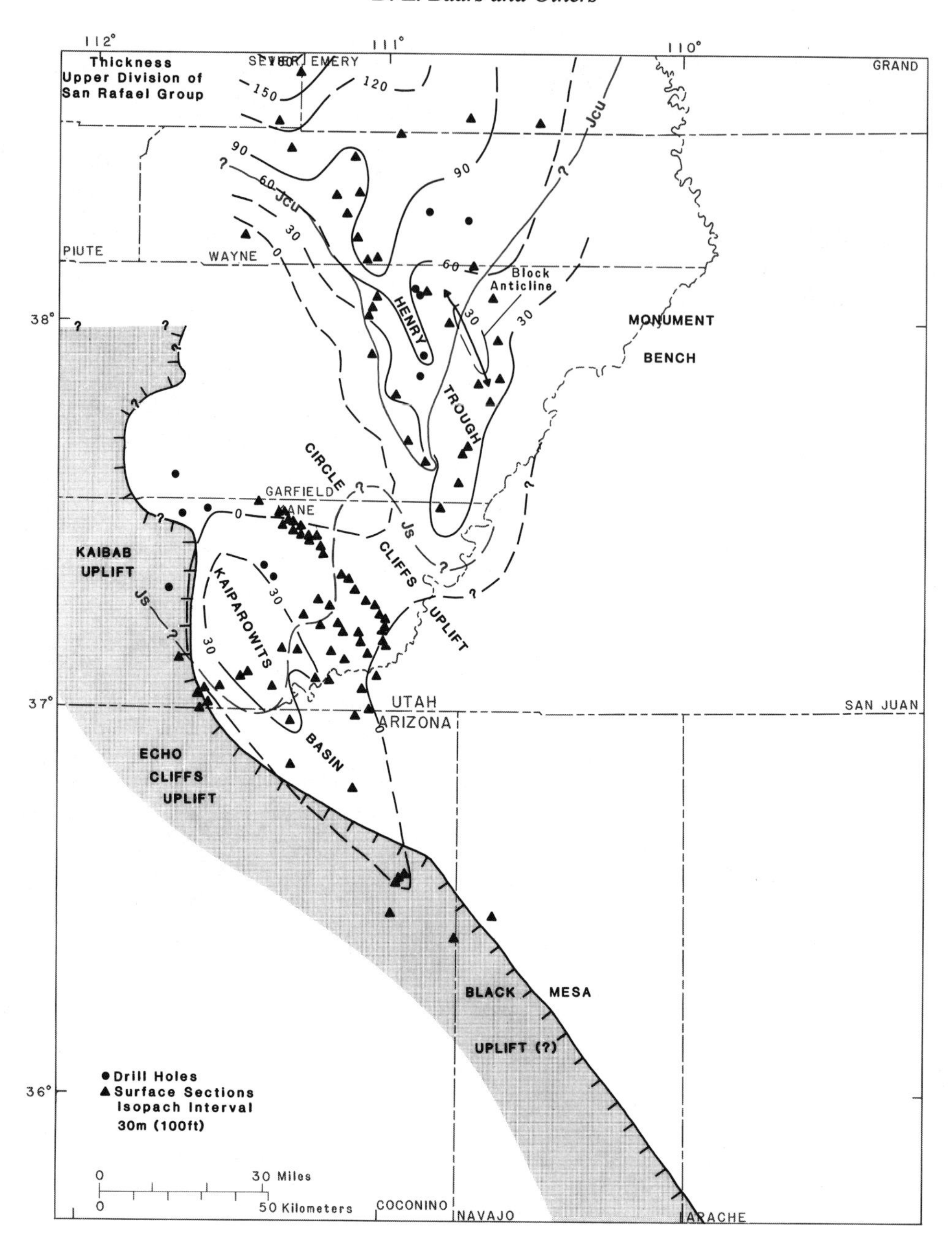

Figure 23. Isopach map of the upper division of the San Rafael Group. The red line marks the southern extent of the Curtis Formation. The dashed red line indicates the southern extent of Summerville Formation and red beds in the lower part of the Romana Sandstone in the Kaiparowits region. The shaded area is where the Dakota Formation bevels the upper division.

red mudstone bed as much as 3 m thick at the base of the Romana in much of the Kaiparowits region is a tongue of the Summerville deposited during the greatest extent of the Summerville seaway.

Crossbedding studies in eolian beds indicate that the regional wind regime changed during deposition of the upper division. Eolian crossbeds in older Jurassic rocks dip generally southwest, south, or southeast (Poole, 1962), whereas eolian crossbeds in the Romana Sandstone and overlying Morrison Formation mostly dip east or northeast. The change in directions apparently reflects passage of this part of the North American continent into a latitudinal belt with different prevailing wind directions (Peterson and Caputo, 1986).

Tectonically, the upper division marks a dramatic change in the structure of the region (Fig. 23). Chert and quartzite pebbles and cobbles in the Curtis, upper Summerville, and Romana that were derived from Paleozoic rocks farther west suggest the beginning of tectonic activity in the Sevier orogenic belt. Structural lows evolved in the Kaiparowits and Henry Mountains regions, and positive movement began in the Circle Cliffs, Echo Cliffs–Kaibab, and Monument regions. Each of the positive structural elements influenced sediment distribution such that the Curtis and Summerville tend to be restricted to the structural lows. Southeastward truncation of the Romana Sandstone suggests that Black Mesa may have been a positive area, perhaps as an early manifestation of the positive structure that was there in the Late Jurassic.

UPPER JURASSIC SERIES

Morrison Formation

Lower Division. The lower division of the Morrison Formation includes the Salt Wash and Tidwell Members in most of the three basin areas and correlative beds in the lower part of the Recapture Member in southeastern Black Mesa (Fig. 19). The Salt Wash consists largely of fluvial sandstone or conglomerate and thin mudstone beds deposited in fluvial, overbank floodplain, and local lacustrine environments (Peterson, 1984). The Tidwell contains mudstone, sandstone, and thin limestone lenses; it also contains gypsum in the northern part of the Henry Mountains region. The Tidwell was deposited distally or laterally to the Salt Wash alluvial complex in large lakes, mudflats, small eolian dune fields, and evaporative environments; fluvial deposits are rare. The lower Recapture consists of sandstone and mudstone deposited in mudflat, lacustrine, and possibly fluvial environments that were distal to the Salt Wash alluvial complex.

The Henry Mountains and Kaiparowits regions continued as structural basins, and the Circle Cliffs region persisted as an uplift during deposition of the lower division of the Morrison Formation (Peterson, 1984). Thickening of the lower division away from the Monument Uplift indicates that this structure was a broad anticlinal structure rather than the structural bench that it had been earlier. The lower division thins and pinches out by onlap onto paleotopographic and paleostructural highs, one in the vicinity of Black Mesa and the other about 30 km northwest of the Henry Basin (Peterson, 1986b).

Upper Division. The upper division of the Morrison consists solely of the Brushy Basin Member in the Henry Mountains and northern Kaiparowits regions (Fig. 19). This member consists largely of smectitic mudstone, thin bentonite beds, and some sandstone. The Fiftymile Member of the Morrison is a facies of the Brushy Basin present only in the southeastern Kaiparowits region. There, the lower part of the Fiftymile consists largely of mudstone and scarce sandstone; the upper part consists largely of sandstone (Peterson, 1988). The upper part of the Recapture Member of the Morrison, consisting largely of overbank, mudflat, and lacustrine mudstone and eolian and fluvial sandstone, currently is recognized as the sole representative of the upper division in the Black Mesa region. A fluvial unit formerly called the Westwater Canyon Member of the Morrison at Black Mesa is now thought to be in the uppermost part of the Recapture, as it is finer grained and also contains more feldspar pebbles and fewer pebbles of volcanic materials than the Westwater Canyon farther east and northeast.

The Brushy Basin, lower Fiftymile, and most of the upper Recapture were deposited in a broad lowland region containing large mudflats, lakes, small dune fields, and scarce streams. Fluvial sandstones in the upper Fiftymile and uppermost Recapture comprise an alluvial complex that prograded northeastward across the Black Mesa and southeastern Kaiparowits regions toward mudflat and lacustrine areas represented by the Brushy Basin Member. Abundant smectitic clay and bentonite beds in the Brushy Basin are alteration products of volcanic ash carried into the region by winds from the west and southwest. Evidence of a large playa lake has been found in northwestern New Mexico (Bell, 1981, 1986; Turner-Peterson, 1985), and studies in progress show that it extended into the northeastern part of the study area east of the Monument Uplift (C. E. Turner-Peterson, oral communication, 1986).

The tectonic history of the upper division of the Morrison is poorly understood because much of the stratigraphic record was removed in the Early Cretaceous. Isopach trends indicate subsidence in the Henry Mountains region and upwarping on the Monument Uplift. Continuation of the structural high at Black Mesa is suggested by onlap of the Recapture Member in this area (Peterson, 1986).

LOWER CRETACEOUS SERIES

Cedar Mountain Formation

The Cedar Mountain Formation (Fig. 19) consists of interbedded fluvial sandstone and overbank mudstone. The sediment was derived from the Sevier highlands farther west and deposited, probably as an eastward thinning wedge, in the northwestern Henry Mountains region and northern Kaiparowits region (Craig, 1981). Early Cretaceous structural activity in the region is poorly known; however, beveling of Early Cretaceous strata away from the crest of the Monument region suggests uplift on this structure prior to deposition of the Dakota Formation (Peterson, 1986).

UPPER CRETACEOUS AND PALEOGENE ROCKS

Upper Cretaceous rocks contain transgressive-regressive deposits of continental and marine strata deposited on the west side of a broad epicontinental seaway that extended from the Gulf of Mexico northward to the Arctic Ocean. Emphasis here is on depositional environments using the Kaiparowits region as the primary guide because it contains the most complete section in the region (Fig. 24). Moreover, it is the only one of the three basins that has been examined in sufficient detail to evaluate growth on pre-Laramide Late Cretaceous structures.

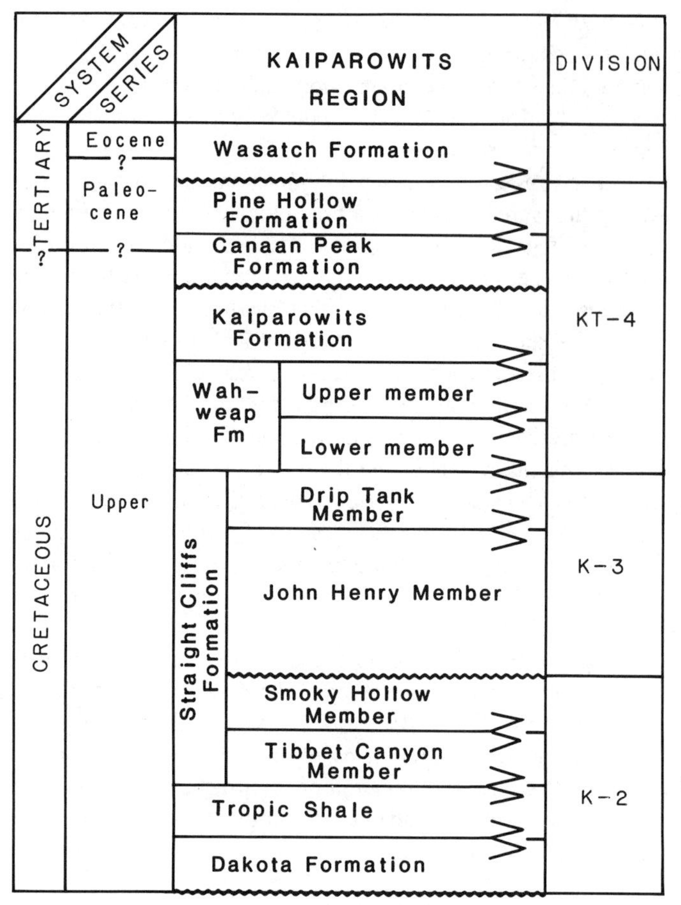

Figure 24. Upper Cretaceous and lower Tertiary formations in the Kaiparowits region.

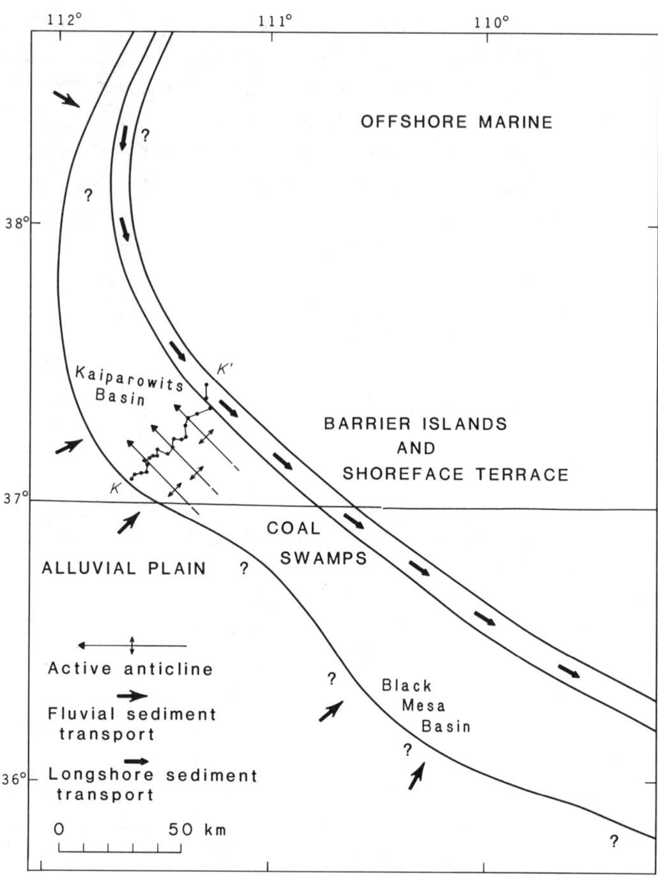

Figure 25. Distribution of depositional environments during deposition of the Christensen coal zone in the John Henry Member of the Straight Cliffs Formation. The distribution of depositional environments was similar throughout deposition of the Straight Cliffs Formation. K–K′, line of stratigraphic section in Figure 26. For clarity, the active synclines are not shown.

Lower Division

The lower or K-2 division of the Upper Cretaceous (Fig. 24) was deposited during a single transgressive-regressive cycle across all three basins. The shoreline was irregular and advanced mainly westward and southwestward during the transgression (Repenning and Page, 1956; Peterson, 1969b). In contrast, the shoreline was fairly linear during regression (Fig. 25), extending southeastward across the Kaiparowits and Black Mesa regions and retreating to the northeast across the Henry Mountains region.

The Calico Bed (Fig. 26-A; Peterson, 1969a) and similar widespread fluvial sandstone beds at higher stratigraphic levels are intimately related to major regressions of the Late Cretaceous strandline in the three-basin region. They suggest a brief pulse of uplift in the source region, probably the result of thrusting, and the subsequent outpouring of large quantities of detritus that accompanied the major shoreline regressions.

Sedimentologic studies indicate a complex structural history for Upper Cretaceous rocks of the Kaiparowits region (Peterson, 1969b). The interpretation is based largely on the distribution of coal, which was deposited under conditions that were especially sensitive to small amounts of crustal movements when the supply of clastic sediment was low. Subsidence early in the depositional history of the lower division is shown by the distribution of coal beds in the Dakota Formation, which occur in small synclines in the structurally deepest part of the Kaiparowits Basin (Peterson, 1969b). Beds in the upper part of the division show repeated movement on folds, and occasionally on some faults, by the thinning or absence of coal zones over paleoanticlines and upthrown fault blocks (Fig. 26-A; Peterson, 1969b). Basinal downwarping followed by uplift and erosion is shown by truncation of beds on the northeastern and southwestern limbs of the Kaiparowits Basin (top of Fig. 26-A).

Middle Division

Sedimentary rocks of the K-3 or middle division (Figs. 24,

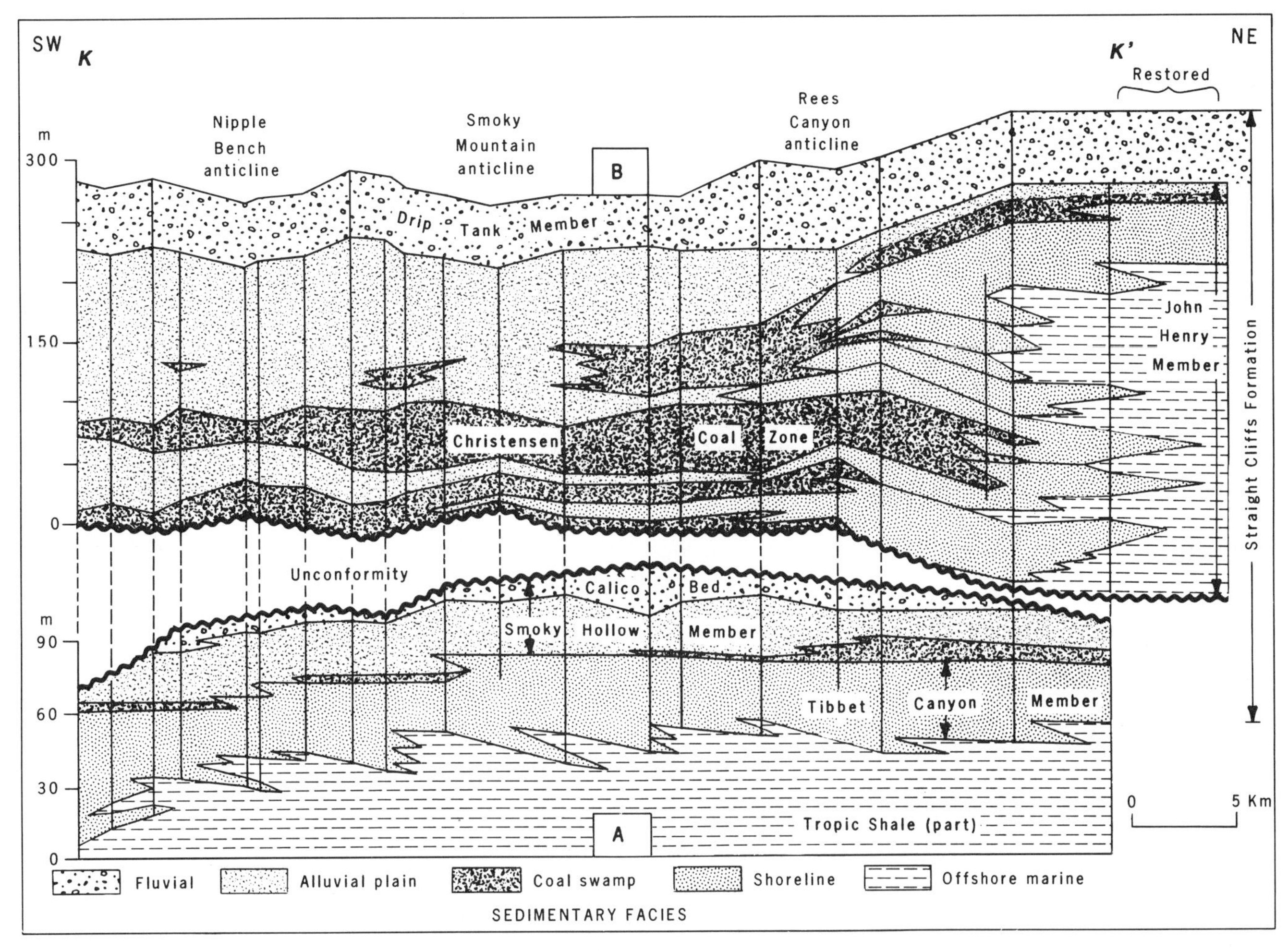

Figure 26. Stratigraphic sections showing part of the K-2 division (A) and the K-3 division (B) in the Kaiparowits region. The sections trend normal to Late Cretaceous shorelines. Note the different vertical scales for the two sections. The line of section is shown on Figure 25.

26-B) came from source terranes southwest and northwest of the Kaiparowits region and were deposited during two transgressive-regressive cycles (Peterson and Kirk, 1977). Northeast-flowing streams originating in the southwestern terrane carried sand and small chert pebbles (<2 cm long) to the region. During deposition of the John Henry Member (Fig. 26), the streams deposited most of their load on an alluvial plain southwest of the shoreline; they rarely extended across the narrow belt of coastal swamps and lagoons to empty directly into the seaway. Thus, John Henry streams supplied little sediment to the shoreline. At the shoreline, southeastward-flowing longshore currents supplied large quantities of sand, including large chert and quartzite pebbles (as much as 8 cm long). Thus, the brief transgressions and regressions in the middle of the John Henry (Fig. 26-B; Peterson and others, 1980) are related more to sediment supplied by longshore currents from a source farther northwest than to sediment supplied by fluvial processes from a source to the southwest. In Black Mesa, the sedimentological history of the middle division appears similar to that of the Kaiparowits region (Peterson and Kirk, 1977). The regression near the end of John Henry time marks the end of marine deposition in the three-basin region (Molenaar, 1983). The division is capped by the Drip Tank Member, which is another widespread fluvial sandstone unit that accompanied a major strandline regression.

The Kaiparowits region probably was a structural platform, folded into shallow anticlines and synclines and bordered on the northeast by a trough of unknown extent (Peterson, 1969b). Three actively growing, northwest-plunging anticlines and synclines were present and influenced the thickness and distribution of the extensive coal deposits in the John Henry Member (Fig. 26-B). The shoreline sandstone beds tend to pinch out landward on or near the crests of paleoanticlines. In addition, several growth faults were active during the early stages of deposition of the John Henry Member. Tectonic activity in the region is not known to have occurred during deposition of the Drip Tank, which may have taken place too quickly to have been influenced by slow growth on the folds.

Upper Division

The upper or K-4 division of latest Cretaceous (Campanian-Maastrichtian) and Paleocene(?) age includes strata deposited during the Laramide Orogeny. The division consists of sediment that originated in highland source regions west and southwest of the three-basin region. Thickening of the Canaan Peak and Pine Hollow Formations in the northwestern Kaiparowits region (Bowers, 1972) marks the last episode of Laramide downwarping of the Kaiparowits Basin. These formations are locally truncated by the flat-lying Wasatch Formation, which marks the end of Laramide tectonic activity in the region.

ECONOMIC RESOURCES

Jurassic and Cretaceous strata in the Henry, Kaiparowits, and Black Mesa regions contain appreciable quantities of uranium, coal, and gypsum. Uranium occurs in fluvial sandstone beds of the Salt Wash Member of the Morrison Formation in the eastern part of the Henry Basin (Peterson, 1980). Large quantities of subbituminous to low-rank bituminous coal are present in Upper Cretaceous rocks of all three basins. Appreciable quantities of gypsum are present in the Morrison and Carmel Formations, but a lack of nearby railways makes it unlikely that the mineral will be mined in the near future.

CONCLUSIONS

Clastic sedimentation by marine and nonmarine processes dominated the three-basin region during the Jurassic and Cretaceous. Sediment distribution and type was at least partly governed by slight but recognizable movement on the major structures. A major change in structural setting occurred in late Middle Jurassic time, coinciding with formation of the J-3 unconformity. Laramide deformation largely consisted of reactivation of preexisting folds, many of which had been active in the pre-Laramide Cretaceous and Jurassic.

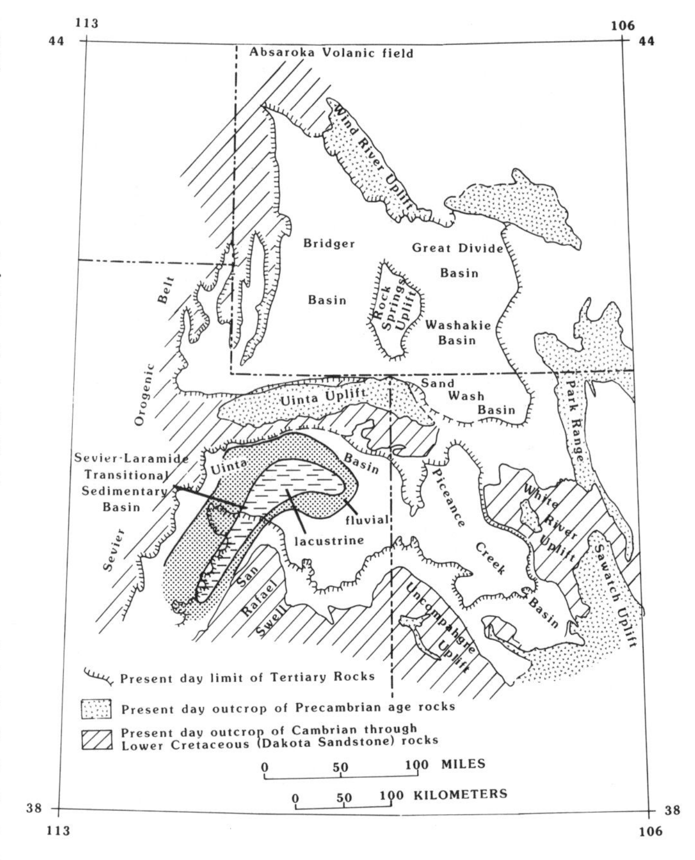

Figure 27. Map showing locations of Precambrian basement outcrops, the Uinta and Piceance Creek Basins, and the basins that make up the greater Green River basin in Utah, Colorado and Wyoming. The approximate extent of the Sevier-Laramide transition sedimentary basin during Late Cretaceous Maastrichtian time is also shown (from Fouch and others, 1983). Key is shown on figure 29.

EARLY CENOZOIC HISTORY OF THE UINTA AND PICEANCE CREEK BASINS, COLORADO AND UTAH

Ronald C. Johnson

The Uinta and Piceance Creek Basins are closely associated structural and sedimentary basins that were actively subsiding during the Laramide Orogeny from latest Cretaceous through about the end of Eocene time. The Uinta Basin is bounded on the west by the Sevier orogenic belt, an area of active uplift and eastward thrusting from possibly as early as Late Jurassic through the early Cenozoic (Armstrong, 1968). Laramide uplifts define the remaining boundaries of both the Uinta and Piceance Creek Basins. The Uinta Basin is bounded on the north by the Uinta Uplift and on the south by the San Rafael Anticline and the Uncompahgre Uplift. The Piceance Creek Basin is bounded on the north by the Uinta Uplift and the Axial Basin Anticline, on the east by the White River Uplift, and on the south by the Sawatch and Uncompahgre uplifts (Fig. 27).

Both basins are highly asymmetrical, with the structural trough of the Uinta Basin lying just south of the Uinta Uplift and the structural trough of the Piceance Creek Basin lying just west of the White River Uplift (Fig. 28). The two basins are separated structurally by the Douglas Creek Arch, a broad north-south trending anticline. During the early Cenozoic, the arch acted as a hingeline between the adjacent subsiding basins, and in essence, the flanks of the arch are also flanks of the adjacent basins (Johnson and Finn, 1985; 1986). A fairly thin interval of early Ceno-

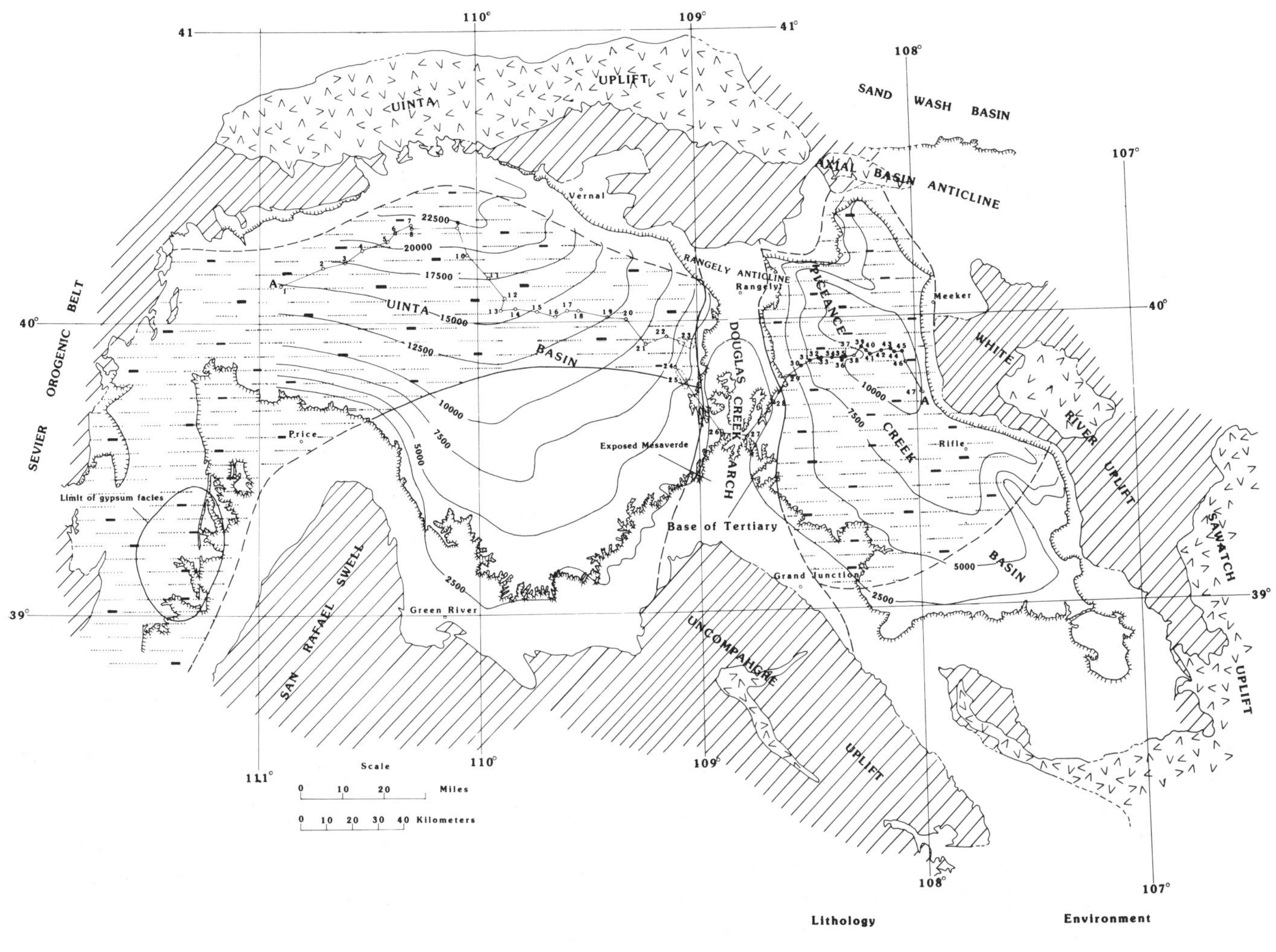

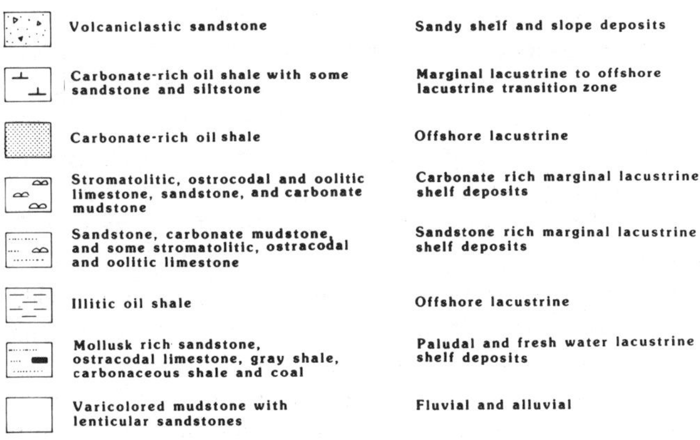

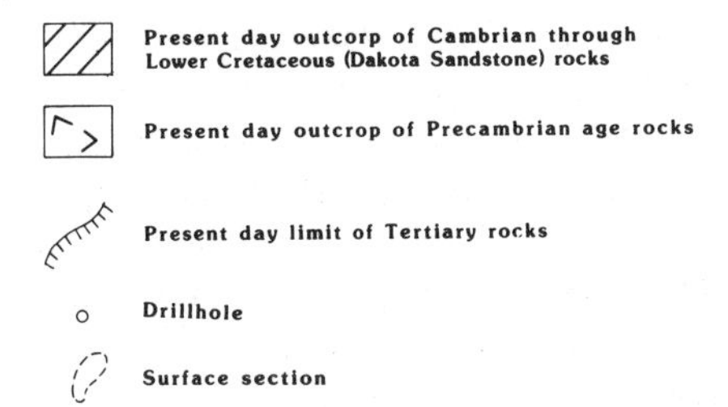

Figure 28. Map showing the distribution of lithologies and interpretations of environments of deposition during the maximum extent of late Paleocene age Lake Flagstaff in the Uinta Basin and its equivalent in the Piceance Creek Basin. The isopachs are the approximate original thickness in feet of early Cenozoic sedimentary rocks prior to later periods of erosion. The area labeled "Exposed Mesaverde" on the Douglas Creek Arch is where the Cretaceous-Tertiary unconformity had not, at this time, been onlapped by early Tertiary sedimentary rocks. The accompanying key is for Figures 28 to 32.

TABLE 1. LIST OF DRILLHOLES AND MEASURED SECTIONS USED IN CROSS SECTION

1. Shell Ute no. 1-16D9
2. Mountain Fuel Cedar Rim no. 1
3. Mountain Fuel Cedar Rim no. 3
4. Phillips Stewart B 1
5. Shell Murdock no. 1
6. Shell Bleazard no. 1-18B4
7. Shell Brotherson no. 1-2B4
8. Shell Brotherson no. 1-11B4
9. Gulf Reese Estate no. 1-10B2
10. Sinclair Senor Mortenson no. 1
11. Gulf Valley Wash no. 1
12. Union of California Gov. no. 1 691
13. Davis Parette Bench no. 5
14. Pan American Parette Bench no. 2
15. Shamrock Uteland Mine no. 1
16. Continental no. 22-1
17. CIG Natural Buttes no. 72-18-9-21
18. CIG Bitter Creek no. 45-15-9-21
19. Belco Chapita Wells no. 5
20. Continental Chapita no. 2
21. El Paso Southman Canyon no. 4
22. Moab Gem no. 2
23. Hells Hole measured section
24. Continental Evacuation Creek no. 22-1
25. Evacuation Creek measured section
26. Baxter Pass measured section
27. Douglas Pass measured section
28. Brushy Point measured section
29. Cathedral Bluffs measured section
30. American Minerals Ryan Creek no. 3
31. American Minerals Ryan Creek no. 2
32. TOSCO no. 1-722
33. Equity Sulfur Creek no. 7
34. Stuarco Sulfur Creek Govt. no. 1
35. Equity Sulfur Creek Govt. no. 1
36. Sinclair Bradshaw no. 1
37. Equity Sulfur Creek no. 5
38. Equity Johnson no. 2
39. Mobil Piceance Creek no. 68-11
40. Mobil Piceance Creek no. 54-13
41. Mobil Piceance Creek no. 62-19G
42. General Petroleum Piceance Creek no. 84
43. Mobil Piceance Creek no. T73-13G
44. General Petroleum Piceance Creek no. 28-19G
45. Indian Wells Piceance Crek no. 16-2-95
46. General Petroleum Piceance no. 45-33
47. Fourteenmile Creek measured section

zoic rocks is preserved along the crest of the arch (Fig. 28), indicating that at times base level was higher than the crest of the arch. Isopach maps for the Piceance Creek Basin and the eastern half of the Uinta Basin suggest that the early Cenozoic subsidence histories of the two basins were significantly different (Johnson and Finn, 1985; 1986). The Paleocene and early Eocene section is thickest in the Piceance Creek Basin, whereas the middle and late Eocene section is thickest in the eastern half of the Uinta Basin. These findings may help explain some of the shifts in sedimentary environments that occurred in the two basins during the early Cenozoic. Rocks from the margins of the two basins have been largely removed by recent erosion. The present-day outcrop of the Dakota Sandstone, shown on the lithologic and facies maps, is probably a good approximation of the original extent of the two sedimentary basins.

The history of the two basins is divided into six time periods: (1) prebasin to basin formation, (2) early basin sediments, (3) development of late Paleocene age lacustrine and paludal systems, (4) latest Paleocene through early Eocene alluvial period, (5) early Eocene fresh-water lakes, and (6) early to late Eocene saline Lake Uinta. An east-west cross section through both basins (Fig. 29; Table 1) and maps showing distribution of lithologies and interpretations of environments of deposition for four different time periods are presented here. Sources used for the Uinta Basin include Dane, 1955; Picard, 1959; Roberts, 1964; Cashion, 1967; Fouch and Cashion, 1979; and Fouch, 1981; those used for the Piceance Creek Basin include Bradley, 1931; Donnell, 1961; Johnson and May, 1978; Johnson, 1979a, 1979b, 1979c, 1985; Johnson and others, 1979a, 1979b, and 1979c; and Johnson, 1985. These works were supplemented by a considerable amount of subsurface information and unpublished surface sections measured by the author.

PREBASIN TO BASIN FORMATION PERIOD

The Sevier orogenic belt, an area of eastward thrust-faulting from Late Jurassic through the early Cenozoic, borders the Uinta Basin on the west. A trough or foreland basin formed to the east of the thrust belt, and rapid downwarping in the foreland basin during Cretaceous time created an epeiric seaway that extended from the Arctic Ocean to the Gulf of Mexico. Fanglomerates were deposited in the foreland basin just east of the Sevier orogenic belt (Fouch and others, 1983). These graded to the east, first into finer grained fluvial and paludal sediments, then into marginal marine sandstone, and finally into marine shale.

During the Late Cretaceous, the seaway was gradually filled by a series of sediment pulses from the west, and by Late Cretaceous time (late Campanian or early Maastrichtian), the shoreline had retreated to east of the Piceance Creek Basin. Following the regression, nonmarine sediments of the Mesaverde Group were deposited over the area of the Uinta and Piceance Creek Basins. These Mesaverde rocks today are as much as 1,300 m thick; however, the original thickness may have been much greater because an undetermined amount of the Mesaverde was removed by erosion during latest Cretaceous through early Tertiary time.

The Laramide Orogeny began during latest Cretaceous time in the foreland basin east of the Sevier orogenic belt and is in part synchronous with thrusting along the Sevier orogenic belt. The Laramide Orogeny was characterized by large-scale, basement-involved uplift and downwarping along reverse and high-angle thrust faults, and contrasts sharply with the thin-skinned decollement thrusting that characterized the largely earlier Sevier Orogeny (Armstrong, 1968). The Laramide Orogeny affected a broad area of the foreland basin from southern Montana through New

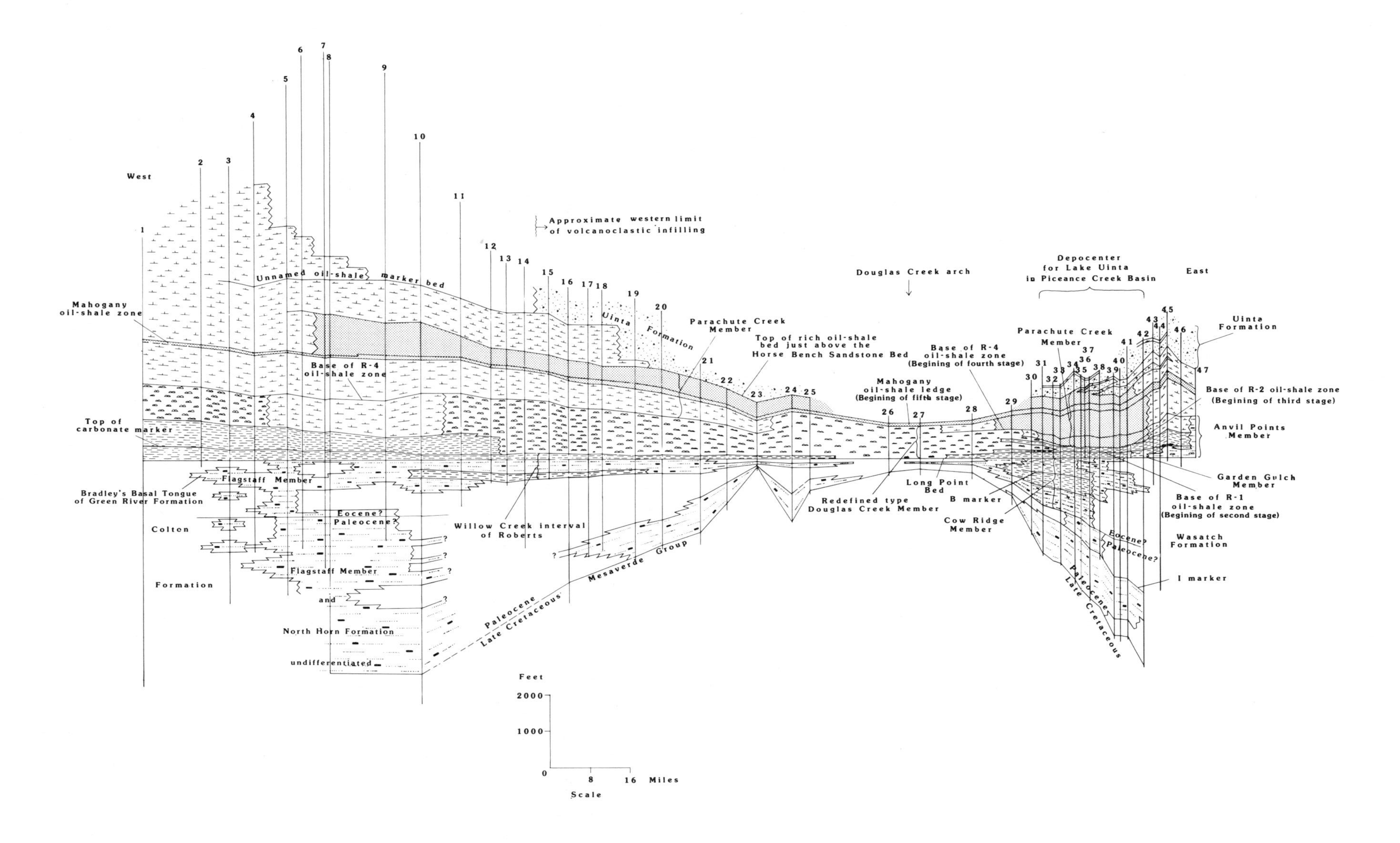

Figure 29. Cross section showing early Cenozoic rocks in the Uinta and Piceance Creek Basins. The location of the cross section is shown on Figure 28. The drillholes and measured sections used are listed in Table 1.

Mexico and formed many much smaller basins and uplifts. The transition occurred gradually, and the foreland basin continued to exist for a time as a fairly continuous structural and sedimentary basin punctuated by several isolated early Laramide uplifts. As the orogeny developed, the Laramide uplifts became major sources of sediments, and subsidence trends and surface drainage patterns within the affected area of the foreland basin were radically altered.

Dating of the onset of structural growth on Laramide uplifts is generally imprecise, but many of the uplifts that define the Uinta and Piceance Creek Basins began to rise prior to the end of the Cretaceous and prior to the end of Mesaverde deposition. One of the oldest uplifts appears to be the Sawatch Uplift southeast of the Piceance Creek Basin. A porphyry sill in that uplift was dated as 72.2 ± 2.2 Ma (Obradovich and others, 1969; Tweto, 1975), older than much of the Mesaverde that is preserved along the trough of the adjacent Piceance Creek Basin. This uplift may have been the source of igneous rock fragments and angular feldspar grains reported in the upper part of the Mesaverde in the southeastern Piceance Creek Basin near the Sawatch Uplift (Hansley, 1981). The effects of this uplift on Late Cretaceous depositional patterns in the foreland basin appear to have been confined to a fairly narrow area adjacent to the uplift.

Structural relationships on the north flank of the Uinta Uplift suggest that uplift began prior to the end of the Cretaceous, as more than 2,550 m of Cretaceous strata were removed by erosion, prior to burial by the Paleocene Fort Union Formation (Hansen, 1965). The Uncompahgre Uplift, south of the Uinta and Piceance Creek Basins, is part of the much larger San Juan–San Luis Uplift. Unconformities in the San Juan Basin to the south suggest that elements of this uplift may have begun to rise during the late Campanian, and that almost certainly there was activity by the Maastrichtian (Dane, 1946). The San Rafael Anticline, south of the Uinta Basin, probably began to rise during latest Campanian time (Fouch and others, 1983).

Widespread beveling of Upper Cretaceous rocks began during early stages of the Laramide Orogeny, producing an unconformity that extends over probably all of the Piceance Creek Basin (Johnson and May, 1978; 1980) and all but the western part of the Uinta Basin (Fouch and Cashion, 1979; Fouch and others, 1983). This unconformity, where present, conveniently divides foreland basin rocks from early Cenozoic rocks deposited in the Laramide, Uinta, and Piceance Creek Basins. The maximum thickness of Upper Cretaceous rocks removed at the unconformity appears to have been considerable. The interval from the Castlegate Sandstone (a Campanian age marginal marine unit near the base of the Mesaverde) to the unconformity thins from about 2,130 m along the trough of the Piceance Creek Basin to about 760 m over the crest of the Douglas Creek Arch (Johnson and Finn, 1985; 1986); much of this thinning appears to be due to truncation of Campanian and Maastrichtian age rocks. The interval maintains a near constant thickness of 600–700 m in the eastern half of the Uinta Basin. The isopach map does not cover the western part of the Uinta Basin where continuous deposition occurred during much of the erosional episode. The isopach map appears to outline a broad area of uplift and beveling during the early stages of the Laramide Orogeny that extended over the Douglas Creek Arch area and the eastern half of the Uinta Basin (Johnson and Finn, 1986).

Continuous deposition occurred in the western part of the Uinta Basin during much of the lacuna and to the south in a sedimentary basin bounded by the Sevier orogenic belt on the west, the rising San Rafael Anticline on the southeast, and the rising Uinta Uplift on the northeast (Fig. 27). The basin has been called the Lake Flagstaff sedimentary basin of late Paleocene age by Stanley and Collinson (1979). The existence of this sedimentary basin has been variably attributed to downwarping during the final stages of the Sevier Orogeny (Spieker, 1946), to block faulting and downwarping during the early stages of the Laramide Orogeny (La Rocque, 1960), and to a combination of both (Stanley and Collinson, 1979). The basin will here be referred to as a transitional basin that was partly Sevier and partly Laramide in origin.

The North Horn Formation was deposited in this transitional sedimentary basin from Late Cretaceous (Maastrichtian) to late Paleocene time. The Maastrichtian part of the North Horn is correlative with the upper Mesaverde Group, preserved beneath the Cretaceous-Tertiary unconformity along the trough of the Piceance Creek Basin. According to Fouch and others (1983), the basin contained several warm, shallow, alkaline lakes fed by centrally draining streams.

EARLY BASIN SEDIMENTS

During the Paleocene, the transitional sedimentary basin spread eastward along the south flank of the rising Uinta Uplift and southeast along the east flank of the San Rafael Anticline, and by late Paleocene time, the Cretaceous-Tertiary unconformity had been onlapped to within a few kilometers of the crest of the Douglas Creek Arch. Conglomerate and conglomeratic sandstones, with pebbles consisting mainly of varicolored chert, silicified limestone, and quartzite, commonly compose the first unit found above the unconformity in the Uinta Basin east of the San Rafael Anticline (Fouch and Cashion, 1979; Francyzk and Nichols, 1986). This unit, informally called the conglomeratic beds at Dark Canyon (Fouch and others, 1983), is generally less than 30 m thick and is Paleocene in age where dated (Franczyk and Nichols, 1986). The unit can be traced on outcrop to within a few kilometers of the crest of the Douglas Creek Arch. Overlying the Dark Canyon beds is the largely fluvial Wasatch Formation of late Paleocene to early Eocene age (Fouch and others, 1983).

In the Piceance Creek Basin, sediments first began to accumulate on the unconformity surface in early to middle Paleocene time and spread westward, eventually burying the crest of the Douglas Creek Arch in early to middle Eocene time. Westward onlapping of Paleocene sediments on the unconformity has been noted in the southwestern part of the basin (Johnson and May, 1978) and is also probably responsible for the westward thinning of the lower member of the Fort Union Formation along the White River in the northern part of the basin (Hail, 1973, 1974).

A thin unnamed basal conglomerate or conglomeratic sandstone, similar to the beds at Dark Canyon in the Uinta Basin, was the first unit deposited above the Cretaceous-Tertiary unconformity throughout most of the Piceance Creek Basin. Pebbles and cobbles in this unit consist of varicolored chert, quartzite, and silicified limestone. The unit is time transgressive and varies in age from early to middle Paleocene in the eastern part of the basin, near the structural axis (Hail, 1973; Newman, 1974), to late Paleocene where it wedges out a few kilometers east of the crest of the Douglas Creek arch (Johnson and May, 1978, 1980). The pebbles and cobbles are similar in composition to but are, on the average, somewhat larger than those found in scattered lenses in the upper 150 m of the underlying Mesaverde in the southern part of the Piceance Creek Basin (Johnson and May, 1978; 1980) and may have been derived in part from that source.

A Paleocene fluvial sequence of sandstone and varicolored shale, derived largely from local Laramide uplifts, occurs above the basal conglomerate. This unit is generally included in the Fort Union Formation in the northern part of the basin (Hail, 1973; Pipiringos and Rosenlund, 1977) and in the Wasatch Formation in the southern part (Donnell, 1969). Volcaniclastic pebbles and cobbles, thought to be derived from the Sawatch Uplift (Tweto, 1975), are common in the lower part of this sequence in the southeastern and eastern areas of the basin. This unit, originally called the "Ruby beds" (Hills, 1890; Eldridge, 1894) and later included in the Wasatch Formation, is about 750 m thick at Ruby Peak in the southeasternmost part of the basin (Eldridge, 1894). It thins to the north to about 100 m at Rifle Gap (Johnson, 1982) and is a maximum of 20 m thick near Meeker in the northeastern part of the basin (Pipiringos and Rosenlund, 1977). The volcaniclastic conglomeratic sequence thins and grades into finer grained fluvial rocks toward the west (Johnson and others, 1979b, 1979c) and is not present along the western margin of the basin.

DEVELOPMENT OF LATE PALEOCENE AGE LACUSTRINE AND PALUDAL SYSTEMS

During the late Paleocene, the small lakes in the transitional sedimentary basin west of the San Rafael Anticline expanded and coalesced to form one large lake, Lake Flagstaff, that expanded to within a few kilometers of the crest of the Douglas Creek Arch before the end of the Paleocene (Figs. 28 and 29). The Flagstaff Limestone was deposited in this lake. It was originally described on the Wasatch Plateau southwest of the Uinta Basin by Spieker and Reeside (1925) and Spieker (1946) and traced along outcrop into the western Uinta Basin by Spieker (1949). The Flagstaff Limestone was correlated into the subsurface in much of the western Uinta Basin by Ryder and others (1976) and occurs in the subsurface in the eastern Uinta Basin (Fig. 29).

On the Wasatch Plateau the Flagstaff Limestone had been divided into as many as seven distinct units (Gill, 1950); most recently it was divided into three members, in ascending order: the Ferron Mountain Member, the Cove Mountain Member, and the Musinia Peak Member (Stanley and Collinson, 1979). The Ferron Mountain and Musinia Peak Members, which represent high stands of Lake Flagstaff, consist of highly fossiliferous limestone with abundant gastropod, pelecypod, ostracode, and charophyte remains that indicate shallow, well-oxygenated fresh water. The Cove Mountain Member represents a more restricted stage and consists of nonfossiliferous dolomicrite, mudstone, sandstone, and bedded and nodular gypsum.

In the western Uinta Basin, the Flagstaff Limestone was recently defined as a member of the Green River Formation (Fouch, 1976). Three general facies were recognized by Ryder and others (1976) going from lake margin to lake center: (1) interbedded sandstone, siltstone, gray calcareous claystone, algal coal, and oncolitic-bearing grain-supported carbonate; (2) highly fossiliferous gray mud and grain-supported limestone; and (3) dark-gray mud-supported limestone. Calcite is the dominant carbonate mineral. Some anhydrite nodules occur in the Flagstaff Member near the structural trough of the Uinta Basin (Fouch, 1981), which may correlate with the restricted Cove Mountain phase of Lake Flagstaff on the Wasatch Plateau.

A somewhat similar late Paleocene sequence of gray shale, carbonaceous shale, thin coal beds, thin fossiliferous limestones, and lenticular sandstones is found in the Piceance Creek Basin (Figs. 28, 29). This sequence appears to represent more paludal and less lacustrine depositional environments than the Flagstaff Member of the Uinta Basin. It overlies the largely fluvial unit in the central and eastern parts of the basin but directly overlies the thin, discontinuous basal Tertiary conglomerate in the west, where the fluvial unit has lapped out on the Cretaceous-Tertiary unconformity. These paludal-lacustrine rocks in turn lap out on the Cretaceous-Tertiary unconformity a few kilometers east of the crest of the Douglas Creek Arch (Figs. 28 and 29) (Johnson and May, 1978; 1980; Johnson, 1985). This widespread unit extends to near the boundaries of the Piceance Creek Basin. It is part of the Atwell Gulch Member of the Wasatch Formation in the southern part of the basin (Donnell, 1969) and part of the Fort Union Formation in the northern part (Hail, 1973; Pipiringos and Rosenlund, 1977).

During most of the Paleocene, the Piceance Creek Basin is thought to have been a separate hydrologic basin from the Uinta Basin. The presence of cobbles derived from the Sawatch uplift in the Paleocene section of the Sand Wash Basin to the north suggests that the Piceance may have drained in that direction. However, during maximum expansion of the paludal and lacustrine systems in the Uinta and Piceance Creek Basins during the late Paleocene, the two systems were separated by no more than a 25-km wide area near the crest of the Douglas Creek Arch. It is possible that the two basins were at least periodically connected hydrologically by that time.

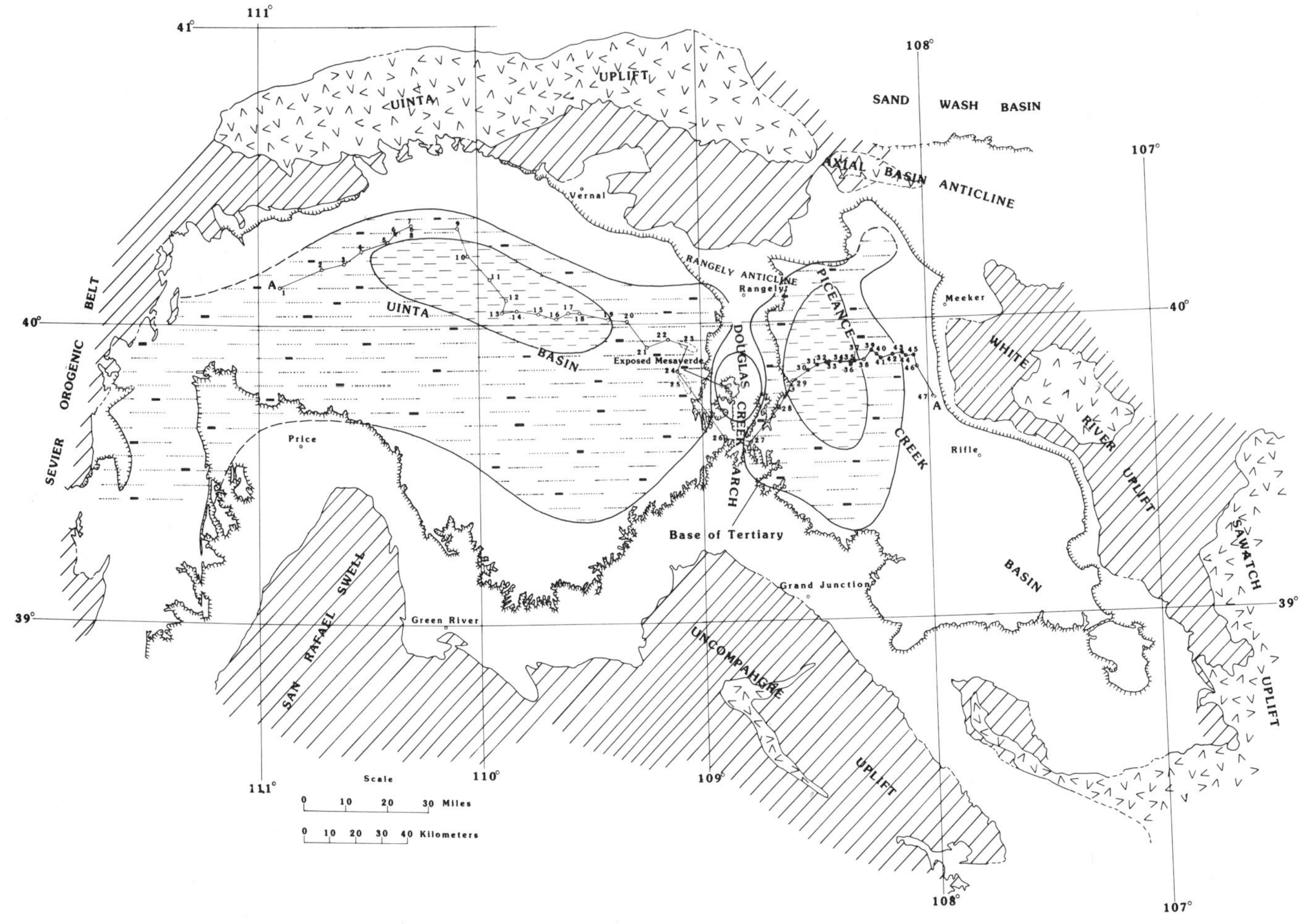

Figure 30. Map showing the distribution of lithologies and interpretations of environments of deposition during maximum extent of early Eocene freshwater lakes. The area labeled "Exposed Mesaverde" on the Douglas Creek Arch is where the Cretaceous-Tertiary unconformity had not, at this time, been onlapped by early Tertiary sedimentary rocks.

LATEST PALEOCENE THROUGH EARLY EOCENE ALLUVIAL PERIOD

Before the end of the Paleocene, wedges of varicolored mudstone and sandstone of predominantly fluvial origin began to prograde from the basin margins inward, eventually replacing the upper Paleocene limestones, shales, and coals in all but relatively small areas near the structural troughs of both basins and adjacent to the Sevier orogenic belt. Rocks from this interval are included in the Colton Formation in the western and central Uinta Basin, and in the Wasatch Formation in the eastern Uinta and Piceance Creek Basins. Infilling began in latest Paleocene time around most of the Uinta Basin (Ryder and others, 1976) and in the northern Piceance Creek Basin (Hail, 1973; Pipiringos and Rosenlund, 1977), and in latest Paleocene to early Eocene time in the southwestern Piceance Creek Basin (Johnson and May, 1978). The expansion of the alluvial environments during this period has been traditionally attributed to renewed elevation of the surrounding Laramide uplifts. Isopach maps of Johnson and Finn (1986) for the Piceance Creek Basin indicate an acceleration of subsidence beginning near the end of the Paleocene.

EARLY EOCENE FRESHWATER LAKES

After the late Paleocene to early Eocene influx of clastics, the area of freshwater lacustrine deposition in both basins began to expand again until, during maximum transgression, the two lakes extended to within a few kilometers of the crest of the Douglas Creek Arch (Figs. 29 and 30). They may have connected across the crest of the arch for a relatively brief period of time. Unfortunately, rocks of this age have been removed by erosion between Douglas Pass and Rangely, where the connection would have been. Nonetheless, it is likely that there was some sort of hydrologic connection between the two basins during much of the early freshwater stage. Rocks of this interval in the Piceance Creek Basin were originally included in the lower part of the

Douglas Creek Member of the Green River Formation (Bradley, 1931) but were recently removed from the Douglas Creek and renamed the Cow Ridge Member of the Green River Formation (Johnson, 1984). In the Uinta Basin, Bradley (1931) named a tongue of this freshwater unit the basal tongue of the Green River Formation where it crops out in Indian Canyon. Roberts (1964) included these rocks, as well as a section of younger rocks, in his Willow Creek interval (Fig. 29).

In the Piceance Creek Basin, marginal lacustrine rocks from this freshwater interval consist mainly of mollusc- and ostracod-rich sandstone and limestone, gray shale, and carbonaceous shale with thin coal beds. Carbonaceous shale and coal are the dominant lithologies along the southwestern margin of the basin, while sandstone is most abundant along the northern and eastern margins. The wedge-edge of the Cow Ridge Member adjacent to the crest of the Douglas Creek Arch consists mainly of ostrocodal limestone. Laterally persistent low-grade oil shales, with Fischer assay oil-yield values ranging from about 5 to 15 gallons per ton, were deposited in the north-central part of the Piceance Creek Basin in an area of offshore lacustrine deposition (Figs. 29 and 30).

In the Uinta Basin, contemporaneous lacustrine rocks are similar to those in the Piceance Creek Basin, but the marginal lacustrine rocks include minor oolites, oncolites, and stromatolites, and offshore lacustrine rocks appear to contain less kerogen. As in the Piceance, the wedge-edge of this interval just west of the crest of the Douglas Creek Arch consists mainly of ostracodal limestone. Ryder and others (1976) divided the lacustrine rocks from this freshwater period in the western part of the Uinta Basin into three facies that represent deposition from lake margin to lake center: (1) ostracodal and oolitic grainstones with horizontal and low-angle cross stratification and minor sandstone; (2) gray, fossiliferous, mud-supported carbonate; and (3) brown, kerogenous, ostracode-bearing, mud-supported carbonate. The third, or kerogen-rich, facies was deposited in an offshore lacustrine depocenter in the north-central part of the basin. The intermediate facies is commonly absent and, for simplicity, is included with the marginal lacustrine facies on Figures 29 and 30.

At maximum extent, the marginal lacustrine facies was as much as 30 km wide in the Piceance Creek Basin and as much as 65 km wide in the Uinta Basin. These broad shallow-water shelves covered a much larger area than the offshore kerogen-rich shale facies. Near the end of the freshwater lacustrine stage, variegated fluvial deposits filled most of the marginal lacustrine areas, and both lakes retreated toward their areas of offshore deposition. The lakes appear to have remained fresh during this regression, indicating that outflow occurred at least periodically. An isopach map of rocks from the Cretaceous-Tertiary unconformity to the end of the freshwater lacustrine period (Fig. 31) indicates that the Douglas Creek Arch acted as a hingeline between the two subsiding basins throughout this time interval.

The return to widespread freshwater lacustrine deposition in both basins could be considered the beginning of Lake Uinta, and indeed Bradley (1931) considered these rocks as Lake Uinta deposits. However, Bradley (1931, pl. 3) clearly believed that the two freshwater systems formed one lake across the crest of the Douglas Creek Arch, a connection that later detailed mapping and detailed stratigraphic studies indicate could only have occurred for a fairly brief time during maximum transgression, if at all. If we concur with Bradley and consider these freshwater lacustrine sequences as deposits of Lake Uinta, then there were two Lake Uintas, possibly connected by some sort of river system throughout most, if not all, of the early freshwater stage. The two lakes appear to have been similar to Lake Lumen, the early freshwater phase of Lake Gosiute (Pipiringos, 1962), which existed at approximately the same time in the greater Green River Basin of Wyoming and northwesternmost Colorado.

THE DEVELOPMENT OF SALINE LAKE UINTA

At about the beginning of middle Eocene time, the freshwater lakes in both basins began to expand and deepen during the Long Point transgression (Johnson, 1984) and were connected across the crest of the Douglas Creek Arch to form one continuous lake (Figs. 29 and 31). This lake (Lake Uinta of Bradley, 1931) existed as a single body until it was infilled near the end of the Eocene. A mollusc-rich basal transgressive bed, the Long Point Bed, marks this transgression in exposures throughout much of the Piceance Creek Basin (Johnson, 1984). Freshwater molluscs, common throughout the underlying freshwater lacustrine sequence, are rarely found more than 5 m above the basal transgressive bed in both basins, indicating that Lake Uinta remained at least brackish and probably saline throughout its remaining history.

Johnson (1985) suggested that the evolution of Lake Uinta from two relatively small freshwater lakes to one much larger saline lake was caused by a rising spillpoint for the two basins. At first, the rising of the spillpoint would have acted as a dam causing the lake to expand; however, increased evaporation from the newly expanded surface area would not allow expansion to continue indefinitely. Eventually the point was reached where evaporation and loss through leakage equaled water supplied to the lakes. As the spillpoint continued to rise beyond this critical point, the outlet would have been lost, and the two basins would have evolved into a closed hydrologic system.

Although Lake Uinta extended across the crest of the Douglas Creek Arch, isopach maps indicate that the arch was still active (Fig. 32). The slow rate of subsidence over the arch may have been largely responsible for the area of shallow marginal lacustrine deposition that persisted along the crest of the arch throughout much of Lake Uinta's history (Dyni, 1981). This shoal on the crest of the arch created two areas of deeper oil-shale deposition, one in each basin (Fig. 32), and strongly influenced circulation in the lake. Only the Piceance Creek Basin depocenter accumulated saline minerals, suggesting that the shallow crest of the arch prohibited deep water brine movement between the two areas of oil-shale deposition (Dyni, 1981; Johnson, 1985).

Marginal lacustrine rocks deposited after the Long Point

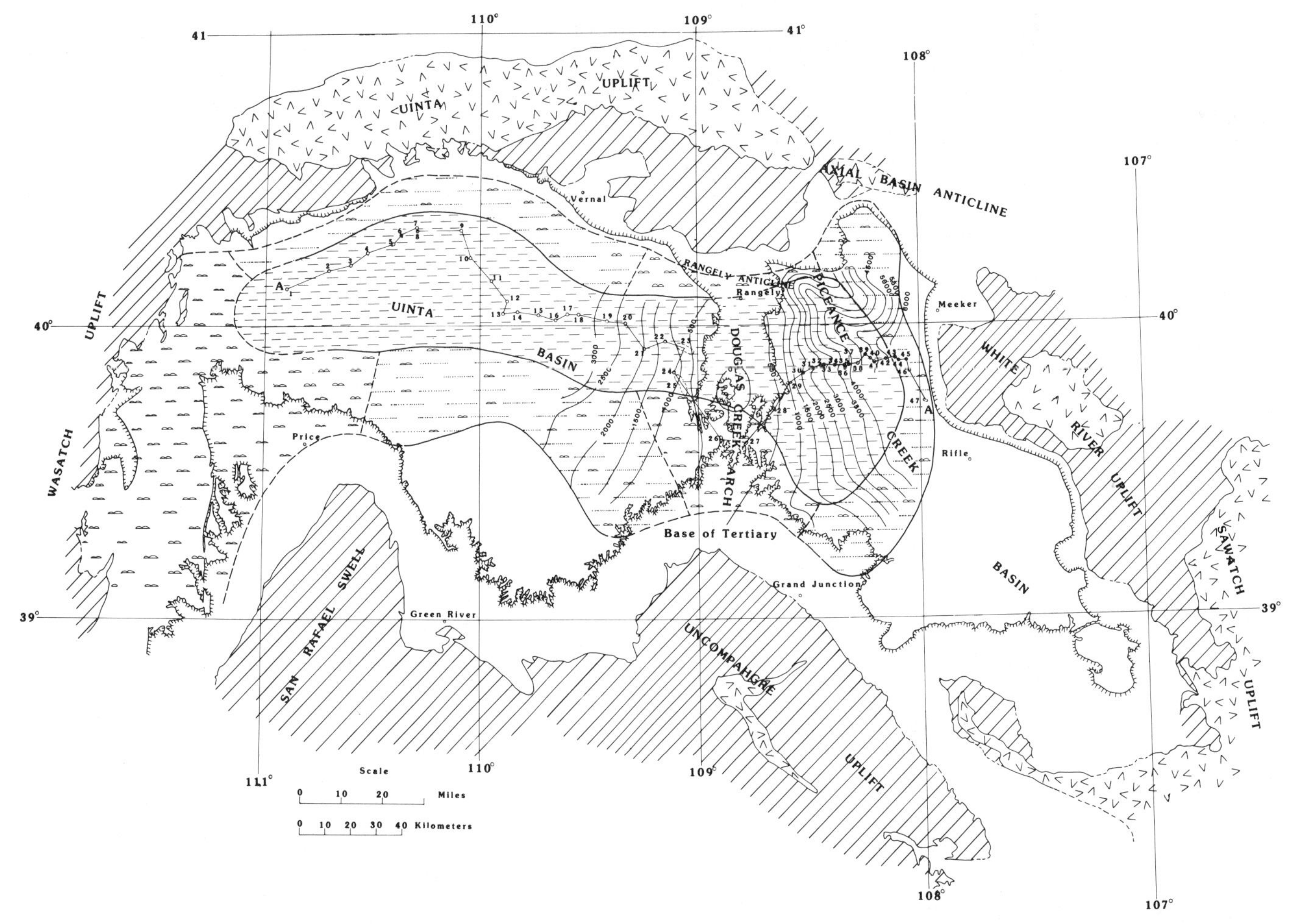

Figure 31. Map showing the distribution of lithologies and interpretations of environments of deposition just after the Long Point transgression. The isopach is the thickness in feet of rocks from the Cretaceous-Tertiary unconformity to the base of the Long Point Bed or its equivalent.

transgression are heterogeneous and highly variable but far more carbonate rich than the marginal lacustrine rocks from the preceding freshwater period. The more recent marginal lacustrine rocks are divided into two general facies: (1) a sandy unit consisting mainly of fairly persistent, mostly ripple-laminated sandstone with some carbonate-rich mudstone and stromatolitic, ostracodal, and oolitic limestone, and (2) a carbonate-rich unit consisting mainly of stromatolitic, ostracodal, and oolitic limestone and carbonate-rich mudstone with some ripple-laminated sandstone. The carbonate unit was dominant in areas with low rates of clastic influx such as the southwest Piceance Creek Basin and the southwest Uinta Basin. The sandy facies was dominant in areas with high rates of clastic influx such as the south-central and northern parts of the Uinta Basin and along the eastern margin of the Piceance Creek Basin.

In the Piceance Creek Basin, sandy marginal lacustrine units have been included in the Douglas Creek Member (Bradley, 1931; Waldron and others, 1951; Donnell, 1961; Snow, 1970), the Basal Sandstone Member (Hail, 1972, 1973, 1974), and the Anvil Points Member (Donnell, 1961; O'Sullivan, 1974; Pipiringos and Johnson, 1975, 1976; Johnson, 1975; 1977) of the Green River Formation. Carbonate-rich marginal lacustrine rocks are generally included in the Douglas Creek Member. In the Uinta Basin, marginal lacustrine rocks have been included in the Douglas Creek Member (Cashion, 1967), the black shale facies (Picard, 1955), and the Willow Creek interval (Roberts, 1964).

Oil shale was the dominant lithology in the two areas of offshore deposition. Several persistent rich and lean oil-shale zones appear to be time-stratigraphic units (Cashion and Donnell, 1972). The lower six zones are numbered zero through six, with the R-0 rich oil-shale zone overlying the Long Point Bed. All six rich zones occur in the Piceance Creek Basin, whereas only the R-0 and the R-4 through R-6 zones are recognized in the Uinta Basin. The rich zones are separated by lean oil-shale zones labeled L-1 through L-5. The Mahogany oil-shale zone occurs above the R-6 zone and is by far the most widespread and most economi-

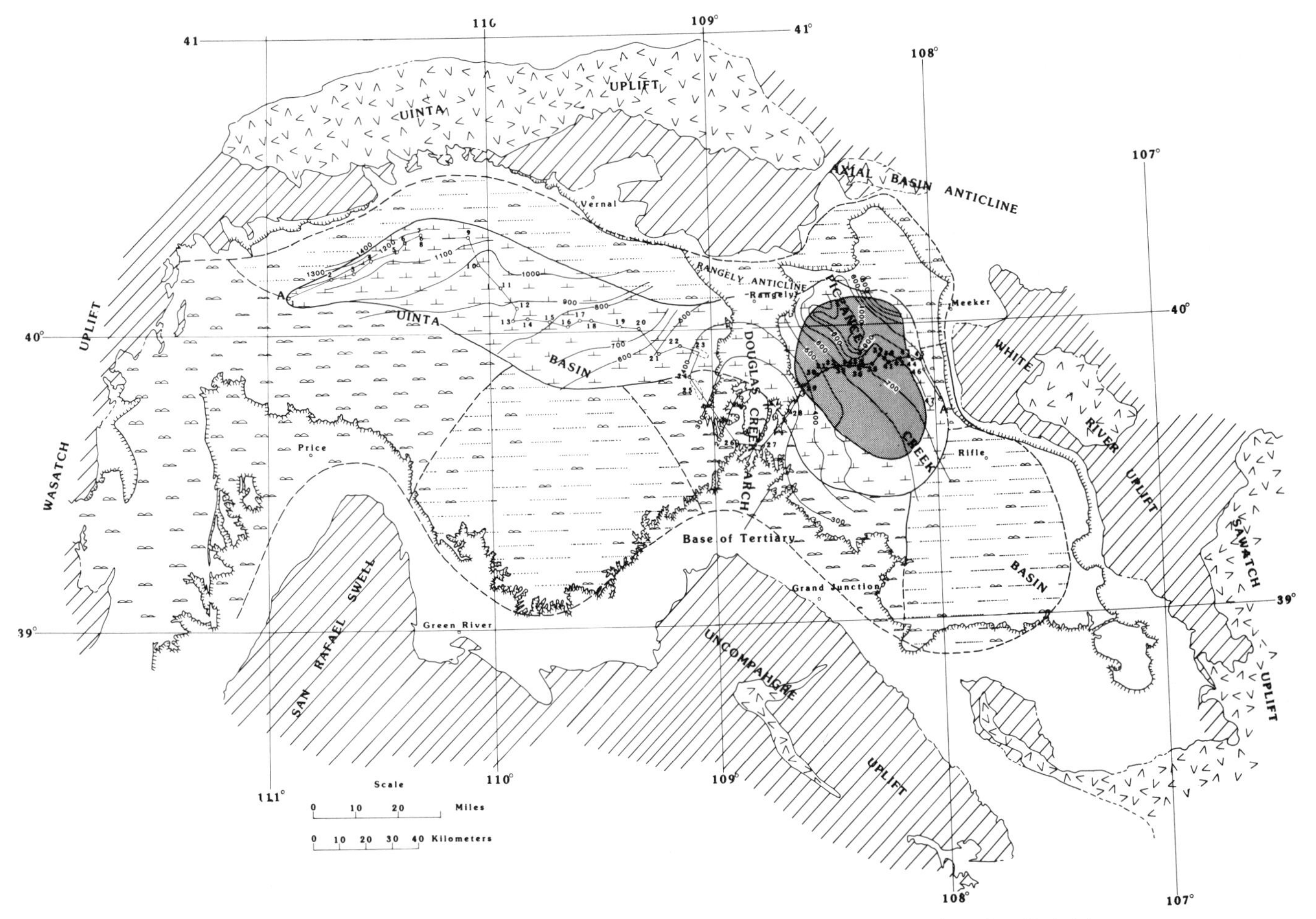

Figure 32. Map showing the distribution of lithologies and interpretations of deposition of the R-4 rich oil-shale zone. The isopach shows the thickness in feet of the interval from the base of the R-4 zone to the base of the Mahogany oil-shale zone.

cally important oil-shale zone in both basins. Several less important oil-shale zones occur above the Mahogany zone (Pitman and Donnell, 1973).

Variations in the composition of oil shales deposited in the Piceance Creek Basin seem to reflect changes in water chemistry of Lake Uinta (Dyni, 1974; Robb and Smith, 1974). The first oil-shale zone deposited, the R-0 zone, is fairly low-grade, averaging about 20 gallons per ton, and is illitic. In contrast, the illitic oil shales in the overlying R-1 oil-shale zone average as much as 40 gallons per ton. A shift from illitic to dolomitic oil shale occurs in the next zone, the R-2 oil-shale zone, and the remaining oil shales of the Green River Formation are all dolomitic. The sodium bicarbonate mineral nahcolite first occurs in the R-2 and R-3 zones as disseminated crystals, but nahcolite is not found in the lower part of the R-4 zone. The base of the R-4 corresponds to a slight transgression that may have diluted the brines. Figure 32 shows Lake Uinta after this transgression. Bedded nahcolite and bedded halite are present in the R-5 and L-5 zones, while only sparse nodules of nahcolite occur in the oil shales above the L-5 zone. Johnson (1985) correlated many of these oil-shale zones with their marginal lacustrine equivalents and constructed paleogeographic maps for several time periods (Figs. 30-32) The older illitic oil shales are usually mapped as the Garden Gulch Member in the Piceance Creek Basin and included in the black-shale facies in the Uinta Basin (Picard, 1955). The younger dolomitic oil shales are mapped as the Parachute Creek Member in both basins.

Lake Uinta has been described as both a deep, chemically stratified lake (Bradley, 1963; Bradley and Eugster, 1969) and a shallow playa lake (Eugster and Surdam, 1973). Both models explain some of the sedimentary features observed, and some aspects of both models may ultimately be adopted. The paleogeographic reconstructions by Johnson (1985) show that Lake Uinta had very broad, shallow, marginal playa-like shelves that produced bicarbonate-rich brines that accumulated in offshore areas of oil-shale deposition in the Piceance Creek Basin; however, large-scale clinoform bedding has been identified in rocks that were deposited between the marginal shelves and the area of oil-shale deposition (Johnson, 1981; 1985). Maximum relief on the clinoform bed sets is over 400 m. It seems unlikely that these large-scale features can be explained unless the area of oil-shale deposition was as much as several hundred meters deep during much of the history of Lake Uinta.

Lake Uinta began to expand during the latter part of R-6 oil-shale deposition and reached its maximum size during deposition of the Mahogany oil-shale bed (Fig. 29). During maximum transgression, the area of rich oil-shale deposition covered almost the entire marginal shelves. The actual extent of the lake during this brief period is not known because most of the marginal lacustrine rocks have been eroded; however, Lake Uinta may have extended to near the margins of the sedimentary basins. Volcaniclastics derived from the Absaroka volcanic field in northwestern Wyoming entered Lake Uinta for the first time in the northern part of the Piceance Creek Basin during deposition of the Mahogany bed. This indicates that the greater Green River Basin was draining south into Lake Uinta by this time, and the new source of water may have caused the transgression (Surdam and Stanley, 1979). The volcaniclastics prograded southward, filling the Piceance Creek Basin part of Lake Uinta. Infilling of Lake Uinta continued, finally pushing the lake considerably west of the crest of the Douglas Creek Arch and flushing the brines into the area of oil-shale deposition in the Uinta Basin, where saline mineral precipitation continued. Lake Uinta persisted as a saline lake in the central and western part of the Uinta Basin until near the end of the Eocene, when it was filled by sediments derived largely from local Laramide sources. As much as 1,825 m of largely fluvial rocks of the upper Eocene to lower Oligocene Duchesne River Formation presently overlie the Green River Formation near the trough of the Uinta Basin.

Johnson and Finn (1986) estimated about 4,000 m of subsidence in the Uinta Basin as compared to about 1,500 m of subsidence in the Piceance Creek Basin in post-Mahogany time. Almost 60 percent of the total Laramide subsidence in the Uinta Basin and only about 35 percent of the total subsidence in the Piceance Creek Basin is post-Mahogany. This difference may in part explain why Lake Uinta persisted in the Uinta Basin long after the Piceance Creek Basin was filled. The results may better define periods of movement on Laramide uplifts, if it is assumed that uplift was linked to subsidence in adjacent basins. This suggests that the Uinta Uplift north of the Uinta Basin was very active during the late Eocene. In contrast, movement on the White River Uplift east of the Piceance Creek Basin was apparently diminishing during that time. Subsidence ceased, and widespread erosional surfaces developed across both basins during the early Oligocene.

GREATER GREEN RIVER BASIN

R. T. Ryder

INTRODUCTION

The greater Green River Basin encompasses four basins (Great Divide, Green River, Sand Wash, and Washakie) and four intrabasin uplifts (Cherokee Arch, Moxa Arch, Rock Springs Uplift, and Wamsutter Arch) on the strongly compressed craton of southwest Wyoming and adjacent Colorado and Utah (Fig. 33; Plate 2). The Great Divide Basin of this paper is the Red Desert Basin of Woodward (Plate 2).

Recently published seismic profiles used in conjunction with deep drill holes have been extremely valuable in resolving the geometry of basement-involved structures at the margins of the greater Green River Basin. These data convincingly show, for example, that basement uplifts flanking three sides of the basin are the product of horizontal compression rather than vertical uplift (Hamilton, 1978; Bally and Snelson, 1980). Seismic profiles and deep drill holes also have increased our understanding of the Wyoming-Idaho thrust belt along the western side of the greater Green River Basin, even though the decollement style of deformation of the thrust belt already had been identified from surface geologic maps (Royse and others, 1975). Subsidence and sedimentation during active stages of basin development are intimately associated with structures adjacent to the basin perimeter; for that reason a discussion of the geometry and history of movement of these structures, supported whenever possible by published seismic data, constitutes the first part of this section. Intrabasin uplifts also are discussed.

Fragmentation of the craton that in large part controlled the structural configuration of the present-day greater Green River Basin and its sedimentary deposits is Laramide in age (Campanian through late Eocene). However, prior to Laramide deformation and sedimentation, the region now occupied by the greater Green River Basin was a foreland basin (Late Jurassic through earliest Eocene) adjacent to the Wyoming-Idaho thrust belt, and before that it was a continental shelf (Cambrian through Middle Jurassic). The stage following the Laramide, which was characterized by basin filling, extensional faulting, and partial collapse of several basement uplifts, further modified the structural configuration of the basin. The second part of this chapter summarizes the nature and evolution of the sedimentary record of the greater Green River Basin in terms of the shelf, foreland-basin, Laramide, and post-Laramide stages.

GEOMETRY AND AGE OF BASIN-MARGIN AND INTRABASIN UPLIFTS

Northern Margin

Three major uplifts—the Gros Ventre, Wind River, and Granite Mountains—bound the 325-km-long northern margin of the greater Green River Basin (Fig. 33).

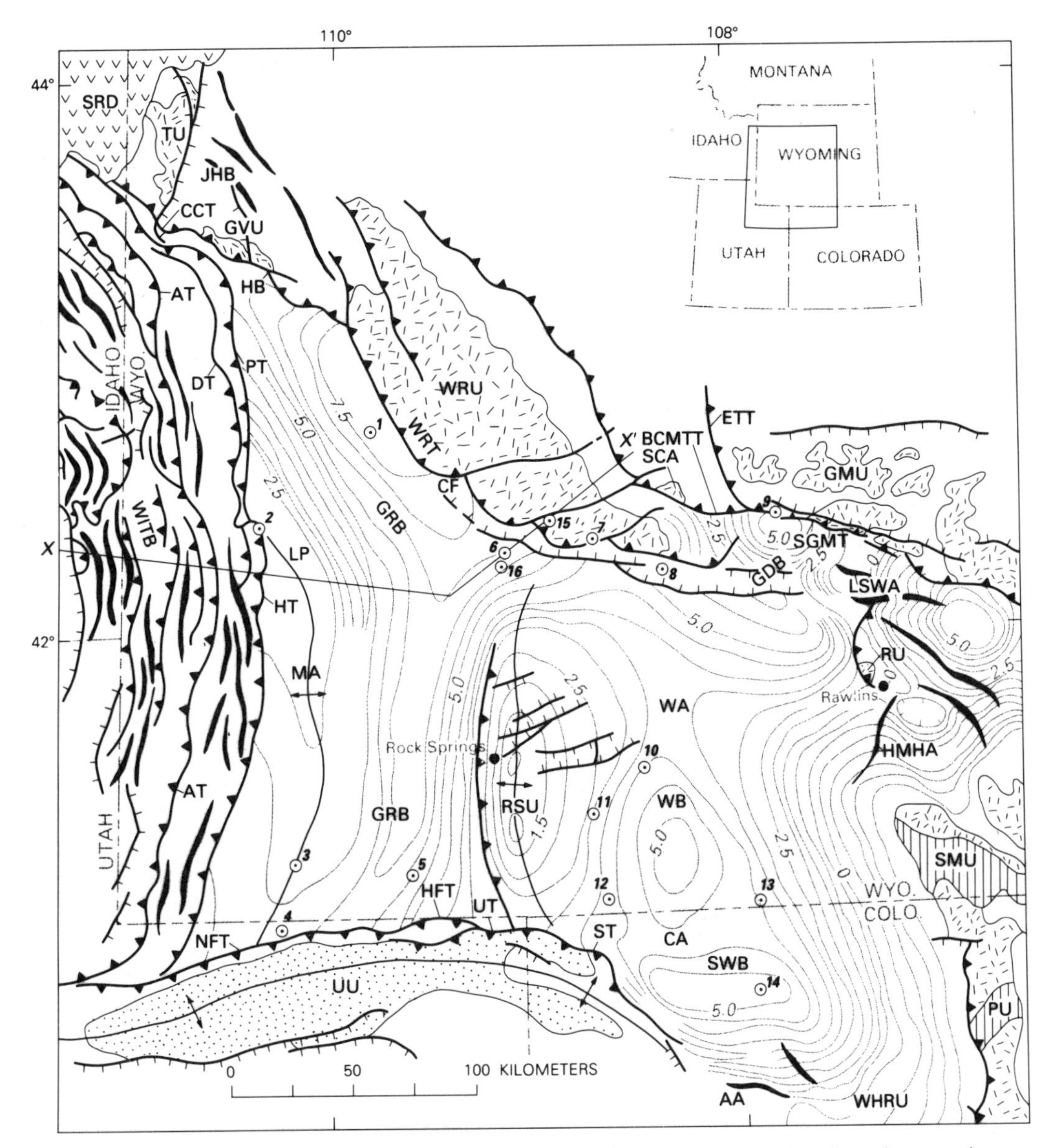

Figure 33. Tectonic map of southwestern Wyoming and adjacent states showing the major tectonic elements of the greater Green River Basin. The map is taken from King (1969). Minor modifications have been added based on the work of Blackstone (1979), Love and Christiansen (1985), and Reynolds (1968). Structure contours, in kilometers below mean sea level, are drawn on top of Precambrian basement rocks. Except along the Moxa Arch, structural contours are not redrawn to conform with new drill-hole data. Data relating to drill holes 1 to 16 on the map are presented in Table 2. Major tectonic features are identified as follows: AA, Axial Arch; AT, Absaroka thrust fault; BCMTT, Beaver Creek and Mormon Trail thrust faults; CA, Cherokee Arch; CCT, Cache Creek thrust fault; CF, Continental Fault; DT, Darby thrust fault; ETT, Emigrant Trail thrust fault; GDB, Great Divide Basin; GMU, Granite Mountains Uplift; GRB, Green River Basin; GVU, Gros Ventre Uplift; HB, Hoback Basin; HFT, Henry's Fork thrust fault; HMHA, Hatfield and Miller Hill Anticlines; HT, Hogsback thrust fault; JHB, Jackson Hole basin; LSWA, Lost Soldier and Wertz Anticlines; MA, Moxa Arch; LP, La Barge Platform; NFT, North Flank thrust fault; PT, Prospect thrust fault; PU, Park Uplift; RSU, Rock Springs Uplift; RU, Rawlins Uplift; SCA, Sweetwater Crossing Anticline; SGMT, South Granite Mountains thrust fault; SMU, Sierra Madre Uplift; SRD, Snake River downwarp; ST, Sparks thrust fault; SWB, Sand Wash Basin; TU, Teton Uplift; UT, Uinta thrust fault; UU, Uinta Uplift; WA, Wamsutter Arch; WB, Washakie Basin; WHRU, White River Uplift; WITB, Wyoming-Idaho thrust belt; WRT, Wind River thrust fault; and WRU, Wind River Uplift. Basement rocks are identified as follows: random-dash pattern, basement rocks of Archean age; vertical-line pattern, basement rocks of Early Proterozoic age; stippled pattern, basement rocks of Middle Proterozoic age. Tertiary volcanic rocks are identified by a random V pattern. Line of geologic cross section (Fig. 34) is X–X'.

TABLE 2. NAME AND LOCATION OF DRILL HOLES IDENTIFIED IN FIGURE 33

Name	Location	Total Depth	Age of Rocks at Total Depth
1. El Paso No. 1 Wagon Wheel	Sec.5,T.30N,R.108W. Sublette Co., WY	19,000 ft (5.8 km)	Late Cretaceous (Santonian)
2. Mobil No. 22-19G Tip Top Unit	Sec.19,T.28N.,R.113W. Sublette Co., WY	15,435 ft (4.7 km)	Cambrian
3. Mountain Fuel Supply No. 1 Butcher Knife	Sec.29,T.15N.,R.112W. Uinta Co., WY	18,843 ft (5.7 km)	Mississippian
4. CIG Exploration No. 1 Cook	Sec.17,T.3N.,R.15E. Summit Co. UT	21,786 ft (6.6 km)	Mississippian
5. Tom Brown No. 1 Current Creek	Sec.20,T.14N.,R.108W. Sweetwater Co., WY	19,250 ft (5.9 km)	Triassic
6. Superior Oil No. 1 Current Creek	Sec.27,T.27N.,R.103W. Sublette Co., WY	20,521 ft (6.25 km)	Late Cretaceous (Santonian-Coniacian)
7. Amoco No. 1 South Pass	Sec.17,T.27N.,R.100W. Fremont Co., WY	22,970 ft (7.0 km)	Late Jurassic
8. Mountain Fuel Supply No. 1 Cyclone Rim	Sec.14,T.26N.,R.96W. Sweetwater Co., WY	16,002 ft (4.9 km)	Late Cretaceous
9. Sinclair No. 1 Cooper Creek	Sec.19,T.28N.,R.90W. Fremont Co., WY	12,225 ft (3.7 km)	Late Cretaceous (Turonian)
10. Texaco No. 15 Table Rock	Sec.2,T.18N.,R.98W. Sweetwater Co., WY	17,339 ft (5.3 km)	Pennsylvanian
11. Champlain No. 1 Brady	Sec.2,T.16N.,R.101W. Sweetwaater Co., WY	16,502 ft (5.0 km)	Cambrian
12. Mountain Fuel Supply No. 1 Kinney	Sec.18,T.13N.,R.99W. Sweetwater Co., WY	18,636 ft (5.7 km)	Mississippian
13. Phillips No. 8 Baggs	Sec.10,T.12N.,R.92W. Carbon Co., WY	16,248 ft (5.0 km)	Precambrian
14. Humble No. 1 Lay Creek	Sec.13,T.8N.,R.93W. Moffat Co., CO	16,381 ft (5.0 km)	Devonian
15. American Quasar No. 1 Skinner Federal	Sec.32,T.28N.,R.101W. Fremont Co., WY	15,040 ft (4.6 km)	Cretaceous
16. Rainbow Resources No. 1-34 Pacific Creek Federal	Sec.34,T.27N,R.103W. Sublette Co., WY	25,764 ft (7.85 km)	Mississippian

Archean granite of the Gros Ventre Uplift has overridden the northern margin of the Green River Basin along the Cache Creek thrust fault and subsidiary faults (Dorr and others, 1977; Love, 1982). No seismic lines that cross the Cache Creek Fault have been published, but by analogy to the adjacent Wind River thrust fault (Fig. 33; Zawislak and Smithson, 1981; Allmendinger and others, 1983), the Cache Creek Fault probably dips northward between 30 and 40 degrees. At its northwestern end, the Cache Creek fault system passes south of the Teton Uplift and continues into east-central Idaho where it disappears under the volcanic rocks of the Snake River downwarp (Fig. 33; Love, 1982; Love and Christiansen, 1985). Southeastward, the Cache Creek Fault system branches into several splay faults and merges with the Wind River thrust fault (Dorr and others, 1977; Love and Christiansen, 1985). The Gros Ventre Uplift and adjoining Teton Uplift emerged as a contiguous structure in latest Cretaceous and early Paleocene time, separating the Green River Basin from the Jackson Hole Basin to the north (Love, 1982). Stratigraphic studies of lower Tertiary synorogenic deposits in the northwesternmost part of the Green River Basin (commonly known as the Hoback Basin) by Dorr and others (1977) indicate that the Gros Ventre Uplift and Cache Creek fault system were most active in middle Paleocene to middle early Eocene time, and that accelerated growth of these structures progressed from northwest to southeast.

The COCORP seismic profile across the southeastern end of

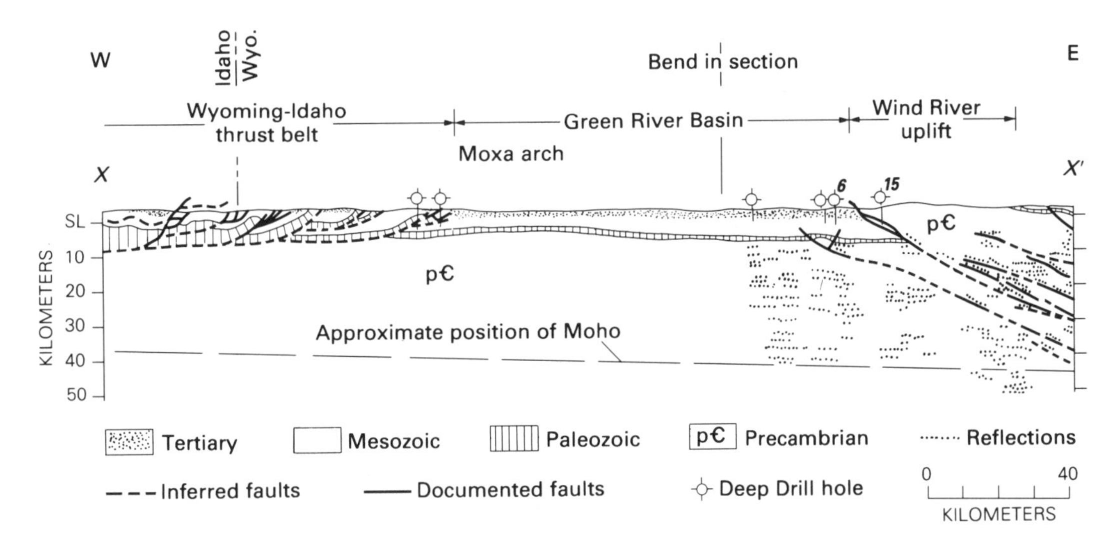

Figure 34. Cross section through the Wyoming-Idaho thrust belt, the Green River Basin, and the Wind River Uplift. Line of section is located on Figure 33. The section is from Bally and Snelson (1980). Drill holes 6 and 15 are identified in Table 2.

the Wind River Uplift shows that the Archean basement rocks of the uplift have overridden sedimentary rocks of the Green River Basin approximately 18–20 km along the Wind River thrust fault (Fig. 34; Allmendinger and others, 1983). Approximately 12 km of vertical structural relief is indicated on the basement. This thrust fault is buried along most of its 260-km length except for an exposed branch at the northern part of the uplift where Archean granite is juxtaposed against Mesozoic sedimentary rocks (Fig. 33; Love and Christiansen, 1985). Near the southeastern part of the Wind River Uplift, the thrust probably changes from a northwesterly trend to a northeasterly trend and terminates near the junction of the Beaver Creek and Mormon Trail thrust fault system and the Emigrant Trail thrust fault (Fig. 33). The Wind River Uplift probably began in latest Cretaceous time (Love and others, 1963), was greatly accelerated in Paleocene and early Eocene time as evidenced by arkosic sandstone deposits up to 2.5 km thick along the northern margin of the Green River Basin (Shaughnessy and Butcher, 1974), and ended in middle Eocene time as indicated by the folding of syntectonic sediments of that age that overlie the Wind River thrust fault (Steidtmann and others, 1983; Zeller and Stephens, 1969). The uplift was later modified by large down-to-the-north normal faults, which include the Continental Fault, located in front of and subparallel to the Wind River thrust fault (Fig. 33; Zeller and Stephens, 1969).

Near the southeastern plunge of the Wind River Uplift, the margin of the Great Divide Basin shifts abruptly northward about 25 km to the Beaver Creek and Mormon Trail thrust fault system and the South Granite Mountains thrust fault (Fig. 33; Love, 1970). These faults form the southern margin of the Sweetwater Crossing Anticline and Granite Mountains Uplift, respectively, and judging from the configuration of the nearby Emigrant Trail thrust fault (Berg, 1961), they have carried Archean granite at least several kilometers over sedimentary rocks of the Great Divide Basin. Pebbles in the Lance Formation suggest that the Granite Mountains were uplifted initially in Maastrichtian time (Reynolds, 1976). By Paleocene time, the granite core of the Granite Mountains Uplift was exposed and contributed moderate amounts of feldspathic debris to deposits in the Great Divide Basin (Love, 1970). The rate of uplift of the Granite Mountains Uplift and Sweetwater Crossing Anticline was accelerated greatly in earliest Eocene time, causing the deposition of as much as 1.2 km of arkosic sandstone in the Great Divide Basin and major movement along the South Granite Mountains, Emigrant Trail, Beaver Creek, and Mormon Trail thrust faults (Love, 1970). In late Eocene time and again in Miocene through Pliocene time, the Granite Mountains Uplift and Sweetwater Crossing Anticline were modified by large, generally down-to-the-north normal faults (Love, 1970; Reynolds, 1968).

Eastern Margin

From north to south, the 250-km-long eastern margin of the greater Green River Basin is bordered by the Lost Soldier and Wertz Anticlines, Rawlins Uplift, Hatfield and Miller Hill Anticlines, Sierra Madre Uplift, and Park Uplift (Fig. 33; Love and Christiansen, 1985; Tweto, 1979a).

Multiple unconformities, abrupt facies changes, and depositional thinning indicate a complex growth history beginning in early Campanian time and lasting into the middle Paleocene for the ancestral Lost Soldier Anticline (Reynolds, 1976). A final stage of folding in late Eocene time divided the ancestral structure into the present-day northwest-trending Lost Soldier and Wertz Anticlines (Reynolds, 1976).

The compact Rawlins Uplift, with Archean basement rocks exposed in its core, has overridden sedimentary deposits of the Great Divide basin along an east-dipping thrust fault system located along the western margin of the uplift (Gries, 1983; Love and Christiansen, 1985). Vertical structural relief as much as 3 km is documented between the basement rocks of the uplift and the adjacent basin (Knight, 1951). The southeast-facing Hatfield and Miller Hill anticlinal trend plunges northeastward beneath the Rawlins Uplift and crosses the nose of the Sierra Madre Uplift at a high angle (Fig. 33).

The Sierra Madre and Park Uplifts are the dominant structures along the eastern border of the greater Green River Basin. The western flank of the Park Uplift is bounded by a major thrust fault that has displaced Early Proterozoic gneissic rocks and Archean granitic rocks over the eastern margin of the Sand Wash Basin (Tweto, 1979a). Recent drilling indicates that basement rocks of Archean and Early Proterozoic age of the Sierra Madre Uplift probably are also detached along an east-dipping thrust fault system, now buried beneath Miocene strata of the Washakie Basin (Gries, 1981). By Paleocene time, these uplifts were supplying granite detritus to the adjacent Sand Wash and Washakie Basins, and the supply continued into Eocene time (Tweto, 1975; Colson, 1969). Pennsylvanian and Early Permian growth also is indicated for the Sierra Madre and Park Uplifts by adjacent arkosic deposits of that age (Mallory, 1972a) and by the truncation, along the flanks of the uplifts, of Pennsylvanian and older rocks beneath Upper Permian and Triassic rocks (Ritzma, 1951).

Southern Margin

Most of the 375-km-long southern margin of the greater Green River Basin is bounded by the east-west–trending Uinta uplift and its southeast-trending extension, the Axial Arch (Fig. 33). The northern flank of the White River Uplift bounds the remainder of the southern margin of the basin (Fig. 33).

From west to east, the North Flank, the Uinta, and the Sparks thrust faults separate the Uinta Uplift from the adjacent greater Green River Basin (Fig. 33; Hansen, 1965; Ritzma, 1971; Stokes and Madsen, 1961). The Henry's Fork thrust fault, located approximately 10 km north of the Uinta Fault, is probably the eastern extension of the North Flank Fault (Fig. 33). The structural axes of the Moxa Arch, the Green River Basin, and the Rock Springs Uplift are truncated by these faults, whereas thrust faults in the Utah sector of the Wyoming-Idaho thrust belt swing westward and merge tangentially with these faults (Fig. 33). A seismic line across the eastern end of the buried North Flank thrust fault, supplemented by data from a deep hole drilled through the northern margin of the Uinta Uplift, demonstrates that Middle Proterozoic quartzite and shale of the uplift have been thrust north approximately 12–14 km over younger sedimentary rocks of the Green River Basin (Clement, 1983). Vertical structural relief on the Precambrian basement is approximately 8 km. Bally and Snelson (1980) applied the term structural inversion to the process that transformed the precursor Middle Proterozoic basin into the Uinta Uplift.

An angular unconformity between Paleocene and late Campanian strata along the northern side of the Uinta Fault, several kilometers west of the Colorado-Utah border, indicates that the Uinta Uplift and perhaps the Uinta thrust fault began to develop in latest Cretaceous time (Hansen, 1965). In this same area, thick locally derived conglomeratic deposits of late Paleocene and (or) early Eocene age are cut by the Uinta Fault, suggesting that accelerated growth of the Uinta Uplift and thrust faulting occurred then. The youngest age limit on thrust faulting along the northern margin of the Uinta Uplift is more tightly constrained along the Henry's Fork Fault. Here, thick, southerly derived conglomerate deposits equivalent to middle Eocene lake beds are interrupted by an intraformational unconformity that truncates the Henry's Fork Fault (Anderman, 1955). The North Flank thrust fault and subsidiary faults, which form the large "triangle zone" interpreted by Clement (1983), do not seem to cut beds younger than Paleocene; however, substantial Eocene activity along these faults is implied by southward depositional thinning and multiple unconformities recorded in Eocene strata overlying the faulted margin of the uplift. The Uinta Uplift experienced several stages of extension between late Eocene and Pliocene time that led to the partial collapse of the uplift along large normal faults (Hansen, 1965; Sales, 1983).

The complexly faulted Axial Arch, flanked by basement-involved thrust faults with opposing directions of transport, maintains the same approximate trend as the Uinta Uplift (Stone, 1975). The adjoining northern flank of the White River Uplift is marked by numerous north- to northeast-plunging anticlinal folds underlain by thrust faults that have involved Precambrian basement rocks (Livesey, 1985; Stone, 1975; Tweto, 1979a). The White River Uplift was active in Paleocene time, whereas the Axial Arch was not active until Eocene time (McDonald, 1972, 1975).

Western Margin

The Wyoming-Idaho thrust belt forms the 325-km-long western margin of the greater Green River Basin. The thrust belt is characterized by thin sheets of sedimentary cover rocks that have been detached from the underlying gently west-dipping Precambrian basement rocks and imbricated along east-directed thrust faults (Figs. 33, 34). The Absaroka thrust fault, the dominant structure of the Wyoming-Idaho thrust belt, extends uninterrupted from beneath the Snake River downwarp in east-central Idaho to the northern flank of the Uinta Uplift in Utah (Fig. 33; Blackstone, 1979; Crittenden, 1974). The Prospect thrust fault marks the approximate leading edge of the northern part of the thrust belt, whereas the Hogsback thrust fault marks the approximate leading edge of the southern part (Fig. 33). The meeting of these faults in an area of complex structure resulted from a transverse ramp in the footwall of the Prospect Fault (Royse and others, 1975). Although the decollement style of deformation in the Wyoming-Idaho thrust belt had been well documented in earlier field studies (Rubey and Hubbert, 1959; Armstrong and

Oriel, 1965), seismic profiles and deep drill holes have greatly increased our perspective of surface and buried structures; for example, seismic and drill-hole data demonstrate that the western margin of the Green River Basin and the associated Moxa Arch extend for approximately 20 km beneath the Prospect thrust sheet (Royse and others, 1975). A similar amount of overlap exists between the western margin of the Green River Basin and the Hogsback thrust sheet (Royse and others, 1975).

Several of the major thrust sheets appear to be associated with a synorogenic conglomerate that was derived from, and later was overridden by, the leading edge of the sheet. This structural and stratigraphic evidence, plus the age of deposits overlying the subcrop trace of the thrust faults indicates that thrust faulting in this belt appears to have progressed from west to east (Armstrong and Oriel, 1965; Royse and others, 1975). Dates assigned to the following thrusts illustrate the point: thrust faults in the western part of the belt are assigned a latest Jurassic to Early Cretaceous age, the Absaroka thrust fault had middle Campanian and Maastrichtian phases of growth, and the Prospect and Hogsback Faults are assigned middle early Eocene and pre–latest Paleocene dates, respectively (Armstrong and Oriel, 1965; Oriel and Armstrong, 1966; Royse and others, 1975). These dates also indicate that the easternmost faults of the thrust belt grew simultaneously with the Gros Ventre and Uinta Uplifts, a conclusion supported by the eastward-facing curvature of the thrust belt and by overlapping structures and intertonguing synorogenic deposits of thrust-belt and basement-uplift origin (Crittenden, 1974; Dorr and others, 1977).

The sense of motion along several thrust faults in the belt was reversed in late Miocene and early Pliocene time. Extension with collapse produced half-graben structures filled with thick nonmarine deposits (Dorr and others, 1977; Royse and others, 1975).

Intrabasin Uplifts

The doubly plunging Rock Springs Uplift, with rocks as old as Santonian age exposed in its core, is the most conspicuous uplift within the greater Green River Basin (Fig. 33). This uplift extends from the southeastern part of the Wind River Uplift to near the eastern end of the Uinta Uplift and separates the Green River Basin on the western side of the Great Divide and the Washakie Basin on the eastern side. Westward-facing asymmetry and curvature of the uplift probably are caused by east-west–oriented compression and by the buried east-dipping thrust fault that Garing and Tainter (1985) recognized along the western margin of the uplift. Smaller buried east-dipping thrust faults offset the basement along the eastern flank of the uplift (Garing and Tainter, 1985). Northeast-trending transverse normal faults cut across the crest and flanks of the uplift. An angular unconformity between Paleocene rocks and Upper Cretaceous rocks as younger as Maastrichtian age defines the time of major uplift (Roehler, 1983). Probably the thrust fault along the western flank of the uplift also was formed in latest Cretaceous time, because its subcrop trace is buried beneath Paleocene rocks (Love and Christiansen, 1985). Intermittent growth of the Rock Springs Uplift must have continued at least through the middle Eocene, because lacustrine rocks of that age are gently tilted by the uplift and are cut by transverse normal faults (Roehler, 1978).

The Moxa Arch is another prominent uplift in the greater Green River Basin (Stockton and Hawkins, 1985). This uplift is buried beneath uppermost Cretaceous and lower Tertiary rocks along its entire length, from the northern end of the Hogsback thrust fault to the Uinta Uplift (Fig. 33). Drill hole and seismic data indicate that the arch plunges south and disappears beneath the Uinta Uplift. The northern end of the arch, commonly referred to as the La Barge Platform, is overridden by the Wyoming-Idaho thrust belt. Drill-hole data also indicate that the arch is convex east in plan view. Angular unconformities recognized in subsurface stratigraphic studies indicate that the arch experienced initial uplift and truncation in early to middle Turonian time, followed by a second period of major uplift and truncation in late Campanian time (Merewether and others, 1984; Roehler, 1965). Stratigraphic studies by Wach (1977) indicate that the Moxa Arch was only mildly active in latest Cretaceous and early Tertiary time. Times of pre-Laramide crustal instability recorded on the Moxa Arch (and on the Lost Soldier Anticline, discussed in the section "Eastern Margin") are the result of crustal "conditioning" prior to major crustal fragmentation in the region (Bally and Snelson, 1980).

Two subtle east-west–trending uplifts, the Wamsutter and the Cherokee Arches, divide the eastern half of the greater Green River Basin into three subbasins (Fig. 33). The Wamsutter Arch, the larger of the two uplifts, separates the Great Divide Basin on the north from the Washakie Basin on the south; the Cherokee Arch separates the Washakie Basin on the north from the Sand Wash Basin on the south. Judging from isopach maps of early Tertiary rocks across the uplifts and the age of the youngest rocks involved in the uplifts, the Wamsutter and Cherokee Arches appear to have developed in Eocene time (McDonald, 1975). However, Weimer (1966) suggested that the western part of the Wamsutter Arch had a history of tectonic growth going back to early Late Cretaceous time.

SEDIMENTARY RECORD

Thickness Patterns

Seismic and deep drill-hole data indicate that the sedimentary sequence in the greater Green River Basin commonly exceeds a thickness of 6 km and locally exceeds 10 km in the deepest parts of the basin. Key drill holes are located on Figure 33 and listed in Table 2.

The COCORP seismic profile over the southeastern end of the Wind River Uplift suggests that the thickness of the sedimentary rocks in the adjacent Green River Basin ranges between 9 and 12.5 km (Fig. 34; Allmendinger and others, 1983; Basham and Martin, 1985; Zawislak and Smithson, 1981). A 7.85-km-deep drill hole located near the COCORP seismic line (drill hole No. 16) encountered Mississippian rocks at total depth. Gries

(1983) estimated, on the basis of seismic data, that the thickness of the sedimentary cover of the Green River Basin in front of the central part of the Wind River Uplift is approximately 10 km. The deepest drill hole in this part of the Green River Basin (drill hole No. 1) stopped at 5.8 km in rocks of probable late Santonian age (Shaughnessy and Butcher, 1974). On the basis of seismic data, Garing and Tainter (1985) suggested that the sedimentary rocks in the southern end of the Green River Basin attain a maximum thickness of approximately 7.5 km. This estimate seems reasonable in view of a nearby deep drill hole (drill hole No. 5) in which Triassic rocks were encountered at a total depth of 5.9 km.

The thickness of the sedimentary cover over the Moxa Arch increases southward from about 4.9 km near the northern end of the Hogsback thrust (drill hole No. 2) to about 6 km near the southern end of the arch (drill hole No. 3) and attains a maximum of about 7 km near the southern end of the arch (drill hole No. 4) (Clement, 1983; Marzolf, 1965).

Drilling in the deepest part of the Great Divide Basin in front of the Wind River thrust fault has reached uppermost Jurassic rocks at a total depth of 7 km (drill hole No. 7) and Campanian rocks at a total depth of 4.9 km (drill hole No. 8). A realistic total thickness of the sedimentary rocks in this part of the Great Divide Basin is between 8 and 9 km. The sedimentary section in the shallower northern part of the Great Divide Basin is estimated to be about 5.5 km thick. This estimate is based on a 3.7-km-deep hole (drill hole No. 9) drilled through the southern margin of the Granite Mountains Uplift and into upper Turonian rocks at total depth and on a geologic cross section through this drill hole by Love (1970).

Four deep holes drilled on the basin flanks (drill holes 10–13) and a regional seismic line (Garing and Tainter, 1985) define the thickness of the sedimentary sequence in the Washakie Basin. From north to south, drill holes 10–12 on the western side of the basin are 5.3, 5.0, and 5.7 km deep, respectively, and end in Pennsylvanian, Cambrian, and Mississippian rocks at their respective total depths (Roehler, 1977, 1978). These drill holes located on anticlinal structures indicate that the 6-km depth to Precambrian basement rocks near the southern end of the Rock Springs Uplift, estimated by Gries (1981), is probably correct. According to B. E. Law (written communications, 1985), the deepest part of the Washakie Basin is located about 30 km northeast of drill hole No. 12 where the sedimentary section is probably between 10 and 11 km thick. Drill hole No. 13, located on the Cherokee Arch south of the regional seismic line, reached Precambrian basement rocks at 5 km.

Using seismic data, Livesey (1985) predicted that the depth to Precambrian basement rocks is about 4.5 km in the eastern part of the Sand Wash Basin. The deepest hole in the Sand Wash Basin (drill hole No. 14), located approximately 60 km northwest of the source of the seismic data, reached Devonian rocks at a total depth of 5 km.

Shelf Stage of Sedimentation

From Middle Cambrian through Middle Jurassic time, the area occupied by the present-day greater Green River Basin and adjacent basement uplifts was part of the Rocky Mountain shelf as defined by Peterson (1977). Marine carbonate rocks were the dominant shelf deposits in Paleozoic time, but in Triassic and Jurassic time, nonmarine rocks derived from the craton occupied much of the eastern part of the shelf. Numerous unconformities interrupted shelf sedimentation (Fig. 35). Isopach maps indicate that the shelf deposits thicken gradually westward toward the shelf margin near the present-day Wyoming and Idaho border, beyond which the deposits thicken abruptly into the Cordilleran miogeosyncline (Armstrong and Oriel, 1965; Peterson, 1977). The total thickness of sedimentary rocks deposited on the shelf during the shelf stage ranges from about 2.3 km in the western part of the Green River Basin to about 0.9 km near the eastern part of the Great Divide Basin (Armstrong and Oriel, 1965; Keller and Thomaidis, 1971; Marzolf, 1965; Peterson, 1977).

Intermittent carbonate sedimentation alternating with uplift on the northwestern flank of the Transcontinental Arch during Cambrian through Mississippian time produced four east- to southeast-thinning stratigraphic sequences bounded by unconformities. The thickest of these sequences—composed, in ascending order, of the Flathead Sandstone, Gros Ventre Formation (shale and limestone), and Gallatin Limestone—records the initial transgression of the Paleozoic sea over Precambrian basement rocks (Lochman-Balk, 1972). The zero edge of this sequence along the northern and western margins of the Sierra Madre Uplift marks the former position of the Transcontinental Arch. A major Sag (Colorado Sag) in the Transcontinental Arch permitted the Cambrian seaway to swing into northwestern Colorado and to deposit the Lodore Sandstone and Peerless Formation. The next two sequences deposited across the northwestern flank of the Transcontinental Arch were the Bighorn Dolomite of Middle and Late Ordovician age and the Darby Formation (limestone and shale) of Middle and Late Devonian age. The zero edge of these sequences trends northeasterly across the greater Green River Basin from near the western end of the Uinta Uplift to the southeastern end of the Wind River Uplift (Armstrong and Oriel, 1965; Baars, 1972; Foster, 1972). The Manitou Limestone of Early Ordovician age and the Chaffee Formation (limestone and sandstone) of Late Devonian and Mississippian(?) age were deposited in the northwestern part of the Colorado Sag in the region now occupied by the Sand Wash Basin. The fourth sequence, composed of the Madison Limestone of Mississippian age, the equivalent Leadville Limestone, and the Humbug and Doughnut Formations, extends across the entire greater Green River Basin, except along the western side of the

Figure 35. Stratigraphic correlation chart for Phanerozoic and Precambrian rocks of the greater Green River Basin. The chart is modified from charts published by the Wyoming Geological Association Stratigraphic Nomenclature Committee (1969) and Walker and Wiloth (1975). Radiometric age (in Ma) is taken from the Geological Time Scale compiled by Palmer (1983). Time scale on figure is not linear.

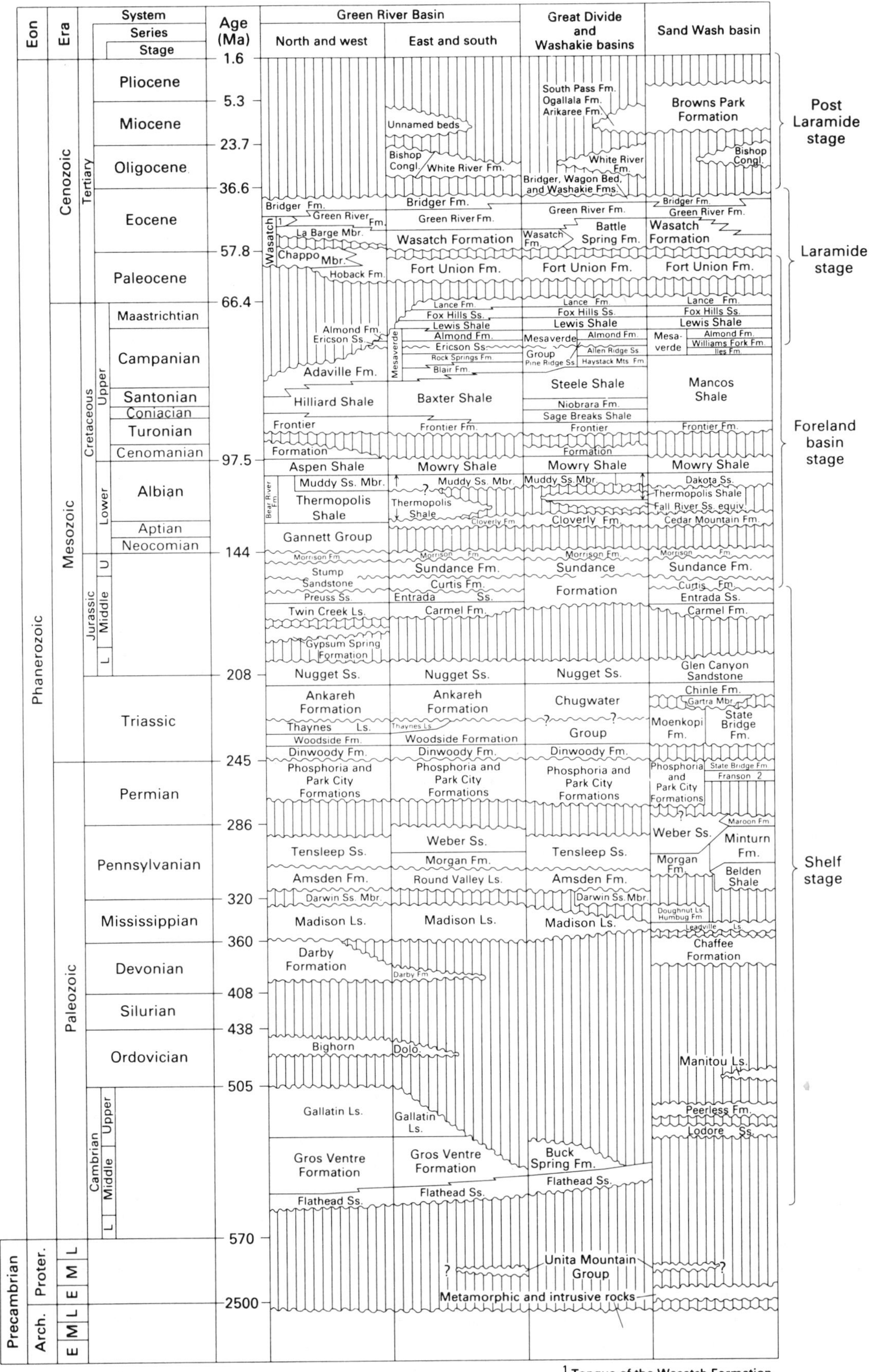

[1] Tongue of the Wasatch Formation

[2] Member of the Park City Formation

Sierra Madre and Park uplifts where Mississippian rocks have been removed by Pennsylvanian tectonism (Craig, 1972; Sando, 1979). Also included in this sequence is a lower tongue of the Madison Limestone that has been assigned a latest Devonian age (Baars, 1972; Sando, 1979).

Following a period of epeirogenic uplift, erosion, and karstification of the Madison Limestone, clean quartzose sandstone of the Darwin Sandstone Member of the Amsden Formation was deposited in latest Mississippian time over most of the shelf in southwestern Wyoming (Maughan, 1979). Subsequent erosion before the deposition of the remainder of the Amsden Formation in Early and Middle Pennsylvanian time left only remnants of its once widespread Darwin Sandstone Member (Maughan, 1979). The red shale and limestone units of the Amsden Formation (upper part) and the equivalent Round Valley Limestone, which rest unconformably on the Darwin Sandstone and the Madison Limestone, were deposited in the shallow water of the advancing Pennsylvanian sea (Maughan, 1979). Emergence of the ancestral Sierra Madre and Park Range Uplifts, beginning in late Middle Pennsylvanian time, shed arkosic sandstone and red shale of the Maroon and Minturn Formations westward onto the shelf now occupied by the Sand Wash Basin (Mallory, 1972a). North and northwestward, these deposits merged with marine limestone and quartzose sandstone of the Morgan Formation and the Weber Sandstone (Mallory, 1972a). The Weber Sandstone and the equivalent Tensleep Sandstone, both largely of eolian origin, blanketed much of southwestern Wyoming and adjacent states in late Middle to Late Pennsylvanian and Early Permian time (Mallory, 1972a; Rascoe and Baars, 1972).

The Rocky Mountain shelf in southwestern Wyoming, emergent during most of Early and Middle Permian time, was transgressed in Late Permian time and covered by deposits of the Phosphoria and Park City Formations. The Phosphoria, characterized by chert, phosphatic shale, and phosphorite, is confined largely to the miogeosyncline west of the present-day greater Green River Basin; whereas the Park City, characterized by carbonate rock, occupies the shelf area (Maughan, 1975; Peterson, 1977). However, two phosphatic shale members (Meade Peak and Retort Shale Members) and a chert member (Tosi Chert Member) of the Phosphoria extend eastward across most of the greater Green River Basin and intertongue with carbonate rock members (Grandeur and Franson Members) of the Park City Formation (Maughan, 1975; Peterson, 1977). In the Sand Wash Basin, the Upper Permian sequence includes a red bed unit named the State Bridge (lower part) Formation (Rascoe and Baars, 1972).

Early Triassic shelf deposits of southwestern Wyoming are characterized by two major intertonguing facies: (1) a western facies of shallow-marine limestone and (2) an eastern facies of nonmarine red shale and sandstone (Koch, 1976; MacLachlan, 1972). The limestone units are named the Dinwoody Formation and the Thaynes Limestone; the red beds are named the Woodside Formation, the Chugwater Group (lower part), the Ankareh (lower part), the State Bridge (upper part), and the Moenkopi Formations. Lower Triassic rocks appear to rest unconformably on Permian rocks and, in turn, are overlain unconformably by Upper Triassic nonmarine red beds of the Ankareh (upper part), Chugwater Group (upper part), and Chinle Formations (MacLachlan, 1972). The deposition of the widespread eolian Nugget Sandstone and its equivalent Glen Canyon Group in Late Triassic and Early Jurassic time maintained the dominance of nonmarine rocks across the shelf (MacLachlan, 1972; Peterson, 1972; Pipiringos and O'Sullivan, 1978).

After a prolonged period of uplift, westward tilting, and erosion, marine conditions were restored along the western margin of the shelf in early Middle Jurassic time. The result was the deposition of red beds, limestone, and evaporite of the Gypsum Spring Formation (Peterson, 1972; Pipiringos and O'Sullivan, 1978). The deposition of the Gypsum Spring was followed by a second cycle of uplift, erosion, and marine invasion in Middle Jurassic time that led to the deposition of the marine Twin Creek Limestone and the equivalent Carmel and Sundance (lower part) Formations (Peterson, 1972; Pipiringos and O'Sullivan, 1978; Imlay, 1982). The shelf stage ended in upper Middle Jurassic time with the deposition of the marine to partly marine Preuss Sandstone over the Twin Creek Limestone, the deposition of the eolian Entrada Sandstone over the Carmel Formation, and the continued deposition of the marine Sundance Formation (lower part (Pipiringos and O'Sullivan, 1978; Imlay, 1982).

Foreland-basin Stage of Sedimentation

From Late Jurassic through Late Cretaceous time, the present-day greater Green River basin and the adjacent basement uplifts were located in a foreland basin that was initially generated and then sustained by the Wyoming-Idaho thrust belt and later modified by basement tectonics. Westerly derived terrigenous clastic sediments filled a moatlike depression adjacent to the evolving Wyoming-Idaho thrust belt that migrated eastward in time, leaving a Lower Cretaceous depocenter with as much as 2.5 km of strata in east-central Idaho and a Cenomanian through Santonian depocenter with as much as 2.7 km of strata in southwesternmost Wyoming (Armstrong and Oriel, 1965; Jordan, 1981). Two additional depocenters, Cenomanian through Santonian and Campanian through Maastrichtian in age and containing as much as 3 km of strata, were formed in south-central Wyoming near the future sites of the northern Green River, the Great Divide, the Washakie, and the Sand Wash Basins (Weimer and Haun, 1960; Cross and Pilger; 1978; Jordan, 1981).

Researchers are debating whether or not the present-day greater Green River Basin was part of a foreland basin. Jordan (1981) restricted the foreland basin to the narrow, north-trending belt of maximum Cretaceous deposition coincident with the present-day Wyoming-Idaho thrust belt. Cretaceous depocenters in the greater Green River Basin east of the Moxa Arch are not considered to be part of a foreland basin by Jordan (1981) because their origin is attributed to mechanisms other than loading by the Wyoming-Idaho thrust belt, such as subcrustal loading by a shallowly subducted plate (Cross and Pilger, 1978) and (or) loading by thrust-faulted basement uplifts of the Central Rocky Mountain province.

In this chapter the broader definition of foreland basin (foredeep) from Bally and Snelson (1980) is used. Their definition of foreland basin not only includes the thick moatlike deposits adjacent to the thrust belt but also the updip extension of these deposits above the basement ramp that typically dips gently toward and beneath the thrust belt. Although the characteristic basement ramp of a classical foreland basin is absent beneath the present-day greater Green River Basin, Bally and Snelson (1980) and Royse and others (1975) assumed that it was present before severe dislocation of the basement in Laramide time. Thus, Bally and Snelson (1980) recognized the greater Green River Basin and adjacent basins in the Central Rocky Mountain province as part of a foreland basin dominated by basement tectonics.

Marine terrigenous clastic rocks of the Stump Sandstone (lower part) and the equivalent Curtis Formation mark the beginning of the foreland basin stage (Fig. 35; Peterson, 1972). Dominated by sandstone derived from tectonic uplands in the developing thrust belt and local emergent areas on the craton, these units of late Middle Jurassic age rest unconformably on the Entrada Sandstone and the equivalent Preuss Sandstone. According to Pipiringos and O'Sullivan (1978), the Stump (lower part) and the Curtis Formations were subjected to uplift and erosion, removed in parts of northwest Colorado and south-central Wyoming, and conformably overlain by the upper parts of the Sundance and Stump Formations, which are Late Jurassic in age. Following another period of regional emergence, Jurassic sedimentation ended in the vicinity of the present-day greater Green River Basin with the deposition of nonmarine conglomerate, conglomeratic sandstone, and variegated shale of the Morrison Formation (Peterson, 1972; Pipiringos and O'Sullivan, 1978; Imlay, 1982).

The Lower Cretaceous Gannett Group marked the beginning of Cretaceous sedimentation. This unit extends eastward from the Wyoming-Idaho thrust belt, where it is composed of nonmarine conglomerate, red shale, and limestone that rest unconformably on marine deposits of the Stump, to the western margin of the Green River Basin, where it is comprised of finer grained deposits that presumably rest unconformably on the Morrison (Pipiringos and O'Sullivan, 1978). Units equivalent to the Gannett Group in the central and eastern parts of the greater Green River Basin consist of unconformity-bound conglomerate and red shale sequences, generally no thicker than 45 m, assigned to the Cedar Mountain and the Cloverly Formations (McGookey and others, 1972; Young, 1975).

The Bear River Formation of Early Cretaceous age succeeds the Gannett Group in the western part of the greater Green River Basin. Three subdivisions of this sequence and equivalent strata farther eastward—a lower sandstone and mudstone unit, a middle black shale unit, and an upper sandstone unit—correlate, respectively, with the Fall River Sandstone of the Black Hills area and the Thermopolis Shale and its Muddy Sandstone Member of north-central Wyoming (McGookey and others, 1972; Oriel, 1969). The black claystone unit with local sandstone beds, described by Oriel (1969) beneath the base of the Bear River probably belongs to the uppermost Gannett Group or to the Cloverly. The Fall River Sandstone equivalent and Thermopolis Shale document the arrival of the Western Interior seaway into the area now occupied by the greater Green River Basin. After uplift of at least the eastern part of the greater Green River Basin, the Thermopolis Shale was subjected to subaerial erosion. The ensuing deposition of the nearshore marine Muddy Sandstone Member at the top of the Thermopolis marked the initial advance of the Western Interior seaway that culminated with the deposition of the widespread Mowry Shale of late Albian age and its equivalent, the Aspen Shale (McGookey and others, 1972; Cobban and Reeside, 1952a).

Throughout Late Cretaceous time, the Western Interior seaway either covered or was peripheral to the present-day greater Green River Basin. Tectonically active uplands in the adjacent thrust belt supplied an enormous quantity of terrigenous clastic sediments to the seaway, where they became segregated into four major lithofacies. From west to east for any given time, these lithofacies change from (1) sandstone and mudstone of alluvial plain origin to (2) sandstone, carbonaceous mudstone, and coal of lagoonal-paludal origin to (3) sandstone of nearshore marine origin to (4) shale of offshore marine origin. Moreover, Upper Cretaceous rocks in the basin can be subdivided into three major regressive events and two major transgressive events. The regressive events are identified by significant eastward migration away from the source area of the nearshore marine sandstone units and associated nonmarine units, whereas the transgressive events are identified by significant westward migration of the offshore marine shale units.

The Frontier Formation records the first major Upper Cretaceous regressive event in the area now occupied by the greater Green River Basin (Cobban and Reeside, 1952b). In the western part of the area, an unconformity divides the Frontier into a lower part of Cenomanian to early and earliest middle Turonian age and an upper part of latest middle and late Turonian to early Coniacian age, which contain nonmarine and marine components (Cobban and Reeside, 1952b; Merewether and others, 1984). Erosion accompanying this unconformity has removed the older part of the Frontier Formation from the Moxa Arch, the southern part of the Green River Basin, and the Sand Wash Basin, thus permitting the youngest part of the Frontier, which is entirely marine in these localities, to rest directly on the Mowry Shale (Merewether and others, 1984). This unconformity also extends into the eastern part of the Great Divide and the Washakie Basins, but here the lower part of the Frontier still is preserved (Merewether, 1983).

Following the first major transgression of the Western Interior seaway recorded by the Coniacian and Santonian parts of the Hilliard, the Baxter, the Steele (Cody used by some geologists), and the Mancos Shales, a second major regressive event occurred in which the rocks of the Mesaverde Group and equivalent parts of the Hilliard, the Baxter, the Steele, and the Mancos Shales were deposited across the greater Green River Basin (McGookey and others, 1972; Roehler, 1983; Weimer, 1960). Shoreline sandstone and equivalent lagoonal-paludal deposits of the Adaville, Rock Springs, and Blair Formations become progressively younger to the east. The more or less orderly eastward progradation of the shoreline was interrupted in early late Campanian time by regional arching over much of present-day Wyoming, including the Moxa Arch, the ancestral Lost Soldier Anticline, and the future site of the Rock Springs Uplift (Roehler, 1965; Gill and Cobban, 1966; Reynolds, 1976; Jordan, 1981). Westerly derived fluvial sandstone and conglomeratic sandstone of the Ericson Sandstone (upper part) and Pine Ridge Sandstone were deposited above the unconformity (Reynolds, 1976; Miller, 1977). The upper part of the Williams Fork Formation of lagoonal-paludal origin in the Sand Wash Basin is equivalent to the Ericson Formation (upper part) and the Pine Ridge Sandstones (Miller, 1977). The second regressive event was completed with the deposition of the lower part of the Almond Formation of paludal-lagoonal and nearshore margin origin across the Rock Springs Uplift and the adjacent Great Divide, Washakie, and Sand Wash Basins (McGookey and others, 1972; Miller, 1977; Roehler, 1983; Weimer, 1960).

The second major Upper Cretaceous transgressive event across the future site of the greater Green River Basin occurred in early Maastrichtian time and resulted in the deposition of the upper part of the Almond Formation and the Lewis Shale (Roehler, 1965; Weimer, 1960). The Lewis Shale was deposited across the future site of the Rock Springs Uplift and around the flanks of the uplifted ancestral Lost Soldier Anticline (Reynolds, 1976) before succumbing to a third major regressive sequence marked by the Fox Hills Sandstone of nearshore-marine origin and the overlying Lance Formation of continental origin. By the end of late Maastrichtian time, the Western Interior seaway had disappeared from southwestern Wyoming and adjacent states, leaving the area to subaerial processes.

Laramide tectonism, which was first manifested by depocenters of middle to Upper Cretaceous strata in the central and eastern parts of the foreland basin and by the uplift and erosion of parts of the basin in Campanian time, was reactivated in latest Cretaceous and Paleocene time. At this time, the crust was deformed into well-defined basement uplifts, including the Rock Springs Uplift (Roehler, 1983), and the cores of most of the emerging basement uplifts adjacent to the greater Green River Basin were eroded to Precambrian basement rocks. Terrigenous clastic sediments derived from these emerging basement uplifts formed the Fort Union Formation of Paleocene age. This event thus marked a significant shift in the style of sedimentation from a foreland basin with a dominant western source to discrete subbasins of the greater Green River Basin with local sources. Thrust faulting in the adjacent Wyoming-Idaho thrust belt began to wane in Paleocene time but continued to supply detritus to a foreland-basin sequence called the Hoback Formation and the Chappo Member of the Wasatch Formation (Dorr and others, 1977; Oriel, 1969). The foreland basin stage ended in latest Paleocene to earliest Eocene time, the age of final movement on the frontal thrust sheets of the Wyoming-Idaho thrust belt (Armstrong and Oriel, 1965; Oriel and Armstrong, 1966; Oriel, 1969).

Laramide Stage of Sedimentation

As implied in the previous discussion, the Laramide stage of sedimentation overlapped in time with the foreland basin stage (Fig. 35). However, the amount of overlap between these stages depends on where the lower limit of the Laramide stage is placed. Using the criterion of the first deposition of feldspathic rock, the appearance of coarse feldspathic sandstone above the unconformity at the base of the Pine Ridge Sandstone indicates that the Laramide stage began in late Campanian time (M. W. Reynolds, written communication, 1986). Reactivation of the late Campanian phase of uplift and erosion across the northern margin of the present-day greater Green River Basin in Paleocene time produced the moderately feldspathic nonmarine beds of the Fort Union Formation. Sandstone and conglomeratic sandstone of fluvial origin; sandstone, siltstone, and gray-green shale of floodplain origin; and coal and carbonaceous shale of paludal origin are the major lithologic components of the Fort Union (Love, 1970; McDonald, 1972, 1975). The Fort Union is reported to be 0.9 km thick in the southern Green River Basin (Clement, 1983), 1.2 km thick in the Great Divide Basin (McDonald, 1972), 1.3 km thick in the Washakie Basin (McDonald, 1975), and 0.9 km thick in the Sand Wash Basin (McDonald, 1975).

Early Eocene time brought an even greater period of crustal instability to the region. The Fort Union Formation was uplifted throughout the region, strongly upturned and truncated along the margins of the basement uplifts, and covered by sandstone and variegated shale of the Wasatch Formation and equivalent Battle Spring Formation (Love, 1970; McDonald, 1972, 1975; Reynolds, 1976). In contrast to the Wasatch and Battle Spring Formations in the northern Green River Basin and in the Great Divide, Washakie, and Sand Wash Basins, which were derived from a granitic terrane, the Wasatch of the southern and western parts of the Green River Basin was derived from a sedimentary terrane (Oriel, 1962; Hansen, 1965).

By middle Eocene time, structural and topographic relief had developed to the extent that the greater Green River Basin probably became a topographically closed basin in which lacustrine beds of the Green River Formation and equivalent fluvial beds of the Wasatch and the Battle Spring Formations accumulated (Bradley, 1964). The Green River Formation consists of a basal lacustrine and paludal phase named the Luman Tongue, two expanded lacustrine phases named the Tipton Shale and Laney Shale Members, and a restricted lacustrine phase named the Wilkins Peak Member (Bradley, 1964; Surdam and Stanley, 1980). The richest oil shale beds in the Green River Formation of Wyoming are present in the Tipton and Laney Shale Members, and trona deposits are present in the Wilkins Peak Member (Bradley, 1964; Eugster and Hardy, 1975; Surdam and Stanley,

1980). In late middle to late(?) Eocene time, the lacustrine system of the Green River was replaced by tuffaceous fluvial deposits of the Bridger, the Wagon Bed, and the Washakie Formations. Eocene deposits are reported to be 2.6 km thick in the southern Green River Basin (Clement, 1983), 1 km thick in the Great Divide basin (McDonald, 1972), and 2.3 km in the Washakie and the Sand Wash Basins (McDonald, 1975).

Post-Laramide Stage of Sedimentation

The post-Laramide stage of sedimentation was characterized by the infilling of the greater Green River Basin and the readjustment of basin-margin uplifts by extensional tectonics. The stratigraphic record is incomplete but suggests that integrated drainage was reestablished in Oligocene time and formed a large eastward-sloping alluvial cone flanked by basement uplifts. Alluvial fans bordered the uplifts, forming the Bishop Conglomerate that graded basinward into fine-grained deposits of the White River Formation (Fig. 35). Partial to total collapse of the basement uplifts surrounding the greater Green River Basin in Miocene and Pliocene time (Hansen, 1965; Love, 1970; Reynolds, 1976; Sales, 1983) formed coarse-grained deposits of the Browns Park Formation and fine-grained deposits of the equivalent South Pass, Ogallala, and Arikaree Formations. Some units of the White River, the South Pass, the Ogallala, and the Arikaree Formations have an important volcanic component. Major fault scarps in Quaternary deposits along the south flank of the Granite Mountains (M. W. Reynolds, written communication, 1986) indicate that parts of the greater Green River Basin margin have been active recently.

AXIAL BASINS OF THE NORTHERN AND CENTRAL RIO GRANDE RIFTS

Charles E. Chapin

The Rio Grande Rift is a north-trending break in the continental lithosphere that separates the Colorado Plateau on the west from the interior of the craton on the east. From near Socorro, New Mexico, to the vicinity of Leadville, Colorado, a distance of approximately 550 km, the Rio Grande Rift consists mainly of four axial basins aligned in a right-stepping echelon pattern between uplifted and outward-tilted shoulders. From south to north, these basins are the Albuquerque, Española, San Luis, and upper Arkansas Basins (Fig. 36). Rifting continues north of Leadville to near the Wyoming border but as a wide zone of normal faulting without discrete axial basins (Tweto, 1979b). South of the Albuquerque Basin, the rift bifurcates and widens into a series of parallel basins and intrarift tilted-block uplifts (Chapin, 1971) that resembles the Basin and Range province. This southern third of the rift is not discussed here. Regional syntheses of the Socorro to Leadville portion of the Rio Grande Rift are available in Chapin (1971), Cordell (1978), Kelley (1979), Tweto (1979b), and Baldridge and others (1984).

ONSET OF RIFTING

Burke (1980, p. 42) defines rifts as "elongate depressions overlying places where the entire thickness of the lithosphere has ruptured in extension." Mohr (1982, p. 243) utilizes morphology and surface structure alone to define rift valley (or just "rift") as "a parallel-sided, down-faulted valley, tens of kilometers in width and at least a few hundred kilometers in length." Mohr also states that "graben and rift valleys are commonly asymmetric about their longitudinal axes, and where a faulted margin faces a warped margin, a so-called half graben is developed." He defines a rift system as "a set of rift valleys united through closely parallel, en-echelon or anastomosing links." Certainly the Rio Grande Rift has fit these definitions for the past 10 m.y. The question is, "Was the Rio Grande Rift an intracontinental rift during the early stages of its development?"

Lipman and Mehnert (1975) place the beginning of rifting at about 26 Ma (older radiometric ages cited herein have been adjusted for current IUGS constants) on the basis of an angular unconformity along the west side of the San Luis Basin. The unconformity was cut on middle Tertiary regional ash-flow sheets (31–27 Ma) and overlain by basalt flows as old as 25–27 Ma (Fig. 36) and alluvial-fan deposits of the Los Pinos Formation (herein considered part of the Santa Fe Group). This age is in excellent agreement with the 26–27 Ma age of the Amalia Tuff erupted from the Questa Caldera and interbedded in the Los Pinos Formation (Fig. 36) along the southwest side of the San Luis Basin (Lipman and others, 1986; Manley, 1984). In the upper Arkansas Basin, Tweto (1979b) places the beginning of rifting at younger than the 30-Ma Gribbles Park Tuff deposited in east-trending paleovalleys that crossed the site of the upper Arkansas Basin (Epis and Chapin, 1975). A rhyolitic vent at Nathrop, Colorado, that cuts faults of the graben border system is approximately the same age. In the Española Basin, the oldest volcanic rocks interbedded with basin-fill sediments are mafic flows dated at 24.6 to 25.8 Ma (Fig. 36) and interbedded with the Picuris Formation and the Jarita Member of the Los Pinos Formation (Bachman and Mehnert, 1978; Baldridge and others, 1980). For the Albuquerque Basin, the best evidence for the age of the oldest synrift fill comes from the southern margin, where a 27-Ma basaltic andesite flow at Cerritos de los Minas is interbedded with basal sedimentary deposits of the Popotosa Formation (Machette, 1978), and where a 22 Ma basalt flow is interbedded in strongly tilted lower basin fill along the southeast flank of the basin (Bachman and Mehnert, 1978). In the Mulligan Gulch Graben, 30 km southwest of the Albuquerque Basin, 26-Ma dacite flows at Arroyo Montosa are interbedded with lower Santa Fe fanglomerates (Osburn and Chapin, 1983).

Important evidence for the beginning of rifting is also provided by the ages of the youngest middle Tertiary ash-flow sheets that crossed the rift prior to the development of sufficient topographic relief to prevent their crossing. Precise $^{40}Ar/^{39}Ar$ dating (McIntosh and others, 1986) indicates that four major ash-flow sheets were emplaced across the Rio Grande Rift at the south end

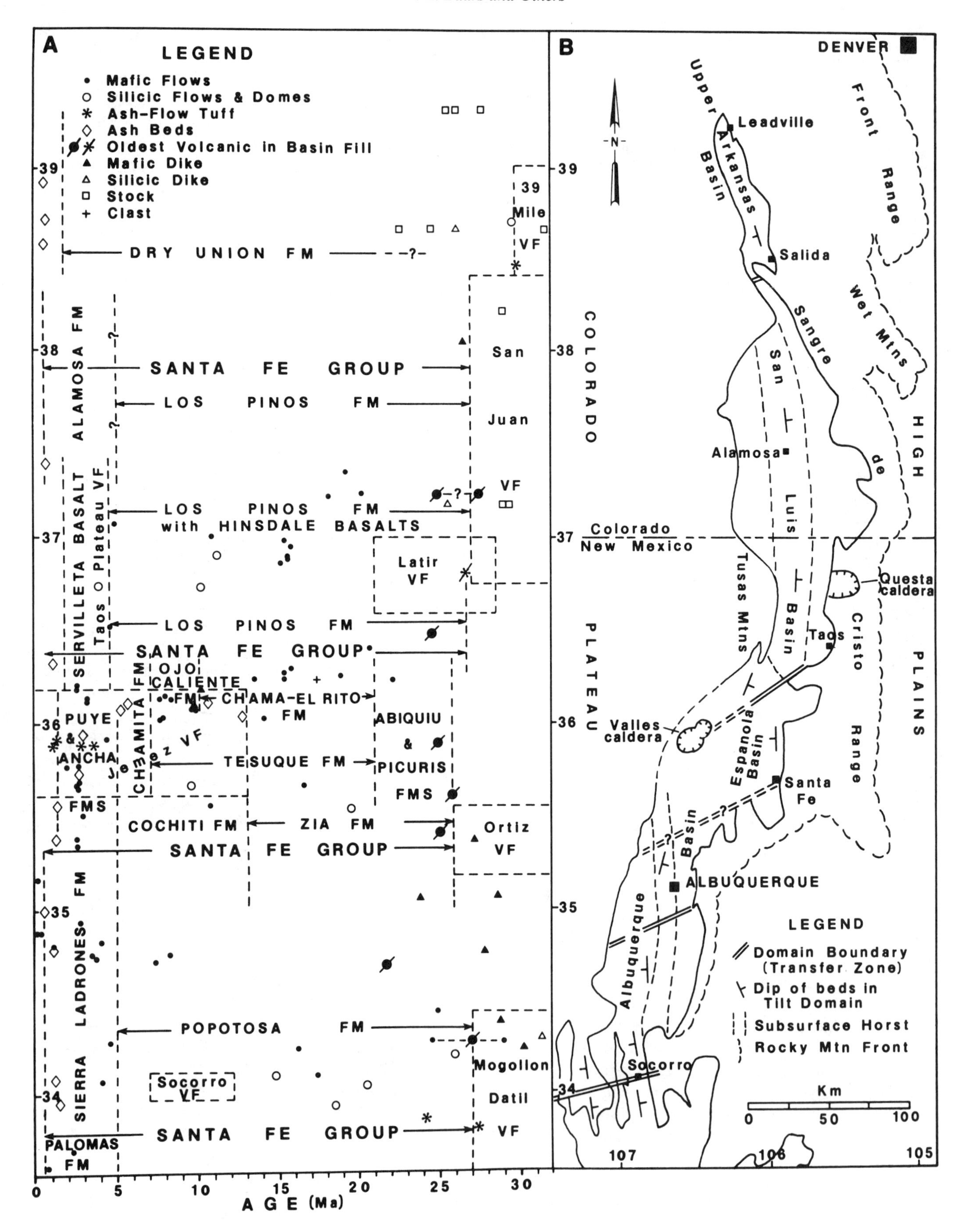
A
LEGEND
Mafic Flows
Silicic Flows & Domes
Ash-Flow Tuff
Ash Beds
Oldest Volcanic in Basin Fill
Mafic Dike
Silicic Dike
Stock
Clast
DRY UNION FM
39 Mile VF
ALAMOSA FM
SANTA FE GROUP
LOS PINOS FM
San Juan VF
SERVILLETA BASALT
Taos Plateau VF
LOS PINOS FM with HINSDALE BASALTS
Latir VF
LOS PINOS FM
SANTA FE GROUP
OJO CALIENTE FM
CHAMA-EL RITO FM
CHAMITA FM
PUYE & ANCHA FMS
Jemez VF
TESUQUE FM
ABIQUIU & PICURIS FMS
COCHITI FM
ZIA FM
Ortiz VF
SANTA FE GROUP
SIERRA LADRONES FM
POPOTOSA FM
Socorro VF
Mogollon Datil VF
SANTA FE GROUP
PALOMAS FM
AGE (Ma)
B
DENVER
Upper Arkansas Basin
Leadville
Salida
Front Range
Wet Mtns
Sangre de Cristo Range
San Luis Basin
Alamosa
COLORADO PLATEAU
HIGH PLAINS
Colorado
New Mexico
Tusas Mtns
Questa caldera
Taos
Valles caldera
Espanola Basin
Santa Fe
ALBUQUERQUE
Albuquerque Basin
Socorro
LEGEND
Domain Boundary (Transfer Zone)
Dip of beds in Tilt Domain
Subsurface Horst
Rocky Mtn Front
Km

of the Albuquerque Basin in the short interval between 28.8 and 27.4 Ma. Large volumes of basaltic andesite lavas are interbedded with these ash-flow sheets to form a distinctly bimodal suite of rocks. The truly catastrophic eruption rates during this brief 1.4-m.y. interval, after a 3.4-m.y. hiatus in volcanism (McIntosh and others, 1986), suggest a relationship to the onset of rifting. Angular unconformities between successive ash-flow sheets, separated by wedge-shaped accumulations of mafic lavas (Chamberlin, 1983), indicate that rapid extension and domino-style faulting began during the early part of the 28.8–27.4 Ma interval. Rapid distension of the crust and infilling of strike valleys with mafic lavas prevented development of significant topographic relief and permitted successive ash flows to cross the rift. This precise dating of the onset of rifting at the south end of the Albuquerque Basin agrees well with the less-precise conventional K-Ar dating of the beginning of rifting in the San Luis and Upper Arkansas Basins, as discussed previously. Also, the 26–27 Ma Questa caldera on the southeast side of the San Luis Basin was intensely deformed concurrent with caldera formation by closely spaced listric normal faults in a style and timing very similar to that at the south end of the Albuquerque Basin (Lipman, 1983). The broad, shallow, early rift basin in the Questa area allowed the Amalia Tuff to be deposited as far west as the Tusas Mountains on the west side of the present San Luis Basin and as far east as the High Plains (Lipman, 1983).

The beginning of rifting coincided with an interval of intense pyroclastic volcanism from caldera centers, both within and outside the early rift basins. If these volcanic rocks are proven to be rift related, then the Rio Grande Rift is not so devoid of volcanic rocks as some workers have presumed (Baldridge and others, 1984). Elston and Bornhorst (1979) conclude that the Rio Grande Rift did not begin until after a mid-Miocene lull in volcanic activity that ended about 15 Ma. They assign the copious late Oligocene–early Miocene pyroclastic activity and basaltic andesite volcanism to a period of back-arc extension. Regardless of the cause of extension, however, sedimentary basins were accumulating bolson-type sediments along the entire length of the Rio Grande Rift by 25–27 Ma, as demonstrated by radiometric dating of interbedded volcanic rocks. The same timing has been demonstrated for the beginning of sedimentation in fault-bounded basins along the projected trend of the Rio Grande Rift in northern Colorado (Izett, 1975; Tweto, 1979b), as well as in the southern Rio Grande Rift (Chapin and Seager, 1975; Seager and others, 1984). Throughout its history, the Rio Grande Rift was opening as a north-trending zone of extension marked by a linear chain of sedimentary basins between the Colorado Plateau on the west and the High Plains on the east. Thus, using any reasonable definition of a continental rift, it seems difficult to deny that the Rio Grande Rift was a structural and sedimentological entity from late Oligocene to present. The causes, rates, and styles of extension have varied during the long history of the Rio Grande Rift. However, a perusal of literature on continental rifts around the world reveals that such variations are common to many rifts (Mohr, 1982; Williams, 1982). Furthermore, as pointed out by Williams (1982, p. 197), broad downwarped troughs, wider than the present rift structures, marked initial stages in the evolution of the Kenya, Rhine, and Biakal Rifts.

This rather lengthy discussion of the beginning of rifting is necessary in order to understand the stratigraphic and structural evolution of the rift and what constitutes the fill of the rift basins. The title of this volume, *Sedimentary Cover—North American Craton: U.S.,* necessitates the question, What part of the sedimentary cover is contained within and related to axial basins of the northern and central Rio Grande Rift? Bryan (1938, p. 107) stated: "This series of basins forms a structural depression that is referred to in this paper as the Rio Grande depression" and (p. 205) "The main body of sedimentary deposits of the Rio Grande depression, from the north end of the San Luis Valley to and beyond El Paso, is considered to be of the same general age and to belong to the Santa Fe Formation." In raising the Santa Fe Formation to group status, Spiegel and Baldwin (1963, p. 39) stated: "Therefore, the Santa Fe Group is here considered to be a broad term including sedimentary and volcanic rocks related to the Rio Grande trough, with a range in age from middle(?) Miocene to Pleistocene(?)." Since publication of these papers, the upper Arkansas Basin has been recognized as part of the Rio Grande rift (Van Alstine, 1968; Chapin, 1971), and better dating has extended the age of basin-fill sediments from the middle Miocene(?) of Spiegel and Baldwin back to late Oligocene. Even though the late Oligocene regional ash-flow sheets are probably related to the onset of rifting, they are arbitrarily excluded from

Figure 36. (A) Latitude versus age plot of radiometric ages of igneous rocks and stratigraphic nomenclature for axial basins of the northern and central Rio Grande Rift. Major volcanic fields are shown as time-space boxes without plotting individual ages except for rocks that constrain boundaries of synrift sedimentary units. References and details of the radiometric ages are available in a data bank at the New Mexico Bureau of Mines and Mineral Resources. Older K-Ar ages have been adjusted for the current IUGS constants using the table of Dalrymple (1979). Mafic flows include some andesites. Stratigraphic nomenclature is inadequate and conflicting in many areas. Most synrift sedimentary units were deposited on coalescing alluvial fans and thus have gradational, interfingering boundaries with adjacent units. Obsolete stratigraphic terms have been deleted where possible; terms such as tuff and sand have been changed to formation to better reflect lithologic variability. All unit boundaries are dashed because future mapping and dating will necessitate many modifications. Santa Fe Group is used in the broad, regional sense of Bryan (1938) and Spiegel and Baldwin (1963). (B) Sketch map showing axial basins of the northern and central Rio Grande Rift (same scale as A). All basins are half grabens with asymmetry changing across complex transfer zones that separate domains of west-tilted versus east-tilted strata. Most transfer zones developed where the rift broke across pre-existing lineaments. Major prerift subsurface horsts are shown to underlie the Albuquerque (Baars, 1982) and San Luis (Tweto, 1979b; Gries, 1985) Basins. Modified from Tweto (1978) and Woodward and others (1978).

the Santa Fe Group because most of them are not interbedded with basin-fill sediments, and they provide a convenient and practical stratigraphic boundary for the base of the Santa Fe Group.

BASIN GEOMETRY

As defined by the present distribution of basin-fill sediments, the axial basins in the northern and central Rio Grande Rift range from 80 to 240 km in length and from 2 to 95 km in width, with an average width of approximately 50 km. Basin-fill deposits range up to 5.5 km in thickness. The basins are asymmetric half grabens, hinged down on one side with major fault boundaries on the opposing side. The sense of asymmetry shifts back and forth from basin to basin and occasionally within a basin (Fig. 36). The transverse boundaries between east-tilted versus west-tilted basins, or basin segments, are scissorslike shear zones that are essentially small-scale continental transform fault zones (sometimes called transfer faults). The best documented of these structures are the Socorro transverse shear zone (Chapin and others, 1978) south of the Albuquerque Basin and the Embudo fault zone (Muehlberger, 1979) between the San Luis and Española Basins.

The transverse structures formed where the north-trending rift broke across pre-existing northeast- or west-northwest–trending lineaments of probable Precambrian ancestry (Fig. 36). Thus, the length of the axial basins was determined primarily by the spacing of major lineaments. Structurally complex constrictions formed at the lineament crossings; basin-fill sedimentary deposits are relatively thin on these constrictions and are commonly strongly deformed. The transverse structures tend to leak magmas and to be conduits for ascending hydrothermal fluids; they also act as sills that constrict and alter groundwater flow patterns (Bryan, 1938; Anderholm, 1983; Coons and Kelly, 1984). As shown on Figure 36, synrift magmatism has been most continuous where the rift crosses the Morenci Lineament in the Socorro area and the broader Jemez lineament in northern New Mexico.

Structural relief, as determined by differences in elevation of Precambrian rocks cropping out in mountain ranges bordering the rift and the Precambrian-Phanerozoic interface beneath rift basins, ranges up to 12 km and averages about 5 km. For example, the top of the Precambrian basement lies at depths as much as 8,700 m below sea level in the Albuquerque Basin (Black, 1982), yet Precambrian rocks crop out as high as 3,300 m above sea level in the adjacent Sandia Mountains. Except for the Upper Arkansas Basin, structural relief is generally greater along the east side of the Rio Grande Rift (Chapin, 1971).

Synrift sedimentary deposits range in thickness from a wedge edge on the hinged sides of basins to as much as 5.5 km in the deeper portions. Major variations in thickness also occur within basins across intrabasin horsts. A cross section of the San Luis Basin constructed from oil tests (Tweto, 1979b) indicates that the synrift sedimentary fill varies from 2,600 m in the eastern (deep) side of the half-graben, to 500 m above a longitudinal intrabasin horst (Fig. 36), to 0 to 1,600 m in the western (hinged) side of the basin. However, interpretation of a recent seismic reflection line in the same general area by Gries (1985) indicates that approximately one-half to two-thirds of the Tertiary section in the eastern (deep) side of the San Luis Basin is Paleocene and Eocene in age, and thus is prerift. This reduces the maximum thickness of synrift sediments to approximately 1,500 m. The Paleogene (mainly Eocene) sediments were deposited during late Laramide wrench faulting in which the Colorado Plateau was translated approximately 100 km north-northeast, relative to the High Plains and frontal ranges of the southern Rocky Mountains (Chapin and Cather, 1981; Chapin, 1983). Eocene sediments are known to underlie the Noegene fill of the San Luis, Española, and Albuquerque Basins and probably underlie the upper Arkansas Basin as well.

Late Cenozoic extension reactivated many of the Laramide wrench faults, both as high-angle normal faults and as oblique-slip faults with a left slip sense of motion. The least principal horizontal stress direction for the southwestern United States was oriented WSW-ENE (Zoback and others, 1981) oblique to the northerly trend of the rift during approximately two-thirds of rift time (30–10 Ma). Thus, a minor amount of left slip during rifting is to be expected (Kelley, 1979; Withjack, 1984). Prerift strike-slip movement, however, was dextral (Chapin and Cather, 1981). High-angle, deeply penetrating faults that originated as wrench faults, some as long ago as the Precambrian (Tweto, 1979b, 1980b), have exerted an important influence on the location and trend of the Rio Grande Rift and on the size and shape of the depocenters in which synrift sediments accumulated.

Longitudinal horsts that originated prior to rifting underlie both the San Luis and Albuquerque Basins. The Alamosa Horst in the San Luis Basin is approximately 180 km long by 20 to 30 km wide and shows structural relief on the Precambrian-Phanerozoic interface of several kilometers (Tweto, 1979b; Keller and others, 1984; Gries, 1985). The horst apparently developed during Laramide wrench faulting, as Paleogene sedimentary deposits thin across the structure (Tweto, 1979b; Gries, 1985). Late Oligocene regional ash-flow sheets from the San Juan volcanic field pinch out against the horst, which apparently acted as a barrier to their eastward dispersal, at least at the latitude of the seismic line published by Gries (1985). Sedimentary units of the Santa Fe Group also thin across the structure, which split the San Luis Basin longitudinally into two narrow, highly elongate subbasins until it was finally buried by clastic sediments of the Santa Fe Group.

The longitudinal horst beneath the Albuquerque Basin is approximately 145 km long by 11 to 26 km wide and has structural relief on the Precambrian-Phanerozoic interface of several kilometers (Baars, 1982). In contrast with the Alamosa Horst, the horst beneath the Albuquerque Basin is not visible on the regional gravity map of Cordell and others (1982). It apparently had relatively little influence on thicknesses of Mesozoic and Cenozoic sedimentary units, although Mississippian and Pennsylvanian rocks thin dramatically across it (Baars, 1982). Minor rejuvenation of the horst during rifting produced a low-amplitude anticlinal structure visible on the cross section of the northern Albuquerque Basin published by Black and Hiss (1974). Several recent oil and gas tests have been drilled along the trend of the buried horst. Baars (1982, p. 156) makes the point that "rifting of the Rio Grande system recurred over basement rifting long in place."

SYNRIFT SEDIMENTARY DEPOSITS

Synrift sedimentary and volcanic deposits of the Rio Grande Rift are collectively known as the Santa Fe Group (Fig. 36). For a review of the long and often controversial usage of the term Santa Fe, see Hawley (1978) and Tedford (1981, 1982). Present practice is to use Santa Fe as a group term that includes all the synrift basin fill, both volcanic and sedimentary, ranging in age from late Oligocene to Quaternary, but excluding deposits that postdate entrenchment of the Rio Grande in middle Pleistocene time. Such broad usage of the term Santa Fe is obviously advantageous in the San Luis and Albuquerque Basins where most of the basin fill is not exposed and in the upper Arkansas Basin where the older basin fill is not exposed and the younger fill has not been subdivided nor studied in detail. In the Española Basin, where numerous studies have been published and the basin fill has been subdivided and dated in much detail, the various formations and their members make up the Santa Fe Group. Eventually, as more detailed stratigraphic studies are published and map coverage becomes more complete, the Santa Fe Group will probably be elevated to supergroup status, and some of the thicker and more complex formations will become groups.

Sedimentary fill consists mainly of broad, coalescing alluvial-fan deposits that grade basinward into fine-grained alluvial-plain and playa deposits in the early closed basins of the Rio Grande Rift. After through drainage was established in early Pliocene time (Bachman and Mehnert, 1978), the bordering piedmont deposits interfingered with axial-river sands and gravels. In some areas, broad pediments have been cut on tilted fault-block uplifts; the sedimentary veneers on these pediments are also included in the Santa Fe Group when of appropriate age. Where dissected and exposed, the sedimentary deposits, from mountain front basinward, variously consist of pediment gravels; proximal, medial, and distal fan deposits (also known as piedmont-slope and alluvial-plain deposits); and playa deposits or axial-river sands and gravels, depending on whether drainage was closed or open. Eolian dune sands are also present in some basins, and freshwater lacustrine deposits are present in the Albuquerque Basin and in the Plio-Pleistocene Alamosa Formation that underlies much of the San Luis Basin.

Strata of the Santa Fe Group are typically poorly indurated and pinkish buff to gray in color. However, older basin fills in areas that have experienced diagenetic reactions induced by alkaline, saline basin waters are very well indurated and reddish brown to brick red in color (Chapin and Lindley, 1986). The compositions of clasts vary markedly from one alluvial fan or piedmont slope to another depending upon the source regions. For example, in the Española Basin, the Tesuque Formation in the eastern half of the basin consists mostly of pinkish-buff, arkosic sandstones and siltstones derived from basement rocks exposed in the Sangre de Cristo Range (Galusha and Blick, 1971; Manley, 1979), whereas the laterally equivalent Los Pinos, Chama–El Rito, and Ojo Caliente Formations in the northern and western portions of the basin contain abundant volcanic detritus derived from the San Juan and Latir volcanic fields (Kelley, 1978; Manley, 1981; May, 1984; Ekas and others, 1984). Similar relationships are present in other basins.

The last 10 m.y. of rift history was a time of profound change, both tectonically and sedimentologically. The region was uplifted an average of 1,100 m (Axelrod and Bailey, 1976), more or less concurrently with a change in the direction of regional extension in the southwestern United States from WSW-ENE to E-W or WNW-ESE (Zoback and others, 1981). Block faulting greatly increased topographic relief across basin margins and narrowed the basins by uplifting their edges, causing stripping of uplifted proximal and medial alluvial-fan deposits and juxtaposing relatively fine grained distal fan deposits against the faulted margins of some basins. Increased runoff from newly elevated alpine regions integrated the drainage to form the ancestral Rio Grande prior to 4.5 Ma. Well-sorted, axial-river deposits of sand and gravel accumulated concurrently with relatively poorly sorted piedmont-slope deposits shed from the basin margins and graded to the throughgoing river. These contrasting facies intertongued in a complex manner as a result of changes in climate, periods of aggradation versus incision of the river, avulsion of the axial-river channel, basin subsidence, and progradation rates of opposing piedmont slopes.

Axial river deposits and their intertonguing piedmont facies are known as the Camp Rice and Palomas Formations in southern New Mexico (Hawley, 1975; Lozinsky and Hawley, 1986), the Sierra Ladrones Formation in the Socorro area and in most of the Albuquerque Basin (Machette, 1978), and the Puye Formation in the western Española Basin (Griggs, 1964; Smith and others, 1970). Terminology for deposits of similar age in the northern Albuquerque Basin varies from upper Buff Member of the Santa Fe Formation (Bryan and McCann, 1937), to Ceja Member of the Santa Fe Formation (Kelley, 1977), to various informal terms (Tedford, 1982), or simply upper Santa Fe Group. The Ancha Formation (Spiegel and Baldwin, 1963) of the Española Basin, and the Servilleta Formation (Lambert, 1966) of the southern San Luis Basin, have a similar early Pliocene to middle Pleistocene age and depositional setting. Lipman and Mehnert (1979) have redefined the Servilleta to include only the distinctive olivine tholeiite basalts of the Taos Plateau and have relegated the interbedded sedimentary rocks to correlative upper Santa Fe units, such as the Ancha Formation. The interbedded lacustrine and fluvial deposits of the Alamosa Formation that underlie much of the San Luis Basin have been dated at 0.6 to 0.9 Ma in outcrop (Rogers, 1984) and probably extend back to early Pliocene in age. In the Mapco-Amoco well, 29 km northeast of Alamosa (Burroughs, 1981), a 625-m section of Alamosa Formation was penetrated, and a Mio-Pliocene pollen age was obtained from underlying sediments of the Santa Fe Group undifferentiated (Gries, 1985). The Alamosa Formation should also be included in the Santa Fe Group using the broad definition of Spiegel and Baldwin (1963). The correlation charts of Hawley (1978) provide a concise summary of stratigraphic nomenclature and an excellent regional perspective on correlations and ages of units.

Post–Santa Fe deposits, laid down after initial incision of the Rio Grande at about 0.5 Ma, consist of alluvium on a stepped sequence of valley-border surfaces, terrace deposits, eolian sands, paleosols, basalt flows, glacial moraines and outwash gravels, arroyo and river valley alluvium, and—in closed basins like the northern San Luis Basin—alluvial-fan, playa, and eolian deposits. Alternate incision and aggradation by the Rio Grande during glacial and interglacial stages formed a complex inner-valley stratigraphy. Summaries of Quaternary stratigraphy are available by Hawley and others (1976) and Gile and others (1981) for New Mexico, and by Scott (1975) for central Colorado. In addition, Hawley (1978) contains a wealth of information on late Cenozoic stratigraphy for the full length of the rift.

Stratigraphic units of the upper Santa Fe Group (Camp Rice, Palomas, Sierra Ladrones, Puye, Ancha, Alamosa, etc.) are commonly bounded above and below by unconformities. The upper unconformities separate these units from a complex of late Pleistocene and Holocene deposits that postdate incision of the basin fill by the Rio Grande. The lower unconformities separate these units from older portions of the basin fill that are usually more strongly tilted and better indurated. These lower unconformities developed during the strong pulse of epeirogenic uplift and block faulting that occurred between approximately 7 and 4 Ma (Chapin, 1979). The oldest dated axial-river deposits of the ancestral Rio Grande are 4.1 to 4.6 Ma (Bachman and Mehnert, 1978). Thus, the unconformities at the base of upper Santa Fe units usually separate axial river and intertonguing floodplain and piedmont deposits from older closed-basin deposits. Where basins are still closed, as in the northern San Luis Basin and parts of several basins in the southern Rio Grande Rift, sedimentation has been relatively continuous throughout the Neogene, and the upper limit of the Santa Fe Group is difficult to establish.

The Upper Arkansas Basin is not in the Rio Grande drainage, so its history differs somewhat from that of other rift basins. Here, basin-fill sediments consist of the Dry Union Formation (Santa Fe Group), heavily mantled in the western half of the basin by Pleistocene alluvial deposits, which include glacial moraines and outwash gravels. Lenses of Bishop ash (0.7 Ma) and Pearlette Type-O ash (0.6 Ma) have been found in the lower Pleistocene alluvium (Scott and others, 1975). The Pleistocene deposits rest on extensive pediments carved across gently tilted beds of the Dry Union Formation. The time of formation of these pediments is poorly constrained but is probably related to the late Miocene and Pliocene period of epeirogenic uplift.

TECTONIC EVOLUTION AND STRUCTURAL GEOLOGY OF THE RATON BASIN, COLORADO AND NEW MEXICO, AND HUERFANO PARK, COLORADO

Ira S. Merin, John R. Everett, and Peter R. Rose

INTRODUCTION

Intense deformation during the Laramide Orogeny created the Sangre de Cristo Uplift and adjacent Raton Basin and Huerfano Park (Fig. 37). The Raton Basin is an asymmetrical basin with a north-trending axial deep that is parallel and adjacent to the Sangre de Cristo Range. In the northern part of the Raton Basin, this axial deep, the La Veta Syncline (Johnson and others, 1958), is a north-trending structure that turns northwestward and passes into Huerfano Park. Sediments along the western edge of the La Veta Syncline are extensively deformed adjacent to the Sangre de Cristo Range; the eastern flank of the La Veta Syncline dips gently westward and is only mildly deformed by open folds and faults with small displacements.

The resource potential of the Raton Basin is very promising. Extensive deposits of high to medium volatile bituminous coal are present in the Vermejo and Raton Formations (Fig. 38) (Tremain, 1980; Pillmore, 1969; Harbour and Dixon, 1959). Additionally, the coal beds of the Vermejo Formation contain extensive quantities of methane (e.g., up to 500 cubic feet of gas per ton of coal, according to Danilchik and others, 1979) and constitute a viable exploration objective for gas lying at drilling depths of less than 1,000 m throughout most of the basin (Tremain, 1980). Drilling activity has shown that oil and gas are present in the basin; however, commercial production of hydrocarbons has not been established to date (Dolly and Meissner, 1977; Speer, 1976; McCaslin, 1981; 1983; 1984). Rose and others (1984) suggest the presence of an extensive basin-centered gas accumulation. Commercial deposits of CO_2 have been discovered in Huerfano Park (Renfro, 1979). In addition, the west flank of Raton Basin and Huerfano Park contains an 8- to 15-km-wide fringe of folded and thrusted sediments with numerous closed and untested structures.

GEOLOGIC EVOLUTION

The principal tectonic events responsible for forming the structures of the Raton Basin and Huerfano Park are the Pennsylvanian-Permian orogenic event related to the Ancestral Rockies deformation, the Laramide Orogeny, and Neogene extension.

Pennsylvanian-Permian

Pennsylvanian and Early Permian deformation and subsidence in the Raton Basin region are part of the broader deformation of the Ancestral Rockies, an event believed to be related to

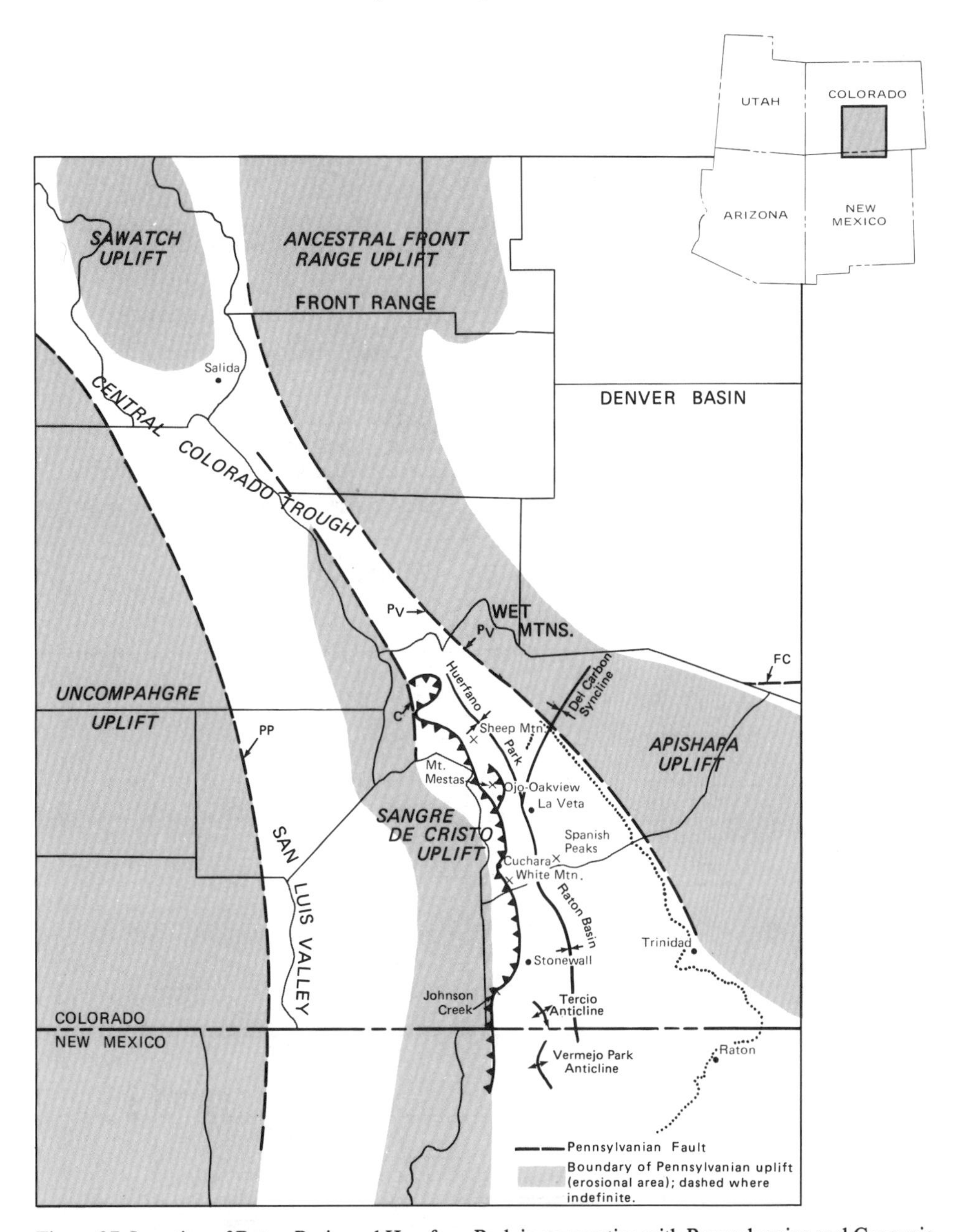

Figure 37. Location of Raton Basin and Huerfano Park in perspective with Pennsylvanian and Cenozoic tectonic features. Pennsylvanian features shown with oblique lettering; Cenozoic features with roman lettering. Sawatch and Sangre de Cristo Uplifts interpreted as Pennsylvanian tectonic features at general locations of similar Cenozoic features. Pennsylvanian faults: C, Crestone; FC, Freezout Creek; PP, Pecos-Picuris; PV, Pleasant Valley. Cenozoic features: Huerfano Park; Raton basin; Del Carbon Syncline; Tercio Anticline; Vermejo Park Anticline; Wet Mountain Valley graben; thrust symbol marks Sange de Cristo Uplift thrust front. Outcrop of Trinidad Formation (dotted) shows general outline of Raton Basin (modified after DeVoto, 1980b).

plate interaction along the southern and eastern margins of North America (Kluth and Coney, 1981). Suturing of the southern margin of North America with other continental masses has been inferred to be the result of right-lateral coupling as South America moved westward relative to North America (Kluth and Coney, 1981; Dewey, 1982). According to these interpretations, the Ancestral Rockies developed when that portion of the craton located between the Cordilleran Orogen and the Ouachita/Marathon Orogen was wrenched and pushed northwestward as the convergence progressed (Kluth and Coney, 1981). This produced foreland-type deformation characterized by block faulted mountains (e.g., Uncompahgre Uplift, Ancestral Front Range) and rap-

Era	Age	Series or Stage	Tectonic Activity	Stratigraphic Formation	Lithology	Thickness	Hydrocarbon Shows		Potential Source Rocks
CENOZOIC	RECENT			ALLUVIUM, DUNES, LANDSLIDES, SOIL ZONES		0-200'			
	PLEISTOCENE PLIOCENE			OGALLALA FM		200-500'			
	MIOCENE OLIGOCENE			DEVILS HOLE FM WALL MTN TUFF INTRUSIONS, DIKES, SILLS INTRUDE ENTIRE SECTION		0-1500'			
	EOCENE	UPPER	LATE LARAMIDE	FARASITA FM		0-1200'			
				HUERFANO FM		0-2000'			
		MIDDLE LOWER		CUCHARA FM		0-5000'			
	PALEOCENE	UPPER	EARLY LARAMIDE	POISON CANYON FM *		0-2500'	☼		
		LOWER		RATON FM * (basal)		0-2075'	● ☼		+
MESOZOIC	CRETACEOUS	MAASTRICHTIAN		VERMEJO FM		0-360'	☼		+
				TRINIDAD SS		0-255'	● ☼		
		CAMPANIAN	SEVIER	PIERRE SH		1300-2900'	● ☼		+
		SANTONIAN & CONIACIAN		NIOBARA: SMOKEY HILL MARL FT HAYES LMS		900' 0-55'	● ☼		+
		TURONIAN & CENOMANIAN		BENTON: CARLILE SH GREENHORN LMS GRANEROS SH		165-225' 20-70' 175-400'	● ☼ ☼ ● ☼		+ + + +
		ALBIAN & APTIAN		DAKOTA SS PURGATOIRE FM		140-200' 100-150'	● ☼	C	+
	JURASSIC			MORRISON WANAKAH ENTRADA		150-400' 30-100' 40-100'	● ☼ ●	C	+
	TRIASSIC	UPPER		DOCKUM GROUP		0-1200'			
PALEOZOIC	PERMIAN	GUADALUPE & LEONARD	ANCESTRAL ROCKIES	BERNAL FM SAN ANDRESS LMS GLORIETA SS YESO FM		0-125' 10-20' 0-200' 200-400'	●	He	+
		WOLFCAMP		SANGRE DE CRISTO FM *		700-5300'	●		
	PENNSYLVANIAN	VIRGIL & MISSOURI		MAGDALENA GROUP: MADERA FM		4000-6000'	● ☼		+
		DES MOINES & ATOKA		DEER CREEK *			● ☼		
	MISSISSIPPIAN	MORROW		KEBER					+
	DEVONIAN			TERERRO FM		40-50'			
				ESPIRITU SANTO FM		25'			
	PRE-CAMBRIAN			MAFIC GNEISS METAQUARTZITC GROUP GRANITE & GRANITE GNEISS		7000' ? 5000' ? 4000' ?			

IM 0722

idly subsiding fault-bounded troughs (e.g., Paradox Basin, Central Colorado Trough).

During this period of tectonic activity, the sites of the future Raton Basin and Huerfano Park were located in the Central Colorado Trough between the Uncompahgre and the Apishapa–Ancestral Front Range basement uplifts (Fig. 37; De Voto, 1980b). Two smaller uplifts of unknown areal extent, the Sawatch Uplift and the Ancestral Sangre de Cristo Uplift, were present in this trough.

Coarse-grained and poorly sorted arkose, conglomerate, and siltstone of the Sangre de Cristo Formation and Magdalena Group (Fig. 38), which were deposited in the trough, reflect erosion of the rising adjacent uplifts. Periods of intense tectonic activity alternated with quiescent periods that resulted in deposition of fossiliferous limestone, red shale, black shale, and siltstone.

Much of the present-day shape of the Raton Basin and Huerfano Park was inherited from faults active during this period. The Crestone Fault, located along the east flank of the northern Sangre de Cristo Mountains, and the Pleasant Valley Fault, located along the southwestern edge of the Ancestral Front Range–Apishapa Uplift, mimic present-day basin boundaries (Fig. 37). Activity along these faults was intense near Salida, Colorado, during the Missourian and Virgilian, resulting in deposition of boulders of granite and gneiss and abrupt facies changes adjacent to the faults (De Voto, 1972b; De Voto and Peel, 1972). Presumably this intense activity also occurred farther to the south along the east flank of the Sangre de Cristo Mountains, adjacent to the Raton Basin; however, such features would be difficult to distinguish from faults active during Laramide deformation. The abrupt thinning of the Sangre de Cristo Formation in the subsurface of the Raton basin on the west side of the Apishapa Uplift (Shaw, 1958) suggests that the Pleasant Valley Fault extended to at least the vicinity of Trinidad, Colorado. Cretaceous and early Tertiary strata are folded into a monocline along this portion of the northeastern boundary of the Raton Basin, implying that reactivation of the Pleasant Valley Fault occurred during the early Tertiary (Laramide Orogeny).

Additionally, reactivation of northeast-trending faults may have contributed to the shape of the Raton Basin. The southeast boundary of the Raton Basin parallels the northeast trend of the Colorado mineral belt and other northeast-trending Precambrian faults (see Warner, 1978). A basement shear zone (Lewis and others, 1969), expressed in aeromagnetic data (Zietz and Kirby, 1972), parallels the Colorado mineral belt and apparently separates the Raton Basin from Huerfano Park.

Figure 38. Stratigraphic column, Raton Basin and Huerfano Park. In "Hydrocarbon Shows" column, C and He refer to commercial quantities of CO_2 and helium respectively; solid circles indicate oil; open circles indicate gas. In "Stratigraphic Formation" column, * refers to presence of conglomerates containing pebbles of crystalline rock (modified after Dolly and Meissner, 1977).

Late Cretaceous to Eocene Activity

The Laramide Orogeny was a two-phased orogenic event (Chapin and Cather, 1981; Gries, 1983b). Based on the orientation of Laramide-age dikes and the orientation of Laramide-age compressional structures, Chapin and Cather (1981) showed that the orogeny began with early Laramide maximum compressive stress oriented east-northeast during Late Cretaceous to Paleocene time and ended with late Laramide maximum compression oriented northeast in Eocene time. Livaccari and Keith (1984, 1985) produced a series of paleotectonic maps (Figs. 39 and 40) showing that the plate motion data of Engebretson and others (1984) are compatible with the orientation of Laramide-age structures and dikes, supporting Chapin and Cather's idea of a dual-phase Laramide Orogeny marked by stress regimes of different orientation. In general, north-south to northwest-trending compressional structures formed during the early phase of the Laramide Orogeny, and most northwest to east-west–trending compressional structures formed during the late phase of the Laramide Orogeny.

Early Laramide activity resulted in deposition of at least 1.6 km of clastic sediments in the Raton Basin during the Maastrichtian and Paleocene. The initial effects of the Laramide Orogeny may be seen as early as late Campanian, being recorded as coarse siltstone units in the upper Pierre Shale (Baltz, 1965). However, the first intense period of orogenic activity did not occur until late Maastrichtian to early Paleocene when arkose and conglomerates, bearing clasts of sedimentary and crystalline rocks (Wood and others, 1957) of the Raton Formation, were deposited. This intense period of tectonic activity continued into and intensified during the Paleocene; it is recorded by the coarse arkose and conglomerate, bearing pebbles of granite and gneiss, of the Poison Canyon Formation. The presence of fresh feldspar in the Poison Canyon Formation (Wood and others, 1957; Johnson, 1959) and pebbles 6 to 8 cm in diameter (larger than those in the Raton Formation, Johnson, 1959) implies that this period of activity was more intense than the earlier period. Certainly the locus of uplift was closer.

The composition, facies changes, and unconformable relations of the Raton and Poison Canyon Formations indicate sources with exposed Precambrian rocks in highlands southwest and northwest of the Raton Basin (Tweto, 1975). The northwestern source was evidently the Wet Mountains, a rejuvenated element of the late Paleozoic Ancestral Front Range. The southwestern source was the Laramide-age San Luis Uplift (referred to below as the San Luis–Sangre de Cristo Uplift), now buried beneath the San Luis Valley. This feature coincided in location with the southern (i.e., San Luis) portion of the late Paleozoic Uncompahgre–San Luis highland. Evidence that the present Sangre de Cristo Mountains could not have been the source (i.e., that the Laramide San Luis–Sangre de Cristo Uplift did not extend topographically this far east) includes the following: (1) the eastern slopes and crest of the Sangre de Cristo Mountains are mantled with Upper Paleozoic sedimentary rocks

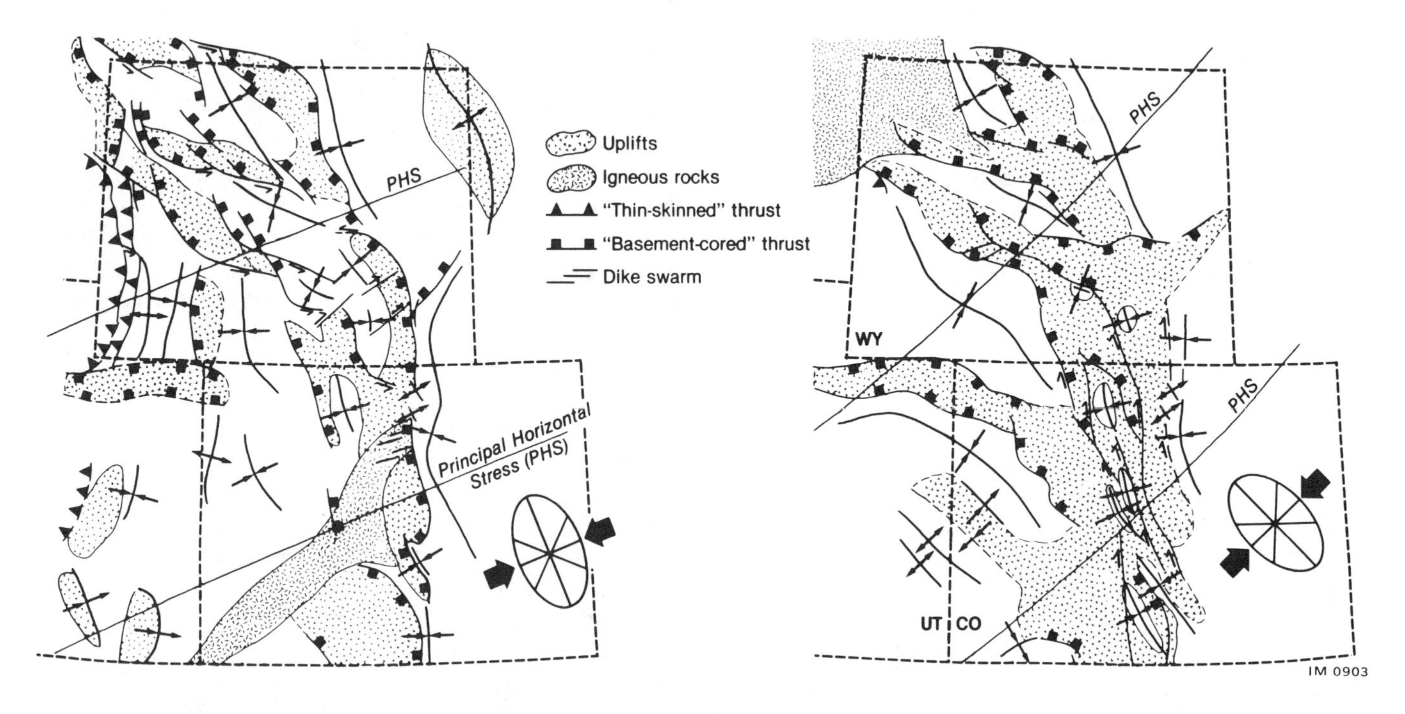

Figure 39. Paleotectonic map of Colorado and Wyoming showing the orientation of the principal horizontal stress (PHS), uplifts, and structures active during the early Laramide Orogeny, Maastrichtian to Paleocene time (simplified from Livaccari and Keith, 1985).

Figure 40. Paleotectonic map of Colorado and Wyoming showing the orientation of the principal horizontal stress (PHS), uplifts, and structures active during the late Laramide Orogeny, Eocene time (simplified from Livacarri and Keith, 1985).

of dissimilar composition to those of the Maastrichtian and Paleocene sedimentary rocks in the Raton Basin, (2) pebbles of crystalline rocks in the Paleocene Poison Canyon Formation are coarser than those in the Paleozoic conglomerates, and (3) feldspar grains present in the Poison Canyon Formation are fresh, indicating close proximity to a crystalline source (Tweto, 1975). Apparently, the Sangre de Cristo Mountains did not become a source of sediment until Eocene time.

Late Laramide activity resulted in the deposition of at least 2.6 km of sediments (Cuchara, Huerfano, and Farasita Formations) during the Eocene. The Cuchara Formation truncates various formations from the Poison Canyon down to the Pierre and contains red beds virtually identical to those of the Pennsylvanian-Permian Sangre de Cristo Formation (Baltz, 1965). Scott and Taylor (1975) reported that the Huerfano Formation also contains red beds virtually identical to the Sangre de Cristo Formation. This implies that these upper Paleozoic sediments along the Sangre de Cristo Mountains were the source (Baltz, 1965) and that during Cuchara and Huerfano deposition (Eocene) the Laramide San Luis–Sangre de Cristo Uplift had expanded topographically to the east to include these mountains (Tweto, 1975). The upper part of the Huerfano Formation intertongues with the Farasita Formation in Huerfano Park, implying that they are partially contemporaneous (Johnson, 1959; Johnson and others, 1958; Scott and Taylor, 1975; Tweto, 1975). The Farasita contains boulder conglomerates with clasts up to 2 to 6 m in diameter, principally of crystalline rocks and lesser quantities of Jurassic and Permian rocks (Johnson, 1959). This material reflects close proximity to an exposed crystalline terrain along the Wet Mountains (Scott and Taylor, 1975).

During the Eocene, north-south–trending right-slip deformation occurred along the eastern margin of the Colorado Plateau (Kelley, 1955b; Chapin and Cather, 1981; Chapin, 1983). Chapin (1983) speculated that faults along the eastern and western margins of the San Luis–Sangre de Cristo and Wet Mountain Uplifts accommodated some of this right-slip movement, producing a series of en echelon, north-northwest–trending, right-slip faults and northwest-trending thrusts. Eocene right-slip along these uplifts may have occurred either along new high-angle faults or along the preexisting, early-Laramide, west-dipping thrusts or a combination of the two. Either way, the overprinting of early Laramide thrusts with late Laramide right-lateral, oblique-slip faults produced a complex structure resulting from strongly convergent wrenching. Seismic data (Ovellette and Boucher, 1983; Rose and others, 1984; Applegate and Rose, 1985) across the west margin of Raton Basin and Huerfano Park show that the west flank of these basins exhibits a complexly thrusted and folded wedge of sediment. Additionally, this complex structure continues farther to the west as indicated by the numerous north-northwest- to north-south–trending faults that

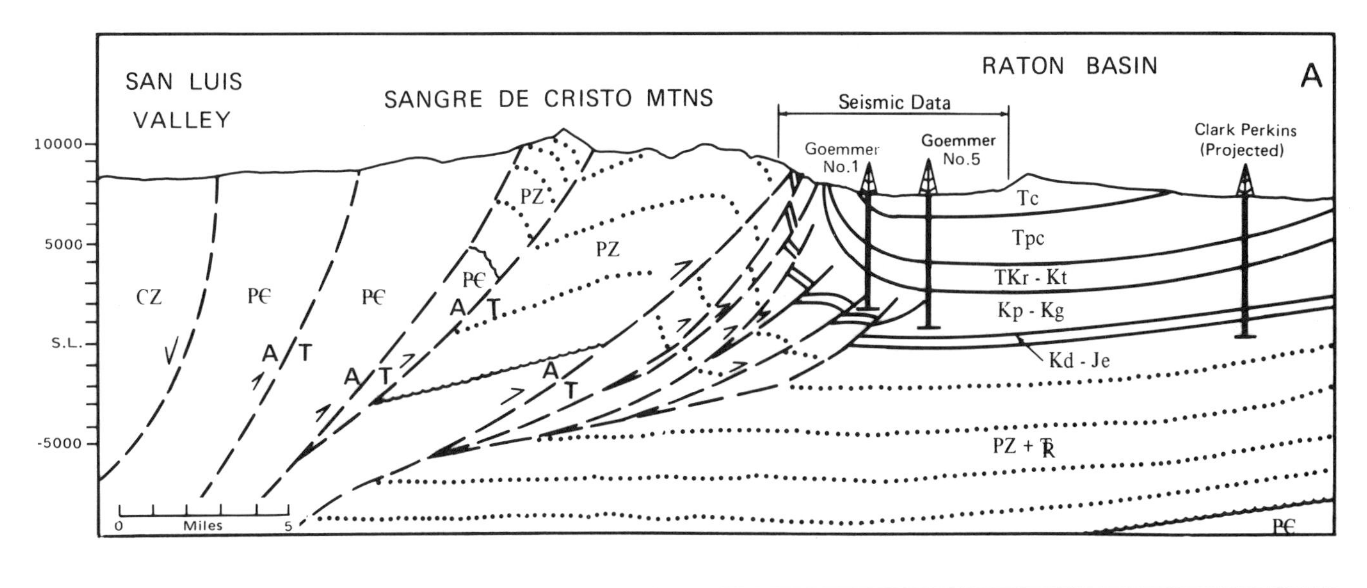

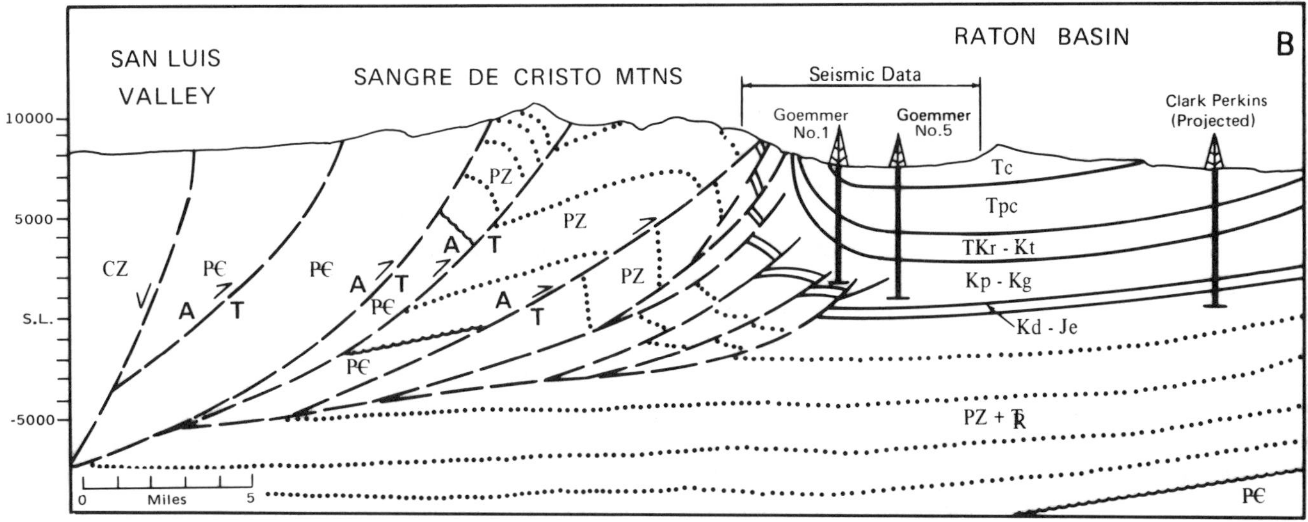

Figure 41. Schematic regional cross sections illustrating two possible styles of structures that may have resulted from late Laramide wrenching: A, a positive flower structure, and B, a series of low-angle, oblique-slip thrust faults. The cross sections trend east-west approximately along the southern edge of T29S. Faults are solid where confirmed by seismic data (Rose and others, 1984; Applegate and Rose, 1985). Angle of dip of thrusts was modeled after analogous features mapped by Gries (1981).

can be mapped on Landsat imagery and aerial photography near the crest of the Sangre de Cristo Mountains (see Fig. 42).

Strongly convergent wrenching, such as that believed to have occurred along the Sangre de Cristo Mountains during the Eocene, may result in a positive flower structure (Lowell, 1972), low-angle detached oblique-slip thrusts (Crouch and others, 1984), or structures intermediate between these extremes. Unfortunately, the published seismic data for this area do not cross the Sangre de Cristo Mountains, limiting us to speculation on the precise structural style present in that region. One view is that the deformed wedge along the west flank of Raton Basin and Huerfano Park is the eastern margin of a positive flower structure, located along the Sangre de Cristo Mountains. Such a feature may have nucleated on, and disrupted, the lower part of an earlier thrust zone (Fig. 41a). An alterative view (Fig. 41b) is that the complex structures along the west flanks of these basins are the result of low-angle, oblique-slip, thrusts. These low-angle, oblique-slip thrusts may have been cut by extensional faults along the east flank of the San Luis Valley during Neogene time. If the early Laramide thrusts dip at about 30°, as shown by Gries (1981) to be the case elsewhere, then a positive flower structure nucleated along early Laramide thrusts might be difficult to distinguish from low-angle oblique-slip thrusts.

Because the zone of decoupling along most wrench faults is

tens of kilometers wide, these faults probably affected both the Sangre de Cristo Mountains and the adjacent Raton Basin. Tercio anticline, a north-northwest–trending asymmetric anticline along the western edge of Raton Basin (Fig. 42), has the proper orientation and vergence for north-trending right-slip movement and supports the idea that such tectonic activity affected the basin.

Chapin (1983) also speculated that a north-northwest–trending, right-slip zone exists along the southwestern edge of the Wet Mountains. This zone, continuing to the southeast to mark the northeastern edge of the Raton Basin, roughly corresponds to the monocline defining the northeast flank of the Raton Basin and to a fault that was active during the Pennsylvanian and Permian (the Pleasant Valley Fault; De Voto, 1980b). This implies that the northeast flank of the Raton Basin was formed by Laramide reactivation of an older fault zone.

Rio Grande Rifting

Post-Laramide orogenic activity is related to the Rio Grande rifting event. During the late Eocene–early Oligocene, after Laramide tectonic activity ceased and before Oligocene volcanism began, a low relief landscape formed as a result of erosion of the Laramide uplifts and filling of adjacent basins (Epis and others, 1980). Widespread volcanism occurred during the Oligocene and Miocene, producing numerous igneous plugs (e.g., Spanish Peaks), dikes, sills, and volcanic debris that covered this post-Laramide surface (Taylor, 1975; Epis and others, 1980). The earliest volcanic activity, 35–36 Ma, produced the Wall Mountain Tuff (Epis and others, 1980).

During the Neogene, the Sangre de Cristo Mountains and the Wet Mountains were uplifted as horsts, and the adjacent San Luis Valley and Wet Mountain Valley–Huerfano Park were downdropped as grabens. The subsidence of Wet Mountain Valley–Huerfano Park actually began during the Laramide Orogeny and was rejuvenated in the Neogene along previously existing Laramide thrusts (Lindsey and others, 1983). The San Luis Valley was part of the San Luis–Sangre de Cristo Uplift during the early phase of the Laramide Orogeny. Downdropping of the San Luis Valley occurred during late-Laramide wrench faulting and again during the Neogene (Gries, 1985). This activity probably occurred along existing Laramide faults such as the late Laramide oblique right-slip thrust faults, leaving only the leading edge of the Laramide uplift preserved as the Sangre de Cristo Mountains (Sales, 1983; Lindsey and others, 1983). Prior to the collapse of the San Luis Basin, the Sangre de Cristo Mountains were merely the flank of the broader, early Laramide San Luis–Sangre de Cristo Uplift that shed sediments across the Sangre de Cristo Mountains into the Raton Basin.

The elevation difference between peaks of the Sangre de Cristo Mountains and the Wet Mountain Valley floor is about 3 km. This represents a minimum value for the vertical throw on this fault, because no Neogene sediments are present in the Sangre de Cristo Mountains (Taylor, 1975). Evidence of a similar nature was used by Epis and others (1980) to determine that the vertical throw between the west flank of the Sangre de Cristo Mountains and the San Luis Valley is on the order of 7 km.

STRUCTURAL GEOLOGY

The structures present in the Raton Basin and Huerfano Park were formed primarily during the Laramide Orogeny or are remnants of older features reactivated by the Laramide Orogeny. These basins are marked by gently west-dipping lower Tertiary–age sediment, except along the west flank, where lower Tertiary– through Pennsylvanian-age sediments are intensely deformed into a fold and fault belt 8–15 km in width (Fig. 42). Even though the structures of these two basins were formed during the same tectonic event, each basin appears to be characterized by different structural styles that we believe are a reflection of slightly different responses to Laramide compression. Furthermore, the Raton Basin may be divided into two subareas: a Northern Raton Basin and a Southern Raton Basin.

Huerfano Park

Huerfano Park is a northwest-trending basin bounded on both sides by faults of significant magnitude. Right-slip faults that separate the Wet Mountains from the basin define the northeast boundary of Huerfano Park (Chapin, 1983). The southwest flank of Huerfano Park is an overthrust zone composed of three very lobate low-angle thrust salients, each of which shows considerable horizontal shortening (Johnson, 1959). The salients are defined by east-west- to northeast-trending tear faults that merge into north-northwest- to northwest-trending thrusts. The southern edge of the southernmost salient is a northeast-trending tear fault passing through the Oakview-Ojo area; the fault approximately corresponds to a northeast-trending basement shear zone (see Lewis and others, 1969). It is clear from seismic data that this tear fault has a thrust component (Applegate and Rose, 1985); however, because the seismic events are discontinuous where the seismic line crosses the tear fault, the angle of dip of the fault is unclear. More clearly shown by this same seismic line is that considerable horizontal shortening occurs along thrusts of the southernmost salient, located along the western portion of the seismic line. The northernmost salient is a large imbricate thrust package that has at least 7.5 km of horizontal movement (Johnson, 1959) and places Pennsylvanian (Madera Formation) through Late Cretaceous (Pierre Shale) sediments in thrust contact with Paleocene sediments (Poison Canyon Formation). A klippe of Paleocene (Poison Canyon Formation) and Eocene (Cuchara Formation) sediment was probably part of the middle salient and also has a minimum of 7.5 km of horizontal movement (Johnson, 1959). East of the middle salient are outcrops of Pierre Shale and Poison Canyon, Cuchara, and Huerfano Formations that are folded in a syncline-anticline couplet. A thrust fault probably cuts the anticline at depth.

The structural front of the Huerfano Park overthrust zone trends north-northwest to northwest. Theoretically, faults with

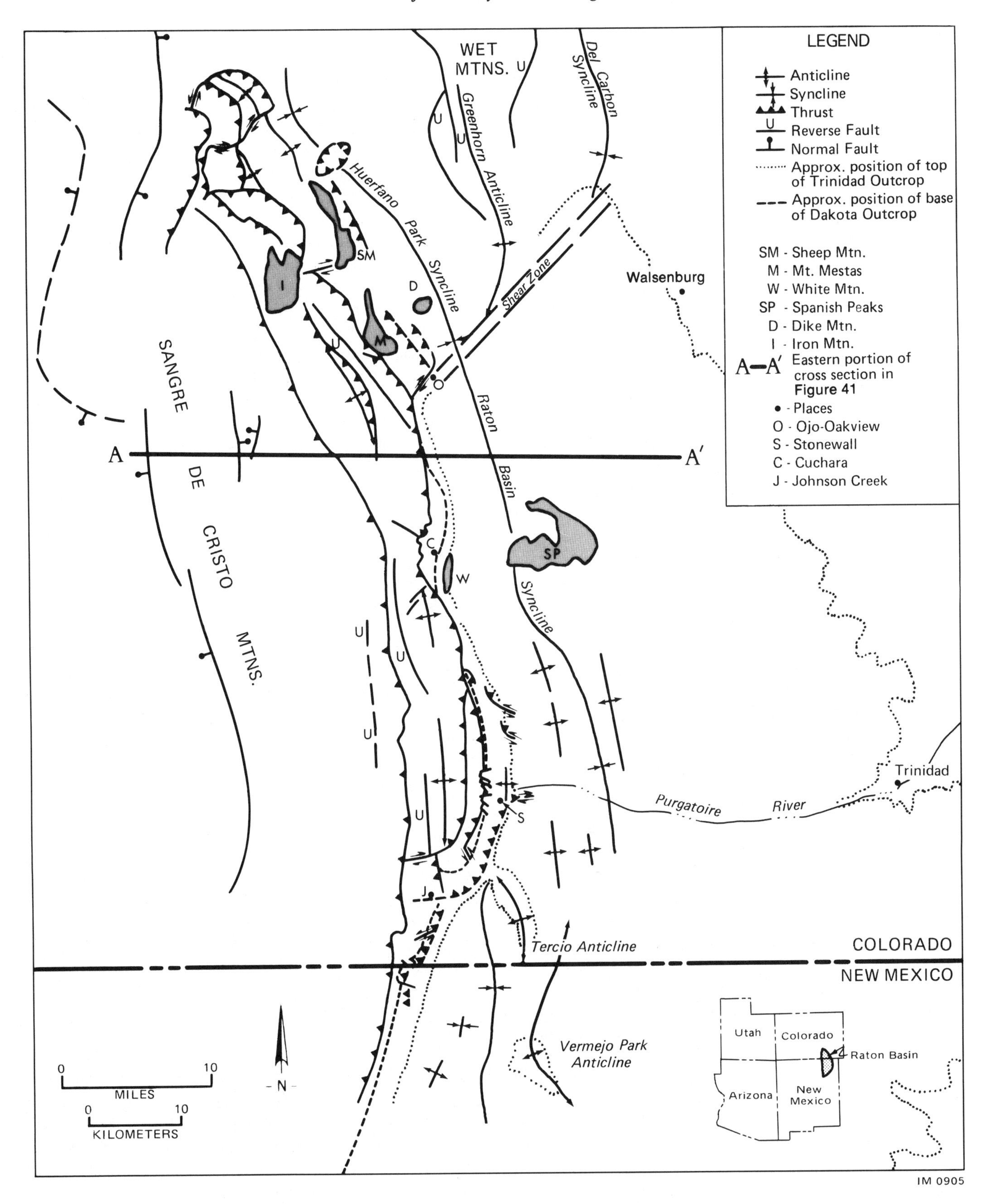

Figure 42. Generalized tectonic map of Raton Basin and Huerfano Park.

this orientation would have oblique thrust movement during both the early and late Laramide orogenic pulses. Therefore, folds and thrusts formed in this area during the early Laramide Orogeny would probably continue to grow as folds and thrusts during the late Laramide Orogeny. This suggests that this area should exhibit considerably more horizontal shortening than the Raton Basin to the south.

Northern Raton Basin

The western boundary of the Northern Raton Basin is an overthrust zone with two gently lobate thrust salients. The northern salient extends north from White Mountain (about 2 km east of Cuchara) to just south of the Ojo-Oakview area, where it terminates against the northeast-trending right-lateral shear that defines the southern limit of Huerfano Park. The northern and southern salients of the Northern Raton Basin meet in the vicinity of White Mountain. The southern salient is terminated on the south by an east-northeast–trending fault along Johnson Creek (near the north end of the Tercio Anticline).

The salients of the Northern Raton overthrust zone consist of Pennsylvanian through Cretaceous sediments imbricately stacked in north-south–trending thrusts and folds that gently curve to the north-northwest or south-southwest. Along the eastern margin of these salients, highly folded and thrusted Pennsylvanian and Permian rocks lie in thrust contact with apparently gently deformed Mesozoic and younger strata. The Mesozoic and younger rocks form a belt of east-dipping outcrops that range from vertical near the fault contact with Paleozoic rocks to nearly horizontal in the basin center. Seismic data reveals that the Mesozoic rocks below the upper Pierre are allochthonous—cut at depth by a series of imbricate thrusts (Rose and others, 1984). The thrusts and attendant folds have produced at least 4 km of horizontal shortening (Fig. 41). Most of the crumpling is confined to the Graneros through Pierre interval. This appears similar to conditions encountered in drill holes elsewhere in the basin (Speer, 1976). The amount of horizontal shortening within the Paleozoic strata cannot be accurately determined because of poor stratigraphic control; however, we believe it is on the order of several kilometers.

The southern salient appears to be more intensely deformed and exhibits greater horizontal shortening than the northern salient. Evidence for this includes tear faults associated with thrusts, outcrop width, and folds in the basin adjacent to the southern salient. The width of the Mesozoic outcrop belt of the southern salient is about twice that of the northern salient, even though dips of these strata are similar in both salients. Because no regional stratigraphic north-south thickness changes occur in this region, this difference must be accounted for by thrust faults that structurally thicken the Mesozoic section of the southern salient. A small north-south–trending anticline near Stonewall, Colorado, probably marks a thrust at depth. Because this fold is too small to account for the apparent difference in deformation between the northern and southern salients, we believe other thrusts (previously unrecognized) must be present in the Mesozoic section.

Numerous mapped tear faults that splay from the thrust that brings the Sangre de Cristo Formation in contact with Jurassic strata are present in the southern salient. However, few such features are mapped in the northern salient (Johnson and others, 1958; Wood and others, 1956). These tear faults clearly offset the Dakota Formation. Previous mapping shows these faults to be almost randomly oriented. Aerial photography and Landsat Thematic Mapper imagery show that west-northwest–trending stream valleys pass through Dakota outcrops that are offset left-laterally, whereas east-northeast–trending valleys typically pass through Dakota outcrops that are offset right-laterally. Additionally, slickensides occur in Dakota outcrops that correspond in position and relative displacement (i.e., left-slip) to topographic offset of the Dakota outcrop about 3.0 km northwest of Stonewall. These slickensided surfaces are vertical and strike west-northwest, and the near horizontal slickensides suggest a west-northwest–trending left-slip tear fault. Slickensided surfaces, dipping about 25° west, found in an adjacent Dakota outcrop, have slickensides that indicate thrust movement. This implies that thrusting and strike-slip movement occurred approximately contemporaneously, and these tear faults are splays in the major thrust. These splays take up horizontal compression, causing additional minor horizontal movement of strata in the footwall. This suggests the presence of compressional features (thrusts or folds) in front of the major thrust of the southern salient.

Several gentle folds are present in the basin adjacent to the southern salient. The apparent offset of the synclinal axis of the basin and the termination of some of these folds along the basinward projection of the tear faults cutting the thrusts of the southern salient suggest that these folds overlie thrust faults at depth. These features also suggest that horizontal shortening across this part of the basin exceeds that of the northern salient.

The structural front of the Northern Raton Basin overthrust zone trends north-south. This direction parallels thrust faults formed during the early Laramide Orogeny and oblique right-slip faults formed during the late Laramide Orogeny. This implies that north-south–trending thrusts and folds that formed during early deformation subsequently may have either become refolded to trend northwest to north-northwest or been decapitated by similarly trending thrust faults.

Southern Raton Basin

The Southern Raton Basin has a nonlobate, north-northeast–trending fault and fold belt along the west flank of the basin south of Johnson Creek, Colorado. This area contains fewer thrusts and folds and is narrower than the two northern overthrust areas. Basinward of this belt is the major north-northwest–trending, asymmetric Tercio Anticline. The amount of horizontal shortening across this belt is unknown.

Theoretically, the north-northeast–trending structural front would be a right oblique-slip zone during both the early and late stages of the Laramide Orogeny. The Tercio Anticline is oriented properly for a compressional fold related to right oblique-slip along north-northeast–trending faults.

SUMMARY

The Raton Basin and Huerfano Park inherited their configuration from tectonic events that reactivated older zones of crustal weakness. The events responsible for forming the structures present in these basins include Neogene extension, the Late Cretaceous to early Tertiary Laramide Orogeny, and Pennsylvanian-Permian orogenic activity related to the Ancestral Rockies deformation. These two basins are characterized by gently west-dipping lower Tertiary–age sediment except along the west flank, where lower Tertiary– through Pennsylvanian–age sediments are intensely deformed into a fold and fault belt 8 to 15 km in width.

Raton Basin and Huerfano Park were both formed during the Laramide Orogeny; however, both basins and the northern and southern subareas of the Raton Basin are characterized by different structural styles that reflect different responses to Laramide compression. The style of thrusts along the southwest flank of Huerfano Park is very lobate and shows a large amount of horizontal movement. By comparison, the thrusts along the west flank of the Northern Raton Basin are mildly lobate and have moderate horizontal movement, and those in the Southern Raton Basin are straight with minor horizontal movement. Most Laramide thrusts with lobat plan-view geometry have principally dip-slip rather than strike-slip displacement and have a low-angle of dip, whereas Laramide thrusts with greater strike-slip than dip-slip movement are straighter and have a steeper dip (Berg, 1983). This supports the concept that the thrusts in the Huerfano Park overthrust are low-angle and have a large component dip-slip, whereas those of the other two areas have both dip-slip and strike-slip components and a comparatively steeper dip.

ACKNOWLEDGMENT

This paper benefited from thorough and critical reviews by Frank Kottlowski, John W. Rold, Allison R. (Pete) Palmer, Charles E. Chapin, and Orville Russell, and from discussion with Charles E. Chapin and Richard F. Livaccarri on Laramide tectonics. We express our sincere gratitude to Earth Satellite Corporation for permission to publish this data and to Dorothy B. McCabe and David L. Redecke for preparation of the illustrations.

SOUTH PARK

Richard H. DeVoto

South Park, an intermontane structural and sedimentary basin in central Colorado, has a complex Tertiary history. The Laramide and Tertiary structural activity represents a rejuvenation of structural features that created a similar structural-sedimentary basin in the Late Paleozoic (De Voto, 1980b).

LATEST CRETACEOUS–PALEOCENE

The Laramide (latest Cretaceous–Paleocene) in central Colorado was a time of uplift of basement blocks, in many places along bounding faults; erosion in the uplifted mountainous areas and deposition in the intermontane South Park Basin; compression, folding, and thrusting of the sedimentary sequence in the synclinal area of South Park; and associated igneous activity. The Sawatch and Front Range Uplifts were initiated in the latest Cretaceous (±70 Ma), the Cretaceous sea withdrew from the area, the broad expanse of South Park was tilted eastward on the east flank of the Sawatch Anticline, and extensive erosion occurred on the uplifts and within South Park in the latest Cretaceous and probably into the early Paleocene. A major quartz latite to andesitic volcanic pile was extruded in the area of northwestern South Park during this time. Several intrusives ranging in age from 70 to 60 Ma in the Leadville area represent the subvolcanic intrusives, and up to 350 m of andesitic flows, flow breccias, tuffs, and tuffaceous sediments that occur stratigraphically at the base of the South Park Formation in South Park represent the preserved extrusive remnants of this volcanism (Fig. 43).

The South Park Formation includes up to 3,000 m of, in ascending sequence, the basal andesitic volcanic and volcaniclastic unit (Reinecker Ridge Volcanic Member), a mixed-lithology conglomerate unit, a tuffaceous unit (Link Spring Tuff Member), and an upper arkosic unit. The basal South Park beds rest with low angular discordance on Upper Cretaceous strata. They fill a major 3- to 5-km-wide paleovalley and rest 1,500 m stratigraphically below the top of the Cretaceous sequence, with 20° angular discordance at the southwesternmost exposures of the South Park beds 10 km north of Hartsel. The mixed-lithology conglomerate unit, consisting of up to 1,800 m of conglomerate, sandstone, and tuffaceous mudstone, overlaps the Reinecker Ridge andesitic unit and fills the paleovalley. The conglomerate clasts display the result of erosional unroofing of the Sawatch uplift to the west, with a stratigraphic progression upward of cobbles of Reinecker Ridge Andesite, Laramide porphyries of the Leadville area, blocks of Dakota Sandstone (Cretaceous) and Pennsylvanian red beds, lower and middle Paleozoic quartzites and dolomites, and Precambrian gneiss, granite, pegmatite, and quartzite (Wyant and Barker, 1976). Thus, most of the lower South Park strata were derived from erosion of the western part of South Park and the Sawatch Uplift area to the west.

The Link Spring Tuff Member includes up to 210 m of

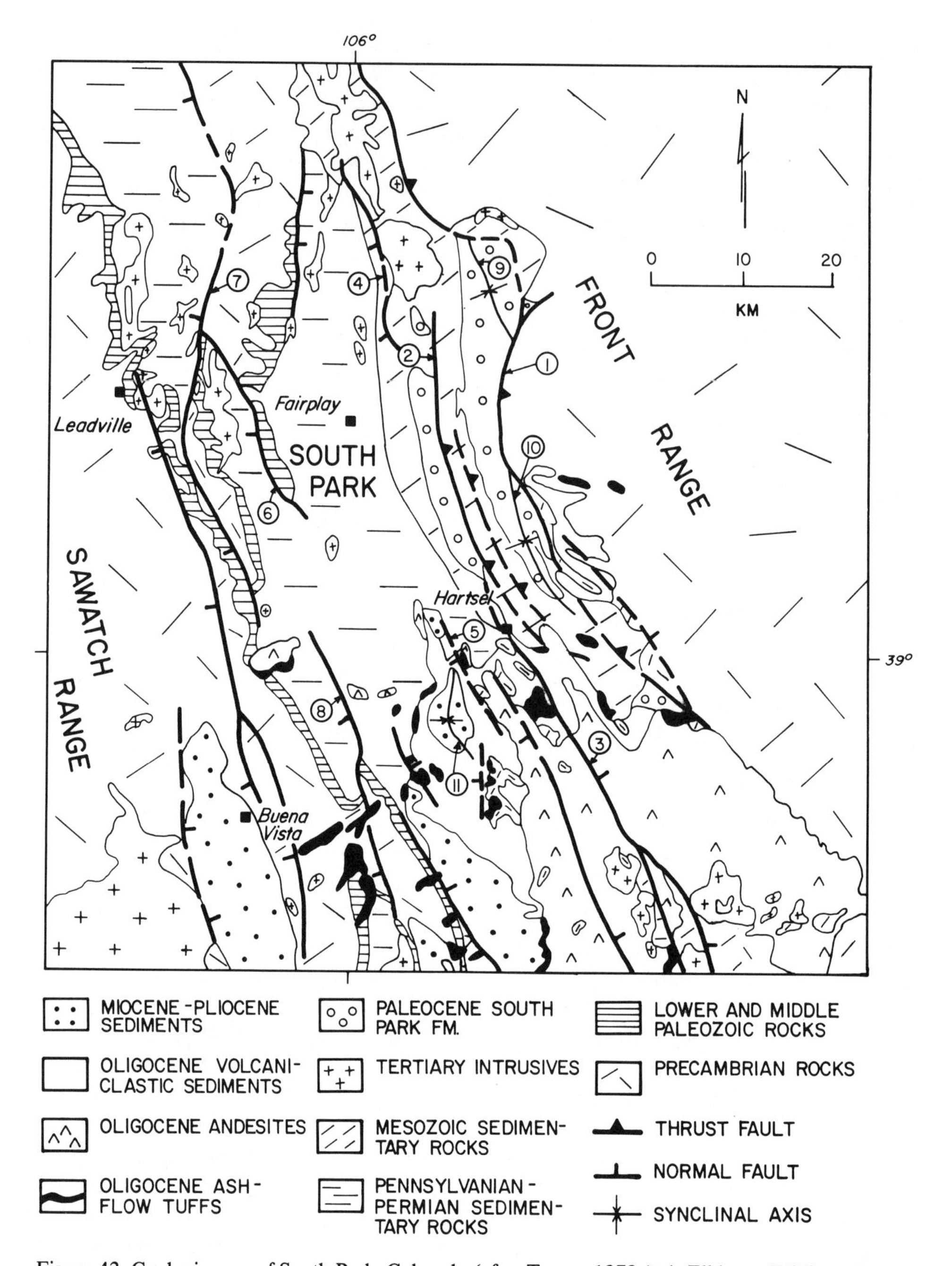

Figure 43. Geologic map of South Park, Colorado (after Tweto, 1979a). 1: Elkhorn–Williams Range fault—Elkhorn Fault along most of eastern South Park becomes Williams Range Fault to the north of South Park; 2: South Park–South Platte thrust complex; 3: Santa Maria–Currant Creek Fault; 4: Boreas Pass Fault; 5: Agate Creek Fault; 6: London Fault; 7: Mosquito-Weston Fault; 8: Trout Creek Fault; 9: Michigan Hill Syncline; 10: San Isabel Syncline; 11: Antero Syncline.

waterlaid tuff, minor andesite flow breccia, and tuffaceous sandstone. Age dating of biotite in this unit indicates a Paleocene age (58.4 Ma) (Wyant and Barker, 1976).

As much as 600 m of arkosic sandstone and conglomerate, apparently derived from erosion of Precambrian rocks from the core of the Front Range Uplift to the east, occur stratigraphically below the Link Spring Tuff Member in the area of the San Isabel Syncline (Fig. 43) (Sawatzky, 1967). The arkosic unit in the area of the Michigan Hill Syncline includes up to 1,070 m of arkosic sandstone, conglomerate, and mudstone stratigraphically above the Link Spring Tuff Member. These arkoses to the north include several boulder beds containing blocks of Precambrian granitic and metamorphic rocks up to 3 m across (Wyant and Barker, 1976).

The lack of conspicuous angular discordance within the 3,000 m of South Park strata suggests that the South Park beds accumulated as the Sawatch Range was progressively uplifted and eroded in the latest Cretaceous and Paleocene. The boulder beds in the arkosic unit suggest proximity to a high-relief erosional area of the Front Range to the east, probably induced by Paleocene displacement on the Elkhorn–Williams Range Fault. Major compression, folding, and thrusting of the South Park strata, however, did not occur until after deposition of the entire South Park sequence. The lack of lacustrine sediments in the South Park strata and the presence of volcanic detritus in Paleocene strata south of Canon City suggest that the Paleocene drainage in South Park flowed south or southeastward out of South Park (Scott, 1975; Epis and others, 1980).

Westward thrusting along the Elkhorn–Williams Range Thrust on the western margin of the Front Range Uplift has caused Precambrian rocks to be thrust over South Park and older strata and the synclinal axis of the South Park Basin, the Michigan Hill–San Isabel Syncline (Fig. 43) (Sawatzky, 1967, 1972). Folding and overturning of the South Park strata occurred in front of the thrust to form the strongly asymmetric synclinal nature of South Park. At least two associated westward-directed thrusts broke the Cretaceous and Paleocene strata in front of the Elkhorn Thrust. Other north-northwest–trending folds developed in the sedimentary sequence in South Park at this time also. This major westward-directed thrusting, compression, and folding probably occurred late in the Paleocene and early in the Eocene, after deposition of the entire South Park sequence and prior to a prolonged period of Eocene erosion.

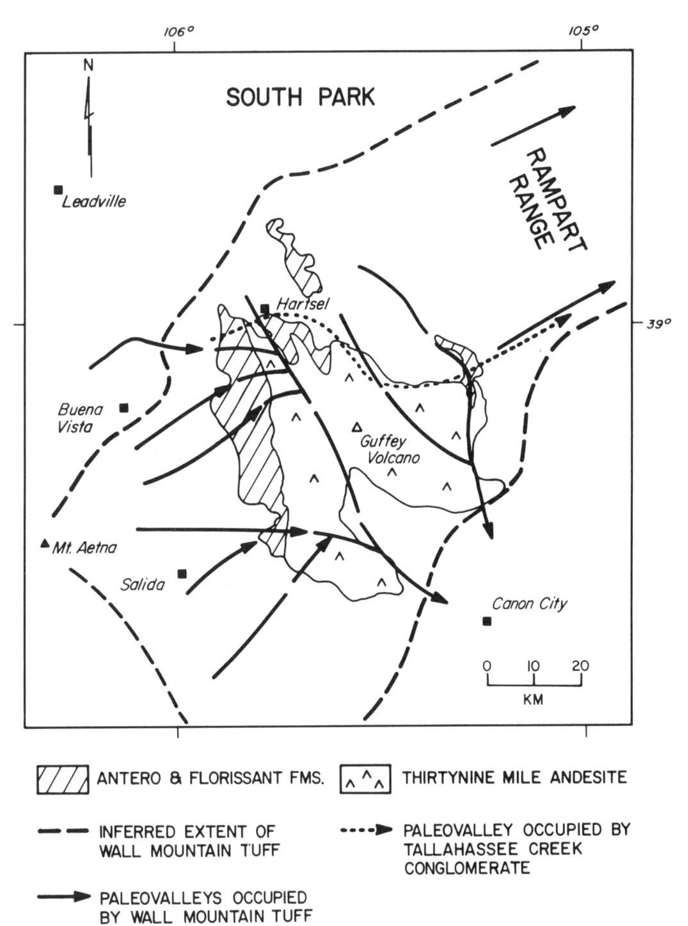

Figure 44. Oligocene prevolcanic valleys and major volcanic features, southern South Park, Colorado (after Epis and others, 1980).

EOCENE

Remnants of a widespread, relatively low relief, late Eocene surface of erosion occur extensively in southern South Park and on the southern Front Range (Epis and Chapin, 1975; Epis and others, 1980). Most of the Eocene was a time of relative tectonic inactivity, deep erosional dissection of Laramide structures, and mountain pedimentation. The only Eocene sediments (with the possible exception of the uppermost South Park beds) that occur in South Park are a thin sequence of arkosic sands that occur in paleovalleys cut on the late Eocene surface in southern South Park. To the south of South Park, up to 300 m of this late Eocene arkosic Echo Park Alluvium occupies north-northwest–trending paleovalleys in grabens or half grabens (Epis and Chapin, 1975). Thus, drainage from South Park throughout the Eocene was to the south along these paleovalleys, particularly the valley controlled by the Currant Creek Fault (Fig. 43).

OLIGOCENE

A complex sequence of volcanic rocks and volcaniclastic sediments was deposited during the Oligocene in southern South Park. Rhyolitic ash flows of the Wall Mountain Tuff (34–36 Ma) were erupted from a caldera near Mt. Aetna in the southern Sawatch Range (Fig. 44); they filled paleovalleys and overrode interfluve divides, covering an area of at least 9,500 km^2 that extended across southern South Park and the Rampart Range (southern Front Range) to Castle Rock in the Denver Basin (De Voto, 1971, 1972a; Epis and others, 1980; Epis and Chapin, 1968, 1974). The early Oligocene paleovalleys that were filled with 100 to 200 m of the welded ash flows have undergone topographic inversion, as the Wall Mountain Tuff now forms ridges on the landscape.

Up to 200 m of volcanic conglomerate and sandstone of the early Oligocene Tallahassee Creek Conglomerate overlie partially eroded sections of the Wall Mountain Tuff in southern South Park. The Tallahassee Creek Conglomerate occupies many of the same paleovalleys that were partially filled by the Wall Mountain Tuff as well as a major eastward drainage that extended from Buena Vista to Florissant, entirely across southern South Park, and beyond across the Front Range (Fig. 44) (De Voto, 1971, 1972a; Epis and others, 1980).

Beginning about 34 Ma, the Thirtynine Mile Andesite was erupted from many small vents in southeastern South Park, disrupting and damming the eastward- and southeastward-flowing drainages. Growth of the large composite Guffey Volcano over a period of two million years (32–34 Ma) caused the accumulation of a thick sequence of laharic, flow, and pyroclastic breccias that covered an area of at least 3,000 km^2 (Epis and others, 1980). The volcanic damming of the drainages and the eruption of intermediate (dacitic to andesitic) ash flows from a western source (probably the Mt. Aetna Caldera) caused a thick sequence (up to 600 m) of volcaniclastic and lacustrine sediments of the Antero Formation to accumulate in lakes immediately peripheral to and updrainage from the andesite field in southern South Park (De Voto, 1971, 1972a; Epis and others, 1980; Smith, 1982). Facies changes and structural relationships indicate that subsidence of the Antero Syncline and normal faulting of the Agate Creek Fault occurred during deposition of the Antero beds in the Oligocene (De Voto, 1971; Smith, 1982). Thus, the accumulation of the thick Antero lake-bed sequence was principally the result of volcanic damming of drainages but was augmented by structural activity as well.

MIOCENE-PLIOCENE

Major tectonic activity, uplift and erosion of mountain blocks, and deposition in adjacent grabens occurred in central Colorado throughout the Miocene and Pliocene. Much of the structural relief on the normal faults (Fig. 43) and most of the present topographic relief of the mountainous areas in central Colorado are due to this period of Late Tertiary tensional tectonic activity (Scott, 1975; Epis and others, 1980). The displacements occurred along north-northwest–trending faults, many of which had undergone earlier late Paleozoic and/or early Tertiary offset (De Voto, 1980b).

Up to 300 m of conglomerate, sandstone, and siltstone of the Wagontongue and Trump Formations of late Miocene age were deposited in the actively downdropped, fault-bounded Antero Basin and syncline block in southwestern South Park (De Voto, 1971, 1972a; Smith, 1982). Up to 1,400 m of Miocene and Pliocene coarse-grained alluvial sediments occupy the Arkansas River Graben near Buena Vista southwest of South Park (Fig. 43) (Epis and others, 1980).

The late Miocene sediments of South Park (Trump Formation) rest with angular unconformity on the Oligocene Antero beds. The Miocene strata are also folded into the Antero Syncline. Thus, block faulting was recurrently active throughout the Miocene and Pliocene. This Late Tertiary vertical block faulting offset the extensive late Eocene erosion surface and the Oligocene volcanic units and created most of the present relief of the mountains of central Colorado (Epis and others, 1980). Erosion has dominated the late Pliocene and Quaternary history of South Park.

SEDIMENTARY ROCKS OF THE DENVER BASIN

Bruce F. Curtis

INTRODUCTION

The Denver Basin, the easternmost and one of the largest of the Rocky Mountain basins, underlies an area of about 155,000 km^2, situated mainly in eastern Colorado but extending into southeastern Wyoming, western Nebraska, and northwesternmost Kansas (Fig. 45). The basin is unusual because its western flank involves Cordilleran foreland structure, whereas its eastern border is structurally similar to the stable craton of the Mid-Continent. Likewise, many of the sedimentary rocks of the basin display marked west-to-east facies changes worthy of close study. This section presents a brief overview of the sedimentary strata of the Denver Basin—their nature, their evolution, and their importance to human activities.

STRUCTURAL OUTLINE OF THE DENVER BASIN

Structure of the Margins

The steeply inclined western flank of the Denver Basin abuts against the Wet Mountains and southern Front Range Uplifts along a zone of related boundary faults (Figs. 45 and 46). A considerable body of evidence indicates that these are reverse faults whose surfaces dip westward through the sedimentary section at angles varying from low to high. Questions as to whether these faults are listric in character, steepen with depth, or are sigmoid in cross section are not completely resolved; however, there is increasing acceptance of the view that the attitudes of the faults not only may correspond to any of these patterns, but may change substantially within distances of a few kilometers (Tweto, 1983; Hansen, 1984; Chapin and Cather, 1983; Jacob, 1983; Stone, 1985a; R. A. Johnson, personal communication, 1984). Many of the faults seem to have early ancestry, related ultimately to features of the Precambrian basement, and motion along them, not always with the same sense of relative displacement, is evidenced at various times in geologic history. Trends of a number of Denver Basin flank faults that had Laramide and late Cenozoic motion were also apparent in Pennsylvanian time. Thus, interpretations of regional stress fields that shaped the Denver Basin in Laramide time are fraught with great uncertainty.

A few kilometers north of the 40th parallel, in Boulder County, Colorado, the trend of the western flank of the Denver Basin swings rather abruptly from north-northwest to nearly north. Also from the same location northward to approximately the Colorado-Wyoming line, the structural pattern of the basin flank differs from that farther south, displaying a steep eastward slope that is complexly modified by smaller superposed folds and faults (Anderman and Ackman, 1963). Some of the folds are nearly symmetrical, whereas others are steeper either toward the west or east. Some have faulted flanks, and many are thought to overlie deep faults (Fig. 46). Geologists (e.g., Warner, 1956) have noted that this stretch of complex basin margin architecture lies where the mountain front is intersected by a broad, northeasterly trending belt of faulted basement rocks above which sedimentary beds in the basin display structural disturbance and unusual variations in thickness. Of various names applied to this feature (e.g., Siouxia, Sioux Uplift or Arch, Colorado Lineament), "Transcontinental Arch" seems most appropriate for this discussion.

In Wyoming, the basin flank bordered by the Laramie Range assumes the character of a steep monoclinal flexure complicated by reverse faulting and a few flank folds (Anderman and Ackman, 1963; G. G. Anderman, personal communication, 1984; D. L. Blackstone, Jr., personal communication, 1984). Only a few faults had been recognized at scattered surface locations until recent seismic studies brought indications that additional, probably more important, faults are completely hidden beneath the Tertiary cover (R. A. Johnson, personal communication, 1984). As the details of these crustal breaks are not generally known, the western border of the basin is depicted on Fig. 45 as a single fault zone, even though it surely is not that simple.

Near 42° latitude, in southern Platte County, Wyoming, the Laramie Range intersects an uplifted structural feature that trends northeasterly to form the northwestern flank of the Denver Basin. This feature, called the Hartville Uplift or Arch, has been interpreted as a product of Laramide, northwesterly directed, reverse faulting, followed by normal faulting with downdropping to the southeast in late Cenozoic time (Droullard, 1963). Although some geologists have suggested that late Paleozoic tectonic activity occurred at the Hartville Arch location, the stratigraphic and geophysical evidence of it is not particularly compelling. The arch merges to the northeast with anticlinal folds that plunge off the Black Hills Uplift. The northern tip of the Denver Basin is formed by the crumpled monoclinal southern flank of the Black Hills Uplift. This feature is mainly a product of Laramide deformation, although irregularities in ancient sedimentary patterns have been interpreted as reflecting earlier deformation in the Black Hills area (Momper, 1963; Wilson, 1978; Weimer and others, 1982).

Only a slight structural saddle separates the Black Hills Uplift from a gently arched trend that extends far southeastward through western Nebraska into Kansas (Fig. 45). This upwarp, called the Chadron Arch in its northern portion and the Cambridge Arch farther southeast, forms the gently sloping northeast flank of the Denver Basin. The arch undoubtedly has Pennsylvanian ancestry, as shown by the absence of early Paleozoic rocks and the onlap of Pennsylvanian beds on its flanks (see Fig. 50). Substantial renewal of arching, with some faulting in post–Upper Cretaceous–pre-Oligocene time, is revealed in the subcrop pattern beneath flat-lying Tertiary beds, especially at the northern end of the arch (see Fig. 54).

The Las Animas Arch, which forms the southeastern flank of the Denver Basin, extends about 300 km through southeastern Colorado into northwestern Kansas to join the Cambridge Arch in a broad, structurally featureless area where the sinuous courses of the mapped basement contours are determined more by ancient topography on the crystalline rocks than by tectonic deformation (Fig. 45). The Las Animas Arch trend probably received occasional mild uplift in late Paleozoic time, as suggested by thinning of Pennsylvanian and Permian beds over its crest (Rascoe, 1978). In Laramide time the arch achieved essentially its present form as downwarp of the Denver Basin furnished westward slope to then gently eastward-dipping Cretaceous beds and imparted a slight northeasterly plunge to the crest of the arch.

The southern border of the Denver Basin lies along zones of faulting with displacements of 300 to 1,600 m (Fig. 45). These fault trends, known from well information, have been projected somewhat speculatively into zones bordering the Wet Mountains and southern Front Range (Tweto, 1980b), but the style of faulting is not documented. The faults clearly were active in late Paleozoic time when the Apishapa Uplift south of them was elevated and became a sedimentary source area throughout Pennsylvanian and Early Permian time (Fig. 50). Renewed, considerably gentler uplift in Laramide time followed the outline of the earlier feature in at least a general way.

It is apparent that the form of the Denver Basin as a downwarp on the cratonic shelf began to appear at least as far back as Pennsylvanian time. However, some of the bordering features first emerged in Late Cretaceous and early Cenozoic time, and other inherited uplifts were rejuvenated then, so that the basin is most accurately considered to be of Laramide origin.

Intra-basin Structure

The most intriguing aspect of structure within the Denver Basin is the loosely defined, northeasterly trending, faulted belt called the Transcontinental Arch in the preceding discussion. Many faults in this trend, which crosses the central Denver Basin, have been shown by Tweto (1980b) and are displayed in Figure 46. Others, such as northeasterly trending faults of small displacement lying on the southwestern flank of the Cambridge Arch, have received more causal mention by geologists (Anderman and Ackman, 1963), and undoubtedly many additional ones remained undefined or unrecognized. The faulted belt has been considered to be part of a lengthy fracture trend that extends some 1,100 km from northwestern Arizona to the vicinity of the Chadron-Cambridge Arch, and may possibly stretch as much as 1,200 km farther northeast (Warner, 1956, 1978, 1980). The southeastern border of the Transcontinental Arch is rather well delineated by shear zones exposed in the Colorado mountains, by the change in mountain front structure at about 40°north latitude, and by folds and faults within the basin. However, the northwestern fringe of the arch trend is so obscure that various accounts treat the belt as being anywhere from 100 to 200 km wide. In the latter interpretation (preferred by the author), the northwestern

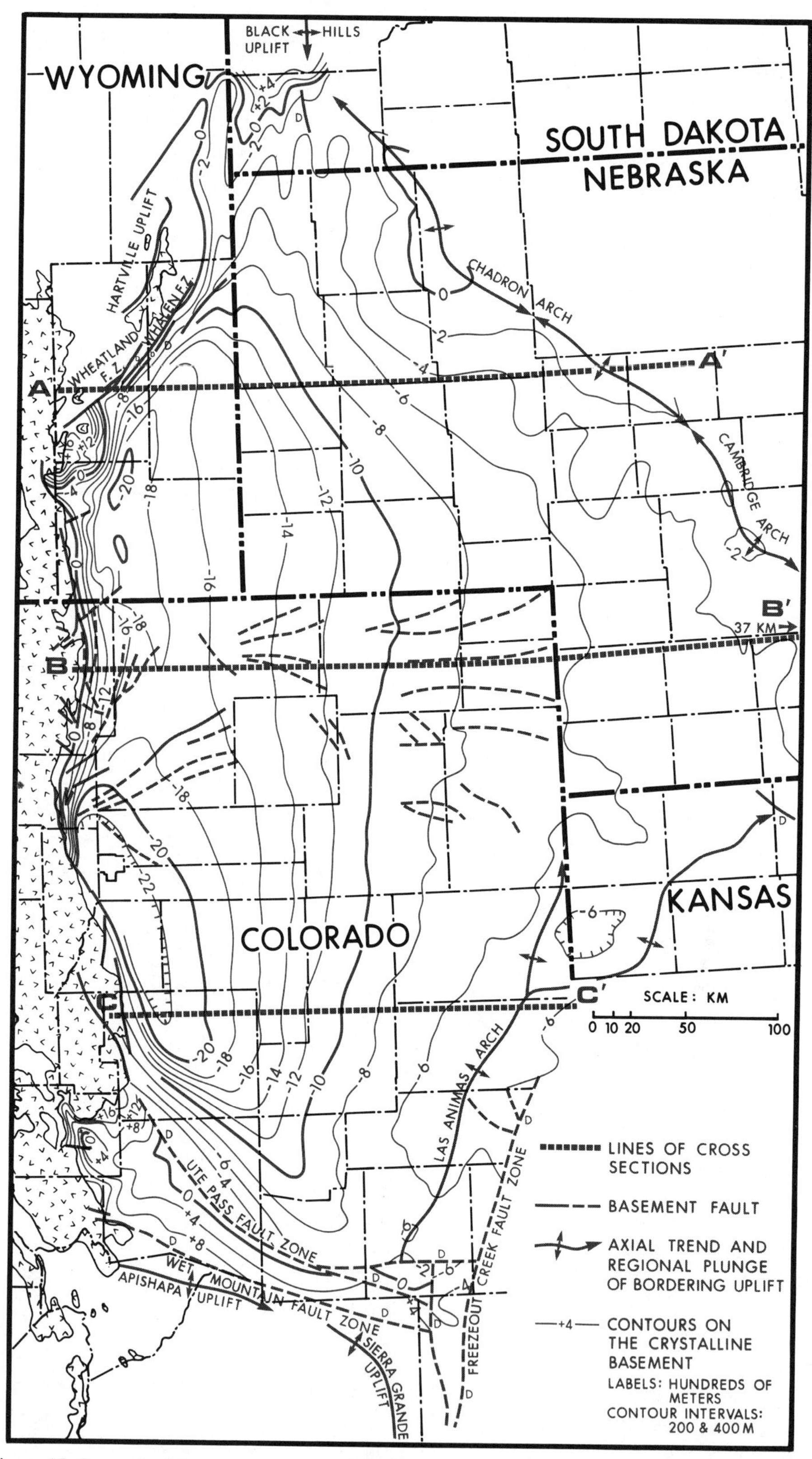

Figure 45. Generalized basement structure of the Denver Basin. Based upon information from Anderman and Ackman, 1963; MacLachlan and Kleinkopf, 1969; Davis and Weimer, 1976; Tweto, 1979a, 1980b, 1983; D. L. Blackstone, Jr., personal communication, 1984; R. A. Johnson, personal communication, 1984; and from geophysical and geological subsurface records. Cross-section lines are those of Figure 46.

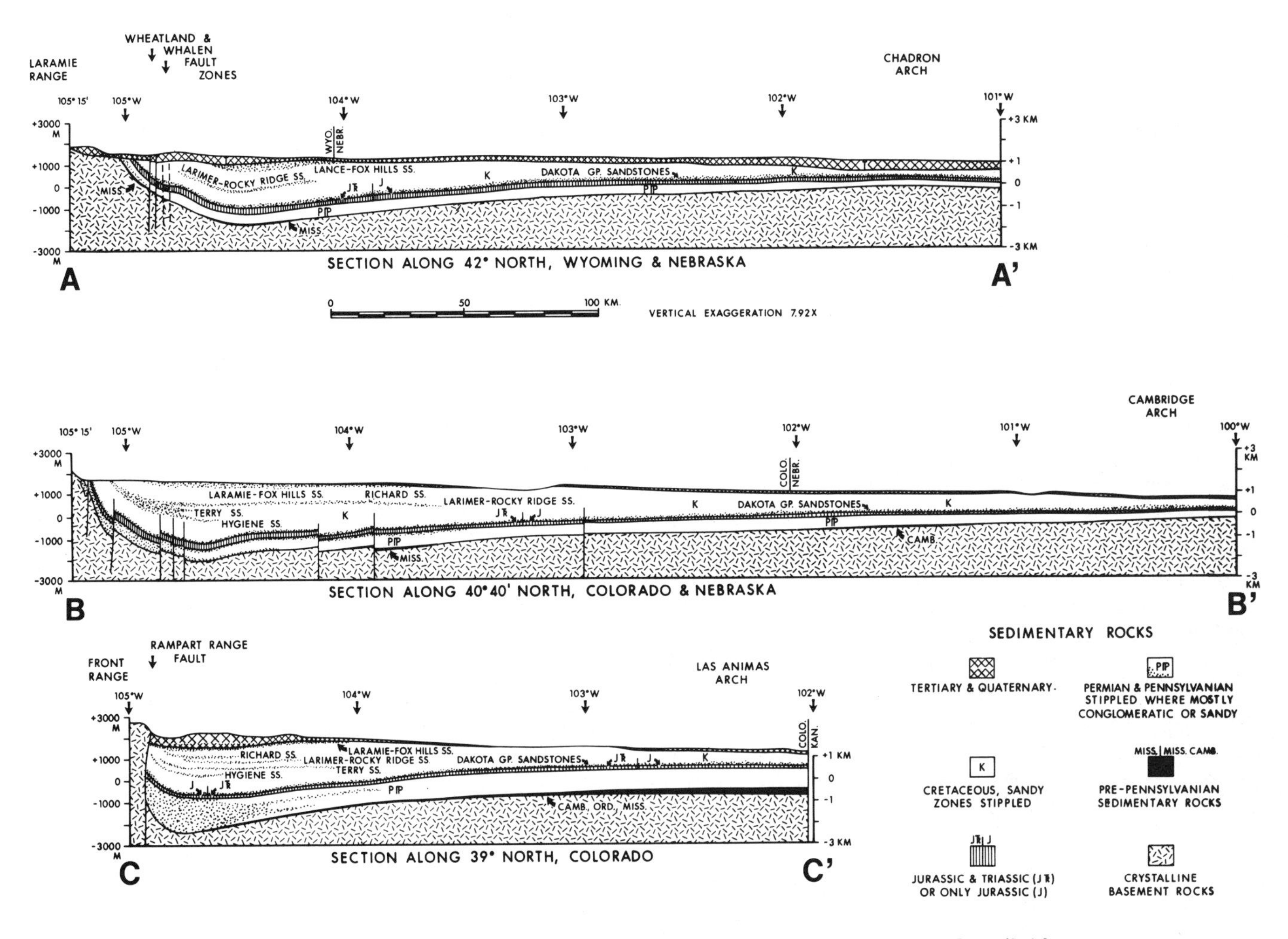

Figure 46. Sections across the Denver Basin along lines of latitude shown in Figure 45. Compiled from work of Anderman and Ackman, 1963; Martin, 1965; Rocky Mountain Association of Geologists Research Committee, 1976; Tweto, 1983; and subsurface data.

margin of the trend lies approximately along the Hartville Arch; thus the Transcontinental Arch appears to underlie practically all of the northern Denver Basin. Although the tectonic classification of this arch is debated, most geologists tend to view it as a wrench trend crossing the craton (Stone, 1969, 1985a; Warner, 1978; Anderman and Ackman, 1963), while noting that its tectonic effects on the sedimentary section appear to be mostly those attributable to vertical components of movement (Weimer, 1984; Sonnenberg and Weimer, 1981). Some of the irregularities in thickness and distribution of sedimentary strata that have been attributed to unusual mobility of the basement along the Transcontinental Arch trend are pointed out in the following sections.

Other intra-basin structural features worthy of note include (1) the Greeley Arch, a structural saddle expressed in the Cretaceous beds, which separates the deepest parts of the basin above part of the Transcontinental Arch trend; (2) some anticlines (many of which are petroleum productive) with essentially north-south axes lying along the mountain front where it intersects the Transcontinental Arch; (3) anticlinal noses that plunge southward off the southern flank of the Black Hills; (4) strongly asymmetric to monoclinal folds above deeper faults of various trends lying on the western flank of the basin (Hansen and Crosby, 1982; Stone, 1985a); (5) southwesterly trending low-amplitude folds that lie along the southwest flank of the Chadron-Cambridge Arch (imperfectly depicted in Figure 45 because of its large contour interval); (6) northwesterly trending faults deep in the basin (one of which, lying just northeast of Denver, became famous when earthquakes followed injection of waste fluids into fractured basement rock at the fault location) (Hollister and Weimer, 1968); and (7) the Goshen Uplift, a broad doming of Oligocene beds that is made evident by Cretaceous inliers in southern Goshen Conty, Wyoming (see Fig. 55), but is represented at depth (Fig. 45) only by slight flattening of contoured slope because it was imposed upon pre-existing regional syncline structure.

THE SEDIMENTARY ROCK SEQUENCE

Pre-Pennsylvanian Rocks (Sauk, Tippecanoe, and Kaskaskia Sequences)

Dolomitic limestones, rather pure quartz sandstones, and some shales are present between the crystalline basement and the Pennsylvanian beds in much of the Denver Basin (Figs. 46–49). Collectively these rocks, representing the Sauk, Tippecanoe, and Kaskaskia sequences of Sloss (1963), constitute a bit less than 1 percent of the total sedimentary cover.

Cambrian sandstones extend as thin tongues into the northernmost part of the Denver Basin. They also appear in outcrops along the southern Front Range margin and eastward in the subsurface. The Cambrian Peerless Shale has been identified in the subsurface in the southern part of the Denver Basin.

Lower Ordovician Manitou (Arbuckle Group) carbonates occur rather irregularly in outcrops along the southwestern border of the Denver Basin and extend southeastward from there in the subsurface. The middle Ordovician Harding (Simpson) Sandstone and the upper Ordovician Fremont (Viola) Dolomite apparently are represented in the subsurface of the basin only close to their somewhat erratic outcroppings on the edge of the Cañon City embayment (the narrow southwesternmost arm of the basin) where they are as thick as 30 and 90 m respectively. The Harding, but not the Fremont, occurs in the outcrop closest to the Fig. 45 section line C–C′. Consequently the Fremont is not shown in the right-hand column of the stratigraphic chart (Fig. 47). Both formations are thought to have extended into the areas of their present outcrops from the Mid-Continent and to have become isolated by later erosion.

Uppermost Devonian strata are now thought to be represented in the northernmost tip of the Denver Basin as the basal part of the mostly Mississippian Guernsey Limestone. Otherwise, Silurian and Devonian rocks are absent, although they occur within 25 and 60 km of the basin's western margin, thus strongly suggesting their former presence within the basin area.

The Mississippian rocks in the basin are believed to form a physically continuous body whose age differs from place to place (Fig. 49). Use of the name Madison for the Mississippian beds in the subsurface of the central Denver Basin is debatable (Fig. 47), but it is common in the absence of any generally accepted alternative term. The Mississippian age assignment for the Williams Canyon Limestone is supported by DeVoto (1980a), although others differ in their interpretations (e.g., Ross and Tweto, 1980).

Pennsylvanian, Permian and Triassic Rocks (Absaroka Sequence)

Pennsylvanian-age rocks are present throughout the Denver Basin and generally grade from conglomerates and sandstones on the west to finer clastics and carbonates typical of marine and marginal marine conditions farther east (Fig. 47). The map in Figure 50 represents the situation in Desmoinesian time when uplift of a source area bordering the basin on the south and southwest apparently was at its maximum. An eastern, and less prominent, source area included the Chadron Arch during at least part of Pennsylvanian time. The thickest Pennsylvanian sections, of more than 1,200 m, accumulated near the southern end of the present Front Range (DeVoto, 1980b).

The Permian rocks represent depositional conditions that evolved from those of Pennsylvanian time; hence, beds of the two systems are not easily separable. In general the Permian strata contain fewer carbonates, finer grained red clastics, and substantial amounts of evaporites (Fig. 51). Most stratigraphic names used for the Pennsylvanian and Permian in the subsurface of the eastern Denver Basin (Fig. 47) result from projection (with a fair degree of confidence) of the terminology of the well-known Kansas sections.

In the western Denver Basin, the Permian rocks pass upward into lower Triassic beds of similar character. Red shales and siltstones containing scattered gypsum deposits are typical of these strata, which extend southward in a long tongue almost the length of the basin and constitute the upper part of the Lykins Formation. The two lobes of Upper Triassic rocks, called the Jelm Formation and the Dockum Group, are similar terrestrially deposited sandstones; however, there is no evidence that they ever were joined as parts of a single deposit (Fig. 52).

Together, the Pennsylvanian, Permian, and Triassic rocks of the Denver Basin represent the deposits of the Absaroka sequence of Sloss (1963). They constitute about 16 percent of the basin's total sedimentary section.

Jurassic, Cretaceous and Early Cenozoic Rocks (Zuni Sequence)

Everywhere within the Denver Basin, Jurassic rocks lie unconformably on the Triassic or Permian formations beneath them (Fig. 47). The oldest Jurassic beds of medium-grained pink- to buff-colored sandstone, appearing in the northern part of the basin, have been correlated with the Entrada Sandstone of Utah and the lower part of the Sundance Formation of the Black Hills. A separate wedge of similar sandstone, nearly equivalent in age, extends into the southernmost Denver Basin, where it is called Exeter or Entrada (Fig. 53). The succeeding layers of limy shales, with some sandstones and evaporites that lie beneath the Morrison Formation over most of the basin, are called the Ralston Creek Formation or "Lower Morrison." These beds are considered, in Figure 53, to be Oxfordian in age and thus approximately equivalent to the upper Sundance or Redwater Shale of Wyoming, following the interpretation of Peterson (1972). However,

Figure 47. Correlation of Denver Basin sedimentary strata. Thicknesses shown in meters. Adapted from Pearl, 1976; Love and others, 1980; J. D. Love, personal communication, 1985.

PERIOD/EPOCH	NORTHERN DENVER BASIN (VICINITY OF SECTION A-A′)		CENTRAL DENVER BASIN (VICINITY OF SECTION B-B′)		SOUTHERN DENVER BASIN (VICINITY OF SECTION C-C′)	
QUATERNARY	eolian and alluvial deposits	sands, gravels and silts 0–100 m.	alluvial and eolian deposits	silts, sands, and gravels 0–120 m.	eolian and alluvial deposits	silts and sands, 0–40 m.
TERTIARY: PLIOCENE						
TERTIARY: MIOCENE	OGALLALA FORMATION	gravels, sands, silts, clays 0–200 m.	OGALLALA FORMATION	gravels, sands, silts 0–180 m.	OGALLALA FORMATION	gravels, silts, sands 0–45 m.
	ARIKAREE FM.	silty sandstone 0–300 m.	ARIKAREE FM.	conglomerate sandstone and siltstone 0–45 m.		
TERTIARY: OLIGOCENE	WHITE RIVER GP.	siltstone and some sandstone 0–180 m.	WHITE RIVER GP.	fluvial sandstone and siltstone 0–200 m.	CASTLE ROCK FM.	conglomerate sandstone 0–90 m.
					WALL MOUNTAIN TUFF	tuffstone 0–6 m.
TERTIARY: EOCENE						
TERTIARY: PALEOCENE					DAWSON ARKOSE	arkosic, conglomerate sandstone, claystone, and carbonaceous, silty sandstone 0–750 m.
					ARAPAHOE FM.	ss., cgl. 0–70 m.
CRETACEOUS: UPPER	LANCE FM.	sandstone, shale, thin coals 0–500 m.	LARAMIE FM.	coal bearing silty sandstones 0–30 m.	LARAMIE FM.	siltstone, sandstone, coals 0–170 m.
	FOX HILLS SS.	silty ss. 0–60 m.	FOX HILLS SS.	silty ss. 0–45 m.	FOX HILLS SS.	silty sandstone 0–90 m.
	PIERRE SHALE < HYGIENE < ZONE	lenticular > silty sss. > gray silty shale 80–1700 m.	PIERRE SHALE < RICHARD < LARIMER < ROCKY RIDGE < TERRY < HYGIENE	len- tic- ular silty sss. > gray silty shale 300–2500 m.	PIERRE SHALE < HYGIENE < ZONE	gray shale sandy siltstones 0–10 m. > 150–1900 m.
	NIOBRARA SHALE: SMOKY HILL SHALE	chalky sh. 70–90 m.	NIOBRARA FM.: SMOKY HILL SH.	chalky sh. 70–100 m.	NIOBRARA FM.: APISHAPA SH.	chalky sh. 100–190 m.
	NIOBRARA SHALE: FT. HAYS LS.	limestone 30–35 m.	NIOBRARA FM.: FT. HAYS LS.	limestone 6–12 m.	NIOBRARA FM.: TIMPAS LS.	limestone 10–40 m.
	CODELL SANDSTONE	sandstone 0–30 m.	CODELL SANDSTONE	silty sandstone 0–6 m.	CODELL SANDSTONE	silty sandstone 0–8 m.
	CARLILE SHALE	silty shale 35–100 m.	CARLILE SHALE	silty shale 12–30 m.	CARLILE SHALE	silty shale 20–25 m.
	GREENHORN LIMESTONE	limestone, shale 10–30 m.	GREENHORN LIMESTONE	limestone, shale 60–85 m.	GREENHORN LIMESTONE	limestone, shale 55–65 m.
	GRANEROS SHALE	dark shale 45–85 m.	GRANEROS SHALE	dark shale 50–65 m.	GRANEROS SHALE	dark gray shale 30–70 m.
	"D" SS.	sandstone 0–35 m.	"D" SS.	sandstone 0–30 m.	"D" SS.	
CRETACEOUS: LOWER	MOWRY SHALE	gray shale 0–45 m.	DAKOTA GROUP: HUNTSMAN SH.	gray shale 0–12 m.	DAKOTA GROUP: HUNTSMAN SHALE	sh. 0–6 m. > ss. 0–6 m.
	MUDDY SANDSTONE	sandstone 15–45 m.	DAKOTA GROUP: "J" SANDSTONE	sandstone 15–30 m.	DAKOTA GROUP: "J" SANDSTONE	sandstone 20–75 m.
	SKULL CREEK SHALE	shale 20–50 m.	DAKOTA GROUP: SKULL CREEK SH.	gray shale 30–60 m.	DAKOTA GROUP: SKULL CREEK SH.	shale 18–30 m.
	CLOVERLY FM.	ss., sh., cgl. 15–180 m.	DAKOTA GROUP: PLAINVIEW-LYTLE SS.	sandstone 15–60 m.	DAKOTA GROUP: LYTLE SANDSTONE	sandstone 10–35 m.
JURASSIC	MORRISON FORMATION	sandstone, variegated shale, conglomerate 50–85 m.	MORRISON FORMATION	conglomerate, sandstone and shale 25–75 m.	MORRISON FORMATION	silty sandstones and claystones 55–100 m.
	SUNDANCE FM. UPPER	gray-green shaly, limy ss. 0–45 m.	RALSTON CREEK FORMATION	sandstone, evaporites, limestone 10–60 m.	"LOWER MORRISON"	siltstone, limestone evaporites 20–35 m.
	SUNDANCE FM. LOWER	sandstone, shale 0–30 m.	ENTRADA SANDSTONE	well sorted sandstone 0–40 m.		
TRIASSIC	JELM FORMATION	sandstone 0–15 m.	JELM FORMATION	light red ss. 0–40 m.	DOCKUM GROUP	red shaly ss. 0–30 m.
	CHUGWATER FORMATION	red beds 0–150 m.	LYKINS FORMATION	red-brown siltstone and shale 0–95 m.	LYKINS FM.	red sh. 0–70 m.
PERMIAN	GOOSE EGG FORMATION	red shaly sandstones, evaporites and thin limestones 150–250 m.	LYKINS FORMATION: TALOGA FM.		LYKINS FM.: TALOGA FM.	red beds and evaporites 50–120 m.
			FORRELL LS. / DAY CREEK DOL.	sandy ls., dol. 2–20 m.	DAY CREEK DOL.	
			WHITEHORSE FM.	red silty shales, limestones, evaporites 20–75 m.	WHITEHORSE FM.	
			DOG CR. SH.		DOG CR. SH.	shaly ss. and evaporites 70–100 m.
			MINNEKAHTA LS.-BLAINE FM.		LYONS SS.: BLAINE FM.	ss. 30–60 m.
			LYONS SS.-NIPPEWALLA GP.	ss., shaly ss., evaps. 6–40 m.	NIPPEWALLA GP.	
			OWL CANYON FM.-SUMNER GP.	red sls.,ss., evaps. 30–75 m.	FOUNTAIN FM: SUMNER GP.	sandy shale and evaporites 220–270 m.
			INGLESIDE LS.-CHASE GP.	ls.,ss.,evaps. 30–100 m.	CHASE GP.	
			COUNCIL GROVE AND ADMIRE GPS.	limestone, sandstones, evaporites 45–135 m.	COUNCIL GROVE AND ADMIRE GPS.	
PENNSYLVANIAN	HARTVILLE FORMATION	red to gray shales, sandstones carbonates and evaporites 280–400 m.	FOUNTAIN FM: WABAUNSEE GP.	Fountain sandstone, siltstone and conglomerate 270 m. on west changes eastward to limestone, evaporites and sandstone 200–300 m.	FOUNTAIN FM: WABAUNSEE GP.	arkosic conglomerate 950–1350 m.; limestones and shaly sandstones 400–600 m.
			SHAWNEE GP.		SHAWNEE GP.	
			DOUGLAS GP.		DOUGLAS GP.	
			LANSING GP.		LANSING GP.	
			KANSAS CITY GP.		KANSAS CITY GP.	
			PLEASANTON GP.		PLEASANTON GP.	
			MARMATON GP.		MARMATON GP.	
			CHEROKEE GP.		CHEROKEE GP.	
			ATOKA GP.		ATOKA GP.	
			MORROW GP.		MORROW GP.	
MISSISSIPPIAN	GUERNSEY FORMATION	limestone 0–45 m.	MADISON FORMATION	limestone 0–30 m.	BEULAH LS. / HARDSCRABBLE LS.	indicated limestones = 60 m. on west side of basin; change eastward to include all Mississippian 100 m.
					WILLIAMS CANYON LS.	
DEVONIAN						
SILURIAN						
ORDOVICIAN					SIMPSON GP.	sandstone 0–14 m.
					MANITOU LS. / ARBUCKLE GP.	limestone and dolomite 20–85 m.
CAMBRIAN UPPER					PEERLESS (BONNETERRE) FM.	dol., ss., sh. 9–20 m.
			REAGAN (LAMOTTE) SS.	ss. and dol. 0–25 m.	SAWATCH (REAGAN) SS.	sandstone 18–25 m.
CAMBRIAN LOWER AND MIDDLE						
PRECAMBRIAN ERA	PRECAMBRIAN BASEMENT CRYSTALLINE ROCKS					

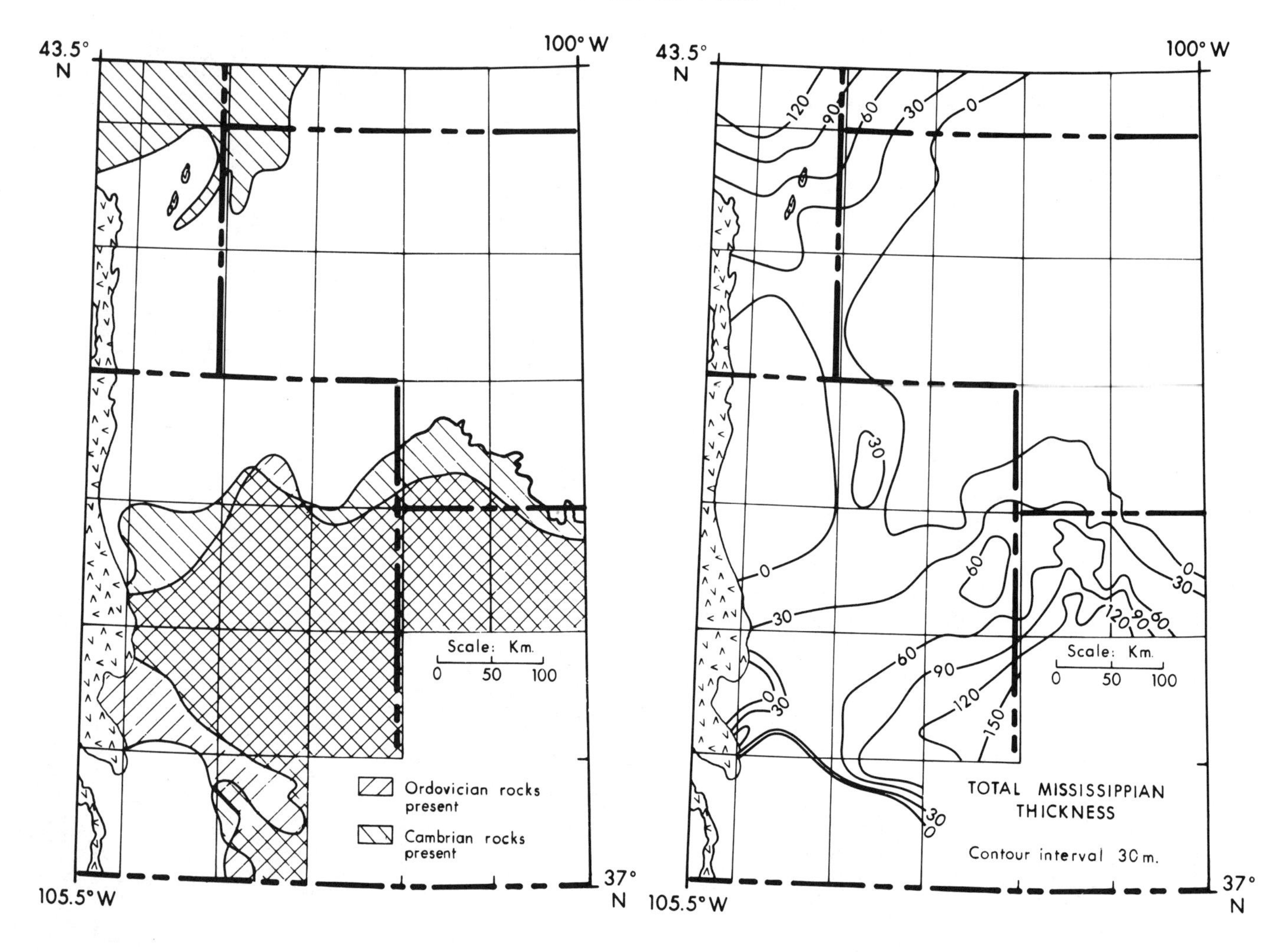

Figure 48. Present distribution of lower Paleozoic rocks. Modified from Ross and Tweto, 1980; Sonnenberg and Weimer, 1981; Anderman and Ackman, 1963.

Figure 49. Isopach map of Mississippian rocks. Modified from DeVoto, 1980a; Sonnenberg and Weimer, 1981; Maughan, 1963; Craig, 1972.

some geologists consider these rocks to be Kimmeridgian in age and correlative with the lower part of the Morrison Formation west of the basin. If this latter interpretation is accepted, the upper Sundance equivalents extend southward in the Denver Basin only to about the latitude of the Colorado-Wyoming boundary. Overlying conglomeratic sandstones, varicolored shales, and a few limestones, constituting most or all of the Morrison Formation, are present over the entire basin as the uppermost Jurassic beds.

Above the often obscure unconformity at the top of the Morrison Formation lies a sequence of alternating sandstones and shales that may be physically correlated but are not everywhere contemporaneous in age. The sandstones of this Lower Cretaceous and lowermost Upper Cretaceous section, called the Dakota Group, are petroleum productive in the central Denver Basin; consequently, the subsurface stratigraphic details of the Dakota rocks have become well known (Fig. 47).

The superjacent sequence of the Graneros, Greenhorn, Carlile, and Niobrara Formations (Fig. 54) consists of dark shales, calcareous shales, and limestones, except for the thin and discontinuous Codell Sandstone Member of the Carlile, which is bounded by surfaces of unconformity. Weimer and Sonnenberg (1983) distinguish one body of Codell that lies in the southeastern Denver Basin from another, much larger, correlative mass in the northwest. The two are separated by a lacuna belt about 70 to 80 km wide trending northeasterly across the basin from the southern Front Range through the northeast corner of Colorado. Petroleum production is being developed in the south part of the northern sandstone body.

The thick body of gray shales called the Pierre is notably less calcareous than shales lower in the Cretaceous section. Tongues of generally fine grained and clayey sandstones that play out eastward within the Pierre Shale have aroused some interest as objectives for petroleum exploration (Fig. 47).

Upwardly the Pierre Shale grades into the Fox Hills Sand-

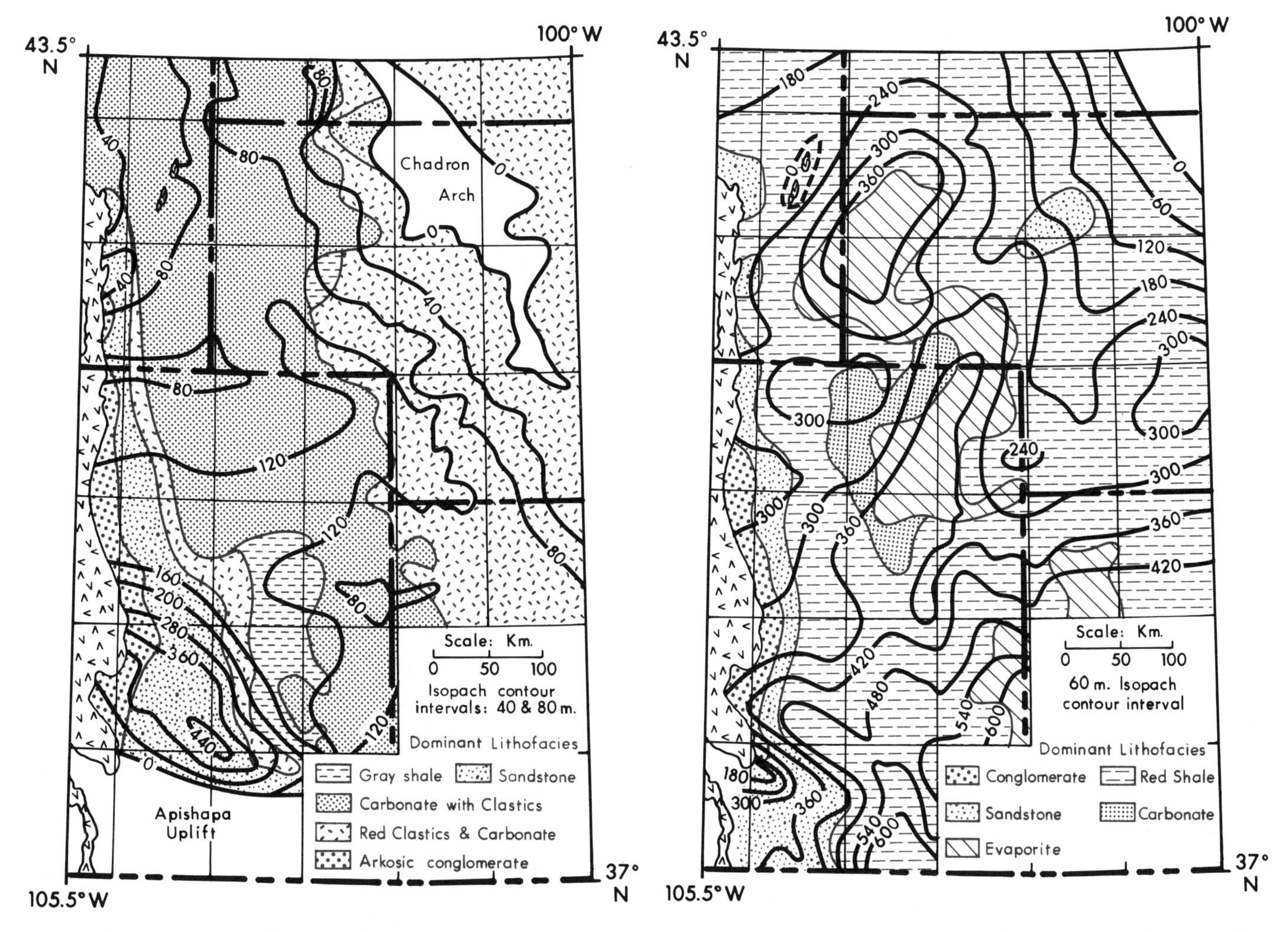

Figure 50. Lithofacies and isopach map of Des Moines age beds. Comprises middle portions of Hartville and Fountain Formations or Cherokee and Marmaton Groups of Kansas. Modified from Mallory, 1972a; Martin, 1965.

Figure 51. Isopach and dominant lithofacies map of Permian age rocks. Adapted from Martin, 1965; Rascoe and Baars, 1972.

stone, which contains zones of thinly interbedded sandstone, siltstone, and shale and also some fine grained massive sandstone beds. The Fox Hills in turn grades upward into coal-bearing sandstones and shales of the Laramie Formation or the equivalent Lance Formation of the northern Denver Basin. As a result of early Tertiary erosion of their former edges, the Fox Hills and Laramie beds are present only over an area lying within the central part of the basin (Fig. 55). The constriction in the outline of this area of occurrence coincides, at least approximately, with the southern border of the Transcontinental Arch and also with the location where the important Cache la Poudre and South Platte Rivers flow easterly across these formations.

Terrestrially deposited strata containing the Cretaceous-Tertiary boundary lie above the Laramie beds in the southern Denver Basin (Fig. 55). On the west these rocks of the Arapahoe and Dawson Formations are mostly coarse conglomerates and sandstones, but they change eastward to include a great deal of sandy siltstone, claystone, and lignitic material. All of them bear a complex facies relationship to strata called the Denver Formation in the northern part of their area of occurrence (Bryant and others, 1981). Undoubtedly these formations once extended some distance east, north, and south beyond their present erosionally determined edges, but whether that projection might be of the order of tens or of hundreds of kilometers is not known. Recent dating information indicates that the Arapahoe and Dawson beds record more or less continuous deposition in the Denver Basin from latest Cretaceous into Eocene time—corresponding to the span of Laramide Orogeny. Consequently, local geologic history is most readily understood by adapting Sloss's (1963) original sequence definitions to include beds of late Paleocene and early Eocene ages within the Zuni sequence of the Denver Basin. Then the widespread late Eocene erosion surface, or unconformity, may be regarded as the boundary between the Zuni and Tejas sequences in this region (Fig. 47) (Tweto, 1975).

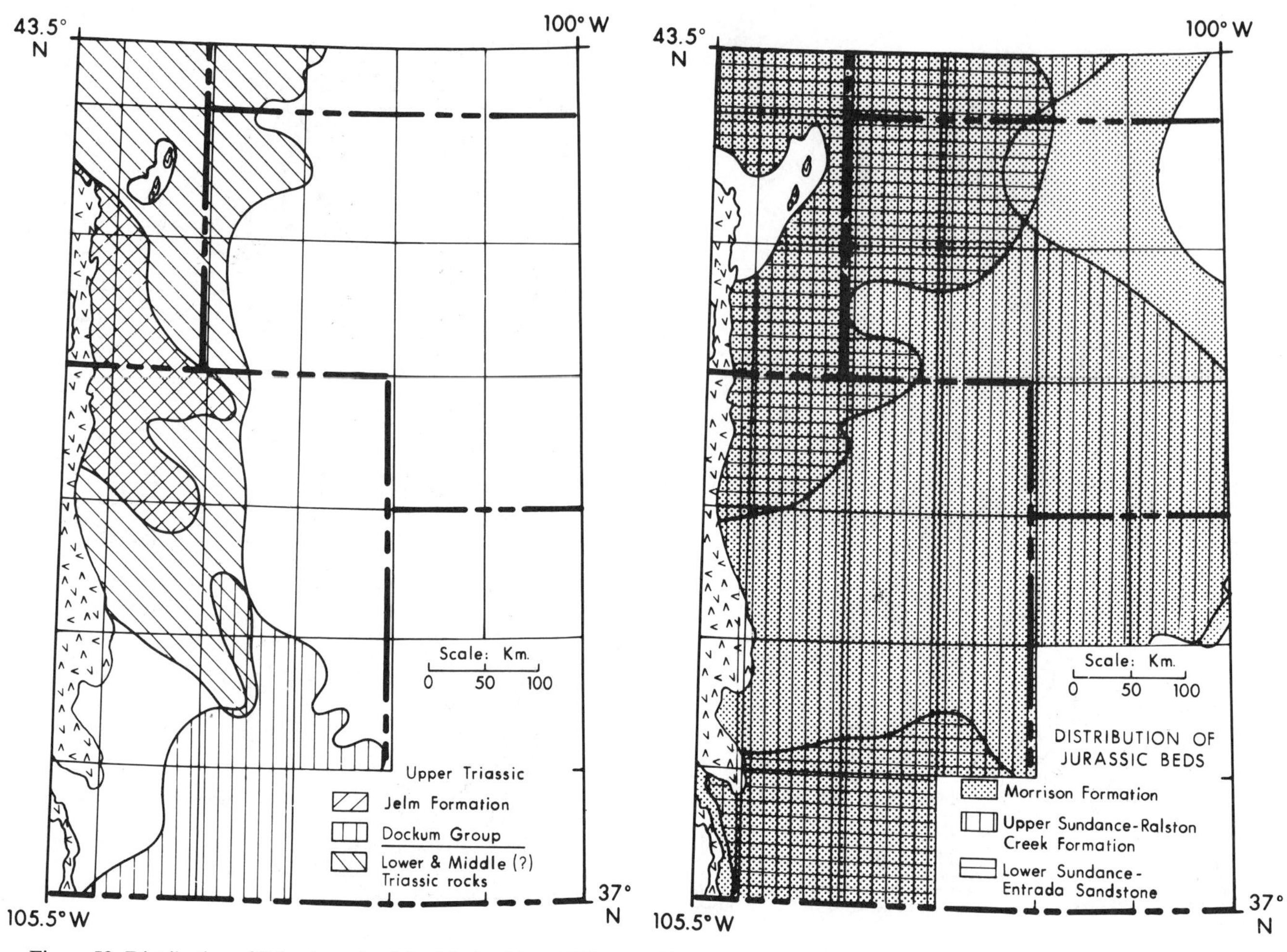

Figure 52. Distribution of Triassic rocks. After MacLachlan, 1972.

Figure 53. Areas of occurrence of Jurassic sedimentary rocks. Modified from Curtis, 1963; Martin, 1965; Peterson, 1972.

Post-Eocene Rocks (Tejas Sequence)

Deposits of the Castle Rock Conglomerate cap buttes and mesas scattered through the areas indicated for Oligocene beds in the southwestern Denver Basin (Fig. 55). The vitric Wall Mountain Tuff lies at the base of the Castle Rock in its westerly outcrops, and clasts of the welded tuff are included in the conglomerates above it. Oligocene deposits widely present through the central-eastern and northern Denver Basin consist mainly of gray and brown siltstones with some channeled coarse sandstones. Within these beds of the White River Group, lower (Chadron) and upper (Brule) formations can be distinguished in many parts of the Denver Basin.

In roughly the northern half of the Denver Basin, moderately consolidated gray to brown conglomeratic sandstones, siltstones, and claystones of the Arikaree Formation unconformably overlie the White River beds. The Arikaree is now thought to contain some Oligocene deposits in its lowest part, but most of it is considered Miocene in age, as are the overlying, very widespread beds of the Ogallala Formation. The variably consolidated, cross-bedded and channeled conglomerates, sandstones, and siltstones of the Ogallala generally become finer grained eastward across the basin, and in places they include some lenticular limestones, probably of algal origin. Large clasts of volcanic rocks are contained in the western outcroppings of the Ogallala, and ash beds are everywhere common within the formation.

Quaternary deposits are extensively represented in the Denver Basin as alluvial material bordering modern and ancient stream courses and as surficial deposits of eolian sands, silts, and loess. An especially large Holocene dune field called the Sand Hills lies in northwestern Nebraska (Fig. 58).

EVOLUTION OF THE SEDIMENTARY COVER

Sauk, Tippecanoe, and Kaskaskia Sequences (Early Paleozoic)

The carbonate and mature clastic rocks of the Sauk, Tippecanoe, and Kaskaskia sequences in the Denver Basin

(Figs. 47–49) lie on what appears to be a surface of very low relief cut on the Precambrian crystalline basement. The nature of these beds bespeaks deposition in a warm climate on a passive cratonic shelf where slight changes in relative sea levels favored, variously: slow accumulation of carbonates, deposition of clastics carried from distant and from local low insular sources, or exposure to periods of nondeposition and erosion. Silurian rocks doubtless were once present, as evidenced by blocks found nearby in diatremes at about 41°N, 105½°W. The Transcontinental Arch trend probably was slightly but recurrently elevated above nearby areas, as suggested by the thinning of Mississippian rocks there and the absence of any earlier Paleozoic beds (Fig. 48 and 49). Late in Mississippian time, low sea levels and slight elevation of the shelf resulted in erosion over the entire area of the present Denver Basin. Where Mississippian carbonates are present, their contact with overlying Pennsylvanian beds is a karstic surface of unconformity.

Absaroka Sequence (Pennsylvanian, Permian, and Triassic)

Beginning early in Pennsylvanian time, the tectonic and sedimentation patterns of the Denver Basin area changed drastically. Fault block mountains were raised in the approximate locations of the Apishapa, Wet Mountains, and southern Front Range Uplifts (Fig. 47). These highlands extended northwesterly, with diminished elevation, into central Wyoming. The site of the Chadron-Cambridge Arch was upwarped early in Pennsylvanian time and stripped to basement over part of its crest (Fig. 50). Much more subtle tectonism along the present Las Animas Arch trend is reflected in the Absaroka sequence sedimentary patterns (Rascoe, 1978).

This new regional framework may well have developed from an exterior continental collision, as pictured by Kluth and Coney (1981); whatever the ultimate cause, it is clear that many of the major structural elements defining the present Denver Basin had their inceptions in Pennsylvanian time. The actively subsiding basin site received sediments mainly from the strongly elevated lands. Therefore the resulting deposits generally are thickest within about 200 km of their western and southwestern source areas; they gradually thin eastward onto the craton, and change from alluvial fan deposits on the west to sediments of coastal plains and shallow fluctuating seas farther east.

A similar, but evolving, depositional pattern is apparent in the later Absaroka sequence beds. As uplift of the western and southwestern highlands diminished and finally ended, only a few areas of low residual hills remained. The basin east of them gradually was filled with increasingly fine grained sediments, as the seas hesitatingly withdrew and the climate became quite dry. All this resulted in evaporite deposits of restricted marine basins (Fig. 51), in reddish siltstones that accumulated on broad mud flats (Figs. 51 and 52), and finally in the coarser alluvial sediments that lie at the top of the Absaroka sequence beds (Fig. 52).

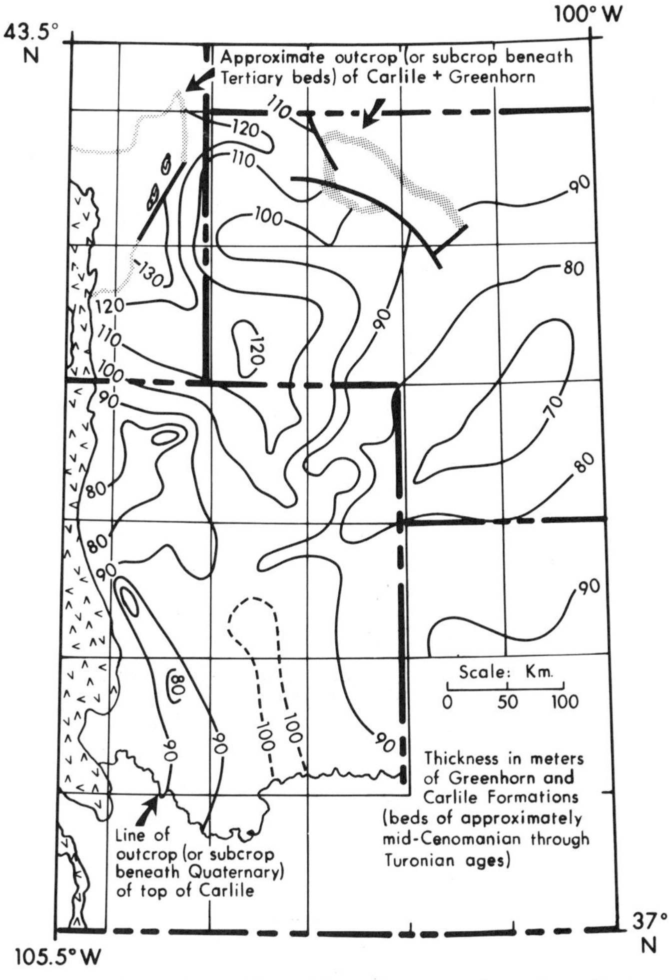

Figure 54. Isopach map of two Upper Cretaceous formations. Contour interval 10 m. Compiled from subsurface data.

Zuni Sequence (Jurassic, Cretaceous, and Early Tertiary)

The cratonic shelf at the site of the Denver Basin was emergent during Early Jurassic time when an erosional surface formed on Triassic and Permian rocks (Fig. 47). The directly overlying Zuni sequence beds indicate that marine transgression from the northwest barely reached the northernmost part of the present basin, where "lower" Sundance nearshore sandstones change southeastward into shore and dune deposits. A lobe of terrestrially deposited sandstone accumulated concomitantly near the southern margin of the basin site (Fig. 53). Broader transgression from the north and west is recorded by the "upper" Sundance (Ralston Creek) beds, which contain restricted marine and tidal flat deposits over most of the Denver Basin area. Late in Jurassic time the site of the Denver Basin was part of an extensive regional alluvial plain where the Morrison Formation accumulated (Fig. 53).

The Dakota Group and younger Zuni sequence beds record the development and later decline of a huge troughlike foreland

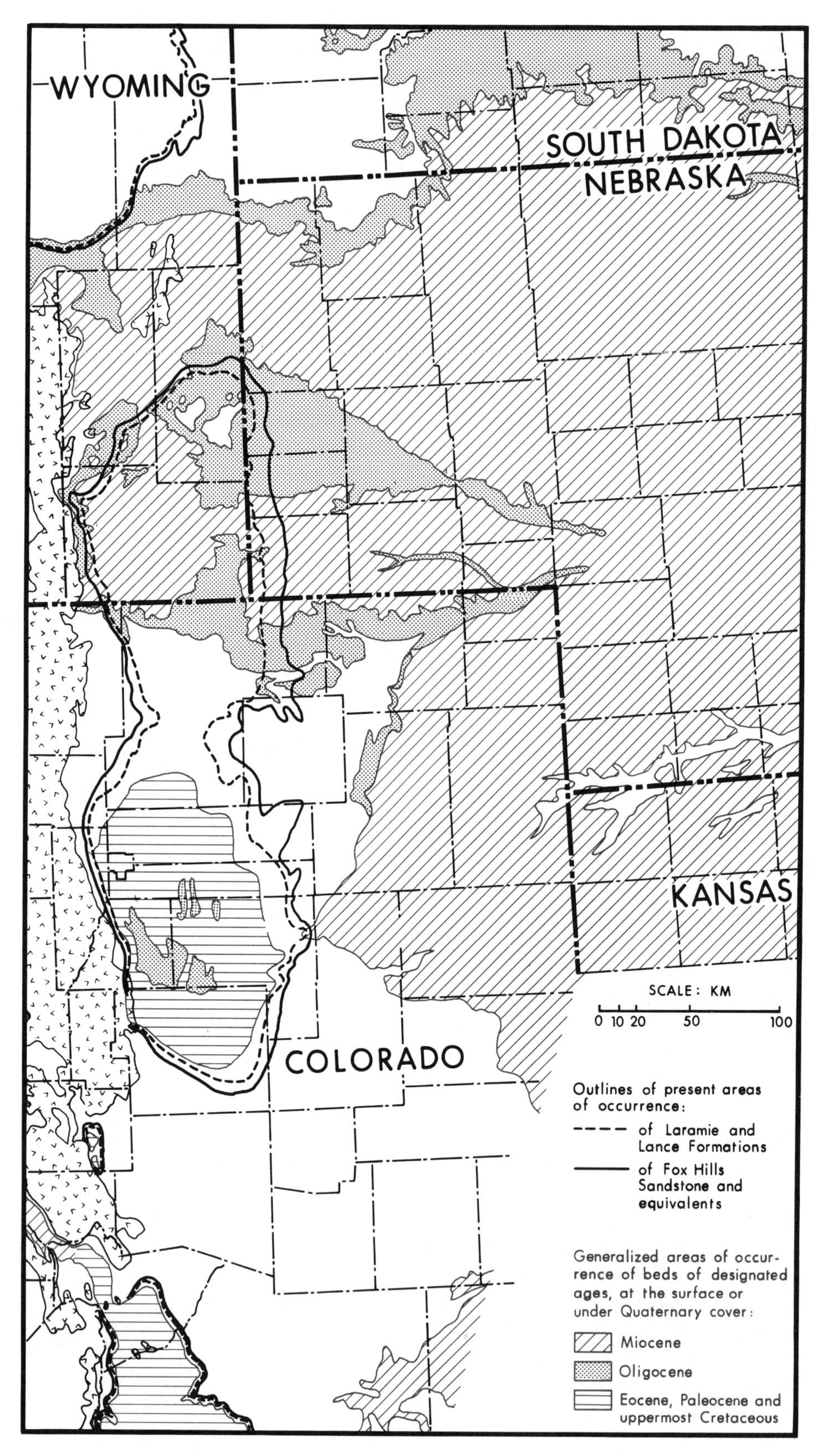

Figure 55. Late Cretaceous and Tertiary deposits. Compiled from state geologic maps of the U.S. Geological Survey and subsurface records.

depression, called the Western Interior Basin, within which lay the site of the Denver Basin. At its maximum, the vast downwarp extended across the craton from the Gulf of Mexico to the arctic regions; it received large volumes of sediment from rising Cordilleran mountains to the west, lesser amounts from cratonic lands to the east, and minor supplies from small internal uplifts. The lowermost sandstones of the Dakota Group are mainly fluvial deposits of an extensive stream system whose main river flowed northward within the nascent Western Interior Basin and was fed, in most of the Denver Basin area, by northeastward-flowing tributaries (McGookey and others, 1972). Succeeding beds record general marine advances from the north corresponding to the Skull Creek and Huntsman Shales and marine regressions correlative with the "J" and "D" sandstones. These oil-productive sandstones have been interpreted as coastal plain, deltaic, and nearshore deposits of variable age that are partly or entirely the lateral equivalents of marine shales elsewhere (MacKenzie and Poole, 1962; McGookey and others, 1972; Weimer, 1984).

Inundation of the Western Interior Trough following Dakota Group deposition brought the Denver Basin site under marine water that had fairly high organic productivity and depth probably less than 200 m. The seas received limited fine clastic material from distant terrigenous sources, and at times the supply of such debris dwindled sufficiently that carbonate became an important to dominant depositional constituent. Although the resulting sedimentary beds, extending upward through the Niobrara Formation (Fig. 47), tend regionally to be fairly uniform in thickness (Fig. 54), they give evidence of small differential elevations of underlying basement blocks, of incidents of submarine scour, and possibly of one episode of emergence of the sea floor (Weimer, 1984).

The succeeding mass of Pierre Shale accumulated in about 16 million years, but it constitutes approximately 47 percent of the entire sedimentary section in the central Denver Basin. Tongues of sandstone, typical of shelf deposits, project eastward into the beds of marine silty shale (Fig. 47).

A withdrawal of sea waters is recorded in the general upward progression to deposits of nearshore, shoreline, deltaic, coastal plain, paludal, and fluvial environments displayed in the Fox Hills, Laramie-Lance, Arapahoe, and Dawson beds. In explaining this emergence, the relative importance of falling sea level and of basin filling attendant upon diminishing subsidence remains a subject of debate. Nevertheless, the paludal and alluvial (coastal?) plain character of beds remaining in the Denver Basin is not inconsistent with the speculation that a narrow (Cannonball) seaway may have covered the eastern part of the basin as late as Paleocene time, while fluvial deposits accumulated in the western part (Figs. 47 and 55).

During latest Cretaceous through early Eocene time, the Wet Mountains and southern Front Range obviously were persistently uplifted along their ancient Pennsylvanian trend, but dating of Laramide events in the area of the northern Denver Basin is less certain, as no beds of those ages are known there (Figs. 47 and 55). Incomplete evidence suggests, however, that the northern Front Range–Laramie Range, the Hartville Arch, and the Black Hills are late Laramide features without Pennsylvanian ancestry and with little or no structural expression or relief until their principal tectonic elevation in late Paleocene and early Eocene time. In fact, it seems probable that all of the area of the Denver Basin lying north of the southeastern margin of the Transcontinental Arch was regionally uplifted at approximately this time, while the southern part of the basin was downwarped as the upper Dawson deposits accumulated. It is further inferred, from very sketchy evidence, that the features forming the eastern and southern boundaries of the Denver Basin also developed as late Laramide elements. The Chadron-Cambridge Arch and the Apishapa Uplift were rejuvenated along inherited Pennsylvanian trends, and the relative downwarp of the southern Denver Basin brought expression to the Las Animas Arch through a mild reversal of regional dips.

Tejas Sequence (Later Tertiary and Quaternary as defined here)

During a late Eocene period of tectonic dormancy, a very extensive east-sloping, fluvially smoothed surface developed over the region of the southern Denver Basin. This primarily erosional surface crosses uplifted rocks ranging in age from Precambrian on the basin edges to early Tertiary within the basin (Scott, 1982). Evidence of possible extension of such a plain into and adjoining the northern Denver Basin is ambiguous, but at least it is reasonable to suppose that the widespread White River Group was deposited upon a northward continuation of this same late Eocene surface (Figs. 47 and 55).

A shift from the late Eocene erosional regime to one of fluvial deposition in Oligocene time probably was caused mainly by change to a dry climate; however, increased volcanism and minor uplift in the sediment source region west of the Denver Basin doubtless also played a role (Clark, 1975; Scott, 1975). The White River beds, which cover a very large area with their nearly flat-lying fluvial sediments, have been interpreted as deposits of eastward flooding, braiding streams carrying material from both distant and nearby basement uplifts and from volcanic centers (Fig. 55). Some of the alluvial deposits were aggraded to lap on to adjoining hilly source areas, such as the Front and Laramie Ranges, Hartville Arch, and Black Hills, where stream courses crossed them.

Continuation of these conditions is reflected in the succeeding Arikaree and Ogallala fluvial deposits. These beds cover an even larger area than the White River Group and commonly contain even more volcanic sediment (Figs. 47 and 55).

Fluvial deposition ceased in Pliocene time when uplift of the entire region initiated the present condition of general erosion. In the northern Denver Basin the Tertiary deposits were domed by the Goshen uplift early in the Pliocene (Fig. 55), and this was followed by normal faulting nearby in late Pliocene time when the Hartville Arch was displaced upward at least 200 m with respect to the basin southeast of it (Fig. 45). Some incomplete evidence suggests that the Laramie Range may have been slightly lifted and tilted eastward in Pliocene time, causing superposed incision of some streams, and beheading of others, that had carried Miocene sediments across the site of the range. Uplift and block faulting of the Wet Mountains and Front Range appear to have begun late in Miocene time and increased greatly in the Pliocene, when these ranges were differentially elevated 500 to 1,200 m with respect to the Denver Basin. Stripping of the Miocene and most of the Oligocene deposits from the southwestern part of the basin followed.

Diminished uplift and cooler, moister climatic conditions in Quaternary time brought about complex changes in the drainage within the Denver Basin. Fanlike alluvial deposits left near the mountain borders and floodplain gravels deposited along stream courses later were incised, and new deposits of similar nature were left at progressively lower levels, while eolian loess and sand deposits accumulated over sizable areas in the eastern part of the basin. These youngest portions of the sedimentary cover are important as aquifers and as sand and gravel resources.

MINERAL RESOURCES IN THE SEDIMENTARY ROCKS OF THE DENVER BASIN

Petroleum

Oil was first discovered in the Denver Basin in 1862 at the Florence Field in the Cañon City embayment. Some 2 million metric tons have since been produced from that field's fractured reservoir in the Pierre Shale (Figs. 47 and 56). The Boulder Field, discovered in 1901, was the first of several productive areas found by surface mapping and later geophysical work on anticlines bordering the mountain front in the west central part of the basin (Fig. 56). Most of the modest production of these fields has come from sandstones in the Cretaceous beds and in lesser quantities from eolian Lyons Sandstone in the Permian section. Small anticlines lying farther east near the basin axis also produce oil from the Lyons (Fig. 56).

By far the most important oil and gas production has come from the "D" and "J" sandstones of the Dakota Group on the eastern flank of the Denver Basin. Following a 1949 discovery in Cheyenne County, Nebraska, hundreds of oil and gas accumulations were found, chiefly within the area designated "D" and "J" fairway on Figure 56. They are attributable primarily to stratigraphic variations in the "D" and "J" sandstones and secondarily to irregular structural features of low relief. Exploration in the fairway has increased recently following discoveries of oil in the Pennsylvanian and Permian beds in western Nebraska.

West of the fairway, on the trend of the Transcontinental arch and near the axis of the Denver Basin, the Wattenberg and associated fields produce gas from "J" sandstone of low permeability as well as oil and gas from the Codell Sandstone, Niobrara Formation, and the sandstones within the Pierre Shale. Many of the reservoirs are partly or wholly dependent upon secondary or fracture permeability for fluid production.

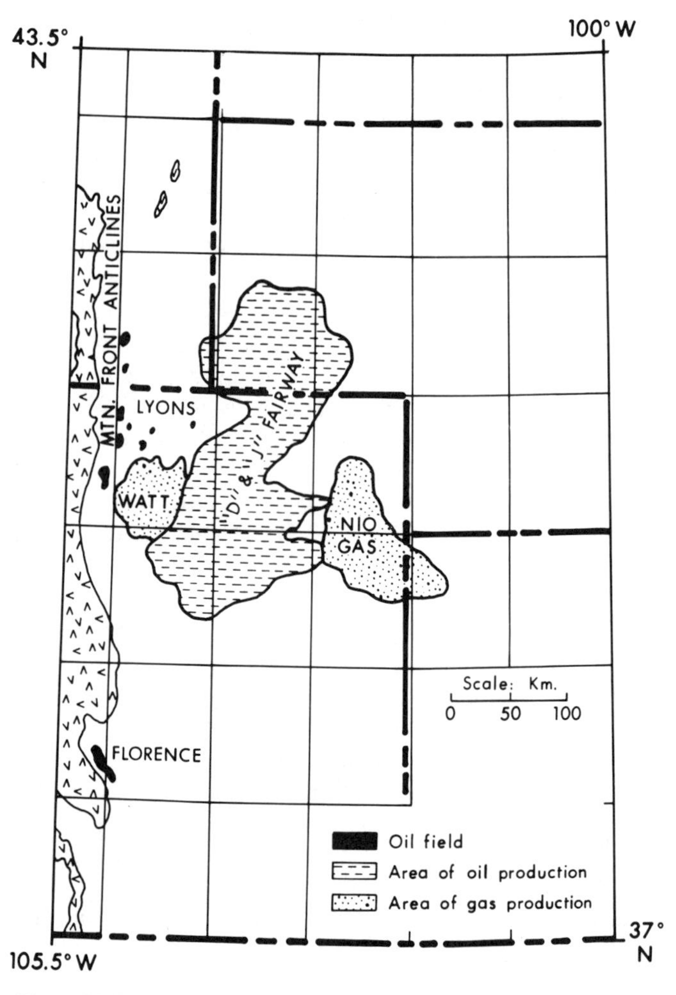

Figure 56. Petroleum resources of the Denver Basin. NIO GAS = area of gas fields in Niobrara Chalk. WATT = area of Wattenberg and associated fields.

An area of gas production from Niobrara chalk of low permeability but high porosity was discovered on the eastern side of the basin in 1919 (Fig. 50). However, the shallow, structurally controlled accumulations were not outlined and developed until suitable economic conditions evolved in the period from 1975 to 1983.

A cluster of fields that produce oil from Paleozoic beds lies on the easternmost edge of the Denver Basin, where the Las Animas and Cambridge Arches merge. In this area, not shown on Figure 56, the principal reservoir rocks are derived from ancient carbonate banks within the Pennsylvanian Lansing and Kansas City Groups (Fig. 47).

Uranium

Roll-front–type uranium deposits that are currently noncommercial have been found in sandstones of the Fox Hills and

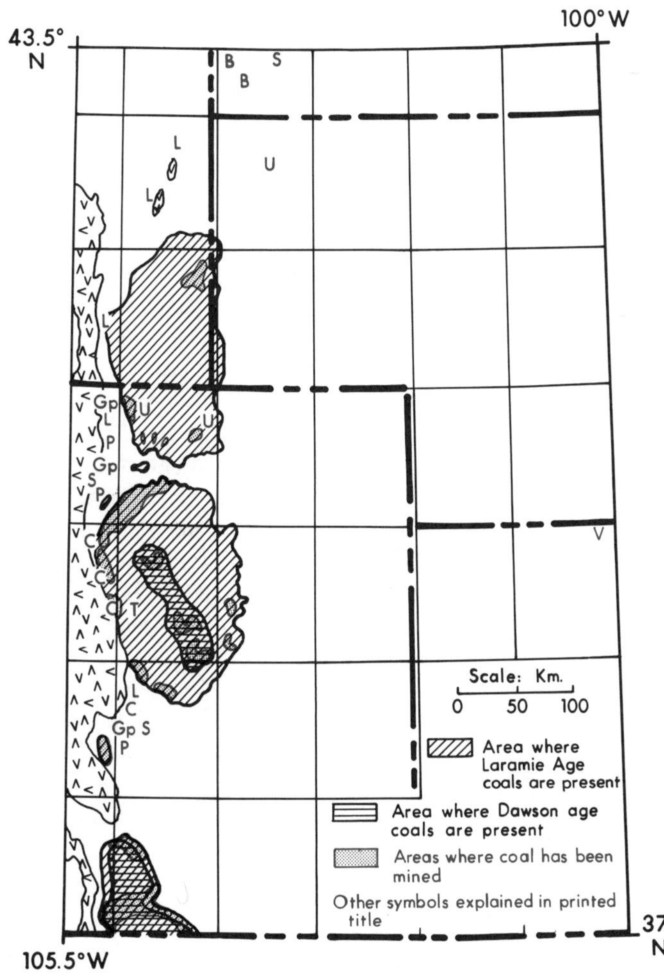

Figure 57. Mineral resources in the sedimentary rocks of the Denver Basin. Symbols in red: B = coals in Lakota of Black Hills; Gp = gypsum deposits; P = Portland cement sites; L = lime sources; C = high quality clays; S = sandstone quarries; T = tuffstone quarry; V = volcanic ash; U = uranium mining. Compiled from G. B. Glass, personal communication, 1985; R. Bretz, personal communication, 1985; R. Burchett, personal communication, 1985; Hansen and Crosby, 1982; del Rio, 1960; Soister, 1978; and other widely scattered sources.

Lance Formations close to where these beds pass beneath the eroded edge of Oligocene strata in northern Colorado (Figs. 55 and 57). Ground waters percolating eastward in the White River and underlying Cretaceous beds are believed to have carried the uranium to the depositional sites. Similar uranium deposits in channel sandstones of the basal White River Group in northwestern Nebraska are under consideration for exploitation (Fig. 57).

Coal

Although there is virtually no coal mining at present within the Denver Basin, a very substantial reserve remains. Subbituminous B and C coals in the Laramie-Lance formations constitute the major resource that has been mined at the locations shown in Figure 57. The most important of these localities are the Foothills-Boulder-Weld Field on the northwest and the Colorado Springs Field on the southwest, of the large southerly area of occurrence. Past production from these two fields has been about 107 million and 14 million metric tons respectively. Additionally, in the small Cañon City structural embayment (the southwesternmost corner of the Denver Basin), 40 million metric tons of high volatile bituminous C coal has been mined from the Vermejo Formation—an approximate stratigraphic equivalent of the Laramie.

Little is known about the Lance Formation coals that lie in the northern large area of Laramie-age beds shown on Figure 57. These seams, presumably of subbituminous C rank, are thought to be about a third thinner and more discontinuous than those farther south, as indicated by sparse data derived from sporadic mining along the southern edge of this north area, from occasional noncommercial mining in the Goshen Hole area of Wyoming, and from well records.

Another large resource of uncertain economic potential is that in the lower Dawson and equivalent beds within the area shown on Figure 57. In the past, lignite beds within a zone as much as 150 m thick were mined from the Dawson-Denver on a very small scale. Recently Soister (1978) estimated that more than 9 billion metric tons of the lignite could be exploited by underground gasification techniques.

Other Materials

The sedimentary beds of the Denver Basin contain resources of gypsum (in the Lykins Formation), Portland cement rock (in the Niobrara Formation), limestone (in the Manitou, Madison-Guernsey, Hartville, Ingleside, Forelle, and Fort Hays strata), clay and shale (principally in the Dakota, Laramie, Graneros, and Pierre Formations), building stone (mainly in the Lyons, Dakota, Sawatch, and Fountain Formations and the Wall Mountain Tuff), and volcanic ash (in the Ogallala Formation and Brule and Arikaree strata). Figure 57 shows the most important quarry sites for these materials. Denver Basin sedimentary deposits also provide many local sources of sand and gravel and of crushed rock.

Groundwater

Aquifers of great importance to the economy of the Denver Basin are contained in the Tertiary and Quaternary beds of the High Plains and in the Laramie, Fox Hills, Arapahoe, Denver, and Dawson Formations in the central part of the basin (Romero, 1976) (Fig. 58). Other formations serve as more local groundwater sources on the basin margins, a prominent example being the Cloverly Group sandstones on the southern flank of the Black Hills.

ACKNOWLEDGMENTS

John P. Lockridge did initial work on this section, provided information for the discussions of petroleum and uranium, ar-

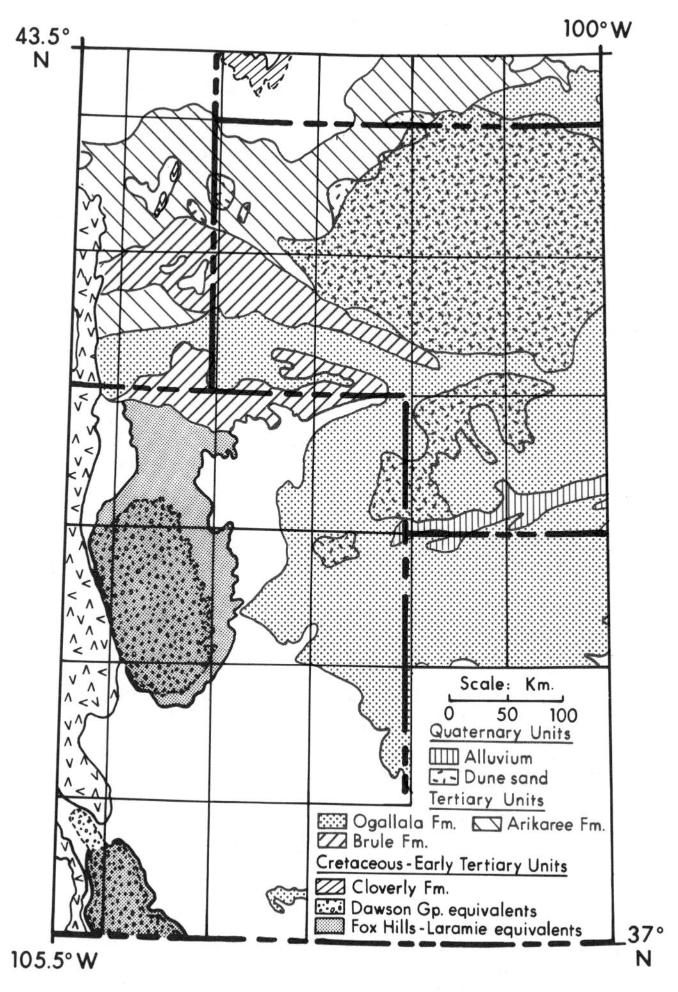

Figure 58. Stratigraphic components of the principal bedrock aquifers of the Denver Basin area. Adapted from: Gutentag and others, 1984; Robson, 1983; Repplier and others, 1980; R. Bretz, personal communication, 1985.

ranged the drafting, and maintained a lively interest in the progress of the study. Many other colleagues generously helped the author, especially Richard Bretz (South Dakota Geological Survey), Ray Burchett and Harold DeGraw (Nebraska Conservation and Survey Division), Donald L. Blackstone, Jr. and Roy A. Johnson (University of Wyoming), Gary B. Glass (Wyoming Geological Survey), and J. D. Love (U.S. Geological Survey).

GEOLOGY OF THE WIND RIVER BASIN, CENTRAL WYOMING

J. D. Love

The Wind River Basin of central Wyoming is a major downwarp shaped like a parallelogram (Fig. 59), completely surrounded by mountains or anticlinal uplifts. It has been the major source of uranium ore in Wyoming and also contains 113 oil and gas fields.

BASEMENT ROCKS

Few drill holes have reached Precambrian rocks beneath the Wind River Basin, so interpretation of the kinds and ages of the basement rocks is largely based on projections across the basin from exposures in the Wind River Range on the west, the Owl Creek and Washakie Ranges on the north, the Bighorn Mountains on the northeast, and the Granite Mountains on the south. On the Casper Arch, several drill holes passed through nearly 2,700 m of Precambrian rocks before reaching the overridden Cretaceous section (Gries, 1983a). Classification, correlation, and lateral extent of Precambrian rock types are presented elsewhere (Love and Christiansen, 1985). The oldest rocks are metasedimentary and metavolcanic rocks in the Granite Mountains, dated at more than 3.2 Ga. Other metasedimentary rocks in the Wind River Range are slightly younger but still older than 2.8 Ga; metagabbro in the same range is dated at about 2.6 Ga. Amphibolite in the Owl Creek Mountains is older than 2.7 Ga.

The oldest plutonic rock is quartz monzonite, more than 2.9 Ga in the Bighorn Mountains. Granites in the Wind River, Owl Creek, and Granite Mountains are about 2.6 Ga. All these rocks are Archean, and their general trend is northeast or east, so it seems likely that the entire Wind River Basin is floored by Archean rocks.

SEDIMENTARY ROCKS

Table 3 summarizes the data on Phanerozoic sedimentary rocks in the Wind River Basin. All periods are represented except Silurian. Formation names, descriptions, correlations, and areas of outcrops are given in Love and Christiansen (1980; 1985). The regional geology is described in many papers in the Geologic Atlas of the Rocky Mountain Region (Mallory, 1972c). The nature of these rocks is especially important in this basin because they contain large resources of groundwater, oil and gas, and uranium and lesser amounts of coal.

The outcrop area of Paleocene and lower Eocene rocks that mantle a major part of the basin interior is about 9,600 km^2. The basin is asymmetric: The long axis trends generally westward, and the deepest part is near the northern margin (Fig. 59). Broad outcrops of Mesozoic strata are present on the east and west sides, and narrower outcrops of Paleozoic rocks are along the west and south margins.

TECTONIC OVERVIEW

The tectonic pattern of the Wind River Basin as it is today (Fig. 59) is chiefly the result of folding and thrust-faulting during the Laramide revolution, which in this basin began in latest Cretaceous time and terminated between early and middle Eocene times. This Laramide pattern was superimposed on an older,

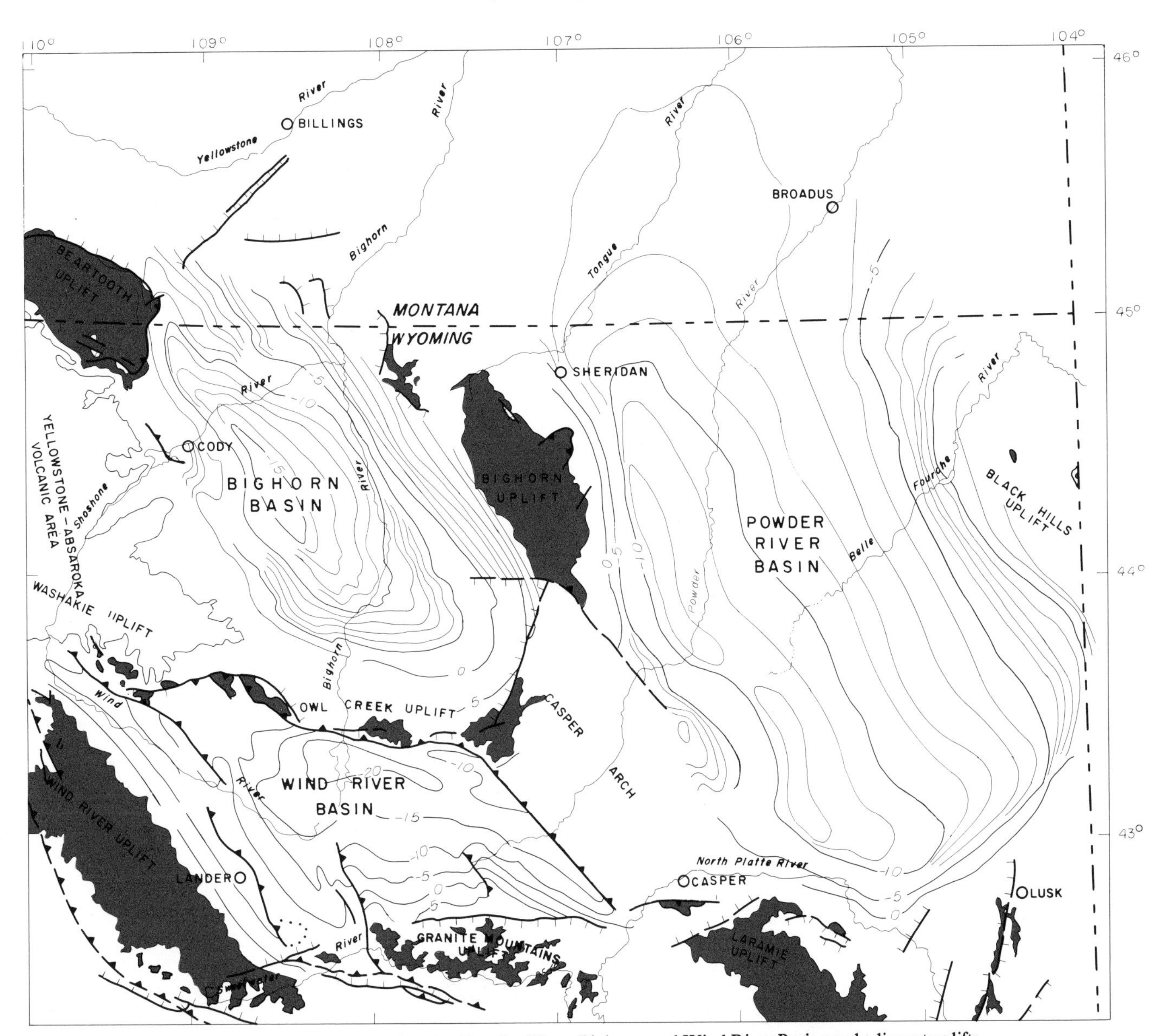

Figure 59. Tectonic map showing Powder River, Bighorn, and Wind River Basins and adjacent uplifts. Precambrian areas on uplifts are patterned. Contours are on the Precambrian surface and at intervals of 5,000 feet. Interbasin areas and uplifts are not contoured. Thrust and reverse faults have sawteeth on upper plate; normal faults have hachures on downdropped plate. Details of structures along the west margins of the Powder River and Bighorn Basins and the north margin of the Wind River Basin are not shown. References to these are given in the individual basin descriptions. Modified from Bayley and Muehlberger, 1968.

broader pattern of broad-scale gentle tilting, warping, and erosion (Love, 1951, p. 1108; Keefer and Van Lieu, 1966), which had occurred in Mesozoic and Paleozoic times. The most intense deformation in and around the basin occurred during earliest Eocene time. The major folding and thrust-faulting at that time was accompanied by emplacement of large detachment masses of Paleozoic and Mesozoic rocks from the Washakie Range in the northwest part of the basin. After middle or late Miocene time there was regional northward tilting, excavation of the Wind River Basin, exhumation of the buried flanking Laramide mountains, isostatic rise of the southern basin margin, downfaulting of the south end of the Bighorn Mountains and eastern part of the Owl Creek Mountains, and northward diversion of Wind River. Cross sections of structures throughout the Wind River Basin have been presented elsewhere (Keefer, 1970; Love, 1978; Ray and Keefer, 1985; Gries, 1983a, 1983b). These structures are so

TABLE 3. PALEOZOIC, MESOZOIC, AND TERTIARY ROCKS WITHIN AND ALONG THE MARGINS OF THE WIND RIVER BASIN
(For formation names, correlations, and areas of outcrops, refer to Love and Christiansen, 1980; 1985).

Geologic Age	Thickness ft (m)	Description
Miocene	0-900 (0-274)	White tuffaceous sandstone, thin white limestone, vitric tuff, gray claystone; conglomerates at top and base.
Oligocene	0-600 (0-183)	Claystone and siltstone, white to pale pink, massive, blocky, tuffaceous. Thin beds of vitric tuff; locally derived conglomerates near top and base.
Late and Middle Eocene	0-2,500 (0-762)	Along east and south margins: claystone, siltstone, and sandstone, green, yellow, and gray, tuffaceous. Along northwest margin: upper part is olive drab to green hard slabby tuff; lower part is highly variegated soft bentonitic tuff and claystone. Some strata are highly seleniferous and gray to brown.
Early Eocene	0-9,000 (0-2,743)	Variegated claystone, siltstone, sandstone, and conglomerate; very coarse conglomerate locally at base and top. These strata are the chief host for uranium in the southeastern part of the basin. In the deepest part of the basin there is 2,500 ft (762 m) of black and gray shale near base. In the northwestern part there are large detachment masses of Paleozoic and Mesozoic rocks near base.
Paleocene	0-8,000 (0-2,438)	Gray and brown, generally noncalcareous claystone, shale, and sandstone with some lenticular conglomerates and a few coal beds. In the northeastern part of the basin, the upper section contains a brown and gray shale sequence 2,500 ft (762 m) thick, possibly marine.
Late Cretaceous	5,600-8,000 (1,707-2,438)	From top down: gray and buff sandstone, gray to black shale, and thin coal beds (thickness 0-6,000 ft; 0-1,829 m); white to brown soft tuffaceous sandstone, thin carbonaceous shale and yellow bentonite, grading eastward into marine shale (thickness 800-1,300 ft; 244-396 m); white to gray sandstone, carbonaceous shale, and coal (thickness 550-2,000 ft; 168-610 m); gray marine shale with thin sandstones (thickness 3,000-4,700 ft; 914-1,433 m); gray marine sandstones and brown to black shale (thickness 500-1,000 ft; 152-305 m).
Early Cretaceous	525-1,250 (160-381)	Marine and brackish-water siliceous shale, bentonite beds, and tan sandstone; nonmarine sandstone and variegated claystone near base; basal cliff-forming conglomerate.
Jurassic	300-850 (91-259)	Uppermost 100-200 ft (30-61 m): nonmarine gray sandstone and variegated claystone; middle 200-400 ft (61-122 m); marine glauconitic green shale and sandstone and red shale; basal 250 ft (76 m); red shale, gypsum, and anhydrite wedging out eastward in southeastern half of basin.
Jurassic(?) and Triassic	0-600 (0-183)	Red cross-bedded sandstone (Nugget Sandstone) wedging out eastward; absent in eastern half of basin.
Triassic	800-1,450 (244-442)	Red shale, siltstone, and silty sandstone; widespread thin gray limestone (Alcova Limestone) in upper part, associated with thick red sandstones; olive drab dolomitic siltstone at base in western half of basin, intertonguing eastward with red shale, dolomite, limestone, and anhydrite.
Permian	150-350 (46-107)	Tan cherty dolomite, quartzitic sandstone, green and black shale, and phosphorite beds in western half of basin, intertonguing eastward with red shale, purple limestone and dolomite, and anhydrite.
Pennsylvanian	400-700 (122-213)	Upper part chiefly white to tan sandstone; lower part interbedded sandstone, cherty dolomite, limestone, and red and green shale and claystone.
Mississippian	300-900 (91-274)	Blue-gray limestone and dolomite, coarsely crystalline in part, cavernous, cherty; some red shale and widespread sandstone near top.
Devonian	0-250 (0-76)	Upper part: yellow, gray, and pink dolomitic siltstone, shale, and dolomite; lower part: fetid brown hard dolomite and tan limestone; present only in western one-third of basin and wedges out eastward.
Silurian	Absent	
Ordovician	0-400 (0-122)	Gray dolomite, hard, massive, siliceous; forms cliffs on outcrops; gray marine sandstone locally at base; wedges out eastward at about middle of basin.
Cambrian	200-1,200 (61-366)	Widespread blue-gray marine limestone, mottled with yellow splotches; flat-pebble limestone conglomerates; green micaceous shale that is glauconitic in part, becoming sandy toward base. Widespread basal dull red sandstone in contact with Precambrian rocks.

varied that no single cross section of the basin is representative of the tectonic pattern as a whole.

The difference in altitude on the Precambrian surface between the deepest part of the basin (7,600 m below sea level) and that on the Owl Creek Mountains (2,400 m above sea level), 13 km to the northeast of the deep, is about 10,000 m. A comparison of the amounts of mountain-basin structural relief involving the Wind River Basin (Keefer and Love, 1963) and other basins and mountains in Wyoming, and the magnitude of Cenozoic sedimentation in these basins, is presented elsewhere (Love, 1960).

GEOLOGIC HISTORY

Only those Phanerozoic events of regional or possible economic significance are discussed here. More detailed descriptions of the geology of the region are presented elsewhere (Mallory, 1972c).

Archean to Middle Cambrian Hiatus

The Middle Cambrian sea advanced eastward across a gently rolling surface, partly on Archean bedrock and partly on regolith that had accumulated during portions of two billion years of missing Earth history in central Wyoming. The regolith was derived from a great variety of Archean rocks, and some of them, especially along the southwestern margin of the Wind River Basin, contain considerable gold. Archean rocks along the south flank of the Owl Creek Mountains contain significant amounts of tungsten, copper, beryllium, and other elements. No detailed studies have been made of the composition of the regolith around or within the basin, with respect to concentrations of economic mineral deposits.

Paleozoic Events

Cambrian was a time of crustal stability in central Wyoming, but after the last Cambrian sea retreated westward, the region was tilted northwestward, and some erosion occurred prior to advance of the Ordovician sea across at least the western half of the basin. In this sea the largely nonclastic Bighorn Dolomite was deposited. The original extent of the dolomite is not known. A tectonic uplift somewhere in the region is indicated by small clasts of fresh gray granite in the dolomite along the western margin of the basin, but their source has not yet been located. These clasts suggest a nearby uplift from which more than 300 m of Cambrian rocks had been eroded (or not deposited), prior to Ordovician deposition.

There is no record of what happened in the Wind River Basin during Silurian time, but prior to Middle Devonian there was regional northwestward tilting and erosion that removed some, or perhaps all, of the Ordovician rocks from the eastern half of the basin. Devonian marine strata are present in only the western one-third of the basin. Their original eastern extent is not known. Central Wyoming probably was again tilted northwestward prior to deposition of Mississippian strata. They were deposited entirely across the Wind River Basin and, in the eastern half, probably incorporated regolith that may have accumulated during all or part of the 75-m.y. time interval between Late Ordovician and Early Mississippian. No studies have been made of the remnants of this regolith with respect to concentrations of economic mineral deposits.

Pennsylvanian strata are present throughout the Wind River Basin. The Pathfinder uplift (Mallory, 1972a), southeast of the basin, was a major source of sand to the Pennsylvanian strata of the basin during the older half of Pennsylvanian time. Younger Pennsylvanian strata were eroded away from the western part of the basin prior to Permian deposition, probably as a result of broad north-south arching of that region (Love, 1954). Permian strata consist of cherty dolomites, shales, and phosphorites in the western part of the basin; they intertongue eastward with red beds, limestones, and anhydrite. West of this facies change, the Permian rocks contain high-sulfur oil and gas, but to the east they are barren.

Mesozoic Events

Marine Lower Triassic rocks, overlain by red beds, were deposited in the western part of the Wind River Basin. Red bed deposition continued from Permian throughout much of Triassic time in the eastern part of the basin. Intertonguing with, and overlying the main part of the Triassic sequence, is a soft, porous red sandstone wedge (Nugget Sandstone) that thickens southwestward to as much as 180 m but is eroded from the eastern part of the basin. The age of these strata is controversial. A regional unconformity is present between them and Middle Jurassic rocks. An impact structure was emplaced between Triassic and Middle Jurassic time near the northeastern margin of the basin (Love and Christiansen, 1985). Middle Jurassic was a time of deposition of anhydrite, gypsum, red beds, and thin dolomite beds throughout most of the Wind River Basin. Evaporite deposition was followed by a marine invasion and then retreat in Late Jurassic time. The last deposit during the Jurassic was a widespread nonmarine sequence of lenticular sandstones and dully variegated claystones (Morrison Formation).

There is no conspicuous regional unconformity between the nonmarine uppermost Jurassic and nonmarine Lower Cretaceous strata. Throughout all but the latest part of Cretaceous time, muddy seas repeatedly advanced westward and then retreated, leaving thick sequences of intertonguing sandstones and shales. Sandstones are thicker and more abundant in the western part of the basin. There is a hiatus of regional extent near the base of Upper Cretaceous rocks (within the Frontier Formation), but no conspicuous unconformities developed until late in Cretaceous time. Then the Granite Mountains along the south margin of the Wind River Basin, the Wind River Range along the west side, and the Washakie Range along the northwest margin began to rise and to be eroded. They provided clastic debris to the basin as it was sinking, and as much as 1800 m of youngest Cretaceous strata accumulated. Marine conditions became increasingly restricted late in Cretaceous time.

Cenozoic Events

The Wind River Basin continued to sink and the bordering mountains to rise during Paleocene and early Eocene time. Fluvial conditions persisted except for early late Paleocene time, when the central part of the basin was flooded by a large lake or an arm of the Cannonball Sea (Keefer, 1965). Nearly 1,800 m of carbon-rich silt and clay was deposited in this body of water. All the adjacent mountain ranges were eventually stripped to their Precambrian cores, and as much as 5,000 m of sediment, mostly clastic, was deposited in the deepest part of the basin. All major thrust faults along the mountain margins and within the basin were developed, chiefly in earliest Eocene time; faulting continued with lesser frequency and intensity through the remainder of early Eocene time. During earliest Eocene time, the Owl Creek and Washakie ranges rose so rapidly that large masses of both Paleozoic and Mesozoic rocks slid or were thrust southward and southwestward into the basin and then were buried by additional lowest Eocene strata. The main drainage flowed eastward out of the basin throughout early Eocene time, and the sandstone strata deposited along it contain shallower groundwater resources throughout much of the basin.

During middle and late Eocene time the Wind River Basin was almost entirely filled with volcaniclastic debris that was blown and washed southeastward from the Yellowstone-Absaroka volcanic area (Fig. 59; Love and others, 1963, fig. 8). Volcanic eruptions began near the end of early Eocene time and terminated about the end of late Eocene time. The entire Wyoming region was tilted westward during this interval, thereby establishing internal drainage and the development of large lakes. In Oligocene time, eastward drainage was reestablished, and volcanic debris from an unknown distant source was deposited along with locally derived clastics. This cycle of deposition continued through at least part of middle Miocene time, and more than 300 m of strata, chiefly tuffaceous sandstones, accumulated. By then, the mountains bordering the Wind River Basin were largely buried (Love, 1970).

Probably after middle Miocene time, regional uplift and northward tilting of the northern half of the Wind River Basin took place. Wind River was diverted northward from the central part of the basin and established a superimposed course across the Owl Creek Mountains. The regional uplift started the reexcavation of the Wind River Basin and exhumation of the adjacent mountains. This has continued to the present time. Many hundred cubic kilometers of sediment were eroded out of the basin, probably mostly in the last ten million years. In places where rates of erosion can be measured, chiefly in the western half of the basin, the rate of basin excavation during the last three million years has been about 0.3 m in 1,600 years.

Because of the vast weight of debris removed from the basin in a relatively short time, there has been conspicuous isostatic rebound, especially along the southern margin of the Wind River Basin where even the youngest Cenozoic strata are tilted southward, away from the basin. Large normal faults developed within the Granite Mountains near the south margin of the Wind River Basin and dropped the core of that uplift more than one thousand meters. Another normal fault system along the interface between the basin and the Bighorn and Owl Creek mountains dropped these mountain cores, at least locally, about 600 m (Love, 1978). The relation of isostatic rebound to the normal faults has not been determined.

ECONOMIC GEOLOGY

Between 1954 and 1983, uranium was the chief mineral resource in the Wind River Basin. During this brief span, about twice as much uranium was produced from this basin as from all other areas in Wyoming. Cumulative production was approximately 25 million tons of ore with an average grade of about 0.017 percent U_3O_8 (R. E. Harris, written communication, 1985). Thus, the yield would be approximately 8.5 million pounds of uranium oxide. Most of the ore came from lower Eocene strata along the southern margin of the Wind River Basin (Gas Hills area, not shown on Fig. 59) and a minor amount from Precambrian and middle Eocene rocks along the south flank of the Owl Creek Mountains.

Oil and gas are also major economic mineral resources in the Wind River Basin. The first oil field (Dallas) in Wyoming, discovered in 1883 near the western margin of the basin, is still productive. There are 113 named and producing fields in and adjacent to the basin. Cumulative production is estimated to be more than 377,400,000 barrels of oil and about 1.2 trillion cubic feet of gas (Barlow and Haun, 1978; updated through 1983). The basin also contains coal resources estimated at approximately 1 billion tons. Cumulative coal production through 1973 (the last year of recorded production) was about 4 million tons (Glass and Roberts, 1978).

Water is another major resource in the Wind River Basin. The chief aquifer system is composed of Paleozoic rocks that are present throughout the subsurface area of the basin. Smaller, less continuous, and shallower individual aquifers are present in Jurassic, Cretaceous, Paleocene, and Eocene strata (Richter, 1981). These are characterized by large variations in both quantity and quality of water because of lateral changes in lithologies. The petroleum industry is the largest user of groundwater.

The composition of the preserved remnants of regolith that accumulated during parts of 2 billion years after Archean rocks were formed and prior to deposition of the oldest Cambrian strata, and during the 140-m.y. time interval between Late Cambrian and Early Mississippian, has not been studied with respect to possible concentrations of economic mineral deposits.

GEOLOGY OF THE BIGHORN BASIN, NORTHERN WYOMING AND SOUTHERN MONTANA

J. D. Love

The Bighorn Basin of northern Wyoming and southern Montana has mountain uplifts on the eastern, southern, and western margins (Plate 2; Fig. 59). The basin is of economic interest because it has 149 producing oil fields, at least 1.8 billion tons of coal, many million tons of bentonite, and large groundwater resources.

BASEMENT ROCKS

Very few drill holes have reached Precambrian basement rocks within the Bighorn Basin. The present interpretation of the kinds and ages of basement rocks is based on projections across the basin from exposures in the Beartooth Mountains to the west, the Bighorn Mountains to the east, and the Owl Creek Mountains to the south (see correlation, classification, and lateral extent of rock types in Love and Christiansen, 1980). The oldest rocks are metasedimentary strata in the Beartooth Mountains, dated at more than 3.4 Ga. These were cut by granite with an age of about 2.7 Ga. Gneisses in the Bighorn Mountains, on the opposite side of the basin, are about 3 Ga and are cut by quartz monzonite with an age of 2.9 Ga. At the south end of the basin, an amphibolite complex is older than 2.8 Ga, and this is cut by granite with an age of 2.6 Ga. The metamorphic rocks trend northeast, so presumably much of the Bighorn Basin is underlain by them and by younger granitic plutons, probably all of Archean age.

SEDIMENTARY ROCKS

Table 4 summarizes the data on sedimentary rocks in the Bighorn Basin. All periods are represented except the Silurian. The lithology of these rocks is especially important in this basin because they contain large resources of groundwater, coal, oil and gas, and bentonite.

The outcrop area of Paleocene and lower Eocene rocks within the major part of the structural basin is about 9,300 km^2. The basin is asymmetric and trends northwest; the deepest part is near the western margin, about 16 km southeast of Cody (Fig. 59). Broad outcrops of Mesozoic strata surround all but the northwest margin of the basin, and extensive areas of Paleozoic strata are exposed on the east and south flanks of the basin.

TECTONIC OVERVIEW

The tectonic pattern of the Bighorn Basin is mainly the result of folding and thrust-faulting during the Laramide Revolution, which here began in Late Cretaceous time and terminated during early Eocene time. This Laramide tectonic pattern was superimposed on that representing less violent older Mesozoic and Paleozoic events, chiefly broad-scale gentle tilting, warping, and erosion. Tectonic maps in more detail than that shown on Figure 59 and cross sections of many structures throughout the Bighorn Basin have been presented elsewhere (Blackstone, 1985, 1986; Pierce, 1978, and references cited therein; Stone, 1985b; Ver Ploeg, 1985; Parker, 1986). These structures are so varied that no single cross section of the basin is representative of the tectonic pattern as a whole.

During late early and early middle Eocene time, large-scale detachment masses of Paleozoic and Mesozoic rocks slid into the western part of the basin from rising uplands to the west (Pierce, 1973; 1978). Regional northward tilting, rapid basin excavation, and isostatic adjustment as a result of this excavation probably occurred after middle Miocene time.

The Precambrian surface along the trough line of the basin is about 7,010 m below sea level. In a small, separate, structurally low area (not shown on the generalized Fig. 59) about 24 km north of Cody, the Precambrian surface is 6,400 m below sea level, but on the Beartooth Mountains 16 km to the west, the surface is approximately 3,000 m above sea level. Hence, the structural relief along the western basin along the western basin margin is about 9,400 m.

GEOLOGIC HISTORY

Only those Paleozoic and later events of regional or possible economic significance are discussed here. More detailed descriptions of the geology of the region are presented elsewhere (Mallory, 1972c).

Archean to Middle Cambrian Hiatus

The Middle Cambrian sea advanced eastward across a gently rolling surface, partly on bedrock and partly on regolith that had accumulated during portions of two billion years of missing Earth history in this region. Concentrations of regolith were derived from a great variety of Archean rocks, and some of them undoubtedly furnished concentrations of insoluble minerals. Monazite, gold, and other minerals have been observed in regolith on the Bighorn Mountains along the east side of the Bighorn Basin. No detailed studies have been made of the composition of the regolith here or elsewhere in the region with respect to concentrations of economic mineral deposits.

Paleozoic Events

The Cambrian Period was a time of crustal stability in this part of northern Wyoming, but after the last Cambrian sea retreated westward, the southeastern part of the basin was tilted northwestward, and some erosion occurred prior to advance of the Ordovician sea eastward across the Bighorn Basin. Dolomite containing very little clastic debris was deposited. There is no record of what happened in the basin during Silurian time, but prior to Middle Devonian there was northward tilting of northern Wyoming and erosion that removed most of the Ordovician

TABLE 4. PALEOZOIC, MESOZOIC, AND TERTIARY ROCKS WITHIN AND ALONG THE MARGINS OF THE BIGHORN BASIN
(For formation names, correlations, and areas of outcrop, refer to Love and Christiansen, 1980; 1985)

Geologic Age	Thickness ft (m)	Description
Miocene	0-200 (0-61)	White tuffaceous sandstone and thin, locally derived conglomerate.
Oligocene	0-200 (0-61)	Light-gray, tan, and pale-green tuffaceous bentonitic claystone and siltstone and white lenticular limestone; locally derived lenticular conglomerate at base.
Late and middle Eocene	0-6,000 (0-1,830) (west side, 0-850; [0-259])	Upper part is gray and white coarse volcaniclastic conglomerate; middle part is green tuff; lower part is bentonitic variegated claystone and tuff; locally contains coal, oil shale, and uranium phosphate.
Early Eocene	2,000-3,000+ (610-914)	Claystone and siltstone, variegated to drab, interbedded with brown and gray lenticular sandstones. Tongues of locally derived conglomerate extend eastward into the basin; tongues of gold-bearing quartzite conglomerate from a distant source to the west are present near base. Thin oil shales are locally present in upper part; very little volcanic debris present.
Paleocene	1,000-10,000 (305-3,050)	Gray to buff sandstone, drab to dully variegated claystone and siltstone, and coal beds, one of which is as much as 20 ft (6 m) thick. Gold-bearing quartzite clast conglomerates in western part.
Late Cretaceous	3,000-8,000 (914-2,438)	From top down: golden-brown sandstone, white tuffaceous sandstone interbedded with thin carbonaceous shale and yellow bentonite (combined thickness about 2,500 ft; 762 m); light-gray sandstone interbedded with carbonaceous shale and some major coal beds (thickness 800-1,200 ft; 244-366 m); gray marine shale (1,500-3,000 ft; 457-1,067 m); and a basal marine sandstone, carbonaceous shale, and thin lenticular conglomerates (500-700 ft; 152-213 m).
Early Cretaceous	800-1,300 (244-396)	Marine and brackish-water black siliceous shale, bentonite beds, and tan sandstone; nonmarine sandstone and variegated claystone near base.
Jurassic	275-700 (84-213)	Uppermost part: nonmarine gray sandstone and variegated claystone; remainder marine glauconitic green shale and sandstone and red shale; basal thick gypsum beds and red shale throughout basin.
Triassic	600-820 (183-250)	Red shale, siltstone, silty sandstone; widespread thin gray limestone bed (Alcova Limestone) in upper part in southern half of basin, associated with thick red sandstones; olive-drab dolomitic siltstone at base in western half of basin, intertonguing eastward with red shale and anhydrite.
Permian	50-250 (15-76)	Tan dolomite, quartzitic sandstone, chert, and green and black shale in western part, intertonguing eastward with red shale and anhydrite. Thin phosphorite beds in southwestern part; thick gray sandstone at base along southeastern margin of basin.
Pennsylvanian	250-600 (76-183)	Upper part chiefly white to tan sandstone; lower part interbedded sandstone, cherty dolomite, limestone, and red and green shale and claystone.
Mississippian	500-1,000 (152-305)	Blue-gray limestone and dolomite, coarse-grained, cavernous, cherty; some red shale and a widespread sandstone near top.
Devonian	0-500 (0-152)	Upper part; yellow and gray dolomitic siltstone, shale, and dolomite; black shale locally at top. Lower part: fetid brown hard dolomite and tan limestone; absent in southeastern part of basin.
Silurian	Absent	
Ordovician	0-450 (0-137)	Gray dolomite and dolomitic limestone, hard, massive, siliceous; forms cliffs on outcrops; absent along southeastern margin of basin.
Cambrian	900-1,300 (274-396)	Widespread blue-gray limestone mottled with yellow splotches, flat-pebble limestone conglomerates, green micaceous shale, glauconitic in part, becoming sandy toward base. Widespread basal dull red sandstone in contact with Precambrian rocks.

sequence from the southeastern part of the basin. Devonian strata are present only in the northern part of the Bighorn Basin but probably were once present throughout the area.

It is likely that northern Wyoming was tilted northwestward and Devonian rocks eroded away from the southeastern part of the basin prior to Mississippian deposition. Mississippian strata were deposited entirely across the Bighorn Basin, and in the southeastern part probably incorporated regolith that may have accumulated during all or part of the 75-m.y. time interval between Late Ordovician and Early Mississippian. No studies have been made of this regolith with respect to concentrations of economic mineral deposits.

Pennsylvanian marine strata are present throughout the Bighorn Basin. At the base and at the top are widespread sandstones that are fair to good aquifers in places. During the latter part of Pennsylvanian time, the Bighorn Basin was uplifted and some of the strata were eroded away, but the configuration of this uplift is not well known. Permian strata consist of porous and permeable cherty dolomites in the western part of the basin, intertonguing eastward with red beds and anhydrite. From the area of these facies changes westward, the Permian rocks contain high-sulfur oil and gas.

Mesozoic Events

Red bed deposition continued throughout much of Triassic time in the Bighorn Basin, but conditions fostering evaporite deposition terminated in Early Triassic time. The upper part of the Triassic section and all the Lower Jurassic(?) rocks, which are present in areas to the south and southwest, have been eroded away from the Bighorn Basin region. An unconformity of regional extent is present between Triassic and Middle Jurassic rocks throughout the Bighorn Basin. Middle Jurassic again was a time of deposition of anhydrite, gypsum, and red beds. Evaporite deposition was followed by a marine invasion and then retreat in Late Jurassic time. An area of thinning of marine Upper Jurassic rocks along the southeastern margin of the basin suggests development of a gentle northeast-trending arch at that time. The last deposition in the Jurassic was a widespread nonmarine sequence of lenticular sandstones and dully variegated claystones (Morrison Formation).

No conspicuous regional unconformity exists between the nonmarine uppermost Jurassic and nonmarine Lower Cretaceous strata. Throughout the remainder of Cretaceous time, mud-laden seas advanced westward and then retreated, then advanced and retreated over and over again, leaving thick sequences of intertonguing sandstone and shale. Sandstones are thicker and more abundant in the western part of the basin than in the eastern part. There are no conspicuous unconformities, although minor ones are present and more common in the western part of the basin. Areas of deposition of marine and brackish-water strata became smaller late in Cretaceous time, and widespread peat swamps developed. Near the end of Cretaceous time, the seas retreated permanently and the Washakie Range (Fig. 59), southwest of the Bighorn Basin, rose and was eroded to its Paleozoic core. The widespread unconformity between Cretaceous and Paleocene strata was reduced in magnitude eastward across the basin.

Cenozoic Events

Beginning in Paleocene time and continuing through the early Eocene, the Bighorn Basin, especially the western part, was involved in the Laramide revolution. The Beartooth, Bighorn, Owl Creek, and Washakie Ranges rose, were thrust-faulted in various directions, and were stripped to their Precambrian cores. They contributed an enormous volume of clastic debris to the Bighorn Basin, which was formed at this time. Most of the oil- and gas-bearing anticlines developed in earliest Eocene time, concurrently with several thrust-fault systems within the deep western part of the basin. Somewhat later but still largely in early Eocene time, huge detachment masses slid eastward into the basin from the highlands to the west, along a north-south marginal area at least 56 km long. Among the more notable of these are the Heart Mountain and South Fork thrust masses.

In Paleocene time, much of the drainage was eastward, but after the Bighorn and Owl Creek Mountains rose and blocked this system, drainage was diverted northward out of the newly formed basin. These paleodrainage systems and the strata deposited along them are directly related to the shallower groundwater resources throughout much of the basin.

During middle and late Eocene time, the Bighorn Basin was almost entirely filled with volcaniclastic debris that was blown and washed eastward from the Yellowstone-Absaroka volcanic area (Fig. 59; McKenna and Love, 1972). Volcanic eruptions began near the end of early Eocene time and terminated about the end of late Eocene time. Remnants of Oligocene and lower Miocene strata are preserved on the Bighorn Mountains and are presumed to have extended across the Bighorn Basin. During Oligocene time, some clasts of volcanic rock from the Absaroka volcanic field were carried eastward across the south end of the buried Bighorn Mountains and deposited in the Powder River Basin (Love, 1952).

Probably after early or middle Miocene time there was regional uplift and northward tilting of the Bighorn Basin region. The Wind River–Bighorn River was superimposed across the Owl Creek Mountains at the south margin of the basin and the Bighorn River across the northwestern part of the Bighorn Mountains. The regional uplift started the reexcavation of the Bighorn Basin and exhumation of the adjacent mountains; this process has continued to the present time. Many hundred cubic kilometers of sediment were eroded out of the basin, probably mostly in the last 10 m.y. In the few places where rates of erosion can be measured, the rate of stream downcutting during Quaternary time has been about 0.3 m in 1,600–2,000 years.

Because of the vast weight of debris removed from the basin in a relatively short time, there has been some conspicuous isostatic readjustment. This can best be seen on Carter Mountain southwest of Cody, where the upper and middle Eocene strata are tilted westward away from the basin (McKenna and Love, 1972, Fig. 2-A), and on Lysite Mountain at the southeast margin of the Bighorn Basin, where strata of the same age are tilted southward away from the basin. Another consequence of the rapid excavation of the basin, concomitant with rejuvenation of major volcanic activity in the Yellowstone National Park area to the west and probably accompanied by violent earthquakes, was the development of huge debris flows and detachment masses of Eocene rocks along the southwest side of the Bighorn Basin. These Quaternary debris flows are among the largest in the world; they extend 56 km north to south and nearly 48 km eastward onto the basin floor (Bown and Love, 1984). Some of these are similar to the detachment masses that were emplaced during Eocene time in the same region.

ECONOMIC GEOLOGY

Oil and gas are the chief economic mineral resources in the Bighorn Basin. There are 149 named and producing fields. Cumulative oil production is estimated to be nearly 2 billion barrels and gas about 1.8 trillion cubic feet (Barlow and Haun, 1978; updated through 1983). The basin also contains in-place bituminous and subbituminous coal resources of about 1.8 billion tons (R. W. Jones, written communication, 1984). Cumulative coal production through 1983 has been nearly 53 million tons, most of it from Upper Cretaceous rocks.

Another economic resource in the Bighorn Basin is water. Two major bedrock aquifer systems have been identified: Paleozoic strata and Upper Cretaceous–Tertiary strata, undifferentiated (Libra and others, 1981). Ordovician, Mississippian, and Pennsylvanian rocks (Table 4) have the greatest potential for producing large quantities of good water. Cretaceous, Paleocene, and lower Eocene aquifers are characterized by large variations in both quantity and quality of water because of lateral changes in lithologies. The petroleum industry is the largest user of groundwater, with agricultural use second.

In contrast with the large uranium production from the Powder River Basin to the east and the Wind River Basin to the south, that from the Bighorn Basin has been minor—only about 24,000 tons. Most uranium was produced from Paleozoic rocks on the northeast margin of the basin.

Wyoming is the largest producer of bentonite in the world, and almost all of it is mined from Lower Cretaceous rocks on the east side of the Bighorn Basin and in the Powder River Basin. No breakdown by basins is available, but the cumulative production for both basins is about 59 million metric tons through 1982.

GEOLOGY OF THE POWDER RIVER BASIN, NORTHEASTERN WYOMING AND SOUTHEASTERN MONTANA

J. D. Love

The Powder River Basin (Plate 2; Fig. 59) of northeastern Wyoming and southeastern Montana is the largest intermontane structural basin in, and extending northward from, the eastern front of the Rocky Mountains. The basin is notable for several economic reasons. For example, it probably has the fifth largest coal resources in the world. It has more than 500 producing oil and gas fields, and in the last 30 years it has yielded more than 2.5 million pounds of uranium oxide.

BASEMENT ROCKS

Very few drill holes have reached Precambrian basement rocks within the Powder River Basin, so the present interpretation of the kinds and ages of basement rocks is based on projections across the basin from exposures in the Bighorn Mountains to the west and the Black Hills to the east (see correlation, classification, and lateral extent of rock types in Love and Christiansen, 1985; Bayley and Muehlberger, 1968). According to this projection, the oldest rocks are a gneiss complex at least 3 billion years old, extending northeastward from the southern half of the Bighorn Mountains into the basin. The Precambrian rocks in the northern Bighorn Mountains are a quartz diorite and quartz monzonite complex with an age of about 2.9 Ga. Small exposures of Precambrian rocks in the Wyoming part of the Black Hills on the east side of the Powder River Basin are granitic rocks of the 2.6 Ga age group and metasedimentary and metavolcanic rocks of slightly younger age.

SEDIMENTARY ROCKS

Table 5 summarizes the data on sedimentary rocks in the Powder River Basin. All periods of the Phanerozoic are represented except Silurian. The nature of the rocks present is especially important in this basin because they contain large resources of groundwater, coal, oil and gas, uranium, and bentonite.

The outcrop area of Paleocene and lower Eocene rocks within the major part of the basin is about 42,900 km^2. The basin is asymmetric and trends northward; the deepest part is near the western margin (Fig. 59). Broad outcrops of Cretaceous strata are present on the east and southwest sides of the basin, and narrower outcrops of older Mesozoic and Paleozoic rocks are present along the east, south, and west margins.

TECTONIC OVERVIEW

The tectonic pattern of the Powder River Basin as it is today (Fig. 59) is largely the result of folding and thrust-faulting during the Laramide Revolution. The major movements in this basin began near the end of Paleocene time and terminated during early Eocene time. A more detailed tectonic map than that shown in Figure 59, as well as many cross sections of the western margin of the basin, has been published elsewhere (Blackstone, 1981). This Laramide pattern was superimposed on an older pattern of broad-scale gentle tilting, warping, and erosion that had occurred in Mesozoic and Paleozoic times. Later in Eocene time and again in post-Oligocene (perhaps post-middle Miocene) time, the basin

TABLE 5. PALEOZOIC, MESOZOIC, AND TERTIARY ROCKS WITHIN AND ALONG THE MARGINS OF THE POWDER RIVER BASIN
(For formation names, correlations, and areas of outcrop, refer to Love and Christiansen, 1980, and Love and Christiansen, 1985)

Geologic Age	Thickness ft (m)	Description
Miocene	0-600 (0-183)	Light-gray tuffaceous sandstone, thin beds of limestone, and lenticular conglomerates near base.
Oligocene	0-500 (0-152)	Light-gray, pink, and green tuffaceous claystone and siltstone; lenticular conglomerates near base.
Eocene	0-3,000 0-915	Buff arkosic sandstone, lenticular conglomerates, drab to variegated siltstones and claystones, carbonaceous shales, and many thick coal beds near base; contains major deposits of uranium.
Paleocene	0-4,000 (0-1,220)	From top down: thick yellow sandstones and gray and black shale, many thick coal beds; dark-gray shale and concretionary sandstone; gray sandstone, gray and brown carbonaceous shale, and thin coal beds.
Late Cretaceous	6,000-11,000 (1,830-3,353)	Chiefly marine and brackish-water gray and brown sandstone and shale, with sandstones more abundant in upper and lower parts and 3,000-4,000 ft (914-1219 m) of gray shale in middle; many sandstones yield oil and gas.
Early Cretaceous	330-1,000 (100-305)	Marine and brackish-water black shale, bentonite beds, and tan sandstone; some marl and limestone in eastern part; nonmarine sandstone and veriegated claystone near base; thin gold-bearing coal in Black Hills; many sandstones yield oil and gas.
Jurassic	350-650 (107-198)	Uppermost part: nonmarine gray sandstone and variegated claystone; remainder marine red and glauconitic green shale and sandstone; basal sandstone in southern area; basal gypsum and red shale in northern area.
Triassic	150-1,000 (46-305)	Red shale, siltstone, and silty sandstone; some white gypsum beds in lower part; widespread thin gray limestone bed (Alcove Limestone) in upper part in southwestern area.
Permian	75-350 (23-107)	Interbedded red shale and siltstone with thin purple limestone and dolomite and white gypsum; salt beds in some subsurface sections.
Pennsylvanian	600-1,300 (183-396)	Chiefly white to tan sandstone; lower part interbedded with cherty dolomite, limestone, and red and green shale and claystone; widespread thin black radioactive shales; anhydrite and some salt in subsurface in eastern area; yields major amounts of oil and gas.
Mississippian	0-650 (0-198)	Limestone and dolomite, gray, coarse-grained, cavernous, cherty; locally thin sandstone at base.
Devonian	0-100 (0-30)	Yellowish-gray dolomite, limestone, and siltstone; sandy near base; present only in northern part of basin and as an erosion remnant along the southeastern margin.
Silurian	Absent	
Ordovician	0-350 (0-107)	Gray dolomite, hard, massive, siliceous; forms cliffs on outcrops; absent in southern part of basin.
Cambrian	0-600 (0-183)	Green to gray mottled limestone in upper part, green micaceous glauconitic shale in middle, and reddish-brown hard sandstone in lower part; more sandy near eastern margin of basin; thins out near southern margin.

was modified by additional tilting, warping, extensive and rapid basin excavation, isostatic adjustment, and drastic drainage change of at least one river course.

The present difference in altitude on the Precambrian surface between the deepest part of the basin (–3,900 m below sea level) and the adjacent Bighorn Mountains to the west (2,440 m above sea level) is about 6,400 m. South of the town of Sheridan, Wyoming (Fig. 59), where the Bighorn Mountains rise to 3,900 m above sea level, the differential elevation between the mountains and the adjacent basin to the east is about 7,600 m. A comparison of the amounts of mountain-basin structural relief involving the Powder River and other basins and mountains in Wyoming and the magnitude of Cenozoic sedimentation in these basins is presented elsewhere (Love, 1960).

GEOLOGIC HISTORY

Only those Paleozoic and later events of regional or possible economic significance are discussed here. More detailed descriptions of the geology of the region are presented elsewhere (Mallory, 1972c).

Archean to Middle Cambrian Hiatus

A Middle Cambrian sea advanced eastward across a gently rolling surface, partly on bedrock and partly on regolith that had accumulated during portions of two billion years of missing Earth history. Concentrations of regolith were derived from a great variety of Archean rocks.

Paleozoic Events

After the Cambrian sea retreated westward, an Ordovician sea advanced eastward across the Powder River Basin, and dolomite and limestone containing very little clastic debris was deposited. The present wedge edge extends eastward across the southern part of the basin; it is not known how far south the dolomite may originally have been deposited. Inasmuch as Ordovician carbonate rocks are present in diatremes 193 km south of the wedge edge, along the Wyoming-Colorado state line, the dolomite must have extended that far southward across Wyoming. There is no evidence of marginal clastic debris near the present wedge edge in the Powder River Basin.

There is no record of Silurian rocks in the area now occupied by the Powder River Basin, but prior to the Middle Devonian, northeastern Wyoming was tilted northward, and some Ordovician and Cambrian rocks were removed by subsequent erosion. Devonian seas probably advanced eastward across the region. Only remnants of deposits of Devonian dolomite, limestone, shale, siltstone, and sandstone are present. These are preserved in the northeastern part of the basin and in an isolated occurrence of marine strata south of Lusk (Fig. 59) near the southeast margin of the basin.

Prior to the Mississippian, northeastern Wyoming was tilted northwestward, and some Devonian rocks were removed by erosion. Mississippian sediments were deposited entirely across the area now occupied by the Powder River Basin; they probably incorporated regolith accumulated during all or part of the 140-m.y. time interval between Late Cambrian and Early Mississippian.

Pennsylvanian marine rocks were deposited throughout that area. At the base is a widespread sandstone that is a moderately good aquifer in places. Red and green shale and thin, widespread radioactive black shale beds overlie the basal foundations, and dolomite intertongues with thick sandstones near the top of the section. In the eastern part of the area, the upper part of the Pennsylvanian section contains gypsum and anhydrite, which when dissolved out leave cavernous residual breccias. These breccias are porous and permeable and therefore are significant in terms of groundwater and oil and gas reservoir zones. Near the end of Pennsylvanian time, several broad, gentle, northeast-trending folds developed in the northeastern part of the area, and an erosional unconformity separates marine Pennsylvanian rocks from brackish-water and evaporitic Permian rocks. Permian strata consist regionally of red bed and evaporite facies with some halite beds in the western part of the area.

Mesozoic Events

Evaporite and red bed deposition continued through Triassic time but with no salt and with diminishing amounts of gypsum and anhydrite. Upper part of the Triassic section and some Lower Jurassic rocks that are present in areas to the west and southwest were eroded from the area now occupied by the Powder River Basin. Middle Jurassic deposits of anhydrite, gypsum, and red beds occupy the northern third of the area. The original southern extent of these deposits is not known because the region was again tilted northward or northwestward prior to Late Jurassic time, and Middle Jurassic deposits were removed by subsequent erosion. Evaporite deposition was followed by a marine invasion and then retreat in Late Jurassic time. An area of thinning of marine Upper Jurassic rocks in the northern part of the region suggests development of a gentle north-trending arch at that time. The last deposition during the Jurassic was a widespread nonmarine sequence of lenticular sandstones and dully variegated claystones (Morrison Formation). There is no conspicuous regional unconformity between the nonmarine uppermost Jurassic and Lower Cretaceous strata.

Throughout Cretaceous time, muddy seas advanced westward and then retreated, then advanced and retreated over and over again, leaving thick sequences of intertonguing sandstone and shale. Seas retreated eastward late in Cretaceous time, and large coal swamps developed. There is no major unconformity between Cretaceous and Paleocene strata in the Powder River Basin.

Cenozoic Events

Northeastern Wyoming was, for the first time, affected by the Laramide Revolution that had been much more active earlier in areas to the west. The Bighorn and Laramie Mountains rose, were eroded to their Precambrian cores, and contributed much coarse clastic debris to the Powder River Basin. This asymmetric basin, much as it is today, and the lesser anticlines and synclines developed during late Paleocene and early Eocene times. Some thrust faulting occurred along and in the northwestern part of the basin (Blackstone, 1981). Most of the oil- and gas-bearing anticlines were formed at this time, and the lower Eocene clastic debris became the host rocks for subsequently introduced uranium deposits. The Black Hills on the east side of the Powder River Basin rose as a broad domal uplift, probably in Paleocene and early Eocene time. Most of the drainage was to the east and northeast out of the basin at this time, but locally some streams carried clastic debris westward from the Black Hills into the basin. Large coal swamps were formed throughout much of the Powder River Basin.

Drainage was from east to west during middle and late

Eocene time, and there is no record of deposition of the 300 m or more of lacustrine-fluviatile middle and upper Eocene rocks that nearly filled the Bighorn and Wind River Basins to the west and southwest (McKenna and Love, 1972; Love and others, 1963). This nondeposition, westward drainage, and pre-Oligocene erosion were the result of regional westward tilting of eastern and northeastern Wyoming.

At the beginning of Oligocene deposition, eastward drainage was once more established; rivers that flowed eastward from the Yellowstone-Absaroka volcanic terrain across the now-filled Bighorn Basin and across the Bighorn Range deposited volcanic rock clasts from the Absaroka Range in the central Powder River Basin (Love, 1952). Basin filling continued through at least the early part of Miocene and possibly through middle Miocene time (McKenna and Love, 1972). Then there was regional uplift, northward tilting of most of the Powder River Basin, and establishment of the north-flowing Powder River and northeastward-flowing Belle Fourche River. The latter was superimposed across the northern part of the Black Hills Uplift (Fig. 59). At about the same time, there was normal faulting near the southern and southeastern margins of the basin. The regional uplift and perhaps the more localized faulting started the reexcavation of the Powder River Basin and exhumation of the adjacent mountains; this process has continued to the present time. Several hundred cubic kilometers of sediment were eroded out of the basin, probably mostly in the last 10 m.y. In the few places where rates of erosion can be measured, the rate of basin floor excavation in some intervals of Quaternary time has been about 0.3 m per 2,000 years.

Because of the huge weight of debris removed from the basin, there has been some isostatic readjustment, the magnitude of which has not yet been determined. One indication of this adjustment is the major drainage change of the North Platte River. Apparently it flowed northward from the city of Casper (Fig. 59) during part of Quaternary time and carried clastic fragments of distinctive middle Eocene porphyry from the northern part of the Granite Mountains at least as far as a place 32 km north of Casper. Later, because of tectonic movements, the river was diverted eastward along its present course, probably during the last two m.y. The isostatic adjustment was not related to glaciation, for the basin has not been glaciated.

ECONOMIC GEOLOGY

The Powder River Basin is best known for its immense coal deposits, chiefly in Paleocene and lower Eocene strata, that extend from the surface to depths of 1,500 m or more. Many beds of coal are 30 m thick, and one near the western margin is more than 60 m thick. Above a depth of 60 m in Wyoming and 75 m in Montana, the basin contains about 60 billion short tons of coal (Glass, 1984; Sholes and others, 1984), or about 38 percent of the strippable coal reserves in the nation (Energy Information Administration, 1984). By using published conservative estimates of the amount of coal resources identified (mapped and explored) in the basin, it contains almost 283 billion short tons of coal above a depth of 900 m (Glass, 1984; Combo and others, 1949; Averitt, 1975). This constitutes about 16 percent of the nation's total identified coal resources (Averitt, 1975). A recent estimate of 246 billion short tons for only three coal deposits in the Wyoming portion of the basin was made (B. H. Kent, written communication, 1985). Other estimates of the total coal resources of the basin range from 775 billion short tons on federal lands only (B. H. Kent, written communication, 1985) to 1,215 billion short tons (U.S. Geological Survey, 1976). Thus, the Powder River Basin alone would rank fifth in world coal resources, behind Russia, the remainder of the conterminous United States, China, and Alaska.

Total cumulative coal production in the basin through 1983 is about 1,061 million short tons; in 1984, it yielded 164 million tons of coal, about 18.4 percent of the nation's coal production (R. W. Jones, written communication, 1985).

For more than 100 years, oil and gas have been known to exist around the margins of the Powder River Basin. By the end of 1983, 513 named and productive oil and gas fields had been discovered. Cumulative oil production is estimated to be nearly two billion barrels (Barlow and Haun, 1978; updated through 1983) and gas about 1.8 billion MCF (thousand cubic feet). Many of the large and small oil and gas fields were formed in Laramide structures. Other fields occur at facies changes, unconformities, and fault traps, chiefly involving Pennsylvanian, Permian, and Cretaceous rocks.

Uranium is the third major mineral resource in the Powder River Basin, yet it has only been known there since 1951 (Love, 1952). It occurs chiefly in rocks of early Eocene and Early Cretaceous ages, but the ore concentrations are generally thought to have taken place in mid to late Cenozoic time. Cumulative production in the 30 years of development has been approximately 12,700,000 tons of ore (R. E. Harris, written communication, 1985). No good record of grade is available, but the average grade that mills would accept during that time is about 0.01 percent U_3O_8. This would indicate a yield of at least 2.54 million pounds of uranium oxide.

Wyoming is the largest producer of bentonite in the world, and all of it is mined from Cretaceous rocks in the Powder River (chiefly on the Black Hills margin) and Bighorn Basins. No breakdown of production by basins is available, but the cumulative production for both is about 65 million short tons.

Another major economic resource in the Powder River Basin is water. Four principal bedrock aquifer systems have been identified (Feathers and others, 1981): Paleozoic (chiefly Cambrian, Mississippian, and Pennsylvanian systems); Lower Cretaceous; uppermost Cretaceous; and Paleocene–lower Eocene (Table 3 has thicknesses and lithologic descriptions of these rocks). The quality of water in the Paleozoic rocks is generally good except in the northeastern part of the basin where salts and fluorine are more abundant. Cretaceous waters are generally high in salts and fluorine. Paleocene-Eocene waters have the advantage of being shallow but are uneven in volume and quality.

The composition of regolith that accumulated during parts of 2 billion years after Archean rocks were formed and prior to deposition of the oldest Cambrian strata, and the 140-m.y. time interval between Late Cambrian and Early Mississippian, has not been studied with respect to possible concentrations of economic mineral deposits.

REFERENCES

Ahern, J. L., and Mrkvicka, S. R., 1984, A mechanical and thermal model for the evolution of the Williston Basin: Tectonics, v. 3, p. 79–102.

Allmendenger, R. W., and five others, 1982, COCORP profiling across the Rocky Mountain front in southern Wyoming, Part 2, Precambrian basement structure and its influence on Laramide deformation: Geological Society of America Bulletin, v. 93, p. 1253–1263.

Allmendinger, R. W., Brown, L. D., Oliver, J. E., and Kaufman, S., 1983, COCORP deep seismic profiles across the Wind River Mountains, Wyoming, *in* Bally, A. W., ed., Seismic expression of structural styles: American Association of Petroleum Geologist Studies in Geology Series 15, v. 3, p. 3.2.1-29–3.3.1-33.

Anderholm, S. K., 1983, Hydrogeology of the Socorro and La Jencia basins, Socorro County, New Mexico: New Mexico Geological Society, 34th Annual Field Conference, Guidebook, p. 303–310.

Anderman, G. G., 1955, Tertiary deformational history of a portion of the north flank of the Uinta Mountains in the vicinity of Manila, Utah: Wyoming Geological Association Guidebook, 10th Annual Field Conference, p. 130–134.

Anderman, G. G., and Ackman, E. J., 1963, Structure of the Denver-Julesburg basin and surrounding areas, *in* Bolyard, D. W., and Katich, P. J., eds., Geology of the northern Denver basin and adjacent uplifts: Rocky Mountain Association of Geologists, p. 170–175.

Applegate, J. K., and Rose, P. R., 1985, Structure of the Raton basin from a regional seismic line, *in* Gries, R. R., and Dyer, R. C., eds., Seismic exploration of the Rocky Mountain region: Rocky Mountain Association of Geologists, p. 259–266.

Armstrong, F. C., and Oriel, S. S., 1965, Tectonic development of Idaho-Wyoming thrust belt: American Association of Petroleum Geologists Bulletin, v. 49, p. 1847–1866.

Armstrong, R. L., 1968, Sevier orogenic belt in Nevada and Utah: Geological Society of America Bulletin, v. 79, p. 429–458.

—— , 1969, K-Ar dating of laccolithic centers in the Colorado Plateau and vicinity: Geological Society of America Bulletin, v. 80, no. 10, p. 2081–2086.

Atwater, T., 1970, Implications of plate tectonics for the Cenozoic tectonic evolution of western North America: Geological Society of America Bulletin, v. 81, p. 3513–3536.

Averitt, P., 1975, Coal resources of the United States, January 1, 1974: U.S. Geological Survey Bulletin 1412, 131 p.

Axelrod, D. E., and Bailey, H. P., 1976, Tertiary vegetation, climate, and altitude of the Rio Grande depression, New Mexico–Colorado: Paleobiology, v. 2, p. 235–254.

Baars, D. L., 1962, Permian System of the Colorado Plateau: American Association of Petroleum Geologists Bulletin, v. 46, p. 149–218.

—— , 1966, Pre-Pennsylvanian paleotectonics; Key to basin evolution and petroleum occurrences in Paradox basin, Utah and Colorado: American Association of Petroleum Geologists Bulletin, v. 50, p. 2082–2111.

—— , compiler, 1972, Devonian System, *in* Mallory, W. W., ed., Geological atlas of the Rocky Mountain region: Rocky Mountain Association of Geologists, p. 90–99.

—— , 1976, The Colorado Plateau aulacogen, key to continental scale basement rifting: 2nd International Conference on Basement Tectonics Proceedings, p. 157–164.

—— , 1982, Paleozoic history of the Albuquerque trough; Implications of basement control on Rio Grande rift: New Mexico Geological Society, 33rd Annual Field Conference, Guidebook, p. 153–157.

—— , 1985, Regional stratigraphy of Permian "eolian" sandstones, southeastern Utah [abs.]: Abstracts and Program, Society of Economic Paleontologists and Mineralogists, 1985 Mid-Year Meeting, p. 8.

Baars, D. L., and See, P. D., 1968, Pre-Pennsylvanian stratigraphy and paleotectonics of the San Juan Mountains, southwestern Colorado: Geological Society of America Bulletin, v. 79, p. 333–350.

Baars, D. L., and Stevenson, G. M., 1982, Subtle stratigraphic traps in Paleozoic rocks of Paradox basin, *in* Halbouty, M. T., ed., Deliberate search for the subtle trap: American Association of Petroleum Geologists Memoir 32, p. 131–158.

—— , 1984, The San Luis uplift, an enigma of the Ancestral Rockies: The Mountain Geologist, v. 21, no. 2, p. 57–67.

Bachman, G. O., and Mehnert, H. H., 1978, New K-Ar dates and the late Pliocene and Holocene geomorphic history of the central Rio Grande region, New Mexico: Geological Society of America Bulletin, v. 89, p. 283–292.

Baker, A. A., 1933, Geology and oil possibilities of the Moab District, Grand and San Juan Counties, Utah: U.S. Geological Survey Bulletin 841, 95 p.

Baker, A. A., Dane, C. H., and Reeside, J. B., Jr., 1936, Correlation of the Jurassic formations of parts of Utah, Arizona, New Mexico, and Colorado: U.S. Geological Survey Professional Paper 183, 166 p.

Baldridge, W. S., Damon, P. E., Shafiqullah, M., and Bridwell, R. J., 1980, Evolution of the central Rio Grande rift, New Mexico; New potassium-argon ages: Earth and Planetary Science Letters, v. 51, p. 309–321.

Baldridge, W. S., Olsen, K. H., and Callender, J. F., 1984, Rio Grande rift; Problems and perspectives: New Mexico Geological Society, 35th Annual Field Conference, Guidebook, p. 1–12.

Bally, A. W., and Snelson, S., 1980, Realms of subsidence, *in* Facts and principles of world petroleum occurrence: Canadian Society of Petroleum Geologists Memoir 6, p. 9–94.

Baltz, E. H., 1965, Stratigraphy and history of Raton basin and notes on San Luis basin, Colorado–New Mexico: American Association of Petroleum Geologists Bulletin, v. 49, no. 11, p. 2041–2075.

—— , 1967, Stratigraphy and regional tectonic implications of part of Upper Cretaceous and Tertiary rocks, east-central San Juan Basin, New Mexico: U.S. Geological Survey Professional Paper 552, 100 p.

Barlow, J. A., and Haun, J. D., 1978, Oil and gas production, reserves, and resources in Wyoming: Minerals Division, Wyoming Department of Economic Planning and Development, 18 p.

Bartleson, B. L., 1968, Stratigraphy and petrology of the Gothic Formation, Elk Mountains, Colorado [Ph.D. thesis]: Boulder, University of Colorado, 216 p.

—— , 1972, Permo-Pennsylvanian stratigraphy and history of the Crested Butte–Aspen region: Colorado School of Mines Quarterly, v. 67, no. 4, p. 187–248.

Basham, W. L., and Martin, W. F., 1985, Seismic line across the Wind River thrust fault, Wyoming, *in* Gries, R. R., and Dyer, R. C., eds., Seismic exploration of the Rocky Mountain region: Rocky Mountain Association of Geologists and Denver Geophysical Society, p. 59–66.

Bayley, R. W., and Muehlberger, W. R., 1968, Basement rock map of the United States: U.S. Geological Survey, scale 1:2,500,000.

Bell, T. E., 1981, A Jurassic closed basin in the Morrison Formation: Geological Society of America Abstracts with Programs, v. 13, p. 406.

—— , 1986, Deposition and diagensis in the Brushy Basin Member and upper part of the Westwater Canyon Member of the Morrison Formation, San Juan Basin, New Mexico, *in* Turner-Peterson, C. E., Santos, E. S., and Fishman, N. S., eds., A basin analysis case study; The Morrison Formation, Grants uranium region, New Mexico: American Association of Petroleum Geologists Studies in Geology, no. 22, p. 77–91.

Berg, R. R., 1961, Laramide tectonics of the Wind River Mountains, *in* Wilmoth, G. J., ed.: Wyoming Geological Association Guidebook, 16th Annual Field Conference, p. 70–80.

—— , 1983, Geometry of the Wind River thrust, Wyoming, *in* Lowell, J. D., ed., Rocky Mountain foreland basins and uplifts: Rocky Mountain Association of Geologists, p. 257–262.

Bissell, H. J., 1964, Lithology and petrography of the Weber Formation in Utah and Colorado, *in* Guidebook of the geology and mineral resources of the Uinta basin: Intermountain Association of Petroleum Geologists, p. 67–91.

Bissell, H. J., and Childs, O. E., 1958, The Weber Formation of Utah and Colorado, *in* Curtis, B. F., ed., Symposium on Pennsylvanian rocks of Colorado and adjacent areas: Rocky Mountain Association of Geologists, p. 26–30.

Black, B. A., 1982, Oil and gas exploration in the Albuquerque Basin: New Mexico Geological Society, 33rd Annual Field Conference, Guidebook, p. 313–324.

Black, B. A., and Hiss, W. L., 1974, Structure and stratigraphy in the vicinity of the Shell Oil Company Santa Fe Pacific no. 1 test well, southern Sandoval County, New Mexico: New Mexico Geological Society, 25th Annual Field Conference, Guidebook, p. 365–370.

Blackstone, D. L., Jr., compiler, 1979, Tectonic map of the overthrust belt western Wyoming, southeastern Idaho, and northeastern Utah showing current oil and gas drilling and development: Geological Survey of Wyoming, 1 sheet, scale 1:316,800.

—— , 1981, Compression as an agent in deformation of the east-central flank of the Bighorn Mountains, Sheridan and Johnson Counties, Wyoming: Contributions to Geology, University of Wyoming, v. 19, no. 2, p. 105–122.

—— , 1985, South Fork detachment fault, Park County, Wyoming; Geometry, extent, source: Contributions to Geology, University of Wyoming, v. 23, no. 2, p. 47–62.

—— , 1986, Foreland compressional tectonics, southern Bighorn Basin, Wyoming: Geological Survey of Wyoming Report of Investigations no. 34, 32 p.

Blakey, R. C., and four others, 1983, Paleogeography of Middle Jurassic continental, shoreline, and shallow marine sedimentation, southern Utah, *in* Reynolds, M. W., and Dolly, E. D., eds., Mesozoic paleogeography of the west-central United States: Rocky Mountain Section, Society of Economic Paleontologists and Mineralogists Symposium 2, p. 77–100.

Boggs, S., Jr., 1966, Petrology of the Minturn Formation, east-central Eagle County, Colorado: American Association of Petroleum Geologists Bulletin, v. 50, p. 1399–1422.

Bowers, W. E., 1972, The Canaan Peak, Pine Hollow, and Wasatch Formations in the Table Cliff region, Garfield County, Utah: U.S. Geological Survey Bulletin 1331-B, p. B1–B39.

Bown, T. M., and Love, J. D., 1984, Comment *on* 'Catastrophic debris-avalanche from ancestral Mount Shasta volcano, California': Geology, v. 13, no. 1, p. 79–80.

Bradley, W. H., 1931, Origin and microfossils of the oil shale of the Green River Formation of Colorado and Utah: U.S. Geological Survey Professional Paper 168, 58 p.

—— , 1963, Paleolimnology, *in* Frey, D. G., ed., Limnology in North America: Madison, University of Wisconsin Press, p. 621–652.

Bradley, W. H., 1964, Geology of Green River Formation and associated Eocene rocks in southwestern Wyoming and adjacent parts of Colorado and Utah: U.S. Geological Survey Professional Paper 496-A, 86 p.

Bradley, W. H., and Eugster, H. P., 1969, Geochemistry and paleolimnology of the trona deposits and associated authigenic minerals of the Green River Formation of Wyoming: U.S. Geological Survey Professional Paper 496-B, 71 p.

Brewer, J. A., and four others, 1982, COCORP profiling across the Rocky Mountain front in southern Wyoming, Part 1, Laramide structure: Geological Society of America Bulletin, v. 93, p. 1242–1252.

Brill, K. G., Jr., 1944, Late Paleozoic stratigraphy, west-central and northwestern Colorado: Geological Society of America Bulletin, v. 55, p. 621–656.

—— , 1952, Stratigraphy in the Permo-Pennsylvanian zeugogeosyncline of Colorado and northern New Mexico: Geological Society of America Bulletin, v. 63, p. 809–880.

—— , 1958, The Belden Formation, *in* Curtis, B. F., ed., Symposium on Pennsylvanian rocks of Colorado and adjacent areas: Rocky Mountain Association of Geologists, p. 102–105.

Brown, W. G., 1983, Sequential development of the fold-thrust model of foreland deformation, *in* Lowell, W. D., ed., Rocky Mountain Foreland Basins and Uplifts: Rocky Mountain Association of Geologists, p. 57–64.

Bryan, K., 1938, Geology and ground-water conditions of the Rio Grande depression in Colorado and New Mexico, *in* Rio Grande joint investigations in the upper Rio Grande Basin in Colorado, New Mexico, and Texas: U.S. Natural Resources Planning Board, Washington, D.C., v. 1, pt. 2, p. 197–225.

Bryan, K., and McCann, F. T., 1937, The Ceja del Rio Puerco; A border feature of the Basin and Range Province in New Mexico, Part I, Stratigraphy and structure: Journal of Geology, v. 45, p. 108–828.

Bryant, B., McGrew, L. W., and Wobus, R. A., 1981, Geologic map of the Denver 1° × 2° quadrangle, north-central Colorado: U.S. Geological Survey Miscellaneous Investigations Series, Map I-1163.

Burke, K., 1980, Intracontinental rifts and aulacogens, *in* Continental tectonics: Washington, D.C., National Academy of Sciences, p. 42–49.

Burroughs, R. L., 1981, A summary of the geology of the San Luis Basin, Colorado–New Mexico, with emphasis on the geothermal potential for the Monte Vista graben: Colorado Geological Survey, Special Publication 17, 30 p.

Camfield, P. A., and Gough, D. I., 1977, A possible Proterozoic plate boundary in North America: Canadian Journal of Earth Sciences, v. 14, p. 1229–1238.

Cape, C. D., McGeary, S., and Thompson, G. A., 1983, Cenozoic normal faulting and the shallow structure of the Rio Grande rift near Socorro, New Mexico: Geological Society of America Bulletin, v. 95, p. 3–14.

Cashion, W. B., 1967, Geology and fuel resources of the Green River Formation southeastern Uinta Basin, Utah and Colorado: U.S. Geological Survey Professional paper 548, 48 p.

Cashion, W. B., and Donnell, J. R., 1972, Chart showing correlation of selected key units in the organic-rich sequence of the Green River Formation, Piceance Creek Basin, Colorado and Uinta Basin, Utah: U.S. Geological Survey Oil and Gas Investigations Chart, OC-65.

Cater, F. W., and Elston, D. P., 1963, Structural development of salt anticlines of Colorado and Utah, *in* Backbone of the Americas: American Association of Petroleum Geologists, Memoir 2, p. 152–159.

Chamberlain, R. M., 1983, Cenozoic domino-style crustal extensions in the Lemitar Mountains, New Mexico; A summary: New Mexico Geological Society, 34th Annual Field Conference, Guidebook, p. 111–118.

Chapin, C. E., 1971, The Rio Grande rift, part I, Modifications and additions: New Mexico Geological Society, 22nd Annual Field Conference, Guidebook, p. 191–201.

—— , 1979, Evolution of the Rio Grande rift: a summary, *in* Reicker, R. E., ed., Rio Grande rift; Tectonics and magmatism, Washington, D.C., American Geophysical Union, p. 1–5.

—— , 1983, An overview of Laramide wrench faulting in the southern Rocky Mountains with emphasis on petroleum exploration, *in* Lowell, J. D., ed., Rocky Mountain foreland basins and uplifts: Rocky Mountain Association of Geologists, p. 169–179.

Chapin, C. E., and Cather, S. M., 1981, Eocene tectonics and sedimentation in the Colorado Plateau–Rocky Mountains area, *in* Dickinson, W. R., and Payne, W. D., eds., Relations of tectonics to ore deposits in the southern Cordillera: Arizona Geological Society Digest, v. 14, p. 173–198.

—— , 1983, Eocene tectonics and sediments in the Colorado Plateau Rocky Mountain area, *in* Lowell, J. D., ed., Rocky Mountain foreland basins and uplifts: Rocky Mountain Association of Geologists, p. 33–56.

Chapin, C. E., and Lindley, J. I., 1986, Potassium metasomatism of igneous and sedimentary rocks in detachment terranes and other sedimentary basins; Economic implications: Arizona Geological Society Digest, v. 16, p. 118–126.

Chapin, C. E., and Seager, W. R., 1975, Evolution of the Rio Grande rift in the Socorro and Las Cruces areas: New Mexico Geological Society, 26th Annual Field Conference, Guidebook, p. 297–321.

Chapin, C. E., and four others, 1978, Exploration framework of the Socorro geothermal area, New Mexico, *in* Chapin, C. E., and Elston, W. E., eds., Field guide to selected cauldrons and mining districts of the Datil-Mogollon volcanic field, New Mexico: New Mexico Geological Society Special Publication 7, p. 115–129, map in pocket.

Chronic, J., and Stevens, C., 1958, Pennsylvanian paleogeography in the McCoy area, Eagle County, *in* Curtis, B. F., ed., Symposium on Pennsylvanian rocks of Colorado and adjacent areas: Rocky Mountain Association of Geologists, p. 86–90.

Clark, J. R., 1975, Controls of sedimentation and provenance of sediments in the Oligocene of the Central Rocky Mountains, *in* Curtis, B. F., ed., Cenozoic history of the southern Rocky Mountains: Geological Society of America Memoir 144, p. 95–117.

Clement, J. H., 1983, North flank of the Uinta Mountains, Utah, *in* Bally, A. W., ed., Seismic expression of structural styles: American Association of Petroleum Geologists Studies in Geology Series 15, v. 3, p. 3.2.2.29–3.2.2.32.

Cobban, W. A., and Reeside, J. B., Jr., 1952a, Correlation of the Cretaceous formations of the Western Interior of the United States: Geological Society of America Bulletin, v. 63, p. 1011–1044.

——, 1952b, Frontier Formation, Wyoming and adjacent areas: American Association of Petroleum Geologists Bulletin, v. 36, p. 1913–1961.

Colson, C. T., 1969, Stratigraphy and production of the Tertiary formations in the Sand Wash and Washakie basins, *in* Barlow, J. A., Jr., ed.: Wyoming Geological Association Guidebook, 21st Annual Field Conference, p. 121–128.

Combo, J. X., Brown, D. M., Pulver, H. F., and Taylor, D. A., 1949, Coal resources of Montana: U.S. Geological Survey Circular 53, 28 p.

Coney, P. J., and Reynolds, S. J., 1977, Cordilleran Benioff zones: Nature, v. 270, p. 403–406.

Coons, L. M., and Kelly, T. E., 1984, Regional hydrology and the effect of structural control on the flow of ground water in the Rio Grande trough, northern New Mexico: New Mexico Geological Society, 35th Annual Field Conference, Guidebook, p. 241–244.

Cordell, L., 1978, Regional geophysical setting of the Rio Grande rift: Geological Society of America Bulletin, v. 89, p. 1073–1090.

Cordell, L., Keller, G. R., and Hildenbrand, T. G., 1982, Bouguer gravity map of the Rio Grande rift: U.S. Geological Survey Geophysical Series Map GP-949, scale 1:1,000,000.

Craig, L. C., 1972, Mississippian System, *in* Mallory, W. W., ed., Geologic atlas of the Rocky Mountain region: Rocky Mountain Association of Geologists, p. 100–110.

——, 1981, Lower Cretaceous rocks, southwestern Colorado and southeastern Utah, *in* Wiegand, D. L., ed., Geology of the Paradox basin: Rocky Mountain Association of Geologists, p. 195–200.

Crittenden, M. D., Jr., 1974, Regional extent and age of thrusts near Rockport Reservoir and relation to possible exploration targets in northern Utah: American Association of Petroleum Geologists Bulletin, v. 58, p. 2428–2435.

Cross, T. A., and Pilger, R. H., Jr., 1978, Tectonic controls of Late Cretaceous sedimentation, Western Interior, USA: Nature, v. 274, p. 653–657.

Crouch, J. K., Bachman, S. B., and Shay, J. T., 1984, Post-Miocene compressional tectonics along the central California margin, *in* Crouch, J. K., and Bachman, S. B., eds., Tectonics and sedimentation along the California margin: Pacific Section, Society of Economic Paleontologists and Mineralogists, v. 38, p. 37–54.

Crowley, K. D., Ahern, J. L., and Naeser, C. W., 1985, Origin and epeirogenic history of the Williston Basin; Evidence from fission-track analysis of apatite: Geology, v. 13, p. 620–623.

Curtis, B. F., 1963, Jurassic stratigraphic relationships in the northern Denver basin, *in* Bolyard, D. W., and Katich, P. J., eds., Geology of the northern Denver basin and adjacent uplifts: Rocky Mountain Association of Geologists, p. 111–118.

Dallmus, K. F., 1958, Mechanics of basin evolution and its relation to the habitat of oil in the basin, in Week, L. G., ed., Habitat of Oil: American Association of Petroleum Geologists, p. 883–931.

Dalrymple, G. R., 1979, Critical tables for conversion of K-Ar ages from old to new constrants: Geology, v. 7, p. 558–560.

Dane, C. H., 1946, Stratigraphic relations of Eocene, Paleocene, and latest Cretaceous of eastern San Juan Basin, New Mexico: U.S. Geological Survey Oil and Gas Investigation Preliminary Chart 24.

——, 1955, Stratigraphic and facies relationships of upper part of Green River Formation and lower part of Uinta Formation in Duchesne, Uintah, and Wasatch Counties, Utah: U.S. Geological Survey Oil and Gas Chart OC-52, vertical scale, 1:600; horizontal scale, 1:125,000.

Dane, C. H., Cobban, W. A., and Kauffman, E. G., 1966, Stratigraphy and regional relationships of a reference section for the Juana Lopez Member, Mancos Shale, in the San Juan Basin, New Mexico: U.S. Geological Survey Bulletin 1224-H, 15 p.

Danilchik, W., Schultz, J. E., and Tremain, C. M., 1979, Content of absorbed methane in coal from four core holes in the Raton and Vermejo Formations, Las Animas, Colorado: U.S. Geological Survey Open-File Report 79-762.

Davidson, E. S., 1967, Geology of the Circle Cliffs area, Garfield and Kane Counties, Utah: U.S. Geological Survey Bulletin 1229, 140 p.

Davis, T. L., and Weimer, R. J., 1976, Late Cretaceous growth faulting, Denver basin, Colorado, *in* Epis, R. C., and Weimer, R. J., eds., Studies in Colorado field geology: Professional contributions, Colorado School of Mines, no. 8, p. 280–300.

Del Rio, S. M., ed., 1960, Mineral resources of Colorado, first sequel: Denver, Colorado Mineral Resources Board, 764 p.

DeVoto, R. H., 1971, Geologic history of South Park and geology of the Antero Reservoir Quadrangle, Colorado: Colorado School of Mines Quarterly, v. 66, no. 3, 90 p.

——, 1972a, Cenozoic geologic history of South Park: The Mountain Geologist, v. 9, p. 211–221.

——, 1972b, Pennsylvanian and Permian stratigraphy and tectonism in central Colorado: Colorado School of Mines Quarterly, v. 67, p. 139–185.

——, 1980a, Mississippian stratigraphy and history of Colorado, *in* Kent, H. C., and Porter, K. W., eds., Colorado geology: Rocky Mountain Association of Geologists, p. 57–70.

——, 1980b, Pennsylvanian stratigraphy and history of Colorado, *in* Kent, H. C., and Porter, K. W., eds., Colorado geology: Rocky Mountain Association of Geologists, p. 71–101.

——, 1983, Central Colorado karst-controlled lead-zinc-silver deposits (Leadville, Gilman, Aspen, and others), a Late Paleozoic Mississippi Valley–type district, *in* The genesis of Rocky Mountain ore deposits; Changes with time and tectonics: Denver Region Exploration Geologists Society, p. 51–70.

——, 1985, Sedimentology, dolomitization, karstification, and mineralization of the Leadville Limestone (Mississippian), central Colorado, Field Trip no. 6, *in* Society of Economic Paleontologists and Mineralogists Midyear Meeting field guides: Society of Paleontologists and Mineralogists, p. 6-143–6-180.

——, 1988, Late Mississippian paleokarst, and related mineral deposits, Leadville Formation, central Colorado, *in* James, N. P. and Coquette, P. W. eds., Paleokarst: New York, Springer-Verlag, p. 278–305.

DeVoto, R. H., and Peel, F. A., 1972, Pennsylvanian and Permian stratigraphy and structural history, northern Sangre de Cristo Range, Colorado: Colorado School of Mines Quarterly, v. 67, no. 4, p. 282–320.

DeVoto, R. H., Peel, F. A., and Pierce, W. H., 1971, Pennsylvanian and Permian stratigraphy, tectonism, and history, northern Sangre de Cristo Range, Colorado: New Mexico Geological Society, 22nd Field Conference Guidebook, p. 141–163.

DeVoto, R. H., Bartleson, B. L., Schenk, C. J., and Waechter, N. B., 1986, Late Paleozoic stratigraphy and syndepositional tectonism, northwestern Colorado, *in* Stone, D. S., ed., New interpretations of northwest Colorado geology: Rocky Mountain Association of Geologists, p. 37–50.

Dewey, J. F., 1982, Plate tectonics and the evolution of the British Isles: Journal of the Geological Society of London, v. 139, p. 371–412.

Dickinson, W. R., and Snyder, W. S., 1978, Plate tectonics of the Laramide orogeny: Geological Society of America Memoir 151, p. 355–366.

Dodge, C. N., and Bartleson, B. L., 1986, Minturn Formation, Eagle Basin; An exploration frontier, *in* Stone, D. S., ed., New interpretations of northwest Colorado geology: Rocky Mountain Association of Geologists, p. 113–122.

Dolly, D. E., and Meissner, F. F., 1977, Geology and gas exploration potential, Upper Cretaceous and Lower Tertiary strata, northern Raton Basin, Colorado, *in* Veal, H. K., ed., Exploration frontiers of the central and southern Rockies: Rocky Mountain Association of Geologists, p. 247–270.

Donnell, J. R., 1961, Tertiary geology and oil-shale resources of the Piceance Creek basin between the Colorado and White Rivers, northwestern Colorado: U.S. Geological Survey Bulletin 1082-L, p. 835–891.

—— , 1969, Paleocene and lower Eocene units in the southern part of the Piceance Creek Basin, Colorado: U.S. Geological Survey Bulletin 1274-M, p. M1–M18.

Dorr, J. A., Jr., Spearing, D. R., and Steidtmann, J. R., 1977, Deformation and deposition between a foreland uplift and an impinging thrust belt, Hoback Basin, Wyoming: Geological Society of America Special Paper 177, 82 p.

Driese, S. G., and Dott, R. H., Jr., 1984, Model for sandstone-carbonate "cyclothems" based on upper member of the Morgan Formation (Middle Pennsylvanian) of northern Utah and Colorado: American Association of Petroleum Geologists Bulletin, v. 67, p. 574–597.

Droullard, E. K., 1963, Tectonics of the southeast flank of the Hartville uplift, *in* Bolyard, D. W., and Katich, P. J., eds., Geology of the northern Denver Basin and adjacent uplifts: Rocky Mountain Association of Geologists, p. 176–178.

Dyni, J. R., 1974, Stratigraphy and nahcolite resources of the saline facies of the Green River Formation in northwest Colorado, *in* Energy resources of the Piceance Creek Basin, Colorado: Rocky Mountain Association of Geologist 25th Field Conference Guidebook, p. 111–112.

—— , 1981, Geology and nahcolite deposits and associated oil shales of the Green River Formation on the Piceance Creek Basin, Colorado [Ph.D. thesis]: Boulder, University of Colorado, 144 p.

Eaton, G. P., 1979, Regional geophysics, Cenozoic tectonics, and geologic resources of the Basin and Range Province and adjoining regions, *in* Newman, G. W., and Goode, H. D., eds., Basin and Range Symposium: Rocky Mountain Association of Geologists, p. 11–39.

Ekas, L. M., Ingersoll, R. V., Baldridge, W. S., and Shafiquallah, M., 1984, The Chama–El Rito Member of the Tesuque Formation, Espanola Basin, New Mexico: New Mexico Geological Society, 35th Annual Field Conference, Guidebook, p. 137–143.

Eldridge, G. H., 1894, Anthracite–Crested Butte; Folio 9: U.S. Geological Survey Geologic Atlas of the United States.

Elston, D. P., and Landis, E. R., 1960, Pre-Cutler unconformities and early growth of the Paradox Valley and Gypsum Valley salt anticlines, Colorado, *in* Short papers in the geological sciences: U.S. Geological Survey Professional Paper 400-B, p. B261–265.

Elston, W. E., and Bornhorst, T. J., 1979, The Rio Grande rift in context of regional post–40 m.y. volcanic and tectonic events, *in* Riecker, R. E., ed., Rio Grande rift; Tectonics and magmatism: American Geophysical Union, p. 416–438.

Energy Information Administration, 1984, Coal production—1983: Energy Information Administration DOE/EIA-0118(83), 118 p.

Engebretson, D. C., Cox, A., and Thompson, G. A., 1984, Correlation of plate motions with continental tectonics, Laramide to basin-range: Tectonics, v. 3, n. 2, p. 115–120.

Epis, R. C., and Chapin, C. E., 1968, Geological history of the Thirty-nine Mile Volcanic Field, central Colorado: Colorado School of Mines Quarterly, v. 63, no. 3, p. 51–85.

—— , 1974, Stratigraphic nomenclature of the Thirty-nine Mile volcanic field, central Colorado: U.S. Geological Survey Bulletin 1395-C, 23 p.

—— , 1975, Geomorphic and tectonic implications of the post-Laramide, Late Eocene erosion surface in the Southern Rocky Mountains: Geological Society of America Memoir 144, p. 45–74.

Epis, R. C., Scott, G. R., Taylor, R. B., and Chapin, C. E., 1980, Summary of Cenozoic geomorphic, volcanic, and tectonic features of central Colorado and adjoining areas, *in* Kent, H. D., and Porter, K. W., eds., Colorado geology: Rocky Mountain Association of Geologists, p. 135–156.

Eugster, H. P., and Hardy, L. A., 1975, Sedimentation in an ancient playa-like complex; The Wilkins Peak Member of the Green River Formation of Wyoming: Geological Society of America Bulletin, v. 86, p. 319–334.

Eugster, H. P., and Surdam, R. C., 1973, Depositional environment of the Green River Formation of Wyoming; A preliminary report: Geological Society of America Bulletin, v. 84, no. 4, p. 1115–1120.

Fassett, J. E., ed., 1978, Oil and gas fields of the Four Corners area, v. 1 and 2: Durango, Colorado, Four Corners Geological Society, 727 p.

—— , 1982, Dinosaurs in the San Juan Basin, New Mexico, may have survived the event that results in creation of an iridium-enriched zone near the Cretaceous/Tertiary boundary, *in* Silver, L. P., and Schultz, P. H., eds., Geological implications of impacts of large asteroids and comets on the earth: Geological Society of America Special Paper 190, p. 435–447.

—— , 1983, Stratigraphy and oil and gas production of Northwest New Mexico updated through 1983, *in* Oil and gas fields of the Four Corners area, v. 3: Durango, Colorado, Four Corners Geological Society, p. 849–863.

—— , 1985, Early Tertiary paleogeography and paleotectonics of the San Juan Basin area, New Mexico and Colorado, *in* Flores, R. M., and Kaplan, S. S., eds., Cenozoic paleogeography of the west-central United States, Rocky Mountain Paleogeography Symposium 3: Rocky Mountain Section, Society of Economic Paleontologists and Mineralogists, p. 317–334.

Fassett, J. E., and Hinds, J. S., 1971, Geology and fuel resources of the Fruitland Formation and Kirtland Shale of the San Juan Basin, New Mexico and Colorado: U.S. Geological Survey Professional Paper 676, 76 p.

Fassett, J. E., Lucas, S. G., and O'Neill, F. M., 1986, Dinosaurs, pollen and spores, and the age of the Ojo Alamo Sandstone, San Juan Basin, New Mexico, *in* The Cretaceous-Tertiary boundary in the San Juan and Raton Basins, New Mexico and Colorado: Geological Society of America Special Paper 209, p. 17–34.

Feathers, K. R., Libra, R., and Stephenson, T. R., 1981, Occurrence and characteristics of ground water in the Powder River Basin, Wyoming: Water Resources Research Institute, University of Wyoming, v. 1-A, 171 p.

Fetzner, R. W., 1960, Pennsylvanian paleotectonics of Colorado Plateau: American Association of Petroleum Geologists Bulletin, v. 44, no. 8, p. 1371–1413.

Foose, R. M., Wise, D. U., and Garbarini, G. S., 1961, Structural geology of the Beartooth Mountains, Montana and Wyoming: Geological Society of America Bulletin, v. 72, p. 1143–1172.

Foster, N. H., 1972, Ordovician System, *in* Mallory, W. W., ed., Geologic atlas of the Rocky Mountain region: Rocky Mountain Association of Geologists, p. 76–85.

Fouch, T. D., 1976, Revision of the lower part of the Tertiary system in the central and western part of the Uinta Basin, Utah: U.S. Geological Survey Bulletin 1405-C, p. C1–C7.

—— , 1981, Distribution of rock types, lithologic groups, and interpreted depositional environments for some lower Tertiary and Upper Cretaceous rocks from outcrops at Willow Creek–Indian Canyon through the subsurface of Duchesne and Altamont oil fields, southwest to north-central parts of the Uinta Basin, Utah: U.S. Geological Survey Oil and Gas Investigations Chart OC-81, 2 sheets.

Fouch, T. D., and Cashion, W. B., 1979, Distribution of rock types, lithologic groups, and depositional environment for some lower Tertiary and Upper and Lower Cretaceous, and Upper and Middle Jurassic rocks in the subsurface between Altamont oil field and San Arroyo gas field, north-central and northeastern Uinta Basin, Utah: U.S. Geological Survey Open-File Report 79-365, 2 sheets.

Fouch, T. D., and four others, 1983, Patterns and timing of synorogenic sedimentation in Upper Cretaceous rocks of central and northeast Utah, *in* Mesozoic paleogeography of the west-central United States: Rocky Mountain Paleogeography Symposium 2, Rocky Mountain Section, Society of Economic Paleontologists and Mineralogists, p. 305, 336.

Frahme, C. W., and Vaughn, E. B., 1983, Paleozoic geology and seismic stratigraphy of the northern Umcompahgre front, *in* Lowell, J. D., ed., Rocky Mountain foreland basins and uplifts: Rocky Mountain Association of Geologists Guidebook, p. 201–211.

Franczyk, K. J., and Nichols, D. J., 1986, Unconformity-bounded Paleocene conglomerate sequence, southeastern Uinta Basin [abs.]: American Association of Petroleum Geologists, v. 70, no. 8, p. 1039.

Freeman, V. L., 1971, Permian deformation in the Eagle basin, Colorado, *in* Geological Survey Research 1971: U.S. Geological Survey Professional Paper 750-D, p. D80–D83.

Freeman, V. L., and Bryant, B., 1977, Red bed formations in the Aspen region, Colorado, *in* Veal, H. K., ed., Exploration frontiers of the central and southern Rockies: Rocky Mountain Association of Geologists, p. 181–189.

Fryberger, S. G., 1978, Stratigraphy of the Weber Formation (Pennsylvanian and Permian), Dinosaur National Monument and adjacent area, Utah and Colorado [M.S. thesis]: Golden, Colorado School of Mines, 105 p.

—— , 1979, Eolian-fluviatile (continental) origin of ancient stratigraphic trap for petroleum in Weber Sandstone, Rangely oil field, Colorado: The Mountain Geologist, v. 16, no. 1, p. 1–29.

Galton, P. M., 1978, Fabrosauridae, the basal family of ornithischian dinosaurs (Reptilia: Ornithopoda): Paleontologische Zeitschrift, v. 52, p. 138–159.

Galusha, T., and Blick, C., 1971, Stratigraphy of the Santa Fe Group, New Mexico: American Museum of Natural History Bulletin, v. 144, no. 1, 127 p.

Garing, J. D., and Tainter, P. A., 1985, Greater Green River basin regional seismic line, *in* Gries, R. R., and Dyer, R. C., eds., Seismic exploration of the Rocky Mountain region: Rocky Mountain Association of Geologists and the Denver Geophysical Society, p. 233–238.

Gile, L. H., Hawley, J. W., and Grossman, R. B., 1981, Soils and geomorphology in the Basin and Range area of southern New Mexico, Guidebook to the Desert Project: Socorro, New Mexico Bureau of Mines and Mineral Resources Memoir 39, 222 p., 2 maps.

Gill, J. R., 1950, Flagstaff Limestone of the Spring City–Manti area, Sanpete County, Utah [M.S. thesis]: Ohio State University, 209 p.

Gill, J. R., and Cobban, W. A., 1966, Regional unconformity in Late Cretaceous, Wyoming, *in* Geological Survey research 1966: U.S. Geological Survey Professional Paper 550-B, p. B20–B27.

Gilluly, J., and Reeside, J. B., Jr., 1928, Sedimentary rocks of the San Rafael Swell and some adjacent areas in eastern Utah: U.S. Geological Survey Professional Paper 150-D, p. 61–110.

Glass, G. B., 1984, Wyoming description of seams, *in* 1984 Keystone Coal Industry Manual: New York, McGraw Hill, p. 637–663.

Glass, G. B., and Roberts, J. T., 1978, Update on the Wind River coal basin: Wyoming Geological Association Guidebook, 30th Annual Field Conference, p. 363–377.

Gorham, F. D., Jr., 1975, Tectogenesis of the central Colorado Plateau aulacogen: Four Corners Geological Society, 8th Field Conference Guidebook, p. 211–216.

Gries, R. R., 1981, Oil and gas prospecting beneath the Precambrian of foreland thrust plates in the Rocky Mountains: The Mountain Geologist, v. 18, p. 1–18.

—— , 1983a, Oil and gas prospecting beneath Precambrian of foreland thrust plates in the Rocky Mountains: American Association of Petroleum Geologists Bulletin, v. 67, no. 1, p. 1–28.

—— , 1983b, North-south compression of Rocky Mountain foreland structure, *in* Lowell, J. D., ed., Rocky Mountain foreland basins and uplifts: Rocky Mountain Association of Geologists, 1983, p. 9–32.

—— , 1985, Seismic lines in the San Luis Valley, south-central Colorado, *in* Gries, R. R., and Dyer, R. C., eds., Seismic exploration of the Rocky Mountain region: Rocky Mountain Association of Geologists, p. 267–274.

Griggs, R. L., 1964, Geology and groundwater resources of the Los Alamos area, New Mexico: U.S. Geological Survey, Water-Supply Paper 1753, 107 p.

Gutentag, E. D., and others, 1984, Geohydrology of the High Plains aquifer in parts of Colorado, Kansas, Nebraska, New Mexico, Oklahoma, South Dakota, Texas and Wyoming: U.S. Geological Survey Professional Paper 1400-B, 63 p.

Hackman, R. J., and Olson, A. B., 1977, Geology, structure, and uranium deposits of the Gallup 1° × 2° quadrangle, New Mexico and Arizona: U.S. Geological Survey Miscellaneous Investigations Series Map I-981.

Hackman, R. J., and Wyant, D. G., 1973, Geology, structure, and uranium deposits of the Escalante 1° × 2° quadrangle, Utah and Arizona: U.S. Geological Survey Miscellaneous Investigations Series Map I-744.

Hail, W. J., 1972, Preliminary geologic map of the Barcus Creek southeast quadrangle, Rio Blanco County, Colorado: U.S. Geological Survey Miscellaneous Field Investigations Map MF-1189, 2 sheets.

—— , 1973, Geologic map of the Smizer Gulch quadrangle, Rio Blanco and Moffat Counties, Colorado: U.S. Geological Survey Quadrangle Map GQ-1131.

—— , 1974, Geologic map of the Rough Gulch quadrangle, Rio Blanco and Moffat Counties, Colorado: U.S. Geological Survey Geologic Quadrangle Map GQ-1195.

Hamilton, W., 1978, Mesozoic tectonics of the western United States, *in* Howell, D. G., and McDougall, K. A., eds., Mesozoic paleogeography of the western United States: Pacific Section, Society of Economic Paleontologists and Mineralogists, Pacific Coast Paleogeography Symposium 2, p. 33–70.

—— , 1981, Plate tectonic mechanism of Laramide deformation, *in* Boyd, D. W., and Lillegraven, J. A., eds., Rocky Mountain Foreland tectonics: Contributions to Geology, University of Wyoming, v. 19, p. 87–92.

Hansen, W. R., 1965, Geology of the Flaming Gorge area, Utah-Colorado-Wyoming: U.S. Geological Survey Professional Paper 490, 196 p.

—— , 1984, Published letter to The Mountain Geologist Forum: Mountain Geologist, Rocky Mountain Association of Geologists, v. 21, no. 3, p. 73.

Hansen, W. R., and Crosby, E. J., 1982, Environmental geology of the Front Range Urban Corridor and vicinity, Colorado: U.S. Geological Survey Professional Paper 1230, 99 p.

Hansley, P. L., 1981, Mineralogy, diagenesis, and provenance of Upper Cretaceous sandstones from the Ralston Production Company Federal no. 31 well, Piceance Creek Basin, northwestern Colorado: U.S. Geological Survey Open-File Report 81-1295, 21 p.

Harbour, R. L., and Dixon, G. H., 1959, Coal resources, Trinidad-Aguila area, Colorado: U.S. Geological Survey Bulletin 1072 G, p. 445–484.

Hardin, F. R., and Hardin, G. C., 1961, Contemporaneous normal faults of Gulf Coast and their relation to flexures: American Association of Petroleum Geologists Bulletin, v. 45, p. 238–248.

Harshbarger, J. W., Repenning, C. A., and Irwin, J. H., 1957, Stratigraphy of the uppermost Triassic and Jurassic rocks of the Navajo Country: U.S. Geological Survey Professional Paper 291, 74 p.

Hawley, J. W., 1975, Quaternary history of Dona Ana County region, south-central New Mexico: New Mexico Geological Society, 26th Annual Field Conference, Guidebook, p. 139–150.

—— , compiler, 1978, Guidebook to Rio Grande rift in New Mexico and Colorado: New Mexico Bureau of Mines and Mineral Resources Circular 163, 241 p., 2 maps. Scale 1:1,000,000.

Hawley, J. W., Bachman, G. O., and Manley, K., 1976, Quaternary stratigraphy in the Basin and Range and Great Plains provinces, New Mexico and western Texas, *in* Maheny, W. C., ed., Quaternary stratigraphy of North America: Stroudsburg, Pennsylvania, Dowden, Hutchinson and Ross, Inc., p. 235–274.

Haynes, D. D., and Hackman, R. J., 1978, Geology, structure, and uranium deposits of the Marble Canyon 1° × 2° quadrangle, Arizona: U.S. Geological Survey Miscellaneous Investigations Series Map I-1003, scale 1:250,000.

Heylmun, E. B., 1958, Paleozoic stratigraphy and oil possibilities of the Kaiparowits region, Utah: American Association of Petroleum Geologists Bulletin, v. 42, no. 8, p. 1781–1811.

Hills, F. A., Houston, R. S., and Subbarayudu, G. V., 1975, Possible Proterozoic plate boundary in southern Wyoming: Geological Society of America Abstracts with Programs, v. 7, p. 614.

Hills, R. C., 1890, Orographic and structural features of Rocky Mountain geology: Colorado Scientific Society Proceedings, v. 3, p. 362–458.

Hite, R. J., 1960, Stratigraphy of the saline facies of the Paradox Member of the Hermosa Formation of southeastern Utah and southwestern Colorado: Four Corners Geological Society, 3rd Annual Field Conference, Guidebook, p. 86–89.

—— , 1975, An unusual northeast-trending fracture zone and its relations to basement wrench faulting in northern Paradox basin, Utah and Colorado: Four Corners Geological Society, 8th Annual Field Conference, Guidebook,

p. 217–223.
Hite, R. J., and Cater, F. W., 1972, Pennsylvanian rocks and salt anticlines, Paradox basin, Utah and Colorado, *in* Mallory, W. W., ed., Geologic atlas of the Rocky Mountain region: Rocky Mountain Association of Geologists, p. 133–138.
Hoffman, F. M., 1957, Possibilities of Weber stratigraphic traps, Rangely area, northwest Colorado: American Association of Petroleum Geologists Bulletin, v. 41, p. 894–905.
Hollister, J. C., and Weimer, R. J., 1968, Geophysical and geological studies of the relationships between the Denver earthquakes and the Rocky Mountain Arsenal well: Colorado School of Mines Quarterly, v. 63, no. 1, 251 p.
Hoppin, R. A., and Palmquist, J. C., 1965, Basement influence on later deformation; The problem, techniques of investigation, and examples from the Bighorn Mountains, Wyoming: American Association of Petroleum Geologists Bulletin, v. 49, p. 993–1003.
Houston, R. S., 1971, Regional tectonics of the Precambrian rocks of the Wyoming Province and its relationship to Laramide structure: Wyoming Geological Association, 23rd Annual Field Conference, Guidebook, p. 19–28.
Hunt, C. B., Averitt, P., and Miller, R. L., 1953, Geology and geography of the Henry Mountains region, Utah: U.S. Geological Survey Professional Paper 228, 234 p.
Imlay, R. W., 1982, Jurassic (Oxfordian and Late Callovian) ammonites from the western interior region of the United States: U.S. Geological Survey Professional Paper 1232, 44 p.
Ingersoll, R. V., 1982, Triple-junction instability as course for late Cenozoic extension and fragmentation of the western United States: Geology, v. 10, p. 621–624.
Irtem, O., 1977, Stratigraphy of the Minturn Formation (Pennsylvanian) between Glenwood Springs and Craig, Colorado [Ph.D. thesis]: Golden, Colorado School of Mines, 197 p.
—— , 1983, Algal limestones within the Minturn Formation, Meeker to Dotsero area, western Colorado: Colorado School of Mines Quarterly, v. 78, no. 2, 14 p.
Izett, G. A., 1975, Late Cenozoic sedimentation and deformation in northern Colorado and adjoining areas, *in* Curtis, B. F., ed., Cenozoic history of the southern Rocky Mountains: Geological Society of America Memoir 144, p. 179–209.
Jacob, A. F., 1983, Mountain front thrust, southeastern Front Range and northeastern Wet Mountains, Colorado, *in* Lowell, J. D., ed., Rocky Mountain foreland basins and uplifts: Rocky Mountain Association of Geologists, p. 229–244.
Johnson, R. B., 1959, Geology of the Huerfano Park area, Huerfano and Custer Counties, Colorado: U.S. Geological Survey Bulletin, 1071-D, 118 p.
Johnson, R. B., Wood, G. H., Jr., and Harbour, R. L., 1958, Preliminary geology map of the northern part of the Raton Mesa region and Huerfano Park in parts of Las Animas, Huerfano, and Custer Counties, Colorado: U.S. Geological Survey Oil and Gas Investigations Map OM 183.
Johnson, R. C., 1975, Preliminary geologic map, oil-shalc yicld histograms and stratigraphic sections, Long Point quadrangle, Garfield County, Colorado: U.S. Geological Survey Miscellaneous Field Investigations Map MF-688.
—— , 1977, Preliminary geologic map and cross section of "The Saddle" quadrangle, Garfield County, Colorado: U.S. Geological Survey Miscellaneous Field Investigations Map MF-829.
—— , 1979a, Cross Section A-A′ of Upper Cretaceous and lower Tertiary rocks, northern Piceance Creek Basin, Colorado: U.S. Geological Survey Miscellaneous Field Investigations Map MF-1129-A, 2 sheets.
—— , 1979b, Cross section B-B′ of Upper Cretaceous and lower Tertiary rocks, northern Piceance Creek Basin, Colorado: U.S. Geological Survey Miscellaneous Field Investigations Map MF-1129-B, 2 sheets.
—— , 1979c, Cross section C-C′ of Upper Cretaceous and lower Tertiary rocks, northern Piceance Creek Basin, Colorado: U.S. Geological Survey Miscellaneous Field Investigations Map MF-1129-C, 2 sheets.
—— , 1981, Stratigraphic evidence for a deep Eocene Lake Uinta, Piceance Creek Basin, Colorado: Geology, v. 9, no. 2, p. 55–62.
—— , 1982, A measured section of the Late Cretaceous Mesa Verde Group and the lower part of the Wasatch Formation, Rifle Gap, Colorado: U.S. Geological Survey Open-File Report 82-590.
—— , 1984, New names of units in the lower part of the Green River Formation, Piceance Creek basin, Colorado: U.S. Geological Survey Bulletin 1529-I, p. 11–120.
—— , 1985, Early Cenozoic history of the Uinta and Piceance Creek basins, Utah and Colorado, with special reference to the development of Lake Uinta, *in* Flores, R. M., and Kaplan, S. S., eds., Cenozoic paleogeography of the west-central United States, Rocky Mountain Paleogeography Symposium 3: The Rocky Mountain Section, Society of Economic Paleontologists and Mineralogists, p. 247–276.
Johnson, R. C., and Finn, T. M., 1985, Age of the Douglas Creek arch, Colorado and Utah: American Association of Petroleum Geologists Bulletin, v. 69, no. 3, p. 270.
—— , 1986, Cretaceous through Holocene history of the Douglas Creek arch, Colorado and Utah, *in* Stone, D. S., ed., New interpretations of northwest Colorado Geology: Rocky Mountain Association of Geologists, p. 77–95.
Johnson, R. C., and May, F., 1978, Preliminary stratigraphy studies of the upper part of the Mesa Verde Group, the Wasatch Formation, and the lower part of the Green River Formation, DeBeque area, Colorado, including environments of deposition and investigations of palynomorph assemblages: U.S. Geological Survey Miscellaneous Field Studies Map MF-1050, 2 sheets.
—— , 1980, A study of the Cretaceous-Tertiary unconformity in the Piceance Creek Basin, Colorado; The underlying Ohio Creek Formation (Upper Cretaceous) redefined as a member of the Hunter Canyon or Mesa Verde Formation: U.S. Geological Survey Bulletin, 1482-B, 27 p.
Johnson, R. C., Granica, M. P., and Dessenberger, N. C., 1979a, Cross section A-A′ of Upper Cretaceous and lower Tertiary rocks, southern Piceance Creek Basin, Colorado: U.S. Geological Survey Miscellaneous Field Investigations Map MF-1130-A, 2 sheets.
—— , 1979b, Cross section B-B′ of Upper Cretaceous and lower Tertiary rocks, southern Piceance Creek Basin, Colorado: U.S. Geological Survey Miscellaneous Field Investigations Map MF-1130-B, 2 sheets.
—— , 1979c, Cross section C-C′ of Upper Cretaceous and lower Tertiary rocks in the southern Piceance Creek Basin, Colorado: U.S. Geological Survey Miscellaneous Field Investigations Map MF-1130-C, 2 sheets.
Johnson, S. Y., 1987, Sedimentology and paleogeographic significance of six fluvial sandstone bodies in the Maroon Formation, northern Eagle basin, northwest Colorado: U.S. Geological Survey Bulletin 1787-A, 18 p.
Jordan, T. E., 1981, Thrust loads and foreland basin evolution, Cretaceous, western United States: American Association of Petroleum Geologists Bulletin, v. 65, p. 2506–2520.
Katich, P. J., 1958, Stratigraphy of the Eagle Evaporites, *in* Curtis, B. F., ed., Symposium on Pennsylvanian rocks of Colorado and adjacent areas: Rocky Mountain Association of Geologists, p. 106–110.
Keefer, W. R., 1965, Stratigraphy and geologic history of the uppermost Cretaceous, Paleocene, and lower Eocene rocks in the Wind River Basin, Wyoming: U.S. Geological Survey Professional Paper 495-A, 77 p.
—— , 1970, Structural geology of the Wind River Basin, Wyoming: U.S. Geological Survey Professional Paper 495-D, 35 p.
Keefer, W. R., and Love, J. D., 1963, Laramide vertical movements in central Wyoming: Wyoming Contributions to Geology, v. 2, no. 1, p. 47–54.
Keefer, W. R., and Van Lieu, J. A., 1966, Paleozoic formations in the Wind River Basin, Wyoming: U.S. Geological Survey Professional Paper 459-B, 60 p.
Keller, E. R., and Thomaidis, N. D., 1971, Petroleum potential of southwestern Wyoming and adjacent areas, *in* Cram, I. H., ed., Future petroleum provinces of the United States; Their geology and potential: American Association of Petroleum Geologists Memoir 15, p. 656–672.
Keller, G. R., Braile, L. W., and Schlue, J. W., 1978, Regional crustal structure of the Rio Grande rift from surface wave dispersion measurements: International Symposium on the Rio Grande Rift, Program and Abstracts, Los Alamos Scientific Laboratory, p. 47–48.

Keller, G. R., Cordell, L., Davis, G. H., and others, 1984, A geophysical study of the San Luis Basin: New Mexico Geological Society, 35th Annual Field Conference, Guidebook, p. 51–57.

Kelley, V. C., 1951, Tectonics of the San Juan Basin, *in* Guidebook of the south and west sides of the San Juan Basin: New Mexico Geological Society, p. 124–131.

—— , 1955a, Monoclines of the Colorado Plateau: Geological Society of America Bulletin, v. 66, p. 789–804.

—— , 1955b, Regional tectonics of the Colorado Plateau and relationship to the origin and distribution of uranium: University of New Mexico Publications in Geology, no. 5, 120 p.

—— , 1972, New Mexico lineament of the Colorado Rockies front: Geological Society of America Bulletin, v. 83, p. 1849–1852.

—— , 1977, Geology of Albuquerque Basin, New Mexico: Socorro, New Mexico Bureau of Mines and Mineral Resources Memoir 33, 59 p. Map, 1:190,000.

—— , 1978, Geology of Espanola Basin, New Mexico: Socorro, New Mexico Bureau of Mines and Mineral Resources, Geologic Map 48, 1:125,000.

—— , 1979, Tectonics, middle Rio Grande rift, New Mexico, *in* Riecker, R. E., ed., Rio Grande rift; Tectonics and magmatism: American Geophysical Union, p. 57–70.

Kelly, K. E., 1984, Stratigraphy and sedimentology of the Pennsylvanian Coffman Member of the Minturn Formation and Belden Formation, Mosquito Range, Colorado [M.S. thesis]: Golden, Colorado School of Mines, 107 p.

King, P. B., compiler, 1969, Tectonic map of North America: U.S. Geological Survey, 2 sheets, scale 1:5,000,000.

Kluth, C. G., and Coney, P. J., 1981, Plate tectonics of the ancestral Rocky Mountains: Geology, v. 9, p. 10–15.

Knight, S. H., 1951, The Late Cretaceous–Tertiary history of the northern portion of the Hanna basin, Carbon County, Wyoming, *in* Wyoming Geological Association Guidebook, 6th Annual Field Conference, p. 45–53.

Koch, W. J., 1976, Lower Triassic facies in the vicinity of the Cordilleran hingeline, western Wyoming, southeastern Idaho and Utah, *in* Hill, J. G., ed., Geology of the Cordilleran hingeline: Rocky Mountain Association of Geologists, p. 203–218.

Kocurek, G., and Dott, R. H., Jr., 1983, Jurassic paleogeography and paleoclimate of the central and southern Rocky Mountain region, *in* Reynolds, M. W., and Dolly, E. D., eds., Mesozoic paleogeography of the west-central United States: Rocky Mountain Section, Society of Economic Paleontologists and Mineralogists, Rocky Mountain Paleogeography Symposium 2, p. 101–116.

Koelmel, M. H., 1986, Paleotectonic, stratigraphic, and diagenetic history of the Weber Sandstone in the Rangely area, *in* Peterson, J. A., ed., Paleotectonics and sedimentation in the Rocky Mountain region: American Association of Petroleum Geologists Memoir 41, p. 371–376.

Lambert, P. W., 1966, Notes on the late Cenozoic geology of the Taos-Questa area, New Mexico: New Mexico Geological Society, 17th Annual Field Conference, Guidebook, p. 43–50.

Landis, E. R., Dane, C. B., and Cobban, W. A., 1973, Stratigraphic terminology of the Dakota Sandstone and Mancos Shale, west-central New Mexico: U.S. Geological Survey Bulletin 1372-J, 44 p.

Landwehr, W. R., 1967, Belts of major mineralization in western United States: Economic Geology, v. 62, p. 494–501.

Langenheim, R. L., 1952, Pennsylvanian and Permian stratigraphy in the Crested Butte quadrangle, Gunnison County, Colorado: American Association of Petroleum Geologists Bulletin, v. 36, p. 543–574.

—— , 1954, Correlation of Maroon Formation in the Crystal River Valley, Gunnison, Pitkin, and Garfield Counties, Colorado: American Association of Petroleum Geologists Bulletin, v. 38, p. 1748–1779.

LaRocque, A., 1960, Molluscan faunas of the Flagstaff Formation of central Utah: Geological Society of America Memoir 78, 100 p.

Lewis, C. J., Wilde, D. E., and Gerhard, L. C., 1969, Basement structure map of the Raton basin area: Mountain Geologist, v. 6, p. 85–86.

Libra, R., Doremus, D., and Goodwin, C., 1981, Occurrence and characteristics of ground water in the Bighorn Basin, Wyoming: Water Resources Research Institute, University of Wyoming, v. II-A, 114 p.

Lindsey, D. A., Johnson, B. R., and Andriessen, P.A.M., 1983, Laramide and Neogene structure of the northern Sangre de Cristo Range, south-central Colorado, *in* Lowell, J. D., ed., Rocky Mountain foreland basins and uplifts: Rocky Mountain Association of Geologists, p. 219–228.

Lipman, P. W., 1983, The Miocene Questa caldera, northern New Mexico; Relation to batholith emplacement and associated molybdenum mineralization, *in* The genesis of Rocky Mountain ore deposits; Changes in time and tectonics: Denver Region Exploration Geologists Society, p. 133–149.

Lipman, P. W., and Mehnert, H. H., 1975, Late Cenozoic basaltic volcanism and development of the Rio Grande depression in the southern Rocky Mountains: Geological Society of America Memoir 144, p. 119–154.

—— , 1979, The Taos Plateau volcanic field, northern Rio Grande rift, New Mexico, *in* Reicker, R. E., ed., Tectonics and magmatism: Washington, D.C., American Geophysical Union, p. 289–311.

Lipman, P. W., Mehnert, H. H., and Naeser, C. W., 1986, Evolution of the Latir volcanic field, northern New Mexico, and its relation to the Rio Grande rift, as indicated by potassium-argon and fission track dating: Journal of Geophysical Research, v. 91, p. 6329–6345.

Livaccari, R. F., and Keith, S. B., 1984, Tectonic evolution of Sevier-Laramide foreland structures from latest Jurassic through the Eocene [abs.]: American Association of Petroleum Geologist Bulletin, v. 68, no. 4, p. 501.

—— , 1985, Paleotectonic maps, Sevier to Laramide orogeny: [unpublished] Earth Satellite Corporation and Magmachem Exploration.

Livesey, G. B., 1985, Laramide structures of the southeastern Sand Wash basin, *in* Gries, R. R., and Dyer, R. C., eds., Seismic exploration of the Rocky Mountain region: Rocky Mountain Association of Geologists and Denver Geophysical Society, p. 87–94.

Lochman-Balk, C., 1972, Cambrian System, *in* Mallory, W. W., ed., Geologic atlas of the Rocky Mountain region: Rocky Mountain Association of Geologists, p. 60–75.

Love, J. D., 1951, Zones of stratigraphic thinning in relation to oil and gas possibilities in Wyoming [abs.]: American Association of Petroleum Geologists Bulletin, v. 35, no. 5, p. 1108.

—— , 1952, Preliminary report on uranium deposits in the Pumpkin Buttes area, Powder River Basin, Wyoming: U.S. Geological Survey Circular 176, 37 p.

—— , 1954, Tentative diagrammatic correlation of Tensleep, Amsden, Casper, and Hartville formations in Wyoming: Wyoming Geological Association, Guidebook, 9th Annual Field Conference, 1954, map in pocket.

—— , 1960, Cenozoic sedimentation and crustal movements in Wyoming: American Journal of Science, v. 258-A, p. 204–214.

—— , 1970, Cenozoic geology of the Granite Mountains area, central Wyoming: U.S. Geological Survey Professional Paper 495-C, 154 p.

—— , 1978, Cenozoic thrust and normal faulting, and tectonic history of the Badwater area, northeastern margin of Wind River Basin, Wyoming, *in* Wind River Basin Guidebook: Wyoming Geological Association, Guidebook, 30th Annual Field Conference, p. 235–238.

—— , 1982, A possible gap in the western thrust belt in Idaho and Wyoming, *in* Powers, R. B., ed., Geological studies of the Cordilleran thrust belt: Rocky Mountain Association of Geologists, v. 1, p. 247–259.

Love, J. D., and Christiansen, A. C., 1980, Preliminary correlation of stratigraphic units used on $1° \times 2°$ geologic quadrangle maps in Wyoming, *in* Stratigraphy of Wyoming: Wyoming Geological Association Guidebook, 31st Annual Field Conference, p. 279–282, and separate stratigraphic chart.

—— , 1985, Geologic map of Wyoming: U.S. Geological Survey Map, 3 sheets, scale 1:500,000.

Love, J. D., McGrew, P. O., and Thomas, H. D., 1963, Relationship of latest Cretaceous and Tertiary deformation to oil and gas occurrences in Wyoming, *in* Backbone of the Americas: American Association of Petroleum Geologists Memoir 2, p. 196–208.

Love, J. D., Christiansen, A. C., and Sever, C. K., 1980, Geologic map of the Torrington 1° by 2° quadrangle, southeastern Wyoming and western Nebraska: U.S. Geological Survey Miscellaneous Field Studies Map, MF-1184, scale 1:250,000.

Lowell, J. D., 1970, Antithetic faults in upthrusting: American Association of Petroleum Geologists Bulletin, v. 54, p. 1946–1950.
—— , 1972, Spitsbergen Tertiary orogenic belt and the Spitsbergen fracture zone: Geological Society of America Bulletin, v. 83, p. 3091–3102.
Lozinsky, R. P., and Hawley, J. W., 1986, The Palomas Formation of south-central New Mexico; A formal definition: New Mexico Geology, v. 8, p. 73–78.
Machette, M. N., 1978, Geologic map of the San Acacia quadrangle, Socorro County, New Mexico: U.S. Geological Survey, Geologic Quadrangle Map GQ 1415, scale 1:24,000.
MacKenzie, D. B., and Poole, D. M., 1962, Provenance of Dakota Group sandstones of the western interior, *in* Enyert, R. L., and Curry, W. H., III., eds., Symposium on early Cretaceous rocks of Wyoming and adjacent areas: Wyoming Geological Association, p. 62–71.
MacLachlan, J. C., and Kleinkopf, M. D., 1969, eds., Configuration of the Precambrian surface of Colorado: Mountain Geologist, v. 6, no. 4, p. 193–197.
MacLachlan, M. E., compiler, 1972, Triassic System, *in* Mallory, W. W., ed., Geologic atlas of the Rocky Mountain region: Rocky Mountain Association of Geologists, p. 166–176.
Mallory, W. W., 1958, Pennsylvanian coarse arkosic red beds and associated mountains in Colorado, *in* Curtis, B. F., ed., Symposium on Pennsylvanian rocks in Colorado and adjacent areas: Rocky Mountain Association of Geologists, p. 17–20.
—— , 1960, Outline of Pennsylvanian stratigraphy of Colorado, *in* Guide to the geology of Colorado: Rocky Mountain Association of Geologists, p. 23–33.
—— , 1971, The Eagle Valley Evaporite, northwest Colorado; Regional synthesis: U.S. Geological Survey Bulletin 1311-E, p. E1–E37.
—— , 1972a, Regional synthesis of the Pennsylvanian System, *in* Mallory, W. W., ed., Geologic atlas of the Rocky Mountain region: Rocky Mountain Association of Geologists, p. 111–127.
—— , 1972b, Pennsylvanian arkose and Ancestral Rocky Mountains, *in* Mallory, W. W., ed., Geologic atlas of the Rocky Mountain region: Rocky Mountain Association of Geologists, p. 131–132.
—— , ed., 1972c, Geologic atlas of the Rocky Mountain region: Rocky Mountain Association of Geologists, 331 p.
—— , 1977, Regional aspects of the Eagle Valley Evaporite, *in* Veal, H. K., ed., Exploration frontiers of the Central and Southern Rockies: Rocky Mountain Association of Geologists, p. 191–196.
Manley, K., 1979, Stratigraphy and structure of the Espanola Basin, Rio Grande rift, New Mexico, *in* Riecker, R. E., ed., Tectonics and magmatism: American Geophysical Union, p. 71–86.
—— , 1981, Redefinition and description of the Los Pinos Formation of north-central New Mexico: Geological Society of America Bulletin, pt. I, v. 92, p. 984–989.
—— , 1984, Brief summary of the Tertiary geologic history of the Rio Grande rift in northern New Mexico: New Mexico Geological Society, 35th Annual Field Conference, Guidebook, p. 63–66.
Martin, C. A., 1965, Denver basin: American Association of Petroleum Geologists Bulletin, v. 49, no. 11, p. 1908–1925.
Marzolf, J. E., 1965, Description of cores and cuttings of rocks drilled by the Mobil Oil Company in test well 22-19G in the Tip Top unit, Sublette County, Wyoming: U.S. Geological Survey Open-File Report, 2 sheets.
Maughan, E. K., 1963, Mississippian rocks in the Laramie Range, Wyoming, and adjacent areas, *in* Bolyard, D. W., and Katich, P. J., eds., Geology of the northern Denver Basin and adjacent uplifts: Rocky Mountain Association of Geologists, p. 36–40.
—— , 1975, Organic carbon in shale beds of the Permian Phosphoria Formation of eastern Idaho and adjacent states; A summary report, *in* Exum, F. A., and George, G. R., eds.: Wyoming Geological Association, Guidebooks, 27th Annual Field Conference, p. 107–115.
—— , 1979, Pennsylvanian (Upper Carboniferous) System of Wyoming, *in* The Mississippian and Pennsylvanian (Carboniferous) systems in the United States: U.S. Geological Survey Professional Paper 1110-U, p. U16–U33.
May, S. F., 1984, Miocene stratigraphic relations and problems between the Abiquiu, Los Pinos, and Tesuque formations near Ojo Caliente, northern Espanola Basin: New Mexico Geological Society, 35th Annual Field Conference, Guidebook, p. 129–135.
McCaslin, J. C., ed., 1981, Oil and Gas Journal newsletter, July 20, 1981, p. 1.
—— , 1983, More drilling heating up Raton basin play: Oil and Gas Journal, November 14, 1983, p. 229, 231.
—— , 1984, New Mexico basins hosting new wildcats and gas discoveries: Oil and Gas Journal, April 9, 1984, p. 129–131.
McCubbin, D. G., 1969, Cretaceous strike-valley sandstone reservoirs, northwestern New Mexico: American Association of Petroleum Geologists Bulletin, v. 53, no. 10, p. 2114–2140.
McDonald, R. E., 1972, Paleocene and Eocene rocks of the central and southern Rocky Mountain basins, *in* Mallory, W. W., ed., Geologic atlas of the Rocky Mountain region: Rocky Mountain Association of Geologists, p. 243–256.
—— , 1975, Structure and correlation and depositional environments of the Tertiary, Sand Wash and Washakie basins, Colorado and Wyoming, *in* Bolyard, D. W., ed., Deep drilling frontier in the central Rocky Mountains: Rocky Mountain Association of Geologists, p. 175–184.
McFall, C. C., 1955, Geology of the Escalante-Boulder area, Garfield County, Utah [Ph.D. dissertation]: Yale University, 180 p.
McGookey, D. P., and others, 1972, Cretaceous System, *in* Mallory, W. W., ed., Geologic atlas of the Rocky Mountain region: Rocky Mountain Association of Geologists, p. 190–228.
McIntosh, W. D., Sutter, J. F., Chapin, C. E., Osburn, G. R., and Ratte, J. C., 1986, A stratigraphic framework for the eastern Mogollon-Datil volcanic field based on paleomagnetism and high-precision $^{40}Ar/^{39}Ar$ dating of ignimbrites; A progress report: New Mexico Geological Society, 37th Annual Field Conference Guidebook, p. 183–195.
McKenna, M. C., and Love, J. D., 1972, High-level strata containing early Miocene mammals on the Bighorn Mountains, Wyoming: American Museum Novitates No. 2490, 31 p.
Merewether, E. A., 1983, The Frontier Formation and mid-Cretaceous orogeny in the foreland of southeastern Wyoming: The Mountain Geologist, v. 20, p. 121–138.
Merewether, E. A., Blackmon, P. D., and Webb, J. C., 1984, The mid-Cretaceous Frontier Formation near the Moxa arch, southwestern Wyoming: U.S. Geological Survey Professional Paper 1290, 29 p.
Miller, F. X., 1977, Biostratigraphic correlation of the Mesa Verde Group in southwestern Wyoming and northwestern Colorado, *in* Veal, H. K., ed., Exploration frontiers of the central and southern Rockies: Rocky Mountain Association of Geologists, p. 117–137.
Mohr, P., 1982, Musings on continental rifts, *in* Palmason, G., ed., Continental and oceanic rifts: American Geophysical Union and Geological Society of America, Geodynamics Series, v. 8, p. 293–309.
Molenaar, C. M., 1973, Sedimentary facies and correlation of the Gallup Sandstone and associated formations, northwestern New Mexico, *in* Fassett, J. E., ed., Cretaceous and Tertiary rocks of the southern Colorado Plateau: Four Corners Geological Society Memoir, p. 85–110.
—— , 1983, Major depositional cycles and regional correlations of Upper Cretaceous rocks, southern Colorado Plateau and adjacent areas, *in* Reynolds, M. W., and Dolly, E. D., eds., Mesozoic paleogeography of the west-central United States, Rocky Mountain Paleogeography, Symposium 2: Rocky Mountain Section, Society of Economic Paleontologists and Mineralogists, p. 201–224.
Momper, J. A., 1963, Nomenclature, lithofacies, and genesis of Permo-Pennsylvanian rocks in northern Denver Basin, *in* Bolyard, D. W., and Katich, P. J., eds., Geology of the northern Denver Basin and adjacent uplifts: Rocky Mountain Association of Geologists, p. 41–67.
Muehlberger, W. R., 1979, The Embudo fault between Pilar and Arroyo Hondo, New Mexico; An active intracontinental transform fault: New Mexico Geological Society, 30th Annual Field Conference, Guidebook, p. 77–82.
Murray, H. F., 1958, Pennsylvanian stratigraphy of the Maroon trough, *in* Curtis, B. F., ed., Symposium on Pennsylvanian rocks of Colorado and adjacent areas: Rocky Mountain Association of Geologists, p. 47–58.

Newman, K. T., 1974, Palynomorph zones in early Tertiary formations of the Piceance Creek and Uinta Basins, Colorado and Utah, *in* Field Conference, Energy Resources of the Piceance Creek Basin, Colorado: Rocky Mountain Association of Geologists, Guidebook, p. 47–55.

Obradovich, J. D., Muschler, F. E., and Bryant, B., 1969, Potassium-argon ages bearing on igneous and tectonic history of Elk Mountains and vicinity, Colorado; A preliminary report: Geological Society of America Bulletin, v. 80, p. 1749–1756.

Olsen, P. E., and Galton, P. M., 1977, Triassic-Jurassic tetrapod extinctions; Are they real?: Science, v. 197, no. 4307, p. 983–985.

Oriel, S. S., 1962, Main body of Wasatch Formation near La Barge, Wyoming: American Association of Petroleum Geologists Bulletin, v. 46, p. 2161–2173.

—— , 1969, Geology of the Fort Hill quadrangle, Lincoln County, Wyoming: U.S. Geological Survey Professional Paper 594-M, 40 p.

Oriel, S. S., and Armstrong, F. C., 1966, Times of thrusting in Idaho-Wyoming thrust belt, Reply: American Association of Petroleum Geologists Bulletin, v. 50, p. 2614–2621.

Osburn, G. R., and Chapin, C. E., 1983, Nomenclature for Cenozoic rocks of northeast Mogollon-Datil volcanic field, New Mexico: Socorro, New Mexico Bureau of Mines and Mineral Resources, Stratigraphic Chart 1.

O'Sullivan, R. B., 1965, Geology of the Cedar Mesa–Boundary Butte area, San Juan County, Utah: U.S. Geological Survey Bulletin 1186, 128 p.

—— , 1974, Preliminary geologic map of the Segar Mountain quadrangle, Rio Blanco County, Colorado: U.S. Geological Survey Miscellaneous Field Studies Map MF-570, scale 1:24,000.

—— , 1978, Stratigraphic sections of Middle Jurassic San Rafael Group from Lohali Point, Arizona, to Bluff, Utah: U.S. Geological Survey Oil and Gas Investigations Chart OC-77.

—— , 1980a, Stratigraphic sections of Middle Jurassic San Rafael Group and related rocks from the Green River to the Moab area in east-central Utah: U.S. Geological Survey Miscellaneous Field Studies Map MF-1247.

—— , 1980b, Stratigraphic sections of Middle Jurassic San Rafael Group from Wilson Arch to Bluff in southeastern Utah: U.S. Geological Survey Oil and Gas Investigations Chart OC-102.

—— , 1981a, Stratigraphic sections of some Jurassic rocks from near Moab, Utah, to Slick Rock, Colorado: U.S. Geological Survey Oil and Gas Investigations Chart OC-107.

—— , 1981b, Stratigraphic sections of Middle Jurassic Entrada Sandstone and related rocks from Salt Valley to Dewey Bridge in east-central Utah: U.S. Geological Survey Oil and Gas Investigations Chart OC-113.

O'Sullivan, R. B., and Beikman, H. M., 1963, Geology, structure, and uranium deposits of the Shiprock quadrangle, New Mexico and Arizona: U.S. Geological Survey Miscellaneous Geologic Investigations Map I-345, scale 1:250,000.

O'Sullivan, R. B., Repenning, C. A., Beaumont, E. C., and Page, H. G., 1972, Stratigraphy of the Cretaceous rocks and the Tertiary Ojo Alamo Sandstone, Navajo and Hopi Indian Reservations, Arizona, New Mexico, and Utah: U.S. Geological Survey Professional Paper 521-E, 65 p.

Ovellette, R. G., and Boucher, D. A., 1983, Frontal thrust structure, south-central Colorado, *in* Bally, A. W., ed., Seismic expression of structural styles: American Association of Petroleum Geologists Studies in Geology Series 15, v. 3, p. 3.2.2.25–28.

Palmer, A. R., compiler, 1983, Decade of North American Geology 1983 Time Scale: Geology, v. 11, p. 503–504.

Parker, S. E., 1986, Tectono-stratigraphic relations of Upper Cretaceous and lower Tertiary strata of the Bighorn Basin [M.S. thesis]: Laramie, University of Wyoming, 125 p.

Pearl, R. H., 1976, Colorado stratigraphic nomenclature chart, *in* Subsurface cross sections of Colorado: Rocky Mountain Association of Geologists Research Committee, Rocky Mountain Association of Geologists Special Publication no. 2, fig. 2.

Peterson, F. A., 1983, Foreland detachment structures, *in* Lowell, J. D., and Gries, R., eds., Rocky Mountain foreland basins and uplifts: Field Conference—Rocky Mountain Association of Geologists, p. 65–77.

Peterson, Fred, 1969a, Four new members of the Upper Cretaceous Straight Cliffs Formation in the southeastern Kaiparowits region, Kane County, Utah: U.S. Geological Survey Bulletin 1274-J, p. J1–J28.

—— , 1969b, Cretaceous sedimentation and tectonism in the southeastern Kaiparowits region, Utah: U.S. Geological Survey Open-File Report, 259 p.

—— , 1980, Sedimentology as a strategy for uranium exploration; Concepts gained from analysis of a uranium-bearing depositional sequence in the Morrison Formation of south-central Utah, *in* Turner-Peterson, C. E., ed., Uranium in sedimentary rocks; Application of the facies concept to exploration: Rocky Mountain Section, Society of Economic Paleontologists and Mineralogists, p. 65–126.

—— , 1984, Fluvial sedimentation on a quivering craton; Influence of slight crustal movements on fluvial processes, Upper Jurassic Morrison Formation, western Colorado Plateau: Sedimentary Geology, v. 38, no. 14, p. 21–49.

—— , 1986, Jurassic paleotectonics in the west-central part of the Colorado Plateau, *in* Peterson, J. A., ed., Paleotectonics and sedimentation, Rocky Mountain region: American Association of Petroleum Geologists Memoir 41, p. 563–596.

—— , 1988, Stratigraphy and nomenclature of Middle and Upper Jurassic rocks, western Colorado Plateau, Utah and Arizona: U.S. Geological Survey Bulletin 1633-B (in press).

Peterson, Fred, and Caputo, M. V., 1986, The Middle Jurassic Summerville-Curtis interval, a time of transition on the western Colorado Plateau: Geological Society of America Abstracts with Programs, v. 18, no. 5, p. 402.

Peterson, Fred, and Kirk, A. R., 1977, Correlation of the Cretaceous rocks in the San Juan, Black Mesa, Kaiparowits and Henry Basins, southern Colorado Plateau, *in* Fassett, J. E., ed., San Juan Basin III, northwestern New Mexico: New Mexico Geological Society, Guidebook, 28th Field Conference, p. 167–178.

Peterson, Fred, and Pipiringos, G. N., 1979, Stratigraphic relations of the Navajo Sandstone to Middle Jurassic formations, southern Utah and northern Arizona: U.S. Geological Survey Professional Paper 1035-B, B1–B43.

Peterson, Fred, Cornet, B., and Turner-Peterson, C. E., 1977, New data bearing on the stratigraphy and age of the Glen Canyon Group (Triassic and Jurassic) in southern Utah and northern Arizona: Geological Society of America Abstracts with Programs, v. 9, no. 6, p. 755.

Peterson, Fred, Ryder, R. T., and Law, B. E., 1980, Stratigraphy, sedimentology, and regional relationships of the Cretaceous System in the Henry Mountains region, Utah, *in* Goode, H. D., ed., Henry Mountains Symposium: Utah Geological Association, p. 152–170.

Peterson, J. A., 1972, Jurassic System, *in* Mallory, W. W., ed., Geologic atlas of the Rocky Mountain region: Rocky Mountain Association of Geologists, p. 177–189.

—— , 1977, Paleozoic shelf-margins and marginal basins, western Rocky Mountains, *in* Heisey, E. L., and four others, eds., Great Basin, United States: Wyoming Geological Association Guidebook, 29th Annual Field Conference, p. 135–154.

Phoenix, D. A., 1963, Geology of the Lees Ferry area, Coconino County, Arizona: U.S. Geological Survey Bulletin 1137, 86 p.

Picard, M. D., 1955, Subsurface stratigraphy and lithology of the Green River Formation in Uinta Basin, Utah: American Association of Petroleum Geologists Bulletin, v. 39, no. 1, p. 75–102.

—— , 1959, Green River and lower Uinta Formation subsurface stratigraphy in western Uinta Basin, Utah, *in* Williams, N. C., ed., Guidebook to the geology of the Wasatch and Uinta Mountains transition areas: Intermountain Association of Petroleum Geologists, p. 139–149.

Pierce, W. G., 1957, Heart Mountain and South Fork detachment thrusts of Wyoming: American Association of Petroleum Geologists Bulletin, v. 41, p. 591–626.

—— , 1973, Principal features of the Heart Mountain fault and the mechanism problem, *in* DeJong, D., and Scholten, R., eds., Gravity and tectonics: New York, John Wiley and Sons, Inc., p. 457–471.

—— , 1978, Geologic map of the Cody 1° × 2° quadrangle, northwestern Wyoming: U.S. Geological Survey Miscellaneous Field Studies Map MF-

963, scale 1:250,000.

Pierce, W. H., 1969, Geology and the Pennsylvanian-Permian stratigraphy of the Howard area, Fremont County, Colorado [M.S. thesis]: Golden, Colorado School of Mines, 129 p.

Pillmore, C. O., 1969, Geology and coal deposits of the Raton coal field, Colfax County, New Mexico: The Mountain Geologist, v. 6, p. 125–142.

Pipiringos, G. N., 1962, Uranium-bearing coal in the central part of the Great Divide Basin: U.S. Geological Survey Bulletin 1099-A, 103 p.

Pipiringos, G. N., and Johnson, R. C., 1975, Preliminary geologic map of the Buckskin Point quadrangle, Rio Blanco County, Colorado: U.S. Geological Survey Miscellaneous Field Investigations Map MF-651, scale 1:24,000.

——, 1976, Preliminary geologic map and correlation diagram of the White River City quadrangle, Rio Blanco County, Colorado: U.S. Geological Survey Miscellaneous Field Investigations Map MF-736, scale 1:24,000.

Pipiringos, G. N., and O'Sullivan, R. B., 1978, Principal unconformities in Triassic and Jurassic rocks, Western Interior United States; A preliminary survey: U.S. Geological Survey Professional Paper 1035-A, 29 p.

Pipiringos, G. N., and Rosenlund, G., 1977, Preliminary geologic map of the White rock quadrangle, Rio Blanco and Moffat Counties, Colorado: U.S. Geological Survey Miscellaneous Field Investigations Map MF-837, scale 1:24,000.

Pitman, J. K., and Donnell, J. R., 1973, Potential shale-oil resources of a stratigraphic sequence above the Mahogany zone, Green River Formation, Piceance Creek Basin, Colorado: Journal of Research, U.S. Geological Survey, v. 1, no. 4, July-August, 1973, p. 467–473.

Plouff, D., and Pakiser, L. C., 1972, Gravity study of the San Juan Mountains, Colorado: U.S. Geological Survey Professional Paper 800-B, p. B183–B190.

Poole, F. B., 1961, Stream directions in Triassic rocks of the Colorado Plateau: U.S. Geological Survey Professional paper 424-C, p. C139–C141.

——, 1962, Wind directions in late Paleozoic to middle Mesozoic time on the Colorado Plateau: U.S. Geological Survey Professional Paper 450-D, p. D147–D151.

Powell, J. S., 1973, Paleontology and sedimentation of the Kimbeto Member of the Ojo Alamo Sandstone, *in* Fassett, J. E., ed., Cretaceous and Tertiary rocks of the southern Colorado Plateau: Four Corners Geological Society Memoir, p. 111–122.

Prodehl, C., 1979, Crustal structure of the western United States: U.S. Geological Survey Professional Paper 1034, 74 p.

Prodehl, C., and Pakiser, L. C., 1980, Crustal structure of the southern Rocky Mountains from seismic measurements: Geological Society of America Bulletin, v. 91, pt. I, p. 147–155.

Rascoe, B., Jr., 1978, Late Paleozoic structural evolution; The Las Animas arch, *in* Pruit, J. D., and Coffin, P. E., eds., Energy resources of the Denver Basin: Rocky Mountain Association of Geologists, p. 113–127.

Rascoe, B., Jr., and Baars, D. L., 1972, Permian System, *in* Mallory, W. W., ed., Geologic atlas of the Rocky Mountain region: Rocky Mountain Association of Geologists, p. 143–165.

Ray, R. R., and Keefer, W. R., 1985, Wind River Basin, central Wyoming, *in* Gries, R. R., and Dyer, R. C., eds., Seismic exploration of the Rocky Mountain region: Rocky Mountain Association of Geologists and Denver Geophysical Society, p. 201–212.

Reeside, J. B., Jr., 1924, Upper Cretaceous and Tertiary formations of the western part of the San Juan Basin, Colorado and New Mexico: U.S. Geological Survey Professional Paper 134, 70 p.

——, 1944, Maps showing thickness and general character of the Cretaceous deposits in the western interior of the United States: U.S. Geological Survey, Oil and Gas Investigations, Preliminary Map 10, scale 1:13,939,200.

Reeves, F., 1946, Origin and mechanics of the thrust faults adjacent to the Bearpaw Mountains, Montana: Geological Society of America Bulletin, v. 57, p. 1033–1048.

Renfro, J. J., 1979, Sheep Mountain CO_2 production facilities; Conceptual design: Journal of Petroleum Technology, November, 1979, p. 1462–1468.

Repenning, C. A., and Page, H. G., 1956, Late Cretaceous stratigraphy of Black Mesa, Navajo and Hopi Reservations, Arizona: American Association of Petroleum Geologists Bulletin, v. 40, no. 2, p. 255–294.

Repplier, F. N., and others, 1980, Atlas of ground water quality in Colorado: Colorado Geological Survey Map Series no. 16, scale 1:500,000.

Reynolds, M. W., 1968, Geologic map of the Muddy Gap quadrangle, Carbon County, Wyoming: U.S. Geological Survey Geologic Quadrangle Map GQ-771, scale 1:24,000.

——, 1976, Influence of recurrent Laramide structural growth on sedimentation and petroleum accumulation, Lost Soldier area, Wyoming: American Association of Petroleum Geologists Bulletin, v. 60, p. 12–32.

Rice, D. D., 1983, Relation of natural gas composition to thermal maturity and source rock type in San Juan Basin, northwestern New Mexico and southwestern Colorado: American Association of Petroleum Geologists Bulletin, v. 67, no. 8, p. 1199–1218.

Richter, H. R., Jr., 1981, Occurrence and characteristics of ground water in the Wind River Basin, Wyoming: Water Resources Research Institute, University of Wyoming, v. IV-A, 149 p.

Ritzma, H. R., 1951, Paleozoic stratigraphy, north end and west flank of the Sierra Madre, Wyoming-Colorado: Wyoming Geological Association Guidebook, 6th Annual Field Conference, p. 66–67.

——, 1971, Faulting on the north flank of the Uinta Mountains, Utah and Colorado, *in* Renfro, A. R., ed., Wyoming Geological Association Guidebook, 23rd Annual Field Conference: Wyoming Geological Association, p. 145–150.

Robb, W. A., and Smith, J. W., 1974, Mineral profile of oil shales in Colorado core hole no. 1, Piceance Creek Basin, Colorado, *in* 25th Field Conference, Energy Resources of the Piceance Creek Basin, Colorado: Rocky Mountain Association of Geologists Guidebook, p. 91–100.

Roberts, P. K., 1964, Stratigraphy of the Green River Formation, Uinta Basin, Utah [dissertation]: University of Utah, 212 p.

Robson, S. G., 1983, Hydraulic characteristics of the principal bedrock aquifers in the Denver Basin, Colorado: U.S. Geological Survey Hydrologic Investigations Atlas, HA 659.

Rocky Mountain Association of Geologists Research Committee, 1976, Subsurface cross-sections of Colorado: Rocky Mountain Association of Geologists Special Publication no. 2.

Roehler, H. W., 1965, Summary of pre-Laramide Late Cretaceous sedimentation in Rock Springs uplift area, *in* Devoto, R. H., and Bitter, R. K., eds., Wyoming Geological Association Guidebook 19th Annual Field Conference: Wyoming Geological Association, p. 10–12.

——, 1977, Geologic map of the Sand Butte Rim northwest quadrangle, Sweetwater County, Wyoming: U.S. Geological Survey Geologic Quadrangle Map GQ-1362, scale 1:24,000.

——, 1978, Geologic map of the Chicken Creek southeast quadrangle, Sweetwater County, Wyoming: U.S. Geological Survey Geologic Quadrangle Map GQ-1454, scale 1:24,000.

——, 1983, Stratigraphy of Upper Cretaceous and lower Tertiary Rock Springs uplift, Wyoming: U.S. Geological Survey Miscellaneous Investigations Series Map I-1500, 1 sheet.

Rogers, K. L., 1984, A paleontological analysis of the Alamosa Formation, south-central Colorado (Pleistocene, Irvingtonian): New Mexico Geological Society, 35th Annual Field Conference, Guidebook, p. 141–155.

Romero, J. C., 1976, Ground water resources of the bedrock aquifers of the Denver basin, Colorado: Colorado Division of Water Resources Report, 109 p.

Rose, P. R., Everett, J. R., and Merin, I. S., 1984, Possible basin centered gas accumulation, Raton Basin, southern Colorado: Oil and Gas Journal, v. 82, no. 40, p. 190–197.

Ross, R. J., and Tweto, O., 1980, Lower Paleozoic sediments and tectonics in Colorado, *in* Kent, H. C., and Porter, K. W., eds., Colorado geology: Rocky Mountain Association of Geologists, p. 47–56.

Royse, R., Jr., Warner, M. A., and Reese, D. L., 1975, Thrust belt structural geometry and related stratigraphic problems, Wyoming–Idaho–northern Utah, *in* Bolyard, D. W., ed., Deep drilling frontiers of the central Rocky Mountains: Rocky Mountain Association of Geologists, p. 41–54.

Rubey, W. W., and Hubbert, M. K., 1959, Overthrust belt in geosynclinal area of western Wyoming in light of fluid pressure hypothesis, part 2, *of* Role of fluid pressure in mechanics of overthrust faulting: Geological Society of America Bulletin, v. 70, p. 167–206.

Ryder, R. T., Fouch, T. D., and Elison, J. H., 1976, Early Tertiary sedimentation of the western Uinta Basin, Utah: Geological Society of America Bulletin, v. 87, p. 496–512.

Sales, J. K., 1983, Collapse of Rocky Mountain basement uplifts, *in* Lowell, J. D., ed., Rocky Mountain foreland basins and uplifts: Rocky Mountain Association of Geologists, p. 79–97.

Sando, W. H., 1979, Lower part of the Carboniferous, *in* The Mississippian and Pennsylvanian (Carboniferous) Systems in the United States: U.S. Geological Survey Professional paper 1110-U, p. U2–U16.

Sawatzky, D. L., 1967, Tectonic style along the Elkhorn thrust, eastern South Park and western Front Range, Park County, Colorado [M.S. thesis]: Golden, Colorado School of Mines, 206 p.

——, 1972, Structural history of southeastern South Park: The Mountain Geologist, v. 9, p. 223–228.

Schenk, C. J., 1986, Turbidites in the lower part of the Eagle Valley Evaporite, Eagle County, Colorado, and implication for Des Moinesian paleogeography, *in* Stone, D. S., ed., New interpretations of northwest Colorado geology: Rocky Mountain Association of Geologists, p. 123–128.

——, 1988, Sedimentology and stratigraphy of the Eagle Valley Evaporite (Des Moinesian), central Colorado basin [Ph.D. thesis]: Boulder, University of Colorado (in preparation).

Scott, G. R., 1975, Cenozoic surfaces and deposits in the southern Rocky Mountains, *in* Curtis, B. F., ed., Cenozoic history of the southern Rocky Mountains: Geological Society of America Memoir 144, p. 227–248.

——, 1982, Paleovalley and geologic map of northeastern Colorado: U.S. Geological Survey Miscellaneous Investigations Series Map I-1378, scale 1:250,000.

Scott, G. R., and Taylor, R. B., 1975, Post-Paleocene Tertiary rocks and Quaternary volcanic ash of the Wet Mountain Valley, Colorado: U.S. Geological Survey Professional Paper 868, 15 p.

Scott, G. R., Van Alstine, R. E., and Sharp, W. N., 1975, Geologic map of the Poncha Springs quadrangle, Chaffee County, Colorado: U.S. Geological Survey Miscellaneous Field Studies Map MF-658, scale 1:62,500.

Seager, W. R., Shafiqullah, M., Hawley, J. W., and Marvin, R. F., 1984, New K-Ar dates from basalts and the evolution of the southern Rio Grande rift: Geological Society of America Bulletin, v. 95, p. 87–99.

Sharps, S. L., 1955, Correlation of pre-Mancos, post-Weber formations, northwestern Colorado, *in* Intermountain Association of Petroleum Geologists, Guidebook 6th Annual Field Conference, p. 16–17.

——, 1962, Geology of Pagoda quadrangle, northwestern Colorado [Ph.D. thesis]: Boulder, University of Colorado, 364 p.

Shaughnessy, J., and Butcher, R. H., 1974, Geology of Wagon Wheel nuclear stimulation project, Pinedale field, Wyoming: American Association of Petroleum Geologists Bulletin, v. 58, p. 2250–2259.

Shaw, G. L., 1958, Pennsylvanian history and stratigraphy of the Raton basin, *in* Curtis, B. F., ed., Symposium on Pennsylvanian rocks of Colorado and adjacent areas: Rocky Mountain Association of Geologists, p. 74–79.

Sholes, M. A., and four others, 1984, Keystone coal industry manual: New York, McGraw-Hill, p. 551–557.

Sikkink, P.G.L., 1986, Lithofacies relationships and depositional environment of the Tertiary Ojo Alamo Sandstone and related strata, San Juan Basin, New Mexico and Colorado, *in* Fassett, J. E., and Rigby, J. K., Jr., eds., The Cretaceous-Tertiary boundary in the San Juan and Raton basins, New Mexico and Colorado: Geological Society of America Special Paper 209, p. 81–104.

Slack, P. B., 1981, Paleotectonics and hydrocarbon accumulation, Powder River Basin, Wyoming: American Association of Petroleum Geologists Bulletin, v. 65, p. 730–743.

Sloss, L. L., 1963, Sequences in the cratonic interior of North America: Geological Society of America Bulletin, v. 74, no. 2, p. 93–114.

Smith, J. F., Jr., Huff, L. C., Hinrichs, E. N., and Luedke, R. G., 1963, Geology of the Capitol Reef area, Wayne and Garfield Counties, Utah: U.S. Geological Survey Professional Paper 363, 102 p.

Smith, L. B., 1982, Geology and uranium geochemistry of the western margin of the Thirtynine Mile Volcanic Field, Park, Chaffee, and Fremont Counties, Colorado [M.S. thesis]: Golden, Colorado School of Mines, 355 p.

Smith, R. B., 1978, Seismicity, crustal structure, and intraplate tectonics of the interior of the western Cordillera, *in* Smith, R. B., and Eaton, G. P., eds., Cenozoic tectonics and regional geophysics of the western Cordillera: Geological Society of America Memoir 152, p. 111–114.

Smith, R. L., Bailey, R. A., and Ross, C. S., 1970, Geological map of the Jemez Mountains, New Mexico: U.S. Geological Survey Miscellaneous Geologic Investigations Map I-571, scale 1:125,000.

Smithson, S. B., and four others, 1978, Nature of the Wind River thrust, Wyoming, from COCORP deep-reflection data and from gravity data: Geology, v. 6, p. 648–652.

Snow, B. C., 1970, Stratigraphy of basal sandstones on the Green River Formation northeast Piceance Basin, Rio Blanco County, Colorado: The Mountain Geologist, v. 7, p. 3–32.

Soister, P. E., 1978, Geologic setting of coal in Denver Basin, *in* Pruit, J. D., and Coffin, P. E., eds., Energy resources of the Denver Basin: Rocky Mountain Association of Geologists, p. 183–185.

Sonnenberg, S. A., and Weimer, R. J., 1981, Tectonics, sedimentation, and petroleum potential, northern Denver Basin, Colorado, Wyoming, and Nebraska: Colorado School of Mines Quarterly, v. 76, no. 2, 45 p.

Speer, W. R., 1976, Oil and gas exploration in the Raton basin: New Mexico Geological Society Guidebook, 27th Annual Field Conference, Vermejo Park: New Mexico Geological Society, p. 217–226.

Speigel, Z., and Baldwin, B., 1963, Geology and water resources of the Santa Fe area, New Mexico: U.S. Geological Survey Water Supply Paper 1525, 258 p.

Spieker, E. M., 1946, Late Mesozoic and early Cenozoic history of central Utah: U.S. Geological Survey Professional Paper 205-D, p. 117–161.

——, 1949, The transition between the Colorado Plateau and the Great Basin in central Utah: Utah Geological Society Guidebook 4, 106 p.

Spieker, E. M., and Reeside, J. B., 1925, Cretaceous and Tertiary Formations of the Wasatch Plateau: Geological Society of America Bulletin, v. 36, p. 435–454.

Stanley, K. O., and Collinson, J. W., 1979, Depositional history of Paleocene–lower Eocene Flagstaff Limestone and coeval rocks, central Utah: American Association of Petroleum Geologists Bulletin, v. 63, no. 3, p. 311–323.

Stearns, D. W., 1978, Faulting and forced folding in the Rocky Mountain foreland, *in* Matthews, V., III, ed., Laramide folding associated with basement block faulting in the western United States: Geological Society of America Memoir 151, p. 1–36.

Steidtmann, J. R., McGee, L. C., and Middleton, L. T., 1983, Laramide sedimentation, folding, and faulting in the southern Wind River Range, Wyoming, *in* Lowell, J. D., ed., Rocky Mountain foreland basins and uplifts: Rocky Mountain Association of Geologists, p. 161–167.

Stevenson, G. M., and Baars, D. L., 1986, The Paradox; A pull-apart basin of Pennsylvanian age: American Association of Petroleum Geologists Memoir 41, p. 513–539.

Stockton, S. L., and Hawkins, C. M., 1985, Southern Green River basin/Moxa arch, *in* Gries, R. R., and Dyer, R. C., eds., Seismic exploration of the Rocky Mountain region: Rocky Mountain Association of Geologists and Denver Geophysical Society, p. 73–78.

Stokes, W. L., and Madsen, J. H., Jr., compilers, 1961, Northeast quarter of the geologic map of Utah: Utah State Land Board, 1 sheet, scale 1:250,000.

Stone, D. S., 1969, Wrench faulting and Rocky Mountain tectonics: The Mountain Geologist, v. 6, no. 2, p. 67–79.

——, 1975, A dynamic analysis of subsurface structure in northwestern Colorado, *in* Bolyard, D. W., ed., Deep drilling frontiers in the central Rocky Mountains: Rocky Mountain Association of Geologists, p. 33–40.

——, 1977, Tectonic history of the Uncompahgre uplift, *in* Veal, H. K., ed.,

Frontiers of the central and southern Rockies: Rocky Mountain Association of Geologists, p. 23–30.

—— , 1985a, Seismic profiles of the area of the Pierce and Black Hollow fields, Weld County, Colorado, *in* Gries, R. R., and Dyer, R. C., eds., Seismic exploration of the Rocky Mountain region: Rocky Mountain Association of Geologists and Denver Geophysical Society, p. 79–86.

—— , 1985b, Geologic interpretation of seismic profiles, Bighorn Basin, Wyoming, *in* Gries, R. R., and Dyer, R. C., eds., Seismic exploration of the Rocky Mountain region: Rocky Mountain Association of Geologists and Denver Geophysical Society, p. 165–186.

—— , 1986, Seismic and borehole evidence for important pre-Laramide faulting along the Axial arch in northwest Colorado, *in* Stone, D. S., ed., New interpretations of northwest Colorado geology: Rocky Mountain Association of Geologists, p. 19–36.

Surdam, R. C., and Stanley, K. O., 1979, Lacustrine sedimentation during the culminating phase of Eocene Lake Gosiute, Wyoming, Green River Formation: Geological Society of America Bulletin, pt. 1, v. 90, no. 1, p. 93–110.

—— , 1980, The stratigraphic and sedimentologic framework of the Green River Formation, Wyoming: Wyoming Geological Association Guidebook, 31st Annual Field Conference, p. 205–211.

Taranik, J. V., 1974, Stratigraphic and structural evolution of Breckenridge area, central Colorado [Ph.D. thesis]: Golden, Colorado School of Mines, 222 p.

Taylor, R. B., 1975, Neogene tectonism in south-central Colorado, *in* Curtis, B. F., ed., Cenozoic history of the southern Rocky Mountains: Geological Society of America Memoir 144, p. 211–226.

Tedford, R. H., 1981, Mammalian biochronology of the late Cenozoic basins of New Mexico: Geological Society of America Bulletin, pt. I, v. 92, p. 1008–1022.

—— , 1982, Neogene stratigraphy of the northwestern Albuquerque Basin: New Mexico Geological Society, 33rd Annual Field Conference Guidebook, p. 273–278.

Tillman, R. W., 1971, Petrology and paleoenvironments, Robinson Member, Minturn Formation (Des Moinesian), Eagle Basin, Colorado: American Association of Petroleum Geologists Bulletin, v. 55, p. 593–620.

Tremain, C. M., 1980, The coal bed methane potential of the Raton Mesa coal region, Raton basin, Colorado: Colorado Geological Survey Open-File Report 80-4, 48 p.

Turner-Peterson, C. E., 1985, Lacustrine-humate model for primary uranium ore deposits, Grants uranium region, New Mexico: American Association of Petroleum Geologists Bulletin, v. 69, no. 11, p. 1999–2020.

Tweto, O., 1949, Stratigraphy of the Pando area, Eagle County, Colorado: Colorado Scientific Society Proceedings, v. 15, p. 149–235.

—— , 1958, Pennsylvanian stratigraphic section in the Minturn-Pando area, Colorado, *in* Curtis, B. F., ed., Symposium on Pennsylvanian rocks of Colorado and adjacent areas: Rocky Mountain Association of Geologists, p. 80–85.

—— , 1975, Laramide (Late Cretaceous–early Tertiary) orogeny in the southern Rocky Mountains, *in* Curtis, B. F., ed., Cenozoic history of the southern Rocky Mountains: Geological Society of America Memoir 144, p. 1–44.

—— , 1978, Tectonic map of the Rio Grande rift system in Colorado: New Mexico Bureau of Mines and Mineral Resources Circular 163, plate 1, scale 1:1,000,000.

—— , 1979a, Geologic map of Colorado: U.S. Geological Survey, 1 sheet, scale 1:500,000.

—— , 1979b, The Rio Grande rift system in Colorado, *in* Riecker, R. E., ed., Rio Grande rift; Tectonics and magmatism: American Geophysical Union, p. 33–56.

—— , 1980a, Tectonic history of Colorado, *in* Kent, H. C., and Porter, K. W., eds., Colorado geology: Rocky Mountain Association of Geologists, p. 5–9.

—— , 1980b, Precambrian geology of Colorado, *in* Kent, H. C., and Porter, K. W., eds., Colorado geology: Rocky Mountain Association of Geologists, p. 37–46.

—— , 1983, Geologic sections across Colorado: U.S. Geological Survey Miscellaneous Investigations Series Map I-1416, scale 1:500,000.

Tweto, O., and Sims, P. K., 1963, Precambrian ancestry of the Colorado mineral belt: American Association of Petroleum Geologists Bulletin, v. 74, p. 991–1014.

U.S. Geological Survey, 1976, Geological Survey Research, 1976: U.S. Geological Survey Professional Paper 1000, p. 27.

Van Alstine, R. E., 1968, Tertiary trough between the Arkansas and San Luis valleys, Colorado: U.S. Geological Survey Professional Paper 600-C, p. C158–C160.

Ver Ploeg, A. J., 1985, Tectonic map of the Bighorn Basin, Wyoming: Wyoming Geological Survey Open-File Report 85-11.

Wach, P. H., 1977, The Moxa arch, an overthrust model? *in* Heisey, E. L., and four others, eds., Wyoming Geological Association, 29th Annual Field Conference Guidebook, p. 651–664.

Waechter, N. B., and Johnson, W. E., 1986, Pennsylvanian-Permian paleostructure and stratigraphy as interpreted from seismic data, in the Piceance basin, northwest Colorado, *in* Stone, D. S., ed., New interpretations of northwest Colorado geology: Rocky Mountain Association of Geologists, p. 51–64.

Walcott, C. D., 1890, Study of a line of displacement in the Grand Canyon of the Colorado, in northern Arizona: Geological Society of America Bulletin, v. 1, p. 1–86.

Waldron, F. R., Donnell, J. R., and Wright, J. C., 1951, Geology of DeBeque oil-shale area, Garfield and Mesa Counties, Colorado: U.S. Geological Survey Oil and Gas Investigations Map OM-114, scale 1:62,500.

Walker, R. N., and Wiloth, G. J., compilers, 1975, Stratigraphic correlation chart, *in* Bolyard, D. W., ed., Deep drilling frontiers in the central Rocky Mountains: Rocky Mountain Association of Geologists, p. 7.

Walker, T. R., 1972, Bioherms in the Minturn Formation (Des Moinesian age) Vail-Minturn area, Eagle County, Colorado: Colorado School of Mines Quarterly, v. 67, no. 4, p. 249–278.

Warner, L. A., 1956, Tectonics of the Colorado Front Range, *in* 1956 Geological Record: Rocky Mountain Section, American Association of Petroleum Geologists, p. 129–144.

—— , 1978, The Colorado lineament, a middle Precambrian wrench fault system: Geological Society of America Bulletin, v. 89, no. 2, p. 161–171.

—— , 1980, The Colorado lineament, *in* Kent, H. C., and Porter, K. W., eds., Colorado geology: Rocky Mountain Association of Geologists, p. 11–21.

Weimer, R. J., 1960, Upper Cretaceous stratigraphy, Rocky Mountain area: American Association of Petroleum Geologists Bulletin, v. 44, p. 1–20.

—— , 1966, Time-stratigraphic analysis and petroleum accumulations, Patrick Draw field, Sweetwater County, Wyoming: American Association of Petroleum Geologists Bulletin, v. 50, p. 2150–2175.

—— , 1980, Recurrent movement on basement faults; A tectonic style for Colorado and adjacent areas, *in* Kent, H. C., and Porter, K. W., eds., Colorado geology: Rocky Mountain Association of Geologists, p. 23–25.

—— , 1984, Relation of unconformities, tectonics, and sea-level changes, Cretaceous of the western interior, U.S.A., *in* Schlee, J. S., ed., Interregional unconformities and hydrocarbon accumulation: American Association of Petroleum Geologists Memoir 36, p. 7–35.

Weimer, R. J., and Haun, J. D., 1960, Cretaceous stratigraphy, Rocky Mountain Region, U.S.A., *in* Regional paleogeography, Section 12: International Geological Congress, Copenhagen, p. 178–184.

Weimer, R. J., and Sonnenberg, S. A., 1983, Codell Sandstone; New exploration play, Denver basin: Oil and Gas Journal, v. 81, no. 22, p. 119–125.

Weimer, R. J., and 5 others, 1982, Tectonic influences on sedimentation, early Cretaceous, east flank Powder River basin, Wyoming and South Dakota: Colorado School of Mines Quarterly, v. 77, no. 4, 61 p.

Wengerd, S. A., and Matheny, M. L., 1958, Pennsylvanian System of Four Corners region: American Association of Petroleum Geologists Bulletin, v. 42, p. 2048–2106.

Wernicke, B., Spencer, J. E., Burchfiel, B. C., and Guth, P. L., 1982, Magnitude of crustal extension in the southern Great basin: Geology, v. 10, p. 499–502.

Whitaker, R. M., 1975, Upper Pennsylvanian and Permian strata of northeast Utah and northwest Colorado, *in* Bolyard, D. W., ed., Symposium on deep drilling frontiers in the central Rocky Mountains: Rocky Mountain Association of Geologists, p. 75–85.

Williams, L.A.J., 1982, Physical aspects of magmatism in continental rifts, *in* Palmason, G., ed., Continental and oceanic rifts: American Geophysical Union and Geological Society of America, Geodynamics Series, v. 8, p. 193–222.

Williams, P. L., and Hackman, R. J., 1971, Geology, structure, and uranium deposits of the Salina 1° × 2° quadrangle, Utah: U.S. Geological Survey Miscellaneous Geologic Investigations Map I-591, scale 1:250,000.

Wilson, J. M., 1978, Permo-Pennsylvanian of the west-central Nebraska panhandle, *in* Pruit, J. D., and Coffin, P. E., eds., Energy resources of the Denver basin: Rocky Mountain Association of Geologists, p. 129–140.

Wise, D. U., 1963, Keystone faulting and gravity sliding driven by basement uplift of Owl Creek Mountains, Wyoming: American Association of Petroleum Geologists Bullctin, v. 47, p. 586–598.

Withjack, M. O., 1984, Rift deformation produced by combined extension and shearing [abs.]: American Association of Petroleum Geologists Bulletin, v. 68, p. 1838.

Wood, G. H., Jr., Johnson, R. B., and Dixon, G. H., 1956, Geology and coal resources of the Gulnare, Cuchara Pass, and Stonewall area, Huerfano and Las Animas Counties, Colorado: U.S. Geological Survey Coal Investigations Map C-26, 1:31,680.

——, 1957, Geology and coal resources of the Starkville-Weston area, Las Animas County, Colorado: U.S. Geological Survey Bulletin 1051, 68 p.

Woodward, L. A., 1976, Laramide deformation of the Rocky Mountain foreland; Geometry and mechanics, *in* Woodward, L. A., and Northrop, S. A., eds., Tectonics and mineral resources of southwestern North America: New Mexico Geological Society Special Publication No. 6, p. 11–17.

——, 1977, Rate of crustal extension across the Rio Grande rift near Albuquerque, New Mexico: Geology, v. 5, p. 269–272.

Woodward, L. A., and Ingersoll, R. V., 1979, Phanerozoic tectonic setting of Santa Fe Country, *in* Ingersoll, R. V., Woodward, L. A., and James, H. L., eds., Santa Fe Country: New Mexico Geological Society, 30th Annual Field Conference, Guidebook, p. 51–57.

Woodward, L. A., Gibson, G. G., and McLelland, D., 1976, Geologic map and sections of Gallina quadrangle, New Mexico: Socorro, New Mexico Bureau of Mines and Mineral Resources Geological Map 39, scale 1:24,000.

Woodward, L. A., and 6 others, 1978, Tectonic map of Rio Grande rift region in New Mexico, Chihuahua, and Texas: New Mexico Bureau of Mines and Mineral Resources Circular 163, plate 2, scale 1:1,000,000.

Wray, J. L., 1985, Pennsylvanian algal carbonates and associated facies, central Colorado, Field Trip no. 4, *in* Society of Economic Paleontologists and Mineralogists Midyear Meeting Field Guides: Society of Economic Paleontologists and Mineralogists, 29 p.

Wyant, D. G., and Barker, F., 1976, Geologic map of the Milligan Lakes quadrangle, Park County, Colorado: U.S. Geological Survey Map GQ-1343, scale 1:24,000.

Wyoming Geological Association Stratigraphic Nomenclature Committee, 1969, Wyoming stratigraphic nomenclature chart, *in* Barlow, J. A., Jr., ed., Wyoming Geological Association Guidebook, 21st Annual Field Conference: Wyoming Geological Association, p. 7.

Young, R. G., 1975, Lower Cretaceous rocks of northwestern Colorado and northeastern Utah, *in* Bolyard, D. W., ed., Deep drilling frontier in the central Rocky Mountains: Rocky Mountain Association of Geologists, p. 141–147.

Zawislak, R. L., and Smithson, S. B., 1981, Problem and interpretation of COCORP deep seismic reflection data, Wind River Range, Wyoming: Geophysics, v. 46, p. 1684–1701.

Zeller, H. D., and Stephens, E. V., 1969, Geology of the Oregon Buttes area, Sweetwater, Sublette, and Fremont Counties, southwestern Wyoming: U.S. Geological Survey Bulletin 1256, 60 p.

Zietz, I., and Kirby, J. R., 1972, Aeromagnetic map of Colorado: U.S. Geological Survey Geophysical Investigations Map GP-836, scale 1:500,000.

Zoback, M. L., Anderson, R. E., and Thompson, G. A., 1981, Cainozoic evolution of the state of stress and style of tectonism of the Basin and Range province of the western United States: Royal Society of London, Philosophical Transactions, v. A300, p. 407–434.

Manuscript Accepted by the Society December 2, 1986

Printed in U.S.A.

The Geology of North America
Vol. D-2, Sedimentary Cover—North American Craton: U.S.
The Geological Society of America, 1988

Chapter 9

Geology of the Williston Basin (United States portion)

Lee C. Gerhard
Department of Geology, Colorado School of Mines, Golden, Colorado 80401
Sidney B. Anderson
North Dakota Geological Survey, University Station, Grand Forks, North Dakota 58202

INTRODUCTION: GEOGRAPHY AND PHYSIOGRAPHY—GEOLOGICAL ATTRIBUTES

The Williston Basin is an elliptical shaped depression in the western distal Canadian Shield occupying most of North Dakota, northwestern South Dakota, eastern Montana, and parts of southern Manitoba and Saskatchewan, Canada (Fig. 1). This discussion is restricted to the United States portion of the Williston Basin. Glacial or Cenozoic continental deposits mask much of the basin geology. Paleozoic and Mesozoic rocks are exposed infrequently along the basin margin. Therefore, virtually all of the knowledge of this basin is from subsurface geologic data, including cores, samples, well logs, and seismic surveys.

Glaciation during Pleistocene time not only covered most of the basin rocks and structures, but also created a very subdued topography over most of the region. Glacially derived soils form the basis for extensive agricultural land use despite the harsh climate. Fertile glacial Lake Agassiz sediments, in the extreme eastern part of North Dakota, support one of the most productive small grain, potato, and sugar beet farm economies in the country. Less-productive ice-margin deposits support both farming and ranching. Temperatures range from above 100°F to below −40°F, but the northerly latitudes provide long hours of daylight for growing. Areas of exposed Cenozoic rocks in the western part of the state are predominantly ranchlands, although grain farming is also common in river bottoms and where topography is relatively flat. Sharp topographic breaks occur where major drainages cut Cenozoic rocks, exhibited by the Missouri, Little Missouri, Yellowstone, Powder, and Knife Rivers. Badland topography dominates these areas because of the very fine-grained nature of the Cenozoic rocks. Buttes and hillocks are capped by a few sandstone and natural porcellanite beds. Porcellanites formed by subsurface burning of lignite beds are locally referred to as "scoria." These beds are red in color while unaltered rocks are buff and gray. These features provide many rugged topographic areas of scenic interest, several of which have been incorporated into Theodore Roosevelt National Park.

The basin also supports a thriving oil and gas economy

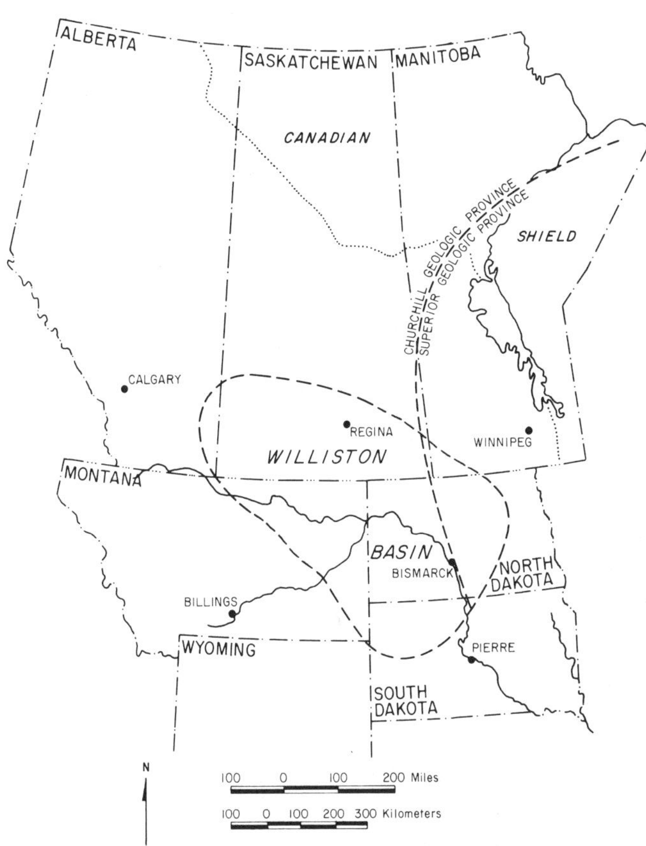

Figure 1. Index map showing location of the Williston Basin and the Superior-Churchill province boundary of the Canadian Shield.

(North Dakota is ranked ninth in oil production in the country). Surface mines supply lignite for electrical power generation. Lignite will also be used for the generation of synthetic natural gas at a plant that is now under construction. Other operations within the basin result in the mining of salt, clay, sand and gravel, and road metal ("scoria").

Rocks deposited during all periods of Phanerozoic time are present in the basin (Fig. 2 and Plate 3). Paleozoic rocks are

Gerhard, L. C., and Anderson, S. B., 1988, Geology of the Williston Basin (United States portion), *in* Sloss, L. L., ed., Sedimentary Cover—North American Craton; U.S.: Boulder, Colorado, Geological Society of America, The Geology of North America, v. D-2.

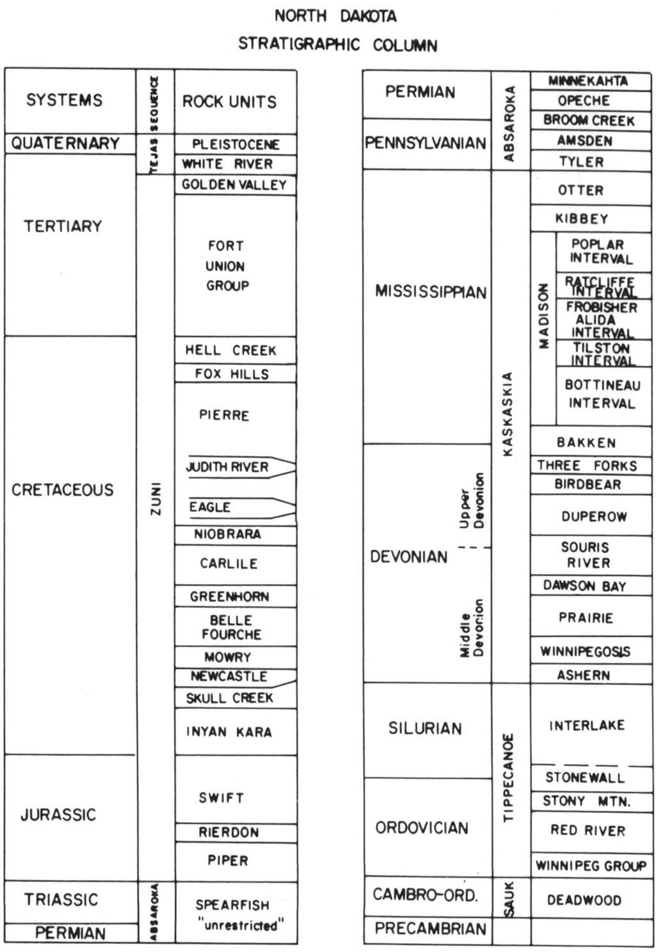

NORTH DAKOTA
STRATIGRAPHIC COLUMN

SYSTEMS	SEQUENCE	ROCK UNITS
QUATERNARY	TEJAS	PLEISTOCENE
	TEJAS	WHITE RIVER
TERTIARY	ZUNI	GOLDEN VALLEY
		FORT UNION GROUP
CRETACEOUS		HELL CREEK
		FOX HILLS
		PIERRE
		JUDITH RIVER
		EAGLE
		NIOBRARA
		CARLILE
		GREENHORN
		BELLE FOURCHE
		MOWRY
		NEWCASTLE
		SKULL CREEK
		INYAN KARA
JURASSIC		SWIFT
		RIERDON
		PIPER
TRIASSIC	ABSAROKA	SPEARFISH "unrestricted"
PERMIAN	ABSAROKA	

SYSTEMS	SEQUENCE	ROCK UNITS
PERMIAN	ABSAROKA	MINNEKAHTA
		OPECHE
		BROOM CREEK
PENNSYLVANIAN		AMSDEN
		TYLER
MISSISSIPPIAN	KASKASKIA	OTTER
		KIBBEY
	MADISON	POPLAR INTERVAL
		RATCLIFFE INTERVAL
		FROBISHER ALIDA INTERVAL
		TILSTON INTERVAL
		BOTTINEAU INTERVAL
		BAKKEN
DEVONIAN	Upper Devonian	THREE FORKS
		BIRDBEAR
		DUPEROW
		SOURIS RIVER
	Middle Devonian	DAWSON BAY
		PRAIRIE
		WINNIPEGOSIS
		ASHERN
SILURIAN	TIPPECANOE	INTERLAKE
ORDOVICIAN		STONEWALL
		STONY MTN.
		RED RIVER
		WINNIPEG GROUP
CAMBRO-ORD.	SAUK	DEADWOOD
PRECAMBRIAN		

Figure 2. Generalized stratigraphic column of the Williston Basin. See plate 3 for additional chronostratigraphic and surface information.

mainly carbonate and evaporites with minor amounts of shale, followed by dominantly clastic Mesozoic and Cenozoic rocks. Structural trends within the basin reflect major directional changes in the structure of the Rocky Mountain belt and are likely associated with an ancient wrench-fault system. Approximately 16,000 feet of sedimentary rocks are present in the deepest part of the basin, southeast of Watford City, North Dakota (McKenzie County). To date, the deepest well drilled in the basin penetrated Precambrian crystalline rocks at 15,340 ft (4,677 m); presently, the deepest oil production is from the Ordovician Red River Formation at 14,343 ft (5,799 m) (Mesa #1-13 Brandvik, Dunn County, North Dakota). Major transgressive-regressive cycles characterize Phanerozoic sedimentation, permitting use of the Sloss (1963) sequence concept, which is herein modified by recognition of a major unconformity within the Kaskaskia sequence. For this reason, stratigraphic units and their significance to deciphering the structural evolution of the Williston Basin are discussed in relation to the sequence concept.

PREVIOUS STUDIES

Several regional studies of the Williston Basin are the basis for many of the ideas presented here. Some of these are the papers of Porter and Fuller (1959), Macauley (1964), Carlson and Anderson (1966), the various papers presented in the Geologic Atlas of the Rocky Mountains (Mallory, 1972b), and Gerhard and others (1982). Recent studies of specific stratigraphic units have been published in two major symposium volumes (Estelle and Miller, eds., 1978; Christopher and Kaldi, eds., 1982) and a number of publications by the North Dakota Geological Survey.

Data for the interpretations are drawn largely from subsurface oil and gas well information; however, extrapolated outcrop information, drainage analysis, and geophysical surveys are also used. Approximately 10,000 subsurface control points are presently available, although relatively few of these extend to the Lower Paleozoic, and many of the data points are concentrated in the major oil- and gas-producing areas. It must be noted by those unfamiliar with subsurface geology that isopach information is particularly useful to the interpretation of ancient tectonic events, especially when combined with analyses of the depositional environments. Many of the cited studies of the North Dakota Geological Survey and others are based upon a thorough study of all of the core control available so as to provide maximum correlation of numerical and lithologic information. In general, lithologic studies relying on single, or purely local, core control have not been used in writing this regional summary.

TECTONIC SETTING

The Canadian Shield rocks extend under the Williston Basin into the Cordilleran orogen. Precambrian rocks have not been systematically studied, but some generalizations can be made. First, gravity, seismic, and geologic mapping indicate that the eastern margin of the basin is a structural declivity congruent with the boundary between Churchill and Superior provinces. The few cores and samples available indicate that the majority of the Precambrian substructure of the basin is of the granulite facies (Peterman and Goldich, 1982) consisting of biotite garnet gneisses and similar rocks. In eastern North Dakota, extremely strong magnetic anomalies on the surface have caused survey lines to vary widely from true directions. Core drilling of the anomalies and detailed magnetic mapping have demonstrated that siliceous specular hematite and lesser amounts of magnetite occur in belts likely associated with the hematitic taconite belts of northern Minnesota.

Among the tectonic elements in the Williston Basin, the eastern hinge-line boundary between the Precambrian basement provinces (Fig. 3), appears to have been one of the most structurally active during Paleozoic sedimentation. Associated structural features are present along the hinge line. Among these are the Newburg Syncline (Fig. 3), resulting from the dissolution of underlying Devonian Prairie Formation salts, and the Burleigh high. No case has yet been made for significant vertical motion along

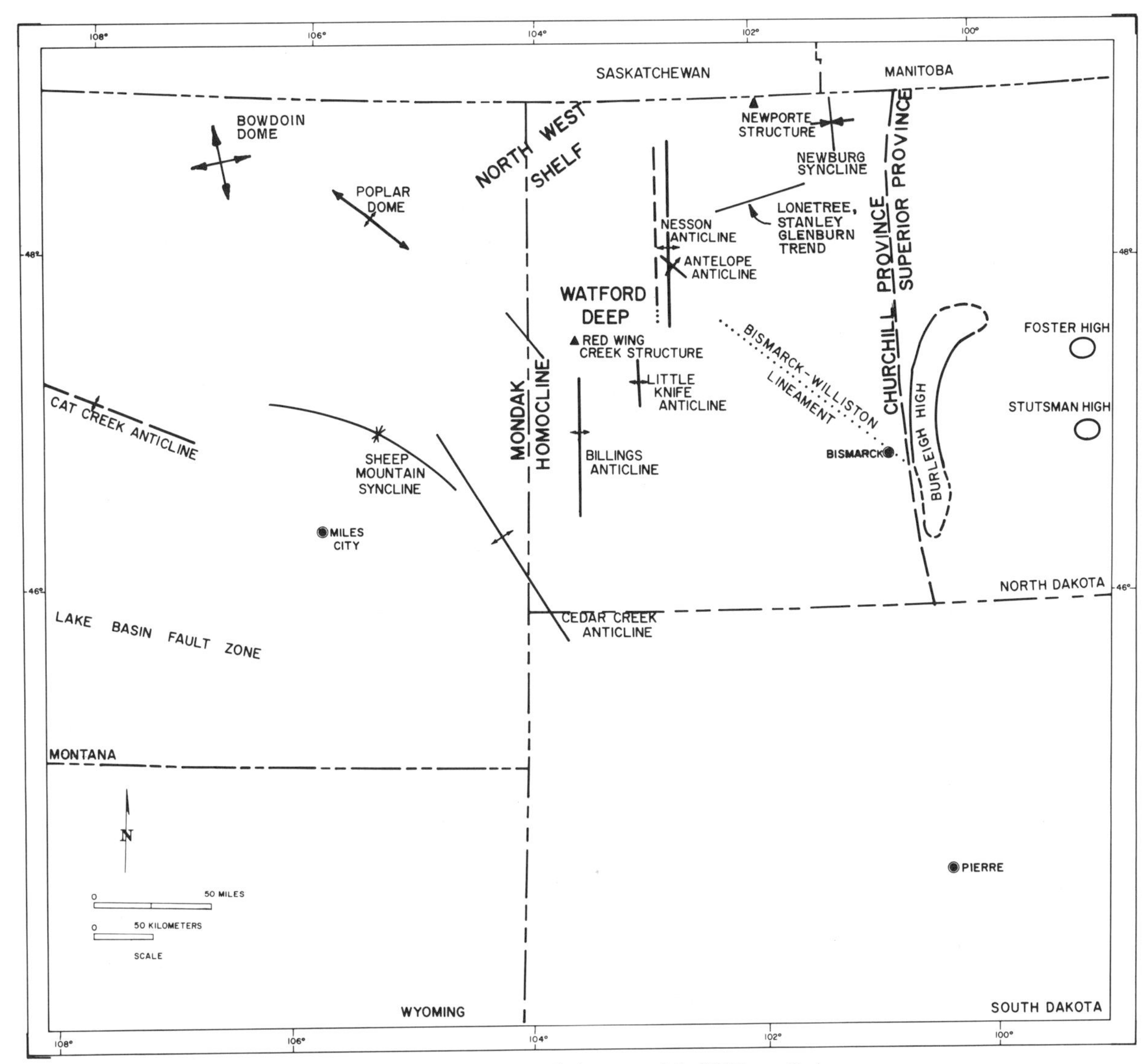

Figure 3. Major structural elements of the Williston Basin.

the hinge except for westward declivity into the basin that appears to have occurred simultaneously with sedimentation. The Williston Basin was a craton-margin or continental-shelf basin at its inception, and has been changed to an intracratonic basin by the deformation of the Cordilleran orogen and crustal additions to the western margin.

Other structural elements that bound the United States portion of the basin are the Sioux Arch, the Black Hills Uplift, the Cedar Creek Anticline, and the Bowdoin Dome (Fig. 4). The western boundary of the basin is indistinct, reflecting the craton-margin origin of the basin and the westward-seaway connections that operated during the Paleozoic and Mesozoic Eras.

Other major structures which had significant control on the basin and its development are the Colorado-Wyoming lineament and the Brockton-Froid-Fromberg fault zone. From exposures in Colorado and Wyoming, the lineament can be traced through southwestern South Dakota and the Sioux Arch into the upper midwest (Fig. 4; Warner, 1978). Warner has demonstrated significant left-lateral shear along this system during Precambrian time. Tweto and Sims (1963) indirectly demonstrated vertical motion during Paleozoic sedimentation. Gerhard (1972), in a study of early Paleozoic tectonics in central Colorado, demonstrated that the shear system had vertical relief during Lower Ordovician time and formed a sill-like barrier separating the eastern interior seaway from the Cordilleran shelf. On the northwestern side of the Williston Basin (U.S. portion), the Brockton-Froid-Fromberg fault zone forms an easily traceable lineament from the Snake River Plain to the Canadian border and beyond. This zone can be identified on LANDSAT imagery, in seismic data, in drainage pattern, and in facies mapping. Gerhard and others (1982) have suggested that this system had a left-lateral origin in Archean time similar to the Colorado-Wyoming Lineament.

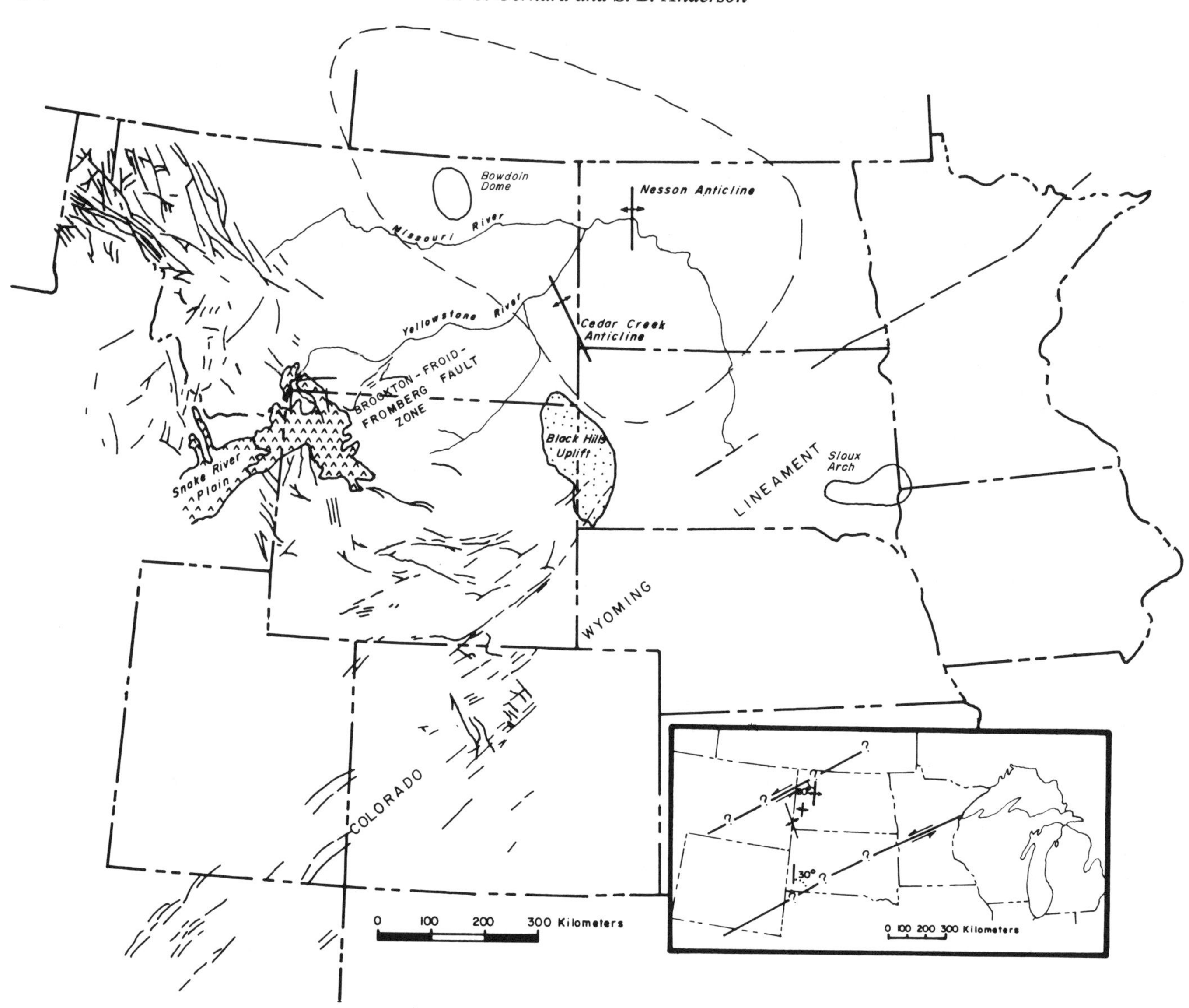

Figure 4. Archean shear systems hypothesized to control the origin of the Williston Basin and other structures shown with present major structural zones of southern, central, and northern Rockies.

Structural grain of the region appears to be related to the offset in the Rocky Mountain chain, indicated by the central Rocky Mountains (Wyoming Rockies), between the north-trending southern and northern Rockies. The central Rockies are characterized by northwesterly orientation. Regional wrench faulting along the Cat Creek and Lake Basin zones (Fig. 3; Thomas, 1974), coupled with the two major shear systems, suggests that structural control of Williston Basin sedimentation is related to large-scale "tears" in the edge of the early Phanerozoic and Precambrian craton.

The two most prominent features in the Williston Basin are the northerly trending Nesson and the northwesterly trending Cedar Creek Anticlines (Fig. 3). These structures have produced large amounts of petroleum and natural gas and have surface expression. Less obvious structures are the Billings Anticline, the Little Knife Anticline, and the Burleigh high.

In contrast with structural knowledge a decade ago, numerous other structures are now known in the basin. In addition to the Cedar Creek Anticline, several northwest-trending features are present. These include the Mondak and Antelope structures and the Bismarck-Williston Lineament. The Mondak, Nesson, Cedar Creek, and Antelope structures each have a fault component.

Subsurface mapping indicates the presence of several northeast-trending lineaments. The Lonetree-Glenburn and the Stanley trends are examples of these. Selective accumulation of hydrocarbons along these trends suggests that they are major structural elements of importance to petroleum trapping and migration. Numerous other smaller structures are detectable with seismic techniques. Many of the seismically identified structures are paleostructures. Many of these structures parallel known trends and may be fault related.

Much of the petroleum exploration in the Williston Basin is governed by the identification of small Lower Paleozoic isopach "thins," usually seen as isochron "thins" in seismic data. Commonly, these cannot be traced upward through Cretaceous rocks, although the Greenhorn to Winnipeg isochron is a common mapping interval. Since many basin petroleum deposits are in part stratigraphically controlled, and velocity problems in surficial sediments and Cenozoic rocks hamper seismic interpretation, isotime and isopach mapping are the accepted tools for exploration decisions, in preference to the structural mapping so important to other areas.

There is little question that sedimentation and structure of the basin are controlled by movement of basement blocks, which were structurally defined during pre-Phanerozoic time. Several demonstrations of sedimentation effects on buried structures have been cited, particularly in studies of the Sauk and Tippecanoe sequences (Gerhard and others, 1982). In some cases, thinning of stratigraphic units is the major effect seen. In others (Carroll, 1979) the effects are early diagenesis, controlling the distribution of effective porosity as a function of position in the depositional environment. Recent mapping on the Nesson Anticline has shown that each of the major sequences ended with an unconformity across the structure, and isopach maps clearly show the effects of block motion. Detailed cross-sections show that motion affected sedimentation as well as final thickness after erosion.

Several writers have attempted to establish a wrench-fault framework for the basin. Kearns and Traut (1979) illustrate satellite-imagery surface lineations which are northwest and northeast trending. Although northwest-trending structures are known to be of significance in the basin, this is one of the few illustrations of northeast trends that appears to control much of the Mississippian, structurally assisted, stratigraphic oil traps in north-central North Dakota. Brown (1978) has used isopach variations in individual stratigraphic units to construct a wrench-fault pattern for the southern part of the basin. Earlier, Thomas (1974) developed a regional theoretical model for wrench-fault tectonics in the Montana and North Dakota portions of the basin. Most recently, Gerhard and others (1982) suggested that the basin origin and development was controlled by Archean left-lateral shear systems, which developed the planes of structural weakness now seen as vertical structures in some cases. This model is used here.

MODEL FOR ORIGIN AND DEVELOPMENT OF THE WILLISTON BASIN

A fully documented geometry of the Archean Canadian Shield in the western part of North America is not available. However, the Colorado-Wyoming lineament appears to be a fault system that is possibly of Archean age and may have been intermittently active since that time. It is postulated that its early history is characterized by left-lateral motion. Geometry of the Black Hills Uplift and other central Rocky Mountain and Williston Basin structures can be attributed to their being caught between the two left-lateral systems (Brockton-Froid-Fromberg on the north, Colorado-Wyoming on the south).

Sedimentation during Sauk and Tippecanoe time took place in a depressed block between the two shear systems. Although later influence of the Brockton-Froid-Fromberg lineament and the Colorado-Wyoming shear zone disrupted that pattern, early Paleozoic sedimentary rocks reflect a "central Rockies depression" in both thickness and facies. The seaway connection to the Cordilleran shelf was through the depression. The eastern terminus of the depression is the hinge line of the Williston Basin, with a hinging occurring because of the mechanical discontinuity caused by a change in rock character at the Churchill-Superior provincial boundary.

All of the mapped structures of the Williston Basin appear congruent with a left-lateral origin. Rejuvenation of ancient planes of weakness have served to establish linearity of more modern structures. Modifications of earlier structural grain may have resulted from Mesozoic and younger plate interactions.

STRATIGRAPHY

General

All periods of the Phanerozoic are represented by sedimentary rocks in the Williston Basin (Fig. 2). As is the case with most stratigraphic assemblages of craton margin or cratonic habit, these rocks can easily be subdivided into transgressive and regressive cycles covering large spans of geologic time. The basic framework of this is the sequence concept (Sloss, 1963). Sloss' cycles, however, have been modified by the recognition of an additional major cycle in the subsurface. Basal units of the early sequences are siliciclastic transgressive sandstones, followed by finer-grained siliciclastics and, most frequently, argillaceous and glauconitic carbonates. Younger sequences lack the glauconite, but the cyclic lithologies are retained. Williston Basin sedimentation in the lower and middle Paleozoic is dominated by carbonates and evaporites, whereas upper Paleozoic and younger rocks are largely siliciclastic with small admixtures of carbonate. Evaporites are especially common in the basin, reflecting the rather low-slope gradients and restricted circulation during Paleozoic sedimentation. Upper Paleozoic sediments are largely fine-grained clastics with evaporites and evaporite-associated carbonates. Mesozoic sedimentation is typical of the upper Great Plains regions with many of the units thinning abruptly in the eastern part of the basin. Cretaceous sedimentation covered the basin and the surrounding region. The post-Cretaceous regression terminates significant marine deposition in the basin. One Paleocene unit is of marine origin (Cannonball Formation); all succeeding stratigraphic units are of terrestrial origin, including the youngest of the carbonate units.

Glacial sediments cover much of the basin and are not dis-

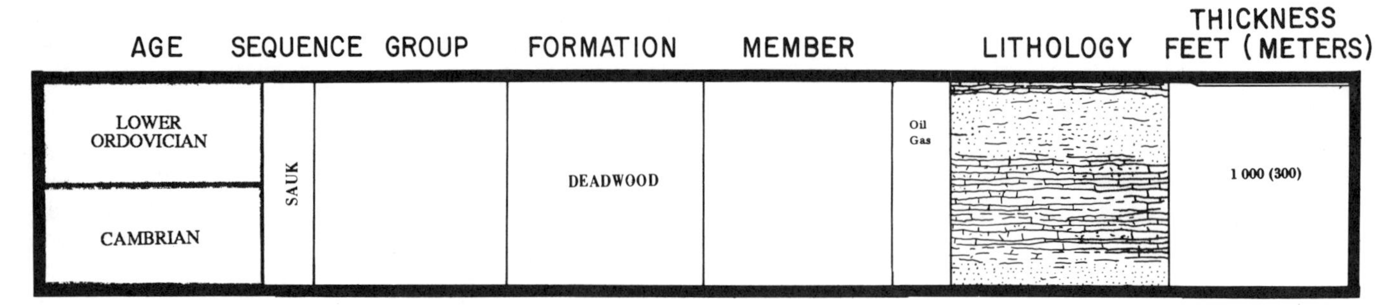

Figure 5. Sauk sequence rocks of the Williston Basin.

cussed in this paper. Although of local importance, they do not shed any light on the geological structure or tectonic history of the basin.

Stratigraphic units are not discussed separately in the following text. Where individual units are of special importance, they are discussed in detail; where not, they are taken as part of the lithologic sequence in which they are found. For details of correlation, age, and subsurface signature, the reader is referred to the Williston Basin COSUNA chart (Ballard, Bluemle, and Gerhard, 1983), to the North Dakota Geological Survey stratigraphic column (Bluemle, Anderson, and Carlson, 1980), and the stratigraphic nomenclature chart (Bluemle, Anderson, and Carlson, 1981; Plate 3).

Sauk Sequence

Cambrian–Lower Ordovician. Sauk sequence sedimentary rocks comprise the Deadwood Formation of Late Cambrian and Early Ordovician age (Fig. 5). This unit nonconformably overlies Precambrian crystalline rocks and is disconformably overlain by Tippecanoe rocks. Transgression of the shallow Deadwood Sea took place eastward over a relatively irregular surface. Precambrian relief associated with the Nesson Anticline is documented, along with several other topographic features. In addition, buried Precambrian highs have not only influenced Sauk sedimentation, they have continued to influence sedimentation of the Tippecanoe-sequence rocks as well. Despite relatively few penetrations of the Deadwood Formation, it is doubtful that its maximum thickness exceeds 1,000 ft (300 m) in the area discussed.

There was no true Williston Basin at the time of Deadwood sedimentation. The Williston Basin area was a large indentation or embayment of the western Cordilleran shelf and was not structurally well defined. In South Dakota, the Sioux Arch was a ridge of quartzite that was both a source of sediment and a barrier to the transgression (see Bunker and others, this volume). Another source of sediment was the Precambrian Shield. Much of the siliciclastic sediment was locally derived from extensively weathered, Precambrian, crystalline rocks.

Deadwood lithology exhibits much of the variation expected in a transgressive cycle. Basal beds are conglomeratic in part and may contain much orthoquartzite. Glauconite is abundant. Middle beds of the Deadwood Formation are largely argillaceous carbonate containing flat pebble conglomerates with abundant glauconite. These are characteristic of Sauk carbonate units elsewhere in the Rocky Mountains. From the present central Williston Basin, these Sauk carbonates begin to grade and thin eastward into more clastic rocks reflecting the proximity of the paleostrandline. The upper beds are clastic, including fucoidal and worm-bored red sandstones (Steece, 1978).

The close of Sauk sequence deposition was marked by minor structural motion on the Nesson Anticline and erosion throughout the Williston Basin (Plate 3). The unconformity that marks the boundary between the Sauk and Tippecanoe sequences was formed during the approximate time of initiation of the Taconic event in the northeastern U.S. Structural changes of a regional nature formed the initial Williston Basin during the hiatus.

Tippecanoe Sequence

Ordovician–Silurian. Tippecanoe sedimentation took place in a basin that was nearly circular, with openings to the southwest through a trough formed between the Colorado-Wyoming and Brockton-Froid-Fromberg shear systems and to the southeast (Fig. 6). Small, local structures in south-central North Dakota were present along a northeast trend, as well as the Burleigh high, which formed just east of the hinge line and parallel to it. The Nesson Anticline was present; isopach information suggests that the Bismarck-Williston lineament was also structurally active (Carroll, 1979). The same isopach data show that the Billings Anticline was well developed. Along the eastern margin of the Red River depositional basin, a large local thickening occupies a round depression which was connected to the main basin. This small sub-basin, unnamed, continued to influence sedimentation well into Mississippian time.

Over 2,500 ft (760 m) of sedimentary rocks were deposited during Tippecanoe time (Fig. 7). The basal transgressive stratigraphic unit is the Winnipeg Formation, comprised of a lower sand and upper shale. The Winnipeg sand produces gas and

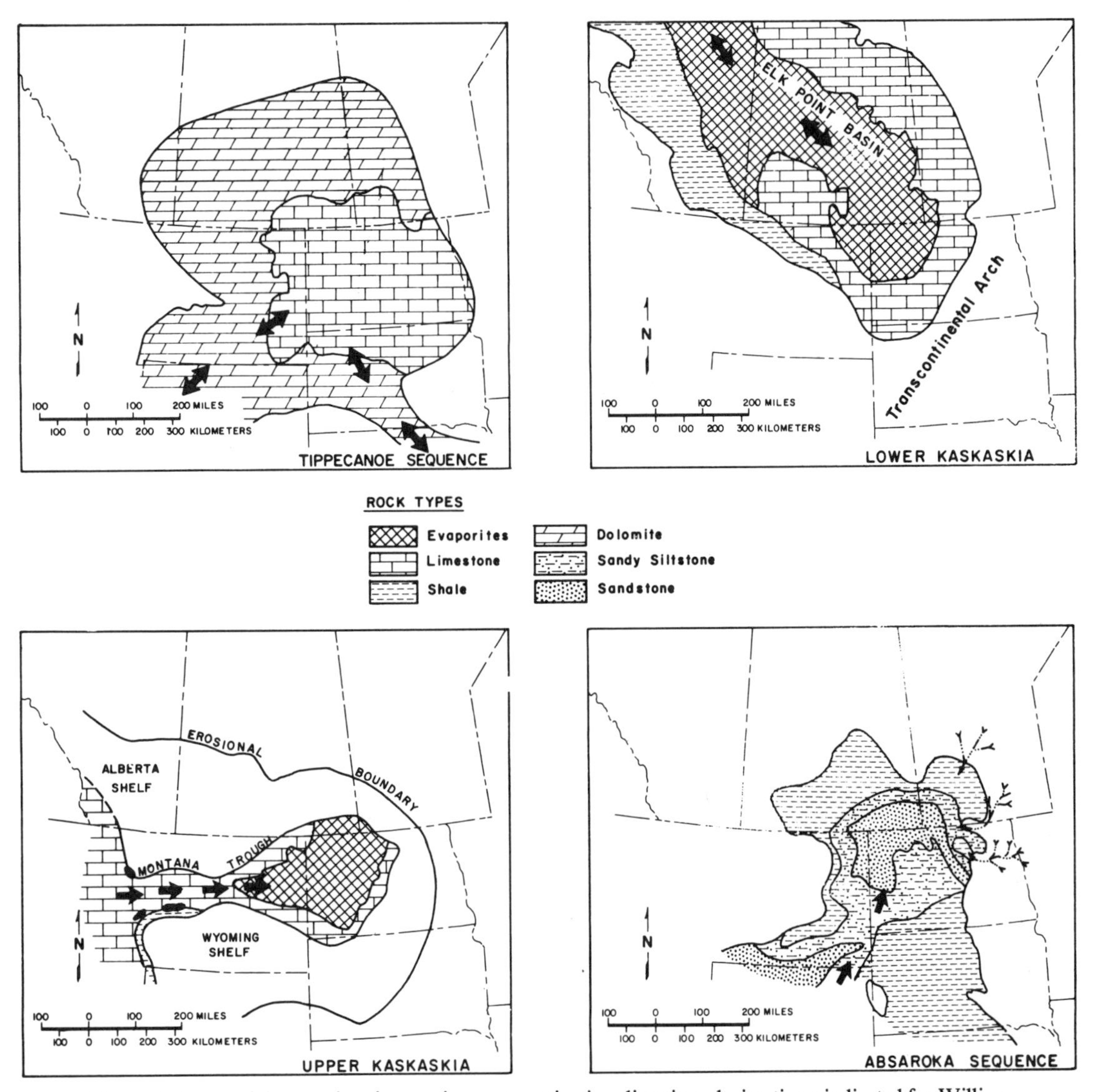

Figure 6. Sequential maps showing marine communication directions during times indicated for Williston Basin. A. Tippecanoe sequence, showing connection with Cordilleran miogeocline on southwest and eastern interior on southeast across Transcontinental Arch. (Modified from Foster, 1972.) B. Lower Kaskaskia sequence communication to Elk Point Basin, Canada. (Modified from Clark and Stern, 1968.) C. Upper Kaskaskia, showing communication of Cordilleran miogeocline through Montana Trough. Dark spots at entrance and along edges of Montana Trough represent bank development. Some bank development is also hypothesized within basin. (Modified from Bjorlie and Anderson (1978.) D. Absaroka sequence, showing flood of clastics from southwest as well as from drainages to northeast and east. Sediment flow probably occurred from farther southeast also. (Modified from Mallory, 1972a, b; and Rascoe and Baars, 1972.)

condensate where it is both clean and relatively free of silica cement, such as in Richardton Field, Stark County, North Dakota. It is surmised that the sand meets these criteria where it crosses ancient structural axes and is the result of winnowing during deposition. Sufficient control to adequately document this assumption is not yet available.

Upper Winnipeg rocks are largely shaly and have long been considered to be major source rocks for lower Paleozoic oil and gas production. These rocks follow the first transgressive facies of the Tippecanoe cycle and are represented by a relatively uniform log character. Over most of its extent, the Winnipeg is considered to be early Mohawkian in age (Ross and others, 1982) and transitional to the beginning phase of major carbonate deposition in the Williston Basin.

Carbonate deposition, characteristic of much of the lower and middle Paleozoic in the Williston Basin, begins in Ordovician time with deposition of the Red River Formation. The Red River is correlative with the Bighorn of Wyoming and the Fremont

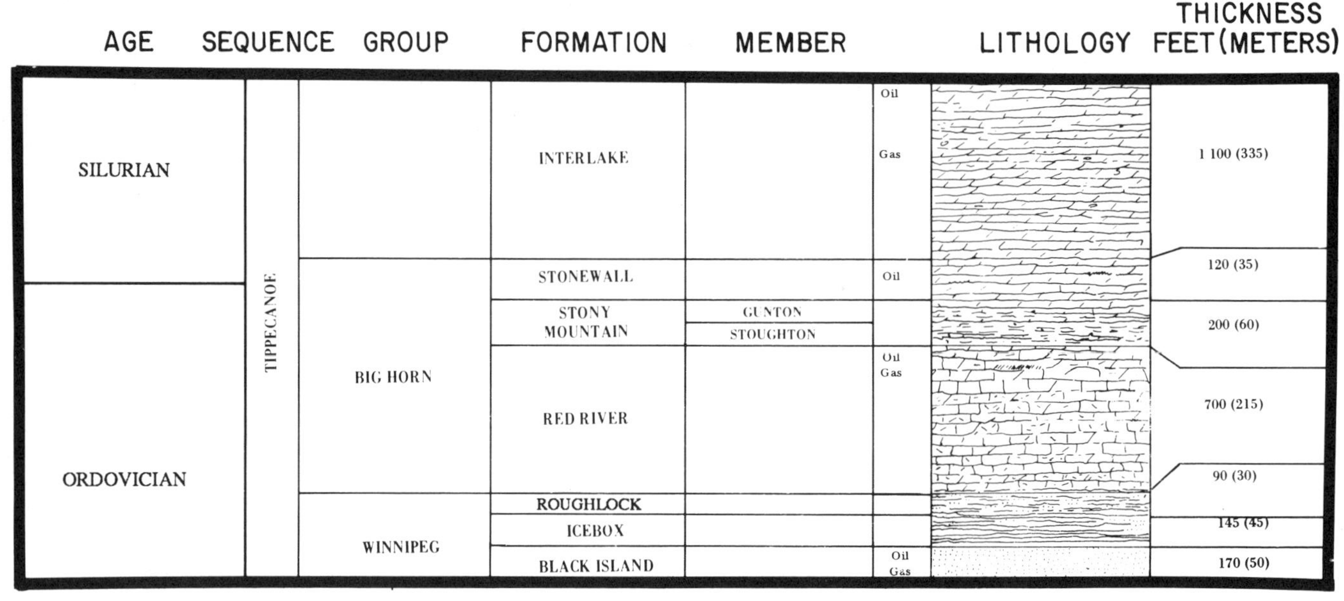

Figure 7. Tippecanoe rocks of the Williston Basin.

Limestone of Colorado. (Abundant fauna indicate a Cinçinnatian age for the unit.) Lower Red River rocks are limestones with dark organic and argillaceous particles. Carroll (1979) characterized these rocks as repetitive cycles of organic-rich, partially dolomitized, fossiliferous limestone and burrowed mudstone. There are generally four "pay" zones recognized in the Red River Formation. The upper part of the lower Red River Formation comprises the lower (D) zone of hydrocarbon production in the basin. Carroll's model of dolomitization for the lower zone and his model for the overlying upper Red River Formation rocks (C, B, and A zones) has generally been adopted by more recent writers. The pervasiveness of dolomitization in the burrowed mudstones of the lower zone increases with depth of the bed, with burrows selectively dolomitized as compared to the matrix. At the base of each bed the matrix is also dolomitized, in some cases creating vuggy porosity. The mechanism he has suggested is gravity-driven brines moving basinward and downward until reaching an impermeable organic-rich limestone layer. The brines have been relatively enriched in magnesium by previous precipitation of gypsum (anhydrite) in the shallow water along the shore or over shoals. In general, lower zone porosity lies shoreward of overlying porosity zones; the better porosity is developed around the peripheries of ancient buried hills. This model is especially relevant to production and exploration in the central part of the basin.

Three lithologic cycles are present in the upper Red River Formation. Each cycle thins successively upward and is comprised of bioturbated skeletal wackestones, packstones, fine-grained dolomitic mudstones, and finally, nodular anhydrite. Fabrics and biotas of the Red River Formation indicate shelf and lagoonal environments. There is no evidence of deep-water environments during deposition of the Red River Formation.

The shape of the basin during Red River deposition was roughly circular (Fig. 8), with a southwesterly connection through the Central Rockies depression to the Cordilleran shelf. Circulation was impaired during deposition, resulting in extensive sheets of anhydrite and anhydritic dolomites. Overall, the unit appears to shoal gradually upwards, with the cycles becoming more closely spaced. A depression on the eastern margin of the basin in north-central North Dakota was connected to the main basin through a narrow passage. Little lithologic data are available from that area, but it is surmised that the Red River in north-central North Dakota is more evaporitic than in the main part of the basin.

Isopach mapping of the Red River Formation indicates that the Nesson Anticline, Billings Anticline, and the various small eastern North Dakota hinge-line highs were active during Red River sedimentation. Isopachs of the Red River are especially important in studying basin evolution because only the basin margins were exposed to erosion.

Overlying the Red River Formation are some thin stratigraphic units (Stony Mountain and Stonewall Formations) that represent the remainder of Ordovician sedimentation. The Stony Mountain Formation is comprised of argillaceous carbonate and shale. The Stonewall Formation is dolostone and limestone and is conformably overlain by the Silurian Interlake Formation.

Together, the Red River, Stony Mountain, and Stonewall Formations are about 1,000 ft (300 m) thick. The overlying and youngest Tippecanoe stratigraphic unit, the Interlake Formation, comprises a maximum of 1,100 ft (335 m) of dolostone and anhydritic dolostones. The classic study of Roehl (1967) described the Interlake Formation in the Cedar Creek Anticline area, and used it as an analog to the Bahamian sedimentologic

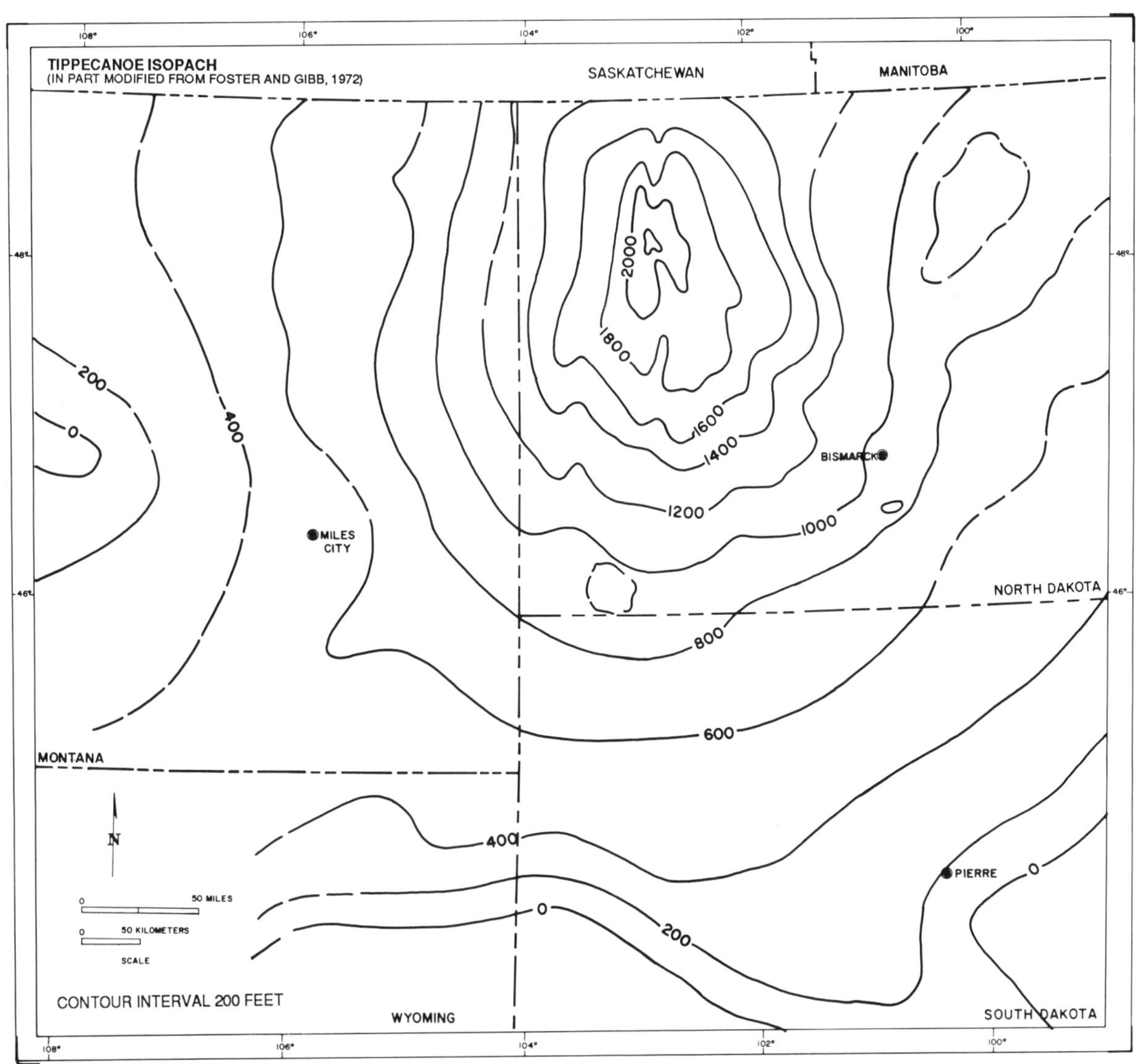

Figure 8. Tippecanoe sequence isopach map of the Williston Basin, showing major southwestern seaway connection, congruent with ancient shear system trends, and minor connection to eastern interior of craton to the southeast.

model. LoBue (1982; personal communication, 1983) has described the lithofacies of the Interlake throughout North Dakota and interpreted its depositional and diagenetic history. The Interlake is lithologically complex and has an intraformational unconformity. At least twelve lithofacies are recognized, although only one or two may occur within any one core. To date, hydrocarbon production is restricted to the upper part of the Interlake Formation, where secondary porosity is developed below and at the post-Tippecanoe unconformity.

Deposition of the Interlake Formation took place in a roughly circular Williston Basin, similar to the basin in Red River time. Littoral flats and shallow sublittoral environments dominate the depositional regime and were laterally, as well as vertically, juxtaposed. Although carbonate muds are abundant throughout the Interlake, rocks in the northeastern part of the basin appear to be somewhat muddier than those in the southwestern and western parts of the basin. Salt beds (halite) are known to occur in local areas in the deeper parts of the basin.

Exposure of the Interlake Formation to erosion during the pre-Kaskaskia regression had several distinct phases, which are preserved as diagenetic events in Interlake rocks. These exposures caused topographic relief and deepened the paleosol development, enhancing porosity and permeability. As Roehl (1967) showed, structural movement on the Cedar Creek Anticline during the pre-Kaskaskia subaerial exposure created well-defined topography in the Interlake. Isopach mapping on the Nesson Anticline indicates that topography was developed there as well.

During restriction of circulation in Interlake time, anhydrite (gypsum) crystals grew in muds in tidal-flat settings. Based on log analysis, halite occurs in numerous areas as discrete beds. In addition, halite is known to occur in the carbonates interstitially or as vug filling locally along the Nesson Anticline.

Pre-Kaskaskia erosion has stripped all Interlake rocks younger than Early Cayugan from the basin. A well-developed karst surface and sedimentary cover was developed during this major intersequence unconformity. It is interesting that the pre-Kaskaskia–post-Tippecanoe unconformity covers the time span of Cayugan through Early Devonian time. That is the approximate time interval of the Acadian event in the middle Appalachians. Middle Devonian rocks are the earliest Kaskaskia rocks in the basin.

Isopach mapping of the entire Tippecanoe sequence (Fig. 8) demonstrates the presence and activity of the Nesson and Antelope Anticlines, Billings Anticline, possibly the Little Knife Anticline, and the Bismarck-Williston lineament. The Cedar Creek Anticline does not greatly influence the overall depositional regime of the Tippecanoe, although it is an active structure by the end of Tippecanoe time. The major opening of the basin to the Cordilleran shelf is through the Central Rockies depression with an additional opening to the interior around the southwestern margin of the Sioux Uplift.

Lower Kaskaskia Sequence

Devonian. Kaskaskia rocks are perhaps the best-known sequence because of their economic importance and because there is more subsurface data available for these rocks than any other sequence (Fig. 9). Tectonic activity along several paleotectonic elements changed the major communication from the Cordilleran shelf (the Central Rockies depression) to a new access route via the Elk Point Basin of the Prairie Provinces of western Canada (Fig. 6). As a result of this change, the lower Kaskaskia rocks have lithologic and stratigraphic affinity with the Canadian Devonian rocks rather than the Devonian rocks of the Cordilleran shelf. No clear consensus exists as to the mechanism for this change. Gerhard and others (1982) attribute the change to the last major uplift along the Transcontinental Arch, which tilted the Williston Basin northward. Growth of the Transcontinental Arch during Early Devonian time caused subaerial erosion within the basin, coinciding with the Acadian event. Subsidence of the Williston Basin continued during lower Kaskaskia time. Sediments thickened northward, juxtaposing thick Tippecanoe and upper Kaskaskia rocks with thin lower Kaskaskia rocks in the central Williston Basin.

A karst surface, represented by siltstones, shales, and some residual breccias, was created on Silurian rocks during the erosional interval that occurred prior to the Devonian deposition. This was reworked in part by the lower Kaskaskia transgression so that the Winnipegosis Formation appears transitional with the Ashern in most logs; cores have demonstrated that a physical disconformity is present but is not easily discernible. The Winnipegosis Formation is one of the most interesting carbonate rock units of the Williston Basin because it contains large pinnacle reefs, which frequently are found to have extensive vuggy porosity. Perrin (1982) presented an extensive core-based study of the Winnipegosis Formation in North Dakota in which she details the lithofacies and depositional history of this unit. She considers the Ashern Formation to be the initial transgression deposit of the Kaskaskia, with a brief hiatus and erosional period marking the top of the Ashern. Renewed transgression provided the setting for deposition of the Winnipegosis. Winnipegosis transgression deposited muddy tidal-flat and subtidal rocks, which are overlain by packstones and grainstones. This is the major transgressive phase of the Winnipegosis that culminates in the stromatoporid pinnacle-reef facies. Although the major reef-forming interval may be categorized as a stromatoporid facies, it is exceedingly more complicated than that. An entire reef complex exists, with well-developed supratidal, patch-reef, lagoonal, pinnacle-reef, shelf, and deeper basin environments.

The final phase of Winnipegosis deposition encompasses a regressive setting with stromatolitic pinnacles and subaerially exposed, weathered dolomites, breccias, and anhydritic beds. Pisolitic beds replace skeletal packstones and grainstones, further substantiating the subaerial nature of last-phase sedimentation and early diagenesis.

An alternate view of the last phase of sedimentation of the Winnipegosis Formation regards the cessation of carbonate formation as a result of increasing salinity in the basin. This is indicated by the fauna, flora, and increasing evaporite content of the marginal sediments. Winnipegosis sedimentation was terminated by the initiation of deposition of halites, potash, and other evaporite minerals of the Prairie Formation (Middle Devonian). The potash-bearing members contain both carnallite and sylvite.

Restricted circulation in the Elk Point Basin is normally regarded as the controlling factor in the deposition of the Prairie Formation (Anderson and Swinehart, 1979). The vast area of the Elk Point Basin, coupled with continued marine influx and shoal environments, may have been able to accomplish the same effects. The evaporites of the Prairie are not everywhere present over the Winnipegosis.

This edge of the Prairie Formation is interpreted to be a solution edge, which controls some of the Mississippian structure that enhances petroleum trapping on the basin's east side. The edge of the Prairie also roughly corresponds to the boundary between the Churchill and Superior provinces of the Canadian Shield. It is conjectural, but the observed relationships indicate that the basin hinge line, Prairie salt solution, and much oil production is controlled by slight tectonic movement along that provincial boundary.

On the west side of the basin in southern Richland County, Montana, Prairie Formation thicknesses vary. In two wells only 1.5 mi apart, a full section of Prairie is present in the eastern well and is absent in the western well. The younger rocks in the western well are thin compared to normal sections in the Upper Devonian. This indicates that structural growth along major fracture trends took place during deposition of lower Kaskaskia

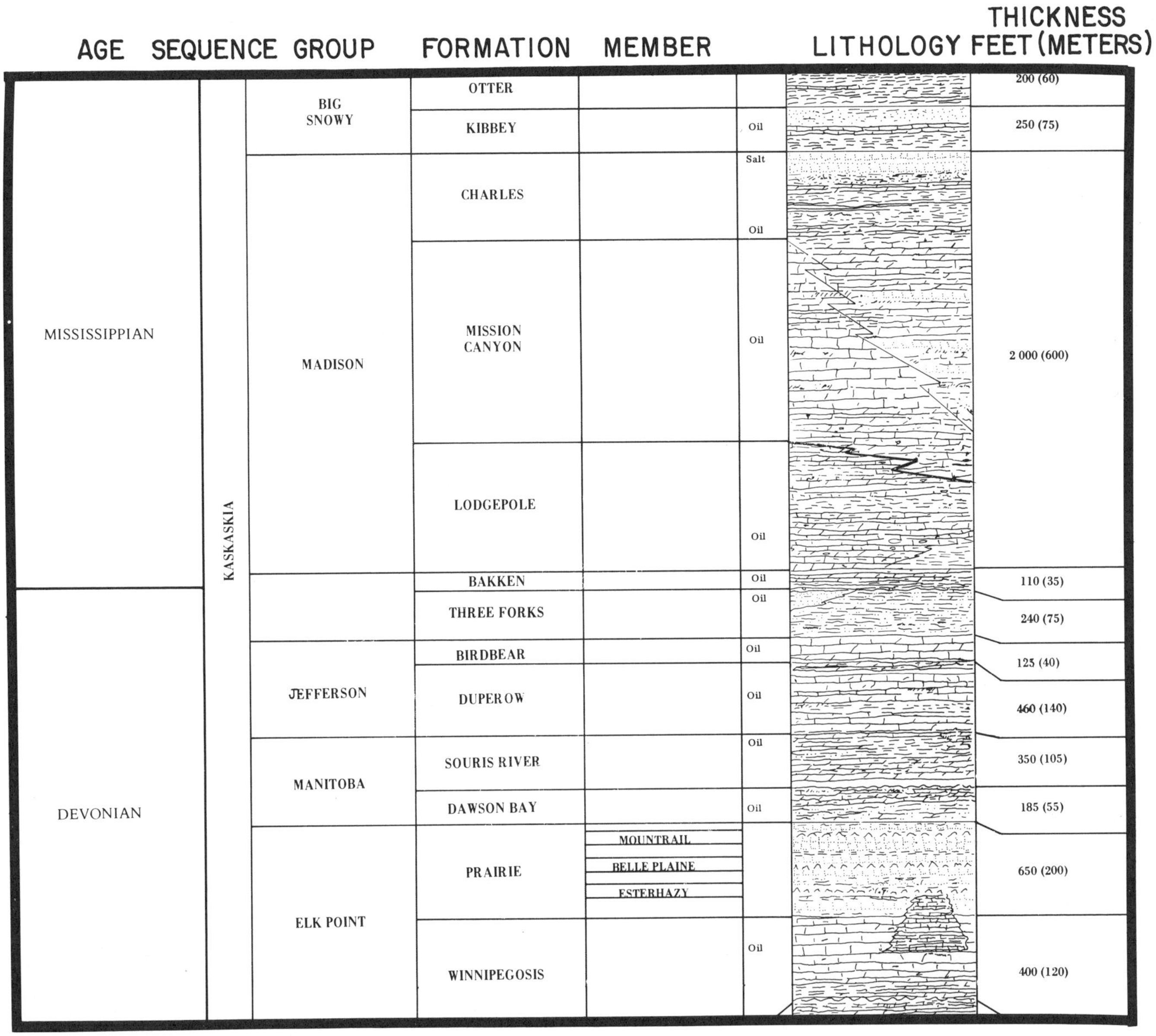

AGE	SEQUENCE	GROUP	FORMATION	MEMBER		LITHOLOGY	THICKNESS FEET (METERS)
MISSISSIPPIAN	KASKASKIA	BIG SNOWY	OTTER				200 (60)
			KIBBEY		Oil		250 (75)
		MADISON	CHARLES		Salt Oil		2 000 (600)
			MISSION CANYON		Oil		
			LODGEPOLE		Oil		
			BAKKEN		Oil		110 (35)
DEVONIAN			THREE FORKS		Oil		240 (75)
		JEFFERSON	BIRDBEAR		Oil		125 (40)
			DUPEROW		Oil		460 (140)
		MANITOBA	SOURIS RIVER		Oil		350 (105)
			DAWSON BAY		Oil		185 (55)
		ELK POINT	PRAIRIE	MOUNTRAIL BELLE PLAINE ESTERHAZY			650 (200)
			WINNIPEGOSIS		Oil		400 (120)

Figure 9. Kaskaskia sequence rocks of the Williston Basin.

rocks. Mondak Field, which crosses the Montana-North Dakota border parallel to the Cedar Creek Anticline, has had its trapping attributed to solution of the Prairie, similar to that in the eastern part of the basin (Kearns and Traut, 1979). Mapping by the present authors suggests instead that there is a depositional edge of the Prairie basinward from the Mondak structure, suggesting that here the limit of evaporites is tectonically controlled by the wrench-fault system previously proposed rather than by salt-solution tectonics. Farther west in Montana, there are several two-stage salt-solution features that are highly productive. Mapping of the theoretical fracture systems and salt thinning by solution establishes a close correlation between the two phenomena and suggests that western solution is largely controlled by fracture systems.

Apparently a second subcycle of transgression-regression ended with the deposition of the Prairie, and a third cycle of carbonates and clastics was deposited before the end of the lower Kaskaskia. Two stratigraphic units, the lower Dawson Bay Formation and upper Souris River Formation, together mark a very shallow phase of sedimentation in the basin. Clastics present are derived from extrabasinal sources. Evaporitic conditions persisted throughout much of both units. To date, only limited hydrocarbon production has been obtained from these units; thus, they have not been extensively cored. Dunn (1982) discusses much of the lithology of the Dawson Bay Formation in Saskatchewan and demonstrates that shelf conditions dominated sedimentation, with bioherms and other skeletal deposits forming a significant part of the rock column. Dean (1982) shows that the Dawson Bay Formation in northwestern North Dakota contains abundant evidence for peritidal deposition. This interpretation is consistent with the tectonic model for lower Kaskaskia deposition.

Cycles of normal and hypersaline waters and related carbonate deposition characterize the remainder of the lower Kaskaskia. The Duperow Formation is characterized by stromatoporid banks and shelf carbonates that are interbedded with peritidal rocks and anhydrite. Several writers have noted that porosity development in the Duperow Formation is largely controlled by postdepositional processes. Burke and Stefanovsky (1982) have shown the variety of minerals and pore-forming processes involved. Prolific rates are obtained where the Duperow is productive. Depositional processes and environments do not have a strong control on porosity.

The Birdbear Formation (Nisku) is largely limestone with some anhydrite. A variety of lithofacies has been described by Halabura (1982). Abundant brachiopods and other invertebrate fossils are present in this unit. Lower Birdbear rocks are characterized by micritic fabrics and the presence of peritidal environmental indicators. Apparent supratidal dolomite and anhydritic carbonates are present. Algal and cryptalgal laminates are abundant. Upper Birdbear lithologies are more characteristic of deeper water and higher energy. Skeletal sands in grain-to-grain contact with interstices or sands with mud-filled pores interbedded with muds are both common. Halabura has also reported oolitic beds and interpreted the environmental complex to be the proximal facies of the deeper-water Nisku rocks to the north and west in Canada. Loeffler (1982) has found similar facies in the Birdbear Formation of North Dakota.

The Three Forks Formation overlies the Birdbear and is predominantly clastic. Sediments are mainly dolomitic and anhydritic siltstones and shales. A regression occurred at the conclusion of Three Forks deposition.

Upper Kaskaskia Sequence

Mississippian. During lower Kaskaskia sedimentation, patterns indicate that the Williston Basin was part of the larger Devonian Elk Point Basin, opening to the Cordilleran seaway to the northwest. Upper Kaskaskia patterns demonstrate that the connection changed from the northwest to almost due west through the Montana Trough (Figs. 6, 10).

The unconformity separating lower and upper Kaskaskia is extensive. The most pronounced erosion appears to have been on the Cedar Creek Anticline, where rocks down to the Silurian Interlake Formation were exposed (Clement, 1976; Gerhard and others, 1982). The eastern margin is represented by an angular unconformity between the Madison Group and stratigraphically lower units down to the Stony Mountain Formation (Upper Ordovician).

General euxinic conditions existed in the central Williston Basin during latest Devonian and earliest Mississippian time. The basal transgressive unit of the upper Kaskaskia is the Bakken Formation, which is comprised of black shales and siltstones (Fig. 11). This stratigraphic unit does not extend over all of the basin, but is restricted to the central portion. Despite its limited geographic extent, the Bakken is of great significance to hydrocarbon production as a source rock for Mississippian oil.

Continued transgression during upper Kaskaskia time sustained shaly limestone and limestone deposition of the Lodgepole Formation (Madison Group). Isolated from the remainder of Lodgepole deposition by carbonate mud bars (Waulsortian mounds), a shallow depression was present along the eastern margin, producing the Carrington shale facies (Bjorlie and Anderson, 1978). Major mounds are exposed in the Big Snowy Mountains. Bjorlie and Anderson suggest that these Waulsortian mounds are probably present along the margins of the Central Montana Trough.

Concurrent with the Carrington facies sediment, Lodgepole sedimentation continued with dark basinal mudstones and wackestones, basin-slope argillaceous skeletal wackestones, packstones, and open-shelf cherty skeletal wackestones and packstones. A partial regressive phase began in mid-Lodgepole time, which caused the shoaling of open-shelf water; oolite bars at the shelf divide muddy, skeletal-pelletal, argillaceous grainstones of the open shelf from shoreward mudstones and skeletal-pelletal wackestones. Cyclic sedimentation of skeletal sediments and oolites has been reported and continued through the upper part of Lodgepole sedimentation, particularly in the eastern portion of the basin. Oolites are not abundant in the upper beds. The regression, which began during Lodgepole sedimentation, is apparently a function of westward progradation of sediments rather than a eustatic sea-level change (Fig. 11).

Belts of sedimentation, including shoreward sabkha sediments and marine carbonates, typify Mission Canyon deposits. Mission Canyon rocks are generally shoaling-upward carbonates followed by anhydrite. However, variations on this theme make stratigraphic study and hydrocarbon exploration difficult.

Muddy skeletal carbonates, packstones, and wackestones, are typical of shelf and basinal sedimentation. Fossils, including corals, are abundant and varied. However, marginal beds interfinger with, and are covered by, anhydrite of apparent sabkha origin and have lateral anhydrite equivalents that appear to be of subaqueous origin. Radial fibrous ooids are abundant. Large coated pisoids and extensive beds of pisolite are especially abundant in the shoaling facies of the Mission Canyon beds on the northeastern and eastern margin. These may also contain many structures otherwise identifiable as having formed in the subaerial or supratidal environment. There is no agreement on the origin of these round, coated grains, although it is likely that the very low depositional slope and highly evaporitic conditions prevailing on the tidal shoals and supratidal zones during Mission Canyon time caused an alteration of environments and a mixture of types of grains.

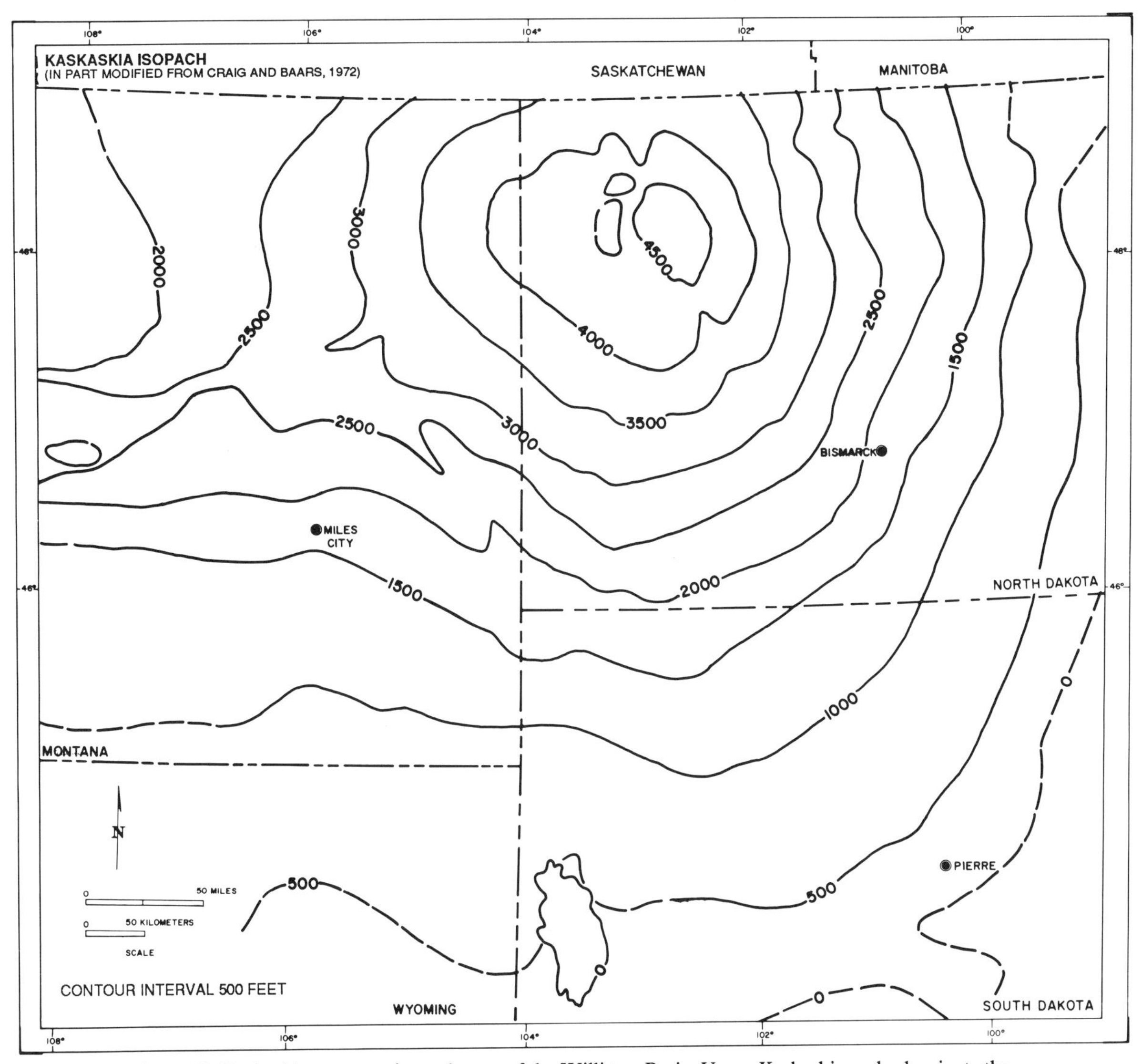

Figure 10. Kaskaskia sequence isopach map of the Williston Basin. Upper Kaskaskia rocks dominate the thickness pattern, obscuring the lower Kaskaskia northwestward tilt of the Williston Basin.

The regression that started during the upper Lodgepole continued slowly throughout Mississippian time, marked by reduction of the area in which carbonates were deposited, while extending the area of evaporite deposition around the periphery of the basin. Although anhydrite is the major evaporite mineral now present, halite beds were also deposited and are used both as correlation markers and as indicators of paleotopography. Madison Group sedimentation was terminated by basin-wide evaporite deposition.

Upper Kaskaskia sedimentation, showing the influence of the Ancestral Rocky Mountain deformation, continued after the end of evaporite deposition in the Charles Formation. Sandstones, shale, and thin limestones form the bulk of the sediments deposited at this time, known as the Big Snowy Group (Chesterian).

Absaroka Sequence

Pennsylvanian-Triassic. Widespread North American tectonism during Late Mississippian and Pennsylvanian time was manifested in the Williston Basin by regional uplift and erosion, lower rates of basin depression, and deposition of siliciclastic sediments from extrabasinal sources. Although several sources are indicated, the major flow of clastics appears to have been from the south (Fig. 6). These sediments were supplied to the basin through a fluvial deltaic system. The basin was not closed; excess sediment was moved westward through the Central Montana Trough. Maximum thickness of Absaroka sequence rocks in the basin is approximately 2,350 ft (715 m). Rocks are predominantly siliciclastic although a few limestones, dolostones, and numerous salt and gypsum beds are present.

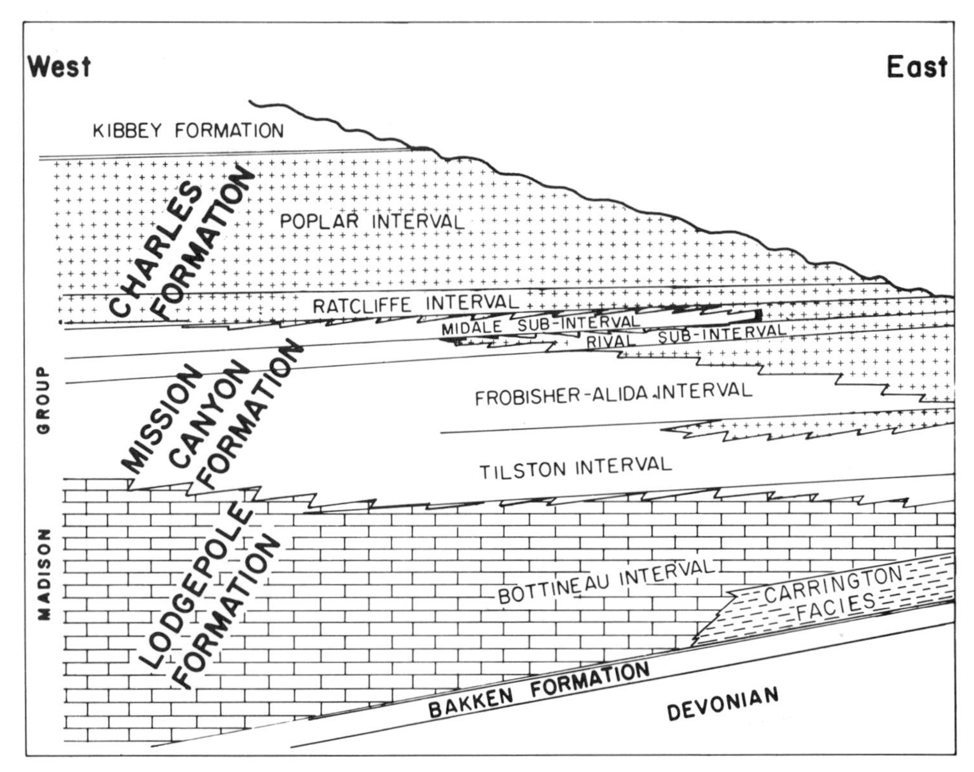

Figure 11. Generalized upper Kaskaskia stratigraphic cross-section from central basin to eastern margin.

Depositional environments are characterized by terrestrial clastics interfingering with marginal marine and evaporite sediments (Fig. 12). Early Absaroka time is characterized by more marine or marginal marine sediments. These become more evaporitic with an overall finer grain size in the upper parts of the Absaroka.

The Tyler Formation has been studied in some detail (Ziebarth, 1964; Grenda, 1978; Sturm, 1983). The fluvial deltaic complex has been well documented in North Dakota. Offshore bars, estuarine environments, and channel complexes have been identified. Paleotopographic prominences associated with major basin structures appear to provide sites for accumulation of winnowed sands that frequently have well-developed porosity. Sturm divided the Tyler into upper and lower stratigraphic units. These are interpreted to be a progradational delta plain, succeeded by a barrier island and bar system. The source of the sediments is either the ancient Sioux Ridge of South Dakota or the Hartville Uplift of eastern Wyoming. Oil and gas production is presently obtained from channel and delta complexes in southwestern North Dakota.

Overlying the Tyler are thinner units of varying lithology that all suggest progradational sedimentation into a basin of increasing salinity. An unconformity above the Broom Creek Formation of the Minnelusa Group records a period of weathering and erosion during the Permian. The remainder of Absaroka deposition includes episodes of hypersalinity in the center of the basin (Opeche Salt, Pine Salt) and fine-grained red beds deposited across the basin. The Triassic Spearfish Formation consists of fine- to medium-grained siliciclastic rocks that fill the basin, prograding eastward across truncated Mississippian carbonates. These rocks are reservoirs for hydrocarbons migrating through the Mississippian to be trapped by permeability barriers in the Spearfish.

Relatively flat topography and clastic sedimentation mark the transition from the Absaroka sequence into the Zuni. By Absaroka time the Williston Basin was truly cratonic, a tectonic setting that has persisted from that time to the present.

Zuni Sequence

Jurassic–Tertiary. Sedimentation of the Zuni sequence in the Williston Basin began with fine- to medium-grained clastic sediments and salt (Fig. 13). Even though most wells penetrate the Zuni sequence, failure to log much of the stratigraphic section above the Cretaceous Pierre has made it difficult to study Cretaceous and Paleocene rocks in the subsurface.

Younger Jurassic rocks (post–middle Piper Formation) contain carbonates, mostly oolitic beds that were deposited on shoal areas within a shallow-marine environment. Although porous through secondary processes, these oolite banks are so far known

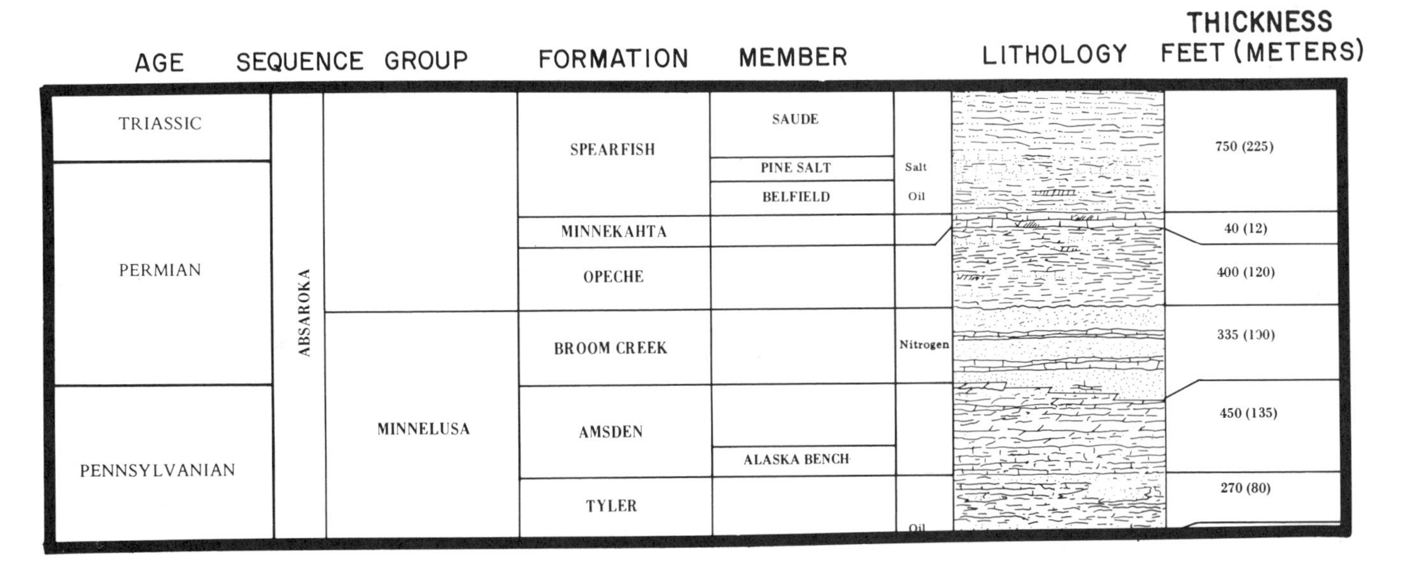

Figure 12. Absaroka sequence rocks of the Williston Basin.

to contain only water. The Swift and Morrison Formations are siltstones, shales, and calcareous sandstones with minor amounts of sandy, freshwater limestone. A basin-wide unconformity that exists between Jurassic and Cretaceous rocks is found at the base of the Inyan Kara Formation. This unconformity represents a time gap of 26 m.y. (Bluemle and others, 1981) and is largely attributed to relatively small emergence of the basin.

Lithologies of the remainder of the Zuni sequence are related to the overall transgressive-regressive cycles of the Cretaceous and earliest Tertiary interior sea in the western U.S., with the transgressive phase culminating in the deposition of the Pierre Shale. The cycle is represented by basal sandstones grading upward into shales with some carbonates and sandstones. Terrestrial deposits began in Cretaceous Hell Creek time and are generally fine-grained sandstones and siltstones.

Paleocene beds are composed of terrestrial paludal and fluvial deposits, in which extensive lignite reserves have been formed. The last marine inundation of the Williston Basin occurred during deposition of the Cannonball Formation of the Fort Union Group. Subsequent sedimentation is entirely terrestrial, consisting of cycles of siltstone, fine-grained sandstone, lignite, and lignitic shales. Rarely, freshwater limestones are present, and a few white pseudoquartzite beds indicate formation of soil zones during deposition of the Fort Union Group.

Upper Zuni sequence rocks contain bentonites and less-altered volcanic ash, possibly derived from the western Laramide Rockies. The relatively moist terrestrial environment, in which these rocks formed, was altered; there is a disconformity between Zuni rocks and Tejas sequence rocks.

Tejas Sequence

Tertiary, Quaternary, and Holocene. Few Tejas rocks are present in the Williston Basin. By Tejas time, the Williston basin became tectonically inactive, although deposition of terrestrial sediments continued (Fig. 14).

Glacial deposits and accompanying erosion have altered the surface of much of the basin. These deposits of ice-contact origin cover much of the northern part of the basin.

The Pleistocene glacial features of the northern part have been well documented in recent years (Clayton and others, 1980; Bluemle, 1977). A virtual plethora of soil stratigraphic names has been proposed for various sediment bodies probed by shallow drilling for water and lignite. Some of the glacial features are classic in exposure and serve as examples of true, continental glaciation, ice-contact features. Small intrusions of Tertiary age are located in the very southern part of the basin near the Black Hills.

MAJOR BASIN STRUCTURES

Nesson Anticline

The Nesson and Cedar Creek Anticlines are the major structures in the basin that have surface expression and are oil productive. Study of subsurface data on the Nesson Anticline has suggested that the structure has been active since the Precambrian (Anderson and others, 1983). The history of the Nesson Anticline began during Precambrian time. Lower Deadwood sediments were deposited around, but not over the crystalline core of the present structure. Upper Deadwood sediments onlapped the top of the structure. Abrupt changes in thickness of rocks across the anticline indicate uplift along a fault on the west side. Detailed mapping of the anticline has revealed that this fault was active throughout the Phanerozoic history of the basin (Fig. 15).

Cedar Creek Anticline

Recent review of the structural history of the Cedar Creek

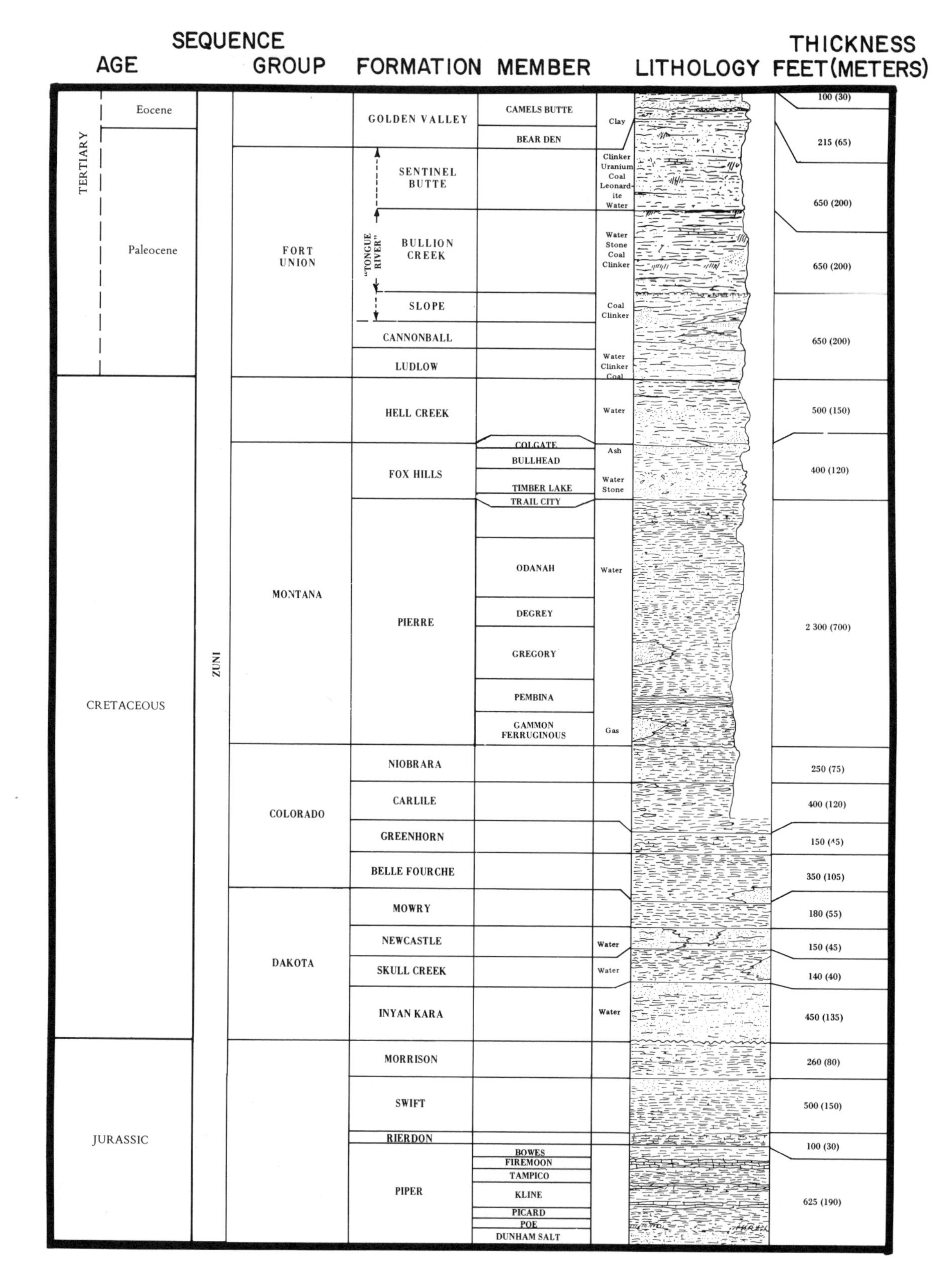

Figure 13. Zuni sequence rocks of the Williston Basin.

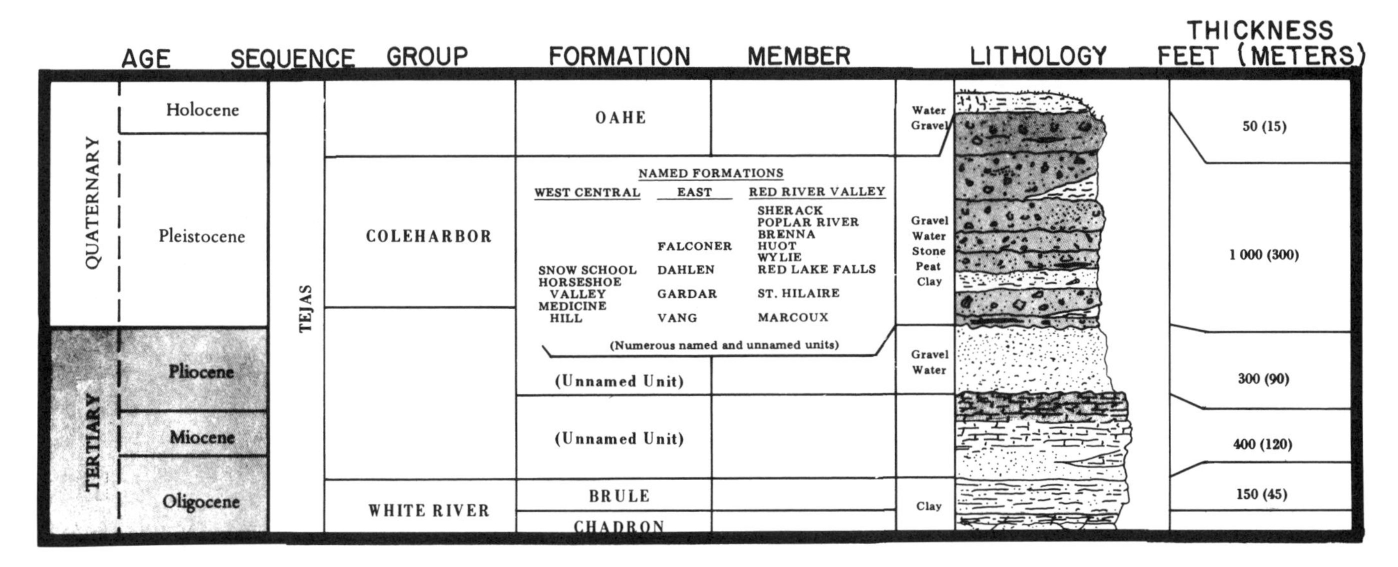

Figure 14. Tejas sequence rocks of the Williston Basin.

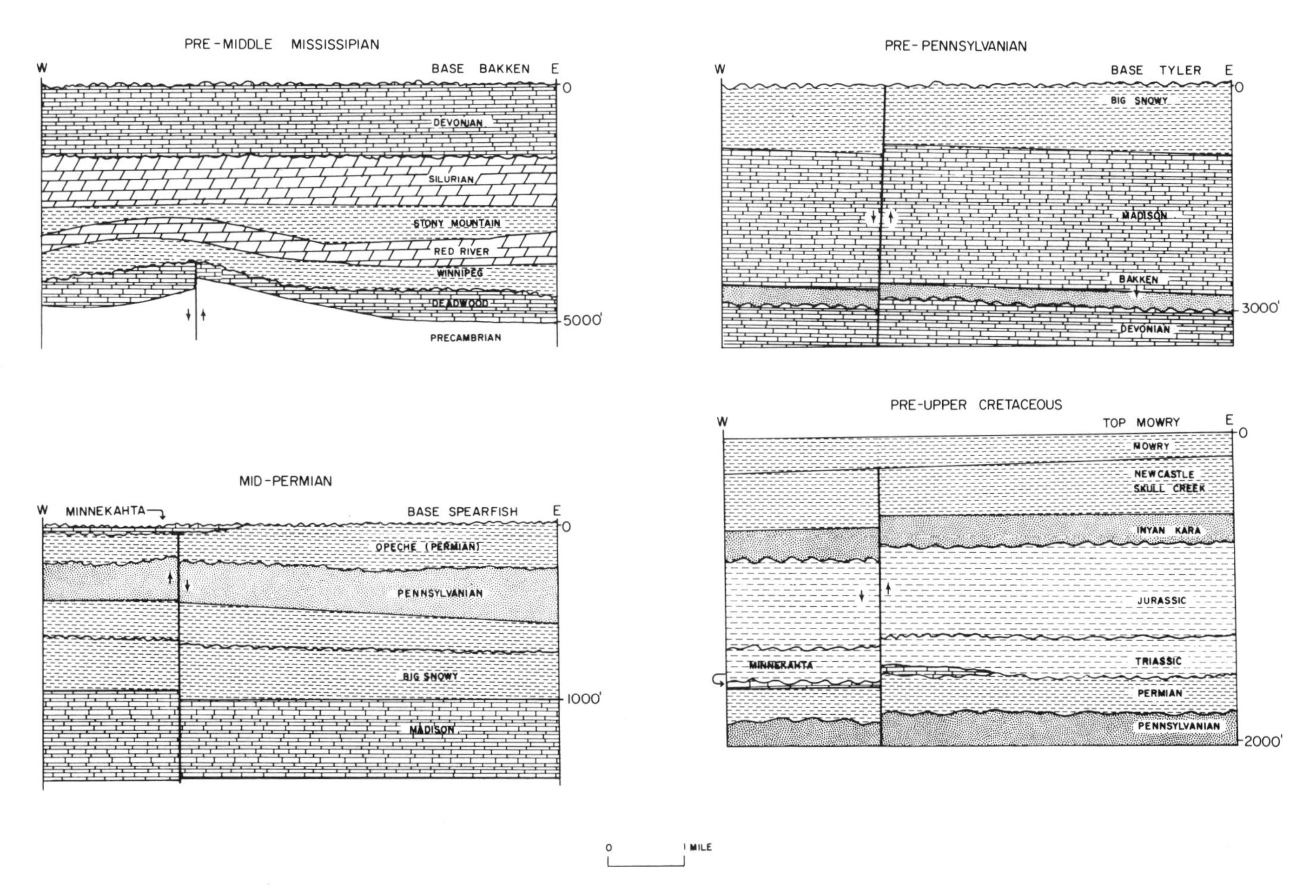

Figure 15. Diagrammatic cross-sections across the Nesson Anticline.

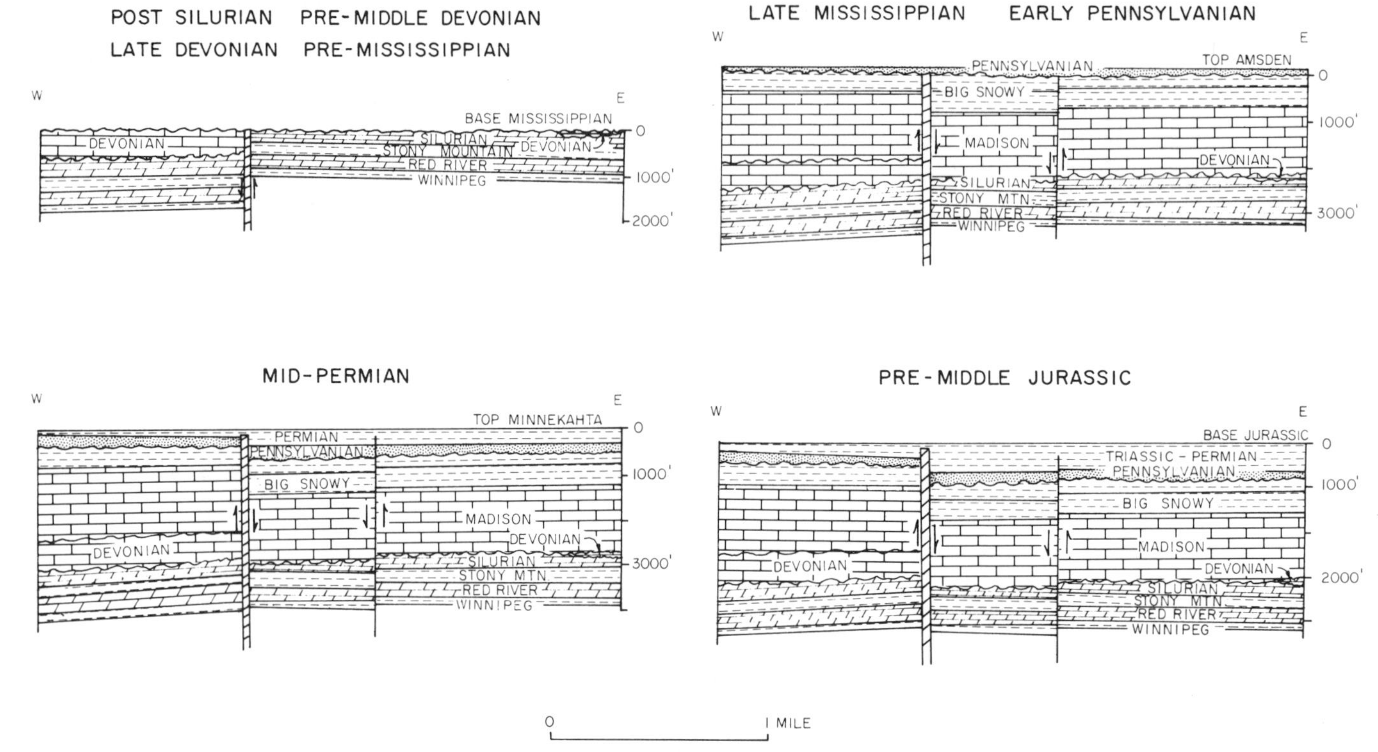

Figure 16. Diagrammatic cross-section across the Cedar Creek Anticline.

Anticline by Clement (1983) demonstrates that the major deformations of the Cedar Creek are congruent with those of the Nesson. The structure marks the southwestern margin of the Williston Basin and therefore has a more pronounced erosional history. The relative intensity of motion and deformation is similar to that of the Nesson. Clement (personal communication) has also demonstrated reversal of motion on faults along the margin of the Cedar Creek Anticline (Gerhard and others, 1982). As Clement points out (1983), the major uplift of the structure took place along deep faults and by drape folding in younger rocks. It appears that the Cedar Creek Anticline has been tectonically more active than the Nesson (Fig. 16).

Red Wing Creek Structure

One of the most controversial and interesting structures in the Williston Basin is the Red Wing Creek structure (Fig. 3), which produces oil from a greatly exaggerated thickness of Mississippian rocks. At the time of its discovery in 1972, this structure was a seismic anomaly that did not fit a pattern common to other Williston Basin structures. Exhibits presented to the North Dakota Industrial Commission during the spacing hearing for the Red Wing Creek Field interpreted this structure as an astrobleme. Succeeding publications by Brennen and others (1975) and Parson and others (1975) presented additional data supporting this hypothesis.

Generally, it appears that sedimentation across the structure followed normal patterns for the Williston Basin through Triassic (Spearfish) time. In the Jurassic a major structural disruption of a very small area occurred (Fig. 17). The discovery well (True Oil #22-27 Burlington Northern) contains a very thick pay section in disturbed Mississippian rocks. The interpretation of the Red Wing Creek structure is based on seismic information and on interpretation of complex data from well logs and samples.

The central part of the structure is an elevated block surrounded by a ring depression. The geometry of the structure is clearly cryptovolcanic and it is difficult to argue against a meteorite impact origin of this complex structure. Bridges (1978) has suggested an intersection of several faults that change orientation through time as an alternate hypothesis (concentricline). It has also been suggested that this structure could result from deep-seated igneous activity (diatreme) or from long-term interaction of adjoining faulted blocks. Detailed analysis of samples, including a search for coesite and other high-pressure silica phases, may determine the validity of the meteorite impact hypothesis.

Newporte Structure

In north-central North Dakota, the Shell #27X-9 Larson well discovered oil in sandstone of the Deadwood Formation in the Newporte structure (Fig. 3; Clement and Mayhew, 1979). Continued development of this field demonstrated the complexi-

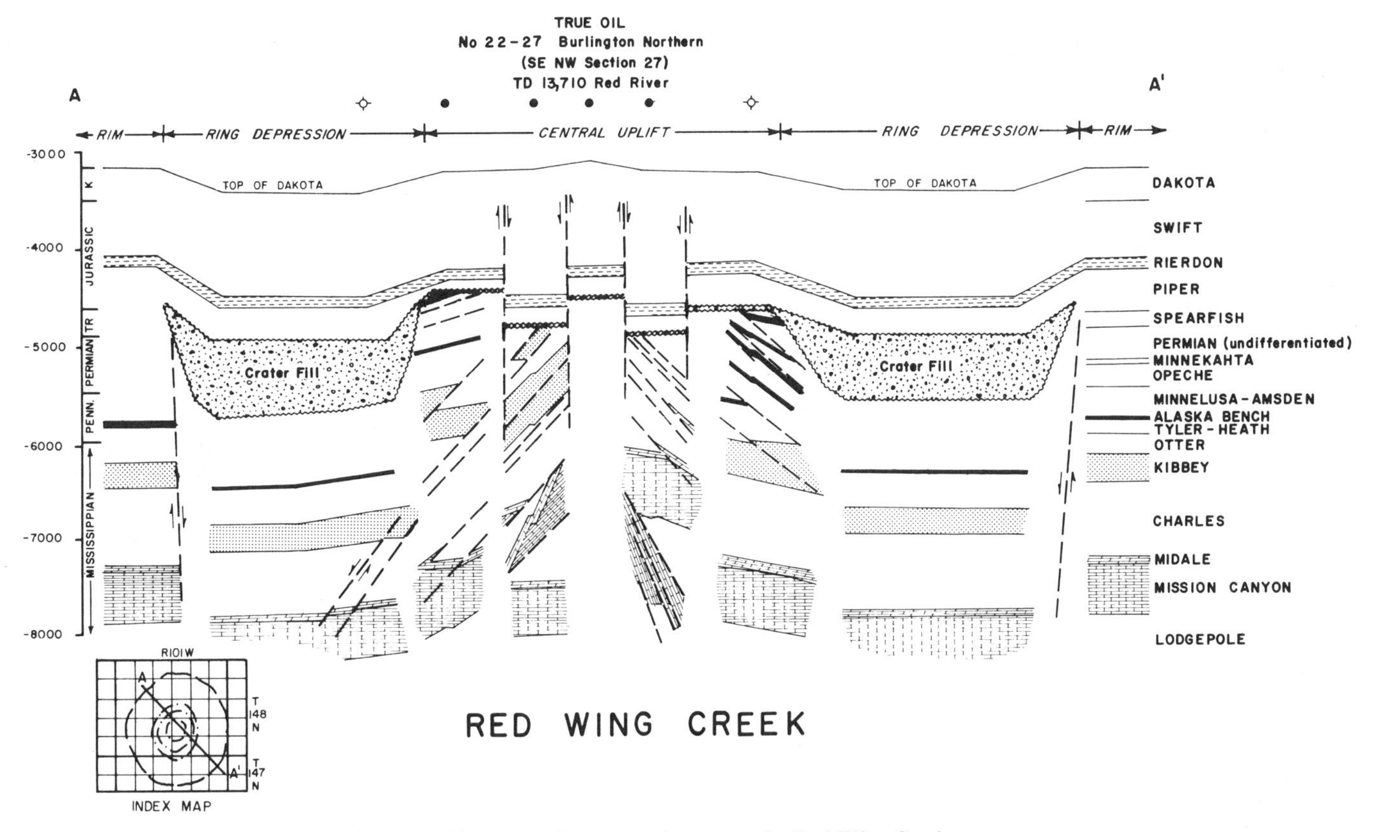

Figure 17. Diagrammatic cross-section across the Red Wing Creek structure.

ties of the Williston Basin. It also includes an oil well producing from Precambrian crystalline rock (Shell #14-34 Mott). Subsurface data suggest that Deadwood Sandstone was deposited between Precambrian hills, and that pre-Winnipeg erosion accompanied by faulting created a small basin of detritus from Deadwood and Precambrian rocks; then Winnipeg sedimentation covered the structure. Donofrio (1981) has suggested that the Newporte structure is possibly a meteorite impact feature.

ENERGY AND MINERAL RESOURCES

Oil and Gas

Oil was discovered in the Williston Basin in Montana on the Cedar Creek Anticline in 1936. Although gas had been produced for some years before, Manitoba's oil production began in 1950, and oil was discovered in North Dakota on the Nesson Anticline in 1951, establishing the Williston Basin as a major oil basin in the Rocky Mountain province. Recent development of deep reservoirs has demonstrated that the basin also contains significant gas reserves.

Several 100-million-barrel oil fields or combined field areas have been found in recent years in the basin, including Little Knife Field (Madison, North Dakota) and the Billings Anticline fields (Duperow and Madison, North Dakota). The Mondak trend may also prove to have large ultimate recoveries (Montana and North Dakota).

More than 10,000 wells have been drilled in the North Dakota portion of the basin. Numerous areas exist that are virtually untested; some deeper horizons, such as the Winnipeg, have only a few penetrations. Volumetric calculation of hydrocarbons produced from source rocks suggests that at least 11 billion barrels of oil have been generated and less than 2 billion have been found. Further search for stratigraphic traps and deliberate application of structural theory to oil migration and trapping problems will add to the discovery rate and volume.

OTHER MINERAL AND ENERGY RESOURCES

The Williston Basin contains two major resources besides hydrocarbons, although present production of either does not begin to compare to the value of the hydrocarbon resource. Lignite is actively mined and is a significant energy and revenue producer in Montana and North Dakota, although most Montana lignite is in the Powder River Basin.

Lignite occurs as part of the Cenozoic rock suite deriving from erosion of the Cretaceous-Paleocene Sevier thrust belt and Elkhorn Volcanics. These rocks, generally called the Fort Union Group, are mostly of terrestrial origin, although the Cannonball

Formation, in the lower part of the Fort Union, is the youngest marine stratigraphic unit in this area.

Since this is low-sulfur but high-water-content fuel, most of the lignite is utilized close to the mining site for electrical generation. Production in North Dakota has doubled in the last five years as new electrical generation plants have gone on stream. The 20-billion-ton reserve and 350-billion-ton resource (possibly as high as 500 billion tons) indicate that lignite will be a major energy source for decades.

Devonian Prairie Formation potash resources are not yet being mined. Approximately 50 billion tons of potash lie south of the international border in North Dakota, with additional resources in Montana (Anderson and Swinehart, 1979). Minor amounts of other minerals are extracted from the basin. Clays (mined locally for brick manufacture), salt (solution mined near Williston, North Dakota), and sand and gravel are the most important. North Dakota is the 22nd-ranked state in sulfur production, but the sulfur is all by-product from processing natural gas.

SUMMARY

Williston Basin evolution in Phanerozoic time begins with Sauk (Upper Cambrian) transgression across a rough Precambrian topographic surface, but without definition of the basin. Tippecanoe sequence rocks show the initial depression of the Williston Basin, with seaway connections west and southwest to the Cordilleran miogeocline, and a probable connection to the midcontinent as well.

During deposition of the Kaskaskia sequence, movement along the Transcontinental Arch tilted the Williston Basin northward, creating a marine connection to the north through the Elk Point Basin during Devonian time, while shutting off the earlier Cordilleran and midcontinent marine connection of the Tippecanoe sequence. During deposition of the upper part of the Kaskaskia sequence (Mississippian), the northern connection was cut off and the Williston Basin reconnected to the Cordilleran sea through the Central Montana Trough.

Pennsylvanian structural disturbance in the Central Rockies disrupted sedimentation patterns of primarily carbonate-dominant lithologies in the basin. Beginning with the Absaroka sequence, clastic sedimentary facies predominate with interspersed evaporites and a few carbonates.

A full marine setting was established during the Cretaceous as part of the Western Interior seaway. Regression accompanying uplift of the Laramide Rockies provided paludal and fluvial environments, which trapped much sediment moving eastward from the Rockies during Paleocene times and assisted in the formation of extensive lignite deposits.

The basin appears to be the result of drag along two left-lateral shear systems. These shear systems formed a depressed block, which now encompasses the Central Rocky Mountains and the southern Williston Basin. Torsion of this block during plate-wide orogenic stress created the vertical structure seen today.

Substantial quantities of lignite, potash, halite, and petroleum are present in the Williston Basin. Lignite mining for electrical generation is a major industry; halite is mined by solution methods from depths of 8,000 ft (2,439 m), but potash resources are not yet being exploited.

Oil and gas production is rapidly expanding in the basin. Very high rates of success in exploration drilling doubled oil production between 1980 and 1984.

REFERENCES

Anderson, S. B., and Swinehart, R. P., 1979, Potash salts in the Williston Basin, U.S.A.: North Dakota Geological Survey Report of Investigation 68, 19 p.

Anderson, S. B., Gerhard, L. C., and LeFever, J. A., 1983, Structural and sedimentologic history of the Nesson anticline [abs.]: American Association of Petroleum Geologists Bulletin, v. 67, p. 1329.

Ballard, W. W., Bluemle, J. P., and Gerhard, L. C., 1983, Correlation of stratigraphic units of North America; northern Rockies/Williston Basin correlation chart: American Association of Petroleum Geologists.

Bjorlie, P. F., and Anderson, S. B., 1978, Stratigraphy and depositional setting of the Carrington Shale facies (Mississippian) of the Williston basin, *in* The economic geology of the Williston basin: Williston basin symposium, Montana Geological Society, p. 165–177.

Bluemle, J. P., 1977, The face of North Dakota; The geologic story: North Dakota Geological Survey Educational Series 11, 73 p.

Bluemle, J. P., Anderson, S. B., and Carlson, C. G., 1980, North Dakota stratigraphic column: North Dakota Geological Survey.

—— , 1981, Williston basin stratigraphic nomenclature chart: North Dakota Geological Survey Miscellaneous Series 61.

Brennan, R. L., Peterson, B. L., and Smith, H. J., 1975, The origin of Red Wing Creek structure; McKenzie County, North Dakota: Wyoming Geological Association Earth Science Bulletin, v. 8, 41 p.

Bridges, L.W.D., 1978, Red Wing Creek field, North Dakota; A concentricline of structural origin, *in* The economic geology of the Williston basin: Williston basin symposium, Montana Geological Society, p. 315–326.

Brown, D. L., 1978, Wrench-style deformational patterns associated with a meridional stress axis recognized in Paleozoic rocks in parts of Montana, South Dakota, and Wyoming, *in* The economic geology of the Williston Basin: Williston basin symposium, Montana Geological Society, p. 17–35.

Burke, R. B., and Stefanovsky, G. L., 1982, Porosity types, geometry, and interpore minerals of the Lower Duperow Formation, Billings nose area, Williston basin, North Dakota, *in* Christopher, J. E., and Kaldi, J., eds., Fourth international Williston basin symposium: Saskatchewan Geological Society Special Publication 6, p. 93–100.

Carlson, C. G., and Anderson, S. B., 1966, Sedimentary and tectonic history of North Dakota part of Williston basin: American Association of Petroleum Geologists Bulletin, v. 49, p. 1833–1846.

Carroll, W. K., 1979, Depositional environments and paragenetic porosity controls, upper Red River Formation, North Dakota: North Dakota Geological Survey Report of Investigation 66, 51 p.

Christopher, J. E., and Kaldi, J., eds., 1982, Fourth international Williston basin symposium: Saskatchewan Geological Society Special Publication 6, 326 p.

Clark, T. H., and Stern, C. W., 1968, Geologic evolution of North America: New York, Ronald Press, 570 p.

Clayton, L., Moran, S. R., and Bluemle, J. P., 1980, Explanatory text to accompany the geologic map of North Dakota: North Dakota Geological Survey Report of Investigation 60, 93 p.

Clement, J. H., 1976, Geologic history; Key to accumulation at Cedar Creek [abs.]: American Association of Petroleum Geologists Bulletin, v. 60, p. 2067–2068.

—— , 1983, Cedar Creek; A significant paleotectonic feature of the Williston basin [abs.]: American Association of Petroleum Geologists Bulletin, v. 67, p. 1332.

Clement, J. H., and Mayhew, T. E., 1979, Newporte discovery opens new pay: Oil and Gas Journal, v. 77, no. 27, p. 165–172.

Dean, K., 1982, Devonian Dawson Bay Formation in northwestern North Dakota, *in* Christopher, J. E., and Kaldi, J., eds., Fourth international Williston basin symposium: Saskatchewan Geological Society Special Publication 6, p. 89–93.

Donofrio, R. R., 1981, Impact craters; Implications for basement hydrocarbon production: Journal of Petroleum Geology, v. 3, no. 3, p. 279–302.

Dunn, C. E., 1982, Geology of the Middle Devonian Dawson Bay Formation in the northern part of the Williston basin, *in* Christopher, J. E., and Kaldi, J., eds., Fourth international Williston basin symposium: Saskatchewan Geological Society Special Publication 6, p. 75–89.

Estelle, D., and Miller, R., eds., 1978, Economic geology of the Williston basin: Williston Basin Symposium, Montana Geological Society, 447 p.

Foster, N. H., 1972, Ordovician System, *in* Geologic atlas of the Rocky Mountain region, U.S.A.: Rocky Mountain Association of Geologists, p. 76–86.

Gerhard, L. C., 1972, Canadian depositional environments and paleotectonics of central Colorado: Quarterly of the Colorado School of Mines, v. 67, p. 1–36.

Gerhard, L. C., Anderson, S. B., LeFever, J. A., and Carlson, C. G., 1982, Geological development, origin, and energy mineral resources of the Williston basin, North Dakota: American Association of Petroleum Geologists Bulletin, v. 66, p. 989–1020.

Grenda, J. C., 1978, Paleozoology of oil well cores from the Tyler Formation (Pennsylvanian) in North Dakota, U.S.A., *in* The economic geology of the Williston basin: Williston Basin Symposium, Montana Geological Society, p. 249–263.

Halabura, S., 1982, Depositional environments of the Upper Devonian Birdbear Formation, Saskatchewan, *in* Christopher, J. E., and Kaldi, J., eds., Fourth international Williston basin symposium: Saskatchewan Geological Society Special Publication 6, p. 113–125.

Kearns, J. R., and Traut, J. D., 1979, Mississippian discoveries revive the Williston basin: World Oil, v. 188, no. 6, p. 52–57.

Loeffler, P., 1982, Depositional environment and diagenesis, Birdbear Formation (Upper Devonian), Williston basin, North Dakota [M.S. thesis]: Grand Forks, University of North Dakota, 268 p.

LoBue, C., 1982, Depositional environments and diagenesis of the Silurian Interlake Formation, Williston basin, western North Dakota, *in* Christopher, J. E., and Kaldi, J., eds., Fourth international Williston basin symposium: Saskatchewan Geological Society Special Publication 6, p. 29–43.

Macauley, G., Penner, D. C., Procter, R. M., and Tisdall, W. H., 1964, Carboniferous, *in* Geologic history of western Canada: Alberta Society of Petroleum Geologists, p. 89–102.

Mallory, W. M., 1972a, Pennsylvanian arkose and the ancestral Rocky Mountains, *in* Geologic atlas of the Rocky Mountain region, U.S.A.: Rocky Mountain Association of Geologists, p. 131–133.

—— , 1972b, Regional synthesis, *in* Geologic atlas of the Rocky Mountain region, U.S.A.: Rocky Mountain Association of Geologists, p. 111–128.

Parson, E. S., Henderson, G. W., and Conti, L. J., 1975, Red Wing Creek field; Cosmic impact structure [abs.]: American Association of Petroleum Geologists Bulletin, v. 59, p. 2197.

Perrin, N. A., 1982, Environments of deposition and diagenesis of the Winnipegosis Formation (Middle Devonian), Williston Basin, North Dakota, *in* Christopher, J. E., and Kaldi, J., eds., Fourth international Williston basin symposium: Saskatchewan Geological Society Special Publication 6, p. 51–67.

Peterman, Z., and Goldich, S. S., 1982, Archean rocks in the Churchill basement of the Williston basin, *in* Christopher, J. E., and Kaldi, J., eds., Fourth international Williston basin symposium: Saskatchewan Geological Society Special Publication 6, p.11–13.

Porter, J. W., and Fuller, J.G.C.M., 1959, Lower Paleozoic rocks of northern Williston basin and adjacent areas: American Association of Petroleum Geologists Bulletin, v. 43, p. 124–190.

Rascoe, B., and Baars, D. L., 1972, Permian System, *in* Geologic atlas of the Rocky Mountain region, U.S.A.: Rocky Mountain Association of Geologists, p. 143–166.

Roehl, P., 1967, Carbonate facies, Williston basin and Bahamas: American Association of Petroleum Geologists Bulletin, v. 51, p. 1979–2033.

Ross, R. J., Jr., and Bergstrom, S. M., eds., 1982, The Ordovician System in the United States: International Union of Geological Sciences, 73 p.

Sloss, L. L., 1963, Sequences in the cratonic interior of North America: Geological Society of America Bulletin, v. 74, p. 93–114.

Steece, F. V., 1978, Deadwood Formation in the Williston basin, South Dakota: Montana Geological Society, 24th Annual Conference, 1978 Williston Basin Symposium, p. 63–71.

Sturm, S. D., 1983, Depositional environments and sandstone diagenesis in the Tyler Formation (Pennsylvanian), southwestern North Dakota: North Dakota Geological Survey Report of Investigation 76, p. 48.

Thomas, G. E., 1974, Lineament-block tectonics; Williston–Blood Creek basin: American Association of Petroleum Geologists Bulletin, v. 58, p. 1305–1322.

Tweto, O., and Sims, P. K., 1963, Precambrian ancestry of the Colorado mineral belt: Geological Society of America Bulletin, v. 74, p. 991–1014.

Warner, L. A., 1978, The Colorado lineament; A middle Precambrian wrench fault system: Geological Society of America Bulletin, v. 89, p. 161–171.

Ziebarth, H. C., 1964, The Tyler Formation of southwestern North Dakota: Third International Williston Basin Symposium, p. 119–126.

MANUSCRIPT ACCEPTED BY THE SOCIETY SEPTEMBER 15, 1986

ACKNOWLEDGMENTS

The writers gratefully acknowledge the assistance and support of James Clement (Shell Oil Company), Cooper Land (Consultant), and Julie LeFever (North Dakota Geological Survey) in preparation of this manuscript. The facilities of the North Dakota Geological Survey were made freely available for manuscript preparation by Don L. Halvorson, State Geologist. Many others contributed with discussions, information, and helpful suggestions to improve this manuscript. So many people's thoughts, ideas, and published data go into regional syntheses that it is impossible to cite them all. We express our appreciation to all of our fellow students of the Williston Basin for their help.

Printed in U.S.A.

The Geology of North America
Vol. D-2, Sedimentary Cover—North American Craton: U.S.
The Geological Society of America, 1988

Chapter 10

Phanerozoic history of the central midcontinent, United States

Bill J. Bunker and Brian J. Witzke
Iowa Geological Survey, 123 North Capitol St., Iowa City, Iowa 52242
W. Lynn Watney
Kansas Geological Survey, 1930 Constant Ave., Campus West, University of Kansas, Lawrence, Kansas 66044
Greg A. Ludvigson
Iowa Geological Survey, 123 North Capitol St., Iowa City, Iowa 52242

INTRODUCTION

The region of the central midcontinent has commonly been termed the "stable interior" of the North American continent. The magnitudes of Phanerozoic crustal deformation in the cratonic interior certainly are very small compared to those known from active continental margins, and the rates of deformation have been generally slower (Schwab, 1976). Nevertheless, the Phanerozoic sedimentary record in the central midcontinent region is replete with evidences of tectonic activity of surprising diversity and pattern. The central midcontinent, as defined for this report, includes Iowa, Kansas, southeastern South Dakota, Nebraska (excluding the panhandle), southern Minnesota, and Missouri north of 37°N latitude.

The Phanerozoic stratigraphic record in the central midcontinent region of North America is divided into six major depositional sequences, each bounded by major interregional unconformities (Sloss, 1963). The structural and stratigraphic development of this region is evaluated utilizing a series of isopach and paleogeologic maps constructed within the general framework of Sloss' (1963) cratonic sequences (Fig. 1).

PRECAMBRIAN BASEMENT FRAMEWORK

Phanerozoic sedimentary rocks in the central midcontinent region are underlain by Archaean and Proterozoic igneous, sedimentary, and metamorphic rocks. The Midcontinent Rift System (MRS), which bisects the midcontinent, is one of several prominent structural features evident in the Precambrian basement complex of this region (Plate 4A). Prominent gravity and magnetic anomalies and drillhole data document extension of the MRS southwestward from Lake Superior beneath the Paleozoic cover into northeastern Kansas (King and Zietz, 1971; Van Schmus and Hinze, 1985) Basalts, related mafic intrusions, and thick sedimentary rock sequences characterize this Keweenawan rift system. Offsets of the MRS near the Nebraska-Kansas and Iowa-Minnesota borders have been interpreted as possible transform faults (Chase and Gilmer, 1973) associated with development of the MRS. The central area of the MRS was uplifted during the Late Proterozoic, dramatically reorganizing the structure and creating a complex fault-bounded horst system along its axis (Craddock, 1972; Anderson and Black, 1982). Phanerozoic faulting and flexuring along the southern margin of the central horst area attest to continued structural activity along Precambrian basement structures associated with the MRS.

The Chadron and Cambridge Arches of Nebraska and the Central Kansas Uplift (Plate 4A) are part of a north-northwest–trending, Precambrian structural high (Black Hills-Central Kansas Uplift, Muehlberger and others, 1967) along the western margin of the study area. This basement feature appears to have remained structurally positive with respect to adjacent areas throughout much of the Phanerozoic. The central area of this feature (i.e., Chadron and Cambridge Arches) cuts across a portion of the north-northeast-trending Transcontinental Arch (Plate 4B), a broad, midcontinent, Phanerozoic structural feature.

The Sioux Ridge, a prominent Precambrian paleotopographic feature, cuts diagonally across the crest of the Transcontinental Arch. The ridge forms an east-west–trending feature occupied by a broad synclinal fold of Proterozoic Sioux Quartzite, which was more resistant to erosion than the flanking, older, Precambrian basement units (Baldwin, 1949; Bunker, 1981). Phanerozoic rock units have onlapped and overstepped the ridge repeatedly, only to be erosionally stripped back during periods of emergence.

The area between the Sioux Ridge and the Black Hills-Central Kansas Uplift served as an intermittent seaway connection (Nebraska Sag, Adler and others, 1971; Plate 4C) between the Williston Basin and the central midcontinent during various intervals in the Paleozoic. At times, seas may have migrated around the northern flank of the Sioux Ridge connecting with the eastern Iowa area.

Bunker, B. J., Witzke, B. J., Watney, W. L., and Ludvigson, G. A., 1988, Phanerozoic history of the central midcontinent, United States, *in* Sloss, L. L., ed., Sedimentary Cover—North American Craton; U.S.: Boulder, Colorado, Geological Society of America, The Geology of North America, v. D-2.

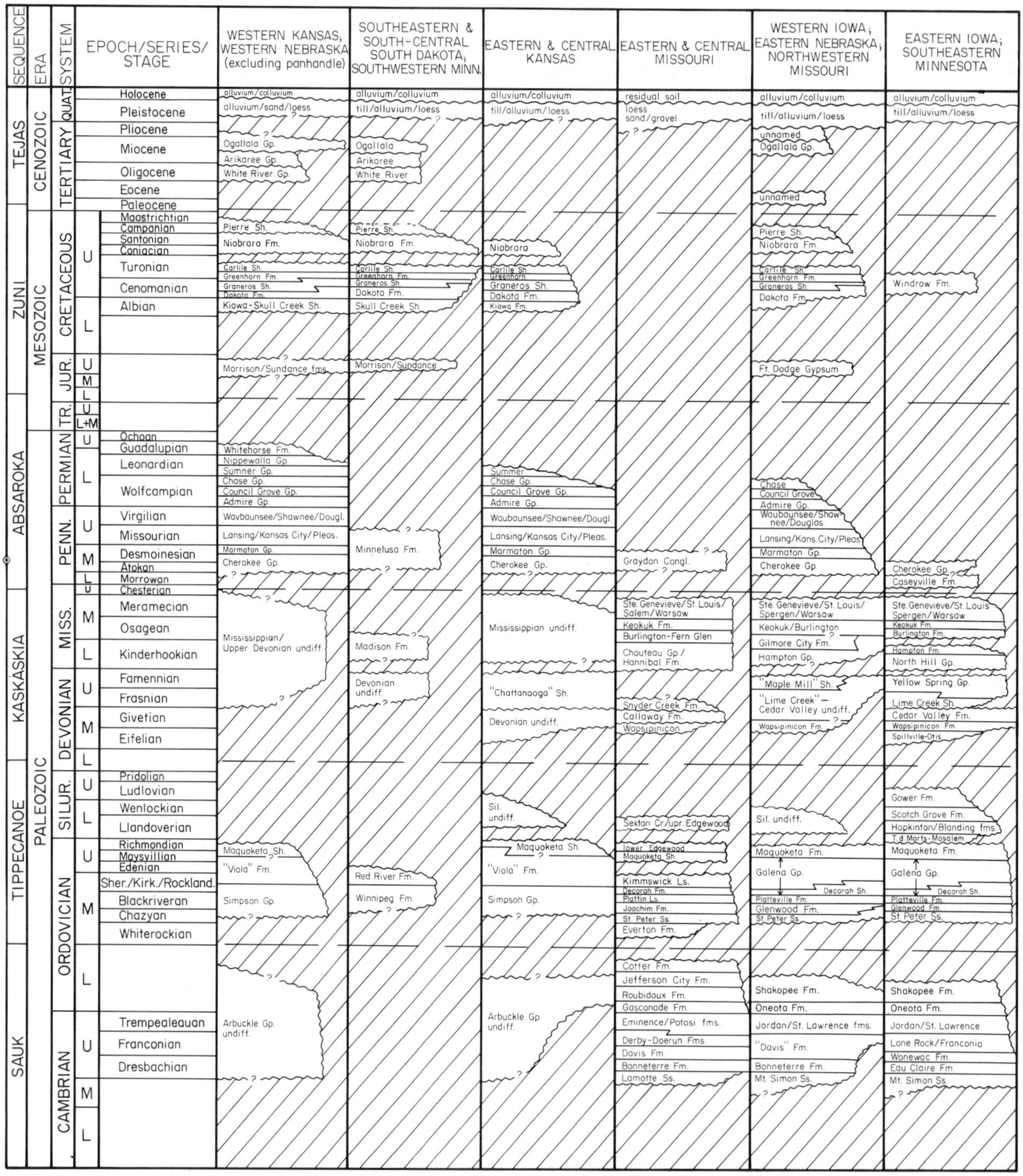

SEQUENCE	ERA	SYSTEM		EPOCH/SERIES/STAGE
TEJAS	CENOZOIC	QUAT.		Holocene
				Pleistocene
		TERTIARY		Pliocene
				Miocene
				Oligocene
				Eocene
				Paleocene
ZUNI	MESOZOIC	CRETACEOUS	U	Maastrichtian
				Campanian
				Santonian
				Coniacian
				Turonian
				Cenomanian
			L	Albian
		JUR.	U	
			M	
			L	
		TR.	U	
			L+M	
ABSAROKA		PERMIAN	U	Ochoan
				Guadalupian
			L	Leonardian
				Wolfcampian
	PALEOZOIC	PENN.	U	Virgilian
				Missourian
			M	Desmoinesian
				Atokan
			L	Morrowan
KASKASKIA		MISS.	U	Chesterian
			M	Meramecian
				Osagean
			L	Kinderhookian
		DEVONIAN	U	Famennian
				Frasnian
			M	Givetian
				Eifelian
			L	
TIPPECANOE		SILUR.	U	Pridolian
				Ludlovian
			L	Wenlockian
				Llandoverian
		ORDOVICIAN	U	Richmondian
				Maysvillian
				Edenian
			M	Sher./Kirk./Rockland.
				Blackriveran
				Chazyan
				Whiterockian
			L	
SAUK		CAMBRIAN	U	Trempealeauan
				Franconian
				Dresbachian
			M	
			L	

WESTERN KANSAS; WESTERN NEBRASKA (excluding panhandle): alluvium/colluvium; alluvium/sand/loess; ?; Ogallala Gp.; Arikaree Gp.; White River Gp.; Pierre Sh.; Niobrara Fm.; Carlile Sh.; Greenhorn Fm.; Graneros Sh.; Dakota Fm.; Kiowa-Skull Creek Sh.; ?; Morrison/Sundance fms.; ?; Whitehorse Fm.; Nippewalla Gp.; Sumner Gp.; Chase Gp.; Council Grove Gp.; Admire Gp.; Wabaunsee/Shawnee/Dougl.; Lansing/Kansas City/Pleas.; Marmaton Gp.; Cherokee Gp.; ?; ?; Mississippian/Upper Devonian undiff.; ?; Maquoketa Sh.; ?; "Viola" Fm.; Simpson Gp.; ?; ?; Arbuckle Gp. undiff.; ?

SOUTHEASTERN & SOUTH-CENTRAL SOUTH DAKOTA; SOUTHWESTERN MINN.: alluvium/colluvium; till/alluvium/loess; ?; Ogallala; Arikaree; White River; Pierre Sh.; Niobrara Fm.; Carlile Sh.; Greenhorn Fm.; Graneros Sh.; Dakota Fm.; Skull Creek Sh.; Morrison/Sundance; ?; Minnelusa Fm.; ?; Madison Fm.; ?; Devonian undiff.; ?; Red River Fm.; Winnipeg Fm.; ?

EASTERN & CENTRAL KANSAS: alluvium/colluvium; till/alluvium/loess; ?; Niobrara; Carlile Sh.; Greenhorn; Graneros Sh.; Dakota Fm.; Kiowa Fm.; Summer; Chase Gp.; Council Grove Gp.; Admire Gp.; Waubaunsee/Shawnee/Dougl.; Lansing/Kansas City/Pleas.; Marmaton Gp.; Cherokee Gp.; ?; Mississippian undiff.; ?; "Chattanooga" Sh.; Devonian undiff.; Sil. undiff.; Maquoketa Sh.; ?; "Viola" Fm.; Simpson Gp.; ?; ?; Arbuckle Gp. undiff.; ?

EASTERN & CENTRAL MISSOURI: residual soil; loess; sand/gravel; ?; Graydon Congl.; ?; Ste. Genevieve/St. Louis/Salem/Warsaw; Keokuk Fm.; Burlington-Fern Glen; Chouteau Gp./Hannibal Fm.; ?; Snyder Creek Fm.; ?; Callaway Fm.; Wapsipinicon; Sexton Cr./upr. Edgewood; lower Edgewood; Maquoketa Sh.; Kimmswick Ls.; Decorah Fm.; Plattin Ls.; Joachim Fm.; St. Peter Ss.; Everton Fm.; Cotter Fm.; Jefferson City Fm.; Roubidoux Fm.; Gasconade Fm.; Eminence/Potosi fms.; Derby-Doerun Fms.; Davis Fm.; Bonneterre Fm.; Lamotte Ss.

WESTERN IOWA; EASTERN NEBRASKA; NORTHWESTERN MISSOURI: alluvium/colluvium; till/alluvium/loess; unnamed; Ogallala Gp.; unnamed; Pierre Sh.; Niobrara Fm.; Carlile Sh.; Greenhorn Fm.; Graneros Sh.; Dakota Fm.; Ft. Dodge Gypsum; Chase; Council Grove; Admire Gp.; Waubaunsee/Shawnee/Douglas; Lansing/Kans. City/Pleas.; Marmaton Gp.; Cherokee Gp.; Ste. Genevieve/St. Louis/Spergen/Warsaw; Keokuk/Burlington; ?; Gilmore City Fm.; Hampton Gp.; "Maple Mill" Sh.; "Lime Creek"–Cedar Valley undiff.; ?; Wapsipinicon Fm.; Sil. undiff.; Maquoketa Fm.; Galena Gp.; Decorah Sh.; Platteville Fm.; Glenwood Fm.; St. Peter Ss.; Shakopee Fm.; Oneota Fm.; Jordan/St. Lawrence fms.; "Davis" Fm.; Bonneterre Fm.; Mt. Simon Ss.; ?

EASTERN IOWA; SOUTHEASTERN MINNESOTA: alluvium/colluvium; till/alluvium/loess; Windrow Fm.; Cherokee Gp.; ?; Caseyville Fm.; Ste. Genevieve/St. Louis/Spergen/Warsaw; Keokuk Fm.; Burlington Fm.; Hampton Fm.; North Hill Gp.; Yellow Spring Gp.; Lime Creek Sh.; Cedar Valley Fm.; Wapsipinicon Fm.; Spillville-Otis; Gower Fm.; Scotch Grove Fm.; Hopkinton/Blanding fms.; T.d Morts-Mosalem; Maquoketa Fm.; Galena Gp.; Decorah Sh.; Platteville Fm.; Glenwood Fm.; St. Peter Ss.; Shakopee Fm.; Oneota Fm.; Jordan/St. Lawrence; Lone Rock/Franconia; Wonewoc Fm.; Eau Claire Fm.; Mt. Simon Ss.; ?

Figure 1. Generalized correlation diagram of Phanerozoic stratigraphic units in the central midcontinent. Largely adapted from GSA COSUNA correlation charts.

SAUK SEQUENCE

Sauk sequence rocks were deposited in the central midcontinent during the Cambrian and Early Ordovician (Fig. 1).

Cambrian. Sauk deposition was initiated over most of the area during the Late Cambrian, although thick, basal Sauk sandstone sequences in portions of Illinois and Iowa probably include older Cambrian strata. The basal Sauk sandstone interval, assigned to the Mt. Simon Sandstone, reaches a thickness in excess of 750 m in northeastern Illinois and ranges from 150 to 450 m across most of eastern and central Iowa. The Mt. Simon thins abruptly against the southeastern margin of the central horst of the MRS in Iowa. Across western Iowa, Missouri, Nebraska, and Kansas, the basal Sauk sandstone sequence (variably assigned to the Lamotte, Reagan, or Mt. Simon sandstones) is markedly thinner (10–100 m) and is locally absent over topographic highs on the Precambrian surface. In general, regional thickness patterns of the basal Sauk sandstone interval appear to be unrelated to that of overlying Sauk strata (Howe and others, 1972).

The Precambrian surface had significant relief prior to burial by Sauk rocks: up to 450 m in the Ozark–St. Francois Mountains area of Missouri (Chenoweth, 1968), up to 500 m across the MRS in central Iowa, and knobs with up to 200 m of relief in Kansas (especially along 38°N). Most of the Precambrian surface in the central midcontinent was buried during the Sauk II interval (Palmer, 1981), although some topographic and structural features remained emergent until later Sauk deposition. Structural patterns developed during deposition of the post-Mt. Simon Sauk interval are reflected on the regional isopach map (Plate 4B). In general, the Sauk sequence thickens southeastward across the study area toward the present-day Illinois Basin and Ozark Uplift area. Northward from the region of maximum Sauk subsidence in the Illinois and "Ozark basin" (Lee, 1943, 1956) areas, the Hollandale Embayment represents an axis of Sauk thickening that trends across eastern Iowa into southern Minnesota. Broad positive structural elements reflected on the Sauk isopach map include the Southeast Nebraska Arch, Transcontinental Arch, and Wisconsin Dome (with its southward extending arch). A broad Precambrian upland area, ancestral to the Central Kansas Uplift (Chenoweth, 1968), occupied the central part of Kansas during the Cambrian. As suggested by Keroher and Kirby (1948), this positive structural feature may not have been buried by Sauk deposits until the Early Ordovician.

The Sauk sequence in the central midcontinent is characterized by a series of depositional cycles (Ostrom, 1970). The first major cycle spread marine and marginal marine deposits across most of the study area during the Dresbachian. Following deposition of the basal Cambrian sandstone, a series of broad marine facies became established in the area by the middle to late Dresbachian; these were: 1) predominantly sandstone in eastern Minnesota and Wisconsin (Eau Claire Formation), 2) mixed facies of siltstone, shale, sandstone, and minor carbonate across eastern Iowa and northeastern Missouri (Eau Claire Formation), and 3) carbonate-dominated facies with some shale and siltstone across central and western Iowa and much of Missouri (Bonneterre Formation). The Wisconsin Dome area and/or nearby regions on the Canadian Shield were the primary clastic source terranes. Potential sources of detrital materials include Keweenawan sediments (especially sandstones) and Precambrian granitic and other igneous/metamorphic rocks. The broad exposed crystalline terrane on the Transcontinental Arch apparently supplied comparatively little clastic material to the adjacent seaway during Sauk deposition.

Keroher and Kirby (1948), Lochman-Balk (1971), and Chenoweth (1968) correlated supposed late Dresbachian carbonate strata (Bonneterre) across eastern Kansas, and suggested that Franconian and early Trempealeauan strata are absent over most of Kansas (Chenoweth, 1968). However, the discovery of Franconian fossils in the basal Cambrian sandstone of southwestern Missouri indicates that the Bonneterre Formation is not present in the area (Kurtz and others, 1975). Instead, the basal Cambrian sandstone-carbonate succession in eastern Kansas may represent a sequence of Franconian sediments that lapped northwestward onto the margins of the ancestral Central Kansas Uplift. However, Dresbachian strata are present in western Kansas (Chenoweth, 1968). Basal Sauk relations are not known with certainty in Nebraska, although Dresbachian units present in western Iowa have been tentatively correlated into eastern Nebraska (Carlson, 1969). The late Dresbachian was marked by a significant regressive event, and a disconformity (Sauk II/Sauk III boundary; Palmer, 1981) apparently formed at the top of the Dresbachian interval over large portions of the central midcontinent.

In a general sense, younger Sauk depositional cycles resemble the Dresbachian pattern, although carbonate-dominated facies are more widespread during the Trempealeauan and Canadian. The Franconian-Trempealeauan cycle was initiated with deposition of a basal sandstone interval in Wisconsin, northern Illinois, Minnesota, and eastern Iowa (Ironton Sandstone, Wonewoc Formation) and eastern Kansas (Lamotte or Reagan Sandstones). Several broad facies tracts characterize subsequent Franconian deposition: 1) primarily glauconitic sandstones in Wisconsin and adjacent areas of Minnesota, Iowa, and Illinois (Lone Rock Formation, Franconia Formation); 2) a mixed facies belt, primarily characterized by siltstone, shale, and dolomite, across much of Iowa and Missouri (Davis Formation); and 3) carbonate-dominated facies in southwestern Iowa, eastern Nebraska, western Missouri, and Kansas (Davis Formation, lower Arbuckle Group). Carbonate-dominated facies expanded across much of Missouri and Iowa during the late Franconian and earliest Trempealeauan (Derby-Doerun Formations).

Most of the central midcontinent area was covered by a carbonate facies during the Trempealeauan (St. Lawrence Formation, Potosi Formation). The late Trempealeauan and earliest Canadian generally marked a period of significant marine offlap during which sandstone units prograded off the Wisconsin Dome area across much of Minnesota, Iowa, and northeastern Missouri (Jordan Sandstone, Momence Sandstone). Coeval carbonate and sandy carbonate facies (Eminence Formation) are present across western Iowa, eastern Nebraska, and much of Missouri, extending westward to the edge of the ancestral Central Kansas Uplift (Keroher and Kirby, 1948). As reported by many workers, maximum regression, generally coincident with the Cambrian-Ordovician boundary, was apparently marked by development of a disconformity over portions of the central midcontinent.

Lower Ordovician. A widespread sandstone and sandy carbonate interval probably represents the basal unit of the succeeding Lower Ordovician depositional cycle across much of the central midcontinent (upper Jordan and/or basal Oneota Formation in eastern Iowa, Minnesota, Wisconsin; Gunter Sandstone in Missouri, western Iowa, eastern Nebraska, and northeastern Kansas). Relatively pure carbonate deposition dominated the remainder of the cycle across most of the study area (Oneota Formation in Minnesota, Wisconsin, Iowa, Illinois; Gasconade Formation in Nebraska, Kansas, Missouri). These deposits apparently onlapped Precambrian granites in southcentral Kansas (Keroher and Kirby, 1948). The final regressive phase of this cycle was marked by development of an unconformity over portions of the midcontinent, especially in the Wisconsin Dome area (Ostrom, 1970) and eastern Kansas (Keroher and Kirby, 1948). The succeeding cycle was intiated with widespread deposition of a basal sandstone and/or sandy carbonate sequence (New Richmond Sandstone in Wisconsin, Iowa, Minnesota, Illinois; Roubidoux Formation in Missouri, Nebraska, and Kansas). Roubidoux sediments buried the Precambrian upland surface in central Kansas (Keroher and Kirby, 1948). Subsequent deposition was dominated by carbonates, although deposition of sandy carbonate and minor sandstone is noteworthy (Shakopee Formation in Wisconsin, Minnesota, Iowa, Illinois; Jefferson City-Cotter Formations in Missouri, Kansas). Later Sauk carbonate and sandy carbonate deposition is documented in southeastern Missouri (Powell and Smithville formations), although equivalent units are absent over most of the central midcontinent due to nondeposition (?) and/or extensive pre-Tippecanoe erosional stripping.

SUB-TIPPECANOE EROSIONAL SURFACE

A prolonged period of erosion, including the Whiterockian and portions of the Chazyan and late Canadian, separated deposition of the Sauk sequence from the overlying Tippecanoe sequence over most of the central midcontinent (Witzke, 1980). The vast sheet of Sauk rocks, primarily dolomite, that was deposited over the area was eroded marginally and around structurally positive features (Wisconsin Arch, Southeast Nebraska Arch). The exposed carbonates were subjected to karstification and valley formation to varying degrees. Prominent karst sinkholes and valleys, up to 200 m deep, are best developed in Kansas (Merriam and Atkinson, 1956), northern Illinois (Buschbach, 1964), northern and western Missouri, eastern Iowa, and Wisconsin. The youngest Sauk stratigraphic units are preserved in the structural depression coinciding with the southern Illinois and "Ozark basin" area. Sauk strata were erosionally stripped from the crest of the Southeast Nebraska Arch, where Precambrian crystalline and clastic rocks form the sub-Tippecanoe surface.

TIPPECANOE SEQUENCE

Middle and Upper Ordovician. In general, the initial Whiterockian phases of Tippecanoe deposition (Fig. 1) were restricted to regions of maximum late Sauk subsidence. Whiterockian sandy carbonates and sandstones (Everton Formation) were deposited on the eroded Sauk surface in southeastern Missouri and southern Illinois. Whiterockian strata also occur as far north as the Oklahoma-Kansas border area (Oil Creek Formation of Simpson Group). However, Whiterockian strata were not deposited over the remainder of the central midcontinent, where erosion remained the dominant process.

As Middle Ordovician seas onlapped the central midcontinent area during the Chazyan, the underlying erosion surface was buried beneath a time-transgressive sheet of St. Peter Sandstone (Dapples, 1955). Sporadically distributed sections of exceptionally thick (greater than 200 m) St. Peter Sandstone represent clastic-filled karst features and valleys on the pre-Tippecanoe erosion surface (Witzke, 1980). The great purity and maturity of the St. Peter quartz arenites over much of the midcontinent suggest that a substantial portion of the St. Peter sand was derived from earlier-generation Sauk and Keweenawan quartz sandstones. The St. Peter shoreline spread onto the margins of the Transcontinental Arch and across most or all of the Southeast Nebraska Arch, apparently by the late Chazyan or early Blackriveran. In those areas (i.e., Nebraska, northwestern Iowa, and southeastern Minnesota), the St. Peter Sandstone is locally argillaceous and silty with interbedded fossiliferous brown and green shales (Witzke, 1980). The St. Peter includes shales and oolitic ironstones across portions of northern Kansas (Leatherock, 1945) and southeastern Nebraska. It extends northwestward from north-central Nebraska (Carlson, 1969) to join with basal Winnipeg clastics in South Dakota. These Middle Ordovician deposits trend directly across the Transcontinental Arch, paralleling the southern margin of the Sioux Ridge along the Nebraska Sag (Plate 4C).

An extensive sandstone body, probably of Blackriveran age, that trends across northern Illinois, southeastern Iowa, and northern Missouri is termed the Starved Rock Standstone (Fraser, 1976). The thin Glenwood Shale unit interfingers with the Starved Rock Sandstone on its northern margin and trends across Iowa, Minnesota, and Wisconsin. Contemporaneous carbonate and evaporite facies (Joachim Formation) are noted south of the Starved Rock Sandstone body in Illinois and eastern Missouri.

Marine carbonate sediments were deposited across the eastern portion of the central midcontinent following St. Peter-Glenwood deposition (Plattin Limestone, Platteville Formation). Correlative strata in the western and northern portions of the study area include significant quantities of terrigenous clastic material, commonly green or brown shales and varying quantities of sandstone, reflecting proximity to source terranes along the Transcontinental Arch and in portions of Kansas (Leatherock, 1945; Sloan, 1972; Witzke, 1980). The overlying Decorah Formation undergoes significant facies variations in the central midcontinent: 1) the Decorah in northwestern Iowa and Minnesota is primarily a shale unit with scattered limestone interbeds; 2) the Decorah in eastern Iowa and Missouri is primarily a carbonate unit with some shale; 3) Decorah or upper Simpson strata in portions of southwestern Iowa, Nebraska, and Kansas include carbonates, shales, and sandstones. The northwestward increase in Decorah shale thicknesses in Iowa and Minnesota identifies the Transcontinental Arch as the source area (Witzke, 1980). Decorah sedimentation was succeeded by "Trentonian" carbonate deposition across the central midcontinent (Galena Group, Kimmswick Limestone, "Viola" Formation). The Decorah-Galena boundary is a diachronous facies transition in Iowa and Minnesota, and the upper Decorah in the region bordering the Transcontinental Arch is a contemporaneous facies to lower Galena carbonates in areas away from the arch (Witzke, 1980).

Continued marine transgression during the late Middle and early Late Ordovician inundated vast areas of the midcontinent. "Viola" carbonates overstepped the Simpson edge in northwestern Kansas (Cole, 1975) and portions of the ancestral Central Kansas Uplift. Large areas of the Transcontinental Arch also were submerged, reducing clastic influx to the epeiric sea and allowing Galena carbonate sedimentation to become more prevalent. Ross (1976) termed this the greatest inundation in North American history. Galena-Viola carbonate strata are generally characterized by skeletal wackestones/packstones and/or fossiliferous dolomites containing varying quantities of chert. For the most part, complex dolomite and limestone facies patterns do not parallel depositional patterns but are secondary diagenetic features (Witzke, 1983a).

A general reorganization of Sauk structural patterns occurred during the Middle Ordovician. The southward thickening of Platteville/Plattin strata, in Illinois and eastern Missouri, resembles Sauk isopach trends, although the northward trending axis of Sauk thickening (Hollandale Embayment) was disrupted by Middle and Late Ordovician uplift of a broad arch trending north from the Ozark region across northern Missouri and southeastern Iowa (Plate 4C) termed the Northeast Missouri Arch (Bunker, 1981). However, the northern extension of the Hollandale Embayment, in northeastern Iowa and southern Minnesota, remained an area of increased subsidence during the Middle Ordovician, where the thickest sequences of Galena Group strata in the central midcontinent are preserved (Witzke, 1983a). Positive structural elements, the Transcontinental and Wisconsin arches, flanked the embayment. Middle Ordovician strata thin southwestward across Iowa, approaching the area of the Southeast Nebraska Arch. Pronounced upwarping of the Ozark Uplift during deposition of the Galena/Kimmswick sequence resulted in erosional beveling of these carbonate strata around the uplift prior to Maquoketa deposition (Templeton and Willman, 1963). The emergence of the Ozark Uplift during Tippecanoe deposition marked the destruction of the older Sauk "Ozark basin." Tippecanoe development of the Chautauqua and Northeast Missouri arches was generally coincident with that of the Ozark Uplift. Viola strata thin toward the Chautauqua Arch in Kansas. Numerous bentonites in the Middle Ordovician sequence of Minnesota, Iowa, Illinois, and Missouri (Kolata, 1983) originated from volcanic sources associated with the newly emerging Taconic Mountains, presaging Upper Ordovician influx of Taconic-derived clastics into the midcontinent.

Carbonate depositional patterns became disrupted during the middle and late Cincinnatian, and Maquoketa shales spread across the central midcontinent area. Lower Maquoketa shale facies in eastern Iowa and Illinois are primarily brown organic-rich shales with interbedded carbonates and phosphorites. This shale facies is replaced to the west and northwest by carbonate-dominated facies. Maquoketa shale-dominated facies progressively overstep Maquoketa carbonate-dominated facies westward in Iowa (Witzke, 1983a). The Viola carbonate sequence in Kansas and adjacent areas is not an exact chronostratigraphic equivalent of the Iowa-Illinois Galena Group carbonate sequence, but includes younger strata (Witzke, 1983a; Adler and others, 1971). The westward spread of shale facies over carbonate facies during Maquoketa deposition in the midcontinent probably represents the distal progradation of clastic sediment from Taconic sources. However, additional clastic sources in the central midcontinent significantly modified this general picture; sand and mud were shed off areas on the Ozark Uplift and Transcontinental Arch (Witzke, 1980).

Structural patterns developed during or after Maquoketa deposition resemble those of the Middle Ordovician, although new features are evident. In particular, the area of the Southeast Nebraska Arch subsided during Maquoketa deposition as evidenced by southwestward thickening of Maquoketa strata in Iowa toward the former arch. The initiation of increased subsidence in the vicinity of the Southeast Nebraska Arch and in eastern Iowa during the late Middle and Late Ordovician marked the early development of the North Kansas and East-Central Iowa basins.

The youngest Ordovician deposits in the midcontinent occur at the top of the Maquoketa interval. Uppermost Ordovician sediments in Iowa include red silty shales with scattered oolitic ironstones, whereas skeletal and oolitic limestones are present in Missouri. An erosional unconformity, with up to 50 m of relief in eastern Iowa, separates Ordovician and Silurian strata over the midcontinent area, which probably developed during a major glacial eustatic drop in sea level (Sheehan, 1978).

Silurian. Outside of the Illinois Basin area, Silurian strata in the central midcontinent are restricted to the East-Central Iowa and North Kansas Basins and the structural sag that connects these two regions (Plate 4D). Silurian carbonate deposition undoubtedly extended far beyond the present-day Silurian edge across much or all of the Transcontinental (Chronic and others, 1969; Colville and Sheehan, 1983) and Chautauqua Arches. However, Silurian strata were removed over vast areas of the midcontinent during a prolonged period of pre-Kaskaskia nondeposition and erosion which lasted some 40 m.y.

The initial transgression of Silurian seas into the midcontinent proceeded from the Illinois area into eastern Iowa, where shaly carbonates were deposited in topographic depressions on the Maquoketa Shale surface during the early Llandoverian. However, Silurian deposition across central Iowa and the North Kansas Basin area did not begin until the middle to late Llandoverian (Carlson and Boucot, 1967; Witzke, 1981), and late Llandoverian strata overlie the Ordovician surface in eastern Missouri (Thompson and Satterfield, 1975). Subsequent Llandoverian-early Wenlockian marine carbonate depositional patterns and biofacies distributions across the midcontinent area were strongly influenced by relative changes in sea level (Johnson, 1980; Witzke, 1983b). Level-bottom marine carbonate environments were locally modified in eastern Iowa during portions of the late Llandoverian and Wenlockian as carbonate mound facies (bioherms) developed on the sea bottom (Witzke, 1983b). Carbonate mound complexes achieved dimensions up to 2.5 km in diameter in Iowa.

A profound change in carbonate depositional patterns occurred during the middle to late Wenlockian in the East-Central Iowa Basin area, as open-marine carbonate environments were replaced by restricted-marine subtidal laminated carbonate environments (Philcox, 1972; Witzke, 1983b). Open-marine epeiric circulation patterns became disrupted as the Late Silurian marine offlap proceeded, creating conditions of increased salinity within the basin. Late Wenlockian-Ludlovian carbonate mound facies containing abundant low-diversity invertebrate faunas interfinger with laminated carbonate facies in eastern Iowa (Witzke, 1983b).

Silurian stratigraphy in the North Kansas Basin area is poorly known, although lithologic and biostratigraphic similarities with eastern Iowa provide the most consistent comparisons. Ireland (1967) proposed a correlation of the North Kansas Basin Silurian section with the Silurian section in Oklahoma, although his correlations are inconsistent with the brachiopod biostratigraphy established in Nebraska (Carlson and Boucot, 1967; Witzke, 1981). The Chautauqua Arch apparently served as an effective barrier separating Silurian carbonate environments in the North Kansas Basin area from Silurian carbonate/clastic environments in Oklahoma. Close similarities between the Silurian sequences in the North Kansas and East-Central Iowa Basins suggest that similar structural and depositional conditions existed in both basins. Maximum thicknesses of Silurian rocks in the central areas of both basins are similar (150 m; Plate 4D), and thickening of individual Silurian rock units towards the center of each basin is documented (Ireland, 1967; Witzke, 1981, 1983b). The absence of Silurian strata in the central midcontinent across the Transcontinental Arch, ancestral Central Kansas Uplift, Northeast Missouri Arch, Chautauqua Arch, and Ozark Uplift suggests that these features were probably positive structural elements during Silurian deposition as well as the subsequent pre-Middle Devonian erosional episode.

PRE-KASKASKIA EROSIONAL EPISODE

Prior to the initial Kaskaskia transgression into the central midcontinent, an extensive period of erosion ensued during which several hundred meters of Tippecanoe, Sauk, and Precambrian rocks were stripped from portions of the continental interior. The pre-Kaskaskia paleogeologic map (Plate 4E) reflects the pattern of basins and arches that had developed prior to Kaskaskia deposition. Because of extensive sub-Kaskaskia erosional stripping, Silurian rocks were preserved only in areas of maximum structural subsidence. The general distribution of Silurian rocks (Plates 4D, 4E) depicts a northeast- to southwest-trending synclinal trough across the central midcontinent, coincident in part to the southeastern margin of the MRS. This depressed area, as discussed previously, has been subdivided into the North Kansas and the East-Central Iowa Basins. Other prominent structural features evident on the pre-Kaskaskia paleogeologic map (Plate 4E) include the Ozark Uplift, Northeast Missouri Arch, Chautauqua Arch, ancestral Central Kansas Uplift, and Transcontinental Arch.

Pre-Kaskaskia paleotopographic relief developed from differential erosional characteristics of truncated pre-Kaskaskia strata, in particular between Silurian carbonates and Upper Ordovician shales. Erosional escarpments of low relief (10–30 m) developed along the pre-Kaskaskia erosional margins of the Silurian, and probably served as effective barriers to open-marine circulation in the initial transgressing Kaskaskia seas (Bunker and others, 1983, 1985).

KASKASKIA SEQUENCE

The Kaskaskia sequence (Fig. 1) in the central midcontinent region includes strata ranging in age from Middle Devonian (Late Eifelian: Hilpman, 1969a, b; Klapper and Barrick, 1983; Bunker and others, 1985) to late Middle Mississippian (Meramecian). The structural framework influencing early Kaskaskia deposition in this region was largely inherited, with some variations, from that which developed during the late Tippecanoe.

Devonian. Collinson and James (1969) considered the Middle Devonian rocks of eastern Iowa and northwestern Illinois to be the southeasternmost transgressive deposits of a vast seaway that extended northwestward into western Canada. Recent biostratigraphic (Klapper and Barrick, 1983) and lithostratigraphic (Bunker and others, 1983, 1985; Witzke and Bunker, 1984) interpretations of the Middle Devonian rocks in north-central Iowa and south-central Minnesota indicate that open-marine carbonate environments (Spillville Formation) characterized this region during the Late Eifelian. These Eifelian units are physically separated from correlative, restricted-marine, carbonate units (Otis Formation) in the East-Central Iowa Basin area (Bunker and others, 1983, 1985) by a paleoescarpment of Silurian carbonate strata in the northern part of the East-Central Iowa Basin. Upper Eifelian rocks are absent elsewhere across central Iowa, although probable upper Eifelian strata are present in the North Kansas Basin area (Hilpman, 1969b). In general, southward-transgressing seas expanded into the central midcontinent area during the Late Eifelian and reoccupied the two major Tippecanoe basinal areas. However, there was apparently no direct Late Eifelian seaway connection between the North Kansas and East-Central Iowa Basins across central Iowa. The Nebraska Sag probably served as a seaway connection between the Williston and the North Kansas Basins during the Middle Devonian onlap. Seaway connections may have been established around the northern flank of the Sioux Ridge with the East-Central Iowa and Williston Basins. Biogeographic similarities between basal Middle Devonian rocks in northeastern Iowa and eastern Wisconsin (Klapper and Barrick, 1983) suggest that seaway connections with the Michigan Basin were also probably established across the area of the Wisconsin Arch. However, extensive post-Devonian erosion has removed Devonian strata from the Transcontinental and Wisconsin Arches, obscuring physical relationships.

Successive expansion of the Middle Devonian seaway, into the central midcontinent region, is recorded by a series of transgressive-regressive depositional cycles (Witzke and Bunker, 1984), which are similar in timing to those noted in Manitoba (Norris and others, 1982), and complimentary to those noted across Euramerica (Johnson and others, 1985). The first cycle, as noted, spread Late Eifelian carbonate (Spillville-Otis formations) environments across portions of the East-Central Iowa and North Kansas Basins. The upper regressive part of this cycle (Wapsipinicon Formation) is characterized by an extensive carbonate and gypsum-anhydrite evaporite sequence (Dorheim and Campbell, 1958; Sendlein, 1964, 1968, 1972), which is preserved in the subsurface of south-central and southeastern Iowa. Brecciated carbonates, textures primarily developed by evaporite solution collapse, characterize this interval in the outcrop area of eastern and northeastern Iowa (Norton, 1920; Bunker and others, 1983, 1985).

Late Givetian-Early Frasnian (Cedar Valley-Shell Rock formations) cyclic sedimentation in the central midcontinent region is similarly characterized by fossiliferous dolomitic intervals at the base of each cycle, recording deposition in open-marine, carbonate, shelf environments during each successive transgressive phase. Laminated, intraclastic, and brecciated carbonates in the upper part of each cycle record deposition in shallow, restricted-subtidal and tidal-flat settings during each regressive phase (Witzke and Bunker, 1984). Evaporites (gypsum-anhydrite) laterally equivalent to each of the regressive phases are present in the subsurface of central Iowa. The lower Cedar Valley Formation represents a widespread transgression (Taghanic Onlap, Johnson, 1970; Klapper and Johnson, 1980), which marked the end of provincialism among brachiopods, corals, and trilobites across the North American continent. The Sangamon Arch in central Illinois, which initially developed during the Middle Devonian (Whiting and Stevenson, 1965), together with the Ozark Uplift and the Chautauqua Arch, formed the southern margin of the merging North Kansas and East-Central Iowa Basins. Cedar Valley carbonate deposition along the southern margin of the developing Iowa Basin was influenced by the influx of Ozark-derived clastics (Hoing Sandstone) across central and northern Missouri and portions of southern Iowa (Fraunfelter 1967; Schumacher, 1976).

Prior to the Late Frasnian transgression across the central midcontinent, a period of erosion and karst development occurred over much of the area. Sinkholes and caverns developed in the Middle Devonian and Silurian carbonates, and stratigraphic leaks of Upper Devonian shale (Independence Shale of eastern Iowa, Uraban, 1972; Klapper, 1975) filled these karst features. Lee (1956) described a pre-Chattanooga Valley (McPherson Valley) with paleotopographic relief of more than 60 m in south-central Kansas, as well as the presence of several Middle Devonian outliers along the western margin of the North Kansas Basin. However, marine and restricted-marine carbonates and evaporites were deposited in central Iowa (Iowa Basin) coincident with erosion in areas to the southeast and west.

Significant changes in structural and depositional patterns occurred prior to and/or during the Late Devonian (Late Frasnian-Famennian) transgression into the central midcontinent. While earlier Middle Devonian marine incursions appear to have spread from northern and eastern seaways, the Late Devonian transgression apparently came from a seaway to the southeast. The Upper Devonian in this region can be grouped into three broad northeast-trending megafacies. The southeasternmost facies (New Albany Shale Group, Cluff and others, 1981) consists predominantly of gray and black to brownish black organic-rich shales. The New Albany Shale Group grades laterally toward the northwest into gray shales, which in turn grade to and are interbedded with carbonates and green shales (Yellow Spring Group, Dorheim and others, 1969). Biostratigraphic investigations of Devonian carbonates in the central midcontinent have also revealed that strata previously classified as Middle Devonian (Collinson, 1967; Adler and others, 1971) are, in part, a Late Devonian carbonate lithosome (Klug and Tynan, 1981; Klug, 1982) as earlier suggested by Carlson (1963). The northwesternmost megafacies is a sparsely fossiliferous carbonate-dominated interval, with abundant solution collapse(?) breccias, which progressively onlaps and oversteps truncated Tippecanoe rocks along the southeastern flank of the Transcontinental Arch. Laminated brown shales, containing a probable Givetian-Frasnian freshwater fish fauna, overlie Cambrian and Precambrian rocks near the Sioux Ridge in northwest Iowa (note outlier on Plate 4F).

The total Devonian isopach map (Plate 4F) does not reflect the individual histories of the separate Middle and Late Devonian structural and depositional regimes. However, the isopach map does suggest a merging of the North Kansas and East-Central Iowa basins into one depositional basin located in central Iowa (i.e., the Iowa Basin), sometime during the late Middle to Late Devonian. This northeast-trending, elongated basinal feature occupies a position coincident to the southeastern margin of the MRS, suggesting probable structural influence on both Middle and Upper Devonian sedimentation. Structural movements along the northwest-southeast-trending Lincoln Fold System in northeastern Missouri and southeastern Iowa locally influenced Devonian depositional and erosional patterns (McQueen and others, 1961) in this area.

Mississippian. Following a brief regressive offlap at the end of the Devonian, Mississippian marine environments transgressed across a region of low erosional relief. The basal sequence of Mississippian rocks (Kinderhookian Series) in the midcontinent is dominated by marine carbonate deposits, although significant quantities of Kinderhookian shale and siltstone occur in southeastern Iowa and northeastern Missouri. Mississippian onlap along the flanks of the Transcontinental Arch and the Cambridge Arch-Central Kansas Uplift has been described for the northern and western parts of the study area (Carlson, 1963, 1979; Goebel and Stewart, 1979). Intermittent seaway connections with the Williston Basin area were probably established through the area of the Nebraska Sag. Similar onlap is also noted around the flanks of the Ozark Uplift (Thompson, 1979), indicating that this area remained a positive structural feature during the Mississippian.

The Mississippian is typically subdivided into four widely recognized series—the Kinderhookian, Osagian, Meramecian, and Chesterian. Many problems involving time-stratigraphic relationships, vertical and lateral facies variations, and the extent of regional and local unconformities have precluded meaningful synthesis of the Mississippian System across the entire central midcontinent area. For example, disconformable relationships between Kinderhookian and Osagean strata in southeastern Iowa attest to late Kinderhookian uplift and erosion. Whether this unconformity is of local or regional significance is not well established. Laudon (1937) used this disconformity in southeastern Iowa to explain the apparent transgressive overlap of crinoid zones within the Burlington Limestone (Osagean). Lithostratigraphic relationships (Harris and Parker, 1964) also suggest that successively younger members of the Burlington progressively overlie younger units of the Kinderhookian series north and west from southeastern Iowa. Parker (1973), however, suggested that a facies relationship may exist between the lower Burlington and portions of the Hampton (formerly considered Kinderhookian) in central Iowa. Recent biostratigraphic investigations (Glenister and Sixt, 1982; Baxter and Brenkle, 1983) have now assigned portions of the Hampton and Gilmore City Formations in central Iowa to the Early Osagean.

Osagean rocks were deposited during a major transgressive episode in the midcontinent. Shallow seas with normal salinities and a diverse assemblage of benthic invertebrates characterized most of this interval. The Central Kansas Uplift was overlapped and the Hugoton Embayment of southwestern Kansas actively subsided (Goebel and Stewart, 1979), connecting southward into the Anadarko Basin.

Meramecian strata in the central midcontinent include carbonates, evaporites, shales, and sandstones deposited in various open-marine and restricted-marine environments during several transgressive-regressive cycles. A major regressive interval during the Meramecian resulted in the development of an erosion surface within the western Kansas sequence (Goebel and Stewart, 1979). Restriction of the Meramecian sea in south-central Iowa permitted development of a gypsum-anhydrite evaporite sequence (Carlson, 1979). Chesterian rocks are generally absent from the region, being restricted to the southern basins. Eroded Chesterian rocks occur within the Hugoton Embayment, and it seems probable that Chesterian deposition may have extended across other portions of the central midcontinent. However, pre-Absaroka erosion has removed any evidence of Chesterian rocks across most of the area.

The present thickness of Mississippian rocks in the midcontinent (Plate 4G) reflects extensive pre-Absaroka uplift and erosion, especially along the trend of the Nemaha Uplift. Nevertheless, the isopach map outlines several areas of increased Mississippian subsidence. 1) The Hugoton Embayment of western Kansas contains the thickest sequence of Mississippian strata in the central midcontinent; however, because of unresolved stratigraphic problems, the thickness patterns shown include some Upper Devonian strata (Hilpman, 1969; Goebel and Stewart, 1979). 2) Mississippian thickening in northeast Kansas and adjacent Missouri reflects development of a shallow basin, which Lee (1946) termed the Ancestral Forest City Basin. 3) The thick Mississippian section in western Iowa is informally included in the "Massena basin." This basin coincides with the western extent of the Late Devonian Iowa Basin. 4) A synclinal depression in northeastern Missouri and southern Iowa is also apparent on the Mississippian isopach map, and is informally termed the "Appanoose trough." This trough is a major structural low associated with the Lincoln Fold System (Plate 4I), indicating active development of Lincoln Fold structures during the Mississippian. Meramecian evaporites are preserved within this feature.

PRE-ABSAROKA EROSION CYCLE

The pre-Absaroka unconformity reflects a profound change in the tectonic framework of the central midcontinent. Renewed uplift of older positive features, and erosional stripping of uplifted strata led to the development of a geologically complex land surface. Early to Middle Pennsylvanian sediments onlap and overstep the erosionally beveled edges of Kaskaskia through Sauk units, and rest on Precambrian basement rocks along the crest of some pre-Absaroka uplifts. The pre-Absaroka paleogeologic map illustrates this complex erosion surface (Plate 4H).

The Nemaha Uplift, a reactivation of the older (Sauk) Southeast Nebraska Arch, is apparent at the center of the map (Plate 4H). Regional beveling removed Kaskaskia and Tippecanoe strata, exposing Precambrian crystalline rocks along the crest of the uplift. Uplift and erosional beveling is also noted along the trend of the Cambridge Arch-Central Kansas Uplift. Lower Tippecanoe and upper Kaskaskia rocks are preserved along the axis of the Nebraska Sag. In east-central Iowa and northwestern Illinois, uplift of the Savanna-Sabula Anticlinal System (Plate 4I) and faulting along the Plum River Fault Zone (Plate 4H) is indicated by regional beveling of Kaskaskia and upper Tippecanoe rocks along the crest of the anticline (Bunker and others, 1985). The East-Central Iowa Basin, developed during Tippecanoe-lower Kaskaskia deposition, had been uplifted and erosionally beveled to the northeast prior to Absaroka deposition.

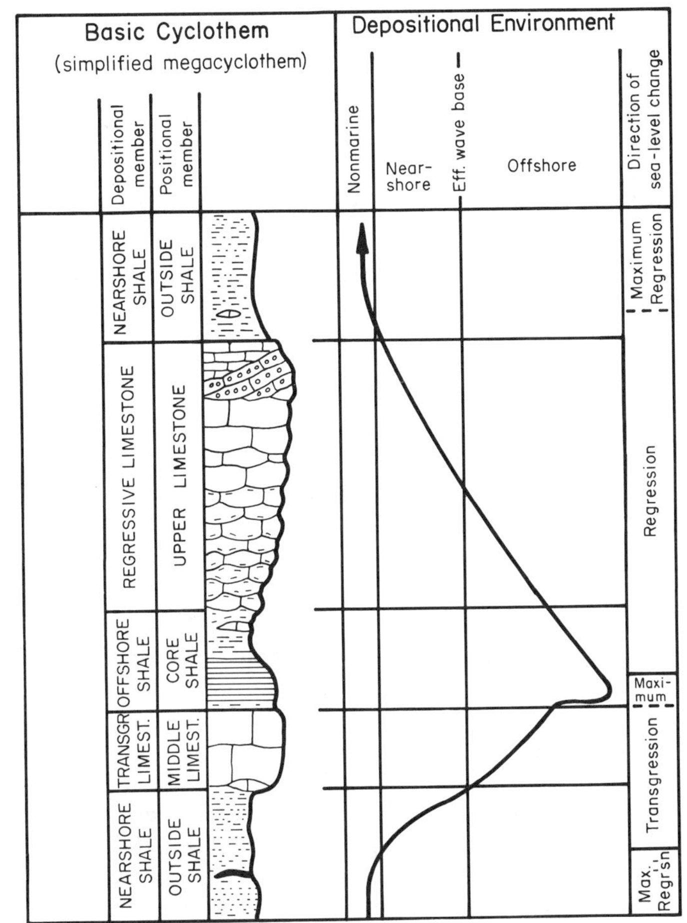

Figure 2. Generalized Upper Pennsylvanian "Kansas cyclothem." Modified from Heckel, 1977, 1980.

Continued presence of the Ozark Uplift is indicated by erosional stripping of Kaskaskia and Tippecanoe strata, so that Pennsylvanian rocks of the region rest upon Lower Ordovician strata.

ABSAROKA SEQUENCE

The Absaroka sequence (Fig. 1) in the central midcontinent consists primarily of strata ranging in age from Middle Pennsylvanian to Late Permian. Triassic strata are present only along the extreme western margin of the study area. Late Chesterian rocks are restricted to the Illinois and Anadarko Basins, but possible stratigraphic leaks have been noted within karstified Kaskaskia rocks in the central midcontinent (Urban, 1971, 1972; Goebel and Stewart, 1979).

Pennsylvanian. Lower Pennsylvanian (Morrowan) rocks in the study area are known from the Hugoton Embayment of southwestern Kansas and the Illinois Basin. Nonmarine Morrowan strata have also been identified in east-central Iowa and northern Illinois (Kosanke and others, 1960; Fitzgerald, 1977) along the southern flank of the Savanna-Sabula Anticlinal System and the crest of the Mississippi River Arch (Bunker and others, 1985). Basal Pennsylvanian strata in the central area of the Forest City Basin have not been dated, and possible Morrowan strata may be included (Wanless, 1975).

Middle Pennsylvanian (Atokan-Desmoinesian) strata (Cherokee and Marmaton groups) cover much of the central midcontinent, but are absent across structural highs along the trend of the Cambridge Arch-Central Kansas Uplift and Nemaha Uplift (Plate 4I). Lower and Middle Pennsylvanian strata are dominated by nonmarine clastic deposits, although marine shale and limestone units occur within the sequence. Clastic sediments of the Cherokee Group are primarily shales and mudstones with locally abundant sandstone and siltstone units. Coal seams, locally thick enough to be of economic interest, occur within the Cherokee Group, especially in the eastern part of the study area. Lower Cherokee strata in the central portion of the Forest City Basin are dominated by black and gray shale with only minor sandstone. However, the upper Cherokee includes beds of coarse-grained arkosic sandstone, possibly derived from Precambrian granites on the crest of the Nemaha Uplift during the late Middle Pennsylvanian (Lee, 1943, 1956). In the Salina Basin, Cherokee strata are dominated by shale and sandstone; coal beds, although present, are less significant than in the basins to the east. Gray silty shales are interstratified with red shale beds, in particular along the flanks of the Central Kansas Uplift (Lee, 1956). Compared to the lower Cherokee, upper Cherokee and Marmaton strata show increasing evidence of marine deposition across the midcontinent, including prominent, cyclic marine shale-limestone units.

Much of subsequent Absaroka sequence deposition was characterized by cyclic patterns (cyclothems) of marine and nonmarine sedimentation, apparently in response to eustatic sea-level fluctuations in the midcontinent. Eustatic fluctuations may relate to the waxing and waning of Gondwanan glaciers during the Late Mississippian through Middle Permian (Crowell, 1978; Heckel, 1980). The idealized "Kansas cyclothem" (Fig. 2), which characterizes many Upper Pennsylvanian cyclic units in the midcontinent, includes four basic components (Heckel, 1977): 1) a thin, basal, transgressive limestone; 2) a phosphatic, black, fissile, offshore marine shale (maximum transgression); 3) a thick regressive limestone; and 4) a sandy nearshore and nonmarine shale locally with sandstone and coal (maximum regression). However, not all cyclothems contain the four basic components. In some cyclothems the basal transgressive limestone is extremely thin to absent and the black marine shale interval directly overlies coal beds or nearshore shale facies ("Illinois cyclothem;" Heckel, 1980). The lack of the black marine shale in some cycles suggests that the magnitudes of eustatic sea-level fluctuations varied. Regional facies variations are noted within individual cyclothems, including carbonate mound and oolite shoal facies (Heckel and Cocke, 1969; Watney, 1980). Regional examination of the Kansas City Group in western Kansas indicates that the entire shelf was exposed repeatedly during late regression in each cycle (Watney, 1984). The configuration of the shelf during deposition of these cycles was controlled by continued though subtle uplift of the Central Kansas Uplift and its extension to the south, the Pratt Anticline. Subtle flexures along the platform were sometimes loci for the formation of ooid shoals, when waves and currents were focused on these areas during the late regressive phases of some cycles (Watney, 1984).

Middle Pennsylvanian faulting along the Humboldt Fault Zone (Plate 4A), apparently contemporaneous with the Ouachita-Marathon Orogeny (Kluth and Coney, 1981), gave rise to the Nemaha Uplift, in part a reactivation of the older (Sauk) Southeast Nebraska Arch. The Nemaha Uplift bisected the region of the late Tippecanoe North Kansas Basin and cut off the southwestern extension of the mid-Kaskaskia Iowa Basin. Up to 320 m of pre-Missourian Pennsylvanian rocks accumulated in the structural depression east of the Humboldt Fault Zone (Plate 4I). This area constitutes the Forest City Basin as defined by Lee (1943, 1946). Maximum differential subsidence of the Forest City Basin was essentially contemporaneous with the rising Nemaha Uplift. The Nemaha Uplift forms the eastern margins of two basins created during this period of increased structural activity: 1) the Salina Basin of north-central Kansas and south-central Nebraska, and 2) the Sedgwick Basin, a structural depression in south-central Kansas that plunges southward to join the Anadarko Basin of Oklahoma. West of the Central Kansas Uplift, the Hugoton Embayment continued to be a region of increased subsidence during the Pennsylvanian.

Eastward thinning of Middle Pennsylvanian strata from the Forest City Basin toward a broad outcrop area in northeastern Missouri indicates the presence of a broad northward-trending arch, apparently the Absaroka expression of the older pre-Kaskaskia Northeast Missouri Arch (Plate 4E). The Mississippi River Arch (Howell, 1935) and the Lincoln Fold System (McQueen and others, 1961) occur along the present day structural crest of this broad feature. The Lincoln Fold, active during the Mississippian, continued as a prominent structural feature during the Middle Pennsylvanian (Searight and Searight, 1961). The Mississippi River Arch, which developed during the Middle Pennsylvanian, apparently formed concurrently with maximum subsidence of the Forest City Basin (Bunker and others, 1982). Pennsylvanian clastic source areas included four general regions: 1) the Ancestral Rocky Mountains to the west, 2) the Marathon-Ouachita Mountains to the south, 3) Canadian Shield sources to the north, and 4) alluvial-deltaic sediments that prograded into the midcontinent from Appalachian sources.

Differential basinal subsidence of the Forest City and Salina Basins decreased during the Late Pennsylvanian. The Nemaha Uplift and Central Kansas Uplift continued to stand in mild positive relief until overlapped by the Kansas City Group (post-Hertha; Lee, 1943, 1946, 1956). Although prominent Middle Pennsylvanian structural features continued to influence Late Pennsylvanian deposition, the Late Pennsylvanian was marked by more subdued structural patterns in the eastern and central portions of the study area. In contrast with Lower and Middle Pennsylvanian units, the widespread lateral continuity of members within the Upper Pennsylvanian cyclothems reflects relative structural stability across much of the midcontinent.

Permian. Permian strata occur across much of Kansas and Nebraska, but are generally absent in the study area east of the Missouri River. Permian strata thicken to the west and southwest in the central midcontinent, reaching thicknesses in excess of 1000 m in portions of southwestern Kansas (Peterson, 1980). Lower Permian (Wolfcampian) strata (Admire, Council Grove, Chase groups) include a series of limestone/shale cyclothemic units similar to those noted in the Upper Pennsylvanian (Mudge and Yochelson, 1962). Gypsum units also occur within the Wolfcampian sequence (Burchett, 1970). Younger Permian strata (Leonardian, Guadalupian), included in the Sumner Group, Nippewalla Group, and Whitehorse Formation in Kansas, are characterized by a thick sequence of shale/sandstone red beds and evaporites (gypsum, anhydrite, and halite) with some carbonate units. More than 40 m of evaporitic deposits occur within the red bed sequence in portions of southwestern Kansas (Merriam, 1963).

Permian structural patterns in the midcontinent strongly influenced Permian deposition. Early Permian subsidence in the Salina Basin and Hugoton Embayment areas generally resembled Pennsylvanian patterns. Structural reorganization of these patterns occurred during the Leonardian-Guadalupian as arching of the Central Kansas Uplift ceased and the areas of the Salina Basin and Central Kansas Uplift began to be tilted as a whole toward the southwest into the Hugoton Embayment (Lee, 1956). In addition, a south-plunging anticline in western Kansas, the Oakley Anticline, developed at that time within the Hugoton Embayment area (Merriam, 1963). The Hugoton Embayment area merged with the Anadarko Basin to the south and extended westward to the Las Animas Arch, which extended northeastward from the Apishipa Uplift in southeastern Colorado into southwestern Nebraska (Rascoe, 1978). This arch had a pronounced effect on sedimentation during the Early Permian, controlling carbonate and evaporite deposition on its flanks. Farther west the relatively shallow Denver Basin spread to the eastern flanks of the ancestral Rocky Mountain source areas in Colorado. Although the youngest strata included in the Absaroka Sequence across most of the central midcontinent are Guadalupian in age, Triassic strata (Dockum Group) are known in a small area of southwestern Kansas. Deposition of the nonmarine Dockum Group was largely controlled by structural settings inherited from relict Late Paleozoic basins (McGowen and others, 1983).

SUB-ZUNI EROSIONAL SURFACE

Upper Jurassic and Cretaceous strata overlie the sub-Zuni erosional surface in the central midcontinent. The Paleozoic sequence is progressively beveled beneath Zuni rocks across northwestern Iowa and northeastern Nebraska in the direction of the Sioux Ridge (Plate 4J), where up to 100 m of relief is developed across strike-oriented Cretaceous valleys (Witzke and others, 1983). Jurassic strata overlie Permian rocks in western

Kansas and central Nebraska, although Jurassic strata truncate the Pennsylvanian sequence and rest directly on Precambrian basement rocks in eastern South Dakota (Plate 4J). Cretaceous strata overstep the eroded Jurassic edge and overlie the Precambrian surface across eastern South Dakota and western Minnesota. Cretaceous outliers in eastern Minnesota, Iowa, and Wisconsin rest on progressively older Paleozoic units in the direction of the Wisconsin Dome. In general, the sub-Zuni erosional surface reflects the position of positive structural and topographic features: 1) Wisconsin Dome, 2) Transcontinental Arch (and the included Sioux Ridge), and 3) an upland area of Precambrian rocks in western Minnesota and eastern South Dakota.

ZUNI SEQUENCE

Jurassic. Zuni deposition in the central midcontinent was initiated during the Late Jurassic, generally coincident with the eastward encroachment of the Sundance Sea. The maximum eastward extent of this epeiric sea is not known, and much of the preserved Jurassic sequence in Kansas, Nebraska, and central South Dakota is apparently of nonmarine origin. Except for structurally preserved outliers in north-central Iowa (Fort Dodge Gypsum), Jurassic strata in the central midcontinent occur entirely in the subsurface. Few stratigraphic studies have been undertaken on these strata and correlations remain uncertain. Nevertheless, the Jurassic of Nebraska-Kansas, which reaches thicknesses in excess of 60 m, shares its closest similarities with the upper Sundance-Morrison interval of Wyoming (Condra and Reed, 1943; Merriam, 1955, 1963). The Kansas Jurassic consists primarily of sandy shale, commonly containing anhydrite and cherty beds in the lower part and limestone beds in the upper part (Merrian, 1955, 1963). Sandstones also occur within the Jurassic interval across Kansas, Nebraska, and South Dakota; sandstone beds are more abundant to the east in Kansas (Merrian, 1955, 1963). As suggested by Brenner (1983), "cratonic siliciclastic sources may have continued supplying nearshore settings along the eastern margin of the epeiric sea as they shifted eastward during the Oxfordian." As such, Jurassic sediments in the central midcontinent probably were deposited as a complex series of nonmarine fluvial and nearshore marine facies. The abundance of anhydrite in the lower part of this interval suggests that tidal-flat or nearshore restricted-marine settings were associated with Late Jurassic transgression along the eastern margin of the seaway.

The economic Fort Dodge Gypsum of Iowa overlies Mississippian and Middle Pennsylvanian strata, and is preserved within a structural depression along the Northern Boundary Fault of the MRS. These Jurassic strata occur over 450 km east of the present Jurassic edge. The Fort Dodge Gypsum contains Late Jurassic palynomorphs (Cross, 1966), and may have been deposited in a marginal marine basin (Bard, 1982). A possible late Jurassic red bed and dolomite interval with minor gypsum (Hallock red beds) also occurs east of the Jurassic edge in northwestern Minnesota (Mossler, 1978). These occurrences suggest that considerable erosional stripping of Jurassic strata occurred prior to burial of the Jurassic edge beneath Cretaceous sediments.

Cretaceous. Cretaceous strata in the central midcontinent were deposited in the eastern margin area of the north-south-trending Western Interior Seaway. Cretaceous sedimentation in this area was controlled by several factors including: 1) the rise and fall of sea level, and hence base level in fluvial systems, during five major transgressive-regressive cycles; 2) relative rates of terrigenous clastic influx from eastern and western source areas; and 3) the distribution of structural and paleotopographic features, most notably the Transcontinental Arch, Sioux Ridge, Precambrian uplands of Minnesota, and Wisconsin Dome (Witzke and others, 1983). Cretaceous deposition in the central midcontinent area was apparently initiated during transgression of the Albian Kiowa-Skull Creek marine cycle. Albian marine shale facies (Kiowa and Skull Creek shales) spread eastward into central Nebraska and Kansas and eastern South Dakota. Correlative nearshore and nonmarine sandstone deposits are known in Kansas and eastern South Dakota, and a portion of the lower Dakota fluvial sandstone sequence in eastern Nebraska and western Iowa may have aggraded as base levels rose during the Kiowa-Skull Creek cycle (Witzke and others, 1983). Eastern-derived, terrigenous, clastic sediments prograded westward across the central midcontinent area during the regressive phase of this marine cycle.

Marine shale facies (Graneros Shale) and offshore pelagic carbonate facies (Greenhorn Formation) spread eastward across the central midcontinent, progressively displacing nearshore and nonmarine facies (Dakota Formation) during the Cenomanian. This marked the initial transgressive phase of the Greenhorn cycle. Nonmarine and marine deposition was initiated across the Precambrian upland surface in western Minnesota during the Cenomanian, although prominent topographic highs, especially the Sioux Ridge, remained emergent through much or all of the Greenhorn cycle. By the Turonian, deposition of Greenhorn carbonates had spread across the western half of the central midcontinent, "reflecting a significant decrease in eastern clastic influx in the offshore areas as shorelines spread eastward to the Wisconsin dome area" (Witzke and others, 1983). Westward progradation of eastern-derived mud and sand (Carlile Shale) during the middle Turonian, followed by marine offlap and subaerial erosion, marked the regressive phase of the Greenhorn cycle. However, a Precambrian granite knob in western Minnesota and eastern South Dakota was buried by lower Carlile strata, suggesting that sea level continued to rise during the early phases of Carlile deposition (Shurr, 1981).

Following a period of late Turonian-early Coniacian erosion, marine carbonate and chalky shale deposition expanded across the western half of the central midcontinent during the Niobrara marine cycle (late Coniacian-Santonian). Eastern source areas supplied clastic material to the nearshore facies of the lower Niobrara Formation, including clastic-dominated facies in western Minnesota and silty-sandy chalk facies in northeast Nebraska and the Sioux Ridge area (Witzke and others, 1983). A unique sequence of clastic rocks overlain by biogenic siliceous strata (Split Rock Creek Formation) is restricted to paleovalleys incised into the flanks of the Sioux Ridge, and correlates, in part, to the Niobrara Formation (Witzke and others, 1983). The low rate of eastern clastic influx, coupled with limited dispersal of western-derived clastics from Cordilleran sources across the axis of the Transcontinental Arch (Rice and Shurr, 1983), permitted widespread development of pelagic carbonate-dominated facies across much of Nebraska, Kansas, and southeastern South Dakota. The Niobrara becomes progressively more shaly northwestward from the arch.

The Pierre Shale of Campanian-Maastrichtian age was deposited disconformably above Niobrara rocks in the central midcontinent, and reaches thicknesses up to 400 m along the western edge of the study area. Members within the Pierre thin eastward to their erosional margin. In general, Pierre clastics were derived from western sources associated with the rising Cordillera, although influx of minor quantities of eastern-derived clastics influenced depositional patterns along the eastern margin of the Western Interior Seaway (Witzke and others, 1983). At times when influx of western clastics was reduced, shaly chalk and chalky shale Pierre facies developed in the eastern portion of the seaway. Lower Pierre strata, including widespread black shale facies, were deposited during the Claggett marine cycle. An erosional unconformity was developed around the Sioux Ridge following Clagget deposition (Witzke and others, 1983). The bulk of Pierre strata was deposited during the succeeding Bearpaw marine cycle. The Cretaceous seaway withdrew from the continental interior during the regressive phase of this cycle.

Zuni deposition in the central midcontinent occurred within two major structural regimes: 1) a broad, relatively stable eastern platform, and 2) a broad hinge-zone area east of the rapidly subsiding basins in the western portion of the Western Interior (Kauffman, 1977). Zuni stratigraphic units thicken westward in the area, reflecting subsidence rates in the midcontinent. Numerous bentonites in the Zuni sequence originated from Cordilleran volcanic sources. In the east, minor structural movements along the Transcontinental Arch and Nemaha Uplift probably influenced Cretaceous deposition (Witzke and others, 1983). Farther west, upwarping along the crest of the Chadron Arch-Cambridge Arch was initiated during or immediately following latest Cretaceous Pierre deposition (Fuenning, 1942; Merrian, 1963). Zuni structural patterns in the midcontinent were modified by later Laramide epeirogenic movement.

SUB-CENOZOIC EROSIONAL SURFACE AND STRUCTURE

The central midcontinent Cretaceous sequence is beveled to the south and east beneath Tertiary strata (Plate 4K). Uplift along the trend of the Cambridge Arch-Chadron Arch was probably a latest Cretaceous-early Tertiary event; up to 400 m of Cretaceous strata were apparently removed from the crest of the Chadron Arch prior to burial by Oliogocene sediments (Plate 4L). As reflected on Cretaceous structure (Fuenning, 1942; Merriam, 1963; Carlson and Reed, 1969) and isopach (Plate 4L) maps, the Cambridge Arch-Chadron Arch trend marks the western margin of a shallow basin that occupies the general area of the Salina Basin and Nebraska Sag (Kennedy Basin). It was apparently during the Tertiary that other broad Laramide epeirogenic features developed, including: 1) broad upwarping along the trend of the Nemaha Uplift (Bunker, 1981), and 2) upwarping along the Las Animas Arch in western Kansas and adjacent Colorado, which separated the Western Kansas Basin from the Denver-Julesburg Basin.

CENOZOIC NONMARINE DEPOSITION

At the onset of Tertiary deposition, the western portion of the central midcontinent was an eastward sloping plain receiving sediments from volcaniclastic material from western sources and epiclastic sediments from the rising Rocky Mountains and Black Hills. This contrasts with the westward-dipping, regional, structural patterns in the Cretaceous. Oligocene bentonitic mudstones, siltstones, and channel sandstones (White River Group) are preserved as far east as north-central Nebraska, and represent the oldest Tertiary deposits in the western portion of the central midcontinent. However, erosional remnants of Lower Eocene strata are known from northeastern Nebraska (M. R. Voorhies, 1984, personal communication). Uppermost Oligocene/Lower Miocene siltstones and sandstones, and volcaniclastic material (Arikaree Group) extend eastward into north-central Nebraska and adjacent South Dakota, and overlie an eroded surface on the White River Group.

Significant erosional downcutting of Arikaree and White River strata preceded deposition of the Ogallala Group. The Ogallala consists primarily of sand, gravel, and silt with beds of limestone and ash. The Ogallala oversteps the Arikaree/White River edge and unconformably overlies Cretaceous strata across Kansas and eastern and southern Nebraska. Although many workers considered the Ogallala to be primarily Pliocene in age, fission-track dating of intercalated volcanic ashes indicates that much of the Ogallala is of Miocene age (Boellstorff, 1978a). The Ogallala apparently includes some Pliocene strata in the western portion of the central midcontinent, but a more complete Pliocene record is preserved in eastern and central Nebraska (Boellstorff, 1978a). The Ogallala is capped by a persistent hard pisolitic limestone over much of its extent (Swineford and others, 1958). This limestone is actually a form of soil caliche, which was originally referred to as the algal limestone (Elias, 1931). Significant erosional downcutting of Ogallala strata, in places incised up to 150 m, preceded Late Pliocene-Pleistocene deposition over much of the western portion of the central midcontinent. Blancan (Late Pliocene) sediments filled some of these valleys, but the deposits were subsequently dissected by later Quaternary drainages.

In general, the Pleistocene sequence in the central midcontinent region includes a complex series of glacial tills, fluvial sands and gravels, paleosols, and lacustrine and aeolian deposits. Numerous Pleistocene erosional events further complicate the stratigraphy. Fission-track dating of volcanic ashes in the sequence helps bracket glacial episodes in the midcontinent (Boellstorff, 1978a, b). Two or more glacial tills are present beneath a 2.2 m.y.-old ash in western Iowa and eastern Nebraska, and represent the oldest Cenozoic glacial deposits known in the U.S. Additional till units occur between 0.7 and 1.2 m.y.-old ashes and above a 0.6 Ma ash (Boellstorff, 1978a, b). These glacial deposits extend across much of Illinois, Iowa, eastern Nebraska, northern Missouri, eastern South Dakota, and northeastern Kansas; their distribution approximates the southern limits of Pleistocene continental glaciation in North America.

Although the classic two-part "Nebraskan" and "Kansas" glacial stage terminology was originally proposed in the area containing this till sequence, new stratigraphic investigations indicate that up to seven or more separate glacial advances are represented in the sequence (Boellstorff, 1978a, b; Hallberg and Boellstorff, 1978). The included glacial till units and associated deposits (paleosols, aeolian/alluvial sediments) are informally lumped together in the pre-Illinoian stage. Eroded pre-Illinoian strata are covered by Illinoian and Wisconsinan glacial and associated deposits across much of the central and northern portions of the central midcontinent. Wisconsinan loess deposits are especially well developed along the Missouri River Valley. The Pleistocene sequence reaches thicknesses in excess of 100 m in portions of western Iowa and eastern Nebraska. West of the limits of glacial deposits in South Dakota, Kansas, and Nebraska, Pleistocene alluvial and aeolian deposits are found in the terraces of incised drainage systems. Widespread Wisconsinan and Holocene aeolian deposits (loess, sand dunes) mantle the terraces and upland bedrock surfaces (Reed and Dreeszen, 1965).

Quaternary depositional and erosional processes have shaped the modern landscape in the central midcontinent. Although the area remains one of relative tectonic stability, limited seismic data indicate that the midcontinent is certainly not quiescent. Late Cenozoic Laramide-related epeirogenic movements probably account for the general eastward tilt of Ogallala strata across the Great Plains (Merriam, 1963). Erosional processes and human activities continue to modify the modern midcontinent landscape.

PHANEROZOIC IGNEOUS ACTIVITY

While the Phanerozoic rock record in the central midcontinent is almost completely comprised of cratonic sedimentary sequences, a suite of alkalic to ultramafic igneous rocks is known in the region from a set of widely scattered dikes, sills, plugs, and diatremes. Isotopic age investigations and field relationships of these rocks show that they were explosively emplaced during several Phanerozoic episodes (Zartman, 1977). Within the study area, Upper Cambrian pyroclastic rocks record explosive submarine volcanism at several localities in southeast Missouri (Wagner and Kisvarsanyi, 1969; Snyder and Gerdemann, 1965; Kisvarsanyi and Hebrank, 1982). In addition, a swarm of Devonian kimberlite and carbonatite diatremes intrude Cambrian strata in the Avon area of southeast Missouri (Kidwell, 1947; Zartman and others, 1967). These Phanerozoic igneous rocks are located close to 38°N latitude, where an east-west-trending axis of recurrent Phanerozoic tectonic activity in eastern and central North America has been described (Snyder and Gerdemann, 1965; Zartman, 1977).

Early Cretaceous kimberlite diatremes intrude Permian strata along the crest of the Nemaha Uplift in northeast Kansas (Brookins, 1970; Brookins and Naeser, 1971). The Elk Creek Carbonatite (Treves and others, 1972a, b; Brookins and others, 1975) intrudes Upper Pennsylvanian strata along the crest of the Nemaha Uplift in southeast Nebraska (R. R. Burchett, 1983, personal communication). Late Cretaceous peridotite dikes intruding Upper Pennsylvanian strata are known along 38°N latitude in southeast Kansas (Knight and Landes, 1932, Zartman and others, 1967; Franks and others, 1971).

In addition to the known areas of explosive igneous activity in the central midcontinent region, a group of cryptoexplosion structures of possible explosive igneous or extraterrestrial impact origin occur within the study area. These include the Crooked Creek (post-Early Ordovician; Hendricks, 1954; Snyder and Gerdeman, 1965) and Decaturville (post-Silurian; Snyder and Gerdemann, 1965; McCracken, 1971) structures located along 38°N latitude in southern Missouri, the Manson Anomaly (latest Cretaceous or Tertiary; Hoppin and Dryden, 1958; Witzke and others, 1983) in northwest Iowa, and the Rock Elm structure (post-Early Ordovician; Cordua, 1985) in southwest Wisconsin.

CONCLUDING REMARKS

The modern structural geology of the central midcontinent, as portrayed on three structural cross sections (Plate 4), reflects a complex Phanerozoic history. Changing patterns of differential crustal movements characterized the region. In comparison to some cratonic basin areas that displayed relatively long-term patterns of unidirectional subsidence (e.g., Michigan and Williston basins), basinal development in the central midcontinent region was a more transitory phenomenon. Areas of maximum basinal subsidence shifted within the region during the Phanerozoic, and reversals in vertical crustal movements through time are apparent at specific localities. The tectonic processes responsible for these complex patterns are poorly understood. The general correspondence of many Phanerozoic structures to Precambrian basement features, such as noted along the MRS in Iowa, suggests that midcontinent Phanerozoic tectonism may be, in part, a reactivation of earlier structures by deep-seated crustal processes.

The Phanerozoic burial histories of three different portions of the Forest City Basin-Nemaha Uplift region are depicted in Figure 3. These subsidence curves were constructed by following the general guidelines discussed by Siever (1983), although no attempt was made to estimate the rates or magnitudes of compactional thinning during the accumulation of the strata. We have,

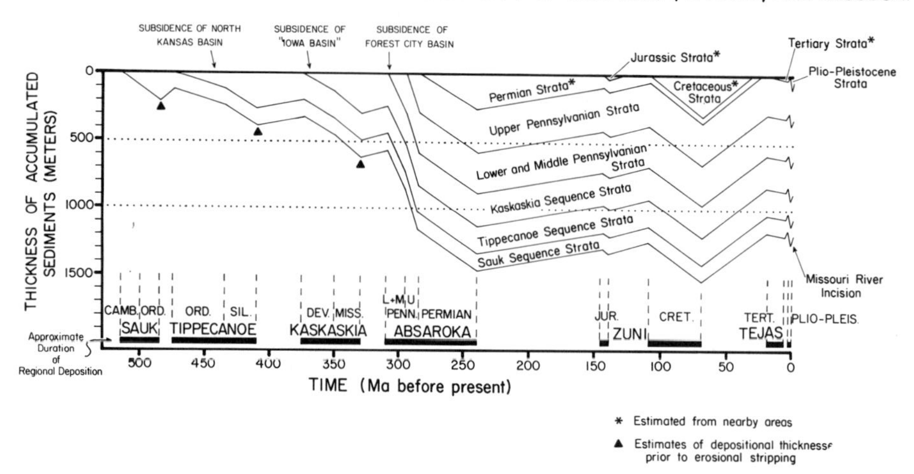

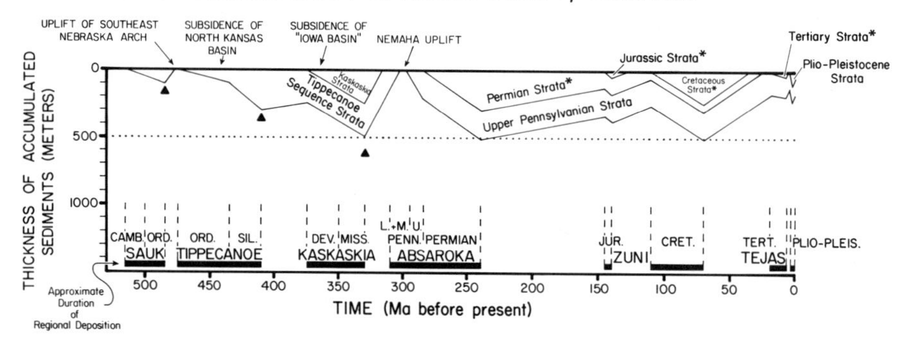

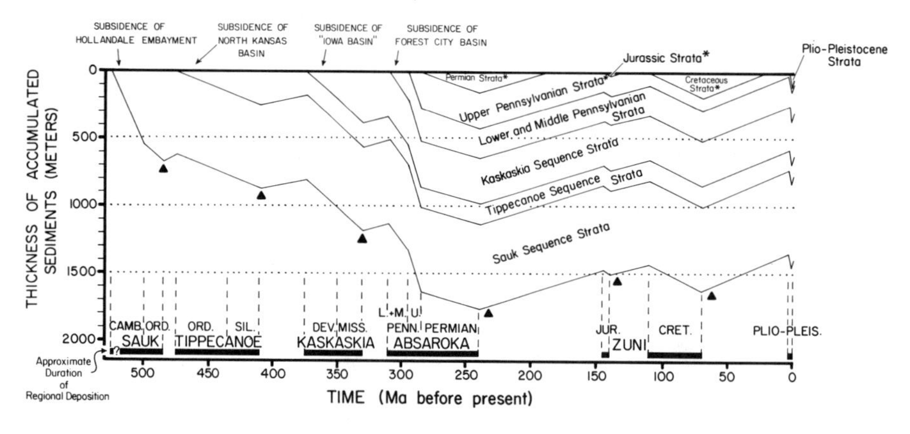

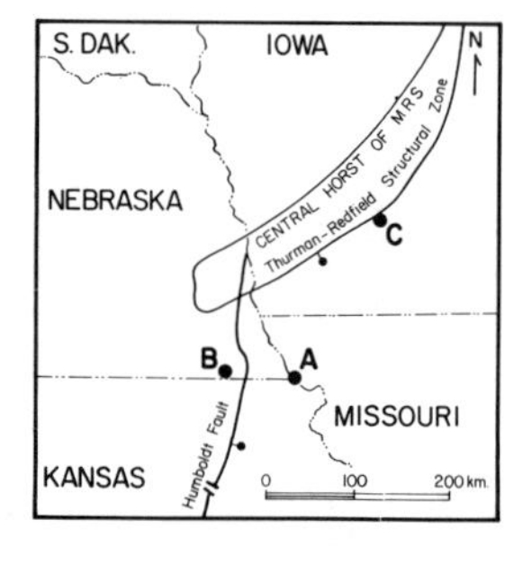

Figure 3. Phanerozoic subsidence curves in the region of the Forest City Basin. A. Center of Forest City Basin at tri-state border of Nebraska, Kansas, and Missouri. B. Crest of Nemaha Uplift in southern Pawnee County, Nebraska. C. Eastern margin of Forest City Basin, in northwest Madison County, Iowa. Based on published data in Carlson, 1970; Burchett, 1982; Bunker, 1982; and Witzke and others, 1983.

however, made an attempt to estimate the magnitudes of erosional exhumation that occurred during extended periods of emergence and/or uplift. The gross timing of depositional and erosional episodes of all three areas were generally similar, but differences in the original depositional thicknesses, chronologies, and magnitudes of post-depositional erosion, and resulting preserved thicknesses illustrate the influence of differential vertical crustal movements on the Phanerozoic stratigraphy of this region.

The Forest City Basin was a relatively short-lived asymmetric fault-bounded sedimentary basin that subsided in synchrony with the ascension of the Nemaha Uplift along the mutually bounding Humboldt Fault Zone during Absaroka Sequence deposition (Fig. 3, A and B). Earlier and subsequent Phanerozoic sedimentation in the area occurred in short-lived depositional basins whose structural geometries were strikingly dissimilar. The resulting superimposition of several discordant structural, depositional, and erosional patterns in the area obviates simple classification of Phanerozoic structural elements in the central midcontinent. Figure 3 shows that the center of the Forest City Basin (Fig. 3, A; Absaroka Sequence), the crest of the Nemaha Uplift (Fig. 3, B; Tippecanoe Sequence, North Kansas Basin), and the eastern margin of the Forest City Basin in central Iowa (Fig. 3, C; Sauk Sequence; Kaskaskia Sequence, Iowa Basin) each have been the locus of maximum subsidence during different Paleozoic intervals. Subsequent Mesozoic and Cenozoic sedimentation in the region was controlled by tectonic and depositional regimes, which were grossly dissimilar to the preceding Paleozoic history.

Paleogeographic syntheses of various Phanerozoic intervals in North America indicate that intraplate tectonism can be grossly correlated with episodes of orogenesis along the continental margins. High-angle faulting along the Humboldt Fault Zone apparently was contemporaneous with high-angle block-faulting in the ancestral Rockies of the western Cordillera, and has been related to collisional tectonics in the Ouachita-Marathon thrust belt (Kluth and Coney, 1981). Likewise, the development of the north-south-trending Mesozoic interior basins of North America have been placed in a larger plate tectonic context (Brenner, 1983; Weimer, 1983). Bevelling of Cretaceous rocks by Cenozoic strata along the Chadron Arch and the western flank of the Nemaha Uplift indicates tectonic activity that was roughly contemporaneous with the development of similarly oriented, but larger-scale Laramide uplifts to the west. These regional relationships indicate that potential exists for integrating the Pennsylvanian-Cenozoic history of the central midcontinent region into a larger, continent-wide tectonic synthesis. The relationships between earlier Paleozoic structural patterns in the central midcontinent and coeval crustal deformation along the continental margins are more obscure, however.

The Phanerozoic crustal dynamics of the central midcontinent region present some of the most intriguing challenges to those interested in mechanically modeling intraplate tectonism. It is still unclear to us which structures and deformational episodes would be best explained in terms of isostatic crust-mantle interactions versus deviatoric stresses in the shallow crust. For example, does the emplacement of Early Cretaceous kimberlites along the axis of the Nemaha Uplift (Brookins and Naeser, 1971) indicate an extensional stress regime or anomalous heat flow during Mesozoic reactivation of the structure (Witzke and others, 1983)? How can the multiple reversals in the direction of vertical crustal movement, which characterized much of this region, be explained? Finally, can the magnitudes of relative vertical crustal movements in the midcontinent be filtered from the stratigraphic record to deduce the magnitudes of Phanerozoic eustatic sea-level changes? Sleep, Nunn, and Chou (1980) have used the mean elevation to the Dakota Formation-Graneros shale contact from 90 well logs in northwest Iowa to deduce the magnitude of eustatic sea-level change since the mid-Cenomanian. Structure contouring of the overlying Turonian Greenhorn Limestone (Bunker, 1981), however, shows that the Cretaceous rocks of the region have experienced relative vertical displacements of greater than 250 m. We are unable to discern any stable benchmark in the region to which other areas might be compared, and so find no reason to suppose that any part of the central midcontinent has remained at a constant elevation relative to other areas for the last 100 m.y.

REFERENCES CITED

Adler, F. J., and 11 others, 1971, Future petroleum provinces of the midcontinent Region 7: American Association of Petroleum Geologists Memoir 15, v. 2, p. 985–1120.

Anderson, R. R., and Black, R. A., 1982, Geological interpretations from geophysical models of the midcontinent geophysical anomaly in southwest Iowa, *in* Van Eck, O. J., Regional tectonics and seismicity of southwestern Iowa: U.S. Nuclear Regulatory Commission, NUREG/CR-2548, p. 27–41.

Andrews, G. W., 1958, Windrow Formation of Upper Mississippi Valley region; A sedimentary and stratigraphic study: Journal of Geology, v. 66, p. 597–624.

Baldwin, W. B., 1949, A preliminary report on the Sioux Quartzite: South Dakota Geological Survey Report of Investigations 63, 34 p.

Bard, G. G., 1982, Petrology and diagenetic features of the Fort Dodge gypsum beds [Ph.D. thesis]: Ames, Iowa State University, 119 p.

Baxter, J. W., and Brenkle, P. L., 1982, Preliminary statement on Mississippian calcareous foraminiferal successions of the midcontinent (U.S.A.) and their correlation to western Europe: Newsletters on stratigraphy, v. 11, no. 3, p. 136–153.

Boellstorff, J., 1978a, Chronology of some Late Cenozoic deposits from the central United States and the Ice Ages: Transactions of the Nebraska Academy of Sciences, v. 6, p. 35–49.

——, 1978b, North American Pleistocene stages reconsidered in light of probable Pliocene–Pleistocene continental glaciation: Science, v. 202, p. 305–307.

Bradbury, J. C., and Atherton, E., 1965, The Precambrian basement of Illinois: Illinois State Geological Survey Circular 382, 13 p.

Brenner, R. L., 1983, Late Jurassic tectonic setting and paleogeography of western interior, North America, *in* Reynolds, M. W., and Dolly, E. D., eds., Mesozoic paleogeography of the west-central United States: Society of Economic Paleontologists and Mineralogists, Rocky Mountain Paleogeography Symposium 2, p. 119–132.

Brookins, D. G., 1970, The Kimberlites of Riley County, Kansas: Kansas State Geological Survey Bulletin, v. 200, 32 p.

Brookins, D. G., and Naeser, C. W., 1971, Age of emplacement of Riley County, Kansas, kimberlites and a possible minimum age for the Dakota Sandstone: Geological Society of America Bulletin, v. 82, p. 1723–1726.

Brookins, D. G., Treves, S. B., and Bolivar, S. L., 1975, Elk Creek, Nebraska, Carbonatite; Strontium geochemistry: Earth and Planetary Science Letters, v. 28, p. 79–82.

Bunker, B. J., 1981, The tectonic history of the Transcontinental Arch and Nemaha Uplift and their relationship to the Cretaceous rocks of the central midcontinent region: Iowa Geological Survey, Guidebook 4, p. 1–23.

—— , 1982, Phanerozoic structural development in the area of the Forest City Basin, southwestern Iowa, *in* Van Eck, O. J., Regional tectonics and seismicity of southwestern Iowa: U.S. Nuclear Regulatory Commission, NUREG/CR-2548, p. 42–61.

Bunker, B. J., Klapper, G., and Witzke, B. J., 1983, New stratigraphic interpretations of the Middle Devonian rocks of Winneshiek and Fayette counties, northeastern Iowa: Geological Society of Iowa, Guidebook 39, 38 p.

Bunker, B. J., Ludvigson, G. A., and Witzke, B. J., 1985, The Plum River Fault Zone and the structural and stratigraphic framework of eastern Iowa: Iowa Geological Survey, Technical Information Series 13, 126 p.

Burchett, R. R., 1969, Geologic bedrock map of Nebraska: Nebraska Conservation and Survey Division, scale 1:1,000,000.

—— , 1970, Occurrence of gypsum in the Johnson Shale (Permian) in Nemaha County, Nebraska: University of Nebraska–Conservation and Survey Division, Resource Report 3, 23 p.

—— , 1982, Thickness and structure maps of the Pennsylvanian and Permian rocks in Nebraska: University of Nebraska–Conservation and Survey Division, Report of Investigations 7, 15 p.

Buschbach, T. C., 1964, Cambrian and Ordovician strata of northeastern Illinois: Illinois State Geological Survey, Report of Investigations 218, 90 p.

—— , 1975, Cambrian System, *in* Willman, H. B., and others, Handbook of Illinois stratigraphy: Illinois State Geological Survey Bulletin 95, p. 34–46.

Carlson, M. P., 1963, Lithostratigraphy and correlation of the Mississippian System in Nebraska: Nebraska Geological Survey Bulletin 21, 46 p.

—— , 1966, Configuration of Precambrian surface in Nebraska: Nebraska Conservation and Survey Division, scale 1:1,000,000.

—— , 1969, Stratigraphic framework of Precambrian and Lower and Middle Paleozoic rocks in the subsurface of Nebraska [Ph.D. thesis]: Lincoln, University of Nebraska, 66 p.

—— , 1970, Distribution and subdivision of Precambrian and Lower and Middle Paleozoic rocks in the subsurface of Nebraska: Nebraska Conservation and Survey Division, Report of Investigations 3, 25 p.

—— , 1979, Nebraska—Iowa region, *in* Craig, L. C., Connor, C. W., and others, Paleotectonic investigations of the Mississippian System in the United States, Part I. Introduction and regional analyses of the Mississippian System: U.S. Geological Survey Professional Paper 1010, p. 107–114.

Carlson, M. P., and Boucot, A. J., 1967, Early Silurian brachiopods from the subsurface of southeastern Nebraska: Journal of Paleontology, v. 41, p. 1121–1125.

Carlson, M. P., and Reed, E. C., 1961, Structure contour map on top of Greenhorn Limestone (Cretaceous): Nebraska Conservation and Survey Division, scale 1:1,000,000.

Chase, C. G., and Gilmer, T. H., 1973, Precambrian plate tectonics—the midcontinent gravity high: Earth and Planetary Science Letters, v. 21, p. 70–78.

Chenoweth, P. A., 1968, Early Paleozoic (Arbuckle) overlap, southern midcontinent, United States: American Association of Petroleum Geologists Bulletin, v. 52, p. 1670–1688.

Chronic, J., McCallum, M. E., Ferris, Jr., C. S., and Eggler, D. H., 1969, Lower Paleozoic rocks in diatremes, southern Wyoming and northern Colorado: Geological Society of America Bulletin, v. 80, p. 149–156.

Cluff, R. M., Reinbold, M. L., and Lineback, J. A., 1981, The New Albany Shale Group of Illinois: Illinois State Geological Survey, Circular 518, 83 p.

Cole, V. B., 1975, Subsurface Ordovician–Cambrian rocks in Kansas: Kansas Geological Survey, Subsurface Geology Series 2, 18 p.

—— , 1976, Configuration of the top of the Precambrian rocks in Kansas: Kansas Geological Survey, Map M-7.

Collinson, C., 1967, Devonian of the north-central region, United States, *in* Oswald, D. H., ed., International symposium on the Devonian System: Calgary, Alberta Society of Petroleum Geologists, v. 1, p. 933–971.

Collinson, C., and James, A., 1969, Conodont zonation of Devonian Cedar Valley Formation and paleogeography of Cedar Valley and Wapsipinicon Formations [abs.]: Geological Society of America Special Paper 121, p. 674.

Colville, V. R., and Sheehan, P. M., 1983, Ordovician–Silurian eustatic sea level changes at Limestone Mountain, Upper Peninsula, Michigan: Geological Society of America Abstracts with Programs, v. 15, p. 256.

Condra, G. E., and Reed, E. C., 1943, The geological section of Nebraska: Nebraska Geological Survey Bulletin 14, 82 p.

Cordua, W. S., 1985, Rock Elm structure, Pierce County, Wisconsin; A possible cryptoexplosion structure: Geology, v. 13, p. 372–374.

Craddock, C., 1972, Keweenawan geology of east-central and southeastern Minnesota, *in* Sims, P. K., and Morey, G. B., eds., Geology of Minnesota; A centennial volume: Minnesota Geological Survey, p. 416–424.

Craig, L. C., and Connor, C. W., coordinators, 1979, Paleotectonic investigations of the Mississippian System in the United States: U.S. Geological Survey Professional Paper 1010, 559 p.

Cross, A. T., 1966, Palynologic evidence of mid-Mesozoic age of Fort Dodge (Iowa) Gypsum [abs.]: Geological Society of America Special Paper 101, p. 46.

Crowell, J. C., 1978, Gondwannan glaciation, cyclothems, continental positioning, and climate change: American Journal of Science, v. 278, p. 1345–1372.

Dapples, E. C., 1955, General lithofacies relationships of St. Peter Sandstone and Simpson Group: American Association of Petroleum Geologists Bulletin, v. 39, p. 444–467.

DeGraw, H. W., 1971, The pre-Oligocene surface in western Nebraska; Its relation to structure and subsequent topographies, *in* Guidebook to the Late Pliocene and Early Pleistocene of Nebraska: Nebraska Conservation and Survey Division, p. 13–21.

Dorheim, F. H., and Campbell, R. B., 1958, Recent gypsum exploration in Iowa: Iowa Academy of Science Proceedings, v. 65, p. 246–253.

Dorheim, F. B., Koch, D. L., and Parker, M. C., 1969, The Yellow Spring Group of the Upper Devonian in Iowa: Iowa Geological Survey, Report of Investigations 9, 30 p.

Elias, M. K., 1931, The geology of Wallace County, Kansas: Kansas Geological Survey Bulletin 18, p. 1–254.

Fitzgerald, D. J., 1977, Petrology of the Pennsylvanian sandstones of Muscatine County, Iowa [M.S. thesis]: Iowa City, University of Iowa, 215 p.

Franks, P. C., Bickford, M. E., Wagner, H. C., 1971, Metamorphism of Precambrian granitic xenoliths in a mica peridotite at Rose Dome, Woodson County, Kansas; Part 2, petrologic and mineralogic studies: Geological Society of America Bulletin, v. 82, p. 2869–2890.

Fraser, G. S., 1976, Sedimentology of a Middle Ordovician quartz arenite carbonate transition in the Upper Mississippi Valley: Geological Society of America Bulletin, v. 87, p. 833–845.

Fraunfelter, G. H., 1967, The stratigraphy of the Cedar City Formation (Middle Devonian) of Missouri: Illinois Academy of Science Transactions, v. 60, p. 9–44.

Fuenning, P., 1942, Thickness and structural study of major divisions of Cretaceous System in Nebraska: American Association of Petroleum Geologists Bulletin, v. 26, p. 1517–1536.

Glenister, B. F., and Sixt, S. C., 1982, Mississippian biofacies–lithofacies trends, northcentral Iowa: Geological Society of Iowa, Iowa Geological Survey, Guidebook 37, 21 p.

Goebel, E. D., and Stewart, G. F., 1979, Kansas, *in* Craig, L. C., Connor, C. W., and others, Paleotectonic investigations of the Mississippian System in the United States; Part I, introduction and regional analyses of the Mississippian System: U.S. Geological Survey Professional Paper 1010, p. 115–123.

Gries, J. P., 1954, Cretaceous rocks of the Williston Basin: American Association Petroleum Geologists Bulletin, v. 38, p. 443–453.

Hallberg, G. R., and Boellstorff, J. D., 1978, Stratigraphic "confusion" in the region of the type areas of Kansan and Nebraskan deposits: Geological Society of America Abstracts with Programs, v. 10, p. 255–256.

Harris, S. E., and Parker, M. C., 1964, Stratigraphy of the Osage Series in southeastern Iowa: Iowa Geological Survey, Report of Investigations 1, 52 p.
Heckel, P. H., 1977, Origin of phosphatic black shale facies in Pennsylvanian cyclothems of midcontinent North America: American Association of Petroleum Geologists Bulletin, v. 61, p. 1045–1068.
——, 1980, Paleogeography of eustatic model for deposition of midcontinent Upper Pennsylvanian cyclothems, *in* Fouch, T. D., and Magathan, E. R., eds., Paleogeography of the west-central United States: Society of Economic Paleontologists and Mineralogists, Rocky Mountain Paleogeography Symposium 1, p. 197–215.
Heckel, P. H., and Cocke, J. M., 1969, Phylloid algal-mound complexes in outcropping upper Pennsylvanian rocks of midcontinent: American Association of Petroleum Geologists Bulletin, v. 53, p. 1058–1074.
Hendricks, H. E., 1954, Geology of the Steelville Quadrangle: Missouri Geological Survey and Water Resources, 2nd Ser., v. 36, p. 52–72.
Hilpman, P. L., 1967, Devonian stratigraphy in Kansas; A progress report, *in* Toomey, D. F., ed., Symposium "Silurian–Devonian rocks of Oklahoma and environs": Tulsa Geological Society Digest, v. 35, p. 88–98.
——, 1969a, Devonian stratigraphy in Kansas; A progress report: Tulsa Geological Society Digest, v. 35, p. 88–98.
——, 1969b, Devonian rocks in Kansas and their epeirogenic significance [Ph.D. thesis]: Lawrence, University of Kansas, 73 p.
Hoppin, R. A., and Dryden, J. E., 1958, An unusual occurrence of Precambrian crystalline rocks beneath glacial drift near Manson, Iowa: Journal of Geology, v. 66, p. 694–699.
Horick, P. J., and Steinhilber, W. L., 1973, Mississippian Aquifer of Iowa: Iowa Geological Survey, Miscellaneous Map Series 3.
Howe, W. B., Kurtz, V. E., and Anderson, K. H., 1972, Correlation of Cambrian strata: Missouri Geological Survey and Water Resources, Report of Investigations 52, 60 p.
Kluth, C. F., and Coney, P. J., 1981, Plate tectonics of the ancestral Rocky Mountains: Geology, v. 9, p. 10–15.
Knight, G. L., and Landes, K. K., 1932, Kansas laccoliths: Journal of Geology, v. 40, p. 1–15.
Koenig, J. W., 1967, The Ozark Uplift and midcontinent Silurian and Devonian stratigraphy, *in* Toomey, D. F., ed., Symposium "Silurian–Devonian rocks of Oklahoma and environs": Tulsa Geological Society Digest, v. 35, p. 119–147.
Kolata, D. R., Huff, W. D., and Frost, J. K., 1983, Correlation of K-bentonites in the Decorah Subgroup of the Mississippi Valley by chemical fingerprinting, *in* Delgado, D. J., ed., Ordovician Galena Group of the Upper Mississippi Valley, deposition, diagenesis, and paleoecology, guidebook for the 13th annual field conference: Great Lakes Section Society of Economic Paleontologists and Mineralogists, p. F1–F15.
Kosanke, R. M., Simon, J. A., Wanless, H. R., and Willman, H. B., 1960, Classification of the Pennsylvanian strata of Illinois: Illinois State Geological Survey, Report of Investigations 214, 84 p.
Kurtz, V. E., Thacker, J. L., Anderson, K. H., and Gerdemann, P. E., 1975, Traverse in Late Cambrian strata: Missouri Geological Survey and Water Resources, Report of Investigations 55, 112 p.
Laudon, L. R., 1937, Stratigraphy of northern extension of Burlington limestone in Missouri and Iowa: American Association of Petroleum Geologists Bulletin, v. 21, p. 1158–1167.
Leatherock, C., 1945, The correlation of rocks of Simpson age in north-central Kansas with the St. Peter Sandstone and associated rocks in northwestern Missouri: Kansas Geological Survey Bulletin 60, pt. 1, p. 1–16.
Lee, W., 1943, The stratigraphy and structural development of the Forest City Basin in Kansas: Kansas Geological Survey Bulletin 51, 142 p.
——, 1946, Structural development of the Forest City Basin of Missouri, Kansas, Iowa, and Nebraska: U.S. Department of the Interior–Geological Survey, Oil and Gas Investigations, Preliminary map 48, 7 sheets, scale 1:1,000,000.
——, 1956, Stratigraphy and structural development of the Salina Basin area: Kansas Geological Survey Bulletin 121, 167 p.
Lochman-Balk, C., 1971, The Cambrian of the craton of the United States, *in* Holland, C. H., ed., Cambrian of the New World: Wiley–Interssciences, p. 79–167.
McCracken, M. H., 1971, Structural features of Missouri: Missouri Geological Survey and Water Resources, Report of Investigations 49, 97 p.
McGowen, J. H., Granata, G. E., and Seni, S. J., 1983, Depositional setting of the Triassic Dockum Group, Texas Panhandle and eastern New Mexico, *in* Reynolds, M. W., and Dolly, E. D., eds., Mesozoic paleogeography of the west-central United States: Society of Economic Paleontologists and Mineralogists, Rocky Mountain Paleogeography Symposium 2, p. 13–38.
McKee, E. D., and Crosby, E. J., and others, 1975, Paleotectonic investigations of Pennsylvanian System in the United States; Part II, Interpretative summary and special features of the Pennsylvanian System: U.S. Geological Survey Professional Paper 853, 192 p.
McQueen, H. A., Hinchey, N. S., and Aid, K., 1961, The Lincoln Fold in Lincoln, Pike, and Ralls counties, northeastern Missouri, *in* Koenig, J. W., Martin, J. A., and Collinson, C. W., 26th regional field conference guidebook, northeastern Missouri and west-central Illinois: Kansas Geological Society, p. 81–85.
Merriam, D. F., 1955, Jurassic rocks in Kansas: American Association of Petroleum Geologists Bulletin, v. 39, p. 31–46.
——, 1963, The geologic history of Kansas: Kansas Geological Survey Bulletin 162, 317 p.
Merriam, D. F., and Atkinson, W. R., 1956, Simpson filled sinkholes in eastern Kansas: Kansas Geological Survey Bulletin 119, p. 61–80.
Miller, B. W., 1971, Petroleum potential of South Dakota: American Association of Petroleum Geologists Memoir 15, v. 1, p. 706–717.
Mossler, J. H., 1978, Results of subsurface investigations in northwestern Minnesota, 1972: Minnesota Geological Survey, Report of Investigations 19, 18 p.
Mudge, M. R., and Yochelson, E. L., 1962, Stratigraphy and paleontology of the uppermost Pennsylvanian and lowermost Permian rocks in Kansas: U.S. Geological Survey Professional Paper 323, 213 p.
Muehlberger, W. R., Denison, R. E., and Lidiak, E. G., 1967, Basement rocks in continental interior of United States: American Association of Petroleum Geologists Bulletin, v. 51, p. 2351–2380.
Munter, J. A., Ludvigson, G. A., and Bunker, B. J., 1983, Hydrogeology and stratigraphy of the Dakota Formation in northwest Iowa: Iowa Geological Survey, Water Supply Bulletin 13, 55 p.
Norris, A. W., Uyeno, T. T., and McCabe, H. R., 1982, Devonian rocks of the Lake Winnipegosis–Lake Manitoba outcrop belt, Manitoba: Geological Survey of Canada Memoir 392, 280 p.
Norton, W. H., 1920, Wapsipinicon breccias of Iowa: Iowa Geological Survey Annual Report v. 27, p. 260–549.
Ostrom, M. E., 1970, Lithologic cycles in Lower Paleozoic rocks of western Wisconsin, *in* Field trip guidebook for Cambrian–Ordovician geology of western Wisconsin: Wisconsin Geological and Natural History Survey, Information Circular 11, p. 10–34.
Palmer, A. R., 1981, Subdivision of the Sauk Sequence, *in* Taylor, M. E., ed., Short papers for the international symposium on the Cambrian System: U.S. Geological Survey Open File Report, p. 81–743.
Parker, M. C., 1973, Mississippian stratigraphy of southeastern Iowa: Iowa Academy of Science Proceedings, v. 80, p. 4–7.
Peterson, J. A., 1980, Permian paleogeography and sedimentary provinces, west central United States, *in* Fouch, T. D., and Magathan, E. R., eds., Paleozoic paleogeography of the west-central United States: Society of Economic Paleontologists and Mineralogists, Rocky Mountain Paleogeography Symposium 1, p. 197–215.
Philcox, M. E., 1972, Burial of reefs by shallow-water carbonates, Silurian of Iowa, U.S.A.: Geologische Rundschau, v. 61, p. 686–708.
Reed, E. C., and Dreeszen, V. H., 1965, Revision of the classification of the Pleistocene deposits of Nebraska: Nebraska Geological Survey Bulletin 23.
Rascoe, B., Jr., 1978, Late Paleozoic structural evolution of the Las Animas Arch, *in* Pruit, J. D., ed., Energy resources of the Denver Basin: Rocky Mountain Association of Geologists, Field Conference Guidebook, p. 113–127.
Rice, D. D., and Shurr, G. W., 1983, Patterns of sedimentation and paleogeog-

raphy across the western interior seaway during time of deposition of Upper Cretaceous Eagle Sandstone and equivalent rocks, northern Great Plains, *in* Reynolds, M. W., and Dolly, E. D., eds., Mesozoic paleogeography of the west-central United States: Society of Economic Paleontologists and Mineralologists, Rocky Mountain Paleogeography Symposium 2, p. 337–358.

Ross, R. J., Jr., 1976, Ordovician sedimentation in the western United States, *in* Bassett, M. G., ed., The Ordovician System: Proceedings of the Palaeontological Association, p. 73–105.

Schumacher, D., 1976, Conodont biofacies and paleoenvironments in Middle Devonian–Upper Devonian boundary beds, central Missouri: Geological Association of Canada Special Paper 15, p. 159–169.

Schwab, F. L., 1976, Modern and ancient sedimentary basins; Comparative accumulation rates: Geology, v. 4, p. 723–727.

Searight, W. V., and Searight, T. K., 1961, Pennsylvanian geology of the Lincoln Fold, *in* Koenig, J. W., Martin, J. A., and Collinson, C. W., 26th annual regional field conference guidebook, northeastern Missouri and west-central Illinois: Kansas Geological Society, p. 155–163.

Sendlein, L.V.A., 1964, Geology of the Sperry Mine, Des Moines County, Iowa [Ph.D. thesis]: Ames, Iowa State University, 79 p.

——, 1968, Geology of the U.S. Gypsum Company's Sperry Mine, *in* Steinhilber, W. L., ed., 2nd Annual Meeting, North-Central Section: Geological Society of America, Field Trip Guidebook, no. 2, 17 p.

——, 1972, Geology of the U.S. Gypsum Sperry, Iowa Mine, Devonian gypsum of Iowa, *in* Proceedings, eighth forum of geology of industrial minerals: Iowa Geological Survey Public Information Circular 5, p. 67–87.

Sheehan, P. M., 1978, The relation of Late Ordovician glaciation to the Ordovician–Silurian changeover in North American brachiopod faunas: Lethaia, v. 6, p. 147–154.

Shurr, G. W., 1981, Cretaceous sea cliffs and structural blocks on the flanks of the Sioux Ridge, South Dakota and Minnesota: Iowa Geological Survey Guidebook 4, p. 25–41.

Siever, R., 1983, Burial history and diagenetic reaction kinetics: American Association of Petroleum Geologists Bulletin, v. 67, p. 684–691.

Sleep, N. H., Nunn, J. A., and Chou, L., 1980, Platform basins: Annual Review of Earth and Planetary Sciences, v. 8, p. 17–34.

Sloan, R. E., 1972, Notes on the Platteville Formation, southeastern Minnesota, *in* Field trip guidebook for Paleozoic and Mesozoic rocks of southeastern Minnesota: Minnesota Geological Survey, Guidebook 4, p. 43–53.

Sloss, L. L., 1963, Sequences in the cratonic interior of North America: Geological Society of America Bulletin, v. 74, p. 93–114.

Snyder, F. G., and Gerdemann, P. E., 1965, Explosive igneous activity along an Illinois–Missouri–Kansas axis: American Journal of Science, v. 263, p. 465–493.

Swineford, A., Bryon, L. A., and Frye, J. C., 1958, Petrology of the Pliocene pisolitic limestone in the Great Plains: Kansas Geological Survey Bulletin 130, part 2, p. 97–116.

Templeton, J. S., and Willman, H. B., 1963, Champlainian Series (Middle Ordovician) in Illinois: Illinois State Geological Survey Bulletin 89, 260 p.

Thompson, T. L., 1979, The Mississippian and Pennsylvanian (Carboniferous) systems in the United States; Missouri: U.S. Geological Survey Professional Paper 1110-N, 22 p.

Thompson, T. L., and Satterfield, I. R., 1975, Stratigraphy and conodont biostratigraphy of strata contiguous to the Ordovician–Silurian boundary in eastern Missouri: Missouri Geological Survey and Water Resources Report of Investigations 57, p. 61–108.

Thwaites, F. T., 1957, Map of the buried Precambrian of Wisconsin: University of Wisconsin and Natural History Survey, scale 1 inch = 38 miles.

Treves, S. B., Smith, R., Carlson, M. P., and Coleman, G., 1972a, Elk Creek carbonatite, Johnson and Pawnee counties, Nebraska: Geological Society of America Abstracts with Programs, v. 4, p. 297.

Treves, S. B., Smith, R., Rinehart, J., Coleman, G., and Carlson, M. P., 1972b, Petrography and mineralogy of the Elk Creek carbonatite, Nebraska: Geological Society of America Abstracts with Programs, v. 4, p. 352.

Urban, J. B., 1971, Palynology and the Independence Shale of Iowa: Bulletin of the American Paleontologist, v. 60, p. 103–189.

——1972, The Independence Shale; A palynologic viewpoint: 36th Annual Tri-State (Iowa, Illinois, and Wisconsin) Geological Field Conference Guidebook, p. 48–52.

Van Schmus, W. R., and Hinze, W. J., 1985, The Midcontinent Rift System: Annual Review of Earth and Planetary Sciences, v. 13, p. 345–383.

Wagner, R. E., and Kisvarsanyi, E. B., 1969, Lapilli tuffs and associated pyroclastic sediments in Upper Cambrian strata along Dent Branch, Washington County, Missouri: Missouri Geological Survey Report of Investigations 43, 80 p.

Wanless, H. R., 1975, Missouri and Iowa, *in* McKee, E. D., and Crosby, E. J., coordinators, Paleotectonic investigations of the Pennsylvanian System in the United States; Part I, Introduction and regional analyses of the Pennsylvanian System: U.S. Geological Survey Professional Paper 853, p. 97–114.

Watney, W. L., 1980, Cyclic sedimentation of the Lansing–Kansas City groups in northwestern Kansas and southwestern Nebraska: Kansas Geological Survey Bulletin 220, 72 p.

——1984, Recognition of favorable reservoir trends in Upper Pennsylvanian cyclic carbonates in western Kansas, *in* Hyne, N. J., ed., Limestones of the mid-continent: Tulsa Geological Society Special Paper 2, p. 201–245.

Weimer, R. J., 1983, Relation of unconformities, tectonics, and sea level changes, Cretaceous of the Denver Basin and adjacent areas, *in* Reynolds, M. W., and Dolly, E. D., eds., Mesozoic paleogeography of the west-central United States: Society of Economic Paleontologists and Mineralogists, Rocky Mountain Paleogeography Symposium 2, p. 359–376.

Whiting, L. L., and Stevenson, D. L., 1965, Sangamon Arch: Illinois State Geological Survey Circular 383, 20 p.

Willman, H. B., and Atherton, E., 1975, Silurian System, *in* Willman, H. B., and others, Handbook of Illinois stratigraphy: Illinois State Geological Survey Bulletin 95, p. 87–104.

Willman, H. B., and Buschbach, T. C., 1975, Ordovician System, *in* Willman, H. B., and others, Handbook of Illinois stratigraphy: Illinois State Geological Survey Bulletin 95, p. 47–87.

Willman, H. B., and Frye, J. C., 1975, Cenozoic Erathem, *in* Willman, H. B., and others, Handbook of Illinois stratigraphy: Illinois State Geological Survey Bulletin 95, p. 206–209.

Willman, H. B., and others, 1976, Geologic map of Illinois: Illinois State Geological Survey, scale 1:500,000.

Witzke, B. J., 1980, Middle and Upper Ordovician paleogeography of the region bordering the Transcontinental Arch, *in* Fouch, T. D., and Magathan, E. R., eds., Paleozoic paleogeography of the west-central United States: Society of Economic Paleontologists and Mineralogists, Rocky Mountain Paleogeography Symposium 1, p. 1–18.

——, 1981, Silurian stratigraphy, petrology, and depositional environments of eastern Iowa [Ph.D. thesis]: Iowa City, University of Iowa, 574 p.

——, 1983a, Ordovician Galena Group in Iowa subsurface, *in* Delgado, D. J., ed., Ordovician Galena Group; Deposition, submarine and later diageneses, and paleoecology: Society of Economic Paleontologists and Mineralogists, Great Lakes Section, 13th Annual Field Conference Guidebook, p. D1–D26.

——, 1983b, Silurian benthic invertebrate associates of eastern Iowa and their paleoenvironmental significance: Wisconsin Academy of Science, Arts, and Letters, v. 71, part 1, p. 21–47.

Witzke, B. J., and Bunker, B. J., 1984, Devonian stratigraphy of north-central Iowa: Iowa Geological Survey Open-File Report 84-2, p. 107–149.

Witzke B. J., Ludvigson, G. A., Poppe, J. R., and Ravn, R. L., 1983, Cretaceous paleogeography along the eastern margin of the western interior seaway, Iowa, southern Minnesota, and eastern Nebraska and South Dakota, *in* Reynolds, M. W., and Dolly, E. D., eds., Mesozoic paleogeography of the west-central United States: Society of Economic Paleontologists and Mineralogists, Rocky Mountain Paleogeography Symposium 2, p. 225–252.

Zartman, R. E., Brock, M. R., Heyl, A. V., and Thomas, H. H., 1967, K-Ar and Rb-Sr ages of some alkalic intrusive rocks from central and eastern United States: American Journal of Science, v. 265, p. 848–870.

MANUSCRIPT ACCEPTED BY THE SOCIETY SEPTEMBER 15, 1986

Printed in U.S.A.

The Geology of North America
Vol. D-2, Sedimentary Cover—North American Craton: U.S.
The Geological Society of America, 1988

Chapter 11

The Permian Basin region

H. N. Frenzel
1118 Mogford, Midland, Texas 79701
R. R. Bloomer
Bloomer and Associates, 310 N. Willis, Abilene, Texas 79603
R. B. Cline
Shell Western Exploration and Production Inc., Box 576, Houston, Texas 77001
J. M. Cys
6236 Wheaton, Fort Worth, Texas 76113
J. E. Galley
Box 1346, Kerrville, Texas 78029
W. R. Gibson
Seeks Oil and Gas, Box 1351, Midland, Texas 79702
J. M. Hills
Department of Geological Sciences, University of Texas, El Paso, Texas 79968
W. E. King and W. R. Seager
New Mexico State University, Las Cruces, New Mexico 88003
F. E. Kottlowski and S. Thompson III
New Mexico Bureau of Mines and Mineral Resources, Socorro, New Mexico 87801
G. C. Luff
Coastal Oil and Gas Corporation, Box 235, Midland, Texas 79702
B. T. Pearson
Box 3508, Midland, Texas 79702
D. C. Van Siclen
4909 Bellaire Blvd., Bellaire, Texas 77401

FOREWORD

H. N. Frenzel

The Permian Basin region, as defined for this chapter, includes all of the Permian basins beneath the high plains of western Texas that lie south of the Red River and Matador Uplifts. It also extends westward into the Southern Rocky Mountain region to include the related Orogrande and Pedregosa Basins in New Mexico. In the central part of the region, surface exposures of the Tertiary Ogallala Formation and other post-Paleozoic strata hide the Paleozoic geology, which includes many anticlinal and faulted structures, sand and shale basins, carbonate banks and reefs marginal to basins, and extensive carbonate shelf areas. On these shelves, carbonates grade into evaporite deposits and eventually into continental clastics. Knowledge of this geology is primarily from wells drilled for oil and gas. Peripheral outcrops of these Paleozoic strata do exist and are important to the interpretation of Permian Basin geology, but many formations and rock facies are known only from the subsurface. With the exception of the two outlying New Mexico basins, petroleum geologists with extensive subsurface experience were selected to write the various parts of this chapter.

GENERAL INTRODUCTION

J. M. Hills and J. E. Galley

The Permian Basin region of west Texas and southern New Mexico lies in the southwest part of the midcontinent craton of North America (Fig. 1). The region comprises all or parts of several physiographic provinces. Physiographic features of the main part of the Permian Basin (Fig. 2) have been largely shaped by erosion, solution, and alluviation during the long period of tectonic stability from the Triassic to the present. The extreme western parts of the region were affected by Tertiary tectonics.

Frenzel, H. N., and 13 others, 1988, The Permian Basin region, *in* Sloss, L. L., ed., Sedimentary Cover—North American Craton; U.S.: Boulder, Colorado, Geological Society of America, The Geology of North America, v. D-2.

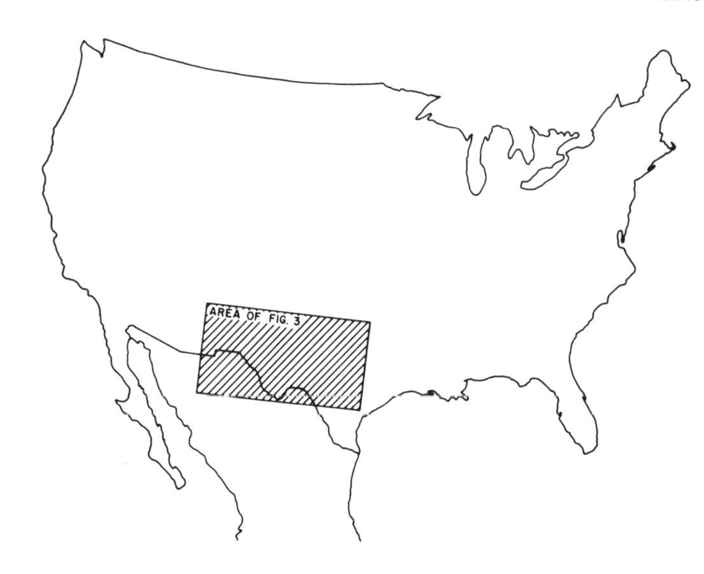

Figure 1. Location of the Permian Basin region.

Because of Mesozoic and Cenozoic cover, in only a few places are the underlying features of the Paleozoic rocks of the main part of the Permian Basin region detectable by surface mapping.

Geologic exploration of the region dates from the Marcy and Pope expeditions in the 1850s in connection with the search for railroad routes through recently annexed territories (Marcy, 1850; Pope, 1855). These were supplemented by the Bartlett and Emory surveys of the Mexico boundary (Bartlett, 1854; Emory, 1857). The ages of the rocks in the region were unknown until the work of the Shumard brothers (Shumard, 1858, 1859) provided convincing evidence of the occurrence of Permian rocks and differentiated them from previously recognized Carboniferous strata. Much later, Girty (1908) gave a detailed account of the upper Permian fossils. P. B. King (1930) and R. E. King (1930) worked out the stratigraphic relations of the surface Paleozoic rocks in the Glass Mountain region and the Guadalupe Mountains (P. B. King, 1948). Early Paleozoic rocks equivalent to those in the Tobosa Basin (Fig. 3) were described by Hill (1900), Richardson (1904), and others. All of these descriptions necessarily were based on outcrops on the periphery of the Permian Basin.

Great advances in knowledge of the subsurface rocks did not begin until the start of oil exploration in the early part of the 20th century. By 1930 the stratigraphic relations within the upper parts of the Permian System were fairly well known (Cartwright, 1930; Lloyd, 1929). Drilling to older Paleozoic rocks did not begin until the 1920s, the first being development of deep production in the Ellenburger (Ordovician) dolomite of the Big Lake Field, Reagan County, in the southern part of the Midland Basin (Bybee and others, 1931). The near cessation of drilling activity during the early 1930s delayed further development of Permian Basin petroleum resources until the middle and late 1930s.

Since that time hundreds of thousands of wells have been drilled in the basin, many of them in the Delaware and Val Verde Basins, to Cambro-Ordovician rocks at depths close to 9,000 m. The continuous flow of geological and geophysical exploration and drilling data has provided a tremendous accumulation of knowledge that has been summarized in many articles (e.g., Udden, 1914; Adams, 1965; Galley, 1958; Hills, 1942, 1972; P. B. King, 1942; and Lloyd, 1929) and cross-sections (Bybee and others, 1931; Cartwright, 1930; Davis and others, 1953; Edwards, 1927; Feldman and others, 1962; Jones and others, 1949; R. E. King and others, 1942; Jones and Matchus, 1984; Roswell Geological Society, 1956; Scobey and others, 1951; Vertrees and others, 1964). The Permian Basin now is probably as well known geologically as any basin of comparable size in the world. Many new stratigraphic units have been established; those that cover a reasonably large area and are accepted by most geologists working in the region are shown in the COSUNA (Correlation of Stratigraphic Units of North America) chart of Hills and Kottlowski, 1983. Of special importance is the paper of Adams and others (1939) on the stratigraphy of the Permian rocks of North America, and the pioneer studies by Dunbar and Skinner (1937) on the Permian fusulinids.

A major basement element of the craton in the Permian Basin region is the Pedernal Massif (Galley, 1958), an ancient, relatively unexplored Precambrian positive region of ill-defined shape and composition at the southern tip of the Transcontinental Arch (Eardley, 1949; "continental backbone" of P. B. King, 1959; Cook and Bally, 1975, p. 41–16). No pre-Mississippian strata exist on the massif; the highest parts, surrounded by Permian and Triassic redbeds, are exposed today in north central New Mexico. The massif underlies much of southern Colorado and much of New Mexico and western Texas. The shores of Paleozoic seas in the southwest part of the North American craton lay on the slopes of the Pedernal Massif, and fingers of that land mass, such as the Texas Arch and the Diablo Uplift, extend, southeastward as peninsulas.

The geologic divisions of the region that are described individually in this chapter are shown on Figure 3. The heart of the Permian Basin is a structural sag consisting of the deep Delaware Basin and the shallower Midland Basin; the two are separated by the Central Basin Uplift, which in Permian time was capped by carbonate reefs and banks and became the Central Basin Platform. The early Paleozoic Tobosa Basin was the forerunner of this major downwarp, lying immediately east of the Diablo Uplift, south of the Matador Uplift, and west of the Texas Arch.

The central Permian Basin is bounded on the south by the front of the Marathon-Ouachita Foldbelt. The linear Val Verde Basin, the southern part of the Kerr Basin, and the Marfa Basin are foreland basins, possibly developed in front of the northwardly advancing Marathon thrust sheets. The Bend Arch (Van Siclen and Bloomer, this chapter) extends northward from the domal Llano Uplift where Precambrian and early Paleozoic rocks are exposed. This feature is the eastern boundary of the Permian Basin (Figs. 2, 3).

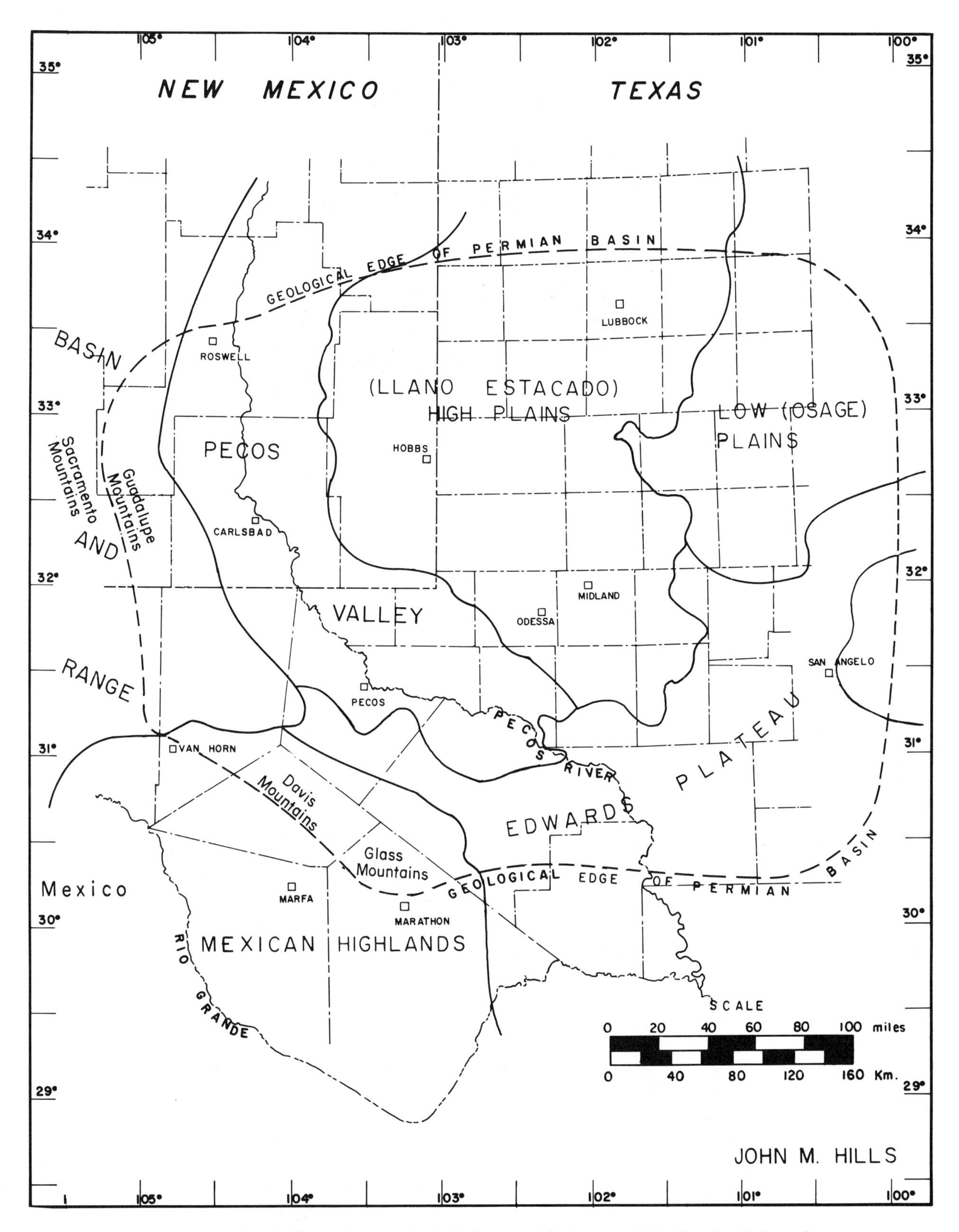

Figure 2. Physiographic provinces and principal geographic features of the Permian Basin region (adapted from Fenneman, 1946).

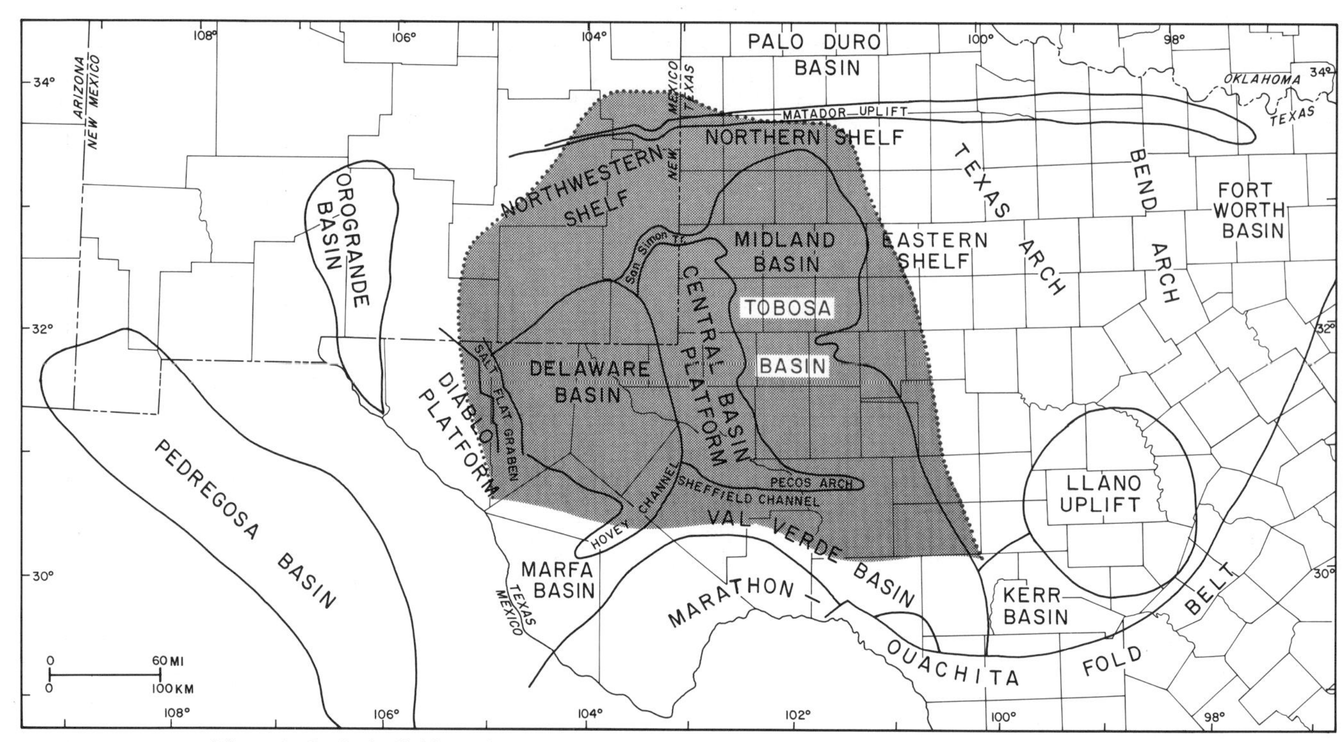

Figure 3. Geologic divisions of the Permian Basin region discussed in this chapter; older Tobosa Basin shown in red.

The Permian Basin is one of several basins, such as the Williston and Denver Basins, that lie in front of younger structures of the Cordilleran province. Although Laramide structures occur in west Texas along the Rio Grande, evidence of tectonic movements of that age in the main Permian Basin region is scanty or nonexistent. However, at the western margin of the Permian Basin, various structures of the Basin-and-Range province occur, such as the Salt Flat Graben (Fig. 3) and the Guadalupe and Sacramento mountain uplifts north and west of the Delaware Basin. Parts of the southwestern edges of the basin have been affected by early and Mid-Tertiary volcanic events, including outpourings of lava and ignimbrites in the Davis Mountains and scattered outcrops of volcanic rocks in the western part of the basin.

The COSUNA charts (Hills and Kottlowski, 1983) give a summary of the stratigraphic units in the Permian Basin region. A generalized chart is given in Figure 4. The Proterozoic rocks underlying the basins are imperfectly known but seem to consist largely of metamorphic and plutonic rocks of cratonic aspect. Sedimentary strata ranging in age from Cambrian through Triassic record the geologic history of the Permian Basin region, the deepest basin of Permian sedimentation in the midcontinent of North America.

The greatest volumes of Paleozoic rocks in the Permian Basin region are of Pennsylvanian or Permian ages and are treated by Cys and Gibson; Van Siclen and Bloomer describe an important subprovince, the Eastern Shelf of the Permian Basin, which overlies the Texas Arch and is representative of the distinctive character of Permian Basin geology; we discuss the Tobosa Basin, which lies beneath the Permian Basin and is the site of post-Ellenburger, pre-Mississippian sedimentary deposition.

Although strictly outside of the geologic boundaries of the Permian Basin, several other related basins are included in this chapter. To the west beyond the Diablo Platform and the Sacramento Mountains is the Orogrande Basin (Fig. 3). While most of this basin lies in New Mexico, it extends southward into El Paso County, Texas. Much further west, another Late Paleozoic basin, the Pedregosa Basin, is located in extreme southwestern New Mexico and adjacent parts of Arizona and Mexico. These basins are discussed by Kottlowski and others. Both of these basins are in a transition zone between the Plains and the Cordilleran geomorphic provinces.

Between the Diablo Platform and the Marathon-Ouachita Foldbelt lies the Marfa Basin. This feature, which may have formed the southwestern part of the Early Paleozoic Tobosa Basin and was later strongly involved in Tertiary deformation and volcanism, is described by Pearson and Luff.

The Kerr Basin is another peripheral basin that lies between the Val Verde Basin (discussed in the section by Cys and Gibson) and the Llano Uplift and north of the Marathon-Ouachita

SYSTEM	SERIES	DELAWARE BASIN	CENTRAL BASIN PLATFORM	EASTERN SHELF	SLOSS SEQUENCES
QUATERNARY	Holocene	Holocene Sand	Holocene Sand	Alluvium	Tejas
TERTIARY	Pliocene	Ogallala	Ogallala	Gravels	
CRETACEOUS	Gulfian Comanchean	Limestone Sand	Limestone	Limestone	Zuni
JURASSIC	Absent				
TRIASSIC		Dockum	Dockum	Dockum	
PERMIAN	Ochoa	Dewey Lake	Dewey Lake	Dewey Lake	
		Rustler	Rustler	Rustler	
		Salado	Salado	Salado	
		Castile			
	Guadalupe	Delaware Mt.: Bell Canyon	Artesia Gr.: Tansill	Artesia Group	
		Cherry Canyon	Yates		
		Brushy Canyon	SevenRivers		
			Queen		
			Grayburg		
		Victorio Peak	San Andres	San Andres	Absaroka
			Glorieta Ss.	San Angelo	
	Leonard	Bone Spring Limestone	Yeso: Clear Fork	Clear Fork	
			Wichita-Abo	Wichita	
	Wolfcamp	Wolfcamp	Wolfcamp	Wolfcamp	
PENNSYLVANIAN	Virgil	Cisco		Cisco	
	Missouri	Canyon		Canyon	
	Des Moines	Strawn	Strawn	Strawn	
	Atoka	Atoka	Atoka	Bend	
	Morrow	Morrow	Morrow		
MISSISSIPPIAN	Chester	Barnett			
	Meramec	Mississippian Limestone	Osage-Meramec	Mississippian Limestone	Kaskaskia
	Osage				
	Kinderhook		Kinderhook		
DEVONIAN	Upper Middle	Woodford	Woodford		
		Thirty one	Thirty one		
SILURIAN	Middle	Wristen	Wristen		Tippecanoe
		Fusselman	Fusselman		
ORDOVICIAN	Upper Middle	Montoya	Montoya		
		Simpson	Simpson		
	Lower	Ellenburger	Ellenburger	Ellenburger	Sauk
CAMBRIAN	Upper	Cambrian	Cambrian Ss.	Cambrian Ss.	
PRE CAMBRIAN		Pre Cambrian	Pre Cambrian	Pre Cambrian	

Figure 4. Principal stratigraphic units of the Permian Basin region. Vertical scale is arbitrary.

Foldbelt in the southeastern part of the region. It is largely covered by Cretaceous beds of the Gulf Coast province. The Paleozoic beds are discussed by Cline.

Transgressive Comanchean (Cretaceous) strata consisting of basal clastics overlain by sequences of shale, marl, and limestone beds, were deposited on the eroded surfaces of Triassic and Paleozoic formations in much of the Permian Basin region but subsequently were eroded from large areas west of the Pecos River. Today, well-bedded Comanchean limestones form the caprock of the Edwards Plateau (Fig. 2) that extends eastward across the southern part of the Permian Basin. Sheets of alluvium, derived in late Tertiary times as outwash of detrital materials from the rejuvenated Rocky Mountain ranges and eastermost mountain blocks of the Basin-and-Range province, cover the Comanchean and older beds in much of the northern half of the basin. These extend eastward to the outcrops of Permian and Pennsylvanian formations in the Low Plains. Thick beds of indurated caliche in the upper layers of the Ogallala Formation (detrital wash) form the caprock of the High Plains. The underlying permeable gravel and sand layers are important aquifers, which are currently being drained for agricultural, industrial, and domestic uses. Bolson deposits of Quaternary age fill intermontane basins west of the Pecos River.

Tectonic events in the history of the Permian Basin region

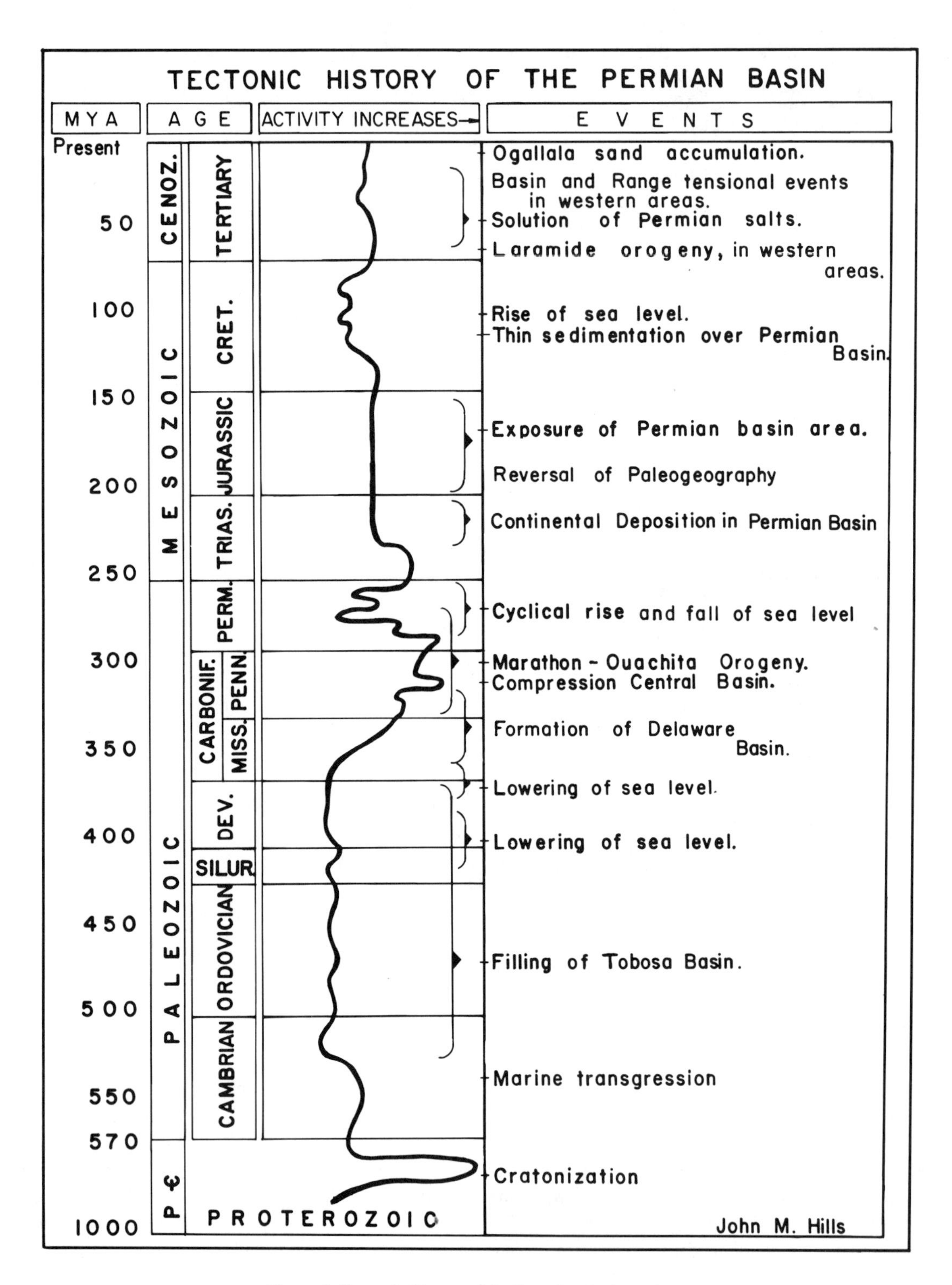

Figure 5. Tectonic history of the Permian Basin region.

are discussed in individual sections of the text and summarized on Figure 5. The general history is one of comparative stability for long stretches of time. Following cratonization in the Precambrian, a long interval of relative quiescence lasted until mid-Mississippian time and was followed by a period of activity culminating in the pulsations of the Marathon-Ouachita Orogeny in late Middle Pennsylvanian and a final pulsation in the Middle of Early Permian time. After that there was another long period of comparative quiescence, broken only by the weak manifestations of Laramide disturbances and the Basin-and-Range events of Tertiary age.

Fluids in the deeper rocks of the Permian basin are largely brines, with salt contents of as high as 100,000 ppm in some of the older rocks of the deeper parts of the basin. On the west and south sides of the basin, hydrodynamic movement of meteoric waters' down dip in permeable beds has diluted these brines to brackish water (Holmquest 1965). Close to the outcrop these waters may even be potable (less than 1,000 ppm dissolved solids).

Locally, structural and stratigraphic traps have filled with important deposits of petroleum and natural gas. In the last 60 years about 20% of the United States production of these hydrocarbons has come from these reservoirs.

THE PRE-PENNSYLVANIAN TOBOSA BASIN

J. M. Hills and J. E. Galley

The Tobosa Basin is an ancient sedimentary basin which was filled chiefly with carbonate and fine-grained clastic sediments until Mississippian time. It ceased to exist as an active tectonic element after deposition of Woodford black shales in Late Devonian time. Its location is the heart of the Permian Basin, for it was the predecessor of the Delaware and Midland Basins.

PRECAMBRIAN

The Proterozoic geography of the basin region is necessarily vague because drilling and geophysical information is scanty. Drilling has reached the Precambrian rocks in many places in the basin, but these rocks have seldom been penetrated more than a few meters. Little is known of the composition of the Proterozoic rocks penetrated and even less of the nature of the deeper ones. Flawn (1956) collected the data then available from drilling and mapped the surface of the Precambrian. A more recent map of the surface configuration is included in Figure 6. Flawn's lithologic determinations show that most of the Permian Basin area is underlain by granitic rocks of the North American Craton. However, in the northern part of the Permian Basin and extending into the Texas panhandle, there are large areas that are underlain by volcanic rocks. Metamorphic rocks in the subsurface, as well as those on the peripheral uplifts in which Precambrian rocks are exposed, indicate that the cratonic crystalline rocks may have been intruded into thick sediments.

Age determinations of Precambrian rocks in the basin are relatively rare but have been made on a few specimens from drill holes and on many from the peripheral uplifts in central New Mexico, the Van Horn area of trans-Pecos Texas, and the Llano uplift of central Texas (Wasserburg and others, 1962; Condie, 1981). The determinations seem to indicate a major metamorphic event about 1.0 to 1.3 Ga.

Gravity work (Cordell and others, 1982; Keller and others, 1980) seems to indicate that the crustal composition beneath the central uplift of the Permian Basin has been strongly affected by major intrusions of basic rocks perhaps of the same age as the metamorphic rocks mentioned above. This interpretation appears to be confirmed by a recent deep hole on the Central Basin Platform, which penetrated 4,400 m of layered gabbro below Permian sedimentary rocks (Hoover and others, 1985). Thick (9–10 km) sedimentary rocks in the Delaware Basin overlying sialic basement seem to account for a gravity low in that area. Seismic work (Stewart and Pakiser, 1962) has shown that the thickness of the crust in the northern part of the basin is somewhat greater than in the areas directly to the east. However, since most of the seismic work has been directed toward investigating the structure of the relatively shallow sedimentary rocks, there is probably much to be learned concerning crustal structure of this region by deeper seismic investigations similar to those of Consortium for Continental Reflection Profiling (COCORP).

CAMBRO-ORDOVICIAN

Because Early and Middle Cambrian sedimentary rocks are absent from the entire Permian Basin region, the area must have been above sea level for hundreds of millions of years from the creation of the intrusive and metamorphic rocks of the late Proterozoic until the deposition of the Late Cambrian Hickory Sandstone Member of the Riley Formation and its stratigraphic equivalents (Hills and Kottlowski, 1983). Being the deposits of a transgressive sea whose shoreline moved northward and northwestward across an eroded surface, these basal sandstones of the Paleozoic sedimentary sequence are oldest in southerly areas and youngest to the north and west. Their combined thickness ranges from less than a meter in parts of southeastern New Mexico to at least 300 m in areas near the Llano Uplift (Galley, 1958). Greatest thicknesses of sandstone accumulated in topographic depressions in the Precambrian surface; sandstone is absent over prominent highs.

In addition to the sandstones, many of which are decidedly ferruginous, the Late Cambrian formations include sandy and glauconitic limestone members, precursors of the succeeding Lower Ordovician Ellenburger Group. In the more southerly, thicker sections near the Llano Uplift, the uppermost Cambrian and lowermost Ordovician carbonates are essentially gradational from one to the other.

The transgressive character of these first Paleozoic deposits is exemplified by the lateral change in facies northwestward, from limestone in the lower Ordovician beds of the Llano area to

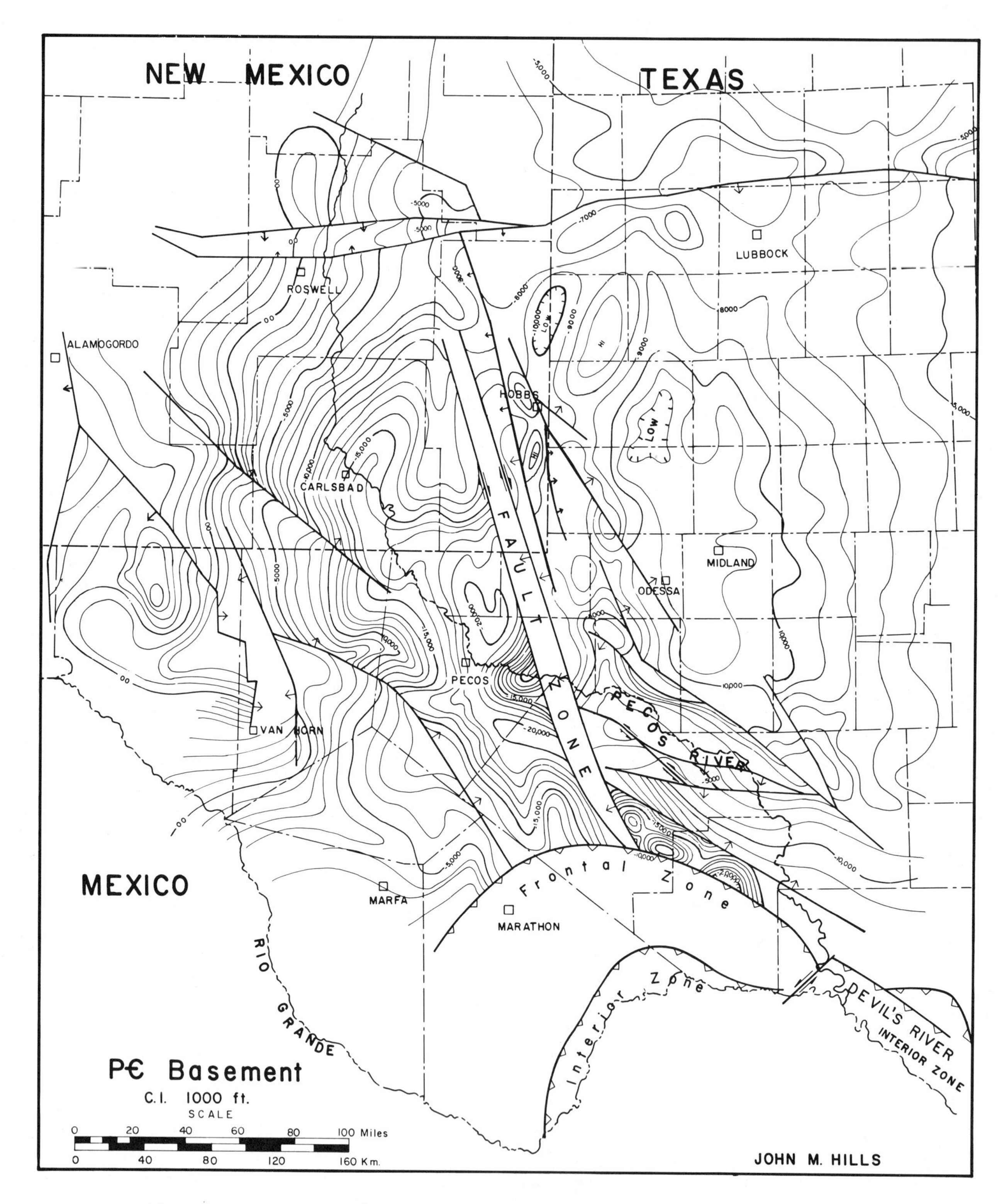

Figure 6. Present surface configuration of the Precambrian basement of the Permian Basin region. Contour interval is 1,000 ft on the top of the Precambrian. Faults are inferred from subsurface data. This surface results from original topography and from tectonic activity during the Phanerozoic.

sandstone in the El Paso area. The Ellenburger Group of limestone formations contains both calcitic and dolomitic members. Deposited in a broad shelf sea extending from the Big Bend and central Texas areas northward, at least to its present pinchout in southeastern New Mexico (Chaves and Roosevelt counties) and the south part of the Texas panhandle (Fig. 7), it attains a thickness of more than 550 m south of the Llano Uplift. At its pinchout the Ellenburger reaches beyond the limits of the overlying Simpson Group, and there is ample evidence (Barnes, 1959) of erosion that removed the upper layers of the Ellenburger in its northern extensions. The correlative Arbuckle Group of dolomite and limestone formations occurs throughout the Anadarko Basin of western Oklahoma and the Texas panhandle, and northward through Kansas and Nebraska. These formations form part of the Sauk sequence of Sloss (1963).

South and east of the Permian Basin the Ellenburger is largely calcitic limestone and contains fossils of Early Ordovician age (Cloud and Barnes, 1948). Northward in the subsurface of the Permian Basin, it has been uniformly dolomitized, and fossils are extremely rare. A few beds of anhydrite are found in central parts of the basin. The dolomite is uniformly mesocrystalline and in many places cherty. Low-grade intergranular porosity is present only locally, but jointing is common, and on anticlinal structures is closely spaced, thus providing adequate porosity for important production of hydrocarbons.

Thick deposits of Ellenburger dolomite in the present Delaware Basin indicate the position of the incipient Tobosa Basin on the shallow Early Ordovician shelf (Figs. 3, 7). Its western edge was the Diablo Uplift, which was already beginning to take shape. Its eastern edge was the ancestral Central Basin Uplift. Farther east the Texas Arch was developing; the depression between it and the Central Basin Uplift was soon to become a part of the Tobosa Basin.

MIDDLE ORDOVICIAN

The restricted distribution of the Simpson Group in the Permian Basin (Fig. 8) represents the fully developed area of the Tobosa Basin in Middle Ordovician time. Accumulation of Simpson strata filled the basin to a depth of more than 600 m as it subsided. Special note is taken of the absence of any thinning in Simpson strata over the position of the ancestral Central Basin Uplift. That feature, which existed in Early Ordovician time and became very active in late Mississippian and later Paleozoic periods, apparently was quiescent during the Middle Ordovician, or it sank at an even greater rate than did the rest of the Tobosa Basin area.

The Simpson Group consists of alternating layers of limestone, sandstone, and dark green shale, the whole overlying a thin basal conglomerate in some places. The regularity of interlayering is remarkably persistent except in its western reaches where a shoreline may have been near. Truncation of uppermost layers as the Texas Arch is approached indicates post-Simpson erosion over that feature, but the group reappears near the Red River and extends across Oklahoma and Kansas without major interruption in the regularity of interbedding. The character of the sandstones, consisting of rounded and frosted quartz grains, persists throughout the midcontinent. The group is absent northwestward in the regions of the Pedernal Massif and the Transcontinental Arch. These beds are the lower part of the Tippecanoe sequence of Sloss (1963).

LATE ORDOVICIAN

The Late Ordovician Montoya Formation overlies the Simpson Group and consists of cherty, finely crystalline carbonates, both dolomitic and calcitic. Detrital grains of Simpson-type quartz sand at the base indicate a possible unconformity between the Montoya and the Simpson, but the duration and degree of erosion are not clear. As in the case of the Simpson, the Montoya in the Permian Basin is restricted to the Tobosa Basin but is present also from the Red River northward, the Texas Arch apparently continuing as a positive element. West of the Diablo Uplift the Montoya is 76 m thick at its type locality north of El Paso, but it is absent over higher parts of the Pedernal Massif and on the Transcontinental Arch. The greatest thickness of Montoya strata is between 150 and 180 m at its depocenter in the Tobosa Basin, equivalent to the southeastern part of the younger Delaware basin. This coincides in position of the depocenter with that of the Simpson Group. Thickening of the Montoya along the east edge of the depocenter is abrupt, suggesting a more rapid rate of subsidence in the depocenter than in the area of the submerged Central Basin Uplift immediately to the east. The presence of a fault zone with downthrow to the west along the belt of thickening has been suggested but not demonstrated. There is no lithologic evidence of a Montoya shoreline at the basin edges, and abrupt thinning there suggests truncation by erosion prior to deposition of the next overlying Silurian rocks.

Montoya carbonates in the area overlying the Central Basin Uplift are predominantly limestone, in conspicuous contrast to predominant dolomite in the area of the depocenter (Galley, 1958), a curious anomaly if the depocenter is assumed to be the area of deepest water. It is to be noted also that the dolomite facies continues to the pinchout of the formation in all directions from the Central Basin Uplift except southward where no pinchout has been observed. An inference that may be drawn is that accumulation of carbonate sediments in the center of the basin proceeded at the same rate as crustal subsidence, thus maintaining shallow-water shelf conditions there.

SILURO-DEVONIAN

The Sylvan gray-green shale, which locally overlies the Montoya in the Tobosa Basin and is generally classified as uppermost Ordovician, appears to be a residual deposit on the Montoya. On this basis it is classified as basal Silurian. The remainder of the section between the Montoya Formation and the Woodford Formation is a sequence of Silurian and Devonian

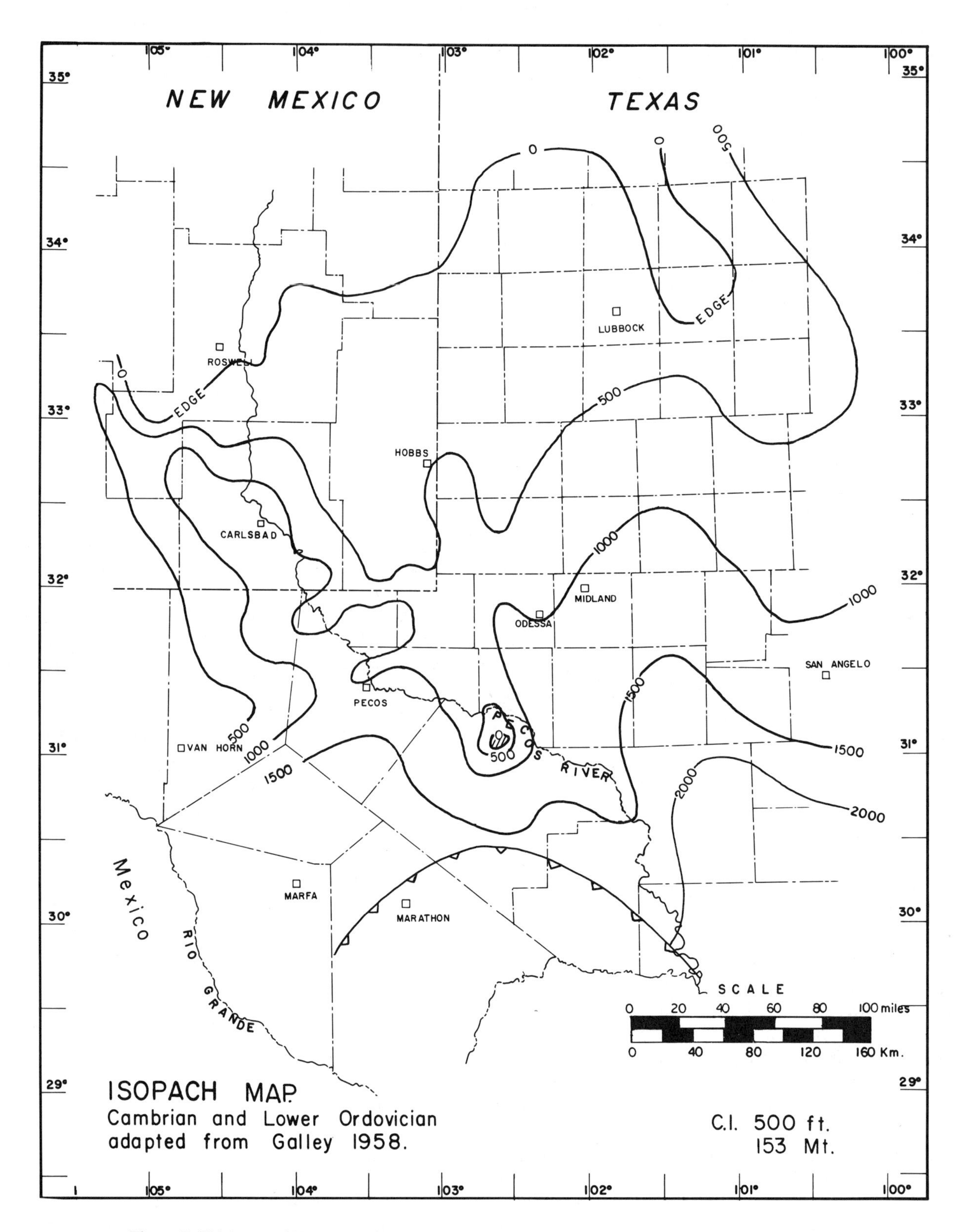

Figure 7. Thickness of Upper Cambrian and Lower Ordovician strata in the Permian Basin; adapted from Galley, 1958. Contour interval is 500 ft.

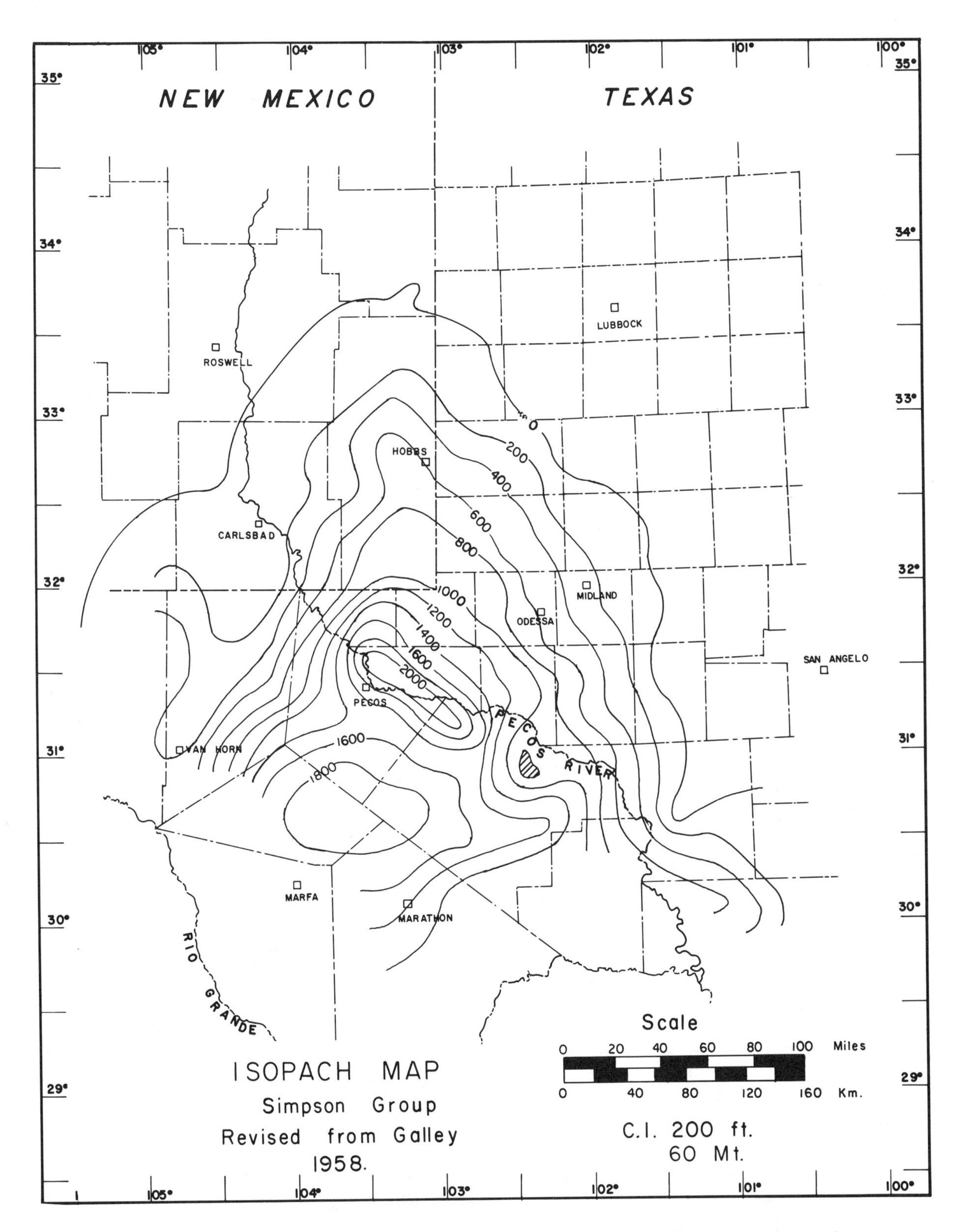

Figure 8. Thickness of the Simpson Group, Middle Ordovician, in the Tobosa Basin. Contour interval is 200 ft (adapted from Galley, 1958).

strata consisting of cherty, medium to coarsely crystalline dolomite and limestone with various amounts of interbedded green shale and inclusions of anhydrite. The proportions of chert relative to carbonates are greatest over the trace of the ancestral Central Basin Uplift, and they increase southward to more than 50 percent (Galley, 1958).

Siluro-Devonian strata are missing on the Bend and Texas arches because of nondeposition or pre-Mississippian erosion, but in Oklahoma they compose the Hunton Group with a thickness of at least 215 m (Shannon, 1962). They are missing also on the Pedernal Massif and the Transcontinental Arch. In the Tobosa Basin the total thickness of the section exceeds 600 m (Fig. 9) and includes the upper part of the Tippecanoe Sequence.

The lowest carbonate beds in the Siluro-Devonian section in the Permian Basin have long been referred to as the Fusselman Formation, which crops out in the El Paso area. It is chiefly dolomitic in the northern and western parts of the Tobosa Basin and calcitic southeastward (McGlasson, 1967). T. S. Jones and E. J. Matchus (1984, personal communication) have pointed out that the dolomite facies is associated with a carbonate shelf of Middle and Upper Silurian age, its edge extending from southern Lea County, New Mexico, to northern and eastern Andrews County, Texas; the shelf facies north of this edge is dolomitic. The subsurface Fusselman is probably equivalent stratigraphically to the lower part of the Fusselman type section in the Franklin Mountains outcrop north of El Paso, and to the Chimneyhill Formation of Oklahoma. It may also be correlative with the middle Caballos Novaculite of the Marathon area, Texas. However, scarcity of fossils makes the correlation uncertain.

Late Silurian was a time of continued carbonate deposition on the shelf and deposition of gray, green, and red shales of the Wristen Formation (Hills and Hoenig, 1979) in the basin to the south. Bioherms originating in the upper part of the Fusselman along the edge of the shelf continued to grow upward. Undifferentiated Silurian limestone beds extend into southern Roosevelt and Chaves Counties, New Mexico (Cys, 1977), and Cochran County, Texas (McGlasson, 1967).

The Lower and Middle Devonian section of the Tobosa Basin comprises limestone strata of the Thirty-one Formation (Hills and Hoenig, 1979), which are cherty in the southwest and central part of the basin. The formation is more siliceous to the south and probably is equivalent to the middle chert of the Caballos Novaculite. It thins westward from 300 m at the type locality in Crane County to less than 30 m in the deeper parts of the Delaware Basin. Middle Devonian beds were eroded and partially truncated in Late Devonian time.

The distribution of dolomitic versus calcitic limestones in the Siluro-Devonian sections is related not to time intervals but rather to lateral facies changes that must have been the result largely of different water depths. As in the Montoya, the area of greatest thickness of Siluro-Devonian carbonates has the highest proportions of dolomite, which suggests maintenance of shallow-water conditions while the basin floor subsided. From the base of the Montoya Formation to the base of the Woodford Shale, the sequence of carbonate strata appears to have been deposited without interruption by any major structural events. Disconformities and minor angularities between unconformable surfaces indicate hiatuses caused by gentle oscillations of sea level. The sequence is essentially a carbonate unit of Late Ordovician through Middle Devonian age, within which formation boundaries in many places are difficult to locate, particularly in the Siluro-Devonian rocks. Paleofaunas in the pre-Mississippian formations are largely mollusks and brachiopods, which presumably favored warm marine waters. The climates of those periods therefore may have been mild, and the seas offshore of the Pedernal landmass must have been rather shallow, conditions that fostered an abundance of marine organisms; thus widespread dolomitization must be blamed for the scarcity of fossils in many areas.

LATE DEVONIAN AND EARLY MISSISSIPPIAN

In Late Devonian time a distinct change came over the sedimentary environments in the Tobosa Basin as well as in much of the continent from the southwestern to eastern United States. Carbonate deposition was greatly diminished, and dark shales unconformably covered the earlier carbonates. The Woodford Formation of the Tobosa Basin, late Devonian (Ellison, 1946, 1950), Early Mississippian (Hollingsworth and Williams, 1951), or perhaps both (Galley, 1958), is highly organic, fine-grained, dark brown to black and very fossiliferous, especially rich in spores and other microfossils including conodonts. In the middle and lower divisions, chert beds are found.

Because the marine invasion that initiated Woodford deposition transgressed a surface composed mostly of carbonate rocks, the basal detrital zone consists only of local patches of thin sandstone associated with light gray shales. All zero edges to the west, north, and east are depositional pinchouts and are marked by nearshore facies of siltstone and fine-grained sandstone. The thickness of the Woodford Formation is less than 60 m over most of the Tobosa Basin, greater thicknesses being limited to a central area around the southeast corner of New Mexico, where maxima over 200 m have been logged (Fig. 10). Strata of the same age and character were deposited in regions as far away to the west as southern New Mexico and southeast Arizona, and to the east from Oklahoma and Kansas continuously to the Appalachian Foldbelt in Pennsylvania (Cook and Bally, 1975). These shales form the basal unit of the Kaskaskia sequence of Sloss (1963).

The origin of the shales of the Woodford Formation, presumably unusual because of the widespread distribution of a formation having such high organic content, has been a subject of speculation and research for many decades. The formation is similar to the correlative Chattanooga Shale. After detailed study of the Chattanooga Shale in the eastern United States, Conant and Swanson (1961) concluded that the most convincing evidence indicates that the formation was deposited in shallow waters, at depths measurable in a few tens rather than hundreds of meters, by a sea transgressing slowly across a flat, low-gradient,

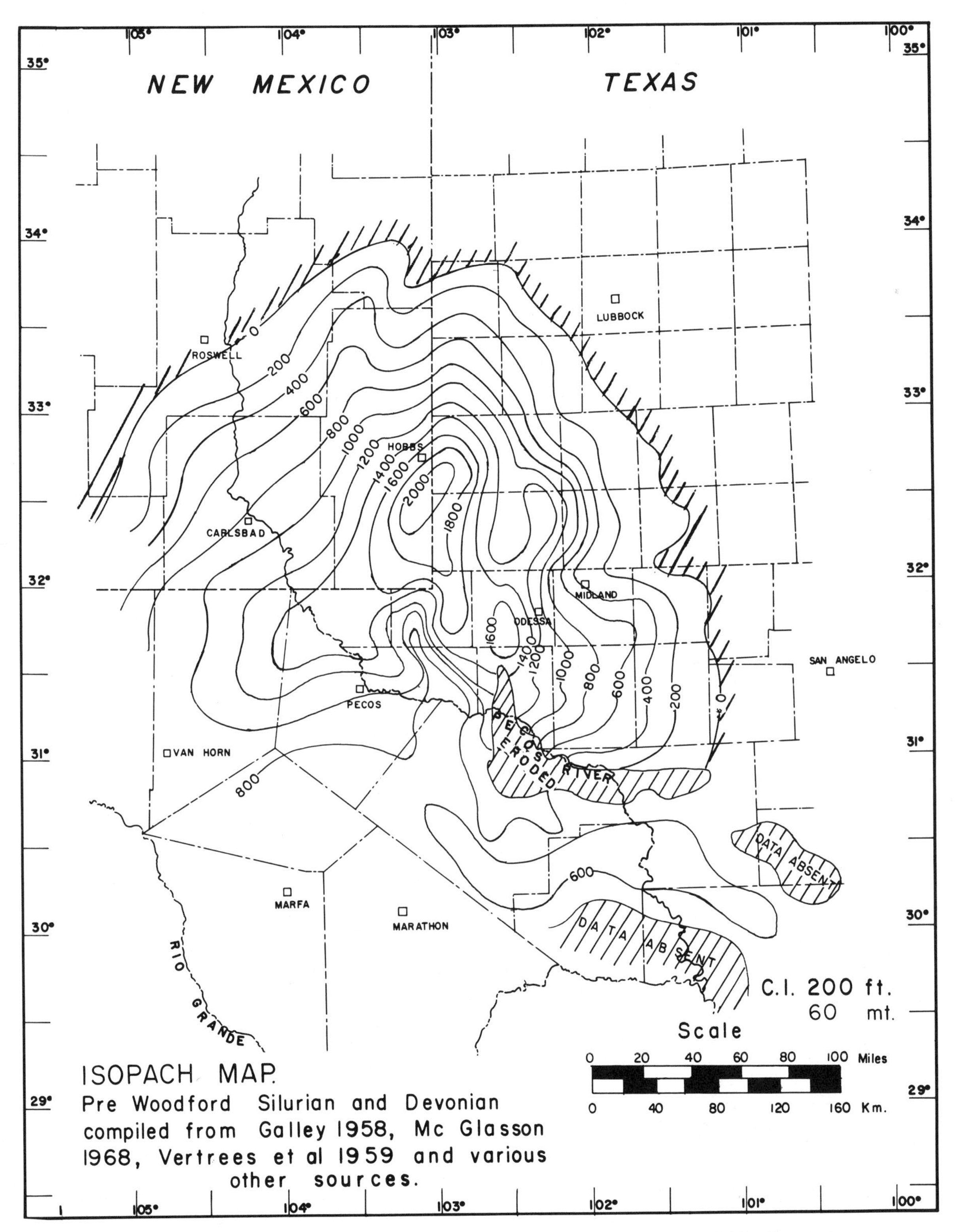

Figure 9. Thickness of pre-Woodford Silurian and Devonian strata in the Tobosa Basin. Contour interval is 200 ft (compiled from Galley, 1958; McGlasson, 1968; Vertrees and others, 1959; and various other sources).

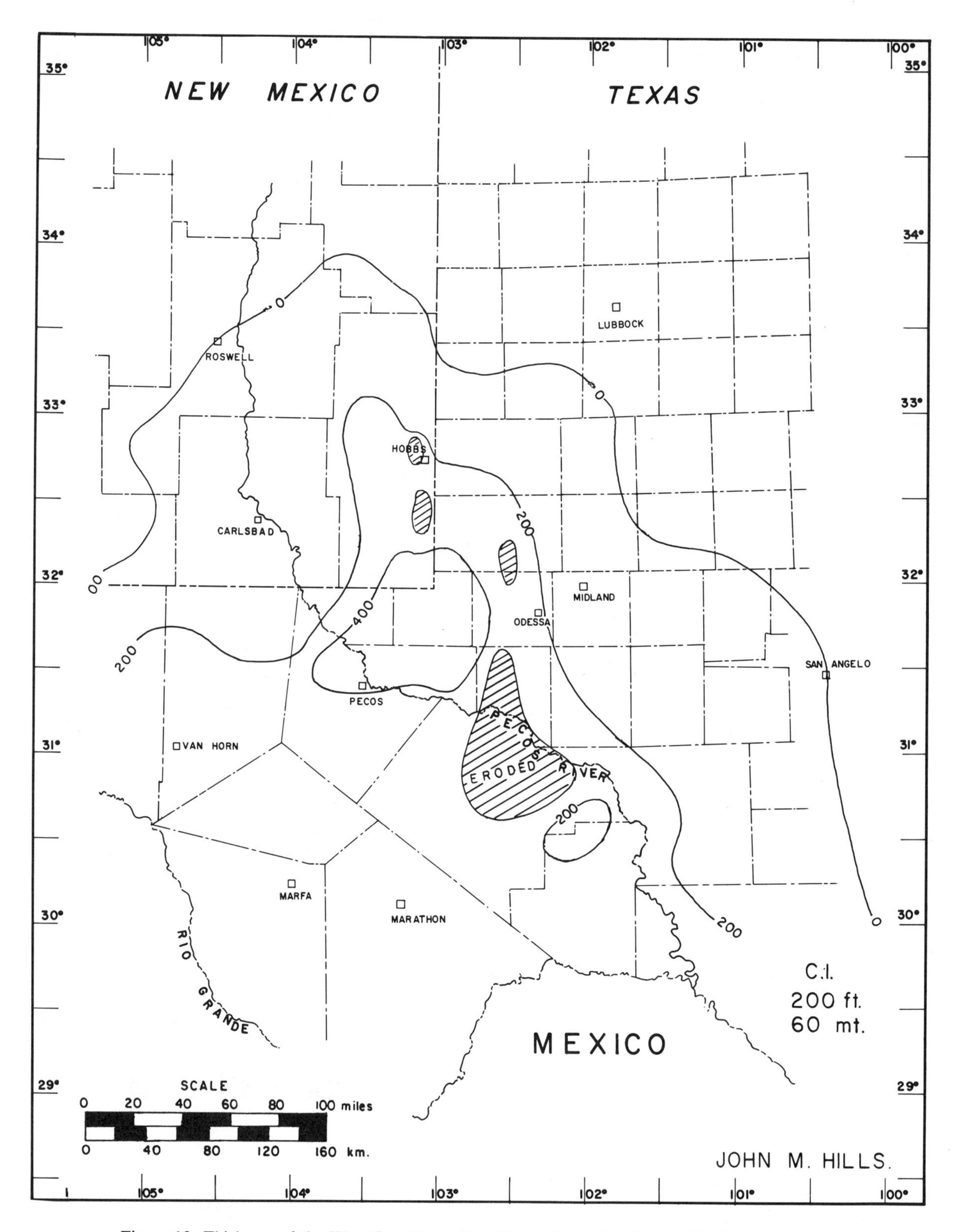

Figure 10. Thickness of the Woodford Formation (Upper Devonian, Lower Mississippian) in the Tobosa Basin. Contour interval is 200 ft. The zero edge is the depositional shoreline except in areas of pre-Kinderhook or later Mississippian erosion in Otero, southernmost Chaves, and Roosevelt Counties, New Mexico, and adjacent counties in Texas. (Compiled from Galley, 1958, and Wright, 1979).

eroded surface of limestone formations. The very fine-grained sediments were transported in suspension by marine currents, and some may have been airborne. Thin interbeds of sandstone are considered to have been derived locally. Abundant fossil plants in the shale are largely species of floating marine algal plants. These may have formed dense mats similar to those in the Sargasso sea of today. Abundant algal mat growths, lack of evidence of violent agitation of the water, and shallow depths, all lead to a presumption of stagnant bottom conditions in which scavengers were absent, and only anaerobic bacteria could survive.

Such an environment would explain the origin of the shales of the Woodford Formation in the Tobosa Basin. The organic content of the shale may have played an important part in the generation of hydrocarbons in the Permian Basin. The anaerobic conditions may have reduced and deposited radioactive compounds in the shale. These have caused the Woodford to become a prominent marker on gamma-ray well logs.

The pinchouts of the Woodford Formation designate the ultimate west, north, and east limits of the Tobosa Basin. When younger Mississippian sediments covered the Woodford Formation, the Tobosa Basin was completely filled and ceased to exist as a separate depositional basin. Since that time it has been a relict structure, although its influence reached into the later Paleozoic history of the Permian Basin.

MISSISSIPPIAN

The change in tectonic and paleogeographic environments near the end of the Devonian Period was apparently the result of only a mild epeirogeny. Early Mississippian sedimentation, like that of the Early Ordovician, was characterized by transgressive deposits of clastics in the south grading eventually into thin carbonates farther north, and finally thicker carbonate deposits covering most of the Permian Basin region. The green-gray Kinderhook Shale was the first formation to be deposited over the Woodford. It spread westward partly up the slope of the Diablo Uplift and eastward where it covered the eroded surface of Ellenburger dolomite in southern sectors of the Texas Arch, and beyond that into Oklahoma.

Osage and Meramec carbonate strata are not dolomitic like most of the pre-Mississippian carbonates, but instead consist generally of finely crystalline, nonporous limestones ranging from pure calcitic limestone to argillaceous and silty, or to siliceous and cherty, limestone in various places. Bioherms developed here and there; some are exposed in outcrops in southern New Mexico (Laudon and Bowsher, 1949), others have been recognized in the subsurface of the Permian Basin and the Bend Arch. Middle Mississippian limestone strata are interbedded to some extent with beds of dark gray and brown shale. Upper Mississippian Chester formations are largely shale, which some geologists believe reach thicknesses of almost a thousand meters in the Delaware Basin (Vertrees and others, 1964). Here they form the lower part of the Absaroka sequence of Sloss (1963). Northward the Chester shales grade to thin-bedded limestones. In northern parts of the Permian Basin the entire Mississippian System is less than 30 m thick (Fig. 11).

Upper Mississippian strata were deposited not only across the relict Tobosa Basin but also over the entire Texas Arch, where they lie on Ellenburger dolomite and even, in its northern extent, directly on Precambrian rocks. Before the deposition of Pennsylvanian strata, however, the Mississippian rocks were attacked by erosion, and today's thicknesses are the result of both onlap at the base of the system and truncation at the top. In places over structural highs, such as those in the Matador alignment (Fig. 11), the entire Mississippian System is absent. Rocks of Chester age are not present on the Pedernal Massif, and the patchy distribution of pre-Chester Mississippian rocks (Cook and Bally, 1975) suggests the action of erosion rather than of deposition in isolated depressions.

The Mississippian Period was transitional between two very different tectonic styles in the evolution of the Permian Basin. From Cambrian time until the Mississippian limestones had been deposited, the Permian Basin region was a relatively stable part of the midcontinent craton. Tectonism was epeirogenic; no great mountains or bathyal deeps were created, and earthquakes must have been few and relatively mild. In Late Mississippian time the earth's crust became restless. Local structures that had remained more or less embryonic began to develop and to assert themselves vigorously. Sedimentary regimens were altered; widespread masses of thick limestone were succeeded by dominantly shale sections, and later by coarser-grained and thicker clastic deposits. Increasingly more violent tectonism became the style and pervaded the geologic environments of the Permian Basin until Middle Permian times. The strata representing this interval form the Absaroka sequence of Sloss (1963).

Structural uplifts bordering and within the Permian Basin, whose prominence began to develop in Late Mississippian time (Galley, 1958) and grew in intensity in Early and Middle Pennsylvanian time, include (1) the Matador Uplift or alignment of highs, where strata of Pennsylvanian age now lie on Precambrian rocks; (2) the Pecos Arch, a relatively small uplift extending eastward from the southern end of the Central Basin Uplift, where Pennsylvanian rocks lie on Cambrian and Ordovician strata; (3) the Central Basin Uplift, where the Permian caprock of the Central Basin Platform lies on Precambrian and folded lower Paleozoic rocks of the eroded uplift itself; and (4) the Diablo Uplift, where Permian beds lie on Precambrian and various Paleozoic rocks. The greatest of all the tectonic episodes that affected the evolution of the Permian Basin was the Marathon-Ouachita Orogeny, whose initial phases occurred during Late Mississippian times, resulting in the accumulation of large thicknesses of flysch deposits (Tesnus Formation) in the Val Verde Basin. All of these Late Mississippian and Pennsylvanian events, and especially the increasingly violent Marathon-Ouachita Orogeny, set the stage for the sedimentary environments that made the Permian Basin the impressive geologic and economic province that it is today.

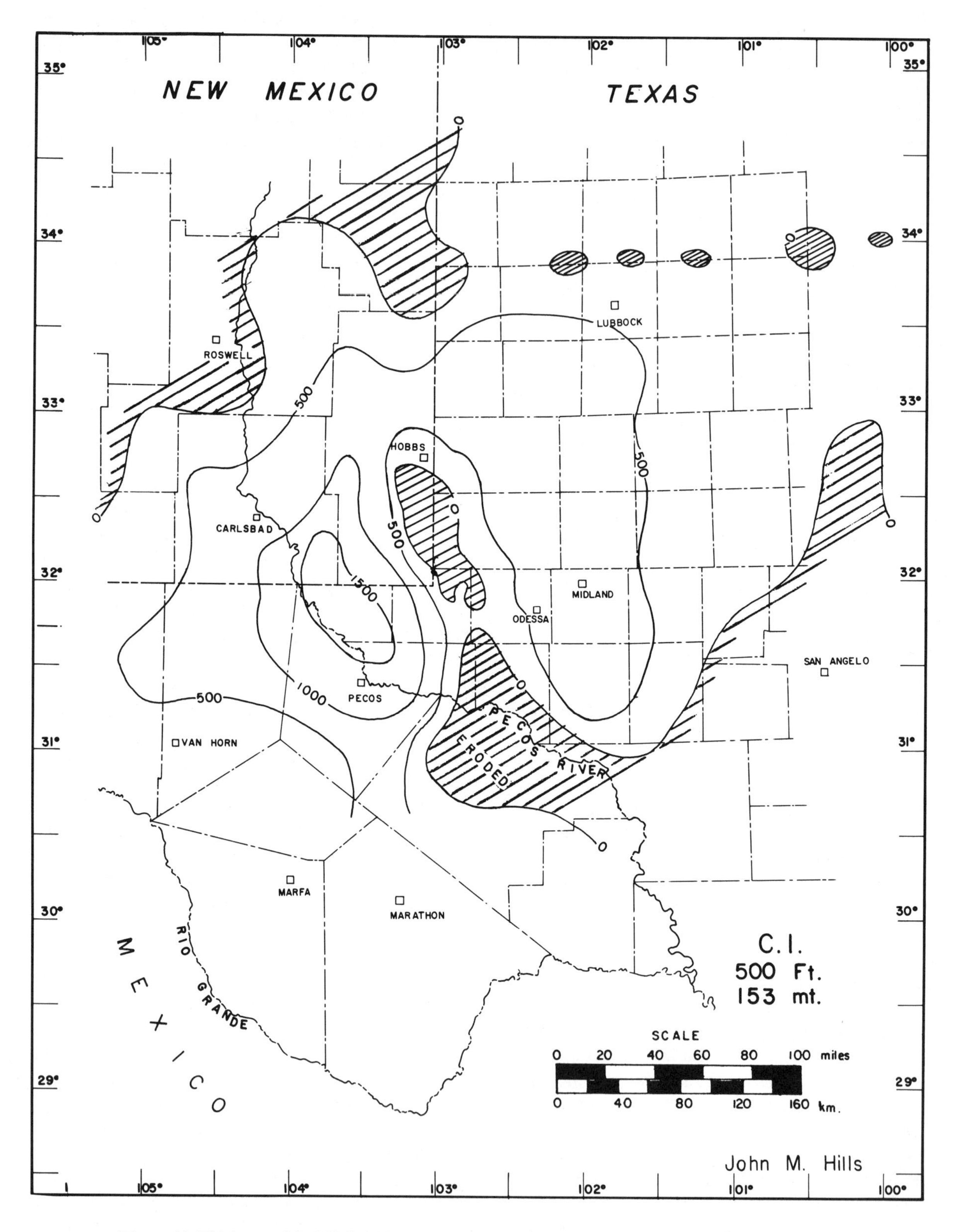

Figure 11. Thickness of the Mississippian System in the Permian Basin. Contour interval is 500 ft. The zero line is the pre-Atoka eroded edge except on the Central Basin Uplift, where it is post-Pennsylvanian. (Compiled from Galley, 1958, and Wright, 1979.)

PENNSYLVANIAN AND PERMIAN GEOLOGY OF THE PERMIAN BASIN REGION

J. M. Cys and W. R. Gibson

PENNSYLVANIAN

The Pennsylvanian rocks of the Central Basin Platform, Diablo Platform, Northwest Shelf, Midland, Delaware, and Val Verde Basins are extremely variable in distribution, thickness, and lithology. The variability results from a continuation of Late Mississippian tectonism into the Pennsylvanian, which caused nondeposition and erosion over positive elements such as the Central Basin Uplift and the Diablo Uplift (Fig. 5). Pennsylvanian strata consist of shales, carbonates, and sandstones deposited in a broad spectrum of depositional environments including sediment-starved basins (Adams and others, 1951), shallow-marine carbonate shelves, fluvial coastal plains, and deltas. Figure 12 depicts the paleogeography during late Pennsylvanian time.

Early Pennsylvanian time marked the breakup of the Tobosa Basin into the present tectonic features of the Permian Basin. The Central Basin Uplift started to form, and the adjacent Midland and Delaware Basins began to subside. The active Diablo Uplift bounded the Delaware Basin on the west and separated it from the Marfa and Orogrande Basins. As the Midland Basin subsided, the Bend Arch structural element separated the Midland Basin to the west from the Fort Worth Basin to the east. To the north, the Matador–Red River Arch arose, separating the Palo Duro Basin from the central part of the Permian Basin. The greatest intensity of tectonism was in the early Pennsylvanian; however, it continued sporadically for the remainder of the Pennsylvanian and into the Early Permian.

Associated with these regional structures, faulting occurred and many of the oil- and gas-producing anticlines of the Permian Basin area were formed. These anticlines are normally asymmetrical, and most of the large ones are faulted. The heights of these structures range from more than 600 m to less than 30 m. The most common structural trend is north northwest–south southeast; however, there are also east–west-trending structures, especially in the Delaware Basin (Fig. 6).

Because of rapid and complex lateral and vertical facies changes, the Pennsylvanian rocks have not been divided into rock-stratigraphic units. Instead, they have been divided into five series based on fusulinid occurrences. From oldest to youngest, these are the Morrow, Atoka, Strawn, Canyon, and Cisco. Lithologic wireline log markers near the fusulinid-defined series boundaries are used for correlation purposes. In some instances, informal lithologic units have been identified in local areas. There is little consensus on lithologic criteria to identify the series boundaries.

Northwest Shelf

The Pennsylvanian System has a total thickness of 840 m to 915 m (Meyer, 1966) on the Northwest Shelf. Lithologically, it can be divided into two sedimentary packages. The Morrow and the Atoka Series form an interbedded sand and shale sequence with a few thin limestones totaling a maximum of 460 m (Meyer, 1966). The dominantly clastic Morrow and Atoka rocks were deposited on a broad shelf; the sediment source was the Pedernal Uplift to the northwest (Meyer, 1966). The sandstones, which form major gas reservoirs, were deposited as alluvial point bars in river channels as well as deltaic sandstone bodies and offshore bars. At the shelf-edge of the developing Delaware Basin, cyclic carbonates are dominant. *Donezella* (algal) bioherms and associated grainstone shoals form hydrocarbon reservoirs along the shelf edge (Mazzullo, 1981).

The Strawn, Canyon, and Cisco rocks are a cyclic sequence of carbonates and shales with a maximum thickness of 460 m (Meyer, 1966). Bioherms and grainstone shoals, both on the shelf and at the shelf edge, form hydrocarbon reservoirs. Most of the oil-producing anticlines and faults on the Northwest Shelf are the result of Pennsylvanian tectonic activity.

Diablo Platform

Throughout most of Pennsylvanian time, the area of the Diablo Platform (Fig. 3) was a stable carbonate shelf. In late Pennsylvanian time, however, local sporadic uplifts (Fig. 12) began in the northern area of the platform, as evidenced by limestone conglomerates in the upper Magdalena Group in the central and northern Hueco Mountains.

In the central part of the Sierra Diablo Mountains, King (1965) described the poorly exposed Pennsylvanian sequence along the basal part of the eastern escarpment as fossiliferous limestone interbedded with some shale. Maximum exposed thickness is approximately 52 m. The sequence dips gently southward and is truncated at an angular unconformity by the overlying Permian Hueco Formation. Fusulinids and cephalopods date the sequence as Atokan and Strawn in age.

In the Hueco Mountains, along the western flank of the Diablo Uplift, 400 m of a continuous sequence of Pennsylvanian rocks is assigned to the Magdalena Group and divided into three informal lithologic units (King and others, 1945). The lower division consists of 150 m of massive, light-colored, cherty limestones containing small organic buildups dominated by *Chaetetes.* The middle division consists of 90 m of late Atokan and early Strawn marl, marly limestone, and shale. The upper division, of late Desmoinesian to early Wolfcampian (Permian) age, is of variable thickness because it is truncated by the Permian Hueco Group in the southern and south-central part of the range; the contact is apparently conformable in the north-central and northern parts of the mountains. In the southern and south-central parts of the mountains, the Magdalena Group consists of fossiliferous, commonly cherty limestones with interbedded marls and scattered limestone and chert-pebble conglomerates. In the south-central area, several thick phylloid algal bioherms occur near the top of the upper division. In the northern and north-central Hueco Mountains, the Magdalena Group is an alternating

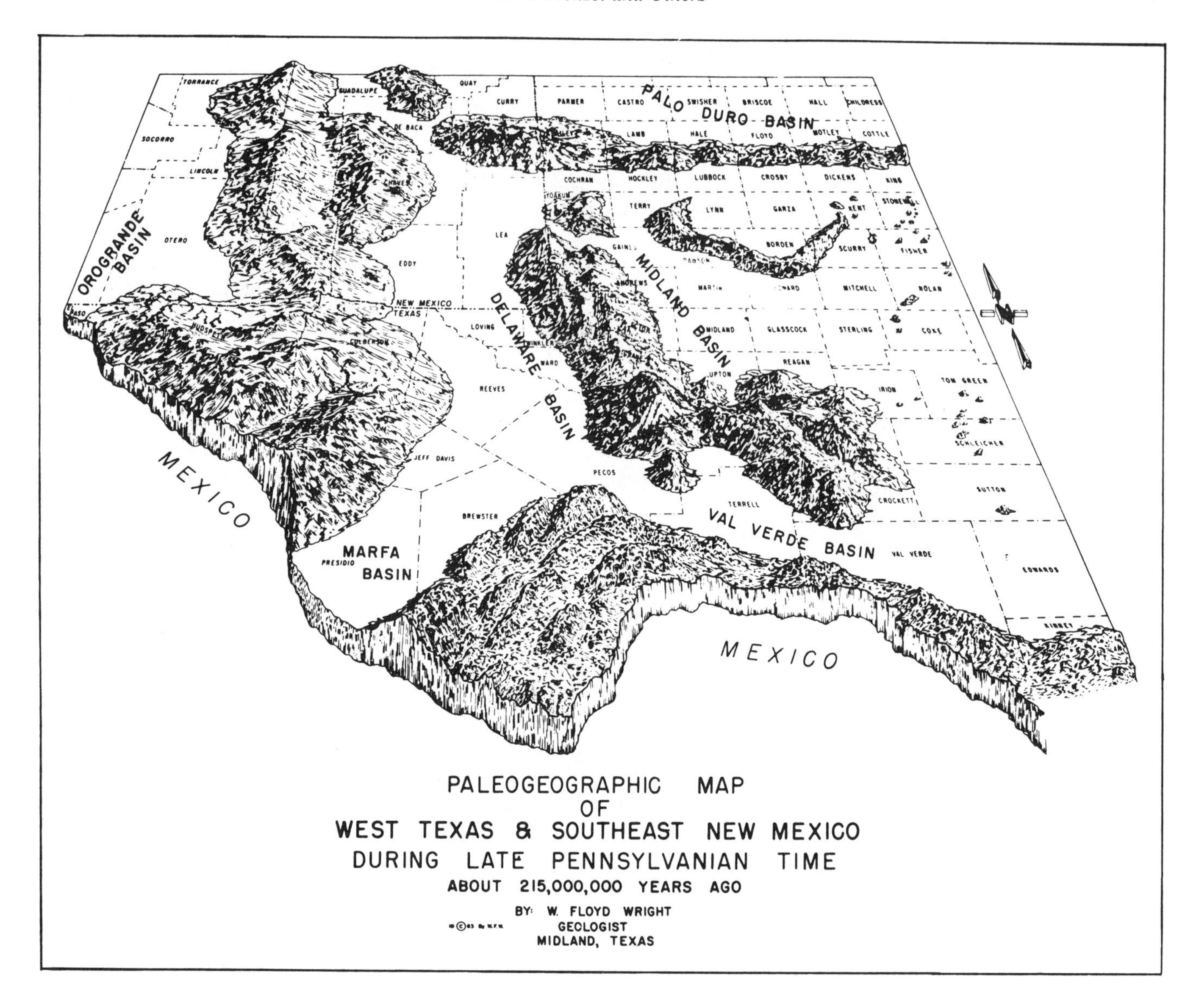

Figure 12. Late Pennsylvanian paleogeographic map of the Permian Basin area (from Wright, 1979).

sequence of chert and limestone pebble conglomerates, medium-bedded to massive limestones and gray to red shales 300 to 360 m thick (Hardie, 1958; Cys, 1983). Williams (1963) identified Wolfcampian (early Permian) fusulinids from the uppermost part of the Magdalena Group in the north central part of the range.

The Diablo Platform was uplifted in late Pennsylvanian and early Permian time, as evidenced by post-early Wolfcampian rocks unconformably juxtaposed on Precambrian rocks.

Central Basin Uplift

The subsurface Central Basin Platform of Permian time was an active high-relief, north-south–oriented, positive element during much of Pennsylvanian time (Fig. 12). During Pennsylvanian time it is properly termed an uplift. It did not become a platform until Permian time. As a result, Pennsylvanian sediment distribution in the vicinity of the uplift is irregular. Atokan and later Pennsylvanian rocks are confined to its flanks as a sequence of interbedded limestones, red, green, and gray shales and scattered conglomerates (Jones, 1953) of highly variable thickness. Limestones are oil and gas productive from structural traps and, in some instances, as stratigraphic traps formed by porous phylloid algal mounds. A typical seismic expression of the Central Basin Uplift and Platform is shown in Fig. 13.

Delaware Basin

The Pennsylvanian System in the subsurface Delaware Basin is generally less than 460 m thick (Vertrees and others,

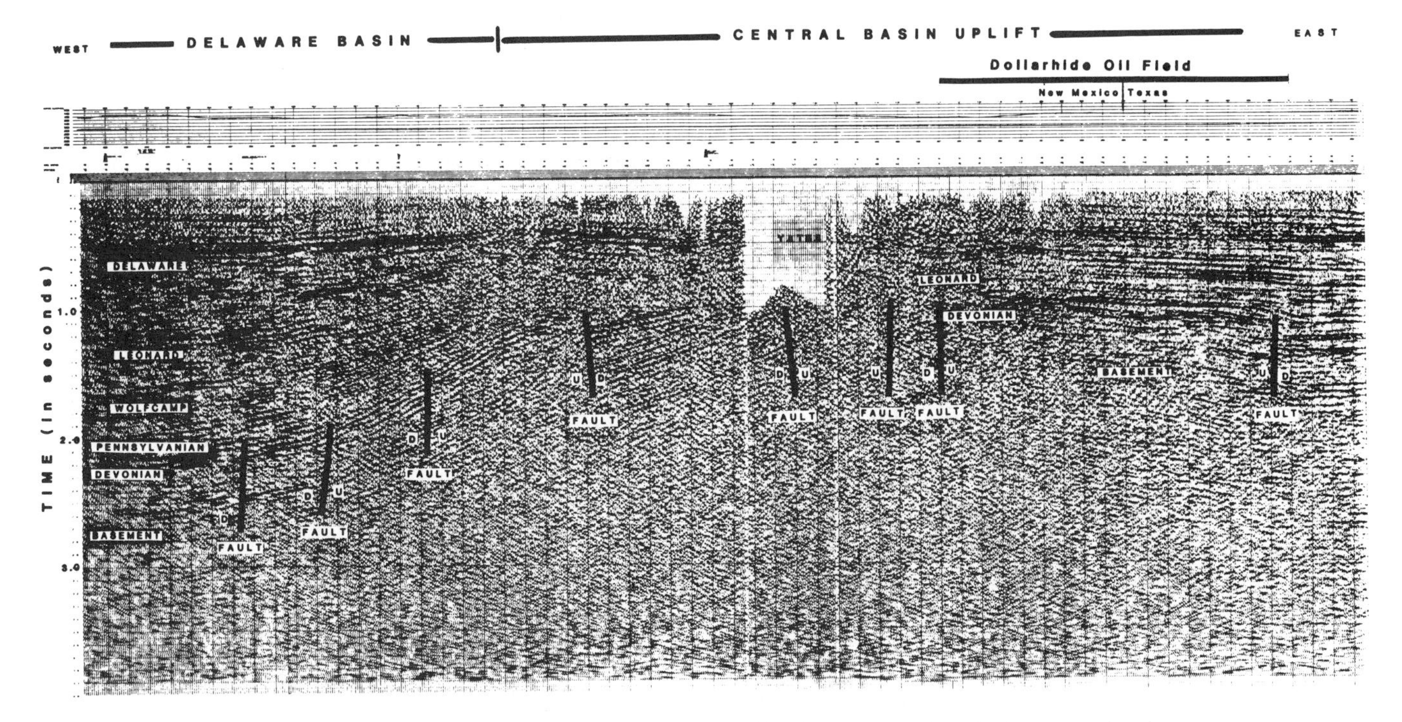

Figure 13. East-west seismic section across the western side of the Central Basin Platform showing the Dollarhide Field area. (Seismic data courtesy of GeoSearch, Inc., Midland, Texas.)

1964). The Pennsylvanian-Mississippian contact in the deep part of the Delaware Basin is typically below 4,250 m; both the upper and lower contacts of the Pennsylvanian System are poorly defined due to the lack of diagnostic fossils and the lack of lithologic contrast within the basinal dark gray shales.

The maximum water depths and rates of subsidence occurred in the eastern part of the basin in a north–south-trend parallel and adjacent to the Central Basin Uplift. The thickest accumulations—predominantly fine-grained terrigenous clastics, especially gray-black shales with some thin interbedded siltstones and sandstones—occurred in this deep basin area of Loving, Ward, and Pecos counties, Texas.

To the west and the northwest, the clastics change facies to shelf carbonates. Gentle regional dip of the Pennsylvanian-Mississippian contact is generally southeast or east. The basinal sandstones are gas-productive in some areas, typically from overpressured reservoirs.

Midland Basin

The subsurface Midland Basin was relatively starved of terrigenous clastic sediments during Pennsylvanian time (Adams and others, 1951). Pennsylvanian rocks are generally less than 300 m thick near the eastern and western shelf margins and less than 150 m thick in the central part of the basin (Davis, 1953; Jones and Matchus, 1984). The strata are predominantly gray to black marine shales with some sands, silts, and carbonates near the basin margins. Although there are several unconformities within the Pennsylvanian strata on and near the shelves, the Midland Basin was an area of tectonic stability during this period. Like the Delaware Basin, the Midland Basin is asymmetric. The Pennsylvanian-Mississippian contact dips gently westward toward the deepest part of the Midland Basin, located in Midland, Martin, and Dawson Counties, Texas. The typical depth to the base of the Pennsylvanian is approximately 3,000 m to 3,500 m below ground level near the basin axis.

The basin is bounded by carbonate shelf areas to the east, south, and west and by the carbonate buildups of the Matador Arch to the north. This east–west structural trend was a Pennsylvanian archipelago, which divided the oil-productive Midland Basin to the south from the relatively unproductive Palo Duro Basin to the north. Pennsylvanian oil is trapped stratigraphically in clastic rocks near shelf margins or hinge lines and also in reef-like carbonate buildups within the Midland Basin.

The most significant Pennsylvanian oil accumulations are in the carbonates of the Horseshoe Atoll, in the northern part of the Midland Basin (Fig. 14). The atoll is formed of a series of Strawn, Canyon, Cisco and, in some places, Wolfcamp biohermal carbonate buildups and their respective talus deposits, as much as 915 m thick (Vest, 1970). These carbonate rocks formed a roughly semicircular atoll on the southern part of the oval lower Strawn Platform.

The Strawn (Desmoinesian) Platform that underlies the biohermal deposts is composed of biostromal carbonate deposits

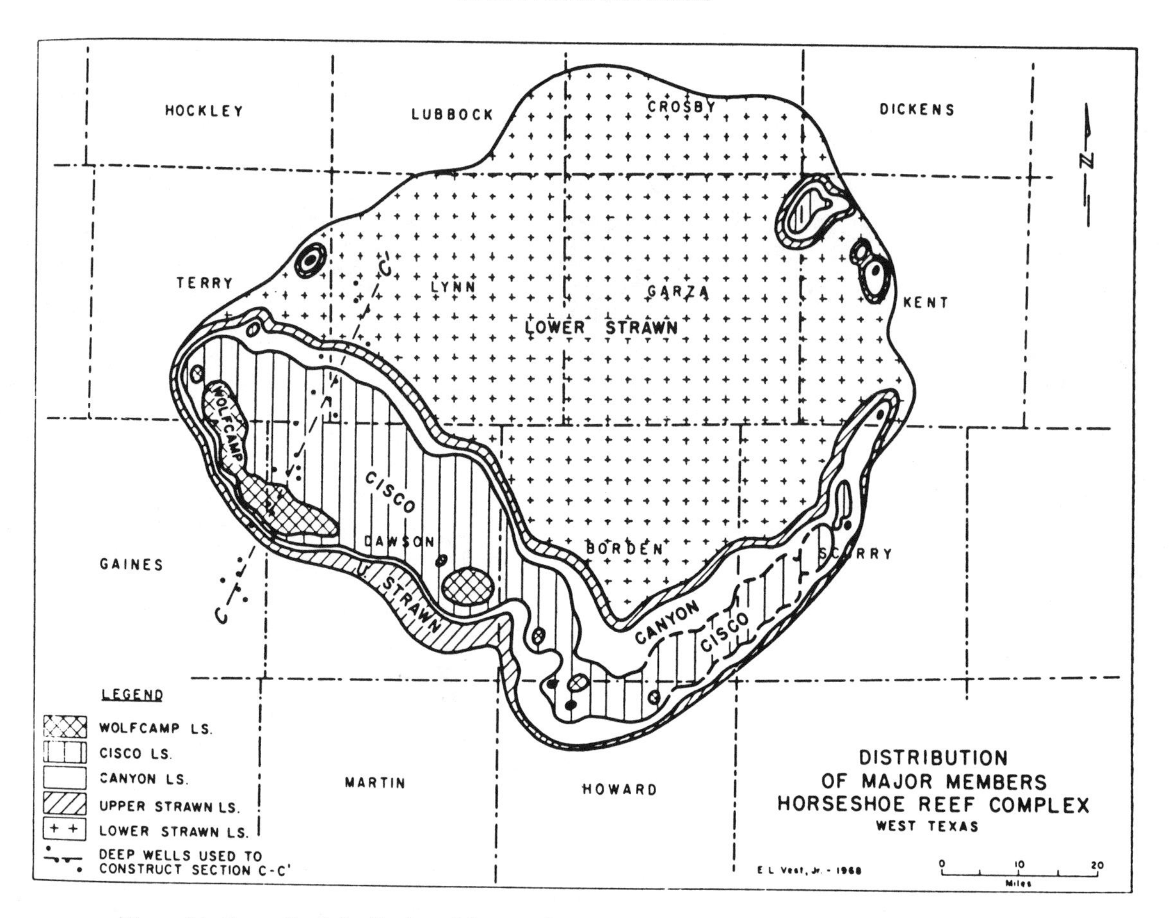

Figure 14. Generalized distribution of Strawn, Canyon, Cisco (Pennsylvanian), and Wolfcamp (Lower Permian) Formations in the Horseshoe Reef complex and location of cross section C-C′ (from Vest, 1970).

up to 230 m thick. It is approximately 145 km east–west by 110 km north–south. The stratigraphic relationship of the lower Strawn biostromes to the overlying Permo-Pennsylvanian carbonate buildup is shown in Figure 15.

Oil production from the Scurry "Reef" Field in the southeastern part of the Horseshoe Atoll has been especially significant. Approximately one third of all of the oil produced from the Strawn, Canyon, and Cisco reservoirs in the Permian Basin has been produced from the Scurry "Reef" (Hartman and Woodward, 1969).

Val Verde Basin

By early Pennsylvanian time, the Val Verde Basin had been separated from the Midland Basin to the north by the Central Basin Uplift (Fig. 12). Vertrees and others (1959) considered the Val Verde Basin to be a southeastern extension of the Delaware Basin in Permo-Pennsylvanian time. The Pennsylvanian sediments deposited in the Val Verde Basin are generally less than 150 m thick and are similar to those of the southern Delaware Basin. The Devils River Uplift defines the southern boundary of the basin. By Strawn time, this area was a developing foredeep basin, receiving sediments from the rising Ouachita Orogen to the south.

PERMIAN

Permian rocks contain the vast majority of the hydrocarbon reserves of west Texas, and the San Andres Formation is the most important oil-producing unit of the Permian Basin area. The Permian Capitan Reef complex, exposed in the Guadalupe Mountains, forms the flanks of the Delaware Basin. It is one of the world's best-exposed ancient reefs. With its classic outcrops, it is an excellent natural laboratory for the study of the complex interrelationships of a shelf-to-basin transition across a shelf margin rimmed by a carbonate buildup (King, 1948; Adams and Frenzel, 1950; Newell and others, 1953; Dunham, 1972; Cys and others, 1977).

The Guadalupe Mountains are also the type locality for the North American Guadalupian Series of the Permian (Adams, and others, 1939). Surface exposures of rocks overlying the Guadalupian rocks in the adjacent Delaware Basin comprise the type locality for the Ochoan Series (Adams and others, 1939; Adams,

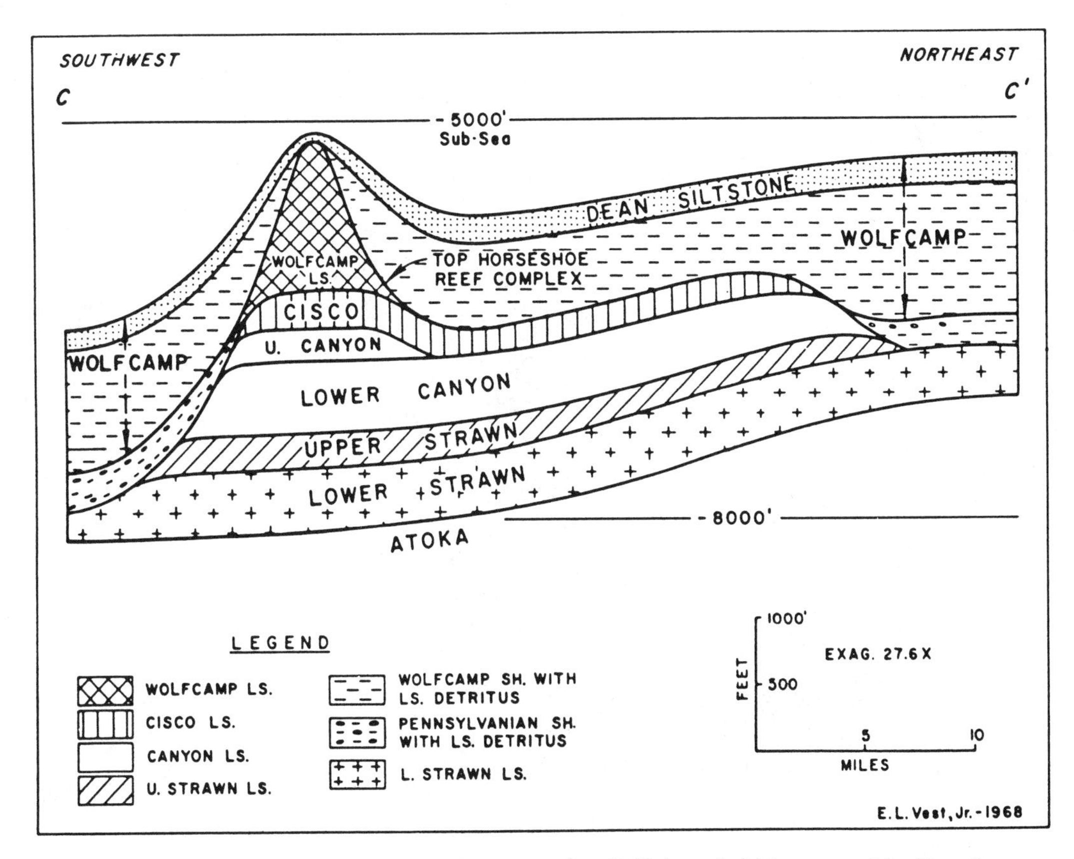

Figure 15. Southwest–northeast schematic cross-section C-C′ through thickest part of the Horseshoe Reef complex (from Vest, 1970).

1944). The type localities of the Wolfcampian and Leonardian Series are located in the Glass Mountains (Adams and others, 1939) on the south margin of the Permian Basin.

The general mode of sediment deposition during the Permian was one of infilling of the Permian Basin region. Figure 16 shows basin paleogeography during late Wolfcampian (early Permian) time. Stratigraphic relationships are shown in Figure 4.

Northwest Shelf

Wolfcampian Time. The Wolfcampian of the Northwest Shelf is a cyclic sequence of limestones and interbedded shales that is assigned to the Hueco Formation. The maximum thickness is 380 m in northwestern Eddy County, New Mexico (Meyer, 1966). Shoaling-upward cycles in the basal Hueco Formation in northern Lea County, New Mexico usually consist of a basal transgressive unit of shale and argillaceous limestones overlain by bioclastic wackestones, phylloid algal wackestones, and bioclastic grainstones (Cys, 1985). Figure 17 shows a typical cycle. The phylloid algal wackestones occur as low-relief bioherms and biostromes (Fig. 18), which are important hydrocarbon reservoirs in the area. The shelf margin is characterized by buildups of tubular foraminifera and the encrusting problematical organism *Tubiphytes.* The buildups are important hydrocarbon reservoirs.

Leonardian Time. In the subsurface on the Northwest Shelf, the lower Leonardian consists of a red-bed sequence assigned to the Abo Formation. It has a conformable and locally interfingering contact with the underlying Hueco Formation. Abo fluvial sandstones in the Chaves County area have recently been drilled for their natural gas accumulations at depths of approximately 915 m to 1,675 m (Gaines and others, 1982).

The evaporite and dolomite content of the Abo increases southward toward the Delaware Basin shelf margin. At the shelf margin, the Abo equivalent Wichita-Albany "formation" is entirely dolomite. (Application of the term "Abo" to this dolomite sequence has been deemed inappropriate by Meyer (1966) and Cys (1975) because the Abo Formation was originally defined as a clastic red-bed sequence and is lithologically distinct from the shelf-edge dolomite.) In New Mexico and in Yoakum and Terry Counties, Texas (Fig. 19), dolomitized organic buildups of the Wichita-Albany "formation" are important hydrocarbon reservoirs. Along the shelf margin of the northern Midland Basin, these buildups are replaced by oolite shoals, which are locally hydrocarbon productive (Mazzullo, 1982).

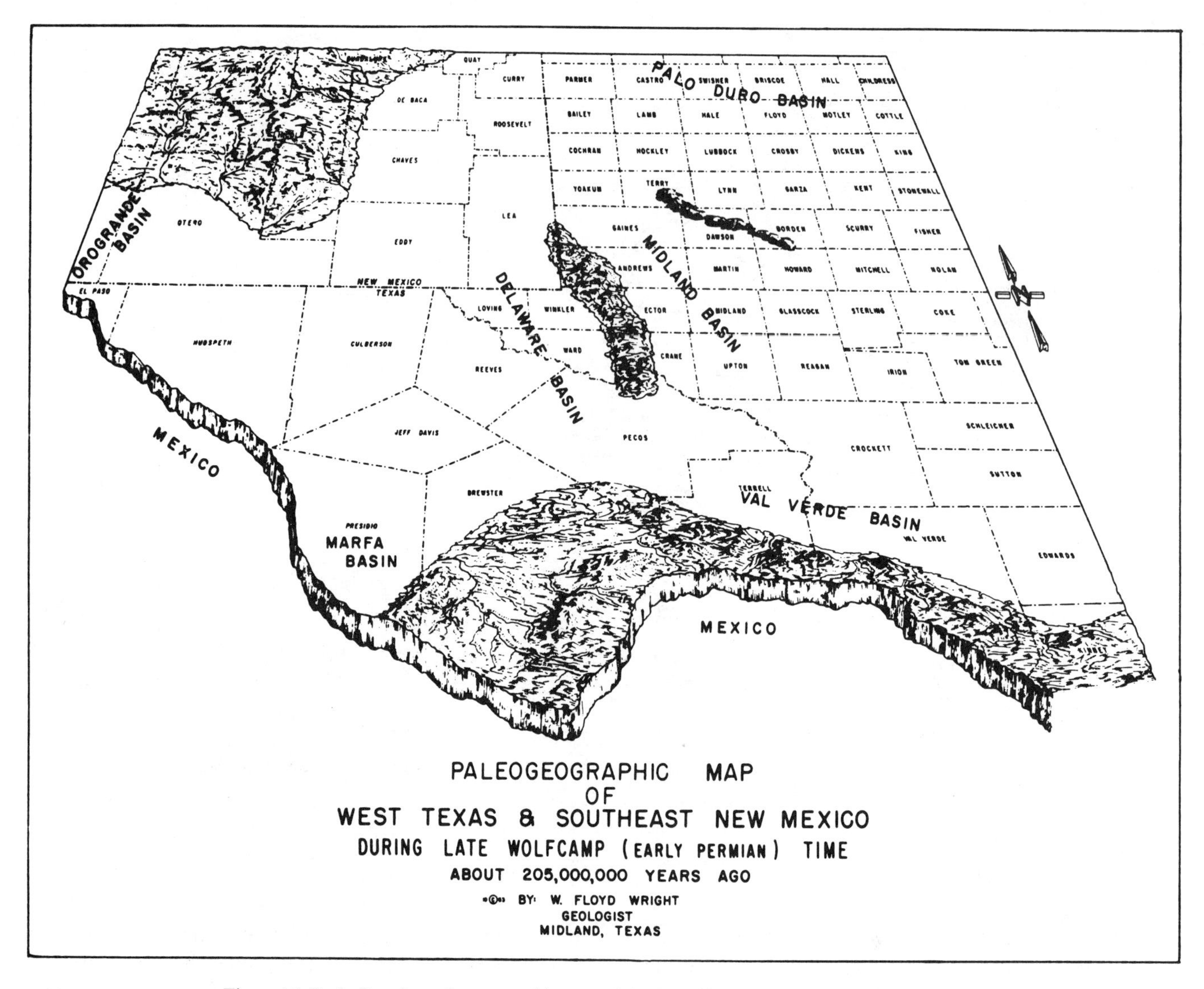

Figure 16. Early Permian paleogeographic map of the Permian Basin area (from Wright, 1979).

Conformably above the Wichita-Albany and Abo Formations are the Yeso Formation in New Mexico and the Clearfork Formation in Texas. The Yeso and Clearfork Formations are complex evaporite and dolomite units with interbedded sandstones in some places. The facies range from sabkha and tidal flat to shelf-edge grainstone barrier bars (Mazzullo, 1982).

The only outcrops of Leonardian strata on the Northwest Shelf occur in the Guadalupe Mountains, along the rugged western scarp. The Victorio Peak Formation consists of 244 m of gray thick-bedded limestone and dolomite. King (1948) and Newell and others (1953) described its lateral transition into the basinal Bone Spring Formation. Newell and others (1953) interpret the Victorio Peak Formation to represent a shelf-edge carbonate bank, but not a true reef.

Guadalupian Time. On the Northwest Shelf, the San Andres Formation is the major hydrocarbon reservoir. The San Andres Formation is 250 m to 425 m thick and consists of cyclic limestones, dolomites, and evaporites deposited on a shallow shelf in environments ranging from open marine to sabkha tidal flat. The hydrocarbons accumulated where porous, dolomitized, open marine and tidal flat limestones pinch out into nonporous anhydritic dolomites, which were deposited in a sabkha environment.

The remaining Guadalupian section is excellently exposed in the Guadalupe Mountains and can be traced into the subsurface of the Northwest Shelf. The middle Guadalupian Goat Seep Limestone crops out in the higher parts of the southern Guadalupe Mountains where it is from 170 to 365 m thick and consists of massive to thick-bedded, occasionally lenticular, gray dolomitic limestone (King, 1948). Basinward, it grades into the

Cherry Canyon Formation; shelfward, to the north, it changes laterally into the Grayburg and Queen Formations, which are thin- to thick-bedded dolomites with some interbedded sandstones.

Conformably overlying the Goat Seep Limestone is the Capitan Formation, which has a maximum thickness of 610 m (King, 1948) and constitutes the world-famous Capitan Reef complex. Lloyd (1929) first recognized the organic reef nature of the Capitan Formation. King (1948) did a masterful job of mapping the geology and describing the basic stratigraphic framework. Expanding on King's (1948) work, Adams and Frenzel (1950), and later Newell and others (1953), refined the stratigraphic framework and presented the barrier reef hypothesis for the Capitan Reef. Dunham (1972), based on comprehensive petrographic work, regarded the Capitan Reef as a downslope mound, with the paleotopographic crest of the Capitan being in the backreef pisolite facies, which he thought formed on subaerially exposed islands as vadose caliche. Most workers now accept the hypothesis that the paleotopographic crest is in the pisolite facies. The exact origin(s) of the pisolites, however, is still debated. Cys and others (1977), Babcock (1974, 1977), and Yurewicz (1976, 1977) have all offered persuasive arguments that the Capitan is a low-relief downslope mound, with its wave-resistant character the result of pervasive submarine cementation. Figure 20 shows a facies mosaic across the Capitan Reef complex.

Capitan deposition began in relatively deep (approximately 100 m) water (Yurewicz, 1976, 1977), which became shallower with time until the crest was in water 30 m deep or less near the end of Capitan deposition (Babcock, 1974, 1977). The primary organic constituents of the Capitan Formation are calcareous sponges, the problematic alga *Archeolithoporella,* and the problematic encrusting organism *Tubiphytes.* The Capitan Reef prograded seaward as it grew upward and outward over its own talus. The talus deposits interfinger with sandstones of the Delaware Mountain Group.

Shelfward, the Capitan grades laterally into the Seven Rivers, Yates, and Tansill Formations, which represent a backreef shelf of dolomites and some interbedded sandstones that change to predominantly evaporites several kilometers from the shelf margin. Average thickness of the Seven Rivers, Yates, and Tansill sequence ranges from 300 m to 490 m.

Ochoan Time. The Ochoan sediments of the Northwest Shelf consist of the Salado, Rustler, and Dewey Lake Formations, which rest unconformably on the underlying Tansill and Capitan Formations. These are evaporitic strata that represent the final stage of Permian Basin filling. They are best represented in the Delaware Basin, but they extend on to the Northwest Shelf, Central Basin Platform, and Midland Basin. These units will be discussed in the Delaware Basin portion of this chapter.

Tectonics. During the Permian, the Northwest Shelf was a stable carbonate platform prograding basinward. In the Guadalupe Mountains, the northeast/southwest-trending Bone Spring "flexure" shows minor syndepositional activity during Leonardian time and coincides with the Victorio Peak shelf edge. The Bone Spring Formation there exhibits numerous complex cut and fill structures as well as diastems.

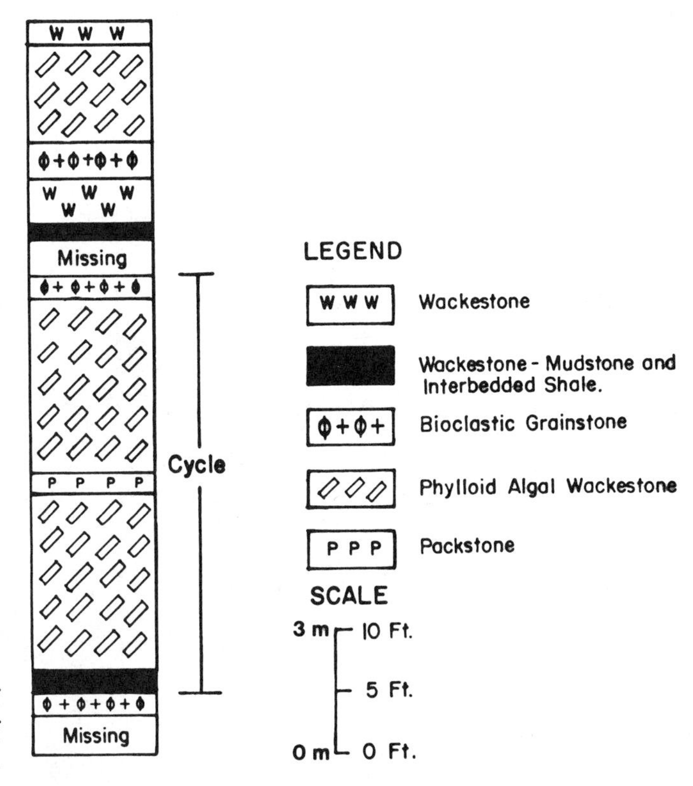

Figure 17. Typical Early Permian shoaling upward carbonate cycle of the Northwest Shelf (after Cys, 1985).

Diablo Platform

Wolfcampian Time. The primary Permian rocks exposed in the Hueco Mountains are limestones and some interbedded shales that are assigned to the Wolfcampian Hueco Group. In ascending order, the Hueco Group consists of the Hueco Canyon Formation, 141 to 195 m of thick-bedded olive-gray limestones and a few interbedded shales; the Cerro Alto Formation, 142 m of gray, thin- to medium-bedded, often marly limestone; and the Alacran Mountain Formation, 190 m of gray, medium- to thick-bedded limestone with a prominent red shale interval (Deer Mountain Red Shale Member) near the middle (Williams, 1963). No higher continuous beds are exposed. In the eastern part of the mountains an "eastern dolomite" facies is exposed. The rocks of this facies are limestone and are younger than the highest part of the Alacran Mountain Formation (Cys, 1975); hence, these Wolfcampian strata are the youngest Permian rocks exposed in the Hueco Mountains.

At the base of the Hueco Canyon Formation is the Powwow Member, which consists of 30 m or less of red shale and some conglomerate. The lower contact is unconformable. As

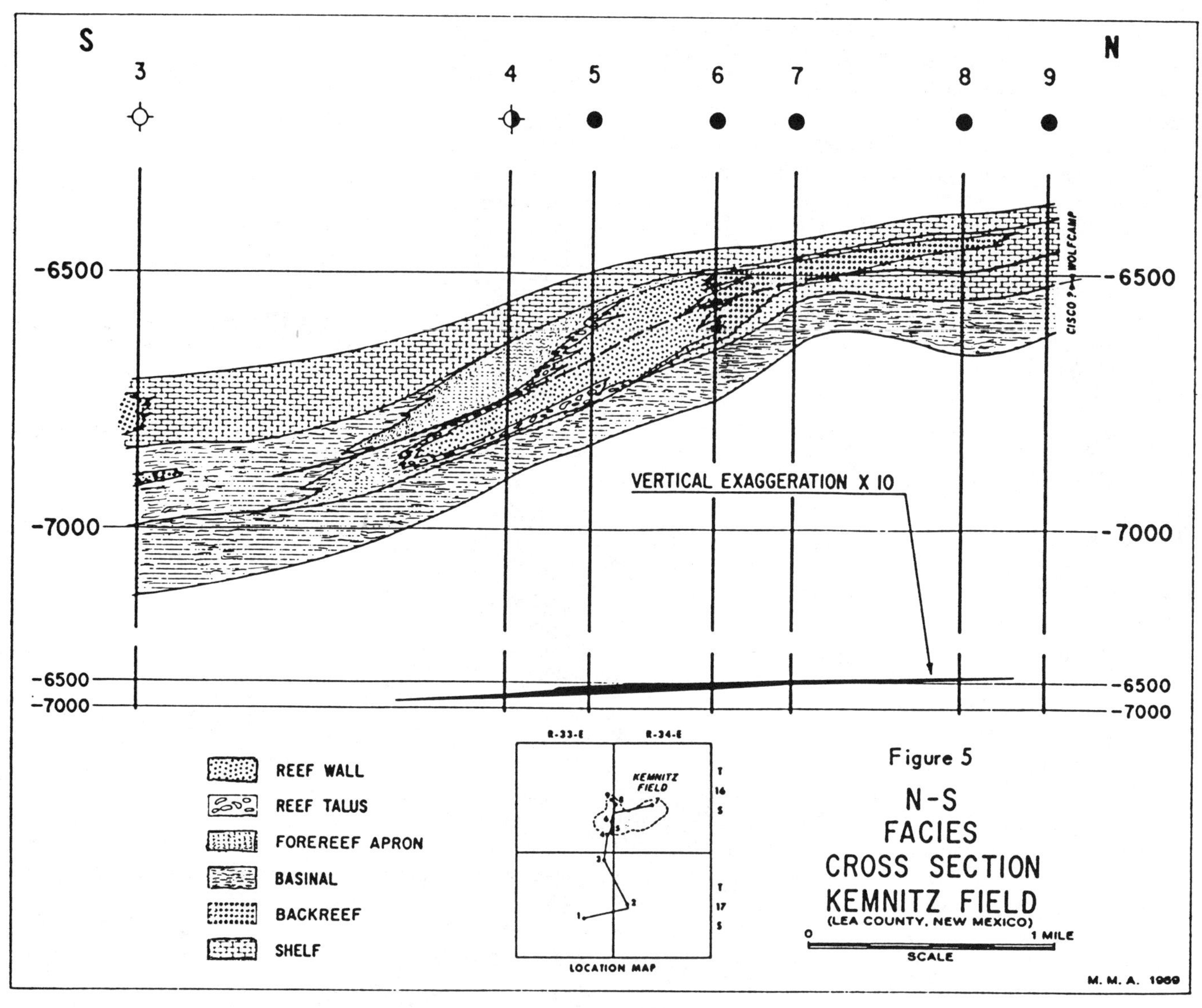

Figure 18. Facies cross-section across Kemnitz Field, a typical Early Permian shelf-edge buildup on the southern margin of the Northwest Shelf, Lea County, New Mexico (from Malek–Aslani, 1970).

mentioned earlier, part of the underlying upper Magdalena Formation is also of Wolfcampian age (Williams, 1963).

The Wolfcampian Hueco Formation rests unconformably on rocks ranging in age from Precambrian to Pennsylvanian in the Sierra Diablo and Apache Mountains. The Hueco Formation consists of 120 to 335 m of medium-bedded gray limestone and basal conglomerate (King, 1965). The conglomerate is assigned to the Powwow Member and has a highly variable thickness with a maximum of approximately 30 m.

Leonardian to Ochoan Time. In the Diablo Platform area, rocks of this age are basically the same as those exposed in the Guadalupe Mountains. In the Sierra Diablo Mountains, the Victorio Peak Formation consists of 460 m of cliff-forming, gray, thick-bedded limestone (King, 1965). Across the Victorio Peak and Babb flexures, the lower part of the Victorio Peak Formation intertongues with the basinal Bone Spring Formation.

In the Apache Mountains, the Capitan Reef complex continues its course around the perimeter of the Delaware Basin but has an east–west trend in contrast to its northeast–southwest trend in the Guadalupe Mountains.

In the Finlay Mountains, a 510- to 535-m-thick sequence of interbedded marlstone, limestone, and calcirudite of Leonardian age, designated the Wilkie Ranch Formation by Cys (1976a), is correlative with the Victorio Peak and Bone Spring Formations in the Sierra Diablo. No other Permian rocks are exposed in these mountains.

Tectonics. The early Wolfcampian was a time of major tectonic activity on the Diablo Platform. Over almost the entire

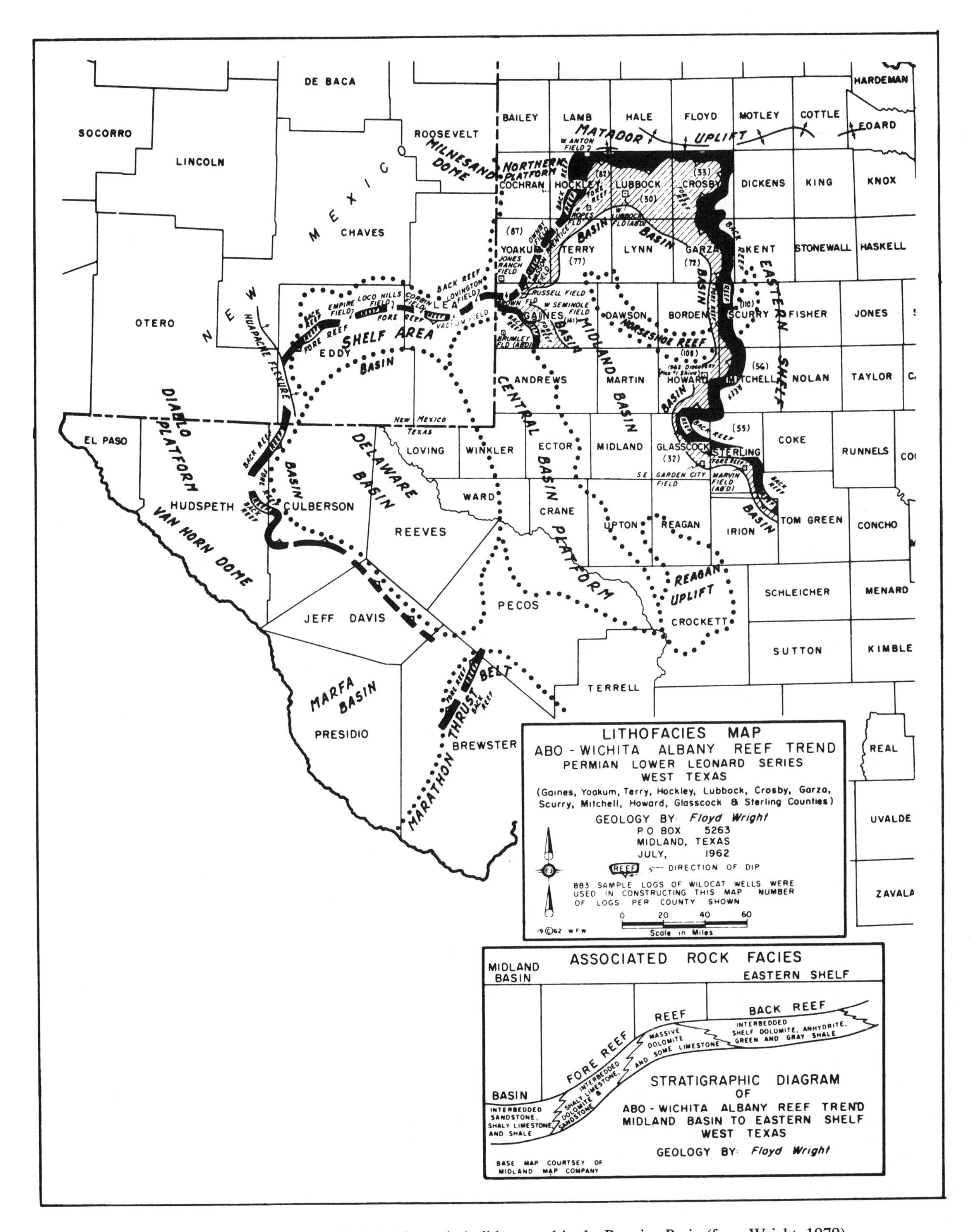

Figure 19. Wichita-Albany (Abo) shelf margin buildup trend in the Permian Basin (from Wright, 1979).

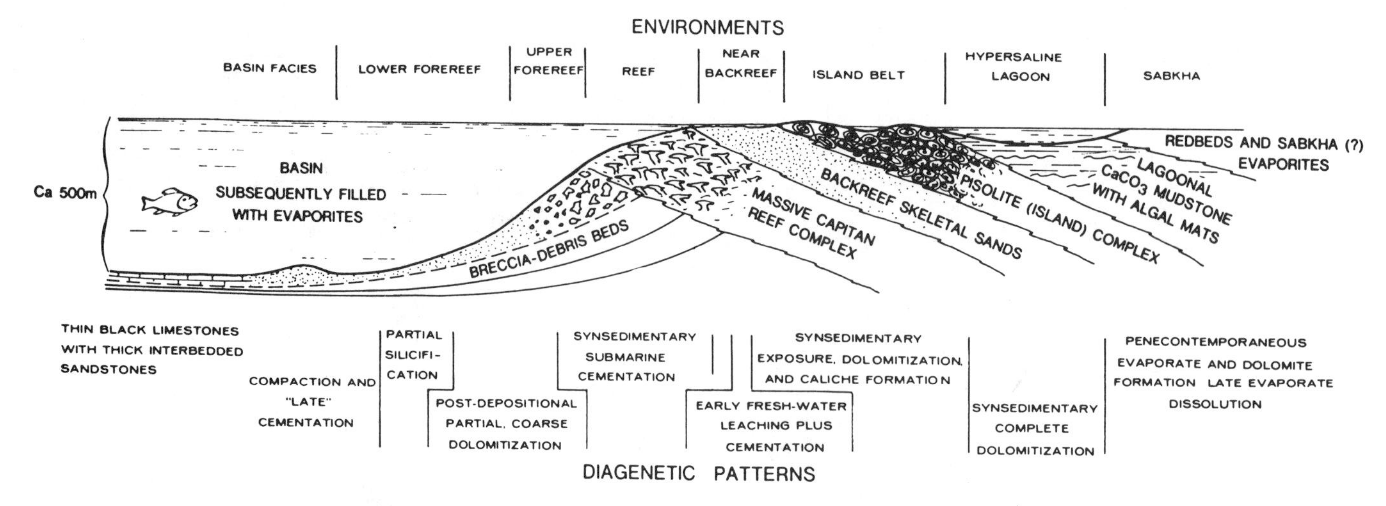

Figure 20. Generalized facies cross-section across the Capitan Reef complex in the Guadalupe Mountains, New Mexico (from Scholle and Halley, 1980).

platform the middle Wolfcampian Hueco Formation rests unconformably on rocks ranging in age from earliest Wolfcampian to Precambrian. The platform then became stable, except for very minor Leonardian tectonic activity on the Babb and Victorio Peak flexures mentioned above. During Leonardian and Guadalupian time the platform carbonates prograded eastward into the Delaware Basin.

Central Basin Uplift and Platform

Wolfcampian Time. The western part of the uplift was emergent during all of Wolfcampian time (Jones, 1953) (Figure 16), More than 300 m of red shale, sand, and local conglomerates assigned to the Powwow Member were deposited on the flanks of this highland area (Jones, 1953). This detrital section interfingers with light-colored limestones that fringe the platform. The whole sequence is assigned to the Hueco Formation and rests unconformably on rocks ranging in age from Precambrian to Pennsylvanian.

Leonardian Time. The lower Leonardian sediments of the Central Basin Platform are mostly light-colored dolomites interbedded with some limestones and a small amount of green to gray shales and anhydrites (Jones, 1953). These are assigned to the Abo Formation or Wichita-Albany unit and are overlain by the Clearfork Formation.

Guadalupian-Ochoan Time. The rocks of the Central Basin Platform are essentially the same as those discussed for the Northwest Shelf (Fig. 21).

Tectonics. The western part of the uplift was a tectonically active positive area during Early Wolfcampian time. After this, the region became a true stable platform on which was built the mass of Permian platform and platform-edge carbonates. The platform carbonates prograded into the Delaware Basin. By Guadalupian time, the Midland Basin had filled to the same level as the platform and Guadalupian backreef sediments are continuous from the platform into the Midland Basin.

Delaware Basin

Wolfcampian Time. Wolfcampian rocks in the Delaware Basin are predominantly dark gray shales with occasional thin limestone or sandstone interbeds. They were deposited in deep marine waters of a rapidly subsiding basin. The basin remained asymmetric throughout lower Permian time; therefore, the thickest Wolfcampian deposits are located in an area of rapid subsidence adjacent to the Central Basin Platform. These strata are a maximum of 4,300 m thick (Feldman and others, 1962) in the southern part of the basin. Wolfcampian rocks thin northward to approximately 1,585 m in Ward County and to less than 610 m in the western Delaware Basin (Wilde, 1976).

The lower Wolfcamp strata in the basin contain large exotic blocks of Mississippian limestone within an area of maximum depositional relief in western Pecos County, Texas. The subsurface interpretation of these blocks explains the repeated upper Paleozoic strata encountered in some of the deep wells adjacent to the western margin of the Central Basin Platform. One major slide block, located in the Coyanosa Field area of northwestern Pecos County, is approximately 26 km long by 15 km wide and is about 610 m thick. It slid westward about 11 km into the Delaware Basin during early Wolfcampian time (Guinan, 1971).

Leonardian Time. Upper Wolfcampian strata grade conformably into the lithologically similar lower Leonardian strata. The Bone Spring Formation is composed predominantly of dark gray, deep marine shales, which grade upward into interbedded shales, quartzose sandstones, and limestones. Rates of sedimentation decreased in Leonardian time because shelf-to-basin deposi-

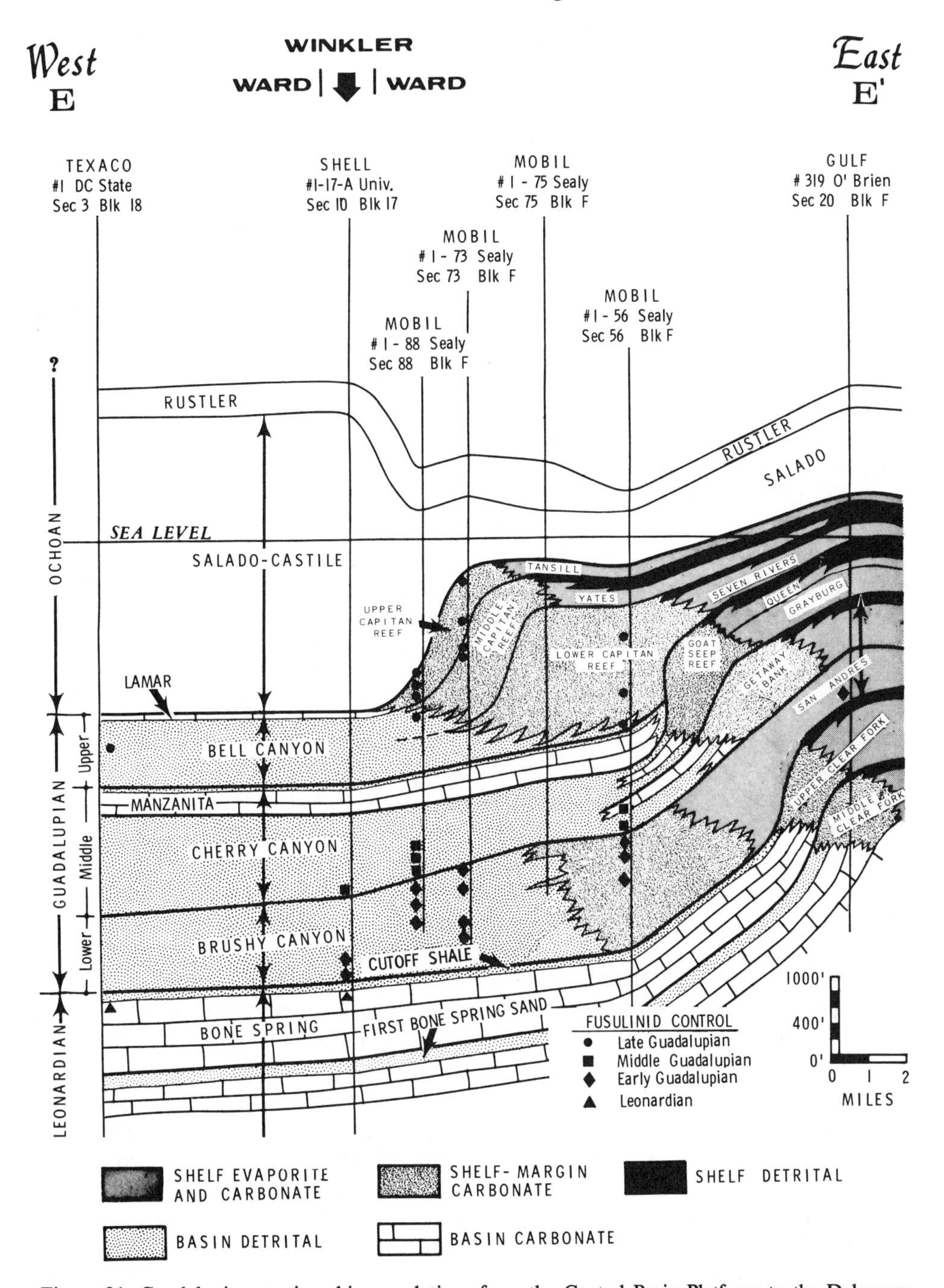

Figure 21. Guadalupian stratigraphic correlations from the Central Basin Platform to the Delaware Basin (from Silver and Todd, 1969).

tional relief had decreased; however, the Delaware Basin continued to be a deep marine basin. The Bone Spring is a maximum of 915 m thick (Hardy, 1976). Shelf-to-basin relationships are shown in Figure 21. The Bone Spring Formation was a significant hydrocarbon-generating source rock.

Guadalupian Time. The contact between the Guadalupian-age Delaware Mountain Group and the underlying Bone Spring Formation appears to be conformable. The Delaware Mountain Group is made up of predominantly light gray, very fine-grained sandstones and siltstones interbedded with shales. These basinal clastics are roughly 900 m thick—half as thick as the facies-equivalent shelf margin carbonates. The group is composed of the Brushy Canyon, Cherry Canyon, and Bell Canyon Formations, in ascending order. The sediments that formed these strata are interpreted to be deep-marine basinal turbidite and density current deposits (Williamson, 1979, Harms, 1974). The

Delaware Mountain Group ranges in thickness from 285 m to a maximum of 1,071 m thick (Cys, 1976b). It contains numerous reservoir quality strata. To date, most of the oil production has been from stratigraphic traps in the Ramsey and Olds Sandstones of the uppermost Bell Canyon Formation. These sandstones are overlain by 6 m of thin limestones of the Lamar Member of upper Guadalupian age.

Ochoan Time. Guadalupian rocks are overlain conformably by the laminated anhydrites and some salt beds of the Ochoan-age Castile Formation (Witt, 1976a). These evaporitic rocks indicate a general lowering of sea level and the final stages of Delaware Basin filling in an arid climate. Terrigenous clastic components of these strata are almost nonexistent. Ochoan strata are up to 1,375 m in this area, of which the Castile Formation represents approximately 640 m (Adams, 1944). This thickness suggests that the maximum water depth was greater than 200 m during Ochoan time.

The laminae of the Castile are alternately light and dark gray to brown in color. Some of the darker laminae may be brown colored due to the inclusion of minor amounts of organic materials. The darker layers are generally more calcitic than the lighter laminae and may represent periods of brine-water intrusions into the hypersaline basin.

The impermeable strata of the Castile Formation function as the seals, which prevented upward migration of much of the oil and gas that accumulated in the underlying Delaware Mountain Group. The Castile Formation is confined to the Delaware Basin and its margins and is overlain by the Salado Formation.

At the end of Castile time, the shallowing hypersaline basin began to spread from the Delaware Basin eastward. By the end of Ochoan time, the Salado had been deposited over the Castile of the Delaware Basin and eastward across the Central Basin Platform, Midland Basin, and Palo Duro Basin. It had also spread northward to include the Northwest Shelf of Lea and Eddy counties in New Mexico. The Salado is a sequence predominantly of halite, with a maximum thickness of 725 m (Witt, 1976b). The Salado in New Mexico contains significant bedded potash deposits. At least three quarters of the U.S. domestic potash production is from underground mines at depths of 244 m to 671 m in the Salado Formation. The mines are located within the Federal Potash Enclave, just east of Carlsbad, New Mexico (Ohl and Eveleth, 1985). The main ores mined are sylvinite and langbeinite.

The Salado has been the focus of intensive scientific research during the 1970s and early 1980s due to its selection by the United States federal government as a formation suited for the storage of transuranic wastes. A Waste Isolation Pilot Plant (WIPP) site between Hobbs, New Mexico and Carlsbad, New Mexico was selected and the excavations of the shafts and underground chambers in the Salado Formation were begun in mid-1983 (Ohl and Eveleth, 1985).

The Salado is overlain by the Rustler Formation, which is composed of up to 120 m of red anhydritic shale, anhydrite, and some dolomite. In the southwestern part of the Delaware Basin the Rustler Formation is mostly limestones and dolomites, which were deposited in more marine conditions than the anhydrites and shales of the northern part of the basin. The basal Rustler Formation is often a red sandstone, especially in the Midland Basin area (Jones, 1953). In places, the contact between the Rustler Formation and the underlying Salado Formation is irregular due to the collapse of beds of the Rustler Formation into cavities formed by the leaching of salt.

The Dewey Lake Formation is the youngest of the Ochoan strata. It is a sequence of red sandstones and shales with minor amounts of gypsum. It is a maximum of 107 m thick in the Delaware Basin (Witt, 1976c) and is unconformably overlain by Triassic red beds of similar lithology.

Midland Basin

Wolfcampian Time. Wolfcampian strata, which conformably overlie Pennsylvanian strata in the Midland Basin, consist mostly of gray shales with interbedded limestones and some sandstones near the basin margins. These rocks attain a thickness of approximately 600 m in the deep parts of the basin (Wilde, 1976).

Some of the Pennsylvanian reefal and carbonate bank deposits of the southern Horseshoe Atoll continued to grow during Wolfcampian time. These stand out as a few pinnacles encased in the dark gray marine shales that covered the rest of the basin area.

Leonardian Time. Leonardian strata, up to 350 m in thickness, were deposited in deep (more than 200 m) marine waters of the Midland Basin. The Dean Sandstone is the lowermost Leonardian formation. This unit is composed of fine-grained quartzose sandstones and siltstones that appear to conformably overlie Wolfcampian shales. It is overlain by the Spraberry Formation, a sequence of fine-grained quartzose sandstones and siltstones interbedded with gray shales and some limestones, forming the Spraberry Trend oil field. More than 140 km long and 40 km wide, this field is a prime example of a stratigraphic (nonstructural) oil accumulation. It is located along the deep basin axis and contains more than 6,650 oil wells, which have produced more than 585 million barrels of oil.

The Dean and Spraberry Formations were deposited during episodes of deep water resedimentation of shelf-derived carbonate debris and during periods of clastic sedimentation via feeder channels and submarine canyons with associated suspension settling of fine-grained sediment (Handford, 1981). The Spraberry Formation is overlain by limestone and shales also of Leonardian age.

Guadalupian Time. The San Andres and Grayburg formations in the Midland Basin were deposited in a variety of facies ranging from normal to restricted shallow-marine hypersaline conditions during Guadalupian time. The basinal carbonate strata of the San Andres Formation range in thickness from a few tens of meters to approximately 300 m. The carbonate strata are thin near the basin center and thicken toward the shelves. The shelf dolomite facies of the San Andres grades into basinal limestones near the shelf edges and then into shales, siltstones, and some sandy limestones near the axis of the Midland Basin (Matchus and Jones, 1984) where the overall thickness of this sequence is up to 820 m.

Ochoan Time. Ochoan rocks in the Midland Basin are mainly anhydrites and salts of the Salado Formation overlain by red shales and sandstones of the Rustler and Dewey Lake Formations. The Ochoan Series is approximately 365 m thick in the Midland Basin. These evaporite strata serve as hydrocarbon seals, which confine oil accumulations in many of the underlying Permian reservoir strata. The Ochoan series of the Midland Basin is quite similar to that of the Delaware Basin, with two exceptions. First, the Castile Formation is absent in the Midland Basin, and second, the Salado, Rustler, and Dewey Lake Formations are thinner in the Midland Basin than their Delaware Basin equivalents. During Late Ochoan time, the Delaware and Midland Basins were no longer separate entities, and the entire region was a large evaporite basin.

Val Verde Basin

The Wolfcampian strata of the Val Verde Basin are similar to the Wolfcampian strata of the southern Delaware Basin, which it joins. The basal Wolfcampian (Permo-Pennsylvanian) contact is conformable, but the upper Wolfcamp-Cretaceous contact is unconformable. Wolfcampian strata in this basin are as much as 4,600 m thick (Wilde, 1976). From 300 to 600 m of Cretaceous carbonates are present over most of the Val Verde Basin. Formations of the Leonardian, Guadalupian and Ochoan Series are only present in the northwestern part of the basin where it adjoins the southern Delaware Basin.

SUMMARY

The Early Pennsylvanian was a time of tectonic activity as the rising Central Basin Uplift separated the Tobosa Basin into the Midland and Delaware Basins with adjoining shelf areas. Tectonic activity sporadically continued through the remainder of the Pennsylvanian with a final major pulse during Wolfcampian (Early Permian) time on the Central Basin Uplift and Diablo Platform.

Because of the above-mentioned tectonic activity, Pennsylvanian rocks are extremely variable in thickness and distribution and exhibit complex lateral and vertical facies changes. Morrow and Atoka rocks are dominantly interbedded sandstones and shales with some local limestones. The Strawn, Canyon, and Cisco are interbedded limestones and shales.

The Permian was characterized by progradation. Carbonate shelves with a broad spectrum of depositional environments—from hypersaline to sabkha to low-energy open-marine shelves, grainstone shoals, and shelf-edge organic buildups—prograded into the surrounding basins. Thick sequences of shales, siltstones, sandstones, and limestones accumulated in the basins by various gravity flow processes. Finally, in Ochoan time, the whole area became an evaporite pan and was filled with a thick evaporite sequence.

EASTERN SHELVES SUBREGION

D. C. Van Siclen and R. R. Bloomer

INTRODUCTION

The name "Eastern Shelves" (Fig. 1) refers to the succession of depositional shelves that, during most of Pennsylvanian and Permian time, prograded westward from the rising Ouachita highlands in eastern Texas, filling first the intervening Fort Worth and Kerr Basins on the east and southeast, and subsequently the "deep" Midland Basin farther west. Although most of the generally 1,200 to 2,800-m-thick stratigraphic section in this subregion was deposited during these two periods, the sedimentary cover also includes perhaps as much as 600 m of early Paleozoic strata, up to 150 m of Late Triassic beds limited to the western edge of the area, "middle" Cretaceous deposits covering most of its southern half and thickening southward to 300 m or more, remnants of Neogene sediments to several tens of meters in thickness along its northern west edge, and thin Pleistocene terrace deposits along the larger valleys.

The present surface formations in the northern three quarters of the Eastern Shelves subregion are predominantly of Permian age, flanked to the southeast by substantial areas of Pennsylvanian strata and along the west by Triassic beds. These systems are overlapped by "middle" Cretaceous strata, which form extensive outliers across the southern central part of the area and cover all the southern quarter except its eastern edge, where early Paleozoic beds crop out on the western flank of the Llano Uplift. Physiographically, the Cretaceous rocks represent the Edwards Plateau Section at the south end of the High Plains, while the exposed late Paleozoic rocks underlie the southwestern Osage Plains Section of the Interior Lowlands. The Cretaceous forms extensive flat-topped mesas and buttes surrounded by steep scarps of massive limestone, while the underlying shaly late Paleozoic beds have been eroded into rolling lowlands separated by numerous fairly low eastward-facing cuestas held up generally by carbonate strata. These cuestas accentuate the rise in the land toward the west, from about 300 m elevation over the Bend Arch to more than 800 m on the edge of the High Plains.

Throughout the Eastern Shelves subregion, Late Cambrian strata rest on a basement of Precambrian crystalline rocks that have experienced only minimal deformation over the last billion years. The margins of this cratonic block, however, were warped and broken to varying degrees by the late Paleozoic Ouachita Orogeny, leaving a series of basins and uplifts around most of its periphery. These bounding structural features include the Matador (Red River) Arch on the north, and to the east, the Bend Arch, which rises southward to form the western edge of the Llano Uplift of central Texas. To the southwest and west the Eastern Shelves pass into the Val Verde and Midland Basins across fairly arbitrary dividing lines that are based mostly on changes of dip and stratigraphy.

Figure 22 and the upper left part of Figure 29 of Johnson

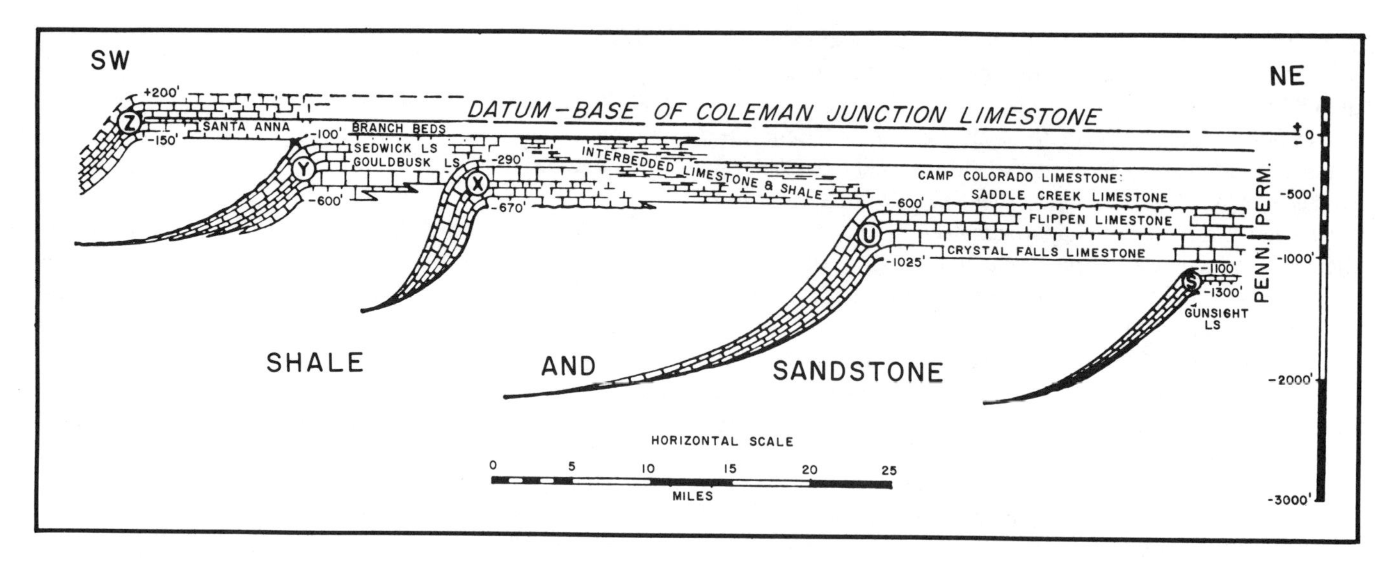

Figure 22. Diagrammatic southwest–northeast stratigraphic section showing the alternation of prominent Late Paleozoic limestone bodies with terrigenous clastic strata in the central-western part of the Eastern Shelves subregion. Environmental conditions responsible for this arrangement included: (1) a relatively deep basin, (2) undergoing long-term subsidence and (3) experiencing eustatic fluctuations in sea level, (4) in a low-latitude location, (5) alongside a rising source of abundant detrital sediments (after Van Siclen, 1958).

and others (this volume) illustrate the Eastern Shelves concept, while the lower right of the figure in Johnson and others shows the nature of the Bend Arch and relationships to the Fort Worth Basin, Ouachita Foldbelt, and overlapping Early Cretaceous strata. Deposition of more than 800 m of fairly continuous carbonate above the westernmost shelf-edge (Fig. 22) created an imposing sedimentary structure of considerable economic importance that has long served as a convenient boundary between the Eastern Shelves and the "deep" Midland Basin. Figure 23 is a single-fold conventional seismic section across this boundary in the extreme northwestern corner of the Eastern Shelves subregion, where moderate progradation allows a number of distinct shelf-edges to be recognized. As defined above, the Eastern Shelves subregion covers about 65,000 km^2.

GEOLOGIC STRUCTURE

The Eastern Shelves subregion is thus a broad area of gentle homoclinal dip between the crest of the Bend Arch and the "deep" Midland Basin. Average regional dip is about one-half degree, almost west in the southwestern half of the area but nearly northwest (and slightly gentler) in the northeastern half. This change in dip direction defines a continuous regional feature that shows up best as a pre-Pennsylvanian isopach thin near the northwestern edge of the Eastern Shelves subregion, where Adams (1965) termed it the Texas Arch (Fig. 1). This inconspicuous arch represents the pre-Permian structural and topographic transition between the Southern Oklahoma Aulacogen and the Tobosa Basin.

The homoclinal westward dip of the Eastern Shelves is rarely interrupted by substantial folds or faults. Most of the folds, which have been explored intensively for petroleum, formed by differential compaction over basement hills or over thickened bodies of limestone or sandstone. Except for compaction anticlines over a few extensive organic reefs or carbonate banks, these are of only local extent. The one extraordinary deformational feature within this subregion is the Fort Chadbourne system of large basement faults (Conselman, 1954). This remarkable line of normal faults strikes very slightly east of north across the entire Eastern Shelves subregion, passing about 15 km east of San Angelo and Sweetwater, Texas. The individual faults may have a slightly more northeasterly strike than the Fort Chadbourne trend, and commonly bound deep asymmetrical grabens with the greatest throw on the western marginal fault. The faults offset the Early Pennsylvanian beds conspicuously, and movement continued throughout that period (Rall and Rall, 1958). These characteristics suggest that the Fort Chadbourne fault system is an antithetic wrench zone related to the more intense deformation of the surrounding regions during the late Paleozoic Ouachita Orogeny.

PALEOZOIC ROCKS AND EVENTS

The entire sedimentary section in the Eastern Shelves subregion was deposited in response to three major tectonic events. Regionally, the Late Cambrian through Devonian strata were laid down under conditions of relatively slow subsidence. When shallow seas submerged the Eastern Shelves subregion late in the

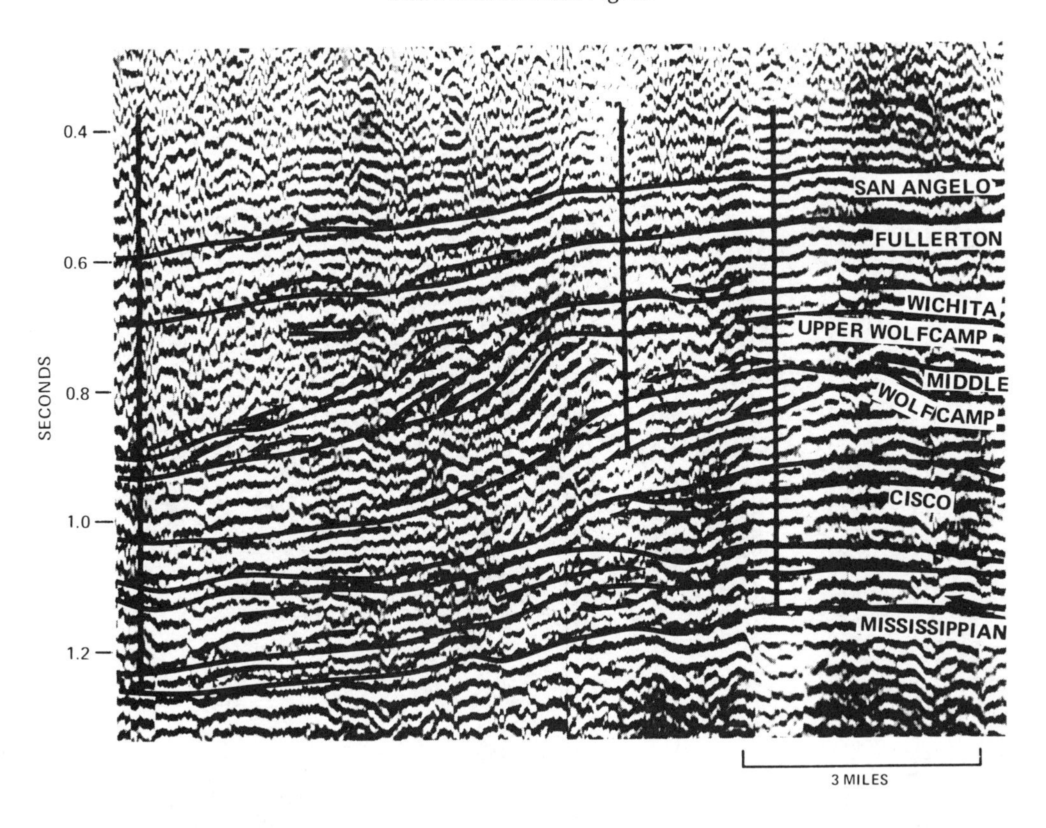

Figure 23. Singlefold conventional seismic section across the series of "superposed" carbonate shelf-edges that mark the western edge of the Eastern Shelves subregion in its extreme northwestern corner, Crosby County, Texas (from Bubb and Hatlelid, 1977).

Cambrian Period the basal sediments were cobbles, gravel, and sand of local derivation, followed by clay from more distant sources. As continuing transgression cleared the seas, an extensive carbonate platform developed before the end of Cambrian time and persisted through Early Ordovician time.

The stratigraphic record from the Middle Ordovician into the Mississippian Period has not been preserved in the Eastern Shelves subregion, due to nondeposition and erosion. However, thin remnants of Silurian and Devonian rocks (including cavern fillings) have been recognized in outcrop in central Texas and are present in the subsurface to the west.

A second major tectonic event was responsible for deposition of the bulk of the stratigraphic section in the Eastern Shelves subregion, including the succession of westward-prograding shelves from which this subregion takes its name. This was the development of the Ouachita Orogen, which caused subsidence along the eastern edge of the Eastern Shelves area, making room for deposition of the Late Mississippian Chappel Limestone, followed by the Barnett Shale as continuing transgression deepened the narrow seaway west of the Ouachita Orogen (Figure 29 of Johnson and others, this volume).

Continued Ouachita overthrusting and the resultant sedimentary loading depressed the Fort Worth Basin, but rapid sedimentation filled it, allowing terrigenous sediments to be carried westward and deposited in the Eastern Shelves area. The basal Pennsylvanian strata in the Eastern Shelves area were predominantly carbonates, frequently resting on a "detrital zone" characterized by the presence of chert pebbles weathered from underlying carbonate sections. The diachronous upper limestone unit of this transgressive section (equivalent to the Marble Falls Limestone of the central Texas outcrop and the Caddo Limestone of the subsurface farther west) formed at least one prominent eastward-facing shelf-edge near the present crest of the Bend Arch (Figure 29 in Johnson and others, this volume), roughly marking the Eastern Shelves' boundary with the Fort Worth Foreland Basin (Namy, 1982, esp. Figure 4). Similar events were taking place in the southern Permian Basin, so that seas transgressing from the south and east finally met in the Eastern Shelves subregion.

Organic reefs (and/or carbonate banks) became established in shallow water along the shore of the advancing sea and over sea-floor hills of various origins, favored perhaps by brief still-stands of relative sea level. Some of these carbonate ridges and knolls built upward persistently in this tropical sea for considerable lengths of time, many surviving into the Late Pennsylvanian, until smothered by the muddy environment of a prograding shelf

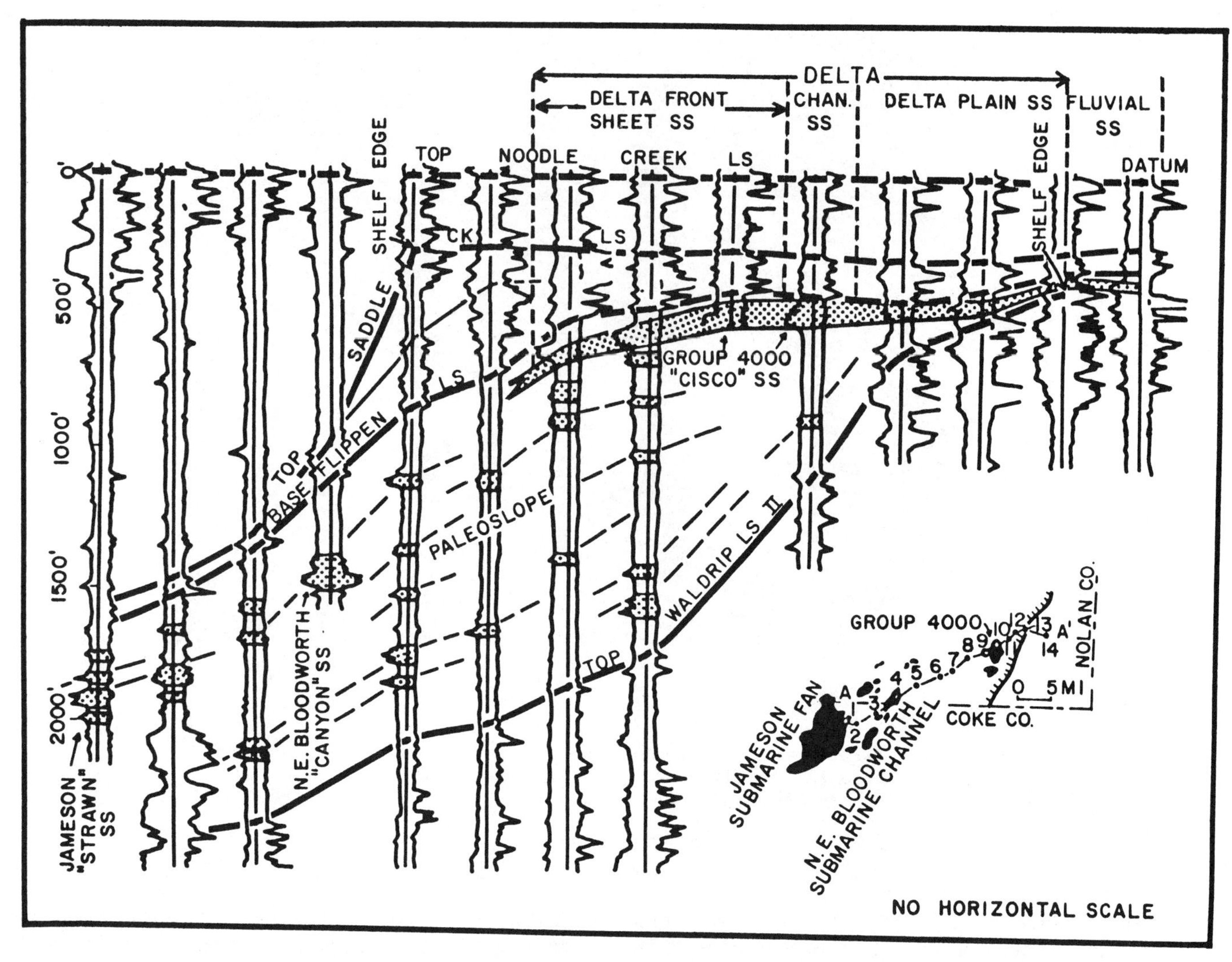

Figure 24. Stratigraphic section across the basal Permian shelf and slope, showing the depositional topography and environments of the productive Cook sandstones, between the Waldrip II and Flippen limestones, in Nolan and Coke Counties, Texas (modified after Bloomer, 1977).

of clastic sediments from the Ouachita Mountains to the east and from the Red River Arch and beyond on the north.

Despite overall subsidence of the Eastern Shelves area, the advance of this terrigenous clastic shelf system was not a continuous process, due to substantial fluctuations in the relative level of the sea. These fluctuations must have been caused in part by local and distant tectonism, but perhaps more importantly, by repeated continental glaciations in the southern hemisphere (Wanless, 1967). Times of lower sea level favored shelf propagation; but when sea level rose above its earlier positions (as by addition of water from melting glaciers) the shelf became submerged and carbonates were deposited widely over the drowned shelf and slope, sometimes thickening along the shelf-edge to form carbonate banks and reefs. Finally, as conditions stabilized, deposition of clastic sediments resumed, constructing a new clastic shelf, adjusted to the higher sea level, over, and commonly beyond, the former one. This buried the transgressive carbonates, which remain as markers that clearly reveal the broad stages of basin filling (Van Siclen, 1958). The form and relationships of the limestone bodies in the upper part of Figure 22 and in Figure 29 (Johnson and others, this volume) illustrate this.

These conditions are typified by the uppermost Pennsylvanian and lowest Permian sediments (Virgil and Wolfcamp Series), which consist of a number of alternating carbonate and clastic shelves and slopes. These formed by terrigenous clastics prograding westward into a sea some hundreds of meters deep, which was differentiated topographically into unda-, clino-, and fondoform environments. During high sea level stages, thin limestone beds with open-marine fossils formed widely in shallow water on the drowned shelves, while over the outer shelf and

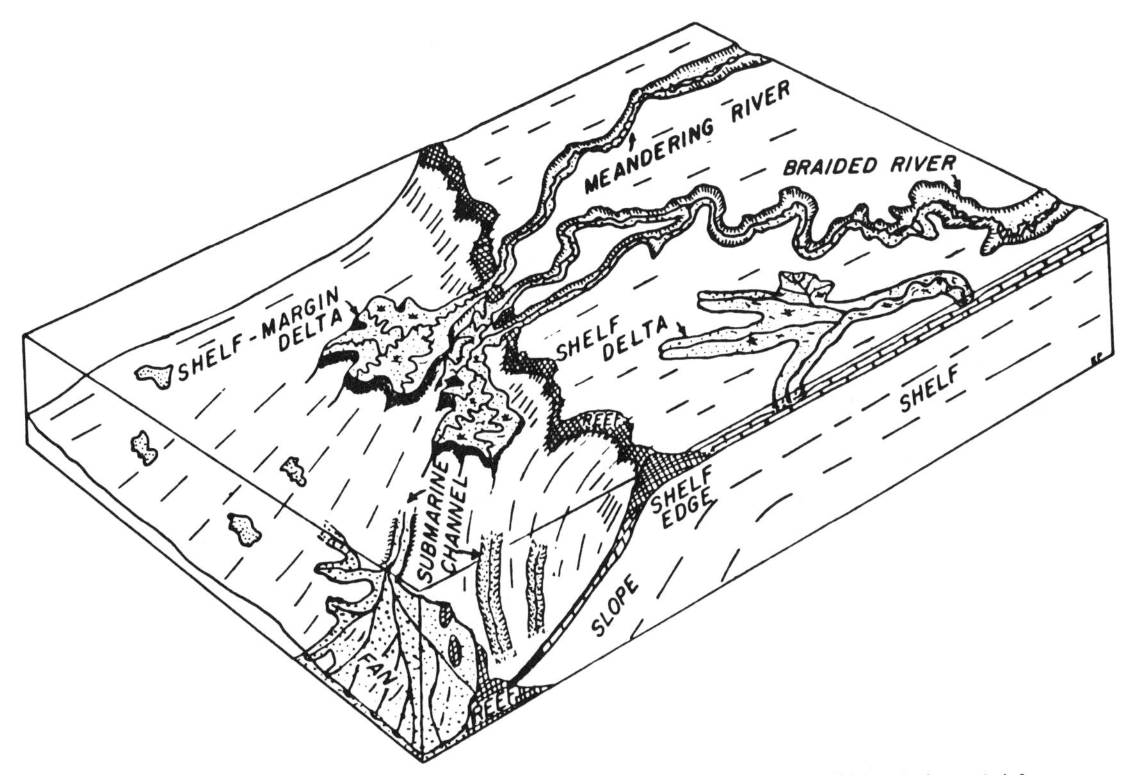

Figure 25. Generalized block diagram of typical eastern shelf and slope conditions during a brief regression at the beginning of the Permian Period, with emphasis on the continuous series of contrasting environments, from fluvial to submarine fan, in which the Cook sandstone bodies were deposited (modified after Bloomer, 1977).

upper slope carbonate deposition steepened the shelf margin to as much as 3° and created shelf-edge barrier reefs and bars. Meanwhile, the subaerial drainage systems farther east continued to supply large amounts of mud and sand, building elongate deltas out onto the shallow shelf and gradually developing an alluvial coastal plain characterized by meandering rivers. Figures 24 and 25 present a well-studied example from the basal Permian. The shelf deposits become younger to the west and are followed from the east by sabkah and alluvial fan deposits, mostly red beds, which characterize the rest of the Permian System (Bloomer, 1977).

KERR BASIN

R. B. Cline

INTRODUCTION

The Kerr Basin of western Texas is one of a series of structural basins that formed along the margin of the Paleozoic craton during the development of the Ouachita Orogeny. The location of the basin centers about Kerr, Bandera, and Real counties (Fig. 3). Bounding elements of the Kerr Basin are the Llano Uplift to the north and east, the frontal zone of the Ouachita thrust belt to the south, and the Edwards Arch to the west. The foreland rocks in the Kerr Basin continue southward under the Ouachita detritus and thrust belt (Fig. 26). The overlying Pennsylvanian foredeep rocks contain both detrital outwash from the Ouachita Orogen and cratonic sources. A post-orogenic Mesozoic–Tertiary sequence overlies the area.

SEDIMENTATION-STRATIGRAPHY

The early Paleozoic sedimentary sequences (Fig. 27) consist of foreland facies dominated by carbonates of the Early Ordovician Ellenburger Group, which overlie carbonates and clastics of the Cambrian Wilberns and Riley Formations; but few wells penetrate the pre-Ellenburger strata.

The Ellenburger Formation of the Kerr Basin is predominantly a fine-grained dolomite but becomes an interbedded dolomite and limestone formation over much of the eastern part of the basin. The Ellenburger grades from the dominantly finely crystalline, occasionally vuggy dolomites, with a suggested tidal-flat origin, to a sequence of cyclic, somewhat cherty limestone and dolomite, suggesting a deeper marine environment eastward in easternmost Kerr and Bandera Counties; limestones dominate the sequence in Kendall and western Comal Counties. The thick-

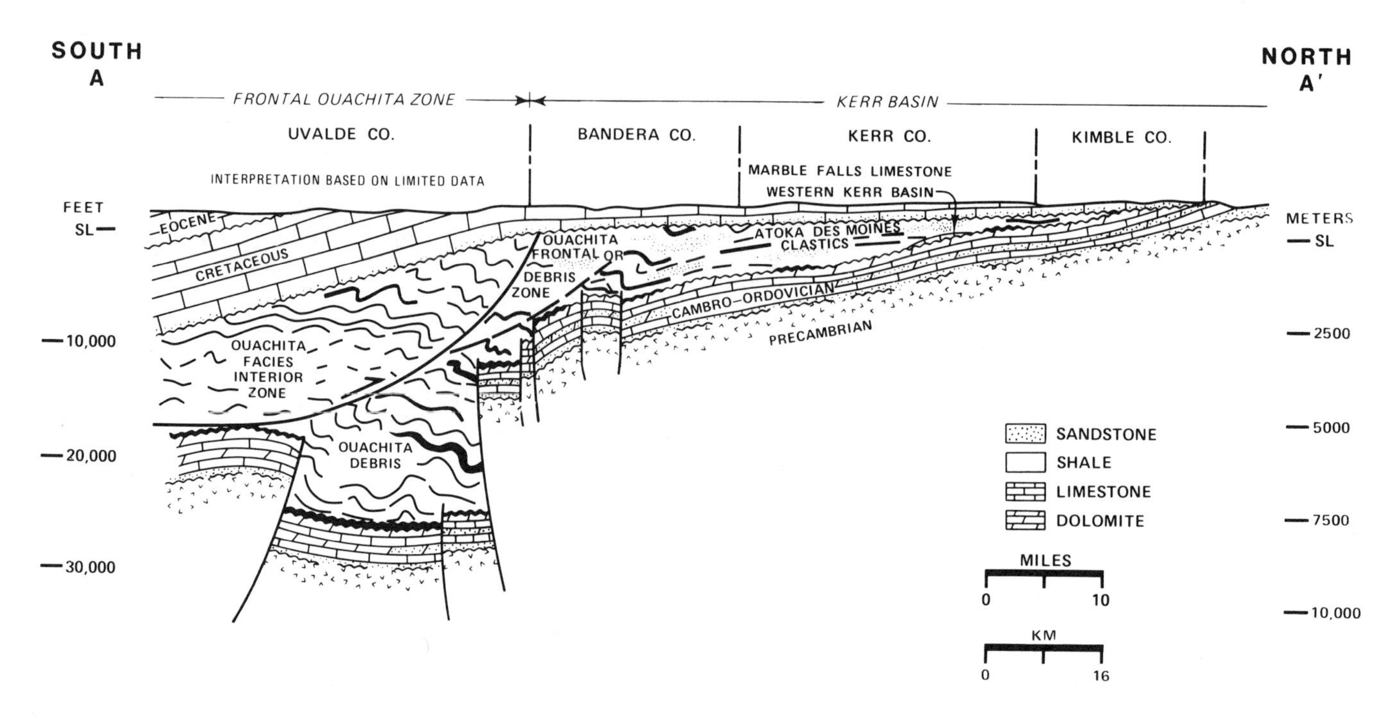

Figure 26. North–south schematic cross-section, Kerr Basin.

ness ranges from zero at the truncated edge on the Llano Uplift to approximately 610 m in the southernmost areas.

Thin remnants of Silurian, Devonian and Mississippian rocks, including richly organic Mississippian Barnett Shale, are preserved in the northern part of the basin. Over most of the basin, these strata have been eroded across the ancestral Texas Arch, and the Morrowan-Atokan Marble Falls Limestone lies unconformably on the Ellenburger Formation. Thick foredeep clastics of Pennsylvanian age, principally Atokan, overlie the Marble Falls. This clastic fill apparently is derived from both northern and southern sources. The detritus from the south contains chert and variegated shale fragments from eroded Ouachita facies. The chert detritus provides stratigraphic markers that can be traced over the southern part of the basin in Uvalde, Bandera, Kendall, and southernmost Kerr Counties. Clastics from the north and east were probably derived from the Llano Uplift during upper Atokan-Desmoinesian time. Clastic facies derived from the north indicate prevalent alluvial and deltaic environments, whereas the clastics derived from the south accumulated in deep-marine environments. Thin organic-rich shales that exist in the Pennsylvanian foredeep sequence generally have a humic aspect.

Lower Cretaceous, Comanchean, strata unconformably overlie the Paleozoic subcrops, which range from lower Paleozoic in the north to upper Paleozoic in the southern part of the basin. Upper Cretaceous and Eocene strata are present in the extreme southern part of the area.

STRUCTURAL-SEDIMENTARY SYNTHESIS

The sediments of the early Paleozoic suggest a passive margin sequence of shallow-margin origin. The Ellenburger, or foreland, facies was structurally depressed in southern Kerr Basin when a northward-migrating foredeep developed in late Mississippian and Pennsylvanian times. The hinterland supplying detritus to this foredeep was the uplifted terrain of the Ouachita orogen that migrated cratonward, loading the area with thrust sheets and supplying a source for foredeep fill. The uplifts in the frontal tectonic zone brought deeply buried foreland facies up to the shallower depths observed today. Deep-burial and high-geothermal gradients probably contributed to the incipient metamorphism of the foreland facies in areas south of and along the trend of the Kerr Basin proper. The loading of the craton by thrust-sheet emplacement and detrital accumulation in lower and mid-Pennsylvanian in the Fort Worth and Kerr Basins might be in part responsible for cratonal bulges such as the Llano Uplift. Morrowian, lower Atokan and older sediments and facies patterns, including eastward-facing Atokan carbonate shelves, can be projected across parts of the Llano Uplift in a southwest–northeast depositional strike. Disruption of this pattern and deposition of the younger Pennsylvanian Marble Falls Limestone on rocks as old as Ordovician suggest post-lower Atokan Uplift of the Llano region. The loading could also be responsible for increased thermal history seen southward within the frontal zone and interior zones.

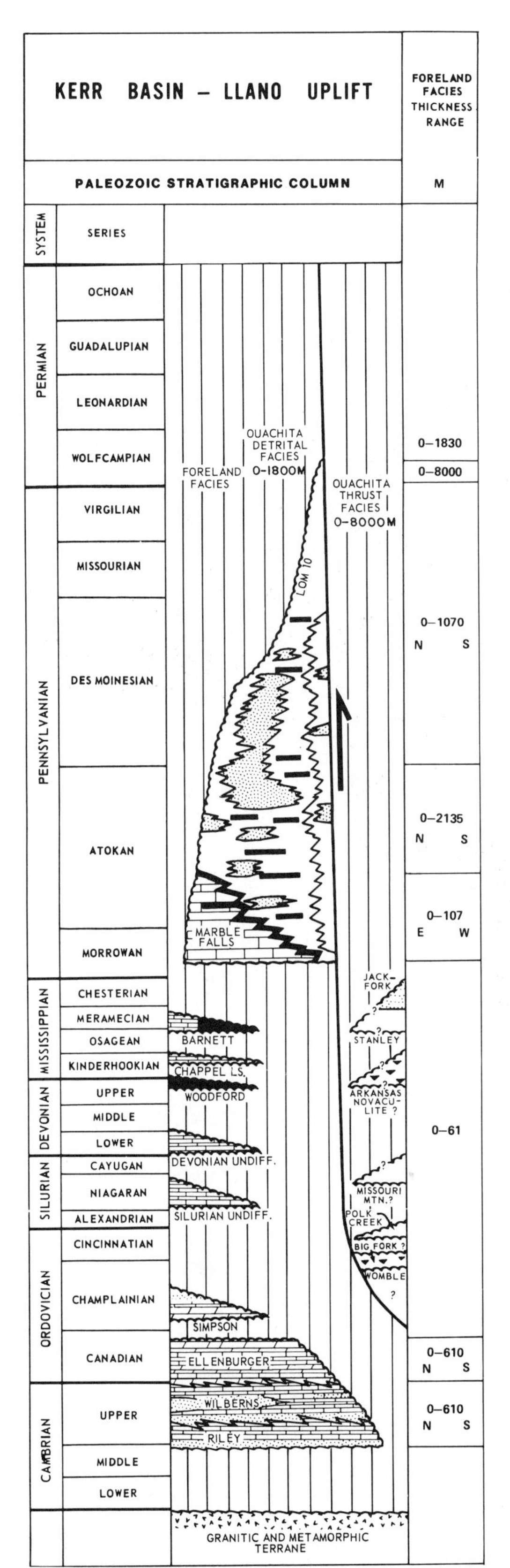

Figure 27. Paleozoic stratigraphic column for the Kerr Basin/Llano Uplift area.

HYDROCARBON POTENTIAL

Liquid hydrocarbons have been recovered from the Ellenburger Formation from wells drilled in the Kerr Basin and natural hydrocarbon gases have been recovered from several wells penetrating the Ellenburger south of the frontal Ouachita thrusts. The Ellenburger carbonates are unmetamorphosed over the central and northern portions. These facies have retained saline fluids and have had liquid hydrocarbon shows indicative of moderate thermal activity. Natural gas shows in the southern areas suggest a more severe thermal history. Marble Falls Limestone penetrations have encountered oil and gas shows including free-phase hydrocarbon recoveries. Other Pennsylvanian horizons have also had hydrocarbon shows reported from cutting samples.

Heavy oil is present in lower Cretaceous sandstones and limestones in portions of the basin.

THE MARFA BASIN

G. C. Luff and B. T. Pearson

LOCATION

The Marfa Basin is the southwesternmost segment of the Permian Basin complex. It is located in that part of West Texas generally known as Trans-Pecos Texas, or the Big Bend region, and encompasses parts of Brewster, Jeff Davis, and Presidio Counties (Fig. 3). It is expressed topographically as an intermontane basin or plateau, with an elevation of 1,500 m near its center at Marfa, bounded by two ranges: the 2,000-m-high Chinati–Sierra Vieja Mountains on the west and the 2,500-m-high Davis Mountains on the east.

GEOLOGICAL FRAMEWORK

In general, the Marfa Basin is the area of sedimentary rocks northwest of the frontal zone of the Ouachita–Marathon Foldbelt, south of the Diablo Platform and northeast of the Rio Grande. It is separated from the Delaware Basin to the northeast by the Hovey Anticline.

The "Paleozoic" Marfa Basin is bounded on the north by eroded uplifts of the Diablo Platform, where Permian carbonates (Hueco) and conglomerates (Pow Wow) lie directly on Precambrian rocks of the Precambrian Texas craton and thrust-faulted, folded Precambrian metamorphic rocks of the Van Horn Mobile Belt (Flawn, 1956). These Precambrian rocks form the basement underlying the Marfa Basin farther south. The basin is limited on the southwest by the uplifted, positive elements of the Chinati Mountains–Sierra Vieja ranges and to the southeast by the Ouachita–Marathon Frontal Zone (Fig. 28), (Flawn and others, 1961).

In the Marfa Basin proper, outcrops older than Cretaceous are scarce, as about 85 percent of the area is covered by Tertiary lavas, tuffs, and continental deposits, so that data supporting strat-

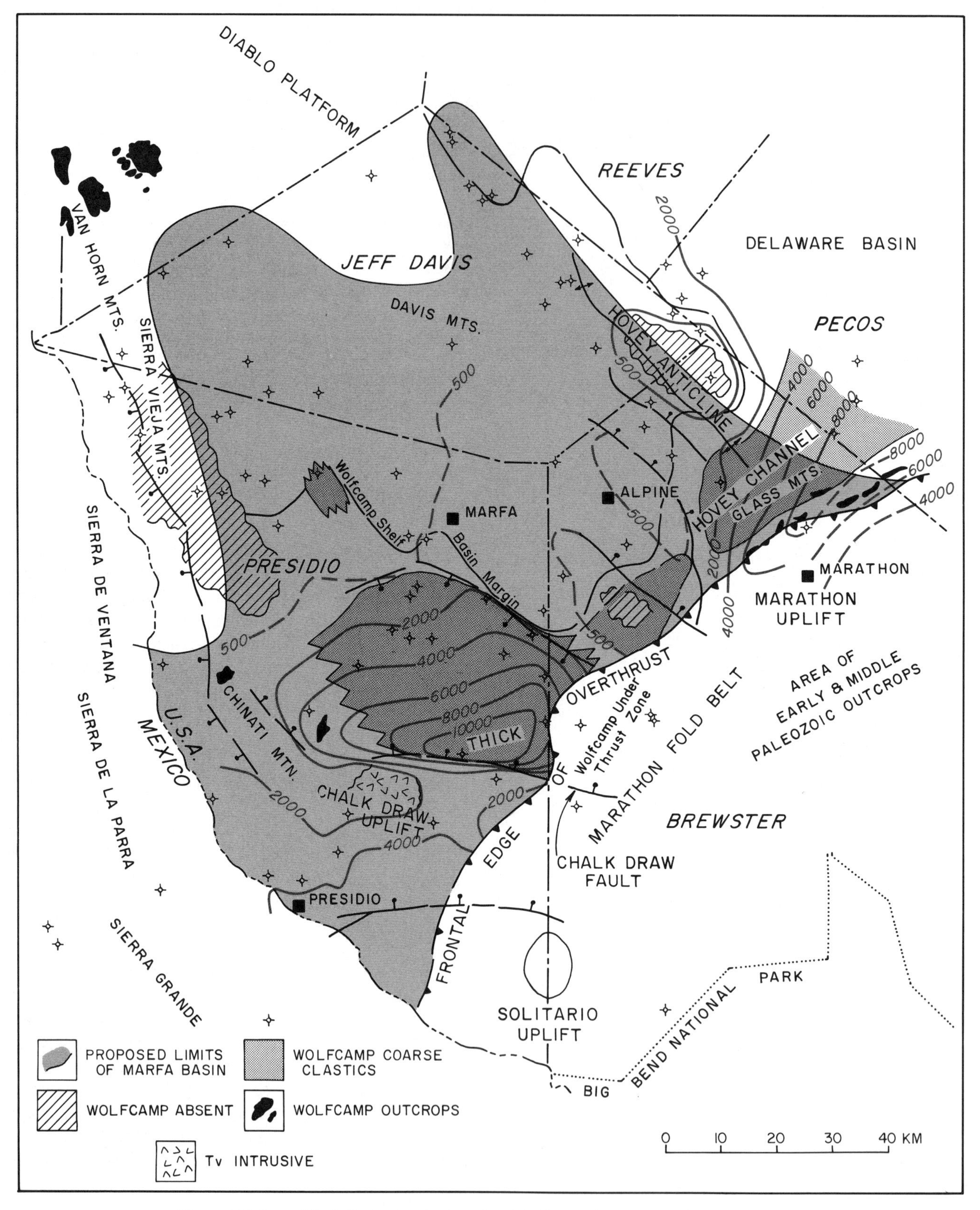

Figure 28. Map of Marfa Basin showing Permian (Wolfcamp) isopachs and facies distribution.

igraphic interpretations have been derived primarily from subsurface well control. Paleozoic rocks are exposed only in uplifted areas on the periphery of the basin. Marine Pennsylvanian rocks are found in the Chinati Mountains in southwestern Presidio County (Amsbury, 1958; Rix, 1953). Permian Wolfcampian, Leonardian, and Guadalupian rocks are also present in the Chinati Mountains, and Upper Permian Leonardian and Guadalupian formations crop out in the Glass Mountains of northern Brewster County (King, P. B., 1930; King, R. E., 1930).

Pre-Permian Paleozoics, which are completely missing from the Davis Mountains, north of Marfa, thicken southwestward across the basin toward the downthrown side of the west-striking Chalk Draw Fault (Pearson, 1981), which has as much as 3,300 m of displacement down to the north. Along this downfaulted southern margin of the Marfa Basin, the total Paleozoic marine section beneath the Tertiary volcanics and continental deposits is 5,000 m thick.

During Paleozoic time the Marfa Basin was characterized by northwest-striking faulted elements of the Diablo Platform on the north. The presence of Silurian rocks at depths of less than 1,000 m in the vicinity of Alpine indicates a shallow Precambrian core and a southerly projection of the Diablo Platform beneath the Tertiary volcanics of the Davis Mountains.

The southeastern boundary of the Marfa Basin is marked by folded and thrust-faulted Cambrian to Pennsylvanian rocks of basinal facies that are exposed at the centers of the Solitario and Marathon Uplifts; Cretaceous rocks lie with great disconformity on these older Paleozoic rocks, and Permian rocks are notably absent. Some older Permian rocks are present beneath the frontal overthrusts of the Ouachita–Marathon Foldbelt, but Permian rocks younger than early Permian (mid-Wolfcampian) are entirely missing (cross-section A-A′, Fig. 28).

EARLY GEOLOGICAL HISTORY AND STRATIGRAPHY

The Marfa Basin is correlative with the Delaware Basin in Paleozoic stages of development, stratigraphy, and lithologies. From Cambrian through early Pennsylvanian time both the Marfa and Delaware Basins were part of the great marine Tobosa Basin, which extended across Texas, New Mexico, and northern Mexico. The broad shallow seas covering the Permian Basin and west Texas resulted in widespread carbonate deposition, which in the Marfa Basin is reflected as an almost continuous sequence of shelf carbonates from early Ordovician through Silurian times. The Paleozoic ocean extended from the Delaware Basin on the east to the Placer de Guadalupe area of Chihuahua, Mexico (Bridges, 1964) and on as far west as the Aldama Platform of western Mexico. As in the Delaware Basin, Woodford Shales were deposited over the area, followed by deposition of Mississippian carbonates. Late Mississippian–Early Pennsylvanian erosion, however, has removed many of the lower Paleozoic units from local uplifts.

The Marfa Basin was affected by the same orogeny that marked the beginning of the Pennsylvanian in adjacent parts of the Permian Basin. Pennsylvanian–Early Permian tectonism formed northwest-trending, fault-bounded uplifts and depressions contemporaneously with similar structures of the Diablo Platform, the Delaware Basin, and the Central Basin Platform. By middle Pennsylvanian time the gently sloping shelf of the western Delaware Basin continued into the Marfa Basin, producing the same shelf-margin-basin sequence of sedimentation. These margins are characterized by carbonate banks and "reef" developments.

Widespread erosion at the beginning of the Permian removed a great deal of the Pennsylvanian section, making reconstruction of the Pennsylvanian history difficult. A thicker section of Pennsylvanian rocks north of the Chalk Draw fault suggests that a depositional depression was beginning to form there in late Pennsylvanian time.

In late Pennsylvanian to mid-Permian time, a trend of block faulting developed in a general northwest–southeast alignment, perhaps as rejuvenation along pre-existing faults or zones of weakness. There is no evidence of faulting in front of (northwest) or parallel to the Ouachita–Marathon Foldbelt.

Early Wolfcampian time actually marks the age of major structural development of the Marfa Basin as a separate entity within the Permian Basin region, when a deep depression formed north of the Chinati–Chalk Draw Uplift and the west-east–trending Chalk Draw Fault (Fig. 28). A downwarping in the central portion of the basin began to receive coarse clastics from surrounding uplifted areas. Continued subsidence allowed deposition of over 3,350 m of Wolfcampian sediments until the "deep" filled and the Wolfcampian depositional environment followed the shelf-margin-basin pattern typical of other units in the Pennsylvanian and Permian (Luff, 1981).

To the southeast, frontal overthrusts of the Ouachita–Marathon Foldbelt moved northwestward over and around buttresses of the Diablo Platform and formed a salient into the relatively depressed Marfa Basin. Well data from Brewster County date the final pulsation of the Ouachita-Marathon movement in the Marfa Basin as Wolfcampian rather than the beginning of the Permian. This overthrusting ended in middle Wolfcampian (Permian) time (Flawn, 1956) and is shown at the southeast end of cross-section A-A′ (Figure 29).

Upthrust Ordovician (Maravillas) limestones and cherts and Devonian (Caballos) novaculites were sources of Permian clastics only along the southeastern basin margin. From south to north, rocks of Ordovician to Pennsylvanian age rapidly change lithology from thick dark clastics, dark limestones, cherts, and novaculites of the Marathon Foldbelt to lighter gray and brown foreland carbonates and thinner cherts and clastics of the Diablo foreland.

Late Permian marine sedimentation in the basin was only slightly influenced by the surrounding uplifted areas of the Diablo Platform and the Ouachita–Marathon Foldbelt. Depositional patterns of the Permian Leonardian and Guadalupian exhibit thick vertical carbonate development, as demonstrated by outcrops in the Glass Mountains. Seismic and well data indicate that

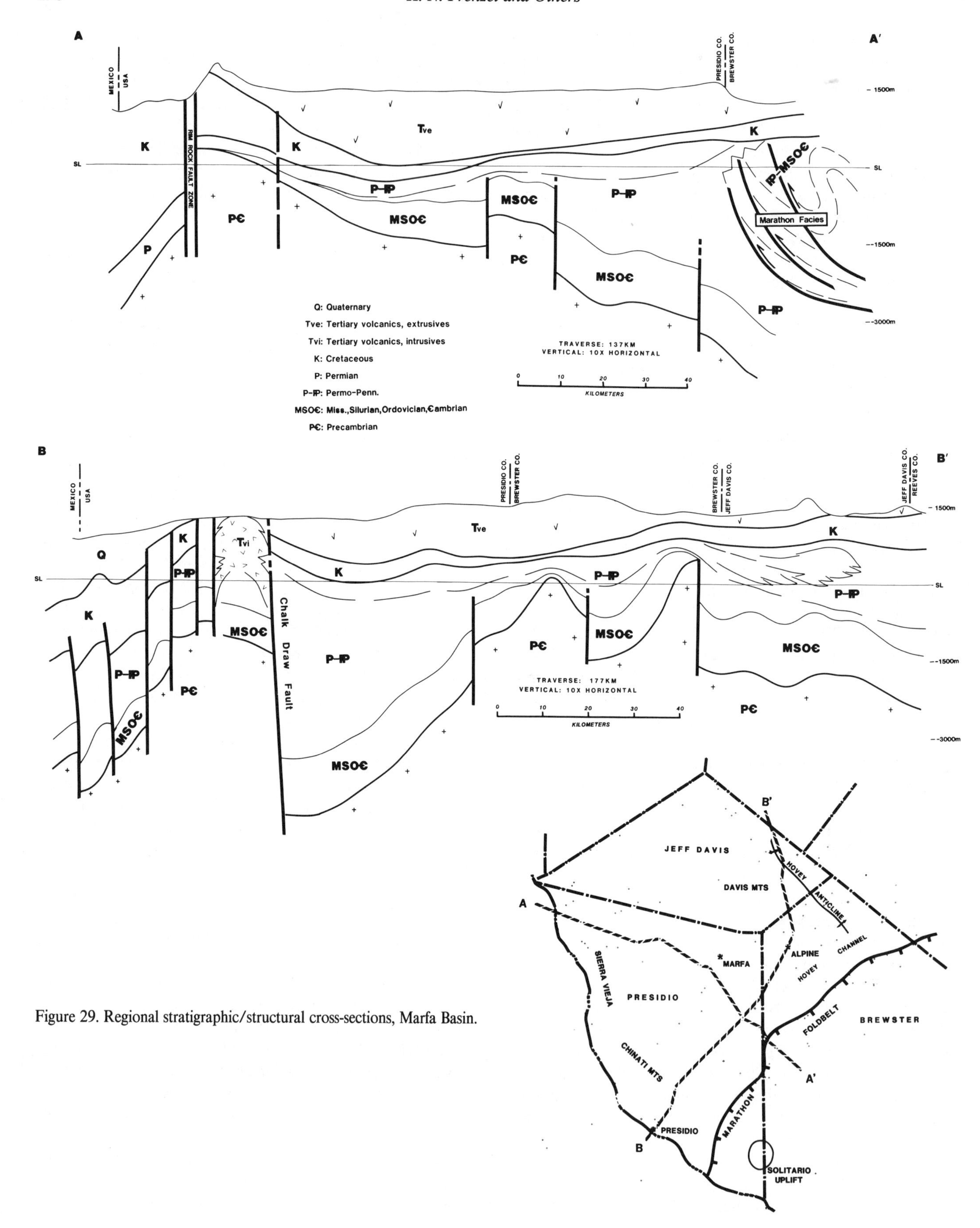

Figure 29. Regional stratigraphic/structural cross-sections, Marfa Basin.

the foldbelt tectonics played relatively little or no part in forming the structural pattern of the basin. They did, however, provide the southern platform edge for development of late Wolfcampian margin carbonates and the shaping of the Hovey Channel.

Permian carbonates were deposited on a shallow shelf between the deeper parts of the Marfa Basin and the Delaware Basin until late Permian (Guadalupian) time, when the Marfa Basin was partly closed off and remained connected to the Delaware Basin only by the Hovey Channel. The channel became a restricted northeast-trending, reef-bounded waterway between landmasses of the Diablo Platform on the northwest and the Ouachita–Marathon Foldbelt on the southeast. The existence of a similar seaway connecting the Marfa Basin to the Permian sea of the Chihuahua region is indicated by Guadalupian reefs in the vicinity of Shafter at the southeast end of the Chinati Mountains.

By the end of Permian-Guadalupian time, the Hovey Channel served as the only direct connection between the Delaware Basin and the Chihuahuan Permian sea. As such it possibly served as a pipeline for the influx of saline waters into the Delaware Basin that replaced evaporating waters and contributed to the concentration of evaporites in the Delaware Basin during Ochoan times. Permian evaporites, which are present in the Delaware Basin on the east and the Malone Mountains area to the northwest, are not present in the Marfa Basin.

The end of the Permian in the Marfa region was again a period of uplift and peneplanation prior to inundation by Cretaceous seas.

LATE HISTORY

After Permian time the Marfa Basin was modified by the development of the Mesozoic Coahuila Platform and the sinking of the adjacent Chihuahua Trough, so that the Cretaceous, which is less than 200 m thick around Marfa and Alpine, becomes 2,300 m thick near the Rio Grande at Presidio. Thick, porous carbonates cropping out at several localities along the Rio Grande indicate hinge-line trends of the Early and Late Cretaceous of the Chihuahua Trough.

Post-Permian deformation decreases in intensity northeastward across the Davis Mountains and finally dies out in the Delaware Basin. Movements during the Laramide orogeny and later periods of Tertiary basin-and-range faulting and Tertiary volcanism transformed the basin into a topographic and structural basin that extends from the Salt Basin on the north to Udden's Sunken Block (Maxwell and others, 1967) on the south. The present overall structural configuration of the Paleozoic rocks of the Marfa Basin is that of a northwest-trending graben, tilted to the southwest, stepping down by faulting from northeast to southwest, covered by 1,000 to 1,500 m of shallow-dipping, normal-faulted Tertiary volcanics and continental deposits. Some recent faulting and seismic activity is evident in the displacement of Quaternary beds and in the 1931 Valentine earthquake.

There is considerable speculation regarding the amount of lateral displacement that may have resulted from structural movements during Laramide and late Tertiary time, the possibility of rifting associated with the Rio Grande Rift, and the relationship of the Marfa Basin to the Basin-and-Range or Cordilleran structural provinces of the western United States and Mexico.

PEDREGOSA AND OROGRANDE BASINS

F. E. Kottlowski, S. Thompson III, W. E. King, and W. R. Seager

The Pedregosa and Orogrande Basins lie mainly in southern New Mexico and extend into adjacent parts of Texas, Mexico, and Arizona. Major sedimentation, as with the Permian Basin, was during late Paleozoic time, mainly during Pennsylvanian time. The present topography is desert country typical of the Mexican Highland and Sacramento sections of the Basin-and-Range physiographic province, where north–south-trending mountain ranges are separated by broad basins.

PEDREGOSA BASIN

The Pedregosa Basin (Zeller, 1965; Kottlowski, 1960, 1963, 1965) lies in the "Cuatro Fronteras Region." As defined by a thickness of 600 m or more of Pennsylvanian strata, it extends from southeastern Arizona, southeastward through southwesternmost New Mexico, the northwesternmost tip of Sonora, and into northwestern and central Chihuahua (Fig. 30). Stratigraphic data are from the several dozen oil tests in the area (Thompson and others, 1978; Thompson, 1982) and superb exposures in the Big Hatchet Mountains of New Mexico (Zeller, 1965; Thompson and Jacka, 1981), Sierra de Palomas in Chihuahua (Diaz and Navarro, 1964; Tovar, 1969), in the Pedregosa Mountains in southeastern Arizona (Epis, 1956), and other ranges.

Structural Geology

The east–southeast-trending Pedregosa Basin lies southwest of the Burro-Florida-Moyotes uplifts and east of the Magdalena Uplift (Navarro and Tovar, 1974). The Florida and Moyotes Uplifts were active areas of erosion during late Paleozoic time, whereas the Burro Uplift and the Graham and Florence Uplifts to the northwest in Arizona may not have developed until Laramide time because upper Cretaceous to Tertiary rocks overlie Precambrian rocks on their crests (Turner, 1962).

The Laramide Orogeny, during late Cretaceous to early Tertiary time, produced locally intense folding and thrusting. Andesitic volcanism covered moderate to large areas, and monzonitic to dioritic intrusions produced mineralization of the region (Elston, 1970; Thompson, 1976). Tertiary volcanic rocks form large parts of some of the mountain ranges; these are Eocene to early Miocene in age and range in composition from rhyolitic ash-flow tuffs to basaltic andesites. Eruptive centers are cauldrons that range in diameter from 8 to 24 km (Elston and others, 1979). Basin-and-range block faulting, occurring mainly in early to late Miocene time, is responsible for much of the present topography.

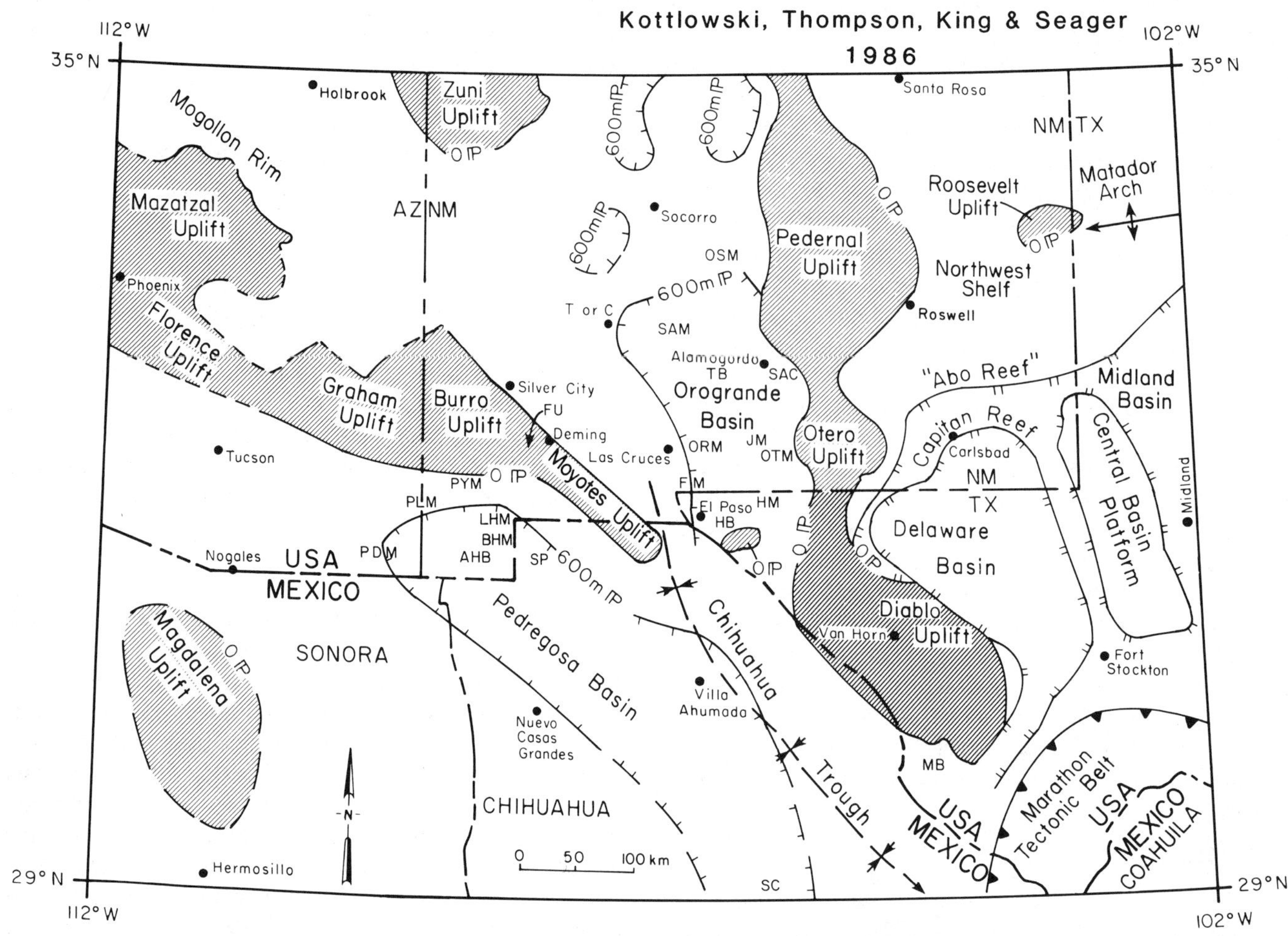

Figure 30. Regional setting of Pedregosa and Orogrande basins. AHB, Alamo Hueco Basin; BHM, Big Hatchet Mountains; FM, Franklin Mountains; FU, Florida Uplift; HB, Hueco Basin; HM, Hueco Mountains; JM, Jarilla Mountains; LHM, Little Hatchet Mountains; MB, Marfa Basin; ORM, Organ Mountains; OSM, Oscura Mountains; OTM, Otero Mesa; PDM, Pedregosa Mountains; PLM, Peloncillo Mountains; PYM, Pyramid Mountains; SAC, Sacramento Mountains; SAM, San Andres Mountains; SC, Sierra del Cuervo; SP, Sierra de Palomas; TB, Tularosa Basin (modified from Greenwood and others, 1977).

Several significant Laramide thrust zones cross the northern part of the Pedregosa area (Zeller, 1975; Drewes, 1978; Woodward and DuChene, 1981). They strike northwest–southeast, with Precambrian or Paleozoic rocks displaced northeastward at least a few kilometers over Lower Cretaceous strata. Thompson (1980) suggested that some major thrust zones may be related to northwest–southeast-trending basement wrench faults, which thus would affect only a smaller part of the Pedregosa Basin and the surrounding margin-to-shelf facies of late Paleozoic rocks.

Rock Sequences

Depositional strike of the Lower Paleozoic rocks was generally east–west across the Pedregosa area. Beginning in Mississippian time, a northwest–southeast-trending axis began to subside to form the Pedregosa Basin (Greenwood and others, 1977).

In most of the Pedregosa area the Bliss Sandstone (Upper Cambrian–Lower Ordovician) overlies the Precambrian granites and granite gneisses. To the west, the Precambrian is overlain by the Bolsa Quartzite (Middle Cambrian) and the Abrigo carbonate rock unit (Upper Cambrian) (see COSUNA chart of Southwest/Southwest Mid-Continent Region; Hayes, 1975). Gradationally overlying the Bliss Sandstone are shallow-marine to intertidal and supratidal limestones and dolostones of the El Paso Formation (Lower Ordovician). Locally, the Bliss is absent, and the El Paso onlaps Precambrian basement highs. The top of the El Paso is truncated by pre-Montoya erosion in much of the region. In the Montoya Group (Middle and Upper Ordovician), the thin basal

Cable Canyon Sandstone grades upward into the main dolostone and chert units of the Montoya.

Overlying the Montoya Group on a gently undulatory disconformity is the Fusselman Dolomite (Lower to Middle Silurian), a shallow-marine shelf deposit with some intertidal to supratidal units. It is overlain with pronounced unconformity by Upper Devonian rocks; its western and northern limits are the results of pre-Middle Devonian erosional truncation, which occurs in about the middle of the Pedregosa Basin area. Prior to deposition of the Upper Devonian rocks, the Fusselman and older Paleozoic formations were tilted down toward the south with their northern and western limits truncated by widespread erosion. Then an extensive blanket of black, shallow-marine shales of the Percha Formation (Upper Devonian) overlapped progressively older Paleozoic beds towards the north. In southeastern Arizona, the Martin Limestone (Middle and Upper Devonian) underlies the Percha Formation.

The contact of the Percha with the overlying Escabrosa Group (Mississippian) is transitional in the Pedregosa area. The Escabrosa consists of dark, argillaceous, cherty limestone and white crinoidol limestone. These shallow-marine deposits are thickest along a northwest–southeast trend, the first evidence of subsidence along the axis of the Pedregosa Basin. The uppermost part of the Escabrosa is transitional into the overlying Paradise Formation (Mississippian-Chesterian), which consists of shallow-marine to coastal deposits of oolitic, skeletal, and arenaceous limestones.

Prior to deposition of the Horquilla Formation (Pennsylvanian-Permian), older rocks were tilted down to the south and their northern limits were truncated by erosion. In places, this pre-Pennsylvanian surface has relief of more than 300 m. Morrowan and Atokan strata consist of deltaic to shallow-marine clastics and dark limestones. During Desmoinesian time, shallow-marine limestones covered most of the area; however, along part of the axis of the southeastward-trending Pedregosa Basin, subsidence exceeded sedimentation, and deep-marine limestones, shales, and sandstones were deposited. This deep-marine depositional area within the Pedregosa was named the Alamo Hueco Basin by Zeller (1965); it is fringed by carbonate-rock reefs. Similar deep-marine sedimentation continued in the Alamo Hueco Basin into Missourian, Virgilian, and Early Wolfcampian times. Thick, porous dolostone units are exposed in the Big Hatchet Mountains along the north side of the basin (Thompson and Jacka, 1981). Transition of shelf-to-basin rocks is shown in outcrops of the central Big Hatchet Mountains. Deep-marine deposits include carbonate turbidites and debris flows interstratified with basinal mudstones and limestones. Schupbach (1973) noted an ancient submarine canyon along the shelf margin, suggesting that proximal parts of the basin may contain significant quantities of sandstone in submarine-fan deposits. The basinal facies extend southeastward into Chihuahua, as indicated by the deep-marine mudstones and limestones described by Wilson and others (1969) and the outcrops in Sierra del Cuervo (Tovar and Valencia, 1974).

Unconformably overlying the Horquilla is the Earp Formation (Upper Wolfcampian–Leonardian). It consists of red-bed facies in the Big Hatchet area and a limestone facies to the southeast similar to the Abo-Hueco transition in the Orogrande Basin area. The Earp is overlain conformably by the Colina-Epitaph (Leonardian) units, a shallow-marine limestone-dolostone transition with some gypsum and clastic deposits in the upper part. Overlying a low-relief disconformity on top of the Epitaph are the Scherrer Sandstone and the Concha-Rainvalley Limestones (Leonardian-Guadalupian), lithostratigraphic correlatives of the Glorieta and San Andres formations of south-central New Mexico.

Triassic-Jurassic sedimentary rocks have not been recognized in the Pedregosa region. Triassic red beds are present to the northwest in northeastern Arizona. Upper marine Jurassic deposits occur to the east in the Chihuahua trough near El Paso and southward. Lower Cretaceous rocks lie unconformably on the Permian in the Pedregosa area (Hayes, 1970); they are as much as 5,000 m thick in the Pedregosa Basin area and consist of Hell-to-Finish red beds, overlain by the U-Bar limestones with rudistids, and then Mojado deltaic to shallow-marine sandstones and mudstones.

During Laramide time, a subaerial erosion surface was developed. It marked the final withdrawal of the seas from the Pedregosa area. In the Burro-Florida Uplift areas, the older rock units were eroded such that Upper Cretaceous or Lower Tertiary nonmarine sedimentary or volcanic rocks locally rest unconformably on Precambrian granite. Middle Tertiary volcanic rocks and local intrusive masses were in place prior to Basin-and-Range faulting. Upper Tertiary-Quaternary clastic deposits filled the graben valleys.

Hydrocarbon Potential

Oil and gas shows have been reported from Paleozoic units in the Pedregosa Basin (Wengerd, 1970F). Petroleum source rocks have been documented in many of the Paleozoic and Cretaceous units (Thompson, 1981); most are gas prone, but significant amounts of oil probably were generated locally. Horquilla shelf-edge dolostones on the margin of the Alamo Hueco (deep-marine) Basin appear to be the most promising petroleum-reservoir units (Thompson, 1980). Other promising reservoirs are shelf dolostones of the Fusselman, fractured-carbonate reservoirs in the El Paso, rudistid limestone buildups of the U-Bar, and deltaic sandstones of the Mojado. Lesser quality reservoirs are seen in the Bliss sandstones, Montoya dolostones, Escabrosa limestones, Paradise limestones, Epitaph dolostones, and Concha carbonates (Greenwood and others, 1977). Rocks beneath the widespread unconformities within the Paleozoic sequence are also exploration targets.

OROGRANDE BASIN

The Orogrande Basin, (Pray, 1959; Thompson, 1942; Wilson, 1967; Kottlowski, 1960, 1963, 1965; Kottlowski and others,

1956), is a late Pennsylvanian–early Permian depositional feature in south-central New Mexico. As arbitrarily outlined by the 600-m isopach thickness of Pennsylvanian rocks, it is roughly 65 km wide and extends from about the southern Oscura Mountains southward for more than 180 km into westernmost Texas. As such, it is almost contiguous with the present day Tularosa and Hueco basins. Superb outcrops of the basinal facies occur in the San Andres, Organ, and Franklin mountains, which bound these present-day structural basins on the west. To the east are the Sacramento Mountains, and their southern extension, Otero Mesa, and the Hueco Mountains.

Geologic Structure

Nearly continuous fault scarps of Quaternary age occur along the eastern edge of the Franklin, Organ (Dunham, 1935), and San Andres Ranges along the west side of the Tularosa and Hueco Basins (Seager, 1980). These are recent expressions of the predominant fault-block structure; both the Tularosa and Hueco Basins are asymmetric, west-tilted grabens, deepest along their western margin, according to available gravity, aeromagnetic, and drill hole data (Mattick, 1967; Bath, 1977; Healey and others, 1978). Generalizing, the Sacramento Mountains, and their northward extension of Sierra Blanca and southern extensions of Otero Mesa and Hueco Mountains, form the eastern wall of the Tularosa-Hueco Grabens; these mountain blocks are uplifted sharply on the west, with the Paleozoic section dipping off to the east. On the west side of the Tularosa-Hueco Grabens, the San Andres–Organ–Franklin chain of ranges is uplifted sharply on the east, and sedimentary beds dip gently to steeply to the west.

In the eastern third of the Tularosa-Hueco basins, raised blocks, particularly on the east side of the Hueco Basin, occur with indication of eastward dip; whether this series of irregular blocks extends from the southern part of the Tularosa Basin toward its northern end is unclear from geophysical data, although obviously the Jarilla Mountains and other isolated blocks west of Alamogordo and Tularosa, where there are areas of Permian outcrops, indicate a complex structure in the eastern part of the basin. These easterly tilted blocks may continue slightly to the west of north to join the southern part of the Oscura Mountains, which are uplifted sharply on their west side and dip off to the east and northeast.

Rock Sequences

Precambrian rock sequences exposed in the San Andres, Organ, and Franklin Mountains are various granitic and metamorphic rocks with locally thick metasediments apparently of two ages, roughly 1.65 Ga and 1 Ga.

Lower Paleozoic rock units are similar to those in the Pedregosa Basin area, the Cambrian-Ordovician Bliss Sandstone, Ordovician El Paso and Montoya groups, Silurian Fusselman Dolomite, and Devonian shales and siltstones (see COSUNA chart of Southwest/Southwest Mid-Continent Region).

Mississippian rocks are conformable with the underlying Devonian strata and are composed of crinoidal debris, calcareous mud, and biohermal mounds in most of the central and northern part of the region; to the south in the southern Sacramento and Franklin Mountains, the crinoidal Lake Valley Formation is missing (Wilson, 1975), probably because of nondeposition. From the southern San Andres and southern Sacramento Mountains southward, the deep marine deposits of black calcareous mud of the Las Cruces and Rancheria Formations accumulated in the deep, starved basin of middle and Late Mississippian time. These basinal deposits buried an irregular karsted and channeled surface formed on the Lake Valley carbonate rocks or on Devonian strata. Withdrawal of the Rancheria seas southward in latest Mississippian time is recorded by the nearshore clastic rocks and oolitic limestones of the Helms Formation.

The basic sedimentary framework changed in Pennsylvanian time in south-central New Mexico; most of the Pennsylvanian structural trends are north–south—at right angles to early Paleozoic trends. Before the Pennsylvanian strata were deposited, the pre-Pennsylvanian beds were tilted down to the south and generally were eroded in the northern part of the area. Erosional remnants of lower Paleozoic rocks are preserved locally. Widespread channels were cut into Mississippian and older strata and filled with debris of the eroded units, particularly of Mississippian cherts.

Morrowan carbonate rocks occur in the Franklin Mountains and onlap the pre-Pennsylvanian surface to the northwest; Morrowan clastics are present in the Sacramento Mountains, whereas the oldest Pennsylvanian strata in the northern San Andres Mountains appear to be of Atokan clastic rocks.

Desmoinesian-Missourian rocks of the Orogrande area are mainly shallow-marine limestones with local lenses of arkosic sandstone and shale derived from the early stages of the uplift of the Pedernal area to the east. In Virgilian time, and locally in late Missourian time, north–south complementary downwarping west of the Pedernal Uplift produced the main development of the Orogrande Basin. The thick, cyclic, mostly clastic deposits of the Panther Seep Formation accumulated in the basin with infilling processes approximately keeping pace with subsidence (Wilson, 1967). Intertidal and supratidal conditions were common as indicated by mud cracks, ripple marks, stromatolites, local occurrences of caliche and gypsum beds, and dolomitized limestone. Fluctuating sea level seems to explain the cyclic repetition of rock types, with coarser clastics forming during periods of low water, and fine-grained clastics or laminated carbonate mud during higher stands of sea level (Wilson, 1967). Maximum thickness of the Virgilian basinal sediments is about 750 m. In Early Permian time, subsidence of the Orogrande Basin continued, resulting in a thick lower Wolfcampian sequence. However, depositional conditions changed as marine limestones and shales dominate the section, with cyclic influxes of terrigenous clastics. In mid-Wolfcampian time, the Pedernal Uplift rose again, and tight folds and vertical faults developed on the western flank. After widespread erosion, nonmarine red-bed clastics of the Abo Formation were spread over the northwest part of the basin area. To the southeast, these redbeds intertongue with shallow-marine limestones of the Hueco Formation. This Abo-Hueco transition seen in the Sacramento (Pray, 1961) and San Andres Mountains takes place within a belt 25 to 50 km wide. Bioherms occur locally near the shorelines. On the east side of the Orogrande Basin, the shaly carbonate-rock Hueco facies grades northeastward, toward the Pedernal Uplift, into red-bed facies within a few mi (Otte, 1959).

In Leonardian time, marine waters covered the area with deposition of the interbedded dolomitic limestones, gypsum beds, sandstones, and siltstones of the Yeso Formation followed by the shallow-marine seas in which the San Andres Formation carbonate rocks were deposited.

Upper Permian (Guadalupian-Ochoan) rocks, as well as those of Triassic and Jurassic age, do not occur in the Orogrande Basin area. The region apparently was uplifted and eroded until Early Cretaceous time when shallow-marine seas from the south spread to deposit the Sarten Sandstone. In Late Cretaceous time, seas from the north and east deposited the Dakota Sandstone and associated overlying beds of the Mancos Shale and younger Cretaceous units. Erosion during Tertiary time left only isolated remnants of the Cretaceous rocks in the region.

Laramide orogeny resulted in deposition of uppermost Cretaceous to Tertiary conglomerates and red beds. In the Organ Mountains (Seager, 1981) and Sierra Blanca regions, masses of andesitic lava flows and tuffs were erupted in Eocene-Oligocene time, followed by thick sequences of rhyolitic ash-flow tuffs and lava flows which, in the Organ Mountains area, were intruded by the Organ batholith of granite to monzodiorite composition.

The graben fill units, as much as 2,750 m thick, are the upper part of the Santa Fe Group and younger Quaternary deposits, generally alluvial fans, piedmont sediments, and lake beds.

REFERENCES CITED

Adams, J. E., 1944, Upper Permian Ochoa series of Delaware Basin, west Texas and southeast New Mexico: American Association of Petroleum Geologists Bulletin, v. 28, no. 11, p. 1596–1625.

——, 1965, Stratigraphic-tectonic development of Delaware Basin: American Association of Petroleum Geologists Bulletin, v. 49, p. 2140–2148.

Adams, J. E., and nine others, 1939, Standard Permian section of North America: American Association of Petroleum Geologists Bulletin, v. 23, p. 1673–1681.

Adams, J. E., and Frenzel, H. N., 1950, Capitan barrier reef, Texas and New Mexico: Journal of Geology, v. 58, p. 289–312.

Adams, J. E., Frenzel, H. N., Rhodes, M. L., and Johnson, D. P., 1951, Starved Pennsylvanian Midland Basin: American Association of Petroleum Geologists Bulletin, v. 35, no. 12, p. 2600–2607.

Amsbury, D. L., 1958, Geologic map of Pinto Canyon area, Presidio County, Texas: Texas Bureau of Economic Geology Quadrangle Map 22, scale 1:63,360.

Babcock, J. A., 1974, The role of algae in the formation of the Capitan Limestone (Permian Guadalupian), Guadalupe Mountains, West Texas and New Mexico [Ph.D. thesis]: Madison, University of Wisconsin, 241 p.

——, 1977, Calcareous algae, organic boundstones, and the genesis of the upper Capitan Limestone (Permian, Guadalupian), Guadalupe Mountains, West Texas and New Mexico, *in* Hileman, M. E., and Mazzulo, S. J., eds., Upper Guadalupian facies, Permian reef complex, Guadalupe Mountains, New Mexico and West Texas, 1977 Field Conference Guidebook: Permian Basin Section, Society of Economic Paleontologists and Mineralogists Publication no. 77-16, v. 1, p. 3–44.

Barnes, V. E., 1959, Stratigraphy of the pre-Simpson Paleozoic subsurface rocks of Texas and southeast New Mexico: University of Texas Publication no. 5924, pt. 1, p. 11–72.

Bartlett, J. R., 1854, Personal narrative of explorations: D. Appleton and Company, v. 1, 2, 412 p. (reprinted 1965 by Rio Grande Press, Chicago).

Bath, G. D., 1977, Aeromagnetic maps with geologic interpretations for the Tularosa Valley, south-central New Mexico: U.S. Geological Survey Open-File Report 77-258, 16 p.

Bloomer, R. R., 1977, Depositional environments of a reservoir sandstone in west-central Texas: American Association of Petroleum Geologists Bulletin, v. 61, no. 3, p. 344–359.

Bridges, L. W., 1964, Stratigraphy, structure, and geologic history of the Mina Plomosa–Placer de Guadalupe Area, *in* Geology of Mina Plomosas–Placer de Guadalupe Area, Chihuahua, Mexico: West Texas Geological Society Publication no. 64-50, p. 50–64.

Bubb, J. N., and Hatlelid, W. G., 1977, Seismic recognition of carbonate buildups, *in* Payton, C. E., ed., Seismic stratigraphy; Applications to hydrocarbon exploration, American Association of Petroleum Geologists Memoir 26, p. 205–212.

Bybee, H. P., and four others, 1931, Detailed cross section from Yates area, Pecos County, Texas, into southeastern New Mexico: American Association of Petroleum Geologists Bulletin, v. 25, p. 1087–1093.

Cartwright, L. D., Jr., 1930, Transverse section of Permian Basin, West Texas and southeast New Mexico: American Association of Petroleum Geologists Bulletin, v. 14, p. 969–981.

Cloud, P. E., Jr., and Barnes, V. E., 1948, The Ellenburger Group of central Texas: Austin, University of Texas Publication no. 4421, 473 p.

Conant, L. C., and Swanson, V. E., 1961, Chattanooga shale and related rocks of central Tennessee and nearby areas: U.S. Geological Survey Professional Paper no. 357, 91 p.

Condie, K. C., 1981, Precambrian rocks of the southwestern United States and adjacent areas of Mexico: Socorro, New Mexico Bureau of Mines and Mineral Resources, Resource Map 13, scale 1:1,460,640.

Conselman, F. B., 1954, Preliminary report on the geology of the Cambrian trend of west-central Texas, *in* Geological Contributions: Abilene Geological Society, p. 10–23.

Cook, T. D., and Bally, A. W., 1975, Stratigraphic atlas of North and Central America, Exploration Department, Shell Oil Company: Princeton, Princeton University Press, p. 272.

Cordell, L., Keller, G. R., and Hildenbrand, T. G., 1982, Bouguer gravity map of the Rio Grande Rift: U.S. Geological Survey, Geological Investigations Map GP-949, scale 1:1,000,000.

Cys, J. M., 1975, New observations on the stratigraphy of key Permian sections of West Texas, *in* Cys, J. M., and Toomey, D. F., eds., Permian exploration, boundaries, and stratigraphy: Midland, West Texas Geological Society and Permian Basin Section, Society of Economic Paleontologists and Mineralogists, p. 22–42.

——, 1976a, Wilkie Ranch Formation, new Permian unit in Finlay Mountains, West Texas: American Association of Petroleum Geologists Bulletin, v. 60, no. 7, p. 1107–1110.

——, 1976, Delaware Mountain Formation, *in* Lexicon of Permian stratigraphic names of West Texas and southeastern New Mexico: Midland, West Texas Geological Society Publication no. 76-66, p. 109–110.

——, 1977, Lower and Middle Paleozoic stratigraphy of the Tatum Basin, New Mexico, *in* The oil and gas fields of southeastern New Mexico: Roswell Geological Society, p. 1–17.

——, 1983, Lower Wolfcampian stratigraphic relationships in the Hueco Mountains, *in* Geology of the Sierra Diablo and southern Hueco Mountains, West Texas: Midland, Permian Basin Section, Society of Economic Paleontologists and Mineralogists, p. 73–75.

——, 1985, Lower Permian phylloid algal mounds, southern Tatum Basin, southeastern New Mexico, U.S.A., *in* Toomey, D. F., and Nitecki, M. H., eds., Paleoalgology, contemporary research, and applications: New York, Springer-Verlag, p. 179–187.

Cys, J. M., and ten others, 1977, Capitan Reef; Evolution of a concept, *in* Hileman, M. E., and Mazzullo, S. J., eds., Upper Guadalupian facies, Permian reef complex, Guadalupe Mountains, New Mexico and West Texas: Permian Basin Section, Society of Economic Paleontologist and Mineralogists, 1977 Field Conference Guidebook, Publication 77-16, v. 1, p. 201–322.

Davis, H. E., and others, 1953, North-south cross section through the Permian Basin of West Texas: Midland, West Texas Geological Society.

Diaz, G., T., and Navarro G., A., 1964, Lithology and stratigraphic correlation of the upper Paleozoic in the region of Palomas, Chihuahua, *in* Geology of Minas Plomosas–Placer de Guadalupe area, Chihuahua, Mexico: West Texas Geological Society Publication 64-50, p. 65–84.

Drewes, H., 1978, The Cordilleran orogenic belt between Nevada and Chihuahua: Geological Society of America Bulletin, v. 89, p. 641–657.

Dunbar, C. O., and Skinner, J. W., 1937, Permian fusulinidae of Texas: University of Texas Bulletin no. 3701, Part 2, p. 517–825.

Dunham, K. C., 1935, Geology of the Organ Mountains: Socorro, New Mexico Bureau of Mines and Mineral Resources Bulletin 11, 272 p.

Dunham, R. J., 1972, Capitan Reef, New Mexico and Texas; Facts and questions to aid interpretation and group discussion: Permian Basin Section, Society of Economic Paleontologists and Mineralogists, no consecutive page numbers.

Eardley, A. J., 1949, Paleotectonic and paleogeologic maps of central and western North America: American Association of Petroleum Geologists Bulletin, v. 33, p. 655–682.

Edwards, E. C., 1927, Stratigraphic position of the Big Lime of West Texas: American Association of Petroleum Geologists Bulletin, v. 33, p. 655–682.

Ellison, S. P., Jr., 1946, Conodonts as Paleozoic guide fossils: American Association of Petroleum Geologists Bulletin, v. 30, p. 93–110.

——, 1950, Subsurface Woodford black shale, West Texas and southeast New Mexico: Austin, Bureau of Economic Geology, University of Texas Report of Investigations, no. 7, 20 p.

Elston, W. E., 1970, Volcano-tectonic control of ore deposits, southwestern New Mexico, *in* Tyrone–Big Hatchet Mountains, Florida Mountains region: New Mexico Geological Society, 21st Field Conference Guidebook, p. 147–153.

Elston, W. E., Erb, E. B., and Deal, E. G., 1979, Tertiary geology of Hidalgo County, New Mexico: New Mexico Geology, v. 1, no. 1, p. 1–6.

Emory, W. H., 1857, Report on the United States and Mexican boundary survey: U.S. 34th Congress, 1st Session, Senate Executive Document 108, 258 p., and House Executive Document 135, 174 p.

Epis, R. C., 1956, Geology of the Pedregosa Mountains, Cochise County, Arizona [Ph.S. thesis]: University of California, 181 p.

Feldman, M. L., and others, 1962, Southwest-northeast cross section Marathon region to Midland Basin: West Texas Geological Society Publication no. 62-47.

Fenneman, N. M., 1946, Physical divisions of the United States: U.S. Geological Survey Map, scale 1:7,000,000.

Flawn, P. T., 1956, Basement rocks of Texas and southeast New Mexico: Austin, Bureau of Economic Geology, University of Texas Ppublicdation no. 5605, 261 p.

Flawn, P. T., Goldstein, A., Jr., King, P. B., and Weaver, C. E., 1961, The Ouachita system: Austin, Bureau of Economic Geology, University of Texas Publication no. 6120, 361 p.

Gaines, R. B., and seven others, 1982, Oil and gas developments in West Texas and eastern New Mexico in 1981: American Association of Petroleum Geologists Bulletin, v. 66, no. 11, p. 1869–1878.

Galley, J. E., 1958, Oil and geology in the Permian Basin of Texas and New Mexico, *in* Weeks, L. G., ed., Habitat of oil, American Association of Petroleum Geologists Special Publication, p. 395–446.

Girty, G. H., 1908, The Guadalupian fauna: U.S. Geological Survey Professional Paper 58, 651 p.

Greenwood, E., Kottlowski, F. E., and Thompson, S., III, 1977, Petroleum potential and stratigraphy of Pedregosa Basin; Comparison with Permian and Orogrande basins: American Association of Petroleum Geologists Bulletin, v. 61, p. 1448–1469.

Guinan, M. A., 1971, More evidence of the slide-block event will follow Delaware Basin drilling: Oil and Gas Journal, July 6, p. 120–127.

Handford, C. R., 1981, Sedimentology and genetic stratigraphy of Dean and Spraberry Formations (Permian), Midland Basin, Texas: American Association of Petroleum Geologists Bulletin, v. 65, no. 9, p. 1602–1616.

Hardie, C. H., 1958, The Pennsylvanian rocks of the northern Hueco Mountains, *in* 1958 Field Trip Guidebook: Midland, West Texas Geological Society, p. 43–45.

Hardy, D. G., 1976, Bone Spring(s) Formation, *in* Lexicon of Permian stratigraphic names of West Texas and southeastern New Mexico: Midland, West Texas Geological Society Publication no. 76-66, p. 51–53.

Harms, J. C., 1974, Brushy Canyon Formation, Texas; Deep-water density current deposit: Geological Society of America Bulletin, v. 85, p. 1763–1784.

Hartman, J. K., and Woodward, L. R., 1969, Future petroleum resources in post-Mississippian strata of north central and West Texas and eastern New Mexico: American Association of Petroleum Geologists Memoir 15, v. 1, p. 752–799.

Hayes, P. T., 1970, Cretaceous paleogeography of southeastern Arizona and adjacent areas: U.S. Geological Survey Professional Paper 658-B, 42 p.

——, 1975, Cambrian and Ordovician rocks of southern Arizona, New Mexico, and westernmost Texas: U.S. Geological Survey Professional Paper 873, 98 p.

Healey, D. L., Wahl, R. R., and Currey, F. E., 1978, Gravity survey of the Rularosa Valley and adjacent areas: U.S. Geological Survey Open-File Report 78-309, 21 p.

Hill, R. T., 1900, The physical geography of the Texas region: U.S. Geological Survey Topographic Folio, no. 3, p. 4, 9.

Hills, J. M., 1942, Rhythm of Permian seas; A paleogeographic study: American Association of Petroleum Geologists Bulletin, v. 26, p. 217–255.

——, 1972, Late Paleozoic sedimentation in West Texas Permian Basin: American Association of Petroleum Geologists Bulletin, v. 56, p. 2303–2322.

Hills, J. M., and Hoenig, M. A., 1979, Proposed type sections for Upper Silurian and Lower Devonian subsurface units in Permian Basin, West Texas: American Association of Petroleum Geologists Bulletin, v. 63, p. 1510–1521.

Hills, J. M., and Kottlowski, F. R., 1983, Correlation of stratigraphic units of North America; southwest/southwestern mid-continent region chart: American Association of Petroleum Geologists.

Hollingsworth, R. V., and Williams, H. L., 1951, Review of subsurface Woodford black shale, West Texas and southeast New Mexico, American Association of Petroleum Geologists Bulletin, v. 35, 627 p.

Holmquist, H. J., 1965, Deep pays in the Delaware and Val Verde basins, *in* Fluids in subsurface environments: American Association of Petroleum Geologist Memoir 4, p. 257–280.

Hoover, J. D., Hills, J. M., Keller, G. R., and Roy, R. F., 1985, A deep drill hole on the central Basin Platform of West Texas [abs.]: Continental Scientific Drilling Workshop, Deep Observation and sampling of Earth's Continental Crust, Houston, Texas, p. 62–63.

Jones, T. S., 1953, Stratigraphy of the Permian Basin of West Texas: Midland, West Texas Geological Society, 63 p.

Jones, T. S., and others, 1949, East-west cross section through the Permian Basin of West Texas: Midland, West Texas Geological Society Publication 49-17.

Jones, T. S., and Matchus, E. J., 1984, East-west cross section through the Permian Basin of West Texas: Midland, West Texas Geological Society Publication 84-79.

Keller, G. R., Hills, J. M., and Djeddi, R., 1980, A regional geological and geophysical study of the Delaware Basin, New Mexico and West Texas:

Trans–Pecos Region, 31st Field Conference, New Mexico Geological Society Guidebook, p. 105–111.

King, P. B., 1930, Geology of the Glass Mountains, Texas, Part I: Austin, University of Texas Bulletin 3038, 167 p.

—— , 1942, Permian of West Texas and southeastern New Mexico: American Association of Petroleum Geologists Bulletin, v. 26, p. 535–763.

—— , 1948, Geology of the southern Guadalupe Mountains, Texas: U.S. Geological Survey Professional Paper 215, 138 p.

—— , 1959, The evolution of North America: Princeton, New Jersey, Princeton University Press, 190 p.

—— , 1965, Geology of the Sierra Diablo region, Texas: U.S. Geological Survey Professional Paper 480, 185 p.

King, P. B., King, R. E., and Knight, J. B., 1945, Geology of the Hueco Mountains, El Paso and Hudspeth Counties, Texas: U.S. Geological Survey Oil and Gas Investigation Preliminary Map 36, 2 sheets, scale 1:63,360.

King, R. E., 1930, Geology of the Glass Mountains, Texas, Part II: Austin, University of Texas Bulletin 3038, 245 p.

King, R. E., and five others, 1942, Resume of the geology of the southern Permian Basin, Texas and New Mexico: Geological Society of America Bulletin, v. 53, p. 539–560.

Kottlowski, F. E., 1960, Summary of Pennsylvanian sections in southwestern New Mexico and southeastern Arizona: Socorro, New Mexico Bureau of Mines and Mineral Resources Bulletin 66, 187 p.

—— , 1963, Paleozoic and Mesozoic strata of western and south-central New Mexico: Socorro, New Mexico Bureau of Mines and Mineral Resources Bulletin 79, 100 p.

—— , 1965, Sedimentary basins of south-central and southwestern New Mexico: American Association of Petroleum Geologists Bulletin, v. 49, p. 2120–2139.

Kottlowski, F. E., Flower, R. H., Thompson, M. L., and Foster, R. W., 1956, Stratigraphic studies of the San Andres Mountains, New Mexico: Socorro, New Mexico Bureau of Mines and Mineral Resources Memoir 1, 132 p.

Laudon, L. R., and Bowsher, A. L., 1949, Mississippian Formations of southwestern New Mexico: Geological Society of America Bulletin, v. 60, p. 1–87.

Lloyd, E. R., 1929, Capitan limestone and associated formations of New Mexico and Texas: American Association of Petroleum Geologists Bulletin, v. 13, p. 645–656.

Luff, G. C., 1981, A brief overview of the Marfa Basin, Marathon–Marfa region of West Texas: Permian Basin Section, Society of Economic Paleontologists and Mineralogists Publication no. 81-20, p. 110–129.

Malek–Aslani, M., 1970, Lower Wolfcampian Reef in Kemnitz Field, Lea County, New Mexico: American Association of Petroleum Geologists Bulletin, v. 54, no. 12, p. 2317–2335.

Marcy, R. B., 1850, Report of a survey for a road from San Antonio to El Paso through Guadalupe Pass: Report of Secretary of War, 31st Congress, 1st session, Senate Executive Document 64, v. 14, p. 14–24.

Mattick, R. E., 1967, A seismic and gravity profile across the Hueco Bolson, Texas, *in* Geological Survey Research 1967: U.S. Geological Survey Professional Paper 575-D, p. D85–D91.

Maxwell, R. A., Lonsdale, J. T., Hazzard, R. T. and Wilson, J. A., 1967, Geology of Big Bend National Park, Brewster County, Texas: University of Texas Publication 6711, 320 p.

Mazullo, S. J., 1981, Facies and burial diagenesis of a carbonate reservoir, Chapman Deep (Atoka) Field, Delaware Basin, Texas: American Association of Petroleum Geologists Bulletin, v. 65, no. 4, p. 850–865.

—— , 1982, Stratigraphy and depositional mosaics of Lower Clearfork and Wichita groups (Permian), northern Midland Basin, Texas: American Association of Petroleum Geologists Bulletin, v. 66, no. 2, p. 210–227.

McGlasson, E. H., 1967, The Siluro–Devonian of West Texas and southeast New Mexico: Tulsa Geological Society Digest, v. 35, p. 148–164.

—— , 1968, Reprint of the above: West Texas Geological Society Field Trip Guidebook, Hueco Mountains, Guadalupe Mountains, and Franklin Mountains: West Texas Geological Society Publication 68-55, p. 34–44.

Meyer, R. F., 1966, Geology of Pennsylvanian and Wolfcampian rocks in southeast New Mexico: Socorro, New Mexico Bureau of Mines and Mineral Resources Memoir 17, 123 p.

Namy, J., 1982, Stratigraphy and hydrocarbon production of the Marble Falls Group, southern part of the Fort Worth Basin, *in* Martin, C. A., ed., Petroleum Geology of the Fort Worth Basin and Bend Arch area: Dallas Geological Society, p. 213–222.

Navarro, G., A., and Tovar, R., J. C., 1974, Stratigraphy and tectonics of the State of Chihuahua, Mexico, *in* Geological Field Trip to the states of Chihuahua and Sinaloa, Mexico: West Texas Geological Society Publication 74-63, p. 87–91.

Newell, N. D., and six others, 1953, The Permian Reef complex of the Guadalupe Mountains region, Texas and New Mexico: San Francisco, W. H. Freeman and Company, 236 p.

Ohl, J. P., and Eveleth, R. W., 1985, New Mexico (Mineral), *in* Minerals Yearbook 1983, v. II: Washington, D.C., U.S. Government Printing Office, p. 375–389.

Otte, C., 1959, Late Pennsylvanian and Early Permian stratigraphy of the northern Sacramento Mountains, Otero County, New Mexico: Socorro, New Mexico Bureau of Mines and Mineral Resources Bulletin 50, 111 p.

Pearson, B. T., 1981, Some structural problems of the Marfa Basin, *in* Pearson, B. T., Marathon–Marfa region of West Texas: Permian Basin Section, Society of Economic Paleontologists and Mineralogists Publication no. 81-20, p. 59–73.

Pope, J., 1855, Report of exploration of a route for the Pacific railroad near the thirty–second parallel of latitude from the Red River to the Rio Grande: U.S. Pacific Railroad Exploration, U.S. Congress 1st Session, Executive Document 129, v. 18, pt. 2, 324 p.

Pray, L. C., 1959, Stratigraphic and structural features of the Sacramento Mountains escarpment, New Mexico: Roswell Geological Society Guidebook 12, p. 87–182.

—— , 1961, Geology of the Sacramento Mountains escarpment, Otero County, New Mexico: Socorro, New Mexico Bureau of Mines and Mineral Resources Bulletin 35, 144 p.

Rall, R. W., and Rall, E. P., 1958, Pennsylvanian subsurface geology of Sutton and Schleicher counties, Texas: American Association of Petroleum Geologists Bulletin, v. 42, no. 4, p. 839–870.

Richardson, G. B., 1904, Report of a reconnaissance in Trans–Pecos Texas north of the Texas and Pacific Railway: Austin, University of Texas Mineral Survey Bulletin 23, 119 p.

Rix, C. C., 1953, Geology of Chinati Peak Quadrangle, Trans–Pecos Texas [Ph.D. thesis]: Austin, University of Texas, 188 p.

Roswell Geological Society, 1956, Cross section of the Permian Basin, Eddy and Lea counties, New Mexico: Roswell, New Mexico, Roswell Geological Society.

Scholle, P. A., and Halley, R. B., 1980, Upper Paleozoic depositional and diagenetic facies in a mature petroleum province (a field guide to the Guadalupe and Sacramento mountains of West Texas and New Mexico): U.S. Geological Survey Branch of Oil and Gas Resources Administrative Report, 195 p.

Schupbach, M. A., 1973, Comparison of slope and basinal sediments of a marginal cratonic basin (Pedregosa Basin, New Mexico) and a marginal geosynclinal basin (southern border of Piemontais geosyncline, Switzerland) [Ph.D. thesis]: Houston, Rice University, 135 p.

Scobey, W. B., and others, 1951, North-south cross section through the Permian Basin of West Texas: West Texas Geological Society Bulletin 51-27.

Seager, W. R., 1980, Quaternary-age fault system in the Tularosa and Hueco Basins, southern New Mexico and West Texas: New Mexico Geological Society Guidebook, 32nd Field Conference, p. 131–135.

—— , 1981, Geology of Organ Mountains and southern San Andres Mountains, New Mexico: Socorro, New Mexico Bureau of Mines and Mineral Resources Memoir 36, 97 p.

Shannon, J. P., Jr., 1962, Hunton Group (Silurian–Devonian) and related strata in Oklahoma: American Association of Petroleum Geologists Bulletin, v. 46, p. 1–29.

Shumard, B. F., 1859, Notice of fossils from the Permian strata of New Mexico,

obtained by the United States expedition under Captain Pope: St. Louis Academy of Science Transactions, v. 1, p. 387–403.
Shumard, G. G., 1858, Observations on the geological formations of the country between the Rio Pecos and the Rio Grande in New Mexico, near the line of the 32nd parallel: St. Louis Academy of Science Transactions, v. 1, p. 273–289.
Silver, B. A., and Todd, R. G., 1969, Permian cyclic strata, northern Midland and Delaware basins, West Texas and southeastern New Mexico: American Association of Petroleum Geologists Bulletin, v. 53, no. 11, p. 2223–2251.
Sloss, L. L., 1963, Sequences in the cratonic interior of North America: Geological Society of America Bulletin, v. 74, p. 93–114.
Stewart, S. W., and Pakiser, L. C., 1962, Crustal structure in eastern New Mexico, interpreted from Gnome shot: Seismological Society of America Bulletin, v. 52, p. 1017–1030.
Thompson, M. L., 1942, Pennsylvanian system in New Mexico: Socorro, New Mexico Bureau of Mines and Mineral Resources Bulletin 17, 92 p.
Thompson, S., III, 1976, Tectonic and igneous effects on petroleum accumulations in southwestern New Mexico: New Mexico Geological Society Special Publications no. 6, p. 122–126.
—— , 1980, Pedregosa Basin's main exploration target is Pennsylvanian dolostone: Oil and Gas Journal, v. 78, p. 202, 207, 210, and 215.
—— , 1981, Petroleum source rocks in exploration wells drilled to Paleozoic and Mesozoic units, Hidalgo and Grant counties, New Mexico: Socorro, New Mexico Bureau of Mines and Mineral Resources Open–File Report 153, 120 p.
—— , 1982, Oil and gas exploration wells in southwestern New Mexico, *in* Powers, R. B., ed., Geologic studies of the Cordilleran thrust belt: Rocky Mountain Association of Geologists, p. 521–536.
Thompson, S., III, and Jacka, A. D., 1981, Pennsylvanian stratigraphy, petrography, and petroleum geology of Big Hatchet Peak section, Hidalgo County, New Mexico: Socorro, New Mexico Bureau of Mines and Mineral Resources Circular 176, 125 p.
Thompson, S., III, Tovar R., J. C., and Conley, J. N., 1978, Oil and gas exploration wells in the Pedregosa Basin: New Mexico Geological Society Guidebook, 29th Field Conference, p. 331–342.
Tovar, J., 1969, Stratigraphic study of the Sierra Santa Rita: New Mexico Geological Society Guidebook, 20th Field Conference, p. 165–170.
Tovar, J., and Valencia, R. J., 1974, Roadlog, Ojinaga to Chihuahua City, *in* Geologic Field Trip through the states of Chihuahua and Silaloa, Mexico: West Texas Geological Society Publication 74-63, p. 7–43.
Turner, G. L., 1962, Deming axis, southeastern Arizona, New Mexico, and Trans–Pecos Texas: New Mexico Geological Society, 13th Field Conference Guidebook, p. 59–71.
Udden, J. A., 1914, The deep boring at Spur: Austin, University of Texas Bulletin, no. 363, 90 p.
Van Siclen, D. C., 1958, Depositional topography; Examples and theory: American Association of Petroleum Geologists Bulletin, v. 42, no. 8, p. 1897–1913.
Vertrees, C. D., Atchison, C. H., and Evans, G. L., 1959, Paleozoic geology of the Delaware and Val Verde basins: West Texas Geological Society Publication 59-43, Guidebook to the Geology of the Val Verde Basin, p. 64–73.
Vertrees, C. D., and others, 1964, Cross section through the Delaware and Val Verde basins from Lea County, New Mexico, to Edwards County, Texas: West Texas Geological Society Publication, no. 64–49.
Vest, E. L., Jr., 1970, Oil fields of Pennsylvanian Permian Horseshoe Atoll, West Texas, *in* Geology of giant petroleum fields: American Association of Petroleum Geologists Memoir 14, p. 185–203.
Wanless, H. R., 1967, Eustatic shifts in sea level during the deposition of Late Paleozoic sediments in the central United States, *in* Elam, J. G., and Chuber, S., eds., Cyclic sedimentation in the Permian Basin: Midland, West Texas Geological Society, p. 41–54.
Wasserburg, G. J., Wetherill, G. W., Silver, L. T., and Flawn, P. T., 1962, Study of the ages of the Precambrian of Texas: Journal of Geophysical Research, v. 67, p. 4021–4047.
Wengerd, S. A., 1970, Petroleum prospects in southwesternmost New Mexico: New Mexico Geological Society, 21st Field Conference Guidebook, p. 91–104.
Wilde, G. L., 1976, Wolfcamp series, *in* Lexicon of Permian stratigraphic names of West Texas and southeastern New Mexico: Midland, West Texas Geological Society Publication 76-66, p. 327–331.
Williams, T. E., 1963, Fusulinidae of the Hueco Group (Lower Permian), Hueco Mountains, Texas: Peabody Museum of Natural History Bulletin 18, 122 p.
Williamson, C. R., 1979, Deep sea sedimentation and stratigraphic traps, Bell Canyon Formation (Permian), Delaware Basin, *in* Sullivan, N. M., ed., Guadalupian Delaware Mountains Group of West Texas and southeast New Mexico: Permian Basin Section, Society of Economic Paleontologists and Mineralogists, p. 39–74.
Wilson, J. L., 1967, Cyclic and reciprocal sedimentation in Virgilian strata of southern New Mexico: Geological Society of America Bulletin, v. 78, p. 805–817.
—— , 1975, Regional Mississippian facies and thickness in southern New Mexico and Chihuahua, Mississippian shelf-edge and basin facies carbonates, Sacramento Mountains and southern New Mexico region: Dallas Geological Society Guidebook, p. 125–128.
Wilson, J. L., Madrid-Solis, A., and Malpica-Cruz, R., 1969, Microfacies of Pennsylvanian and Wolfcampian strata in southwestern United States and Chihuahua, Mexico: New Mexico Geological Society Guidebook, 20th Field Conference, p. 80–90.
Witt, W. J., Jr., 1976, Castile Formation, *in* Lexicon of Permian stratigraphic names of West Texas and southeastern New Mexico: Midland, West Texas Geological Society Publication 76-66, p. 254–256.
—— , 1976b, Salado Formation, *in* Lexicon of Permian stratigraphic names of West Texas and southeastern New Mexico: Midland, West Texas Geological Society Publication 76-66, p. 259–260.
—— , 1976c, Dewey Lake Formation, *in* Lexicon of Permian stratigraphic names of West Texas and southeastern New Mexico: Midland, West Texas Geological Society Publication 76-66, p. 111–112.
Woodward, L. E., and Duchene, H. R., 1981, Overthrust belt of southwestern New Mexico, comparison with Wyoming–Utah overthrust belt: American Association of Petroleum Geologists Bulletin, v. 65, p. 722–729.
Wright, W. F., 1979, Petroleum geology of the Permian Basin: Midland, West Texas Geological Society, 98 p.
Yurewicz, D. A., 1976, Sedimentology, paleoecology, and diagenesis of the massive facies of the lower and Middle Capital Limestone (Permian), Guadalupe Mountains, New Mexico and West Texas [Ph.D. thesis]: Madison, University of Wisconsin, 278 p.
—— , 1977, The origin of the massive facies of the lower and middle Capitan Limestone (Permian), Guadalupe Mountains, New Mexico and West Texas, *in* Hileman, M. E., and Mazzullo, S. J., eds., Upper Guadalupian facies, Permian reef complex, Guadalupe Mountains, New Mexico and West Texas: Permian Basin Section, Society of Economic Paleontologists and Mineralogists, 1977 Field Conference Guidebook, Publication no. 77-16, v. 1, p. 45–92.
Zeller, R. A., 1965, Stratigraphy of the Big Hatchet Mountains, New Mexico: Socorro, New Mexico Bureau of Mines and Mineral Resources Memoir 16, 128 p.
—— , 1975, Structural geology of Big Hatchet Peak quadrangle, Hidalgo County, New Mexico: Socorro, New Mexico Bureau of Mines and Mineral Resources Circular 146, 23 p.

Manuscript Accepted by the Society December 15, 1986

Printed in U.S.A.

The Geology of North America
Vol. D-2, Sedimentary Cover—North American Craton: U.S.
The Geological Society of America, 1988

Chapter 12

Southern Midcontinent region

Kenneth S. Johnson and Thomas W. Amsden
Oklahoma Geological Survey, University of Oklahoma, Norman, Oklahoma 73019
Rodger E. Denison
Mobil Research Laboratory, 13777 Midway Road, Dallas, Texas 75244
Shirley P. Dutton
Bureau of Economic Geology, University of Texas at Austin, Austin, Texas 78713
Arthur G. Goldstein
Department of Geology, Colgate University, Hamilton, New York 13346
Bailey Rascoe, Jr.
5701 Cornell Drive, Bartlesville, Oklahoma 74006
Patrick K. Sutherland
School of Geology and Geophysics, University of Oklahoma, Norman, Oklahoma 73019
Diana Morton Thompson
Chevron U.S.A., P.O. Box 599, Denver, Colorado 80201

INTRODUCTION

Kenneth S. Johnson

The Southern Midcontinent is a complex region characterized by great thicknesses of sediments preserved in a series of major depositional and structural basins separated by orogenic uplifts created mainly during Pennsylvanian time (Plate 5-A). Sedimentary rocks of every geologic system from Precambrian through the Quaternary are preserved within the region, and their diverse lithologies include limestones, dolomites, sandstones, shales, conglomerates, red beds, and evaporites. The strata are a mixture of marine and nonmarine deposits, and generally (except for the red bed–evaporite sequences), they are richly fossiliferous and are well suited to biostratigraphic correlation and interpretation of depositional environments.

Dominant lithologies in most basins of the region are, in ascending order, as follows: a thin transgressive sandstone of Late Cambrian age that covered the basement-rock complex of intrusives, extrusives, and metasediments; overlain by a thick sequence of Late Cambrian through Late Mississippian (Meramecian) carbonates, with minor amounts of sandstone and shale; followed by a thick sequence of terrigenous clastics, with some carbonates, deposited from Late Mississippian (Chesterian) through Early Permian (Wolfcampian) time; then a thick series of red beds and evaporites were deposited during the remainder of the Permian; overlain, in the west only, by Triassic and Jurassic terrestrial red beds; then Cretaceous marine deposits in the south and west; and finally a mantle of Tertiary alluvial-fan, aeolian, and lacustrine sediments in the west.

Owing to the great thickness of strata in most basins of the region, and the many surface and subsurface stratigraphic studies that have been conducted, a plethora of stratigraphic names have been proposed and used in the Southern Midcontinent. The correlation charts and stratigraphic nomenclature used in this chapter are based primarily upon the recently completed COSUNA charts released by the American Association of Petroleum Geologists. The lithology and thickness of selected units are discussed throughout this chapter, and they are summarized in a series of maps (Plates 5 and 6). The thickness and lithology maps show the present distribution and thickness of each mapped unit (not the restored distribution and thickness), and they show the dominant lithology (i.e., mostly limestone, or mostly shale) present in each unit, unless otherwise noted in the explanation.

The Southern Midcontinent contains one of the greatest thicknesses of sedimentary rocks preserved in North America, with as much as 12,000 m of strata in the deep Anadarko Basin. Other major accumulations of sediments are in the Ardmore Basin (about 9,000 m), the southern part of the Arkoma Basin

Johnson, K. S., Amsden, T. W., Denison, R. E., Dutton, S. P., Goldstein, A. G., Rascoe, B., Jr., Sutherland, P. K., and Thompson, D. M., 1988, Southern Midcontinent region, *in* Sloss, L. L., ed., Sedimentary Cover—North American Craton; U.S.: Boulder, Colorado, Geological Society of America, The Geology of North America, v. D-2.

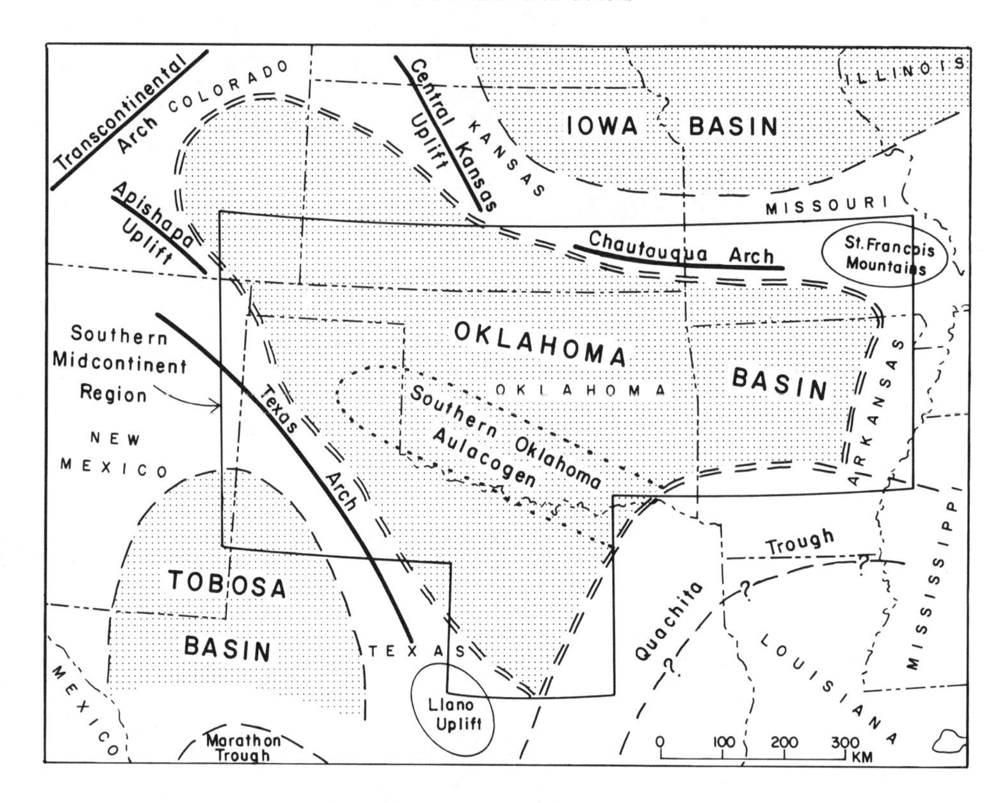

Figure 1. Map of southwestern United States showing outline of Southern Midcontinent region, as discussed in this chapter, and approximate boundary of the Oklahoma Basin and other major features that existed in parts of early and middle Paleozoic time.

(about 8,000 m), the Marietta Basin (about 8,000 m), and the Palo Duro, Dalhart, Hardeman, and Fort Worth Basins (each of which contains 3,000 to 4,000 m). These thick sedimentary packages accumulated along the southern margin of the North American Craton during Paleozoic episodes of subsidence of the Anadarko, Ardmore, and Marietta Basins, and of the foredeep areas north and west of the Ouachita Trough. The west-northwest–trending trough comprising the Anadarko, Ardmore, and Marietta Basins, and the associated uplifts, has been referred to as the Southern Oklahoma Geosyncline (Ham and others, 1964; Ham and Wilson, 1967) or the Southern Oklahoma Aulacogen (Gilbert, 1983; Brewer and others, 1983; see Fig. 1).

The Phanerozoic history of the Southern Midcontinent is characterized by three major periods or phases of tectonics and sedimentation: early and late phases of broad subsidence and epeirogenic movements were interrupted by a middle phase of sharp orogenic uplift and deep subsidence. The early epeirogenic phase is represented by Late Cambrian through Mississippian marine sediments deposited in a broad epicontinental sea, herein called the Oklahoma Basin, that extended across almost all parts of the Southern Midcontinent (Fig. 1). The Oklahoma Basin was a shelf-like area that received a sequence of remarkably thick and extensive marine carbonates interbedded with thinner marine shales and sandstones. These strata are readily correlated though-out the basin. The sediments thicken into protobasins (Anadarko, Ardmore, and others) accentuated later during Pennsylvanian orogenies, and they also were deposited upon and across the present-day major uplifts, from which they were subsequently stripped during Pennsylvanian uplift and erosion. Whereas the Southern Oklahoma Aulacogen was the depocenter for the Oklahoma Basin, the entire basin covered a much greater area: the basin extended from the Ouachita trough on the southeast to a series of broad arches on the north and west that were intermittently uplifted epeirogenically in early and middle Paleozoic time. Adler and others (1971) used the term Oklahoma Basin in referring to the Anadarko Basin and Hugoton Embayment areas during early and middle Paleozoic time, but the use in the current report is for a much larger area.

The middle phase of Phanerozoic development centered on orogenic activity during the Pennsylvanian Period. The broad, shallow-marine Oklahoma Basin was divided by sharply uplifted crustal blocks into a series of well-defined marine basins. The Ouachita trough was destroyed by uplift and northwesterly thrusting of the Ouachita Fold Belt (Plate 5-A). Orogenic activity was limited to folding, faulting, and uplift, and was not accompanied by igneous or metamorphic activity. Sedimentation in the rapidly subsiding basins was dominated by coarse and fine clastics shed from the nearby uplifts, although platform conditions persisted in some of the broad shelf areas away from the uplifts.

The final Phanerozoic phase, from the Permian Period until the present time, consisted of infilling of the basins in the west, and a return to epeirogenic movements. Western basins were filled with Permian red beds and evaporites, and subsequently with Triassic and Jurassic alluvial-deltaic and lacustrine red beds. The last marine incursion occurred when shallow Cretaceous seas spread over the south and west parts of the region. Regional eastward tilting of the Southern Midcontinent during uplift of the Rocky Mountains in Late Cretaceous and Early Tertiary time established the dominant east-flowing river systems, which deposited Late Tertiary alluvial, aeolian, and lacustrine sediments in the west, and Quaternary terraces and alluvial deposits throughout the region.

Owing to the major tectonic and sedimentologic changes that occurred in Pennsylvanian time, the remainder of this chapter is organized in three sections. (1) Pre-Pennsylvanian depositional environments and patterns of sedimentation were laterally persistent across most of the Oklahoma Basin, and thus each of the major pre-Pennsylvanian sedimentary units is discussed for the entire region in the first section. (2) Because Pennsylvanian orogenies subdivided the Southern Midcontinent into a number of major basins and uplifts, each with a somewhat different evolutionary history, the Pennsylvanian and Permian sedimentary rocks are discussed separately for each major basin (and surrounding uplifts) in the second section. (3) The third section contains a discussion of post-Permian rocks and sediments for the entire Southern Midcontinent region, inasmuch as after Permian time the separate major basins ceased having any significant effect upon sedimentation patterns.

The vast size of the Southern Midcontinent, and the complex history of its many basins and uplifts, required the cooperative efforts of eight authors, and each is credited for his/her individual contribution.

PRE-PENNSYLVANIAN ROCKS

REGIONAL STRATIGRAPHIC FRAMEWORK

Kenneth S. Johnson

Pre-Pennsylvanian strata of the Southern Midcontinent are mainly shallow-marine carbonates (limestone and dolomite), interbedded with several sandstone and shale units. These strata are remarkably widespread and laterally persistent throughout most parts of the Oklahoma Basin, reflecting the stability of this part of the craton and the importance of epeirogenic movements during early and middle Paleozoic time. Early studies of the pre-Pennsylvanian units were focused on outcrops in the widely separated Ozark Uplift, Arbuckle and Wichita Mountains, and Llano Uplift areas (Plate 5-A), and thus a different nomenclature for each area was established and ingrained before equivalency of the rock units could be fully established.

In most parts of the Oklahoma Basin, particularly the eastern, central, and southern parts, there is a basal, transgressive sand of late Middle to early Late Cambrian age, variously called the Reagan, Lamott, or Hickory Sandstone. This is overlain by a thick Cambro-Ordovician carbonate section, called the Ellenberger and Moore Hollow Groups in Texas, the Arbuckle Group in Oklahoma and Kansas, and by a series of formation names in Missouri and Arkansas.

The Middle Ordovician Simpson Group and equivalent strata in the eastern part of the region consists of widespread quartzose sandstones and carbonates. The sandstones, named the "Wilcox" sands in many oil fields of Oklahoma, are partly equivalent to the St. Peter Sandstone that extends northward across much of the midwestern United States. The overlying Viola Group carbonates, Sylvan Shale, and Hunton Group (and equivalent) carbonates, ranging in age from Late Ordovician to Middle Devonian, reflect continued shallow-marine shelf deposition in most parts of the region. Several widespread disconformities are due to epeirogenic uplifts.

Lateral persistence of a uniform lithology is best represented by the Upper Devonian–Lower Mississippian organic-rich black shales that stretch across the eastern, central, and southwestern United States. Named the Chattanooga Shale for exposures in Tennessee and the Woodford Shale for exposures in the Arbuckle Mountains, this unit overlies one of the most widespread unconformities of the Southern Midcontinent. Carbonates, cherty carbonates, and some shales typify the remaining Mississippian strata that overlie the Woodford-Chattanooga Shale in the region.

PRECAMBRIAN THROUGH MIDDLE CAMBRIAN

Rodger E. Denison

There are three substantial outcrops of basement rocks in the Southern Midcontinent: the St. Francois Mountains in southeast Missouri and the Wichita and Arbuckle Mountains in southern Oklahoma (Plate 5-B). In addition, several small outcrops of granite are exposed in northeast Oklahoma near the town of Spavinaw. In most of the area the basement is buried beneath Paleozoic cover at depths generally less than 3,000 m. The exceptions are in the Arkoma, Ardmore, and Anadarko Basins where the sedimentary cover is as much as 12,000 m. A large number of wells drilled in search of oil, gas, and other minerals have penetrated the basement in all but the deepest basins.

Detailed petrographic study of samples from these drill holes, together with geochronologic studies on both outcrop and subsurface samples, has led to a reasonably clear understanding of the basement upon which the Paleozoic rocks were deposited. Flawn (1956) in Texas provided the first successful regional study of the buried basement. Later work by Ham and others (1964) in southern Oklahoma, Muehlberger and others (1967) over the entire area, Kisvarsanyi (1979) in Missouri, Bickford and others (1981) in Kansas and Missouri, Denison (1981) in northeast Oklahoma, Denison (1984) in northern Arkansas, and Tweto (1983) in southeast Colorado have provided most of the data on which our understanding is based. A more complete summary of the basement is in Denison and others (1984).

The Precambrian basement in the Southern Midcontinent is characterized by granitic igneous rocks extruded or emplaced at shallow depths (Plate 5-B). Large areas are covered by rhyolites associated with comagmatic epizonal granites. Low-rank metasedimentary rocks account for only a small part of the basement surface, but analysis of deep seismic profiles (Brewer and others, 1981) suggests a possible Precambrian basin in southwest Oklahoma containing as much as 10 km of relatively undisturbed layered rocks. There is little direct evidence to indicate the extent of this basin, but seismic data show similar Precambrian reflections in parts of the Texas Panhandle where most of the basement surface is covered by rhyolites (Budnik, 1984). Massive granitic and metagranitic rocks make up most of the area between the rhyolite fields. Basaltic rocks are common as dikes and sills but are volumetrically not important. Much of the area shown in Texas is believed to be underlain by rocks related to those exposed in the Llano Uplift.

The ages of these Precambrian rocks range from about 1,400 to 1,600 Ma in Kansas and Missouri to near 1,000 Ma in Texas. Neither age nor compositional boundaries of these Precambrian rocks appears to have had a substantial effect on Paleozoic history. In contrast, the two much younger linear basins that strike into the stable interior had a profound influence on later history.

Virtually every aspect of later Paleozoic sedimentary thickness and structural style and direction is strongly influenced by a zone of weakness developed in southern Oklahoma during Cambrian time (Ham and others, 1964; Ham, 1969; Denison, 1982). Rocks emplaced at this time include Early and Middle Cambrian granites, rhyolites, gabbros, and basalts that extend across southern Oklahoma and into the Texas Panhandle: they were emplaced into and upon a sequence of sedimentary rocks referred to as the Tillman Metasedimentary Group. The sequence of Cambrian igneous events in southern Oklahoma is closely constrained and well documented, and this area contains the only suite of crystalline Cambrian basement rocks in the entire central interior of the United States and Canada.

Knowledge of the early history of the present Mississippi Embayment is based almost entirely on geophysical evidence and a few deep drill holes. Howe and Thompson (1984) have summarized the results of recent drilling as well as an interpretation of reflection seismic profiles and a summary of previous work. Only one well is known to have reached crystalline basement in the embayment (Denison, 1984). The well penetrated about 490 m of red arkose before bottoming in a granitic gneiss. The age of the red arkose may be Cambrian. The faults bounding the inferred arkosic basin have been the focus of later Cretaceous intrusions as well as a modern zone of seismicity.

Outside the area of Early and Middle Cambrian igneous activity in southern Oklahoma and the Texas Panhandle, there is no geologic record for the substantial time interval after the final crystallization or metamorphism of the Precambrian basement and before the Cambrian marine transgression. There was extensive erosion to expose rocks that were formed, in some cases, at considerable depth. The surface on which the Cambrian seas transgressed was generally of low relief (about 100 m), supported by a heterogeneous suite of mostly granitic igneous rocks swept clean of any significant debris.

MIDDLE CAMBRIAN THROUGH EARLY ORDOVICIAN

Rodger E. Denison

In Middle and Late Cambrian time seas transgressed to the north and west across the Southern Midcontinent onto what was to become the stable interior of the United States (Plate 5-C). The basement surface was of modest to locally rugged relief composed of largely granitic rocks ranging in age from about 1,600 Ma to 525 Ma. A time-transgressive basal sandstone was deposited except on topographic highs. This sand grades into an overlying succession of shallow-water carbonates deposited more or less continuously until the end of Early Ordovician time. These rocks are well exposed in central Texas, southern Oklahoma, and the St. Francois Mountains and Ozark Uplift of Missouri and Arkansas (Fig. 2) and have been penetrated in all but the deepest basins by numerous drill holes. In the subsurface, virtually any basal sandstone, regardless of age, has been designated with one or another of the basal Cambrian names.

In Oklahoma, the basal sandstone, called the Reagan, is overlain by a bioclastic limestone, the Honey Creek. These two units form the Timbered Hills Group. This is overlain by the Arbuckle Group, which is divided into six limestone units. Two secondary dolomites, formed at the expense of the primary limestones, have also been named (Fig. 2). The lower units are more distinctive lithologically than the upper units, which are identifiable only on the basis of stratigraphic position and fauna. Most of the southern Oklahoma information has been taken from Ham (1969) and work in Donovan (1986).

There are exceptional exposures of Arbuckle Group rocks in the Wichita and western Arbuckle Mountains. The thick, lime-mud–dominated succession of peritidal limestones was deposited on a vast carbonate platform that covered much of the Cratonic interior during the Late Cambrian and Early Ordovician. Although there is only a small, rather disseminated terrigenous

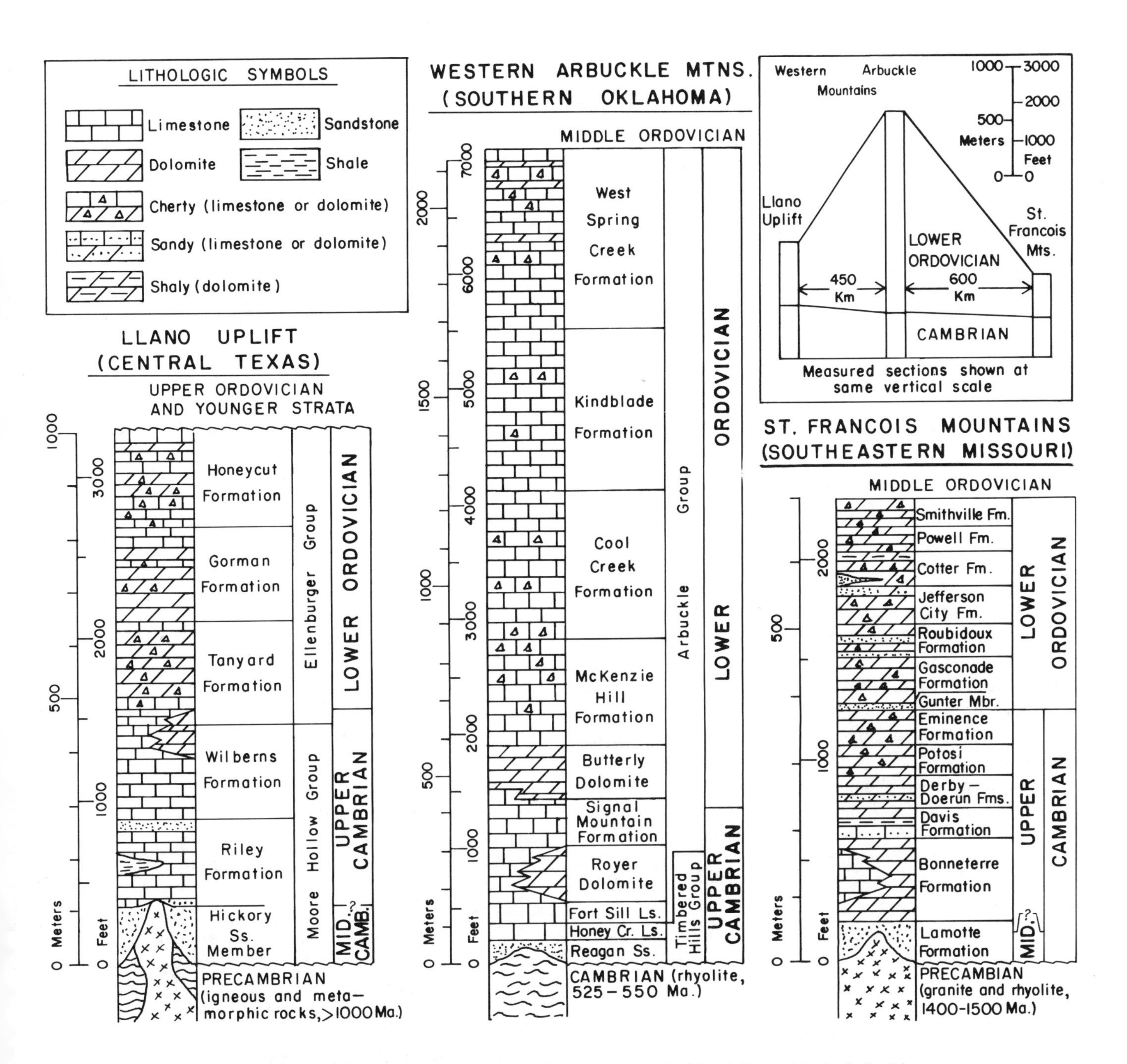

Figure 2. Subdivisions and comparison of lithology (not to scale) of Cambrian and Early Ordovician strata from the principal outcrop areas. The composite St. Francois Mountain section is taken from Hayes and Knight (1961) and Martin and others (1961). The Arbuckle Mountains section in Oklahoma was measured near what is now I-35 by Ham (1969). Age and lithologies in the Llano region are taken from Cloud and Barnes (1948) and Barnes and Bell (1977). Assignment of ages in Oklahoma from Stitt (1977) and Derby (1973) and in Missouri from Kurtz and others (1975) and Ross and others (1982).

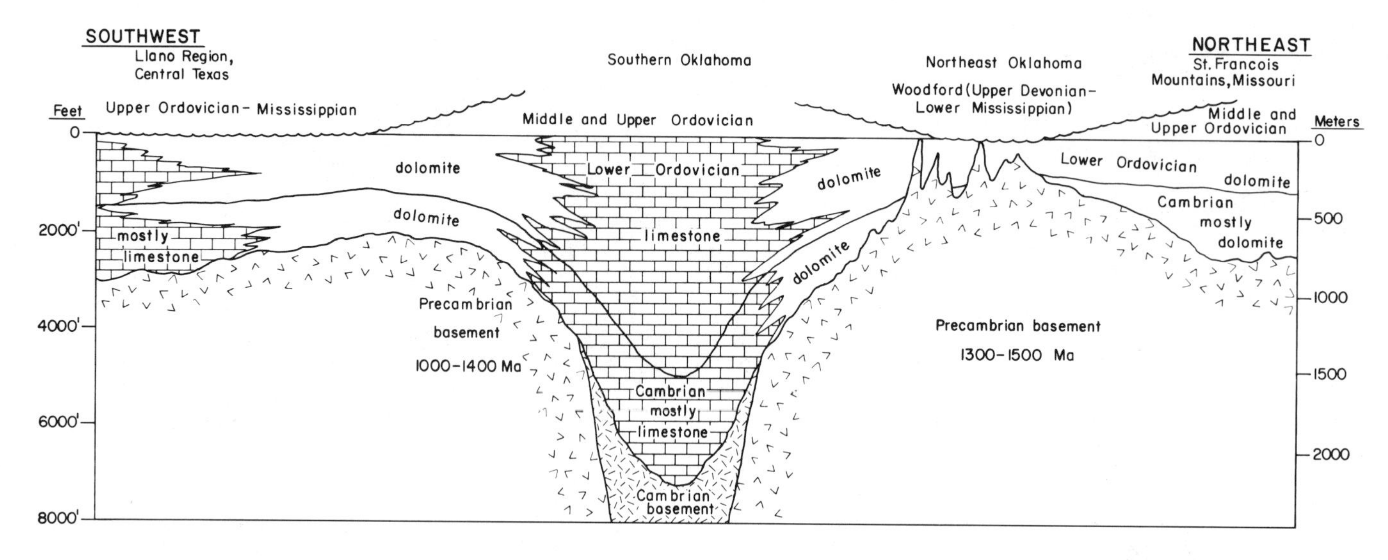

Figure 3. Schematic cross section through the Southern Midcontinent showing restored thickness of Middle Cambrian–Lower Ordovician strata and the generalized distribution of limestone and dolomite (also see Gatewood, 1970).

component in the carbonates, some of the thin sandy zones are remarkably persistent (Ragland and Donovan, 1985a).

The Arbuckle Group is composed almost entirely of limestone and is thickest (more than 2,100 m) in the Southern Oklahoma Aulacogen where the basement rocks are Cambrian in age (Fig. 3). In the eastern Arbuckle Mountains, where the basement is massive Precambrian (about 1,400 Ma) granitic rocks, the Arbuckle is mostly dolomite, contains moderate amounts of quartz sand, and is only about 1,200 m thick. Anhydrite is known from drill holes into the Arbuckle Group in some of the Oklahoma basins (Latham, 1970). Bedded evaporites are not found on the outcrop, but their original presence in outcropping rocks is suggested by collapse features and small anhydrite relicts in chert (Ragland and Donovan, 1985b).

Nearly 1,000 m of Cambrian and Early Ordovician sedimentary rocks are exposed in the Llano Uplift of central Texas. The Middle Cambrian seas transgressed over a surface with as much as 250 m of relief to deposit the basal Hickory Sandstone Member of the Riley Formation (Fig. 2) of the Moore Hollow Group. The younger Cambrian units of the Moore Hollow Group are composed dominantly of peritidal and shallow subtidal carbonate rocks with some significant terrigenous influxes (Barnes and Bell, 1977). Bell and Barnes (in Barnes and others, 1972) have provided a summary of Cambrian sedimentation and nomenclature. The Ordovician Ellenburger rocks are mostly limestone and dolomite with locally abundant chert. Cloud and Barnes (1948) describe the Ellenburger as lime-mud–dominated, shallow-water limestones with extensive dolomite replacement.

In Missouri and northern Arkansas, there are no inclusive terms for rocks equivalent to the Arbuckle and Timbered Hills Groups, but the gross lithologies of the formations (Fig. 2) are quite similar (Hayes and Knight, 1961; Martin and others, 1961). The outcrop pattern is in broad rings around the Precambrian of the St. Francois Mountains. Thickness of the 12 formations composing this time interval must be measured in separate areas. The original thickness of rocks deposited during the Cambrian and Early Ordovician at a given locality is difficult to estimate but is probably in the range of 600 to 900 m. Most of the sequence is dolomite, with lesser amounts of limestone, shale, and sandstone, and locally abundant secondary silica. The environment of deposition, based upon original and relict textures in the carbonates, appears to have been shallow water (Kurtz and others, 1975). The terrigenous content is greater here than in southern Oklahoma.

In the Arkansas part of the Mississippi Embayment, a few wells that have not been carefully studied have penetrated what has been interpreted as the complete, or nearly complete, Cambrian–Lower Ordovician sequence. The section here, if properly correlated, has a drilled thickness in excess of 2,900 m and a composite thickness of nearly 6,400 m, of which 5,000 m is thought to be Cambrian by Howe and Thompson (1984; because of the uncertainty, these data are not shown on Plate 5-C). Howe and Thompson (1984) show much of the Cambrian section to be composed of dark shales and sands, suggesting deeper water than on the adjacent platform.

Correlation between areas of outcropping units is based largely on shelly faunas. In the subsurface this is possible only under fortuitous circumstances where specimens are recovered in cores. In the Ozark region, correlations are made on the basis of the comparison of insoluble residues of outcropping units with those found in the subsurface (e.g., McCracken, 1955). In other areas the correlation is on the uncertain basis of lithology. On the outcrop, where the rocks are best known, the Cambrian–Lower Ordovician stratigraphic section seems to be thickest and most complete. In northeast Oklahoma, where the topography of the basement was most rugged, the older units are apparently missing by nondeposition, indicating this area was a gentle arch during the Late Cambrian (Fig. 3). The basement topography was so rugged here that some of the hills were not covered until the Middle Ordovician (Tarr, 1955).

The Cambrian and Early Ordovician carbonates were deposited over the Southern Midcontinent as part of a vast shallow-water platform that stretched from New York to New Mexico (best shown in Cook and Bally, 1975). The impressive stability of the continental interior following the basal Paleozoic transgression is unmatched during later geologic history. The present thickness of Late Cambrian–Early Ordovician strata (Plate 5-C) is due to the original deposition, modified in some areas by one or more periods of erosion. The earliest erosion period affected the Llano region in pre-Devonian time, the first of seven cited by Cloud and Barnes (1948). Barnes (in Barnes and others, 1959) and Barnes and Bell (1977) also synthesized the regional distribution and thickness of various divisions of the Ellenburger and Moore Hollow Groups in the subsurface of Texas. The pre-Woodford (Late Devonian–Early Mississippian) regional unconformity had a profound influence on the distribution of older rocks in the area, as shown by Tarr and others (1965) in Oklahoma. Deformation in the late Paleozoic caused differential uplift and erosion in Oklahoma and the Texas panhandle to expose basement over large areas. The Cambro-Ordovician rocks have been exposed locally to erosion during several periods from the Pennsylvanian to the present.

MIDDLE ORDOVICIAN THROUGH EARLIEST MISSISSIPPIAN

Thomas W. Amsden

The depositional environment throughout the Oklahoma Basin appears to have been reasonably constant during Middle Ordovician through Early Mississippian time, with shallow carbonate seas supporting rich benthic faunas periodically interrupted by injections of fine to moderately coarse clastics derived from an eastern source. Most of the major lithostratigraphic divisions can be recognized throughout the basin, indicating only moderate lateral variation in sedimentation and paleoenvironment. Deposition was interrupted by two major epeirogenic uplifts, a pre–middle Early Devonian (pre–Frisco-Sallisaw) and a pre–Late Devonian (pre–Woodford-Chattanooga). Both represent broad upwarpings with the maximum uplift on the flanks of the Oklahoma Basin where truncation of the underlying strata was greatest. These uplifts (Figs. 4 and 5) represent large, elongate tectonic features that followed the northwest-southeast structural grain developed in the Precambrian. Little if any folding and faulting accompanied these uplifts, and regional compressive forces were minimal during early and middle Paleozoic time.

In all probability, Middle Ordovician through earliest Mississippian strata were deposited throughout most of the Oklahoma Basin, but their present distribution results from the two episodes of Devonian truncation and, in the case of Woodford-Chattanooga distribution, from post-Woodford truncation. Late Ordovician through Early Mississippian strata are preserved in a small area on the Llano Uplift (just south of the area labeled Llano Uplift on Plate 5-A), and these strata are described briefly at the end of this portion of the chapter.

Middle Ordovician to Lower Mississippian strata in the Oklahoma Basin comprise five major stratigraphic units: Simpson Group, Viola Group, Sylvan-Cason Shale, Hunton Group, and Woodford-Chattanooga Shale (Fig. 4). The Viola and Hunton Groups represent mainly shallow-water carbonate sedimentation interrupted by the Simpson, Sylvan-Cason, and Woodford-Chattanooga depositional episodes when substantial quantities of terrigenous detritus were introduced into the depositional basin. The regional distribution of this detritus indicates it was derived mainly from an eastern source.

This depositional sequence was locally interrupted by numerous diastems and unconformities (Amsden, 1960; Amsden and Sweet, 1983). Two unconformities can be traced over most of this region, and these involve truncation of major proportions. These are the pre–Frisco-Sallisaw (pre–middle Early Devonian) unconformity, during which time strata as old as Late Ordovician were deleted (Fig. 5A), and the pre–Woodford-Chattanooga (pre–Late Devonian) unconformity, which truncated strata as old as the Early Ordovician (Fig. 5B; Jordan, 1965; Amsden, 1975, 1980).

The Simpson Group of Middle Ordovician age includes substantial quantities of sand, silt, and clay-size terrigenous detritus, including well-rounded quartz grains representing multiple-generation sands. This detritus is interspersed with varying amounts of carbonate, especially in the upper part (Bromide), which includes shelly benthic faunas of impressive diversity. Latest Simpson intertidal birdseye limestones occupy most of the western Arkoma Basin (Amsden and Sweet, 1983). Terrigenous detritus, including some graptolitic shale, is present in the lower part of the Viola Group; however, this decreases upward and the strata grade into clean-washed skeletal limestones (Welling Formation = "Fernvale" Formation). The Viola and Simpson Groups are primarily shallow-water sediments with the skeletal limestones representing clean-washed substrate and low turbidity. The combined thickness of both these groups is greatest in southern Oklahoma, where they reach a thickness in excess of 750 m (Plate 5-D). Undoubtedly, deepening of the water occurred from time to time, but the primary factor influencing the paleoenvironment appears to have been the periodic influx of terrigenous sediments, which at times engulfed the substrate.

The Sylvan Shale in the region from western Arkansas into central Oklahoma is a gray shale, noncalcareous in the lower part, becoming increasingly dolomitic in the upper part.

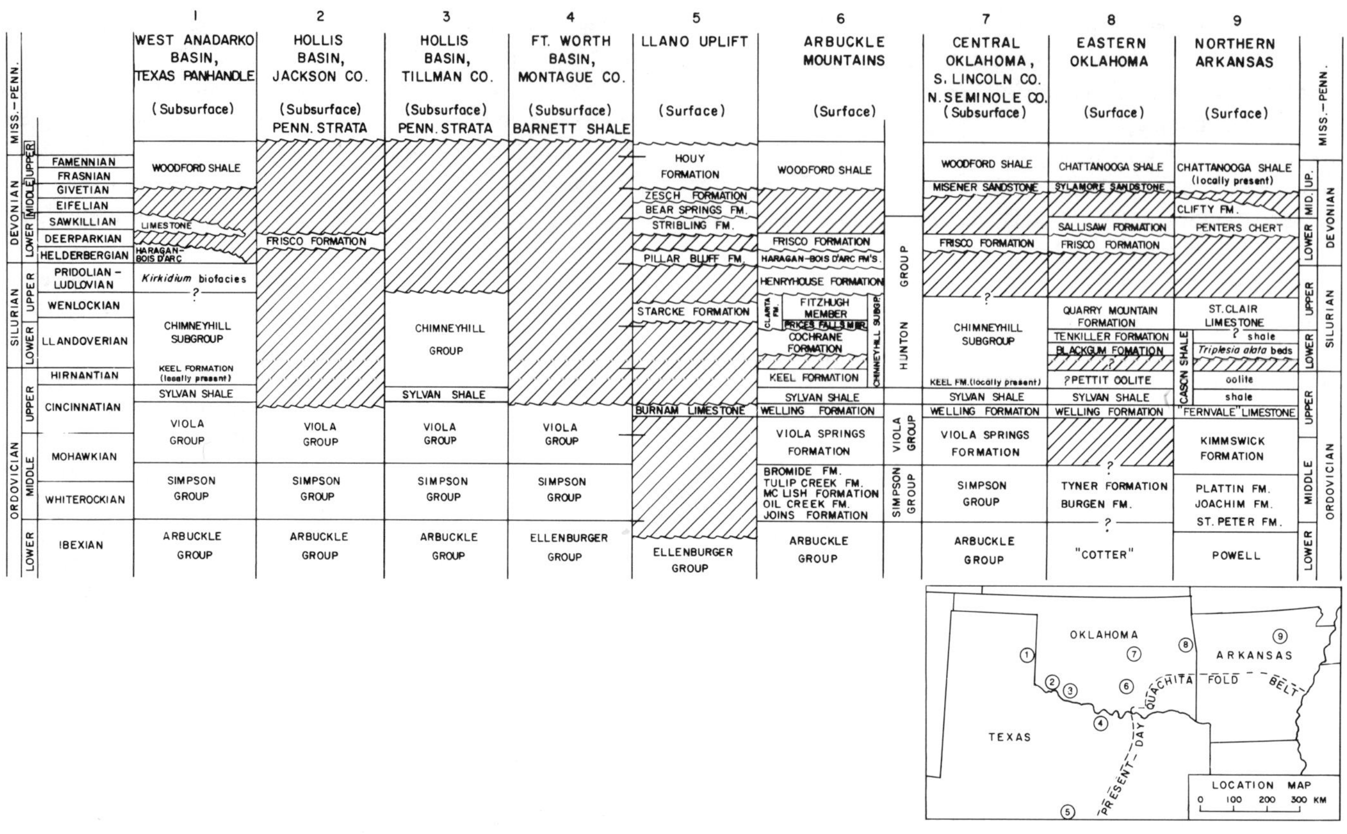

Figure 4. Correlation chart for Ordovician, Silurian, and Devonian strata in the Texas Panhandle, Oklahoma, north Texas, and northern Arkansas. From Amsden (1980) and unpublished data, Amsden and Sweet (1983) and Bozovitch (1963).

Throughout this area the Sylvan includes graptolites and chitinozoans, but no shelly benthic fauna. In western Arkansas it grades into the lower part of the Cason Shale (Fig. 4), which contains much silt-size detritus. In the western part of Oklahoma and the Texas panhandle, the Sylvan Shale becomes increasingly calcareous and locally grades into argillaceous skeletal carbonate (Plate 5-E). The absence of benthic shelly faunas over most of the Sylvan area has commonly been attributed to deep and/or cold water, but the condition can be reasonably explained by assuming that the influx of fine detritus was rapid enough to prevent colonization of the sea floor by benthic organisms. This explanation is in accord with the presence of benthic shelly faunas in the western part of the Sylvan sea, in a region well removed from the source of the terrigenous material. The Sylvan is truncated by erosion around the margins of the Oklahoma Basin and along the Wichita-Amarillo Uplift, but the remaining Sylvan thickens toward the south-central part of the basin where it is in excess of 100 m (Plate 5-E).

The base of the Hunton Group is the Late Ordovician (Hirnantian) Keel oolite, which is a part of an oolitic facies distributed over a wide area in eastern North America, and which represents a shoaling of the water probably related to the eustatic lowering of sea level during North African glaciation. The Keel is overlain by the generally clean-washed skeletal limestones of the Chimneyhill Group. Some silt and clay-size detritus was introduced during Chimneyhill time, but this was largely restricted to the southeastern areas bordering the Ouachita Province (Plate 5-F). The volume of detritus increased sharply in the Late Silurian (Henryhouse) and Early Devonian (Haragan-Bois d'Arc), extending much farther west into the Anadarko Basin. Late Ordovician (Keel), Silurian (Chimneyhill, Henryhouse), and Early Devonian (Haragan–Bois d'Arc) strata in the northern and western part of this region are moderately to heavily dolomitized (Plates 5-F and 5-G), and are a part of the North American Silurian dolomite province, which occupied the continental interior (Amsden, 1975, 1980; Berry and Boucot, 1970). Following deposition of Helderbergian-age strata, the entire area was uplifted and eroded. When deposition resumed, the flood of detritus had ceased, and the Frisco Formation (Deerparkian) is everywhere a clean-washed, low-magnesium limestone.

Latest Early Devonian (Sawkillian) strata are represented by cherts and cherty limestones (Penters and Sallisaw) in the east, grading into clean-washed skeletal limestones in the west (Plate 5-G). The only Middle Devonian strata presently recognized in

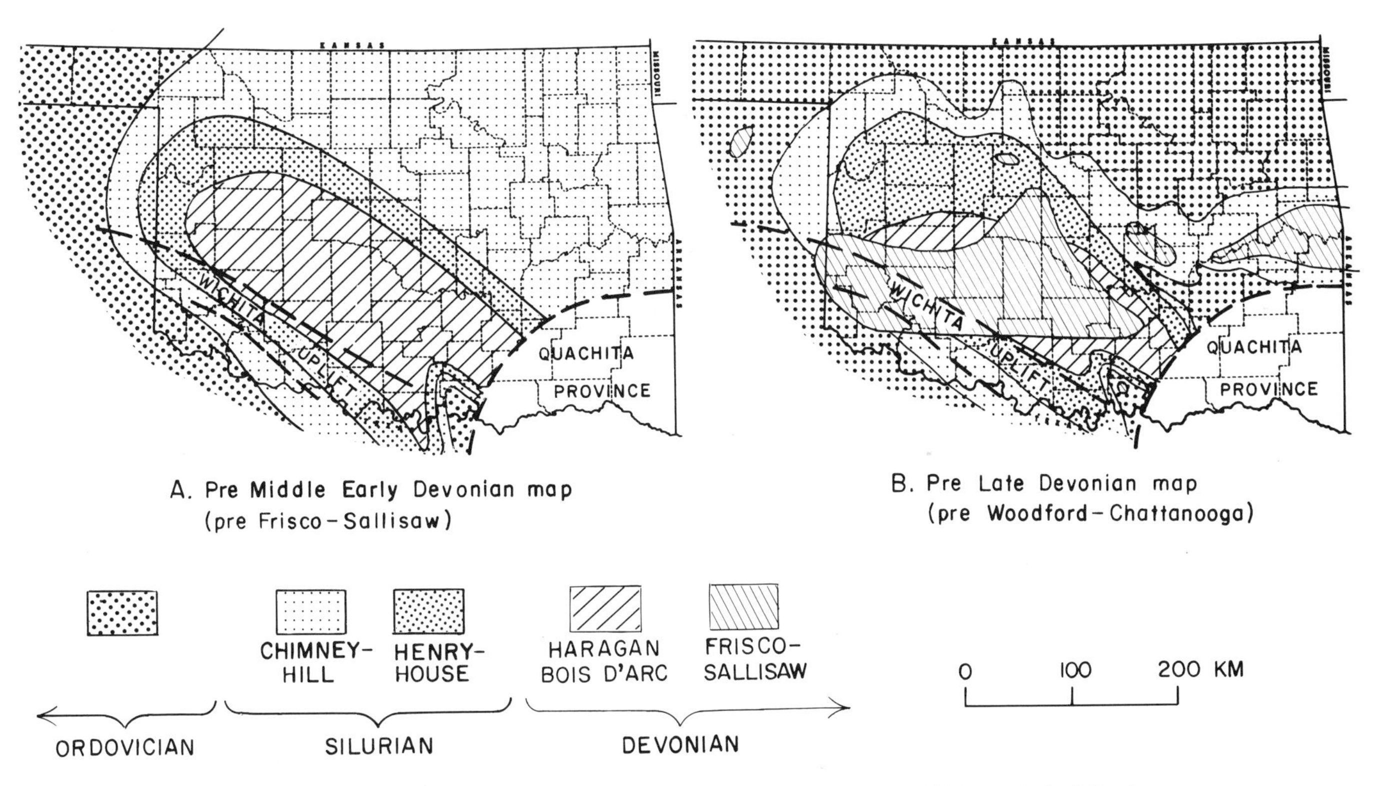

Figure 5. Paleogeologic maps showing inferred distribution of strata in the Oklahoma Basin following two major epeirogenic uplifts. Map A shows pre–middle Early Devonian strata (pre–Frisco-Sallisaw Subcrop map); map B shows pre–Late Devonian strata (pre–Woodford-Chattanooga subcrop map). After Amsden (1980, Figs. 18, 23).

this region are the Clifty Sandstone present in a small area in northwestern Arkansas, and locally the Misener Sandstone in north-central Oklahoma.

Following Early Devonian sedimentation, the area was uplifted and exposed to erosion (Fig. 5B), during which time extensive dissolution and channeling occurred. In late Middle Devonian to early Late Devonian time the Woodford-Chattanooga sea advanced from the east, burying the old erosion surface under a blanket of dark-gray to black, fine silt and clay. Where the advancing sea encountered weathered clastic debris on the underlying eroded surface, this material was incorporated into the basal sediments (Sylamore-Misener Formations). Conditions on the sea floor during deposition inhibited almost all benthic organisms, probably the result of anaerobic conditions and/or a high rate of siltation. Woodford-Chattanooga sediments covered most of the Oklahoma Basin (Plate 5-H).

A sparse representation of middle Paleozoic strata is present in the Llano Uplift region centering around Marble Falls, Burnet County, central Texas. These strata are preserved mainly as infilling or collapse sediments in joints and caves located in Early Ordovician strata of this region. Most formations are represented by very thin sequences of carbonate or cherty carbonate strata whose relationships are not well understood. The stratigraphic units recognized by Seddon (1970) are herein represented in Figure 4. The Burnam Limestone has a conodont fauna of Late Ordovician age. The Starcke (Silurian) and Pillar Bluff (Early Devonian) have a brachiopod and conodont fauna, enabling correlation with equivalent units to the north. The Stribling, Bear Springs, and Zesch Formations consist of sandy and/or cherty limestones that yield brachiopods and conodonts, mainly of Middle Devonian age (Barnes and others, 1947; Seddon, 1970). The Houy Formation comprises several lithologies whose stratigraphic relations are uncertain, but conodont studies indicate a late Middle Devonian to earliest Mississippian age (Seddon, 1970; Kier and others, 1979).

MISSISSIPPIAN SYSTEM

Kenneth S. Johnson

During the Mississippian Period, broad epeirogenic movements continued throughout the Southern Midcontinent. Deposition of shallow-marine limestones and shales was predominant, and in the middle of Mississippian time, cherty limestone deposi-

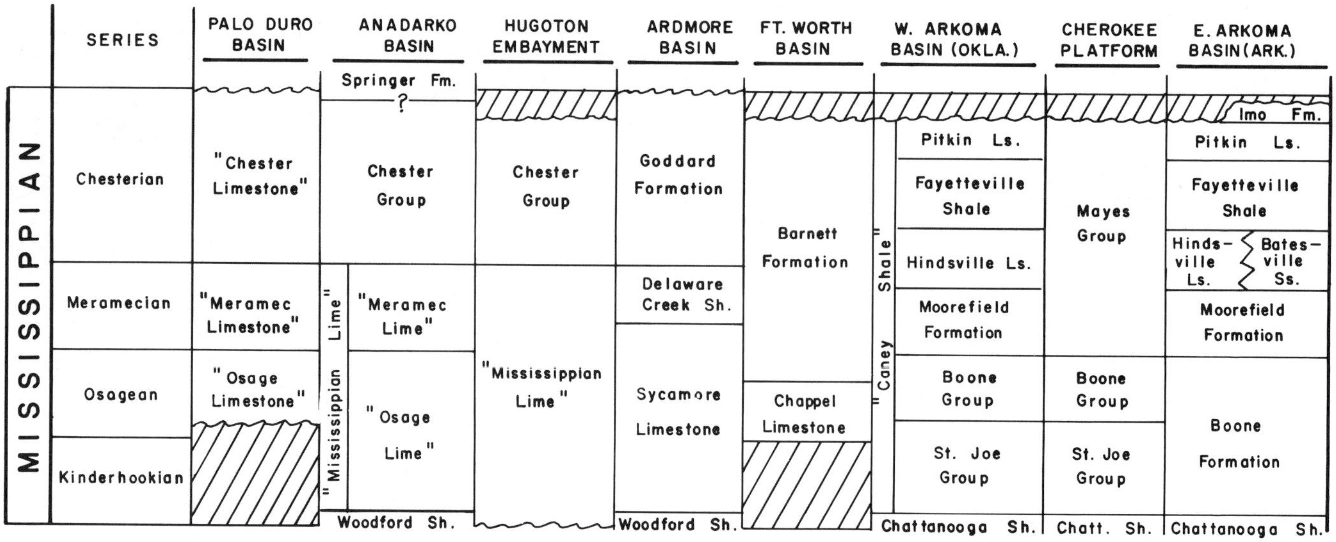

SYSTEM	SERIES	PALO DURO BASIN	ANADARKO BASIN	HUGOTON EMBAYMENT	ARDMORE BASIN	FT. WORTH BASIN	W. ARKOMA BASIN (OKLA.)	CHEROKEE PLATFORM	E. ARKOMA BASIN (ARK.)
			Springer Fm. ?						Imo Fm.
MISSISSIPPIAN	Chesterian	"Chester Limestone"	Chester Group	Chester Group	Goddard Formation	Barnett Formation	Caney Shale: Pitkin Ls.; Fayetteville Shale; Hindsville Ls.	Mayes Group	Pitkin Ls.; Fayetteville Shale; Hinds-ville Ls. / Bates-ville Ss.
	Meramecian	"Meramec Limestone"	"Mississippian Lime": "Meramec Lime"	"Mississippian Lime"	Delaware Creek Sh.		Moorefield Formation		Moorefield Formation
	Osagean	"Osage Limestone"	"Osage Lime"		Sycamore Limestone	Chappel Limestone	Boone Group	Boone Group	Boone Formation
	Kinderhookian						St. Joe Group	St. Joe Group	
			Woodford Sh.		Woodford Sh.		Chattanooga Sh.	Chatt. Sh.	Chattanooga Sh.

Figure 6. Correlation of Mississippian stratigraphic units in the Southern Midcontinent, based upon COSUNA charts (Hills and Kottlowski, 1983; Mankin, 1987).

tion was widespread. Major tectonic features that influenced Mississippian sedimentation include the Oklahoma Basin, the Texas Arch, and the Ozark Uplift; most other regions of the Southern Midcontinent were low-energy shelves or platforms. Correlation of Mississippian stratigraphic units in various parts of the Southern Midcontinent is shown in Figure 6. Principal sources of data on the Mississippian System are by Craig and others (1979), Glick (1979), Frezon and Jordan (1979), Mapel and others (1979), Goebel and Stewart (1979), Ebanks and others (1979), Thompson (1979), Haley and others (1979), Fay and others (1979), Kier and others (1979), and Sable (1979).

Although most of the Woodford-Chattanooga Shale is Late Devonian in age, Mississippian conodonts occur in the top few meters at several localities (Frezon and Jordan, 1979). Commonly, however, the base of the lowest Mississippian (Kinderhookian Series) is placed arbitrarily at the top of the Woodford-Chattanooga Shale. This easily recognized rock-stratigraphic boundary is more indicative of paleotectonic change than is the time-stratigraphic boundary within the upper part of the Woodford-Chattanooga Shale (Frezon and Jordan, 1979).

After withdrawal of the euxinic seas in which the organic-rich black shales of the Woodford-Chattanooga were deposited, the region was inundated in Kinderhookian time (Plate 5-I) by shallow, well-oxygenated, marine waters that originally covered all areas except the Texas Arch and the Llano Uplift. The limestone-shale sequence is less than 30 m thick in the northeast, is typically 15 to 60 m thick in the northwest, and was largely removed from the central and southern parts of Oklahoma by post-Kinderhookian erosion. Sandstone and fine-grained clastics in the northern Texas panhandle are marginal marine deposits on the north flank of the Texas Arch, which itself was inundated by the sea shortly after Kinderhookian time.

The Osagean sea occupied most of the continental interior, and sedimentation in the Southern Midcontinent occurred in aerated, warm, shallow seas that were stirred by currents or waves that gently affected the sea floor. Limestone and cherty limestone were the dominant sediments (Plate 5-J). A rich marine fauna, principally crinoids, flourished in the Osagean sea. The region was affected only by slight epeirogenic movements during the Osagean Epoch. Gentle subsidence or uplift of 100 m or less was enough to cause large areas to be submerged by, or emergent from, the shallow sea. With subsidence of the Texas Arch and Llano Uplift area, marine carbonates were deposited across parts of these provinces. Much of the Osagean rock now present in the Llano Uplift area is preserved as collapse blocks in sinkholes that formed during Meramecian uplift and erosion.

Chert, which occurs in most Osagean rocks, is a replacement of carbonate by silica. Chert generally comprises 10 to 30 percent of the units denoted as cherty limestone or cherty shale on Plates 5-J and 5-K. Principal outcropping cherty beds are in the Boone Formation (or Group) on the flanks of the Ozark Uplift in Arkansas, Missouri, and Oklahoma. The chert locally occurs as interbeds with limestone, as sheetlike masses, and as small- to moderate-sized nodules. Replacement by chert ranges from early diagenetic to postlithification of host limestones, but the source of silica (hydrothermal sources, or spicules from siliceous sponges, or other sources) is uncertain (Giles, 1935; McKnight and Fisher, 1970; Sable, 1979).

Terrigenous clastic sediments are present in many of the Osagean units in the region, but they occur mainly as disseminated clays and silts, and only locally as discrete shales. Carbonates grade southward into shale toward the Ouachita Trough in Arkansas. In the eastern part of the present Arkoma Basin, a tongue of silt and clay extends north-south, reflecting deposition along the seaward extension of a stream system that headed far to the north (Glick, 1979).

The maximum thickness of Osagean strata is in the western part of the Oklahoma Basin where 225 m of cherty limestone is preserved (Plate 5-J). In most other areas, Osagean strata typically range from 50 to 100 m thick. Thickness variations result mainly from post-Osagean erosion. Epeirogenic uplift during Meramecian time caused erosion of Osagean strata deposited in central and southern Oklahoma, and uplift during Meramecian and early Chesterian time caused denudation of Osagean strata in the present Forth Worth Basin area. Removal of Osagean strata along the Wichita-Amarillo Uplift and the Matador Arch resulted mainly from tectonic movements during Pennsylvanian time.

Meramecian marine environments in the Southern Midcontinent resembled those of the Osagean Epoch. Shallow, well-aerated, warm seas were widespread and favored deposition of fossiliferous (commonly crinoidal) and oolitic limestones with some interbeds of claystone, shale, and siltstone (Plate 5-K). Most of the region was a carbonate shelf, and chert formation, similar to the Osagean occurrences, was largely restricted to the northern half of Oklahoma and northwestern Arkansas. The presence of oolitic limestone in many parts of Arkansas, Oklahoma, and Texas indicates that the shallow sea bottom was more agitated by wave and current action during the Meramecian than at other times during the Mississiippian.

In the Arkoma Basin area the Meramecian carbonates grade southward into calcareous shales deposited in and adjacent to the Ouachita Trough. These fine-grained clastics were carried into northeastern Arkansas by streams, and also were transported across the Ouachita Trough from eastern and southeastern sources (Glick, 1979). According to Frezon and Jordan (1979), a depression extended to the west and northwest of the Ouachita Trough across a shoal area into the Oklahoma Basin. The shoal area, which apparently was in the vicinity of the future Arbuckle Mountains, was a high-energy environment where fine-grained clastics were winnowed and transported westward to a low-energy environment in the central part of what would become the Anadarko Basin (Plate 5-K).

There was little or no tectonic activity in the Southern Midcontinent during the Meramecian Epoch. Epeirogenic movements in mid-Meramecian time raised the Ozark region slightly above sea-level, and in late-Meramecian time the Fort Worth Basin area was elevated (Mapel and others, 1979), causing erosion of Meramecian and much of the Osagean strata.

Thickness variations in the Meramecian result mainly from post-Meramecian erosion. In most areas the thickness now ranges from 50 to 150 m; the maximum preserved thickness is nearly 400 m in the western part of the Anadarko Basin. In this area, sedimentation probably was continuous from Meramecian into Chesterian time. Small areas where Meramecian strata are missing—over the Matador Arch, in southern and central Oklahoma, and along the Nemaha Uplift in northern Oklahoma and southern Kansas—result from post-Mississippian tectonic uplift and erosion.

Chesterian seas covered all parts of the Southern Midcontinent, although epeirogenic movements and eustatic changes of sea level caused intermittent emergence of many areas. In much of the region the Chesterian sequence consists of shale with some limestone and marginal sandstones in the lower part, and limestone with some shales in the upper part (Plate 5-L). These strata were deposited in normal marine waters, although the shales were generally laid down in somewhat deeper and more turbid waters than were the limestones.

The Ouachita Trough became an increasingly important factor affecting sedimentation in parts of the Southern Midcontinent during Chesterian time. By now the trough was subsiding sharply along the arcuate trace of the present-day Ouachita Fold Belt. Shales, which reach 1,500 m in thickness within the Ouachita Trough, extend to the north and west into the area of the future Arkoma, Ardmore, and Fort Worth Basins where they interfinger with overlying shelf limestones. The river system in eastern Arkansas broadened to become the Ancestral Mississippi Embayment (Glick, 1979), and this introduced a thick sequence of clastics into the northern part of the Arkoma Basin.

Many of the apparent lithologic changes in Chesterian rocks shown on Plate 5-L do not result from facies variations during deposition. Instead, the changes commonly reflect the amount of post-Mississippian uplift and the depth of erosion around the margins of some of the basins; the upper limestone units have been removed, leaving only the lower shaley units at the basin margins. This is especially evident around the Palo Duro and Hardeman Basins, the Hugoton Embayment area, and the north, west, and south margins of the Fort Worth Basin. Post-Mississippian uplift and erosion along the Nemaha Uplift also account for the absence of Chesterian strata that were originally deposited across the large area that broadens northward in central Oklahoma. Areas where lithologic changes from limestone to shale do occur depositionally are in eastern Oklahoma and the central part of the Anadarko Basin.

The thickest Chesterian rocks in the region are the 600 m of shales in the eastern part of the Arkoma Basin (Plate 5-L), where these strata thicken sharply into the Ouachita basinal facies. Also, about 500 m of shales and limestones are present in the deep Anadarko Basin area. Maximum thicknesses of Chesterian strata in the northeast parts of the Fort Worth and Hardeman Basins are preserved in downthrown blocks on the south side of post-Mississippian faults.

Although the top of the Mississippian System is well marked by a pre-Pennsylvanian unconformity in most parts of the Southern Midcontinent, the Mississippian-Pennsylvanian boundary occurs within the thick sequence of Springer and equivalent shales where sedimentation was uninterrupted in the deep parts of the Anadarko, Ardmore, and Arkoma Basins. These Springer clastics are commonly grouped with the overlying Morrowan strata, from which they are not easily distinguished (McKee and others, 1975), and in this chapter they are described in the separate basin discussions that follow.

Total thickness of the Mississippian is greatest in the western part of the Anadarko Basin (about 1,000 m) and in the southern part of the Arkoma Basin of Arkansas (more than 750 m; Craig and others, 1979). The Mississippian is about 350 m thick in the northeastern part of the Fort Worth Basin, about 250 m thick in the south-central part of the Palo Duro Basin, and 400 m thick in the northern part of the Hardeman Basin. Elsewhere the Mississippian is commonly 100 to 200 m thick. Thickening of the entire Mississippian into the several major basins results partly from depositional thickening toward centers of ancestral basins that were to become predominant in Pennsylvanian time, but also partly from intermittent Mississippian and Pennsylvanian uplift and truncation along the margins of these basins.

PENNSYLVANIAN AND PERMIAN ROCKS

REGIONAL STRATIGRAPHIC FRAMEWORK

Kenneth S. Johnson

Great changes took place in the Southern Midcontinent in Late Mississippian and Pennsylvanian times. Epeirogenic movements and deposition of relatively simple shelf sequences of carbonates (with some shales and sandstones) over vast regions of the Oklahoma Basin, which were typical in the Early and Middle Paleozoic, ended with a series of orogenic movements that subdivided the region into the various tectonic provinces that are so easily recognized today (Plate 5-A). First, an episode of Late Mississippian–Early Pennsylvanian epeirogenic uplift throughout most of the region was accompanied by erosion and development of the pre-Pennsylvanian unconformity that is present in all areas, except where sedimentation was apparently continuous in the deep Anadarko and Ardmore Basins. Then a series of pulses in the Ouachita orogen and the Southern Oklahoma Aulacogen through Early and Middle Pennsylvanian time caused, or contributed to the following: folding and thrusting of the Ouachita Foldbelt; raising of the Amarillo, Wichita, Criner, Arbuckle, Nemaha, Matador, Red River-Electra, and Muenster Uplifts; pronounced downwarping of the Anadarko, Ardmore, Marietta, Arkoma, and Fort Worth Basins; and moderate subsidence of the Hugoton Embayment and the Palo Duro, Hardeman, and Dalhart Basins. Therefore, further discussion of geologic evolution of the Southern Midcontinent must focus on the individual basins into which the region was divided.

Pennsylvanian strata of the region can be characterized as sequences of marine and nonmarine shale, sandstone, conglomerate, and limestone that thicken markedly into the rapidly subsiding basins (Plates 5-M to 6-Q). The thick wedges of terrigenous clastic sediments were shed from nearby uplifts; thinner carbonate sequences were deposited on shallow-water shelf areas distal to the uplifts. Successively younger Pennsylvanian units commonly overlap older units at the margins of the basins and across some of the uplifts. Thin coal beds are abundant in Desmoinesian strata, mainly in the Arkoma Basin and on the Cherokee Platform. Total thickness of Pennsylvanian strata in the various basins is as follows: Anadarko, Ardmore, Marietta (Texas portion), and Arkoma Basins, at least 5,000 m; Fort Worth Basin, 2,000 m; Hardeman Basin, 1,200 m; Palo Duro Basin, 1,000 m; and Dalhart Basin, 750 m. In most of the shelf or platform areas, Pennsylvanian strata are typically 500 to 1,200 m thick.

In Permian time, a fairly well-defined seaway extended north-south from west Texas across the western half of the Southern Midcontinent. Coarse clastics were eroded from the Ouachita lowlands on the east, the ancestral Rocky Mountains (Sierra Grande and Apishapa Uplifts) on the west, and the Amarillo-Wichita Uplift in the center (Plates 6-R to 6-U). Early Permian (Wolfcampian) carbonates and shales, mostly red beds, in the region are overlain by a major evaporite and red-bed sequence (of Leonardian, Guadalupian, and Ochoan age) deposited north of the carbonate reefs that fringed the Midland and Delaware Basins of west Texas and southeast New Mexico. Evaporites (salt and gypsum/anhydrite) thicken into the various basins that continued to subside more than the adjacent uplifts and arches. The maximum thickness of Permian strata in the various basins is about 2,000 m in the Anadarko and Palo Duro Basins, 1,500 m in the Dalhart Basin, and 1,200 m in the Hardeman Basin. Permian strata typically are 600 to 1,200 m thick in nearby shelf or platform areas.

ANADARKO BASIN AND HUGOTON EMBAYMENT

Bailey Rascoe, Jr., and Kenneth S. Johnson

INTRODUCTION

The Anadarko Basin is the deepest interior cratonic basin in the conterminous United States. As much as 12,000 m of Paleozoic sedimentary rocks are present along the axis at the southern margin of this asymmetrical basin, and there may be another 6,000 m of Cambrian-age layered igneous rocks (mainly rhyolite flows and granite sills) beneath the Paleozoic sediments. In the western part of the Anadarko Basin the Paleozoic strata are only 4,000 to 5,000 m thick and they rest directly upon Precambrian igneous and metamorphic rocks (Fig. 7). The Hugoton Embay-

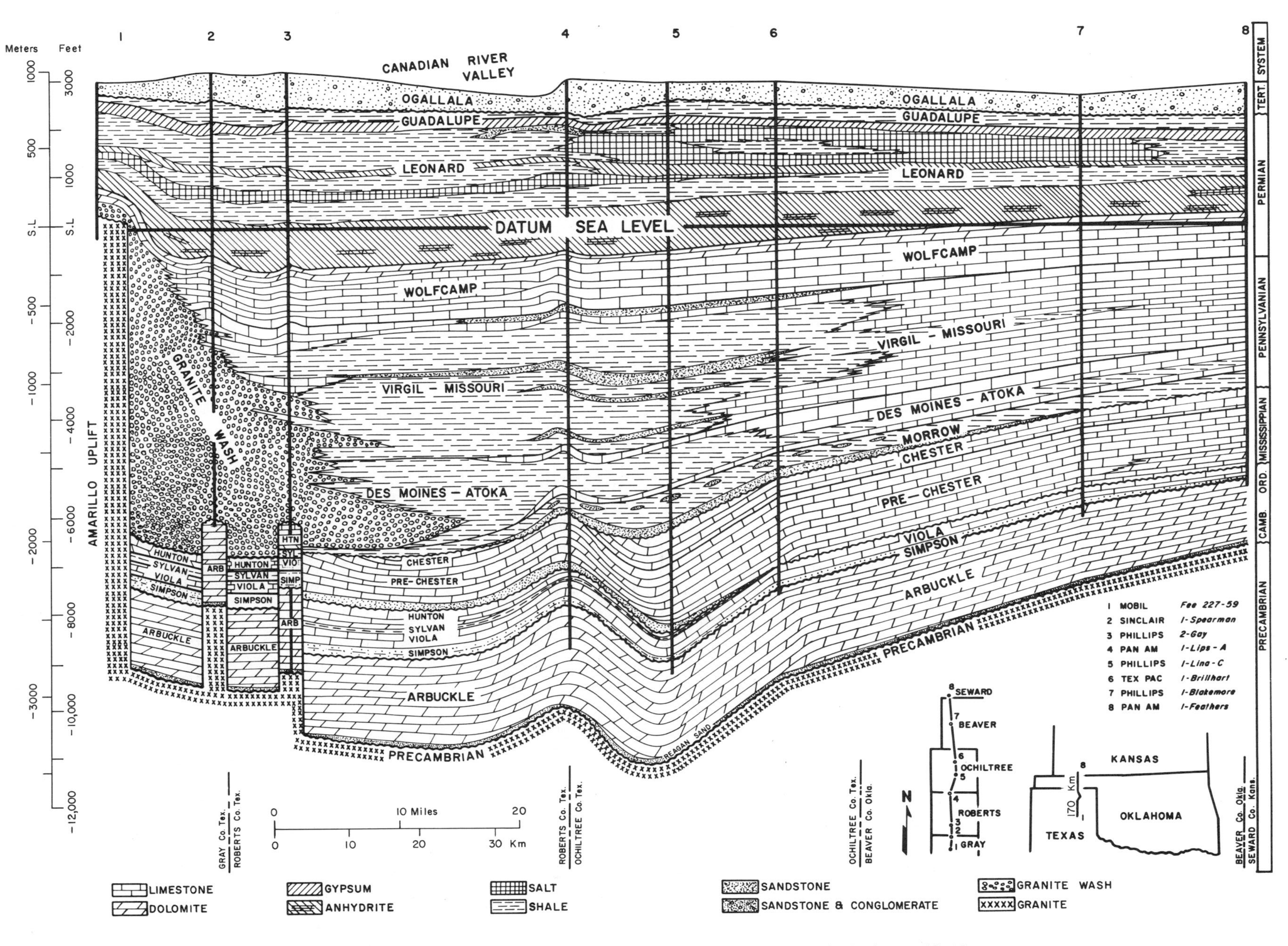

Figure 7. North-south structural cross section in the western part of the Anadarko Basin (modified from Adler and others, 1971).

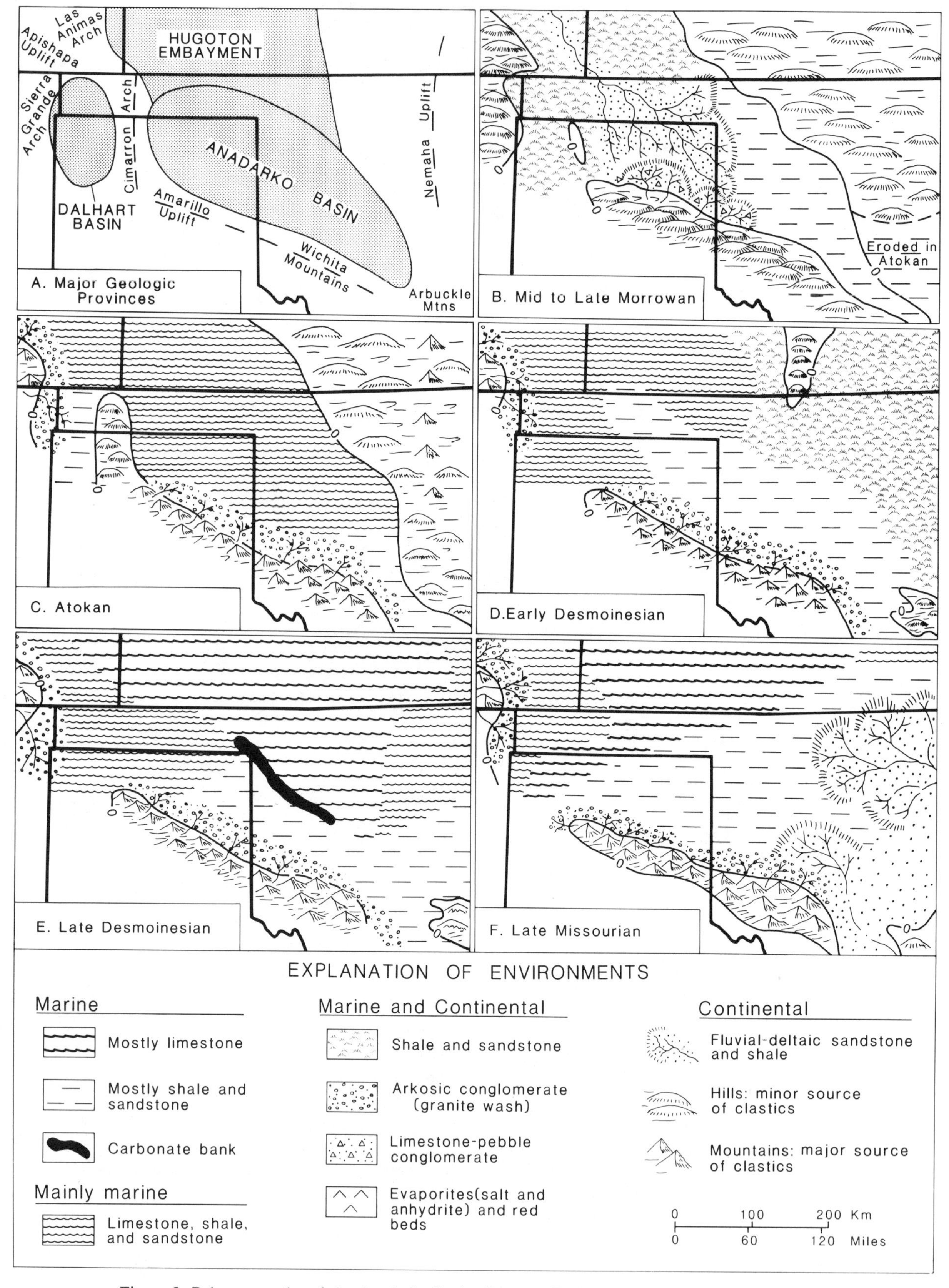

Figure 8. Paleogeography of the Anadarko Basin, Hugoton Embayment, and surrounding areas in Pennsylvanian and Permian times. Maps B through H modified from Rascoe and Adler (1983).

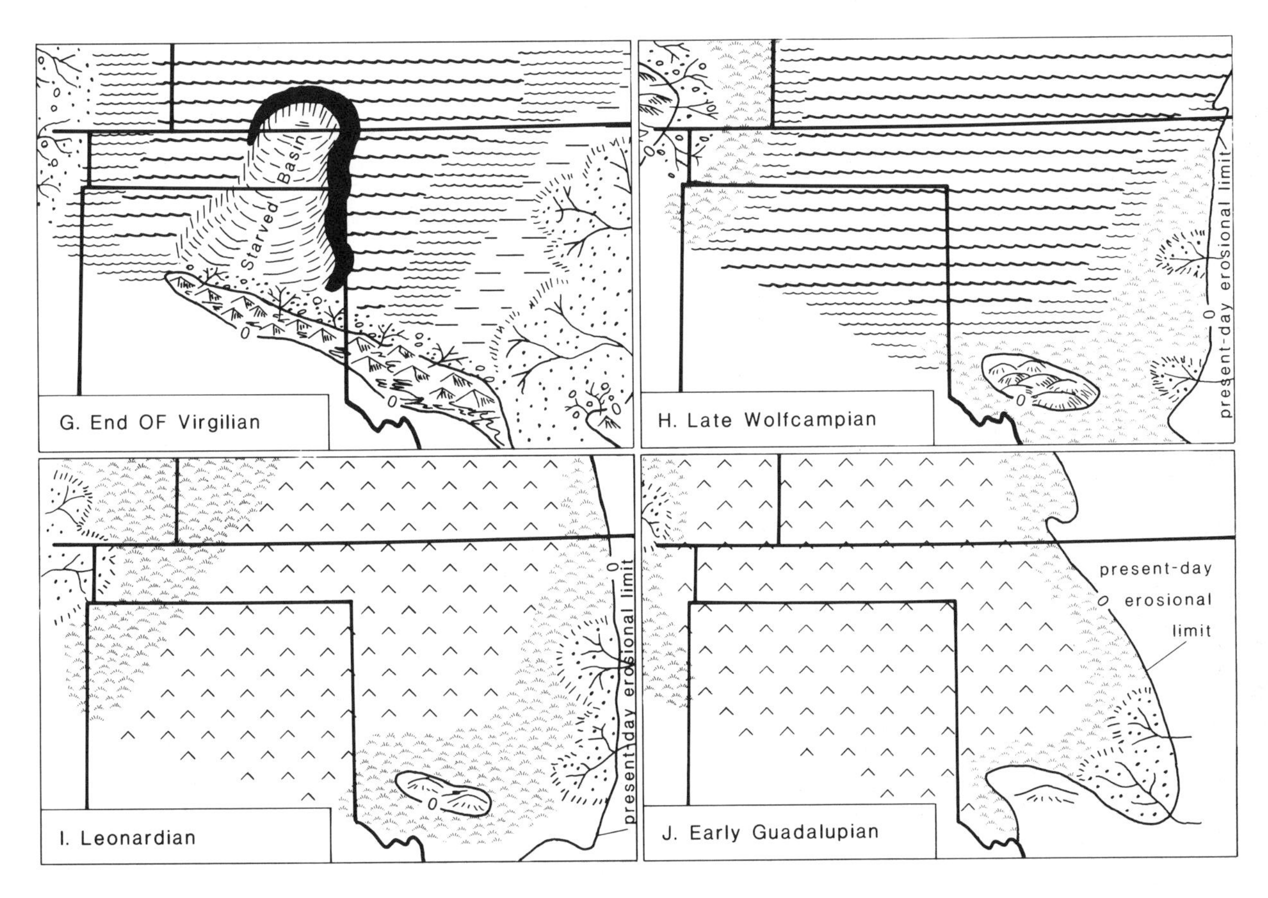

ment, and other areas on the far-northern shelf of the Anadarko Basin, generally contain 2,000 to 3,000 m of Paleozoic sediments resting upon a Precambrian granitic basement.

The Anadarko Basin and adjacent Wichita-Amarillo block subsided together as part of the Oklahoma Basin during early and middle Paleozoic time, and received as much as 4,500 m of pre-Pennsylvanian sediments. The Wichita-Amarillo Uplift was separated from the Anadarko Basin during late Morrowan and early Atokan time by a series of high-angle reverse faults, with displacements of as much as 12,000 m. The basin subsided markedly and received approximately 7,500 m of Pennsylvanian and Permian sediments. The orogenic episodes resulted mainly in block faulting and some folding along the margins of the basin; there was no igneous activity accompanying the tectonism.

The western margin of the Anadarko Basin is formed by the Cimarron Arch, which extends northward from the Wichita-Amarillo Uplift (Fig. 8A). The absence of Atokan and uppermost Morrowan rocks on the Cimarron Arch indicates that this feature was formed during the same tectonic episode as the Wichita-Amarillo Uplift. The Anadarko Basin is bounded on the east by the Nemaha Ridge. The northern shelf of the Anadarko Basin extends across much of western Kansas and is referred to, in that area, as the Hugoton Embayment. The Hugoton Embayment is bounded by the Central Kansas Uplift and the Cambridge Arch on the east and north, and by the Las Animas Arch on the west. Major Pennsylvanian-Permian stratigraphic units of the Anadarko Basin and Hugoton Embayment are given in Figure 9.

PENNSYLVANIAN SYSTEM

Morrowan Series

In most parts of the Anadarko Basin and the Hugoton Embayment, Morrowan sediments were deposited on a surface of eroded Mississippian rocks. Along and adjacent to the axis of the Anadarko Basin, Morrowan beds overlie lithologically similar Springer beds of Late Mississippian and (or) Early Pennsylvanian age. The contact of Morrowan and Springer rocks is typically difficult to determine, and in this area, deposition may have been continuous from Late Mississippian into Early Pennsylvanian time; thus, by convention, Springer strata are often grouped with the Morrowan in regional subsurface mapping (Plate 5-M).

The lower part of the Morrowan Series consists of shallow marine shales, sandstones, and limestones. These sediments, deposited in the transgressing Morrowan sea, onlap the surface of eroded Mississippian rocks and are markedly diachronous from the Anadarko Basin, where they are older, onto the shelf areas to the north and northeast where they are younger. A basal sandstone unit is a sporadic but prominent member of this transgressive sequence. Thickening of this sandstone unit appears to mark

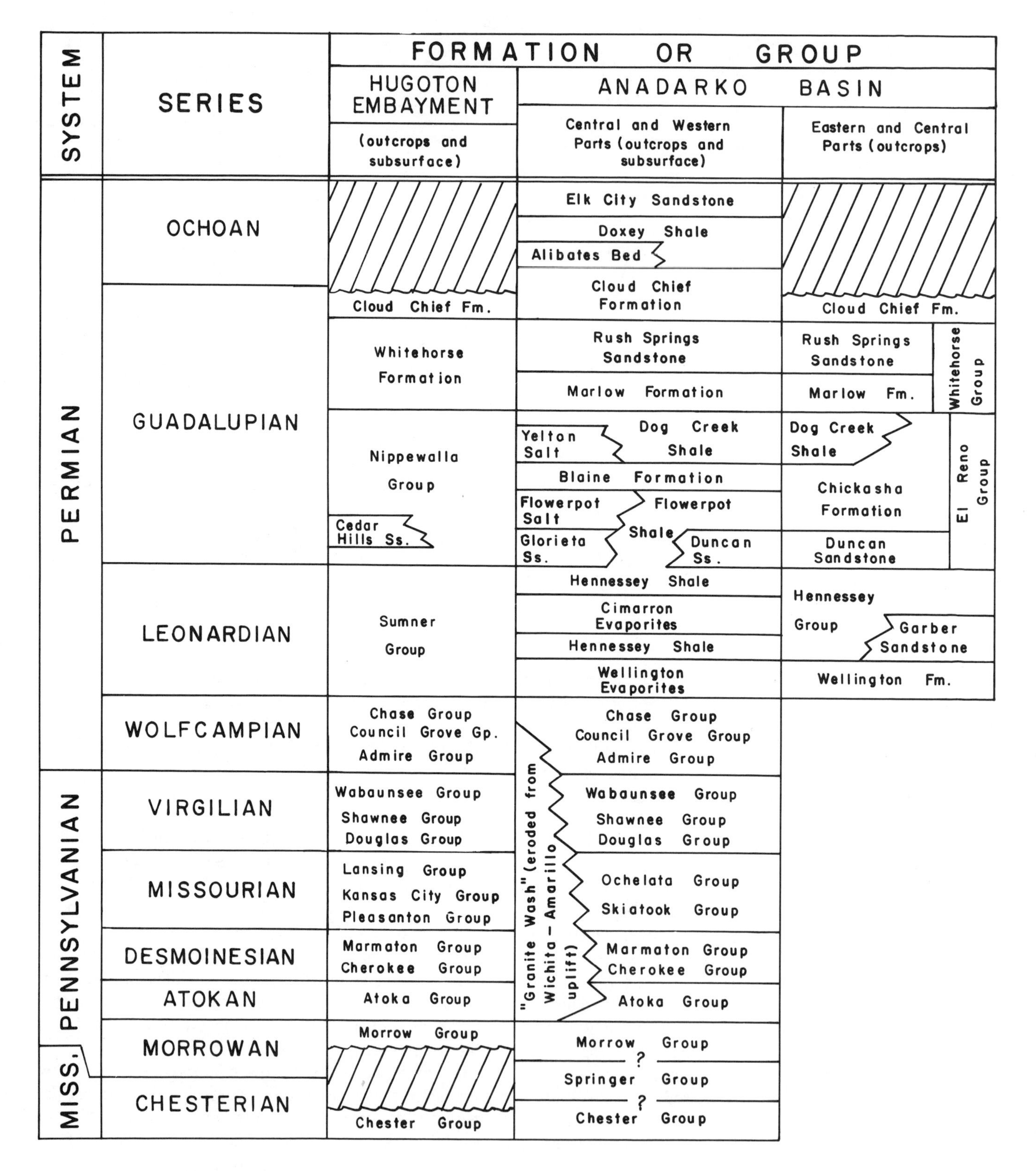

Figure 9. Stratigraphic column of Pennsylvanian and Permian strata in the Anadarko Basin and Hugoton Embayment (modified from Johnson, 1978; and Hills and Kottlowski, 1983).

the positions of ancient shorelines, which formed during still-stands of the Morrowan seas. Carbonate cement is a common constituent of these basal sandstones: it was probably introduced into the marginal marine environment by streams that flowed across the carbonate terrain of exposed Mississippian rocks and emptied into the Morrowan seas (Adams, 1964).

These transgressive lower Morrowan beds are overlain by a sequence that consists principally of shales with lenticular, discontinuous sandstones and minor conglomerates, coals, and thin, dark limestones. This upper Morrowan section is the product of a prograding deltaic phase (Fig. 8B) with the following environmental facies (Swanson, 1979): point-bar and stream-mouth-bar sandstones and conglomerates, prodelta shales, backswamp-marsh shales, and meander-channel-fill sandstones and shales.

Morrowan thickness patterns (Plate 5-M) were strongly influenced by several major tectonic elements: the Anadarko Basin, the Cimarron Arch, and the Dalhart Basin. The Wichita-Amarillo Uplift did not exist at the beginning of Morrowan time, based on two lines of evidence: first, a facies of "granite wash" (arkosic sandstone and/or conglomerate eroded from a nearby granitic terrain) from a southern source, which characterizes Middle and Upper Pennsylvanian units, is not present in the Morrowan section (Fig. 9), thus indicating the absence of the Wichita-Amarillo source area; and second, the intersection of Morrowan thickness patterns with the Wichita-Amarillo Uplift indicates that the formation of that structure followed Morrowan deposition. Shelby (1979) and Evans (1979) have described and mapped upper Morrowan fan-delta chert conglomerates in the Texas panhandle and western Oklahoma. The Wichita-Amarillo Uplift was the source area for these clastics, which were derived by the weathering of cherty limestones and dolomites of Mississippian age. The presence of these chert conglomerates in the upper part of the Morrowan Series dates the initiation of the Wichita-Amarillo Uplift.

Atokan Series

Atokan rocks in the Anadarko Basin and the Hugoton Embayment consist of a cyclic sequence of thin marine limestones and shales (Fig. 8C; Plate 5-N). These sediments compose the so-called "thirteen-finger limestone" of the Atoka Group, and the Atokan age of these rocks was established by the sparse occurrence of *Fusulinella* (Maher, 1948; Lee, 1953). This rock unit is generally 20 to 30 m thick, and the cyclic arrangement of these Atokan limestones and shales resembles the lower Desmoinesian section, which conformably overlies the Atokan.

Along the northern margin of the Wichita-Amarillo Uplift, Atokan limestones and shales grade abruptly into the massive clastic deposits, which consist of granite, limestone, and dolomite fragments. In this "granite wash" sequence the boundaries of the Atokan Series are uncertain; however, the Atokan appears to thicken into this clastic facies so that a maximum thickness of several hundred meters is possible.

In the Anadarko Basin the contact of the Atokan Series with the underlying Morrowan Series ranges from gradational to disconformable. On the northern shelf of the Anadarko Basin, Atokan sediments locally overstep the limit of the Morrowan Series and unconformably onlap rocks of Mississippian age.

Along the southern margin of the Anadarko Basin, pre-Atokan rocks underwent intense folding and faulting as a result of the Wichita orogeny of late Morrowan–early Atokan time. Consequently, Atokan rocks are commonly absent on local positive structures. Erosion of the carbonate and igneous terrain of the emerging Wichita-Amarillo Uplift continued during Atokan time. The Cimarron Arch, which strikes northward across the Texas and Oklahoma panhandles from the western termination of the Wichita-Amarillo Uplift, also experienced uplift in late Morrowan–early Atokan time; as a result Atokan rocks are not present on this arch.

Except for the "granite wash" facies, the relatively thin sequence of Atokan limestone and shale in the Anadarko Basin contrasts markedly with the thick series of Atokan clastic rocks in the Arkoma Basin. These differences are attributed to the rapid subsidence of the Arkoma Basin during Atokan time (in response to downwarping of the adjacent Ouachita geosyncline) compared to the slower, more stable subsidence of the Anadarko Basin. The Anadarko and Arkoma Basins were separated in Atokan time by uplift of a series of narrow, north-trending fault-block mountains along the Nemaha Uplift (Fig. 8C), and broad arching of most of central and northern Oklahoma (also called the Central Oklahoma Arch).

Desmoinesian Series

In the Anadarko Basin and the Hugoton Embayment the Desmoinesian Series consists mostly of cyclic marine limestones and shales; "granite wash" constitutes the Desmoinesian section in a belt adjacent to the Wichita-Amarillo Uplift (Figs. 8D and 8E; Plate 5-O). The lower part of the cyclic marine section is the Cherokee Group, which consists of numerous thin limestones interbedded with shales; on the northern shelf area in Oklahoma, lenticular point-bar and channel-fill sandstones occur in the Cherokee Group. In northwestern Oklahoma and the Texas and Oklahoma panhandles the proportion of limestones to shales is roughly equal. Eastward the increase in limestones in the Cherokee Group is gradual until along the Nemaha Uplift the ratio of limestone to shale is approximately 10:1. The Cherokee is conformably overlain by the Marmaton Group. In the Hugoton Embayment the Marmaton consists of four prominent limestone formations (Lenapah, Altamont, Pawnee, and Ft. Scott, in descending order); in the Anadarko Basin the Marmaton is composed of the Big Limestone (above) and the Oswego Limestone (below). Thickness of the Desmoinesian Series ranges from less than 75 m on the north, to more than 1,500 m along the foredeep of the Anadarko Basin (Plate 5-O).

Desmoinesian rocks conformably overlie the Atokan Series in the Anadarko Basin and the Hugoton Embayment; over this region these strata are very similar, so that the Atokan-

Desmoinesian boundary is difficult to determine. Missourian sediments are mildly disconformable on Desmoinesian strata.

Desmoinesian sediments overstep the limits of the Atokan Series on the northeastern margin of the Anadarko Basin, and onlap pre-Pennsylvanian rocks across the Nemaha Uplift. Thus, the thickness patterns of the Desmoinesian Series reflect the influence of regional positive and negative tectonic elements such as the Nemaha Ridge, the Central Kansas Uplift, the Cimarron Arch, and the Dalhart and Anadarko Basins. The Wichita-Amarillo Uplift continued to rise and to serve as a source of arkosic clastic sediments within the Anadarko Basin.

Missourian Series

The Missourian Series in the Hugoton Embayment and the northern shelf of the Anadarko Basin is predominately carbonate strata with shale interbeds. In the Anadarko Basin the Missourian Series consists of shale and sandstones with minor carbonate units (Fig. 8F; Plate 5-P). Along the western margin of the Hugoton Embayment the Missourian carbonate-shale section is replaced by a clastic sequence of shales, siltstones, and sandstones, commonly red in color. On the northern flank of the Wichita-Amarillo Uplift the Missourian Series is composed of arkosic and carbonate "wash" sediments which were eroded from the uplift (Figs. 7 and 9).

Boundaries of the Missourian Series in the shelf-carbonate section have been well established on faunal and physical evidence. In the basin-clastic section, however, the upper boundary of the Missourian Series has long been a matter of some disagreement. Due to the clastic nature of the section, diagnostic fusulinid evidence is far less plentiful than in the carbonate section to the north; compounding the problem is the difficulty in carrying physical correlations across the zone of lithologic change from the carbonates into the clastics.

The major positive tectonic elements active during Missourian time were: the Arbuckle Uplift, from which conglomeratic debris was eroded; the Wichita-Amarillo Uplift, which was a source area of coarse detritus; the Apishapa Uplift, from which mostly fine-grained clastics were eroded; and the southern part of the Central Kansas Uplift, which influenced the thickness of the Missourian sediments deposited over it. In addition, the northern portion of the Cimarron Arch appears to have influenced the thickness of Missourian rocks in the western Oklahoma Panhandle and southeastern Colorado.

The Missourian cyclic limestone and shale sequence in the Hugoton Embayment is the product of eustatic sea-level changes as described by Heckel (1977) and Watney (this volume). Southward across Kansas toward the Anadarko Basin, the carbonates of the early Missourian thicken at the expense of the shales and form locally thick, linear carbonate banks. Farther south these massive carbonates are replaced abruptly by gray and tan silty shales and fine-grained sandstones; in this section of terrigenous clastics there are several regionally extensive limestone marker beds, including the Checkerboard and Hogshooter Limestones, in the Skiatook Group, and the Dewey and Avant Limestones, in the Ochelata Group. The nature of the transition from shelf carbonates into the Anadarko Basin clastics toward the south is not well understood at this time. Although it is commonly indicated that the transition is "gradational," the existence of interfingering carbonate and clastic lithologies cannot be demonstrated by subsurface well control. It appears that the terrigenous clastics in the Anadarko Basin were not deposited contemporaneously with the thick carbonate deposits, but instead were mainly coeval with shale units of the shelf.

Work in recent years indicates that the Missourian clastic section of the Anadarko Basin is composed of a number of clastic wedges of limited areal extent. Along the eastern shelf of the Anadarko Basin, Galloway and others (1977) have demonstrated the presence of three west-facing clastic wedges or "depositional episodes." In the southern portion of the Anadarko Basin the Medrano Sandstone, in the upper part of the Skiatook Group, is regarded as a delta-front sandstone unit in a clastic wedge that prograded northward. In north-central Oklahoma, work by J. Glenn Cole (personal communication, 1983) demonstrated that the Avant Limestone was deposited on the shelf surface of a northward-prograding clastic wedge, and that on the northward-dipping slope surface, the Avant thins and disappears toward the Kansas shelf: this suggests that in the Anadarko Basin, some (or all) of the carbonate marker beds in the Missourian Series are not necessarily southward projections of carbonates from the Kansas shelf section.

Virgilian Series

The Virgilian Series consists of limestones with shale interbeds in the Hugoton Embayment and adjacent shelf area. In central Oklahoma the Virgilian Series consists of continental to shallow-marine shales, siltstones, and mudstones; to the west, in the Anadarko Basin, the Virgilian is represented by prodelta shales and delta-plain sandstones (Fig. 8G; Plate 6-Q). The shelf carbonates of the Hugoton Embayment grade westward into red shales, siltstones, and some sandstones in southeastern Colorado and northwestern New Mexico. These clastic sediments were probably derived from the Apishapa and Sierra Grande Uplifts, and they represent mixed marine and continental deposits. The Apishapa Uplift was almost completely covered by Virgilian sediment. The Wichita-Amarillo Uplift remained a positive feature during Virgilian time, and coarse detritus was deposited along its northern margin.

In the early part of Virgilian time, shales and sandstones of the Douglas Group and the equivalent Vamoosa Formation were widely deposited over the Hugoton Embayment and the Anadarko Basin. The Endicott Sandstone of the Douglas Group is a prominent unit in the Anadarko Basin across western Oklahoma and the Texas Panhandle; the formation is a fining upward sequence that exhibits a sharp basal contact with the underlying prodelta-shale section. These sediments represent fluvial-deltaic environments, indicating that basin filling was nearly complete.

After Douglas sedimentation the Oread Limestone of the lower Shawnee Group was deposited; the Heebner Shale Member of the Oread Limestone is the most widely distributed of the thin, black, phosphatic deep-water shales in the Upper Pennsylvanian of the Midcontinent, as described by Heckel (1977). The Heebner Shale is believed to mark the time of maximum Pennsylvanian transgression over the Midcontinent, and this marked a fundamental change in sedimentation in the Anadarko Basin: following deposition of the Heebner Shale and the Oread cyclothem, wedges of clastic material derived from the Ouachita source area to the east-southeast accumulated along the eastern margin of the Anadarko Basin and prograded westward. These clastic wedges were deposited during the regressive phases of cyclic sedimentation in response to the lowering of sea level. The transgressive phases, in response to rises in sea level, were marked by the deposition of limestones over the Hugoton Embayment and the adjacent shelf area. These limestones are characterized by shelf-edge carbonate banks that are as much as 100 m thick and consist mainly of fossil hash and oolites. In this manner a constructional shelf was established along the eastern margin of the Anadarko Basin in Virgilian time. Thirteen of these cycles have been delineated in Virgilian time after deposition of the Oread Limestone (Rascoe, 1978). Two of these cycles correlate with the formations that compose the Shawnee Group above the Oread Limestone. The remaining eleven cycles are correlative with formations of the Wabaunsee Group.

PERMIAN SYSTEM

Wolfcampian Series

The Pennsylvanian-Permian boundary, which has not been precisely determined on the basis of floral or faunal evidence in the Anadarko Basin–Hugoton Embayment area, is placed at the top of the Brownsville Limestone of the Wood Siding Formation (top of the Wabaunsee Group) in Oklahoma and southern Kansas. In most of the region the top of the Brownsville Limestone can be readily identified on wireline logs; however, the Brownsville Limestone grades into a clastic facies in southern Oklahoma, and in the northeastern portion of the Texas panhandle the Brownsville Limestone is very difficult to identify.

Tectonic elements present in the Anadarko Basin–Hugoton Embayment region during Late Pennsylvanian time persisted into Early Permian time; however, the influence of these features diminished greatly in the Permian Period. The Wichita-Amarillo Uplift, a significant positive element during Pennsylvanian time, probably was expressed topographically as an archipelago in Early Permian time (Fig. 8H; MacLachlan, 1967). There a terrain of Paleozoic limestone and Precambrian granite was eroded, and clasts of these rocks are found in Wolfcampian sediments on the flank of the uplift (Plate 6-R). The Amarillo Uplift was finally covered by sediments in Wolfcampian time. The Anadarko Basin continued to subside in Early Permian time, but far more slowly than in the Pennsylvanian Period. As a result of continued subsidence, Wolfcampian sediments lie conformably upon the Virgilian strata within the basin, and regional lithofacies patterns of the Wolfcampian and Virgilian Series are quite similar.

In the northeastern Texas panhandle, in an area encompassing the 600 m isopach (Plate 6-R), the Wolfcampian Series consists of massive shelf-margin carbonates and thick sections of slope and basinal shale. These shelf-edge and basinal sediments are unlike the orderly succession of widely traceable shelf limestones and shales to the north in that the vertical and lateral distribution of these rocks is extremely irregular and discontinuous. These rocks infilled the "starved" Anadarko Basin of Late Pennsylvanian age. In the Texas panhandle the Wolfcampian "Brown Dolomite" (upper Council Grove Group and lower Chase Group) covered the Amarillo Uplift. Shelf areas were present to the west, north, and east of the Anadarko Basin. In a northeast-southwest belt across the Midcontinent, Wolfcampian rocks consist mainly of cyclic shallow-marine limestones and shales. These sediments make up the Admire, Council Grove, and Chase Groups; these units and many of their constituent limestone formations can be recognized over Kansas, western Oklahoma, and the Oklahoma panhandle.

Mostly fine-grained clastic red-bed sediments border this belt of limestones and shales on the east and west; these strata represent mixed marine and continental environments, and they are indicative of marine regression from the region. In south-central Oklahoma, clastic sediments considered Wolfcampian in age (and referred to the upper part of the Pontotoc Group) covered all but the highest peaks of the Wichita Uplift. And in the far western part of the region, clastic sediments regarded as Wolfcampian in age are equivalent to part of the "detrital rocks" (McKee and others, 1967, Table I) and the upper part of the Sangre de Cristo Formation of southeastern Colorado and northeastern New Mexico.

Leonardian Series

Leonardian time was marked by continued subsidence of the Anadarko Basin and the Hugoton Embayment, and by continued regression of the sea from the region. As a result, the dominant lithologies of the Leonardian Series are red beds and evaporites deposited in continental and shallowing marine environments (Fig. 8I; Plate 6-S); cyclic patterns of sedimentation suggest that eustatic fluctuations of sea level continued in Leonardian time.

Strata now cropping out on the east side of the Anadarko Basin are the Wellington and Garber Formations and the Hennessey Group. These red-bed clastic units, deposited near the eastern shore of the basin, were derived by erosion of low-land areas that extended across eastern Oklahoma and adjacent areas. The Garber and Wellington outcrops comprise a complex system of interbedded alluvial and deltaic sandstones and shales that are thickest and coarsest in central and south-central Oklahoma. These strata, and shales of the overlying Hennessey, grade westward into two thick evaporite units called the Wellington and

Cimarron evaporites (Jordan and Vosburg, 1963). The Wellington evaporites contain more than 100 to 200 m of interbedded halite, shale, and anhydrite in a unit referred to as the Hutchinson Salt, whereas the Cimarron evaporites consist of more than 200 m of interbedded salt and shale, with a medial anhydrite/dolomite called the Cimarron (Stone Corral) anhydrite. To the west these evaporites grade laterally into sandstones along the western margin of the Permian seaway. The cyclicity of the evaporites is attested by the great lateral continuity and vertical repetition of individual anhydrite, salt, and shale beds 1 to 6 m thick over vast areas of the basin.

The Wichita Uplift continued to have a modest influence on sedimentation in the Anadarko Basin. Clastic debris was shed northward onto the south flank of the basin where it interfingered with the Wellington evaporites, and to a lesser extent with the Cimarron evaporites.

Guadalupian Series

In Guadalupian time the Anadarko Basin continued to subside, and it received as much as 500 m of red beds and evaporites along its depocenter (Fig. 8J; Plate 6-T). The Wichita Uplift subsided at a somewhat slower rate than the basin, and all but the highest mountain peaks were probably buried by fine clastics.

Guadalupian strata in the area are largely unfossiliferous, and thus, age assignments are based primarily upon lithostratigraphic correlation with equivalent strata in the Palo Duro Basin and farther south. The Blaine Formation, consisting of widespread anhydrite, shale, dolomite, and halite interbeds, is equivalent to the upper part of the San Andres Formation in the Palo Duro Basin. Regional correlations show approximate equivalency of the Glorieta, Cedar Hills, and Duncan Sandstones, in various parts of the Anadarko Basin, with the San Angelo Sandstone and basal beds of the San Andres farther to the south. The top of the Guadalupian Series is in the upper part of the Cloud Chief Formation.

An episode of marine regression marked the beginning of Guadalupian time in the Anadarko Basin. Sands entered the basin from the east (Duncan Sandstone), north (Cedar Hills), and northwest (Glorieta), and graded into shales and some salts toward the central and southwest parts of the basin. The Wichita Mountains were essentially buried, and the sources for clastics deposited in the southern and eastern parts of the basin were the lowland areas of eastern Oklahoma and the deeply eroded Ouachita belt of southeastern Oklahoma and northeastern Texas.

A later marine transgression permitted 30 to 60 m of Blaine evaporites to be deposited over all of the Anadarko Basin and Hugoton Embayment area. This in turn was followed by regression that introduced sandstones and scattered thin anhydrites across much of the area in the Whitehorse Group and Cloud Chief Formation. In the eastern part of the basin the Cloud Chief locally contains 30 m of massive anhydrite (gypsum on the outcrop).

Ochoan Series

Ochoan rocks are mainly red-bed sandstones and shales, but they contain some anhydrite and dolomite (Alibates Bed) in the western part of the Anadarko Basin (Plate 6-U). Little is known about Ochoan paleogeography in and around the Anadarko Basin, but it likely was similar to that of late Guadalupian time.

The key to regional correlation and age assignment in the Ochoan is the Alibates Bed. The Alibates is equivalent to the base of the Doxey Shale and to the upper part of the Rustler Formation in the Midland Basin, thus establishing equivalency of the overlying Doxey Shale with the Dewey Lake Formation. It is presumed that the upper part of the Cloud Chief Formation in the Anadarko Basin and Hugoton Embayment is equivalent to the lower Rustler and Salado Formations of the Midland Basin area.

Outcropping Ochoan strata typically contain chaotic structures, collapse features, and other evidence of disturbed bedding due to dissolution of underlying Guadalupian halite beds along the flanks of the Anadarko Basin. Such collapse features also occur in some of the Guadalupian and post-Permian strata.

ARDMORE AND MARIETTA BASINS

Patrick K. Sutherland

INTRODUCTION

The Ardmore and Marietta basins in southern Oklahoma are highly complex structural troughs that have equally complex depositional histories. The basins form a southeastern extension of the Anadarko Basin–Wichita Mountains complex in western Oklahoma. Ham and others (1964) termed this complex the Southern Oklahoma Geosyncline because of its extraordinary scale with as much as 12,200 m of strongly deformed Paleozoic sediments and volcanic flows. They recognized its uniqueness as the deepest sedimentary trough on the North American Craton. This striking feature stands out on tectonic maps of North America; this was probably the fact that led the Soviet geologist Shatski (1946) to apply his newly defined term "aulacogen" to the feature in southern Oklahoma. As used today in the context of plate tectonics, an aulacogen is a long trough of anomalously thick sediments that extends at a high angle onto the foreland of the craton from the continental margin. It has, however, been only during the past ten years that the term Southern Oklahoma Aulacogen has been generally applied in the area. This northwest-trending structural complex is in excess of 484 km in length and about 80 km in width in the area of the Ardmore and Marietta basins. The Upper Cambrian through Upper Pennsylvanian (Missourian) strata in the Ardmore Basin are about 9,100 m thick of which 5,500 m are made up of Mississippian and Pennsylvanian strata (Ham and others, 1964).

STRATIGRAPHIC AND TECTONIC FRAMEWORK

Ham and others (1964) divided the development and filling of the Southern Oklahoma Aulacogen into three stages. The first

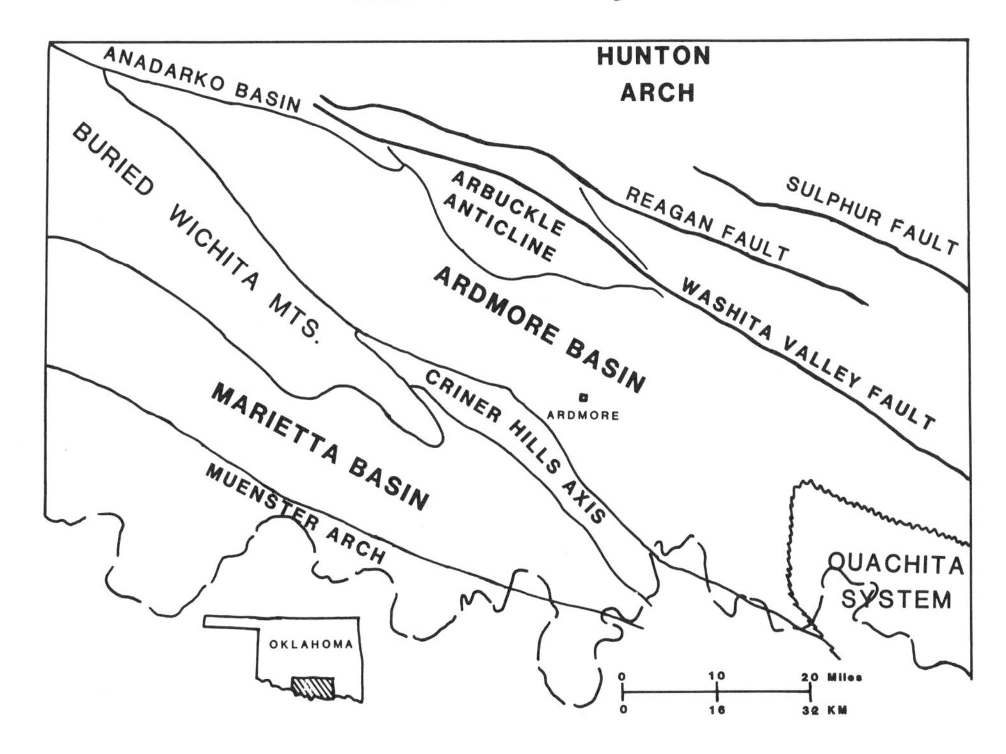

Figure 10. Southern Oklahoma fold belt showing structural provinces (modified from Hicks, 1971; but faults from Jordan, 1962).

(Rifting Stage) was marked during Early and Middle Cambrian time by the development of marginal faults and downwarp associated with both intrusive and extrusive igneous activity. The second (Subsiding Stage), during Late Cambrian through middle Mississippian time, was characterized by subsidence and the accumulation of a thick carbonate sedimentary sequence. The third (Deformation Stage), during the Late Mississippian and Pennsylvanian time, was marked by the development within the aulacogen of major elongated uplifts (e.g., Criner and Wichita Uplifts) in close proximity to long and typically deep basins (e.g., Ardmore, Marietta, and Anadarko Basins; Fig. 10) and the accumulation of a thick clastic sedimentary sequence.

The Southern Oklahoma Aulacogen, as a depositional feature, included also the area of the present Arbuckle Anticline, to the northeast of the Ardmore Basin. The northeastern boundary of the aulacogen was marked by the Washita Valley Fault, which originated in the Cambrian Period during the initial Rifting Stage of aulacogen development. It separates Precambrian granites of the Tishomingo Uplift to the northeast, which underlie the shelf, from Cambrian rhyolites to the southwest, which floor the aulacogen. Ham and others (1964) conclude that this difference in basement rock character affected the type and intensity of folding throughout the Paleozoic and that the older bordering faults associated with the first phase of Cambrian basin filling guided the structural pattern of the late Paleozoic. Thus, the Arbuckle Mountains are divided into two sections along the Washita Valley Fault. To the northeast, where Paleozoic rocks rest upon Precambrian granites, the Paleozoic strata on the craton are relatively thin and are deformed mainly by block faulting. To the southwest of the fault, in the area of the Arbuckle Anticline and Ardmore Basin, the Paleozoic strata are thick, have thick Cambrian volcanic rocks as a floor, and are much more highly deformed (Ham and Wilson, 1967).

During the early and middle Paleozoic, and extending through most of the Mississippian, the Southern Oklahoma Aulacogen was essentially a single broad depositional basin extending from the Washita Valley fault on the northeast to the Muenster Arch on the southwest (Fig. 10). The present division of the aulacogen into complex uplifts and basins is mainly the result of Pennsylvanian orogenies. Three major complex periods of folding and uplift (Deformation Stage) can be differentiated in the area of the Ardmore and Marietta Basins, as follows:

1. Late Mississippian–Early Pennsylvanian uplifts. The Wichita-Criner Orogeny is a term commonly given in the Ardmore Basin area to the time of initial folding of the Criner Hills Uplift, which provided great volumes of early and middle Paleozoic limestone cobbles to the Jolliff Formation (Morrowan) and later to the Bostwick Formation (Atokan). It is generally stated that this orogeny began in Early Pennsylvanian time (Ham and Wilson, 1967) but it, in fact, began in the Chesterian (Late

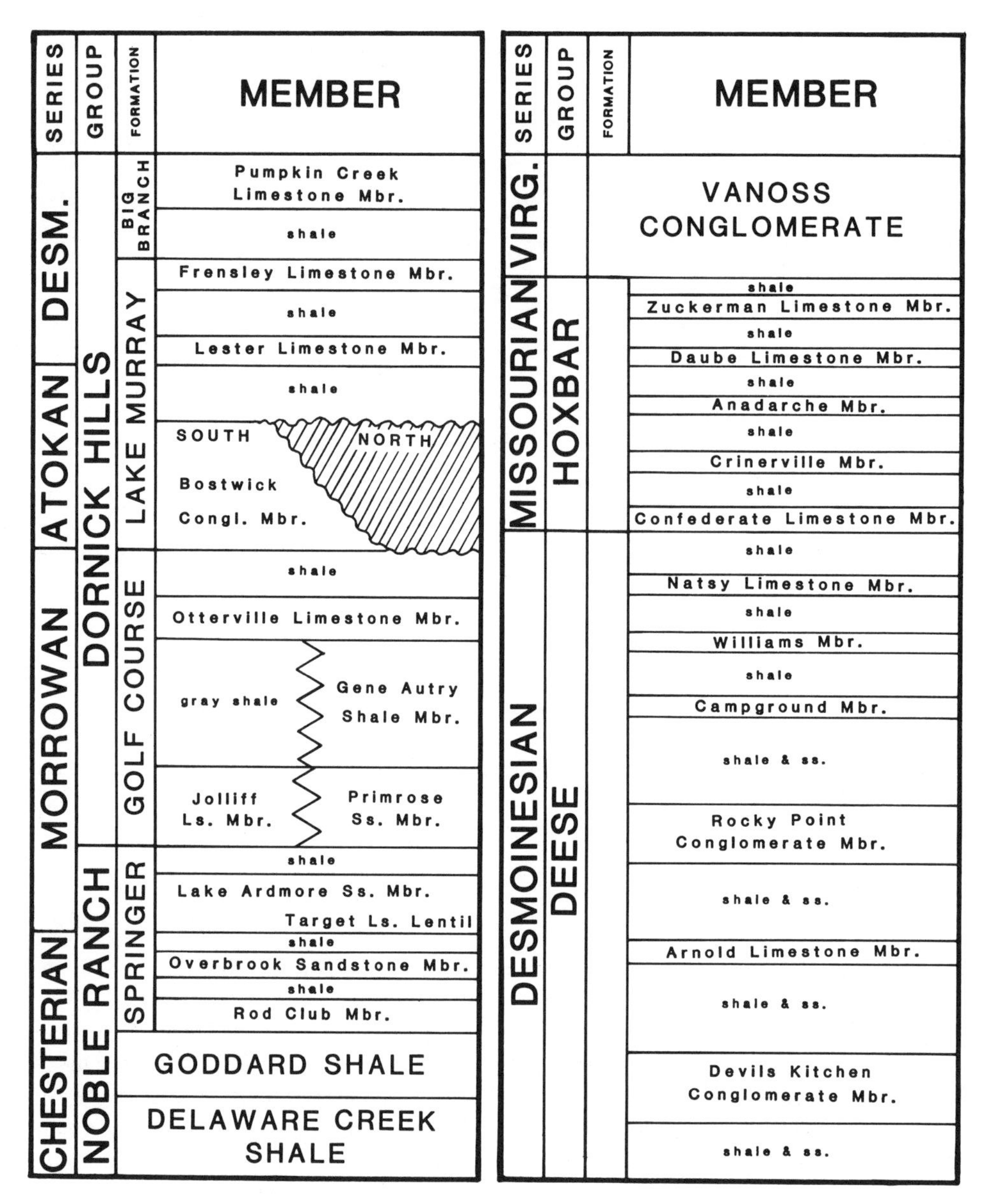

Figure 11. Columnar section of outcropping Pennsylvanian strata in the Ardmore Basin (modified from Tomlinson and McBee, 1959).

Mississippian), at least as early as the deposition of the lower part of the Springer Formation (Fig. 11). The initial folding and erosion along the Criner Hills axis during the Chesterian must have been extremely rapid in that Ordovician limestone pebbles are included in two thin lower Springer conglomerates, one first recorded by Tomlinson and McBee (1959) and the other by Lang (1966, Stop 4). The matrices from these conglomerates contain middle to late Chesterian conodonts (R. C. Grayson, Jr., personal communication, 1985). There was significant further folding during the Morrowan (Jolliff deposition) and even more profound uplift in early Atokan time (Bostwick deposition). In many areas in southern Oklahoma, particularly on crests of truncated uplifts, it is Atokan-aged strata, including conglomerates, that unconformably overlie strata as old as the Early Ordovician Arbuckle Group (Tomlinson and McBee, 1959). It is of interest to point out that the most widely developed and profound Paleozoic unconformity on the North American craton occurs below a surface that varies in age from Morrowan to Early Desmoinesian (Ham and Wilson, 1967).

It was only with the folding that marked the Criner Hills Uplift that the Ardmore and Marietta basins were differentiated for the first time (Ham and others, 1964).

2. Middle Pennsylvanian uplifts. The Criner Hills Uplift, with a northwest strike, shed limestone cobbles in a northeasterly direction in the Ardmore Basin primarily during the Morrowan and Atokan. By Middle Desmoinesian time the chert pebble conglomerates of the Devils Kitchen and Rocky Point reflect the initial uplift of the Ouachita System to the southeast (Fig. 10). On the outcrop in the southern Ardmore Basin area the southeastern source direction, particularly in the Devils Kitchen, is confirmed by the marked increase to the southeast of chert pebble size, percentage of chert pebbles, and thickness of individual chert pebble beds.

3. Late Pennsylvanian uplifts. The final and most profound structural deformation of the Southern Oklahoma Aulacogen is termed the Arbuckle orogeny and was one of strong compression and locally profound uplift, as exemplified by the Arbuckle Anticline. The Ardmore and Marietta basins and the Criner Hills were significantly deformed at this time (Ham and Wilson, 1967). A series of orogenic conglomerates, shed mostly from the Arbuckle Anticline, were deposited in southern Oklahoma during the Middle and Late Virgilian as well as the Early Permian. Those in the Ardmore Basin and around the Criner Hills, generally included in the Vanoss Formation, blanket significant parts of the area and rest with marked angular unconformity upon older rocks that range in age from Precambrian to Late Pennsylvanian (Ham and Wilson, 1967).

Evidence that the Southern Oklahoma Aulacogen was a major subsiding trough during the Late Mississippian and Pennsylvanian is provided by a comparison of thicknesses preserved in the Ardmore Basin with those on the craton to the north. In the Ardmore Basin, Chesterian and Pennsylvanian strata—composed mostly of shales and sandstones plus generally thin limestones—total as much as 5,200 m. Equivalent strata upon the craton, exposed along the northeastern margin of the Arbuckle Mountains, total about 880 m and consist mostly of shales with locally thick limestones and sandstones (Ham, 1969; Morris and Sutherland, 1984). Tomlinson and McBee (1959) provide the only detailed comprehensive description yet published of the stratigraphy and depositional history of the Ardmore Basin.

Little detailed information has been published on the stratigraphy and depositional history of the Late Mississippian and Pennsylvanian strata in the Marietta Basin; the rocks there exist completely in the subsurface are not well dated. The most striking difference in the sequence there compared to that in the Ardmore Basin is the presence of a major pre-Atokan to pre-Desmoinesian unconformity. Across the uplifted parts of the basin, Desmoinesian conglomerates generally rest directly on folded and faulted strata ranging from various formations in the Ordovician to the Devonian Woodford. Locally there are Mississippian shales preserved in the deeper basin areas possibly equivalent to some part of the Delaware Creek and/or the lower Goddard. In the Oklahoma portion of the Marietta Basin there are no strata preserved that are equivalent to the Springer or strata of Morrowan age except on the northeast flank of the basin, in the Enville–southeast Marietta field area in T.7S., Ranges 2 and 3 E. (C. D. Fenstermaker, personal communication, 1986). However, in the Gordonville Trough, a southeastern extension of the Marietta basin in Cook and Grayson Counties, Texas (not shown on Fig. 10), there are beds possibly equivalent to the Springer and also strata of possible Morrowan age (Bradfield, 1969). There are locally nonconglomeratic shales and sandstones that unconformably underlie the Desmoinesian conglomerates, but they rest unconformably on older strata; Bradfield (1959) considers these to be Atokan and possibly early Desmoinesian in age.

DEPOSITIONAL HISTORY

Introduction

In Early and middle Mississippian time the Southern Oklahoma Aulacogen was part of a broad area of starved sedimentation, and strata of these ages are thin and discontinuous. However, this trough clearly became reestablished in Chesterian time as a major subsiding basin. The aulacogen is characterized during Late Mississippian and Early Pennsylvanian time by differentiation of the basin into long, narrow, rapidly folding uplifts and basins (Fig. 10). The uplifts were deeply truncated, producing much detrital material that was deposited in the adjacent basins nearest to the uplifted sources. Deposition within the basins is characterized by marked changes in both facies and thickness as one moves away from the source areas. Some secondary structures within the basins, such as the Caddo Anticline and the Berwyn Syncline in the Ardmore Basin, show evidence of uplift and downwarp during deposition (Tennant, 1981).

Chesterian Series

Chesterian sediments in this area are mostly shales (Delaware Creek and Goddard), and source directions have not been clearly established. For the overlying Springer Formation (Fig. 11), however, paleocurrent data indicate a dominant source direction from the northwest, from the Anadarko Basin. There is also an increase to the northwest of maximum thickness of individual sandstone layers. Only rarely and locally do any Springer sandstone layers on the outcrop in the Ardmore Basin approach or exceed 12 m in thickness. In contrast, in the Sho-Vel-Tum area, located only 40 km to the northwest of the city of Ardmore, individual sandstone intervals commonly exceed 30 m in thickness. The Springer sandstones are parts of several large deltaic complexes that prograded southeastward into the Ardmore Basin. On the outcrop near the city of Ardmore, however, one sees a prodelta facies consisting mostly of shale and representing only the lower parts of deltaic sequences. They generally lack capping remnants of delta front channel or delta plain deposits. The more fully developed deltas are found in the subsurface several miles to the northwest. The Rod Club Sandstone, at the base of the Springer sequence, contains individual sandstone layers less than 3 m thick that consist of graded beds with basal flute casts (Meek, 1983). These layers represent turbidites at the far distal margin of a delta located to the northwest (Meek and others, 1983). The entire Springer sandstone sequence changes facies mostly to basinal shale in the subsurface to the southeast.

The Criner Hills axis began to be uplifted (Wichita-Criner Orogeny) in middle to late Chesterian time, as indicated by the occurrence of thin limestone-pebble conglomerates associated with Rod Club or Overbrook sandstones in the southern part of the Ardmore Basin. These conglomerates, possibly representing submarine debris via gravity flows, are thin and obscure, but the astonishing fact is that they include Ordovician pebbles, which indicates a remarkable rate of uplift and erosion along the Criner Hills axis in late Chesterian time.

Poorly dated shales that presumably correlate with the Delaware Creek and possibly the lower part of the Goddard basinal shales were deposited in the Marietta Basin area before the beginning of uplift on the Criner Hills axis. Tomlinson and McBee (1959, p. 15) postulate that the present area of the Marietta Basin formed part of a positive area, along with the Criner Hills axis, during middle and late Chesterian as well as during Morrowan times.

Morrowan Series

More profound uplift along the Criner Hills axis occurred during Morrowan time as indicated in the southern part of the Ardmore Basin by the occurrence of the Jolliff Limestone cobble conglomerates. The intensity of uplift was greatest along the buried Criner Hills axis south of the present day Criner Hills. In exposures south of Lake Murray the shale intervals between the Jolliff and Otterville and between the Otterville and Bostwick are mostly replaced by limestone cobble conglomerates (Cromwell, 1974). The Jolliff conglomerates accumulated possibly as alluvial outwash fans or as submarine fans (Maley, 1986) along the northeast margin of the uplifting Criner Hills axis but they did not extend across the Ardmore Basin to the northeast. In that area the Primrose Sandstone, with a source from the north and northwest, was being deposited simultaneously (Fig. 11), as indicated by the occurrence of similar conodont faunas in the two units (R. C. Grayson, Jr., personal communication, 1985).

South of Ardmore the base of the Jolliff conglomerates is marked by an unconformity that definitely cuts out the Lake Ardmore (Morrowan) Sandstone Member of the Springer Formation and possibly also the Overbrook (Chesterian) Sandstone Member.

The Morrowan interval in the northern part of the Ardmore Basin above the Primrose includes the Gene Autry Shale and the overlying Otterville Limestone. The Gene Autry was deposited in somewhat deeper water under more restricted conditions. The upper part shows a shallowing upward trend that continued with the deposition of the higher energy, locally shoaling, Otterville Limestone. The Otterville around the Caddo Anticline formed as an oolitic marine sand belt, which supports the conclusion that the Caddo Anticline was being slowly uplifted during the time of Otterville deposition (Maley, 1986). In the Berwyn Syncline, to the northeast of the Caddo Anticline, the Otterville consists of thin layers of fine-grained bioclastic limestone interbedded with shale, representing deposition in slightly deeper, quieter water. On the stable shelf to the northeast, as preserved in the Mill Creek Syncline, more than 120 m of cross-bedded oolitic limestones (Wapanucka Limestone) were deposited simultaneously with deposition of the aulacogen of both the upper Gene Autry Shale and the Otterville Limestone (Sutherland and others, 1982).

Atokan Series

The Wichita-Criner Orogeny intensified in Atokan time with further uplift and the deposition of the highly variable Bostwick Member along the northeast margin of the Criner Hills axis (Fig. 10). The dominant lithologies are limestone and chert pebble conglomerates interbedded irregularly with shales. As with the Jolliff, this unit is thickest, and the limestone cobbles are largest (in excess of 0.5 m) and most numerous at exposures in the southernmost part of the Ardmore Basin, indicating greatest uplift on the Criner Hills axis south of the present day Criner Hills. As with the Jolliff, the limestone cobble conglomerates did not extend northeastward across the Ardmore Basin, suggesting that the cobbles were not carried that far. However, in the northern part of the basin there is a regional unconformity at or near the base of the unnamed shale that overlies the Bostwick in the southern Ardmore Basin (Clopine, 1986; Fig. 11). This shale can be correlated across the entire basin. At several localities in the northeast, this shale—with a thin basal conglomerate and containing the upper Atokan fusulinid *Fusulinella*—rests directly on the Otterville Limestone (Morrowan). Thus, the Bostwick could have been all or partly removed in the northeast by erosion below this late Atokan unconformity.

Desmoinesian Series

By early Desmoinesian time the uplift along the Criner Hills axis had ceased and generally quiescent conditions prevailed across the Ardmore Basin. Detailed facies studies of the Lester, Frensley, and Pumpkin Creek Limestones across the Ardmore Basin show an increase in energy level to the southwest. For example, the Pumpkin Creek Limestone grades from phylloid algal wackestones in the northeast to sandy crinoidal grainstones in the southwest (Tennant, 1981). Much suspended mud was periodically introduced, as was more limited sand. Tennant (1981) records 26 independent paleocurrent flow estimates taken from cross-bedding in sandstone outcrops in this upper Dornick Hills sequence in the northern part of the Ardmore Basin. McGee (1985) made additional paleocurrent measurements in the southern part of the basin. In both areas these indicate a direction of source from the southwest. The sandstones are mostly fine grained and well sorted, and Jacobsen (1959) postulated that these sands came from the erosion of previous Springer sandstones or alternatively from a metamorphic source in the Red River–Muenster Arch, southwest of the Southern Oklahoma Aulacogen (Fig. 10). The latter would require transport across a quiescent Criner Hills and Marietta Basin. Tennant (1981) postulates just such a picture with the development of a deltaic complex prograding from the Muenster Arch, with only the prodelta and delta fringe facies reaching the Ardmore Basin to the northeast. Thus, the general depositional picture is one of alternating transgressions and regressions with the Lester, Frensley, and Pumpkin Creek Limestones representing, in a general way, periods of maximum transgression and intervening unnamed shale and minor sand intervals representing regressive cycles where prodeltaic and terrigenous shelf sediments prograded northeastward out over the carbonate facies (McGee, 1985). The detailed facies are complex; most of the thin limestone layers of the Frensley, in particular, are discontinuous laterally and in part represent deposition between prodelta lobes. The Caddo Anticline and Berwyn Syncline continued to rise and sink in early Desmoinesian time, as indicated by both the high-energy carbonate facies and thinner shale sequences on the Caddo Anticline and the lower energy carbonates and thicker shale sequences in the Berwyn Syncline (Tennant, 1981).

The deposition model just described provides a possible interpretation of the stratigraphic sequence found in the Marietta Basin. In Grayson County, Texas, resting unconformably on older strata, there is a sequence, more than 800 m thick, of gray shales, fine-grained sandstones, and thin limestones. This apparently continuous sequence is included by Bradfield (1959) in the Atokan (middle Dornick Hills), the Lower Desmoinesian (upper Dornick Hills), and the Desmoinesian "Gray Strawn" or "Deese." This interval is poorly dated paleontologically. The possibility is being suggested here that it may be in part equivalent to the Lower Desmoinesian Lester–Frensley–Pumpkin Creek interval in the Ardmore Basin and may be part of a deltaic complex directed toward the northeast from the Muenster Arch. Subsequently, part of this deltaic complex may have been removed by erosion prior to the great influx from the southeast of chert-pebble conglomerates that marked the beginning of deposition of the thick "Red Strawn" or "Deese" in the Marietta Basin and the Devils Kitchen conglomerates in the Ardmore Basin.

In the Ardmore Basin, the first indication of uplift on the Ouachita Foldbelt to the southeast is the occurrence in the southernmost exposure of the Pumpkin Creek Limestone in the Ardmore Basin (on Pumpkin Creek) of common chert granules and a few chert pebbles greater than 1.0 cm in diameter. The chert granules decrease in abundance northward along the outcrop, and chert pebbles have not been observed north of Pumpkin Creek.

The large volume of chert pebbles in the Devils Kitchen Member of the Deese Formation were introduced into the Ardmore Basin in the form of alluvial and/or deltaic deposits in the southern part of the Ardmore Basin. Wood imprints and mud cracks within the interbedded sand facies are common in this area. Cross-bedding transport directions at the Lake Murray spillway are to the west-northwest (Saether, 1976). The conglomerate facies decreases in thickness northwestward with decrease in pebble size. The Rocky Point Member represents a second and similar time of major introduction of chert pebbles from the southeast, but few chert pebbles were carried as far as the northern part of the basin. The upper Deese, above the Rocky Point, has no chert-pebble conglomerates and is more marine in character, with the occurrence of a few thin limestones. The thin sporadic limestone cobble conglomerates (Warren Ranch) in the northern part of the basin, apparently represent slope deposits with a source from the shelf to the north, presumably from the Hunton Arch. Equivalent upper Deese strata in the Marietta Basin are included in the "upper gray shale zone of the 'Strawn' or 'Deese'."

Missourian Series

The depositional history of the Hoxbar Group in the Ardmore Basin is poorly understood, and source directions have not been established. It is predominantly a marine sequence with several named limestone units, but there is one local coal layer (1 m thick) just below the Daube Limestone noted by Tomlinson and McBee (1959). Some of the scattered limestone and chert pebble conglomerates may have come from the uplifting Hunton Arch, on the shelf to the north.

Virgilian Series

The Virgilian Arbuckle Orogeny produced the uplift and sharp folding of the Arbuckle Anticline and simultaneously the sharp folding, compression, and faulting of the Ardmore Basin, Criner Hills axis, and the Marietta Basin. The Collings Ranch Conglomerate, preserved in the central Arbuckle Mountains, was deposited in Middle Virgilian time, followed in Late Virgilian by the deposition of great alluvial fans of limestone-cobble conglomerates (Vanoss Conglomerate) around the north, west, and south margins of the Arbuckle Anticline. In the Ardmore Basin, these limestone-cobble conglomerates were deposited across most of the area and rest unconformably on the steeply truncated margins of all of the Mississippian and Pennsylvanian formations described here, including the Missourian Hoxbar Group (Ham, 1969).

ARKOMA BASIN

Patrick K. Sutherland

STRATIGRAPHIC AND TECTONIC FRAMEWORK

The Arkoma Basin is an arcuate structural feature that extends from the Gulf Coastal Plain in central Arkansas westward 400 km to the Arbuckle Mountains in south-central Oklahoma (Fig. 12). It ranges from 32 to 80 km wide. It is bounded on the north and northwest by the Ozark Uplift and the Northeast Oklahoma Platform. Its southern margin is marked in Oklahoma by the Choctaw Fault and in Arkansas by the Ross Creek Fault.

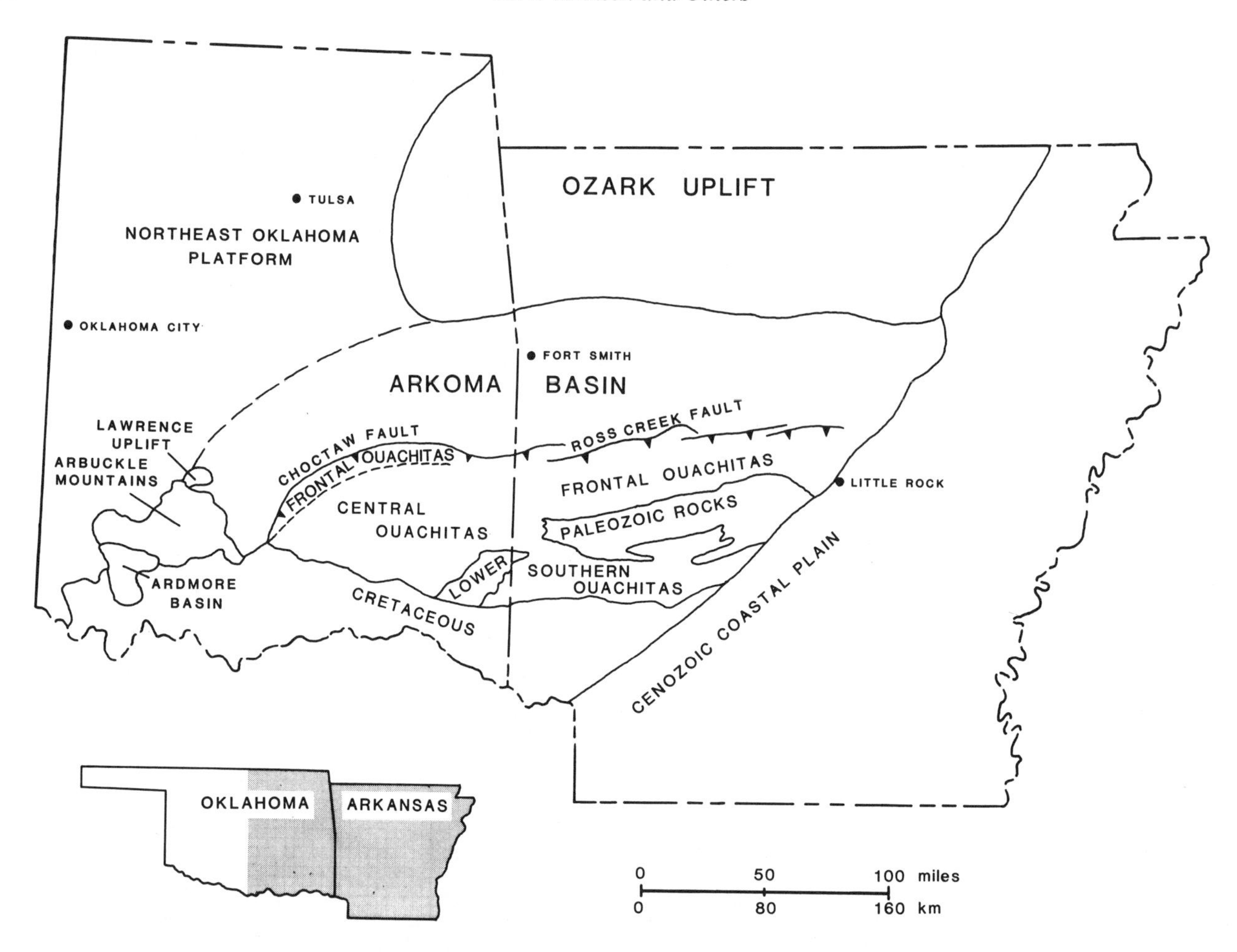

Figure 12. Arkoma Basin of Oklahoma and Arkansas, and surrounding geologic provinces.

Various tectonic models have been proposed for the Ouachita Foldbelt. One of the most widely quoted is a collision model with south-dipping subduction (Wickham and others, 1976; Lillie and others, 1983). There was gradual closure of the Ouachita Trough and the development of a foreland basin followed by uplift and erosion of the Ouachita orogenic belt.

The Arkoma Basin was depositionally part of a stable shelf (here named the Arkoma Shelf) along a passive continental margin during much of its history. Deep-water deposits accumulated in the basin south of the shelf. In Cambrian to Early Mississippian time a thick sequence of shallow-water carbonates was deposited on the shelf, and thinner, deep-water black shales and cherts accumulated in the basin. In middle Mississippian time there was the beginning of deposition of thick turbidites (Stanley Group) in the basin (Ouachita Trough) (Fig. 13), as a result of a gradual sequential narrowing of the Ouachita Trough from east to west (Graham and others, 1976). Turbidites were fed longitudinally westward into the trough. There was no significant change in depositional pattern on the shelf to the north except for the intermittent introduction of terrigenous clastics from the north.

The broad Arkoma Shelf, which included the southern part of the present-day Ozark Uplift as well as the Arkoma Basin, continued as a quiescent and significant depositional feature through Chesterian, Morrowan, and Early Atokan time. In Middle Atokan time the southern margin of the Arkoma Shelf was subjected to flexural bending, caused by continued basin closure, that resulted in sequential northward development of down-to-the-south syndepositional normal faults (Fig. 14). By this process the Arkoma Shelf evolved into an incipient foreland basin. By the end of middle Atokan time, fault movement had ceased and the fault-related basin had been filled with sediment. During the late Atokan, deposition of major deltaic systems prograded both westward and southward over the fault zones (Fig. 14).

In early Desmoinesian time the fully developed foreland basin continued to subside. Deltaic sedimentation, initially from the east but subsequently mostly from the shelf to the north, kept pace with the subsidence, resulting in the thick wedge of shallow-marine and nonmarine late Atokan and early Desmoinesian strata preserved in the present-day Arkoma Basin.

A post-Krebs (Fig. 15) regional unconformity (Oakes, 1953,

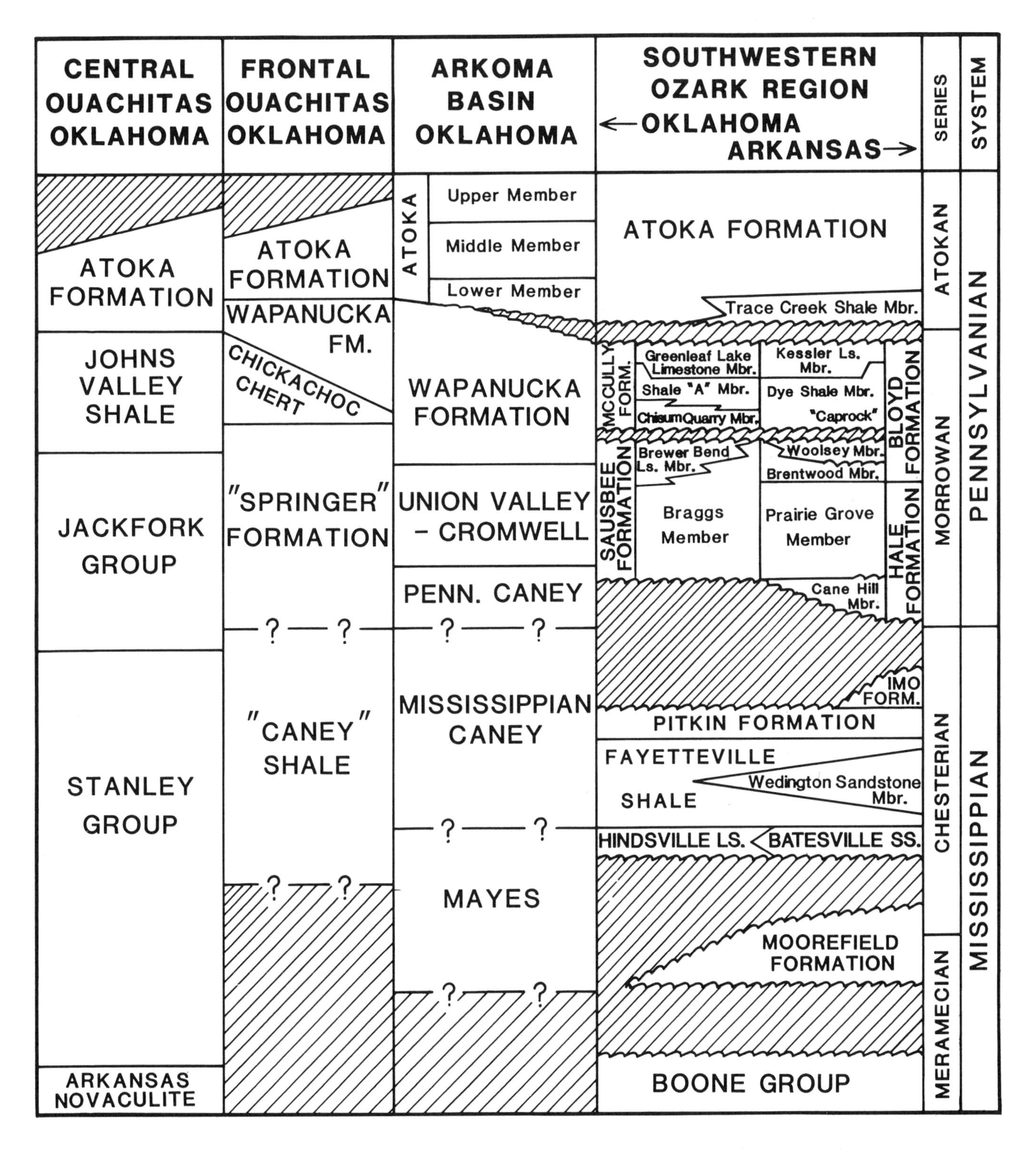

Figure 13. Correlation of Upper Mississippian and Lower Pennsylvanian formations and members across the Arkoma Basin from the Ozark Shelf to the Ouachita Trough.

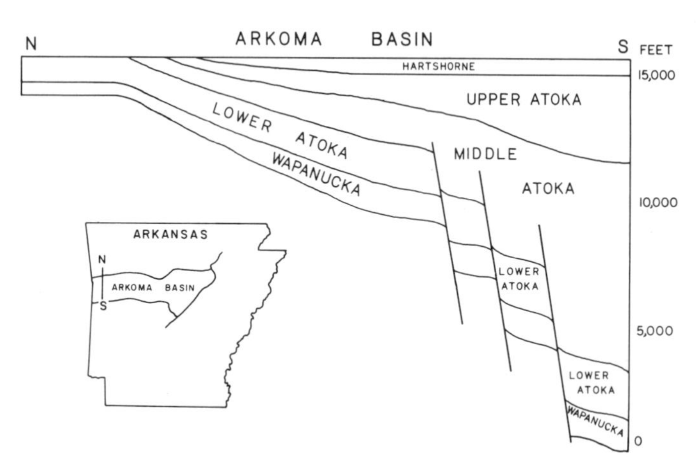

Figure 14. North-south section across Arkoma Basin depicting syndepositional faults that controlled sedimentation during deposition of middle Atoka Formation (modified from Zachry and Sutherland, 1984).

DESMOINESIAN SERIES	MARMATON GROUP	Holdenville Shale
		Wewoka Formation
		Wetumka Shale
		Calvin Sandstone
	CABANISS GROUP	Senora Formation
		Stuart Shale
		Thurman Sandstone
	KREBS GROUP	Boggy Formation
		Savanna Sandstone
		McAlester Shale
		Hartshorne Sandstone

Figure 15. Formations and groups of the Pennsylvanian Desmoinesian Series in the Arkoma Basin and adjacent area to the northwest, in south-central Oklahoma (from Oakes, 1953, 1967, 1977).

1967, 1977) marks the end of major subsidence of the Arkoma Basin, a time of folding and erosion in the Arkoma Basin of both Atoka and Krebs strata, and the sufficient uplift of the Ouachita Foldbelt to expose to erosion the Ordovician and Devonian chert-bearing formations of the core area. By the beginning of deposition of the overlying Cabaniss Group (Thurman Sandstone), chert-pebble conglomerates from a Ouachita source were funneled to the northwest, across the folded Arkoma Basin, into a narrow successor basin to the north of the Hunton Arch. These conglomerates rest unconformably on the Krebs Group (Oakes, 1967, p. 28).

Possible sources for sediments in the Ouachita Trough have been summarized by Thomas (1984). These are based on regional facies distributions, paleocurrent data, and/or sandstone petrography. Two of these source areas are believed to have also provided sediments to the Arkoma Shelf/Basin area to the north. They are (1) the Ouachita Foldbelt that had already been uplifted along the southwest margin of the Black Warrior Basin, which provided lithic arenites that were transported to the northwest along the axis of the sequentially narrowing Ouachita Trough, and (2) sediments transported southwestward through the Illinois Basin, which consisted mostly of quartz arenites. Additional sediments deposited in the Arkoma Shelf/Basin area (first quartz arenites and later sublitharenites) came from the continental interior north of Oklahoma and west of the Ozark Dome, and possibly secondarily, quartz arenites from the Ozark Dome itself in southern Missouri.

The depositional history of the Arkoma Shelf/Basin area can be divided into the following three intervals:

1. Chesterian through Early Atokan. During this interval, terrigenous sediments were derived almost solely from the northeast and north. Quartz arenites were transported southwestward through the Illinois Basin (Potter and Glass, 1958) and from the Ozark Dome and beyond. Some of these terrigenous sediments were transported westward along the Arkoma Shelf in Arkansas, while others were carried further south into the Ouachita Trough.

2. Middle Atokan through early Desmoinesian. During the middle Atokan there was major deposition of lithic arenites, south of the outer shelf margin, in the area where syndepositional faults developed in the southern half of the Arkoma Shelf/Basin as the basin margin shifted northward. These lithic arenites were most likely derived from the southeast, from the erosion of the already uplifted Ouachita orogenic belt to the southwest of the Black Warrior Basin (Graham and others, 1976; Mack and others, 1983; Thomas, 1984). On the Arkoma Shelf there were still quartz arenites with a source from the north. Lithic arenites apparently continued to be derived from the east in Late Atokan time and definitely in the earliest Desmoinesian during the deposition of the Hartshorne Sandstone. The latter was deposited westward along the axis of the newly developed Arkoma Foreland Basin (Houseknecht and others, 1983). Later in the early Desmoinesian, during the deposition of the McAlester, Savanna, and Boggy Formations, most of the terrigenous sediments in the area (mostly sublitharenites) were derived from the craton to the north and west of the Ozark Dome (e.g., Booch and Bartlesville/Bluejacket Sandstones; Bissell and Cleaves, 1986; Visher, 1968).

*3. **Middle Desmoinesian through early Virgilian.*** Beginning with the deposition of the Thurman Sandstone (Cabaniss Group; Fig. 15), chert-pebble conglomerates derived from the newly uplifted core area of the Ouachita Mountains were transported to the northwest. The older Paleozoic rocks in the Ouachita Mountains continued to provide chert-pebble conglomerates, along with muds and sands, to a shallow basin in central Oklahoma, north of the Hunton Arch, through the remainder of the Pennsylvanian.

DEPOSITIONAL HISTORY

Chesterian Series

The Mississippian Chesterian outcrops in the southern Ozark area in Oklahoma and Arkansas consist of interbedded shallow-marine limestones and shales that rarely exceed 200 m in thickness. Included are the upper Moorefield, Hindsville, Fayetteville, Pitkin, and Imo Formations (Fig. 13). Represented are at least two main shoaling-upward sequences. The southern limit of carbonate deposition during Chesterian time occurs in the subsurface 16 to 20 km south of the southern limit of the present outcrop area in Oklahoma (Tulsa Geological Society, 1961; Fig. 16). South of this line, facies change abruptly to shale on a continually deepening, east-trending outer shelf. The terrigenous source for the outer shelf muds is believed to have been to the east because of the marked increase in thickness of the Fayetteville Shale in that direction across northern Arkansas (Ogren, 1968).

The Chesterian interval in the Ouachita Mountains to the south is composed of turbidite deposits of the middle and upper Stanley Group, and the total thickness of that group in the central Ouachitas in Oklahoma is about 3,235 m (Cline, 1960). Sediment transport directions are from the southeast (Niem, 1976; Fig. 16).

Mississippian-Pennsylvanian unconformity

The Late Mississippian was marked by a sea withdrawal from the shelf areas throughout the Southern Midcontinent region. Rascoe and Adler (1983) attribute this emergence to a broad upwarping of the transcontinental arch. A contributing factor may have been the rapid relative sinking of the Ouachita Trough, some upwarping of the Ozark Dome, and a corresponding southward tilt of the Arkoma shelf north of the trough.

During the post-Mississippian emergence and accompanying southward tilt of the shelf, the Chesterian sequence was progressively truncated northward in both Arkansas and Oklahoma, producing a regional angular unconformity at the base of the Pennsylvanian. Both the Pitkin and Fayetteville Formations wedge out northward, and Morrowan strata locally rest directly on the Hindsville Limestone (Fig. 13) in the Oklahoma Ozarks. Regional relief on the unconformity in that area is over 24 m (Sutherland and Henry, 1977b).

Morrowan Series

In the early Morrowan the sea transgressed north from the Ouachita Trough onto the Arkoma Shelf, across the truncated Chesterian surface. The depositional pattern throughout the Morrowan is one of marked lateral changes in facies and thickness. The dominant source direction on the Arkoma Shelf was from the northeast, primarily from the Illinois Basin. Fluvial sandstones and shales (275 m thick) in north-central Arkansas (Glick and others, 1964) change facies westward to a mostly mixed shallow-marine shelf facies (90 m thick) in northwestern Arkansas (Henbest, 1953) and still farther to the west in northeastern Oklahoma to a shallow-marine offshore bank facies (60 m thick; Sutherland and Henry, 1977a).

The Morrowan Series was first described in northwestern Arkansas (Washington County), where it is subdivided into the Hale and Bloyd Formations (Fig. 13). Westward from Washington County, Arkansas, into Oklahoma the names Hale and Bloyd can be used only for a 25-km-wide belt at the eastern margin of Oklahoma. Farther west the lithologic distinction is lost, as there is an overall marked westward increase in percentage of limestone and a corresponding decrease in the percentage of sandstone. Sutherland and Henry (1977a) subdivided this carbonate sequence into the Sausbee and McCully Formations (Fig. 13) on the basis of regional disconformity at the top of the Sausbee. This break coincides in northwestern Arkansas with a regional disconformity at the base of the Dye Shale Member of the Bloyd Formation (Fig. 13).

The Morrowan carbonate facies in the southern Ozarks of northeastern Oklahoma extends southward for about 16 km in the subsurface of the Arkoma Basin south of the southern limit of the present outcrop area in Oklahoma. Farther south there is a distinct decrease in the percentage of limestone in the Morrowan interval below the Wapanucka Limestone, and an increase in the percentage of shale and sandstone (Fig. 17).

In the central and southern part of the Arkoma Basin the typical Morrowan sequence begins with the Pennsylvanian Caney, which cannot be subdivided in most places from the underlying Mississippian Caney. Most subsurface workers therefore use the base of the overlying Cromwell Sandstone as a marker for the base of the Pennsylvanian (Fig. 13).

The Cromwell Sandstone is a subsurface term, and the unit extends throughout most of the Arkoma Basin in Oklahoma as a fine- to medium-grained subrounded calcareous sandstone overlain in most areas by a thin arenaceous limestone. This interval is

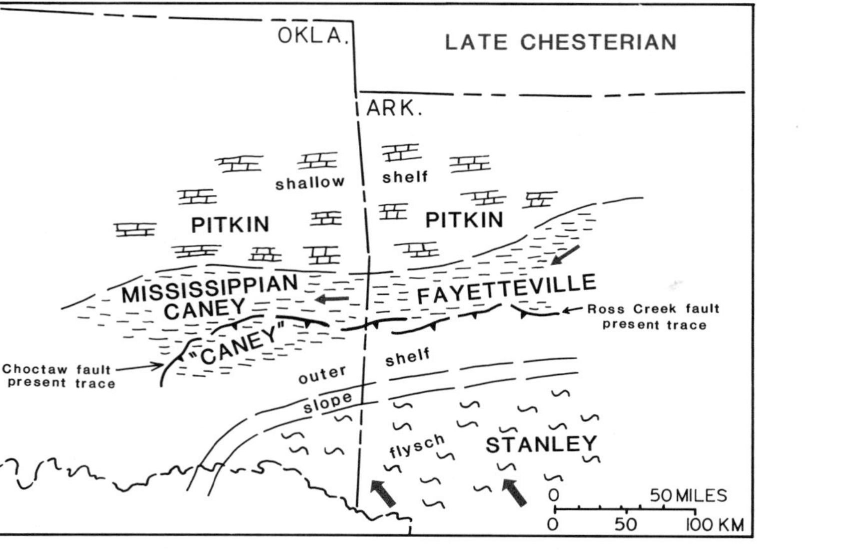

Figure 16. Late Chesterian paleogeographic map. Large and small red arrows indicate major and minor directions of sediment transport. Compare with Figure 12 for location. Dashed traces of present-day locations of Choctaw and Ross Creek Faults are given for reference on Figures 16 to 20. Sources include Tulsa Geological Society (1961), Niem (1976), Glick (1979), Thomas (1985), Hanford (1986).

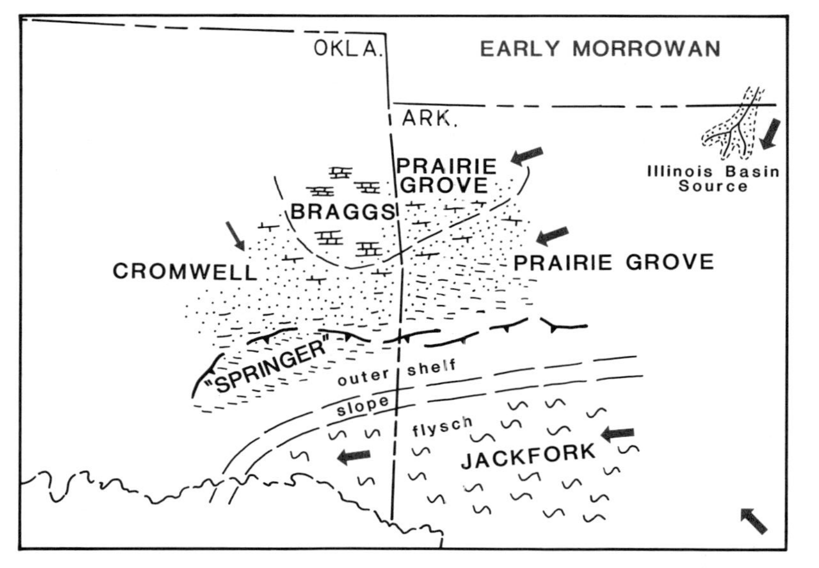

Figure 17. Early Morrowan paleogeographic map. Sources include Morris (1971), Sutherland and Henry (1977a), Foshee (1980), Jefferies (1982), and Moiola and Shanmugam (1984).

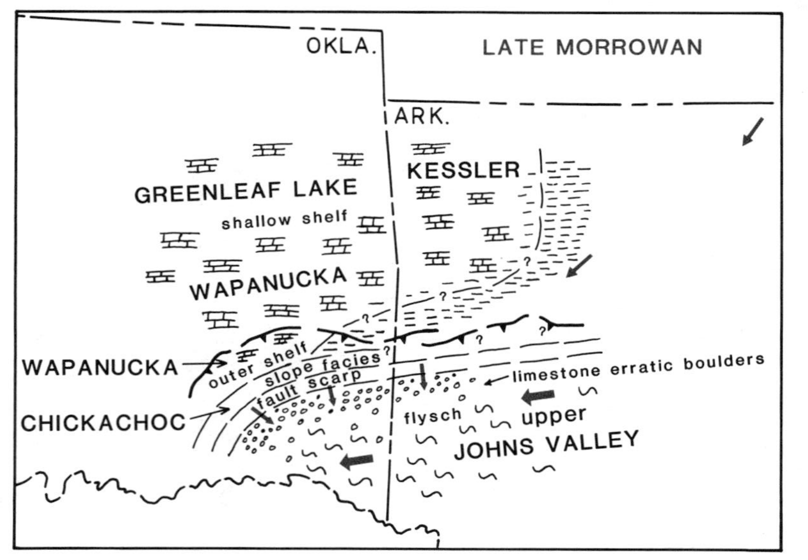

Figure 18. Late Morrowan paleogeographic map. Sources include Shideler (1970), Sutherland and Henry (1977a), and Grayson (1979).

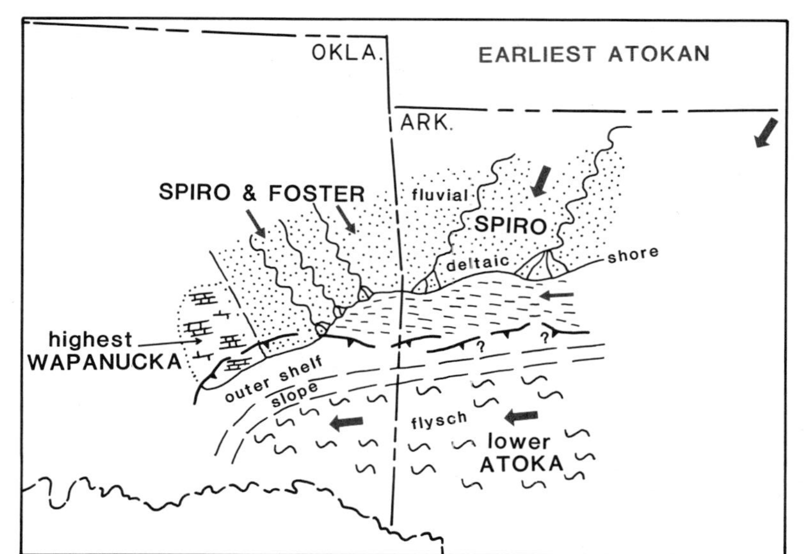

Figure 19. Earliest Atokan paleogeographic map. Sources include Lumsden and others (1971), Grayson (1979), Sutherland and Manger (1979), and Parker (1981).

equivalent to the Prairie Grove Member of the Hale Formation in Arkansas and to the Braggs Member of the Sausbee Formation in the southwestern Ozarks in Oklahoma (Fig. 13). It has not as yet been recognized in the frontal Ouachitas where apparently equivalent outer shelf shales are included in the "Springer" Formation (Fig. 17). The Cromwell is equivalent to the Union Valley, which crops out on the Lawrence Uplift at the north end of the Arbuckle Mountains in Oklahoma (Figs. 12, 13). In the subsurface of Oklahoma the Cromwell consists of multiple, discontinuous, calcareous sandstones separated by thin shales. The Cromwell is over 35 m in thickness across a broad western area in Oklahoma and exceeds 60 m locally (Jefferies, 1982; Fig. 17).

In the subsurface in Arkansas the equivalent Prairie Grove Formation reaches a maximum recorded thickness of 87 m. In addition to the overall thickening to the east there is a general coarsening in size of quartz grains and an increase in the number of sandstone intervals, indicating a quartz sand source to the northeast (Foshee, 1980).

Depositional patterns were complex on the Arkoma Shelf in middle Morrowan time (not shown on a figure). The Brentwood and upper Braggs Limestones were deposited on the inner shelf. This was followed by gradual regression, causing emergence and development of nonmarine environmental conditions in northwestern Arkansas (Woolsey), followed by erosion and the development of the middle Morrowan disconformity (Fig. 13). This was followed by rapid transgression and deposition in the late Morrowan of Kessler Limestone in Arkansas and the Greenleaf Lake Limestone in northeastern Oklahoma. With gradual shallowing on the outer shelf, deeper water shales in the lower part of the Wapanucka Formation gave way to the deposition of the overlying Wapanucka Limestone. Oolitic grainstones and carbonate mudstones were deposited across the entire shelf area (Fig. 18), followed by regional emergence and erosion.

The Wapanucka Formation crops out along Limestone Ridge in the frontal Ouachitas. In this area it was deposited on the outermost shelf. The highest limestone in the Wapanucka is of Atokan age and grades eastward along the ridge into the Spiro Sandstone (Grayson, 1979; Fig. 19; see following section). The total thickness of the Wapanucka on Limestone Ridge is typically about 88 m. Basinward (southward), closely spaced successive fault blocks show an equivalent slope facies (Chickachoc Chert), as much as 206 m thick, that consists of thick shale, thinner spiculites, and a few very thin spiculiferous limestones (Sutherland and Grayson, 1977; Fig. 18). Grayson (1979) has been able to correlate the shelf and slope facies using conodonts.

In the Ouachita Trough to the south, turbidite deposition continued during Morrowan time with the deposition of the Jackfork Sandstone, typically about 1,675 m thick (Fig. 17), and the Johns Valley Shale, as much as 265 m thick (Cline, 1960; Fig. 18). Transport directions recorded for the Jackfork Sandstone on the outcrop in both Arkansas and Oklahoma are to the west parallel to the basin axis (Moiola and Shanmugam, 1984; Fig. 17). Owen and Carozzi (1986) give evidence that part of the source for the upper Jackfork sandstones was to the southeast of Arkansas, from the orogenic belt southwest of the Black Warrior Basin. The Johns Valley is well known for its great variety of erratic limestone boulders that possibly originated from fault scarps on the shelf margin to the north (Shideler, 1970; Fig. 18).

Atokan Series

The Atokan Series is mainly represented by the Atoka Formation in most of the Arkoma Basin of Arkansas and Oklahoma and by the upper part of the Wapanucka Formation in the western segment of the frontal Ouachitas in Oklahoma and in the southwestern part of the Arkoma Basin (Fig. 19). The Atoka Formation ranges in thickness from 305 to 396 m along the northern margin of the Arkoma Basin in Arkansas (Zachry, 1983) to about 6,400 m at the southern margin of the basin in Arkansas just north of the Ross Creek Fault (Haley, 1982).

A regional unconformity separates the Atoka Formation from underlying strata of Morrowan age in all areas except along the southern part of the Arkoma Shelf and in the basin to the south. The sea was displaced from the shelf to the north primarily by a southward tilting of the Morrow surface (as an aspect of the subsidence of the Ouachita Trough), and extensive subaerial erosion resulted. Progressively older strata were eroded northward on this inclined surface. Stream valleys with a relief of at least 35 m were cut into the post-Morrowan surface (Lumsden and others, 1971).

The usage in this paper of informal lower, middle, and upper members of the Atoka Formation follows that of Buchanan and Johnson (1968) and Zachry (1983). In this usage the lower, middle, and upper intervals are based on the depositional history of each in response to the structural history of the basin during Atokan time. The middle Atoka was deposited during the development of syndepositional faults and was characterized by marked increases in thickness of section on the downthrown (south) side of these faults. The lower Atoka predates such faulting, and the upper Atoka postdates it. The development of the syndepositional faults was not synchronous and it appears that the southernmost syndepositional faults became active earliest and that active faulting migrated northward with time (Houseknecht, 1986). The middle Atoka makes up the major part of the thickness of the formation in the southern part of the basin.

Deposition of the lower Atoka in Oklahoma, with a source from the north or northwest, was initiated by the development of meandering fluvial systems and small deltas on the eroded surface of the underlying Wapanucka (Foster sand channels; Fig. 19). This was followed by a rapid northward transgression of a coastal sand complex (Spiro) to form a blanket sand unit (Lumsden and others, 1971). The maximum combined Spiro-Foster thickness recorded by Jefferies (1982) in Oklahoma is 62 m.

Depositional history of the Spiro in Arkansas was similar but with a much greater volume of sand being introduced from the northeast. Meandering fluvial systems that graded into deltas

(Fig. 19) were terminated by a broad regional marine transgression to the north. This caused the final redistribution of sand and produced the unit's sheet-like geometry (Parker, 1981). In Arkansas the name Spiro refers to the entire fluvial, deltaic, and sheet sand package that reaches a maximum recorded thickness of 89 m (Parker, 1981). To the south and southeast the Spiro changes facies to more basinal shale deposits (Fig. 19).

The lower Atoka in Arkansas ranges in thickness from 150 m adjacent to the northern margin of the present Arkoma Basin to approximately 305 m in the south (Zachry, 1983). The interval is composed of seven to eight sandstone units separated by units of shale and includes the Spiro at the base and the Sells at the top. The sandstones above the Spiro are replaced by shale westward in Oklahoma. Individual sandstone units above the Spiro (range from 6 to 60 m) show uniformity of thickness of individual sand units well into the southern part of the Arkoma Basin and are not affected by growth faulting (Zachry, 1983). A general increase in sandstone unit thickness occurs to the northeast in Arkansas, and intervals of shale are thinner, indicating a source to the northeast (Fig 19).

The lower Atoka sandstones in Arkansas, from the Spiro through the Sells, have the character of mature sandstones (quartz arenites), with a change in sandstone character occurring above the Sells in the central to southern part of the basin (D. L. Zachry, personal communication, 1987). The Spiro Sandstone in Oklahoma also has the character of a mature sandstone (Lumsden and others, 1971).

Beginning approximately with the deposition of the middle Atoka, the southern margin of the Arkoma Shelf was subjected to flexural bending, caused by continued basin closure, that resulted in the development of large east-trending normal faults (Fig. 20). The middle Atoka in the northern parts of the present Arkoma Basin in Arkansas is comparatively thin and is characterized by four to five sandstone units separated by intervals of shale (Zachry, 1983). These mature sandstones (quartz arenites) accumulated in delta and tidal flat systems that prograded southwestward across the northern part of the Arkoma Shelf from sources to the north and northeast.

In the middle Atoka south of the Mulberry Fault (Fig. 20) the sandstones have the composition of lithic arenites that contain 75 to 90 percent quartz, 5 to 25 percent metamorphic lithic fragments, and up to 8 percent feldspar (Houseknecht, 1986). The source of the latter two constituents cannot be from either the Ozarks or the Illinois Basin, and they most likely came from the southeast, from erosion of the already uplifted Ouachita Orogenic Belt, on the southwest margin of the Black Warrior Basin (Thomas, 1984). Thus, during the deposition of the middle Atoka, such sediments were carried westward along the axis of the incipient foreland basin that replaced the outer shelf at least as far north as the Mulberry Fault (Houseknecht, 1986; Fig. 20).

The middle Atoka interval in Oklahoma is composed dominantly of shale with a few thick sandstone units. It is best developed in the southern part of the basin and displays marked increases in thickness on the downthrown sides of east-trending syndepositional normal faults (Fig. 20). A major sandstone within this interval is the Red Oak, which is confined to the south side of the San Bois Fault (Fig. 20). Vedros and Visher (1978) believe that the Red Oak accumulated in a submarine-fan environment and that sediment was supplied by way of a submarine canyon cut into the scarp of a normal fault to the north. Houseknecht and Kacena (1983) propose that the normal fault blocks may have resembled half grabens in cross section and that these may have acted as sediment dispersal conduits that funneled sediment westward, parallel to the fault, from shallow to deeper portions of the slope. They state that sand-body geometry on the south side of the growth fault suggests a channel system and may not be associated with a submarine-fan complex.

Middle Atoka sandstones such as the Red Oak have been interpreted by many authors (e.g., Vedros and Visher, 1978) as having been deposited in "deep water" as a result of significant structural relief associated with the syndepositional faults. Houseknecht (1986) alternatively postulates that the Red Oak and similar sandstones in the area were deposited below wave base but at comparatively shallow-water depths (possibly in the order of 100 m), and that deposition possibly occurred on a gently dipping, muddy slope lacking a bathymetrically distinct shelf-slope-rise geometry.

In the southern part of the Arkoma Basin in Arkansas, Stone and McFarland (1981, Stop 16) describe strata exposed at Blue Mountain Dam as representing deep-water sediments. These strata consist of turbidites that show southward transport, which they interpret as upper submarine fan channels that partly dissect a probable slope facies. Still farther to the south in the Ouachita trough, turbidites moved westward on the deep basin plain (Fig. 20).

At the west margin of the Arkoma Shelf, Sutherland (1984) records that the initial uplift on the Hunton Arch is indicated by the occurrence of limestone pebbles and cobbles at the base of the Atoka Formation; these can be derived only from the underlying Wapanucka Limestone. Grayson (1984) has identified middle Atokan conodonts from the lower part of the Atoka Formation in that area, which crops out today on the northeast flank of the Arbuckle Mountains.

Upper Atoka strata are not cut by the normal faults that produced the thick sediment fill characteristic of the middle Atoka in the southern Arkoma Basin (Fig. 14). Upper Atoka sediments were deposited in a fully developed Arkoma Foreland Basin. Shallow-water sedimentation kept pace with continued subsidence. In the northern part of the basin, deltaic systems prograded southward. Paleocurrent directions have not been described for upper Atoka sandstones, but indirect evidence farther south in Arkansas suggests possible westward transport along the axis of the newly developed foreland basin in a manner similar to that found in the overlying Hartshorne Sandstone (lower Desmoinesian; Fig. 21).

The upper Atoka is up to 915 m in thickness in the south-

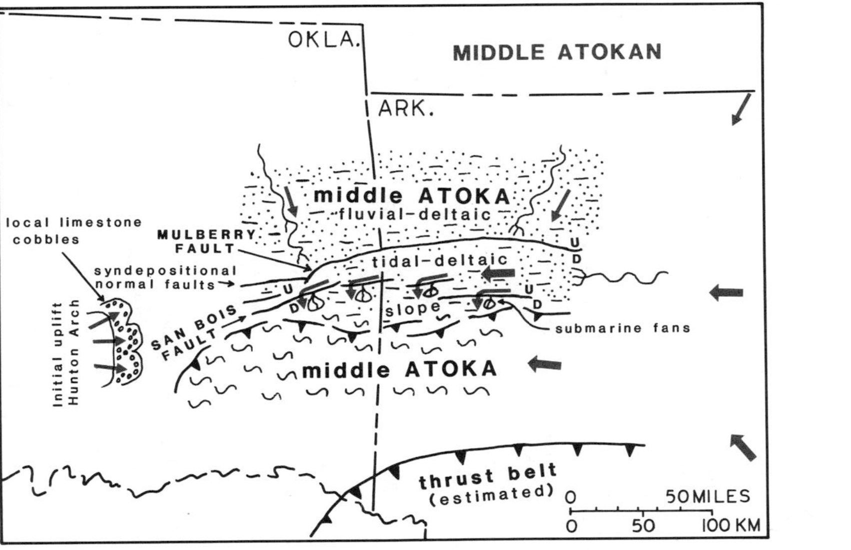

Figure 20. Middle Atokan paleogeographic map. Sources include Koinm and Dickey (1967), Vedros and Visher (1978), Houseknecht and Kacena (1983), Zachry (1983), Grayson (1984), Sutherland (1984), Zachry and Sutherland (1984), and Houseknecht (1986).

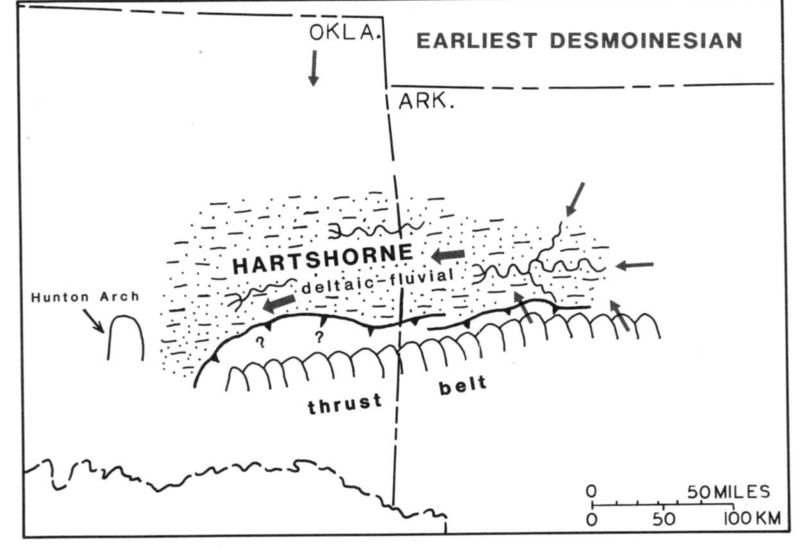

Figure 21. Earliest Desmoinesian paleogeographic map. Sources include Houseknecht and others (1983).

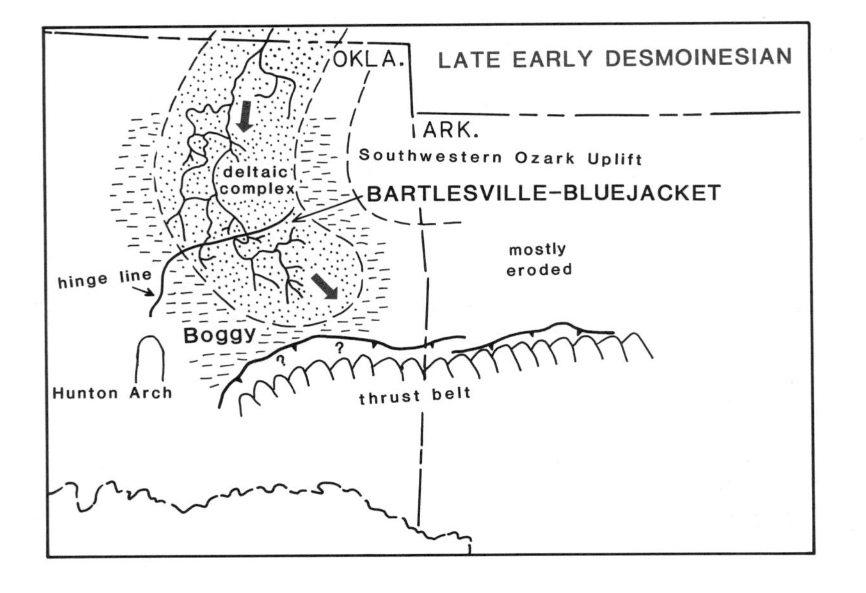

Figure 22. Late early Desmoinesian paleogeographic map. Sources include Weirich (1953), Visher (1968), and Visher and others (1971).

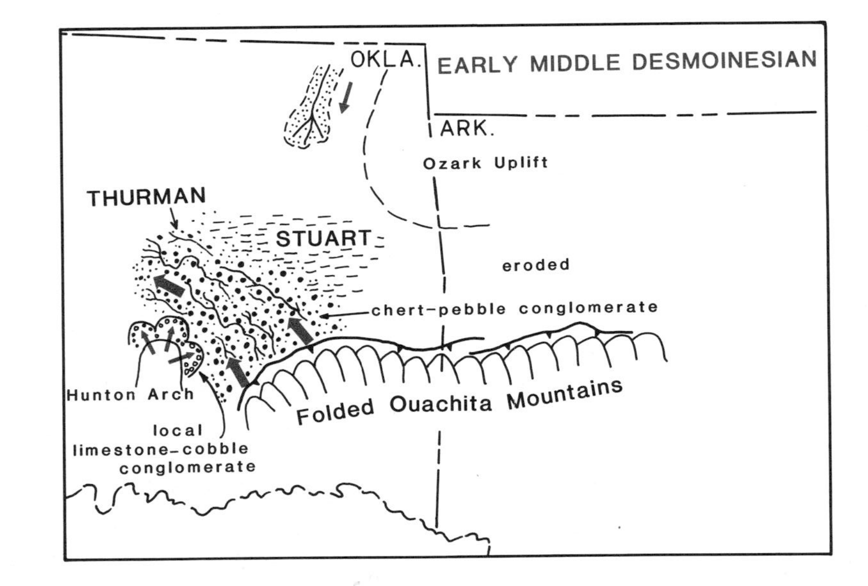

Figure 23. Early middle Desmoinesian paleogeographic map. Sources include Oakes (1948, 1953, 1967, 1977) and Jones (1957).

central part of the Arkoma Basin in Arkansas (Zachry, 1983). The interval in Oklahoma is similar to that in Arkansas but is composed predominantly of shale with thin discontinuous sands (Zachry and Sutherland, 1984).

Desmoinesian Series

The Desmoinesian Series in the Arkoma Basin and adjacent area to the northwest consists of the Krebs, Cabaniss, and Marmaton Groups (Fig. 15). Only the Krebs is preserved across the present-day basin, and units above the Hartshorne Sandstone have been mostly removed by erosion in Arkansas. The Cabaniss and Marmaton Groups crop out along the northwest margin of the basin in Oklahoma.

The Krebs Group consists of the Hartshorne, McAlester, Savanna, and Boggy Formations (Fig. 15). The Hartshorne Sandstone was deposited in high-constructive, tidally influenced deltaic systems that prograded from east to west, coinciding approximately with the present-day axis of the Arkoma Basin. Fluvial sediments in Arkansas came from the northeast, southeast, and possibly east (Houseknecht and others, 1983; Fig. 21). The sediments from the southeast apparently came from the developing Ouachita Foldbelt, but such a source is not recorded westward in Oklahoma (Fig. 21). In Arkansas, the Hartshorne displays a single progradational deltaic sequence, but in Oklahoma there are two such complexes with associated coal beds (Houseknecht and others, 1983; Fig. 21).

The only confirmed direction of source for the overlying McAlester, Savanna, and Boggy Formations (Fig. 15) is from the shelf to the north (Visher, 1968; Weirich, 1953). All of these units show marked thickening southward into the subsiding foreland basin (Fig. 22). The depositional pattern was nonmarine to deltaic, with rapid deposition of sands, muds, and thin coals. In the Morrowan and early Atokan, terrigenous sediments from the north were mostly quartz arenites (e.g., Spiro Sandstone). By the time of deposition of the McAlester Formation, sands from the north had changed mostly to sublitharenites (Bissell and Cleaves, 1986).

McAlester to Boggy depositional patterns were complex and included several major northwestward transgressions followed by regressive southward progradations of fluvial/deltaic systems across the shelf and into the Arkoma Basin. The largest such deltaic complex was developed during deposition of the lower part of the Boggy Formation and contains the Bartlesville-Bluejacket Sandstone Member of the Boggy (Visher, 1968; Visher and others, 1971; Fig. 22). This system extended southward across eastern Kansas and eastern Oklahoma and entered the Arkoma Basin as an extensive fluvial-dominated deltaic complex. The source was from the continental interior north of Oklahoma. The Boggy Formation, including the Bartlesville-Bluejacket, reached a thickness of 610 m in the Arkoma Basin (south of hinge line of Weirich, 1953; Fig. 22) compared to a thickness of less than 150 m on the shelf 80 km to the north (Weirich, 1953).

The initial uplift of the Ozark Mountains in northeastern Oklahoma is suggested by the arcuate pattern and the apparent southwestward deflection of the Bartlesville-Bluejacket deltaic complex in northeastern Oklahoma (Fig. 22).

Uplift, folding, and erosion of the foreland basin occurred following the deposition of the Krebs Group. Boggy and pre-Boggy rocks are more complexly folded and faulted than are post-Boggy strata and there is a conspicuous difference in strike between the two (Oakes, 1967, p. 30). Shales and fine-grained sandstones of the Boggy, below the unconformity, change abruptly to coarser sandstones and chert-pebble conglomerates in the overlying Thurman Sandstone (Oakes, 1953). Some southwest-trending faults, probably associated with the Ozark Uplift to the northeast, cut the Boggy but do not cut post-Boggy rocks (Oakes, 1967).

The Krebs is the only part of the Desmoinesian that was deposited during major subsidence of the Arkoma Foreland Basin before initial folding of the area. The Krebs is 2,195 m thick in the basin, compared to 240 m thick on the shelf 80 km to the north (Fig. 22).

The Thurman Sandstone, at the base of the Cabaniss Group, documents a marked change in depositional setting in the Arkoma Basin area. The Thurman chert-pebble conglomerates are the first indication of significant uplift of the Ouachita Foldbelt that exposed the Ordovician and Devonian cherts of the core area to erosion (Fig. 23). The conglomerates of the Thurman were deposited in a narrow successor basin, northeast of the Hunton Arch, in which the depocenter had shifted northwestward, farther onto the craton, from that of the late Atokan–early Desmoinesian foreland basin. The conglomerates of the Thurman are thickest and coarsest in western Pittsburg County, and the unit thins to the southwest onto the flank of the Hunton Arch and to the northeast. The Thurman is missing northeast of the Canadian River where it is overlapped by the Stuart Shale (Oakes, 1967; Fig. 23). The Thurman chert debris, derived from the Ouachita Foldbelt in Oklahoma, was transported toward the northwest as part of a fluvial system feeding deltaic and shallow-marine environments (Jones, 1957; Fig. 23).

During the deposition of the remainder of the Desmoinesian Cabaniss and Marmaton Groups (Fig. 15) the successor basin (termed the Arkoma seaway by Bennison, 1984) continued to receive terrigenous sediments, including some chert-pebble conglomerates, from the erosion of the Ouachita Foldbelt. The narrow basin was about 80 km in diameter immediately northeast of the Hunton Arch during the deposition of the Marmaton Group (Bennison, 1984, Fig. 4). A terrigenous Marmaton sequence reaches a thickness in this basin of 457 m compared to a more predominantly carbonate sequence on the shelf 80 km to the north, at Tulsa, Oklahoma, of 244 m.

The Ouachita Mountains continued through the remainder of the Pennsylvanian and into the Permian as the primary source of terrigenous sediments in central Oklahoma. These sediments bypassed the Hunton Arch (Fig. 23), which provided limestone cobbles only locally along the southern margin of the basin.

PALO DURO, HARDEMAN, AND DALHART BASINS

Shirley P. Dutton and Arthur G. Goldstein

INTRODUCTION

The Palo Duro, Dalhart, and Hardeman Basins are shallow, intracratonic basins located in the Texas panhandle and adjacent areas (Plate 5-A). These three basins (including the Hollis Basin, which is regarded as a distinct part of the Hardeman Basin by some workers) contain rocks of Precambrian to Quaternary age, but most of the fill is upper Paleozoic. They first developed as structural basins in Early Pennsylvanian time in conjunction with the uplift of surrounding basement blocks (Nicholson, 1960). Prior to this time they responded to the broad epeirogenic downwarps and arching seen elsewhere on the western side of the preexisting Oklahoma Basin. The Palo Duro Basin is asymmetrical and deepens to 3,000 m just north of the Matador Arch, whereas the Dalhart and Hardeman Basins are more symmetrical and are deepest in their centers.

TECTONIC AND STRATIGRAPHIC FRAMEWORK

The western part of the Oklahoma Basin was quiescent during early and middle Paleozoic time. Then the Palo Duro, Dalhart, and Hardeman Basins were formed, each bounded by crustal blocks uplifted during the Pennsylvanian Period in a thick-skinned (basement-involved) deformation (Goldstein, 1982). Each uplift is bounded on one side by a major fault that commonly shows evidence of having existed before late Paleozoic deformation. Northwest-trending faults are the dominant structures within the basins. Most fault movement occurred during the Pennsylvanian and Permian, but some faults cut the entire Permian System and affect the thickness and distribution of Triassic sediments (Dutton and others, 1982).

Post-Mississippian depositional history (Fig. 24) of these basins may be divided into four major stages (Handford and others, 1980): (1) formation of the basins between tectonic blocks, and subsequent deposition of basement-derived, fan-delta arkoses around uplifts flanking the basins (Fig. 25); (2) planation and burial of the uplifts and infilling of the deep basins with shelf-margin carbonate and basinal facies through Early Permian time (Fig. 26); (3) encroachment of continental red-bed facies from sources to the north and west, and deposition of thick Lower to Upper Permian evaporites in shallow-marine to subaerial environments (Fig. 25, Plates 6-S, 6-T, and 6-U); and (4) marine retreat during Late Permian time and development of a Triassic lacustrine basin caused by continental rifting and drainage reversal.

The subsidence and deposition that occurred during the Mississippian, Pennsylvanian, and Permian Periods account for almost all the sediments of the three basins. Peak subsidence occurred in Meramecian (middle Mississippian), Atokan (Early Pennsylvanian), and Missourian (Late Pennsylvanian) time (Goldstein, 1984). Regional control on subsidence is suggested by the burial of the bounding uplifts in Early Permian time (Fig. 25 and Plate 6-R). Both the Amarillo Uplift and the Matador Arch are now buried by 600 to 1,800 m of post-Pennsylvanian sediment, although they were subaerially exposed during Pennsylvanian time.

SYSTEM	SERIES	FORMATIONS AND GROUPS: PALO DURO BASIN		FORMATIONS AND GROUPS: DALHART BASIN	
PERMIAN	Ochoan	Post-San Andres interval	Dewey Lake Fm.	Post-Blaine red beds	Dewey Lake Fm.
			Alibates Formation		Alibates Fm.
			Salado Formation		Undifferentiated
	Guadalupian		Tansill Formation		
			Yates Formation		
			Seven Rivers Fm.		
			Queen/Grayburg Fm.		
		San Andres Formation		Blaine Formation	
	Leonardian	Clear Fork Group	Glorieta Formation	Glorieta Formation	
			Upper Clear Fork Fm.	Clear Fork Fm.	
			Tubb Formation	Undifferentiated Tubb – Wichita red beds	
			Lower Clear Fork Fm.		
			Red Cave Fm.		
		Wichita Group			
	Wolfcampian	Wolfcamp Group (Undifferentiated)			
PENNSYLVANIAN	Virgilian	Cisco Group			
	Missourian	Canyon Group			
	Desmoinesian	Strawn Group			
	Atokan	Bend Group			
	Morrowan				
MISS.	Chesterian	"Chester Limestone"			

Figure 24. Stratigraphic chart of Pennsylvanian and Permian strata in the Palo Duro and Dalhart Basins (after Nicholson, 1960; McGillis and Presley, 1981).

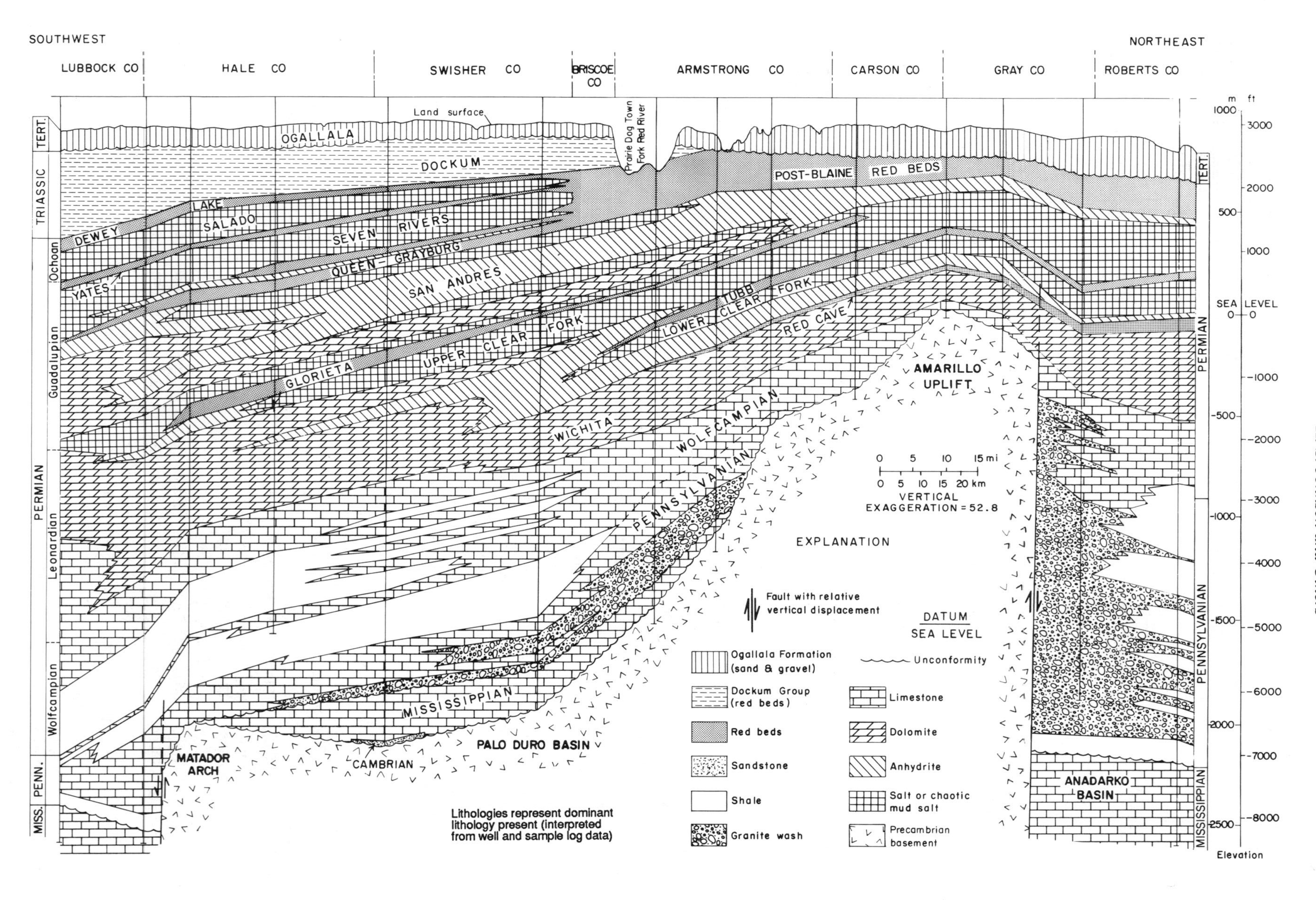

Figure 25. Structural cross section extending north-south through the Palo Duro Basin (after Dutton and others, 1982).

PENNSYLVANIAN SYSTEM AND WOLFCAMPIAN SERIES

Sedimentation during Pennsylvanian and Wolfcampian time was influenced by basin structure and subsidence, and by faulting in the surrounding uplifts. Some basement highlands remained exposed throughout the Pennsylvanian (Plates 5-M to 5-P and 6-Q). Pennsylvanian rocks in the Palo Duro, Dalhart, and Hardeman Basins include the following groups (Fig. 24): Bend (Morrowan and Atokan Series), Strawn (Desmoinesian Series), Canyon (Missourian Series), and Cisco (Virgilian Series).

Depositional conditions were generally similar during the Late Pennsylvanian and the beginning of the Permian (Fig. 26). Wolfcampian rocks record the transition in the three-basin area from relatively deep basins to restricted carbonate platforms (Handford, 1980). Basement uplifts remained emergent at the beginning of the Early Permian, but were finally covered by shallow-marine deposits at the end of Wolfcampian time (Fig. 27).

Pennsylvanian and Wolfcampian strata can be divided into four depositional systems (Figs. 26 and 27): (1) fan-delta system, (2) carbonate shelf and shelf-margin system, (3) high-constructive delta system, and (4) slope and basin system. Each system is characterized by a distinctive facies assemblage, vertical sequence, spatial distribution, and geophysical log signature (Handford and Dutton, 1980; Handford and others, 1981).

Fan-delta system

During the Pennsylvanian and Early Permian, a large volume of coarse arkosic sediment (granite wash) was eroded from the basement uplifts that rimmed the Palo Duro, Dalhart, and Hardeman Basins. The volume of granite wash was greatest in the Desmoinesian, which suggests that the major movement of the Amarillo Uplift occurred during this time (Goldstein, 1984). By Late Pennsylvanian and Wolfcampian time, the extensively eroded highland areas no longer supplied as much clastic sediment to the basins, and granite-wash deposition was confined to the flanks of the uplifts (Fig. 27, Plates 5-P, 6-Q, and 6-R).

Sequences of granite wash in the Texas panhandle are interpreted to be fan-delta deposits (Dutton, 1980). Faults commonly bound thick, proximal fan-delta deposits, indicating that structural displacement occurred during deposition. Fan-delta sequences in the Palo Duro Basin are commonly capped by shallow-marine limestones that were deposited on abandoned fan surfaces following compaction and subsidence. Interbedded granite wash and limestone indicate repeated episodes of fan-delta progradation into a shallow carbonate-shelf environment.

Shelf and shelf-margin system

Carbonate shelf and shelf-margin complexes developed seaward of the fan-delta systems (Figs. 26 and 27). Shelf margins retreated landward during the Pennsylvanian because of combined effects of subsidence and clastic sedimentation (Dutton, 1980). The early Wolfcampian shelf-edge position marked the maximum retreat (Fig. 27). In middle to late Wolfcampian time, the basin closed as shelf margins prograded basinward and southward toward the Midland Basin (Fig. 27; Handford and Dutton, 1980).

Shelf-margin deposits in the Texas panhandle are probably similar to other Late Paleozoic carbonate buildups in the Midcontinent that were formed by encrusting and sediment-baffling organisms, especially phylloid algae (Wilson, 1975). These carbonate buildups probably did not form a wave-resistant framework.

Shelf-margin deposits in the Palo Duro Basin are commonly dolomitized (Fig. 26). The dolomite is most likely a diagenetic replacement mineral because it crosses apparent bedding or facies boundaries.

High-constructive delta system

Elongate to lobate sandstone bodies in the Hardeman and southeastern Palo Duro basins delineate a system of westward prograding, high-constructive deltas (Figs. 26 and 27). The sediment source was probably the Wichita Mountains in Oklahoma. Deltaic progradation had reached as far west as Cottle County in the Hardeman Basin by Late Pennsylvanian time (Fig. 27; Frezon and Dixon, 1975). Clastics entering the Palo Duro Basin generally remained confined to the shelf, but in a few areas they were transported through the shelf margin into the basin.

Basin and slope system

Missourian, Virgilian, and Wolfcampian rocks along the basin axes consist of silty shales, dark micritic limestones, and thin sandstones that were deposited in slope and basinal environments (Figs. 26 and 27). Pulses of sediment probably entered the basin through passes between carbonate buildups along the shelf margins. Between these depositional episodes the basin was essentially starved. Clastic detritus and carbonate debris from the shelf margins were carried by turbidity currents and debris flows into the basin through submarine channels.

LEONARDIAN SERIES THROUGH OCHOAN SERIES

By late Wolfcampian time, the shelf margins had migrated to the southern edge of the Palo Duro Basin. The Palo Duro, Dalhart, and Hardeman Basins became a low-relief, back-shelf environment. Post-Wolfcampian Permian strata are composed of evaporites and red beds (Fig. 25, Plates 6-S to 6-U) that record deposition in a range of marine shelf to supratidal environments along an arid coastline (Presley and McGillis, 1982). Evaporite depositional systems in the Palo Duro Basin interfingered to the south with a shallow-marine shelf and shelf-margin system in the northern Midland Basin.

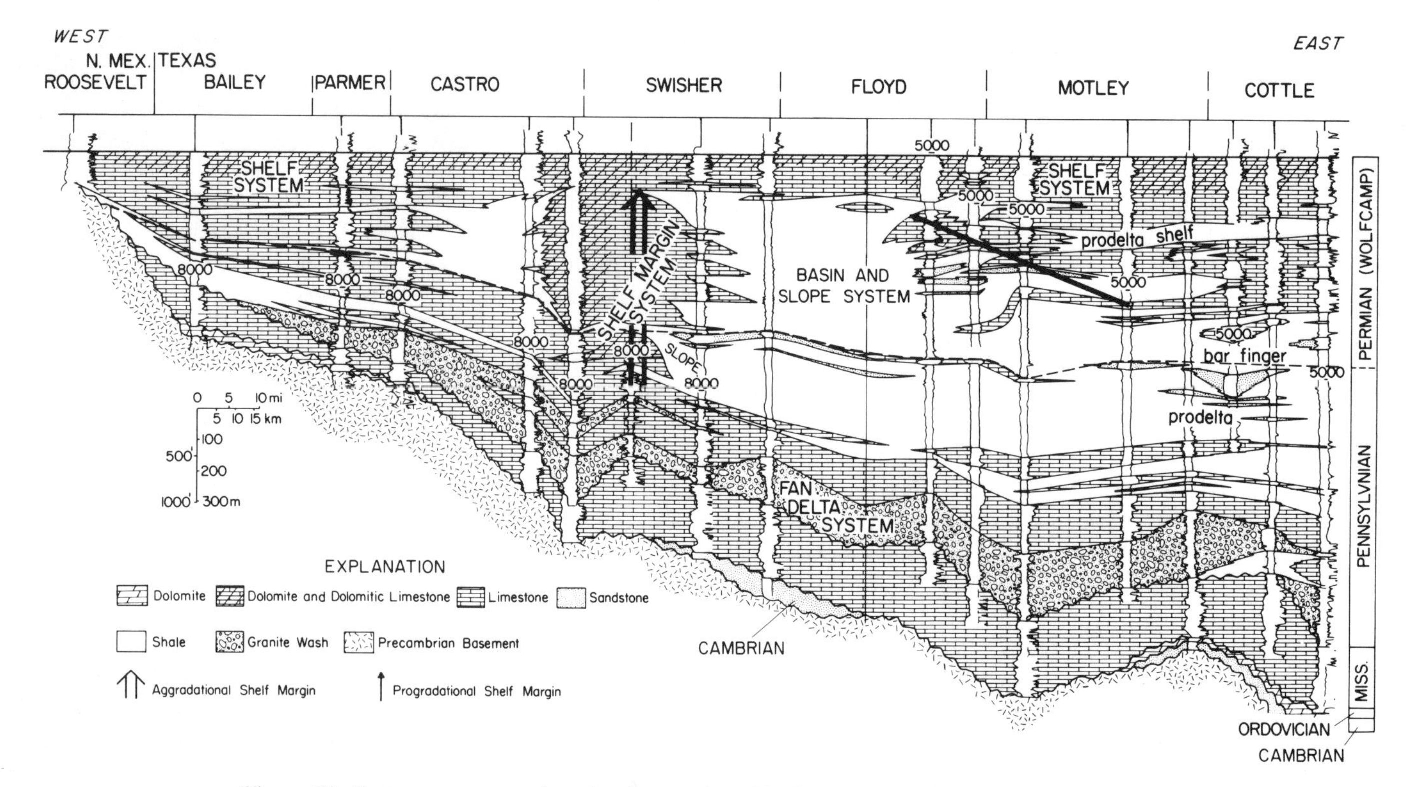

Figure 26. East-west cross section showing stratigraphic framework and depositional systems of Pennsylvanian-Wolfcampian strata and underlying rocks in the Palo Duro and Hardeman Basins (after Handford and Dutton, 1980). Datum is top of Wolfcampian Series. Depths are in feet below ground surface.

The evaporite section can be divided into five major genetic units (Figs. 24 and 25): (1) the Wichita and Red Cave, (2) the lower Clear Fork and Tubb, (3) the upper Clear Fork and Glorieta, (4) the San Andres, and (5) the post–San Andres (Presley, 1980). Each genetic unit except the post–San Andres records a major basinward (southerly) facies shift through time (Presley, 1980). These genetic sequences can be subdivided into secondary cycles that record more localized episodes of transgression and regression. The youngest unit, the post–San Andres, is primarily a grouping of similar facies, and it does not exhibit a broadly regressive character (Presley, personal communication, 1983).

Wichita–Red Cave genetic unit

The Wichita Group marks the beginning of evaporite deposition in the Palo Duro Basin (Fig. 25). Strata in the Wichita Group were deposited in a coastal sabkha that was bordered on the south by the deep Midland Basin and on the west and northwest by an alluvial-fan plain (Handford, 1979). Dolomite and anhydrite were deposited in the Texas panhandle, whereas bedded salt was deposited farther landward in Oklahoma and Kansas. The Wichita sabkha deposits pass into terrigenous red beds in the Dalhart Basin and east-central New Mexico.

The Wichita Group is conformably overlain by and laterally equivalent to the Red Cave Formation (Clear Fork Group; Fig. 24). The Red Cave consists of red-bed clastics and interbedded carbonate-evaporite members (Fig. 25). The clastics were deposited in mud-rich, coastal to continental sabkhas and, farther landward, a wadi-plain system (Handford and Fredericks, 1980; Handford and others, 1981). The Red Cave clastics interfinger to the south with anhydrite and dolomite that were deposited in carbonate-evaporite coastal sabkhas (Handford and Fredericks, 1980).

Lower Clear Fork–Tubb genetic unit

Lower Clear Fork strata were deposited in coastal-evaporite and carbonate environments that were similar to Wichita environments (Handford and others, 1981). However, because of a southerly migration of evaporite environments, bedded salt was deposited in the northern Palo Duro Basin for the first time (Fig. 25). A southerly facies tract in the lower Clear Fork consists of: (1) terrigenous red beds deposited on an alluvial-eolian plain

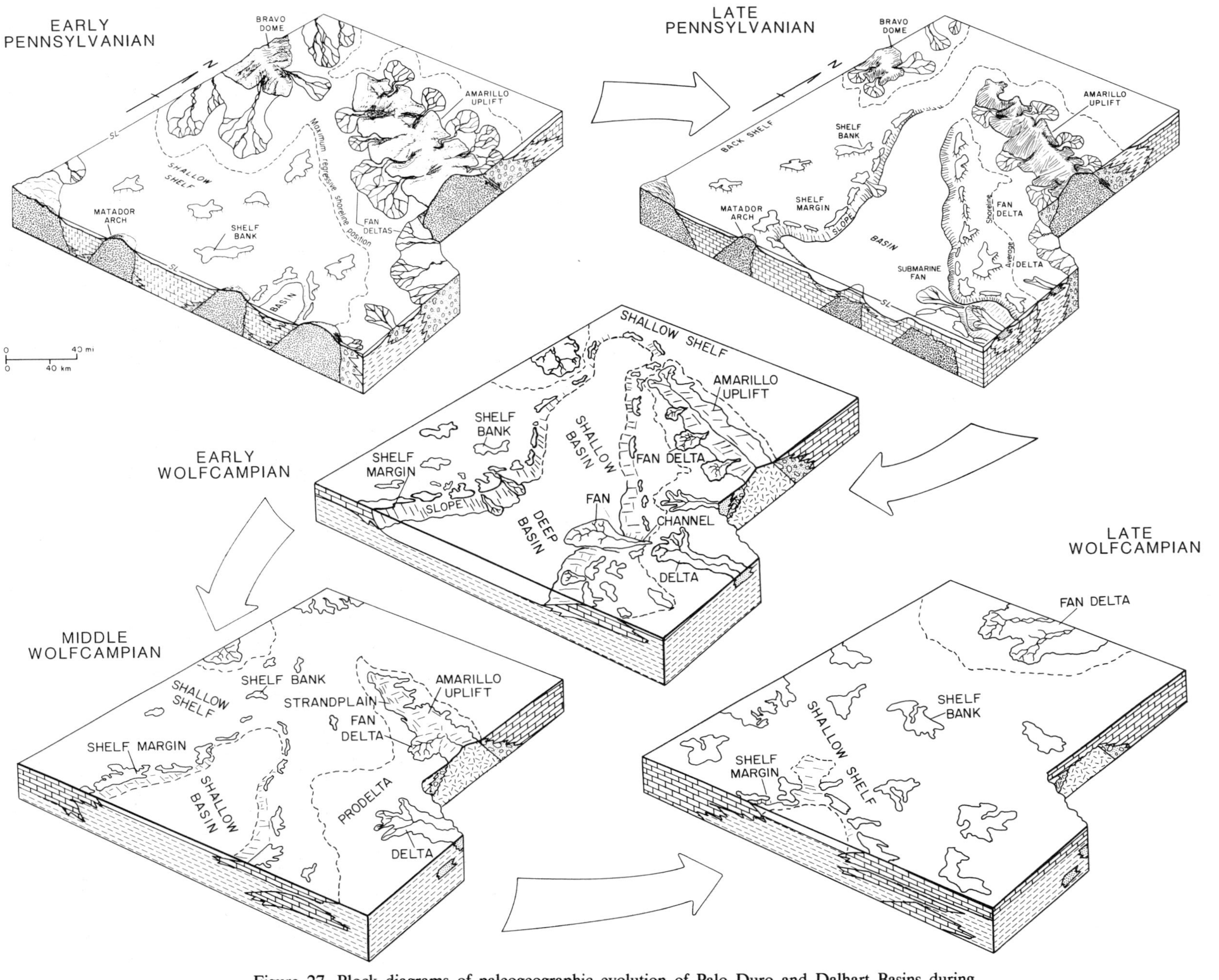

Figure 27. Block diagrams of paleogeographic evolution of Palo Duro and Dalhart Basins during Pennsylvanian and Wolfcampian time (after Handford and Dutton, 1980).

and inner-sabkha mud flat; (2) chaotic mudstone-halite formed in a saline mud flat; (3) banded to massive halite and laminated anhydrite deposited in inner-sabkha salt pans; and (4) dolomite and nodular to massive anhydrite deposited in marginal-sabkha and inner-shelf environments (Handford, 1981).

The overlying Tubb Formation (Fig. 25) contains two facies assemblages: (1) laterally persistent red beds, and (2) evaporites and carbonates that interfinger updip with the red beds (Presley, 1980). Each red bed records a southerly migration of siliciclastic mud-flat environments. Subsequent transgression of the mud flats permitted the resumption of carbonate-evaporite sedimentation.

Upper Clear Fork–Glorieta genetic unit

The upper Clear Fork Formation (Fig. 25) consists of a north-south facies tract similar to that of the lower Clear Fork (Presley and McGillis, 1982). A common facies in the updip part of the upper Clear Fork and other evaporite units is chaotic mudstone-salt, which varies from salt containing intercrystalline mudstone to mudstone with scattered salt crystals. This facies is interpreted as a saline mud-flat deposit in which halite crystals grew displacively near the surface (Presley and McGillis, 1982). The other evaporites in the upper Clear Fork probably formed in salt-pan environments basinward of the saline mud flats. Terrigenous red beds were deposited updip of the evaporites in mid-flat and wadi-plain environments (Presley, 1981).

In Glorieta time, terrigenous-clastic environments migrated basinward. Glorieta clastic units in the Palo Duro Basin are primarily mudstones and siltstones that were deposited in mud flats (Fig. 25). Glorieta clastics in the Dalhart Basin are eolian sheet sandstones (Presley and McGillis, 1982).

San Andres genetic unit

The San Andres Formation marks return to coastal evaporite and carbonate sedimentation in the Palo Duro Basin; the formation is characterized by thick (up to 100 m) carbonate-anhydrite-halite cycles (Fig. 25; Presley, 1981). Thick, salt-pan halite beds in the San Andres Formation are commonly banded into layers of dark halite with disseminated mud and other impurities and purer, clear salt layers (Presley and McGillis, 1982). San Andres deposits in the Palo Duro Basin contain less terrigenous sand or mud than do other Upper Permian formations. However, the Hardeman Basin was the site of terrigenous clastic deposition in early San Andres time, as deltas of the San Angelo Formation prograded into the area from the east (Smith, 1974). In later San Andres time, sabkha and tidal-flat environments developed in the Hardeman Basin (Smith, 1974).

Post–San Andres genetic unit

Post–San Andres strata (Figs. 24 and 25) are composed predominantly of salt and terrigenous red beds deposited in salt pans, saline mud flats, and eolian/continental sabkha environments (Presley, 1980, 1981). In the Seven Rivers and Salado Formations, mudstones interfinger basinward with massive salts (Fig. 25). Salt-bearing facies in the post–San Andres formations extend south of the Palo Duro Basin into the Midland and Delaware Basins. These deposits accumulated during the last stages of the regional Permian regression and record the final Paleozoic marine incursions in the Texas panhandle (McGillis and Presley, 1981).

FORT WORTH BASIN

Diana Morton Thompson

INTRODUCTION

The Fort Worth Basin, north-central Texas, is a Paleozoic foreland basin that was downwarped during Early Pennsylvanian time in response to the same tectonic stresses that produced the Ouachita Thrust Belt. It is bounded on the east by the Ouachita Thrust Belt, on the north by the Red River–Electra and Muenster Arches, on the south by the Llano Uplift, and on the west by the Concho Platform–Bend Arch (Plate 5-A). The basin strikes north-south and is asymmetric to the east.

In its deepest part the Fort Worth Basin contains a maximum known thickness of 4,000 m of Paleozoic rocks unconformably overlain by outcropping Cretaceous strata (Turner, 1957). The majority of the basin fill is Pennsylvanian in age (Figs. 28 and 29) with Cambrian, Ordovician, Mississippian, and Permian rocks also present.

The Paleozoic sedimentary sequence reflects a general transition from a passive continental margin to an actively subsiding basin. As part of the greater Oklahoma Basin, the Fort Worth area was subjected to epeirogenic movements and predominantly carbonate deposition during the early and middle Paleozoic. The Fort Worth Basin was then filled during Pennsylvanian time by terrigenous clastics derived from the adjacent uplifts. Pennsylvanian terrigenous sequences offlap to the west, indicating a progressive westward shift of depocenters as high-constructive fan-delta and delta systems prograded across the basin. The distribution of depositional environments and facies was highly variable with carbonate slope, shelf, and deltaic environments existing contemporaneously (Fig. 29).

TECTONIC AND STRATIGRAPHIC FRAMEWORK

In Pre-Pennsylvanian time the area that later became the Fort Worth Basin lay on the eastern side of the Texas Arch in the southern part of the Oklahoma Basin. The Texas Arch acted as a stable cratonic element throughout much of the Paleozoic, while the Fort Worth Basin area subsided intermittently to receive approximately 1,500 m of early to middle Paleozoic carbonates and other marine sediments. Recognition of the regional extent and longevity of this part of the Oklahoma Basin implies that present structural divisions between the Fort Worth Basin and adjacent

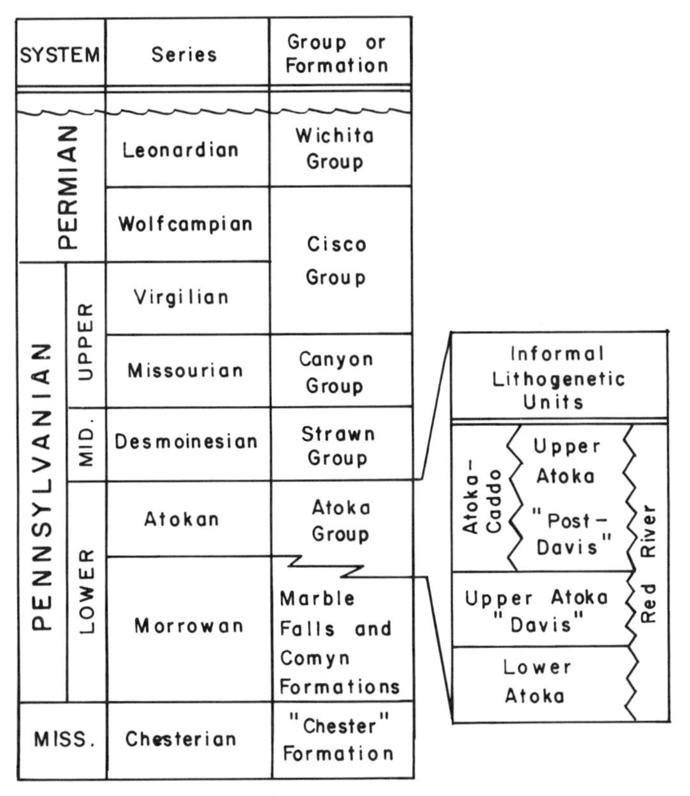

Figure 28. Generalized stratigraphy of the Fort Worth Basin, including a classification of the subsurface Atoka Group. Modified from Thompson (1982).

basins, such as the Hardeman and Ardmore Basins, are strictly Pennsylvanian features that resulted from reactivation of old fault systems during Ouachita deformation. With the advent of the Ouachita Orogeny, the regional structural elements surrounding the Fort Worth Basin were formed. Those structures are: the Ouachita Thrust Belt, the Red River–Electra–Muenster Arches, the Bend Arch, and the Llano Uplift (Plate 5-A).

The Ouachita Thrust Belt is a 2,000-km-long band of deformed Paleozoic rocks that marks the southern margin of the North American Craton (Hatcher and others, 1988). The thrust belt is exposed only in the Ouachita Mountains of Oklahoma-Arkansas, and in the Marathon Uplift of Texas. Most of the belt, including that portion just east of the Fort Worth Basin, is covered unconformably by Cretaceous and Cenozoic strata.

The Red River–Electra Arch, on the north side of the Fort Worth Basin, is composed of a series of discontinuous fault blocks that strike west-northwest. The faults are believed to have been controlled by basement fractures that were initiated in the Late Precambrian and reactivated during Ouachita deformation (Ham and others, 1964).

The Muenster Arch, to the northeast of the basin, consists of a northwest-trending uplifted block of Cambrian to Mississippian age sedimentary rocks mantling Precambrian igneous and metamorphic basement. The southwest flank of the Muenster Arch, adjacent to the Fort Worth Basin, is bounded by a series of faults that exhibit displacement down to the southwest. Total displacement across these faults is estimated to be 1,500 m (Flawn and others, 1961).

The Bend Arch is a broad structural element west of the Fort Worth Basin. It is observed in the subsurface as an elongate ridge that strikes north-south and plunges to the north. The Bend Arch represents the hingeline between the subsiding Fort Worth Basin to the east and the stable Concho Platform to the west. The hingeline formed in Late Mississippian and Early Pennsylvanian time as stresses generated by Ouachita deformation created the Fort Worth Basin. It did not act as a physical barrier to sediment transport or as a sediment source.

The Llano Uplift is located at the south end of the Bend Arch. It is a structural dome that was stabilized by massive granitic intrusions in late Precambrian time (Flawn and others, 1961). During the Early–Middle Pennsylvanian the area was uplifted and faulted into a series of horsts and grabens that strike parallel to the leading edge of the Ouachita Thrust Belt. The Llano Uplift was an intermittent positive feature throughout the Paleozoic and even exerted an influence on Cretaceous facies distribution. Today, it is a topographic basin floored by Precambrian igneous and metamorphic rocks, and rimmed by outcrops of Cambrian, Ordovician, Devonian, Mississippian, Pennsylvanian, and Cretaceous rocks.

As the Fort Worth Basin subsided, Atokan-age intrabasinal faults developed in response to extensional deformation. Faulting and deposition were contemporaneous, with individual faults acting as subsidiary hingelines. In the central part of the basin, faults strike northeast, subparallel to the western edge of the Ouachita Thrust Belt, and coincide with trends of major faults in the Llano Uplift. These faults are inferred to exhibit normal displacement, and most are downthrown to the southeast. However, they also form an en-echelon series of horsts and grabens. Near the northern margin of the basin the faults strike subparallel to the Electra and Muenster Arches and are downthrown toward the center of the basin (Thompson, 1982).

The generalized stratigraphy of the Fort Worth Basin, lithologic descriptions of major units, and a partial list of the more recent workers are summarized in Kier and others (1979) and Thompson (1982). Nomenclature has been extrapolated from outcrop work in the Llano Uplift and Eastern Shelf areas; however, these terms are not always applicable in the subsurface. Because of time-transgressive facies and facies changes, many discrepancies exist, particularly within the Pennsylvanian.

PENNSYLVANIAN SYSTEM

Morrowan Series

During Morrowan time the Fort Worth Basin was inundated by transgressive seas, indicating renewed subsidence. The

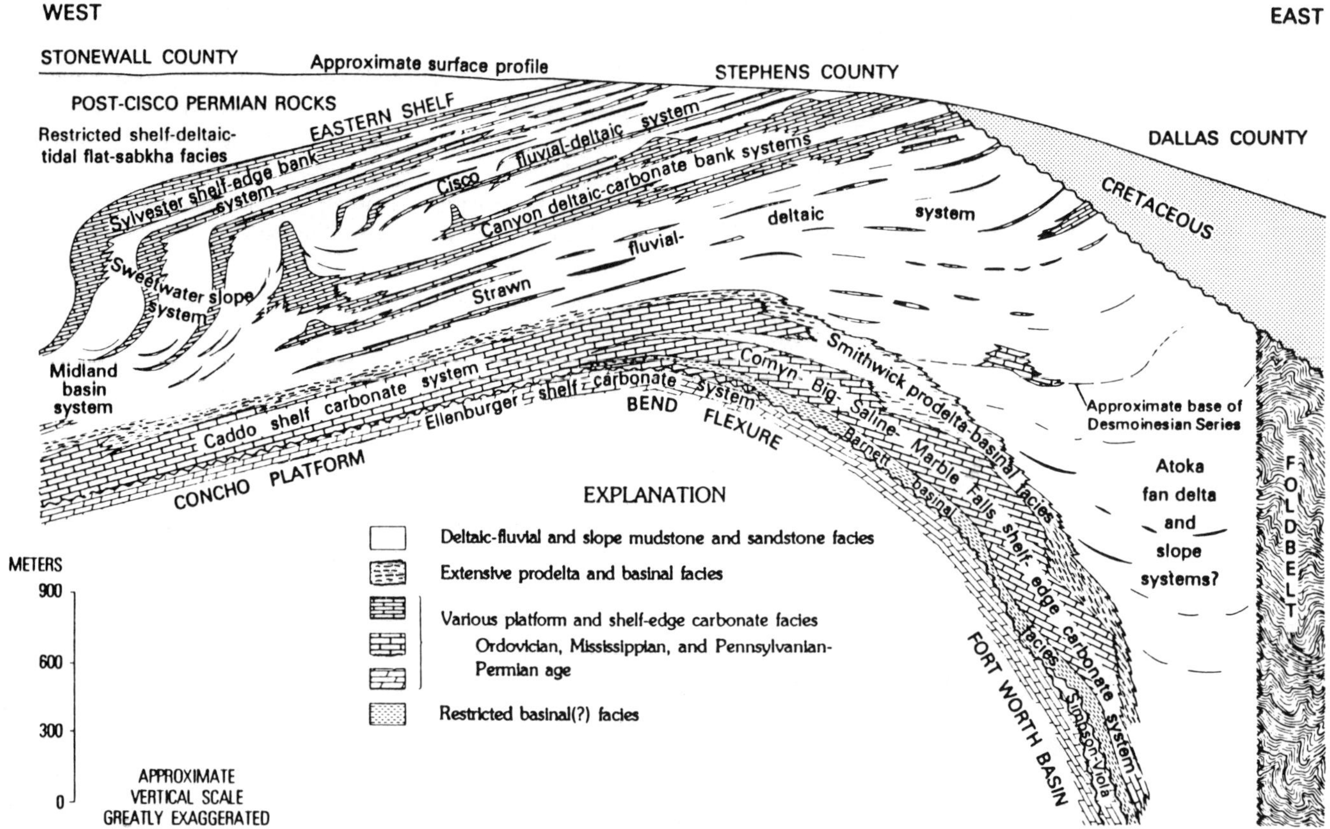

Figure 29. Schematic cross section illustrating sedimentary fill of the Fort Worth Basin and offlap of depositional systems across the Concho Platform (Eastern Shelf). From Kier and others (1979).

resulting carbonates compose the Marble Falls Formation. On the Llano Uplift, a Bahama-like carbonate platform developed that was periodically emergent and submergent. Algal bioherms and oolite shoal facies predominated (Kier, 1980). In the Fort Worth Basin a deeper water facies characterized by black shale and spiculitic limestone was deposited (Plate 5-M), with more than 150 m of strata still preserved in the subsurface.

Springer-age rocks have not been recognized within the Fort Worth Basin area. This may be due to either nondeposition or post-Springer pre-Morrowan erosion. The author favors erosion because Mississippian rocks are irregularly truncated within the basin, and a break in deposition is also observed in the Hardeman and Palo Duro Basins. This suggests broad regional uplift, perhaps associated with movement of the Amarillo-Wichita Mountains.

Atokan Series

Although the Ouachita orogeny was initiated earlier, clear evidence of major structural movement is not recorded in sediments of the Fort Worth Basin until the Atokan Epoch (Plate 5-N). During Atokan time the Ouachita Thrust Belt was emplaced, and reactivation of old faults along with residual positive features gave the Fort Worth Basin distinct geologic and geographic boundaries. Atokan rocks consist of interbedded shales, conglomerates, sandstones, and thin limestones. Chert conglomerates derived from the Ouachita Thrust Belt were shed into the rapidly subsiding basin, along with arkosic granite washes derived from the Red River and Electra Arches, and minor amounts of sediment derived from the Muenster Arch and Llano Uplift. Subsidence was greatest along the leading edge of the Ouachita Thrust Belt and at the southern end of the Muenster Arch where approximately 1,800 m of sediments (or 50 to 75 percent of the Paleozoic basin fill) were deposited (Plate 5-N). Principal studies of Atokan strata in the area include Ng (1979), Lovick and others (1982), and Thompson (1982).

Atokan stratigraphic relationships are highly complex, and there are significant variations in the interpretation of the subsurface data. In the northern half of the basin, five packages of sedimentation can be defined (Fig. 28): (1) the lower Atoka lithogenetic unit, (2) the upper Atoka "Davis" lithogenetic subunit, (3) the upper Atoka "Post-Davis" lithogenetic subunit,

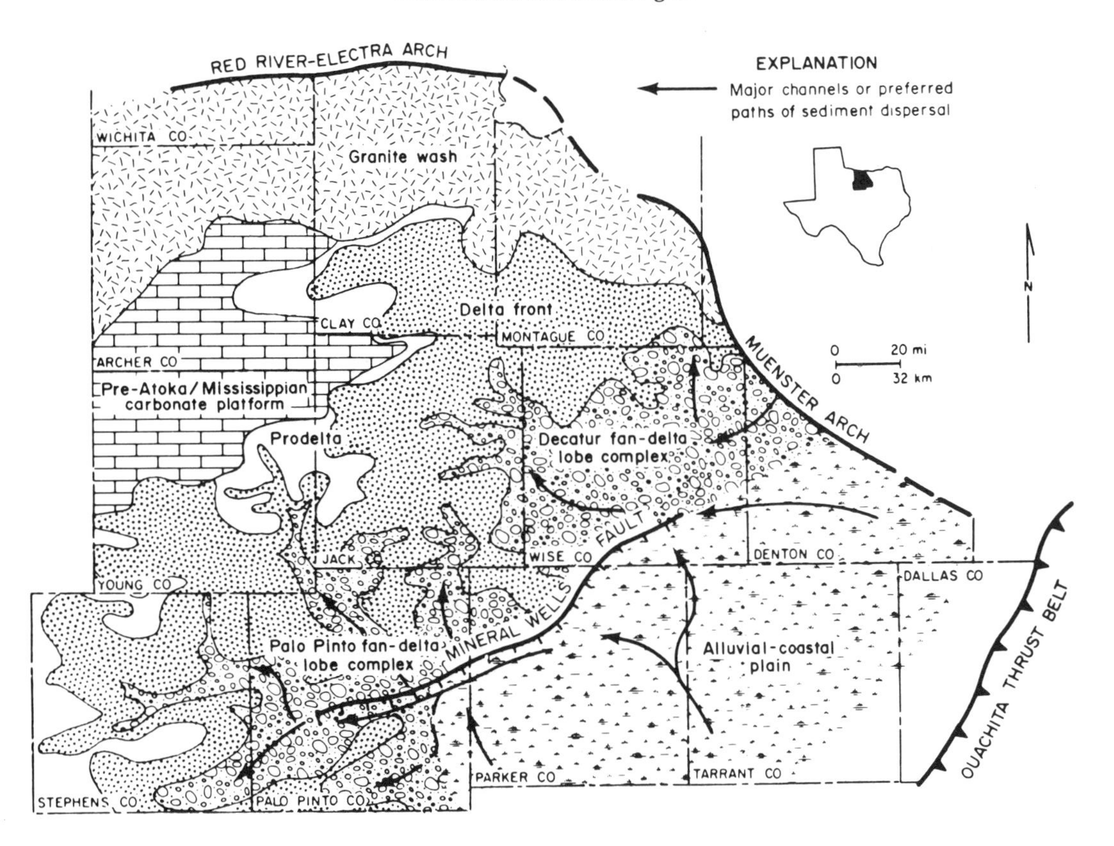

Figure 30. Distribution of lower Atoka facies, northern Fort Worth Basin, Texas. Similar depositional systems are inferred to have existed in the southern half of the basin. From Thompson (1982).

(4) the Red River lithogenetic unit, and (5) the Atoka-Caddo lithogenetic unit (Thompson, 1982).

The lower Atoka unit is interpreted to be a fluvially dominated fan-delta system, characterized by a highly digitate sandstone geometry, extensive interfingering of contemporaneously deposited terrigenous and carbonate strata, progradational facies sequences, and fault-controlled facies distribution (Fig. 30). This lithogenetic unit reflects the initial transition from a shallow marine to a terrigenous-dominated environment as chert-rich sediment was rapidly shed from the tectonically active Ouachita Thrust Belt and locally from the Muenster Arch.

The upper Atoka "Davis" subunit is interpreted to be a system of coalesced wave-dominated deltas, characterized by a thick, strike-oriented sandstone geometry, absence of significant contemporaneous limestone deposits, a predominance of coastal barrier facies, and electric log patterns that suggest concurrent progradation and aggradation. This lithogenetic subunit reflects a period of tectonic quiescence that allowed coastal marine processes to dominate fluvial processes and terrigenous sediment input.

The upper Atoka "Post-Davis" subunit is interpreted to be a thin, poorly integrated, fluvially dominated fan-delta system, similar to the lower Atoka unit. It is characterized by a narrow, highly digitate sandstone geometry and three east-west–trending axes of deposition along which delta lobe complexes prograded into the basin (Fig. 31). This lithogenetic subunit reflects renewed Ouachita tectonic activity that resulted in sporadic influx of sands and gravels rich in chert and metamorphic and volcanic rock fragments derived from the Ouachita Thrust Belt and Muenster Arch.

The Red River lithogenetic unit is characterized by thick, stacked sequences of arkosic conglomerate derived from the Red River–Electra Arch. The granite wash was deposited in a series of alluvial fans and fan deltas that existed throughout Atokan time and reflects continual uplift of the Red River–Electra Arch. The Atoka-Caddo lithogenetic unit roughly corresponds to the "Caddo reef" facies and is contemporaneous with upper Atoka deposition. It is interpreted to be a series of algal buildups that grew along the western margin of the basin out of the intense influence of terrigenous sedimentation and along the axis of the Bend Arch.

Desmoinesian Series (Strawn Group)

The Strawn Group of Desmoinesian age consists mainly of

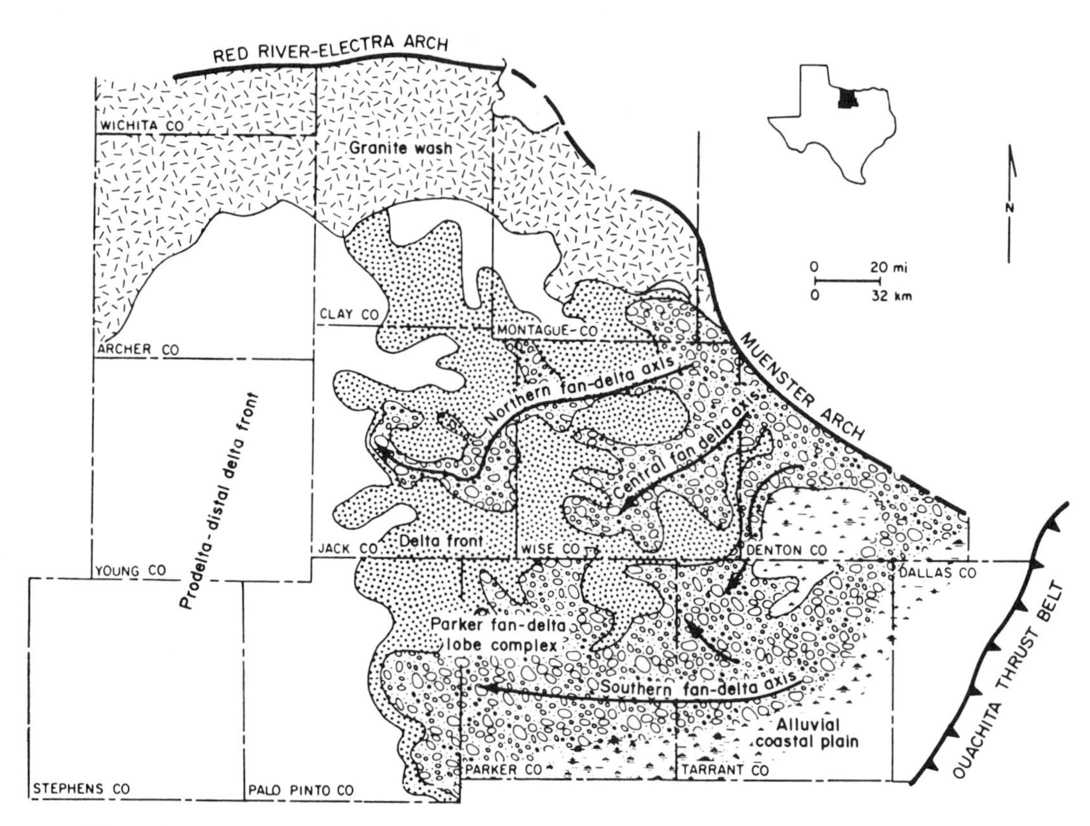

Figure 31. Distribution of upper Atoka "post-Davis" facies, northern Fort Worth Basin, Texas. Similar depositional systems are inferred to have existed in the southern half of the basin. From Thompson (1982).

shales and other fine-grained clastics, and it is more than 1,200 m thick in the northern part of the Fort Worth Basin (Plate 5-O). Two styles of deposition are present in the Strawn Group: (1) the lower Strawn, believed to be genetically related to the Atoka Group, and (2) the middle and upper Strawn, believed to be related to the overlying Canyon Group (Brown and others, 1973).

There is a lack of published information on the lower Strawn. This may be due to scattered well control in the eastern part of the basin, and due to the limited distribution of the unit. Lower Strawn depositional systems are assumed to be similar to those of the upper Atoka "Post-Davis" and imply continuing tectonic uplift to the north and east coupled with basin subsidence.

Middle and upper Strawn rocks are known from outcrops on the east side of the basin and from subsurface work on the Concho Platform to the west. By middle Strawn deposition, subsidence rates were substantially reduced, and high-constructive fluvial-deltaic systems were able to prograde over the older basin fill out onto the Concho Platform. Eight transgressive-regressive cycles are present in the middle and upper Strawn (Cleaves, 1982). Cyclicity was controlled by avulsion of major delta complexes, moderate subsidence within the Fort Worth Basin, and subsidence of the Midland Basin, which resulted in progressive regional tilting to the west.

Major Strawn depositional features present in the Fort Worth Basin are: (1) the Perrin Delta complex, a major site of deposition for the remainder of the Pennsylvanian; (2) the Eastland Delta complex; (3) the Bowie Delta complex; and (4) the Thurber Embayment (Fig. 32; Cleaves, 1982). The Perrin and Eastland Delta complexes are characterized by high-constructive elongate and lobate deltas, incised valley fill systems, multilateral sandstone distribution reflecting low subsidence rates, and chert-rich sediment derived from the Ouachita Thrust Belt. The Bowie Delta complex is characterized by high-constructive elongate deltas, thin prodelta facies, multistoried sandstone distribution reflecting contemporaneous subsidence in the northern end of the basin, and coarse conglomerates derived from the Red River–Electra Arch and possibly the Arbuckle Mountains. The Thurber Embayment, a major deltaic embayment, is characterized by sheet sandstones, bituminous shale, and coal.

Adjacent to the Llano Uplift, rocks of Desmoinesian age contain pebbles of the Marble Falls Formation within prodelta shale. This implies the emergence of the Llano Uplift and extensive transport of prodelta shale from westward-building Strawn deltas (Kimberly, 1961; Turner, 1970; Watson, 1980).

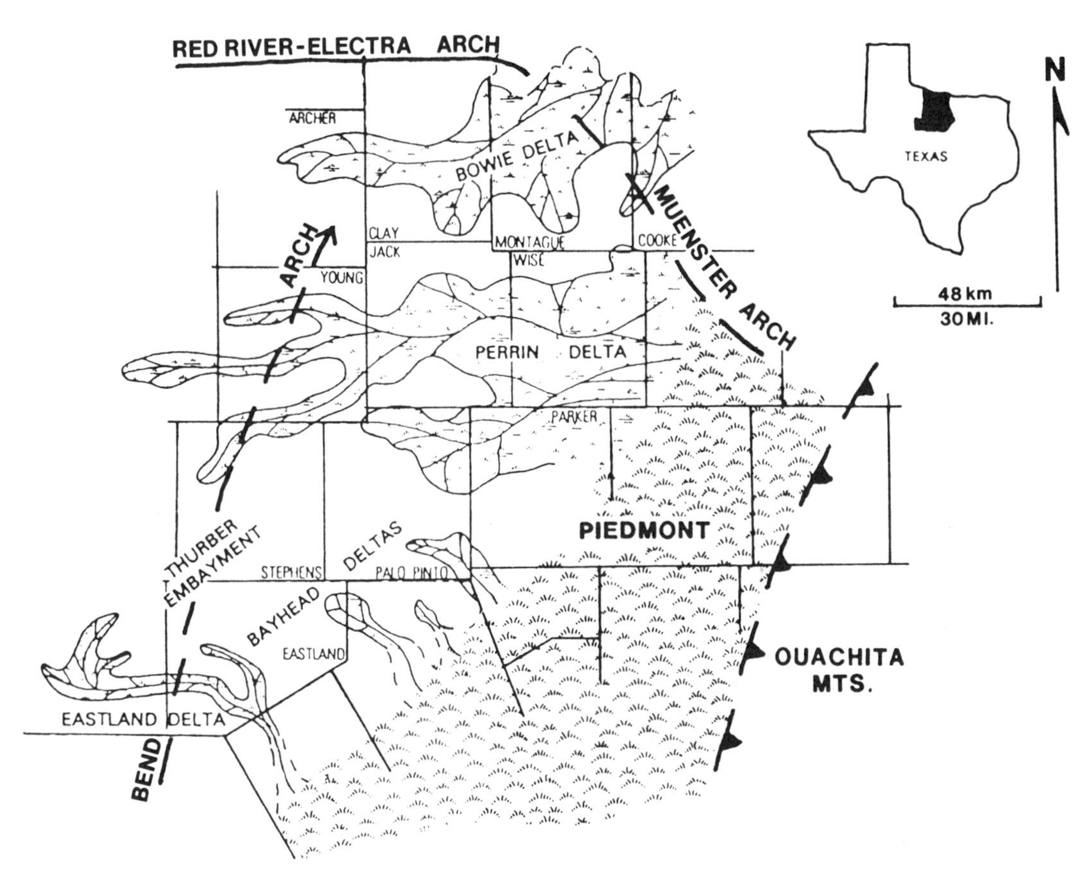

Figure 32. Distribution of Strawn Group depositional systems in the Fort Worth Basin, north-central Texas. Similar types of systems are inferred to have existed in the southern part of the basin. Modified from Cleaves (1975, 1982) and Kier and others (1979).

Missourian Series (Canyon Group)

Rocks of the Canyon Group crop out in the northwest corner of the Fort Worth Basin and are present in the shallow subsurface to the west (Fig. 29 and Plate 5-P). The tectonic and depositional style during Missourian time was similar to that of late Desmoinesian, and nearly 900 m of strata, mostly shales, are still preserved. During Canyon deposition, high-constructive deltas continued to prograde over the older basin fill and across the Concho Platform farther west (Fig. 29). A facies tract of piedmont-fluvial-deltaic-carbonate shelf and shelf edge became well established. However, slow subsidence and sedimentation rates resulted in an increase in destructive facies.

Three cycles of delta building and destruction are present in the Canyon (Erxleben, 1975). Cycles reflect periods of delta progradation, avulsion, abandonment, destruction, marine encroachment, and subsequent deposition of shelf carbonates.

Major Canyon depositional features present in the Fort Worth Basin are: (1) the Perrin Delta complex, (2) the Henrietta Fan-Delta complex, and (3) the Red River, Chico Ridge, and Winchell carbonate bank ("build-up") systems (Fig. 33; Erxleben, 1975). The Perrin Delta is characterized by the same depositional processes established during Strawn deposition. It reflects progradation of sediment interpreted to have been derived from the Ouachita Thrust Belt and from rocks of the Atoka and Strawn Groups that were uplifted along the eastern edge of the basin. The Henrietta Fan-Delta complex is characterized by thick, stacked sequences of arkosic and lithic conglomerate interpreted to have been derived from the Ouachita Thrust Belt and from the Wichita Uplift. Subsidence and sedimentation rates in the northern end of the basin were comparable, and a low-relief fan-delta plain facies with thin coal beds developed. Canyon carbonate-bank systems had depositional relief but were composed of nonframework-building organisms, such as phylloid algae, bryozoa, and crinoids (Erxleben, 1975; Wermund, 1975; Cherng, 1982). The Red River bank formed on the fault blocks of the Red River–Electra Arch. The Winchell and Chico Ridge banks grew on top of abandoned delta lobes. The relief on these banks was sufficient to deflect the prograding Perrin Delta complex.

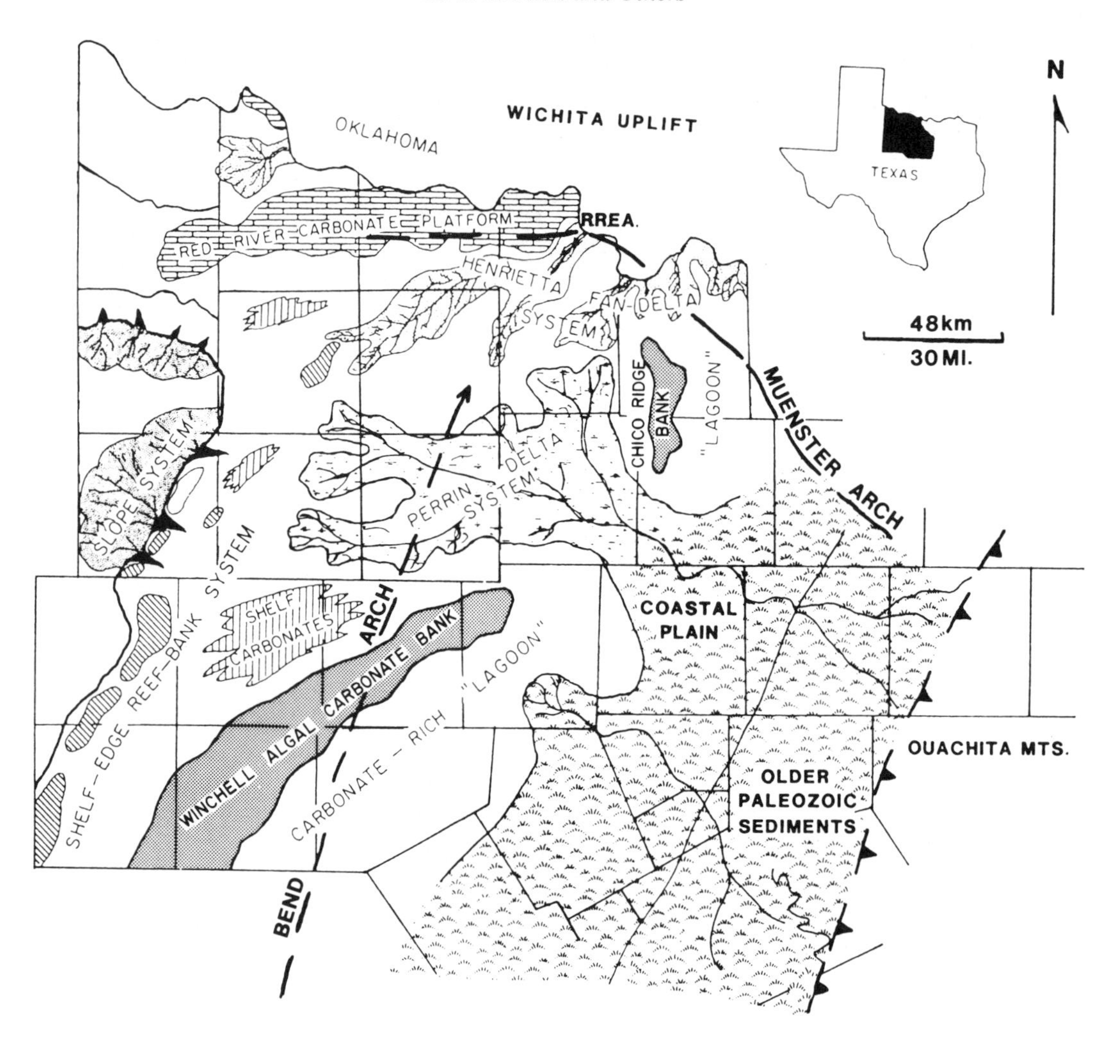

Figure 33. Distribution of Canyon Group depositional systems, north-central Texas. Similar types of systems are inferred to have existed in the southern part of the basin. RREA, Red River–Electra Arch. Modified from Erxleben (1975).

Virgilian Series (Lower part of Cisco Group)

Rocks of Virgilian age, in the lower part of the Cisco Group, are restricted to the northwestern corner of the Fort Worth Basin (Plate 6-Q). Up to 450 m of strata are preserved in the far northern part of the basin. Like the underlying Canyon Group, they consist of mixed terrigenous and carbonate depositional systems, with shales dominant. Although extensive systems of fluvial-deltaic clastics and shelf-edge carbonates exist downdip farther to the west, only the updip fluvial and deltaic facies are preserved in outcrop and shallow subsurface of the Fort Worth Basin (Turner, 1957; Galloway and Brown, 1972).

PERMIAN SYSTEM

Permian rocks are restricted to the northwest corner of the basin and consist of nearly 500 m of alluvial-deltaic and marginal-marine sandstones, mudstones, and shales of Wolfcampian and earliest Leonardian age (Plates 6-R and 6-S). These strata make up the upper portion of the Cisco Group and also the Wichita Group. The clastics consist of detritus from the Ouachita Thrust Belt, which by this time had been eroded down to low mountains or hills of only moderate relief. Clastic units are red, brown, and gray, and they grade westward across the Concho Platform into thick carbonates of the Hardeman Basin and the Eastern Shelf of the Midland Basin.

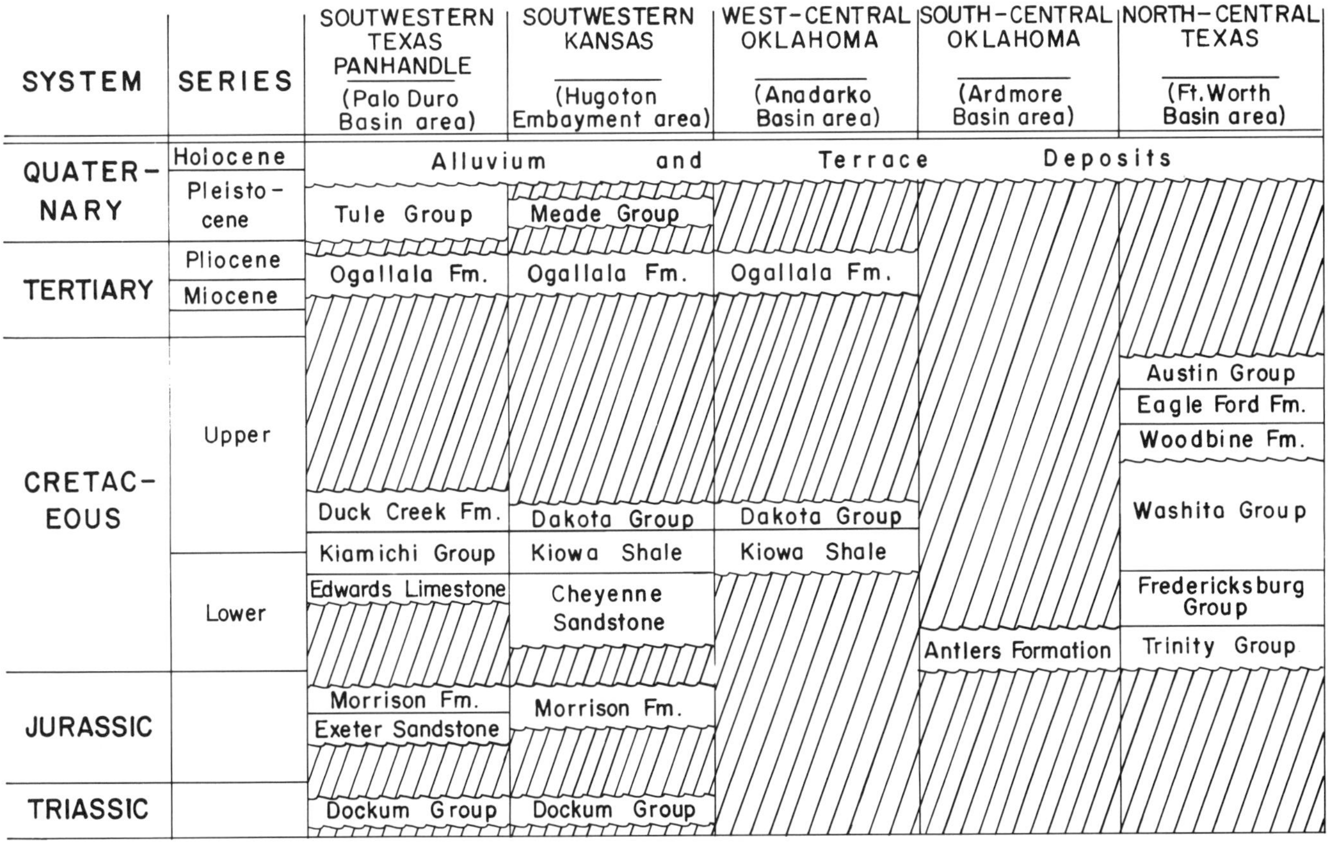

SYSTEM	SERIES	SOUTWESTERN TEXAS PANHANDLE (Palo Duro Basin area)	SOUTWESTERN KANSAS (Hugoton Embayment area)	WEST-CENTRAL OKLAHOMA (Anadarko Basin area)	SOUTH-CENTRAL OKLAHOMA (Ardmore Basin area)	NORTH-CENTRAL TEXAS (Ft. Worth Basin area)
QUATERNARY	Holocene	Alluvium and Terrace Deposits				
	Pleistocene	Tule Group	Meade Group			
TERTIARY	Pliocene	Ogallala Fm.	Ogallala Fm.	Ogallala Fm.		
	Miocene					
CRETACEOUS	Upper	Duck Creek Fm.	Dakota Group	Dakota Group		Austin Group Eagle Ford Fm. Woodbine Fm. Washita Group
	Lower	Kiamichi Group Edwards Limestone	Kiowa Shale Cheyenne Sandstone	Kiowa Shale	Antlers Formation	Fredericksburg Group Trinity Group
JURASSIC		Morrison Fm. Exeter Sandstone	Morrison Fm.			
TRIASSIC		Dockum Group	Dockum Group			

Figure 34. Correlation of post-Permian stratigraphic units in the Southern Midcontinent, based upon COSUNA Charts (Hills and Kottlowski, 1983; Mankin, 1987).

POST-PERMIAN ROCKS

Kenneth S. Johnson

Post-Permian rocks of the Southern Midcontinent include remnants of Triassic, Jurassic, Cretaceous, Tertiary, and Quaternary strata (Fig. 34). Deposition of these sediments was not influenced by the tectonic provinces that were so pronounced and critical to Pennsylvanian and Permian sedimentation.

TRIASSIC SYSTEM

Triassic strata, limited to the western part of the Southern Midcontinent (Plate 6-V), are a red-bed sequence referred to the Dockum Group of Middle and Late Triassic age (Hills and Kottlowski, 1983). These strata were described on a regional basis by McKee and others (1959), and more detailed studies were conducted later by McGowan and others (1979) and Dutton and Simpkins (1986).

The Dockum Group was deposited in a closed continental basin in fluvial, deltaic, and lacustrine environments (McGowan and others, 1979). The thickness of the Dockum ranges from less than 75 m around the now-eroded periphery of the basin to more than 600 m in the southwest (Plate 6-V). This unit has been subdivided informally into the lower Dockum and upper Dockum. The lower Dockum Group consists of fine- to coarse-grained quartzose sandstones and conglomerates deposited in braided and meandering streams, in alluvial fans and fan deltas, and in high-constructive lobate deltas (McGowan and others, 1979; Dutton and Simpkins, 1986). These coarse-grained sediments are more abundant around the periphery of the basin but they are interbedded with mudstones, and they grade laterally into mudstones deposited in lacustrine and prodelta environments toward the central parts of the basin.

The upper Dockum Group is more mud rich than the lower part. Thick sandstone units are not extensive in the upper

Dockum, except in the southern part of the region, because the fluvial-deltaic systems at this time were generally small (McGowen and others, 1979).

JURASSIC SYSTEM

Late Jurassic strata consist of sandstones and shales of the Exeter Sandstone and overlying Morrison Formation in the northwestern part of the Southern Midcontinent (Fig. 34 and Plate 6-W). They are a westward-thickening wedge of fluvial sediments derived by erosion of low plains areas that lay to the east. The Exeter is fine- to medium-grained, cross-bedded sandstone that is white, light brown, and orange (Barnes, 1984). The overlying Morrison consists of red-brown, green, and gray mudstone, sandstone, and minor limestone (Stovall, 1943); fossil dinosaur bones are common (Stovall, 1938).

CRETACEOUS SYSTEM

Lower and Upper Cretaceous shales, sandstones, and limestones are present around the perimeter of the Southern Midcontinent (Fig. 34 and Plate 6-X). The Cretaceous seaway extended across most of the western half of the region during the last great marine inundation of the western interior of the United States. Formation of the Rocky Mountains during the Laramide orogeny of Late Cretaceous and Early Tertiary time imparted an eastward and southeastward tilt to the entire region and eventually caused withdrawal of the Cretaceous sea.

Cretaceous strata in north-central Texas are considered part of the Gulf Coastal Plain, but they overlap the area of the Fort Worth Basin (Fig. 29). A series of transgressive deposits were spread northward from the ancestral Gulf of Mexico (Scott and others, 1978; Corwin, 1982) and they interfinger with alluvial and deltaic complexes to the north and northwest. Deposition was cyclic, with alternation of terrigenous and shallow-marine facies reflecting eustatic sea-level changes or a combination of irregular subsidence and sedimentation rates. Eventually, the encroaching seas submerged all of north-central Texas and then spread northwestward across western Oklahoma, the Texas panhandle, and the Western Interior. A major unconformity is well exposed throughout north-central Texas and southern Oklahoma where Cretaceous strata rest on rocks ranging in age from Precambrian through Permian.

Cretaceous strata in the western part of the Southern Midcontinent are isolated erosional remnants of widespread sediments deposited in the interior seaway. More than 150 m of predominantly marine sandstones, shales, and limestones were deposited in the area (Merriam, 1963). Southward thinning of the Cretaceous results from broad uplift and truncation before the overlying Tertiary sediments were deposited. Small outliers of Cretaceous marine strata in western Oklahoma consist of chaotic blocks and masses (typically 10 to 500 m across) of limestone and shale that have collapsed some 50 to 100 m, due to dissolution of underlying Permian salts, and are now juxtaposed against Permian rocks exposed at the land surface.

TERTIARY SYSTEM

The Ogallala Formation, of Miocene-Pliocene age, is widespread in the High Plains area of the Southern Midcontinent (Plate 6-Y). It is a buff- and light-colored sequence of interbedded fluvial and windblown sediments deposited upon an erosional surface cut into Permian red beds (and locally into Triassic, Jurassic, or Cretaceous strata). The Ogallala was deposited by low-gradient alluvial systems that flowed to the east and southeast from the Rocky Mountains (Seni, 1980). Medial-fan facies, which occur in the northern and northwestern Texas panhandle, include sand and gravel deposited in narrow stream channels separated by interchannel mud and silt (Seni, 1980).

Whereas early work indicated that the Ogallala was almost entirely fluvial in origin (Seni, 1980), more recent studies show that the fluvial deposits are confined to the lower part of the Ogallala and that the middle and upper parts of the Ogallala are mainly windblown deposits (Gustavson and Holliday, 1985). Ogallala deposition apparently ended between 3.5 and 2.4 Ma, which would be in the late Pliocene (Gustavson and Finley, 1985). Beds of calcrete or silcrete, commonly called caliche, are at or near the top of the Ogallala, and they form the cap rock of the High Plains.

QUATERNARY SYSTEM

Quaternary sediments are alluvial, eolian, and lacustrine deposits derived from rivers and streams flowing eastward and southeastward across all parts of the region. The Tule Group and Meade Group consist in part of sediments deposited in local subsidence basins formed by subsurface dissolution of Permian salt in the Texas and Oklahoma panhandles and southwestern Kansas (Gustavson and Finley, 1985). The Tule Group includes, in ascending order, the Blackwater Draw, Double Lakes, and Tahoka Formations.

Younger Pleistocene and Holocene sediments consist of terrace deposits and alluvium along and near the present-day streams and rivers of the region. Typically these deposits are 3 to 15 m thick, although locally they reach 30 m along the major rivers.

REFERENCES

Adams, W. L., 1964, Diagenetic aspects of lower Morrowan Pennsylvanian sandstones, northwestern Oklahoma: American Association of Petroleum Geologists Bulletin, v. 48, p. 1568–1580.

Adler, F. J., and 11 others, 1971, Future petroleum provinces of the Mid-Continent, Region 7, *in* Cram, I. H., ed., Future petroleum provinces of the United States; Their geology and potential: American Association of Petroleum Geologists Memoir 15, v. 2, p. 985–1120.

Amsden, T. W., 1960, Hunton stratigraphy; Stratigraphy and paleontology of the Hunton Group in the Arbuckle Mountains region, part 4: Oklahoma Geological Survey Bulletin 84, 311 p.

——, 1975, Hunton Group (Late Ordovician, Silurian, and Early Devonian) in the Anadarko Basin of Oklahoma: Oklahoma Geological Survey Bulletin 121, 214 p.

——, 1980, Hunton Group (Late Ordovician, Silurian, and Early Devonian) in the Arkoma Basin of Oklahoma: Oklahoma Geological Survey Bulletin 129, 136 p.

Amsden, T. W., and Sweet, W. C., 1983, Upper Bromide Formation and Viola Group (Middle and Upper Ordovician) in eastern Oklahoma: Oklahoma Geological Survey Bulletin 132, 76 p.

Barnes, V. E., 1984, Dalhart sheet: The University of Texas at Austin, Bureau of Economic Geology Geologic Atlas of Texas, scale 1:250,000.

Barnes, V. E., and Bell, W. C., 1977, The Moore Hollow Group of central Texas: The University of Texas at Austin, Bureau of Economic Geology Report of Investigation 88, 169 p.

Barnes, V. E., Cloud, P. E., Jr., and Warren, L. E., 1947, Devonian rocks of central Texas: Geological Society America Bulletin, v. 58, p. 125–140.

Barnes, V. E., and 6 others, 1959, Stratigraphy of pre-Simpson Paleozoic subsurface rocks of Texas and southeast New Mexico: The University of Texas at Austin, Bureau of Economic Geology Publication 5924, v. I, 294 p.

Barnes, V. E., and 6 others, 1972, Geology of the Llano region and Austin area, field excursion: The University of Texas at Austin, Bureau of Economic Geology Guidebook 13, 154 p.

Bennison, A. P., 1984, Shelf to trough correlations of Late Desmoinesian and Missourian carbonate banks and related strata, northeast Oklahoma, *in* Hyne, N. J., ed., Limestones of the Mid-Continent: Tulsa Geological Society Special Publication 2, p. 93–126.

Berry, W.B.N., and Boucot, A. J., 1970, Correlation of the North American Silurian rocks: Geological Society of America Special Paper 102, 289 p.

Bickford, M. E., Harrower, K. L., Hoppe, W. J., Nelson, B. K., Nusbaum, R. L., and Thomas, J. J., 1981, Rb-Sr and U-Pb geochronology and distribution of rock types in the Precambrian basement of Missouri and Kansas: Geological Society of America Bulletin, v. 92, p. 323–341.

Bissell, C. R., and Cleaves, A. W., 1986, Depositional systems and sandstone diagenesis in the McAlester Formation (surface and subsurface) of east-central Oklahoma: Geological Society of America Abstracts with Programs, v. 18, no. 3, p. 212.

Bozovitch, S., 1963, Pre-Mississippian carbonate rocks in the Hollis Basin of Oklahoma [M.S. thesis]: Norman, University of Oklahoma, 82 p.

Bradfield, H. H., 1959, Petroleum geology of Grayson County, Texas, *in* Petroleum geology of southern Oklahoma: American Association of Petroleum Geologists, v. 2, p. 53–100.

——, 1969, Stratigraphy of deeper Marietta Basin in Oklahoma and Texas [abs.]: American Association of Petroleum Geologists Bulletin, v. 52, p. 193–194.

Brewer, J. A., Good, R., Oliver, J. E., Brown, L. D., and Kaufman, S., 1983, COCORP profiling across the southern Oklahoma aulacogen: overthrusting of the Wichita Mountains and compression within the Anadarko Basin: Geology, v. 11, p. 109–114.

Brown, L. F., Jr., Cleaves, A. W., II, and Erxleben, A. W., 1973, Pennsylvanian depositional systems in north-central Texas—a guide for interpreting terrigenous clastic facies in a cratonic basin: The University of Texas at Austin, Bureau of Economic Geology Guidebook 14, 122 p.

Buchanan, R. S., and Johnson, F. K., 1968, Bonanza gas field—a model for Arkoma Basin growth faulting, *in* Cline, L. M., ed., A guidebook to the geology of the western Arkoma Basin and Ouachita Mountains, Oklahoma: Oklahoma City Geological Society Guidebook, p. 75–85.

Budnik, R. T., 1984, Seismic reflection evidence of a later Proterozoic basin in the Texas Panhandle: Geologic Society of America Abstracts with Programs, v. 16, no. 2, p. 79.

Cherng, J. C., 1982, Depositional environment and diagenesis of the Chico Ridge Limestone bank (Upper Pennsylvanian) north-central Texas, *in* Martin, C. A., ed., Petroleum geology of the Fort Worth Basin and Bend Arch area: Dallas Geological Society, p. 35–47.

Cleaves, A. W., II, 1975, Upper Des Moinesian–lower Missourian depositional systems (Pennsylvanian), north-central Texas [Ph.D. thesis]: The Univerity of Texas at Austin, 256 p.

——, 1982, Upper Strawn depositional systems of north-central Texas, *in* Martin, C. A., ed., Petroleum geology of the Fort Worth Basin and Bend Arch area: Dallas Geological Society, p. 49–95.

Cline, L. M., 1960, Late Paleozoic rocks of the Ouachita Mountains: Oklahoma Geological Survey Bulletin 85, 113 p.

Clopine, W. W., 1986, The lithostratigraphy, biostratigraphy, and depositional history of the Atokan Series (Middle Pennsylvanian) in the Ardmore Basin, Oklahoma [M.S. thesis]: Norman, University of Oklahoma, 161 p.

Cloud, P. E., and Barnes, V. E., 1948, The Ellenburger Group of central Texas: University of Texas Publication 4621, 473 p.

Cook, T. D., and Bally, A., eds., 1975, Stratigraphic atlas of North and Central America: Princeton, New Jersey, Princeton University Press, 272 p.

Corwin, L. W., 1982, Stratigraphy of the Fredericksburg Group north of the Colorado River, Texas: Waco, Texas, Baylor Geological Studies Bulletin 40, 64 p.

Craig, L. C., and 24 others, 1979, Paleotectonic investigations of the Mississippian System in the United States: U.S. Geological Survey Professional Paper 1010, part I, p. 1–369; part II, p. 371–559; part III, 15 plates.

Cromwell, D. W., 1974, The stratigraphy and environment of deposition of the lower Dornick Hills Group (Lower Pennsylvanian), Ardmore Basin, Oklahoma [M.S. thesis]: Norman, University of Oklahoma, 138 p.

Denison, R. E., 1981, Basement rocks in northeast Oklahoma: Oklahoma Geological Survey Circular 84, 84 p.

——, 1982, Geologic cross section from the Arbuckle Mountains to the Muenster Arch, southern Oklahoma and Texas: Geological Society of America, Map and Chart Series MC-28R, 8 p., scale 1:250,000.

——, 1984, Basement rocks in northern Arkansas, *in* McFarland, J. B., and Bush, W. V., eds., Contributions to the geology of Arkansas, Volume II: Arkansas Geological Commission Miscellaneous Publication 18, p. 33–49.

Denison, R. E., Lidiak, E. G., Bickford, M. E., and Kisvarsanyi, E. B., 1984, Geology and geochronology of Precambrian rocks in the central interior region of the United States: U.S. Geological Survey Professional Paper 1241-C, 20 p.

Derby, J. R., 1973, Lower Ordovician-Middle Ordovician boundary in western Arbuckle Mountains, Oklahoma, *in* Regional geology of the Arbuckle Mountains, Oklahoma: Oklahoma Geological Survey Special Publication 73-3, p. 24–26.

Donovan, R. N., ed., 1986, The Slick Hills of southwestern Oklahoma; Fragments of an aulacogen?: Oklahoma Geological Survey Guidebook 24, 112 p.

Dutton, A. R., and Simpkins, W. W., 1986, Hydrogeochemistry and water resources of the Triassic Lower Dockum Group in the Texas Panhandle and eastern New Mexico: The University of Texas at Austin, Bureau of Economic Geology Report of Investigations 161, 51 p.

Dutton, S. P., 1980, Depositional systems and hydrocarbon resource potential of the Pennsylvanian System, Palo Duro and Dalhart Basins, Texas Panhandle: The University of Texas at Austin, Bureau of Economic Geology Geological Circular 80-8, 49 p.

Dutton, S. P., Goldstein, A. G., and Ruppel, S. C., 1982, Petroleum potential of the Palo Duro Basin, Texas Panhandle: The University of Texas at Austin, Bureau of Economic Geology Report of Investigations 123, 87 p.

Ebanks, W. J., and 6 others, 1979, Kansas, *in* The Mississippian and Pennsylvanian (Carboniferous) Systems in the United States: U.S. Geological Survey Professional Paper 1110-Q, 30 p.

Erxleben, A. W., 1975, Depositional systems in Canyon Group (Pennsylvanian System) north-central Texas: The University of Texas at Austin, Bureau of Economic Geology Report of Investigations 82, 75 p.

Evans, J. L., 1979, Major structural and stratigraphic features of the Anadarko Basin, *in* Pennsylvanian sandstones of the Mid-Continent: Tulsa Geological Society Special Publication 1, p. 97–113.

Fay, R. O., 1965, Geology and mineral resources of Woods County, Oklahoma: Oklahoma Geological Survey Bulletin 106, 189 p.

Fay, R. O., Friedman, S. A., Johnson, U. S., Roberts, J. F., Rose, W. D., and Sutherland, P. K., 1979, Oklahoma, *in* The Mississippian and Pennsylvanian (Carboniferous) Systems in the United States: U.S. Geological Survey Professional Paper 1110-R, 35 p.

Flawn, P. T., 1956, Basement rocks of Texas and southeast New Mexico: The University of Texas at Austin, Bureau of Economic Geology Publication 5605, 261 p.

Flawn, P. T., Goldstein, A. G., Jr., King, P. B., and Weaver, C. E., 1961, The Ouachita System: The University of Texas at Austin, Bureau of Economic Geology Publication 6120, 401 p.

Foshee, R. R., 1980, Lithostratigraphy and depositional systems of the Bloyd and Hale Formations (Pennsylvanian), in the western Arkoma Basin of Arkansas [M.S. thesis]: Fayetteville, University of Arkansas, 169 p.

Frezon, S. E., and Dixon, G. H., 1975, Texas Panhandle and Oklahoma, *in* Paleotectonic investigations of the Pennsylvanian System in the United States; Part I, Introduction and regional analyses of the Pennsylvanian System: U.S. Geological Survey Professional Paper 853-J, p. 177–194.

Frezon, S. E., and Jordan, L., 1979, Oklahoma, *in* Craig, L. C., and 24 others, Paleotectonic investigations of the Mississippian System in the United States; Part I, Introduction and regional analysis of the Mississippian System: U.S. Geological Survey Professional Paper 1010-I, p. 147–159.

Galloway, W. E., and Brown, L. F., Jr., 1972, Depositional systems and shelf-slope relationships in Upper Pennsylvanian rocks, north-central Texas: The University of Texas at Austin, Bureau of Economic Geology Report of Investigations 75, 62 p.

Galloway, W. E., Yancey, M. S., and Whipple, A. P., 1977, Seismic stratigraphic model of depositional platform margin, eastern Anadarko Basin, Oklahoma: American Association of Petroleum Geologists Bulletin, v. 61, p. 1437–1447.

Gatewood, L. E., 1970, Oklahoma City Field; Anatomy of a giant: American Association of Petroleum Geologists Memoir 14, p. 233–254.

Gilbert, M. C., 1983, Timing and chemistry of igneous events associated with the southern Oklahoma aulacogen: Tectonophysics, v. 94, p. 439–453.

Giles, A. W., 1935, Boone chert: Geological Society of America Bulletin, v. 46, p. 1815–1878.

Glick, E. E., 1979, Arkansas, *in* Craig, L. C., and 24 others, Paleotectonic investigations of the Mississippian System in the United States; Part I, Introduction and regional analysis of the Mississippian System: U.S. Geological Survey Professional Paper 1010-H, p. 125–145.

Glick, E. E., Frezon, S. E., and Gordon, M., Jr., 1964, Witts Springs Formation of Morrowan age in the Snowball quadrangle, north-central Arkansas: U.S. Geological Survey Bulletin 1194-D, 16 p.

Goebel, E. D., and Stewart, G. F., 1979, Kansas, *in* Craig, L. C., and 24 others, Paleotectonic investigations of the Mississippian System in the United States; Part I, Introduction and regional analysis of the Mississippian System: U.S. Geological Survey Professional Paper 1010-G, p. 115–123.

Goldstein, A. G., 1982, Regional and plate tectonic setting of the Palo Duro Basin, *in* Gustavson, T. C., and 21 others, Geology and geohydrology of the Palo Duro Basin, Texas Panhandle: The University of Texas at Austin, Bureau of Economic Geology Geological Circular 82-7, p. 6–10.

—— , 1984, Tectonic controls on late Paleozoic foreland subsidence, south-central United States: Journal of Geology, v. 92, p. 217–222.

Graham, S. A., Ingersoll, R. V., and Dickinson, W. R., 1976, Common provenance for lithic grains in carboniferous sandstones from Ouachita Mountains and Black Warrior Basin: Journal of Sedimentary Petrology, v. 46, p. 620–632.

Grayson, R. C., Jr., 1979, Stop descriptions—fifth day, *in* Sutherland, P. K., and Manger, W. L., eds., Mississippian-Pennsylvanian shelf-to-basin transition, Ozarks and Ouachitas, Oklahoma and Arkansas: Oklahoma Geological Survey Guidebook 19, p. 67–76.

—— , 1984, Morrowan and Atokan (Pennsylvanian) conodonts from the northeast margin of the Arbuckle Mountains, southern Oklahoma, *in* Sutherland, P. K., and Manger, W. L., eds., The Atokan Series and its boundaries: Oklahoma Geological Survey Bulletin 136, p. 41–63.

Gustavson, T. C., and Finley, R. J., 1985, Late Cenozoic geomorphic evolution of the Texas Panhandle and northeastern New Mexico—case studies of structural controls on regional drainage development: The University of Texas at Austin, Bureau of Economic Geology Report of Investigations 148, 42 p.

Gustavson, T. C., and Holliday, V. T., 1985, Depositional architecture of the Quaternary Blackwater Draw and Tertiary Ogallala Formations, Texas Panhandle and eastern New Mexico: The University of Texas at Austin, Bureau of Economic Geology Open-File Report OF-WTWI-1985-23, 60 p.

Gustavson, T. C., and 21 others, 1982, Geology and geohydrology of the Palo Duro Basin, Texas Panhandle, a report on the progress of nuclear waste isolation feasibility studies (1981): The University of Texas at Austin, Bureau of Economic Geology Geological Circular 82-7, 212 p.

Haley, B. R., 1982, Geology and energy resources of the Arkoma Basin, Oklahoma and Arkansas: University of Missouri at Rolla, UMR Journal, no. 3, p. 43–53.

Haley, B. R., Glick, E. E., Caplan, W. M., Holbrook, D. F., and Stone, C. G., 1979, Arkansas, *in* The Mississippian and Pennsylvanian (Carboniferous) Systems in the United States: U.S. Geological Survey Professional Paper 1110-O, 14 p.

Ham, W. E., 1969, Regional geology of the Arbuckle Mountains, Oklahoma: Oklahoma Geological Survey Guidebook 17, 52 p.

Ham, W. E., and Wilson, J. L., 1967, Paleozoic epeirogeny and orogeny in the central United States: American Journal of Science, v. 265, p. 332–407.

Ham, W. E., Denison, R. E., and Merritt, C. A., 1964, Basement rocks and structural evolution of southern Oklahoma: Oklahoma Geological Survey Bulletin 95, 302 p.

Handford, C. R., 1979, Lower Permian depositional systems, *in* Dutton, S. P., Finley, R. J., Galloway, R. J., Gustavson, T. C., Hanford, C. R., and Presley, M. W., Geology and geohydrology of the Palo Duro Basin, Texas Panhandle: The University of Texas at Austin, Bureau of Economic Geology Geological Circular 79-1, p. 26–38.

—— , 1980, Lower Permian facies of the Palo Duro Basin, Texas; Depositional systems, shelf-margin evolution, paleogeography, and petroleum potential: The University of Texas at Austin, Bureau of Economic Geology Report of Investigations 102, 31 p.

—— , 1981, Coastal sabkha and salt pan deposition of the lower Clear Fork Formation (Permian), Texas: Journal of Sedimentary Petrology, v. 51, no. 3, p. 761–778.

—— , 1986, Facies and bedding sequences in shelf-storm-deposited carbonates—Fayetteville Shale and Pitkin Limestone (Mississippian) Arkansas: Journal of Sedimentary Petrology, v. 56, p. 123–137.

Handford, C. R., and Dutton, S. P., 1980, Pennsylvanian–Lower Permian depositional systems and shelf-margin evolution, Palo Duro Basin, Texas: American Association of Petroleum Geologists Bulletin, v. 64, no. 1, p. 88–106.

Handford, C. R., and Fredericks, P. E., 1980, Facies patterns and depositional history of a Permian sabkha complex—Red Cave Formation, Texas Panhandle: The University of Texas at Austin, Bureau of Economic Geology Geological Circular 80-9, 38 p.

Handford, C. R., Presley, M. W., and Dutton, S. P., 1980, Depositional and tectonic evolution of a basement bounded, intracratonic basin, Palo Duro Basin, Texas [abs.]: American Association of Petroleum Geologists Bulletin, v. 64, no. 5, p. 717.

Handford, C. R., Dutton, S. P., and Fredericks, P. E., 1981, Regional cross sections of the Texas Panhandle; Precambrian to mid-Permian: The University of Texas at Austin, Bureau of Economic Geology Cross Sections.

Hatcher, R. D., Jr., Thomas, W. A., and Viele, G. W., eds., The Appalachian-Ouachita Orogen in the United States: Boulder, Colorado, Geological Society of America, The Geology of North America, v. F-2 (in press).

Hayes, W. C., and Knight, R. D., 1961, Cambrian system in the stratigraphic succession in Missouri: Missouri Geological Survey, v. XL, second series, p. 14–20.

Heckel, P. H., 1977, Origin of phosphatic black shale facies in Pennsylvanian cyclothems of Mid-Continent North America: American Association of Petroleum Geologists Bulletin, v. 61, p. 1045–1068.

Henbest, L. G., 1953, Morrow Group and lower Atoka Formation of Arkansas: American Association of Petroleum Geologists Bulletin, v. 37, p. 1935–1953.

Hicks, I. C., 1971, Southern Oklahoma fold belt, *in* Cram, I. H., ed., Future petroleum provinces of the United States; Their geology and potential: American Association of Petroleum Geologists Memoir 15, v. 2, p. 1070–1077.

Hills, J. M., and Kottlowski, F. E., coordinators, 1983, Correlation of stratigraphic units in North America—southwest/southwest Mid-Continent correlation chart: American Association of Petroleum Geologists.

Houseknecht, D. W., 1986, Evolution from passive margin to foreland basin; The Atoka Formation of the Arkoma Basin, south-central U.S.A., *in* Allen, P. A., and Homewood, P., eds., Foreland Basins: Oxford, Blackwell Scientific Publications, International Association of Sedimentologists Special Publication 8, p. 327–345.

Houseknecht, D. W., and Kacena, J. A., 1983, Tectonic and sedimentary evolution of the Arkoma foreland basin, *in* Houseknecht, D. W., ed., Tectonic-sedimentary evolution of the Arkoma Basin: Society of Economic Paleontologists and Mineralogists Midcontinent Section Guidebook, v. 1, p. 3–33.

Houseknecht, D. W., Zaengle, J. F., Steyaert, D. J., Matteo, A. P., Jr., and Kuhn, M. A., 1983, Facies and depositional environments of the Desmoinesian Hartshorne Sandstone, Arkoma Basin, *in* Houseknecht, D. W., ed., Tectonic-sedimentary evolution of the Arkoma Basin: Society of Economic Paleontologists and Mineralogists Midcontinent Section Guidebook, v. 1, p. 53–82.

Howe, J. R., and Thompson, T. L., 1984, Tectonics, sedimentation, and hydrocarbon potential of the Reelfoot rift: Oil and Gas Journal, v. 82, no. 46, p. 179–190.

Jacobsen, L., 1959, Petrology of Pennsylvanian sandstones and conglomerates of the Ardmore Basin: Oklahoma Geological Survey Bulletin 79, 144 p.

Jefferies, B. K., 1982, Stratigraphy and depositional patterns of the Union Valley, Wapanucka, and Lower Atoka formations [M.S. thesis]: Fayetteville, University of Arkansas, 102 p.

Johnson, K. S., 1978, Stratigraphy and mineral resources of Guadalupian and Ochoan rocks in the Texas Panhandle and western Oklahoma, *in* Austin, G. S., compiler, Geology and mineral deposits of Ochoan rocks in Delaware Basin and adjacent areas: New Mexico Bureau of Mines and Mineral Resources Circular 159, p. 57–62.

Johnson, K. S., and Gonzales, S., 1978, Salt deposits of the United States and regional geologic characteristics important for storage of radioactive wastes: Report to Union Carbide Corporation, Office of Waste Isolation, Oak Ridge National Laboratories, Y/OWI/SUB-7414/1, 188 p.

Johnson, K. S., and 6 others, 1979, Geology and earth resources of Oklahoma; An atlas of maps and cross sections: Oklahoma Geological Survey Educational Publication 1, 8 p.

Jones, J. G., 1957, Geology of the Ashland-Kiowa area, Pittsburg County, Oklahoma [M.S. thesis]: Norman, University of Oklahoma, 126 p.

Jordan, L., 1962, Geologic map and section of the pre-Pennsylvanian rocks in Oklahoma: Oklahoma Geological Survey Map GM-5, scale 1:750,000.

—— , 1965, Frisco Formation (Devonian) in borehole, Jackson County, Oklahoma: Oklahoma Geology Notes, v. 25, p. 20–27.

Jordan, L., and Vosburg, D. L., 1963, Permian salt and associated evaporites in the Anadarko Basin of the western Oklahoma-Texas Panhandle region: Oklahoma Geological Survey Bulletin 102, 76 p.

Kier, R. S., 1980, Depositional history of the Marble Falls Formation of the Llano region, central Texas, *in* Geology of the Llano region, central Texas: West Texas Geological Society Guidebook to the Annual Field Trip, p. 59–75.

Kier, R. S., Brown, L. F., Jr., and McBride, E. F., 1979, Texas, *in* The Mississippian and Pennsylvanian (Carboniferous) Systems in the United States: U.S. Geological Survey Professional Paper 1110-S, 45 p.

Kimberly, J. E., 1961, Sedimentology of the Smithwick Formation, Burnet County, Texas [M.S. thesis]: The University of Texas at Austin, 95 p.

King, P. B., 1969, Tectonic map of North America: U.S. Geological Survey, scale 1:5,000,000.

King, P. B., and Beikman, H. M., compilers, 1974, Geologic map of the United States: U.S. Geological Survey, scale 1:2,500,000.

Kisvarsanyi, E. B., 1979, Geologic map of the Precambrian of Missouri: Missouri Department of Natural Resources, Geological Survey, Contribution of Precambrian Geology, Map 7, scale 1:1,000,000.

Koinm, D. N., and Dickey, P. A., 1967, Growth faulting in McAlester Basin of Oklahoma: American Association of Petroleum Geologists Bulletin, v. 51, p. 710–718.

Kurtz, V. E., Thacker, J. L., Anderson, H. A., and Gerdemann, P. E., 1975, Traverse in Late Cambrian strata from the St. Francois Mountains, Missouri, to Delaware County, Oklahoma: Missouri Department of Natural Resources, Geological Survey Report of Investigations 55, 118 p.

Lang, R. C., III, 1966, Road Log; First day of conference, *in* Pennsylvanian of the Ardmore Basin, southern Oklahoma: Ardmore Geological Society Guidebook, p. 42–45.

Latham, J. W., 1970, Healdton field, Carter County, Oklahoma: American Association of Petroleum Geologists Memoir 14, p. 255–276.

Lee, W., 1953, Subsurface cross section from Meade County to Smith County, Kansas: Kansas State Geological Survey Oil and Gas Investigations 9, 23 p.

Lillie, R. J., and 7 others, 1983, Crustal structure of Ouachita Mountains, Arkansas—model based on integration of COCORP reflection profiles and regional geophysical data: American Association of Petroleum Geologists Bulletin, v. 67, p. 907–931.

Lovick, G. P., Mazzine, C. G., and Kotila, D. A., 1982, Atokan clastics—depositional environments in a Foreland basin, *in* Martin, C. A., ed., Petroleum geology of the Fort Worth Basin and Bend Arch area: Dallas Geological Society, p. 193–211.

Lumsden, D. N., Pittman, E. D., and Buchanan, R. S., 1971, Sedimentation and petrology of Spiro and Foster sand (Pennsylvanian), McAlester Basin, Oklahoma: American Association of Petroleum Geologists Bulletin, v. 55, p. 254–266.

Mack, G. H., Thomas, W. A., and Horsey, C. A., 1983, Composition of Carboniferous sandstones and tectonic framework of southern Appalachian–Ouachita orogen: Journal of Sedimentary Petrology, v. 53, p. 931–946.

MacLachlan, M. E., 1967, Oklahoma, *in* Paleotectonic investigations of the Permian System in the United States: U.S. Geological Survey Professional Paper 515-E, p. 81–92.

Maher, J. C., 1948, Subsurface geologic cross section from Baca County to Yuma County, Kansas: Kansas State Geological Survey Oil and Gas Investigations 6, 11 p.

Maley, M. P., 1986, Depositional history of the Upper Morrowan (Pennsylvanian) strata of the Ardmore Basin, Oklahoma [M.S. thesis]: Norman, University of Oklahoma, 206 p.

Mankin, C. J., coordinator, 1987, Correlation of stratigraphic units in North America—Texas-Oklahoma tectonic region correlation chart: American Association of Petroleum Geologists.

Mapel, W. J., Johnson, R. B., Bachman, G. O., and Varnes, K. L., 1979, Southern Midcontinent and Southern Rocky Mountains region, *in* Craig, L. C., and 24 others, Paleotectonic investigations of the Mississippian System in the United States; Part I, Introduction and regional analysis of the Mississippian System: U.S. Geological Survey Professional Paper 1010-J, p. 161–187.

Martin, J. A., Knight, R. D., and Hayes, W. C., 1961, Ordovician System in the stratigraphic succession in Missouri: Missouri Geological Survey, v. XL,

second series, p. 20–32.
McCracken, E., 1955, Correlation of insoluble residue zones of upper Arbuckle of Missouri and southern Kansas: American Association of Petroleum Geologists Bulletin, v. 39, no. 1, p. 47–59.
McGee, D. T., 1985, Lithostratigraphy and depositional history of the upper Dornick Hills Group (Early Desmoinesian, Pennsylvanian) of the Ardmore Basin, Oklahoma [M.S. thesis]: Norman, University of Oklahoma, 348 p.
McGillis, K. A., and Presley, M. W., 1981, Tansill, Salado, and Alibates Formations; Upper Permian evaporite/carbonate strata of the Texas Panhandle: The University of Texas at Austin, Bureau of Economic Geology Geological Circular 81-8, 31 p.
McGowen, J. H., Granata, G. E., and Seni, S. J., 1979, Depositional framework of the Lower Dockum Group (Triassic), Texas Panhandle: The University of Texas at Austin, Bureau of Economic Geology Report of Investigations 97, 60 p.
McKee, E. D., and 9 others, 1956, Paleotectonic maps of the Jurassic System: U.S. Geological Survey Miscellaneous Geologic Investigations Map I-175, 6 p. plus maps.
McKee, E. D., and 6 others, 1959, Paleotectonic maps of the Triassic System: U.S. Geological Survey Miscellaneous Geologic Investigations Map I-300, 33 p. plus maps.
McKee, E. D., and 16 others, 1967, Paleotectonic maps of the Permian System: U.S. Geological Survey Miscellaneous Geologic Investigations Map I-450, scale 1:5,000,000.
McKee, E. D., and 17 others, 1975, Paleotectonic investigations of the Pennsylvanian System in the United States: U.S. Geological Survey Professional Paper 853, part I, 349 p.; part II, 192 p.; part III, 17 plates.
McKnight, E. T., and Fischer, R. P., 1970, Geology and ore deposits of the Picher field, Oklahoma and Kansas: U.S. Geological Survey Professional Paper 588, 165 p.
Meek, F. B., III, 1983, The lithostratigraphy and depositional environments of the Springer and lower Golf Course Formations (Mississippian-Pennsylvanian) in the Ardmore Basin, Oklahoma [M.S. thesis]: Norman, University of Oklahoma, 212 p.
Meek, F. B., III, Elmore, R. D., and Sutherland, P. K., 1983, Progradational sequence in the Springer Formation, Ardmore Basin, Oklahoma: American Association of Petroleum Geologists Bulletin, v. 67, p. 512.
Merriam, D. F., 1963, The geologic history of Kansas: State Geological Survey of Kansas Bulletin 162, 317 p.
Moiola, R. J., and Shanmugam, G., 1984, Submarine fan sedimentation, Ouachita Mountains, Arkansas and Oklahoma: Gulf Coast Association of Geological Societies Transactions, v. 34, p. 175–182.
Morris, R. C., and Sutherland, P. K., 1984, The Mississippian-Pennsylvanian boundary in the Ouachita Mountains, Arbuckle Mountains and Ardmore Basin, Arkansas and Oklahoma, *in* Sutherland, P. K. and Manger, W. L., eds., Biostratigraphy: Ninth International Congress on Carboniferous Stratigraphy and Geology, Compte Rendu, v. 2, 630 p.; Carbondale, Illinois, Southern Illinois University Press.
Muehlberger, W. R., Denison, R. E., and Lidiak, E. G., 1967, Basement rocks in the continental interior of the United States: American Association of Petroleum Geologists Bulletin, v. 51, p. 2351–2380.
Ng, D.T.W., 1979, Subsurface study of Atoka (Lower Pennsylvanian) clastic rocks in parts of Jack, Palo Pinto, Parker, and Wise Counties, north-central Texas: American Association of Petroleum Geologists Bulletin, v. 63, p. 50–66.
Nicholson, J. H., 1960, Geology of the Texas Panhandle, *in* Aspects of the geology of Texas, a symposium: The University of Texas at Austin, Bureau of Economic Geology Publication 6017, p. 51–54.
Niem, A. R., 1976, Patterns of flysch deposition and deep-sea fans in the lower Stanley Group (Mississippian), Ouachita Mountains, Oklahoma and Arkansas: Journal of Sedimentary Petrology, v. 46, p. 633–646.
O'Connor, H., and McClain, T., 1982, Ogallala aquifer study in Kansas—geohydrology: Kansas Water Office, 99 p.
Oakes, M. C., 1948, Chert River, an inferred Carboniferous stream of southeastern Oklahoma: Oklahoma Academy of Sciences Proceedings, v. 28, p. 70–71.
——, 1953, Krebs and Cabaniss Groups, of Pennsylvanian age, in Oklahoma: American Association of Petroleum Geologists Bulletin, v. 37, p. 1523–1526.
——, 1967, Geology and petroleum of McIntosh County, Oklahoma; Part I, Geology and mineral resources of McIntosh County: Oklahoma Geological Survey Bulletin 111, p. 1–49.
——, 1977, Geology and mineral resources (exclusive of petroleum) of Muskogee County, Oklahoma: Oklahoma Geological Survey Bulletin 122, 78 p.
Ogren, D. E., 1968, Stratigraphy of Upper Mississippian rocks of northern Arkansas: American Association of Petroleum Geologists Bulletin, v. 52, p. 282–294.
Owen, M. R., and Carozzi, A. V., 1986, Southern provenance of upper Jackfork Sandstone, southern Ouachita Mountains; Cathodoluminescence petrology: Geological Society of America Bulletin, v. 97, p. 110–115.
Parker, D. C., 1981, Regional lithostratigraphic and depositional systems of the basal Atoka sandstones in the Arkoma Basin of Arkansas [M.S. thesis]: Fayetteville, University of Arkansas, 161 p.
Potter, P. E., and Glass, H. D., 1958, Petrology and sedimentation of the Pennsylvanian sediments in southern Illinois—a vertical profile: Illinois State Geological Survey Report of Investigations 204, 60 p.
Presley, M. W., 1980, Upper Permian salt-bearing stratigraphic units, *in* Gustavson, T. C., and 21 others, Geology and geohydrology of the Palo Duro Basin, Texas Panhandle: The University of Texas at Austin, Bureau of Economic Geology Geological Circular 80-7, p. 12–23.
——, 1981, Middle and Upper Permian salt-bearing strata of the Texas Panhandle; Lithologic and facies cross sections: The University of Texas at Austin, Bureau of Economic Geology Cross Sections.
Presley, M. W., and McGillis, K. A., 1982, Coastal evaporite and tidal-flat sediments of the upper Clear Fork and Glorieta Formations, Texas Panhandle: The University of Texas at Austin, Bureau of Economic Geology Report of Investigations 115, 50 p.
Ragland, D. A., and Donovan, R. N., 1985a, The Thacher Creek Member; Basal unit of the Cool Creek Formation in southern Oklahoma: Oklahoma Geology Notes, v. 45, p. 84–91.
——, 1985b, The Cool Creek Formation (Ordovician) at Turner Falls in the Arbuckle Mountains of southern Oklahoma: Oklahoma Geology Notes, v. 45, p. 132–148.
Rascoe, B., Jr., 1978, Sedimentary cycles in the Virgilian Series (Upper Pennsylvanian) of the Anadarko Basin, Parts 1 and 2: Oklahoma City Geological Society Shale Shaker, v. 28, p. 123–131, p. 144–149.
Rascoe, B., Jr., and Adler, F. J., 1983, Permo-Carboniferous hydrocarbon accumulations, Mid-Continent, U.S.A.: American Association of Petroleum Geologists Bulletin, v. 67, p. 979–1001.
Ross, R. J., Jr., and 27 others, 1982, The Ordovician System in the United States: International Union of Geological Sciences Publication 12, 73 p.
Sable, E. G., 1979, Eastern interior region, *in* Craig, L. C., and 24 others, Paleotectonic investigations of the Mississippian System in the United States; Part I, Introduction and regional analysis of the Mississippian System: U.S. Geological Survey 1010-E, p. 59–106.
Saether, O. M., 1976, Depositional history of the Devils Kitchen sandstones and conglomerates in the Ardmore Basin, southern Oklahoma [M.S. thesis]: Norman, University of Oklahoma, 95 p.
Scott, R. W., Fee, D., Magee, R., and Laali, H., 1978, Epeiric depositional models for the Lower Cretaceous Washita Group, north-central Texas: The University of Texas at Austin, Bureau of Economic Geology Report of Investigations 94, 23 p.
Seddon, G., 1970, Pre-Chappel conodonts of the Llano Region, Texas: The University of Texas at Austin, Bureau Economic Geology Report of Investigations 68, 69 p.
Seni, S. J., 1980, Sand-body geometry and depositional systems, Ogallala Formation, Texas: The University of Texas at Austin, Bureau of Economic Geology Report of Investigations 105, 36 p.
Shatski, N. S., 1946, The Great Donets Basin and the Wichita System, *in* Com-

parative tectonics of ancient platforms: SSSR, Akademii Nauk, Izvestia, Geologicheskii Seriya 1, p. 5–62 (in Russian).

Shelby, J. M., 1979, Upper Morrow fan-delta deposits of Anadarko Basin [abs.]: American Association of Petroleum Geologists Bulletin, v. 63, p. 2119.

Shideler, G. L., 1970, Provenance of Johns Valley boulders in late Paleozoic facies, southeastern Oklahoma and southwestern Arkansas: American Association of Petroleum Geologists Bulletin, v. 54, p. 789–806.

Smith, G. E., 1974, Depositional systems, San Angelo Formation (Permian), north Texas; Facies control of red-bed copper mineralization: The University of Texas at Austin, Bureau of Economic Geology Report of Investigations 80, 74 p.

Stitt, J. H., 1977, Late Cambrian and earliest Ordovician trilobites, Wichita Mountains area, Oklahoma: Oklahoma Geological Survey Bulletin 124, 79 p.

Stone, C. G., and McFarland, J. D., III, 1981, Field guide to the Paleozoic rocks of the Ouachita Mountains and Arkansas Valley provinces, Arkansas: Arkansas Geological Commission Guidebook 81-1, 140 p.

Stovall, J. W., 1938, The Morrison of Oklahoma and its dinosaurs: Journal of Geology, v. 46, p. 583–600.

——, 1943, Stratigraphy of the Cimarron Valley (Mesozoic rocks), *in* Schoff, S. L., and Stovall, J. W., eds., Geology and ground water resources of Cimarron County, Oklahoma: Oklahoma Geological Survey Bulletin 64, p. 43–119.

Sutherland, P. K., 1984, Unroofing of the Arbuckle Mountains, Oklahoma, during the time of folding; An update: Geological Society of America Abstracts with Programs, v. 16, p. 115.

Sutherland, P. K., and Grayson, R. C., Jr., 1977, Basinward facies changes in the Wapanucka Formation (Lower Pennsylvanian), Indian Nation Turnpike, Ouachita Mountains, Oklahoma: Oklahoma Geology Notes, v. 37, p. 39–44.

Sutherland, P. K., and Henry, T. W., 1977a, Carbonate platform facies and new stratigraphic nomenclature of the Morrowan Series (Lower and Middle Pennsylvanian), northeastern Oklahoma: Geological Society of America Bulletin, v. 88, p. 425–440.

——, 1977b, Stratigraphy of uppermost Mississippian and Lower Pennsylvanian strata in northeastern Oklahoma, *in* Sutherland, P. K., and Manger, W. L., eds., Upper Chesterian-Morrowan stratigraphy and the Mississippian-Pennsylvanian boundary in northeastern Oklahoma and northwestern Arkansas: Oklahoma Geological Survey Guidebook 18, p. 41–48.

Sutherland, P. K., and Manger, W. L., eds., 1979, Mississippian-Pennsylvanian shelf-to-basin transition, Ozark and Ouachita regions, Oklahoma and Arkansas: Oklahoma Geological Survey Guidebook 19, 81 p.

Sutherland, P. K., Archinal, B. E., and Grubbs, R. K., 1982, Morrowan and Atokan (Pennsylvanian) stratigraphy in the Arbuckle Mountains area, Oklahoma, *in* Sutherland, P. K., ed., Lower and Middle Pennsylvanian stratigraphy in south-central Oklahoma: Oklahoma Geological Survey Guidebook 20, p. 1–17.

Swanson, D. C., 1979, Deltaic deposits in the Pennsylvanian upper Morrow Formation of the Anadarko Basin, *in* Pennsylvanian sandstones of the Mid-Continent: Tulsa Geological Society Special Publication 1, p. 115–168.

Tarr, R. S., 1955, Paleogeological maps at the base of Woodford and Hunton isopachous map of Oklahoma: American Association of Petroleum Geologists Bulletin, v. 39, p. 1851–1858.

Tarr, R. S., Jordan, L., and Rowland, T. L., 1965, Geologic map and section of pre-Woodford rocks in Oklahoma: Oklahoma Geological Survey Map GM-9, scale 1:750,000.

Tennant, S. H., 1981, Lithostratigraphy and depositional environments of the upper Dornick Hills Group (Lower Pennsylvanian) in the northern part of the Ardmore Basin, Oklahoma [M.S. thesis]: Norman, University of Oklahoma, 291 p.

Thomas, W. A., 1984, Carboniferous tectonic framework of the continental margin of southeastern North America: Ninth International Congress on Carboniferous Stratigraphy and Geology, Compte Rendu, v. 3, p. 291–302.

——, 1985, The Appalachian-Ouachita connection; Paleozoic orogenic belt at the southern margin of North America: Annual Review Earth and Planetary Sciences, v. 13, p. 175–199.

Thompson, D. M., 1982, The Atoka Group (lower-Middle Pennsylvanian), northern Fort Worth Basin, Texas; Terrigenous depositional systems, diagenesis, reservoir distribution, and quality: The University of Texas at Austin, Bureau of Economic Geology Report of Investigations 125, 62 p.

Thompson, T. L., 1979, Missouri, *in* The Mississippian and Pennsylvanian (Carboniferous) Systems in the United States: U.S. Geological Survey Professional Paper 1110-N, 22 p.

Tomlinson, C. W., and McBee, W., Jr., 1959, Pennsylvanian sediments and orogenies of Ardmore district, Oklahoma, *in* Petroleum geology of southern Oklahoma: American Association of Petroleum Geologists, v. 2, p. 3–52.

Tulsa Geological Society, 1961, Stratigraphy of the Arkoma Basin: Tulsa Geological Society Digest, v. 29, p. 55–88.

Turner, G. L., 1957, Paleozoic stratigraphy of the Fort Worth Basin, *in* Conselman, F. B., ed., Study of the Lower Pennsylvanian and Mississippian rocks of the northeast Llano Uplift: Abilene and Fort Worth Geological Societies Joint Field Trip Guidebook, p. 57–77.

Turner, N. L., 1970, Carboniferous stratigraphy of western San Saba County, Texas [Ph.D. thesis]: The University of Texas at Austin, 377 p.

Tweto, O., 1983, Las Animas Formation (upper Precambrian) in the subsurface of southeastern Colorado: U.S. Geological Survey Bulletin 1529-G, 14 p.

Vedros, S. G., and Visher, G. S., 1978, The Red Oak Sandstone; A hydrocarbon-producing submarine fan deposit, *in* Stanley, D. J., and Kelling, G., eds., Sedimentation in submarine canyons, fans, and trenches: Stroudsburg, Pennsylvania, Dowden, Hutchinson and Ross, Incorporated, p. 292–308.

Visher, G. S., 1968, Depositional framework of the Bluejacket-Bartlesville Sandstone, *in* Visher, G. S., ed., Geology of the Bluejacket-Bartlesville Sandstone, Oklahoma: Oklahoma Geological Society Guidebook, p. 32–44.

Visher, G. S., Saitta, S., and Phares, R. S., 1971, Pennsylvanian delta patterns and petroleum occurrences in eastern Oklahoma: American Association of Petroleum Geologists Bulletin, v. 55, p. 1206–1230.

Watson, W. G., 1980, Paleozoic stratigraphy of the Llano Area—a review, *in* Geology of the Llano region, central Texas: West Texas Geological Society Guidebook to the Annual Field Trip, p. 28–51.

Weeks, J. B., and Gutentag, E. D., 1981, Bedrock geology, altitude of base, and 1980 saturated thickness of the High Plains Aquifer in parts of Colorado, Kansas, Nebraska, New Mexico, Oklahoma, South Dakota, Texas, and Wyoming: U.S. Geological Survey Hydrologic Investigations Atlas HA-0648, 2 sheets, scale 1:2,500,000.

Weirich, T. E., 1953, Shelf principle of oil origin, migration, and accumulation: American Association of Petroleum Geologists Bulletin, v. 37, p. 2027–2045.

Wermund, E. G., 1975, Upper Pennsylvanian limestone banks, north-central Texas: The University of Texas at Austin, Bureau of Economic Geology Geological Circular 75-3, 60 p.

Wickham, J. S., Roeder, D. R., and Briggs, G., 1976, Plate tectonics model for the Ouachita fold belt: Geology, v. 4, p. 173–176.

Wilson, J. L., 1975, Carbonate facies in geologic history: New York, Springer-Verlag, 471 p.

Wise, O. A., and Caplan, W. M., 1979, Silurian and Devonian rocks of northern Arkansas: Arkansas Geological Commission Information Circular 25, 14 p.

Zachry, D. L., 1983, Sedimentary framework of the Atoka Formation, Arkoma Basin, Arkansas, *in* Houseknecht, D. W., ed., Tectonic-sedimentary evolution of the Arkoma Basin: Society of Economic Paleontologists and Mineralogists Midcontinent Section Guidebook, v. 1, p. 34–52.

Zachry, D. L., and Sutherland, P. K., 1984, Stratigraphy and depositional framework of the Atoka Formation (Pennsylvanian) Arkoma Basin of Arkansas and Oklahoma, *in* Sutherland, P. K., and Manger, W. L., eds., The Atokan Series (Pennsylvanian) and its boundaries—a symposium: Oklahoma Geological Survey Bulletin 136, p. 9–17.

MANUSCRIPT ACCEPTED BY THE SOCIETY DECEMBER 14, 1987

Printed in U.S.A.

The Geology of North America
Vol. D-2, Sedimentary Cover—North American Craton: U.S.
The Geological Society of America, 1988

Chapter 13

Michigan Basin

James H. Fisher
Department of Geological Sciences, Michigan State University, East Lansing, Michigan 48824
Michael W. Barratt
Patrick Petroleum Company, Jackson, Michigan 49204
John B. Droste
Department of Geology, Indiana University, Bloomington, Indiana 47405
Robert H. Shaver
Indiana Geological Survey and Department of Geology, Indiana University, Bloomington, Indiana 47405

INTRODUCTION

Limit of Study Area

The Michigan Basin (Fig. 1) is bordered on the west by the Wisconsin Highland and to the north and east of Lake Huron by the Canadian Shield. The Algonquin Arch is a major Precambrian feature trending northeastward in Ontario but becoming almost east–west at the border of Michigan. The northwest Ohio area was a platform during the early part of the Paleozoic, and the Findlay Arch is a post-Silurian structure that developed on this platform. The Findlay Arch plunges northeastward and dies out in the western end of Lake Erie. Contrary to many previously published maps, the Findlay Arch is separated from the Algonquin Arch both in trend and in time of formation. The Kankakee Arch was present by Early Ordovician time as a low-relief feature. Droste and others (1975) have described the detailed geologic history of this area and identified an early to middle Paleozoic feature in the area of the present Kankakee Arch, which they named the Wabash Platform. The Wisconsin Highland is a persistent positive area, which was definitely present by Late Cambrian time.

Areas omitted from this study are Manitoulin Island, the Bruce Peninsula, and southwestern Ontario, which are covered in a separate publication by Stott and Aitken (in preparation). Also omitted is the Door Peninsula of Wisconsin, which because of its isolation from the rest of the Michigan Basin is better related to the strata adjacent to the Wisconsin Highland. Northeastern Illinois is included in the chapter on the Illinois Basin.

In order to avoid repetition, it is understood that statements in this chapter regarding the Michigan area, refer to the Lower Peninsula of Michigan. Comments regarding the Upper Peninsula will be specifically noted as such. The names of counties are shown on the Precambrian structure map (Fig. 5) for reference purposes. It should be noted in all isopach maps that the contoured area of each sequence includes the partially eroded section of the sequence.

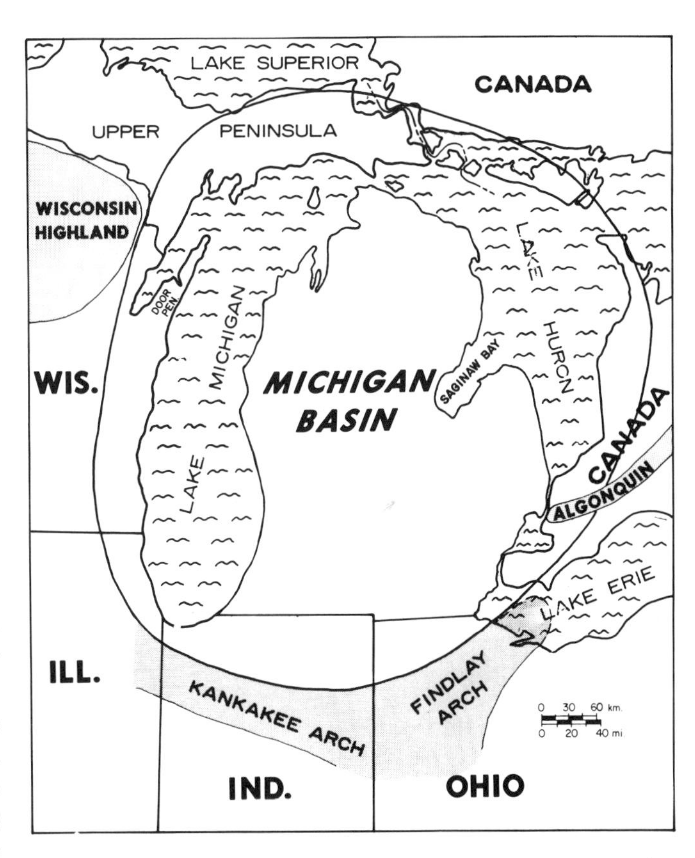

Figure 1. Location of the Michigan Basin and the major surrounding tectonic features.

Fisher, J. H., Barratt, M. W., Droste, J. B., and Shaver, R. H., 1988, Michigan Basin, *in* Sloss, L. L., ed., Sedimentary Cover—North American Craton; U.S.: Boulder, Colorado, Geological Society of America, The Geology of North America, v. D-2.

Physiography—Resume of Glaciation

The Michigan Basin covers approximately 198,387 km^2 in portions of eastern Wisconsin, northeastern Illinois, northern Indiana and Ohio, southwestern Ontario, and nearly all of the state of Michigan. The basin area has been subjected to numerous episodes of glaciation during Pleistocene time (Dorr and Eschman, 1970). Ice advances and retreats account for a complex physiography of glacial lake deposits, moraines, drumlins, eskers, outwash plains, and glacial drift that ranges from 0 to 305 m in thickness. Maximum drift thicknesses are located in Emmet, Otsego, Wexford, and Grand Traverse Counties. Although glacial material was deposited with each major ice advance and retreat, the surface deposits are late Wisconsinan (<35 ka). Three major ice lobes, following the Lake Michigan, Lake Huron–Saginaw Bay, and Lake Erie Basins, advanced and retreated independently, depositing a series of end moraines and outwash plains (Fig. 2). The final retreat of ice from the area occurred approximately 9.5 ka. Rebound of the Lake Superior region apparently is still going on at a rate of 0.15 m per 100 years. Bedrock configurations for southern Michigan, northern Indiana, and northern Ohio have been mapped by Johnson and Keller (1972) and by Schneider and Keller (1970).

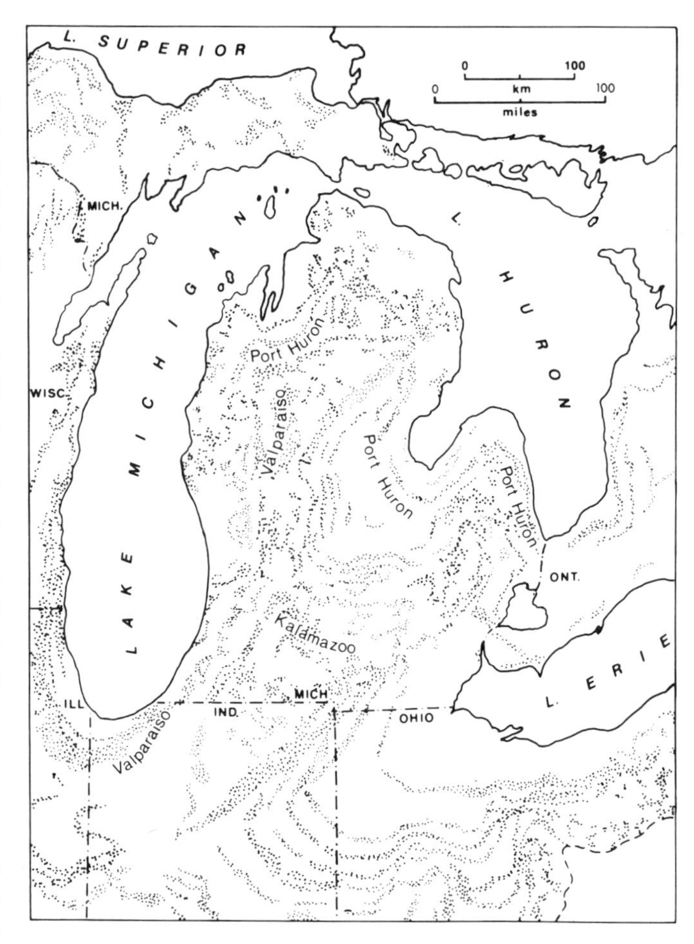

Figure 2. Major glacial moraines of the Michigan Basin (after Leverett and Taylor, 1975).

Sources of Data

A number of studies have been made in the outcrop areas surrounding the Michigan Basin. The majority of these investigations have relied on paleontology for age determinations. Unfortunately, these marginal sections are atypical of the central basin, in thickness and, in many places, in lithology. Further, the identification of megafossils in well cuttings is most uncommon. Studies of microfossils in the central basin are almost nonexistent. Therefore, a dichotomy has developed in the naming of outcrop formations as opposed to central basin units. In view of this, the ages assigned to some of the central basin formations are at best tentative.

Regional studies require an adequate distribution of data. Well samples and cores in Michigan do not meet this requirement. They are, however, an excellent source of supplementary information. Written lithologic well descriptions submitted to the State of Michigan are of variable quality. Some of the older descriptions are very detailed and are quite accurate, while recent descriptions tend to be of poorer quality. Therefore, regional correlations in Michigan are based very heavily on wire line logs, and the formation nomenclature used in every day work is that of the central basin. The data utilized in this report were derived largely from wire line logs supplemented by written lithologic well descriptions plus occasional core studies. Approximately 37,600 wells have been drilled in Michigan in the search for oil and gas. Not all of these wells were logged by wire-line logging methods; most of them do, however, have written lithologic descriptions. Gamma-ray logs for some wells are not available because the operators chose to keep them confidential.

STRUCTURE

Major Gravity Trends

The gravity map (Fig. 3) is dominated by the mid-Michigan gravity high (Hinze and others, 1975) consisting of a series of strong positive anomalies, the flanks of which are well defined on the gravity map by closely spaced contours (see also Hinze and Braile, this volume). The anomaly begins in southeastern Michigan and trends northwest to the central part of the state where it changes direction abruptly and extends northward through Grand Traverse Bay and on into the Upper Peninsula. It has been suggested that the mid-Michigan gravity high represents a rift zone partially filled with Keweenawan basalts (Hinze and others, 1975). Rudman and others (1965) consider the Michigan rift to be an arcuate extension by way of Lake Superior of the midcontinent rift zone that reaches from Minnesota to Kansas. Burke and Dewey (1973) proposed the concept of a triple junction in the eastern part of Lake Superior with the Kapuskasing Fault Zone trending northeast toward Hudson Bay as the third arm. An

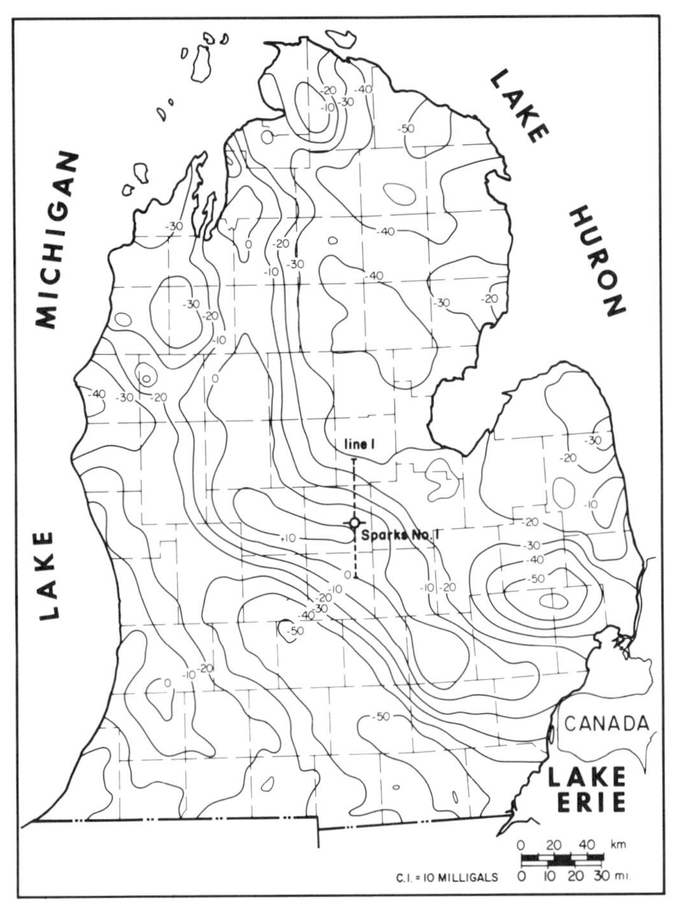

Figure 3. Bouguer gravity anomaly map of the Southern Peninsula of Michigan (after Hinze and others, 1971). Dotted line shows COCORP seismic line 1 through the location of the Sparks well.

alternative third arm trending north from Lake Superior through Lake Nipigon in Canada has been suggested by Halls (1978). The mid-Michigan rift is parallel or subparallel to many of the fold structures in the Phanerozoic sediments. This parallelism may indicate that Keweenawan rifting occurred along pre-existing faults or zones of weakness in the older Precambrian basement. It should be noted that the rift zone is not an early forerunner of the Michigan Basin since it does not represent the site of future depocenters.

The northwest end of a second gravity high is apparent in southwestern Michigan. Rudman (1965) has traced this high as far southeast as eastern Kentucky.

Seismic Cross-section

The deepest well in Michigan is the McClure-Sparks (5,323 m) located in Gratiot County (sec. 8, T10N-R2W) on the northeast flank of the mid-Michigan gravity high (Fig. 3). This well encountered a normal Phanerozoic section to 3,711 m (12,176 ft) at which point a section of Precambrian red beds 1,612 + m (5,290+ ft) thick was penetrated. Several gabbro intrusives were encountered near total depth, but the well did not reach the igneous or metamorphic basement. The Precambrian sediments consist of shale and siltstone, predominantly red with occasional bands of green color, and some interbedded sandstones. Fowler and Kuenzi (1978) have characterized these beds as turbidites accumulating in a rift zone. Van der Voo and Watts (1978) have dated the red beds as Keweenawanan on the basis of paleomagnetic evidence. In terms of color, lithology, and thickness, the most likely correlative unit is the Freda Formation of Keweenawan age in the Upper Peninsula.

The consortium for continental reflection profiling (COCORP) ran three seismic lines in the vicinity of the Sparks well (Brown and others, 1982). The north–south line (Figs. 3 and 4) shows the relatively flat Paleozoic rocks immediately underlain by red beds draped in a graben-like pattern. The deeper, highly reflective beds (some of which are enhanced by heavy lines) may be lava flows. The thick series of red bed turbidites deposited in a down-warped and possibly faulted trough lends credence to the rift concept. If the deeper, highly reflective beds are Keweenawan lava flows, they would add further support to the hypothesis of Keweenawan rifting.

Other proprietary seismic surveys in the Michigan Basin indicate a basement that is highly faulted; however, most of these faults die out upward and generally are not observed above the Dundee Limestone (Middle Devonian) in the central basin area. In some of the largest faults in southeastern Michigan, fault displacement can be traced upward well into the Mississippian section.

Basin Tectonics

There are no Precambrian basement control wells other than the Sparks well in the central part of the Michigan Basin. Most of the Precambrian tests are scattered about the periphery of the basin, and the Precambrian structure map (Fig. 5) is, therefore, highly simplified and heavily based on projections from fairly deep wells. A portrayal of this sort, constructed from very limited data, gives a false impression of the nature of the Precambrian surface. On the basis of seismic surveys and structural features in the basin, the Precambrian surface may be interpreted as a complexly faulted series of horst and graben structures. The deepest part of the Precambrian igneous-metamorphic basement is located in northern Gladwin County.

More dense well coverage in southeastern Michigan permits the delineation of a few of the larger faults (Fig. 5). The Howell and Lucas-Monroe structures are high-angle normal faults with displacements of more than a hundred meters. The Sanilac feature is a high-angle reverse fault. The presumed Albion-Scipio Fault lies along a linear oil field of the same name. This field occurs in a narrow dolomitized zone approximately 56 km long and 0.8–1.6 km wide. The assumption is that the linear dolomitization pattern is fault controlled, and because no vertical offset is

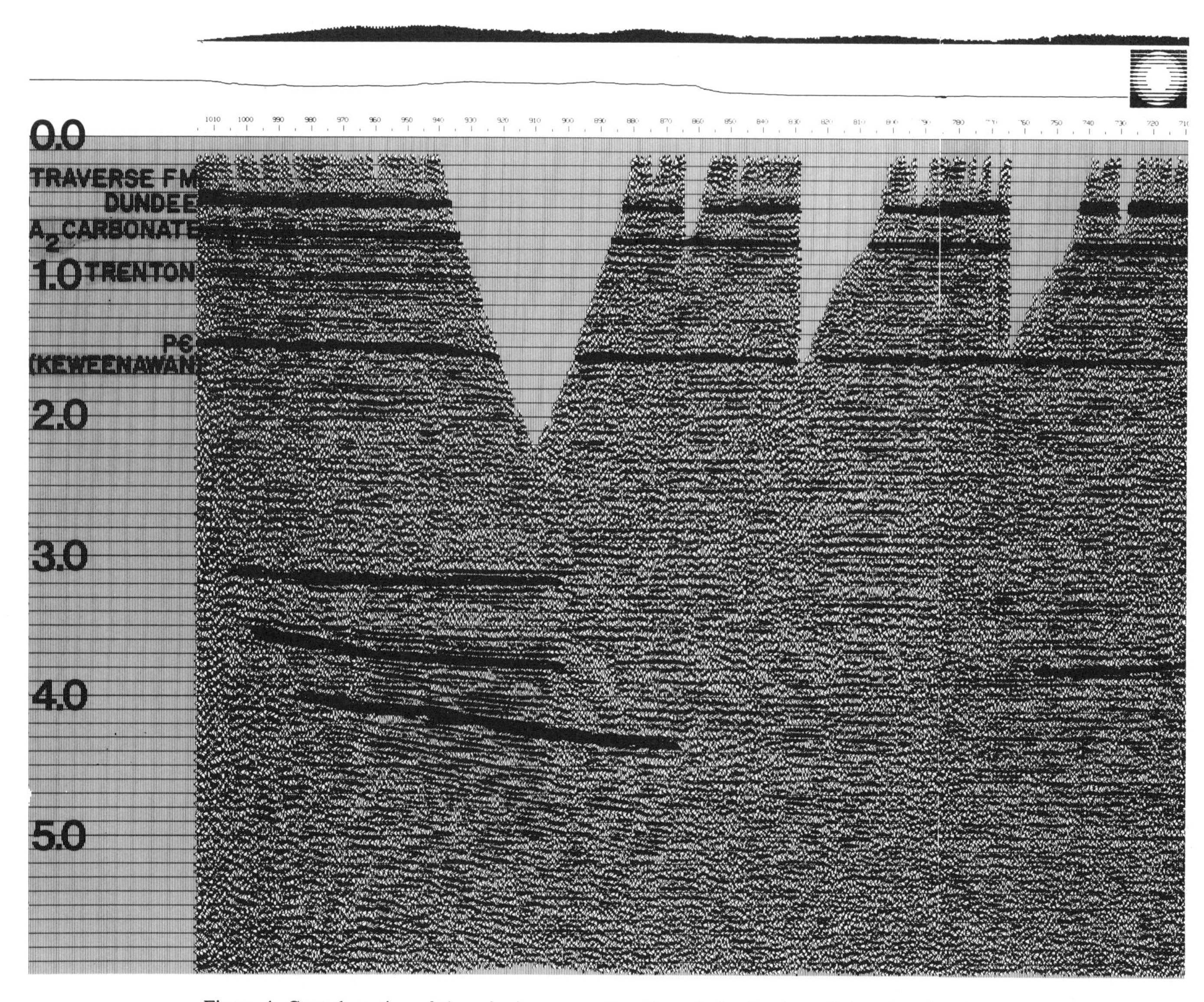

Figure 4. Central portion of the seismic cross-section through the Sparks well location showing a possible graben structure. (Some Phanerozoic formations and some of the possible Precambrian lava flows have been enhanced by darker lines.)

observed as deep as Lower Ordovician rocks, the further assumption is made that this structure is a strike-slip fault. A detailed gravity survey over the field (Hinze and Merritt, 1969) indicates the possibility that the basement is upthrown on the northeast side of the fault. Apparently this scarp was covered by Cambrian sediments, providing a fairly level surface for the deposition of Ordovician strata that were then fractured by a later strike-slip movement. Well cores show fracturing and brecciation, but offsets are measured in inches. The suggestion often is made that the displacement on the Albion-Scipio Fault is left lateral (Harding, 1974), but little field evidence exists to support this conclusion.

Detailed studies in southeastern Michigan (Mescher, 1980; Fisher, J. A., 1981), particularly along the Lucas-Monroe Fault, have shown that basement faulting has been intermittently active throughout early Paleozoic time and that the peak of activity was reached at the end of the Mississippian Period. In some examples, structural inversion of faults can be demonstrated. Note that all four of the faults shown in Fig. 5 are upthrown on the northeast side.

The structural grain of the folds in the basin is to the northwest (Fig. 6), and the largest folds are in the southeastern

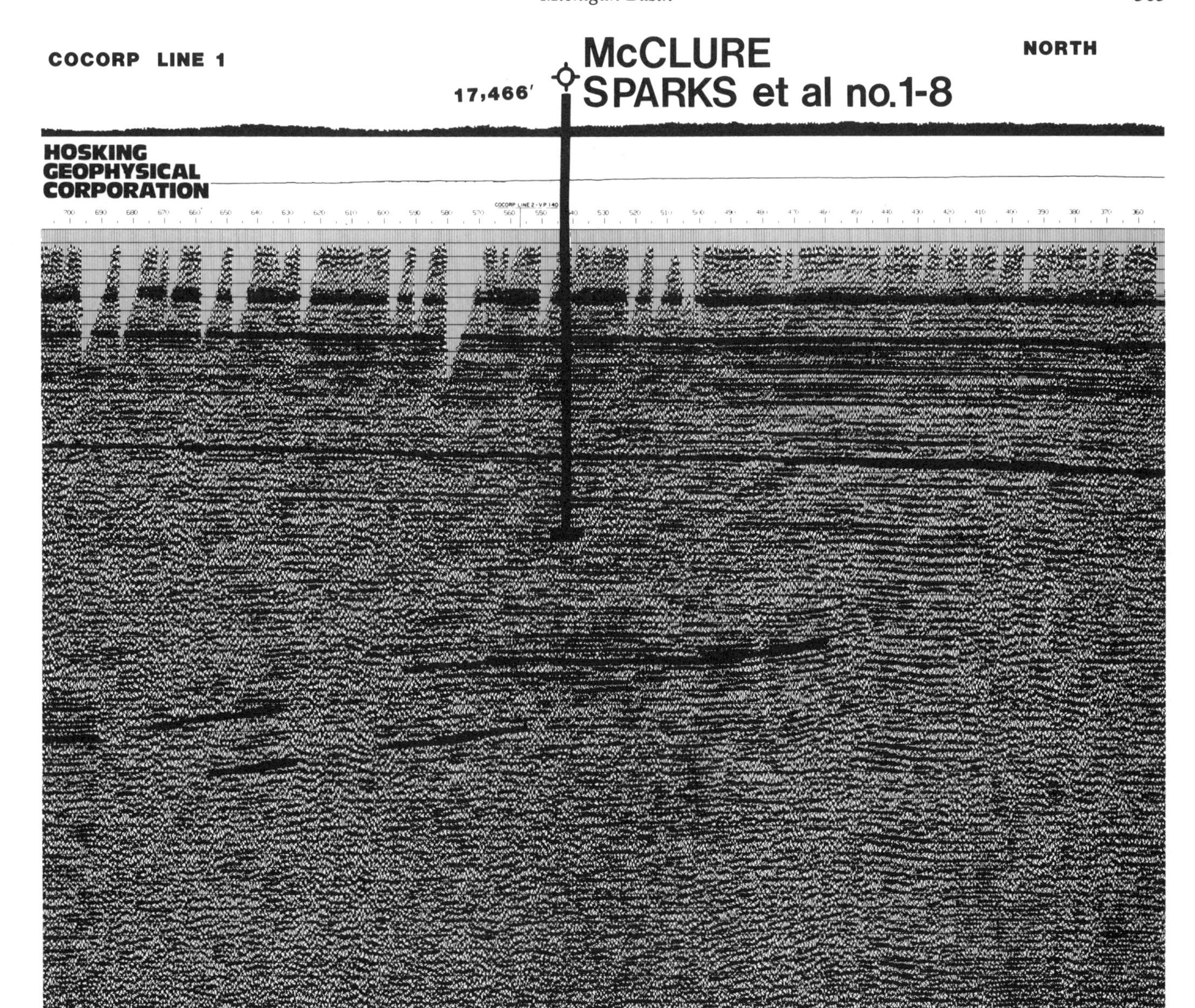

and east-central portions of the basin. One of the largest anticlines is the West Branch Anticline in Ogemaw County, which has a closure in excess of 122 m. Some of the folds are asymmetrical but there is no preferred pattern of asymmetry that could be related to compressional stress. Although the center of the basin is highly folded, the margins of the basin are relatively undeformed. The faults were not drawn in the Lucas-Monroe and Sanilac areas in Figure 6 in order to show the degree of contour distortion, which justifies faults in those areas.

At various times during the Paleozoic Era, two sectors on the basin margin have acted as stable platform areas. The first of these, the St. Clair Platform, occupies parts of Macomb and St. Clair counties (Fig. 7). The second, the Allegan Platform, is located in the general area of Ottawa, Barry, and Allegan Counties. At various times, sediments exhibit rapid thickening off the edges of these platforms into the center of the basin. This thickening is particularly evident during Salina time. Faulting at the basinward margins of these platforms is possible but there is no evidence to support this, other than abrupt changes in thickness and lithology that occur in some formations along platform margins. The platforms seem at times to have been more resistant to basin sag than other parts of the basin margin.

The Canadian Shield north of Lake Huron has a well-developed rectilinear pattern of jointing and faulting, and it is reasonable to assume that the same sort of pattern prevails in the Precambrian rocks beneath the Michigan Basin sediments. The structures observed in the basin are primarily the product of vertical movements of basement blocks, activated by regional

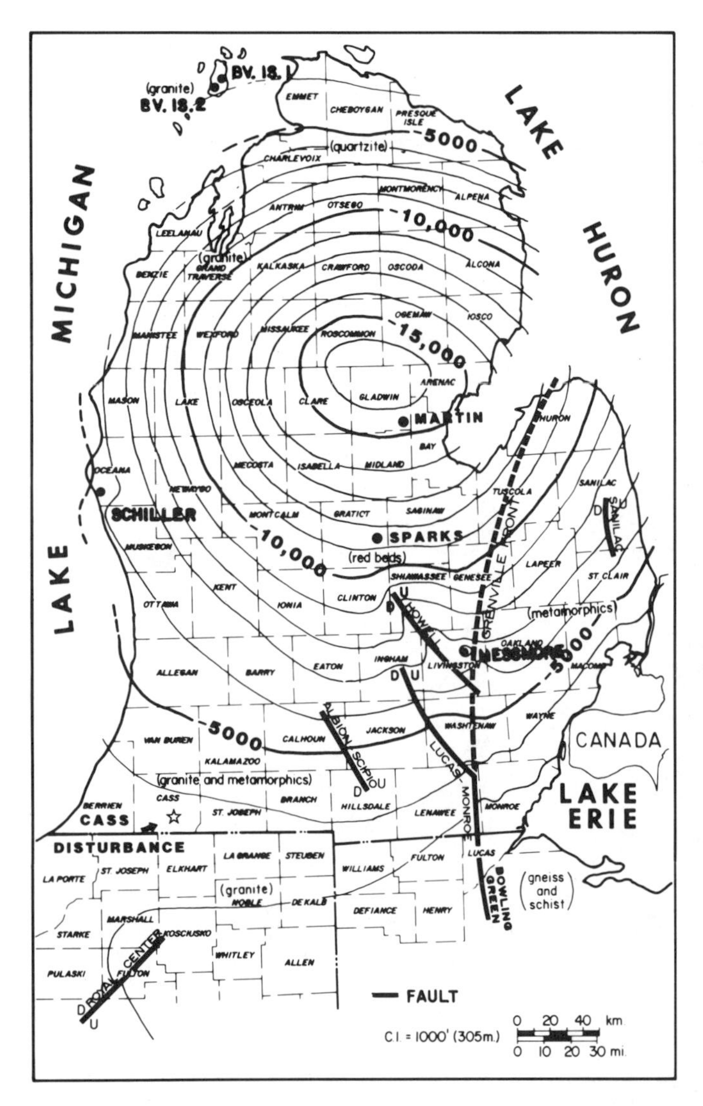

Figure 5. Structure contour map on the Precambrian showing the major faults, the Grenville Front, the Cass Disturbance, and the generalized basement rock types.

stresses (Fig. 7). Similar occurrences have been described in the Williston Basin (Thomas, 1974). There are two wells that reach the Precambrian that support this concept. Basement in the Mobil-Messmore well (sec. 11, T3N-R5E) (Fig. 5) in east-central Livingston County is 250 m high in reference to the regional Precambrian trend. The increase in elevation is approximately equivalent to the thickness of the Eau Claire Formation and Mount Simon Sandstone, which are absent in this well, demonstrating that a Precambrian high was present during the Late Cambrian marine invasion or shortly thereafter.

The basement in the second well, Amoco-Schiller (sec. 10, T13N-R18W) (Fig. 5) in the southwest corner of Oceana County, has an inferred Precambrian top that is 230 m higher than the regional Precambrian structure. There is a question as to whether this well reached the Precambrian, and the samples have been misplaced and are no longer available. The brief sample description of 332 m of Precambrian describes a sandstone that is similar to the overlying Mount Simon. The Precambrian top is based on the gamma-ray log that goes off scale at 1,874 m and has a series of off-scale curves interspersed with more normal sandstone responses, from there to total depth, 2,207 m. The off-scale curves are not a scale change phenomenon. The overlying Mount Simon is 101 m thick, which is anomalously thin compared to regional trends. We have used the log Precambrian top, which accounts for the displacement of the contours (Fig. 5).

Cass Cryptoexplosion Structure

Recent drilling in Cass County has outlined an area of complex high-angle reverse faulting similar to features in the central United States that are interpreted as cryptovolcanic structures. The two critical wells are the Hallwell-Lawson and the Hallwell-Hawkes, both in sec. 28, T7S-R14W. The top of the Cambrian in the Hawkes well is at least 370 m higher than the top in the Lawson, only 1.1 km to the west. The faulting occurred during Late Ordovician time. The Cass feature is very similar to the cryptoexplosion structure at Glasford, Illinois (Buschbach and Ryan, 1963).

GEOLOGIC EVOLUTION

Basement Provinces

The Phanerozoic sediments reach as far west as Marquette in the Upper Peninsula of Michigan. West of there, a complex portion of the Precambrian Shield has been the subject of many investigations. Excellent summaries of this area are presented in two GSA memoirs (Wold and Hinze, 1982; Medaris, 1983).

The nature of the basement in the Lower Peninsula is much more obscure. With one minor exception, the basement tests are all on the periphery of the basin; however, some generalizations can be made. The trend of the Grenville metamorphic front (Fig. 5) is southwesterly from Ontario, skirting the east end of Manitoulin Island and entering Michigan at Saginaw Bay along the western edge of Huron County. From there it generally trends south and parallels part of the Lucas-Monroe Fault (Bowling Green Fault in Ohio). Lidiak and others (1966) reported gneiss and schist in the Precambrian basement in northern Ohio just east of the Bowling Green Fault, but found granite to the west in northern Indiana. This is in agreement with basement rock types (Fig. 5) found in southeastern Michigan, which consist mostly of granite gneiss and schist. Southwestern Michigan is underlain by granite with some areas of metamorphosed sediments. In northeastern Michigan, quartzite is the dominant rock type; however, one well report lists a chloritic greenstone below about 60 m of quartzite. There are three Precambrian tests in northwestern Michigan; two on Beaver Island (Fig. 5) encountered granite and granite wash; one in central Grand Traverse County bottomed in granite. All three of these wells lie along the western edge of the

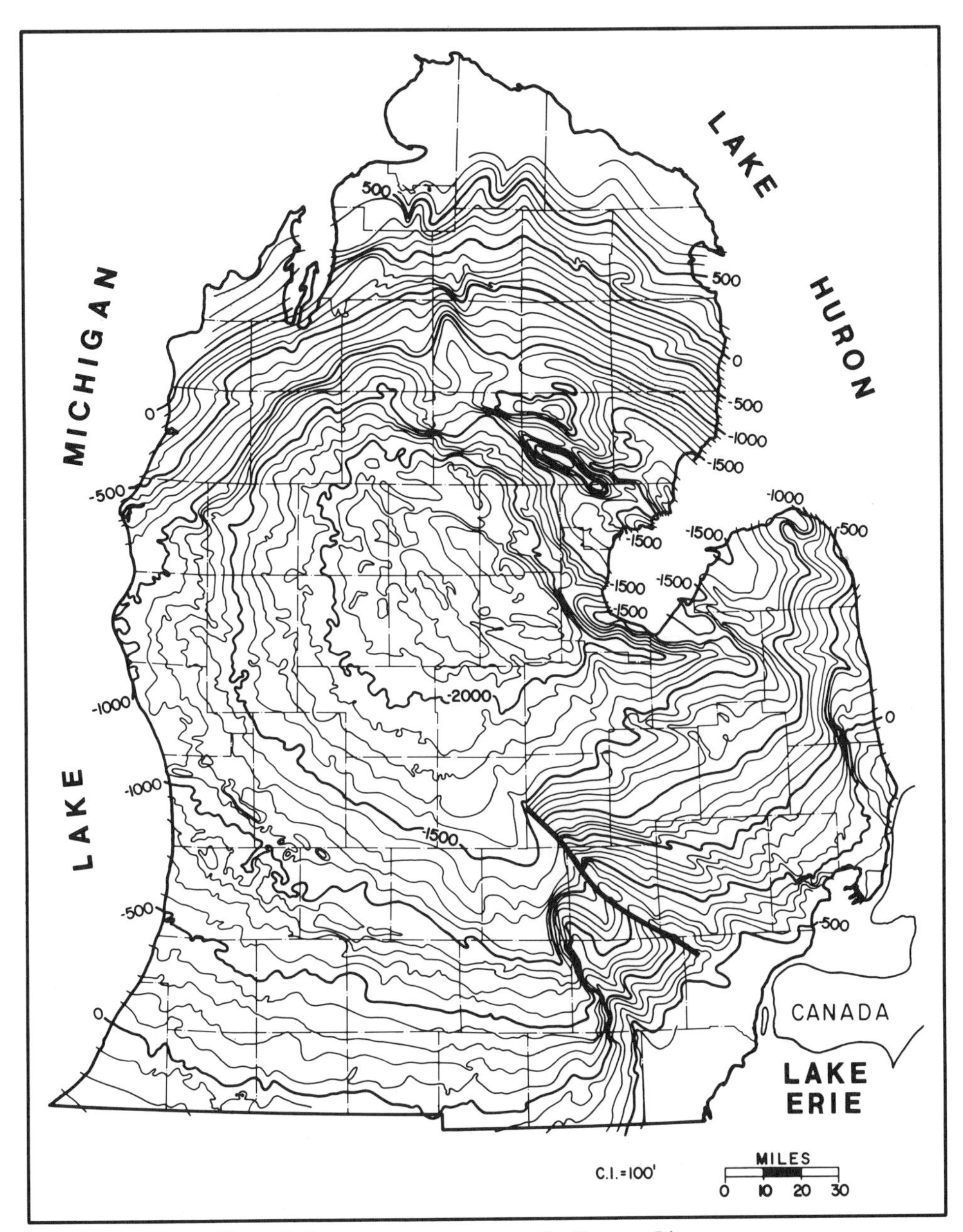

Figure 6. Structure contour map on the Traverse Limestone.

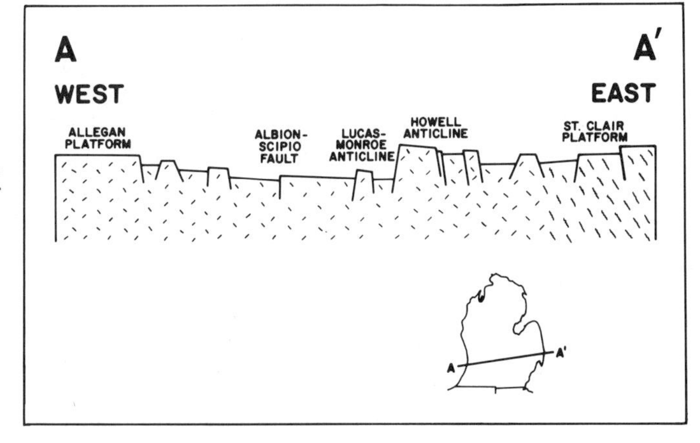

Figure 7. Diagrammatic cross-section of the Precambrian basement.

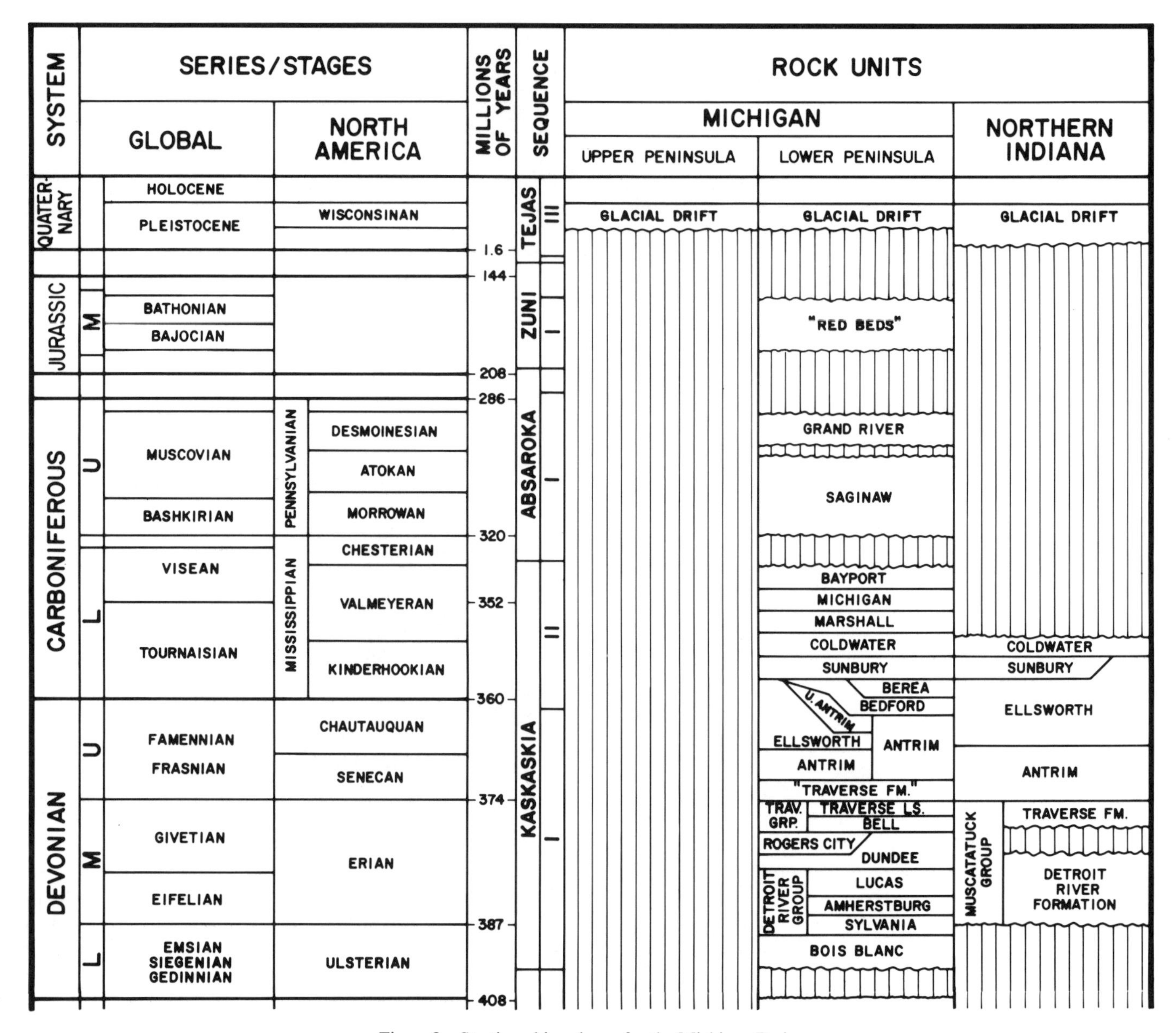

Figure 8a. Stratigraphic column for the Michigan Basin.

mid-Michigan gravity anomaly. A well in east-central Livingston County, the Messmore (Fig. 5), near the southern end of the anomaly, encountered quartzite in the Precambrian. The Sparks well (Fig. 5) is the only Precambrian test on the gravity anomaly feature and is the only well that has encountered Keweenawan sedimentary red beds.

Granite wash (detritus), where reported, is considered to be Precambrian in age. Across southern Michigan the range in thickness of granite wash is up to 10 m. In eastern Michigan there is a regular pattern of thinning in the granite wash, from 63 m in southeastern Huron County to zero in southern St. Clair County. In the northwest part of the basin, on Beaver Island (Fig. 5), in two wells 6.5 km apart, the westernmost well reported 6.4 m of wash but the eastern well was still in wash after drilling 247 m below the Mount Simon. The "granite wash" reported in the Beaver Island wells is sandy, arkosic, coarse material that appears to be the distal portion of a granite wash accumulation.

Stratigraphy

The stratigraphic nomenclature used in this chapter (Fig. 8) conforms, in general, to the standard subsurface terminology currently used in the Michigan Basin. The ages of the subsurface units have been related most closely to the northern Indiana

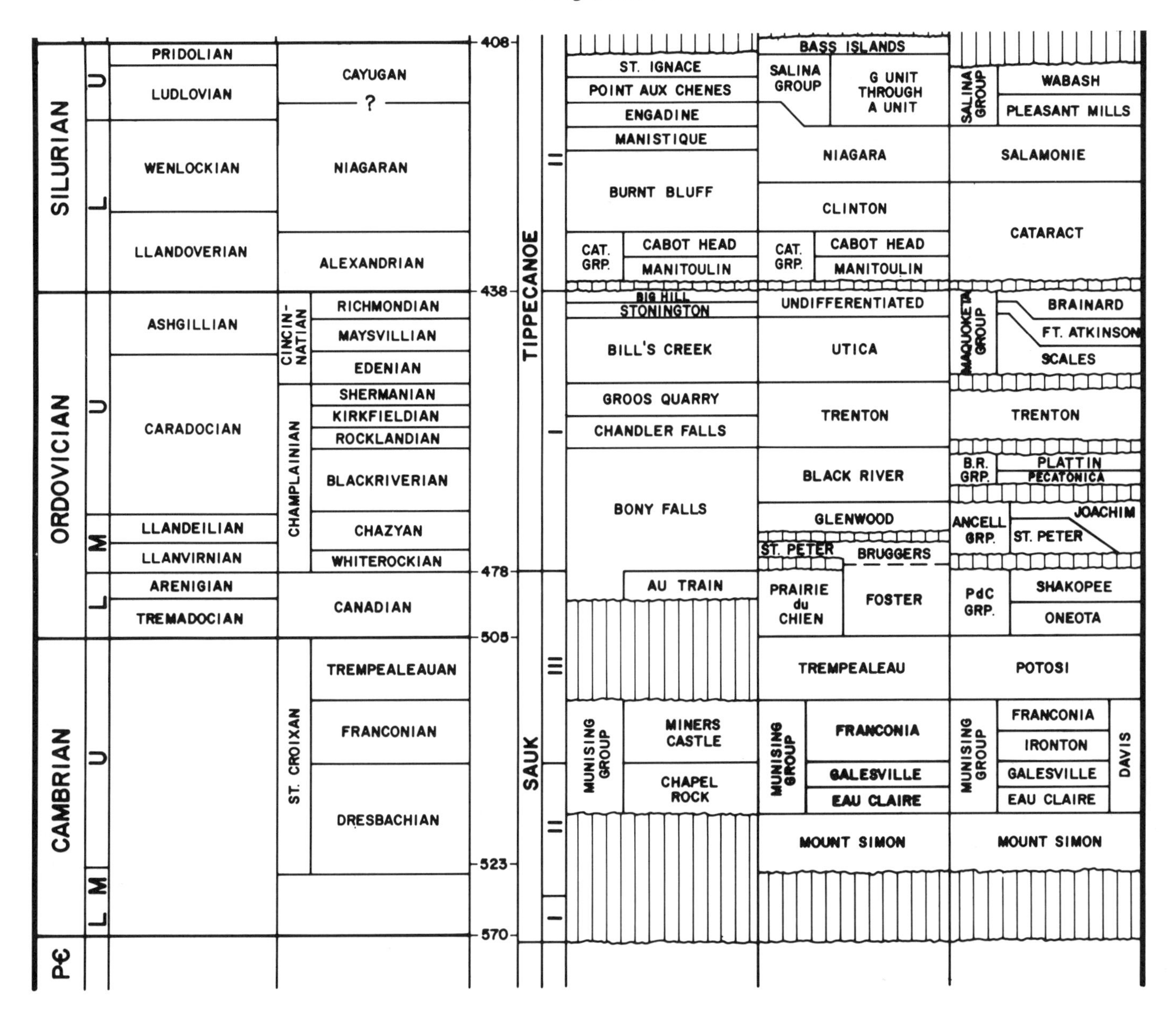

Figure 8b. Stratigraphic column for the Michigan Basin.

section, which are based on both macro- and micropaleontology. The stratigraphic sequences employed in the isopach maps are based on those developed by Sloss (1963). The general geology of the Bruce Peninsula and southwestern Ontario on the Canadian side of the Michigan Basin has been described in detail by Sanford and Brady (1955), Liberty and Bolton (1971), Beards (1967), and Brigham (1971).

The log characteristics of the various stratigraphic units are shown in Figure 9. The unlabeled lines across the vertical columns mark 500-ft (152 m) intervals. The tick marks indicate 100-ft (30.5-m) intervals. The location of the Martin well, which has the thickest Phanerozoic section drilled to date, is shown in Figure 5. The well began in glacial drift, which directly overlies the Saginaw (Pennsylvanian), and did not reach the base of the Mount Simon at the well total depth of 4,834 m (15,859 ft). The approximate depth of the Precambrian is estimated to be 4,998 m. A southwest–northeast cross-section (Plate 7) shows the distribution and thickness of units along a line passing through the Martin well.

Sauk II

The Sauk II sequence in Michigan extends from the base of the Mount Simon to the top of the Galesville (Fig. 10). The

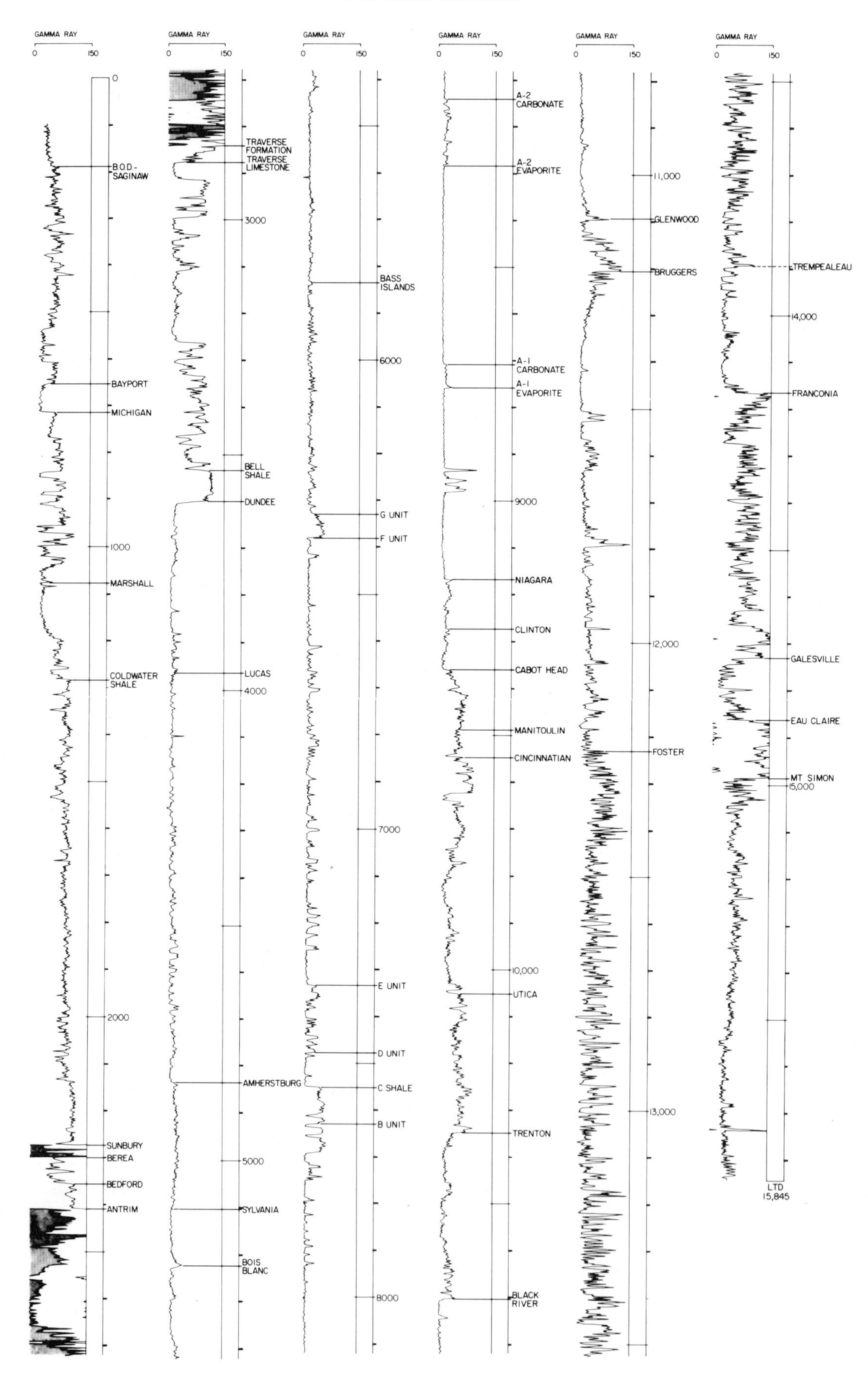
GAMMA RAY
0
150
B.O.D.-SAGINAW
BAYPORT
MICHIGAN
1000
MARSHALL
COLDWATER SHALE
2000
SUNBURY
BEREA
BEDFORD
ANTRIM
TRAVERSE FORMATION
TRAVERSE LIMESTONE
3000
BELL SHALE
DUNDEE
LUCAS
4000
AMHERSTBURG
5000
SYLVANIA
BOIS BLANC
BASS ISLANDS
6000
G UNIT
F UNIT
7000
E UNIT
D UNIT
C SHALE
B UNIT
8000
A-2 CARBONATE
A-2 EVAPORITE
A-1 CARBONATE
A-1 EVAPORITE
9000
NIAGARA
CLINTON
CABOT HEAD
MANITOULIN
CINCINNATIAN
10,000
UTICA
TRENTON
BLACK RIVER
11,000
GLENWOOD
BRUGGERS
12,000
FOSTER
13,000
TREMPEALEAU
14,000
FRANCONIA
GALESVILLE
EAU CLAIRE
MT SIMON
15,000
LTD 15,845

Mount Simon is a sandstone, shaly near the top and arkosic at the base. The Eau Claire Formation consists predominantly of shales and sandstones, which are often glauconitic. The Galesville is a sandstone, usually white, which grades eastward into a sandy dolomite. An embryonic Michigan Basin was present at this time, with the Sauk II depocenter corresponding to the Precambrian structural low. The greater portion of the interval consists of Mount Simon, which is estimated to be 460 m thick in Gladwin County. The major part of the thickening to the southwest in Michigan is also due to the increase in the Mount Simon, which reaches 460 m in that area and continues to thicken into an 850-m depocenter just southwest of Chicago in northeastern Illinois (Buschbach, 1964). The thickening in the Chicago area is unusual because this area does not serve as a depocenter for any of the succeeding formations. The situation of a double depocenter during Mount Simon time, one in northeastern Illinois and one in Michigan, one stabilizing while the other in Michigan continued to subside, presents a difficult problem in the geophysical modeling of the origin of the Michigan Basin. The overlying Eau Claire thickens into a depocenter in southern Illinois and Indiana (Becker and others, 1978).

Note that the reversal between the 1,000-ft contour lines (Fig. 10) parallels the southwest side of the mid-Michigan gravity anomaly. Whether this relationship has any geologic significance is unknown at the present time. The pattern of contour lines in southeastern Michigan indicates that the Lucas-Monroe and Howell Anticlines, and the Algonquin Arch were present. Sanford and Quillian (1959) described a progressive overlap of Cambrian units on the western end of the Algonquin Arch.

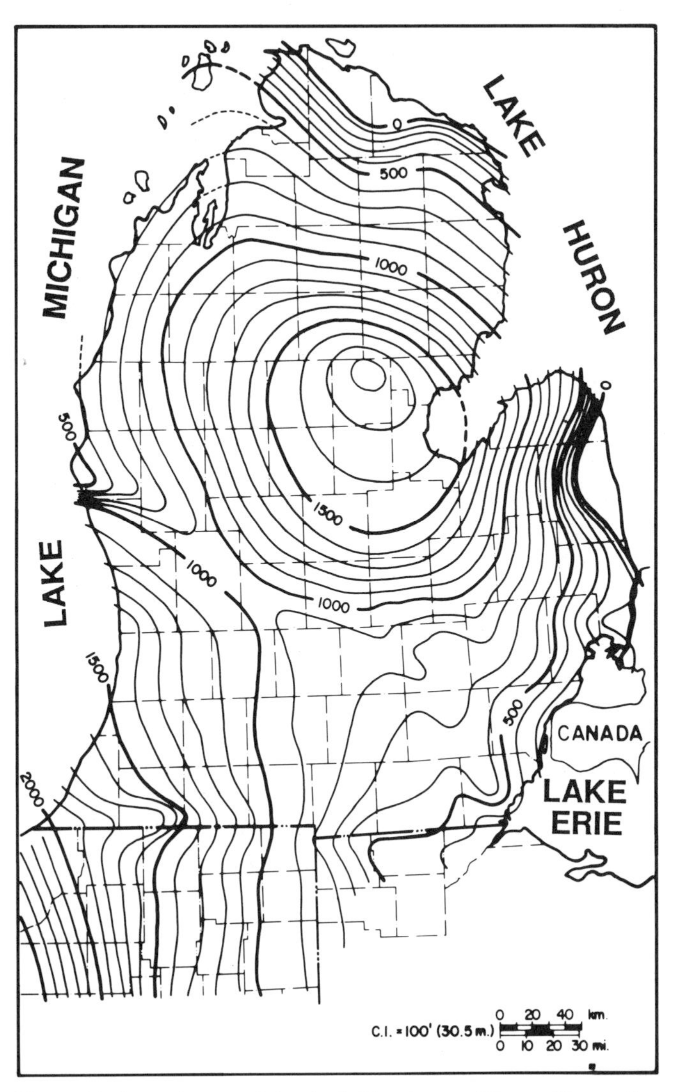

Figure 10. Isopach map of the Sauk II sequence (Mount Simon, Eau Claire, Galesville).

Sauk III

The Sauk III sequence for the midwestern United States (Fig. 11) extends from the top of the Galesville to the major unconformity at the base of the Whiterockian and Chazyan rocks. The lower unit of the Sauk III sequence is the Franconia Formation, which consists of shaly sandstones and dolomites, most of which are glauconitic. The overlying Trempealeau Formation is a rather uniform dolomite. The top of the Trempealeau in southwestern Michigan lies immediately below a shaly zone that is mostly red or pink. This top becomes increasingly difficult to determine basinward. It is questionable in the Sparks well (Fig. 5) and very questionable in the Martin well where it is shown by a dashed line on the gamma ray log (Fig. 9). The Prairie du Chien Group consists of a series of tan to white dolomites and some zones of oolitic chert. There is an unconformity at the top of the Prairie du Chien in southwestern Michigan in the shelf area; however, this unconformity becomes difficult to trace into the central basin where a thick facies of dark shaly, dolomitic siltstone is encountered. In Michigan, the upper boundary of the Sauk III sequence presents a problem. A good deal of confusion has existed concerning pre-Glenwood (Middle Ordovician) stratigraphy in Michigan since the drilling of the Brazos-State Foster well (sec. 28, T24N-R2E) in Ogemaw County in 1964. This well encountered 296 m of clean sandstone below the Glenwood, underlain by 460+ m of black to dark gray, shaly, dolomitic siltstone containing variable amounts of nodular anhydrite. This section was unlike any pre-Glenwood section known in Michigan at that time; the Michigan State Geological Survey, with limited data available, classified it as Cambrian. Current deep drilling has provided a more complete view of the pre-Glenwood units, and a recent conodont study by Repetski and Harris (1981) placed the great bulk of the shaly dolomitic siltstone in the Foster well in the Lower Ordovician. According to the conodont data, the base of

Figure 9. Geophysical log markers shown on the gamma-ray log curve of the Hunt-Martin No. 1 well.

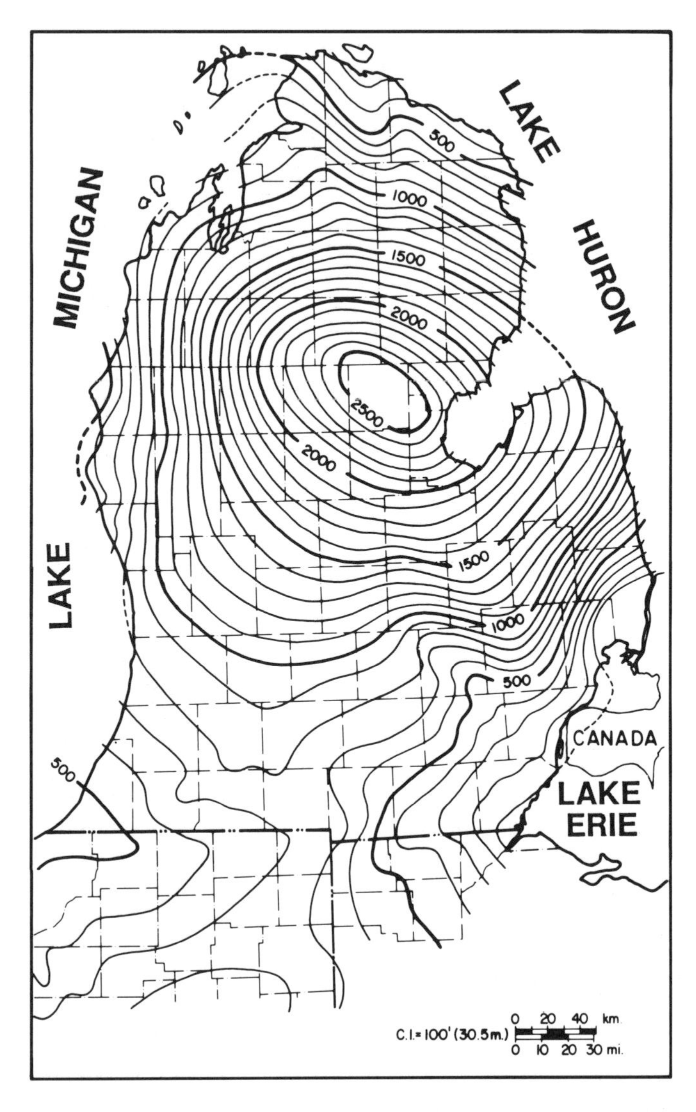

Figure 11. Isopach map of the Sauk III sequence (Franconia, Tempealeau, Prairie du Chien).

the Middle Ordovician would lie within the upper part of the dolomitic siltstone and somewhere in the interval 3,548 to 3,638 m. This entire section has been logged and cored and shows no readily apparent unconformity. The contact of the dolomitic siltstone with the overlying sandstone, while fairly sharp in the Foster well, is transitional in most wells. The top of the sandstone has an irregular surface, shows evidence of secondary porosity in the upper portion, and is overlain by a variety of rock types, all suggesting an unconformable surface. The unconformity at the top of the sandstone is either post-Whiterockian or post-Chazyan.

Fisher and Barratt (1985) suggested that the name Foster Formation be applied to the dark shaly dolomitic siltstone seen in the Brazos–State Foster well from 3,498 m (11,478 ft) to the total depth of 3,961 m (12,996 ft). The age assigned, based on the section examined by Repetski and Harris (1981), is Canadian and Whiterockian. No paleontological study was made of the upper 48 m (159 ft). We have also proposed the name Bruggers Sandstone for the overlying sandstone as seen in the Dart (JEM)–Bruggers 3 well (sec. 7, T24N-R6W) in Missaukee County from 3,094 to 3,451 m (10,152 to 11,322 ft), a total of 357 m (1,170 ft). The tentative age assigned is either Whiterockian or Chazyan (Fig. 8). These correlations and age assignments are contrary to the correlations of the Foster well by Bricker and others (1983). There is a possibility that while the rest of the Midwest was undergoing erosion at the end of Early Ordovician time, transitional sedimentation occurred in the Michigan Basin. Therefore, the placement of the upper boundary of the Sauk III sequence at the base of the Bruggers is arbitrary and a matter of expendience for mapping purposes.

The depocenter of the Sauk III sequence also corresponds closely to the Precambrian structural low. The Howell Anticline was present during Sauk III time. Both the Allegan and the St. Clair Platforms are evident at this time. The geographic extent and the magnitude of deposition during the Sauk III interval, or more specifically, during the deposition of the Foster, indicates the beginning of the Michigan Basin in its present form.

Tippecanoe I

The Tippecanoe I sequence (Fig. 12) extends from the base of the Bruggers to the top of the Upper Ordovician. The Bruggers is a white sandstone that is fairly uniform in the central basin but becomes interbedded with carbonate rocks and shales toward the southern shelf margins, for example, in the Sparks well (Fig. 5). The sandstone in the central basin is generally clear to white; in some places it has a greenish tinge due to included chlorite and glauconite. The sorting is bimodal with coarse, rounded, frosted grains interspersed with fine, angular to subangular grains. Pyrite and feldspar occur in some zones and make up 1 to 2% of the sample. The sandstone is highly bioturbated, suggesting a shallow-water environment. The Bruggers is the deepest productive zone in the Michigan Basin, producing gas at a depth of 3,430 m.

The St. Peter Sandstone has a patchy distribution in southwestern Michigan from Berrien to Hillsdale County. It is a sandstone composed of clear to white quartz grains that are generally rounded. The formation commonly grades upward from coarse to fine-grained sandstone. The St. Peter averages 6 m in thickness but its occurrence is extremely erratic. It may be 18 m thick in one well and absent in an adjacent well. In one well in southern Cass County (Mannes–Smith 1, sec. 20, T7S-R14W), the Prairie du Chien and most of the Trempealeau are absent and the St. Peter Sandstone is 164 m thick. Similar occurrences have been reported from northeastern Illinois (Buschbach, 1964; Willman and others, 1975) and from northwestern Indiana (Droste and others, 1982). These abrupt variations in thickness may be related to karst solution features developed on the underlying erosion surface of the Prairie du Chien carbonate rocks. The question of the lateral relationship of St. Peter and the Bruggers Sandstones is unresolved.

The overlying Glenwood, in southern Michigan is generally a green sandy shale, containing some interbeds of sandstone and limestone. This interval thickens toward the central basin and includes some dolomite beds. It is not certain that all of the units in this thickened zone belong in the Glenwood. Some workers who believe that the Bruggers is older than Chazyan consider that some of the lenticular sandstones above the Bruggers unconformity are part of the St. Peter. The problem is unresolved at present and the writers have included these miscellaneous rock types in the thickened basinal sequence in the Glenwood.

The Black River is a micritic limestone commonly containing nodules of brown chert. Mapping indicates that the Michigan area was an embayment open to the southeast during Black River deposition. The Trenton is primarily limestone and has a widespread dolomite cap. Isopach maps of the Trenton demonstrate that the basin had returned to its circular form by Trenton time. The Albion–Scipio Field, which has yielded 120 million barrels of oil and 210 billion cubic feet of gas, produces from a dolomitized fracture zone in Trenton and Black River Limestones.

The overlying Cincinnatian Series of rocks is usually undivided, with the exception of the Utica Shale, which is a black shale at the base of the series. Lilienthal (1978) prepared an extensive series of cross-sections of the basin along with lithologic descriptions of the formations. He subdivided the Cincinnatian above the Utica into five separate units, numbered one through five from the base upward. These units are alternations of limestone and shale and can be traced over most of the basin. The top of the Ordovician is marked by a major unconformity. Gamma ray log cross sections by Nurmi (1972) and Lilienthal (1978) graphically illustrate the loss of section at the top of the Cincinnatian from the center of the basin outward. Sanford (1961) has described the Ordovician strata in southwestern Ontario.

Tippecanoe I strata reach a maximum thickness over the Precambrian structural low. The Allegan Platform was well developed at this time. The thickening to the southeast is largely attributable to the Black River and the Utica.

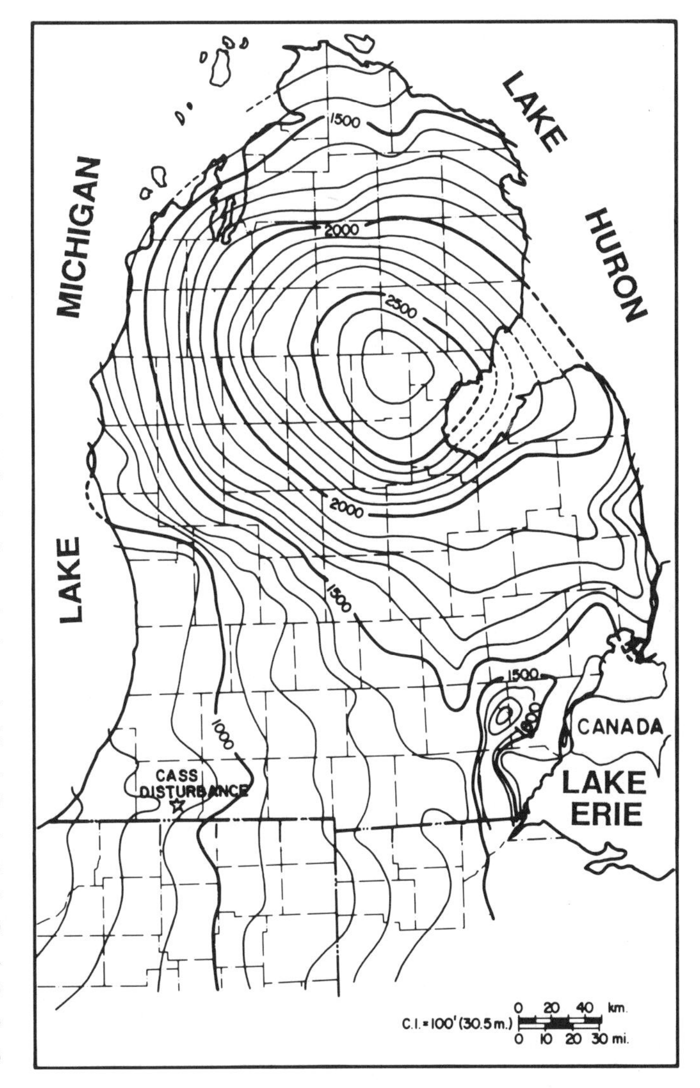

Figure 12. Isopach map of the Tippecanoe I sequence (Middle and Upper Ordovician).

Tippecanoe II

The Tippecanoe II sequence is equivalent to all of the Silurian (Fig. 13) and is bounded above and below by major unconformities. The isopach interval includes the Salina Group, approximately 1,000 m thick in the central basin, which tends to mask the pattern of the Niagara barrier reef, 180 m thick, which encircles the basin. The Niagara of the central basin is generally less than 30 m in thickness.

The Manitoulin at the base of the Tippecanoe II sequence is a gray dolomite of variable thickness. It is overlain by the Cabot Head green and red shales with some limestone stringers. There is a reciprocal relationship in thickness pattern between the Cabot Head and the Manitoulin. The Clinton in the southern half of the basin is very thin, 2–12 m and consists of a thin carbonate rock overlain by shale. In the northern part of the basin it is a cherty dolomite that attains a thickness of 120 m in Antrim County.

The Niagara barrier reef is primarily dolomite. The thin basinal facies is a dark gray carbonaceous limestone except in a few areas where pink crinoidal limestones occur. The point of steepening, or the foot of the barrier reef, shown by the dashed line in Figure 13, is approximately at the 200-ft isopach line (Fisher, J. H., 1973). Pinnacle reefs occur in a belt 16 to 24 km wide, basinward from this line. While the barrier reef, north and south, exhibits the same thickness, pinnacle reef height varies from 90 to 110 m in the southern area to 90 to 180 m in the northern trend (Mantek, 1973; Ells, 1967). The Silurian pinnacle reef trend currently produces the bulk of the oil and gas in Michigan.

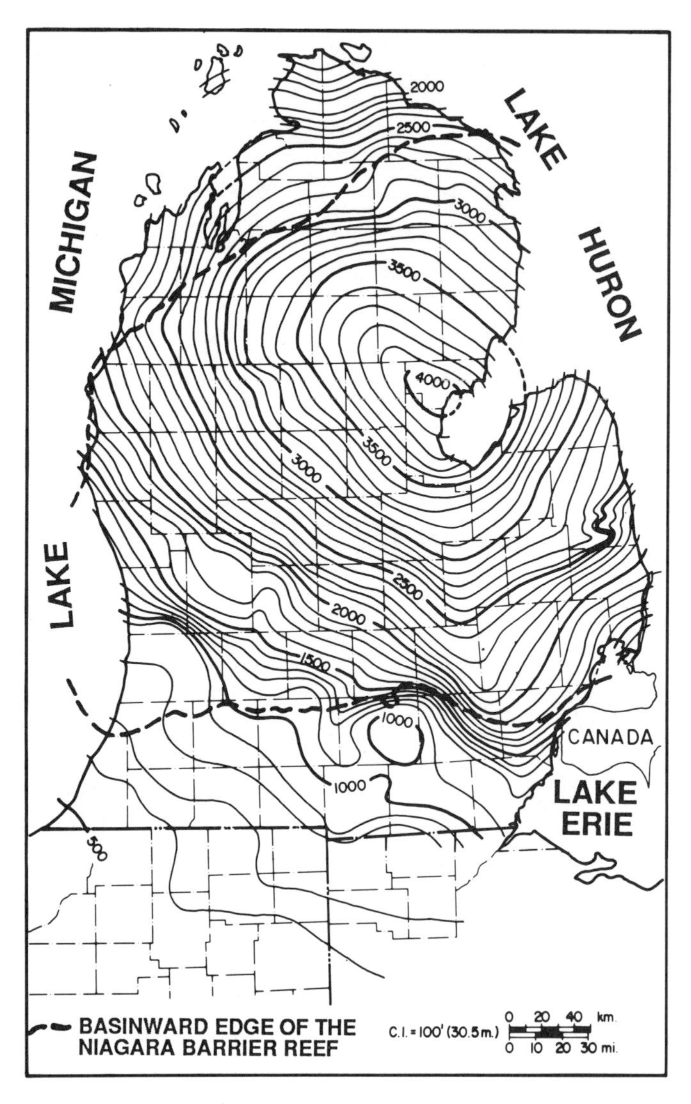

Figure 13. Isopach map of the Tippecanoe II sequence (Silurian).

Controversy exists concerning the time of formation of these pinnacle reefs. Gill (1975) contends that the pinnacle reefs are entirely Niagara in age, while Mesolella and others (1974) argue for two stages of growth, one in the Niagara and a later major resumption of growth during A-1 carbonate time. Shaver and others (1978) and Shaver and Sunderman (1983) believe that reef growth in the Fort Wayne Bank of northern Indiana was continuous through Niagara and Salina time, implying that there was no drawdown of sea level in the Michigan Basin during that interval.

Conodont zonations have been established in northern Indiana (Rexroad, 1980; Rexroad and Droste, 1982) for rock units stratigraphically equivalent to the Manitoulin, Cabot Head, Clinton, and Niagara Formations.

The Salina Group in Michigan was first subdivided by Landes (1945a). Present usage is slightly modified from that of Landes. The Salina Group from the base upward, consists of A-1 evaporite, A-1 carbonate, A-2 evaporite, A-2 carbonate, followed by units B through G. The A evaporite units are thick salts in the central basin that grade into thin anhydrites around the basin margin. The A-1 salt in the central basin contains numerous interbeds of sylvite (Elowski, 1980). The B, D, and F units are salts with minor amounts of shale, anhydrite, and dolomite. At the basin margins, there is a facies change in these units into a relatively thin assemblage of the minor components (Jannsens, 1977; Droste and Shaver, 1982). The C and G units are dominantly shales. The E unit is composed of shale and dolomite with some anhydrite stringers. The Bass Islands (formerly unit H in the Landes classification) has been removed from the Salina Group. The Bass Islands is mostly dolomite with minor shale, salt, and anhydrite stringers.

The widespread controversy over deep versus shallow-water evaporites extends to the evaporite deposits of the Michigan Basin. Detailed discussions can be found in Dellwig and Evans (1969), Schmalz (1969), Sloss (1969), Nurmi and Friedman (1977), Droste and Shaver (1977), and Matthews and Egleson (1974).

The depocenter for Tippecanoe II shifted slightly eastward from previous centers in Gladwin County. The Allegan Platform was well defined. There are several embayments of A-1 salt along the edge of the Allegan Platform (Vary and Elenbaas, 1968). Salt flowage in this area created doming where the salt thins markedly and changes to anhydrite. Selective leaching has also created salt islands, and differential compaction over these islands has created oil and gas productive domed structures. The A-2 salt edge lies slightly basinward of the A-1 salt edge in the Allegan area. Elsewhere in the southern basin, the A-2 salt edge is located well to the south of the A-1 salt edge. The St. Clair Platform is not defined in Figure 13; however, the A-1 evaporite transition from basin salt to anhydrite marks the northern edge of the platform. In the northern basin the isopach lines marking the A-1 and A-2 salt edges are intertwined along the face of the barrier reef, with the A-2 tending to be a little more northward. There is a major deviation of the A-2 in the Grand Traverse Bay area where a major salt embayment extends to the northern end of the bay.

After Niagara time the Michigan Basin subsided more rapidly in the northern reef area compared to the southern reef zone. An isopach map of the Salina Group shows 1,500 ft (457 m) of strata along the basinward crest of the barrier reef to the north in comparison to 500 ft (152 m) in a similar position along the southern reef. The A evaporites also reach a higher position on the face of the northern barrier reef than the same evaporites in the southern reef area.

Regional correlations of the Salina Group from New York to the Michigan Basin have been established by Rickard (1969).

Kaskaskia I

The Kaskaskia I sequence (Fig. 14) is equivalent to the Devonian System of Michigan. The basal unit of Kaskaskia I is the Bois Blanc Formation, a cherty dolomite ranging from 0 to 244 m in thickness (Lilienthal, 1978). According to Gardner (1974), the Bois Blanc has a facies relationship with the lower part of the overlying Sylvania Sandstone. The term Sylvania is sometimes applied to rather widespread sandy dolomites in southwestern and northwestern Michigan. The writers prefer to restrict the term Sylvania to the sandstone unit in central and southeastern Michigan as described by Landes (1951). The Sylvania is a major source of brine for the chemical industry in Midland County. The Sylvania is the basal formation of the Detroit River Group in Michigan and Ohio but is not separately recognized in Indiana. It is overlain by the black limestone and dark brown dolomite of the Amherstburg Formation. The Lucas Formation consists of salt, dolomite, anhydrite, and sandstone; it is 305 m thick, of which 122 m is salt, Landes (1951). The Lucas is an important producer of oil and gas. The top of the Detroit River Group is difficult to define on gamma-ray logs; it is usually picked arbitrarily at the first appearance of bedded anhydrite below the Dundee Limestone. The stratigraphy of the Detroit River rocks in northern Indiana has been described by Doheny and others (1975).

The Dundee is predominantly a buff to brown crystalline limestone in the central basin and grades westward into coarse bioclastic shelf limestone. The unit is mostly dolomite in the western and southwestern areas and contains some anhydrite. The term Rogers City Limestone is applied in some areas to a darker phase in the upper Dundee but this zone is difficult to correlate over any appreciable distance. The Dundee is one of the most prolific oil formations in the Michigan Basin.

The Bell Shale, consisting of gray fossiliferous shale, is the basal unit of the Traverse Group. It is absent in southwestern Michigan (Gardner, 1974), and the Dundee directly overlies Detroit River carbonates. In northern Indiana, it is difficult to separate these units, and there is probably some Dundee equivalent in the upper part of what is identified as Detroit River.

The Traverse Limestone in western Michigan is a carbonate shelf facies—mostly gray limestone but with some secondary dolomite near the top. In a small area on the extreme west side of the state, anhydrite occurs about 60 m below the top of the formation, indicating a back reef restricted environment. Many small bioherms, up to 9 m in height, occur in western and northern Michigan. The Traverse Limestone to the east and southeast has a progressive increase in shale content. The change to shale occurs from the base upward and from the top of the formation downward until in southeastern Michigan it becomes mostly shale containing a few limestone stringers. In Indiana, however, the term Traverse Formation is assigned to this same unit with some modification in its stratigraphic content.

The upper formation of the Traverse Group is the so-called "Traverse formation," or as it is also sometimes identified, the Traverse shale. It is a gray shale with interbedded limestones, which are more abundant near the base. The upper gray shales are commonly interbedded with the overlying black Antrim shales, which creates an artificial placement of the top of the "Traverse formation" below the last off-scale curve on the gamma ray log (Ells, 1978). The "Traverse formation" is Late Devonian in age (Ehlers and Kesling, 1970), in contrast to the remainder of the Traverse Group which is Middle Devonian. Where the Traverse Limestone becomes increasingly shaly toward the east, it becomes increasingly difficult to determine the base of the "Traverse formation" and the two units are then combined under the term Traverse. Considering the age of the "Traverse formation" and the degree of interbedding with the overlying Antrim, it seems reasonable to consider this gray shale as a basal facies of the Antrim and to drop the name "Traverse formation," which would agree with the practice already in use in Indiana. There, the "Traverse formation" of Michigan is simply the lower part of the Antrim Shale.

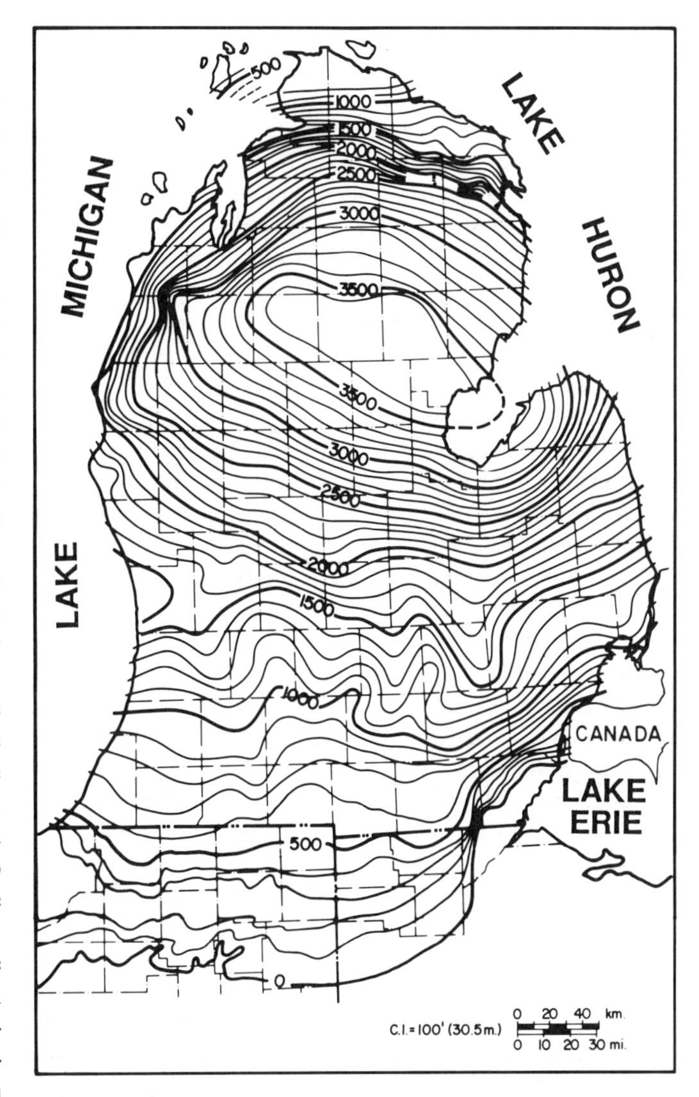

Figure 14. Isopach map of the Kaskaskia I sequence (Devonian).

Janssens (1970) and Sparling (1983) have described the stratigraphy of the Middle Devonian of northern Ohio. Orr

(1971) has identified conodonts from the Middle Devonian of the Michigan Basin. Rickard (1984) established correlations of the Lower and Middle Devonian of the Lake Erie region.

The Antrim is part of a widespread body of black shale in the central United States. It is a black, fissile, pyritic, organic-rich shale containing an abundance of spores and a number of argillaceous limestone stringers. In southeastern Michigan several tongues of gray shale are present in the formation. In western Michigan the black Amtrim grades upward into a brown-black facies, which Ells (1978) placed in the overlying gray-green Ellsworth Shale. The changes in color are due to decreasing organic content and greater oxidation of the shale. The change can be identified on gamma-ray logs by the extreme off-scale pattern of black Antrim compared to the slightly off-scale pattern of brown-black shale, and to the on-scale pattern of the gray-green Ellsworth. The Ellsworth Shale, in western Michigan, extends upward to the base of the Coldwater. Eastward, black shale appears at the top of the Ellsworth and increases in thickness to an area in the central basin where some workers arbitrarily switch terms and identify the entire section as the Antrim. An east–west cross-section through the central basin (Fisher, J. H., 1980) supports the interpretation that the Ellsworth is a transgressive-regressive unit originating in the Wisconsin area. The Ellsworth may be correlative with the green shales of the Maple Mill Formation in central Iowa and the Saverton Shale of northeastern Missouri. As the influx of Ellsworth sediments waned, the black shale environment encroached westward, partially covering the Ellsworth (Fig. 8). The New Albany Shale of Indiana is equivalent to the Ellsworth and Antrim Shales of Michigan (Lineback, 1970).

The Bedford is a gray prodelta shale unit. It represents a deposit in extension of a west-trending distributary of a large delta system created by a river flowing south from Canada with the major delta distributaries in Ohio and Pennsylvania (Pepper and others, 1954). In Ohio the Devonian-Mississippian boundary occurs within the Bedford (DeWitt, 1970). The boundary is uncertain in Michigan and is placed at the Bedford-Berea contact. In western Michigan the boundary is in the upper part of the Ellsworth.

The isopach map of Kaskaskia I (Fig. 14) shows a large elongate depocenter that includes the centers of sedimentation of earlier sequences. Shelf conditions tended to prevail in southwestern Michigan.

Kaskaskia II

Erosion at the end of Mississippian time thinned the Kaskaskia II sequence over the structural highs in the Michigan Basin that were strongly uplifted at this time. Only a few of these thins have been shown (Fig. 15) by small isolated closures over some of the major anticlines. The Howell and Lucas-Monroe Anticlines were positive features at this time as were the Allegan and St. Clair Platforms. The depocenter was once again fairly broad but was tending toward a north–south orientation.

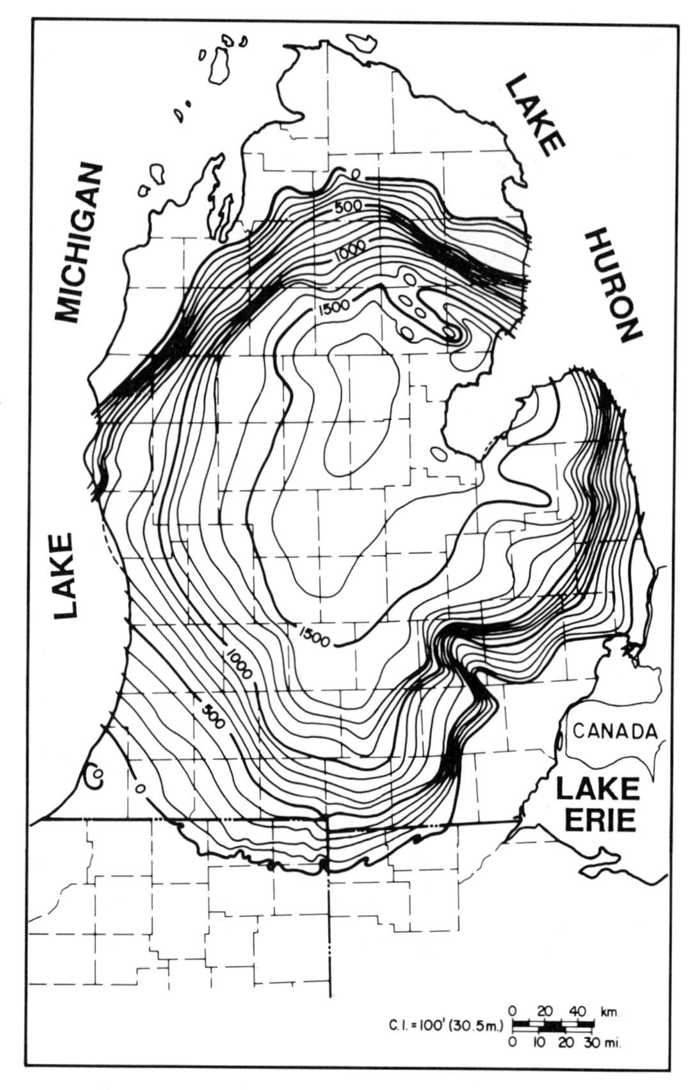

Figure 15. Isopach map of the Kaskaskia II sequence (Mississippian).

The basal unit, the Berea Sandstone, is a light gray, fine-grained sandstone and siltstone interbedded with gray shales. The coarsest grain sizes in the Berea, and also the thickest sections, are found in Sanilac and Lapeer Counties. The Berea is a deltaic sand deposit on the distal portion of a delta plain and is restricted to the eastern part of the Michigan Basin.

The black Sunbury shale is 43 m thick in Huron and Sanilac counties in eastern Michigan and thins to zero along the western edge of Michigan. It is similar in appearance to the black shale of the Antrim. At the base of the overlying Coldwater Shale there is a distinctive red shale (3 to 6 m) thick, containing thin layers of reddish limestone or dolomite. It is a prominent marker bed over most of Michigan, although it is best developed in the western half of the state. The total Coldwater is about 305 m thick in the central basin. It is mainly a gray shale with some maroon and

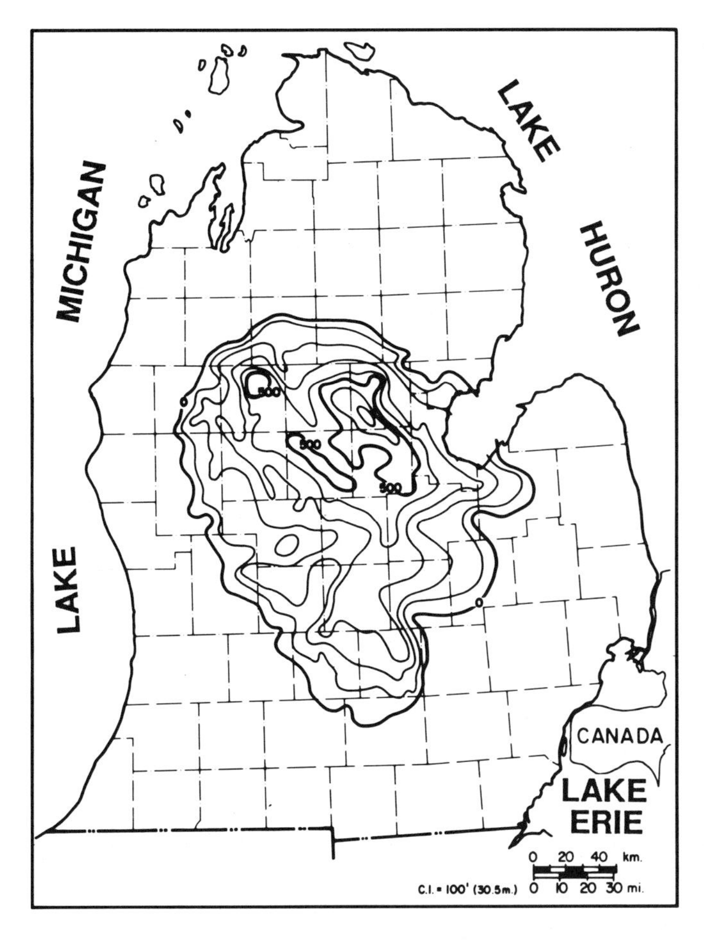

Figure 16. Isopach map of the Absaroka I sequence (Pennsylvanian).

green shales in the upper part. Several thin limestones occur in the Coldwater in the western part of the basin, while locally in the eastern basin, siltstones are present in the lower and upper portions of the unit. There is a transition zone from the upper Coldwater into the Marshall. The dominant lithology of the Marshall is sandstone, mostly gray to brown, although some red sandstones occur. Minor amounts of shale are present, especially in the lower part. In most places the formation is capped by a thin, sandy dolomite.

The Michigan Formation is composed mainly of gray shales with sandstones in the lower portion that are similar to those of the Marshall. A thin but widespread brown limestone is present near the middle of the unit. A group of three anhydrite beds separated by shales occurs in the upper part of the Michigan Formation. The uppermost formation of the Kaskaskia II sequence is the Bayport Limestone, a gray to tan, cherty, lithographic limestone. The unit varies in thickness due to post-Mississippian erosion and is absent from some areas of the central basin.

Absaroka I

The Pennsylvanian System of Michigan (Fig. 16) reaches a maximum thickness of 230 m. It is divided into two units. The lowermost, the Saginaw Formation, consists of a basal sandstone generally less than 30 m thick overlain by both marine and nonmarine units, mainly interbedded lenticular shales, thin coals, and thin limestones (Wanless and Shideler, 1975). The workable coals average 0.6 to 1.2 m thick, but have little horizontal extent (Cohee, 1950). There are a number of unconformities in this sequence, indicating rapid fluctuations in sedimentation. The coals are low-grade bituminous. About 46 million tons of coal were produced between 1835 and 1950, but currently there is no commercial production of coal in Michigan. The overlying Grand River Formation is essentially a sandstone. Ells (1979) and Vugrinovich (1984) have described the Pennsylvanian stratigraphy of the Michigan Basin.

Pennsylvanian well information is poor. The interval is usually cased off in drilling, and samples from this zone frequently are not collected. Many of the wire-line logs stop short of the Pennsylvanian section. As a result, few attempts have been made to establish correlations based on radioactivity logs.

Zuni I

The so-called "red beds" (Fig. 17) of Middle Jurassic age consist of red sandstones and shales and some intercalated gypsum. Patterns in more detailed isopach maps (not shown here) suggest deposition in stream-eroded valleys cut into the subjacent Pennsylvanian rocks. A Kimmeridgian age was determined by a study of the spore and pollen flora (Shaffer, 1969). Recent reevaluation of this flora by Aureal Cross (personal communication, 1984) indicates a Bajocian-Bathonian age.

Lower Paleozoic of the Upper Peninsula of Michigan

The Northern Michigan Highland, an east–west-trending feature centered in the Upper Peninsula, was present during Late Cambrian time. From the Straits of Mackinac just north of Emmet County, Cambrian rocks thin from 122 m to less than 30 m along the crest of this positive area. Northward from the axis of the Highland, Cambrian rocks thicken to about 75 m along the Lake Superior shoreline. The northern section is entirely sandstone (Hamblin, 1958), while the southern section more nearly resembles the sandstone-shale-carbonate Cambrian section of the northern Michigan Basin.

The Northern Michigan Highland was apparently buried by Cambrian sediments since the contours of isopach maps of the Ordovician show no deviation in that area from the regular pattern of northward thinning. The maximum thickness of the Ordovician at the Straits of Mackinac is about 215 m. Both the Foster and the Bruggers, as well as the other standard Ordovician basinal units with the exception of the St. Peter, are present in the immediate area of the Straits. Northward the unit distinctions are not as clear and other names are substituted (Fig. 8b).

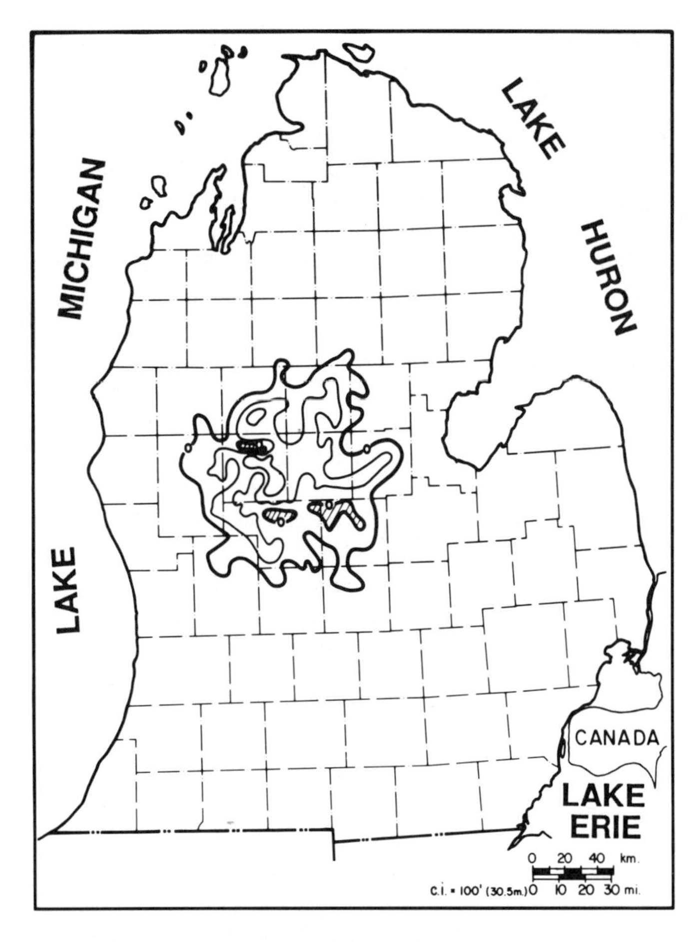

Figure 17. Isopach map of the Zuni I sequence (Jurassic).

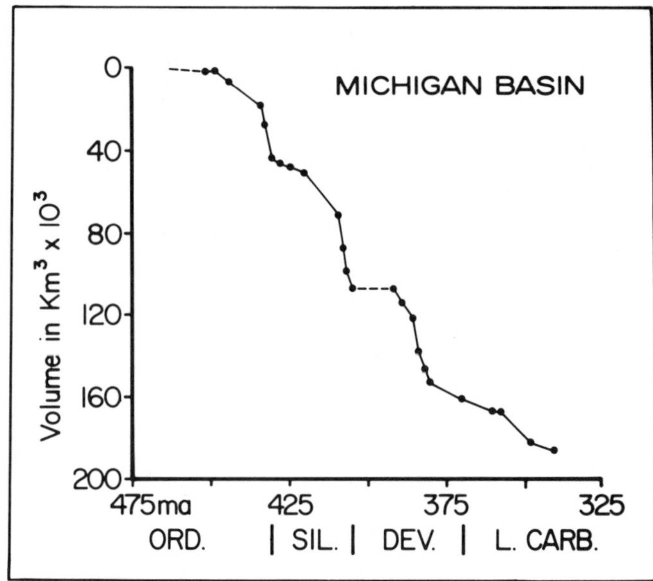

Figure 18. Middle Paleozoic subsidence curve of the Michigan Basin (Sloss, 1982).

The Silurian is present only along the southern margin of the Upper Peninsula (Landes and others, 1945b; Ehlers, 1973). It consists of about 75 m of Niagaran dolomite (Fig. 8b) overlain by 150 to 185 m of the Point aux Chenes Formation, a series of interbedded salts and dolomites which becomes shaly in the upper portion. The upper unit, the Saint Ignace Formation, consists of about 120 m of dolomite, some of it sandy.

The Mackinac Breccia in the Straits area is a collapse breccia (Landes and others, 1945b) created by the solution of salt beds in the Point aux Chenes Formation. Limestone blocks as high stratigraphically as the Detroit River unit are present.

TECTONIC-STRATIGRAPHIC SUMMARY

Depositional Patterns

The North American sequences (Sloss, 1963), while generalizing several diverse stratigraphic events, do show where the bulk of the sediments are concentrated for given sequences (Fig. 18). Certain thick formations or groups establish the general depocenter of a sequence, while other formations in the same sequence may have widely differing centers of sedimentation. The depocenters in Sauk II were largely controlled by the Mount Simon Sandstone forming the Gladwin County center and the ephemeral center in northeastern Illinois. The overlying Eau Claire Formation, thickening southwestward into southern Illinois does not fit either of these patterns. The Tippecanoe I depocenter is dominated by the Bruggers Sandstone, but isopach maps of Black River or of Utica rocks show that these formations thicken to the southeast toward Ohio. The depocenter for Tippecanoe II was controlled by the thick Salina Group, but the thickest part of the Clinton is to the northwest in the Grand Traverse Bay area. These deviations from sequence depocenters are relatively small, but they are significant. They demonstrate clearly that, although a major depocenter persisted in Gladwin County through Kaskaskia I time, many widely scattered subsidence centers did at certain times supplant the main depocenter. The later centers for Kaskaskia II, Absaroka I, and Zuni I deposition are broader and shifted to the southwest. Clearly the Michigan Basin did not subside about a given point throughout time, nor was the subsidence uniform, considering the major unconformities present in the sedimentary sequence.

Sauk II deposition was dominated by sandstones, but this changed during Sauk III time to a carbonate and shale sequence. With the exception of the incursion of the Bruggers sand, this carbonate and shale environment persisted until the end of Niagara deposition, when large amounts of evaporites were added to the carbonate-shale units. After the Traverse Limestone was deposited, the sedimentational pattern of the Michigan Basin changed to shale with increasing amounts of sandstone occurring by Middle Mississippian time, probably in response to the rising Appalachian Mountains to the east. A major erosional truncation of the sediments occurred at the beginning of the Pennsylvanian Period and was followed by the deposition of a typical Carboniferous coal sequence.

Studies of vitrinite reflectance of outcrop samples of Pennsylvanian coals at Michigan State University (Moyer, 1982) indicate that these coal beds were at one time overlain by a minimum of 900 m of younger rocks. Other studies of the thermal alteration index (TAI) of organic material in the basin by Cercone (1984) indicate approximately the same added thickness. The missing section is probably Pennsylvanian in age. Geothermal gradients in the basin during the Paleozoic Era were higher than present gradients (Cercone, 1984; Nunn and others, 1984), which explains the occurrence of oil at shallow depths (183 m) in some Michigan fields.

Deformational History

It is reasonable to assume that the pattern of highly fractured Precambrian rocks observed north of Lake Superior extends southward beneath the Michigan Basin. The profusion of faults seen in proprietary seismic surveys in the basin supports this concept. Detailed studies of the major faults in southeastern Michigan indicate that repeated movements occurred along these faults throughout the early part of the Paleozoic and that reversals took place commonly along some of these faults (Mescher, 1980; Fisher, J. A., 1981). Major uplift of most of the basin fault blocks occurred at the end of the Mississippian Period. Most of the faults die out upward and are not evident above the Dundee. A few of the larger faults penetrate upward into the Mississippian rocks.

The majority of the folds in Michigan are located near the center of the basin. This trend is northwest. If these folds were compressional in origin, this would require a deforming stress oriented northeast–southwest. Theoretically, this stress acting on a large basin should produce folds along the basin margins with appropriate patterns of asymmetry. No major compressional force from these directions is known during the history of the basin; the major folds are in the basin center, while the basin margins are relatively undeformed; no consistent pattern of asymmetry is present in the folded areas. Therefore, some origin for these folds other than compressional folding must be sought. Anticlines in the basin do become tighter at depth, and to some this may suggest compaction over old Precambrian highs. The overall thickness of Phanerozoic sediments in the central basin area, (5,000 to 5,200 m), makes it unlikely that compaction is a major factor in creating the folds seen at shallow depths in the central basin. Folds in the Michigan Basin, therefore, may be controlled by vertical tectonics. The horst and graben blocks of the Precambrian basement, moving along high-angle normal faults and an occasional high-angle reverse fault, moved up or down in response to regional stresses affecting large portions of the North American continent. Ells (1969), in the cross-section of what he has named the Washtenaw Anticlinorium, indicates a fault-block origin for the Howell and Lucas-Monroe anticlines.

The persistance of the St. Clair and Allegan platforms demonstrates that portions of the basin margin at times resisted subsidence, and the uniform thickening of sediments from basin margin to basin center that is so often depicted for the Michigan Basin is not always present. Note that the very thick Foster Formation and Bruggers Sandstone formations lie entirely basinward from the margins of these platforms. The formation of pinnacle reefs suggests that the pinnacles were created in a down-faulted area along the front of the Niagara barrier reef.

The ultimate cause of the sinking of the Michigan Basin is uncertain. Haxby and others (1976) suggested that mantle diapirs penetrated the lower crust and converted the gabbro to eclogite. Cooling of the dense eclogite then brought about the sinking of the basin. Sloss (1982) preferred "variation in the rate of flow of melt from continental to ocean asthenosphere" as a mechanism for basin formation. Sleep and others (1980a) suggested a thermal contraction theory for the origin of platform basins such as the Michigan Basin. All of the mechanisms encounter difficulties in accounting for the interruptions of sedimentation during the development of major unconformities and for the random distribution of the depocenters of certain formations (i.e., Eau Claire, Black River, Clinton), and particularly the double depocenter of the Mount Simon Sandstone (northeastern Illinois and central Michigan).

Comparison with Other Basins

The Michigan, Williston, and Illinois Basins are large, ovate basins with depocenters rather than depositional axes. They are largely carbonate basins with a marked absence of thick arkoses. They are located in an arcuate pattern adjacent to the Canadian Shield, and the positive areas surrounding these basins generally are of low relief. In contrast to this, the Anadarko and Denver Basins, along the margin of the craton, are smaller elongate basins that are highly asymmetrical, with the depositional axes along the seaward side of the basin. They contain abundant clastics, including arkose, derived from uplifted fault blocks along their seaward margins. Much of the subsidence in the Anadarko and Denver Basins occurred during Atokan and Desmoinesian time after the major subsidence of the Michigan Basin had been completed. The tectonic basins along the plate margins are different in origin than the sag basins represented by the Michigan, Illinois, and Williston Basins. It is suggested that the sinking of the Michigan Basin is associated with some event, perhaps a chain of simultaneous hot spots in the mantle occurring along an arcuate line including the Illinois and Williston Basins. Whatever the nature of this event, it must include some mechanism to account for shifts in depocenters within a given basin and the interruptions of sedimentation indicated by unconformities. The double Mount Simon basin in the Michigan-Illinois area illustrates one aspect of the problem. Sloss (1972) noted that the Moscow Basin has patterns similar to intracratonic basins such as the Michigan Basin. Studies of basin patterns of other continents (Kingston, 1983) may aid in solving the problem of why intracratonic basins subsided.

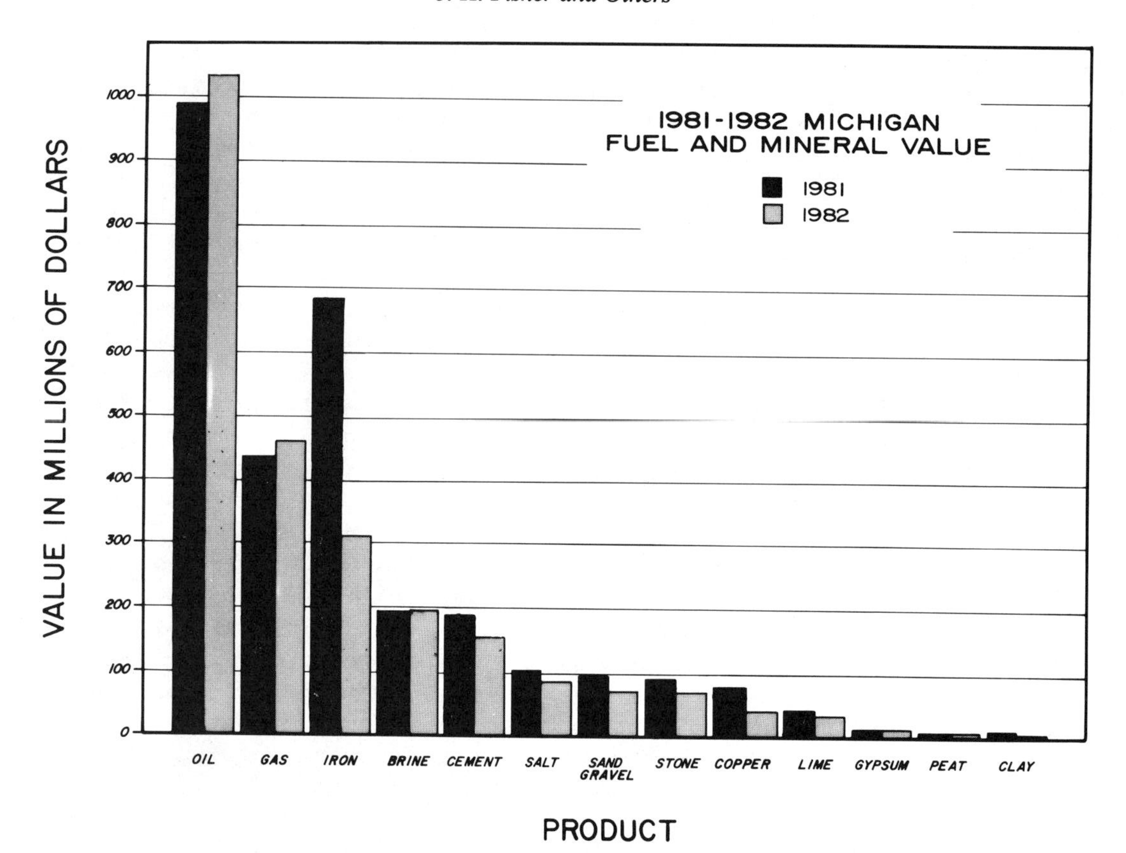

Figure 19. Graph of the dollar value of Michigan's fuel and mineral resources.

ENERGY, MINERAL, AND WATER RESOURCES

The Michigan Basin and the adjoining Upper Peninsula of Michigan contain a variety of energy and mineral resources. Situated in the center of the Great Lakes area, Michigan has an abundant supply of fresh water, an important commodity at the present time and an invaluable resource for the future.

Oil and gas production revenues now account for the two highest resource values. Iron ore, brine, cement, salt, sand and gravel, stone, copper, lime, gypsum, peat, and clay account for the remainder of the mineral resources (Fig. 19). Detailed information is available in the Atlas of Michigan (Sommers, 1977).

REFERENCES CITED

Barratt, M. W., 1981, Recent exploration activity in Michigan: Ontario Petroleum Institute, Twentieth Annual Conference, Technical Papers, no. 6, p. 1–31.

Beards, R. J., 1967, Guide to the subsurface Paleozoic stratigraphy of southern Ontario: Ontario Department of Energy and Resource Management, Paper 67-2, 19 p.

Becker, L. E., Hreha, A. J., and Dawson, T. A., 1978, Pre-Knox (Cambrian) stratigraphy in Indiana: Indiana Geological Survey Bulletin 57, 72 p.

Bricker, D. M., Milstein, R. L., and Reszka, C. R., Jr., 1983, Selected studies of Cambro-Ordovician sediments within the Michigan Basin: Michigan Geological Survey Report of Investigation, no. 26, 54 p.

Brigham, R. J., 1971, Structural geology of southwestern Ontario and southeastern Michigan: Ontario Department of Mines and Northern Affairs, Paper 71-2, 110 p.

Brown, L., Jensen, L., Oliver, J., Kaufman, S., and Steiner, D., 1982, Rift structure beneath the Michigan Basin from COCORP profiling: Geology, v. 10, p. 645–649.

Burke, K., and Dewey, J. F., 1973, Plume-generated triple junctions; Key indicators in applying plate tectonics to old rocks: Journal of Geology, v. 81, no. 4, p. 406–433.

Buschbach, T. C., 1964, Cambrian and Ordovician strata of northeastern Illinois: Illinois Geological Survey Report of Investigations 218, 90 p.

Buschbach, T. C., and Ryan, R., 1963, Ordovician explosion structure at Glasford, Illinois: American Association of Petroleum Geologists Bulletin, v. 47, no. 12, p. 2015–2022.

Catacosinos, P., 1973, Cambrian lithostratigraphy of Michigan Basin: American Association of Petroleum Geologists Bulletin, v. 57, no. 12, p. 2404–2418.

Cercone, K. R., 1984, Thermal history of the Michigan Basin: American Association of Petroleum Geologists Bulletin, v. 68, no. 2, p. 130–136.

Chase, C. G., and Gilmer, T. H., 1973, Precambrian plate tectonics; The midcontinent gravity high: Earth and Planetary Science Letters, v. 21, p. 70–78.

Cohee, G. V., 1947, Lithology and thickness of the Traverse Group in the Michigan Basin: U.S. Geological Survey Oil and Gas Investigations, Preliminary Chart no. 28.

Cohee, G. V., Burns, R. N., Brown, A., Brant, R. A., and Wright, D., 1950, Coal resources in Michigan: U.S. Geological Survey, Circular 77, 56 p.

Cohee, G. V., Macha, C., and Holk, M., 1951, Thickness and lithology of Upper Devonian and Carboniferous rocks in Michigan: U.S. Geological Survey Oil and Gas Investigations, Chart OC 41 (in five sheets), Sheet 4.

Dellwig, L. F., and Evans, R., 1969, Depositional processes in Salina salt of Michigan, Ohio, and New York: American Association of Petroleum Geologists Bulletin, v. 53, no. 4, p. 949–956.

Department of Geology, Western Michigan University, 1981, Hydrogeology for underground injection control in Michigan: Kalamazoo, Michigan, Department of Geology, Western Michigan University, Part I.

DeWitt, W., Jr., 1970, Age of Bedford Shale, Berea Sandstone, and Sunbury Shale in the Appalachian and Michigan Basins, Pennsylvania, Ohio, and Michigan: U.S. Geological Survey Bulletin 1294-G, p. G1–G11.

Doheny, E. J., Droste, J. B., and Shaver, R. H., 1975, Stratigraphy of the Detroit River Formation (Middle Devonian) of northern Indiana: Indiana Geological Survey Bulletin 53, 86 p.

Dorr, J. A., and Eschman, D. F., 1970, Geology of Michigan: Ann Arbor, Michigan, The University of Michigan Press, 476 p.

Droste, J. B., and Shaver, R. H., 1977, Synchronization of deposition, Silurian reef-bearing rocks on Wabash Platform of cyclic evaporites of Michigan Basin: American Association of Petroleum Geologists Studies in Geology, no. 5, p. 93–109.

—— , 1982, The Salina Group (Middle and Upper Silurian, of Indiana: Indiana Geological Survey Special Report 24, 41 p.

—— , 1983, Atlas of Early and Middle Paleozoic paleogeography of the southern Great Lakes area: Indiana Geological Survey Special Report 32, 32 p.

Droste, J. B., Shaver, R. H., and Lazor, J. D., 1975, Middle Devonian paleogeography of the Wabash Platform, Indiana, Illinois, and Ohio: Geology, v. 3, p. 269–272.

Droste, J. B., Abdulkareem, T. F., and Patton, J. B., 1982, Stratigraphy of the Ancell and Black River groups (Ordovician) in Indiana: Indiana Geological Survey (Occasional Paper 36, 15 p.)

Ehlers, G. M., 1973, Stratigraphy of the Niagaran Series of the Northern Peninsula of Michigan: University of Michigan Museum of Paleontology, Papers on Paleontology, no. 3, 200 p.

Ehlers, G. M., and Kesling, R. V., 1970, Devonian strata of Alpena and Presque Isle counties, Michigan: Michigan Basin Geological Society, Annual Field Excursion, 130 p.

Ells, G. D., 1958, Notes on the Devonian-Silurian in the subsurface of southwest Michigan: Michigan Geological Survey Progress Report, no. 18, 55 p.

—— , 1967, Michigan's Silurian oil and gas pools: Michigan Geological Survey Report of Investigations 2, 49 p.

—— , 1969, Architecture of the Michigan Basin: Michigan Basin Geological Society Annual Field Excursion, p. 60–88.

—— , 1978, Stratigraphic cross-sections extending from Devonian Antrim Shale to Mississippian Sunbury Shale in the Michigan Basin: U.S. Department of Energy, Contract No. EX-76-C-01-2346, FE-2346-30, Springfield, Virginia, National Technical Information Service.

—— , 1979, The Mississippian and Pennsylvania (Carboniferous) systems in the United States—Michigan: U.S. Geological Survey, Professional Paper 1110-J, p. J1–J17.

Elowski, R. C., 1980, Potassium salts—Potash of the Salina A-1 Evaporite in the Michigan Basin: Michigan Geological Survey Report of Investigation 25, 15 p.

Fisher, J. A., 1981, Fault patterns in southeastern Michigan [M.S. thesis]: East Lansing, Michigan State University, 80 p.

Fisher, J. H., 1973, Petroleum occurrence in the Silurian reefs of Michigan: Ontario Petroleum Institute, Twelfth Annual Conference, Technical Papers, no. 9, p. 1–10.

—— , ed., 1977, Introduction to reefs and evaporites—Concepts and depositional models: American Association of Petroleum Geologists Studies in Geology, no. 5, p. vi–vii.

—— , 1980, Stratigraphy of the Upper Devonian–Lower Mississippian of Michigan: U.S. Department of Energy, Contract No. AC20-76LC10153, FE 2346-80, Springfield, Virginia, National Technical Information Service.

Fisher, J. H., and Barratt, M. W., 1985, Exploration in Ordovician of central Michigan Basin: American Association of Petroleum Geologists Bulletin, v. 12, p. 2065–2076.

Fowler, J. H., and Kuenzi, W. D., 1978, Keweenawan turbidites in Michigan: Journal of Geophysical Research, v. 83, no. B12, p. 5833–5843.

Gardner, W. C., 1974, Middle Devonian stratigraphy and depositional environments in the Michigan Basin: Michigan Basin Geological Society Special Papers, no. 1, 138 p.

Gill, D., 1975, Discussion *to* Reply *on* Cyclic deposition of Silurian carbonates and evaporites in Michigan Basin: American Association of Petroleum Geologists Bulletin, v. 59, no. 3, p. 538–542.

Halls, H. C., 1978, The late Precambrian North American rift system; A survey of recent geological and geophysical investigations: *in* Ramberg, I. B., and Nueman, E. R., eds., Tectonics and geophysics of continental rifts: Dordrecht, Holland, D. Reidel Publishing Company, p. 111–123.

Hamblin, W. K., 1958, The Cambrian sandstones of Michigan: Geological Survey of Michigan Publication 51, 146 p.

Harding, T. P., 1974: Petroleum traps associated with wrench faults: American Association of Petroleum Geologists Bulletin, v. 58, no. 7, p. 1290–1304.

Haxby, W. F., Turcotte, D. L., and Bird, J. M., 1976, Thermal and mechanical evolution of the Michigan Basin: Tectonophysics, v. 36, p. 57–75.

Hinze, W. J., and Merritt, D. W., 1969, Basement rocks of the Southern Peninsula of Michigan: Michigan Geological Society Annual Field Excursion Guidebook, p. 28–59.

Hinze, W. J., Kellogg, R. L., and Merritt, D. W., 1971, Gravity and aeromagnetic anomaly maps of the Southern Peninsula of Michigan: Michigan Geological Survey Report of Investigations 14, 15 p.

Hinze, W. J., Kellogg, R. L., and O'Hara, N. W., 1975, Geophysical studies of basement geology of Southern Peninsula of Michigan: American Association of Petroleum Geologists Bulletin, v. 59, no. 9, p. 1562–1584.

Huh, J. M., Briggs, L. I., and Gill, D., 1977, Depositional environments of pinnacle reefs, Niagara and Salina groups, northern shelf, Michigan Basin: American Association of Petroleum Geologists Studies in Geology, no. 5, p. 1–21.

Janssens, A., 1970, Middle Devonian formations in the subsurface of northwestern Ohio: Ohio Geological Survey Report of Investigations 78, 22 p.

—— , 1973, Stratigraphy of the Cambrian and Lower Ordovician rocks in Ohio: Ohio Geological Survey Bulletin 64, 197 p.

—— , 1977, Silurian rocks in the subsurface of northwestern Ohio: Ohio Geological Survey Report of Investigations 100, 96 p.

Johnson, G. H., and Keller, S. J., 1972, Geological map of the 1° × 2° Fort Wayne Quadrangle, Indiana, Michigan, and Ohio, showing bedrock and unconsolidated deposits: Indiana Geological Survey Regional Geological Map 8.

Kingston, D. R., Dishroon, C. P., and Williams, P. A., 1983, Global basin classification system: American Association of Petroleum Geologists Bulletin, v. 67, no. 12, 2175–2193.

Landes, K. K., 1945a, The Salina and Bass islands rocks in the Michigan Basin: U.S. Geological Survey Oil and Gas Investigations, Preliminary Map no. 40.

—— , 1951, Detroit River Group in the Michigan Basin: U.S. Geological Survey Circular 133, 23 p.

Landes, K. K., Ehlers, G. M., and Stanley, G. M., 1945b, Geology of the Mackinac Straits region and subsurface geology of northern Southern Peninsula: Michigan Geological Survey Publication 44, Geological Series 37, 204 p.

Liberty, B. A., and Bolton, T. E., 1971, Paleozoic geology of the Bruce Peninsula area, Ontario: Geological Survey of Canada Memoir 360, 61 p.

Lidiak, E. G., Marvin, R. F., Thomas, H. H., and Bass, M. N., 1966, Geochronology of the midcontinent region, United States: Journal of Geophysical Research, v. 71, no. 22, p. 5427–5438.

Lilenthal, R. T., 1978, Stratigraphic cross-sections of the Michigan Basin: Michigan Geological Survey Report of Investigations 19, 36 p., plates 1–89.

Lineback, J. A., 1970, Stratigraphy of the New Albany Shale in Indiana: Indiana Geological Survey Bulletin 44, 73 p.

Mantek, W., 1973, Niagaran pinnacle reefs in Michigan: Ontario Petroleum Institute, Twelfth Annual Conference, Technical Papers, no. 1, p. 1–32.
Matthews, R., and Egleson, G. C., 1974, Origin and implications of a mid-basin potash facies in the Salina salt of Michigan: The Northern Ohio Geological Society, Fourth Symposium on Salt, v. 1, p. 15–33.
Medaris, L. G., Jr., ed., 1983, Early Proterozoic geology of the Great Lakes region: Geological Society of America Memoir 160, 141 p.
Mescher, P. K., 1980, Structural evolution of southeastern Michigan—Middle Ordovician to Middle Silurian [M.S. thesis]: East Lansing, Michigan State University, 120 p.
Mesolella, K. J., Robinson, J. D., McCormick, L. M., and Ormiston, A. R., 1974, Cyclic deposition of Silurian carbonates and evaporites in Michigan Basin: American Association of Petroleum Geologists Bulletin, v. 58, no. 1, p. 34–62.
Moyer, R. B., 1982, Thermal maturity and organic content of selected Paleozoic formations—Michigan Basin [M.S. thesis]: East Lansing, Michigan State University, 62 p.
Newcombe, R. B., 1933, Oil and gas fields of Michigan: Michigan Geological Survey, Publication 38, Geological Series 32, 293 p.
Nunn, J. A., Sleep, N. H., and Moore, W. E., 1984, Thermal subsidence and generation of hydrocarbons in Michigan Basin: American Association of Petroleum Geologists Bulletin, v. 68, no. 3, p. 296–315.
Nurmi, R. D., 1972, Upper Ordovician stratigraphy of the Southern Peninsula of Michigan [M.S. thesis]: East Lansing, Michigan State University, 48 p.
Nurmi, R. D., and Friedman, G. M., 1977, Sedimentology and depositional environments of basin-center evaporites, Lower Salina Group (Upper Silurian), Michigan Basin: American Association of Petroleum Geologists Studies in Geology, no. 5, p. 23–52.
Oray, E., Hinze, W. J., and O'Hara, N., 1973, Gravity and magnetic evidence for the eastern termination of the Lake Superior Syncline: Geological Society of America Bulletin, v. 84, no. 8, p. 2763–2780.
Orr, R. W., 1971, Conodonts from Middle Devonian strata of the Michigan Basin: Indiana Geological Survey Bulletin 45, 110 p.
Owens, G. L., 1967, The Precambrian surface of Ohio: Ohio Geological Survey Report of Investigations, no. 64, 8 p.
Pepper, J. F., deWitt, W., Jr., and Demarest, D. F., 1954, Geology of Bedford Shale and Berea Sandstone in the Appalachian Basin: U.S. Geological Survey Professional Paper 259, 111 p.
Repetski, J., and Harris, A., 1981, Report on referred fossils, Lower and Middle Ordovician, Ogemaw County, Michigan: USGS Paleontology and Stratigraphy Branch, written communication.
Rexroad, C. B., 1980, Stratigraphy and conodont paleontology of the Cataract Formation and the Salamonie Dolomite (Silurian) in northeastern Indiana: Indiana Geological Survey Bulletin 58, 83 p.
Rexroad, C. B., and Droste, J. B., 1982, Stratigraphy and conodont paleontology of the Sexton Creek Limestone and the Salamonie Dolomite (Silurian) in northwestern Indiana: Indiana Geological Survey Special Report 25, 29 p.
Rickard, L. V., 1969, Stratigraphy of the Upper Silurian Salina Group, New York, Pennsylvania, Ohio, Ontario: New York State Museum and Science Service, Map and Chart Series 12, 57 p. (plus maps and charts).
—— , 1984, Correlation of the subsurface Lower and Middle Devonian of the Lake Erie region: Geological Society of America Bulletin, v. 95, p. 814–828.
Rudman, A. J., Summerson, C. H., and Hinze, W. J., 1965, Geology of basement in midwestern United States: American Association of Petroleum Geologists Bulletin, v. 49, no. 7, p. 894–904.
Sanford, B. V., 1961, Subsurface stratigraphy of Ordovician rocks in southwestern Ontario: Geological Survey of Canada Paper 60-26, 54 p.
Sanford, B. V., and Brady, W. B., 1955, Paleozoic geology of the Windsor-Sarnia area, Ontario: Geological Survey of Canada Memoir 278, 65 p.
Sanford, B. V., and Quillian, R. G., 1959, Subsurface stratigraphy of Upper Cambrian rocks in southwestern Ontario: Geological Survey of Canada Paper 58-12, 29 p.
Schmalz, R. F., 1969, Deep water evaporite deposition: American Association of Petroleum Geologists Bulletin, v. 53, no. 4, p. 798–823.
Schneider, A. F., and Keller, S. J., 1970, Geological map of the 1° × 2° Chicago Quadrangle, Indiana, Illinois, and Michigan, showing bedrock and unconsolidated deposits: Indiana Geological Survey, Regional Geologic map 4.
Shaffer, B. L., 1969, Palynology of the Michigan "Red Beds" [Ph.D. thesis]: East Lansing, Michigan State University, 250 p.
Shaver, R. H., 1974, Silurian reefs of northern Indiana; Reef and inter-reef macrofaunas: American Association of Petroleum Geologists Bulletin, v. 58, no. 6, p. 934–956.
Shaver, R. H., and nine others, 1978, The search for a Silurian reef model; Great Lakes area: Indiana Geological Survey Special Report 15, 36 p.
Shaver, R. H., and five others, 1983, Silurian reef and inter-reef strata as responses to a cyclical succession of environments, southern Great Lake area (Field trip 12), *in* Field trips in midwestern geology: Geological Society of America, Indiana Geological Survey, and Indiana University, Department of Geology, Bloomington, Indiana, v. 1, p. 141–196.
Sleep, N. H., Nunn, J. A., and Chou, L., 1980a, Platform basins: Annual Reviews of Earth and Planetary Science, v. 8, p. 17–34.
Sleep, N. H., and Sloss, L. L., 1980b, The Michigan Basin: American Geophysical Union, Geodynamics Series, v. 1, p. 93–98.
Sloss, L. L., 1963, Sequences in the cratonic interior of North America: Geological Society of America Bulletin, v. 74, p. 93–114.
—— , 1969, Evaporite deposition from layered solutions: American Association of Petroleum Geologists Bulletin, v. 53, no. 4, p. 776–789.
—— , 1972, Synchrony of Phanerozoic sedimentary-tectonic events of the North American craton and the Russian Platform: 24th International Geological Congress, Montreal, sec. 6, p. 24–32.
—— , 1982, Subsidence of continental margins; The case for alternatives to thermal contraction: American Geophysical Union, Geodynamics Series, v. 6, p. 197–200.
Sommers, L. M., 1977, Atlas of Michigan: East Lansing, Michigan State University Press, 242 p.
Sparling, D. R., 1983, Conodont biostratigraphy and biofacies of lower Middle Devonian limestone, north-central Ohio: Journal of Paleontology, v. 57, p. 825–864.
Stonehouse, H. B., 1969, The Precambrian around and under the Michigan Basin: Michigan Basin Geological Society Annual Field Excursion, p. 15–27.
Summerson, C. H., 1962, Precambrian in Ohio and adjoining areas: Ohio Geological Survey Report of Investigations, no. 44, 16 p.
Thomas, G. E., 1974, Lineament-block tectonics, Williston–Blood Creek Basin: American Association of Petroleum Geologists Bulletin, v. 58, no. 7, p. 1305–1322.
Van Der Voo, R., and Watts, D. R., 1978, Paleomagnetic results from igneous and sedimentary rocks from the Michigan Basin borehole: Journal of Geophysical Research, v. 83, no. B12, p. 5844–5848.
Vary, J. A., Elenbaas, J. R., and Johnson, M. A., 1968, Gas in Michigan Basin: American Association of Petroleum Geologists, Memoir 9, v. 2, p. 1761–1697.
Vugrinovich, R., 1984, Lithostratigraphy and depositional environments of the Pennsylvanian rocks and the Bayport Formation of the Michigan Basin: Michigan Geological Survey Report of Investigation 27, 33 p.
Wanless, H. R., and Shideler, G. L., 1975, *in* Paleotectonic investigations of the Pennsylvanian system in the U.S., part I; Introduction and regional analysis of the Pennsylvanian System: U.S. Geological Survey Professional Paper 853-B, p. 9–15.
Willman, H. B., and seven others, 1975, Handbook of Illinois stratigraphy: Illinois Geological Survey Bulletin 95, 261 p.
Wold, R. J. and Hinze, W. J., 1982, Geology and tectonics of the Lake Superior Basin: Geological Society of America Memoir 156, 280 p.

ACKNOWLEDGMENTS

The Michigan portion of this report was prepared by Fisher and Barratt. The Indiana and Ohio part of the Michigan Basin was provided by Droste and Shaver. The Michigan writers have benefited greatly from discussions with William Roth, Glenn Sleight, Garland Ells, and Donald Eschman.

Financial support for Fisher was provided by the Exxon Foundation and for Barratt by the Patrick Petroleum Company.

MANUSCRIPT ACCEPTED BY THE SOCIETY DECEMBER 2, 1986

Printed in U.S.A.

The Geology of North America
Vol. D-2, Sedimentary Cover—North American Craton: U.S.
The Geological Society of America, 1988

Chapter 14

Illinois Basin region

Charles Collinson, Michael L. Sargent, and James R. Jennings
Illinois State Geological Survey, 615 East Peabody Drive, Champaign, Illinois 61820

INTRODUCTION

The Illinois Basin is a spoon-shaped structure located mostly in central and southern Illinois, southwestern Indiana, and western Kentucky (Figs. 1 and 2). Its long dimension extends northwest-southeast for 600 km, and its maximum width exceeds 320 km. The greatest thickness of sedimentary fill is in southern Illinois and western Kentucky (in the Rough Creek Graben), where a maximum of approximately 7,000 m has been reported (Anderson, 1984; Bertagne and others, 1986; Hester, 1984; Schwalb, 1983). This section thins both depositionally and erosionally to less than 450 m on arches and domes that surround the basin. The Paleozoic sedimentary fill ranges in age from Early or Middle Cambrian to very Early Permian. Permian rock, however, is present only in a small fault block in the subsurface of western Kentucky. Pennsylvanian bedrock is present directly below surficial material over most of the basin (Fig. 2) and is surrounded by broad outcrop belts of Mississippian strata on the western, eastern, and southern sides. Silurian, Ordovician, and Cambrian rocks dominate the northern slopes. Devonian and Silurian rocks form broad outcrop belts in northeastern Illinois, as well as northern and eastern Indiana. Relatively narrow outcrop belts of lower and middle Paleozoic rocks flank the Ozark Uplift to the southwest.

Various boundaries have been suggested to separate the Illinois Basin (in the restricted sense) from the surrounding arches (Fig. 2). If the generally used –500-ft (–150-m) contour on top of the Ordovician Ottawa (top of Trenton and equivalents) Megagroup (megagroups are formally called supergroups in some areas outside Illinois) is chosen (Fig. 2), the Illinois Basin has an area of about 105,000 km^2 and contains some 311,000 km^3 of sediment (Bell and others, 1964). These estimates do not include more recently discovered, thick Lower and Middle Cambrian (Schwalb, 1982; Houseknecht and Weaverling, 1983) rift-fill sediments in the New Madrid Rift Complex (Braile and others, 1982b). Their areal extent and thickness are poorly understood.

Paleozoic rocks are separated by major unconformities into the Sauk, Tippecanoe, Kaskaskia, and Absaroka sequences (Sloss, 1963, 1982). Although relatively unimportant in the stratigraphy of the basin region, strata of the Zuni and Tejas sequences (Cretaceous, Tertiary, and Quaternary) overlap the southern edge of the Illinois Basin in the Mississippi Embayment, and a few outliers are present farther north. A mantle of Pleistocene glacial drift covers the central and northern parts (Figs. 3 and 4).

Paleozoic rock of the basin is estimated to include a somewhat larger volume of carbonates than volume of siliciclastics. Nearly one-third of the total is shale and siltstone, while nearly one-fifth is sandstone. The carbonates are mainly dolomite in the lower Paleozoic and mainly limestone in the upper Paleozoic. Chert is a widely distributed subordinate constituent of both the limestones and the dolomites. During all except possibly the closing epochs of Paleozoic time, the basin was open to the south (Potter and Pryor, 1961), and most stratigraphic units thicken southward toward the Reelfoot Rift area (Figs. 4 and 5). The dominant source of siliciclastics in the Illinois Basin was the Canadian Shield area to the north (Potter and Pryor, 1961), although Appalachian orogenic belts supplied clays, most notably during Middle to Late Ordovician and Late Devonian times. Erosion of sedimentary rocks from surrounding arches also made a contribution.

STRUCTURAL SETTING

The Illinois Basin was located approximately 300 km north of the margin of the craton during the Paleozoic. The basin was a broad trough that was open to the south during most of this time. The basin was formed by the combination of broad downward flexing in the north and development of a deep, rapidly subsiding graben complex in the south. Rates of vertical movement varied from time to time and place to place, but they generally had a net downward trend throughout the Paleozoic. The surrounding arches also subsided but had diverse histories and smaller net downward movements. The southern seaway was open until sometime after the end of the Pennsylvanian when the Pascola Arch was uplifted (Fig. 5).

The Wisconsin Arch (Figs. 1 and 5), an extension of the Canadian Shield, borders the Illinois Basin on the north and is overlapped from the south by lower Paleozoic rocks. Much of the

Collinson, C., Sargent, M. L., and Jennings, J. R., 1988, *in* Sloss, L. L., ed., Sedimentary Cover—North American Craton; U.S.: Boulder, Colorado, Geological Society of America, The Geology of North America, v. D-2.

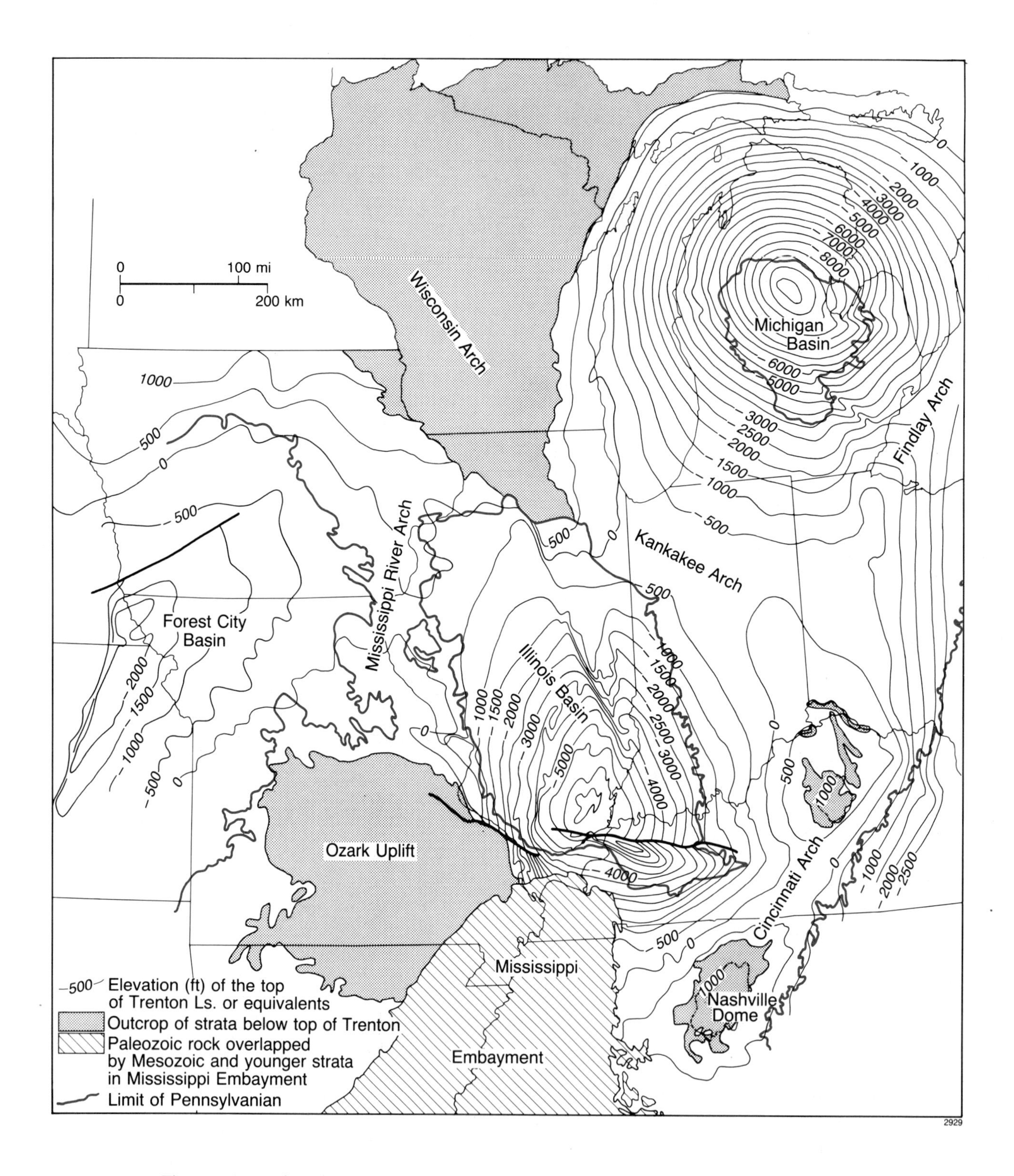

Figure 1. Map of north-central United States showing major basins, arches and domes. Structure contours after Cohee and others (1962).

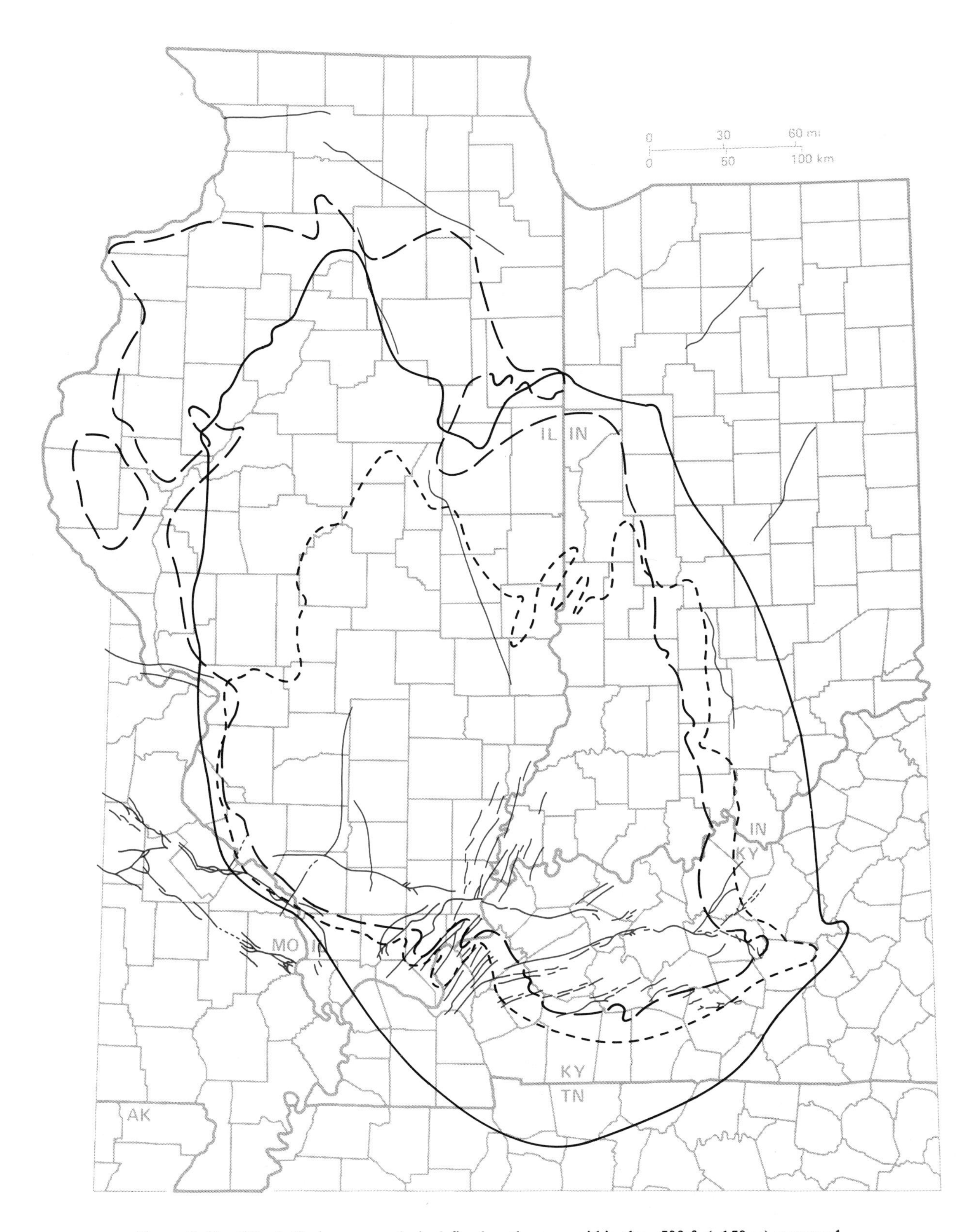

Figure 2. The Illinois Basin commonly is defined as the area within the –500-ft (–150-m) structural contour (solid line) on top of the Ottawa Megagroup (Figs. 3, 10, top of the Middle Ordovician Champlainian Series). Also, it has been defined as the area underlying Pennsylvanian rocks (long dashes), or the area underlying Chesterian rocks (short dashes). Faults are shown in red. (In part after Bristol and Buchsbach, 1971).

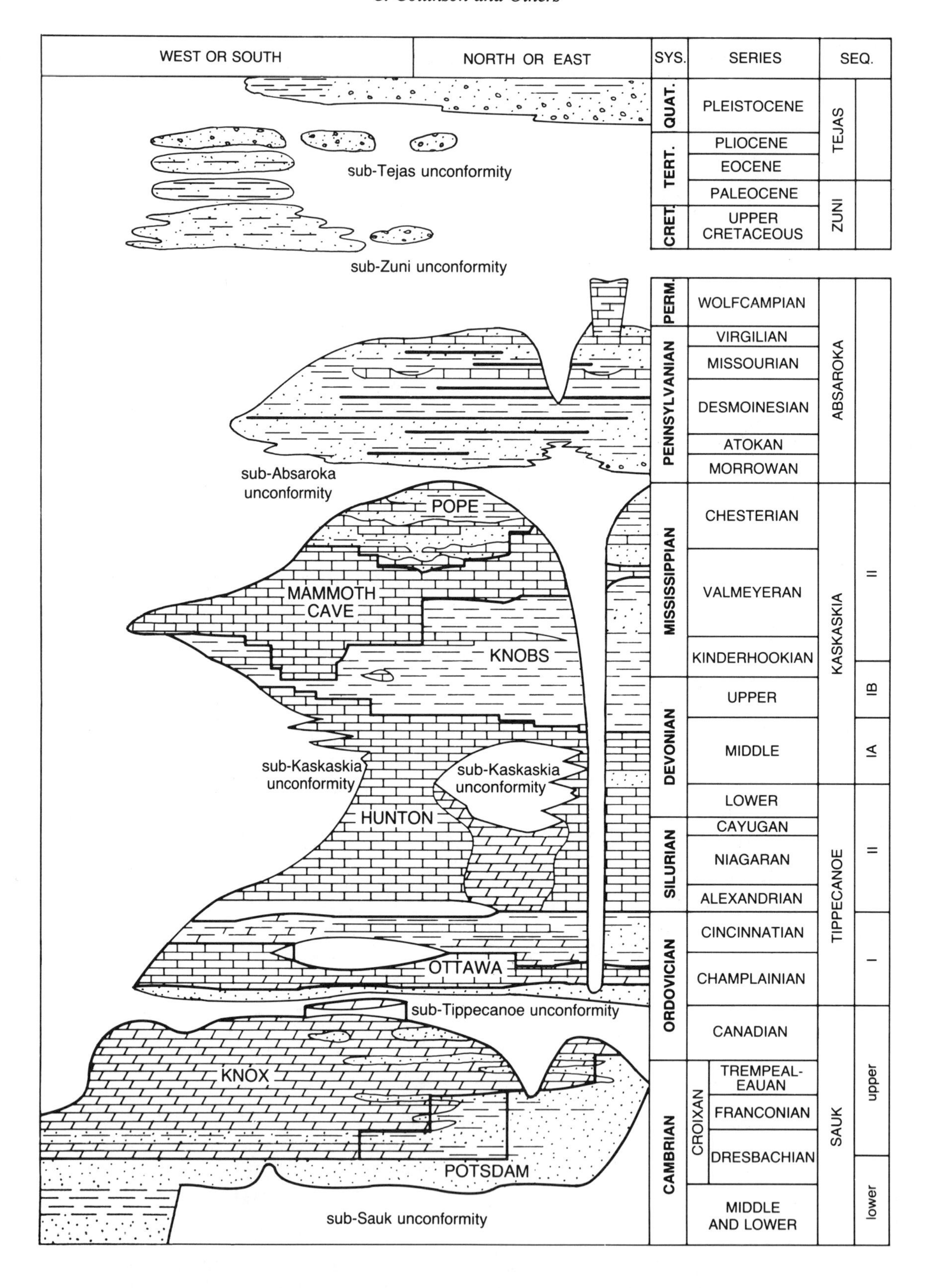

Figure 3. Diagrammatic cross section showing the relationships of megagroups (shown in capital letters), series, and sequences to bounding unconformities and to lithology (modified from Willman and others, 1975, Fig. 14).

Paleozoic cover, once thicker and more extensive than now, has been removed by erosion. Contributions of sediment from the Canadian Shield during much of the Paleozoic, therefore, came from areas well to the north of the present limit of Paleozoic rocks.

The first clear indication of the forthcoming separation of the Illinois and Michigan Basins was the development of the Kankakee Arch (Atherton, 1971) in northern Indiana at the close of Canadian (Early Ordovician) time. Its crest was near the northern border of Indiana, a little north of the location of its crest later in the Paleozoic.

The Cincinnati Arch (Figs. 1 and 5), which adjoins the basin on the east, is the positive structure that separates the Appalachian Basin from the Illinois and Michigan Basins. The arch is low and broad with several well-recognized components, including the Nashville Dome, which lies at its southern end in Tennessee. A relatively positive area in the vicinity of the Cincinnati Arch has generally existed since Precambrian time, as indicated by thinning of most units toward this region (e.g., Figs. 7, 9, 11, 16, 18, 26, 30, and 32). A notable exception is the Upper Ordovician Maquoketa Group and equivalents (Fig. 22).

The Mississippi River Arch (Howell, 1935; Edmund and Anderson, 1967; Anderson and others, 1982), which has also been called the Northeast Missouri Arch (Lee, 1943; Bunker and others, 1985), separates the Illinois Basin from the Forest City Basin (Figs. 1 and 5). The axis of the arch lies near the present course of the Mississippi River, forming a saddle between the Wisconsin Arch to the north and the Ozark Uplift to the south. This general region was at times a positive area at least as far back as early Paleozoic time (Bunker and others, 1985). The Mississippi River Arch was present during the deposition of basal Pennsylvanian strata (Howell, 1935; Lee, 1943). Underlying beds are truncated (Branson, 1962; Edmund and Anderson, 1967). Subsequent erosion has completely removed Pennsylvanian rocks from the axis of the arch, separating the areas of Pennsylvanian strata in Illinois from those in Iowa (Fig. 1).

In the area of the Ozark Uplift (Figs. 1 and 5), which borders the basin on the southwest, Late Cambrian sediments were deposited on a Precambrian surface having a relief of at least 600 m. Paleozoic sedimentation was punctuated by several episodes of uplift (McCracken, 1971; Tikrity, 1968). Later tectonic movements have kept the Ozark Uplift a positive structure.

The Pascola Arch (Fig. 5) closes the Illinois Basin on the south (Atherton, 1971) and lies between the Ozark Uplift and the Nashville Dome. The arch is a relatively recent feature, inasmuch as the Illinois Basin was a trough open to the south through most of Paleozoic time. The arch includes beveled rocks as young as Pennsylvanian around its flanks. Erosion on its crest has cut deeply into St. Croixan (Upper Cambrian) strata (Bond and others, 1971). The time of uplift of the arch is obscure, but it may have begun at the same time as the Ouachita orogeny near the close of the Paleozoic Era. Upper Cretaceous sediments of the Mississippi Embayment overlap the Pascola Arch and are related to Mesozoic subsidence in the embayment.

Widespread faulting in the extreme southern part of the Illinois Basin included the development of a rapidly subsiding rift complex, which was most active during the very early Paleozoic (Braile and others, 1982a, 1982b, 1986; Houseknecht and Viele, 1984). The Rough Creek–Shawneetown and Pennyrile Fault Systems (Fig. 5) are key structural elements in this complex (Soderberg and Keller, 1981). The Rough Creek–Shawneetown Fault System was interpreted by Heyl (1972) to be part of his 38th Parallel Lineament, which he regards as extending from West Virginia to Colorado.

PRECAMBRIAN

Paleozoic rocks range in thickness from 450 m in the north to more than 7,000 m in the south (Anderson, 1984; Atherton, 1975; Schwalb, 1983), and only about 50 deep test wells have penetrated the entire Paleozoic succession in the Illinois Basin region. Thus, information on composition and topography of the Precambrian basement is sparse. Fortunately, geophysical surveys (Bertagne and others, 1986; Braile and others, 1984; Buschbach, 1983; Heigold, 1976; Hester, 1984; Hildenbrand and others, 1982; Howe and Thompson, 1984; Sexton and others, 1986) have greatly added to knowledge of the area. In addition, excellent Precambrian exposures are located in the nearby Baraboo Range of southern Wisconsin and in the Ozark Uplift of southeastern Missouri, exposures that provide opportunity to examine analogous features. Thus, relatively new data have made possible a substantial reinterpretation of the configuration of the Precambrian surface in the southern part of the Illinois Basin (Fig. 6).

Precambrian rocks in the basement differ considerably from each other in age and composition (Fig. 13; Bickford and others, 1986; Lidiak and others, 1966; Shaver and others, 1985; Sims and Peterman, 1986; Zietz and others, 1966). In northern and western Illinois, wells drilled to the basement have encountered granite, monzonite, granodiorite, rhyolite, and felsite. These rocks are part of the Eastern Granite–Rhyolite Province that extends from Iowa and Missouri eastward to Ohio. These rocks range from 1.42 to 1.5 billion years old (Bickford and others, 1986). South of this province, the basement appears to be younger where dates are available (Bickford and others, 1986; Thomas and others, 1984). The Precambrian basement in southwestern Illinois consists of granite, rhyolite, rhyolite porphyry, and granodiorite, whereas the basement in southeastern Illinois consists of granite and rhyolite. Rock recovered from one well in Lawrence County, southern Indiana, is basaltic in composition (Biggs and others, 1960). A boring south of the Rough Creek Fault System in south-central Kentucky encountered syenite (Hoppe and others, 1983).

The surface of the Precambrian basement in the Illinois Basin exhibits considerable relief, and its configuration has affected the Paleozoic section that covers it (Atherton, 1975; Buschbach, 1983; Sargent and Buschbach, 1985). In the Ozarks, where it has been exposed by erosion of Cambrian strata, the Precambrian surface shows at least 600 m of relief. The Precambrian surface in Wisconsin is likewise very rugged. In

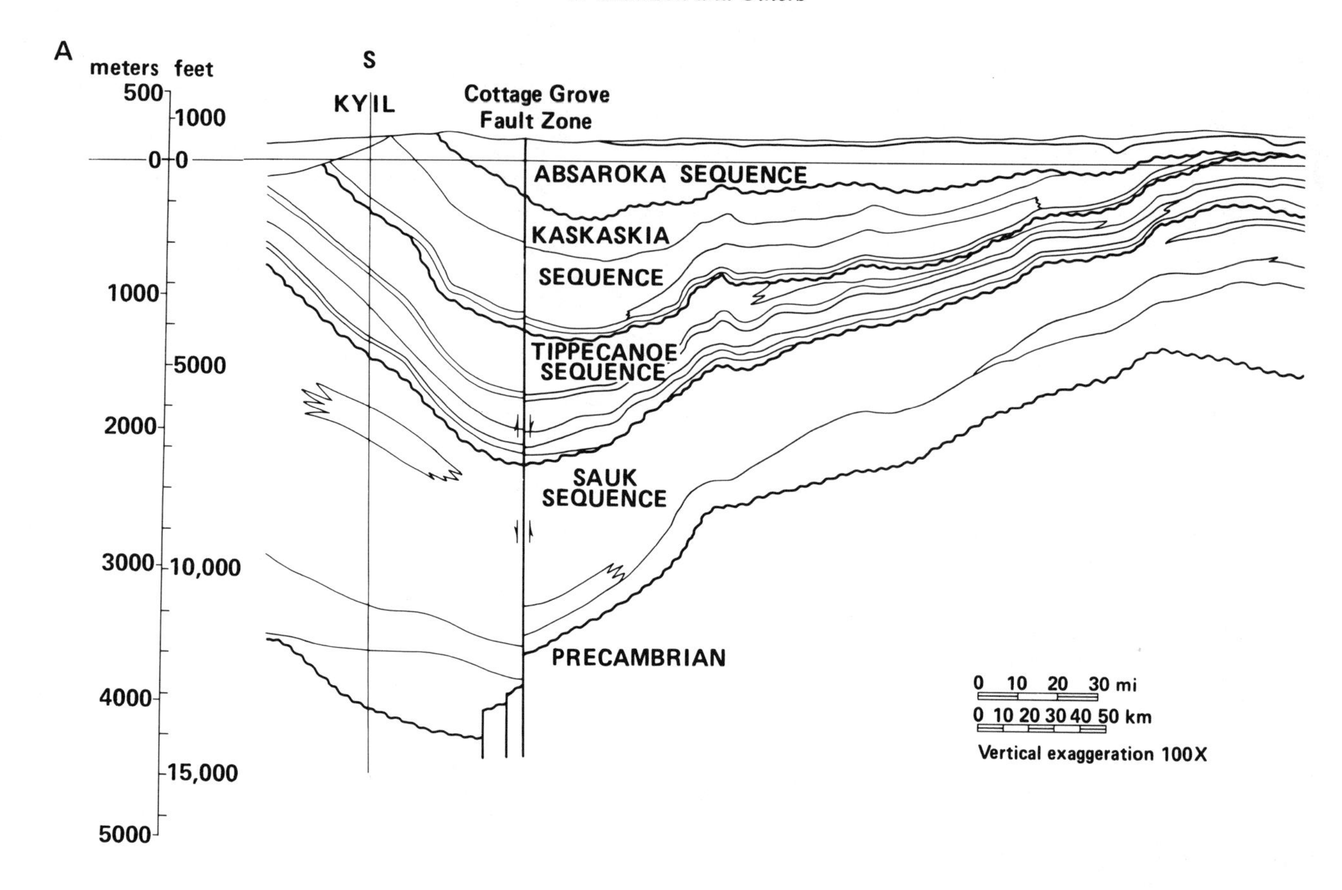

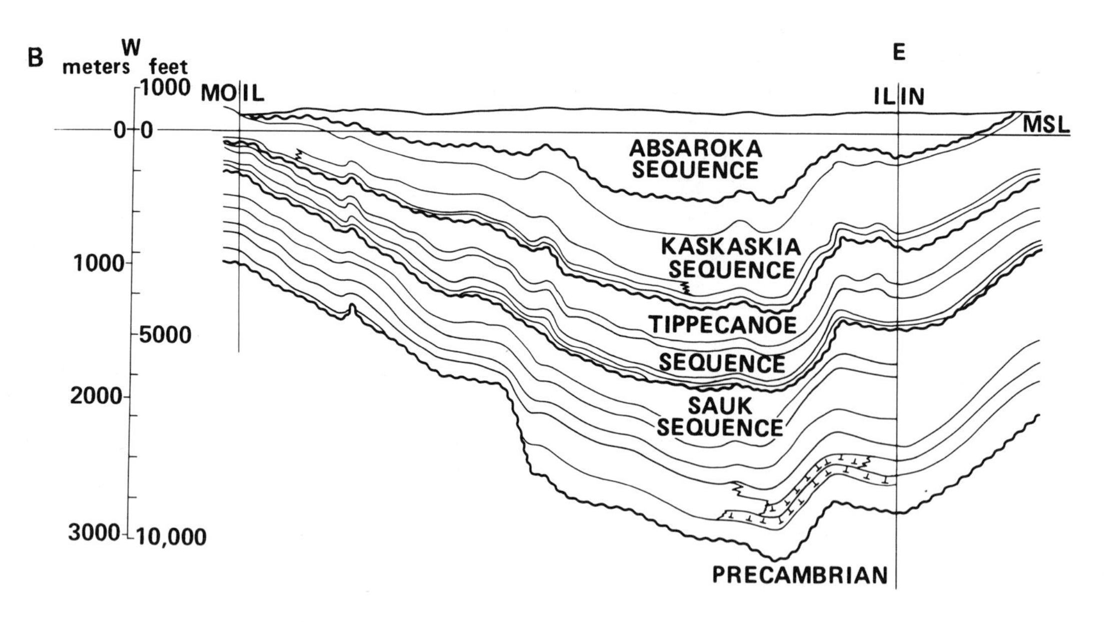

Figure 4 (continued on facing page). Geologic cross sections of the Illinois Basin showing thickening of sequences and their relationships to regional and local structures. Note the truncation of sequences in the Mississippi Embayment area on the south end of cross section A.

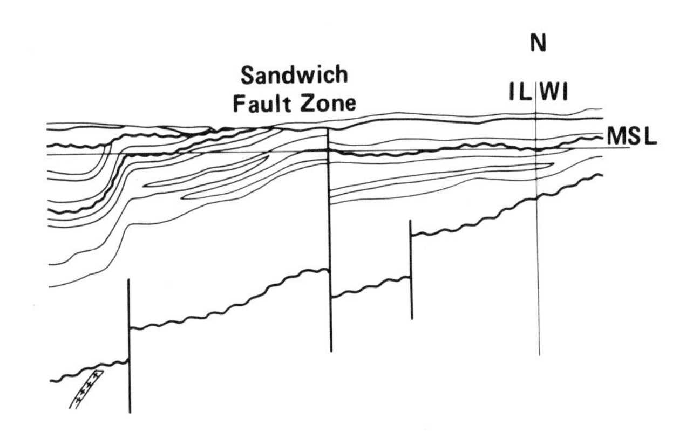
N
Sandwich
Fault Zone
IL WI
MSL

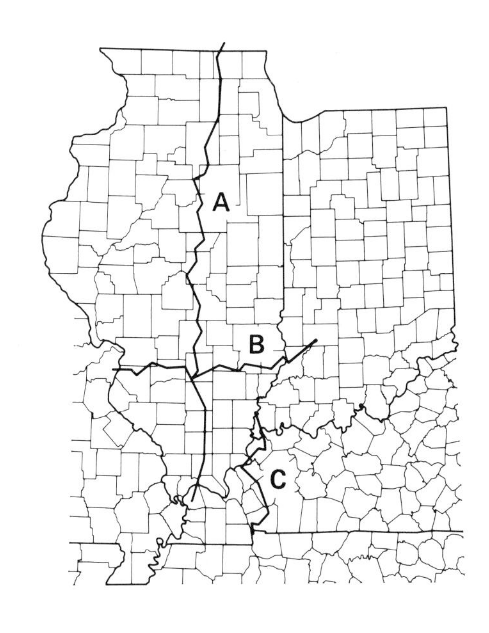
A
B
C

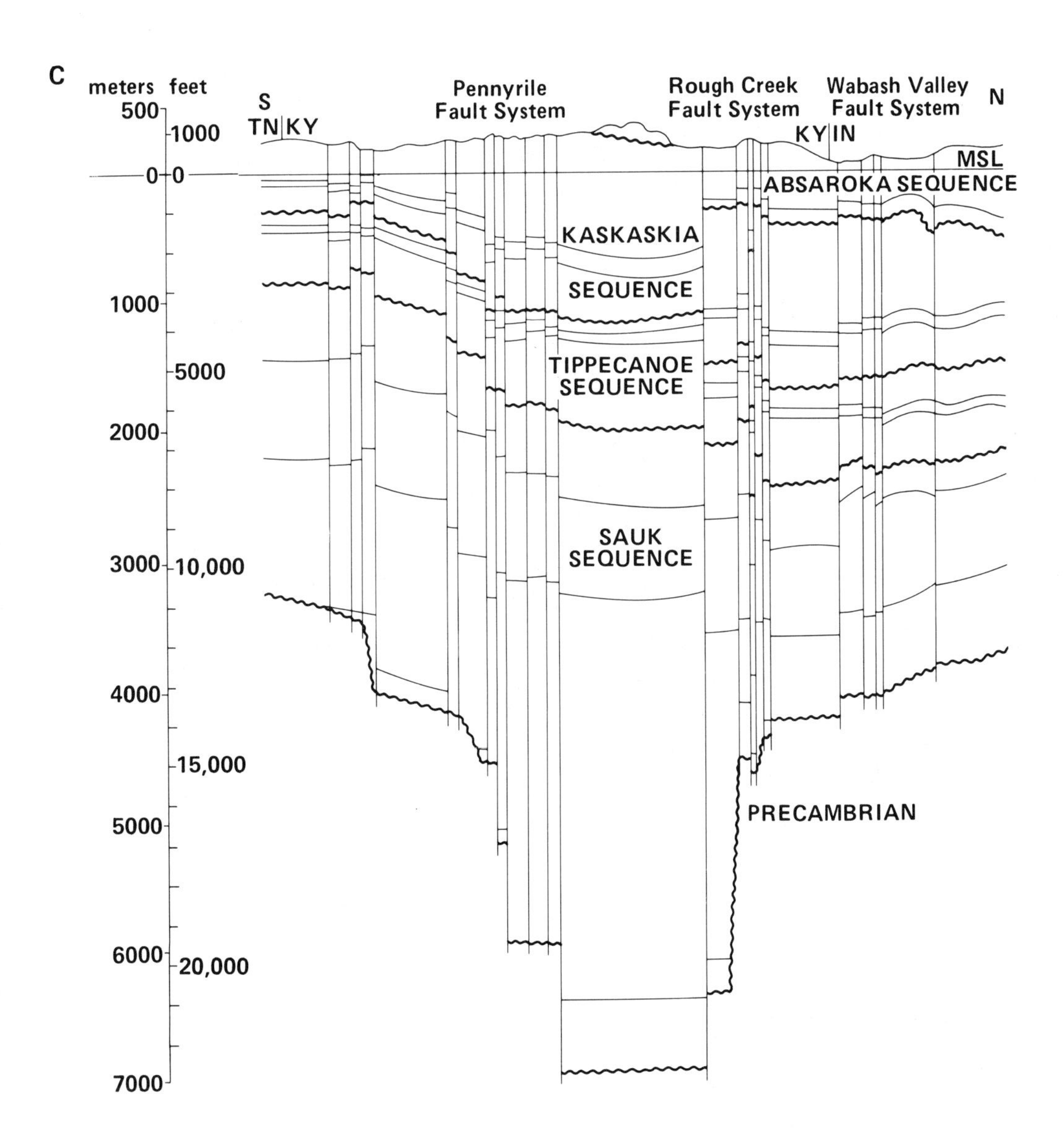
C
meters feet
500
1000
0 0
1000
5000
2000
3000 10,000
4000
15,000
5000
6000 20,000
7000
S
TN KY
Pennyrile
Fault System
Rough Creek
Fault System
Wabash Valley
Fault System
N
KY IN
MSL
ABSAROKA SEQUENCE
KASKASKIA
SEQUENCE
TIPPECANOE
SEQUENCE
SAUK
SEQUENCE
PRECAMBRIAN

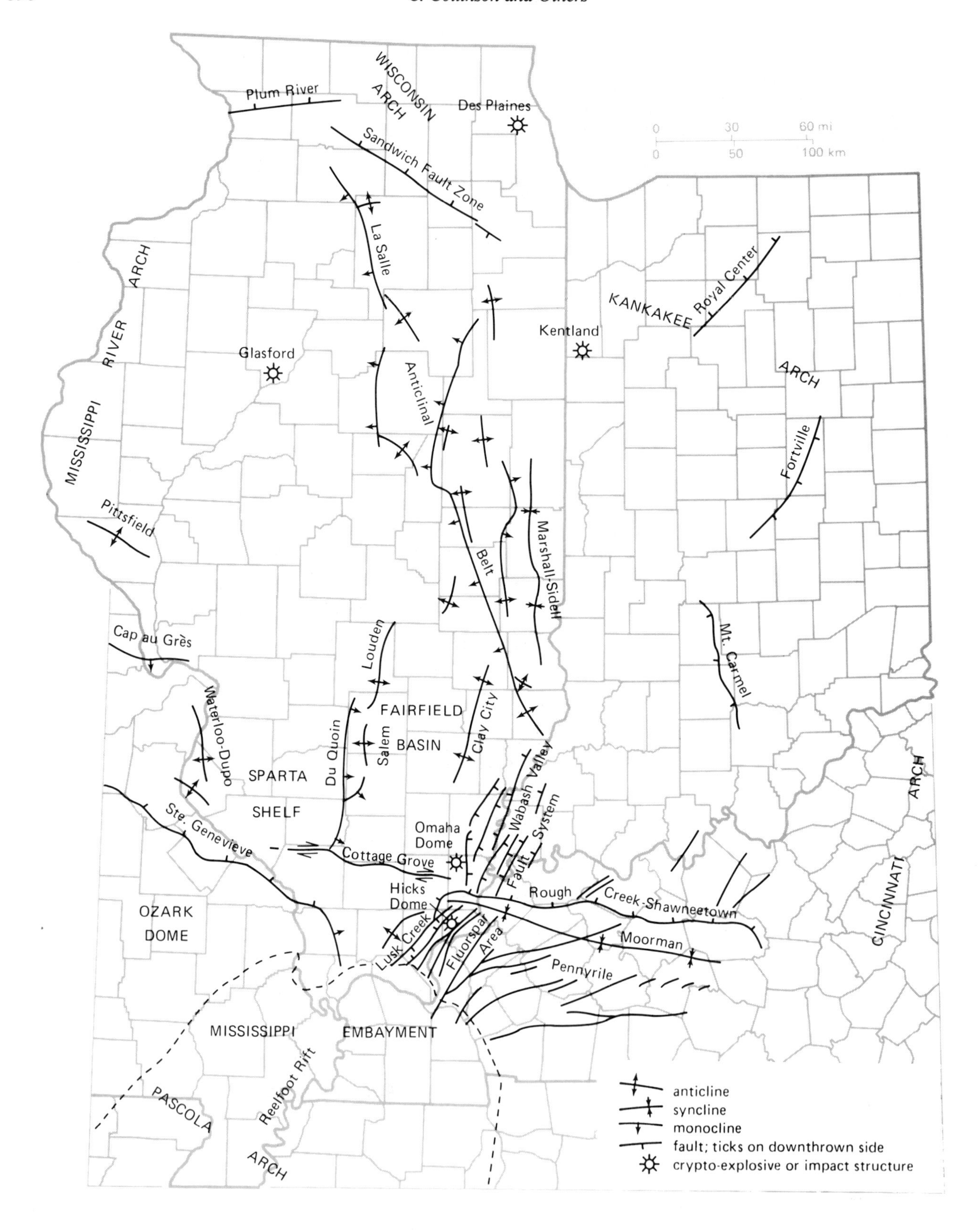

Figure 5. Structural features of the Illinois Basin area (prepared by W. John Nelson and Janis D. Treworgy; modified from Treworgy, 1981).

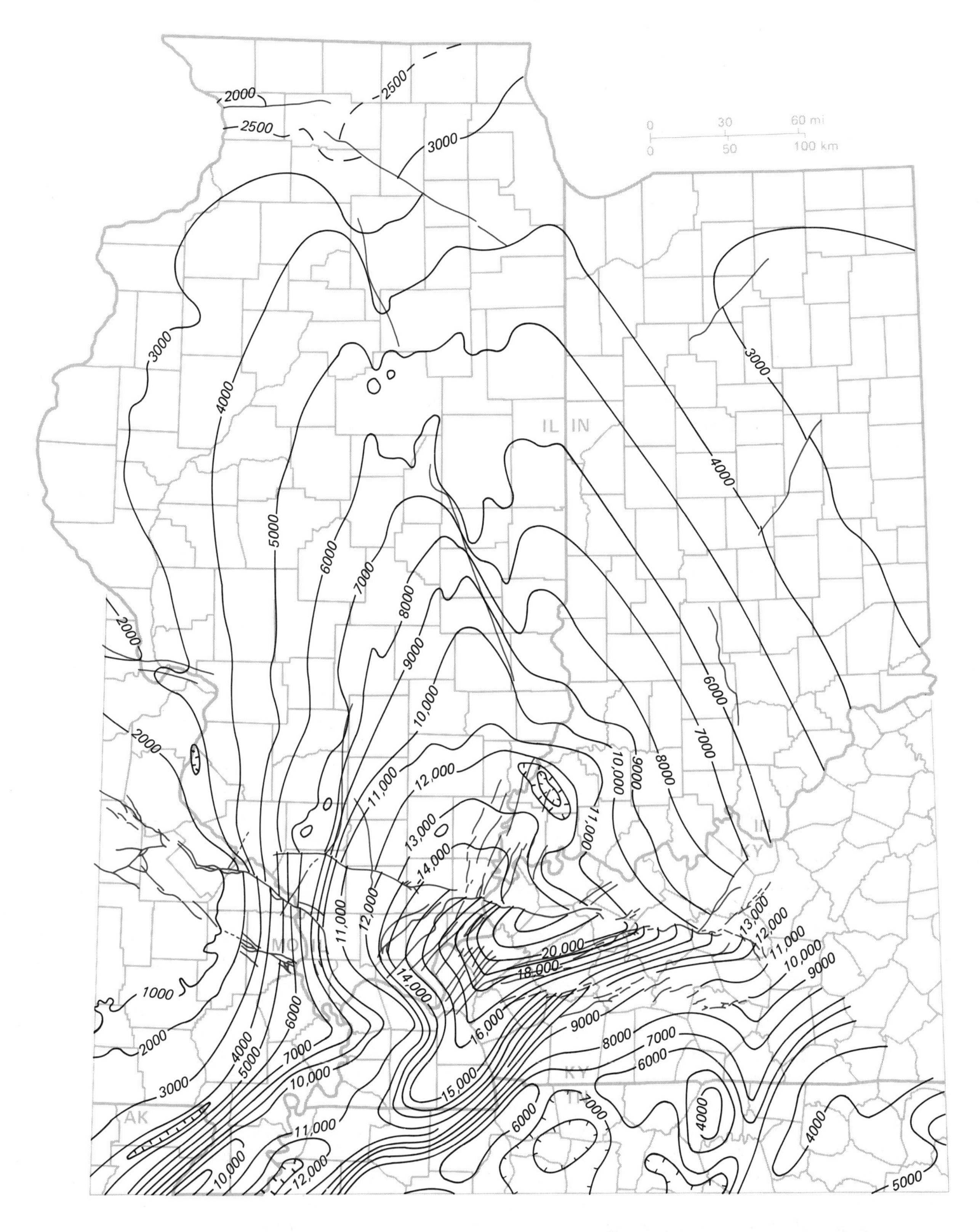

Figure 6. Elevation below sea level (in feet) of the Precambrian basement in the Illinois Basin and adjacent areas. Faults are shown in red. Compiled and interpreted from Sargent and Buschbach (1985), Buschbach (1983, and unpublished maps), Braile and others (1984), Hildenbrand and others (1982), Howe and Thompson (1984), Rudman and others (1965), and Stearns and Reesman (1986).

western Illinois, the Precambrian surface exhibits 240 m of relief between wells only 13 km apart (Atherton, 1971). The Sparta Shelf (Fig. 5), a feature with its origin in the crystalline basement rocks of southwestern Illinois, stood 230 to 300 m above the lower area to the east prior to deposition of Paleozoic sediment (Schwalb, 1982, 1983). In addition, buried Precambrian hills forming the cores of structures such as the Clay City Anticlinal Belt (Fig. 5) verify that the terrain of the basement was rugged. Commonly the Cambrian Mt. Simon Sandstone is missing over Precambrian topographic high points, indicating relief in the neighborhood of 100 m.

Basement rocks obtained from wells are surprisingly fresh, including both the upper Precambrian rocks and detrital grains in the basal arkose of the overlying Mt. Simon Sandstone. However, most test wells have been drilled on Precambrian paleotopographic highs from which weathered rock would have been eroded. Low areas may not have been sampled.

Many major structures found in Phanerozoic rocks are related to features in the Precambrian basement (Figs. 5 and 6). Such major structures as the La Salle Anticlinal Belt, the Fairfield Basin, the Reelfoot Rift (Ervin and McGinnis, 1975), and the Rough Creek Graben are clearly related to major structural features of the basement. Even small structures, such as the Pittsfield Anticline and the Cap au Grès Faulted Flexure, can be traced to basement features. Important structural features of the Precambrian basement (Fig. 5) include the Ozark Uplift, Cincinnati Arch, Sparta Shelf, Clay City Anticline, and the Wisconsin Arch. Movement has occurred in all of these areas since Precambrian time. In addition, the Rough Creek–Shawneetown, Cottage Grove, and Pennyrile fault systems may reflect reactivation of early faults (Nelson and Lumm, 1984; Schwalb, 1982).

The Ozark Uplift (Figs. 1, 5, and 6) is one of the oldest features in the region. In Precambrian time, the ancestral St. Francois Mountains were built by volcanism, broken by faulting, and then eroded before Precambrian sedimentation began (Kisvarsanyi, 1981). Stinchcomb (1976) has identified at least one Precambrian marine incursion in sediments of southeastern Missouri. Tolman and Robertson (1969) indicated that the southeastern Missouri portion of the Ozark Uplift dates from late Precambrian time. After various stages of uplift, the area subsided and was buried by Cambrian sediments.

Schwalb (1983) suggested that near the end of Precambrian time a rift probably opened, extending 160 km northeastward from northeastern Arkansas. This rift is called the New Madrid Rift Complex (Braile and others, 1986), and includes the Rough Creek Graben and Reelfoot Rift. Step faults presumably lined the edges of central grabens where total displacements probably exceed 150 m. The rifting action evidently caused a rotational movement on its eastern end, creating zones of weakness that opened the rift complex, which extends east and south from southern Illinois (Fig. 5). The Rough Creek–Shawneetown Fault System marks the northern edge of the Rough Creek Graben, whereas the Pennyrile Fault System marks the southern edge (Soderberg and Keller, 1981). The Reelfoot Rift appears to be bounded on the northwest by an extension of the Lusk Creek Fault, and on the southeast by an unnamed fault buried beneath Mississippi Embayment sediments (Kolata and Nelson, 1988). The development of the New Madrid Rift Complex represented the beginning of Illinois Basin. The rift complex apparently persisted, continually subsiding, throughout the Paleozoic (Schwalb, 1983) and received the oldest sediments of the Sauk sequence (Fig. 7).

SAUK SEQUENCE

The profound unconformity on the Precambrian basement, representing as much as 1 billion years of time, marks the base of the Sauk sequence (Sloss and others, 1949). The Sauk includes all strata between the deeply eroded Precambrian surface and the widespread unconformity (sub-Tippecanoe) beneath the Ordovician St. Peter Sandstone and Everton Dolomite (Figs. 7 to 13). The sequence represents a single major marine transgression, broken by minor emergences on the margins of the Illinois Basin.

As the product of the earliest of the central cratonic transgressions, the Sauk includes a vast accumulation of siliciclastics derived from deeply eroded Precambrian rocks and consequently represents the greatest volume of sediment in any named sequence. Over most of the basin, the Sauk sequence can be subdivided into two lithic units or megagroups (Fig. 3). The siliciclastic Potsdam Megagroup (Emmons, 1838; Swann and Willman, 1961; Droste and Patton, 1985) is the lower unit. The carbonate Knox Megagroup (Safford, 1869; Swann and Willman, 1961) is the higher one. Willman and others (1975) exclude transitional beds from either megagroup, but we include transitional beds in the Potsdam. The Potsdam Megagroup consists mainly of sandstone and underlies the entire area. It ranges in thickness from more than 1,200 m in northeastern Illinois to about 300 m in Iowa and Missouri. Eastward, its thickness exceeds 300 m almost to Ohio (Bond, 1971; Bond and others, 1971; Becker, 1978). In western Kentucky (in the Rough Creek Graben), it may exceed 3,300 m (Bertagne and Leising, 1988). In northern Illinois, all Cambrian rocks are included in the Potsdam. These strata become more calcareous southward. In central Illinois, at least half of the Cambrian section is sufficiently dolomitic to be included in the Knox Megagroup. In western Kentucky (Figs. 7 to 13), the Cambrian section, with the exception of the Mt. Simon and lower part of the Eau Claire, is virtually all carbonate and therefore is assigned to the Knox along with the Ordovician Prairie du Chien Group (Swann and Willman, 1961).

The oldest datable Sauk sediments thus far encountered are thick Middle Cambrian shales in the Rough Creek Graben (Fig. 13) of western Kentucky (Schwalb, 1969, 1982, 1983). Otherwise, the Mt. Simon Sandstone (Upper Cambrian) is the oldest Paleozoic unit. Except for small areas where Precambrian highs protrude through the Mt. Simon, it is present throughout the area north of the rift complex (Fig. 7).

The Mt. Simon consists of fine to coarse, partly pebbly, friable (in areas where it is not deeply buried) sandstone, which is

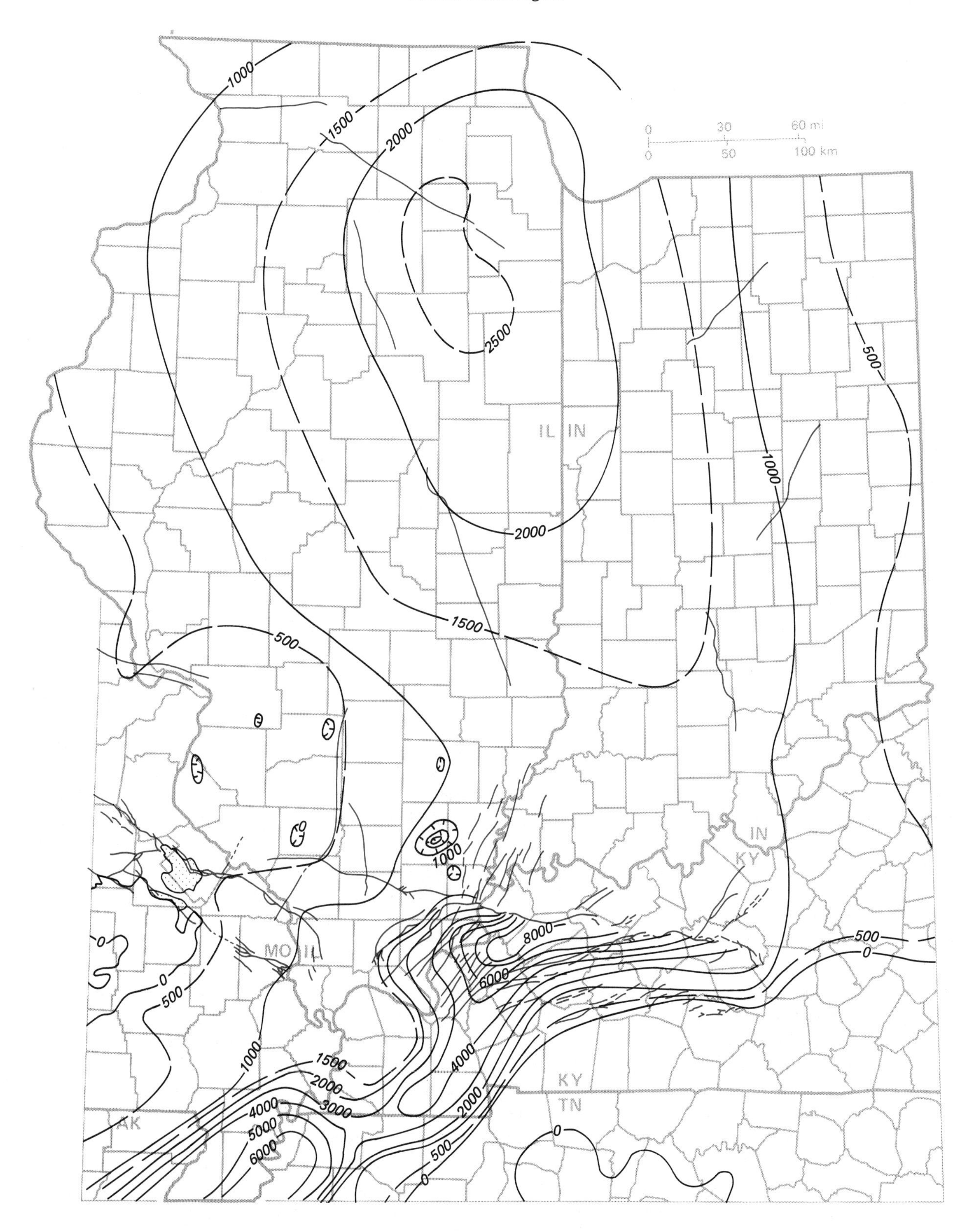

Figure 7. Thickness (in feet) of the Mt. Simon Sandstone and the pre-Mt. Simon formations (Sauk subsequences I and II in the Illinois Basin). Faults are shown in red. Stipple indicates outcrop areas. Some isopach lines are terminated at the Shawneetown–Rough Creek fault system. Thickness is believed to change greatly in a short distance, but dat are not sufficient to delineate the exact pattern.

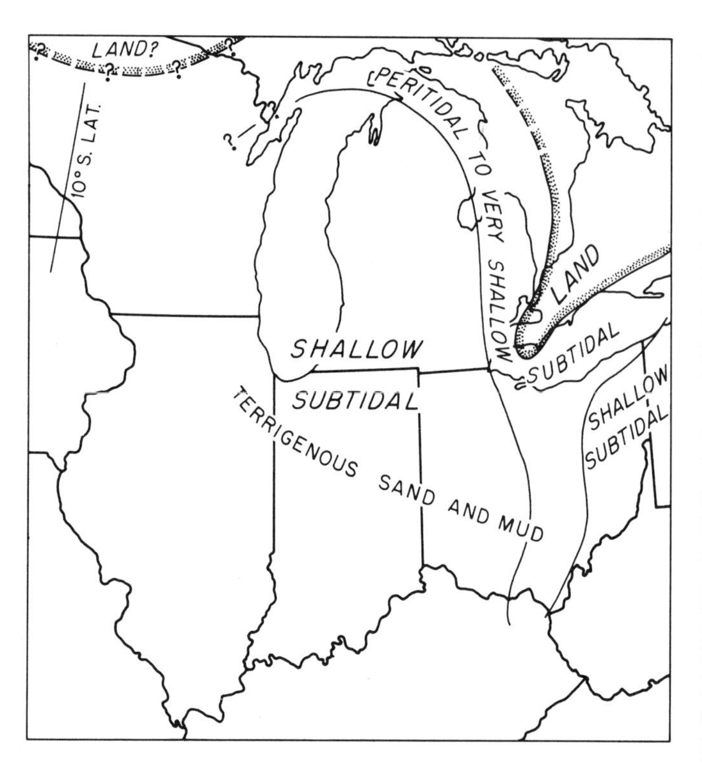

Figure 8. Paleogeography of the early Dresbachian Stage (Mt. Simon Sandstone) showing broad shallow seas covering the entire basin area (from Droste and Shaver, 1983, Fig. 4).

poorly sorted (Buschbach, 1975). In the deeper parts of the basin area, porosity has generally been reduced by grain rotation and sliding, and development of sutured contacts by pressure solution (Hoholick and others, 1984). The Mt. Simon was largely derived from igneous/metamorphic source areas in the Canadian Shield far to the north and the Ozark Uplift to the west. It contains debris from the old basement surface as well as recycled sands from older siliciclastic formations. A basal arkosic part is as much as 100 m thick (Templeton, 1951). The Mt. Simon (Fig. 8), as well as the entire Sauk, is interpreted as being deposited in a subtidal sea (Droste and Shaver, 1983).

During deposition of the Mt. Simon, two depocenters were present in the Illinois Basin area (Fig. 7). One was in northeastern Illinois where more than 900 m of sand accumulated. The second was located in western Kentucky where more than 2,400 m of sediment, much of it pre–Mt. Simon, accumulated in the Rough Creek Graben (Schwalb, 1982). The Mt. Simon represents a shoreward subtidal platform facies; whereas Eau Claire sandy carbonates represent a more offshore facies. The overlying Knox mainly represents a shallow to somewhat deeper subtidal seaward facies, which includes purer parts of the dolomitic Eau Claire, the Derby-Doerun Member of the Franconia Formation, the Potosi Dolomite, the dolomitic facies of the Eminence Formation, and the Canadian (lower Ordovician) Prairie du Chien Group (Figs. 10 and 12).

In the eastern part of the Illinois Basin, that part of the Knox Megagroup that overlies the Eau Claire is referred to the Davis Formation and the Knox Supergroup (Figs. 13 and 14; Droste and Patton, 1985; Shaver and others, 1986). However, in the rift complex, all but the Everton are referred to a Knox Group, which includes in ascending order: Elvins Formation, Potosi Dolomite, Eminence Dolomite, Gunter Sandstone, Gasconade Dolomite, Roubidoux Formation, Jefferson City Dolomite, and Cotter Dolomite.

Most deformation after the close of the Middle Cambrian consisted of broad regional warping and differential sinking, which produced gentle arches, domes, and basins. The St. Francois Mountains continued to be a positive feature from Precambrian time, while the Illinois Basin began subsiding. The old Canadian Shield area on the north was at the high side of the paleoslope that dipped to the south and to the southwest. The Illinois Basin was open through the area now occupied by the Mississippi Embayment into the Black Warrior and Arkoma Basins to the south. By middle St. Croixan time (time of Franconia deposition), the outlines of the northern half of the Illinois Basin were much as we know them today, and the northeastern Illinois depocenter ceased to exist.

Transport of siliciclastics from the northern shield areas continued throughout deposition of the Sauk sequence. Following Mt. Simon deposition, however, there was a trend toward more carbonate and less siliciclastic deposition, beginning with the middle part of the Eau Claire in the central part of the Illinois Basin. Carbonate deposition extended progressively farther northward until well into Late Cambrian time when the Potosi Dolomite covered most of the area. Pulses of siliciclastics from the north and northwest continued to enter the basin in Early Ordovician time, and formed the Gunter and New Richmond Sandstones. Deposition was virtually continuous throughout most of the Illinois Basin until a minor regression of the sea followed deposition of the Eminence Formation (Buschbach, 1975). In Early Ordovician time, similar local regressions occurred between deposition of the Gunter Sandstone and the Oneota Dolomite as well as between the Oneota and the New Richmond Sandstone (Willman and Buschbach, 1975). The rift complex continued to sink throughout Cambrian and Early Ordovician time. As a result, more than 4,900 m of sediments accumulated in the Rough Creek Graben. Nearly as much probably accumulated in the Reelfoot Basin (Figs. 7 to 13).

Thicknesses of lower Sauk, Mt. Simon, and older strata, give evidence of movement in the New Madrid Rift Complex at least as early as Middle Cambrian time (Fig. 7). Thicknesses of upper Sauk show evidence of continued movement in the complex following Mt. Simon deposition (Figs. 8 to 10). The Sparta Shelf and Clay City Anticlinal Belt (Fig. 5) both affected deposition through St. Croixan and Canadian time. Likewise, there is evidence of St. Croixan or earlier subsidence in the vicinity of Ste. Genevieve, Missouri (Figs. 5 and 9). The thickness map of lower

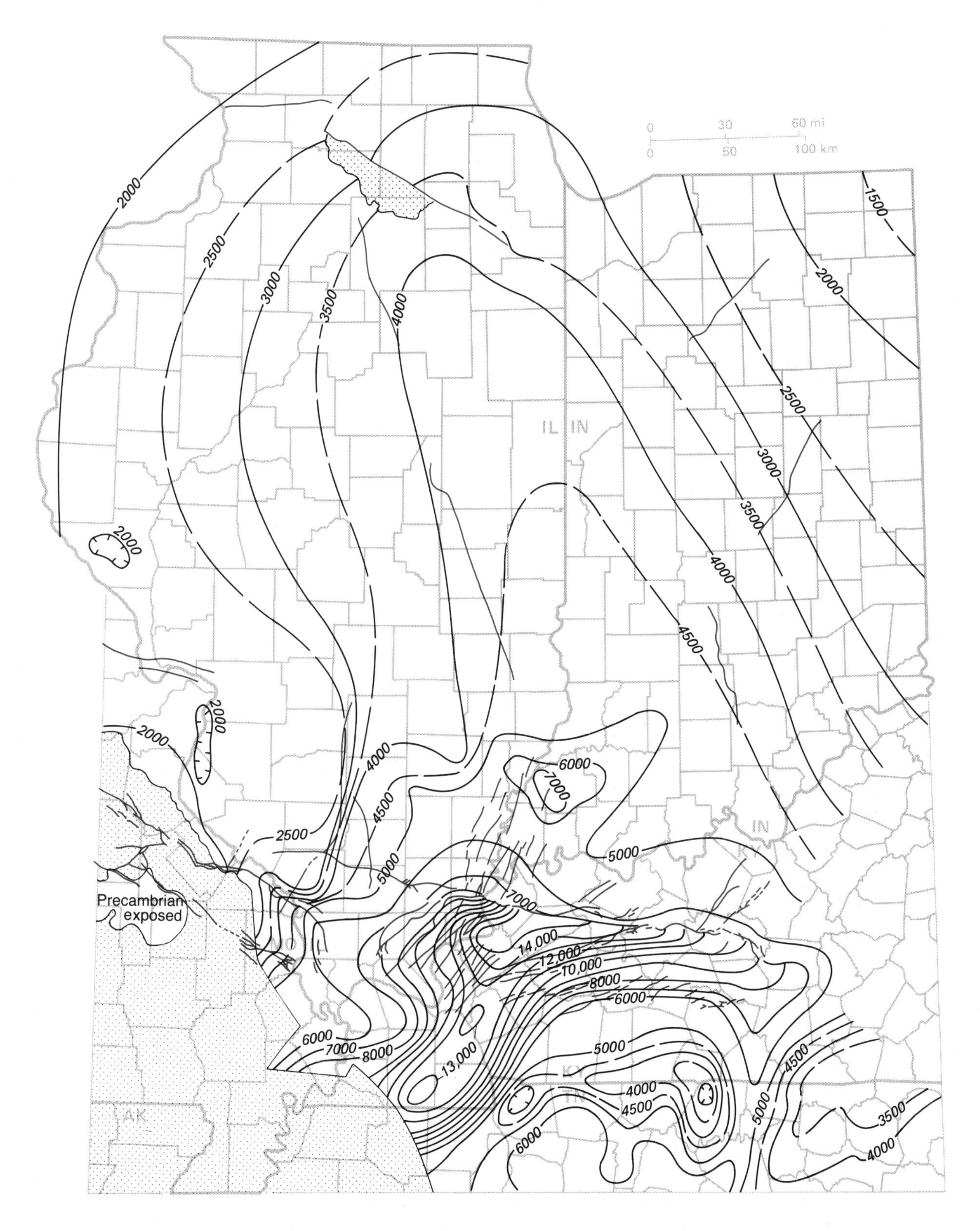

Figure 9. Thickness (in feet) of the Sauk sequence in the Illinois Basin. Faults are shown in red. Stipple represents areas of outcrop and areas eroded post-Ordovician. Two main centers of deposition are shown. Map shows 500 ft contours in areas less than 5,000 ft and shows 1,000 ft contours in areas greater than 5,000 ft (modified from Sargent, 1988).

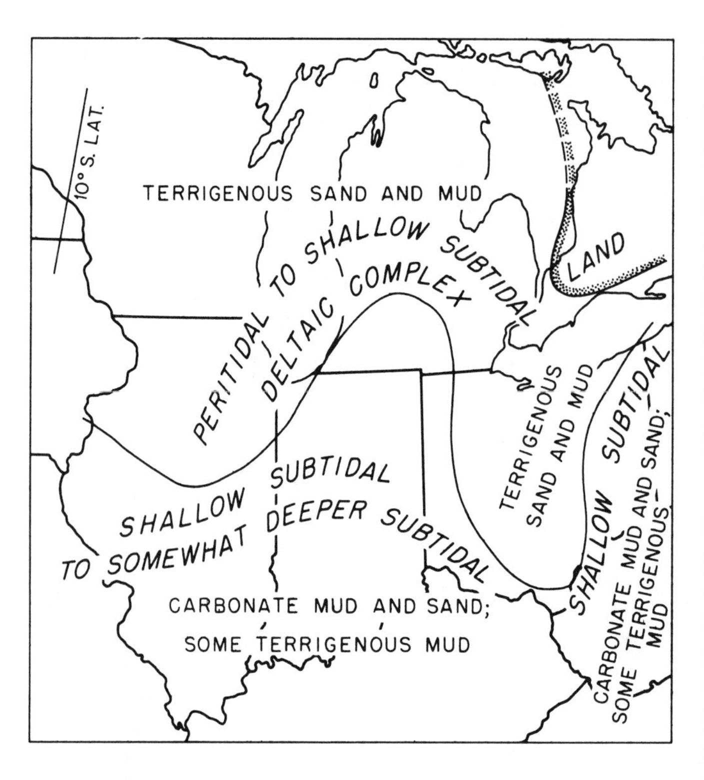

Figure 10. Paleogeography of the Franconian Stage (Franconia Formation) showing seas in the basin area (from Droste and Shaver, 1983, Fig. 7).

Sauk strata (Fig. 7) indicates that there was a positive area near the position of the Nashville Dome. A thin area of upper Sauk lies in about the same position. The Kankakee Arch in northern Indiana and northeastern Illinois rose at the end of Canadian time, beginning the separation of the Illinois Basin from the Michigan Basin. As a result, Sauk deposition stopped in this area, although it may have continued without interruption in the southern part of the basin. As the sea withdrew, a long period of subaerial exposure began. Northern Illinois was deeply eroded, and karst topography developed on the exposed carbonate formations (Buschbach, 1964). The resultant unconformity marks the top of the Sauk sequence (Figs. 3 and 13).

TIPPECANOE SEQUENCE

The Tippecanoe sequence, of Sloss and others (1949) and Sloss (1963, 1982) (Figs. 14, 15, 24, and 26), overlies the sub–St. Peter/Everton (sub-Champlainian) unconformity. In the area where the Everton Dolomite is more than 100 m thick (Fig. 16), nearly continuous sedimentation may have occurred from Canadian through Champlainian time. This area contains the most nearly complete stratigraphic succession in the basin, and no strata are known to be missing. Furthermore, no faunal breaks have been identified. This is probably because there are no cores that sample the Canadian and basal Champlainian section where the Everton is thick, and because faunas are sparse. The sequence is bounded at the top by the sub–Middle Devonian (sub–Grand Tower and Wapsipinicon) unconformity. Beyond the limits of Lower Devonian rocks (Fig. 28), the top of the Tippecanoe is coincident with the top of Silurian strata. Within the areal extent of the Lower Devonian and especially in parts of western Kentucky, the Tippecanoe-Kaskaskia contact occurs between Lower and Middle Devonian carbonate units. In this area, the distinction between these sequences is not always clear, and deposition may have been continuous. The Tippecanoe is subdivided into Tippecanoe subsequences I and II by the basin-wide unconformity between Ordovician and Silurian rocks (Figs. 14 and 24).

The Shakopee Dolomite (below the Everton Dolomite) is Canadian in age, except for some possibly Whiterockian strata present near its top at a few sites (Shaw, 1987). Evidence (Norby and others, 1986; Rexroad and others, 1982) indicates that the Everton Dolomite is a little younger and belongs entirely to the early Champlainian (Whiterockian). The St. Peter Sandstone, which occurs above the Everton, is no older than middle Chazyan. These relationships suggest that the sea withdrew and erosion occurred in part of the basin both before and after Everton deposition. Distribution of the Everton is apparently not related to the distribution of either the overlying St. Peter Sandstone or the underlying Shakopee Dolomite. It may instead represent sediment from a distinct marine incursion, one that covered the basin as far north as central Illinois. Because the Everton overlies the basal unconformity, it is included with other Middle Ordovician formations in the Tippecanoe sequence; however, it is assigned to the underlying Knox Megagroup because of its predominantly carbonate lithology.

Tippecanoe I (Figs. 14 and 15) includes the Everton Dolomite (where it is present) or the St. Peter Sandstone at the bottom (Figs. 16 to 19), the carbonate rocks of the Platteville and Galena Groups (Black River Group; Trenton, Kimmswick, and Lexington Limestones) in the middle (Figs. 20 and 21), and the siliciclastic rocks of the Upper Ordovician Maquoketa Group (Figs 22 and 23) at the top. Formations in the Tippecanoe I, which overlapped eroded Sauk beds below, were slightly eroded and then were overlapped in turn by succeeding Tippecanoe II formations. Carbonates of the Silurian, together with cherts and carbonates of the Lower Devonian, constitute the Tippecanoe II subsequence (Figs. 24 to 26).

Tippecanoe I began with a partial transgression of the area that resulted in deposition of the Everton Dolomite (Figs. 16 and 17) in the deeper part of the Illinois Basin (Droste and Shaver, 1983; Sargent and Norby, 1986). Erosion continued on exposed Sauk strata over most of the craton, however. Following deposition of the Everton, another regression of the sea ensued that affected at least the margins of the Illinois Basin. Afterward, there was a major transgression marked by deposition of St. Peter Sandstone (and in places by its facies equivalent, the Dutchtown Limestone; Figs. 18 and 19), which is believed to represent a

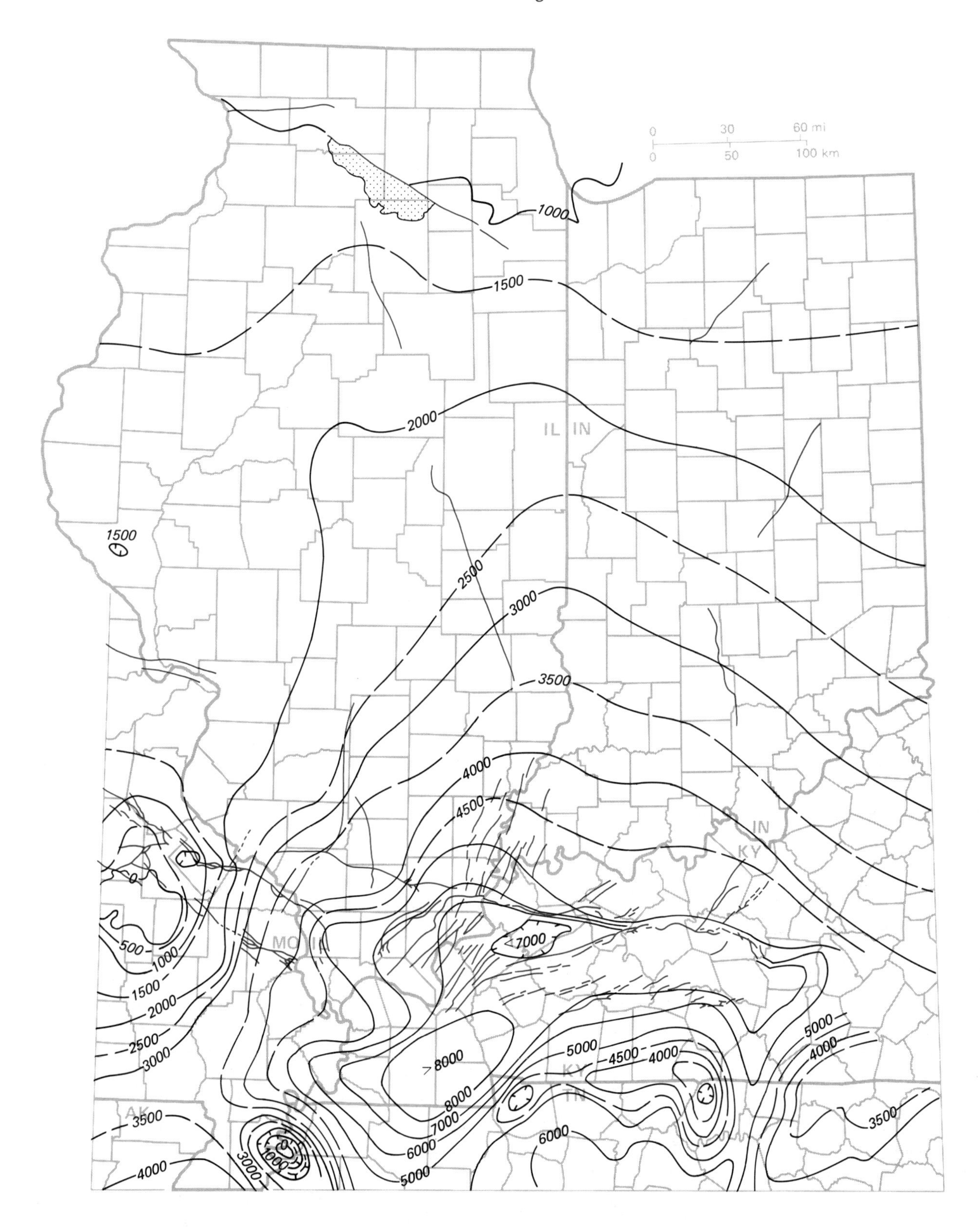

Figure 11. Thickness (in feet) of the upper Sauk sequence in the Illinois Basin. Stipple shows outcrop area. Faults are shown in red. Some isopachs terminate at faults that were active during sedimentation. Eroded Sauk not indicated in southern part of basin (modified from Sargent, 1985, 1988).

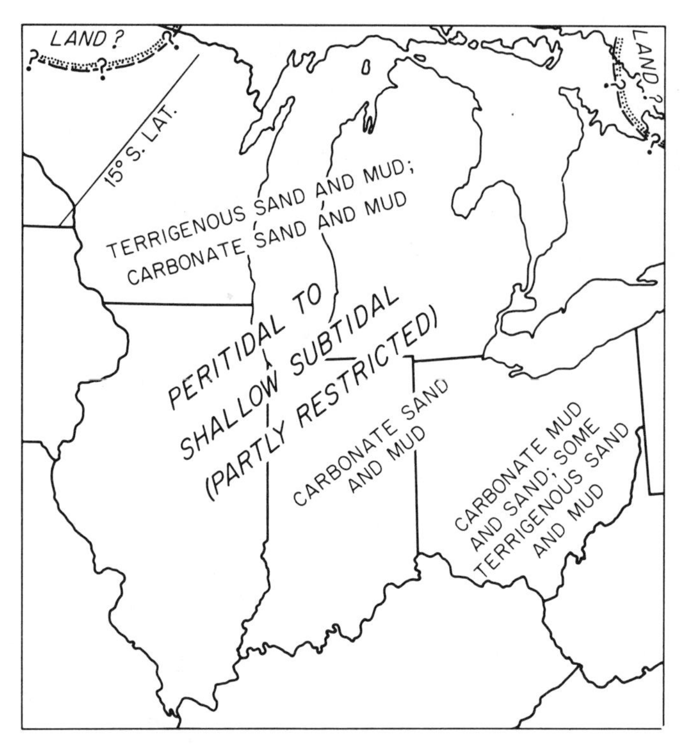

Figure 12. Paleogeography of the middle part of the Canadian Series (Prairie du Chien, upper Knox; from Droste and Shaver, 1983, Fig. 11).

nearshore blanket of marine sand (Fraser, 1976). The St. Peter covers the Illinois Basin uniformly except where it thickens by filling irregularities in the underlying, deeply weathered, karst topography (Buschbach, 1964). The succeeding rocks consist of Champlainian biogenic carbonates with minor components of terrigenous clay, silt, and sand (Figs. 20 and 21). These were transported from the nearby Ozark Uplift as well as from the distant Transcontinental Arch to the west and Canadian Shield to the north. Although deposition was generally very uniform over large areas of the Illinois Basin during this time, minor unconformities occur throughout the subsequence. They can be attributed to eustatic drops in sea level and to gentle local and regional warping. For example, there is a hiatus represented in early Blackriveran rocks (between the Ancell Group and Platteville Group). Also, at the end of Blackriveran time, seas again receded from the margins of the basin, leaving a minor erosional surface on the top of Platteville rocks (Willman and Kolata, 1978). In much of the basin, a hardground is developed at the top of Platteville. Clear seas soon returned, however, and carbonate deposition resumed, although with an increasing component of fine siliciclastics from the west (Witzke, 1980). As dominantly carbonate deposition ended with the Galena Group, there again was regression and erosion during Maysvillian time that affected all but the deeper basin areas (Willman and Kolata, 1978). K-bentonites from Appalachian sources occur in the Galena and Platteville Groups (Kolata and others, 1986). The overlying Maquoketa Group (Figs. 22 and 23) resulted from a massive influx of fine terrigenous siliciclastics from Appalachian sources. Those sources continued to be significant throughout the remainder of the history of the Illinois Basin (e.g., Gray, 1972). Following deposition of the Maquoketa, the sea withdrew from the Illinois Basin, except possibly from the extreme southern part. The cause of this withdrawal has been regarded as a glacio-eustatic drop in sea level (Berry and Boucot, 1973). This withdrawal left the basin subaerially exposed for a short interval, and the soft Maquoketa sediments were eroded to a surface of low relief. When seas returned, siliciclastic sediment of the Maquoketa was incorporated into the lower Silurian strata throughout much of the area (Gray, 1972; Mikulic and others, 1985). The top of the Maquoketa is the upper limit of the Tippecanoe I subsequence.

The Tippecanoe II subsequence, which includes the entire Silurian as well as the Lower Devonian (Figs. 24 to 26), was deposited following a complete transgression of the Illinois Basin by a shallow sea (Willman and Atherton, 1975).

As deposition of the Tippecanoe II subsequence began, the Early Silurian seas were relatively slow in returning to the upper flanks of the basin area. Even during middle Alexandrian time, large areas of central Illinois, as well as most of Indiana, continued to be subaerially exposed (Fig. 25).

Later, during Early Silurian time, there was widespread submergence, bringing renewed deposition dominated by biogenic carbonates. An interruption in sedimentation occurred at the end of Alexandrian time, causing a slight disconformity at the top of the Alexandrian Series (Mikulic, 1977), but erosion was minor. Widespread carbonate deposition resumed during the Niagaran. Grainstones, oolitic carbonates, and reefs with accompanying flank beds were deposited (Lowenstam, 1949, 1950, 1952, 1957; Shaver and others, 1978; Droste and Shaver, 1980, 1987).

Beginning in early Niagaran time (St. Clair Limestone/Salamonie Dolomite), carbonate banks and buildups (reefs) became widespread. The shallow, basin-fringing Terre Haute Bank (Droste and Shaver, 1980, 1983) extended around the northern, northeastern, and eastern flanks of the Illinois Basin. In the deeper central part, argillaceous, silty limestone of the Moccasin Springs Formation was deposited. During late Niagaran and Cayugan time, reefs were both numerous and widely distributed over the Illinois Basin area (Fig. 27).

The most notable rock unit of the Cayugan Series is the silty, siliceous Bailey Limestone. Deposition of the Bailey apparently continued uninterrupted (Collinson and Atherton, 1975) into Devonian time in the deep part of the Illinois Basin (Fig. 24). On the flanks of the basin, the Bailey contains coarse carbonates with numerous reefs.

In Early Devonian time, the seas withdrew sufficiently to expose the upper flanks of the Illinois Basin in western and north-

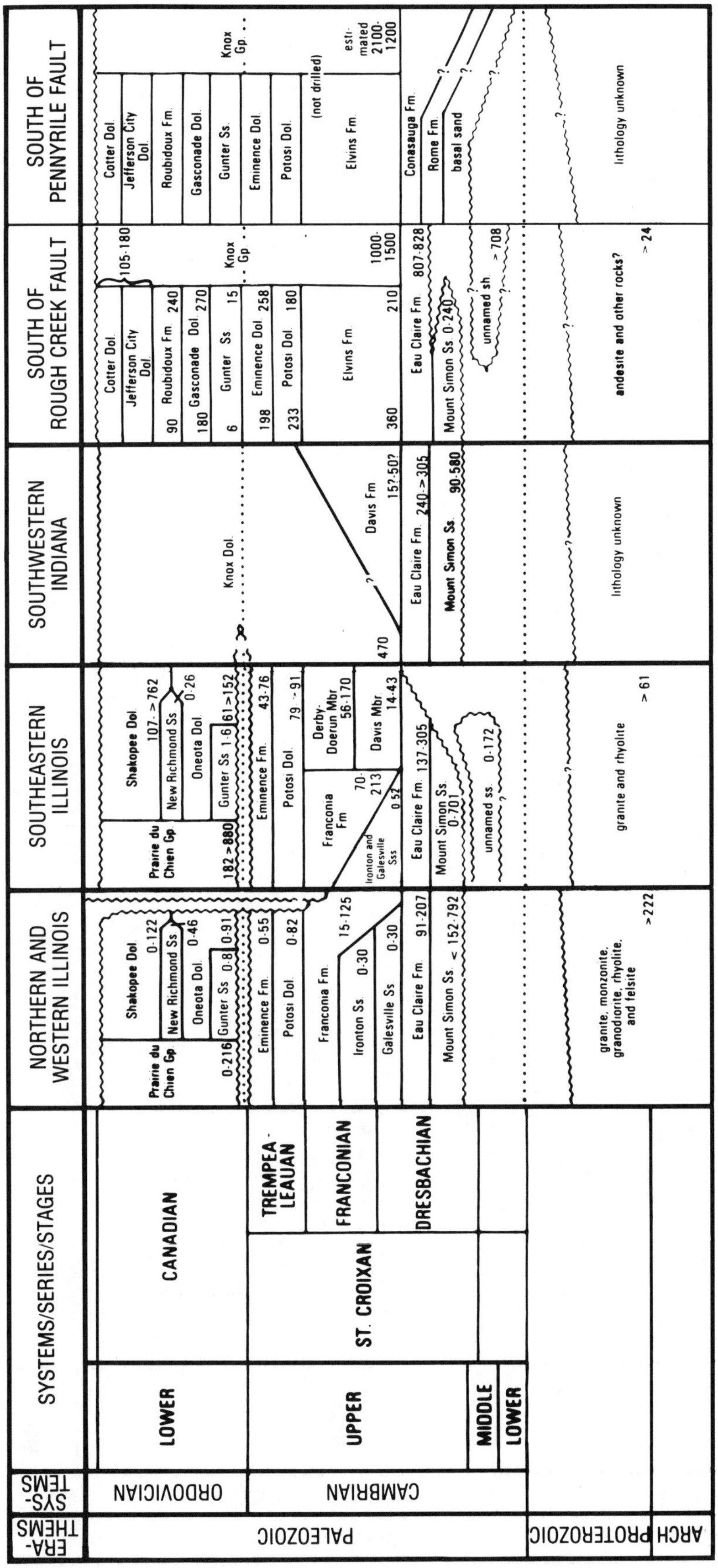

ERATHEMS	SYSTEMS	SYSTEMS/SERIES/STAGES			NORTHERN AND WESTERN ILLINOIS	SOUTHEASTERN ILLINOIS	SOUTHWESTERN INDIANA	SOUTH OF ROUGH CREEK FAULT	SOUTH OF PENNYRILE FAULT
PALEOZOIC	ORDOVICIAN	LOWER	CANADIAN		Prairie du Chien Gp.: Shakopee Dol. 0-122	Prairie du Chien Gp.: Shakopee Dol. 107->762	Knox Dol.	Cotter Dol.; Jefferson City Dol. 105-180	Cotter Dol.; Jefferson City Dol.
					New Richmond Ss.	New Richmond Ss.		90 Roubidoux Fm. 240	Roubidoux Fm.
					Oneota Dol. 0-46	Oneota Dol. 0-26		180 Gasconade Dol. 270	Gasconade Dol.
					Gunter Ss. 0-8; 0-216; 0-91	Gunter Ss. 1-6; 182->880; 61->152		6 Gunter Ss. 15; Knox Gp.	Gunter Ss.; Knox Gp.
	CAMBRIAN	UPPER	ST. CROIXAN	TREMPEALEAUAN	Eminence Fm. 0-55	Eminence Fm. 43-76		198 Eminence Dol. 258	Eminence Dol.
					Potosi Dol. 0-82	Potosi Dol. 79-91		233 Potosi Dol. 180	Potosi Dol.
				FRANCONIAN	Franconia Fm. 15-125; Ironton Ss. 0-30; Galesville Ss. 0-30	Franconia Fm. 70-213; Derby-Doerun Mbr. 56-170; Davis Mbr. 14-43; Ironton and Galesville Sss. 0-52	470 ? Davis Fm. 15?-50?	Elvins Fm. 360 210; 1000-1500	(not drilled) Elvins Fm.; estimated 2100-1200
				DRESBACHIAN	Eau Claire Fm. 91-207	Eau Claire Fm. 137-305	Eau Claire Fm. 240->305	Eau Claire Fm. 807-828	Conasauga Fm.
					Mount Simon Ss. < 152-792	Mount Simon Ss. 0-701	Mount Simon Ss. 90-580	Mount Simon Ss. 0-240	Rome Fm. ?
		MIDDLE				unnamed ss. 0-172 ?		unnamed sh. ? >708	basal sand ?
		LOWER							?
PROTEROZOIC					granite, monzonite, granodiorite, rhyolite, and felsite >222	? granite and rhyolite >61	? lithology unknown	? andesite and other rocks? >24	? lithology unknown
ARCH									

Figure 13. Units of the Precambrian and the Sauk sequence in the Illinois Basin. Modified from COSUNA correlation chart (Shaver and others, 1985). Wavy lines represent unconformities.

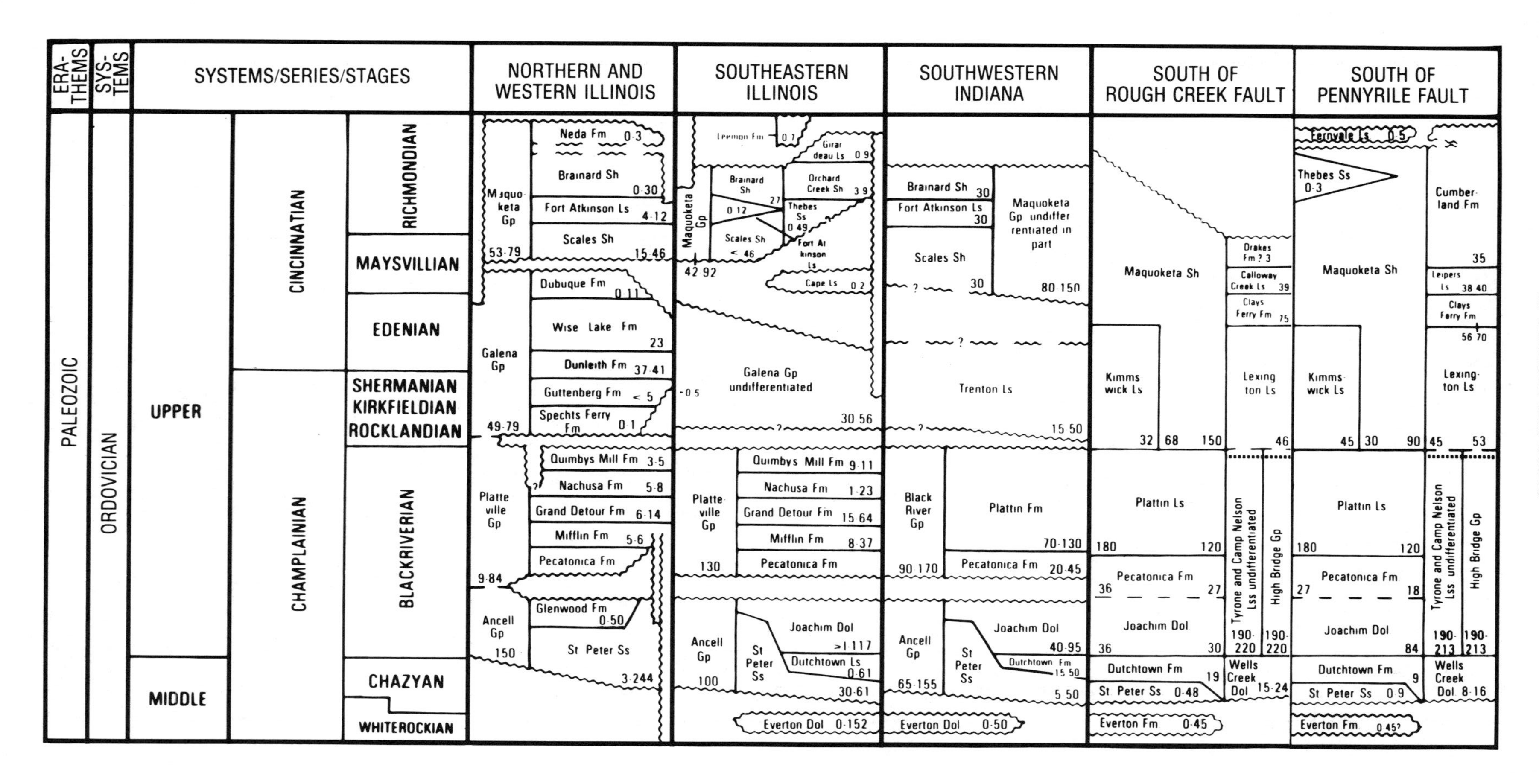

Figure 14. Stratigraphic units of the Tippecanoe I subsequence in the Illinois Basin. Modified from COSUNA correlation chart (Shaver and others, 1985). Wavy lines represent unconformities. The use of "middle" Ordovician and "upper" Ordovician in the text is informal and does not correspond to the above usage.

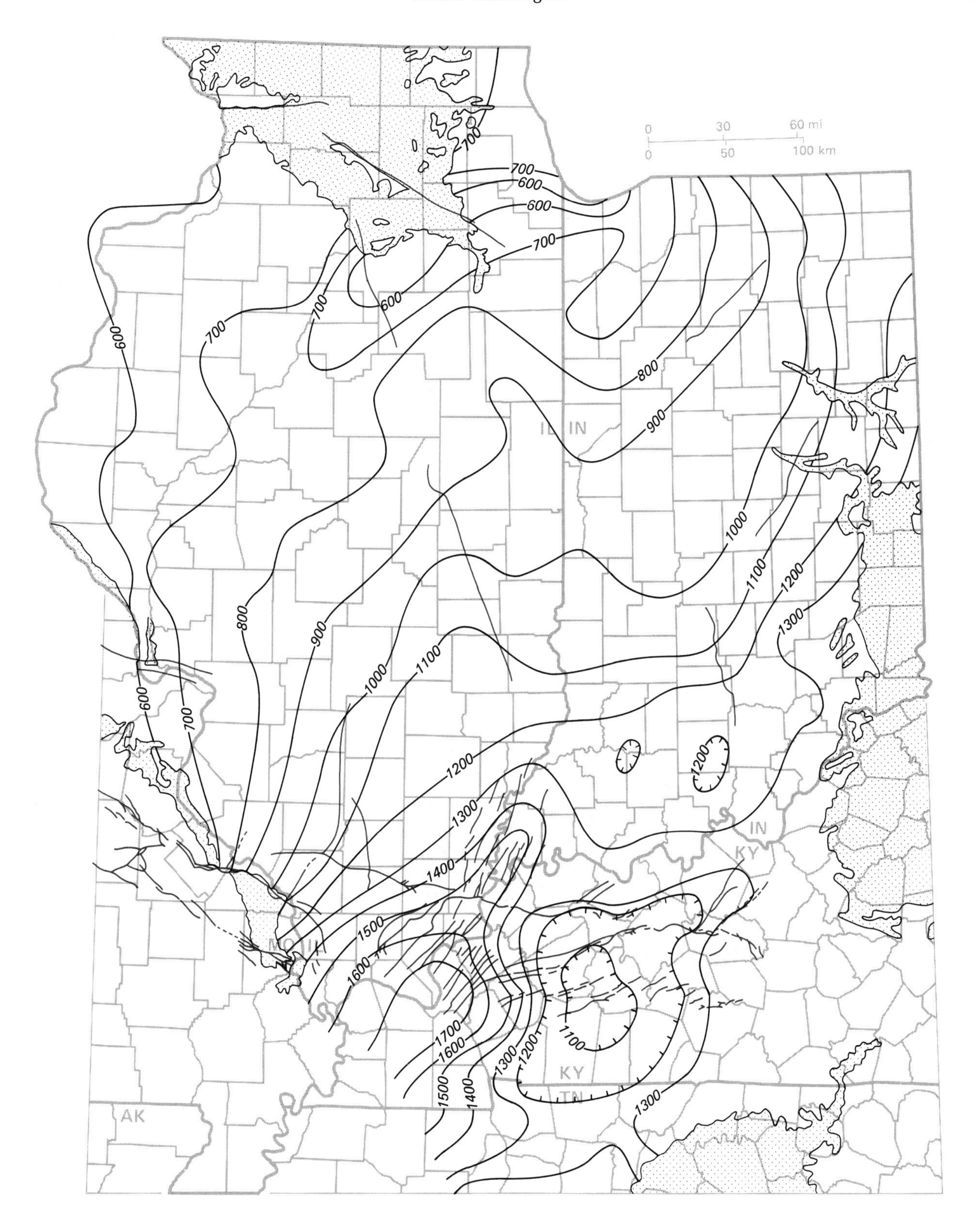

Figure 15. Thickness (in feet) of the Tippecanoe I subsequence (St. Peter Sandstone or Everton Dolomite up through the Maquoketa Group). Faults are shown in red. Stipple represents areas of outcrop or where the subsequence is unconformably overlain by rocks younger than Silurian.

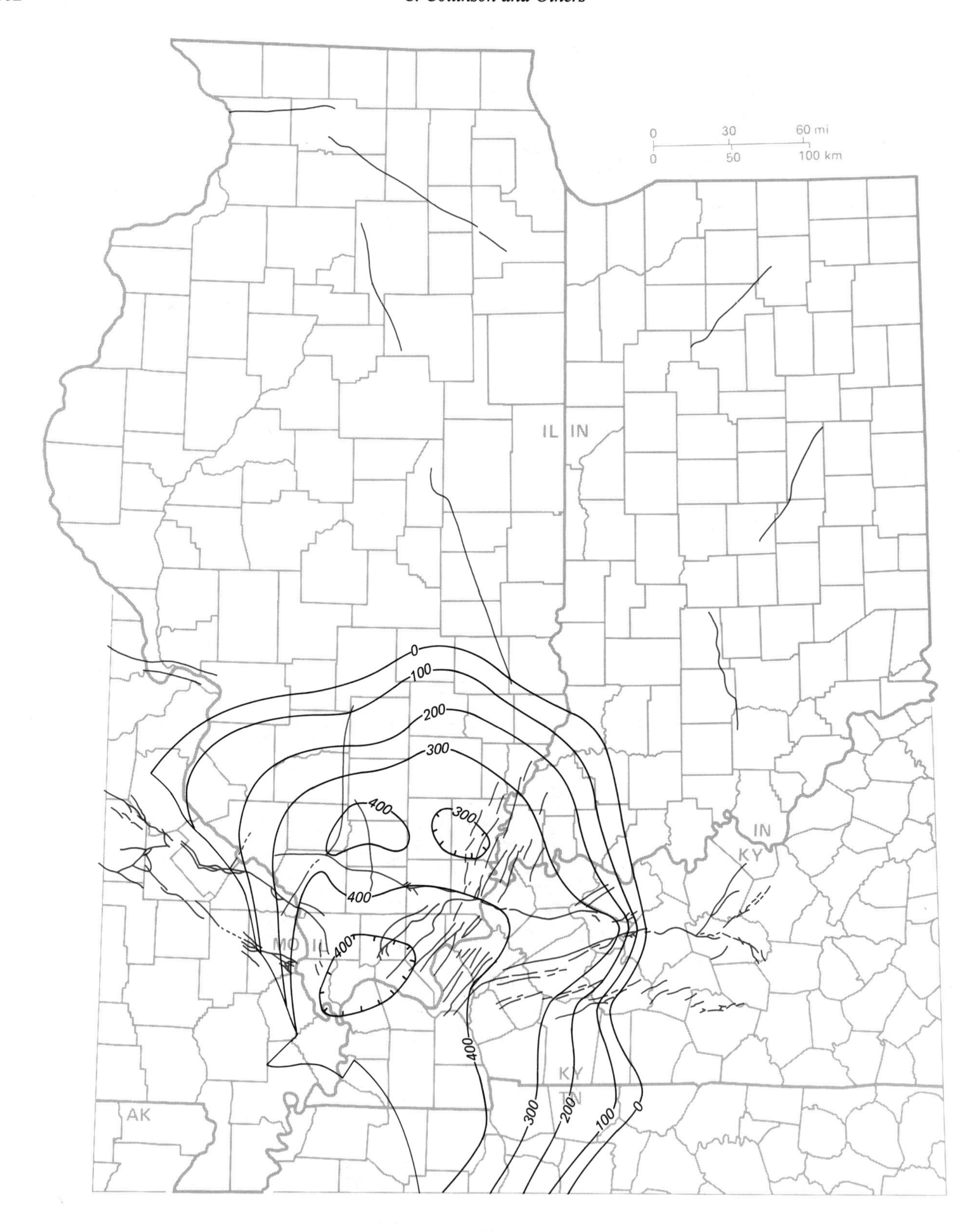

Figure 16. Thickness (in feet) of the Everton Dolomite, a Whiterockian unit isolated by unconformities but showing the broad generalized form of the Illinois Basin at the beginning of Tippecanoe deposition (data from Sargent and Norby, 1986; modified from Droste and Shaver, 1983). Faults are shown in red.

ern Illinois as well as in central and northeastern Indiana (Collinson and others, 1967a; Rogers, 1972). In the central part of the basin, great quantities of silica were deposited in the upper part of the Bailey Limestone (Lower Devonian), Grassy Knob Chert, and Clear Creek Chert. Much smaller amounts of silica were deposited in the Flat Gap Limestone and Backbone Limestone (Figs. 24 and 28). All but possibly the deepest part of the basin became emergent at the end of Early Devonian (Ulsterian) time. In the central area, the Dutch Creek Sandstone of the overlying Kaskaskia sequence appears largely conformable with the underlying Clear Creek Chert of the Tippecanoe sequence.

KASKASKIA SEQUENCE

The Kaskaskia sequence in the Illinois Basin (Figs. 29 to 36) is marked at the base by the Lower Devonian/Middle Devonian unconformity, which is widely developed over the eastern United States. The top of the Kaskaskia is placed at the unconformity beneath the Pennsylvanian System. Erosion that precedes deposition of the Kaskaskia was widespread and deep. In some places it completely removed Niagaran (middle Silurian) rocks (Collinson, 1967). Temporally associated with the unconformity, there was tilting of the Sparta Shelf (Fig. 5), early movement in the Ste. Genevieve Fault Zone (Fig. 5), uplift of the Sangamon Arch (Fig. 30), and uplift of both north (Freeman, 1951) and south sides of the Rough Creek Graben (Figs. 4 and 30). The Kankakee Arch (Fig. 5) on the northeast was broad and low.

The Kaskaskia sequence (Figs. 29 and 33) is divided into Subsequences I and II at the Devonian-Mississippian boundary, which may be a hiatus in some areas. It is largely conformable in the Illinois Basin area, where commonly it can be identified only by paleontological means. The Kaskaskia subsequence I (Fig. 29) is subdivided into IA and IB. The boundary between them is placed at an erosional hiatus that developed during brief exposure of all except the deeper part of the Illinois Basin at the end of Middle Devonian time. There is a resultant erosional surface on top of the Cedar Valley, Lingle, North Vernon, and Sellerburg Limestones. The erosional surface is overlain by a discontinuous lag sand (Collinson and others, 1967a), the Sylamore Sandstone in northeastern Missouri, western Illinois, and central Illinois (Fig. 29).

The oldest Kaskaskia sediments were sands (i.e., Dutch Creek Sandstone) swept into the central basin largely from its exposed slopes, and carbonates (both dolomites and limestones, e.g., Wapsipinicon Limestone, Geneva Dolomite Member of the Grand Tower Limestone) were deposited in shallow—in places evaporitic—seas that transgressed onto the coastal flats along the flanks of the Illinois Basin (Figs. 30 and 31). Conditions of deposition began to change shortly thereafter with the introduction of clay from the Acadian orogen. The terrigenous clastics were deposited as local lenses in Middle Devonian limestone units (e.g., Cedar Valley Limestone, Lingle Formation, North Vernon Limestone, and Sellersburg Limestone; Collinson and others, 1967a, 1967b). Banks of carbonate sand and mud (Grand Tower Limestone and Jeffersonville Limestone) were deposited across south-

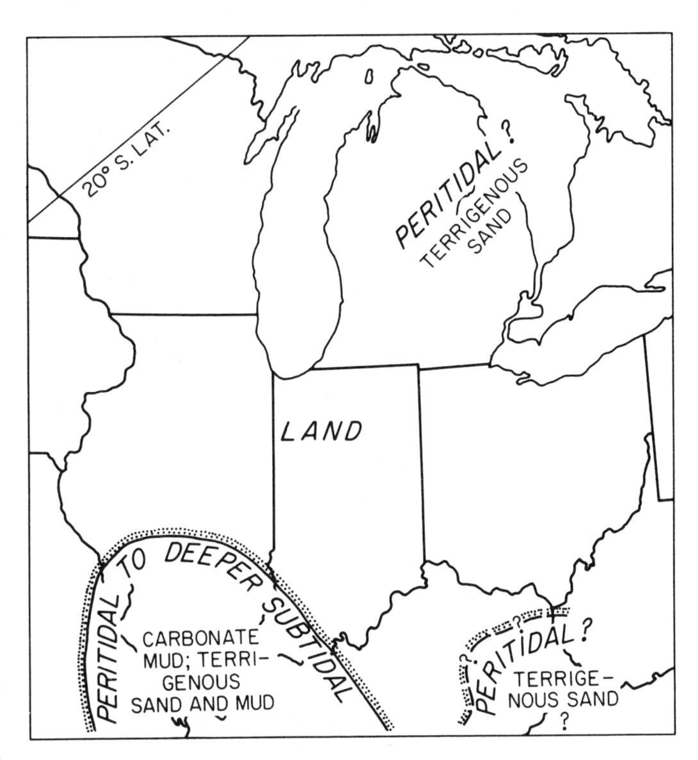

Figure 17. Paleogeography of the Whiterockian Stage (Everton Dolomite; from Droste and Shaver, 1983, Fig. 14).

ern Illinois, southern Indiana, and western Kentucky. The Tioga Bentonite Bed, an ash bed derived from Appalachian volcanism, occurs in this interval over a broad region (Collinson and others, 1967a). In southeastern Missouri, Devonian igneous intrusives are indicative of tectonic activity adjacent to the Illinois Basin (Tarr and Keller, 1933; Zartman and others, 1967).

The influx of clay from the Acadian orogen increased with time. In addition, minor sands were contributed from the nearby Ozark Uplift. The New Albany Group (mostly shale; Figs. 29 and 32) was deposited during Late Devonian through Early Mississippian (Kinderhookian) time (Collinson and others, 1967a; Collinson and Atherton, 1975). Black laminated shales, characteristic of high plant productivity (much of it apparently algal) combined with low-oxygen conditions, covered most of the central midcontinent. Black shales gave way to gray shales that buried the Sangamon Arch (Cluff and others, 1981; Lineback, 1968). Greatest thicknesses of the New Albany (120 to 150 m) are centered in the southern part of the Illinois Basin. The New Albany also thickens northwestward into eastern Iowa.

The Kaskaskia I and II subsequences of Sloss (1982) are very difficult to separate in the Illinois Basin. Consequently, the thicknesses of Pope and Mammoth Cave Megagroups (Figs. 3 and 33) have been combined to approximate the thickness of the Kaskaskia II subsequence. The New Albany Group (Figs. 29 and

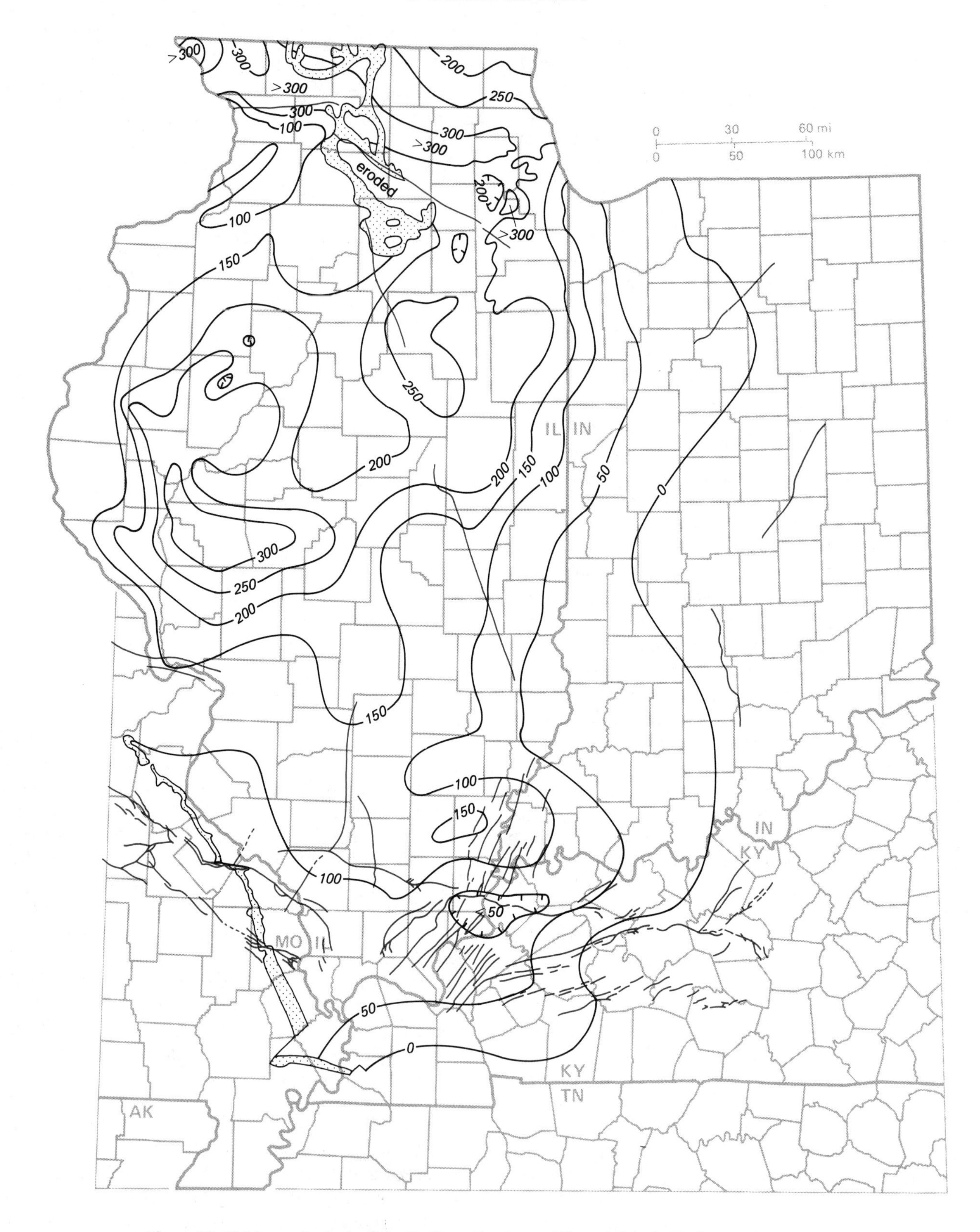

Figure 18. Thickness (in feet) of the St. Peter Sandstone (Champlainian). Faults are shown in red. Stipple indicates outcrop areas and pre-Cretaceous subcrop. Some closely spaced contours are deleted in northern Illinois (data from Sargent and Norby, 1986; and modified from Buschbach, 1964; Willman and Buschbach, 1975).

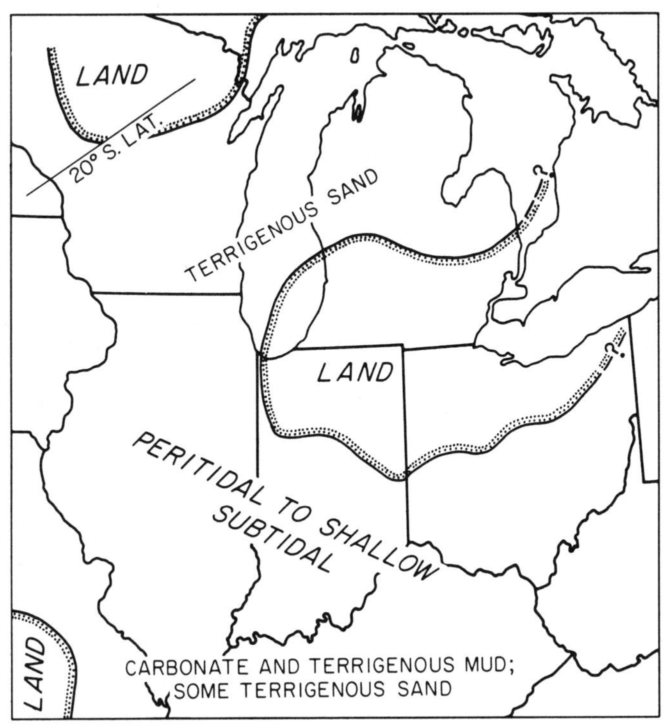

Figure 19. Paleogeography of the Chazyan Stage (St. Peter Sandstone and Dutchtown Limestone; from Droste and Shaver, 1983, Fig. 17).

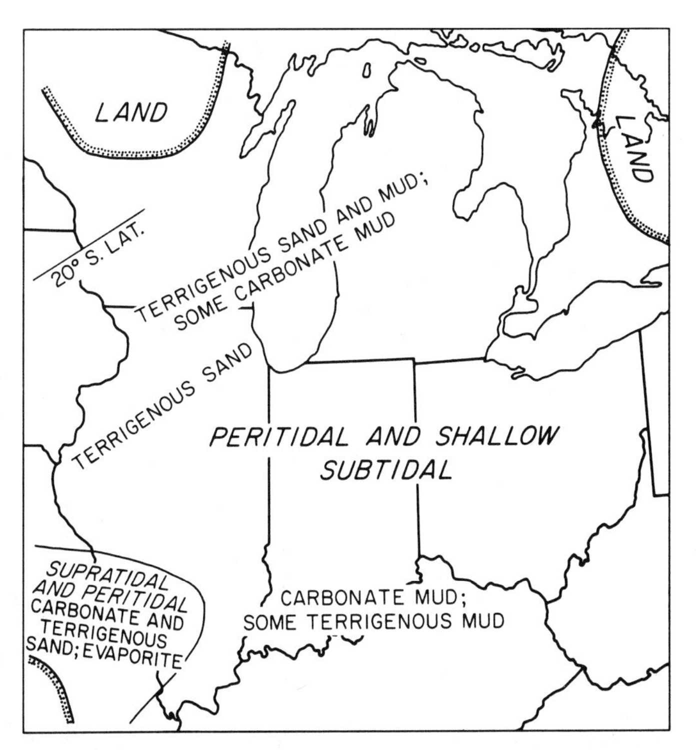

Figure 20. Paleogeography of the Blackriveran Stage (Platteville/Black River Group; from Droste and Shaver, 1983, Fig. 19).

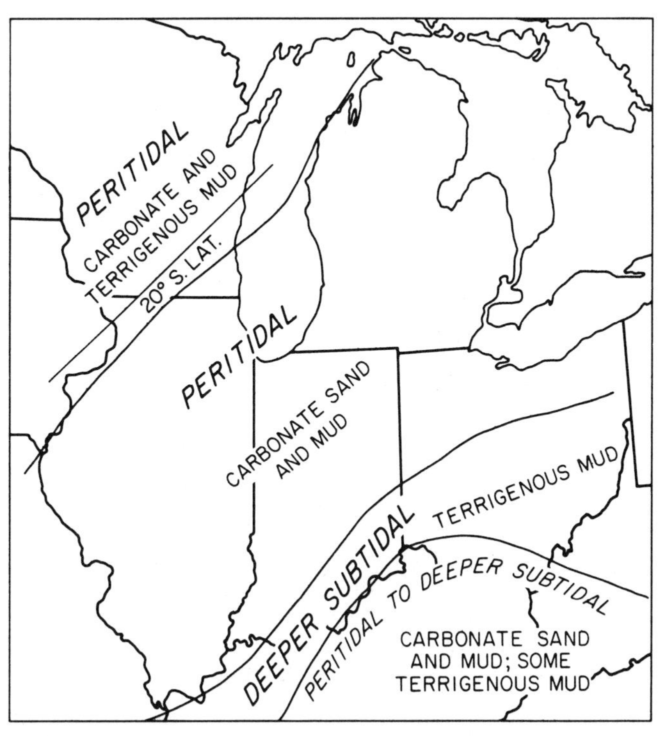

Figure 21. Paleogeography of the Kirkfieldian Stage (Galena–Trenton) showing peritidal seas throughout the basin area (from Droste and Shaver, 1983, Fig. 23).

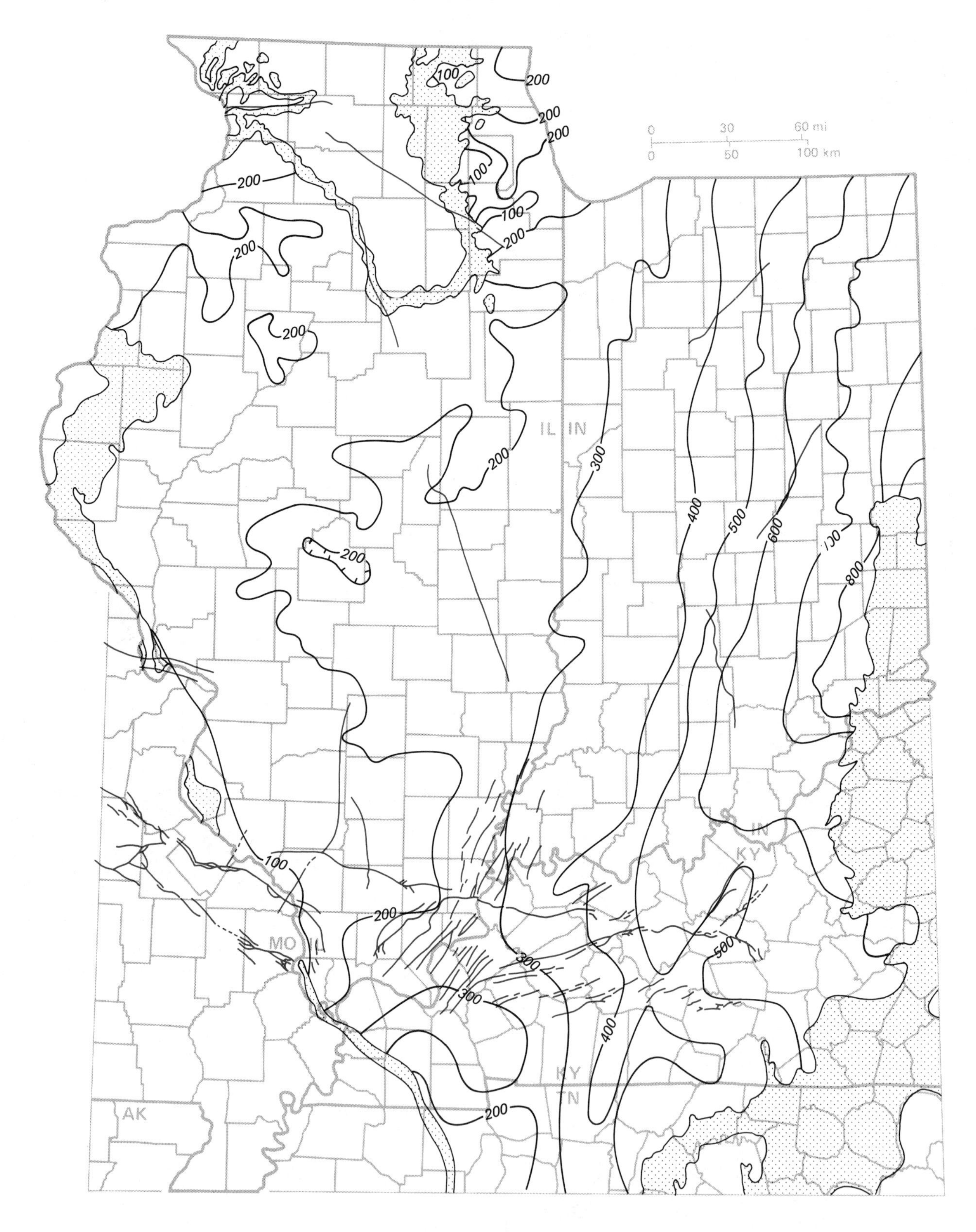

Figure 22. Thickness (in feet) of the Maquoketa Shale Group in the Illinois Basin (based on Willman and Buschbach, 1975; Kolata and Graese, 1983; and Schwalb, personal communication). Stippling indicates outcrop areas and areas where the Maquoketa is unconformably overlain by rocks younger than Silurian.

32), which is part of both Kaskaskia I and II, has been illustrated separately from the Middle Devonian Series (Figs. 29 and 30), which contains most of the Kaskaskia I strata.

At the end of the New Albany deposition (Kinderhookian time), a relatively thin carbonate unit (the Chouteau Limestone of Missouri and Illinois, the Rockford Limestone of Indiana and Kentucky; see Fig. 34) was deposited on the margins of the central Illinois Basin in relatively deep water. The sea remained clear during early Valmeyeran time; nevertheless, the Illinois Basin continued to sink, leaving much of it in a sediment-starved condition (Lineback, 1966; Lane, 1978). The starved area extended southwestward at least as far as Oklahoma. During this time, the relatively thin Springville Shale (Fig. 35) slowly accumulated in the deep basin area (Atherton and others, 1975; Cluff and others, 1981). A large carbonate bank of crinoidal limestone formed on the northwestern and western edge of the Illinois Basin. The Fern Glen, Burlington, and Keokuk Formations represent parts of this accumulation. The bank extended as far west as the Transcontinental Arch in Kansas. Named the Burlington Bank (Lane, 1978), it towered (Fig. 35) over the depths of the starved basin. The Fort Payne Chert, consisting of calcareous and siliceous mudstone derived from eastern sources, slowly filled the southeastern part of the basin. Beds of the most distal part of the Borden Delta (Fig. 35) were building westward across Kentucky and Indiana simultaneously with the development of the Burlington Bank (Lineback, 1981). In middle Valmeyeran time, Borden siliciclastics filled the deep portion of the basin with more than 180 m of silty sediment. The siliciclastic sediment thinned out onto the eastern part of the Burlington Shelf; this westernmost deposit of the Borden Delta is called the Warsaw Shale. The Warsaw was succeeded by sandy, bryozoan-crinoid–rich beds of the Sonora and Salem Formations. To the east the Borden Siltstone was succeeded by the Fort Payne Formation, the bryozoan-crinoid–rich Ullin Limestone, and the Salem Limestone as well as by the Sanders Group of Indiana. These bryozoan-crinoid–rich limestones are indicative of clear seas with good circulation. The Ullin Limestone, derived from biogenic carbonates in the eastern shallows, filled much of the southern part of the Illinois Basin (Fig. 34), returning it to shallow-water conditions. The paleoslope was to the southwest.

The succeeding St. Louis Limestone, a widespread unit, is mainly a cherty limestone with interbedded calcareous mudstone. These lithologies formed mainly in the deeper areas. The St. Louis also contains biogenic grainstone, fine-grained dolomite, and anhydrite (Atherton and others, 1975). Algal mats, lime mudbanks, gypsum, and bryozoan-crinoidal carbonates represent deposits from shallow supratidal, intertidal, and subtidal environments, respectively, along margins of the basin.

The overlying Ste. Genevieve Limestone is equally widespread. It represents combinations of calcareous mudstones and grainstones in cyclical successions that include major occurrences of oolitic limestone and (in the upper part) small amounts of quartz sandstone. The lime mud of the Ste. Genevieve was mainly deposited in deeper seaways and protected embayments

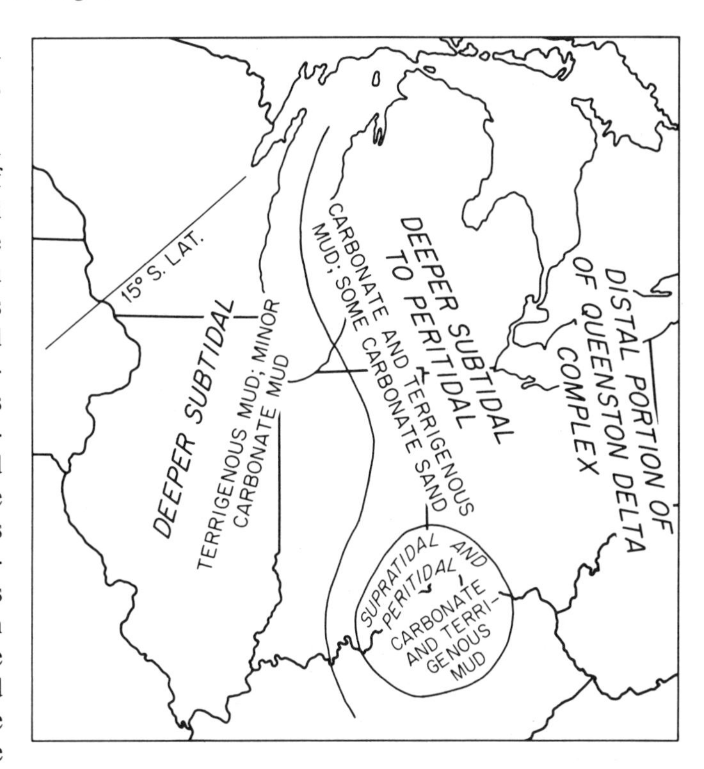

Figure 23. Paleogeography of late Richmondian Stage (Maquoketa Group) showing encroachment of terrigenous muds from the northeast (from Droste and Shaver, 1983, Fig. 28).

where wave energy was low. Oolite was deposited on carbonate shoals and adjacent deeper waters, forming, where porous, the important oil-production McClosky pay zones of the Illinois Basin. The Aux Vases Sandstone overlies the Ste. Genevieve and, unlike Late Mississippian sands generally, contains a significant component from a westerly source (Potter and Pryor, 1961).

A major change in the regional pattern of sedimentation took place at approximately the end of the Valmeyeran Epoch, and for the remainder of the Mississippian, the Michigan River System of Swann (1963, 1968) transported sediments into the Illinois Basin from the northeast. This river (Fig. 36), its source far to the northeast, introduced sediments that were deposited in shifting prograding delta complexes. Orogenic activity in the Appalachian tectonic belts provided the source for most Chesterian terrigenous clastic sediment, according to Potter and Pryor (1961). Swann (1963), however, suggested that the primary source area was the eastern part of the Canadian Shield. Combinations of intermittent subsidence of the basin, lateral shifts in delta systems, and eustatic sea-level changes have been invoked to account for the cyclic succession of limestone, sandstone, and shale, which makes up the Chesterian (Swann, 1964). During the furthest progradation of each Late Mississippian delta complex, large areas of the Illinois Basin were filled. The thin carbonaceous

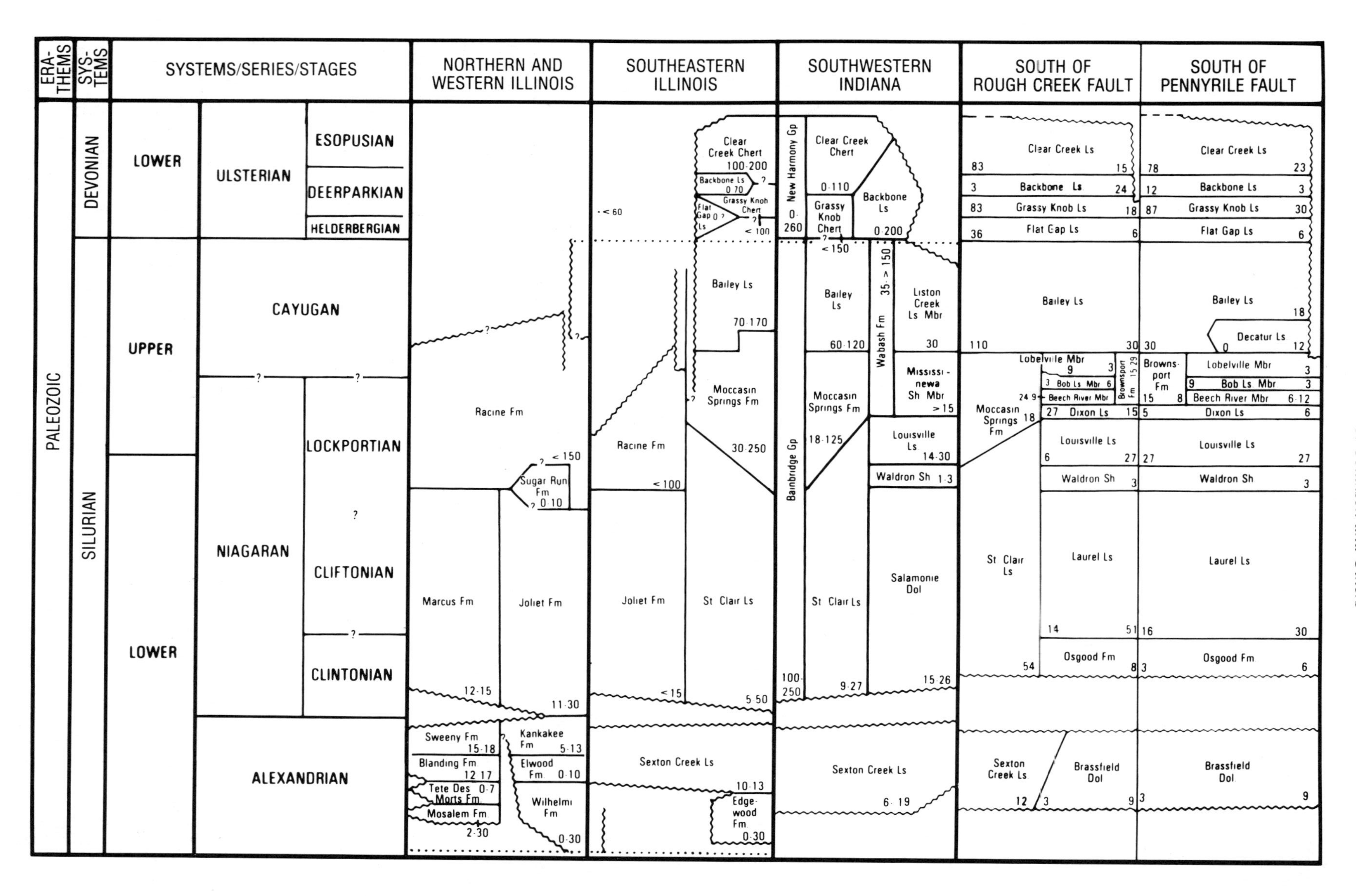

Figure 24. Stratigraphic units of the Tippecanoe II subsequence in the Illinois Basin. Modified from COSUNA correlation chart (Shaver and others, 1985). Wavy lines represent unconformities.

beds that were deposited on top of the delta platforms (Jennings, 1984) foreshadowed the deposition of economically important coals of the Pennsylvanian. Marine limestone is much more important in the Late Mississippian, however, and some Chesterian units are substantially like strata in the Valmeyeran. Chesterian deposition was terminated by a major retreat of the sea, and even the southernmost part of the Illinois Basin was for a time emergent. The basin area was exposed to erosion, and some streams cut deep valleys into its surface. Caves and sinkholes were formed in the exposed Ordovician, Silurian, Devonian, and Mississippian carbonate rocks, especially in northern and western Illinois.

There was major tectonic activity involving movements along the Ste. Genevieve Fault Zone, ancestral Rough Creek–Shawneetown Fault System, Du Quoin Monocline, Waterloo-Dupo Anticline, Cap au Grès Faulted Flexure (Lincoln Fold and Troy-Brussels Syncline), Mississippi River Arch, Salem and Louden Anticlines, and La Salle Anticlinal Belt (Fig. 5; e.g., Nelson and Lumm, 1984, 1985; Krausse and Treworgy, 1979).

ABSAROKA SEQUENCE

In the Illinois Basin, the Absaroka sequence is represented almost entirely by strata of the Pennsylvanian System (Hopkins and Simon, 1975; Palmer and Dutcher, 1979a, 1979b; Shaver and others, 1986), although a small amount of the lowermost part of the Permian System is preserved in a small fault block in Kentucky (Fig. 37). Absaroka strata overlie a great post-Kaskaskia unconformity, and a considerable chronostratigraphic gap exists, especially in the northern part of the Illinois Basin. They underlie an even greater unconformity at the base of the Cretaceous System. Maximum thickness of strata representing the Absaroka sequence exceeds 900 m in the Fairfield Basin and may be as much as 1,370 m in the Moorman Syncline near its southern limit (Fig. 38). Coalification studies indicate that an additional thickness of hundreds of meters may have been lost to erosion (Damberger, 1971, 1974). Furthermore, Pennsylvanian rocks may have once extended north to the Illinois-Wisconsin state line, inasmuch as Pennsylvanian strata are present in downfaulted blocks of the Des Plaines Disturbance northwest of Chicago (Fig. 5; Emrich and Bergstrom, 1962). Today, however, the main body of rock of the Pennsylvanian System in the Illinois Basin lies about 64 km south-southwest of the Des Plaines Disturbance.

A number of tectonic events took place immediately before, or during, earliest Absaroka deposition (Wanless, 1975). The La Salle Anticlinal Belt (Fig. 5) rose in the northern part of the Illinois Basin, while erosion stripped away its top and streams breached its southern end (Howard, 1979). Nevertheless, it remained a positive feature throughout the time of Absaroka deposition and defined the eastern margin of the Fairfield Basin. The Mississippi River Arch (Fig. 1) was present during Early Pennsylvanian time. Folding along the Cap au Grès Faulted Flexure in western Illinois occurred during this interval as did the renewed activity of the Du Quoin Monocline, which partitions the Sparta Shelf from the

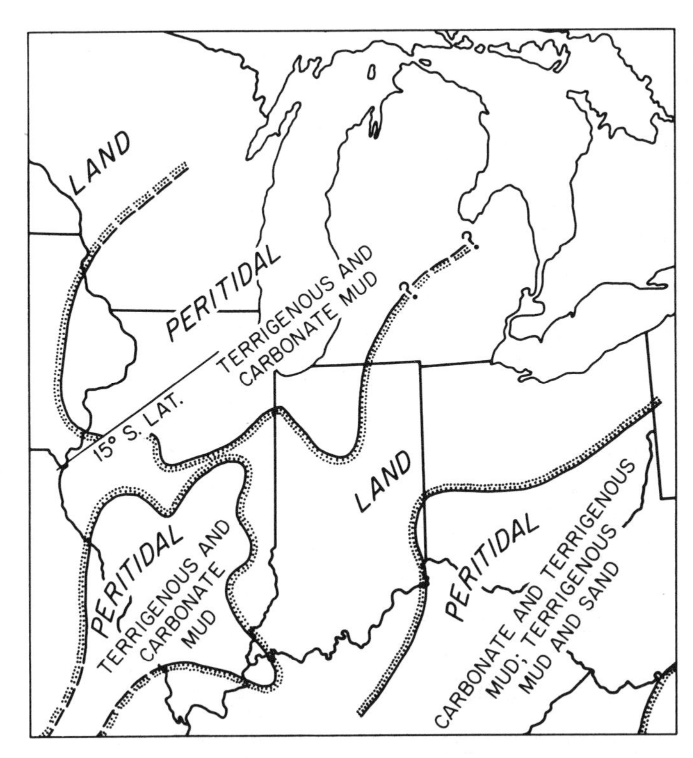

Figure 25. Paleogeography of middle Alexandrian (Early Silurian) time showing widespread emergence and broad shallow seas in the central part of the basin (from Droste and Shaver, 1983, Fig. 30).

Fairfield Basin (Krausse and Treworgy, 1979; Fig. 5). The Ozark Uplift was active during Early Pennsylvanian time, although it later became a shoal and received Pennsylvanian sediment (Tikrity, 1968).

As a result of multiple vertical crustal movements and whatever eustatic drop in sea level may have taken place, the entire area of the basin lay exposed at the beginning of the Pennsylvanian. Erosion cut valleys as much as 32 km wide and 130 m deep (Howard, 1979) before the Pennsylvanian sediments covered the area. The trend of the paleovalleys indicates a southwesterly paleoslope, as does cross-bedding in sediments above and below (Bristol and Howard, 1971; Howard, 1979; Potter and Pryor, 1961). In the southern part of the Illinois Basin, Pennsylvanian strata overlie Late Mississippian rocks. Northward, successively older sub-Pennsylvanian units are truncated. On the crest of the La Salle Anticlinal Belt in northern Illinois, Pennsylvanian rocks directly overlie the Ordovician St. Peter Sandstone (e.g., Hopkins and Simon, 1975).

Subsidence of the Illinois Basin in Early Pennsylvanian time allowed sedimentation to resume. The first areas to begin accumulating Pennsylvanian sediments lie mainly near the southern end of the Illinois Basin, the region lowest on the paleoslope. Although there is a hiatus at the base of the system even in this

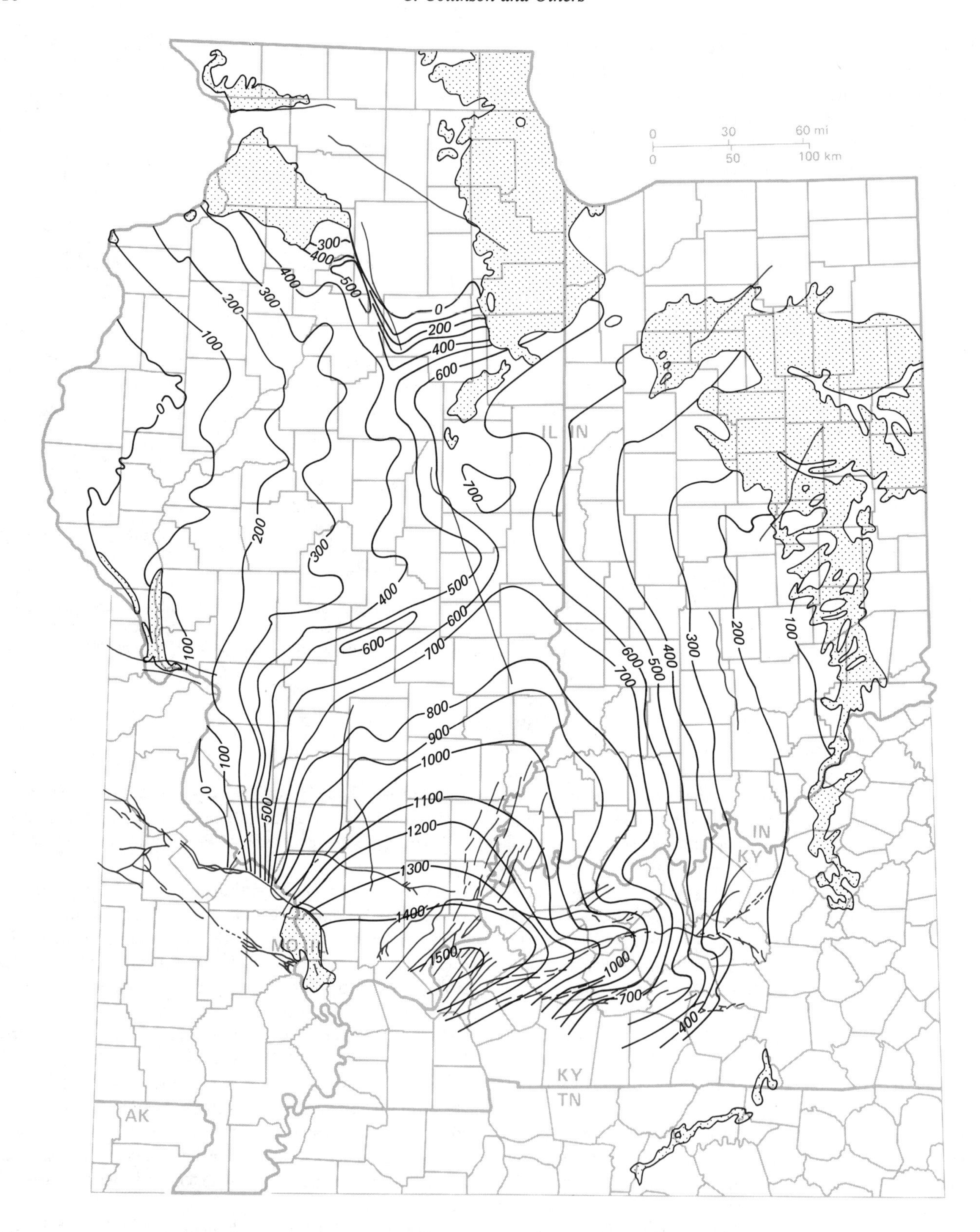

Figure 26. Thickness (in feet) of the Tippecanoe II subsequence, which extends from the base of the Silurian (Alexandrian) to the top of the Lower Devonian (Ulsterian; Fig. 24). Faults are shown in red. Stipple represents areas of outcrop or where the subsequence is unconformably overlain by rocks younger than Devonian.

region, Pennsylvanian sediment had begun to accumulate by mid-Morrowan time (Jennings and Fraunfelter, 1986; Peppers, 1984). Strata that fill sinkholes of sub-Pennsylvanian paleokarst (e.g., Edmund and Anderson, 1967) were also deposited quite early in the period (Peppers, 1984).

As the Illinois Basin continued to subside during the Pennsylvanian, strata were deposited progressively farther north. Where the northern limit of the Pennsylvanian lies on the La Salle Anticlinal Belt, the strata at the base of the system fall within the Desmoinesian Series (Middle Pennsylvanian in age). Differential subsidence also resulted in a much thinner sequence of later Desmoinesian and Missourian rocks in the northern part of the basin than toward the south (e.g., Hopkins and Simon, 1975).

Deposition of early Absaroka strata was dominated by a massive influx of coarse terrigenous siliciclastics. Indeed, the presence of quartz granules and pebbles is helpful in distinguishing basal Absaroka strata from otherwise similar siliciclastics of the Chesterian. Early Absaroka strata, referred to the Caseyville or Mansfield Formations (Fig. 37), locally exceed 150 m, but are highly variable in thickness partly because they were deposited on an irregular surface.

A few Caseyville (and equivalent) strata were deposited under marine conditions, as indicated by the presence of marine invertebrate fossils (e.g., Devera and others, 1987; Jennings and Fraunfelter, 1986). The gradient of the paleoslope and the climate of the area, however, did not permit a return to clear-water open-marine sedimentation of the sort that was widespread during the Mississippian in the Illinois Basin. Discontinuous coals or carbonaceous zones were also deposited at various horizons. Facies relationships among these beds are complex. The absence of widely traceable beds (e.g., limestones representing a single marine transgression) and the scarcity of invertebrate fossils have made determining ages for Lower Pennsylvanian strata difficult. Nevertheless, biostratigraphic information (Devera and others, 1987; Jennings, 1974, 1984; Jennings and Fraunfelter, 1986; Peppers, 1984; Shaver and others, 1986) is consistent with a Morrowan age for all Caseyville and equivalent strata.

The overlying Atokan strata of the Abbott, Tradewater, and Mansfield-Brazil Formations are lithologically transitional between Caseyville (and lower Mansfield) Formation and younger units (Hopkins and Simon, 1975; Shaver and others, 1985). Whereas a relatively steep gradient on the paleoslope caused the incorporation of large quantities of coarse siliciclastics into the Caseyville Formation, there was a progressive reduction of the gradient of the paleoslope from the Lower to the Middle Pennsylvanian. As a consequence, finer siliciclastics are typical of the Desmoinesian, Missourian, and Virgilian deposits. An increased amount of feldspar is present in Atokan (Abbott, Tradewater, Mansfield-Brazil) and later sandstones as compared with the dominantly quartzose sandstones of the Chesterian and Morrowan (Caseyville-Mansfield). This greater amount of feldspar in sandstones of the Atokan is related to a source area in the Canadian Shield to the northeast (Potter and Pryor, 1961). The Atokan strata also include more named coal beds, although they tend

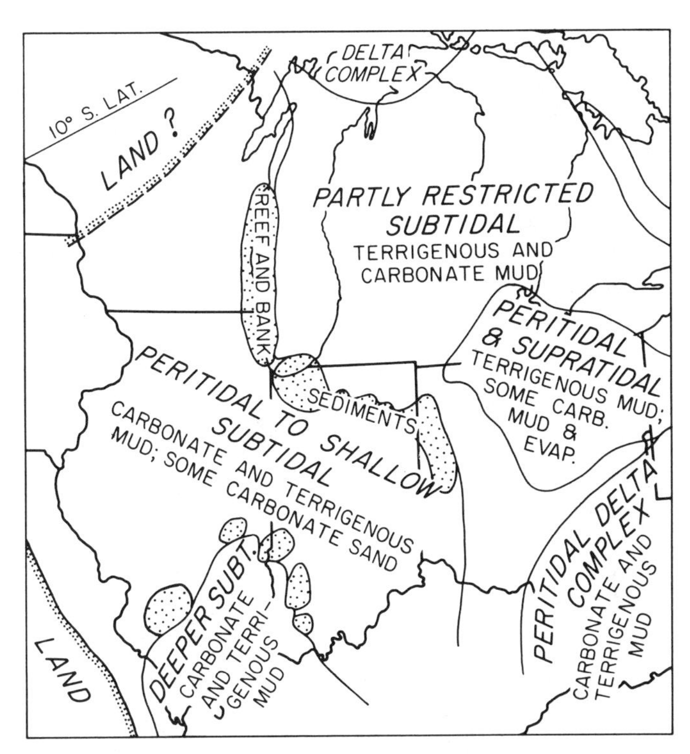

Figure 27. Paleogeography of early Cayugan time (Late Silurian). Stippled areas indicate reef tracts and carbonate banks (from Droste and Shaver, 1983, Fig. 38).

to be thin, and have better-developed limestones than do earliest Pennsylvanian sediments.

The Desmoinesian Series (Kewanee Group and lower part of the McLeansboro Group, Hopkins and Simon, 1975; Staunton Formation, Carbondale Group, and lower McLeansboro Group, Shaver and others, 1985; upper Tradewater, Carbondale Formation, and lower Sturgis Formation, Shaver and others, 1985) contains significant deposits of limestone, sandstone, claystone, gray shale and siltstone, black shale, and coal. Fine terrigenous siliciclastics dominate the Desmoinesian quantitatively, but economically, coal is by far the most significant rock type. Virtually all of the mineable coal in the Illinois Basin was deposited during this time. In contrast to the abrupt lateral facies changes characteristic of the Morrowan and Atokan (Caseyville and Abbott), the lateral continuity of many individual beds (many of which are less than 1 or 2 m thick) of the Desmoinesian is remarkable. This continuity results partly from the fact that the very slight gradient of the paleoslope allowed small vertical shifts in sea level, from either eustatic or tectonic causes, to shift the environment of large areas between terrestrial and marine. Deltaic complexes prograded rapidly across the exceedingly shallow shelf. More than 30 cycles of transgression followed by deltaic progradation are recognizable in the Desmoinesian Series alone. At times, nearly the entire Illinois Basin was inundated by marine waters; whereas, at

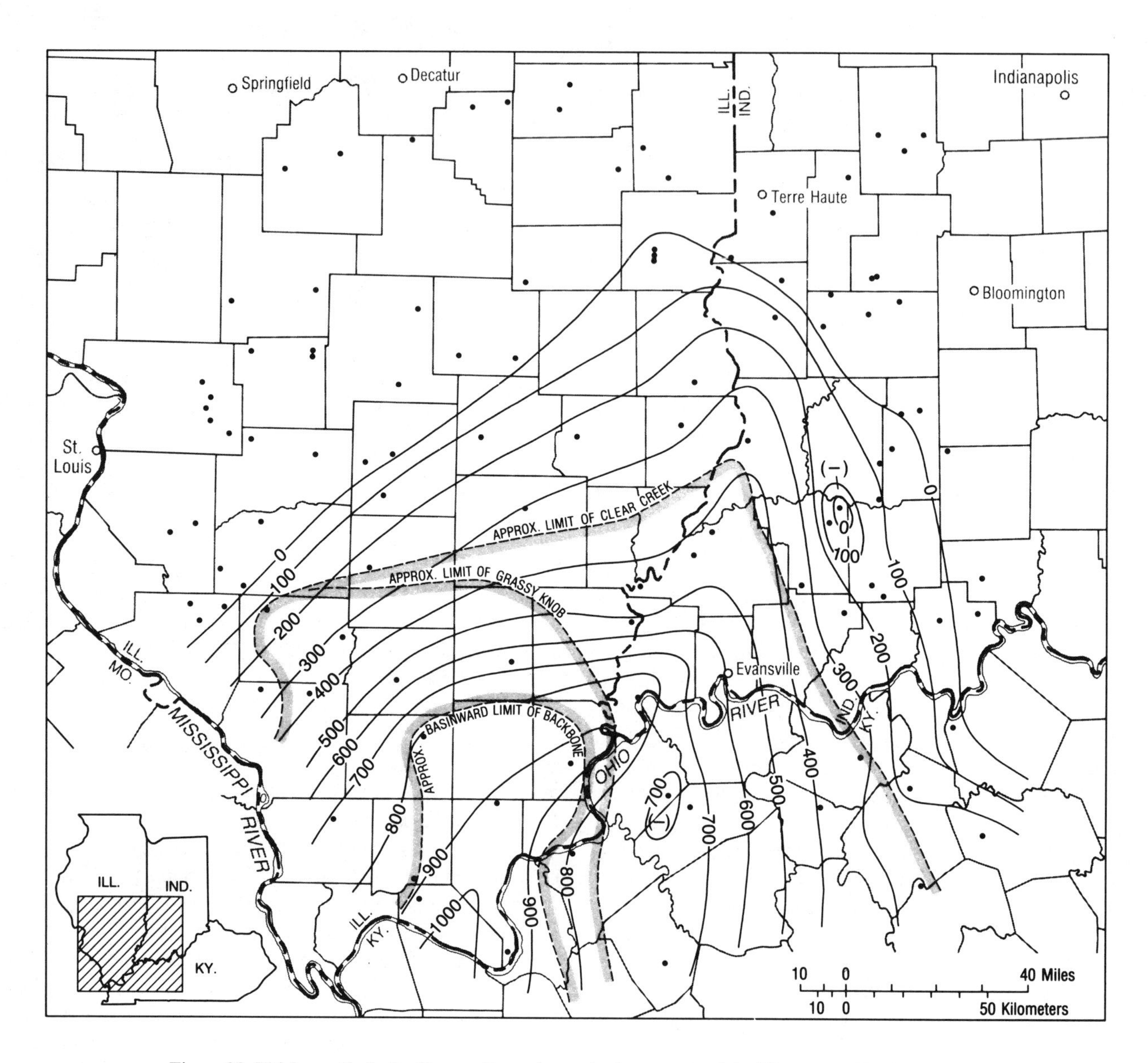

Figure 28. Thickness (in feet) of Lower Devonian rocks (upper part of the Tippecanoe II subsequence) showing their restricted distribution resulting from post-Tippecanoe erosion over all but the most central part of the basin (from Droste and Shaver, 1987, Fig. 6).

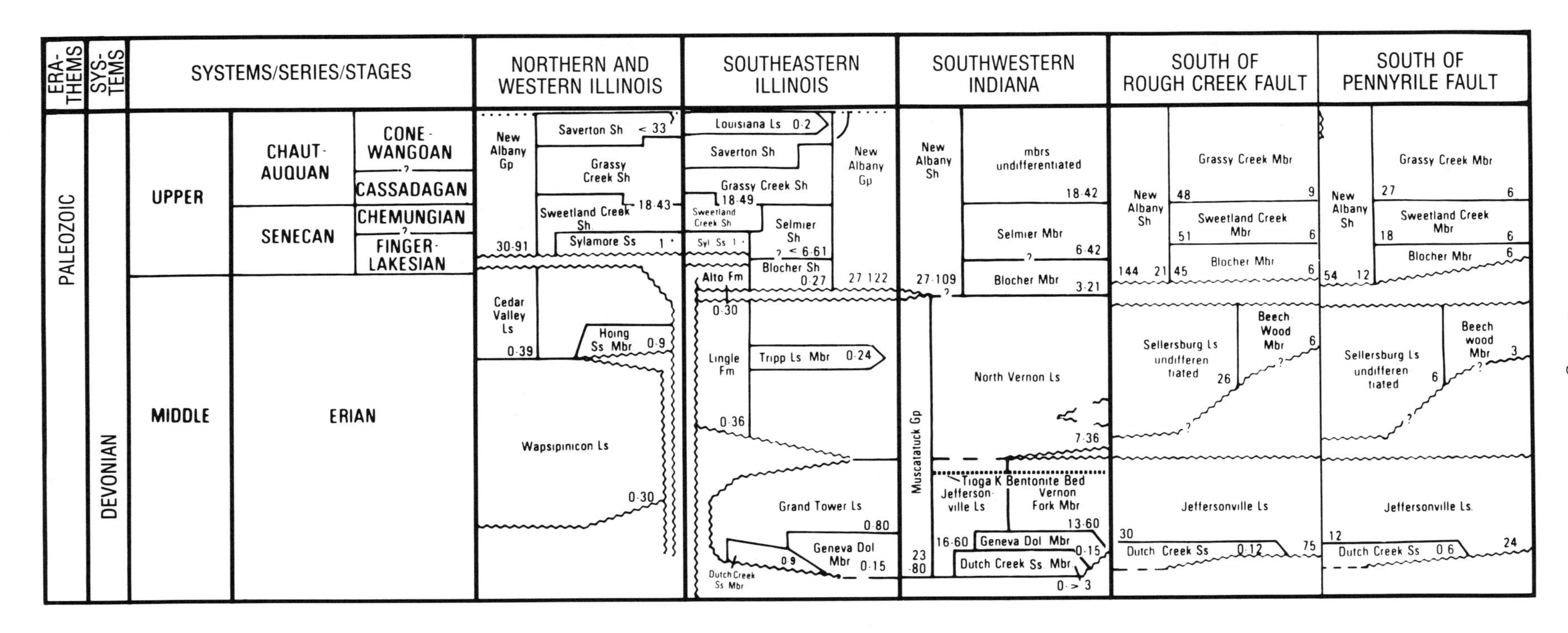

Figure 29. Stratigraphic units of the Kaskaskia I subsequence. Modified from COSUNA correlation chart (Shaver and others, 1985). Wavy lines represent unconformities.

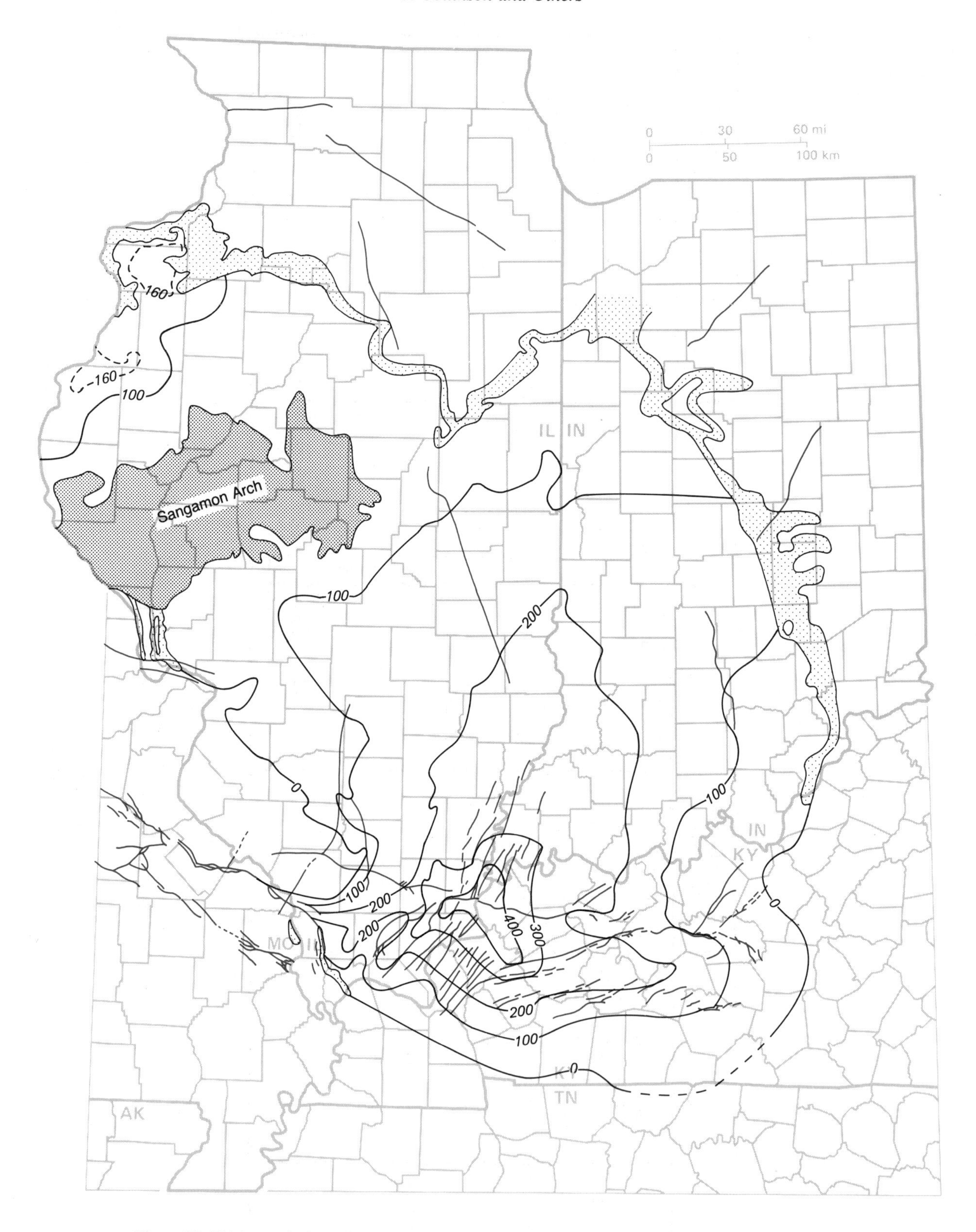

Figure 30. Thickness (in feet) of the lower part of the Kaskaskia I subsequence (Middle Devonian only) showing the gentle slopes and evidence of slight subsidence in the Illinois Basin. Faults are shown in red. Light stipple represents outcrop areas and places where rocks younger than Devonian overlie the formations unconformably. Denser stipple represents area where Middle Devonian strata were not deposited on the Sangamon Arch.

other times, it was nearly filled by deltaic sediment allowing development of swamp forests over most of its area. Of particular interest are paleochannels that existed contemporaneously with peat from the swamp forests (Gluskoter and Hopkins, 1970; Treworgy and Jacobson, 1985). The rivers that occupied these channels meandered for many kilometers through the swamp forests of the delta platform, and the channels were ultimately filled with terrigenous siliciclastics. The rivers remained active even as the sea encroached on the swamp forests, and they deposited gray shale or siltstone above the coal in nearby regions. In areas more distant from these paleochannels, black shales or marine limestone may directly overlie the coal (Fig. 39).

As in the Desmoinesian, fine siliciclastic deposition predominated during Missourian and Virgilian time (Late Pennsylvanian), yielding many shales (gray and black), siltstones, and claystones. Beds of the Missourian and Virgilian (upper part of the McLeansboro Group and Sturgis Formation, Hopkins and Simon, 1975; Shaver and others, 1985), however, contain more limestones and correspondingly fewer (and thinner) coals than the underlying beds. Limestone deposition reached its maximum in mid-Missourian time, and limestone constitutes as much as 10 percent of that section. The proportion of sandstone increases in upper Missourian to Virgilian rocks, but sandstone does not reach the proportion present in Morrowan and Atokan strata. Nevertheless, some channel-fill sandstone is as much as 24 m thick (Hopkins and Simon, 1975). Black shales, limestones, and coals (although they are thin) of the Missourian and Virgilian show lateral continuity nearly equivalent to the continuity of strata of the Desmoinesian, in contrast to the lenticular character of Morrowan-Atokan strata. The Missourian and Virgilian have a combined thickness of over 245 m in the Fairfield Basin where the top is an erosional surface. In the Moorman Syncline, where the entire section is present below Permian beds, their combined thickness reaches 550 m.

Deposition of Absaroka rocks continued, apparently without interruption, into Early Permian (Wolfcampian) time, at least in the vicinity of the Rough Creek–Shawneetown fault zone. In this area, 120 m of Permian strata named the Mauzy Formation (Kehn and others, 1982) have been penetrated. Judging from the maturity of coals (Damberger, 1971, 1974) and degree of compaction of Desmoinesian claystones (Altschaeffl and Harrison, 1959), Pennsylvanian rocks appear to have been buried by many hundred meters of rock. The maximum thickness of the Pennsylvanian in the Illinois Basin (with emphasis on the section overlain by Permian strata) indicates that much of this rock must have been Permian in age. Rocks of the Mauzy Formation are dominantly marine and contain about 25 percent limestone.

Effective subsidence of the Illinois Basin ceased some time after deposition of Wolfcampian sediments. Only much later, when the Mississippian Embayment formed during Late Cretaceous time, did any of the area of the Illinois Basin again subside significantly. A hiatus of about 200 million years is thus present.

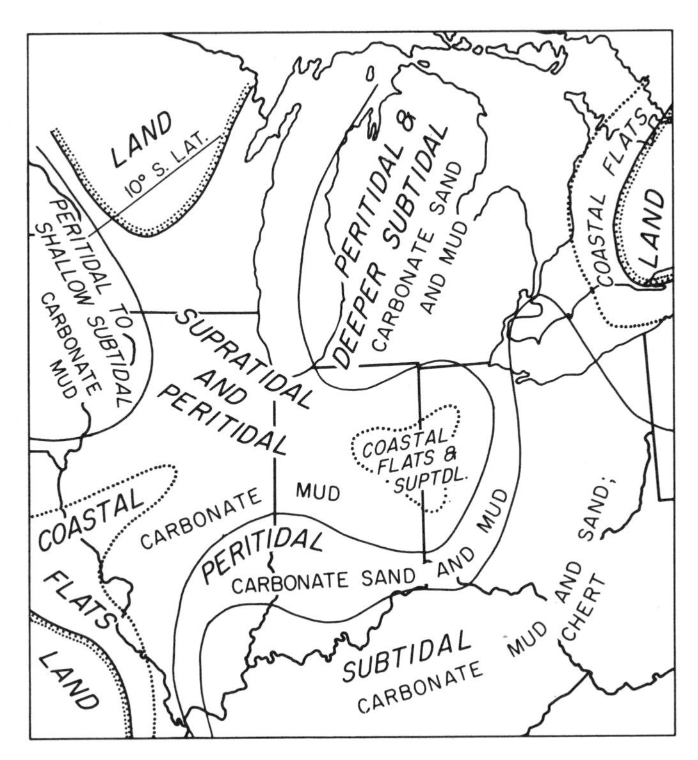

Figure 31. Paleogeography during part of Middle Devonian time showing emergent coasts and shallow coastal flats (from Droste and Shaver, 1983, Fig. 45).

POST-ABSAROKA

A number of tectonic events occurred in the Illinois Basin after the close of Absaroka deposition. Tectonism occurred in the deepest parts of the Illinois Basin in both Late Pennsylvanian and Permian times. There were major movements on the Cottage Grove Fault System (Nelson and Krausse, 1981), the Wabash Valley Fault System, Hicks Dome, the Fluorspar Area Fault Complex, the Rough Creek–Shawneetown Fault System, and the Pennyrile Fault System (Nelson and Lumm, 1984). Igneous intrusives in southern Illinois are Permian in age and help to date other tectonic activity in the region (Zartman and others, 1967; Nelson and Lumm, 1984). The Pascola Arch was uplifted by about 4,000 m, closing the southern margin of the Illinois Basin (Atherton, 1971; Buschbach and Atherton, 1979). This movement may have begun even by Late Pennsylvanian time (Houseknecht, 1983) and certainly had begun by the Triassic (Buschbach and Atherton, 1979; Kolata and Nelson, 1988; Schwalb, 1982). The result was a positive area reaching from the Nashville Dome almost to the Ozark Uplift. By Late Cretaceous time the arch had been deeply eroded, reduced to base level, and

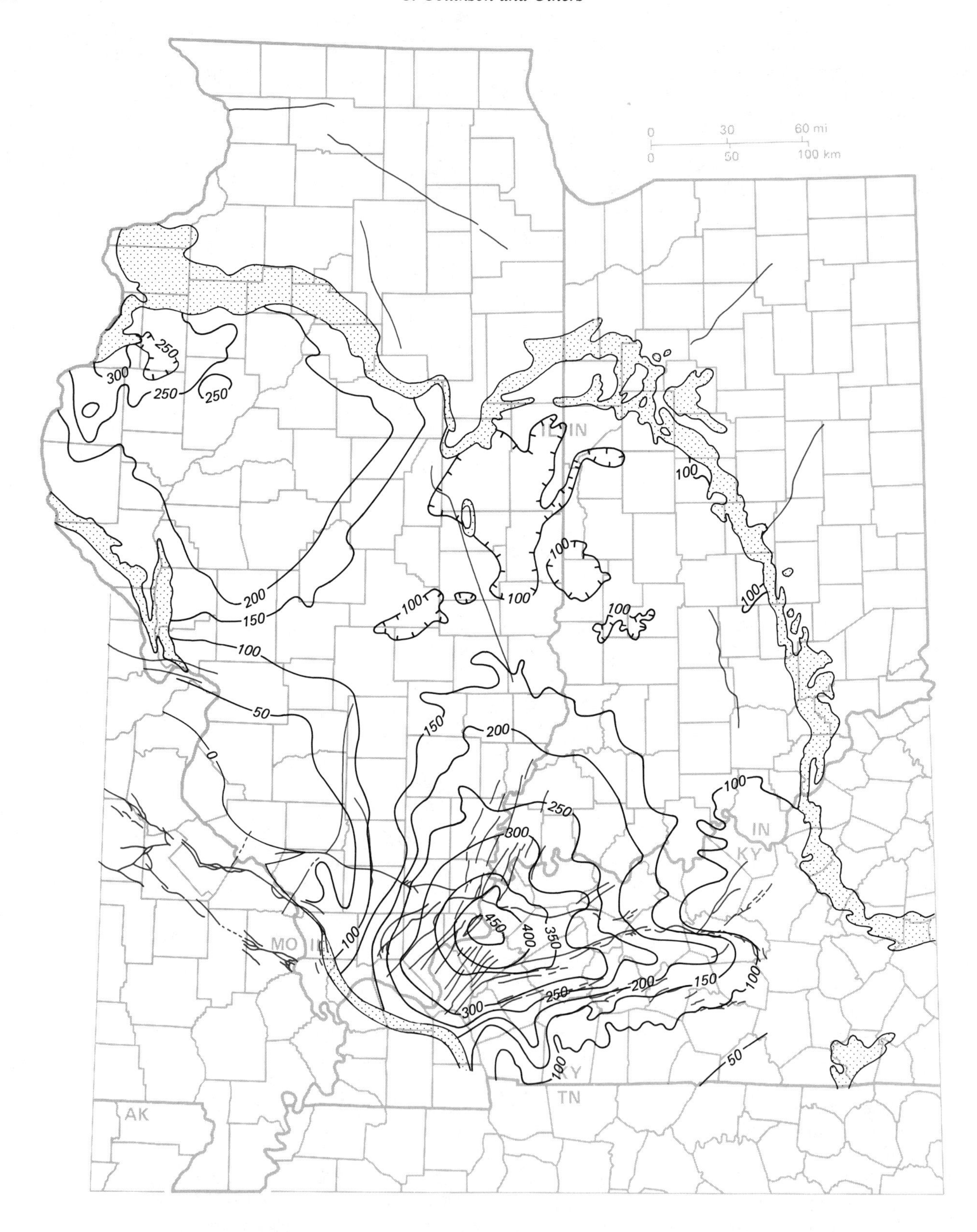

Figure 32. Thickness (in feet) of the upper part of Kaskaskia I and lower part of Kaskaskia II subsequences (New Albany Group from Lineback, 1968) showing evidence of renewed subsidence in the Moorman Syncline area. Faults are shown in red. Stippled areas represent the New Albany outcrop or places where the New Albany is unconformably overlain by rocks younger than Mississippian.

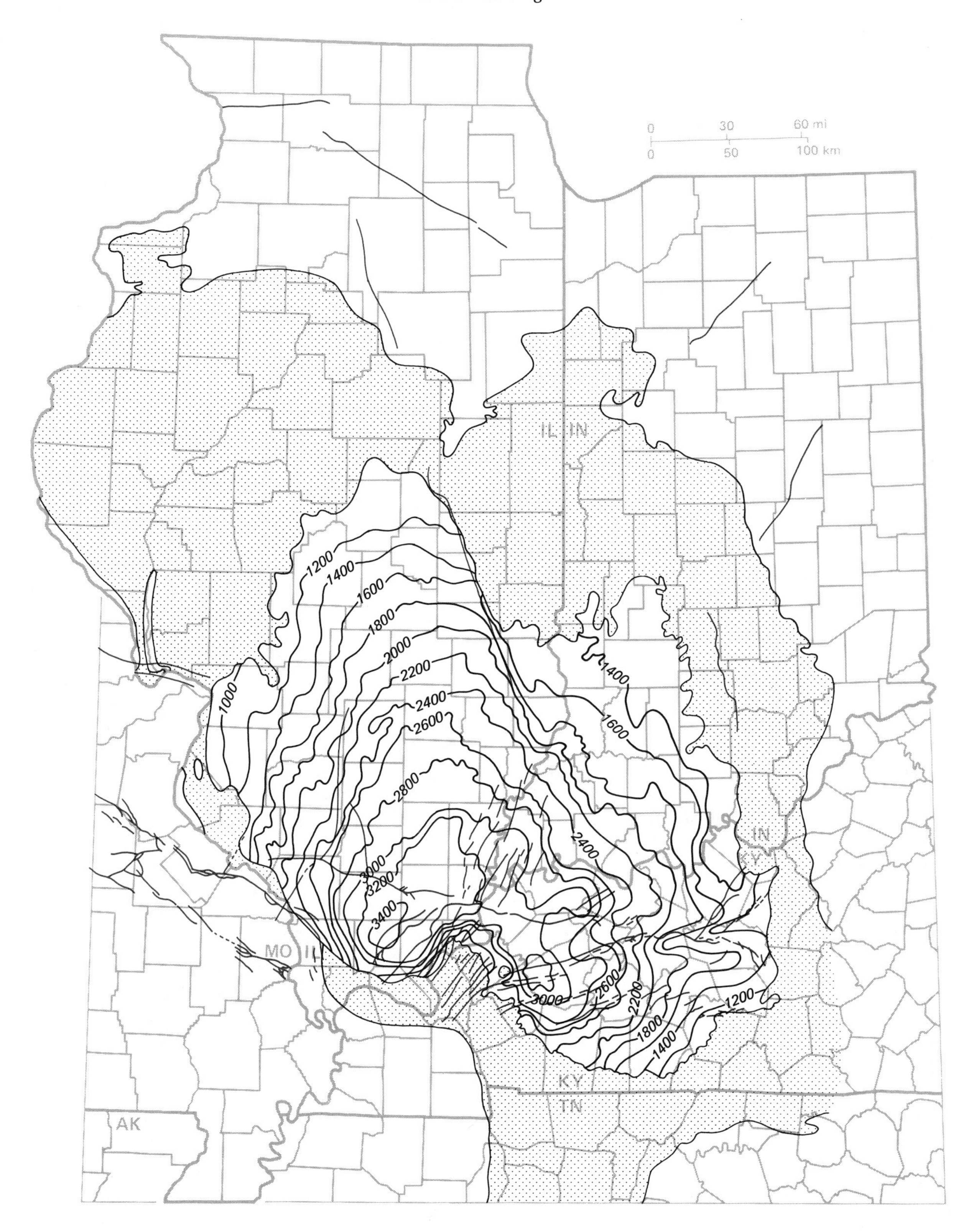

Figure 33. Thickness map (in feet) of the Kaskaskia II subsequence showing early development of the Fairfield Basin and continued subsidence in the Moorman Syncline. Faults are shown in red. Stipple indicates areas of outcrop and subcrops of the Kaskaskia II beyond the erosional limit of the Chesterian Series.

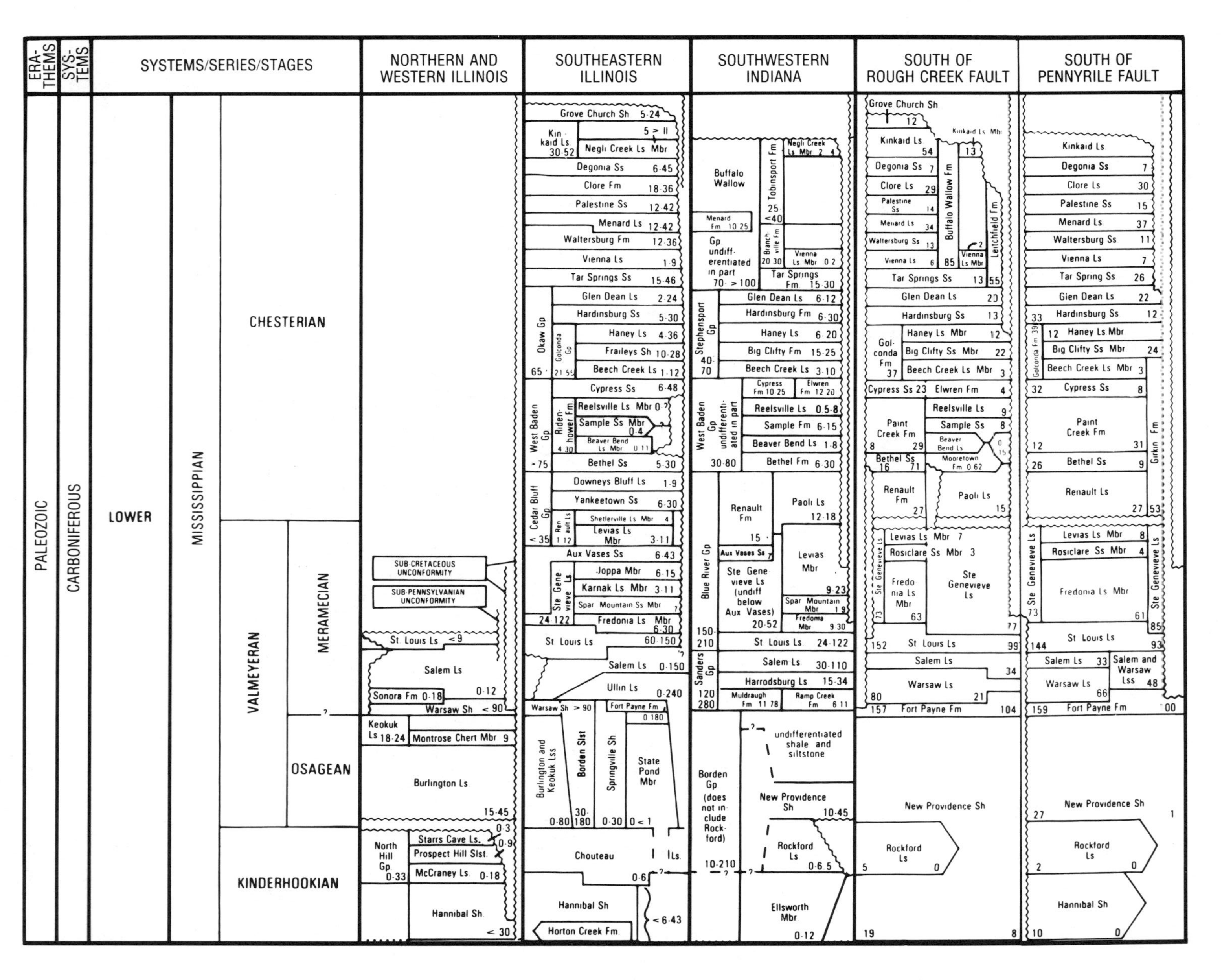

Figure 34. Chronostratigraphic units of the Kaskaskia II subsequence. Modified from COSUNA correlation chart (Shaver and others, 1985). Wavy lines represent unconformities.

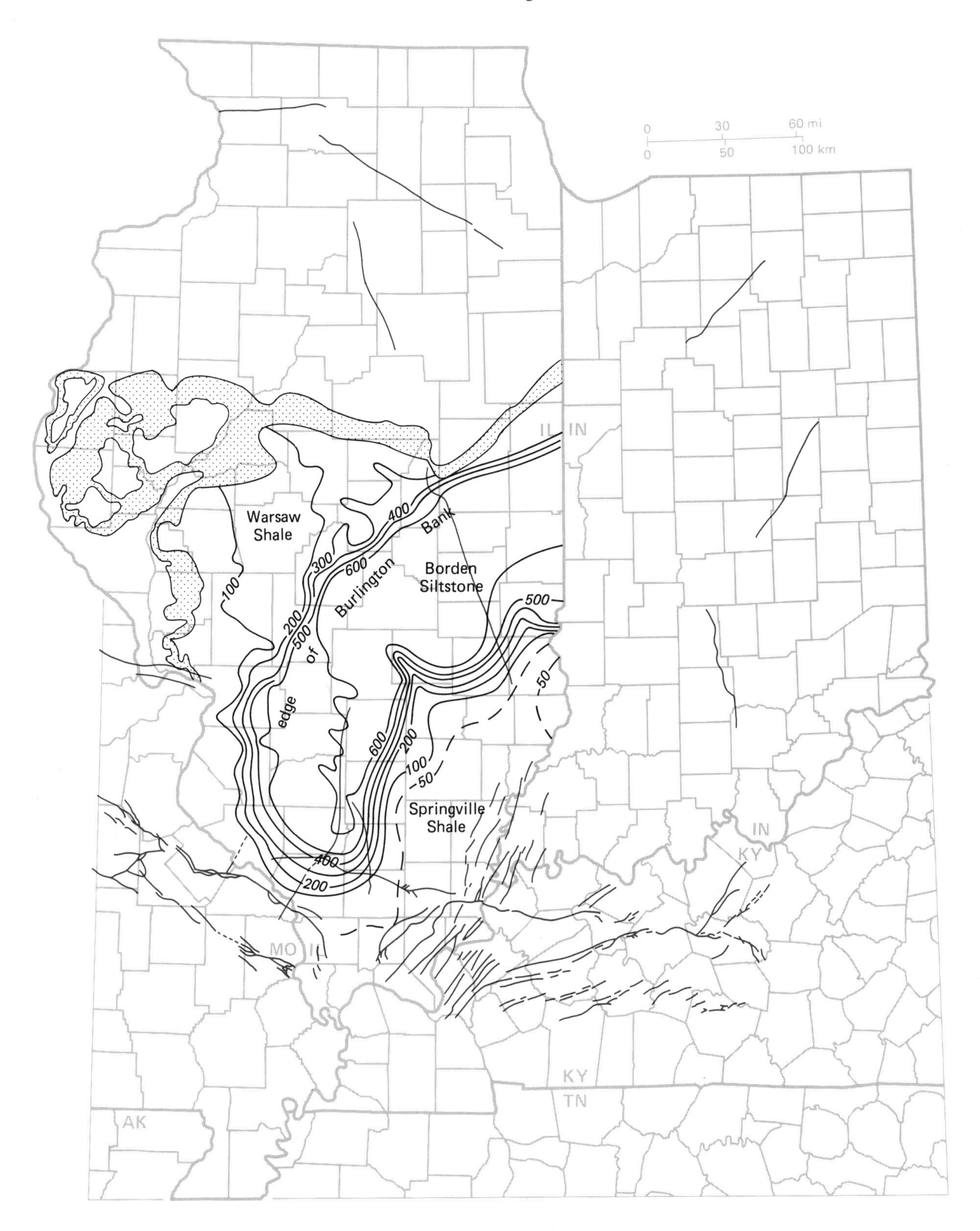

Figure 35. Thickness (in feet) of the Borden Siltstone Delta (part of Kaskaskia II) and the adjacent Warsaw and Springville Shales in Illinois. The Borden filled the starved basin situated southeast of the Burlington Bank (after Lineback, 1966). Faults are shown in red. Stipple indicates areas of outcrop or areas where the Warsaw Shale is unconformably overlain by rocks younger than Mississippian.

had subsided as part of the Mississippi Embayment to be covered by Late Cretaceous sediments.

The Zuni sequence consists mainly of Late Cretaceous and early Paleocene rocks that overlap the truncated edges of Paleozoic rocks at the upper end of the Mississippi Embayment in southernmost Illinois, southeastern Missouri, and westernmost Kentucky (Fig. 3). In addition, scattered Cretaceous outliers are present in western Illinois. The Zuni attains a maximum thickness of about 600 m in the extreme southern part of the Illinois Basin area (Willman and Frye, 1975).

The Cretaceous part of the Zuni sequence is the thickest and most widely distributed part, but includes only Late Cretaceous sediment. It pinches out northward, and at its furthest extent, only the Little Bear Soil and overlying gravels of the Tuscaloosa Formation are present. The McNairy Formation overlies the Tuscaloosa and is a relatively thick succession of fine sand, carbonaceous silts, clay, and lignite. In turn, it is overlain by shallow-marine glauconitic silts and silty clays of the Owl Creek Formation. All of the formations have unconformable top and bottom contacts, indicating repeated fluctuations in sea level relative to the land surface.

In western Illinois, the Cretaceous nearshore marine Baylis Formation, which reaches 30 m in thickness, consists of sand, clayey sand, and gravel (Willman and Frye, 1975). The formation probably was once continuous westward to Iowa and Kansas, based on the lithologic similarity of Cretaceous strata in that region (Willman and Frye, 1975).

The Paleocene Clayton Formation unconformably overlies the Cretaceous strata and consists of marine, glauconitic, micaceous silty clay. It is overlain by the Porters Creek Formation, also Paleocene, which is a massive marine clay, gray to black in color.

The Tejas sequence includes the Eocene Wilcox Formation and Pliocene Grover, Mounds, and Luce Gravels, as well as various other unnamed gravels. Deposition of these gravels took place on an emergent craton, and erosion proceeded for more than a million years before Pleistocene glaciers moved southward, covering most of the Illinois Basin with a blanket of Quaternary deposits.

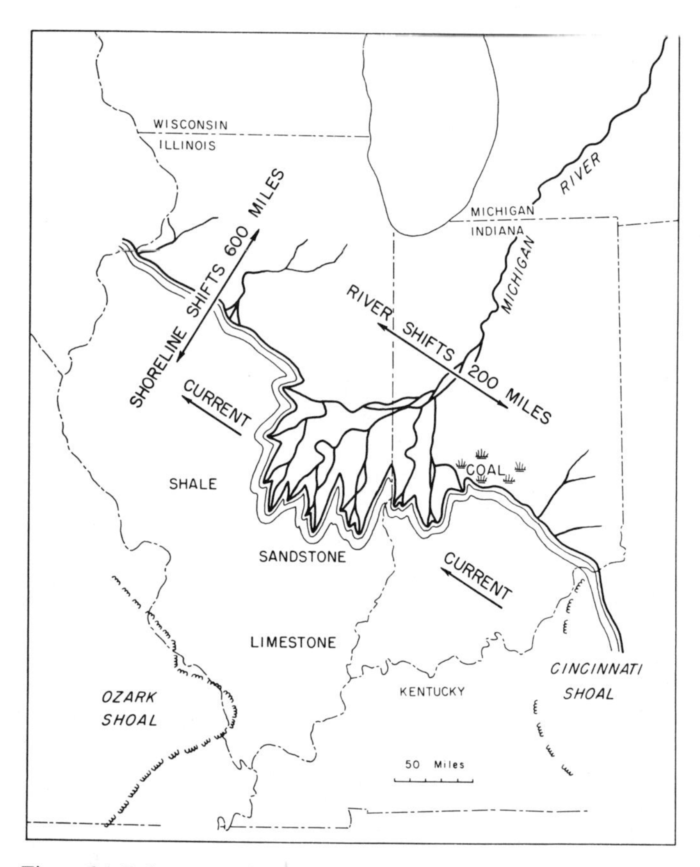

Figure 36. Paleogeography of the Illinois Basin during part of the Late Mississippian. The Michigan River system of Swann (1963) had its source to the northeast and carried sediments that were deposited as a series of prograding delta complexes in the basin.

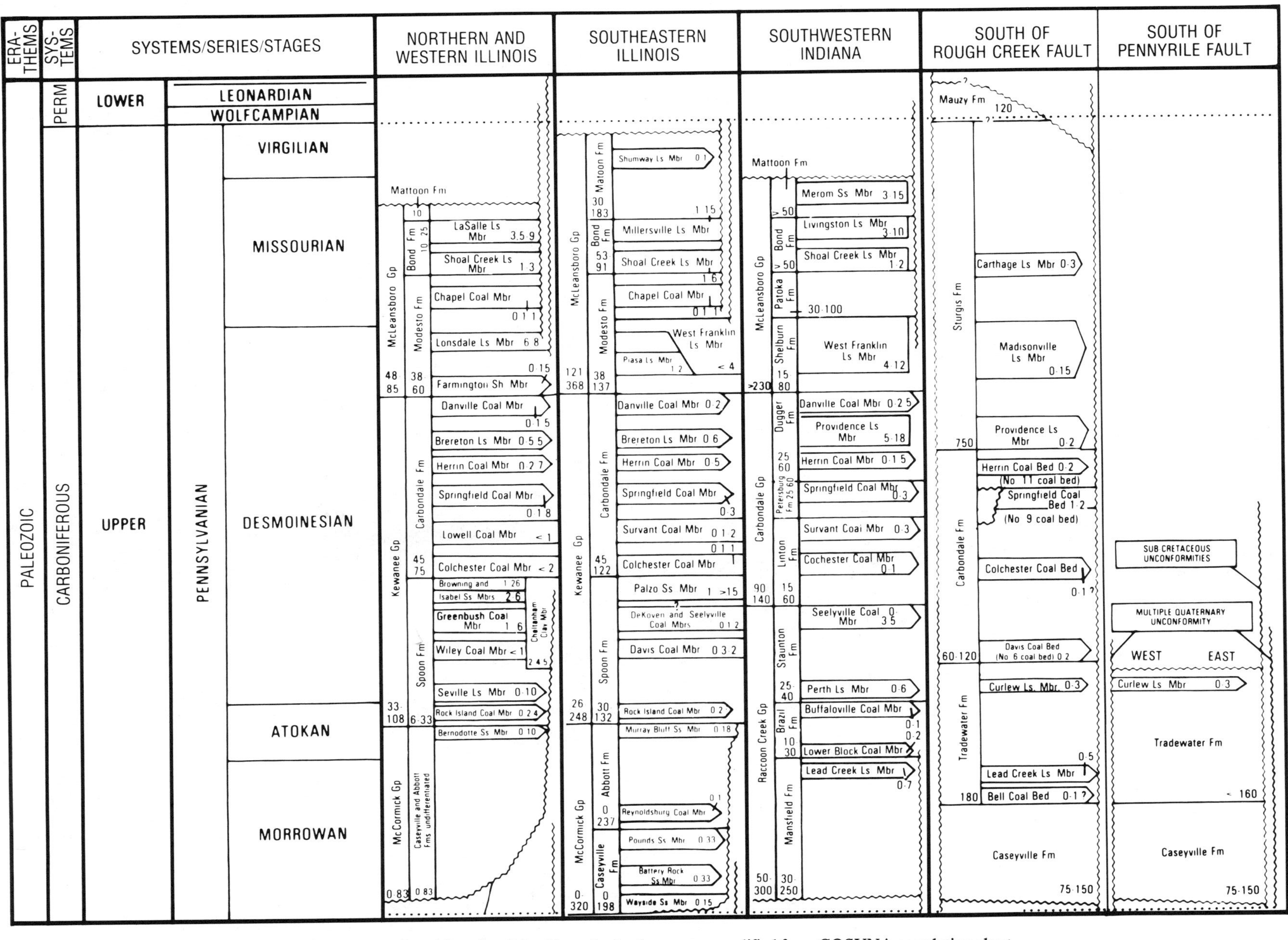

Figure 37. Stratigraphic units of the Absaroka I subsequence modified from COSUNA correlation chart (Shaver and others, 1985). Wavy lines represent unconformities. The position of the Morrowan-Atokan boundary is uncertain and it is placed at a different position by Hopkins and Simon (1975).

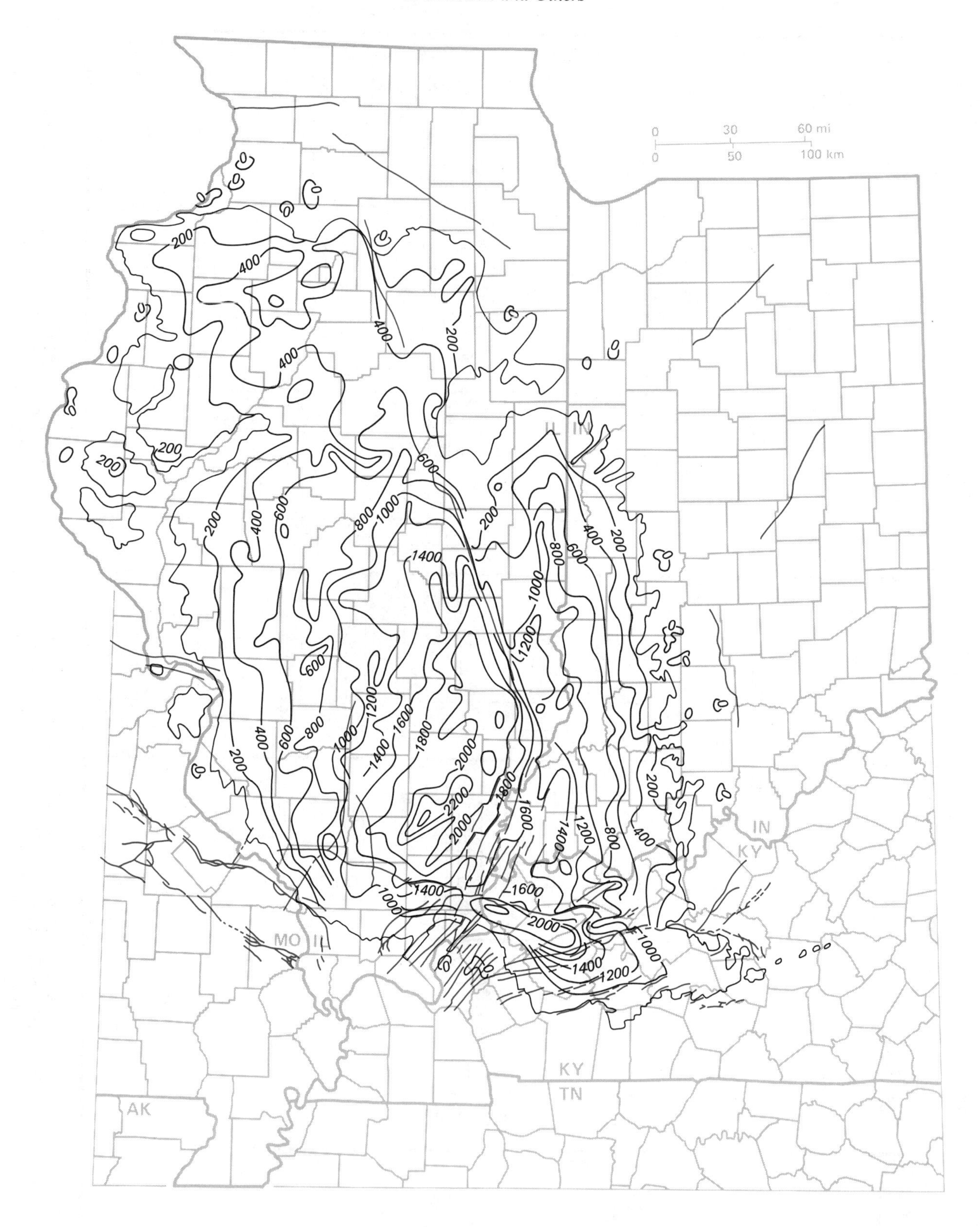

Figure 38. Thickness (in feet) of the Absaroka sequence (Pennsylvanian and Permian) in the Illinois Basin (modified from Wanless, 1975).

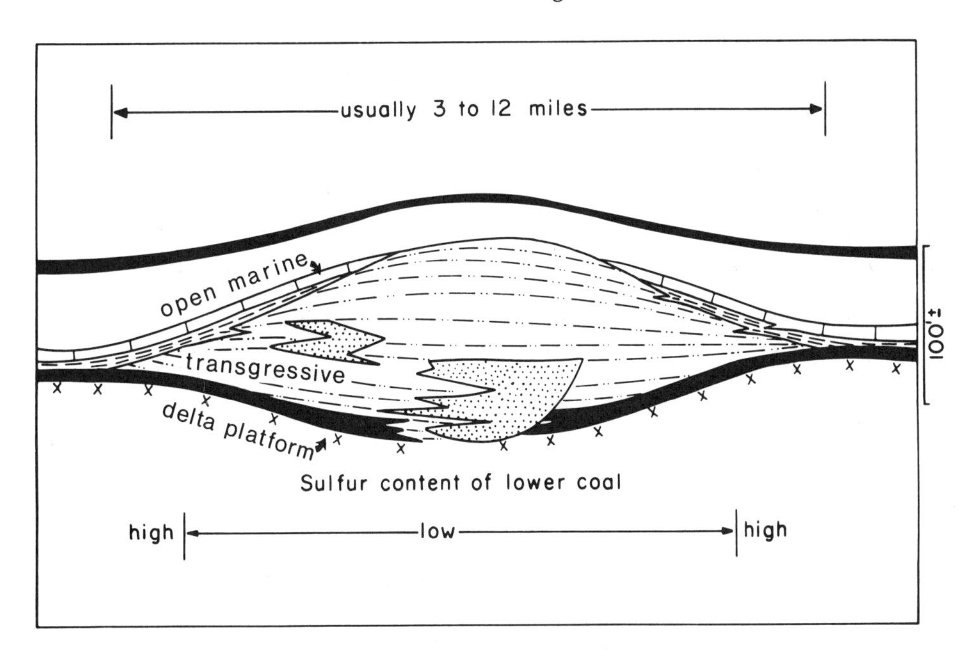

Figure 39. Diagrammatic cross section along depositional strike in the vicinity of a paleochannel showing distribution of limestone (brick pattern), sandstone (stipples), gray silty shale (dashes and stipples), black shale (superposed dashes), claystone (x's), and coal (black). Open space is unspecified lithology (modified from Gluskoter and Hopkins, 1970).

REFERENCES

Altschaeffl, A. G., and Harrison, W., 1959, Estimation of a minimum depth of burial for a Pennsylvanian underclay: Journal of Sedimentary Petrology, v. 29, p. 178–185.

Anderson, R. C., Tyler, R. D., and Hammer, W. R., 1982, The Mississippi River Arch, Quad-Cities region: 46th Annual Tri-State Field Conference, Augustana College, Rock Island, Illinois, 63 p.

Anderson, W. H., 1984, Phanerozoic cross section along 88° W. Longitude, western Kentucky; U.S. Geological Survey strategic and critical minerals assessment program: Kentucky State Geological Survey Open-File Report, 1 chart, scale 1:500,000.

Atherton, E., 1971, Tectonic development of the Eastern Interior Region of the United States, *in* Bond, D. C., chairman, Background materials for symposium on future petroleum potential of NPC Region 9: Illinois State Geological Survey Illinois Petroleum 96, p. 29–43.

—— , 1975, Precambrian rocks, *in* Willman, H. B., and 7 others, Handbook of Illinois stratigraphy: Illinois State Geological Survey Bulletin 95, p. 32–33.

Atherton, E., Collinson, C., and Lineback, J. A., 1975, Mississippian System, *in* Willman, H. B., and 7 others, Handbook of Illinois stratigraphy: Illinois State Geological Survey Bulletin 95, p. 123–162.

Becker, L. E., Andrew, H. J., and Dawson, T. A., 1978, Pre-Knox (Cambrian) stratigraphy in Indiana: Indiana Department of Natural Resources Geological Survey Bulletin 57, 72 p.

Bell, A. H., Atherton, E., Buschbach, T. C., and Swann, D. H., 1964, Deep oil possibilities of the Illinois Basin: Illinois State Geological Survey Circular 368, 38 p.

Berry, W.B.N., and Boucot, A. J., 1973, Glacio-eustatic control of Late Ordovician-Early Silurian platform sedimentation and faunal changes: Geological Society of America Bulletin, v. 84, p. 275–284.

Bertagne, A. J., Pisasale, E. T., and Leising, T. C., 1986, Interpretation of recent seismic data from a frontier hydrocarbon province; The western Rough Creek Graben, southern Illinois and western Kentucky [abs.]: American Association of Petroleum Geologists Bulletin, v. 70, no. 5, p. 564–565.

Bertagne, A. J., and Leising, T. C., 1988, Interpretation of seismic data from the Rough Creek Graben, western Kentucky and southern Illinois, *in* Leighton, M. W., ed., Interior cratonic sag basins: American Association of Petroleum Geologists Petroleum Basin Series, v. 1 (in press).

Bickford, M. E., Van Schmus, W. R., and Zietz, I., 1986, Proterozoic history of the midcontinent region of North America: Geological Society of America Geology, v. 14, no. 6, p. 492–496.

Biggs, M. E., Blakely, R. F., Rudman, A. J., 1960, Seismic velocities and synthetic seismogram computed from a continuous velocity log of a test well to the basement complex in Lawrence County, Indiana: Indiana Department of Natural Resources Geological Survey Report of Progress no. 21, 15 p.

Bond, D. C., chairman, 1971, Background materials for symposium on future petroleum potential of NPC Region 9 (Illinois Basin, Cincinnati Arch, and northern part of Mississippi Embayment): Illinois State Geological Survey Illinois Petroleum 96, 63 p.

Bond, D. C., and 12 others, 1971, Possible future petroleum potential of Region 9; Illinois Basin, Cincinnati Arch, and northern Mississippi Embayment, *in* Cram, I. H., ed., Future petroleum provinces of the United States; Their geology and potential: American Association of Petroleum Geologists Memoir 15, v. 2, p. 1165–1218.

Braile, L. W., Hinze, W. J., Keller, G. R., and Lidiak, E. G., 1982a, The northeastern extension of the New Madrid Seismic Zone, *in* McKeown, R. A., and Pakiser, L. C., eds., Investigations of the New Madrid, Missouri, earthquake region: U.S. Geological Survey Professional Paper 1236L, p. 173–184.

Braile, L. W., Keller, G. R., Hinze, W. J., and Lidiak, E. G., 1982b, An ancient rift complex and its relation to contemporary seismicity in the New Madrid Seismic Zone: Tectonics, v. 1, p. 225–237.

Braile, L. W., Hinze, W. J., Sexton, J. L., Keller, G. R., and Lidiak, E. G., 1984, Tectonic development of the New Madrid Seismic Zone: U.S. Geological Survey Open-File Report 84-770, p. 204–233.

Braile, L. W., Hinze, W. J., Keller, G. R., Lidiak, E. G., and Sexton, J. L., 1986, Tectonic development of the New Madrid Rift Complex, Mississippi Embayment, North America: Tectonophysics, v. 128, p. 1–21.

Branson, C. C., 1962, Pennsylvanian System of the Mid-Continent, *in* Branson, C. C., ed., Pennsylvanian System in the United States: American Association of Petroleum Geologists, p. 431–460.

Bristol, H. B., and Buschbach, T. C., 1971, Structural features of the Eastern Interior Region of the United States, *in* Bond, D. C., chairman, Background materials for symposium on future petroleum potential of NPC Region 9 (Illinois Basin, Cincinnati Arch, and northern part of Mississippi Embayment): Illinois State Geological Survey Illinois Petroleum 96, 63 p.

Bristol, H. M., and Howard, R. H., 1971, Paleogeologic map of the sub-Pennsylvanian Chesterian (Upper Mississippian) surface in the Illinois Basin: Illinois State Geological Survey Circular 458, 14 p.

Bunker, B. J., Ludvigson, G. A., and Witzke, B. J., 1985, The Plum River fault zone and the structural and stratigraphic framework of eastern Iowa: Iowa Geological Survey Technical Information Series no. 13, 126 p.

Buschbach, T. C., 1964, Cambrian and Ordovician strata of northeastern Illinois: Illinois State Geological Survey Report of Investigations 218, 90 p.

—— , 1975, Cambrian System, *in* Willman, H. B., and 7 others, Handbook of Illinois stratigraphy: Illinois State Geological Survey Bulletin 95, p. 34–46.

—— , 1983, New Madrid seismotectonic study: U.S. Nuclear Regulatory Commission NUREG/CR-3173, 197 p.

Buschbach, T. C., and Atherton, E., 1979, History of the structural uplift of the southern margin of the Illinois Basin, *in* Palmer, J. E., and Dutcher, R. R., eds., Depositional and structural history of the Pennsylvanian System of the Illinois Basin; Part 2, Invited papers: Illinois State Geological Survey Guidebook 15a, p. 112–115.

Cluff, R. M., Reinbold, M. L., and Lineback, J. A., 1981, The New Albany Shale Group of Illinois: Illinois State Geological Survey Circular 518, 58 p., 4 pls.

Cohee, G. V., and 15 others, 1962, Tectonic map of the United States: American Association of Petroleum Geologists, sheets 1 and 2, scale 1:2,500,000.

Collinson, C., 1967, Devonian of the north-central region, United States, *in* International Symposium on the Devonian System: Alberta Society of Petroleum Geologists, v. 1, p. 933–939.

Collinson, C., and Atherton, E., 1975, Devonian System, *in* Willman, H. B., and 7 others, Handbook of Illinois stratigraphy: Illinois State Geological Survey Bulletin 95, p. 104–123.

Collinson, C., Becker, L. E., James, G. W., Koenig, J. W., and Swann, D. H., 1967a, Illinois Basin, *in* International Symposium on the Devonian System: Alberta Society of Petroleum Geologists, v. 1, p. 940–962.

Collinson, C., Carlson, M. P., Dorheim, F. H. and Koenig, J. W., 1967b, Central Iowa Basin, *in* International Symposium on the Devonian System: Alberta Society of Petroleum Geologists, v. 1, p. 963–971.

Damberger, H. H., 1971, Coalification pattern of the Illinois Basin: Economic Geology, v. 66, p. 488–494.

—— , 1974, Coalification patterns of the Pennsylvanian coal basins of the eastern United States: Geological Society of America Special Paper 153, p. 53–74.

Devera, J. A., Mason, C. E., and Peppers, R. A., 1987, A marine shale in the Caseyville Formation (Lower Pennsylvanian) in southern Illinois: Geological Society of America Abstracts with Programs, v. 19, p. 196.

Droste, J. B., and Patton, J. B., 1985, Lithostratigraphy of the Sauk Sequence in Indiana: Indiana Department of Natural Resources Geological Survey Occasional Paper 47, 24 p.

Droste, J. B., and Shaver, R. H., 1980, Recognition of buried Silurian reefs in southwestern Indiana; application to the Terre Haute Bank: Journal of Geology, v. 88, p. 567–587.

—— , 1983, Atlas of early and middle Paleozoic paleogeography of the southern Great Lakes area: Indiana Department of Natural Resources Geological Survey Special Report 22, 32 p.

—— , 1987, Upper Silurian and Lower Devonian stratigraphy of the central Illinois Basin: Indiana Department of Natural Resources Geological Survey Special Report 39, 19 p.

Edmund, R. W., and Anderson, R. C., 1967, The Mississippi River Arch: 31st Annual Tri-State Geological Guidebook, Augustana College, Rock Island, Illinois, 64 p.

Emmons, E., 1838, Report of the Second Geological District of the state of New York: New York Geological Survey 2nd Annual Report, p. 186–252.

Emrich, G. H., and Bergstrom, R. E., 1962, The Des Plaines Disturbance, northeastern Illinois: Geological Society of America Bulletin, v. 73, no. 8, p. 959–968.

Ervin, C. P., and McGinnis, L. D., 1975, Reelfoot Rift; Reactivated precursor to the Mississippi Embayment: Geological Society of America Bulletin, v. 86, no. 9, p. 1287–1295.

Fraser, G. S., 1976, Sedimentology of a Middle Ordovician quartz arenite; Carbonate transition in the Upper Mississippi Valley: Geological Society of America Bulletin, v. 87, no. 6, p. 833–845.

Freeman, L. B., 1951, Regional aspects of Silurian and Devonian stratigraphy [Ky.]; Part 1: American Association of Petroleum Geologists Bulletin, v. 35, no. 1, p. 1–61; Parts 1–2, Kentucky Geological Survey, series 9, Bulletin 6, 565 p.

Gluskoter, H. J., and Hopkins, M. E., 1970, Distribution of sulphur in Illinois coals, *in* Smith, W. H., Nance, R. B., Hopkins, M. E., Johnson, R. G., and Shabica, C. W., eds., Depositional environments in parts of the Carbondale Formation—western and northern Illinois; Francis Creek Shale and associated strata and Mazon Creek biota: Illinois State Geological Survey Guidebook Series 8, p. 89–95.

Gray, H. H., 1972, Lithostratigraphy of the Maquoketa Group (Ordovician) in Indiana: Indiana Department of Natural Resources Geological Survey Special Report 7, 31 p.

Heigold, P. C., 1976, An aeromagnetic survey of southwestern Illinois: Illinois State Geological Survey Circular 495, 28 p.

Hester, N. C., 1984, Origin and evolution of the Cambrian age Moorman Trough of Kentucky: Geological Society of America Abstracts with Programs, v. 15, no. 3, p. 145.

Heyl, A. V., Jr., 1972, The 38th Parallel Lineament and its relationship to ore deposits: Economic Geology, v. 16, p. 879–894.

Hildenbrand, T. G., Kane, M. F., and Hendricks, J. D., 1982, Magnetic basement in the upper Mississippi Embayment region; A preliminary report, *in* McKeown, F. A., and Pakiser, L. C., eds., Investigations of the New Madrid, Missouri, earthquake region: U.S. Geological Survey ProfessionalPaper 1236, p. 39–53.

Hoholick, J. D., Metarko, T., and Potter, P. E., 1984, Regional variations of porosity and cement; St. Peter and Mount Simon Sandstones in Illinois Basin: American Association of Petroleum Geologists Bulletin, v. 68, no. 6, p. 753–764.

Hopkins, M. E., and Simon, J. A., 1975, Pennsylvanian System, *in* Willman, H. B., and 7 others, Handbook of Illinois stratigraphy: Illinois State Geological Survey Bulletin 95, p. 163–201.

Hoppe, W. J., Montgomery, C. W., Van Schmus, W. R., 1983, Age and significance of Precambrian basement samples from northern Illinois and adjacent states: Journal of Geophysical Research, v. 88, no. B9, p. 7276–7286.

Houseknecht, D. W., 1983, Tectonic-sedimentary evolution of the Arkoma Basin: Society of Economic Paleontologists and Mineralogists, Midcontinent Section, v. 1, 119 p.

Houseknecht, D. W., and Viele, G. W., 1984, Early Paleozoic tectonic and sedimentary history of the northern Mississippi Embayment: Geological Society of America Abstracts with Programs, v. 16, no. 3, p. 146.

Houseknecht, D. W., and Weaverling, P. H., 1983, Early Paleozoic sedimentation in Reelfoot Rift [abs.]: American Association of Petroleum Geologists Bulletin, v. 67, no. 9, p. 1456.

Howard, R. H., 1979, The Mississippian-Pennsylvanian unconformity in the Illinois Basin; Old and new thinking, *in* Palmer, J. E., and Dutcher, R. R., eds., Depositional and structural history of the Pennsylvanian System of the Illinois Basin; Part 2, Invited papers: Illinois State Geological Survey Guidebook Series 15a, p. 34–43.

Howe, J. R., and Thompson, T. L., 1984, Tectonics, sedimentation, and hydrocarbon potential of the Reelfoot Rift: Oil and Gas Journal, v. 82, p. 179–190.

Howell, J. V., 1935, The Mississippi River Arch: Kansas Geological Society Guidebook, 9th Annual Field Conference, p. 386–389.

Jennings, J. R., 1974, Lower Pennsylvanian plants of Illinois; I, A flora from the Pounds Sandstone Member of the Caseyville Formation: Journal of Paleontology, v. 48, p. 459–473.

——, 1984, Distribution of fossil plant taxa in the Upper Mississippian and Lower Pennsyulvanian of the Illinois Basin; Neuvième Congrès International de Stratigraphie et de Géologie du Carbonifère, Compte Rendu: Carbondale, Southern Illinois University Press, v. 2, p. 301–312.

Jennings, J. R., and Fraunfelter, G. H., 1986, Macropaleontology of the upper boundary of the type Mississippian: Illinois State Academy of Science Transactions, v. 79, p. 253–261.

Kehn, T. M., Beard, J. G., and Williamson, A. D., 1982, Mauzy Formation, a new stratigraphic unit of Permian age in western Kentucky: U.S. Geological Survey Bulletin 1529-H, p. H73–H86.

Kisvarsanyi, E. B., 1981, Geology of the Precambrian St. Francois Terrane, southeastern Missouri: Missouri Division of Geology and Land Survey Contribution to Precambrian Geology No. 8, Report of Investigations 64, 58 p.

Kolata, D. R., and Graese, A. M., 1983, Lithostratigraphy and depositional environment of the Maquoketa Group (Ordovician) in northern Illinois: Illinois State Geological Survey Circular 528, 49 p., 2 maps in pocket.

Kolata, D. R., and Nelson, W. J., 1988, Tectonic history of the Illinois Basin, *in* Leighton, M. W., ed., Interior cratonic sag basins: American Association of Petroleum Geologists Petroleum Basin Series, v. 1 (in press).

Kolata, D. R., Frost, J. K., and Huff, W. D., 1986, K-bentonites of the Ordovician Decorah Subgroup, upper Mississippi Valley; Correlation by chemical fingerprinting: Ilinois State Geological Survey Circular 537, 30 p.

Krausse, H.-F., and Treworgy, C. G., 1979, Major structures of the southern part of the Illinois Basin, *in* Palmer, J. E., and Dutcher, R. R., eds., Deposition and structural history of the Pennsylvanian System of the Illinois Basin; Part 2, Invited papers: Illinois State Geological Survey Guidebook 15a, p. 115–119.

Lane, H. R., 1978, The Burlington Shelf (Mississippian, north-central United States): Geologica et Paleontologica, v. 12, p. 165–176.

Lee, W., 1943, The stratigraphy and structural development of the Forest City Basin in Kansas: State Geological Survey of Kansas Bulletin 51, 140 p.

Lidiak, E. G., Marvin, R. F., Thomas, H. N., and Bass, M. N., 1966, Geochronology of the Midcontinent region; Part 4, Eastern area: Journal of Geophysical Research, v. 71, p. 5427–5438.

Lineback, J. A., 1966, Deep-water sediments adjacent to the Borden Siltstone (Mississippian) delta in southern Illinois: Illinois State Geological Survey Circular 401, 48 p.

——, 1968, Subdivisions and depositional environments of New Albany Shale (Devonian-Mississippian) in Indiana: American Association of Petroleum Geologists Bulletin, v. 52, p. 1291–1303.

——, 1981, The eastern margin of the Burlington-Keokuk (Valmeyeran) carbonate bank in Illinois: Illinois State Geological Survey Circular 520, 24 p.

Lowenstam, H. A., 1949, Niagaran reefs in Illinois and their relation to oil accumulation: Illinois State Geological Survey Report of Investigation 145, 36 p.

——, 1950, Niagaran reefs of the Great Lakes area: Journal of Geology, v. 58, p. 430–487.

——, 1952, Some new observations on Niagaran reefs in Illinois: Illinois Academy of Science Transactions, v. 45, p. 100–107.

——, 1957, Niagaran reefs in the Great Lakes area: Geological Society of America Memoir 67, v. 2, p. 215–248.

McCracken, M. H., 1971, Structural features of Missouri: Missouri Geological Survey and Water Resources Report of Investigations no. 49, 99 p.

Mikulic, D. G., 1977, A preliminary revision of the Silurian stratigraphy of southeastern Wisconsin, *in* Nelson, K. G., ed., Geology of southeastern Wisconsin: 41st Annual Tri-State Field Conference Guidebook, p. A6–A34.

Mikulic, D. G., Sargent, M. L., Norby, R. D., and Kolata, D. R., 1985, Silurian geology of the Des Plaines River valley, northeastern Illinois: Illinois State Geological Survey Guidebook 17, 56 p.

Nelson, W. J., and Krausse, H.-F., 1981, The Cottage Grove Fault System in southern Illinois: Illinois State Geological Survey Circular 522, 65 p.

Nelson, W. J., and Lumm, D. K., 1984, Structural geology of southeastern Illinois and vicinity: Illinois State Geological Survey Contract/Grant Report 1984-2, 127 p.

——, 1985, The Ste. Genevieve Fault Zone, Missouri and Illinois: Illinois State Geological Survey Contract/Grant Report 1985-3, 93 p.

Norby, R. D., Sargent, M. L., and Ethington, R. L., 1986, Conodonts from the Everton Dolomite (Middle Ordovician) in southern Illinois: Geological Society of America Abstracts with Programs, v. 18, no. 3, p. 258.

Palmer, J. E., and Dutcher, R. R., eds., 1979a, Depositional and structural history of the Pennsylvanian System of the Illinois Basin; Part 1, Road log and description of stops: Illinois State Geological Survey Guidebook Series 15, 116 p.

——, eds., 1979b, Depositional and structural history of the Pennsylvanian System of the Illinois Basin; Part 2, Invited papers: Illinois State Geological Survey Guidebook Series 15a, 158 p.

Peppers, R. A., 1984, Comparison of miospore assemblages in the Pennsylvanian System of Illinois with those in the Upper Carboniferous of Western Europe; Neuvième Congrès International de Stratigraphie et de Géologie du Carbonifère, Compte Rendu: Carbondale, Southern Illinois University Press, v. 2, p. 483–502.

Potter, P. E., and Pryor, W. A., 1961, Dispersal centers of Paleozoic and later clastics of the upper Mississippi Valley and adjacent areas: Geological Society of America Bulletin, v. 72, no. 8, p. 1195–1249.

Rexroad, C. B., Droste, J. B., and Ethington, R. L., 1982, Conodonts from the Everton Dolomite and the St. Peter Sandstone (lower Middle Ordovician) in a core from southwestern Indiana: Indiana Department of Natural Resources Geological Survey Occasional Paper 39, 13 p.

Rogers, J. E., 1972, Silurian and Lower Devonian stratigraphy and paleobasin development; Illinois Basin, central U.S. [Ph.D. thesis]: University of Illinois at Urbana-Champaign, 144 p.

Rudman, A. J., Summerson, C. H., and Hinze, W. J., 1965, Geology of basement in midwestern United States: American Association of Petroleum Geologists Bulletin, v. 49, no. 7, p. 894–904.

Safford, J. M., 1869, Geology of Tennessee: Nashville, Tennessee, S. C. Mercer, 550 p.

Sargent, M. L., 1985, Lithofacies and isopach mapping of the Sauk Sequence in Illinois omitting the Mount Simon Sandstone and older strata: Geological Society of America Abstracts with Programs, v. 17, no. 5, p. 325.

——, 1988, Sauk Sequence, *in* Leighton, M. W., ed., Interior cratonic sag basins: American Association of Petroleum Geologists Petroleum Basin Series, v. 1 (in press).

Sargent, M. L., and Buschbach, T. C., 1985, Map of the morphology of the top of Precambrian crystalline rocks in Illinois: Geological Society of America Abstracts with Programs, v. 17, no. 5, p. 324.

Sargent, M. L., and Norby, R. D., 1986, Distribution of the Whiterockian Everton Dolomite in Illinois: Geological Society of America Abstracts with Programs, v. 18, no. 4, p. 322.

Schwalb, H. R., 1969, Paleozoic Geology of the Jackson Purchase Region, Kentucky: Kentucky Geological Survey, Series X, Report of Investigation 10,

40 p.
——, 1982, Paleozoic geology of the New Madrid area: U.S. Nuclear Regulatory Commission NUREG/CR-2909, 61 p.
——, 1983, Paleozoic geology of the New Madrid area, *in* New Madrid seismotectonic study, summaries of activities from 1977 through 1981: U.S. Nuclear Regulatory Commission NUREG/CR-3173, p. 42–68.
Sexton, J. L., Braile, L. W., Hinze, W. J., and Campbell, M. J., 1986, Seismic reflection profiling studies of a buried Precambrian rift beneath the Wabash Valley Fault Zone: Geophysics, v. 51, no. 3, p. 640–660.
Shaver, R. H., and 9 others, 1978, The search for a Silurian reef model; Great Lakes area: Indiana Department of Natural Resources Geological Survey Special Report 15, 36 p.
Shaver, R. H., and 31 others, 1985, Midwestern basins and arches region correlation chart, *in* Childs, O. E., Steele, G., and Salvador, A., directors, Lindberg, F. A., ed., Correlation of stratigraphic units in North America Project: American Association of Petroleum Geologists COSUNA Chart Series.
Shaver, R. H., and 16 others, 1986, Compendium of Paleozoic rock-unit stratigraphy in Indiana; A revision: Indiana Department of Natural Resources Geological Survey Bulletin 59, 203 p.
Shaw, T. H., 1987, Lower Ordovician conodonts from the subsurface of north-central Illinois: Geological Society of America Abstracts with Programs, v. 19, no. 4, p. 244.
Sims, P. K., and Peterman, Z. E., 1986, Early Proterozoic Central Plains orogen; A major buried structure in the north-central United States: Geology, v. 14, no. 6, p. 488–491.
Sloss, L. L., 1963, Sequences in the cratonic interior of North America: Geological Society of America Bulletin, v. 74, p. 93–114.
——, 1982, The Midcontinent Province; United States, *in* Palmer, A. R., ed., Perspectives in regional geological synthesis: Geological Society of America DNAG Special Publication 1, p. 27–39.
Sloss, L. L., Krumbein, W. C., and Dapples, E. C., 1949, Integrated facies analysis, *in* Longwell, C. R., chairman, Sedimentary facies in geologic history: Geological Society of America Memoir 39, p. 91–123.
Soderberg, R. K., and Keller, G. R., 1981, Geophysical evidence for deep basin in western Kentucky: American Association of Petroleum Geologists Bulletin, v. 65, p. 226–234.
Stearns, R. G., and Reesman, A. L., 1986, Cambrian to Holocene structural and burial history of the Nashville Dome: American Association of Petroleum Geologists Bulletin, v. 70, p. 143–154.
Stinchcomb, B. L., 1976, Precambrian algal stromatolites and stromatolitic limestones in the St. Francois Mountains of southeast Missouri: Missouri Geological Survey Report of Investigation no. 61, p. 122–131.
Swann, D. H., 1963, Classification of Genevievian and Chesterian (Late Mississippian) rocks of Illinois: Illinois State Geological Survey Report of Investigation 216, 91 p.
——, 1964, Late Mississippian rhythmic sediments of Mississippian Valley: American Association of Petroleum Geologists Bulletin, v. 48, no. 5, p. 637–658.
——, 1968, A summary of geologic history of the Illinois Basin, *in* Geology and petroleum production of the Illinois Basin; A Symposium: Cooperative Publication of the Illinois, Indiana, and Kentucky Geological Societies, p. 3–21.
Swann, D. H., and Willman, H. B., 1961, Megagroups in Illinois: American Association of Petroleum Geologists Bulletin, v. 45, p. 471–483.
Tarr, W. A., and Keller, W. D., 1933, A post-Devonian igneous intrusion in southeastern Missouri: Journal of Geology, v. 41, p. 815–823.
Templeton, J. S., 1951, The Mount Simon Sandstone in northern Illinois: Illinois Geological Survey Circular 170, p. 151–159.
Thomas, J. J., Shuster, R. D., and Bickford, M. E., 1984, A terrane of 1,350- to 1,400-m.y.-old silicic volcanic and plutonic rocks in the buried Proterozoic of the mid-continent and in the Wet Mountains, Colorado: Geological Society of America Bulletin, v. 95, p. 1150–1157.
Tikrity, S. S., 1968, Tectonic genesis of the Ozark Uplift [Ph.D. thesis]: St. Louis, Missouri, Washington University, 196 p.
Tolman, C. F., and Robertson, F., 1969, Exposed Precambrian rocks in southeast Missouri: Missouri Geological Survey and Water Resources Report of Investigation 44, 68 p.
Treworgy, C. G., and Jacobson, R. J., 1985, Paleoenvironments and distribution of low-sulphur coal in Illinois; Neuvième Congrès International de Stratigraphie et de Géologie du Carbonifère, Compte Rendu: Carbondale, Southern Illinois University Press, v. 4, p. 349–359.
Treworgy, J. D., 1981, Structural features in Illinois; A compendium: Illinois State Geological Survey Circular 519, 22 p., plus map, scale 1:690,000.
Wanless, H. R., 1975, Illinois Basin region, *in* McKee, E. D., and Crosby, E. J., eds., Paleotectonic investigations of the Pennsylvanian System in the United States; Part 1, Introduction and regional analyses of the Pennsylvanian System: U.S. Geological Survey Professional Paper 853, p. 71–95, Plate 11.
Willman, H. B., and Atherton, E., 1975, Silurian System, *in* Willman, H. B., and 7 others, Handbook of Illinois stratigraphy: Illinois State Geological Survey Bulletin 95, p. 87–104.
Willman, H. B., and Buschbach, T. C., 1975, Ordovician System, *in* Willman, H. B., and 7 others, Handbook of Illinois stratigraphy: Illinois State Geological Survey Bulletin 95, p. 47–87.
Willman, H. B., and Frye, J. C., 1975, Mesozoic Erathem, *in* Willman, H. B., and 7 others, Handbook of Illinois stratigraphy: Illinois State Geological Survey Bulletin 95, p. 201–206.
Willman, H. B., and Kolata, D. R., 1978, The Platteville and Galena Groups in Illinois: Illinois State Geological Survey Circular 502, 75 p.
Willman, H. B., and 7 others, 1975, Handbook of Illinois stratigraphy: Illinois State Geological Survey Bulletin 95, 170 p.
Witzke, B. J., 1980, Middle and upper Ordovician paleogeography of the region bordering the Transcontinental Arch, *in* Fouch, T. D., and Magathan, E. R., eds., Paleozoic paleogeography of west-central United States; West-Central United States Paleography Symposium 1: Society of Economic Paleontologists and Mineralogists, p. 1–18.
Zartman, R. E., Brock, M. R., Heyl, A. V., and Thomas, H. H., 1967, K-Ar and Rb-Sr ages of some alkalic intrusive rocks from central and eastern United States: American Journal of Science, v. 265, no. 10, p. 848–870.
Zietz, I., King, E. R., Geddes, W., and Lidiak, E. G., 1966, Crustal study of a continental strip from the Atlantic Ocean to the Rocky Mountains: Geological Society of America Bulletin, v. 77, p. 1427–1448.

MANUSCRIPT ACCEPTED BY THE SOCIETY NOVEMBER 25, 1987

ACKNOWLEDGMENTS

We thank Heinz H. Damberger, Russell J. Jacobson, Dennis R. Kolata, W. John Nelson, Rodney D. Norby, C. Brian Trask, Janis D. Treworgy, David A. Williams, and Allen D. Williamson for making their unpublished data and manuscripts available to us. We thank Allison R. Palmer and Laurence L. Sloss for reviewing the manuscript, and wish to note our special appreciation to Robert H. Shaver for his assistance in review. We thank Sandra K. Stecyk, Pamella K. Foster, and Barbara J. Stiff for help in drafting the figures and Joanne E. Klitzing for typing the manuscript.

Printed in U.S.A.

The Geology of North America
Vol. D-2, Sedimentary Cover—North American Craton: U.S.
The Geological Society of America, 1988

Chapter 15

The Appalachian Basin

Robert C. Milici
Virginia Division of Mineral Resources, Charlottesville, Virginia 22903
Wallace de Witt, Jr.
U.S. Geological Survey, MS 955, Reston, Virginia 22092

INTRODUCTION

The central and part of the southern Appalachian Valley and Ridge and Plateau are underlain by an elongate basin that extends from the Great Lakes southwestward to a low broad arch that lies along the southern Tennessee border (Fig. 1). This arch, a branch of the Cincinnati Arch, separates the Appalachian Basin (sensu stricto) from the Black Warrior Basin in Alabama and Mississippi (Thomas, this volume). In its narrower dimension, the Appalachian Basin extends from an eastern edge buried beneath Piedmont thrust sheets (Cook and Oliver, 1981), westward to the crest of the Cincinnati Arch.

The basin is ovoid; its deepest part lies in eastern Pennsylvania. The strata that fill the basin are only a few thousand meters thick on the western basin margin and thicken to about 13,000 m to the north and east (Colton, 1970). Thickness variations reflect the change in composition of basin-filling strata, from strata dominated by limestone and dolomite in the south and west to strata dominated by quartz sand, silt, and clay to the north and east.

The last major synthesis of Appalachian Basin stratigraphy is that published by Colton (1970). This chapter is a generalized summary of the stratigraphic development and filling of the Appalachian Basin, from its inception in the late Precambrian to its deformation at the end of the Paleozoic. Variations in stratigraphic nomenclature reflect the multiplicity of sources used in the compilation. Supplementary tables (Tables 1–7) of stratigraphic terminology contain the names of units not discussed in the text. The stratigraphic nomenclature used in this chapter may not be internally consistent, nor is it necessarily in agreement with the nomenclature currently in use by the Virginia Division of Mineral Resources or by the U.S. Geological Survey (USGS). This situation reflects the general inconsistency of the literature used in the synthesis.

The first sediments of the Appalachian Basin, generally late Precambrian to Cambrian in age, were deposited on Precambrian crystallines of Grenville age (800$^+$ to 1,000 Ma) (Bartholomew and others, 1984). Across much of the region, these basement rocks are known only from a few widely scattered oil and gas tests. On the eastern margin of the basin, however, extensive exposures of the igneous and metamorphic rocks of this late Proterozoic terrane occur along the Blue Ridge Anticlinorium. In places, paleotopographic relief of several hundred meters on the Precambrian crystalline terrane (Reed, 1955; Gathright, 1976; Allen, 1967) indicates that the terrane had experienced a long period of weathering and erosion prior to latest Precambrian subsidence.

The late Precambrian sequence that was deposited along the eastern margin of the Appalachian Basin records the initial fracturing of the North American protocontinent, the separation of sialic crustal masses thus formed, and the generation of an ancient ocean, Iapetus (Wehr and Glover, 1985). On the western side of the Blue Ridge Anticlinorium in Virginia, late Precambrian basalts of the Catoctin Formation welled up through feeder dikes and spread laterally in great subaerial sheets (Reed, 1955), in some places directly upon crystalline basement, in other places upon a veneer of arkosic alluvial sediments (Table 1). Similarly, to the south in Virginia, Tennessee, and North Carolina, the Bakersville dike swarm records the earliest rifts stages of Iapetus (Goldberg and others, 1986). On the eastern side of the anticlinorium, thick sequences of siliciclastic sediment appear to have poured out over the continental margin into an adjacent ocean basin before eruption of the Catoctin. In some places, these sediments contain mafic igneous rocks that have oceanic affinities (Conley, 1987). Elsewhere along the Blue Ridge, thick clastic sequences of late Precambrian age are interpreted to have filled pull-apart basins or grabens that formed coincidental to the opening of Iapetus (Wehr and Glover, 1985).

The opening of Iapetus, submergence of the North American craton, and development of an early Paleozoic passive margin are recorded by the thick wedge of Cambrian and Lower Ordovician strata that accumulated on the slowly subsiding shelf (Fig. 2). In general, basal strata are siliciclastics that were derived from Precambrian crystalline rocks, whereas overlying beds in this sequence are dominated by carbonate strata that accumulated far from ancient cratonic shorelines.

Milici, R. C., and de Witt, W., Jr., 1988, The Appalachian Basin, *in* Sloss, ed., Sedimentary Cover—North American Craton; U.S.: Boulder, Colorado, Geological Society of America, The Geology of North America, v. D-2.

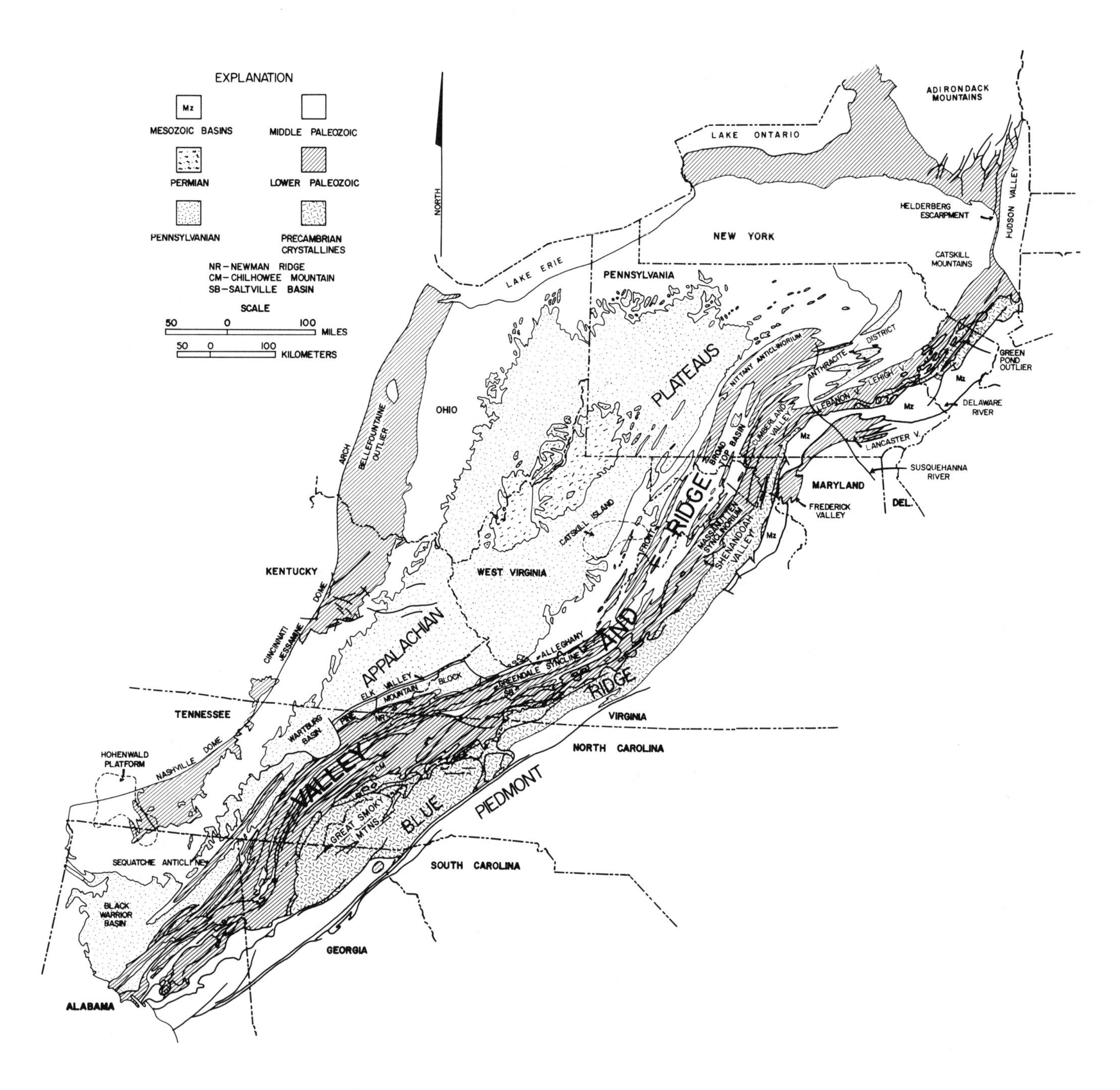

Figure 1. Generalized geologic map of the Appalachian Basin.

TABLE 1. PRECAMBRIAN AND LOWER CAMBRIAN STRATIGRAPHIC NOMENCLATURE IN THE APPALACHIAN BASIN.

AGE	GREAT SMOKY MOUNTAINS	NORTHEASTERN TENNESSEE	NORTHERN VIRGINIA, MARYLAND	EASTERN PENNSYLVANIA		SOUTHEASTERN NEW YORK
EARLY CAMBRIAN	SHADY DOLOMITE	SHADY DOLOMITE	TOMSTOWN DOL	LEITHSVILLE	VINTAGE	STISSING
	CHILHOWEE GROUP: HELENMODE FM	ERWIN: HELENMODE MBR	ANTIETAM FM (ARABY FM (LOWER PART))	HARDYSTON FM	ANTIETAM FM	POUGHQUAG QUARTZITE
	CHILHOWEE GROUP: HESSE SANDSTONE	ERWIN: HESSE QUARTZITE MBR				
	CHILHOWEE GROUP: MURRAY SHALE	ERWIN: MURRAY SHALE MBR				
	CHILHOWEE GROUP: NEBO SANDSTONE	ERWIN: NEBO QUARTZITE MBR				
	CHILHOWEE GROUP: NICHOLS SHALE	HAMPTON FM	HARPERS FM (ARABY FM (LOWER PART))		HARPERS FM	
CAMBRIAN ?	CHILHOWEE GROUP: COCHRAN FM	UNICOI FM	WEVERTON FM ("LOUDOUN" FM) (ARABY FM (LOWER PART))		CHICKIES FM	
PRECAMBRIAN	OCOEE SERIES	MT ROGERS FM GRENVILLE BASEMENT	CATOCTIN FM SWIFT RUN FM GRENVILLE BASEMENT	GRENVILLE BASEMENT		GRENVILLE BASEMENT

TABLE 2. CAMBRIAN STRATIGRAPHIC NOMENCLATURE IN THE APPALACHIAN BASIN.

AGE	EASTERN TENNESSEE AND SOUTHWESTERN VIRGINIA	NORTHERN VIRGINIA AND WESTERN MARYLAND	FREDERICK VALLEY, MARYLAND	VALLEY AND RIDGE, PENNSYLVANIA	EASTERN PENNSYLVANIA	OHIO	NEW YORK
LATE CAMBRIAN	COPPER RIDGE DOLOMITE / CONOCOCHEAGUE DOLOMITE	CONOCOCHEAGUE FORMATION	FREDERICK LIMESTONE	BEEKMANTOWN (PART)	BEEK.	KNOX DOL (PART)	LITTLE FALLS DOLOMITE
				GATESBURG FM / CONOCOCHEAGUE GROUP	CONOCOCHEAGUE GROUP (CONESTOGA FORMATION)	KERBEL FM	GALWAY FM
						"CONASAUGA" FM	
	MAYNARDVILLE	ELBROOK DOLOMITE		WARRIOR FM / ELBROOK FORMATION		"ROME" FORMATION	POTSDAM SANDSTONE
	NOLICHUCKY SHALE (CONASAUGA SHALE / ELBROOK DOLOMITE)			?		MOUNT SIMON SANDSTONE	
MIDDLE CAMBRIAN	MARYVILLE LIMESTONE (CONASAUGA SHALE / HONAKER FM / ELBROOK DOLOMITE)		ARABY FORMATION (PART)	PLEASANT HILL FORMATION / ELBROOK FORMATION	BUFFALO SPRINGS (LEITHSVILLE FORMATION / CONESTOGA FORMATION)		
	ROGERSVILLE SHALE				ZOOKS CORNER		
	RUTLEDGE LIMESTONE				LEDGER FM		
EARLY CAMBRIAN	ROME ("WAUTAUGA") FORMATION	WAYNESBORO FORMATION		WAYNESBORO FORMATION	KINZERS FM		
	SHADY DOLOMITE	TOMSTOWN DOLOMITE		TOMSTOWN FORMATION	VINTAGE FM (LEITHSVILLE FORMATION / ?)		

TABLE 3. ORDOVICIAN STRATIGRAPHIC NOMENCLATURE IN THE APPALACHIAN BASIN.

AGE	MIDDLE AND LATE ORDOVICIAN	EARLY ORDOVICIAN
TENNESSEE AND SOUTHWESTERN VIRGINIA	CHICKAMAUGA LS (SUPERGROUP); STONES RIVER; LEIPERS; SHELLMOUND; SEQUATCHIE; JUNIATA FM; INMAN FM; REEDSVILLE SH; MARTINSBURG FORMATION; NASHVILLE; BAYS FM; OTTOSEE; CHAPMAN RIDGE; HOLSTON; SEVIER; LENOIR LS	KNOX GROUP (PART); c MASCOT DOL; KINGSPORT FM l; CHEPULTEPEC DOLOMITE; JONESBORO LIMESTONE
VIRGINIA VALLEY AND RIDGE, CENTRAL AND SOUTHERN	SEQUATCHIE FM; REEDSVILLE SHALE; TRENTON; EGGLESTON; MOCCASIN; BAYS; WITTEN; WARDELL; BENBOLT; ROCKDELL; WASSUM; CHATHAM HILL; RICH V; LIBERTY HALL; LINCOLNSHIRE; EFFNA; LINCOLNSHIRE; ELWAY LS; BLACKFORD FM; LENOIR; KNOBS; PAPERVILLE	c BEEKMANTOWN FORMATION l; CHEPULTEPEC DOL; STONEHENGE LS
NORTHERN VIRGINIA, MARYLAND AND CUMBERLAND VALLEY PENNSYLVANIA	JUNIATA FM; OSWEGO SS; MARTINSBURG SHALE; LIBERTY HALL/LANTZ MILLS; LINCOLNSHIRE LS; NEW MARKET; CHAMBERSBURG LS; ST PAUL	PINESBURG STATION DOL; ?; c ROCKDALE RUN FORMATION l; b STONEHENGE LIMESTONE; BEEKMANTOWN GROUP
CENTRAL PENNSYLVANIA	JUNIATA FM; OSWEGO; BALD EAGLE; REEDSVILLE SHALE; ANTES SH; COBURN FM; SALONA FM; NEALMONT FM; LINDEN, HALL FM; SNYDER FM; HATTER FM; LOYSBURG FM	BELLEFONTE DOLOMITE; c AXEMAN LS c; NITTANY l DOLOMITE; b LARKE DOL; STONEHENGE LS b; BEEKMANTOWN GROUP
EASTERN PENNSYLVANIA	MARTINSBURG FORMATION; HERSHEY LS; MYERSTOWN; JACKSONBURG; ANNVILLE FM	ONTELAUNEE FORMATION; ?; EPLER FORMATION; RICKENBACH DOL; STONEHENGE LS; BEEKMANTOWN GROUP
OHIO	QUEENSTON FORMATION; SHALE AND LIMESTONE (UNNAMED); TRENTON LIMESTONE; BLACK RIVER GROUP; GLENWOOD SHALE	KNOX DOL (PART)
CENTRAL AND WESTERN NEW YORK	QUEENSTON SHALE; LORRAINE: OSWEGO, PULASKI, WHETSTONE GULF; SCHENECTADY; FRANKFORT SHALE; UTICA; HILLIER; TRENTON GROUP; DOLGEVILLE; BLACK RIVER GP	CHUCTANUNDA DOLOSTONE; TRIBES HILL FORMATION; BEEKMANTOWN GP

c- CERATOPEA, l-LECANOSPIRA, b-BELLEFONTIA

THE CAMBRIAN

Basal clastics, the first transgression

Chilhowee Group. The Chilhowee Group and its equivalents consist almost entirely of siliciclastic sequences that are impure and arkosic at the base and contain clean orthoquartzites at the top (Table 1). They lie along the western flank of the Blue Ridge Anticlinorium throughout the length of the Appalachian Basin, and on both the eastern and western flanks of the anticlinorium in northern Virginia, Maryland, and southern Pennsylvania. The eastward extent of these siliciclastics beneath crystalline thrust sheets of the Piedmont and Blue Ridge Mountains and the westward extent of siliciclastics beneath Valley and Ridge thrust sheets are unknown. Sparse data indicate that lower and middle parts of the group were deposited along the eastern basin margin.

The Chilhowee Group unconformably overlies a variety of formations along its outcrop in the southern and central Appalachian Blue Ridge (Table 1). In northern Georgia and Tennessee, it overlies the late Precambrian Ocoee Supergroup. In northeastern Tennessee and southwestern Virginia, the Mount Rogers Volcanic Group (late Precambrian) lies beneath the Chilhowee. In central and northern Virginia, the Chilhowee overlies either allochthonous crystalline basement or Catoctin Greenstone (late Precambrian), a relationship that persists northeastward into Maryland and Pennsylvania. In eastern Pennsylvania and New Jersey, thin Upper Cambrian sandstones constitute the basal clastics, which there overlie authochthonous crystalline basement.

The Chilhowee Group is subdivided differently along the length of the Appalachians and the age of the group is in question. Cambrian fossils were collected from uppermost beds by Keith (1903) in the Cranberry Quadrangle of Tennessee and North Carolina. Walcott (1891) reports Cambrian fossils from correlative beds in eastern Tennessee, and Butts (1940) mentions Chilhowee fossils in his report on Virginia. Similar fossils of Cambrian age were found in the Chilhowee of Maryland and southern Pennsylvania (Walcott, 1886). The stratigraphic position of the base of the Cambrian is in doubt (King, 1949; Laurence, 1963; Laurence and Palmer, 1963). A recent fossil discovery by Simpson and Sundberg (1987) in Virginia, however, suggests that most, if not all, of the Chilhowee is Cambrian. The USGS designates the lower, unfossiliferous part of the Chilhowee as Cambrian(?).

The Chilhowee Group overlies Precambrian strata with a moderate unconformity in its type area in eastern Tennessee (King, 1964; Neuman and Nelson, 1965), where it is about 920 m thick. There, in the lower part, it consists of cross-bedded, pebbly arkose, siltstone, shale, and massive quartzite. This se-

TABLE 4. SILURIAN STRATIGRAPHIC NOMENCLATURE IN THE APPALACHIAN BASIN.

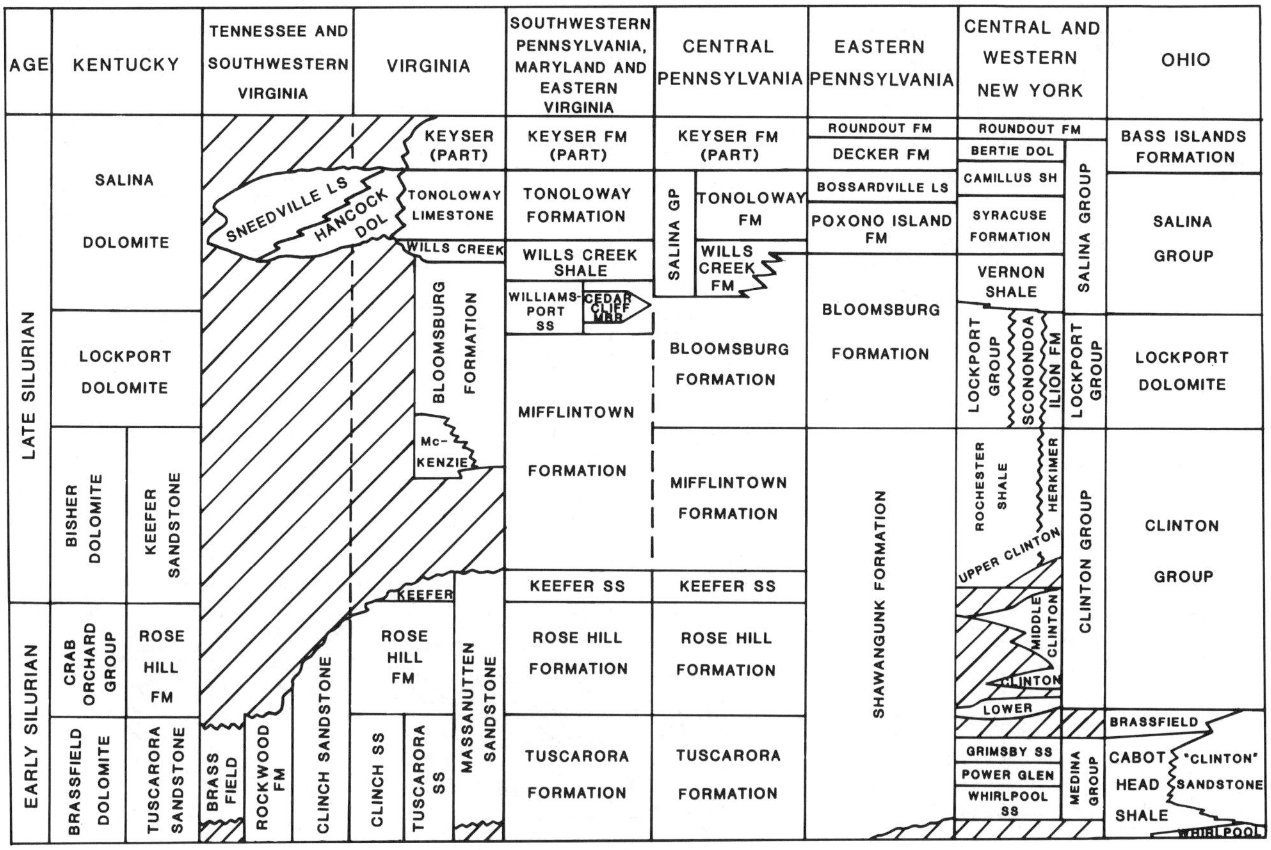

AGE	KENTUCKY	TENNESSEE AND SOUTHWESTERN VIRGINIA	VIRGINIA	SOUTHWESTERN PENNSYLVANIA, MARYLAND AND EASTERN VIRGINIA	CENTRAL PENNSYLVANIA	EASTERN PENNSYLVANIA	CENTRAL AND WESTERN NEW YORK	OHIO
LATE SILURIAN	SALINA DOLOMITE	SNEEDVILLE LS; HANCOCK DOL	KEYSER (PART); TONOLOWAY LIMESTONE; WILLS CREEK	KEYSER FM (PART); TONOLOWAY FORMATION; WILLS CREEK SHALE; WILLIAMS-PORT SS; CEDAR CLIFF MBR	KEYSER FM (PART); SALINA GP: TONOLOWAY FM, WILLS CREEK FM	ROUNDOUT FM; DECKER FM; BOSSARDVILLE LS; POXONO ISLAND FM	ROUNDOUT FM; SALINA GROUP: BERTIE DOL, CAMILLUS SH, SYRACUSE FORMATION, VERNON SHALE	BASS ISLANDS FORMATION; SALINA GROUP
	LOCKPORT DOLOMITE		BLOOMSBURG FORMATION	MIFFLINTOWN FORMATION	BLOOMSBURG FORMATION	BLOOMSBURG FORMATION	LOCKPORT GROUP; SCONONDOA; ILION FM; LOCKPORT GROUP	LOCKPORT DOLOMITE
	BISHER DOLOMITE; KEEFER SANDSTONE		Mc-KENZIE		MIFFLINTOWN FORMATION	SHAWANGUNK FORMATION	CLINTON GROUP: ROCHESTER SHALE; HERKIMER; UPPER CLINTON	CLINTON GROUP
			KEEFER	KEEFER SS	KEEFER SS		MIDDLE CLINTON	
EARLY SILURIAN	CRAB ORCHARD GROUP; ROSE HILL FM	CLINCH SANDSTONE	ROSE HILL FM; MASSANUTTEN SANDSTONE	ROSE HILL FORMATION	ROSE HILL FORMATION		CLINTON; LOWER	
	BRASSFIELD DOLOMITE; TUSCARORA SANDSTONE	BRASS FIELD; ROCKWOOD FM	CLINCH SS; TUSCARORA SS	TUSCARORA FORMATION	TUSCARORA FORMATION		MEDINA GROUP: GRIMSBY SS, POWER GLEN, WHIRLPOOL SS	BRASSFIELD; CABOT HEAD SHALE; "CLINTON" SANDSTONE; WHIRLPOOL

quence is overlain by silty and sandy shale that, in places, contains beds of feldspathic sandstone and *Scolithus*-bearing quartzite; argillaceous to silty and laminated shale; siltstone, fine-grained glauconitic sandstone, and quartzite; and, at the top (Keith, 1903), micaceous shale interbedded with layers of fossiliferous dolomitic sandstone.

In northeastern Tennessee, the Chilhowee Group averages 1,220 m in thickness but locally is as much as 2,290 m thick (King and Ferguson, 1960). In some places, the Chilhowee overlies Precambrian granitic and metamorphic basement rocks. In other localities, it overlies stratified sedimentary rocks and lavas of the Mount Rogers Volcanic Group. The basal Chilhowee consists generally of arkosic and conglomeratic sandstones, quartzite, and some shales and siltstones that locally contain lenses and beds of amygdaloidal basalt up to 30 m thick. The basal Chilhowee is overlain by shale, siltstone, and *Scolithus*-bearing quartzite and arkosic sandstone, followed by interbedded quartzite, siltstone, and shale. The uppermost beds consist of calcareous and dolomitic shale, siltstone, and sandstone.

In southwestern Virginia, Butts (1940, p. 30) measured about 800 m of conglomerate, sandstone, arkosic sandstone, and some shale in a section that contains three lava flows and assigned the section to the basal formation in the Chilhowee Group. The contact between the Chilhowee and the Precambrian basement is obscure. Butts (1940, p. 31) describes the basement as "red rock with decayed feldspathic inclusions."

The middle formation of the Chilhowee consists of about 610 m of sandstones, shale, and slate (Miller, 1944). In one place, Miller (1944) measured 235 m of interbedded sandstone, quartzite, and shale in the uppermost formation of the Chilhowee Group, which is generally 460 to 610 m thick in southwestern Virginia. In general, the uppermost beds of the Chilhowee are glauconite-bearing or shaly sandstone similar to those in Tennessee.

In the Elkton area of northern Virginia, Grenville crystallines are overlain in places by 30 m of pre-Chilhowee, late Precambrian arkosic quartzites and conglomerates (King, 1950). Slate and pyroclastic rocks commonly are interbedded with these coarser-grained siliciclastics, and, in places, these strata intertongue with the overlying Catoctin Greenstone (late Precambrian). The Catoctin Greenstone consists of basaltic flows, some of which have columnar joints, as well as tuffaceous phyllite, slate, and pyroclastic rocks, and generally ranges from 100 to 300 m thick. In places, it pinches out entirely, so that the overly-

TABLE 5. DEVONIAN STRATIGRAPHIC NOMENCLATURE IN THE APPALACHIAN BASIN.

AGE	EASTERN TENNESSEE AND SOUTHWESTERN VIRGINIA	NORTHERN VIRGINIA AND WEST VIRGINIA	WESTERN MARYLAND	CENTRAL PENNSYLVANIA	NORTHERN OHIO	NORTHWESTERN NEW YORK
LATE DEVONIAN	CHATTANOOGA SHALE: BIG STONE GAP SHALE (PART) MIDDLE GRAY UNIT LOWER BLACK UNIT	HAMPSHIRE FORMATION CHEMUNG FORMATION BRAILLER FM	HAMPSHIRE FORMATION FOREKNOBS FM / CHEMUNG SHERR FM BRAILLER FM HARRELL SH	CATSKILL FORMATION TRIMMERS ROCK LS BRAILLER HARRELL SHALE	OHIO SHALE: CLEVELAND MBR CHAGRIN SHALE HURON MBR OLENTANGY SHALE ?	OSWAYO CATTARAUGUS FM SALAMANCA CGL PANAMA CGL CHADAKOIN FM NE & WESTFIELD SHS PERRYSBURG FM JAVA FM WEST FALLS FM SONYEA FM GENESEE FM
MIDDLE DEVONIAN	LOWER BLACK UNIT TIOGA ASH BED	MILLBORO SHALE HUNTERSVILLE CHERT / NEEDMORE SHALE	MAHANTANGO FORMATION MARCELLUS SH NEEDMORE SH	TULLY LIMESTONE MAHANTANGO FORMATION MARCELLUS SHALE SELINSGROVE LIMESTONE NEEDMORE SHALE	OLENTANGY SHALE ? PROUT LIMESTONE PLUM BROOK SHALE ? DELAWARE LIMESTONE COLUMBUS LIMESTONE	MOSCOW FORMATION LUDLOWVILLE FORMATION SKANEATELES FORMATION MARCELLUS FORMATION TIOGA ONONDAGA LS
EARLY DEVONIAN 3	WILDCAT VALLEY SANDSTONE				BOIS BLANC LIMESTONE	BOIS BLANC FM
EARLY DEVONIAN 2	WILDCAT VALLEY SANDSTONE	ORISKANY SANDSTONE LICKING CREEK LS	ORISKANY SANDSTONE SHRIVER CHERT MANDATA SHALE	OLD POINT FORMATION: RIDGELEY MBR SHRIVER MBR MANDATA SH MBR	ORISKANY SS	
EARLY DEVONIAN 1		CORRIGANVILLE HEALING SPRINGS SS NEW CREEK LIMESTONE KEYSER LS (PART)	CORRIGANVILLE LS NEW CREEK LIMESTONE KEYSER LS (PART)	OLD POINT FORMATION: LOWER MEMBER	? LS OF HELDERBERG AGE ?	

1. HELDERBERG, 2. DEERPARK, 3. ONESQUETHAW

ing Chilhowee lies directly upon Grenvillian basement rocks. King (1950) concludes that the Catoctin and older Precambrian rocks were tilted and eroded before Chilhowee deposition.

In northern Virginia and Maryland, the Chilhowee Group, about 915 m thick, consists of spotted, purple or red, pyroclastic slate at the base (King, 1950; Whitaker, 1955) that, in places, contains a thin bed of amygdaloidal lava. Beds of quartzite, conglomerate, arkose, sandstone, siltstone, and phyllite overlie these slates. Above these pyroclastic slates are argillaceous phyllites, siltstones, and sandstones (King, 1950; Whitaker, 1955). The uppermost unit of the Chilhowee in northern Virginia and Maryland generally is composed of *Scolithus*-bearing sandstone and quartzite. In some places in Maryland, the upper Chilhowee is absent, and lower formations are overlain either by the Tomstown Dolomite or by the Frederick Limestone (Whitaker, 1955).

In eastern Pennsylvania, southeast of the Gettysburg Mesozoic basin, the Chilhowee Group overlies Precambrian crystallines unconformably. In some places there is a basal formation that consists of 75 to 150 m of conglomerate, pebbly quartzite, arkosic quartize, and a few beds of black slate. The upper part of the unit generally is cleaner, consisting of 60 to 120 m of *Scolithus*-bearing quartite.

About 305 m of mildly metamorphosed quartzose gray-green phyllite overlie the basal Chilhowee in eastern Pennsylvania (Stose and Stose, 1944, p. 14). This unit is overlain gradationally by 30 to 60 m of quartzite that, in its upper part, is calcareous and ferruginous and contains Lower Cambrian fossils. The contact with the overlying dolomite is gradational.

Platform equivalents. The Chilhowee thins significantly to the northeast, north, and west, generally by overstepping its basal beds. To the northeast, in east-central Pennsylvania and adjacent New Jersey, the Hardyston Quartzite of Early Cambrian age overlies Precambrian gneiss without an intervening section of Cambrian(?) formations. The thickness of the Hardyston is as much as 240 m (Aaron, 1969). Its lower part consists of arkosic sandstones and conglomerates that appear to have been derived from underlying crystallines, whereas the upper part is clean calcareous quartzite that contains Early Cambrian fossils in New Jersey. This latter unit, too, grades upward into an overlying impure dolomite. Aaron (1969) concludes that the lower part of the Hardyston originated as alluvium on a moderate-relief Precambrian terrane; inundation of that terrane cleaned the sand and deposited the upper part of the formation in shallow marine environments. Similarly, an approximate equivalent in southeast-

TABLE 6. MISSISSIPPIAN STRATIGRAPHIC NOMENCLATURE IN THE APPALACHIAN BASIN.

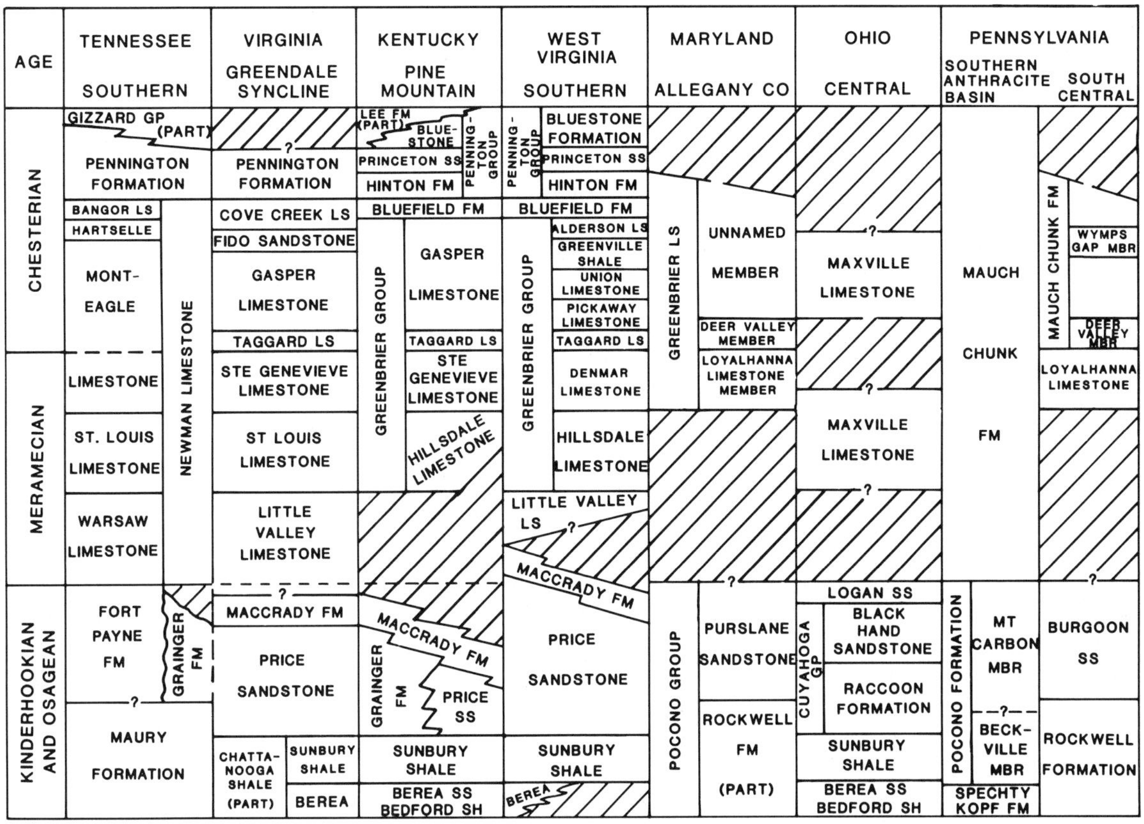

AGE	TENNESSEE SOUTHERN	VIRGINIA GREENDALE SYNCLINE	KENTUCKY PINE MOUNTAIN	WEST VIRGINIA SOUTHERN	MARYLAND ALLEGANY CO	OHIO CENTRAL	PENNSYLVANIA SOUTHERN ANTHRACITE BASIN	PENNSYLVANIA SOUTH CENTRAL
CHESTERIAN	GIZZARD GP (PART) PENNINGTON FORMATION BANGOR LS HARTSELLE MONT-EAGLE LIMESTONE (NEWMAN LIMESTONE)	? PENNINGTON FORMATION COVE CREEK LS FIDO SANDSTONE GASPER LIMESTONE TAGGARD LS STE GENEVIEVE LIMESTONE	LEE FM (PART) BLUE-STONE PRINCETON SS HINTON FM (PENNINGTON GROUP) BLUEFIELD FM GREENBRIER GROUP: GASPER LIMESTONE TAGGARD LS STE GENEVIEVE LIMESTONE	BLUESTONE FORMATION PRINCETON SS HINTON FM (PENNINGTON GROUP) BLUEFIELD FM GREENBRIER GROUP: ALDERSON LS GREENVILLE SHALE UNION LIMESTONE PICKAWAY LIMESTONE TAGGARD LS DENMAR LIMESTONE	GREENBRIER LS: UNNAMED MEMBER DEER VALLEY MEMBER LOYALHANNA LIMESTONE MEMBER	? MAXVILLE LIMESTONE ?	MAUCH CHUNK FM	MAUCH CHUNK FM: WYMPS GAP MBR DEER VALLEY MBR LOYALHANNA LIMESTONE
MERAMECIAN	LIMESTONE ST. LOUIS LIMESTONE WARSAW LIMESTONE	ST LOUIS LIMESTONE LITTLE VALLEY LIMESTONE ?	HILLSDALE LIMESTONE	HILLSDALE LIMESTONE LITTLE VALLEY LS ?		MAXVILLE LIMESTONE ?		?
KINDERHOOKIAN AND OSAGEAN	FORT PAYNE FM GRAINGER FM ? MAURY FORMATION	MACCRADY FM PRICE SANDSTONE CHATTA-NOOGA SHALE (PART) SUNBURY SHALE BEREA	MACCRADY FM GRAINGER FM PRICE SS SUNBURY SHALE BEREA SS BEDFORD SH	MACCRADY FM PRICE SANDSTONE SUNBURY SHALE BEREA	POCONO GROUP: PURSLANE SANDSTONE ROCKWELL FM (PART)	LOGAN SS CUYAHOGA GP: BLACK HAND SANDSTONE RACCOON FORMATION SUNBURY SHALE BEREA SS BEDFORD SH	POCONO FORMATION: MT CARBON MBR ? BECK-VILLE MBR SPECHTY KOPF FM	BURGOON SS ROCKWELL FORMATION

ern New York, the Poughquag Quartzite, consists of about 40 m of feldspathic sandstones and siltstones that were deposited in and near the Early Cambrian transgressive shoreline.

Along the northern margin of the Adirondacks, the basal clastic formation, the Potsdam Sandstone, ranges from 30 to 100 m thick. The formation thickens to the east to as much as 500 m in and adjacent to the Champlain rift basin (Swartz, 1948; Fisher, 1977). In the type area, the upper part is relatively pure quartzite, whereas lower beds are feldspathic and contain pebbles and cobbles of quartz, quartzite, and a few pieces of granite (Chadwick, 1920; Swartz, 1948). Swartz (1948, p. 1537) concludes that the irregular nature of the Potsdam reflects deposition on a surface marked by 30 to 60 m of local relief and in valleys up to several miles wide.

In the northeastern Appalachian Basin, quartzites and sandstones are not present everywhere at the base of the Paleozoic. In the Mohawk Valley, Cambrian dolomite overlies Grenville crystallines (Fisher, 1977, Plate 3), and northward along Lake Ontario, the entire Cambrian section is overlapped so that Ordovician limestones are in contact with basement.

Siliciclastic strata of the Mount Simon Sandstone overlie basement crystallines in the subsurface of Ohio, along the Cincinnati arch (Janssens, 1973). Their thickness, too, appears to be related to the local relief of that Precambrian surface, and, in places, they are absent. Elsewhere, the Mount Simon ranges generally from 60 to 120 m thick. The formation generally consists of sandstone that grades downward into conglomeratic sandstone or sandy conglomerate. Janssens (1973) suggests that the paucity of fossils and the absence of glauconite indicate a nonmarine origin for the lower beds; the cleaner nature of the upper part of the formation resulted from the reworking of the sandstone sediments during the earliest marine transgression.

Harris (1964) correlates Cambrian strata from surface sections in Tennessee and Virginia into the subsurface of Virginia and adjacent parts of eastern Kentucky. In general, he shows that the Rome Formation is a time transgressive unit of Early Cambrian age in Tennessee and of Middle Cambrian age in eastern Kentucky where it overlies Precambrian crystalline basement.

Lower Cambrian carbonate formations

Inundation of the Paleozoic craton during deposition of the Chilhowee Group moved sources of siliciclastic sediment progressively farther to the north and west, away from the eastern

TABLE 7. PENNSYLVANIAN STRATIGRAPHIC NOMENCLATURE IN THE APPALACHIAN BASIN.

AGE	EASTERN TENNESSEE	SOUTHWESTERN VIRGINIA	EASTERN KENTUCKY	WEST VIRGINIA		MARYLAND	OHIO	PENNSYLVANIA WESTERN	PENNSYLVANIA EASTERN
LATE PENNSYLVANIAN				DUNKARD GP (PART)		DUNKARD GP (PART)	DUNKARD GP (PART)	DUNKARD (PART)	LLEWELLYN FORMATION
LATE PENNSYLVANIAN			MONONGAHELA FORMATION	MONONGAHELA GROUP		MONONGAHELA GROUP	MONONGAHELA GROUP	MONONGAHELA GROUP	LLEWELLYN FORMATION
LATE PENNSYLVANIAN			CONEMAUGH GROUP	CONEMAUGH GROUP		CONEMAUGH GROUP	CONEMAUGH GROUP	CONEMAUGH GROUP	LLEWELLYN FORMATION
MIDDLE PENNSYLVANIAN			BREATHITT FORMATION	ALLEGHENY GROUP		ALLEGHENY GROUP	ALLEGHENY GROUP	ALLEGHENY GROUP	LLEWELLYN FORMATION
MIDDLE PENNSYLVANIAN	CROSS MTN FM		BREATHITT FORMATION	POTTSVILLE GROUP	KANAWHA FORMATION	POTTSVILLE GROUP	POTTSVILLE GROUP	POTTSVILLE GROUP	POTTSVILLE FM
MIDDLE PENNSYLVANIAN	VOWELL MTN	HARLAN FM	BREATHITT FORMATION	POTTSVILLE GROUP	KANAWHA FORMATION	POTTSVILLE GROUP	POTTSVILLE GROUP	POTTSVILLE GROUP	POTTSVILLE FM
MIDDLE PENNSYLVANIAN	REDOAK MTN	WISE FM	BREATHITT FORMATION	POTTSVILLE GROUP	KANAWHA FORMATION	POTTSVILLE GROUP	POTTSVILLE GROUP	POTTSVILLE GROUP	POTTSVILLE FM
MIDDLE PENNSYLVANIAN	GRAVES GAP FM	GLADEVILLE SANDSTONE	BREATHITT FORMATION	POTTSVILLE GROUP	KANAWHA FORMATION	POTTSVILLE GROUP	POTTSVILLE GROUP	POTTSVILLE GROUP	POTTSVILLE FM
MIDDLE PENNSYLVANIAN	INDIAN BLUFF FM	GLADEVILLE SANDSTONE	BREATHITT FORMATION	POTTSVILLE GROUP	KANAWHA FORMATION	POTTSVILLE GROUP	POTTSVILLE GROUP	POTTSVILLE GROUP	POTTSVILLE FM
MIDDLE PENNSYLVANIAN	SLATESTONE FM		BREATHITT FORMATION	POTTSVILLE GROUP	KANAWHA FORMATION	POTTSVILLE GROUP	POTTSVILLE GROUP	POTTSVILLE GROUP	POTTSVILLE FM
EARLY PENNSYLVANIAN	CROOKED FORK GROUP	NORTON FM / NEW RIVER FM	BREATHITT FORMATION / LEE FORMATION (PART)	POTTSVILLE GROUP	NEW RIVER FORMATION	POTTSVILLE GROUP	POTTSVILLE GROUP	POTTSVILLE GROUP	POTTSVILLE FM / MAUCH CHUNK FM
EARLY PENNSYLVANIAN	CRAB ORCHARD MOUNTAINS GROUP	LEE FM (PART) / NEW RIVER FM / POCAHONTAS	LEE FORMATION (PART)	POTTSVILLE GROUP	NEW RIVER FORMATION	POTTSVILLE GROUP	SHARON CGL	POTTSVILLE GROUP	MAUCH CHUNK FM
EARLY PENNSYLVANIAN	GIZZARD GROUP (PART)	POCAHONTAS / BLUESTONE (PART)	LEE FORMATION (PART)	POTTSVILLE GROUP	POCAHONTAS FM	POTTSVILLE GROUP		POTTSVILLE GROUP	MAUCH CHUNK FM

margin of the Appalachian Basin. The siliciclastics were generally displaced by deposition of carbonates, thereby forming the oldest calcareous formations of the southern and central Appalachians. In the southern Appalachians, these rocks are known as the Shady Dolomite, whereas to the north in the central Appalachians, they are called Tomstown (Table 2).

Shady Dolomite. In Tennessee, in and near the Great Smoky Mountains, the Shady Dolomite consists of crystalline, gray dolomite (Neuman and Nelson, 1965). In general, the formation is poorly exposed because it is covered either by jasperoid-rich residuum or by colluvium and alluvium from nearby mountains. Shaly beds in the upper part of the formation mark the transition of the Shady into the overlying Rome Formation. The Shady and its residuum are host to a variety of mineral deposits of past and potential economic value, including iron, manganese, lead, and zinc, although none are mined currently. Residual "buckfat" clays may be up to 75 m thick, and irregularly shaped dolomite pinnacles 3 to 30 m high commonly are exposed in old abandoned hydraulic mines (King and Ferguson, 1960, p. 51).

In southwestern Virginia, the Shady generally consists of 600 to 1,300 m of dolomite and limestone. In some eastern exposures in southwestern Virginia (Stose and Jonas, 1938), the unusual amount of siliciclastics and breccia beds suggested to Currier (1935) and others that this Shady sequence represents a nearshore facies that was thrust westward upon the more common carbonate beds of the Shady. Rodgers (1968), however, suggested that these beds accumulated on the Paleozoic shelf edge and were subsequently thrust to their present location. This interpretation was modified by Pfeil and Read (1980) and Read and Pfeil (1983), who identified Shady shelf, shelf edge, foreslope, and slope facies that accumulated in a marginal sea setting, rather than facing the higher-energy regime of an open ocean.

Tomstown Dolomite. The Tomstown Dolomite, a northern equivalent of the Shady, is about 305 to 460 m thick in northern Virginia and consists mostly of gray dolomite and some shale (King, 1950; Edmundson, 1945). As in Tennessee, residuum above these carbonate rocks in the Elkton area is a tough, waxy, "buckfat" clay that contains a variety of impure siliceous materials together with potentially ore-grade iron and manganese oxides. These ores were extensively mined in the Elkton area from about 1836 to 1941 (King, 1950).

The Tomstown thins northeastward along the Great Valley

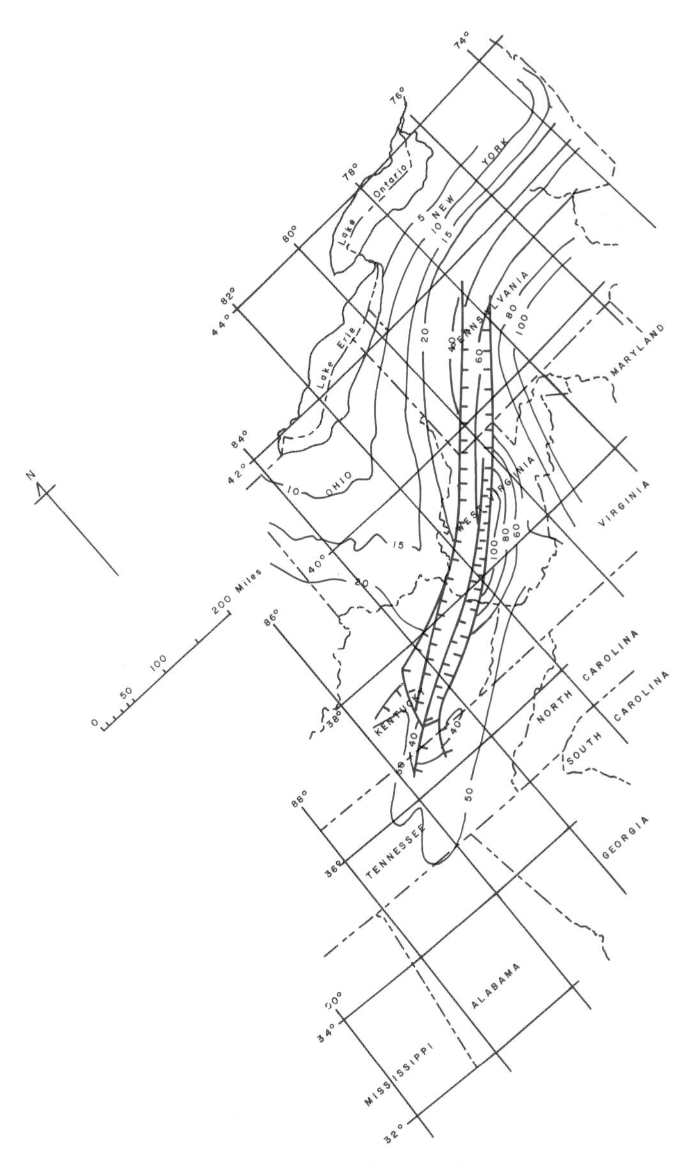

Figure 2. Thickness of Cambrian and Lower Ordovician rocks in the Appalachian Basin. Isopachs are in feet; ticks are on downthrown side of fault. (From Harris, 1975, 1978.)

into Maryland, where it consists of 150 m of interbedded limestone and dolomite (Reinhardt and Wall, 1975). Reinhardt and Wall (1975) determined that the principal lithofacies of the Tomstown represents accumulations on a stable carbonate platform margin, in lagoon-bay and intertidal to supratidal flat environments.

In its type area in the southern Cumberland Valley of Pennsylvania and along the eastern side of the Great Valley, the Tomstown consists of about 305 to 610 m of dolomite, together with some arenaceous and crystalline or banded limestone (Stose, 1906, 1932; Root, 1968, p. 7, 8).

Cambrian platform margin, Maryland and Pennsylvania

East of the Triassic basins, in Frederick Valley, Maryland, Lower and Middle Cambrian phyllites and metasiltstones of the Araby Formation are correlated with part of the Chilhowee and the Tomstown and Waynesboro Formations of the Great Valley by Reinhardt (1974). He also includes the Tomstown Dolomite of Jonas and Stose (1936) with the overlying Frederick Limestone, and concludes that this Lower Cambrian to Lower Ordovician sequence represents a progressive shallowing from deep-water dolomite at the base to shallow-water high-energy limestones at the top.

In eastern Pennsylvania, basin, slope, and shelf deposits of the Vintage, Kinzers, and Ledger Formations are of Early and Middle Cambrian age (Gohn, 1976). The Vintage consists of 150 to 245 m of knotty or finely banded limestone and dolomite layers (Jonas and Stose, 1930; Meisler and Becher, 1968). The overlying Kinzers is about 45 to 155 m of shale, impure dolomite, spotted limestone, and megabreccia (Reinhardt, 1977). In places, it is highly fossiliferous, containing fossils of Early and Middle Cambrian age (Kauffman and Campbell, 1969; Palmer, 1971; Berg and others, 1983). The Ledger consists generally of massive dolomite that, in places, contains some zones of relatively pure limestone. Its thickness ranges from 30 to 365 m. The Ledger is overlain by argillaceous limestone, limestone conglomerates, and calcareous sandstones of the Conestoga Formation above an extensive unconformity (Jonas and Stose, 1930; Stose and Jonas, 1923, 1939; Berg and others, 1983). Spectacular limestone conglomerates are described in the Conestoga by Jonas and Stose (1930) and Stose and Stose (1944). They thought that the Conestoga overlies older beds with marked unconformity. Subsequently, Rodgers (1968) interpreted the Conestoga as a deep-water deposit that lapped over contemporaneous shallow-water deposits, thus marking the edge of the lower Paleozoic carbonate shelf. The limestone congomerates in his interpretation represent debris flows from the edge of the bank into the deeper basin.

Cambrian regressive deposits and formations of the Rome Trough

Rome Trough. Cambrian transgression of the southeastern United States was interrupted during Early Cambrian time by regional uplift and eastward progradation of siliciclastic sediments derived from the craton interior. This regional regression resulted at first in the deposition of the Rome Formation and its siliciclastic equivalents and then, in the southern Appalachians, of the overlying Conasauga Shale. To the east and north, Conasauga Shale interfingers with limestone and dolomite units, and in eastern Tennessee and Virginia these beds grade laterally into dolomite.

The uplift that produced the Cambrian regression also appears to have caused enough crustal extension beneath the Appalachian Basin to have formed a regionally extensive intracratonic

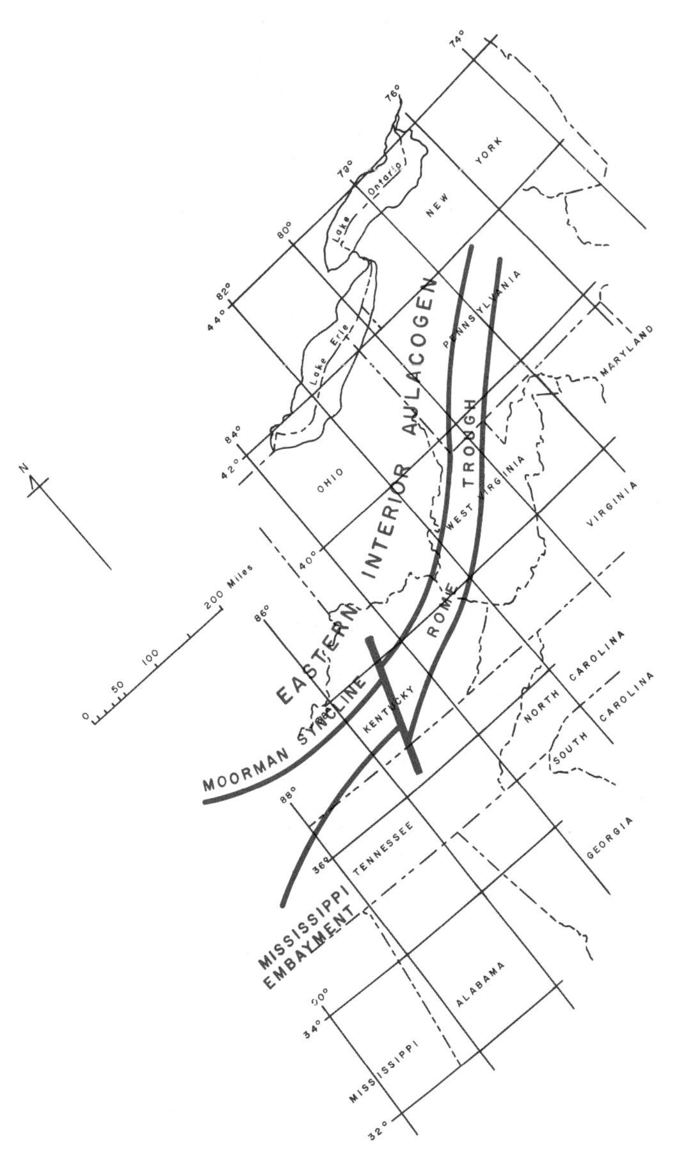

Figure 3. Probable extent of the Eastern Interior aulacogen and its component parts. (From Harris, 1978, Fig. 1.)

graben system from Pennsylvania to the Mississippi Embayment (Harris, 1975). This system, called the Eastern Interior aulacogen by Harris (1978), consists of two parts. The eastern part, called the Rome Trough by McGuire and Howell (1963), extends from Pennsylvania through West Virginia to eastern Kentucky (Fig. 3). The western part of the aulacogen, the Moorman Syncline, is beyond the area of the Appalachian Basin, extending into Kentucky from the northern part of the Mississippi Embayment (Fig. 2). The subsurface connection between the two parts of the aulacogen has not yet been demonstrated by drilling, so they may indeed be discontinuous, rather than constituting a single elongate extensional basin (Harris, 1978).

The Rome Trough was formed before medial Cambrian. Donaldson and others (1975) identified fossils of Middle Cambrian age from cores taken in West Virginia near the Kentucky border. About 490 m of strata lie between the fossiliferous zone and basement, and Harris (1978) speculates that this unfossiliferous zone might be Lower Cambrian. The graben was active during much of the early Paleozoic, as is evidence by overthickening of strata lying over the trough at least into the Silurian. Harris (1978) divides the tectonic history of the Rome Trough into an early graben stage, a transitional stage, and a later downwarping stage. The early graben stage is marked by a thickening of Cambrian and Lower Ordovician strata in the deeper part of the trough by almost 2,000 m when compared with the same strata adjacent to the trough. Differences in sedimentary accumulations were smaller during the Middle and Late Ordovician; Harris (1978) describes a 460-m movement on the southeastern border fault in Kentucky that occurred during that time (Fig. 4).

Isopachs of Silurian strata are deflected westward across the trough, indicating that it was still active (Fig. 5). As Harris (1978) points out, the Silurian was a time of regional subsidence of the Appalachian Basin; the thickest area of sediment accumulation was near the center line of the graben.

Rome Formation. In general, Lower and Middle Cambrian siliciclastics of the Appalachian Basin constitute a regressive–transgressive sequence between the Shady and younger Cambrian carbonate formations. The maximum extent of the regression was during deposition of the upper part of the Rome Formation, which, in eastern Tennessee, is characterized by abundant beds of sandstone. In general, the Rome contains progressively less sandstone eastward and northward toward the Blue Ridge of northeastern Tennessee and Virginia, where it is more calcareous and called the Watauga phase by Rodgers (1953).

The Rome is underlain by thrust faults everywhere in the western strike belts of eastern Tennessee; therefore basal beds of the Rome Formation are absent. Farther to the east, along the toe of the Blue Ridge, all the Rome or equivalent Waynesboro Formation is brought to the surface above Shady or Tomstown, although these formations commonly are complexly folded and faulted, and unbroken sections are lacking.

In the southern Appalachian Valley and Ridge, the Rome ranges up to about 305 m thick. In some places in Tennessee, the formation may be broadly divided vertically into a lower shale unit, a relatively thin dolomite unit, interbedded sandstones, siltstones, and shales, and an upper sandstone unit. In other places, the upper Rome consists of interbedded sandstone, siltstone, and dolomite beds, with the top of the formation placed above the highest conspicuous sandstone (Mixon and Harris, 1971; Milici, 1973; Harris and Milici, 1977).

In eastern Tennessee, the formation represents a montage of littoral depositional environments, ranging from flasered tidal-flat beds to offshore subtidal sandbars (Harvey and Maher, 1948; Milici, 1973). Although not abundant, fossils such as trilobites (*Olenellus*) and small phosphatic-shelled braciopods are common

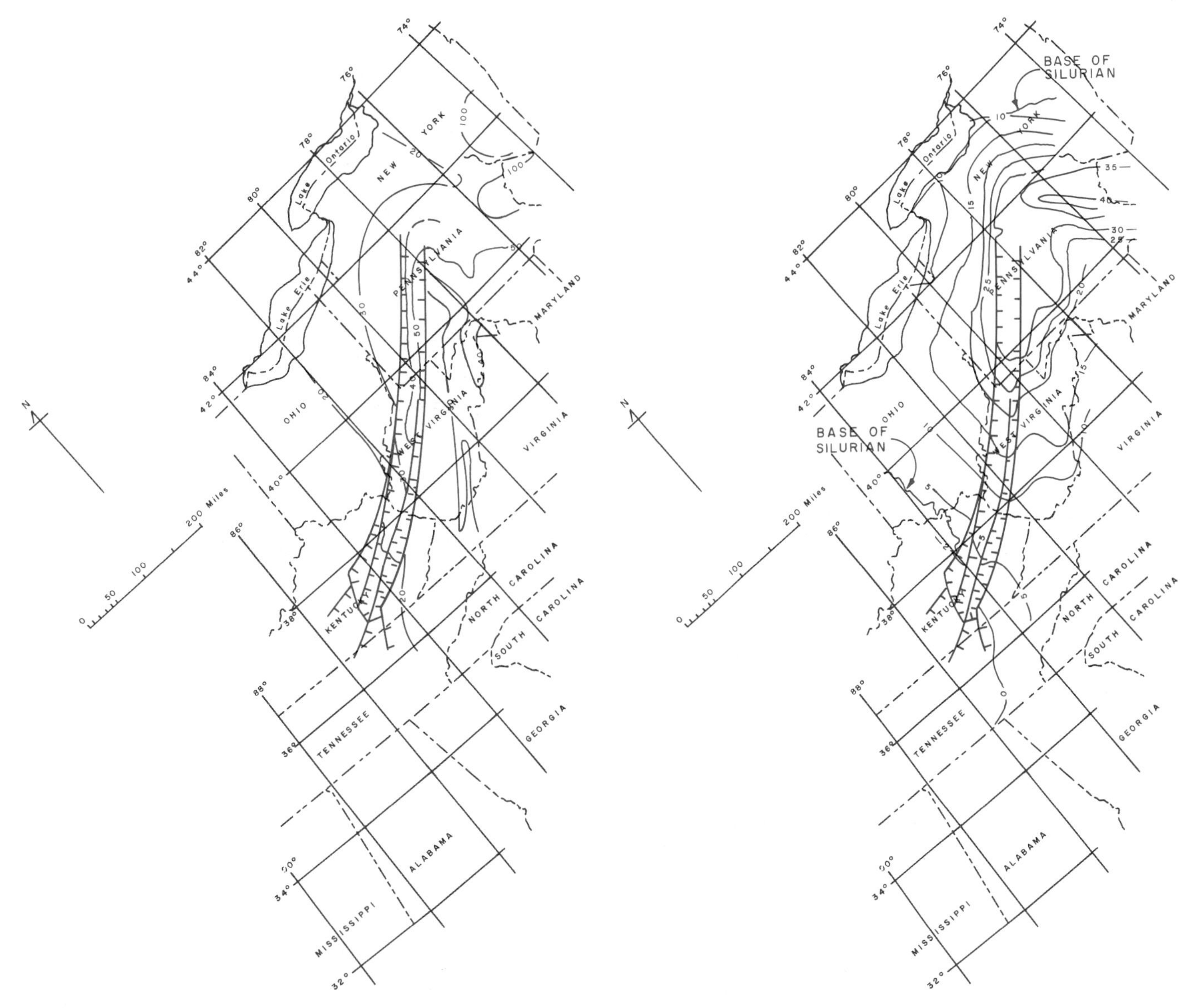

Figure 4. Thickness of Middle and Late Ordovician rocks in the Appalachian Basin. Isopachs are in feet; ticks are on downthrown side of fault. (From Harris, 1978.)

Figure 5. Thickness of Silurian rocks in the Appalachian Basin. Isopachs are in feet; ticks are on downthrown side of fault. (From Harris, 1978.)

in subtidal beds. In Virginia, the Rome and its correlative, the Waynesboro, have been classified as Lower and Middle Cambrian. Rodgers and Kent (1948), however, removed the Middle Cambrian shale beds in eastern Tennessee and placed them with the Conasauga Group. These shale beds appear to grade laterally into the base of Honaker or Elbrook Dolomites in northeastern Tennessee and adjacent southwestern Virginia, so that the Rome in that region may also be entirely Lower Cambrian (Rodgers, 1953, p. 52).

Waynesboro Formation. The central Appalachian equivalent of the Rome, the Waynesboro Formation, extends from west-central Virginia northeastward into Pennsylvania. The Waynesboro consists of about 610 m of red and green shale, dolomite, and lesser amounts of limestone. The Waynesboro thins into Maryland, where it is about 185 m thick and may be divided into two units—an upper shale and sandstone unit and a lower dolomite and shale unit that contains only a few beds of sandstone (Cloos, 1951). In general, the Waynesboro was deposited in peritidal environments similar to those of the Rome but is dominated by carbonate, rather than by siliciclastic, sediments.

Conasauga Group. Regional submergence of the Appalachian Basin during the medial and Late Cambrian resulted in a general confinement of Conasauga siliciclastic deposition to the southern and western parts of the Appalachian Basin. In general, the Conasauga exhibits three phases in eastern Tennessee and adjacent Virginia—a southwestern phase consisting largely of shale, a central phase consisting of interbedded shale and limestone formations, and a northeastern phase consisting largely of dolomite (Butts, 1940; Rodgers, 1953). The group is about 610 m thick.

Considering the distribution of siliciclastics and carbonates in the area of the Appalachian Basin, Conasauga muds must have been derived from a southwestern source. The facies change of Conasauga shale and limestone formations into Honaker and Elbrook dolomites takes place in the northeasternmost counties of Tennessee and in the southernmost counties in the Virginia Valley and Ridge (Rodgers, 1953).

Vertical alternation of carbonate and shale facies within the Conasauga Group represents relatively minor regressive (shale) and transgressive (carbonate) episodes in the southern part of the Appalachian Basin. The vertical transition of the Nolichucky Shale to the Maynardville Formation records the end of significant Cambrian shale deposition in the southern Appalachians.

In eastern Tennessee near Knoxville, Nolichucky shales and limestones apparently were deposited in lagoonal environments (Milici and others, 1973). There, some of the limestones are algally laminated, some are stromatolitic, and others contain oolites and intraclasts. The deposits containing oolites and intraclasts are interpreted to represent washover beds that were derived from adjacent areas near oolite-rich shoals. Nolichucky shales in this intrashelf basin grade laterally into bordering limestones and dolomites, both to the north and east in eastern Tennessee and southeastern Virginia (Markello and Read, 1981, 1982). Carbonate strata were deposited on the east near the shelf edge in tidal flat, shoal, and ramp environments. Nolichucky muds accumulated on the west and south in an adjacent shallow-water basin that contains storm-generated carbonate conglomerates (washovers). Algal carbonates are abundant and are represented by beds of stromatolites, thrombolites, and cryptalgalaminites (Oder and Bumbarner, 1961). The Nolichucky intrashelf basin ultimately was filled and capped by Copper Ridge and Conococheague carbonate tidal flat deposits (Markello and Read, 1981).

Middle and Upper Cambrian carbonate formations, central Appalachians

Elbrook Dolomite. In the central Appalachians, the Elbrook Dolomite extends from the Valley and Ridge of Virginia and Maryland into Pennsylvania. In northwestern Virginia, these carbonate strata consist of 335 to 370 m of platy or laminated dolomite overlain by 490 to 550 m of fine-grained dolomite interbedded with algal limestone, flat pebble conglomerate, dolomitic siltstone, and a few thin sandstones. These beds are arranged in 2- to 3-m cycles that are interpreted to represent deposition in subtidal to supratidal environments (Gathright and others, 1978a, b, c, d).

The Elbrook ranges from 610 to 915 m thick in the Cumberland Valley region of Pennsylvania (Stose, 1909, p. 5; Fauth, 1968; Root, 1968) where it consists of thin-bedded limestone and calcareous shale. In the Lancaster Valley of Pennsylvania, Meisler and Becher (1968) divide the Elbrook Formation into the Zooks Corner and Buffalo Springs Formations (Table 2). The sequence consists of about 460 m of sandy and silty dolomite at the base, overlain by an interbedded limestone and dolomite sequence. To the northeast in the Lehigh Valley division of the Great Valley, basal Cambrian siliciclastics are overlain by the Leithsville Formation, 245 to 275 m of magnesian, shaly and dolomitic limestones, and dolomite, which has some beds of sericitic shale (Howell and others, 1950; Drake, 1965).

The basal Knox Group (Upper Cambrian part) and equivalents

The basal formation of the Knox Group in eastern Tennessee and southwestern Virginia, the Copper Ridge Dolomite, consists of 245 to 335 m of dark, crystalline dolomite with beds of lighter gray, fine-grained dolomite (Rodgers, 1953). The dolomite contains hydrocarbons, and freshly broken rock yields a fetid odor. Stromatolites are common in some beds, giving the dolomite a knotty appearance. Upon weathering, the formation yields a soil rich in chert that has commonly replaced oolites and cryptozoon masses.

In northeastern Tennessee and throughout much of the valley of Virginia, the Copper Ridge Dolomite is replaced laterally by the Conococheague Limestone in eastern strike belts. The limestone is 670 m thick in northeastern Tennessee (King and Ferguson, 1960). Read (1983) describes cyclic deposits within the Conococheague in southwestern Virginia that record repetitions of tidal flat submergence followed by deposition in progressively shallowing conditions. The cycles generally consist of ooid and skeletal sands at the base, which reflect an initial submergence and transgression. The sands are overlain by shallow subtidal ribbon carbonates and thrombolites, which in turn are overlain by tidal flat laminates and cryptalgalaminites. In places, the latter group contains silicified evaporites.

Similar lithologies within the Massanutten Synclinorium of Virginia are described by Gathright and others (1978a, b, c). Within the synclinorium, the lower part of the Conococheague consists of siliceous dolomite interbedded with thin quartz sandstones and some shale. The upper part is cyclic and is interpreted also to represent successive episodes of submergence and deposition resulting in progressively shallowing subtidal to supratidal environments. In some localities, the upper member contains quartz sands that are thought to have been blown southeastward by Cambrian winds from remote parts of the uncovered crystalline craton onto the marginal carbonate platform.

Demicco (1982) recognizes four major lithofacies within the Conococheague in western Maryland: cyclic, thrombolite-grainstone, ribbon rock, and quartz arenite. Cyclic facies represent the lateral progradation of tidal flats over shallow near-shore lagoonal deposits. Thrombolite-grainstone facies are deposits of shelf lagoon environments and record the growth of algal patch reefs among actively migrating shoals. Ribbon-rock facies represent shallow-subtidal to low-intertidal deposits, whereas quartz arenites were deposited in a mud-dominated environment of the shelf lagoon.

The Conococheague ranges from about 580 to 915 m thick in eastern Pennsylvania, including about 90 m of basal orthoquartzites and arenaceous limestones (Wilson, 1952; Cloos, 1951, p. 55–56; Root, 1968; MacLachlan and others, 1975; Berg and others, 1983; Meisler and Becher, 1968; Drake, 1965; MacLachlan, 1979). Much of the Conococheague is laminated limestone, in which flat-pebble conglomerates, ripple marks, and mud cracks are common. In many places, cycles are common and are composed of intraformational conglomerate; mechanical, stromatolitic and oolitic limestones; limestone interbanded or interlaminated with dolomite; and thin dolomite beds. In other places, the formation consists chiefly of micritic, algal limestone and some interbedded skeletal limestone and dolomite.

Northwestward in the Valley and Ridge of Pennsylvania, frosted and rounded sand grains increase in Conococheague equivalents; locally, sand constitutes more than 50 percent of some beds (Wilson, 1952, p. 310). There, the Gatesburg Formation is about 580 to 610 m thick (Wilson, 1952; Berg and others, 1983) and consists of massive dolomite, interbedded orthoquartzite and dolomite, interbedded limestone and dolomite, and chertified oolites and stromatolites.

In New York, Upper Cambrian dolomites exhibit a time-transgressive but partly equivalent interfingering facies relations with underlying sandstones. The dolomites range from about 610 m thick in southeastern New York to a feather edge along the southern shore of Lake Ontario, where they are entirely removed by erosion and where Middle Ordovician rocks directly overlie Precambrian basement (Rickard, 1973; Zenger, 1981).

In the Mohawk Valley of New York, Upper Cambrian strata are predominantly dolomite, but sandstone and sandstone-dolomite mixtures are common (Zenger, 1981). These strata are well known for their abundant algal stromatolites and for their doubly terminated quartz crystals ("Herkimer County diamonds"). The basal part of the unit overlies Precambrian crystalline rock and contains thin basal conglomerates and sandstone. The contact with the overlying Ordovician formation is conformable and, in places, gradational.

Cambrian subsurface sequences, central western Appalachian Basin

The transgressive relationship of Cambrian strata between the Valley and Ridge of Tennessee and Virginia into the subsurface of Kentucky and Ohio was demonstrated by Calvert (1962) and Harris (1964). In general, stratigraphic units of the southern Appalachian Valley and Ridge become progressively younger to the west, although they are lithologically persistent over a wide area. Janssens (1973) divided Cambrian beds in Ohio into a basal Mount Simon Sandstone, units he identified as the "Rome" and "Conasauga", and the Kerbel Formation, and the lower part of the Knox Dolomite. Westward in Ohio, beds below the Knox grade laterally into glauconitic siltstone and sandstones. Because many Appalachian stratigraphers do not agree with the correlations of the Rome and Conasauga from the Appalachians into Ohio, they are herein placed between quotation marks.

In Ohio, the "Rome" Formation ranges from 58 to 220 m thick and consists primarily of gradations between dolomite and sandstone. Some of the sandy dolomite is oolitic and pelletal, especially toward the base of the formation (Janssens, 1973).

The "Conasauga" Formation, 12 to 134 m thick in Ohio, consists of three lithofacies—a southern lithofacies of interbedded red and green shale, siltstone, sandstone, and limestone; an eastern lithofacies of sandy dolomite interbedded with sandstone; and a northern lithofacies consisting of glauconitic siltstone and sandstone. Basal "Conasauga" beds grade northward in Ohio into the "Rome", whereas the uppermost Conasauga grades into the Kerbel (Janssens, 1973, p. 15).

The Kerbel Formation of Janssens (1973) consists of fine- to coarse-grained sandstones, up to 52 m thick, that overlie basal sandstones or the Conasauga and in turn are overlain by the Knox Dolomite in the Ohio subsurface. Janssens (1973) concludes that the Kerbel is of deltaic origin, having a northern source of terrigenous clastic sediment.

The Knox Dolomite, which consists of a mixture of dolomite, sandstone, and some limestone, ranges from about 425 m thick in southern Ohio to a featheredge along the shores of Lake Erie, where it is truncated by the post-Knox unconformity.

The Warrior and Gatesburg Formations of the western Pennsylvania subsurface are approximately equivalent to the "Conasauga" and Knox of Ohio (Wagner, 1966, Fig. 2). The Warrior is a 122-m-thick sequence of argillaceous dolomite that lies between the basal Cambrian sandstones and overlying sandy dolomites. The Warrior thins northwestward to where it grades laterally into sandstone in Canada but thickens to the southwest into the Appalachian Basin. In the subsurface of northwestern Pennsylvania, the Gatesburg is between 200 and 230 m thick (Wagner, 1966, Plate 3). The formation consists dominantly of a very fine to finely crystalline dolomite, some oolitic, and includes sandy zones containing fine- to coarse-grained, rounded and frosted quartz sand grains. The Gatesburg is truncated by a regional unconformity to the northwest into Ontario.

THE ORDOVICIAN

The Knox Group (Lower Ordovician part) and equivalents

Knox Group, upper part. The Knox Group is a sequence of carbonate shelf deposits some 915 m thick that was deposited during the Late Cambrian and Early Ordovician in the southern Appalachians. Subdivision of the Knox Dolomite in eastern Tennessee was encouraged by the search for Mississippi Valley–type zinc deposits during the 1930s and 1940s, and the results of studies by the USGS and other workers were summarized by Rodgers (1943, 1953) and by Bridge (1956).

Because of the great depth of weathering of Knox carbonate rock in the southern Appalachians, much of the geologic mapping of the group depends on recognition of insoluble key or marker beds or silicified fossils in the residuum, that is, distinctive sandstone and chert lithologies or *Lecanospira*-bearing chert. In contrast, a detailed bed-by-bed stratigraphy for the upper part of the Knox Group has resulted from many years of exploratory core drilling in the Mascot–Jefferson City zinc district in eastern Tennessee (for example, see Harris, 1969, Fig. 4).

Bridge (1956) divided the Cambrian and Ordovician Knox Group into five formations in central eastern Tennessee. These dominantly dolomite facies grade eastward into limestone (Rodgers, 1953). Subdivision of the four Ordovician formations in the upper part of the Knox Group in eastern Tennessee proved a difficult task for field geologists who in many instances combined units where they could not recognize key beds or chertified guide fossils in residuum. Because of this difficulty and inconsistencies in mapping, Harris (1969) revised and simplified the stratigraphy of the upper part of the Knox Group by dividing it into the Chepultepec Dolomite, the Kingsport Formation, and the Mascot Dolomite (Table 3). He pointed out that the sandstone zone at the Cambrian-Ordovician boundary is readily mappable, whereas formations of the Knox previously defined by fossils are obscure.

The Chepultepec Dolomite, lowermost Ordovician formation of the Knox Group, is generally from 152 to 245 m thick in eastern Tennessee. In western strike belts, the Chepultepec Dolomite is chiefly light-colored, finely crystalline dolomite, having a conspicuous zone of arenaceous dolomite 2.5 to 3 m thick at the base. It becomes more calcareous eastward, where it consists of slightly dolomitic limestone interbedded with dolomite (Bridge, 1956, p. 38). The overlying Kingsport consists of 60 to 105 m of medium to coarsely crystalline dolomite and limestone. The Mascot Dolomite is 75 to 205 m thick, depending on the depth of erosion beneath the pre–Middle Ordovician unconformity, and is dominantly light gray, finely crystalline dolomite.

In northeastern Tennessee, the eastern limestone facies of the Ordovician part of the Knox Group is called the Jonesboro Limestone. It consists of about 300 to 365 m of limestone and dolomite, and includes some zones of sandy and shaly limestone (King and Ferguson, 1960).

Butts (1940) mapped the basal Ordovician Chepultepec Dolomite in southwestern Virginia; it consists of 150 to 195 m of relatively pure limestone interbedded with magnesian limestone and dolomite. In this area, the overlying Beekmantown consists of 150 to 365 m of dolomite and limestone; it becomes much thicker, however, to the northeast. *Lecanospira* occurs in a lower fossil zone, whereas *Ceratopea* is common in an upper fossil zone within the Beekmantown (Butts, 1940, p. 116–117).

Beekmantown Group. Edmundson and Nunan (1973) and Rader and Biggs (1975) applied Sando's (1957) nomenclature for Lower Ordovician strata in Maryland for use in northern Virginia, elevating the Beekmantown to group status and lowering its base to include Chepultepec equivalents. In northern Virginia, the basal Beekmantown formation consists of 185 to 245 m of crinkly laminated limestone, bioclastic limestone, banded limestone, well-bedded limestone, and siliceous laminated algal limestone. In general, the middle formation comprises about 730 m of limestone and dolomite. Mottled and laminated dolomite and algal structures are common. Some limestones in the lower part of the Beekmantown Group contain *Lecanospira*, whereas others in the upper part are characterized by *Ceratopea* (Edmundson and Nunan, 1973, p. 35–39).

In some places in northern Virginia, the uppermost formation of the Beekmantown Group consists of 90 to 120 m of fine-grained dolomite (Edmundson and Nunan, 1973, p. 39–40). In other places it is absent, apparently removed by erosion; as a result, Middle Ordovician limestones overlie the Beekmantown unconformably. Carbonate conglomerates, in beds from a few centimeters to 6 m thick, were mapped by Rader and Biggs (1975) along the basal contact with the Middle Ordovician limestone.

In western Maryland, the Beekmantown Group is about 1,100 m thick, is similarly subdivided, and consists at the base of about 230 m of thin-bedded limestone, silty limestone, massive algal beds, and stromatolitic bioherms, and includes zones of fossiliferous, oolitic, flat-pebble conglomerate (Sando, 1958). The middle formation of the Beekmantown consists of silty and algal limestone in its lower part and dolomite and dolomitic limestone in its upper part; it is generally about 315 to 780 m thick (Sando, 1957). In western Maryland, the upper part of the group consists primarily of laminated or mottled, cherty dolomite and it ranges from about 113 to 150 m thick.

The Beekmantown Group in Pennsylvania ranges in age from uppermost Lower Cambrian to lowermost Middle Ordovician (Berg and others, 1983). The basal formation of the Beekmantown consists of about 300 m of limestone in the southern part of the Great Valley of Pennsylvania. This limestone formation thins to the north and east to about 75 m in Lehigh Valley by regional interfingering with overlying and underlying dolomite formations (Hobson, 1963, P. 58).

The overlying formations, chiefly composed of interbedded limestone and dolomite, thicken from about 245 m near the Schuylkill River southwestward into 700 m of strata containing algal limestone, stromatolites, fragmental limestones, and pebble conglomerates. The uppermost Beekmantown unit in central eastern Pennsylvania is about 230 m thick and consists chiefly of dolomite but includes some beds of shale and limestone. Its contact with the overlying limestones is sharp and may be disconformable. In some localities in the northeastern part of the Great Valley, Trenton-age rocks unconformably overlie the Beekmantown (Hobson, 1963, p. 23; Berg and others, 1983).

In central Pennsylvania, the Beekmantown Group consists of five formations. Lowermost Ordovician limestones generally range from 150 to 305 m thick and generally grade northwestward into dolomite facies and northward into interbedded limestones and dolomites (Wagner, 1966). The overlying formation is composed almost entirely of dolomite about 365 m thick (Spelman, 1966), which, in some localities contains abundant quartz sand and silt grains. *Lecanospira* and other fossils occur in oolitic dolomite beds in the lower part of this unit. Above the dolomite in central Pennsylvania is a limestone unit that ranges generally from 55 to 122 m thick, although in places it is absent. It grades into dolomite laterally to the southwest of the Nittany Anticlinorium (Wagner, 1966). In central Pennsylvania, the uppermost formation of the Beekmantown consists of about 610 m of dolomite. Lower beds of this unit thin to the north and west, and in western Pennsylvania, western New York, and northeastern Ohio, underlying formations are missing. They apparently were eroded and thinned because of Ordovician uplift along the Cincinnati arch (Wagner, 1966, p. 25).

Lower Ordovician strata in the Mohawk Valley of New York consist of 30 to 38 m of limestone and dolomite at the base; then interbedded shale and sandy limestone; then dolomitic limestone and dolomite; and then silty and sandy, fossiliferous, oolitic calcarenites, flat-pebble conglomerates, and dolomite beds. The overlying unit is about 6 to 12 m thick, consists entirely of dolomite, and in its upper part is very siliceous (Fisher, 1954, 1965).

Pre–Middle Ordovician unconformity. Except in the deeper parts of the Appalachian Basin, the Lower–Middle Ordovician boundary is a marked regional unconformity (Bridge, 1955; Milici, 1973; Harris and Repetski, 1983). This unconformity resulted from regional differential uplift of the Lower Ordovician carbonate platform related to the first collisional events of the Taconic Orogeny, when the early Paleozoic passive continental margin became tectonically active. Local relief on the unconformity is commonly several tens of meters and may be more. On a regional basis, several hundred meters of section appear to have been removed by erosion toward the basin margins in Alabama and Ohio where Middle Ordovician strata directly overlie rocks of Late Cambrian age. Similarly, in New York, pre–Middle Ordovician beds are progressively overstepped to the north, to where Middle Ordovician strata lie directly upon basement crystallines along the southern shore of Lake Ontario. By the Middle Ordovician, before the first great influx of siliciclastics from eastern sources, the general outline of the Appalachian Basin had formed. Thinner deposits around the basin margin to the south, west, and north underwent extensive erosion, and thicker Lower Ordovician deposits on the eastern margin in northern Virginia, Maryland, and Pennsylvania graded continuously into overlying Middle Ordovician beds (Harris and Repetski, 1983). Structurally, the basin was a gently inclined monocline, extending from the crest of the Cincinnati Arch eastward to the Iapetus coastline. Formation of the structural basin, however, was not complete until the passive margin became active and Taconic tectonic lands were thrust up along the eastern margin of the continent.

The Middle and Upper Ordovician shelf deposits

Middle and Upper Ordovician strata of the Appalachians were deposited as tectonic lands were thrust up along the eastern continental margin, shedding siliciclastic flysch northward, westward, and southward over carbonate sediments. At first, the siliciclastics were trapped by foreland basins that developed along the platform margin. Ultimately, however, they filled these basins and spread blanket like across the carbonate shelf (Fig. 4).

The erosion surface. The beds above the pre–Middle Ordovician unconformity in the southern Appalachians are commonly silty dolomites and limestones that contain angular or subrounded clasts of chert and dolomite derived from erosion of subjacent beds. Chert fragments seldom exceed a few centimeters across, but carbonate blocks in places are 30 cm or more in diameter. In some places, pinnacles of Knox protrude upward into overlying Middle Ordovician strata (for example, see Milici, 1969). In other places, extensive terra rossa deposits consisting of red and green shales and argillaceous dolomite compose the basal Middle Ordovician beds. The shales and dolomite commonly overlie a basal conglomerate and appear to fill broad depressions on the pre–Middle Ordovician surface (e.g., Milici and Smith, 1969). Elsewhere, channel-fill conglomerates are composed of chert, quartz pebbles, and sand as much as 60 m thick (Rodgers and Kent, 1948, p. 32) or there are deep sinkholes bearing peculiar fossils (Laurence, 1944; Bridge, 1955; Caster and Brooks, 1956).

Tectonic lands. Conglomerates composed of reworked Cambrian and Ordovician limestones, siltstones, and sandstones are widespread in the Ordovician shale deposits along the toe of the Blue Ridge from southwestern Virginia to northwestern Georgia (Kellberg and Grant, 1956). At least one such deposit, in Virginia, contains clasts of granite gneiss, indicating that the nearby Precambrian crystalline basement had been unroofed and eroded (Rader and Gathright, 1986).

These conglomerates are substantial evidence that segments of the platform margin, including Grenville crystalline basement, had been thrust upward upon the shelf and eroded, thereby providing the sediments that filled rapidly subsiding foreland basins to the west. These tectonic lands, Blountia (Kay, 1951) and Taconica (Kay, 1937), were marginal to the subsiding Appalachian miogeocline and provided terrigenous sediments thereto throughout most of the remainder of the Paleozoic.

The Middle and Upper Ordovician geologic history of the Appalachians records the progressive westward encroachment of siliciclastics upon the carbonate platform. According to Colton (1970), shale depocenters at first formed in two places along the length of the eastern margin of the Appalachian Basin (Blountian phase of the Taconic Orogeny, Rodgers, 1953, p. 94), but later in the Ordovician, terrigenous clastics entered the Appalachian Basin from the northeast and migrated southwestward and westward, forming the great Queenston delta. Middle and Upper Ordovician strata in the Appalachian Basin may be divided into three broad groupings—accumulations on the stable platform, deposits of the platform margin or unstable shelf, and basinal facies composed primarily of dark shales and turbidites.

Stones River and Nashville Groups and equivalents. Ordovician stable shelf deposits, dominantly limestones, extend from the Cincinnati Arch eastward into the western part of the Valley and Ridge. The deposits are exposed along the arch on two domes, the Nashville in Tennessee and the Jessamine in Kentucky and southernmost Ohio. To the east they are brought to the surface by the Sequatchie anticline in Tennessee and Alabama and by westernmost thrust sheets and anticlines of the Valley and Ridge.

The exposed part of the Stones River Group in central Tennessee is 110 m thick (Bassler, 1932; Wilson, 1949). Basal beds of the group, about 122 m thick, are known from subsurface studies in central Tennessee (Bentall and Collins, 1945). In general, the group consists of two major lithofacies, relatively pure, massively bedded, commonly cherty limestone containing massive types of hydrozoans and corals, and thinly bedded, rarely cherty limestone containing partings of silt and clay, fragile ramose and ribbon-like bryozoans, and brachiopods as the dominant fauna (Wilson, 1949, p. 67).

Four bentonite beds (T-1 through T-4), from the upper part of the Stones River Group in the Tennessee central basin, are described by Wilson (1949). The lower two beds are generally thin and difficult to map. The upper two beds are generally each about a meter thick. All the beds except T-4 are widely distributed in central Tennessee. Bed T-4 is present in only a few places in the eastern part of the Nashville Basin but is widely distributed in the Sequatchie Anticline (Milici, 1969), in northwestern Georgia (Milici and Smith, 1969), and in the western part of the Tennessee Valley and Ridge (Rodgers, 1953, p. 93; Wilson, 1979, p. 26). The bentonites are considered by some to be isochronous ash falls. Regional stratigraphic relationships, however, show that almost everywhere T-3 lies at an environmental boundary between subtidal limestones below and more argillaceous and in part desiccated limestones above, which suggests that its position, too, is environmentally controlled. The thicker bentonites in the Stones River may be the result of erosion of ash falls that accumulated on tidal flats and were subsequently redeposited and preserved in nearby subtidal environments.

The Stones River Group is exposed fully in Sequatchie Valley, Tennessee, where it is about 260 m thick. There, the basal unit overlies the post-Knox unconformity and contains lithologic units typical of basal Middle Ordovician deposits. In northwestern Georgia, the Stones River Group is about 290 m thick and generally contains the same formations as in Sequatchie Valley (Milici and Smith, 1969).

Although the Stones River Group is about 510 m thick along the eastern part of the Valley and Ridge in Tennessee near Chattanooga, it is divisible there into its typical suite of formations (Wilson, 1979). The lower part of the group appears to thin eastward across the Valley and Ridge in the Chattanooga area, grading latrally into more argillaceous and silty limestones, siltstones, and shales (Wilson, 1979).

The Stones River Group apparently is a little older in the Valley and Ridge than in the Nashville Basin of Tennessee (Ross and others, 1982); this suggests that depositional environments and associated lithofacies transgressed westward and upward onto the Cincinnati Arch. Apparently, the Cincinnati Arch was intermittently active during deposition of the Stones River; consequently unconformities are found within and at the top of the Group (Wilson, 1949, 1962). Intermittent uplift on the arch continued at least to the Devonian (Wilson, 1962), when black mud blanketed the truncated edges of numerous upwarped formations in central Tennessee (Wilson, 1949, Plate 2).

The Nashville Group consists of three argillaceous limestone formations in Tennessee and northwestern Georgia, and each of these may be further divided into several lithologic members (Wilson, 1949). In general, the group ranges between 76 and 135 m thick in Tennessee; most of this variation is caused by a pronounced thickening of the upper part eastward toward the platform margin. The group is generally more argillaceous than the Stones River and, in many places, contains numerous invertebrate fossils. Phosphate is an important constituent of some of those limestones, and central Tennessee is a source of relatively small amounts of high-grade phosphate.

Rodgers (1953) mapped four informal units within the Chickamauga Limestone in the three western Valley and Ridge strike belts of eastern Tennessee. Units 1, 2, and 3 apparently correlate with the Stones River. Bentonites T-3 and T-4 lie in the upper beds of unit 3 and extend eastward into grayish red calcareous mudstones of the Moccasin Formation (Rodgers, 1953, Fig. 4). Chickamauga unit 4, overlying the bentonite-bearing strata but below the Reedsville Shale (Upper Ordovician), correlates with the Nashville Group.

The two thick bentonites lie at the base and within the upper members of the Eggleston Formation (Upper Ordovician) in the Powell Valley anticline in southwestern Virginia (Rosenkrans, 1941; Miller and Brosgé, 1954) (Table 3). The Eggleston and beds below, down to the top of the Knox Group, generally correlate with the Stones River, whereas the overlying Trenton is in the same stratigraphic position as the Nashville Group in central Tennessee. In general, the Ordovician formations below the Eggleston are relatively pure limestones of variable lithologies, some cherty and some fossiliferous, that are interbedded with yellowish-gray and greenish-gray–weathering calcareous mudstones. These mudstones grade southeastward into Moccasin-type calcareous red beds, which contain abundant evidence of subareal exposure.

The Trenton Limestone in the oil district of southwestern Virginia consists of about 168 to 185 m of fossiliferous limestones, some coquinal; numerous partings or thin beds of gray shale are also present. This unit is the principal producer of oil in southwestern Virginia, apparently from highly fractured reservoirs. The formation overlies the Eggleston with a sharp but conformable contact and is overlain with apparent conformity by the Reedsville Shale (Miller and Fuller, 1954; Miller and Brosgé, 1954).

The exposed Ordovician carbonate strata of the Jessamine Dome in Kentucky consist of about 122 m of limestone and dolomite of the High Bridge Group, and the more argillaceous units of the Lexington Limestone (Cressman, 1973, p. 9; Cressman and Noger, 1976; Ross and others, 1982). Basal Middle Ordovician beds are not exposed; about 90 m of the oldest unit are exposed and contain fucoidal calcilutites (micrite pelmicrite, biopelmicrite) similar to those within the Stones River Group in Tennessee. Terra rossa deposits at the base of the unit overlie the Knox Dolomite unconformably in the subsurface (Wells Creek Dolomite of drillers). Next above are about 15 to 60 m of finely crystalline calcareous dolomite and dolomitic calcilutite formations. These formations contain a variety of lithologic features that indicate they were deposited on low tidal flats or in adjacent shallow marine lagoons (Cressman, 1973; Cressman and Noger, 1976).

Like the Stones River of Tennessee, the High Bridge Group contains two extensive bentonites. The upper, the "mud cave" of drillers, is preserved locally at the top of the group. The lower, the "pencil cave," occurs 4 to 6 m below the top of the High Bridge and is an excellent marker bed for regional correlation.

The overlying Lexington Limestone ranges from 55 to about 105 m thick. It consists mostly of limestone, some interbedded with shale. The formation correlates generally with the Nashville Group of common usage in the group's type area in Tennessee. Ross and others (1982) include the Inman and Leipers Formations, in our opinion erroneously, within the Nashville Group. According to them, the lower beds of the Lexington correlate with the uppermost Stones River strata.

Basal beds of the Lexington unconformably overlie those of the High Bridge, in some places directly upon its uppermost bentonite and in other places upon limestone. Cressman (1973) estimates that a maximum of 3 m of upper High Bridge was eroded before deposition of basal Lexington beds, which, in some places, contain scattered fragments of reworked limestone. In general, the Lexington is a subtidal deposit interpreted by Cressman (1973) to have been deposited in normal marine environments. Some beds, however, appear to have been deposited in lagoonal-intertidal to shallow-subtidal environments. The upper part of the Lexington Limestone grades both upward and laterally northward into marine formations composed of interbedded limestone and shale. In contrast with the Nashville dome, little or no movement occurred on the Jessamine dome during the Middle and Upper Ordovician (Cressman, 1973, p. 55).

High Bridge and Lexington equivalents persist into the subsurface in Ohio where they are called Black River and Trenton (Stith, 1979). Stith (1979) correlates the Black River with the High Bridge Group in Kentucky, and from the cores that he studied, recognizes supratidal, intertidal, and subtidal depositional environments. At the turn of the century, the Trenton (or Lexington) of northwestern Ohio was a major producer of oil and natural gas generally west of or upon the axis of the Cincinnati arch (Landes, 1970); it is still producing, although in relatively small amounts.

Middle and Upper Ordovician carbonate strata range from about 460 m thick in the Nittany anticlinorium of central Pennsylvania to about 200 m in the northwestern part of the state. A lower carbonate unit is generally limestone, with beds of dolomite increasing toward the base. The middle unit has considerably less dolomite, whereas the overlying unit consists largely of limestones. The uppermost unit is a black shale (Antes, Utica) of variable thickness, generally thinning westward (Wagner, 1966).

Rickard (1973) and Fisher (1977, 1982) show similar facies relationships for middle Ordovician strata across New York, from relatively thin Black River and Trenton carbonate sequences overlain by thick Utica Shale on the east, to relatively thick carbonate sequences overlain by thin shale on the west. Isopach maps of the Black River and Trenton sequences show that a shallow synclinal trough filled with abnormally thick Black River and Trenton carbonate strata extends from north of the Adirondacks south and west, across eastern Lake Ontario and into central Pennsylvania (Rickard, 1973, pl. 7, 8). Black River carbonates generally overlie Precambrian crystalline basement in the northern and western parts of the Appalachian Basin. To the south and east, they unconformably overlie progressively younger Cambrian and Lower Ordovician carbonates in the basin trough, where deposition was continuous (Rickard, 1973).

The Black River Group contains three formations in the Black River Valley (Fisher, 1965). The lowermost is about 45 m thick, and its basal beds consist of dolomitic arkose grading vertically into dolostone and dolomitic shale where the unit overlies gneissic basement. It grades further upward and laterally into white calcilutites of the middle unit, which in turn grade into black, cherty calcilutites of the upper formation (Fisher, 1977, Fig. 3). The middle formation is generally about 12 m thick in the Black River Valley (Kay, 1937), but is thinner and patchy in the Mohawk Valley (Fisher, 1965, 1977). Overlying limestones in the Black River and Mohawk Valleys range from about 3 to 12 m thick. In general, the Black River Group consists of carbonate-rock facies that represent several major environments, including supratidal and intertidal mudflats, shallow-subtidal or low-intertidal muds, bioclastic muds and sands, *Tetradium*-formed subtidal wave baffles, and subtidal bioclastic muds and sands (Textoris, 1968; Walker and LaPorte, 1970).

The Trenton Group, a term that has been used by some (e.g., Kay, 1937) to include equivalent thick shales and sandstone formations of eastern New York, consists of as much as 245 m of generally subtidal limestones in New York (Rickard, 1973). Current usage (Rickard 1973; Fisher, 1977, Plate 4) restricts the term to the limestone facies, which is consistent with the usage of "Trenton" throughout the Appalachian Basin.

In the Mohawk Valley, the Trenton is generally less than 26 m thick but can reach a maximum thickness of 122 m (Fisher, 1977, Plate 4). Overlying Utica shales thicken reciprocally, ranging from 488 m in the eastern Mohawk Valley—where they grade eastward into Snake Hill shales and siltstones—to 229 m thick in the northwestern part of the Valley. In the Black River Valley, Trenton limestones about 150 m thick are overlain unconformably by 61 m of Utica Shale. The transition from limestone deposits to shale is laterally and vertically gradational (Kay, 1937; Fisher, 1977). Shelf-edge limestones contain a diminutive bottom fauna, whereas interbedded black shales contain graptolites.

Platform margin and basinal facies

In general, Middle and Upper Ordovician strata along the eastern side of the Appalachian Basin record the change from a divergent to a convergent continental margin tectonic environment. The Middle and Upper Ordovician carbonate formations of the stable carbonate platform grade eastward and have a general increase of siliciclastic content, first into the more shaly and silty formations of the platform margin and then into more basinal shales and turbidites (Rodgers, 1953; Neuman, 1955). In general, basinal deposits accumulated contemporaneously with these shelf carbonates in two depocenters, one in northwestern Georgia and adjacent parts of Alabama and Tennessee, and the other centered in eastern Pennsylvania, Maryland, and northern Virginia (Colton, 1970). These depocenters appear to have been inherited from those of the Cambrian and Ordovician carbonate platform (Read, 1980). As these basins enlarged during the Ordovician, the platform margin was buried by eastward-derived siliciclastic sediments (e.g., see Rodgers, 1953, Fig. 4). The relationship of carbonate units to more basinal facies in Tennessee is described by Walker and others (1983) and Ruppel and Walker (1984); the following discussion is largely from their papers.

The initial Ordovician submergence of the exposed Lower Ordovician carbonate platform (Knox) was followed, in some places, first by reworking of terra rossa deposits and then by deposition of peritidal to subtidal inner platform carbonate strata (the Lenoir Formation) in the central and eastern parts of the Valley and Ridge of Tennessee. Ensuing rapid subsidence on the east inundated the Lenoir, forming a basin as much as 700 to 1,000 m deep in which hemipelagic sediments accumulated (Shanmugam and Lash, 1982). Several million years later the platform margin was displaced about 25 km westward, submerging more of the eroded Knox terrane and resulting in the development of an outer (eastern) marginal reef complex (Walker and Ferrigno, 1973).

Contemporaneously, distal turbidites derived from the uplifted Taconic orogen spread westward and southward (Jones and Dennison, 1970) and began to accumulate in the eastern part of the basin. Carbonate strata in the Valley and Ridge west of the reef tract record a progressive trend from supratidal to subtidal environments. Next above are widespread deposits of higher-energy shallow-water mud mound and sand bank environments; these, in turn, are overlain by siliciclastic strata. Ultimately, the eastern basin filled and was capped by terrigenous tidal-flat muds and sands, which grade westward across the platform margin into mixed supratidal to shallow subtidal carbonate and siliciclastic sediments.

Foundering of the shallow-water deposits was followed by progressive onlapping of the Martinsburg Shale (Upper Ordovician) on the carbonate platform. In Tennessee, the Martinsburg consists of approximately 215 to 300 m of calcareous shale interbedded with very fossiliferous limestone and siltstone (Rodgers, 1953).

A similar transition from shallow carbonate shelf to deep basinal facies took place in southwestern Virginia (Read 1980, 1982). Accumulations of transgressive limestone are thin because of the rapid downwarping of the shelf margin on the eastern side of the miogeocline. These are overlain by thick shales, the distal equivalents of submarine fans derived from tectonic uplands to the southeast. The deepening basin encroached on emergent Lower Ordovician carbonate strata from the southwest, and an onlap sequence consisting of basin and slope black shales and limestones, downslope carbonate buildups, shelf-edge limestones, shallow carbonate buildups, shallow subtidal platform deposits, and peritidal and shoal deposits was formed (Read, 1980, 1982). The buildups, ranging up to 60 km wide and 250 m thick, are carbonate banks that formed by rapid accumulation of skeletal material, baffling action of pelmatozoans and bryozoans, and widespread marine cementation, which resulted in the formation of resistant structures (Read, 1980). With continued submergence, the basin extended toward northern Virginia, and fossiliferous shaly limestones that were deposited along the platform margin grade eastward into basinal shales. Next, an offlap sequence developed as shallow platform sequences prograded southeastward over the deeper water and basinal facies. Carbonate deposition was restricted as the basin filled with siliciclastics. The siliciclastics grade northeastward, first into marine shales and limestones and then into marine limestones.

As the southern basin filled, shale deposition was restricted to northern Virginia, subsequently the site for accumulation of thick Martinsburg flysch deposits. In southwestern Virginia, the Martinsburg Formation overlies peritidal beds and accumulated on an open marine platform. Kreisa (1981), describing the stratigraphy and sedimentary structures of the Martinsburg, shows that the formation, which is generally progradational, has easterly derived siliciclastics migrating progressively westward over more calcareous sediments. To the east, the formation is about 400 m thick, thinning westward to about 280 m. Kreisa (1981) concludes that, within the Martinsburg, sequences of whole-fossil packstones overlain by laminated siltstone, very fine sandstone, and shale represent storm-generated deposits. In southwestern Virginia and Tennessee, the Martinsburg is overlain by peritidal Upper Ordovician red beds of the Sequatchie or Juniata, except in the few places where they have been removed by erosion. At those places, Silurian sandstones unconformably overlie the Martinsburg (Kreisa and Springer, 1987).

The Ordovician basin-to-shelf transition in northern Virginia lies generally across the Massanutten synclinorium (Rader and Henika, 1978). Post-Beekmantown subsidence resulted in the deposition of New Market peritidal limestones. To the west, the New Market is overlain by calcareous mudstones and cherty calcarenites. In some places to the east, where the basin deepened abruptly, carbonate strata are overlain by black pelagic slates and calcareous turbidites. Contemporaneous with basinal deposition, slope, shelf, and shoal deposits accumulated in the more stable westerly area. These shallower deposits were overrun as Martinsburg black muds and siliciclastic turbidites spread westward, ultimately filling the subsiding basin. The Martinsburg is overlain unconformably by Silurian sandstones in the Massanutten Synclinorium.

Middle and Upper Ordovician stratigraphic sequences in the western part of the Valley and Ridge of Virginia and adjacent West Virginia are similar to those in southwestern Virginia. The reader is referred to Kay (1956), Perry (1972), Weed (1982), and Rader (1982) for specific information.

In Maryland and in the adjacent Cumberland Valley of Pennsylvania, Middle and Upper Ordovician carbonate strata consist of the upper part of the St. Paul Group and the Chambersburg Limestone (Table 3). Neuman (1951) divides the St. Paul Group into two formations. The lower unit is about 50 m thick in its type area and consists largely of calcilutite and fossiliferous granular limestone. Its boundary with the underlying Beekmantown is indefinite and occurs within a zone up to 15 m thick of interbedded limestone and dolomite. The contact with the overlying limestones is conformable and more easily distinguished, especially where granular limestones are overlain by calcilutites. The upper unit of the St. Paul is primarily a gray calcilutite, ranging from about 27 m in northern Virginia to a little more than 215 m in southern Pennsylvania. The lower unit thickens in the same direction, from 34 m in southern Maryland to 210 m in Pennsylvania, so that the maximum measured thickness for the group is about 425 m (Neuman, 1951, p. 319). The St. Paul Group apparently formed in depositional settings similar to those of modern carbonate tidal flats, marshes, lakes, and subtidal environments under rainy, subtropical to tropical conditions (Mitchell, 1982). The basal calcilutite is overlain by the Chambersburg Limestone (Cloos, 1951; Neuman, 1951; Root, 1968). The Chambersburg is an argillaceous, fossiliferous, cobbly-weathering limestone about 225 m thick that accumulated on a shallow shelf. The siliciclastic content of the Chambersburg increases near the top of the formation, where it grades into the overlying Martinsburg Formation (Mitchell, 1982).

In the Lebanon Valley, Pennsylvania, the Beekmantown is overlain by thick-bedded limestone that contains zones of high-purity calcium carbonate. The formation ranges up to 76 m thick. Its lower contact with the Beekmantown is conformable, whereas its upper contact is disconformable (Prouty, 1959; MacLachlan, 1967). The overlying formation consists of 60 to 75 m of well-bedded dense limestone. In some places, the limestone contains thin beds of shale, and in its upper half, a 15-m-thick zone contains bentonite beds. This unit grades upward into as much as 300 m of impure shaly limestone.

The Beekmantown is overlain by the Jacksonburg Formation in the Lehigh Valley of Pennsylvania and in adjacent parts of New Jersey (Miller, 1937; Sherwood, 1967). The formation is about 300 m thick in Pennsylvania. Miller (1937, p. 1706) observed an angular unconformity between the Jacksonburg and underlying carbonate strata, marked by a basal conglomerate consisting of angular pebbles and cobbles of the underlying Beekmantown. Sherwood (1964) reports that the basal conglomerate zone is as much as 30 m thick in New Jersey and describes the Jacksonburg-Beekmantown contact as "abrupt but conformable" in Pennsylvania. The contact of the Jacksonburg with the overlying Martinsburg is conformable and gradational (Sherwood, 1964, p. 19). In extreme northern New Jersey, the entire pre-Silurian sequence is absent, and Silurian conglomerates overlie Precambrian rocks unconformably (Miller, 1937, p. 1701). Both Miller (1937) and Sherwood (1964) report bentonite beds from within the Jacksonburg Formation.

The Jacksonburg grades upward into the overlying Martinsburg Formation. In eastern Pennsylvania and western New Jersey, the Martinsburg consists of three members—a lower thin-bedded slate, 1,220 m thick; a middle unit, containing graywackes, about 850 m thick; and an upper thick-bedded slate unit, 915 to 1,830 m thick (Drake and Epstein, 1967). The Martinsburg is unconformably overlain by the Shawangunk Conglomerate (Silurian) in eastern Pennsylvania, the Tuscarora (Silurian) in the central Valley and Ridge, the Juniata Formation (Upper Ordovician) in southern Pennsylvania, and, in many places, by the Bald Eagle or Oswego Sandstones. In the Lancaster Valley, Martinsburg equivalents consist of about 610 m of dark-colored fissile shale, which constains beds of arkosic and vitreous sandstones (Jonas and Stose, 1930; Berg and others, 1983).

Blanket deposits

Reedsville Shale. Upper Ordovician shales generally overlie middle Ordovician shales and limestones from central Pennsylvania to the western Valley and Ridge of Tennessee. Compared with the Martinsburg, the Reedsville is a relatively thin clay-shale unit, ranging from about 760 m thick in Pennsylvania to 30 m thick in Tennessee. In central and western Pennsylvania, the Reedsville overlies the Ordovician Antes graptolitic black shale, which is at least 15 m thick (Pierce, 1966, p. 19; Berg and others, 1983). The contact with the overlying Bald Eagle Sandstone is gradational.

The Antes-Reedsville sequence represents environments that range from mud deposition in a relatively deep, stagnant basin to progressively higher-energy distal and proximal turbidites deposited on steeper slopes nearer the eastern basin margin (Pierce, 1966). In the central Appalachians, the upper part of the Reedsville and lower part of the overlying Oswego or Bald Eagle appear to have been deposited by several small coalescing deltas (Horowitz, 1966). Thompson (1970a) interprets the upper 15 to 30 m of Reedsville as being deposited in a shallow marine, nearshore environment, primarily because of its abundant, diverse fossil assemblage. The upper part of the formation contains an *Orthorhynchula*-bearing fossiliferous unit, 15 to 60 m thick, which is overlain in different places by the Bald Eagle, Oswego, Massanutten, Juniata, or Sequatchie Formations (Bretsky, 1969). Pierce (1966), however, considers the *Orthorhynchua* zone to be the basal part of the Bald Eagle Sandstone.

The Martinsburg and Reedsville appear to have developed in different tectonic settings. Based on a study of Martinsburg sedimentary structures, McBride (1962) and Stephens and Wright (1981) determined that the formation consists of turbidites, interbedded pelagic shales, and radiolarian chert. Paleocurrent trends in the Martinsburg are generally from southwest to northeast along the length of the trough, indicating that much of the basin fill was derived from erosion of southern highlands. In contrast, graywacke units are coarser and more immature to the northeast, suggesting that these siliciclastics at least were derived from erosion of more northerly areas.

The upper part of the Martinsburg and the Reedsville exhibit an entirely different facies distribution. The Reedsville is not confined to marginal foreland basins, but spreads instead blanket like over much of the Appalachian Basin. Bretsky (1969) illustrates a major siliciclastic source area on the northeast that grades southward and westward through alluvial and flood plain environments into marine limestones and shales. As a result, upper Reedsville shelf sediments exhibit a pattern that is generally similar to the one that subsequently developed during deposition of the Upper Ordovician Queenston delta. The Antes–Reedsville–Bald Eagle sequence of Pennsylvania grades northward into the Utica Shale and overlying formations of the Lorraine Group in New York (Fisher, 1977, Plate 1).

Queenston delta. Uppermost Ordovician strata of the Appalachian region compose the great Queenston delta, which grades westward from continental red beds of the Juniata Formation to lower delta plain, grayish red shales and siltstones of the Queenston Formation (Dennison, 1976). Farther west, the Queenston is replaced by Upper Ordovician carbonate and shale along the Cincinnati arch.

Overlying the Reedsville or Martinsburg in Pennsylvania is a sequence of siliciclastics, called Bald Eagle (gray-green) and Juniata (red), which is about 610 m thick. Thompson (1970a) divides these units into lithofacies that represent marginal marine environments at the base and grade upward into fluvial deposits. In these rocks, the color change from gray-green to red is to a large degree secondary, does not reflect depositional environments, and is not a precise stratigraphic boundary (Thompson, 1970b). To the south and west, the Juniata is replaced first by calcareous red and green mud flats of the Sequatchie Formation and then by marine gray limestones and shales of the Shellmound Formation in southernmost Tennessee and adjacent Alabama (Milici and Wedow, 1977). Milici and Wedow (1977) interpret Upper Ordovician deposits in southern Tennessee as being part of a major regressive-transgressive cycle that began in the Ordovician and ended in the Silurian.

Distal equivalents. To the west, in the Nashville Basin of Tennessee, Upper Ordovician strata consist of up to 75 m of argillaceous, fossiliferous limestone and shale that are divided into four formations (Wilson, 1949). In the Blue Grass region of Kentucky, Upper Ordovician strata are divided into nine limestone, shale, and dolomite formations, which have an aggregate thickness of about 220 m (Weir and others, 1984). Weir and others (1984, p. E92) conclude that these Upper Ordovician lithofacies of central Kentucky accumulated on an extensively developed open-marine shelf "of shallow to moderate depth but locally in intertidal and supratidal environments."

The Taconic orogeny

The Taconic Orogeny began in the Middle Ordovician, when the passive eastern margin of the Appalachian Basin became tectonically active. Rodgers (1971, p. 1170) ascribes four general phases or events to this orogeny: "(1) disconformity on the carbonate bank (perhaps angular unconformity at its eastern or 'internal' margin); (2) severe early deformation in a more easternly or 'internal' volcanic zone; (3) gravity slides from the latter into the former area, following an 'inversion of relief'; and finally (4) widespread deformation (generally with some metamorphism), especially strong in the more westerly or 'external' zones. . . ." The Taconic Orogeny has been attributed to the collision of an island arc system with the North American continent (e.g., see Wehr and Glover, 1985).

THE SILURIAN

Molasse, eastern Appalachian Basin

Silurian siliciclastic molasse and carbonate strata reflect the progressive erosion and lowering of Taconic tectonic highlands, ultimately leading to the development of a restricted evaporite basin of regional extent (Fig. 5).

Shawangunk and Bloomsburg. A thick Lower Silurian clastic sequence in the Appalachian Basin was derived from erosion of the tectonic highlands lifted up during the final phase of the Taconic Orogeny (Table 4). In general, this sequence ranges from about 10 m in southern Tennessee and along the western margin of the Appalachian Basin to more than 730 m in eastern Pennsylvania (Colton, 1970, Fig. 18). The lower part of the sequence in eastern Pennsylvania and northwestern New Jersey is the Shawangunk Formation (Epstein and Epstein, 1972). The formation ranges from about 425 to 640 m thick. It overlies the Martinsburg Formation with angular unconformity, so that progressively older Martinsburg beds are at the contact to the northeast along the outcrop in Pennsylvania and New Jersey (Epstein and Epstein, 1972, Figs. 1 and 2). The Shawangunk, in turn, is overlain gradationally by red beds of the Bloomsburg Formation (Table 4).

Epstein and Epstein (1972) correlate the lower quartzose sandstone and conglomerate members of the Shawangunk with the Tuscarora Sandstone, and the remainder of the formation with the Clinton Group. They conclude that the Shawangunk was deposited in a variety of fluvial environments and environments transitional between marine and continental.

The Bloomsburg Formation is a red-bed sequence that extends across much of the central Appalachians, from eastern Pennsylvania and New Jersey southeastward through Maryland into adjacent parts of Virginia and West Virginia. The thickness of the Bloomsburg ranges from about 8 m in northeastern West Virginia to about 610 m in eastern Pennsylvania (Hoskins, 1961, Plate 8). Like the Shawagunk, the Bloomsburg consists of a variety of lithologic units, generally of grayish red claystone, siltstone, argillaceous sandstone, and small amounts of conglomerate. The Bloomsburg is overlain by the Poxono Island Formation in eastern Pennsylvania, but to the west, red beds are overlain by and grade laterally into the Wills Creek and Mifflintown Formations. Where red beds wedge out in north-central Pennsylvania, they are overlain by formations of the Salina Group. In some places, the Bloomsburg consists of fining-upward cycles as much as 4 m thick that are interpreted to represent the deposits of migrating streams, possibly in a high tidal-flat environment (Epstein and Epstein, 1969, p. 151–152).

Tuscarora and Clinch. In general, the post-Taconic molasse becomes finer grained and more marine to the north, west, and south, away from primary sources near eastern Pennsylvania and northern Virginia. Fluvial current trends in the Shawangunk are generally to the northwest (Epstein and Epstein, 1972, Fig. 23), as are current trends in the Tuscarora from Pennsylvania to southwestern Virginia (Yeakel, 1962; Whisonant, 1977). Lower Silurian sandstone ranges in thickness from a maximum of 150 m in northwestern Virgina and 460 m in eastern Pennsylvania to a minimum of 15 m along the Cincinnati Arch. Along the eastern flank of the Cincinnati Arch, Tuscarora equivalents contain an abundant marine fauna.

Cotter (1983, p. 25) describes the Tuscarora as being deposited "in a variety of shelf, coastal, and fluvial environments that shifted laterally in response to eustatic and continent-wide sea-level changes." Subsequent marine transgression ended Tuscarora deposition and produced the overlying Rose Hill Formation.

The southern Appalachian equivalent of the Tuscarora is the Clinch, which, in its type section on Clinch Mountain, Tennessee, is 95 m thick (Dennison and Boucot, 1974). There, the formation consists of a massive, locally pebbly, lower sandstone member that, like the Tuscarora, contains abundant *Arthrophycus.* The upper 13 m of the formation consists of interbedded sandstones and siltstones in about equal amounts. These are overlain unconformably by about 5 m of fossil-bearing Devonian sandstones.

Rockwood. To the west and south in Tennessee, the Clinch is replaced by the Rockwood Formation, which is composed mostly of interbedded shales, argillaceous to silty limestones, calcarenites, and thin silty sandstones. The Rockwood, which commonly contains several thin, calcareous, hematite-ore zones, thickens from 15 to 18 m in northern Sequatchie Valley eastward to 60 m of siliciclastic strata interbedded with calcarenite in the western Valley and Ridge (Milici and Wedow, 1977). At its easternmost exposure in Tennessee, the formation is about 185 m thick and consists almost entirely of siliciclastic strata (Wilson, 1979).

The Rockwood grades westward into the Brassfield Limestone along the southeastern side of the Nashville Basin in central Tennessee. The Brassfield is about 10 m thick and consists primarily of cherty thin-bedded limestone. The Rockwood and the Brassfield are truncated to the west below the Chattanooga Shale.

Rose Hill and Keefer. In southwestern Virginia, the Clinch Sandstone is overlain by varicolored sandy shales of the Rose Hill Formation in outcrop belts west of Clinch Mountain. The Clinch and the overlying Rose Hill ultimately grade into more shaly and marine facies of the Rockwood Formation to the west and south, from Virginia into the adjacent parts of the Tennessee Valley and Ridge (Miller, 1976). The Rose Hill is thickest in Pennsylvania, ranging up to 240 m. It thins southeastward into West Virginia along the axis of the Appalachian Basin. The Rose Hill generally is considered to have formed in shallow-water near-shore environments or on mudflats marginal to Tuscarora or Keefer beach sands. Basinal shales accumulated within the deeper trough of the Appalachian Basin (Smosna and Patchen, 1978). The Keefer is a widespread sandstone that extends with fairly uniform thickness throughout much of the central Appalachians, ranging up to 12 m thick in western Virginia and more than 35 m thick near the Virginia–West Virginia line (Smosna, 1983). The formation grades westward from shaly sandstone into dolomitic, silty sandstone in central and western West Virginia, and then into sandy dolomite near Ohio. Smosna (1983) interprets the Keefer to have been deposited under shoaling conditions during a decrease in the rate of regional subsidence. Lightner and Whisonant (1983) conclude that Keefer sands were derived largely from the reworking of older Paleozoic deposits and were redeposited in shallow-shelf and intertidal environments.

Folk (1960) concludes that the Tuscarora–Rose Hill–Keefer sequence was deposited during a long period of erosion that followed the Taconic Orogeny. At first, sediments were derived largely from erosion of strata. As this cover was removed, plutonic igneous rocks were exposed and contributed to the provenance of the Silurian deposits. The source land, according to Folk (1960, p. 54–55), was nonvolcanic rather than an island arc, and contained "a sequence of unmetamorphosed sandstones, siltstones, and cherty limestones overlying low-rank metamorphics, hydrothermal veins and plutonic igneous rocks."

McKenzie. In the western part of the Valley and Ridge, in West Virginia, Maryland, and Pennsylvania, the McKenzie consists of 73 to 96 m of shales interbedded with impure limestones. In the subsurface of West Virginia, farther to the west, the McKenzie is a fossiliferous carbonate formation, up to 150 m thick, that overlies either the Rochester Shale or the Keefer Sandstone, and is overlain by either the Williamsport or Salina Formations. In southern Pennsylvania, the Rochester and McKenzie are members of the Mifflintown Formation.

In general, during deposition of the McKenzie in Virginia and West Virginia, carbonate bank and reef facies accumulated to the north, west, and south in New York, Pennsylvania, Ohio, and Kentucky, whereas deposits of the Bloomsburg delta and shallow marine muds and sands were deposited to the southeast generally along the eastern margin of the Appalachian Basin (Patchen and Smosna, 1975).

Williamsport and Wills Creek. Deposition of thin, marine, delta-front Williamsport sands took place during maximal progradation of the Bloomsburg delta (Hoskins, 1961, Fig. 6) when the delta plain had built westward into central Pennsylvania and adjacent parts of Maryland, West Virginia, and Virginia. In general, the Williamsport is less than 15 m thick (Smosna and Patchen, 1978, Fig. 14).

The overlying Wills Creek Formation apparently was formed as a tidal-flat deposit during the final transgression and submergence of the Bloomsburg delta. In general, the formation is a mudrock or a limestone that contains beds of sandstone. In the subsurface, it is 120 m thick or more in West Virginia and nearby Maryland and Pennsylvania, thinning to the southeast into Virginia (Faill and Wells, 1977; Smosna and Patchen, 1978, Fig. 17).

Molasse, northern Appalachian Basin

Medina. In New York, along the northern edge of the Appalachian Basin, Silurian molasse deposits are represented by the Medina and Clinton Groups. Silurian formations extend westward from the southern part of the Adirondack Arch into southwestern Ontario, Canada. In general, the Medina thickens westward along its outcrop belt to about 35 m at the Canadian border (Rickard, 1975, Plate 2). The overlying Clinton attains its maximum thickness of about 100 m about midway along its outcrop belt, thinning abruptly to the west.

In this region, the Medina is composed of sandstones and shales representing a variety of lower delta-plain and associated marine environments (Martini, 1971). Martini (1971) concludes that the Medina formed as a result of the reworking of Bald Eagle and Juniata strata in the Appalachian region, and he correlates the Medina with the Tuscarora. Martini (1971) classifies some beds as Medina that Rickard (1975) includes in the base of the Clinton.

Clinton. Historically, the Clinton is divided into lower, middle, and upper units (Gillette, 1947). The middle Clinton is restricted to a small area in eastern New York; elsewhere, upper Clinton formations overlie lower Clinton beds unconformably (Rickard, 1975, Plate 2).

The base of the Clinton generally is marked by thin sandstones and conglomerates that range from about 2 m thick on the west to 9 m thick on the east (Gillette, 1947; Kilgour, 1963). The lower Clinton becomes thicker to the east, is dominantly shale, and attains a maximum thickness of about 30 m. In this region, it contains two hematitic limestones that have been mined for iron ore.

The middle Clinton in eastern New York consists of two facies. A gray and green shale that ranges up to 27 m thick and grades eastward into red cross-bedded sandstones and conglomerates. In some places, the middle Clinton contains a hematitic iron-ore bed.

Where thickest, the upper Clinton is about 60 m. Limestone and dolomite formations at the base of the upper Clinton in the western area of exposure in New York State grade eastward into the dolomite, which, in turn, intertongues with the shale formation (Rickard, 1975, Plate 2). To the west, the formation becomes dolomitic, pyritic, and silty and ultimately consists of interbedded shale, dolostone, and thin sandstone beds that grade westward into the Rochester Shale (Zenger, 1971; Brett, 1983).

Lockport. In New York, the Lockport Dolomite crops out from near Lake Erie eastward to the Adirondack Arch, along the northern edge of the Appalachian Basin. The formation is about 60 m thick on the west, where it is composed of biohermal carbonate rock, and grades eastward into shale and dolomite (Zenger, 1965; Rickard, 1975).

In New York, Lockport bioherms are generally less than 9 m thick. Most are dolomitized and contain a coral and stromatoporoid fauna. Stromatolites and invertebrate fossils are widespread within the formation. The Lockport was deposited in shallow-water to tidal-flat environments (Zenger, 1965).

Western molasse margin. In the subsurface, from western Virginia through West Virginia to Ohio, Lower Silurian sandstones become thinner and more discontinuous, grading westward from relatively thick Tuscarora into a thinner sandstone and shale sequence called Clinton by Knight (1969). In Ohio, the "Clinton" sandstones consist of channel and bar deposits that generally overlie or grade laterally into the Cabot Head Shale (Pepper and others, 1953). All in turn grade laterally into and are overlain by the Brassfield Limestone in western Ohio (Knight, 1969).

The Cincinnati Arch may have begun as a hinge line during Early Silurian time, foundering perhaps in response to molasse sedimentation in the eastern part of the Appalachian Basin (McDowell, 1983, p. 22–24). Silurian strata along the eastern side of the Cincinnati Arch in Kentucky are divided into two principal units by McDowell (1983). The older Crab Orchard Group ranges up to 63 m thick, whereas the overlying Bisher Dolomite is up to 30 m thick. The lower unit consists of several carbonate and shale formations.

In central and western Ohio, Janssens (1977) shows that "Clinton" sandstones and shales grade westward into about 45 m of Brassfield and Cabot Head shales, dolomites, and limestones that are overlain by about 8 m of dolomite and then by 8 m or less of dolomitic shale and dolomite.

Shelf carbonate and salt deposition

A Silurian and Devonian sequence composed mostly of carbonate rock lies between the post-Taconic molasse and the basal black shales of the Devonian Catskill delta. The carbonate sequence thickens northeastward, from about 10 m in northern Tennessee to more than 975 m in eastern Pennsylvania (Colton, 1970, Fig. 20).

Hancock, Tonoloway, and Keyser. In northeastern Tennessee and adjacent southwestern Virginia, the upper Silurian is represented by the Hancock and Sneedville Formations, which consist of up to 90 m of dolomite and limestone (Rodgers, 1953, p. 103; Miller and Brosgé, 1954, p. 83–86).

The northern Virginia equivalent, the Tonoloway Limestone (Silurian), consists of about 43 to 120 m of limestone and dolomite. In Maryland, the formation thickens to between 120 and 185 m of interbedded limestone and shale; in Pennsylvania, the Tonoloway is composed of argillaceous laminated limestone about 135 m thick (Faill and Wells, 1977).

In Virginia, the overlying Keyser is a nodular limestone and calcareous sandstone unit that ranges from 45 to 76 m thick (Young and Rader, 1974; Butts and Edmundson, 1966; Lesure, 1957). In Maryland, the Keyser consists of about 75 to 90 m of massive and nodular limestones in its lower part and planar-bedded shaly limestone in its upper part (Cloos, 1951; Swartz, 1923). In Pennsylvania, the Keyser generally is composed of 30 to 40 m of nodular, argillaceous limestone and shaly claystone (Hoskins, 1976; Faill and Wells, 1977).

Poxono Island, Bossardville, and Decker. In eastern Pennsylvania and New Jersey, the Bloomsburg is overlain by the Poxono Island Formation, 215 to 245 m of interlaminated and mud-cracked limestone and dolomite, calcareous shale, siltstone, and sandstone. Next above is the Bossardville Limestone, 6 to 12 m of fine-grained argillaceous limestone that, in some places, contains mudcracks with columns as much as 6 m long, overlain by 18 to 21 m of very fine-grained to sublithographic, laminated limestone that, in some places, also contains mud cracks. The Decker Formation consists of 15 to 24 m of arenaceous limestone, calcareous siltstone, and sandstone that grade northwestward into 15 m of crinoidal limestone (Epstein and others, 1967; Berg and others, 1983). Epstein and others (1967) conclude that these beds were deposited on a shallow shelf, in a restricted lagoonal environment, or in barrier-beach to biostromal bank settings.

Salina. The combination of the Bloomsburg delta to the east and carbonate reefs and banks to the west, formed a shallow restricted sea during the Late Silurian that extended from the western part of the Appalachian Basin into the Michigan Basin (Alling and Briggs, 1961). Salt-bearing strata range up to 915 m thick in the center of the Michigan Basin and up to 460 m thick in the axial part of the Appalachian Basin, astride the New York–Pennsylvania boundary. Limestone, dolomite, anhydrite, and halite display regular concentric patterns around the Appalachian salt basin (Smosna and Patchen, 1978; Rickard, 1969).

The principal salt-bearing formation of the Salina Group attains a maximum thickness of 460 to 490 m in the heart of the Appalachian salt basin. The salt-bearing interval is 365 to 396 m thick. Lithofacies exhibit a generally concentric pattern around the basin; dolomite and shale grade inward into anhydrite, dolomite, and shale and then into salt-bearing beds (Rickard, 1969, Plate 7 and 8).

Rondout and Bass Islands. The Salina Group is overlain by limestone, shale, and dolomite of the Rondout and Bass Islands Formations, which range generally from 15 to 45 m thick in New York and Pennsylvania. In some places in New York the units have been removed entirely by erosion, so that Devonian beds lie directly on the uppermost beds of the Salina Group (Rickard, 1969, Plate 12). To the west in Ohio, the Bass Islands Formation consists of up to 76 m of argillaceous and calcareous dolomite (Ulteig, 1964).

THE DEVONIAN

Devonian sediments accumulated in a subsiding foreland basin between the cratonic platform along the Cincinnati Arch on the west and an eastern source area. This landmass, which was uplifted by several pulses of Acadian orogenic activity, shed vast quantities of siliciclastic detritus that at times filled the foreland basin with deltaic sediment that spread westward onto the cratonic platform. Devonian rocks are thickest in eastern Pennsylvania (Fig. 6), where they exceed 3,660 m and thin southwestward to a nondepositional edge in southern Tennessee and in central Alabama (Conant and Swanson, 1961; Glover, 1959).

Early Devonian

Lower Devonian sediments accumulated in a shallow, slowly subsiding basin whose axis extended from southeastern New York into west-central Virginia and contiguous West Virginia. Siliciclastic detritus came largely from eastern source areas. Depending on the relation of exposure to depositional strike, Lower Devonian stratigraphy may appear simple or complex. The basal Helderberg Group of eastern New York (Richard 1962, Laporte, 1967) and the equivalent Keyser Limestone, Licking Creek Limestone, and Shriver Chert in Maryland (Head, 1974) illustrate the stratigraphic complexity (Table 5).

In much of the Appalachian Basin, the Silurian-Devonian boundary lies within a sequence of carbonate rocks and must be located by paleontologic criteria. The Lower Devonian sequence, which is tripartite in much of the eastern half of the basin, contains a lower carbonate shelf and basinal shale sequence, a medial blanket siliciclastic sequence, and an upper shelf carbonate and equivalent basinal chert and shale sequence. Subsidence dominated the northeastern part of the basin, whereas nondeposition and local erosion occurred to the west and southwest.

Lower carbonate shelf sequence. This sequence, characterized by the Helderberg Group of eastern New York, accumulated as a series of related facies ranging from lagoonal to supratidal on the eastern edge of the cratonic platform to shallow neritic farther east in the basin (Rickard, 1962; Anderson, 1971). Rocks range from dolomitic mudrock in the supratidal facies to clean calcarenite in bar and reefal facies. At several places closer to eastern source areas (e.g., in the Green Pond outlier of northern New Jersey or in southwestern Virginia), the carbonate rocks grade laterally into about 100 m of quartzose sandstone (Oliver and others, 1969, 1971). The lower carbonate shelf sequence is about 75 m thick in east-central New York and thickens to about 120 m in the subsurface of northeastern Pennsylvania near the basin's axis. Strata exposed in eastern New York and into the Valley and Ridge of eastern Pennsylvania show a complex of interfingering facies that reflect the interplay of many depositional

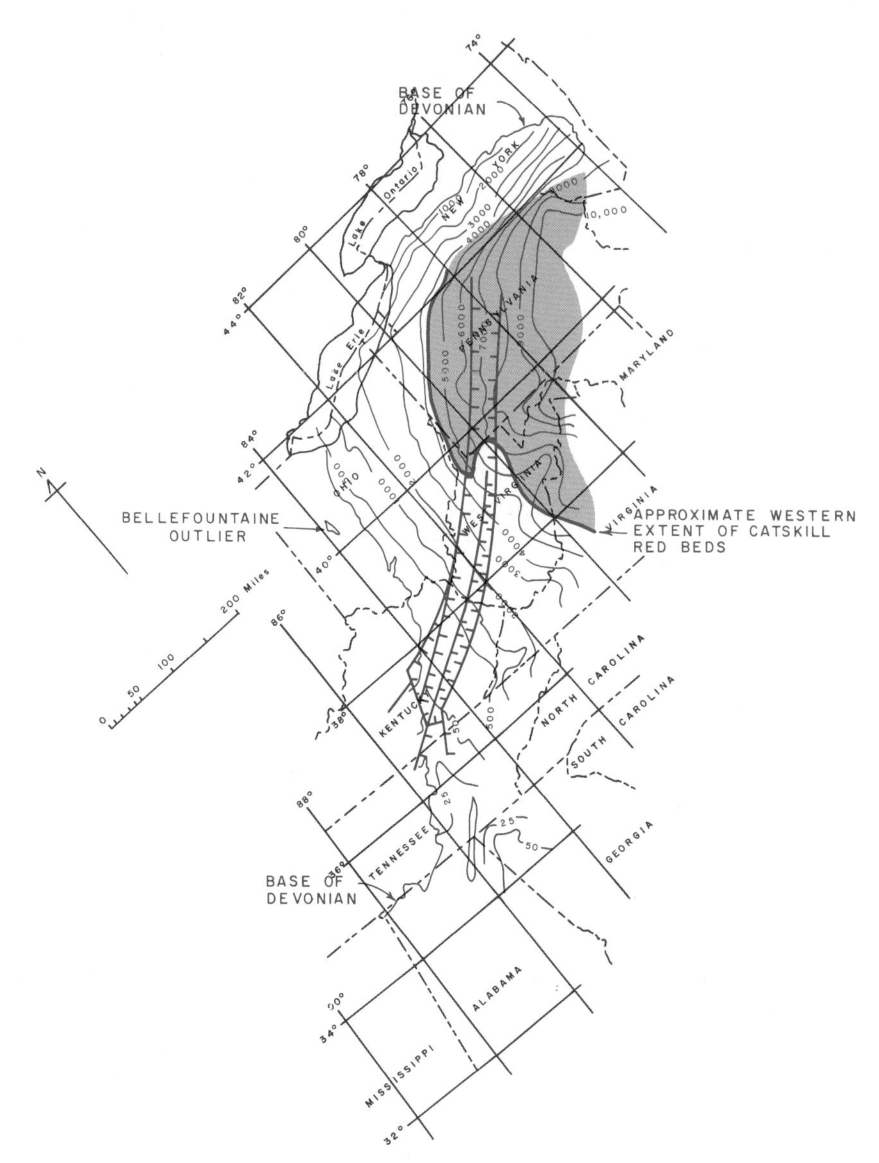

Figure 6. Thickness of Devonian rocks in the Appalachian Basin. Isopachs are in feet; ticks are on downthrown side of fault. (From Harris, 1978.)

environments of the shallow Early Devonian sea (Epstein, and others, 1967). The sequence consists of about 30 m of shaly limestone in central Pennsylvania and grades southward into about 105 m of fossiliferous limestone which has some intercalated chert and dark shale in northwestern West Virginia. The correlation of carbonate units in the Valley and Ridge of West Virginia, Maryland, and Virginia with units of the Helderberg Group in eastern New York has been a long-standing controversy. Bowen (1967) and Head (1974) have simplified the stratigraphy by recognizing locally mappable units in the central Appalachians and expunging nonapplicable New York stratigraphic names.

Near the axis of the foreland basin in western Maryland, the upper part of the lower carbonate sequence consists of as much as 70 m of dark silty calcareous mudrock, silicified siltstone, considerable chert, and a few thin lenses of small-pebble quartzose conglomerate, the Shriver Chert. The Shriver appears to be mainly the deeper water equivalent of the Licking Creek Limestone (Head, 1974).

Lower shelf carbonates thin to the southwest along the Valley and Ridge into central Virginia, where several tongues of quartz sandstone from eastern sources, such as the Clifton Forge, Elbow Ridge, Healing Springs, and Rocky Gap, become intercalated and dominate the sequence. Individual lenticular sandstones may exceed 30 m in thickness. In southwestern Virginia, the Wildcat Valley Sandstone (Miller and others, 1964) represents the total Lower Devonian sequence, resting unconformably on various Silurian units. Helderbergian fossils in the basal part of the Wildcat Valley shows that quartzose clastic sediment accumulated in the shallow sea in Early Devonian time.

Outcrops of Lower Devonian sandstone are scattered along the Valley and Ridge of Tennessee, Georgia, and Alabama. As much as 60 m of fine- to coarse-grained Frog Mountain Sandstone (Kiefer, 1970) occurs locally. Kiefer (1970) suggests that the Frog Mountain was derived from the erosion of Silurian sandstones on the east flank of the Cincinnati Arch and that the Frog Mountain sea deepened eastward away from this western source area. The extent of this seaway is conjectural because Alleghenian thrusting has telescoped, removed, or concealed Devonian rocks to the east.

The wedge of lower carbonate-shelf strata thins westward across the foreland basin from the Valley and Ridge to an erosional edge in eastern Ohio and adjacent Kentucky (Oliver and others, 1971). Drill cuttings from scattered deep wells show that facies patterns in the subsurface are generally similar to those of eastern New York. The rocks become more dolomitic westward across the basin and exhibit features of shallow marine to lagoonal accumulation. A Middle Devonian unconformity truncates the Lower Devonian rocks in the subsurface east of the Cincinnati Arch, suggesting moderate uplift after accumulation of the Lower Devonian strata. The arch was emergent in northern Ohio in earliest Devonian. A shallow valley cut in youngest Silurian dolomite is filled with the Holland Quarry Shale (Carman, 1960), a Lower Devonian brownish black shale that bears an estuarine fauna.

Middle siliciclastic wedge. The middle part of the Lower Devonian sequence, the Oriskany, consists of an extensive sheet of regressive, predominantly fossiliferous, fine- to coarse-grained quartzose marine sandstone that has some lenses of granule to small-pebble conglomerate. It underlies much of the Appalachian Basin from east-central New York to southwestern Virginia (Oliver and others, 1971). The Oriskany sand accumulated in a wave-dominated, high-energy environment in a shallow, well-aerated sea. The Oriskany fauna consists of many thick-shelled, robust fossils.

In outcrop, the Oriskany is more than 110 m thick in parts of Maryland and adjacent West Virginia. It thins both on the surface and in the subsurface to a peripheral depositional edge from west-central New York to northwestern Pennsylvania, southern Ontario, eastern Ohio, eastern Kentucky, and west-central Virginia (Oliver and others, 1971). Locally the Oriskany is absent within the main sheet of sand in western Pennsylvania and adjacent parts of Ohio and New York, either by nondeposition or by post-Oriskany erosion or both. Beds of reworked Oriskany sand in overlying carbonate rocks of the shelf sequence suggest local post-Oriskany erosion.

Outcrops of sandstone containing an Oriskany fauna are scattered along the Valley and Ridge from southwest Virginia to central Alabama. In some places, the Wildcat Valley and the Frog Mountain contain Oriskany fossils, suggesting deposition of quartz sand from several sources.

In the metamorphosed rocks of Alabama's Talladega belt, a 140-m sequence of predominantly silicious, cherty rock, the Jemison Chert (Butts, 1926) also contains an Oriskany fauna. The relation of the Jemison to other Devonian rocks in the Appalachian Basin remains to be resolved because the Talladega belt is an allochthonous sequence that was transported an unknown distance westward by Alleghanian thrusting (Neathery, 1973).

The Oriskany is one of the most important gas-producing rocks of the Appalachian Basin. Its intergranular porosity and permeability have been enhanced by tectonic fracturing under much of the central and eastern parts of the Appalachian Plateau and in the Valley and Ridge. Extensive fracture porosity also makes Oriskany reservoirs ideal for gas storage.

Upper shelf carbonate sequence. Subsidence of the foreland trough and an influx of fine-grained siliciclastic detritus produced a belt of dark shale and mudrock as much as 250 m thick in eastern Pennsylvania (Oliver and others, 1971). Closer to the source area, quartz sandstone accumulated in the Green Pond outlier in New Jersey. Southward along the Valley and Ridge in Maryland, Virginia, and West Virginia, as much as 60 m of medium-gray mudrock composed of scattered beds of black shale and carbonate nodules—the Needmore Shale—accumulated in the deeper part of the basin.

To the west of the main body of Needmore Shale, an extensive tongue of bedded calcareous chert as much as 70 m thick, the Huntersville Chert, underlies much of the eastern half of the Appalachian Plateau from southwestern Virginia to northwestern Pennsylvania. Dark shale and fine-grained sandstone are locally intercalated in the chert. Sponge spicules and thin-shelled fossils indicate accumulation in a protected environment largely below wave base. In southwestern Virginia, the upper part of the Wildcat Valley Sandstone contains a late Early Devonian Schoharie fauna, indicating the accumulation of quartzose siliciclastic detritus from an eastern source throughout the Early Devonian in this area.

To the west and north of the tongue of Huntersville Chert is a broad shelf of shallow-water carbonate rocks, the cherty Onondaga Limestone of New York, and the equivalent Columbus Limestone in Ohio and eastern Kentucky (Oliver and others, 1971). The boundary between shelf and basin facies is gradational and has tongues of one lithology projecting many miles into equivalent facies. The Onondaga, which contains abundant chert in several members, is about 90 to 155 m thick in eastern New York, thins to about 6 m in the subsurface of south-central New York and thickens westward to about 90 m in northeastern Ohio. It contains small reefs in thicker sections in New York (Oliver, 1963) and scattered pinnacle reefs as much as 60 m thick in south-central New York and contiguous Pennsylvania.

The equivalent Columbus Limestone, which contains considerable chert in eastern Ohio, thins from more than 90 m in northwestern Ohio to an erosional distal edge in the subsurface of eastern Kentucky, where it was truncated by Middle Devonian erosion (Oliver and others, 1971). However, the presence of 20 m of Columbus Limestone in the Bellefountaine outlier not far east of the crest of the arch in northwestern Ohio shows that, at least locally, the late Early Devonian sea extended across the arch.

The Acadian orogeny and the great Catskill delta complex

The remainder of the Devonian and the lower Mississippian sequence in the Appalachian Basin records the accumulation of a great thickness of siliciclastic sediment, the Catskill delta complex, that filled the northern part of the foreland trough and spread westward onto the cratonic platform as a result of Acadian orogenic activity to the east. In general, the deltaic sequence prograded westward and southwestward. The complete vertical section shows basinal black or gray shale overlain, in turn, by distal to proximal turbidites, by shallow marine and shoreline sandstones, and, ultimately, by terrestrial deltaic deposits, including some coal-bearing rocks in the Lower Mississippian sequence of Virginia and Pennsylvania. In the basin, an ash fall bed in the Helderberg of New York (Rickard, 1962) and the Tioga ash fall bed or beds indicate the onset of Acadian orogenesis. Tectonism began in the northern part of the basin late in the Early Devonian and spread southward during the Middle Devonian (Faill, 1985). Paleomagnetic data suggest that Acadian tectonism may have been caused by the collision of an ancient landmass, Armorica, with ancestral North America (Laurentia) and the northern European sialic crust (Baltica) (Perroud and others, 1984). In contrast to Taconic and Alleghanian orogenesis, the Acadian produced only moderate local deformation in the Appalachian foreland basin.

Tioga ash fall. One or several closely associated beds of volcanic ash lie within or at the top of the Onondaga Limestone, in the upper part of the Columbus Limestone, or in the basal part of the Marcellus Shale, in the northern part of the Appalachian Basin. In Ebright and others (1949), Fettke named the Tioga for ash beds that he identified in cuttings from deep wells in Tioga County, Pennsylvania. Later the Tioga was found in many places in the northern half of the Appalachian Basin, and its equivalents have been identified in the Illinois and Michigan Basins. Although only a few centimeters thick at most places, the Tioga is an important stratigraphic marker, which may be readily identified by its large biotite content.

Middle Devonian sequence

Catskill delta. During the early Middle Devonian, the foreland basin subsided rapidly, and a euxinic environment spread widely across the northern half of the basin, except in the Green Pond outlier, where silt and silty dark-gray mud accumulated. Black mud, rich in organic detritus, which is the major component of the Marcellus Shale, covered much of the basin. The Marcellus is about 300 m thick in central Pennsylvania and thins to a featheredge below the late Middle Devonian unconformity in the subsurface of eastern Kentucky and adjacent parts of Ohio and West Virginia. Locally, in the subsurface of central Ohio, the Marcellus grades laterally into 10 to 15 m of dark shaly limestone, the Delaware Limestone, which is a western remnant of the once extensive Onondaga-Columbus carbonate shelf.

Prograding coarse-grained siliciclastic sediments, the vangard of the great Catskill delta complex (Willard and others, 1939, p. 371), entered the northeastern part of the Appalachian foreland basin during the Middle Devonian. They appeared first in the Green Pond outlier where several hundred meters of gray Cornwall Shale are overlain by about 760 m of gray shale and sandstone of the Bellevale Sandstone. The Bellevale grades vertically into the coarser-grained pebbly sandstone and conglomerate of the Skunnemunk Conglomerate, which contains beds of red shale and mudrock. This wedge of clastic strata shows a gradational sequence from shallow-marine to terrestrial red-bed facies. Red beds of the Skunnemunk are typical of the Catskill red terrestrial magna facies (Caster, 1934) that dominated much of the northern part of the foreland basin during the Late Devonian. The upper third of the 1,070-m sequence of post-Marcellus Middle Devonian quartzose clastics on the east face of New York's Catskill Mountains contains red beds of the Catskill magnafacies. The Middle Devonian sequence in the Valley and Ridge of central Pennsylvania, the Mahantango Formation, shows a similar vertical increase in coarse-grained siliciclastic rocks, including a scattering of red beds in the upper part in the south-central part of the state (Willard and others, 1939). About half the Mahantango along the Susquehanna River consists of massively bedded gray quartz sandstone. This coarse-grained, near-shore marine facies thins rapidly westward into a finer-grained, deeper-water facies. In west-central Pennsylvania and in central New York, the finer-grained equivalents of the Mahantango consist largely of fossiliferous marine shale and mudrock; some thin but extensive beds of fossiliferous limestone are intercalated in the sequence.

South along the Valley and Ridge from central Pennsylvania to central Virginia, the near-shore facies of the Mahantago Formation, which overlies the Marcellus Shale, thins from about 610 m to less than 60 m, grain size decreases, and the light-gray siliciclastic strata grade into grayish black mudrock similar in appearance to the Marcellus below and to the overlying Upper Devonian black Burket Member of the Harrell Shale. In west-central Virginia, Butts (1940) gave the name Millboro Shale to the 300-m sequence of black beds that includes equivalents of the Marcellus, Mahantango, and Burket. The Millboro thins to the southwest along the Valley and Ridge and adjacent Appalachian Plateau into southwestern Virginia and contiguous West Virginia. It is truncated by a late Middle Devonian unconformity in southwestern Virginia and adjacent eastern Tennessee. Because the exact amount of the Millboro that may be assigned to the Marcellus cannot be determined by existing criteria, the original extent of the Marcellus may only be roughly estimated in the western part of the Appalachian Basin south of the central Ohio.

Near the close of the Middle Devonian, a relatively thin but extensive shallow-water carbonate, the Tully Limestone, accumulated across much of the foreland basin from central New York to southeastern Ohio and adjacent West Virginia (Heckel, 1973). The Tully ranges in thickness from more than 60 m in north-central Pennsylvania to a depositional featheredge along the western and southern periphery of this sheet of carbonate strata. Apparently, more active subsidence in the eastern part of the foreland basin, possibly including some faulting, developed a sediment-trapping trough that retained the flood of siliciclastic detritus, permitting the deposition of the Tully to the west of the sediment trap (Heckel, 1973).

At about Tully time, the western and southern parts of the foreland basin and the adjacent Cincinnati Arch were gently uplifted and eroded to progressively greater depths southward from Ohio to southern Tennessee. Only along the Rome Trough in central Kentucky is a thin sequence of Middle Devonian preserved on the eastern flank of the Cincinnati Arch (Oliver and others, 1971).

Continued subsidence of the northern part of the foreland basin in central New York produced an anoxic environment. This is marked by about 30 m of the black Geneseo Shale Member, the basal unit of the predominantly Late Devonian Genesee Formation.

An erosional hiatus of increasing magnitude to the south and west of the western edge of the Tully Limestone in central New York separates progressively younger Upper Devonian rocks from older Paleozoic strata beneath the unconformity. In north-central Ohio, upper Upper Devonian rocks lie on lower Middle Devonian rocks with apparent concordancy. In southern Kentucky, late Upper Devonian black shales rest upon Upper Ordovician shale and limestone, and in the Valley and Ridge of east Tennessee, the Upper Devonian black Chattanooga Shale lies on Middle Ordovician black shale. The surface of the unconformity is of very low relief, and in most places bedding appears concordant across the unconformity. However, regional stratigraphy indicates that many hundreds of meters of beds were removed in the southern part of the basin during the hiatus represented by the unconformity. Because most of the eroded strata were carbonate or shale, sediment reworked from the cratonic platform cannot be identified positively in the fine-grained younger Devonian strata above the unconformity. However, thinning of individual Upper Devonian shales on the cratonic platform suggests that some uplift occurred locally in parts of the Cincinnati Arch during the Late Devonian.

Upper Devonian sequence

Upper Devonian rocks thin southward from more than 2,750 m in eastern Pennsylvania (Oliver and others, 1971) and are absent on the Hohenwald platform in south-central Tennessee (Conant and Swanson, 1961). From an eastern orogenic source, siliciclastic detritus of the westward-prograding Catskill delta complex filled the northern half of the foreland basin and lapped onto the adjacent cratonic platform. The thousands of meters of deltaic sediment accumulated in a suite of related environments ranging westward from upper and lower delta plain, shoreline, beach, lagoon and bar, high-energy shallow marine, and low-energy turbidite apron, to a deeper-water euxinic basin. In the southern part of the basin remote from sources of coarse-grained clastic detritus, the Upper Devonian sequence is commonly less than 15 m thick and is mainly dark gray to black shale.

The expanding Catskill delta shifted depositional environments westward across the foreland trough. Marine facies were overstepped by more terrestrial strata throughout most of the Late Devonian (Chadwick, 1933; Willard and others, 1939; Rickard, 1964; Oliver and others, 1971). Consequently, the terrestrial red-bed facies, Caster's Catskill magnafacies, is thickest in the northeast and thins to a featheredge to the west and south. In contrast, the basinal black shale facies is thickest in the west central part of the basin and thins to a distal edge in the southern part of the basin (Conant and Swanson, 1961; de Witt and others, 1975). During deposition of the Catskill delta, the westward progradation was temporarily reversed five times by major marine transgressions, which shifted marine facies many tens of kilometers eastward. In Pennsylvania, for example, the basal anoxic Burket facies advanced eastward from near the Ohio state line almost across the state (Wagner, 1963).

Black shale. Most of the Upper Devonian sequence in the western part of the foreland basin consists of black shale rich in organic detritus. The shale consists mainly of clay, with as much as 20 percent organic matter in the southern part of the basin, where the Chattanooga Shale is generally about 10 m thick. Northeastward, particularly along the periphery of the Catskill delta, where black shales interfinger with silty turbidites, the black rocks contain much quartz silt together with moderate clay and organic carbon.

Because diagnostic macrofaunas are absent and because of the strong lithologic similarity of the several Devonian black shales, the age and stratigraphy of these Devonian black shales has been the subject of spirited controversy for more than 125 years (Hass, 1956; Roen, 1981). Recent advances in zonation by use of conodont faunas, and the widespread availability of gamma-ray well logs, which enable correlation of a specific black shale by its radioactive signature, permit geologists to identify and correlate named black shales from type and measured outcrop sections widely throughout the Appalachian Basin (Roen, 1981, 1983; de Witt and Roen, 1985). Of the five regionally extensive Upper Devonian black shales, only the younger Rhinestreet, Huron-Dunkirk, and Cleveland crop out in the western part of the basin. The Middlesex and the older Geneseo-Burket crop out in parts of New York or Pennsylvania and thin westward in the subsurface against older rocks beneath the Middle Devonian unconformity; they are not present in outcrops along the Cincinnati Arch. In contrast, the younger Rhinestreet, Huron-Dunkirk, and Cleveland black shales are well developed in the western part of the basin but grade laterally into coarser-grained gray rocks west of outcrops in the Valley and Ridge of Pennsylvania.

Individual black shales are thickest in the west-central part of the foreland basin, where they are more than 180 m thick

(Harris and others, 1978); locally, individual units may be more than 75 m thick. Regional stratigraphy shows that each unit attains its maximum thickness in a different part of the basin, apparently due to subtle basin tectonics andariations in the supply of sediment.

In northern Ohio, the black Huron and Cleveland Members of the Ohio Shale are more than 200 m thick near Lake Erie. The members thin and coalesce southward into central Kentucky. In south-central Kentucky, the Ohio Shale grades laterally into the upper 10-m-thick Gassaway Member of the Chattanooga Shale. These changes illustrate the thinning of Devonian black shales southward along thc Cincinnati Arch.

Although the black shales are only a small fraction of the total sequence, they are economically important as source beds for oil and gas and locally as reservoir rocks for these hydrocarbons.

Gray shale and turbidite. The marine rocks of the Catskill delta complex largely consist of gray shales, turbidite siltstone, and fine-grained sandstone in the lower prodelta wedge. The near-shore upper part consists of sandstone and siltstone associated with some shale that accumulated in shallow-marine neritic and shoreline environments.

About 600 to 900 m of Brallier turbidites crop out in central Pennsylvania (Lundegard and others, 1978). The turbidite facies lies in the lower part of the Upper Devonian sequence. This facies underlies much of the Appalachian Plateau from southern Virginia to western New York. It grades westward into the dark Chattanooga or Ohio shale sequences by loss of distal turbidites. The Brallier facies progrades westward across the Appalachian Basin and crops out in northern Ohio as the Chagrin Shale, almost at the top of the Devonian sequence (Broadhead and others, 1982).

Shallow marine. The shallow marine facies of the Catskill delta complex consists of about 760 m of fossiliferous, locally pebbly sandstone, siltstone, and intercalated sandy shale in eastern and central Pennsylvania and in adjacent parts of New York, Maryland, Virginia, and West Virginia. Generally, the grain size of units increases vertically and is greatest in bar and beach units in this facies. The shallow marine rocks are overlain by as much as 1,735 m of Catskill red beds (Glaeser, 1963) in eastern Pennsylvania. The shallow marine sequence thins westward with a corresponding decrease in sandstone. In western Pennsylvania and contiguous parts of adjacent states, the thinning distal part of the red Catskill sequence interfingers with the shallow marine facies. A brief transgression near the close of the Late Devonian displaced the facies eastward, and gray shallow marine rocks covered the red-bed facies throughout the western half of Pennsylvania and contiguous parts of New York and West Virgina.

In western Pennsylvania, southeastern New York, and northern West Virginia, the oil- and gas-bearing sandstones of the Venango and Bradford Groups, a series of nearshore sheet-like sands, bars, and associated beach deposits are scattered through about 460 m of greenish gray shale and siltstone. Although these coarse-grained, locally pebbly sandstones constitute less than 10 percent of the shallow marine facies, they are the most important oil-producing rocks in the Appalachian Basin. The first large amounts of oil produced in America were obtained from oil sands of the Venango Group during the early 1860s and since then more than 1.6 billion barrels of oil have been extracted from the Venango and Bradford rocks of the shallow marine facies.

Catskill red beds. The red-bed facies underlies the Valley and Ridge from Pennsylvania to south-central Virginia and much of the northeastern half of the adjacent Appalchian Plateau. Many formation and member names have been applied to parts of the Catskill red beds in parts of these several states. The Catskill facies, which is about 2,750 m thick in eastern Pennsylvania, thins and interfingers with the gray shallow marine facies along the margin of an irregular sheet from western New York to central Virginia. The red beds accumulated as terrestrial deposits associated with anastomosing distributaries on several lobes of the Catskill delta complex. Coarse-grained pebbly channel sandstones alternate with finer-grained overbank and swamp deposits. Evidence of subaerially exposed sediments include desiccation cracks and hail or raindrop impressions. Plant fossils are locally abundant, and a scattering of fossilized freshwater fish have been found in the channel sandstones. Generally, the coarser-grained rocks are more abundant to the east, although pebbly sandstones are intercalated in the red-bed sequence in western Pennsylvania and adjacent West Virginia.

Throughout much of the Late Devonian, sedimentation exceeded subsidence in the foreland basin, and the great Catskill delta complex prograded westward across the northern part of the basin and onto the adjacent part of the unstable cratonic platform. During a moderate transgression near the close of the Devonian Period, as much as 20 m of basinal black shale, the Cleveland Member of the Ohio Shale, overlapped the tongue of gray turbidites in the northwestern part of the basin. To the east as much as 70 m of fossiliferous greenish gray shale and siltstone of the shallow marine facies, the Oswayo Formation (Fettke and Bayles, 1945), overlies the Catskill red-bed facies in west-central Pennsylvania.

THE MISSISSIPPIAN

The general paleogeography and depositional framework that was established during the Late Devonian dominated the Appalachian Basin throughout much of the Early Mississippian (Kinderhookian and Osagean, Fig. 7). In the north, the trough was filled with siliciclastic deltaic deposits and some fringing, marginal shallow-marine sediments. In contrast, shallow-marine to supratidal carbonate sands, silts, and muds blanket much of the southern and central parts of the basin (deWitt and McGrew, 1979, Plate 3B-4B). Locally, prograding delta deposits restricted circulation of the epicontinental sea, and evaporites accumulated in relatively isolated localities. The Late Mississippian (Meramecian) was a time of great marine transgression, when thick sequences of marine carbonates covered much of the western and central parts of the trough. Siliciclastic sediments were largely

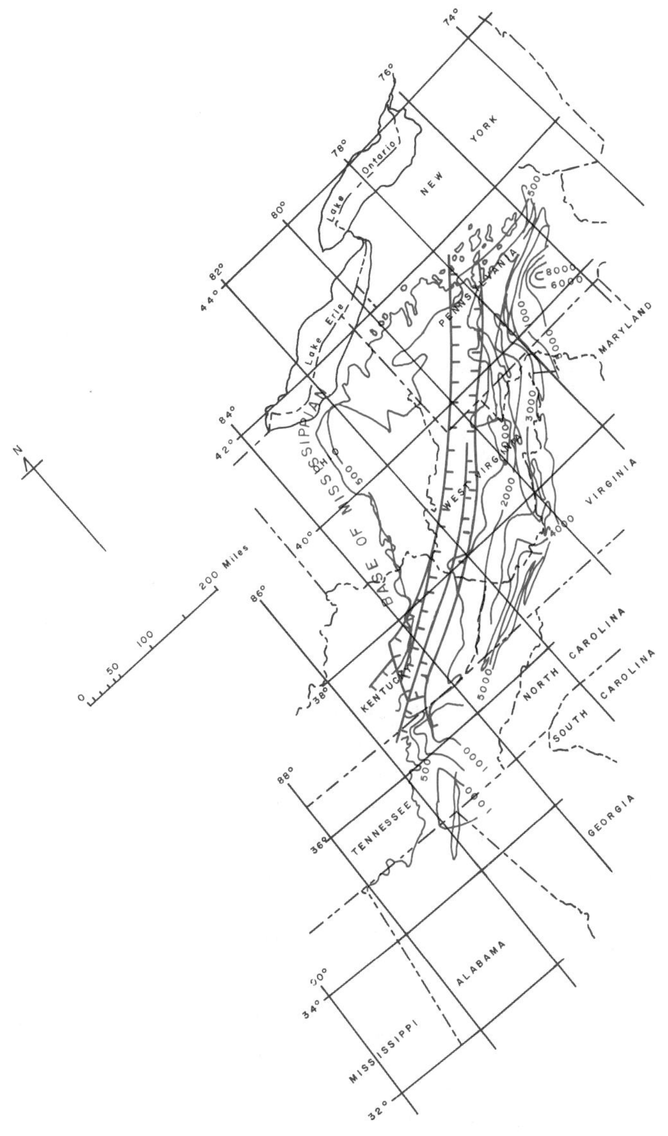

Figure 7. Thickness of Mississippi rocks in the Appalachian Basin. Isopachs are in feet; ticks are on downthrown side of fault. (From Harris, 1978.)

restricted to the northeastern part of the basin. Prograding delta deposits returned to fill the northern part of the trough in Late Mississippian (Chesterian), when thousands of meters of continental red beds accumulated in eastern Pennsylvania. Carbonate strata are succeeded by bar and beach coarse-grained siliciclastic sediments, which are overlain by sediments of the lower and upper delta plain southward from Pennsylvania across West Virginia, eastern Kentucky, and Tennessee into northern Alabama.

Early Mississippian

Delta Systems. Two deltas, the large Pocono delta in eastern Pennsylvania and the smaller Price delta in central Virginia, prograded westward into the northern segment of the Appalachian Basin during the Early Mississippian. The source of siliciclastic detritus that makes up these deltas lay to the east and was uplifted during a part of the Acadian orogeny. In the Anthracite district of eastern Pennsylvania, the Spechty Kopf Formation (Table 6) consists of about 600 m of Early Mississippian red beds, similar to those of the subjacent Catskill Formation, intercalated with gray coarse-grained quartzose rock, similar to those of the overlying Pocono Formation. The Devonian-Mississippian boundary lies within the Spechty Kopf, and its stratigraphic equivalents in western Pennsylvania, Maryland, and northern West Virginia show a similar interfingering of red and gray rocks. East of the Susquehanna River in eastern Pennsylvania, the Pocono Formation at the center of the Pocono delta consists of more than 300 m of quartz-pebble conglomerate. This mass of siliciclastic detritus thins and becomes finer grainer to the west and south. It is about 150 m thick in the Broad Top Basin and about 90 to 110 m thick along the Allegheny Front. In the subsurface, this unit is the well driller's Big Injun sand, which extends west across the basin to central Ohio where its equivalent units are marine. The Big Injun sand is one of the more important oil and gas sands in the Mississippian sequence. In the Valley and Ridge of Maryland, an equivalent sandstone is about 90 m thick. The Big Injun sand thins to the south and fingers out of the sequence in northern Virginia. Farther south, in west-central Virginia, the Cloyd Conglomerate (Butts, 1940), a coarse-grained quartz-pebble conglomerate about 25 m thick, marks the beginning of deposition of the Price deltaic sequence. Above the Cloyd are about 275 to 365 m of the Price Sandstone. Marine sandstone and shale in the lower part grade vertically into terrestrial sandstone with a coal-bed sequence at the top. In some places in the Valley coal fields of west-central Virginia, Price coals, which are as much as 3 m thick and cover a considerable area (Campbell and others, 1925), have been mined extensively since the 1840s. To the southwest in the Valley and Ridge, the Price sandstones and siltstones grade into the greenish gray turbiditic shales, siltstones, and sandstones of the Grainger Formation. In Tennessee, the Grainger is more than 305 m thick in eastern sections where sandstones are abundant, and about 60 m thick to the west where shale is dominant (Milici and others, 1979).

A wide belt of basal Mississippian marine delta-front sands fringe the Pocono and Price deltas on the west, and, in the subsurface, anastomosing tongues of well-sorted, porous sand extend westward across Pennsylvania into eastern Ohio and eastern West Virginia, where they have produced natural gas abundantly in places (Pepper and others, 1954).

At essentially the same time that the basal Pocono sand prograded from the east, clastic detritus from the cratonic Laurentian shield spread southward into the shallow Mississippian sea along the eastern flank of a low shoal marking the site of the

Cincinnati Arch. As much as 45 m of red Bedford mud accumulated in a large spoon-shaped delta, the Red Bedford delta. From north to south, offshore barrier bars and lagoons fringe the delta on the east, and a sand-filled distributary system delineates the long axis of this muddy delta deposit. Uplift to the north and a slight regression of the Mississippian sea initiated channel cutting in the northern part of the Red Bedford delta and on a newly exposed coastal plain in West Virginia (Pepper and others, 1953). A flood of Berea sand from the cratonic shield to the north and from the Price delta to the east filled the channels and prograded across the shallow sea to coalesce in a 30- to 45-m thick sequence of sand and silt in eastern Kentucky (Pepper and others, 1954). In northern Ohio, the Berea fills channels 85 m deep that were scoured into Devonian shales. In contrast, Berea channels in central West Virginia rarely exceed 18 m in depth. Since 1860, the sheet of Berea sand has produced large amounts of oil and gas, and is the most important of the shallow Mississippian and Pennsylvania oil sands of the Appalachian Basin.

An abrupt marine transgression at the close of Berea deposition flooded the restricted basin, producing an anoxic environment and a widespread sheet of black mud as much as 18 m thick—the Sunbury—which is the youngest of the extensive Devonian-Mississippian black source-rock shales. The Sunbury blanketed the western half of the basin from northern Tennessee to northern Ohio. It thins to a few inches and merges into the top of the Chattanooga Shale in south-central Kentucky and eastern Tennessee (Roen, 1984), where its presence can be demonstrated only by its characteristic *Siphonodella* conodont fauna.

Clearing seas dispersed the Sunbury anoxic environment. The Cuyahoga Group and Logan Sandstone of Ohio and the Borden Formation of eastern Kentucky, a 120- to 185-m sequence of siliciclastic sediment ranging from gray mud to coarse-grained pebbly sandstone, accumulated next, as the western marine phase of the Pocono-Price delta complex. In general, grain-size of Cuyahoga rocks and their equivalents decreases southward. In outcrop in southern Ohio and in adjacent northern Kentucky, these rocks are fine-grained sandstone and siltstone turbidites (Moore and Clarke, 1970). In outcrop in eastern Kentucky, the Borden consists of about 120 m of shale intercalated with several sheets of turbidites (Rice and others, 1979). These dominantly shaly strata represent the deeper-water distal part of the large Pocono-Price delta complex.

In northeastern Tennessee, the gray or greenish gray shales of the Borden Formation in the Appalachian Plateau and the Grainger Formation in the Valley and Ridge thin abruptly along the front of the Borden delta (Milici and others, 1979) and underlie the slightly younger cherty carbonate rocks of the Fort Payne Formation.

Carbonate platform. Throughout much of eastern Tennessee and adjacent parts of Alabama and northwestern Georgia, less than 1 m of gray shale with phosphate nodules, the Maury Formation, overlies the Chattanooga Shale and is the basal unit in the Mississippian sequence. Overlying the Maury are 30 to 60 m of cherty, silicious, shallow-water limestone and dolomite of the Fort Payne Formation. In some places, quartz geodes in the Fort Payne apparently resulted from silica replacing nodular anhydrite that had accumulated in tidal-flat and lagoonal environments (Milici and others, 1979). In northeastern Tennessee adjacent to the Borden front, the lower part of the Fort Payne contains biohermal mounds as much as 25 m thick of relatively local extent. These small mounds, which contain oil and gas in Morgan and Scott Counties, Tennessee, have produced about 4 million barrels of oil in the past 25 years (Milici and others, 1979). To the southeast in Georgia and Alabama, the Fort Payne consists largely of cherty limestone or cherty dolomite. In eastern exposures in the Valley and Ridge, a dark gray calcareous shale and argillaceous calcareous mudstone facies about 50 m thick makes up much of the Fort Payne (Thomas and Cramer, 1979). The eastern increase in clayey mud suggests a clastic source lay to the east.

The Fort Payne marks the beginning of the last major marine transgression in the Appalachian Basin. The initial flooding inundated the platform on the west flank of the basin during the Early Mississippian, whereas clastic deltaic rocks dominated the northern half of the basin. In much of the western and central parts of the basin, the Fort Payne is the precursor of widespread deposition of carbonate sediment during the Middle and early Late Mississippian.

Red beds and evaporites. The uppermost Lower Mississippian of the Price delta, the Maccrady Formation, consists of red shale, siltstone, and sandstone, and locally much dolomite, anhydrite, gypsum, and halite. In outcrop in the Valley and Ridge of west-central and southwestern Virginia, the Maccrady is a sequence of terrestrial to marine red beds about 12 to 490 m thick (Butts, 1940; Cooper, 1963). In a small fault-bound structure near Saltville, Smyth County, the Maccrady is more than 600 m thick and contains large amounts of salt, gypsum, and anhydrite (Cooper, 1966). Although the Maccrady appears to have been deposited mainly in a restricted basin, locally the red beds in western Virginia contain marine fossils. Maccrady red rocks extend south in the Valley and Ridge into northeastern Tennessee, where they form the basal 18 to 22 m of the Newman Limestone in the Greendale syncline of the Clinch Mountain belt (Milici and others, 1979). The Maccrady extends into the subsurface of southern West Virginia and adjacent eastern Kentucky. In many places, well drillers identify the Maccrady as the Red Injun sand, which is as much as 30 m thick. The red rock thins to the north and west, and a few meters of purplish red shale in the Borden Formation in east-central Kentucky (Weir and others, 1966) represent the western tip of the red Maccrady in the central part of the Appalachian Basin.

Late Mississippian (Meramecian and Chesterian)

Shallow seas flooded much of the foreland basin and adjacent unstable cratonic platform along the Cincinnati arch during the Meramecian. Shallow-neritic to supratidal carbonate sediments blanketed much of the basin, except along its northeastern and eastern sides, where siliciclastic deltaic detritus accumulated in smaller deltas, near the site of the Pocono-Price delta complex (deWitt and McGrew, 1979, Plate 5B) and farther south in the Valley and Ridge.

Pennsylvania. In the Anthracite district of eastern Pennsylvania, about 1,825 to 2,440 m of Meramecian and Chesterian terrestrial red beds, the Mauch Chunk Formation, separate the underlying Pocono conglomerates from the overlying Pottsville conglomeratic sandstones. The red-bed sequence shows many attributes of subaerial fluvial deposition, including channel-filling sandstones and fine-grained overbank and swamp deposits. Mudcracked shales, raindrop impressions, and mudstone-clast conglomerates are abundant. Plant fossils occur locally in the Mauch Chunk.

To the southwest, in the Broadtop Basin of central Pennsylvania, a 2- to 8-m reddish gray to greenish gray argillaceous, sparsely fossiliferous limestone lies at the base of the Mauch Chunk red beds (Butts, 1945); its sparse fauna appears to be Meramecian. To the west, along the Allegheny Front in south-central Pennsylvania, a 10- to 15-m reddish gray, brown-weathering, strongly cross-bedded, calcareous sandstone or sandy calcarenite, the Loyalhanna, fills much of the Meramecian interval (Adams, 1970). The Loyalhanna sands were derived from cratonic sources to the northwest (Adams, 1970), suggesting a brief pulse of uplift in the northern source area that had previously supplied clastic detritus to the Bedford-Berea delta system.

Virginia. Scant paleontologic data suggest that some of the 300 to 365 m of Maccrady red beds overlying the Price Sandstone in west-central Virginia may be of Late Mississippian age. To the south, in the Greendale Syncline, Meramecian carbonate rocks are more than 765 m thick (deWitt and McGrew, 1979, Plate 5A). The Little Valley Limestone contains some shale and sandstone in the basal part, but most of the unit is shallow-water, yellowish gray, dolomitic calcarenite that includes some intercalated argillaceous calcilutite (Butts, 1940, p. 356–357). The remainder of the Meramecian sequence in the Greendale Syncline consists of shallow-water deposits composed of pelletoid, oolitic, fossiliferous calcarenite intercalated with lagoonal micrite. Chert is abundant in some beds. An 18-m sequence of reddish limestone and shale marks the top of the Middle Mississippian beds in the Greendale area (Butts, 1940). The red beds are a tongue of red-colored rock from an eastern clastic-source area.

West Virginia. During the Late Mississippian, sandy cratonic sediment spread south across the shallow sea to impinge upon the Catskill Island in northeastern West Virginia (Arkle and others, 1979), which effectively stopped the southward spread of quartz sand. South of the island, shallow-water to supratidal carbonate rocks, the lower part of the Greenbrier Limestone, comprise the Meramecian interval along the Allegheny Front in Virginia and in the contiguous subsurface of Virginia and West Virginia. In contrast with equivalent rocks in the Greendale syncline, the Greenbrier Limestone is thinner, averaging about 60 to 90 m thick (Wilpolt and Marden, 1959). The sequence consists of interbedded beach and bar oolitic sands and micritic lagoonal muds. A tidal-flat environment is suggested by nodular and bedded anhydrite in the lower part of the unit. Overlying beds consist of oolitic calcarenite intercalated with calcareous mudstone, and these, too, apparently accumulated in a complex of shallow-neritic, bar and beach, lagoon and tidal-flat environments. The younger beds overstep the older beds northward toward the Catskill Island but did not cover the island during the Middle Mississippian.

Tennessee. Only the lower 300 m of Late Mississippian gray argillaceous limestone, red and gray shale, siltstone and fine-grained sandstone are preserved in the Chilhowee Mountain belt, the most eastern exposures of these strata in Tennessee (Milici and others, 1979). Their considerable content of siliciclastic detritus attests to the proximity of eastern source areas. Similarly in the Greendale Syncline, the lower part of the Newman Limestone consists of about 146 m of interbedded calcareous siltstone, sandstone and silty shale intercalated in fine-grained to oolitic, commonly cherty limestone. The greater component of siliciclastic detritus in the Meramecian strata in the Tennessee segment of the Greendale Syncline suggests closer proximity to the source of detritus or more effective transport to the site of deposition.

To the west and south of the Greendale Syncline, on Newman Ridge, on Pine Mountain, and in Elk Valley, the lower part of the Newman Limestone (Milici and others, 1979) is composed of cherty, light-gray micrite intercalated with oolitic fossiliferous calcarenite. In the Elk Valley section, the basal 6 m of the 125-m section consists of dolomitic, argillaceous limestone with intercalated lenses of coarse-grained quartz sand and jasper-pebble conglomerate. The sequence above contains some greenish gray or grayish red shale interbedded in oolitic or bioclastic calcarenite. These Meramecian carbonates appear to have accumulated in littoral environments of bar, beach, and lagoon.

To the west in Tennessee, marine carbonate rocks blanketed the unstable platform east of the Cincinnati Arch. They crop out along the Sequatchie Anticline and the eastern Highland Rim of the Nashville Dome. The Meramecian sequence includes the lower part of the Monteagle Limestone (Table 6). The upper part of the Monteagle is of Late Mississippian age (deWitt and McGrew, 1979) and is the equivalent of the Gasper Limestone of southwestern Virginia and adjacent eastern Kentucky.

The Warsaw Limestone, about 15 to 55 m thick, is dominantly a sandy, dark-gray, medium- to coarse-grained, cross-bedded calcarenite with some chert, dolomite, and quartz silt. The overlying St. Louis Limestone consists of yellowish gray to brown dolomitic limestone and dolomite about 12 to 55 m thick, that contains abundant spherical to ellipsoidal masses (cannonballs) of bluish gray chert. The St. Louis is characterized by the fossil corals *Lithostrotion proliferum* and *Lithostrotionella castelnaui.* The Monteagle Limestone consists of about 75 m of light- to medium-gray, oolitic, bioclastic limestone intercalated with light-gray micrite. Small amounts of gray shale and yellowish gray dolomite are interbedded locally. Dark chert is scattered throughout the Monteagle. Much of the Monteagle appears to have accumulated on shallow shoals with oolitic sands alternating with calcareous mudrock. Shallow-water high-energy environments lay in close proximity to low-energy lagoonal and mud-flat environments.

Kentucky. To the north in Kentucky, Ettensohn and others (1984) changed the name of the Upper Mississippian Greenbrier to the Slade Formation. In its lower part, the Slade Formation is composed largely of yellowish gray dolomite that has subordinate amounts of calcarenite and calcilutite, greenish gray shale, and dolomitic siltstone. The overlying unit consists of as much as 10 m of abundantly fossiliferous light-gray calcirudite and calcarenite, associated with some light-gray to greenish gray calcilutite and greenish gray shale. Chert nodules occur abundantly, and the guide fossils *Lithostrotion proliferum* and *Lithostrotionella castelnaui* occur in many places. Next above is a unit as much as 28 m thick that consists largely of olive-gray, oolitic, bioclastic calcarenite (Ettensohn and others, 1984, p. 9). The guide fossil *Platycrinites penicillus* indicates correlation with part of the Greenbrier Limestone in Virginia and West Virginia and with the lower part of the Monteagle Limestone in Tennessee. The uppermost member is an extensive sheet-like deposit composed largely of light-gray quartzose pelletoidal to oolitic calcarenite and lesser amounts of olive-gray calcilutite. The unit ranges in thickness from a featheredge to about 30 m.

Ohio. The Meramecian sequence thins northward into east-central Ohio (deWitt and McGrew, 1979, Plate 5-A), where only 2 to 3 m of brownish gray dolomite and dolomitic limestone that has a St. Louis conodont fauna (Scatterday, 1963; Collins, 1979, p. E13, E19) makes up the sequence. This sequence thickens and becomes sandy in southeastern Ohio.

Throughout much of the western part of the basin, deposition was continuous from Meramecian to Chesterian time, and the boundary between the two sequences of carbonate rock is defined paleontologically. Also, subdivision of the deltaic Mauch Chunk red beds of eastern Pennsylvania into Meramecian and Chesterian units is not feasible.

Chesterian

Early in the Chesterian, the transgressing sea reached its maximum and flooded across the foreland basin and western unstable shelf. Later, siliciclastic sediment from the Mauch Chunk delta complex prograded westward, filling the basin and spreading onto the platform. The extensive tectonic source area to the east supplied vast amounts of terrigenous detritus as sedimentation exceeded subsidence in the Appalachian Basin and subaerial sediments displaced marine sediments westward during the closing phase of Mississippian deposition.

Pennsylvania. In the Valley and Ridge of eastern Pennsylvania, the Mauch Chunk Formation consists of 1,830 to 2,750 m of continental siliciclastic red beds. It thins westward to little more than 300 m in the Broadtop Basin and 185 to 250 m along the Allegheny Front in south-central Pennsylvania and contiguous parts of Maryland. Several 3- to 12-m tongues of gray, greenish gray to red, argillaceous, silty, fossiliferous calcarenite are interbedded in the lower part of the argillaceous Mauch Chunk in southwestern Pennsylvania. These limestones are the northern distal edge of the thick wedge of Meramecian and Chesterian carbonate rocks that underlie the Appalachian Plateau and parts of the Valley and Ridge to the south. In western Pennsylvania, the Mauch Chunk thins northward depositionally and was scoured away in part by pre-Pennsylvanian or early Pennsylvanian erosion in much of the northern and west-central parts of the state (Edmunds and others, 1979, p. B21).

Virginia. In the Greendale Syncline, Chesterian rocks are about 1,375 m thick (Butts, 1940; Averitt, 1941). The lower half of the sequence is mainly shallow-marine to supratidal argillaceous, fine- to coarse-textured limestone that contains some shale and quartz sand in the basal part. The amount of shale, silt, and quartz sand increases in the upper half of the carbonate sequence. The limestone is separated into two parts by about 15 m of red, coarse-grained, calcareous quartz sandstone, the Fido Sandstone, which is restricted to the Greendale Syncline and is not present along the adjacent Allegheny Front (Wilpolt and Marden, 1959). The amount of siliciclastic detritus in the limestone and the red color of some beds suggests proximity of the syncline to the eastern source of clastic detritus. The upper half of the Chesterian sequence in the Greendale Syncline, the Pennington Formation, consists of more than 670 m of red, greenish gray and gray shale that is intercalated with white, gray, and red fine- to coarse-grained quartz sandstone. Commonly, the sandstone is ripple marked and cross bedded. A few beds of sandy and silty argillaceous limestone are present, mainly in the lower part of the formation. The Pennington of the Greendale Syncline shows the shift from marine to littoral to terrestrial deposition through a thick sequence of intertonguing marine and subaerially deposited coal-bearing sediment along the front of the Mauch Chunk delta complex.

West Virginia. In northern West Virginia, the Chesterian sequence is relatively thin and consists mainly of limestones in the Greenbrier Group north of the Catskill Island (Arkle and others, 1979). South of this island, the sequence thickens to more than 1,190 m northeast of the local Saltville Basin, which trapped Lower Mississippian evaporites. Southwest along the Allegheny front, Chesterian rocks thin laterally, mainly by a decrease of sediment within stratigraphic units and by gradation of shaly units of the Pennington Group into coarse-grained rocks of the Lee Formation (Englund, 1979) of Mississippian and Pennsylvanian age.

The lower part of the Chesterian sequence south of the Catskill Island is about 185 m thick in the southeastern part of the area and thins to about 30 m near the western tip of Virginia. There, and in the adjacent subsurface of West Virginia and southeastern Kentucky, the sequence consists predominantly of shallow-water to supratidal limestone and some interbedded dolomite and traces of nodular anhydrite. Oolite and bioclastic calcarenite are interbedded with lagoonal and tidal-flat calcilutite (deWitt and McGrew, 1979; Englund, 1979; Milici and others, 1979; Arkle and others, 1979; Sable, 1979; Ettensohn and others, 1984). Siliciclastic shale, siltstone, and sandstone, particularly in the upper part of the sequence, accumulated as near-shore bar and beach, lagoon, and tidal-flat sediments. In southwestern Virginia and adjacent West Virginia, a transitional sequence, consisting of shaly fossiliferous limestone and calcareous gray to red shale that include some interbedded quartz sandstone and impure coal in the upper part, constitutes the Bluefield Formation, which is 75 to 300 m thick along the Allegheny Front. The Bluefield separates the dominantly calcareous Greenbrier Group below from the siliciclastic Pennington Group above. Southward in eastern Tennessee, the transitional Bluefield is included in the upper part of the Newman Limestone (Milici and others, 1979). The Bluefield marks the westward and southwestward progradation of Mauch Chunk siliciclastic sediment into the central part of the Appalachian trough during the middle Late Mississippian, apparently in response to accelerated uplift and erosion of the eastern tectonic source areas.

The Pennington Group of dominantly siliciclastic detritus filled the Appalachian foreland basin and spread westward across the western unstable shelf during the latter part of the Late Mississippian. In southwestern West Virginia and adjacent Virginia, the Pennington is more than 915 m thick. It consists of several extensive sheets of fine- to coarse-grained, locally pebbly, quartzose sandstone in a sequence of gray, red, and black terrestrial shales. Several widespread beds of gray, fossiliferous, argillaceous limestone are intercalated with calcareous fossiliferous shale, mainly in the lower unit (Arkle and others, 1979; Englund, 1979). The Pennington is the shoreline to low-subaerial delta-plain facies of the Mauch Chunk delta complex. The Pennington thins to the north and west under the Appalachian Plateau, and individual members and formations lose their identity in the thinning sequence by loss of clastic detritus and perhaps by erosional unconformity at its top (Arkle and others, 1979).

Kentucky. In eastern Kentucky, the Pennington has been redefined and renamed the Paragon Formation (Ettensohn and others, 1984). There, it consists mainly of gray and red mudrock and shale associated with subordinate amounts of siltstone, sandstone, limestone, and a few coal beds, which accumulated in environments ranging from offshore-bar to low-subaerial delta plain along the western face of the prograding Mauch Chunk delta complex. Bar and lagoon deposits merge eastward into tidal-flat, estuarine-marsh, stream-channel, overbank, and lower delta-plain swamp deposits. The Paragon Formation thins to the north, either by grading laterally into limestone or cut out beneath a pre-Pennsylvanian unconformity in northeastern Kentucky.

Tennessee. In the Valley and Ridge of eastern Tennessee, the Chesterian part of the Newman Limestone is transitional from shallow-marine carbonates to siliciclastic Mauch Chunk detritus that prograded into the central segment of the Appalachian trough.

To the south along the Allegheny Front and in the Sequatchie anticline of east Tennessee, limestone of the Monteagle is overlain by as much as 28 m of ripple- to cross-bedded calcareous quartz sandstone and interbedded silty greenish gray to gray shale—the Hartselle Sandstone (Milici and others, 1979). The Hartselle thins and becomes more shaly to the northeast. Overlying the Hartselle is 25 to 150 m of oolitic to bioclastic, brownish gray to gray, cherty, argillaceous limestone, the Bangor Limestone. It is thickest in southeastern Tennessee and thins to the north and west into northeastern Tennessee (Sedimentation Seminar, 1981) and southern Kentucky. Like the Monteagle, the Bangor was deposited mainly in shallow water, in environments ranging from high-energy bar and shoal to low-energy lagoon and tidal flat.

Overlying the Bangor is the Pennington Formation, a sequence of argillaceous limestone; dolomite; red, gray, or green shale; siltstone; and fine-grained to conglomeratic quartz sandstone 30 to 150 m thick. Pennington rocks grade vertically from shallow-water carbonate strata into beach and tidal-flat deposits. The formation is thickest on the east along the Allegheny front, where it contains the greatest quantity of siliciclastic detritus. Depositional environments recognized in the Pennington of Tennessee include shallow marine, offshore bar, beach, lagoon, tidal flat channel, and overbank splay (Milici, 1974). Tongues of conglomeratic sandstone in the Pennington of northeastern Tennessee are portions of the Lee Formation (Englund and DeLaney, 1966; Englund, 1979), which is a part of the complex, marginal marine, depositional framework in the south central portion of the Appalachian trough.

The boundary between the marine and nearshore rocks of the Pennington and the overlying terrestrial coal-bearing beds of the Gizzard Group of southern Tennessee, is placed arbitrarily at the top of the youngest red bed in the sequence. In northeastern Tennessee, quartz-pebble conglomerates are intercalated vertically and horizontally with red beds, and delineation of the systemic boundary becomes a problem (Milici and others, 1979).

Ohio. In central and south central Ohio, the Chesterian Maxville limestone ranges in thickness from an erosional edge to as much as 55 m. It consists of thinly bedded to massive, pink, brown, olive-gray to grayish black calcilutite and fossiliferous calcarenite. Some moderate-gray to greenish gray shale is interbedded, particularly in the upper part of the sequence. The Maxville was deposited in a shallow-marine environment. In general, the limestone thins to the north, although in places extensive pre-Pennsylvanian erosion appears to have cut deeply into or through the sheet, dissecting it into many blocks of irregular shape and thickness.

PENNSYLVANIAN AND PERMIAN

Mississippian-Pennsylvanian boundary

In general, Carboniferous strata of the Appalachian Basin record the transition from marine carbonate shelf deposits through a sequence of red and green shales, sandstones, and

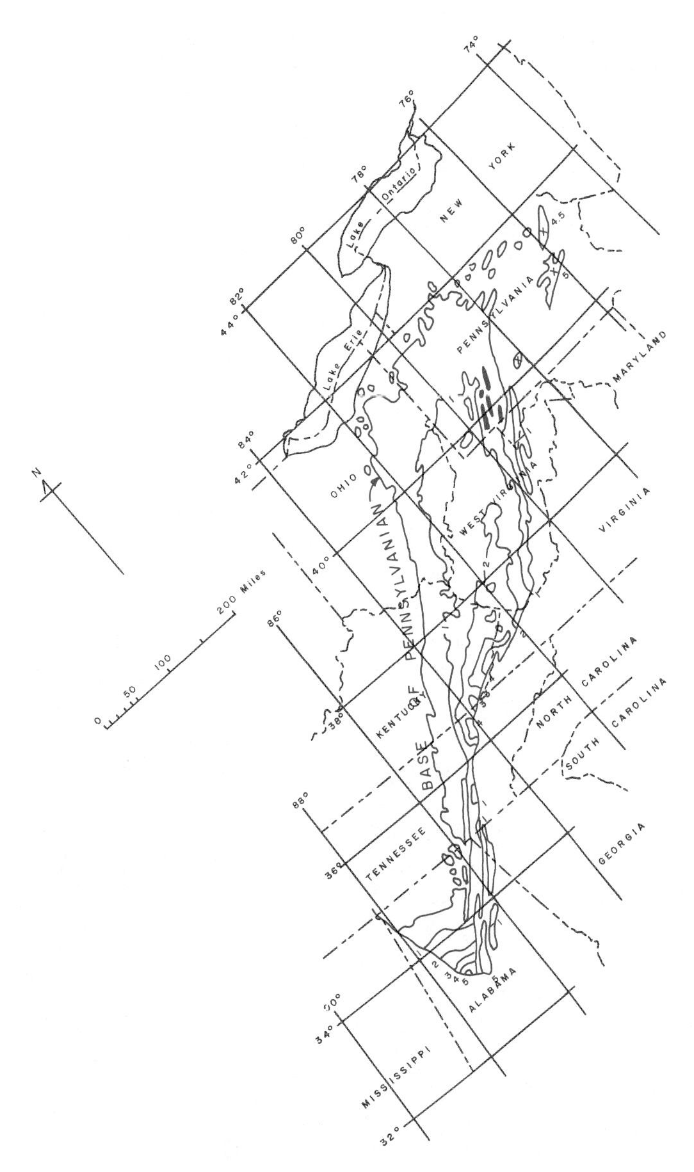

Figure 8. Thickness of Pennsylvanian rocks in the Appalachian Basin. Isopachs are in feet. (From de Witt, 1975.)

minor carbonate rock into coal-bearing shales, siltstones, sandstones, and conglomerates (Fig. 8). Siliciclastic deposition in the upper part of the Mississippian and in the Pennsylvanian reflects the tectonic uplift and erosion of previously deposited Paleozoic strata, as well as of older stratified rocks and Precambrian crystalline basement, which resulted from Alleghanian continental collision (Davis and Erhlich, 1974). During this final phase of sedimentation and tectonism, the basin was bordered on the north by cratonic rocks of the Canadian Shield and the Adirondack Arch, on the east and south by tectonic lands, and on the west by the broad, low, carbonate-covered Cincinnati Arch.

In many places on the eastern side of the Appalachian trough, deposition was continuous from Late Mississippian into Early Pennsylvanian, and the systemic boundary lies within a sequence of similar-appearing rocks. Accordingly, the boundary has been placed arbitrarily within sequences ranging from the southern Anthracite basin in eastern Pennsylvania, to southwestern West Virginia, Virginia, and Tennessee, and to the deep coal basins of Alabama. In contrast, erosion along the northern and western parts of the Appalachian Basin during latest Mississippian or earliest Pennsylvanian time produced a topography that has some relief, so that Pennsylvanian rocks appear to fill paleochannels scoured into or through Mississippian rocks. The systemic boundary is marked by juxtaposed contrasting facies and major differences in fossil fauna and flora (de Witt and McGrew, 1979; Edmunds and others, 1979). In the northeastern edge of the Anthracite district of eastern Pennsylvania and in southwestern New York, coarse-grained, pebbly Pennsylvanian rocks lie on Devonian red beds in the east and on greenish gray fossiliferous rocks in the west. In northeastern Ohio and also in northeastern Kentucky, pebbly Pennsylvanian sandstones fill channels cut into the Lower Mississippian strata. Thus, on the eastern side of the Appalachian Basin, the Mississippian-Pennsylvanian boundary is difficult to locate, whereas along the western periphery of the basin, the boundary is marked in many places by strong contrasts in facies and fossils.

Where the boundary between Carboniferous marine strata and coal-bearing strata is transitional, marine beds grade both laterally and vertically into littoral or deltaic deposits (Ferm and others, 1972; Ferm, 1974; Milici, 1974; Englund, 1979). For example, where coal-bearing strata are interbedded with marine beds containing Mississippian fossils (Milici, 1974), and where thick orthoquartzite lenses of the Lee are interstratified with limestones and shales of the Pennington Formation (Englund, 1968), the boundary is transitional. Elsewhere, in large areas of eastern Kentucky and West Virginia, the nature of the boundary between marine units classified as Mississippian and coal-bearing strata classified as Pennsylvanian is interpreted differently by different workers (compare Ferm and others, 1971; Horne and Ferm, 1976 with Englund, 1979; Englund and Henry, 1979; Rice and others, 1979; Rice, 1984).

Upper Mauch Chunk, Pottsville, and equivalents

During Late Mississippian and Early Pennsylvanian time, siliciclastics were transported into the northern part of the Appalachian Basin from tectonic lands to the east and from cratonic sources to the north (Edmunds and others, 1979). The Mauch Chunk Formation, which attains a maximum thickness of 2,440 to 2,745 m in eastern Pennsylvania, consists generally of red shale, siltstone, and some gray sandstone. Its lower beds intertongue with underlying Mississippian limestones and in the Anthracite district the upper part of the Mauch Chunk is a facies equivalent of Pottsville sandstones and conglomerates (Table 7). Elsewhere in Pennsylvania, the Pottsville–Mauch Chunk relationship is uncertain and has been interpreted in some places as gradational and in other places as erosional. The Mauch Chunk is restricted to the eastern part of the Appalachian Basin in Pennsylvania. In extreme western Pennsylvania and in Ohio, beds below the Mauch Chunk are overlain by Pottsville strata (Arkle, 1974; Edmunds and others, 1979).

Another major area of Carboniferous siliciclastic sedimentation was the Pocahontas delta in parts of southwestern Virginia and adjacent West Virginia, where Pennsylvanian strata are 1,830 m thick. From there, siliciclastic sediments migrated to the west, south, and north. The sediments merged in northern Alabama with sediments from more southerly sources (Ferm, 1974; Hobday, 1974) and in the northern part of the basin with those from both the Canadian Shield and the Pottsville delta (Donaldson, 1974; Edmunds and others, 1979).

Englund (1974) shows that the Pocahontas Formation in Virginia and West Virginia consists of laterally equivalent sandstones of deltaic and marine origin. The Pocahontas ranges up to a maximum thickness of about 230 m in its outcrop area in Virginia (Englund, 1979). It overlies and is interbedded with the Bluestone Formation (Mississippian and Pennsylvanian) and is overlain by the New River Formation (Englund, 1974, 1979). The upper part of the Bluestone, the Pocahontas, and the overlying New River are replaced laterally to the southeast by the Lee Formation, an orthoquartzite-bearing unit about 485 m thick (Miller, 1974; Englund, 1979).

In Virginia, the Lee is overlain by the Norton Formation, which is mostly shale and siltstone, with lesser amounts of sandstone. The Norton increases in thickness from 165 to 600 m northeastward across southwestern Virginia. The overlying Gladeville Sandstone consists of about 15 m of quartzose sandstone. Next above is the Wise Formation, which is about 580 m thick and consists chiefly of immature sandstone, siltstone, shale, and some limestone. The uppermost Pennsylvanian Formation in Virginia, the Harlan Formation, is about 200 m thick and consists mostly of massive sandstones that are overlain by beds of shale, siltstone, and coal (Miller, 1969; Englund, 1979).

In Tennessee, siliciclastics containing orthoquartzites are overlain by sequences containing less-mature sandstones (Milici and others, 1979); the former are widespread over the Cumberland Plateau, whereas the latter are restricted to the Wartburg Basin in the northern Tennessee Plateau. Orthoquartzites and associated finer-grained siliciclastic formations are classified into the Gizzard and Crab Orchard Mountains Groups. The overlying Crooked Fork Group contains a half-dozen interbedded shale and sandstone formations; in turn it is overlain by a half-dozen more formations composed chiefly of shale (American Association of Petroleum Geologists, 1985).

In eastern Kentucky, Pennsylvanian strata are divided into the upper part of the Pennington, Lee, Breathitt, Conemaugh, and Monongahela Formations (Table 7). The upper part of the Pennington Formation contains a Pennsylvanian flora in southeastern Kentucky, and the Mississippian–Pennsylvanian boundary lies below these beds in a gradational zone that contains marine faunas of Mississippian age (Rice and others, 1979). Based on sandstone composition, bed forms, and stratigraphic relationships, Horne and others (1971), Ferm and others (1971), Ferm (1974), and Horne and others (1974) arranged the Newman, Pennington, Lee, and Breathitt Formations in eastern Kentucky into a progradational sequence from offshore bars, barrier islands, lagoons, and delta-plain deposits to those of the alluvial plain. Rice and others (1979, p. F16) rejected this interpretation largely because "the marine rocks that should be associated with a beach or barrier system, are rare in the Lee Formation or intercalated Breathitt strata." Instead, Rice (1984) proposed that Lee orthoquartzites fill southeastward-trending paleovalleys.

Sediments spread from the area of the Pocahontas delta across the Appalachian Basin; marine limestone grades both horizontally and vertically through red and green siliciclastics, first into orthoquartzite beach barrier facies and then into lower and upper delta-plain and alluvial-plain facies (Ferm and Cavaroc, 1969; Ferm, 1974). The Carboniferous sedimentary sequence thins in the direction of progradation, from the margin of the geosyncline onto the more stable shelf. Donaldson (1974) generally concurs with the Ferm-Cavaroc model for New River–Pottsville deposition, pointing out that Early Pennsylvanian quartz-rich sandstones contain marine fossils and "interfinger with underlying marine red beds and overlying coal beds. . . ."

The Ohio section, like that in West Virginia, is divided into the Pottsville, Allegheny, Conemaugh, Monongahela, and Dunkard Groups. The Pottsville averages 78 m thick and consists chiefly of conglomerates, sandstones, shales, coals, and marine limestone beds. A basal Sharon Conglomerate and some of the units above are deposited in channels or valleys cut into underlying Mississippian limestones in the southern part of the state (Collins, 1979) and into the Cuyahoga in the northeastern part of the state.

Allegheny, Conemaugh, Monongahela, and equivalents

Overlying the Pottsville in the Anthracite region is the Llewellen Formation, a nonmarine unit composed of conglomerate, sandstone, siltstone, shale, and coal that ranges up to 1,070 m thick. In the Plateau region to the west, the Pottsville is overlain by the Allegheny Group, about 80 to 100 m thick, which is partly marine in its lower half. The Allegheny, as well as the overlying Conemaugh, Monongahela, and Dunkard Groups, consists largely of sandstone, siltstone, and shale beds associated with coals and some limestones. The Conemaugh Group ranges from 170 to 275 m and, like the Allegheny, contains marine zones in its lower part; its upper part is nonmarine. The overlying Monongahela, with the Pittsburgh coal at its base, ranges from 85 to 115 m thick and is entirely nonmarine in Pennsylvania (Edmunds and others, 1979).

In Ohio, lower delta-plain strata of the Allegheny Group average about 65 m thick and contain much of the state's coal resources. The Conemaugh, 122 m thick, has only a few commercial coal beds and consists largely of sandstones, shales, mud-

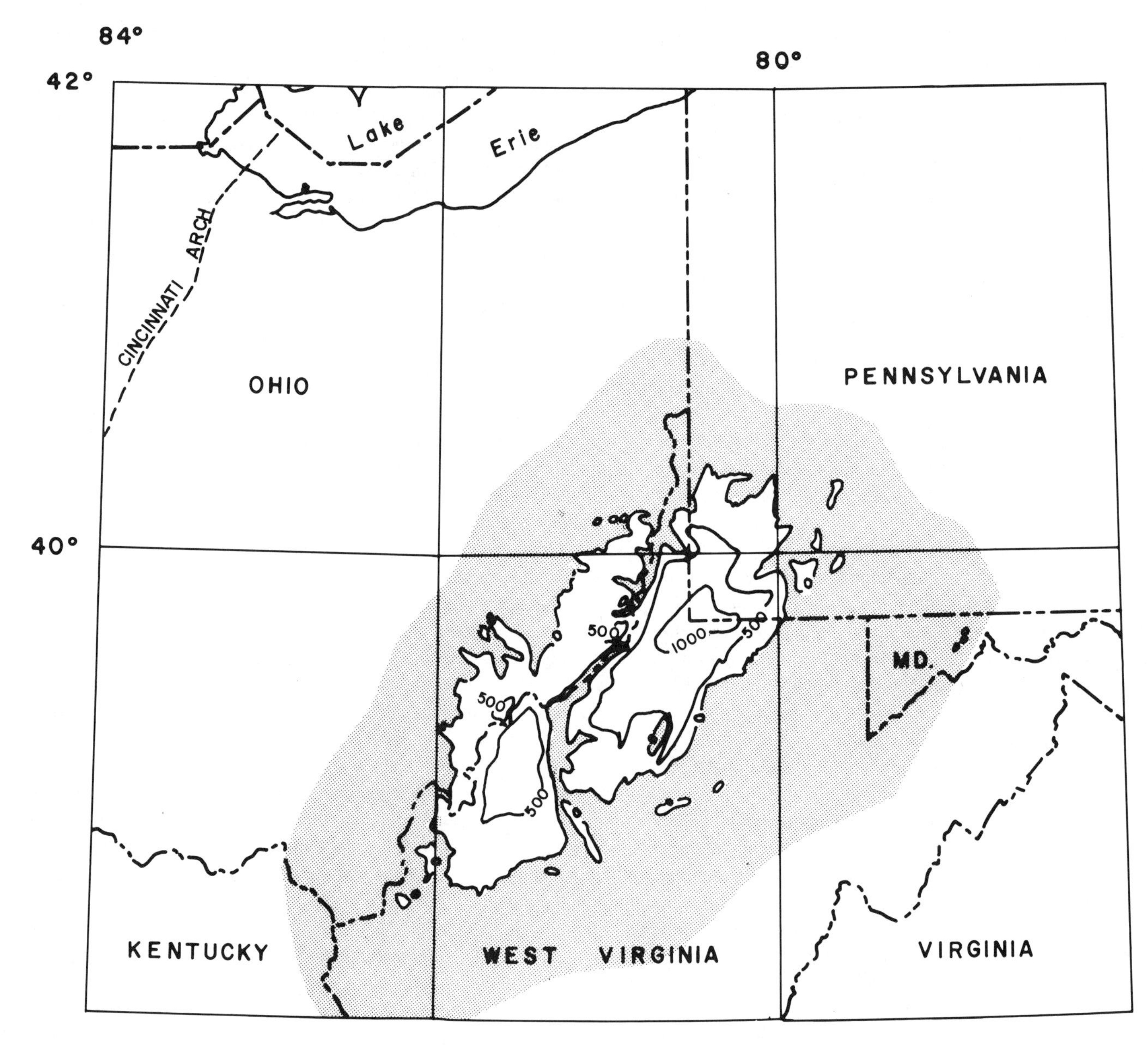

Figure 9. Thickness of Permian rocks in the Appalachian Basin. Isopachs are in feet. (From de Witt, 1975.)

stones, and some marine limestones. Some of the shales and mudstones are red. In its upper part, the Conemaugh becomes more continental in nature and contains freshwater limestone. The overlying Monongahela averages 77 m thick and in Ohio consists of three facies: gray shale with associated limestone and thick coal to the north, red and yellow mudstone and sandstone to the south, and in central eastern Ohio a transitional zone that contains lithologies of both end members.

In West Virginia, Allegheny strata consist of about 30 to 90 m of subgraywacke, shale, and mudstone, associated with several thick coal beds. The Conemaugh Group, about 140 to 260 m thick, consists mostly of red and gray shales and mudstones, and some beds of subgraywacke, marine limestone, and coal. The Monongahela Group is composed of 70 to 120 m of shale, mudstone, subgraywacke, lacustrine limestone, and coal. The Pittsburgh coal bed lies at the base of the group. In some places, the Monongahela Group changes laterally from red beds to lacustrine swamp deposits (Arkle and others, 1979). Donaldson (1974, p. 48–50) interprets Allegheny, Conemaugh, and Monongahela strata to have been deposited by a shallow-water delta adjacent to a strand plain.

In eastern Kentucky, Allegheny equivalents are included in

the upper part of the Breathitt Formation. The Breathitt ranges from 250 to 950 m thick and consists mostly of siltstone, shale, subgraywacke, and some marine limestone. In general, the formation contains more sandstone in its upper part; it includes most of the mineable coal in eastern Kentucky. The overlying Conemaugh and Monongahela Formations generally are not separated in Kentucky. Together, they are about 175 m thick and consist chiefly of siltstone and shale together with some subgraywacke, limestone, and coal. Red and green shales are conspicuous in these units (Rice and others, 1979).

Dunkard

In Pennsylvania and West Virginia, only the lower 335 m of the Dunkard Group are preserved (Fig. 9). The Dunkard consists of impure sandstones, red shales, mudstones, and lacustrine limestones and a few beds of coal. The lower part of the group is classified as Late Pennsylvanian and Early Permian because paleontologic evidence has been viewed as insufficient to definitively place the systemic boundary (Arkle and others, 1979; Berryhill and others, 1971). However, based on palynological data, Clendenning (1975) concluded that the entire Dunkard is Pennsylvanian.

In Ohio, the Dunkard Group consists of about 190 m of variable amounts of sandstone and conglomerate that grade laterally into calcareous shale or mudstone. As in Pennsylvania and West Virginia, red colors locally predominate (Collins, 1979).

Berryhill and others (1971) concluded that these uppermost Pennsylvanian and Permian beds were deposited in an extensive shallow lake or swamp. Water and sediments entered the lake and swamp from highlands to the north, east, and south.

The final phase of Appalachian sedimentation ended when continental collision on a grand scale deformed and uplifted the Appalachian Basin. During the Mesozoic and Cenozoic, following the breakup of Pangaea and the development of the passive Atlantic margin, much of the younger Appalachian stratigraphy was eroded, thereby extensively exposing the older strata and the internal structure of the mountain chain.

SUMMARY

Although the general stratigraphic framework of the Appalachian Basin has been known for many years, its development was placed in a plate tectonic context only during the past two decades. The timing and nature of Late Proterozoic and Paleozoic plate tectonic events in the crystalline hinterland of the Appalachians relate directly to the evolution of the basin.

Tectonic reasons for the Late Proterozoic separation of a protocontinent and formation of the Iapetus Ocean are little understood. The crustal structure of Iapetus, whether continental, transitional, or newly formed oceanic crust, is as yet conjectural. Late Proterozoic–early Paleozoic strata, which are hypothesized to fill extensional basins because of their great thickness and associated mafic igneous rocks, bear little resemblance to Mesozoic strata presumably deposited in a similar tectonic setting.

The Taconic Orogeny, perhaps the result of the collision of an island arc with the eastern edge of the ancient North American continent, initiated emergence of tectonic lands that shed sediment westward into subsiding foreland basins. Post-orogenic molasse was followed by carbonate and salt deposition in platform environments. Acadian continental collision produced the great Catskill delta, its sediments again derived from tectonically formed lands, but this time in the northern part of the mountain chain. As the mountains eroded, carbonate sedimentation replaced siliciclastics throughout most of the basin, until the terminal continental convergence and collision produced the great wedge of Carboniferous siliciclastic strata.

Problems exist at all scales. Regional stratigraphic relationships in the Appalachian Valley and Ridge are not much better understood than they were a generation ago. Interregional correlations between the exposed Paleozoic rocks in the mountains and the subsurface strata along the Cincinnati Arch to the west have not been documented satisfactorily. Currently, little effort is being made to refine systematically the stratigraphy and paleontology of much of the Appalachians, particularly by those preparing detailed geologic quadrangle maps. Certainly, a better understanding of the Appalachian Basin will require an integrated approach, involving stratigraphic and biostratigraphic studies that are closely related to careful and detailed geologic quadrangle mapping. A detailed stratigraphic framework is essential, too, for paleoenvironmental and paleogeographic studies, as well as for petrologic studies, that would better define the nature and development of siliciclastic source areas. The process is iterative. Studies of the tectonics of the continental margin yield data pertinent to the development of the Appalachian Basin, and basin studies better refine the history of lands contiguous to its margin.

REFERENCES CITED

Aaron, J. M., 1969, Petrology and origin of the Hardyston quartzite (Lower Cambrian) in eastern Pennsylvania and western New Jersey, *in* Subitzky, S., ed., Geology of selected areas in New Jersey and eastern Pennsylvania and guidebook of excursions: New Brunswick, New Jersey, Rutgers University Press, p. 23–34.

Adams, R. W., 1970, Loyalhanna Limestone; Cross-bedding and provenance, *in* Fisher, G. W., and others, Studies of Appalachian geology, central and southern: New York, Wiley-Interscience Publishers, p. 98 and Fig. 16.

Allen, R. M., Jr., 1967, Geology and mineral resources of Page County: Virginia Division of Mineral Resources Bulletin 81, 78 p.

Alling, H. I., and Briggs, L. I., 1961, Stratigraphy of Upper Silurian Cayugan evaporites: American Association of Petroleum Geologists Bulletin, v. 45, p. 515–547.

American Association of Petroleum Geologists, 1985, Correlation of stratigraphic units of North America (COSUNA) Project, southern Appalachian region: American Association of Petroleum Geologists correlation chart.

Anderson, E. J., 1971, Interpretation of calcarenite paleoenvironments: Eastern Section of Society of Economic Paleontologists and Mineralogists Field Conference Guidebook, 67 p.

Arkle, T., Jr., 1974, Stratigraphy of the Pennsylvania and Permian systems of the central Appalachians: Geological Society of America Special Paper 148, p. 5–29.

Arkle, T., Jr., and 9 others, 1979, The Mississippian and Pennsylvanian (Carboniferous) systems in the United States; West Virginia and Maryland: U.S. Geological Survey Professional Paper 1110-D, p. D1–D35.

Averitt, P., 1941, The Early Grove gas field, Scott and Washington Counties, Virginia: Virginia Geological Survey Bulletin 56, 50 p.

Bartholomew, M. J., Force, E. R., Sinha, A. K., and Herz, N., eds., 1984, The Grenville event in the Appalachians and related topics: Geological Society of America Special Paper 194, 287 p.

Bassler, R. S., 1932, The stratigraphy of the Central Basin of Tennessee: Tennessee Division of Geology Bulletin 38, 268 p.

Bentall, R., and Collins, J. B., 1945, Subsurface stratigraphy and structure of the pre-Trenton Ordovician and Upper Cambrian rocks in central Tennessee: Tennessee Division of Geology, Oil, and Gas Investigations, Preliminary Chart no. 4.

Berg, T. M., McInerney, M. K., Way, J. H., MacLachlan, D. B., 1983, Stratigraphic correlation chart of Pennsylvania: Pennsylvania Bureau of Topographic and Geologic Survey General Geology Report 75.

Berryhill, H. L., Jr., Schweinfurth, S. P., and Kent, B. H., 1971, Coal-bearing Upper Pennsylvanian and Lower Permian rocks, Washington area, Pennsylvania: U.S. Geological Survey Professional Paper 621, 47 p.

Bowen, Z. P., 1967, Brachiopoda of the Keyser Limestone (Silurian–Devonian) of Maryland and adjacent areas: Geological Society of America Memoir 102, 103 p.

Bretsky, P. W., 1969, Central Appalachian Late Ordovician communities: Geological Society of America Bulletin, v. 80, p. 193–212.

Brett, C. E., 1983, Sedimentology, facies, and depositional environments of the Rochester Shale (Silurian; Wenlockian) in western New York and Ontario: Journal of Sedimentary Petrology, v. 53, no. 3, p. 947–971.

Bridge, J., 1955, Disconformity between Lower and Middle Ordovician series at Douglas Lake, Tennessee: Geological Society of America Bulletin, v. 66, p. 725–730.

—— , 1956, Stratigraphy of the Mascot–Jefferson City zinc district, Tennessee: U.S. Geological Survey Professional Paper 277, 76 p.

Broadhead, R. F., Kepferle, R. C., and Potter, P. E., 1982, Stratigraphic and sedimentologic controls of gas in shale; Example from Upper Devonian of northern Ohio: American Association of Petroleum Geologists Bulletin, v. 66, no. 1, p. 20.

Butts, C., 1926, The Paleozoic rocks, *in* Adams, G. I., Butts, C., Stephenson, L. W., and Cooke, W., Geology of Alabama: Alabama Geological Survey Special Report no. 14, p. 145.

—— , 1940, Geology of the Appalachian Valley in Virginia: Virginia Geological Survey Bulletin 52, part 1, 568 p.

—— , 1945, Description of the Hollidayburg and Huntingdon Quadrangles, Pennsylvania: U.S. Geological Survey Geological Atlas, Folio no. 227, 20 p.

Butts, C., and Edmundson, R. C., 1966, Geology and mineral resources of Frederick County: Virginia Division of Mineral Resources Bulletin 80, 142 p.

Calvert, W. L., 1962, Sub-Trenton rocks from Lee County, Virginia, to Fayette County, Ohio: Ohio Division of Geological Survey Report of Investigations no. 45, 57 p.

Campbell, M. R., and others, 1925, The Valley Coal fields of Virginia: Virginia Geological Survey Bulletin no. 25, 322 p.

Carman, E. J., 1960, The stratigraphy of the Devonian Holland Quarry Shale of Ohio: Fieldiana, v. 14, no. 1, p. 1–5.

Caster, K. E., 1934, The stratigraphy and paleontology of northwestern Pennsylvania; Part 1, Stratigraphy: Bulletin of American Paleontology, v. 21, no. 71, 185 p.

Caster, K. E., and Brooks, H. K., 1956, New fossils from the Canadian–Chazyan (Ordovician) hiatus in Tennessee: Bulletin of American Paleontology, v. 36, no. 157, p. 157–199.

Chadwick, G. H., 1920, The Paleozoic rocks of the Canton Quadrangle: New York State Museum Bulletin 217–18, 60 p.

—— , 1933, Great Catskill delta and revision of Late Devonic succession: Pan-American Geologist, v. 60, no. 2, p. 91–107 and v. 60, no. 5, p. 351.

Clendenning, J. A., 1975, Palynological evidence for a Pennsylvanian age assignment of the Dunkard Group in the Appalachian Basin, Part I, *in* Barlow, J. A., ed., The age of the Dunkard: Proceedings of the First I. C. White Memorial Symposium, West Virginia Geological Economic Survey, p. 195–216.

Cloos, E., 1951, Stratigraphy of sedimentary rocks of Washington County, *in* The physical features of Washington County: Maryland Department of Geology, Mines, and Water Resources, p. 17–94.

Collins, H. R., 1979, The Mississippian and Pennsylvanian (Carboniferous) systems in the United States; Ohio: U.S. Geological Survey Professional Paper 1110-E, p. E1–E26.

Colton, G. W., 1970, The Appalachian Basin; Its depositional sequences and their geologic relationships, *in* Fisher, G. W., Pettijohn, F. J., Reed, J. C., Jr., and Weaver, K. N., eds., Studies of Appalachian geology, central and southern: New York, Interscience Publishers, p. 5–47.

Conant, L. C., and Swanson, V. E., 1961, Chattanooga Shale and related rocks of central Tennessee and nearby areas: U.S. Geological Survey Professional Paper 357, 91 p.

Conley, J. F., 1987, Mafic rocks in the Alligator Back Formation, the upper unit in the Lynchburg Group, in the southwestern Virginia Piedmont; An ophiolite sequence?, *in* Contributions to Virginia geology—V: Virginia Division of Mineral Resources Publication 74, p. 55–68.

Cook, F. A., and Oliver, J. E., 1981, The Late Precambrian–Early Paleozoic continental edge in the Appalachian orogen: American Journal of Science, v. 281, p. 993–1,008.

Cooper, B. N., 1963, Geological excursions in southwestern Virginia; Geological Society of America Southeastern Section Annual Meeting, 1963: Blacksburg, Virginia Polytechnical Institute Engineering Extension Service, Guidebook 2, p. 11–47.

—— , 1966, Geology of the salt and gypsum deposits in the Saltville area, Smyth and Washington Counties, Virginia, *in* Second symposium on salt; Vol. 1, Geology, geochemistry, and mining: Northern Ohio Geological Society, p. 11–34.

Cotter, E., 1983, Shelf, paralic, and fluvial environments and eustatic sea-level fluctuations in the origin of the Tuscarora formation (Lower Silurian) of central Pennsylvania: Journal of Sedimentary Petrology, v. 53, p. 25–49.

Cressman, E. R., 1973, Lithostratigraphy and depositional environments of the Lexington Limestone (Ordovician) of central Kentucky: U.S. Geological Survey Professional Paper 768, 61 p.

Cressman, E. R., and Noger, M. C., 1976, Tidal-flat carbonate environments in the High Bridge Group (Middle Ordovician) of central Kentucky: Kentucky Geological Survey Report of Investigations 18, 15 p.

Currier, L. W., 1935, Zinc and lead region of southwestern Virginia: Virginia Geological Survey Bulletin 43, 122 p.

Davis, M. W., and Ehrlich, R., 1974, Late Paleozoic crustal composition and dynamics in the southeastern United States: Geological Society of America Special Paper 148, p. 171–185.

Demicco, R. V., 1982, Upper Cambrian Conococheague Limestone, *in* Demicco, R. V., and Mitchell, R. W., Facies of the Great American Bank in the central Appalachians, *in* Lyttle, P. J., ed., Central Appalachian geology: American Geological Institute, p. 171–266.

Dennison, J. M., 1976, Appalachian Queenston Delta related to eustatic sea-level drop accompanying late Ordovician glaciation centered in Africa, *in* Bassett, M. G., ed., The Ordovician system: Cardiff, University of Wales Press and National Museum of Wales, p. 107–120.

Dennison, J. M., and Boucot, A. J., 1974, Little War Gap at Clinch Mountain provides standard reference section for Silurian Clinch Sandstone and most nearly complete Devonian section in eastern Tennessee: Southeastern Geology, v. 16, p. 79–101.

de Witt, W., Jr., 1975, Oil and gas data from Upper Paleozoic rocks in the Appalachian Basin; U.S. Geological Survey Map I–917A, scale 1:250,000.

de Witt, W., Jr., and McGrew, L. W., 1979, The Appalachian Basin region, *in* Craig, L. C., and Connor, C. W., eds., Paleotectonic investigations of the Mississippian system in the United States: U.S. Geological Survey Professional Paper 1010–C, part 1, p. 13–48.

de Witt, W., Jr., and Roen, J. B., 1985, Correlation and geographic extent of some Middle and Upper Devonian and Lower Mississippian black shales in the Appalachian Basins: U.S. Geological Survey Bulletin 1605–A, p. A45–A57.

de Witt, W., Jr., Perry, W. J., Jr., and Wallace, L. G., 1975, Oil and gas data from Devonian and Silurian rocks in the Appalachian Basin: U.S. Geological Survey Map I–917B, scale 1:250,000.

Donaldson, A. C., 1974, Pennsylvanian sedimentation of central Appalachians: Geological Society of America Special Paper 148, p. 47–78.

Donaldson, A. C., Heald, M. T., Renton, J. J., and Warshaver, S. M., 1975, Depositional environment of Rome Trough Rocks, Mingo County well, West Virginia [abs.]: American Association of Petroleum Geologists Bulletin, v. 59, p. 1,735.

Drake, A. A., 1965, Carbonate rocks of Cambrian and Ordovician age, Northampton and Bucks Counties, eastern Pennsylvania and Warren and Hunterdon Counties, western New Jersey: U.S. Geological Survey Bulletin 1194–L, 7 p.

Drake, A. A., Jr., and Epstein, J. B., 1967, The Martinsburg Formation (Middle and Upper Ordovician) in the Delaware Valley, Pennsylvania–New Jersey: U.S. Geological Survey Bulletin 1244–H, 16 p.

Ebright, J. R., Fettke, C. R., and Ingham, A. I., 1949, East Fork–Wharton Gas Field, Potter County, Pennsylvania: Pennsylvania Geological Survey, Fourth Series Bulletin M–30, p. 10.

Edmunds, W. E., Berg, T. M., Sevon, W. D., Piotrowski, R. C., Heyman, L., and Rickard, L. V., 1979, The Mississippian and Pennsylvanian (Carboniferous) systems in the United States; Pennsylvania and New York: U.S. Geological Survey Professional Paper 1110–B, p. B1–B33.

Edmundson, R. S., 1945, Industrial limestones and dolomites in Virginia; Northern and central pars of the Shenandoah Valley: Virginia Geologic Survey Bulletin 65, 195 p.

Edmundson, R. S., and Nunan, W. E., 1973, Geology of the Berryville, Stephenson, and Boyce Quadrangles, Virginia: Virginia Division of Mineral Resources Report of Investigations 34, 112 p.

Englund, K. J., 1968, Geology and coal resources of the Elk Valley area, Tennessee and Kentucky: U.S. Geological Survey Professional Paper 572, 59 p.

——, 1974, Sandstone distribution patterns in the Pocahontas Formation of southwest Virginia and southern West Virginia, *in* Briggs, G., ed., Carboniferous of the southeastern United States: Geological Society of America Special Paper 148, p. 31–45.

——, 1979, The Mississippian and Pennsylvanian (Carboniferous) systems in the United States; Virginia: U.S. Geological Survey Professional Paper 1110–C, p. C1–C21.

Englund, K. J., and DeLaney, A. O., 1966, Intertonguing relations of the Lee Formation in southwestern Virginia: U.S. Geological Survey Professional Paper 550–D, p. D47–D52.

Englund, K. J., and Henry, T. W., 1979, The Mississippian–Pennsylvanian boundary in the central Appalachians, *in* Sutherland, P. K., and Manger, W. L., eds., Biostratigraphy: Carbondale, Southern Illinois University Press, p. 330–336.

Epstein, A. G., Epstein, J. B., and Spink, W. J., 1967, Upper Silurian and Lower Devonian stratigraphy of northeastern Pennsylvania, New Jersey, and southeasternmost New York: U.S. Geological Survey Bulletin 1243, 74 p.

Epstein, J. B., and Epstein, A. G., 1969, Geology of the Valley and Ridge province between Delaware water gap and Lehigh gap, Pennsylvania, *in* Subitzky, S., ed., 1969, Geology of selected areas in New Jersey and eastern Pennsylvania and guidebook of excursions: New Brunswick, New Jersey, Rutgers University Press, p. 132–205.

——, 1972, The Shawangunk Formation (Upper Ordovician[?] to Middle Silurian) in eastern Pennsylvania: U.S. Geological Survey Professional Paper 744, 45 p.

Ettensohn, F. R., Rice, C. L., Dever, G. R., Jr., and Chesnut, D. R., 1984, Slade and Paragon Formations; New stratigraphic nomenclature for Mississippi rocks along the Cumberland Escarpment in Kentucky: U.S. Geological Survey Bulletin 1605–B, 37 p.

Faill, R. T., 1985, The Acadian orogeny and the Catskill delta: Geological Society of America Special Paper 201, p. 15–37.

Faill, R. T., and Wells, R. B., 1977, Bedrock geology and mineral resources of the Linden and Williamsport Quadrangles, Lycoming County, Pennsylvania: Pennsylvania Geological Survey Atlas 134ab, Fourth Series, 66 p.

Fauth, J. L., 1968, Geology of the Caledonia Park Quadrangle area, South Mountain Pennsylvania: Pennsylvania Geological Survey Atlas 129a, Fourth Series, 134 p.

Ferm, J. C., 1974, Carboniferous environmental models in eastern United States and their significance, *in* Briggs, G., ed., Carboniferous of the southeastern United States: Geological Society of America Special Paper 148, p. 79–95.

Ferm, J. C., and Cavaroc, V. V., Jr., 1969, A field guide to Allegheny deltaic deposits in the upper Ohio Valley with a commentary on deltaic aspects of Carboniferous rocks in the northern Appalachian Plateau: Ohio Geological Society and Pittsburgh Geological Society Guidebook for the 1969 Spring Field Trip, 21 p.

Ferm, J. C., Horne, J. C., Swinchatt, J. P., and Whaley, P. W., 1971, Carboniferous depositional environments in northeastern Kentucky; Geological Society of Kentucky Guidebook for annual spring field conference, April 1971: Lexington, Kentucky Geological Survey, 30 p.

Ferm, J. C., Milici, R. C., and Eason, J. E., 1972, Carboniferous depositional environments in the Cumberland Plateau of southern Tennessee and northern Alabama: Tennessee Division of Geology Bulletin 33, 32 p.

Fettke, C. R., and Bayles, R. E., 1945, Conemaugh gorge section of Mississippian system southeast of Cramer, Pennsylvania: Pennsylvania Academy of Science Proceedings, v. 19, p. 86–95.

Fisher, D. W., 1954, Lower Ordovician (Canadian) stratigraphy of the Mohawk Valley, New York: Geological Society of America Bulletin, v. 65, p. 71–96.

——, 1965, Mohawk Valley stratigraphy and structure, *in* Guide Book Field Trips in the Schnectady area: New York State Museum and Science Service Educational Leaflet no. 18, p. 1–58.

——, 1977, Correlation of Hadrynian, Cambrian, and Ordovician rocks in New York State: New York State Museum Map and Chart Series n. 25, 75 p.

——, 1982, Synopsis of Ordovician correlations in New York State, *in* Ross, R. J., and others, The Ordovician system in the United States: International Union of Geological Sciences Publication no. 12, p. 45–51.

Folk, R. L., 1960, Petrography and origin of the Tuscarora, Rose Hill, and Keefer Formations, Lower and Middle Silurian of Eastern West Virginia: Journal of Sedimentary Petrology, v. 30, p. 1–58.

Gathright, T. M., II, 1976, Geology of the Shenandoah National Park, Virginia: Virginia Division of Mineral Resources Bulletin 86, 93 p.

Gathright, T. M., II, Henika, W. S., and Sullivan, J. L., III, 1978a, Geology of the Grottoes Quadrangle, Virginia: Virginia Division of Mineral Resources Publication 10, scale 1:24,000.

——, 1978b, Geology of the Mt. Sidney Quadrangle, Virginia: Virginia Division of Mineral Resources Publication 11, scale 1:24,000.

——, 1978c, Geology of the Fort Defiance Quadrangle, Virginia: Virginia Division of Mineral Resources Publication 12, scale 1:24,000.

——, 1978d, Geology of the Crimora Quadrangle, Virginia: Virginia Division of Mineral Resources Publication 13, scale 1:24,000.

Gillette, T., 1947, The Clinton of western and central New York: New York State Museum Bulletin no. 341, 191 p.

Glaeser, J. D., 1963, Catskill reference section and its correlation: *in* Shepps, V. C., Symposium on Middle and Upper Devonian stratigraphy of Pennsylvania and adjacent states: Pennsylvania Geological Survey, Fourth Series, General Geology Report G–39, p. 56–62.

Glover, L., III, 1959, Stratigraphy and uranium content of the Chattanooga Shale in northeastern Alabama, northwestern Georgia, and eastern Tennessee: U.S. Geological Survey Bulletin 1087–E, p. 133–168.

Gohn, G. S., 1976, Sedimentology, stratigraphy, and paleogeography of Lower Paleozoic carbonate rocks, Conestoga Valley, southeastern Pennsylvania [Ph.D. thesis]: Newark, University of Delaware, 315 p.

Goldberg, S. A., Butler, J. R., and Fullager, P. D., 1986, The Bakersville dike swarm; Geochronology and petrogenesis of Late Proterozoic basaltic magmatism in the southern Appalachian Blue Ridge: American Journal of Science, v. 286, p. 403–430.

Harris, A. G., and Repetski, J. E., 1983, Conodonts document continuous to intermittent deposition across the lower-Middle Ordovician boundary; Northern Virginia to Bellefont, PA. [abs.]: Virginia Journal of Science, v. 34, p. 172.

Harris, L. D., 1964, Facies relations of exposed Rome Formation and Conasauga Group of northeastern Tennessee with equivalent rocks in the subsurface of Kentucky and Virginia: U.S. Geological Survey Professional Paper 501–B, p. B25–B29.

—— , 1969, Kingsport Formation and Mascot Dolomite (Lower Ordovician) of east Tennessee, *in* Papers on the stratigraphy and mine geology of the Kingsport and Mascot Formations (Lower Ordovician) of east Tennessee: Tennessee Division of Geology Report of Investigation 23, p. 1–39.

—— , 1975, Oil and gas data from the Lower Ordovician and Cambrian rocks of the Appalachian Basin: U.S. Geological Survey Miscellaneous Investigations Series Map I–917D, scale 1:2,500,000.

—— , 1978, The eastern interior aulacogen and its relation to Devonian shale-gas productions, *in* Second Eastern Gas Shales Symposium Preprint: Morgantown, West Virginia, U.S. Department of Energy, Morgantown Energy Technology Center, p. 55–72.

Harris, L. D., and Milici, R. C., 1977, Characteristics of thin-skinned style of deformation in the southern Appalachians, and potential hydrocarbon traps: U.S. Geological Survey Professional Paper 1018, 40 p.

Harris, L. D., de Witt, W., Jr., and Colton, G. W., 1978, What are possible stratigraphic controls for gas fields in Eastern black shale?: Oil and Gas Journal, v. 76, no. 14, p. 162–165.

Harvey, E. J., and Maher, S. W., 1948, Lithologic and primary structure features in the Rome Formation: Tennessee Academy of Science Journal, v. 23, p. 283–290.

Hass, W. H., 1956, Age and correlaton of the Chattanooga Shale and the Maury Formation: U.S. Geological Survey Professional Paper 286, 47 p.

Head, J. W., 1974, Correlation and paleogeography of upper part of Helderberg Group (Lower Devonian) of central Appalachians: American Association of Petroleum Geologists Bulletin, v. 58, p. 247–259.

Heckel, P. H., 1973, Nature, origin, and significance of the Tully Limestone: Geological Society of America Special Paper 138, 224 p.

Hobday, D. K., 1974, Beach- and barrier-island facies in the Upper Carboniferous of northern Alabama, *in* Briggs, G., ed., Carboniferous of the southeastern United States: Geological Society of America Special Paper 148, p. 209–223.

Hobson, J. P., Jr., 1963, Stratigraphy of the Beekmantown Group in southeastern Pennsylvania: Pennsylvania Geological Survey Bulletin G 37, Fourth Series, 331 p.

Horne, J. C., and Ferm, J. C., 1976, A field guide to Carboniferous depositional environments in the Pocahontas Basin, eastern Kentucky and southern West Virginia: Columbia, University of South Carolina Department of Geology, 129 p.

Horne, J. C., Swinchatt, J. P., and Ferm, J. C., 1971, Lee–Newman barrier shoreline model, *in* Carboniferous depositional environments in northeastern Kentucky: Geological Society of Kentucky Guidebook for Annual Spring Field Conference, p. 5–9.

Horne, J. C., Ferm, J. C., and Swinchatt, J. P., 1974, Depositional model for the Mississippian–Pennsylvanian boundary in northeastern Kentucky: Geological Society of America Special Paper 248, p. 97–114.

Horowitz, D. H., 1966, Evidence for deltaic origin of an Upper Ordovician sequence in the central Appalachians, *in* Shirley, M. L., ed., Deltas in their geologic framework: Houston Geological Society, p. 159–169.

Hoskins, D. M., 1961, Stratigraphy and paleontology of the Bloomsburg Formation of Pennsylvania and adjacent states: Pennsylvania Geological Survey Bulletin G 36, 125 p.

—— , 1976, Geology and mineral resources of the Millersburg 15-minute Quadrangle, Dauphin, Juniata, Northumberland, Perry, and Snyder Counties, Pennsylvania: Pennsylvania Geological Survey Atlas 146, Fourth Series, 38 p.

Howell, B. F., Roberts, H., and Willard, B., 1950, Subdivision and dating of the Cambrian of eastern Pennsylvania: Geological Society of America Bulletin, v. 61, p. 1355–1368.

Janssens, A., 1973, Stratigraphy of the Cambrian and Lower Ordovician rocks in Ohio: Ohio Division of Geological Survey Bulletin 64, 197 p.

—— , 1977, Silurian rocks in the subsurface of northwestern Ohio: Ohio Division of Geological Survey Report of Investigations no. 100, 96 p.

Jonas, A. I., and Stose, G. W., 1930, Geology and mineral resources of the Lancaster Quadrangle: Pennsylvania Geological Survey Topographic and Geologic Atlas 168, Fourth Series, 106 p.

—— , 1936, The reclassification of the Frederick Valley (Maryland) limestones: Geological Society of America Bulletin, v. 47, p. 1657–1674.

Jones, M. L., and Dennison, J. M., 1970, Oriented fossils as paleocurrent indicators in Paleozoic lutites of southern Appalachians: Journal of Sedimentary Petrology, v. 40, p. 642–649.

Kaufman, M. E., and Campbell, L., 1969, Revised interpretation of the Cambrian Kinzers formation in southeastern Pennsylvania: Geological Society of America Abstracts with Programs, v. 1, no. 1, p. 32–33.

Kay, G. M., 1937, Stratigraphy of the Trenton Group: Geological Society of America Bulletin, v. 48, p. 233–302.

Kay, M., 1951, North American geosynclines: Geological Society of America Memoir 48, 143 p.

—— , 1956, Ordovician limestones in the western anticlines of the Appalachians in West Virginia and Virginia northeast of the New River: Geological Society of America Bulletin, v. 67, p. 55–106.

Keith, A., 1903, Description of the Cranberry Quadrangle: U.S. Geological Survey Geologic Atlas, Folio 90, 9 p.

Kellberg, J. M., and Grant, L. F., 1956, Coarse conglomerates of the Middle Ordovician in the southern Appalachian Valley: Geological Society of America Bulletin 67, p. 697–716.

Kiefer, J. D., 1970, Pre-Chattanooga Devonian stratigraphy of Alabama and northwest Georgia [Ph.D. thesis]: Urbana–Champaign, University of Illinois, 175 p.

King, P. B., 1949, The base of the Cambrian in the southern Appalachians: American Journal of Science, v. 247, p. 513–530, 622–645.

—— , 1950, Geology of the Elkton area, Virginia: U.S. Geological Survey Professional Paper 230, 82 p.

—— , 1964, Geology of the central Great Smoky Mountains Tennessee: U.S. Geological Survey Professional Paper 349–C, 148 p.

King, P. B., and Ferguson, H. W., 1960, Geology of northeasternmost Tennessee: U.S. Geological Survey Professional Paper 311, 136 p.

Knight, W. V., 1969, Historical and economic geology of Lower Silurian Clinton Sandstone of northeastern Ohio: American Association of Petroleum Geologists Bulletin, v. 53, p. 1421–1452.

Kreisa, R. D., 1981, Storm-generated sedimentary structures in subtidal marine facies with examples from the Middle and Upper Ordovician of southwestern Virginia: Journal of Sedimentary Petrology, v. 51, p. 823–848.

Kreisa, R. D., and Springer, D. A., 1987, Lithostratigraphy and biostratigraphy of the Martinsburg Formation in southwestern Virginia, with descriptive sections, *in* Contribution to Virginia Geology—V: Virginia Division of Mineral Resources Publication 74, p. 33–54.

Landes, K. K., 1970, Petroleum geology of the United States: New York, Wiley-Interscience, 571 p.

Laporte, L. F., 1967, Carbonate deposition near mean sea-level and resultant facies mosaic; Manlius Formation (Lower Devonian) of New York State: American Association of Petroleum Geologists Bulletin, v. 51, no. 1,

p. 73–101.

Laurence, R. A., 1944, An early Ordovician sinkhole deposit of volcanic ash and fossiliferous sediments in east Tennessee: Journal of Geology, v. 52, p. 235–249.

——, 1963, Rediscovery of the Murray Gap fossil locality, Blount County, Tennessee [abs.]: Tennessee Academy of Science Journal, v. 38, p. 64.

Laurence, R. A., and Palmer, A. R., 1963, Age of the Murray Shale and Hesse Quartzite on Chilhowee Mountain, Blount County, Tennessee: U.S. Geological Survey Professional Paper 475–C, p. C53–C54.

Lesure, F. G., 1957, Geology of the Clifton Forge iron district, Virginia: Bulletin of the Virginia Polytechnic Institute Engineering Experiment Station Series no. 188, 130 p.

Lightner, J., and Whisonant, R. C., 1983, Petrography of some orthoquartzites from the Keefer Formation (Silurian), Montgomery and Giles Counties, southwestern Virginia: Southeastern Geology, v. 24, no. 2, p. 91–99.

Lundegard, P. D., Samuels, N. D., and Pryor, W. A., 1978, The Brallier Formation; Upper Devonian turbidite slope facies of the central and southern Appalachians, *in* Second Eastern Gas Shales Symposium Preprint: Morgantown, West Virginia, U.S. Department of Energy, Morgantown Energy Technology Center/SP–78/6, v. 1, p. 5–23.

MacLachlan, D. B., 1967, Structure and stratigraphy of the limestones and dolomites of Dauphin County, Pennsylvania: Pennsylvania Geological Survey Bulletin G 44, Fourth Series, 168 p.

——, 1979, Geology and mineral resources of the Temple and Fleetwood Quadrangles, Berks County, Pennsylvania: Pennsylvania Geological Survey Atlas 187ab, Fourth Series, 71 p.

MacLachlan, D. B., Buckwalter, T. V., and McLaughlin, D. B., 1975, Geology and mineral resources of the Sinking Spring Quadrangle, Berks and Lancaster Counties, Pennsylvania: Pennsylvania Geological Survey Atlas 177d, Fourth Series, 228 p.

Markello, J. R., and Read, J. F., 1981, Carbonate ramp-to-deeper shale shelf transitions of an Upper Cambrian intrashelf basin, Nolichucky Formation, southwest Virginia Appalachians: Sedimentology, v. 28, p. 573–597.

——, 1982, Upper Cambrian intrashelf basin, Nolichucky Formation, southwest Virginia Appalachians: American Association of Petroleum Geologists Bulletin, v. 66, p. 860–878.

Martini, I. P.,1971, Regional analysis of sedimentology of Medina Formation (Silurian), Ontario and New York: American Association of Petroleum Geologists Bulletin, v. 55, p. 1249–1261.

McBride, E. F., 1972, Flysch and associated beds of the Martinsburg Formation (Ordovician), central Appalachians: Journal of Sedimentary Petrology, v. 32, p. 39–91.

McDowell, R. C., 1983, Stratigraphy of the Silurian outcrop belt on the east side of the Cincinnati arch in Kentucky, with revisions in the nomenclature: U.S. Geological Survey Professional Paper 1151–F, 27 p.

McGuire, W. H., and Howell, P., 1963, Oil and gas possibilities of the Cambrian and Lower Ordovician in Kentucky: Lexington, Spindletop Research Center, 216 p.

Meisler, H., and Becher, A. E., 1968, Carbonate rocks of Cambrian and Orodovician age in the Lancaster Quadrangle, Pennsylvania: U.S. Geological Survey Bulletin 1254–G, 14 p.

Milici, R. C., 1969, Middle Ordovician stratigraphy in central Sequatchie Valley, Tennessee: Southeastern Geology, v. 11, no. 2, p. 111–127.

——, 1973, The stratigraphy of Knox County, Tennessee, *in* Geology of Knox County, Tennessee: Tennessee Division of Geology Bulletin 70, p. 9–24.

——, 1974, Stratigraphy and depositional environments of Upper Mississippian and Lower Pennsylvanian rocks in the southern Cumberland Plateau of Tennessee, *in* Briggs, G., ed., Carboniferous of the southeastern United States: Geological Society of America Special Paper 148, p. 115–133.

Milici, R. C., and Smith, J. W., 1969, The stratigraphy of the Chickamauga Supergroup in its type area: Georgia Geological Survey Bulletin 80, p. 1–35.

Milici, R. C., and Wedow, H., Jr., 1977, Upper Ordovician and Silurian stratigraphy in Sequatchie Valley and parts of the adjacent Valley and Ridge, Tennessee: U.S. Geological Survey Professional Paper 996, 38 p.

Milici, R. C., Brent, W. B., and Walker, K. R., 1973, Depositional environments in upper Conasauga lagoon-fill sequences along I–75 at Copper Ridge, Knox County, Tennessee, *in* Geology of Knox County, Tennessee: Tennessee Division of Geology Bulletin 70, p. 138–148.

Milici, R. C., Briggs, G., Knox, L. M., Sitterly, P. D., and Statler, A. T., 1979, The Mississippian and Pennsylvanian (Carboniferous) systems in the United States; Tennessee: U.S. Geological Survey Professional Paper 1110–G, p. G1–G38.

Miller, M. S., 1974, Stratigraphy and coal beds of Upper Mississippian and Lower Pennsylvanian rocks in southwestern Virginia: Virginia Division of Mineral Resources Bulletin 84, 211 p.

Miller, R. L., 1937, Stratigraphy of the Jacksonburg Limestone: Geological Society of America Bulletin, v. 48, p. 1687–1718.

——, 1944, Geology and manganese deposits of the Glade Mountain district, Virginia: Virginia Geological Survey Bulletin 61, 150 p.

——, 1969, Pennsylvanian formations of southwest Virginia: U.S. Geological Survey Bulletin 1280, 62 p.

——, 1976, Silurian nomenclature and correlations in southwest Virginia and northeast Tennessee: U.S. Geological Survey Bulletin 1405–H, 25 p.

Miller, R. L., and Brosgé, W. P., 1954, Geology and oil resources of the Jonesville district, Lee County, Virginia: U.S. Geological Survey Bulletin 990, 240 p.

Miller, R. L., and Fuller, J. O., 1954, Geology and oil resources of the Rose Hill district; The fenster area of the Cumberland Overthrust Block, Lee County, Virginia: Virginia Geological Survey Bulletin 71, 383 p.

Miller, R. L., Harris, L. D., and Roen, J. B., 1964, The Wildcat Valley Sandstone (Devonian) of southwest Virginia: U.S. Geological Survey Professional Paper 501–B, p. B49–B52.

Mitchell, R. W., 1982, Middle Ordovician St. Paul Group, *in* Demicco, R. V., and Mitchell, R. W., Facies of the Great American Bank in the central Appalachians, *in* Lyttle, P. T., ed., Central Appalachian geology: American Geological Institute, p. 175–216.

Mixon, R. B., and Harris, L. D., 1971, Geology of the Swan Island Quadrangle, Tennessee: U.S. Geological Survey Geologic Quadrangle Map GQ–878, scale 1:24,000.

Moore, B. R., and Clarke, M. K., 1970, The significance of a turbidite sequence in the Borden Formation (Mississippian) of eastern Kentucky and southern Ohio, *in* Lajoie, J., ed., Flysch sedimentology in North America: Geological Association of Canada Special Paper 7, p. 211–218.

Neathery, T. A., 1973, The Talladega front, *in* Carrington, T. J., ed., Talladega metamorphic front: Alabama Geological Society, Eleventh Annual Field Trip Guidebook, p. 1–9.

Neuman, R. B., 1951, St. Paul Group; A revision of the "Stones River" Group of Maryland and adjacent states: Geologic Society of America Bulletin, v. 62, p. 267–324.

——, 1955, Middle Ordovician rocks of the Tellico–Sevier belt, eastern Tennessee: U.S. Geological Survey Professional Paper 274–F, p. 141–178.

Neuman, R. B., and Nelson, W. H., 1965, Geology of the western Great Smoky Mountains, Tennessee: U.S. Geological Survey Professional Paper 349–D, 81 p.

Oder, C.R.L., and Bumgarner, J. G., 1961, Stromatolitic bioherms in the Maynardville (Upper Cambrian) Limestone, Tennessee: Geological Society of America Bulletin, v. 72, p. 1021–1028.

Oliver, W. A., Jr., 1963, Stratigraphy, facies changes, and paleoecology of the Lower Devonian Helderberg Limestones and the Middle Devonian Onondaga Limestone: New York State Geological Assocication Guidebook, Field trip no. 1, p. 11–16.

Oliver, W. A., Jr., de Witt, W., Jr., Dennison, J. M., Hoskins, D. M., and Huddle, J. W., 1969, Correlation of Devonian Rock units in the Appalachian Basin: U.S. Geological Survey Oil and Gas Investigations Chart OC–64.

——, 1971, Isopach and lithofacies maps of the Devonian in the Appalachian Basin: Pennsylvania Geological Survey, Fourth Series, Progress Report 182, 7 plates.

Palmer, A. R., 1971, The Cambrian of the Appalachian and eastern New England regions, eastern United States, *in* Holland, C. E., ed., Cambrian of the New

World: New York, Wiley-Interscience, p. 169–217.

Patchen, D. G., Smosna, R. A., 1975, Stratigraphy and petrology of Middle Silurian McKenzie Formation in West Virginia: American Association of Petroleum Geologists Bulletin, v. 59, p. 2266–2287.

Pepper, J. F., de Witt, W., Jr., and Everhart, G. M., 1953, The "Clinton" sands in Canton, Dover, Mussillon, and Navarre Quadrangles, Ohio: U.S. Geological Survey Bulletin 1003–A, 13 p.

Pepper, J. F., de Witt, W., Jr., and Demarest, D. F., 1954, Geology of the Bedford Shale and Berea Sandstone in the Appalachian Basin: U.S. Geological Survey Professional Paper 259, 61 p.

Perroud, H., Van der Voo, R., and Bonhommet, N., 1984, Paleozoic evolution of the America plate on the basis of paleomagnetic data: Geology, v. 12, p. 579–582.

Perry, W. J., Jr., 1972, The Trenton Group of Nittany anticlinorium eastern West Virginia: West Virginia Geological and Economic Survey Circular 13, 30 p.

Pfeil, R. W., and Read, J. F., 1980, Cambrian carbonate platform margin facies, Shady dolomite, southwestern Virginia, U.S.A.: Journal of Sedimentary Petrology, v. 50, p. 91–116.

Pierce, K. L., 1966, Bedrock and surficial geology of the McConnellsburg Quadrangle, Pennsylvania: Pennsylvania Geological Survey Atlas 109a, Fourth Series, 111 p.

Prouty, C. E., 1959, The Annville, Myerstown, and Hershey Formations of Pennsylvania: Pennsylvania Geological Survey Bulletin G 31, Fourth Series, 47 p.

Rader, E. K., 1982, Valley and Ridge stratigraphic correlations, Virginia: Virginia Division of Mineral Resources Publication 37.

Rader, E. K., and Biggs, T. H., 1975, Geology of the Front Royal Quadrangle, Virginia: Virginia Division of Mineral Resources Report of Investigations 40, 91 p.

Rader, E. K., and Gathright, T. M., II, 1986, Stratigraphic and structural features of Fincastle Valley and Eagle Rock Gorge, Botetourt County, Virginia, *in* Neathery, T. L., ed., Southeastern Section of the Geological Society of America Boulder, Colorado, Geological Society of America, Centennial Field Guide, v. 6, p. 105–108.

Rader, E. K., and Henika, W. S., 1978, Ordovician shelf-to-basin transition, Shenandoah Valley, Virginia, *in* Contributions to Virginia geology, III: Virginia Division of Mineral Resources Publication no. 7, p. 51–65.

Read, J. F., 1980, Carbonate ramp-to-basin transitions and foreland basin evolution, Middle Ordovician sequence, Virginia: American Association of Petroleum Geologists Bulletin, v. 64, p. 1575–1612.

——, 1982, Geometry, facies, and development of Middle Ordovician carbonate buildups, Virginia Appalachians: American Association of Petroleum Geologists Bulletin, v. 66, no. 2, p. 189–209.

——, 1983, Field trip guide to Lower Paleozoic carbonate rocks, Roanoke region, Virginia: Blacksburg, Virginia Polytechnic Institute and State University Department of Geological Sciences, 31 p.

Read, J. F., and Pfeil, R. W., 1983, Fabrics of allocthonous reefal blocks, Shady Dolomite (Lower to Middle Cambrian), Virginia Appalachians: Journal of Sedimentary Petrology, v. 53, p. 761–778.

Reed, J. C., Jr., 1955, Catoctin Formation near Luray, Virginia: Geological Society of America Bulletin, v. 64, p. 871–896.

Reinhardt, J., 1974, Stratigraphy, sedimentology, and Cambro-Ordovician paleogeography of the Frederick Valley, Maryland: Maryland Geological Survey Report of Investigations 23, 74 p.

——, 1977, Cambrian off-shelf sedimentation, central Appalachians: Society of Economic Paleontologists and Mineralogists Special Publication 25, p. 83–112.

Reinhardt, J., and Wall, E., 1975, Tomstown Dolomite (Lower Cambrian), central Appalachian Mountains, and the habitat of *Salterella conulata:* Geological Society of America Bulletin, v. 86, p. 1377–1380.

Rice, C. L., 1984, Sandstone units of the Lee Formation and related strata in eastern Kentucky: U.S. Geological Survey Professional Paper 1151–G, 53 p.

Rice, C. L., Sable, E. G., Dever, G. R., and Kehn, T. M., 1979, The Mississippian and Pennsylvanian (Carboniferous) systems in the United States; Kentucky: U.S. Geological Survey Professional Paper 1110–F, 32 p.

Rickard, L. V., 1962, Late Cayugan (Upper Silurian) and Helderbergian (Lower Devonian) stratigraphy in New York: New York State Museum and Science Service Bulletin 386, 157 p.

——, 1964, Correlation of the Devonian rocks in New York State: New York State Museum and Science Service Geological Survey Map and Chart Series no. 4.

——, 1969, Stratigraphy of the Upper Silurian Salina Group, New York, Pennsylvania, Ohio, Ontario: New York State Museum and Science Service Map and Chart Service no. 12, 57 p.

——, 1973, Stratigraphy and structure of the subsurface Cambrian and Ordovician carbonates of New York: New York State Museum and Science Service Map and Chart Series no. 18, 26 p.

——, 1975, Correlation of the Silurian and Devonian rocks in New York State: New York State Museum and Science Service Map and Chart Series no. 24, 16 p.

Rodgers, J., 1943, Geologic map of the Copper Ridge district, Hawkins, Hancock, and Grainger Counties, Tennessee: U.S. Geologic Survey Strategic Minerals Investigations Preliminary Map.

——, 1953, Geologic map of east Tennessee with explanatory text: Tennessee Division of Geology Bulletin 58, part II, 168 p.

——, 1968, The eastern edge of the North American continent during the Cambrian and Ordovician, *in* Zen, E-an, and others, eds., Studies of Appalachian geology; Northern and maritime: New York, Interscience Publishers, p. 141–149.

——, 1971, The Taconic orogeny: Geological Society of America Bulletin, v. 82, p. 1141–1178.

Rodgers, J., and Kent, D. F., 1948, Stratigraphic section at Lee Valley, Hawkins County, Tennessee: Tennessee Division of Geology Bulletin 55, 47 p.

Roen, J. B., 1981, Regional stratigraphy of the Upper Devonian black shales in the Appalachian Basin, *in* Roberts, T. G., G.S.A. Cincinnati 1981 Field Trip Guidebooks, v. II; Economic geology, structure: Falls Church, Virginia, American Geological Institute, Field Trip No. 3, p. 324–330.

——, 1984, Geology of the Devonian black shales of the Appalachian Basin: Organic Geochemistry, v. 5, no. 4, p. 241–245.

Root, S. I., 1968, Geology and mineral resources of southeastern Franklin County, Pennsylvania: Pennsylvania Geological Survey Atlas, 119cd, Fourth Series, 118 p.

Rosenkrans, R. R., 1941, Stratigraphy of Ordovician bentonite beds in southwestern Virginia: Virginia Geological Survey Bulletin 46, p. 83–111.

Ross, R. J., Jr., and others, 1982, The Ordovician system in the United States: International Union of Geological Sciences Publication 12, 73 p.

Ruppel, S. C., and Walker, K. R., 1984, Petrology and depositional history of a Middle Ordovician carbonate platform: Chickamauga Group, northeastern Tennessee: Geological Society of America Bulletin, v. 95, p. 568–583.

Sable, E. G., 1979, Eastern Interior region, *in* Craig, L. C., and Connor, C. W., eds., Paleotectonic investigations of the Mississippian system in the United States: U.S. Geological Survey Professional Paper 1011, part 1, p. 59–106.

Sando, W. J., 1957, Beekmantown Group (Lower Ordovician) of Maryland: Geological Society of America Memoir 68, 161 p.

——, 1958, Lower Ordovician section near Chambersburg, Pennsylvania: Geological Society of America Bulletin, v. 69, p. 837–854.

Scatterday, J. W., 1963, Stratigraphy and conodont faunas of the Maxville Group [Middle and Upper Mississippian] of Ohio [Ph.D. thesis]: Columbus, Ohio State University, 169 p.

Sedimentation Seminar, 1981, Mississippian and Pennsylvanian section on Interstate 75 south of Jellico, Campbell County, Tennessee: Tennessee Division of Geology Report of Investigations 38, 42 p.

Shanmugam, G., and Lash, G. G., 1982, Analogous tectonic evolution of the Ordovician foredeeps, southern and central Appalachians: Geology, v. 10, p. 562–566.

Sherwood, W. C., 1964, Structure of the Jacksonburg formation in Northampton and Lehigh Counties, Pennsylvania: Pennsylvania Geological Survey Bulletin G 45, Fourth Series, 64 p.

Simpson, E. L., and Sundberg, F. A., 1987, Early Cambrian age for synrift deposits of the Chilhowee Group of southwestern Virginia: Geology, v. 15, p. 123–126.

Smosna, R., 1983, Depositional patterns of a Silurian shelf sand in the central Appalachians: Northeastern Geology, v. 5, p. 100–109.

Smosna, R., and Patchen, D. G., 1978, Silurian evolution of the central Appalachian Basin: American Association of Petroleum Geologists Bulletin, v. 62, p. 2308–2328.

Spelman, A. R., 1966, Stratigraphy of Lower Ordovician Nittany Dolomite in central Pennsylvania: Pennsylvania Geological Survey Bulletin G 47, Fourth Series, 187 p.

Stephens, G. C., and Wright, T. O., 1981, Stratigraphy of the Martinsburg Formation, West of Harrisburg in the Great Valley of Pennsylvania: American Journal of Science, v. 281, p. 1009–1020.

Stith, D. A., 1979, Chemical composition, stratigraphy, and depositional environments of the Black River Group (Middle Ordovician), southwestern Ohio: Ohio Division of Geological Survey Report of Investigations no. 113, 36 p.

Stose, G. W., 1906, The sedimentary rocks of South Mountain, Pennsylvania: Journal of Geology, v. 14, p. 201–220.

—— . 1909, Description of the Mercersburg–Chambersburg district, Pennsylvania: U.S. Geological Survey Geologic Atlas, Folio 170, 19 p.

—— , 1932, Geology and mineral resources of Adams County, Pennsylvania: Pennsylvania Geological Survey Fourth Series, County Report 1, 153 p.

Stose, G. W., and Jonas, A. I., 1923, Ordovician overlap in the Piedmont province of Pennsylvania and Maryland: Geological Society of America Bulletin, v. 34, p. 507–524.

—— , 1938, A southeastern limestone facies of Lower Cambrian dolomite in Wythe and Carroll Counties, Virginia, *in* Contributions to Virginia Geology, II: Virginia Geological Survey Bulletin 51, p. 1–30.

—— , 1939, Geology and mineral resources of York County, Pennsylvania: Pennsylvania Geological Survey Bulletin C 67, Fourth Series, 199 p.

Stose, A. J., and Stose, G. W., 1944, Geology of the Hanover–York District, Pennsylvania: U.S. Geological Survey Professional Paper 204, 84 p.

Swartz, C. K., 1923, Stratigraphic and paleontologic relations of the Silurian strata of Maryland, *in* Maryland Geological Survey, Silurian: Baltimore, Maryland, The Johns Hopkins Press, p. 25–52.

Swartz, F. M., 1948, Tenton and sub-Trenton of outcrop areas in New York, Pennsylvania and Maryland: Pennsylvania Geological Survey Bulletin G 22, Fourth Series, p. 1493–1595.

Textoris, D. A., 1968, Petrology of supratidal, intertidal, and shallow subtidal carbonates, Black River Group, Middle Ordovician, New York, U.S.A., *in* 23rd International Geological Congress Proceedings: Prague, Academia, v. 8, p. 227–248.

Thomas, W. A., and Cramer, H. R., 1979, The Mississippian and Pennsylvanian (Carboniferous) systems in the United States: Georgia: U.S. Geological Survey Professional Paper 1110–A–L, p. H1–H37.

Thompson, A. M., 1970a, Lithofacies and formation nomenclature in Upper Ordovician stratigraphy, central Appalachians: Geological Society of America Bulletin, v. 81, p. 1255–1260.

—— , 1970b, Geochemistry of color genesis in red-bed sequence, Juniata and Bald Eagle Formations, Pennsylvania: Journal of Sedimentary Petrology, v. 40, no. 2, p. 599–615.

Ulteig, J. R., 1964, Upper Niagaran and Cayugan stratigraphy of northeastern Ohio and adjacent areas: Ohio Division of Geological Survey Report of Investigations no. 51, 48 p.

Wagner, W. R., 1963, Correlation of Susquehanna Group in part of northeastern Pennsylvania, *in* Shepps, V. C., Symposium on Middle and Upper Devonian stratigraphy of Pennsylvania and adjacent states: Pennsylvania Geological Survey, Fourth Series, General Geology Report G 39, p. 63–77.

—— , 1966, Stratigraphy of the Cambrian to Middle Ordovician rocks of central and western Pennsylvania: Pennsylvania Geological Survey Bulletin G 49, Fourth Series, 156 p.

Walcott, C. D., 1886, Second contributions to the studies on the Cambrian faunas of North America: U.S. Geological Survey Bulletin 30, 369 p.

—— , 1891, Correlation papers; Cambrian: U.S. Geological Survey Bulletin 81, 447 p.

Walker, K. R., and Ferrigno, K. F., 1973, Major Middle Ordovician reef tract in east Tennessee: American Journal of Science, Cooper Volume 273-A, p. 294–325.

Walker, K. R., and Laporte, L. F., 1970, Congruent fossil communities from Ordovician and Devonian carbonates of New York: Journal of Paleontology, v. 44, p. 928–944.

Walker, K. R., Shanmugam, G., and Ruppel, S. C., 1983, A model for carbonate to terrigenous clastic sequences: Geological Society of America Bulletin, v. 94, p. 700–712.

Weed, E.G.A., 1982, Nealmont Limestone (Middle Ordovician) extending from southwestern Virginia into southwestern New York in the central Appalachians, *in* Stratigraphic notes: U.S. Geological Survey Bulletin 1529–H, p. H135–H136.

Wehr, F., and Glover, L., III, 1985, Stratigraphy and tectonics of the Virginia–North Carolina Blue Ridge; Evolution of a late Proterozoic–Early Paleozoic hinge zone: Geological Society of America Bulletin, v. 96, p. 285–295.

Weir, G. W., Gualtieri, J. L., and Schlanger, S. O., 1966, Borden Formation (Mississippian) in south- and south-central Kentucky U.S. Geological Survey Bulletin 1224–F, p. 15, 31.

Weir, G. W., Peterson, W. L., and Swadley, W. C., 1984, Lithostratigraphy of Upper Ordovician strata exposed in Kentucky: U.S. Geological Survey Professional Paper 1151–E, 121 p.

Whisonant, R. C., 1977, Lower Silurian Tuscarora (Clinch) dispersal patterns in western Virginia: Geological Society of America Bulletin, v. 88, p. 215–220.

Whitaker, J. C., 1955, Geology of Catoctin Mountain, Maryland and Virginia: Geological Society of America Bulletin, v. 66, p. 435–462.

Willard, B., Swartz, F. M., and Cleaves, A. B., 1939, The Devonian of Pennsylvania: Pennsylvania Geological Survey, Fourth Series, Bulletin G–19, p. 141, 199, 371.

Wilpolt, R. H., and Marden, D. W., 1959, Geology and oil and gas possibilities of Upper Mississippian rocks of southwestern Virginia, southern West Virginia, and eastern Kentucky: U.S. Geological Survey Bulletin 1072–K, p. 587–656.

Wilson, C. W., Jr., 1949, Pre-Chattanooga stratigraphy in central Tennessee: Tennessee Division of Geology Bulletin 56, 407 p.

—— , 1962, Stratigraphy and geologic history of Middle Ordovician rocks of central Tennessee: Geological Society of America Bulletin, v. 73, p. 481–504.

Wilson, J. L., 1952, Upper Cambrian stratigraphy in the central Appalachians: Geological Society of America Bulletin, v. 63, p. 275–322.

Wilson, R. L., 1979, The stratigraphy of exposed rocks in Hamilton County, Tennessee, *in* Geology of Hamilton County, Tennessee: Tennessee Division of Geology Bulletin 79, p. 15–37.

Yeakel, L. S., Jr., 1962, Tuscarora, Juniata, and Bald Eagle paleocurrents and paleogeography in the central Appalachians: Geological Society of America Bulletin, v. 73, p. 1515–1540.

Young, R. S., and Rader, E. K., 1974, Geology of the Woodstock, Wolf Gap, Conicville, and Edinburg Quadrangles, Virginia: Virginia Division of Mineral Resources Report of Investigations 35, 69 p.

Zenger, D. H., 1965, Stratigraphy of the Lockport Formation (Middle Silurian) in New York: New York State Museum and Science Service Bulletin no. 404, 210 p.

—— , 1971, Uppermost Clinton (Middle Silurian) stratigraphy and petrology in east-central New York: New York State Museum and Science Service Bulletin no. 417, 58 p.

—— , 1981, Stratigraphy and petrology of the Little Falls dolostone (Upper Cambrian), east-central New York: New York State Museum Map and Chart Series 34, 138 p.

Manuscript Accepted by the Society August 27, 1987

Printed in U.S.A.

The Geology of North America
Vol. D-2, Sedimentary Cover—North American Craton: U.S.
The Geological Society of America, 1988

Chapter 16

The Black Warrior basin

William A. Thomas
Department of Geology, University of Alabama, Tuscaloosa, Alabama 35487

INTRODUCTION

The Paleozoic Black Warrior foreland basin on the southeastern edge of the North American craton in Alabama and Mississippi is similar to other foreland basins along the Appalachian-Ouachita orogen (Fig. 1). The eastern part of the basin is exposed in north-central Alabama, but the western part is covered by Mesozoic-Cenozoic strata of the Gulf Coastal Plain (Mellen, 1947). A regional pre-Mesozoic unconformity truncates Paleozoic strata and dips gently southwestward toward the broad southward-plunging syncline that defines the Mississippi Embayment of the Gulf Coastal Plain (Fig. 1) (Thomas and others, in press).

The oldest sedimentary rocks in the Black Warrior basin are of Cambrian age and rest unconformably on Precambrian crystalline basement rocks. The youngest strata of the Paleozoic succession are of Pennsylvanian age. The Pennsylvanian rocks are unconformably overlain by Cretaceous strata in the western part of the basin and are exposed in the eastern part. The Paleozoic succession includes two distinctly different parts: Cambrian to Early Mississippian shallow-marine shelf facies indicating a passive continental margin, and later Mississippian to Pennsylvanian shallow-marine to deltaic facies reflecting evolution of a foreland basin (Plate 8).

Structurally, the Black Warrior foreland basin is a homocline that dips southwestward away from the craton and extends beneath the front of the Paleozoic Appalachian-Ouachita fold-thrust belt (Fig. 1; Plate 8A). Subsurface data demonstrate continuity of the Paleozoic fold-thrust belt south of the Black Warrior basin from Appalachian outcrops in Alabama, beneath the Gulf Coastal Plain, to the Ouachita Mountains in Arkansas (Fig. 1) (Thomas, 1973, 1985, in press). The north side of the Black Warrior basin is defined by the Nashville dome, a broad intracratonic arch. On the northeast along the Appalachian thrust front, the Black Warrior basin is separated from the Appalachian foreland basin by a low, indistinct arch that plunges southeastward from the Nashville dome. On the west along the Ouachita thrust front, the Black Warrior basin is separated from the Arkoma foreland basin by a southwestward-plunging Paleozoic arch and fault system (Mississippi Valley graben) in the subsurface beneath the Mesozoic-Cenozoic Mississippi Embayment of the Gulf Coastal Plain (Fig. 1; Plate 8A-2) (Thomas and others, in press).

Coal, natural gas, oil, and coalbed methane are produced from various strata in the Black Warrior basin. Coal is obtained from both surface and underground mines in Pennsylvanian strata in the eastern part of the basin, and methane is produced from coal degasification wells in the same area. Natural gas production is mainly from Mississippian sandstones, and oil is produced from some of those reservoirs. Lesser amounts of conventional natural gas are obtained from Pennsylvanian sandstones, and minor discoveries have been made in pre-Mississippian strata. The exposed Mississippian section on the north limb of the basin contains tar sands. A variety of products, including dimension stone, is quarried from Mississippian limestones.

Because much of the Black Warrior basin is in the subsurface beneath Coastal Plain cover and because upper Paleozoic rocks are preserved throughout the deeper part of the basin, stratigraphic and structural data are necessarily derived from deep wells (drill cuttings, cores, and geophysical logs) and geophysical surveys. Distribution of wells is shown on Figure 2, and wells identified by number in the text and illustrations are listed in Table 1. Drill data are abundant through the Mississippian section but are relatively sparse for the older rocks. Outcrop studies are applicable in the eastern part of the basin and in the adjacent Appalachian fold-thrust belt. These data are supplemented by more distant outcrop data from the easternmost exposures in the Ouachita Mountains and Arkoma foreland basin, as well as from the Nashville and Ozark domes. Geophysical data available for use in this synthesis include regional gravity and magnetic maps, as well as proprietary seismic reflection profiles.

STRUCTURAL GEOLOGY

The Black Warrior foreland basin is defined by a homocline having an average dip of less than 2° southwestward from the Nashville dome (Plate 8A). The basin is asymmetric and is characterized by the single long homoclinal limb, by lack of a structu-

Thomas, W. A., 1988, The Black Warrior basin, *in* Sloss, L. L., ed., Sedimentary Cover—North American Craton; U.S.: Boulder, Colorado, Geological Society of America, The Geology of North America, v. D-2.

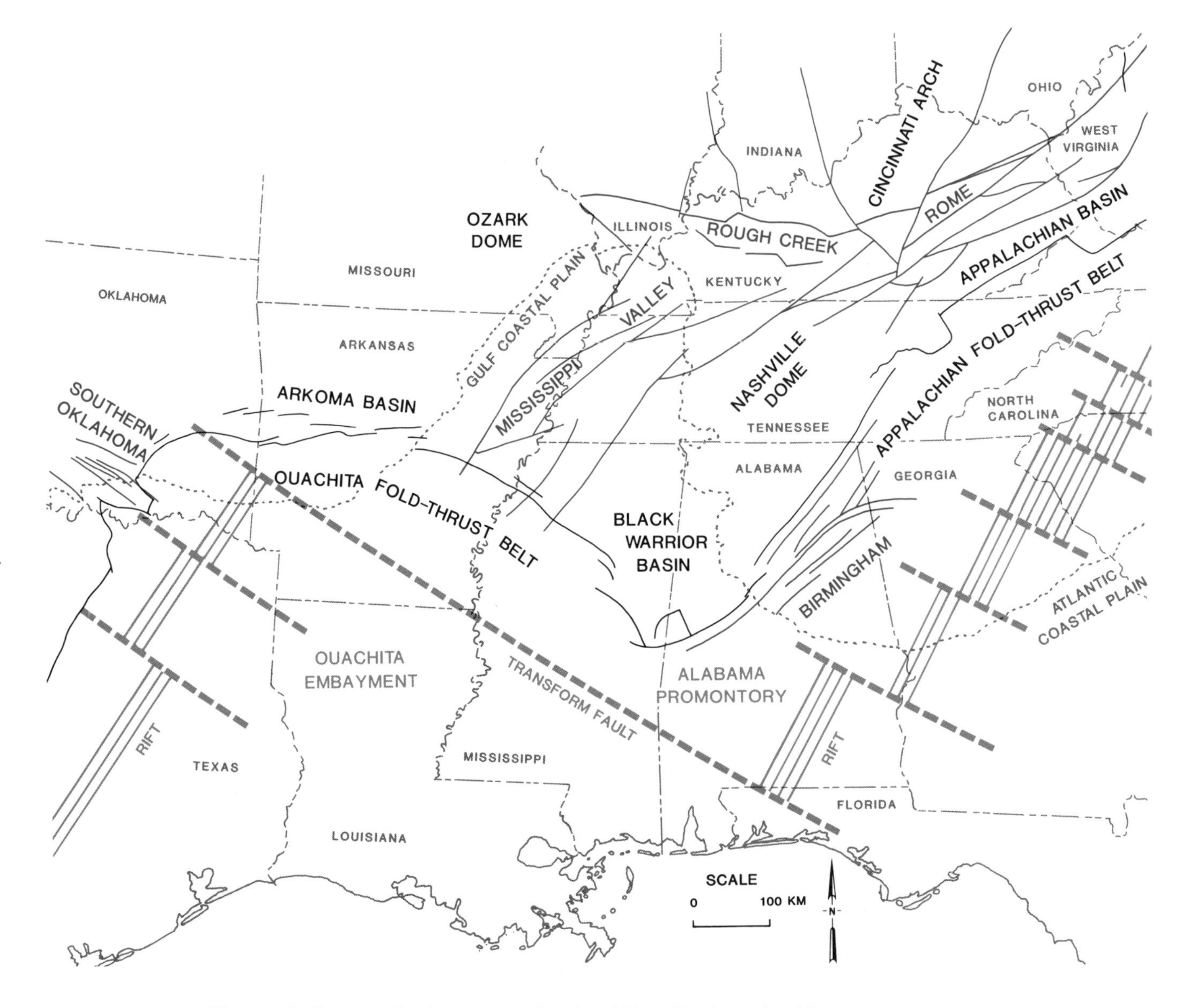

Figure 1. Outline map of regional structural setting of Black Warrior basin with respect to interpreted shape of late Precambrian–early Paleozoic rifted margin (light red), Cambrian basement fault systems (red), and late Paleozoic orogenic belts and cratonic structures (black). Map of Rome trough and Rough Creek graben modified from Harris, 1975; Kulander and Dean, 1978; Soderberg and Keller, 1981.

ral trough marked by opposing dips, and by a down-dip border defined by the frontal structures of the Appalachian-Ouachita fold-thrust belt. The deepest part of the basin is within a northward-concave recess in the front of the Appalachian-Ouachita fold-thrust belt in eastern Mississippi (Plate 8A). The strike of the Black Warrior basin homocline persists southeastward to the structural front of the Appalachian fold-thrust belt. The eastern up-dip part of the basin is defined by the subtle southeastward-plunging nose of the Nashville dome where Mississippian rocks are exposed adjacent to the Appalachian fold-thrust belt in northeastern Alabama (Fig. 2). The oldest rocks exposed on the Nashville dome are Ordovician limestones. Because of the southwestward homoclinal dip from the Nashville dome into the Black Warrior basin, progressively younger rocks are preserved toward the south; and a thick succession of Pennsylvanian rocks is preserved in the deepest part of the Black Warrior basin adjacent to the Appalachian-Ouachita fold-thrust belt (Plate 8A).

The relatively simple southwestward-dipping homocline of the Black Warrior basin is broken by a northwest-trending system of normal faults (Plate 8A). A swarm of faults along the northeast

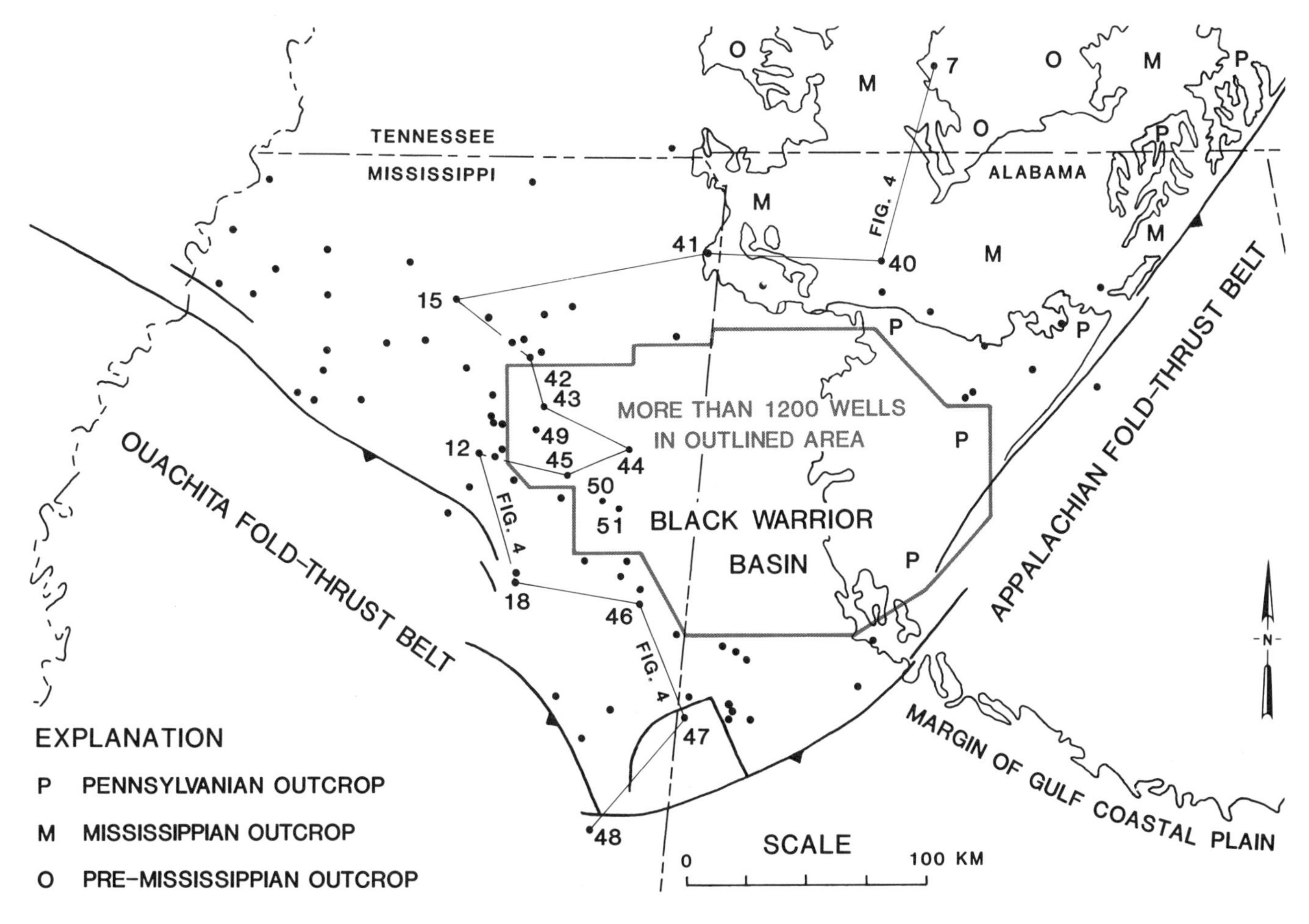

Figure 2. Map of wells (shown by black dots) and outcrop areas of Paleozoic rocks in and around the Black Warrior basin. Within the area of abundant wells (gray outline), only wells in Figures 4 and 5 are shown. Wells in Figures 4 and 5 are identified by number and listed in Table 1. Line of cross section of Figure 4.

side of the system has a cumulative down-to-southwest displacement of approximately 1.5 km. The more southwesterly faults of the system are indicated by seismic data to displace the homocline down-to-southwest as much as 3 km. The faults have a variety of trends and amounts of displacement. Most faults are down-to-southwest, but some are antithetic. Local folds exhibiting low counter-regional dip are associated with some faults. On the southeast, the fault system extends to the front of the Appalachian fold-thrust belt and intersects northeasterly striking Appalachian thrust faults nearly perpendicularly. The fault system extends northwestward entirely across the Black Warrior basin, and the horsetail-like system of faults narrows toward the northwest (Plate 8A). The fault system evidently extends farther west into the Arkoma basin, where a system of large-scale down-to-south normal faults is approximately parallel with the frontal thrust faults of the Ouachitas (Buchanan and Johnson, 1968; Haley, 1982; Houseknecht, 1986). The age of the fault system is not clearly defined. Rocks as young as earliest Middle Pennsylvanian are displaced by the faults in the Black Warrior basin, but rare local stratigraphic anomalies suggest possible fault movement during deposition of the Upper Mississippian and Lower Pennsylvanian rocks. Synsedimentary normal fault movement from Late Mississippian through Middle Pennsylvanian has been documented in the Arkoma basin (Koinm and Dickey, 1967; Buchanan and Johnson, 1968; Haley and Hendricks, 1968; Haley, 1982; Houseknecht, 1986). The Paleozoic normal faults are truncated by Cretaceous Coastal Plain strata, indicating no post-Paleozoic movement (Plate 8A-3).

West of the Black Warrior basin, the Arkoma foreland basin is between the Ouachita fold-thrust belt and the Ozark dome. A structural nose plunges southeastward from the Ozark dome toward the Nashville dome beneath Mesozoic-Cenozoic strata in the Mississippi Embayment of the Gulf Coastal Plain (Thomas, 1985, in press), and wells in northwestern Tennessee (Grohskopf,

TABLE 1. WELLS IDENTIFIED BY NUMBER ON PLATE 8 AND IN THIS CHAPTER

Well	Location	Source of data*
1. U.S. Bureau Mines No. 1 Oliver	Sec. 29, T 22 N, R 11 E, New Madrid Co., Mo.	a,b
2. Strake No. 1 Russell	Sec. 24, T 19 N, R 11 E, Pemiscot Co., Mo.	b,c
3. Benz No. 1 Merritt	Sec. 3, T 4 S, R 1 E, Lake Co., Tn.	d
4. Henderson No. 1 Rice	Sec. 22, T 4 S, R 1 E, Dyer Co., Tn.	a,b
5. Big Chief No. 1 Taylor	Sec. 19, T 5 S, R 6 E, Gibson Co., Tn.	d,e
6. du Pont No. 2 Fee	Sec. 14, T 6 S, R 19 E, Humphreys Co., Tn.	d
7. California No. 1 Beeler	Sec. 4, T 15 S, R 29 E, Giles Co., Tn.	d,e
8. Saga No. 1 Skidmore	Sec. 36, T 7 S, R 1 W, Morgan Co., Al.	f,g
9. Saga No. 1 Hudson	Sec. 16, T 10 S, R 2 E, Blount Co., Al.	f,g
10. Pan American No. 1 Holmes-Davis	Sec. 4, T 21 N, R 9 E, Webster Co., Ms.	c
11. Pan American No. 1 Hodges	Sec. 4, T 22 N, R 10 E, Calhoun Co., Ms.	c
12. Carter No. 1 Crane	Sec. 31, T 13 S, R 1 E, Calhoun Co., Ms.	c
13. Honolulu No. 2 Davis	Sec. 27, T 12 S, R 1 E, Calhoun Co., Ms.	e
14. Salmon No. 1 Patterson	Sec. 12, T 11 S, R 1 E, Pontotoc Co., Ms.	c
15. Pruet & Hughes No. 1 Dunlap	Sec. 18, T 7 S, R 1 W, Lafayette Co., Ms.	e,h
16. Memphis No. 1 Melton	Sec. 12, T 2 S, R 3 E, Tippah Co., Ms.	e
17. Memphis No. 1 Curtis	Sec. 16, T 18 S, R 15 E, Hardin Co., Tn.	d
18. Exxon No. 1 Fulgham	Sec. 33, T 19 N, R 12 E, Oktibbeha Co., Ms.	e
19. McAlester No. A-1 Sudduth	Sec. 6, T 19 N, R 15 E, Oktibbeha Co., Ms.	e
20. Shell No. 1 Gearhiser	Sec. 1, T 18 N, R 17 E, Lowndes Co., Ms.	e
21. Bow Valley No. 1 Bain	Sec. 7, T 18 S, R 16 W, Pickens Co., Al.	f
22. Pruet No. 1 Vann	Sec. 15, T 17 S, R 15 W, Lamar Co., Al.	f
23. Burns No. 20-1 Tomlin	Sec. 20, T 16 S, R 14 W, Lamar Co., Al.	f
24. Warrior No. 23-1 Bynum	Sec. 23, T 14 S, R 13 W, Fayette Co., Al.	f
25. Pine No. 1 Baldwin	Sec. 20, T 13 S, R 11 W, Fayette Co., Al.	c,i
26. Rose No. 2 First National Bank	Sec. 30, T 12 S, R 7 W, Walker Co., Al.	i
27. Texas Eastern No. 1 Horton	Sec. 19, T 10 S, R 4 W, Cullman Co., Al.	c,f
28. Shenandoah No. 1 Smith	Sec. 26, T 9 S, R 2 W, Cullman Co., Al.	f,g
29. Newman No. 1 Nichols	Sec. 2, T 6 S, R 2 E, Madison Co., Al.	i
30. ARCO No. 1 Edgmon	Sec. 6, T 7 N, R 12 W, Faulkner Co., Ar.	j,k
31. Cockrell No. 1 Carter	Sec. 4, T 4 N, R 1 E, St. Francis Co., Ar.	e,j
32. Dow No. 1 Wilson	Sec. 14, T 12 N, R 9 E, Mississippi Co., Ar.	j,l
33. Texaco-Exxon No. 1 Ivy	Sec. 36, T 27 N, R 3 W, Coahoma Co., Ms.	e,h
34. Tipperary No. 1-X Harpole-Campbell	Sec. 8, T 7 S, R 10 W, Quitman Co., Ms.	e
35. Stauffer No. 1 Fee	Sec. 16, T 12 S, R 28 W, Maury Co., Tn.	g
36. Sonat No. 1 Brown	34° 54' 52" N, 85° 28' 32" W, Dade Co., Ga.	g
37. Dow No. 1 Garrigan	Sec. 28, T 15 N, R 10 E, Mississippi Co., Ar.	e,l
38. Henderson No. 1 Markham	Sec. 21, T 2 S, R 1 E, Lake Co., Tn.	b,d
39. Gulf No. 1-A Spinks	Sec. 25, T 2 S, R 13 E, Henry Co., Tn.	a,d
40. Reynolds No. 1 Connors	Sec. 17, T 5 S, R 8 W, Lawrence Co., Al.	i
41. Mississippi No. 1 Southward	Sec. 18, T 5 S, R 11 E, Tishomingo Co., Ms.	c
42. Salmon No. 1 Wilson	Sec. 29, T 9 S, R 3 E, Pontotoc Co., Ms.	c
43. Magnolia No. 1 Warren	Sec. 26, T 11 S, R 3 E, Pontotoc Co., Ms.	c
44. Magnolia No. 1 Pierce	Sec. 22, T 13 S, R 7 E, Monroe Co., Ms.	c
45. Carter No. 1 Pulliam	Sec. 24, T 14 S, R 4 E, Chickasaw Co., Ms.	c
46. Socony Mobil No. 1 Hardy	Sec. 3, T 17 N, R 17 E, Lowndes Co., Ms.	c
47. Sonat No. 1 Hagerman	Sec. 9, T 23 N, R 3 W, Sumter Co., Al.	e,f
48. Amoco-Stack No. 1 Lucky	Sec. 6, T 8 N, R 16 E, Lauderdale Co., Ms.	e
49. Union No. 1 Dinsmore	Sec. 28, T 12 S, R 3 E, Chickasaw Co., Ms.	m
50. Union No. 1 Neal	Sec. 21, T 15 S, R 6 E, Monroe Co., Ms.	e
51. Union No. 1 Watson	Sec. 31, T 15 S, R 7 E, Monroe Co., Ms.	c

*Source of data:

a	Missouri Geological Survey open file	h	Mellen, 1977
b	Grohskopf, 1955	i	McGlamery, 1955
c	sample description by author	j	Denison, 1984
d	Tennessee Division of Geology open file	k	B. R. Haley, unpublished data
e	unpublished industry report	l	Howe, 1985
f	Alabama Geological Survey open file	m	Welch, 1959
g	Neathery and Copeland, 1983		

1955; Tennessee Division of Geology open file) have penetrated Cambrian rocks unconformably below Mesozoic strata. The southeastward-plunging nose of the Ozark dome evidently extends into a southwestward-plunging arch beneath the axis of the Mesozoic-Cenozoic Mississippi Embayment. That arch separates the Arkoma basin from the western end of the Black Warrior basin, and the crest of the arch is paralleled by a system of faults that outline the Mississippi Valley graben (Fig. 1; Plate 8A) (Thomas, 1985). Well data and seismic reflection profiles indicate a basement graben (Howe, 1985). Evidently the Mississippi Valley graben is the southwestern part of a system of basement faults that outline the Rough Creek graben and the Rome trough in the subsurface of Kentucky and West Virginia (Fig. 1) (Woodward, 1961; Harris, 1975; Kulander and Dean, 1978; Webb, 1980; Kane and others, 1981; Soderberg and Keller, 1981; Schwalb, 1982; Thomas, 1985).

Mesozoic and Cenozoic structural evolution of the Gulf Coastal Plain has resulted in southwestward tilting of the Black Warrior basin and Appalachian-Ouachita fold-thrust belt toward the axis of the Mississippi Embayment. Fault systems associated with Mesozoic opening of the Gulf of Mexico crosscut Paleozoic structures of the Appalachian-Ouachita orogenic system, but the Mesozoic faults are restricted to the region south of the Black Warrior foreland basin (Thomas and others, in press).

STRATIGRAPHY

Introduction

The lower part of the Paleozoic sedimentary cover sequence above Precambrian crystalline basement rocks in the Black Warrior basin and surrounding regions consists of a basal clastic unit and an overlying thick, regionally extensive, transgressive carbonate unit that together constitute the Cambrian–Lower Ordovician Sauk sequence (of Sloss, 1963) (Plate 8B). The craton-wide unconformity that regionally marks the top of the Sauk sequence is not distinct in the Black Warrior basin, and the overlying Middle Ordovician consists of carbonate rocks like those in the upper part of the Sauk sequence. The Middle Ordovician through Lower Mississippian succession is dominated by carbonate rocks and chert, but includes some thin clastic units that evidently reflect distant sources. That succession is interrupted by three or more unconformities and is relatively thin in comparison to the underlying Cambrian-Ordovician carbonate sequence and to overlying Mississippian-Pennsylvanian clastic rocks (Plate 8B). The Cambrian through Lower Mississippian strata reflect deposition on a shallow-marine shelf but not in a distinct "basin"; however, the Mississippian-Pennsylvanian rocks constitute a southwestward-thickening clastic wedge that signifies the Black Warrior as a foreland basin (Plate 8B).

Precambrian Basement

Precambrian basement rocks drilled in the region of the Black Warrior basin include granite, rhyolite, granodiorite, and basalt (Plate 8A-2). Northwest of the Black Warrior basin, Precambrian basement rocks exposed on the Ozark dome are rhyolite and granite along with minor intrusions of basalt (Missouri Geological Survey, 1979; Van Schmus and others, 1987). Most basement rocks in the region are lithologically compatible with the Eastern Granite-Rhyolite Province (Van Schmus and others, 1987). Locally, in the Mississippi Valley graben, the basement is granitic and dioritic gneiss (well 32) indicating that the granite-rhyolite province is not totally continuous (Denison, 1984). A possible eastern limit of the granite-rhyolite province is suggested by other lithologic characteristics, such as metamorphic fabric of the basement rocks in well 36 in Georgia (Neathery and Copeland, 1983), low-grade metamorphism of granites in Mississippi (Harrelson and Bicker, 1979), and granodiorite composition in wells 9 and 36.

Ozark dome basement ages (U-Pb zircon) are 1480 Ma (Bickford and Mose, 1975). Radiometric age of granite (well 7) on the Nashville dome is 1120 ± 30 Ma (Rb-Sr) (Wasserburg and others, 1962). Other analyses have yielded anomalously young K-Ar dates that are not meaningful in terms of crystallization ages (well 36, 652 ± 32 Ma; well 8, 752 ± 27 Ma; well 15, 790 ± 70 Ma; well 32, 845 ± 42 Ma) (Riggs, 1976; Neathery and Copeland, 1983; Howe, 1985). The granite in well 7 is within the age range of the Grenville Province (1.0 Ga) suggesting possible southward extent of that province; however, the difference between the Rb-Sr age of the granite in well 7 and the U-Pb zircon age of the Eastern Granite-Rhyolite Province may not be significant (M. E. Bickford, personal communication, 1987).

Basal Clastic Unit

The basal clastic unit of the Sauk sequence exhibits three different styles of distribution: (1) a northwestward-thinning clastic facies along the Appalachian fold-thrust belt; (2) a relatively thin sandstone at the base of a regionally extensive, transgressive sequence northwest of the Appalachians; and (3) locally thick clastic rocks restricted to downthrown basement fault blocks of the Rome–Rough Creek–Mississippi Valley graben system (Fig. 3). Regional stratigraphy indicates northwestward onlap of the Sauk sequence onto the craton (Sloss, 1963). The age of the base of the transgressive sequence decreases northwestward from Early to Late Cambrian, and the age of the top of the basal clastic unit also varies regionally from Early to Late Cambrian (Sloss, 1963; Palmer, 1971). The clastic facies grades upward and, in part, laterally into carbonate facies.

In the eastern part of the Black Warrior basin in Alabama, the section between Upper Cambrian carbonate rocks and Precambrian basement consists of 520 m of strata assigned to the Conasauga (gray and greenish-gray mudstones and limestones) and Rome (red, green, and gray mudstones; varicolored siltstones and sandstones; and limestones) Formations (Fig. 3) (Kidd, 1975; Kidd and Neathery, 1976). Fossils from the Conasauga Formation indicate a Middle Cambrian age. Westward thinning of the basal clastic unit is documented, and westward gradation from a

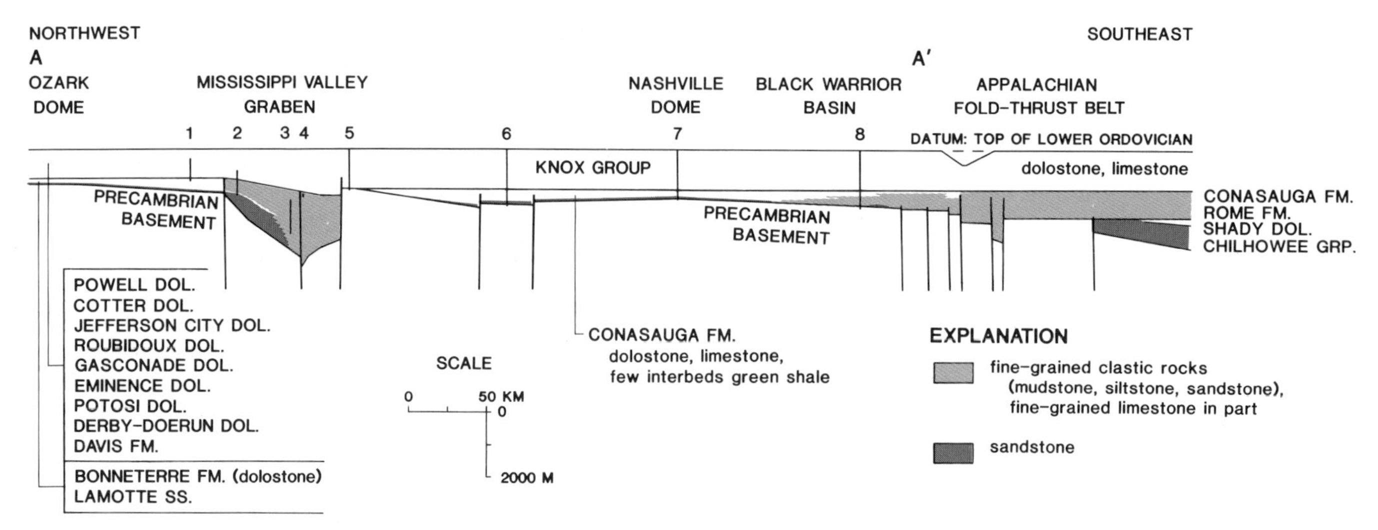

Figure 3. Stratigraphic cross section of Cambrian and Lower Ordovician rocks (Sauk sequence). Line of cross section coincides with the line of cross section A-A′ (Plate 8A-3) and extends southeastward across the Appalachian fold-thrust belt (horizontal dimension palinspastically restored). Wells listed by number in Table 1. Abbreviations: DOL.—Dolostone; FM.—Formation; SS.—Sandstone; GRP.—Group.

fine-grained clastic facies to a carbonate facies is implied (Fig. 3; Plate 8B).

In the western part of the Black Warrior basin in Mississippi, and to the north on the Nashville dome in Tennessee, a relatively thin basal sandstone is overlain by Cambrian carbonate rocks (Fig. 3). The sandstone has a maximum known thickness of less than 80 m; and in some places, possibly on paleotopographic highs, carbonate rocks rest directly on basement. No biostratigraphic data are available to define the precise age of the basal sandstone or of the lower part of the carbonate sequence; however, the lower part of the carbonate succession contains sparse green shale interbeds that have been the basis for lithostratigraphic correlation with the Conasauga Formation to the east (Figs. 3, 4). Northwest of the Black Warrior basin on the Ozark dome, the basal Lamotte Sandstone of Late Cambrian age unconformably overlies basement rocks and is overlain by transgressive Upper Cambrian carbonate rocks; thickness of the sandstone varies greatly because of underlying paleotopography but does not exceed 150 m (Houseknecht and Ethridge, 1978).

The sequence exposed in the Appalachian fold-thrust belt in Alabama southeast of the Black Warrior basin further confirms northwestward transgression of the Sauk sequence. The Lower Cambrian Chilhowee Group (sandstone, conglomerate, and mudstone) is more than 750 m thick; however, the complete thickness is unknown because the lower part is detached at thrust faults (Mack, 1980). The Chilhowee includes fluvial to shallow-marine clastic sediments derived from the craton (Mack, 1980) and is overlain by the transgressive Lower Cambrian Shady Dolostone which is the oldest carbonate component of the Sauk sequence. The Shady is overlain by fine-grained clastic rocks of the Rome and Conasauga Formations, the distribution of which indicates a source on the craton to the northwest (Rodgers, 1968; Palmer, 1971). The clastic facies of the Conasauga grades eastward to a carbonate facies in the Appalachians of Tennessee and Virginia (Palmer, 1971), and the lower part of the Rome grades southeastward into a carbonate facies above the Shady Dolostone in southwestern Virginia (Pfeil and Read, 1980). Similarly, part of the Conasauga clastic facies grades into a carbonate facies in the Appalachians of Alabama, but details of geographic distribution of the two facies are obscured by thrust faults and poor exposure (Palmer, 1971; Thomas and Drahovzal, 1973). Some facies and thickness variations are evidently related to basement faults that are now beneath Appalachian thrust sheets (Thomas, 1986a). The Shady, Rome, and Conasauga aggregate more than 1050 m in the Appalachian fold-thrust belt in Alabama (Copeland and Raymond, 1985), but the sequence thins northwestward across the Black Warrior basin (Fig. 3).

The Rome–Rough Creek–Mississippi Valley graben system, west and north of the Black Warrior basin, contains a clastic sedimentary fill that is locally more than 1 km thick on downthrown fault blocks and is thin or lacking outside the graben system (Fig. 3; Plate 8C-1) (Woodward, 1961; Harris, 1975; Webb, 1980; Kersting, 1982; Schwalb, 1982; Howe, 1985). The fill of the Mississippi Valley graben is characterized by dark-colored mudstones that are in part calcareous and/or silty. The fill also contains very fine-grained sandstone and siltstone, as well as dark-colored fine-grained partly silty to argillaceous limestone. Although the rocks are predominantly dark gray, shades of greenish-gray, red, and brown are also recognizable. A distinct succession of three units in part of the Mississippi Valley graben in Arkansas and Missouri consists of a basal sandstone that is generally arkosic and red, but is quartzose and light colored toward the top (Denison, 1984); a middle unit of light-colored limestone and dolostone; and an upper dark-colored partly calcareous mudstone (Fig. 3). Seismic reflection profiles suggest lateral continuity of the middle carbonate unit over much of the

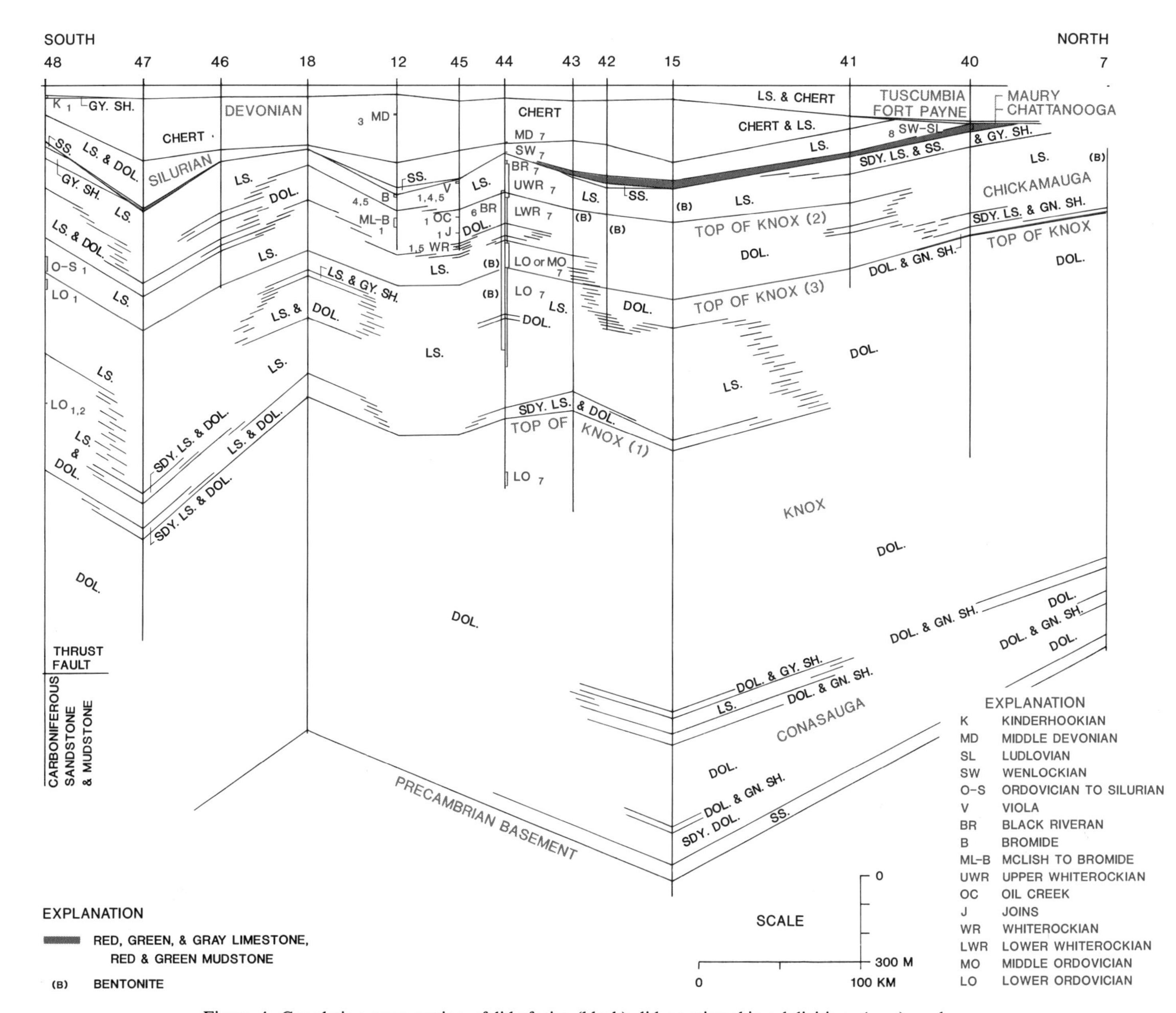

Figure 4. Correlation cross section of lithofacies (black), lithostratigraphic subdivisions (gray), and biostratigraphically controlled subdivisions (red) of Cambrian through Lower Mississippian rocks. Line of cross section shown on Figure 2; wells listed by number in Table 1. Datum: top of Fort Payne–Tuscumbia.

Three alternative correlations of top of Knox Group are: (1) at base of sandy carbonate unit and below thick limestone that contains Middle Ordovician ostracodes (R. W. Harris, *in* Mississippi Geological Society, undated; Thomas, 1972a; J. R. Derby, *in* Ross and others, 1982); (2) at top of highest dolostone (Mellen, 1972; Kidd, 1975); and (3) at top of Lower Ordovician carbonate rocks as indicated by conodonts (J. E. Repetski and L. P. Alberstadt, unpublished data). Lithofacies boundaries between wells 15, 41, and 40 are drawn in accord with "top of Knox (3)" but can be modified to accord with (1) or (2).

Biostratigraphic data and sources of data: 1. conodonts (J. F. Baesemann, G. Klapper, Amoco, unpublished data); 2. gastropods (J. R. Derby, Amoco, unpublished data); 3. trilobites, brachiopods (A. R. Ormiston, Amoco, unpublished data); 4. brachiopods, trilobites, bryozoans (J. R. Derby, Amoco, unpublished data); 5. ostracodes (W. B. Creath, Amoco, unpublished data); 6. ostracodes (R. W. Harris, *in* Mississippi Geological Society, undated); 7. conodonts (J. E. Repetski and L. P. Alberstadt, unpublished data); 8. ostracodes (Lundin and Newton, 1970).

Abbreviations: DOL.—dolostone; SH.—shale, mudstone; LS.—limestone; SS.—sandstone; SDY.—sandy; GY.—gray; GN.—green.

southern part of the graben. The distinct tripartite succession is not recognizable in well records from the northeastern part of the graben, where siltstone, sandstone, limestone, and mudstone apparently are non-systematically distributed, and contacts are gradational both vertically and laterally. In the southernmost well in the graben (well 33), the succession is mostly limestone and dolostone but contains some dark-colored mudstone, dark-colored argillaceous limestone, anhydrite, and a relatively thin basal sandstone (Mellen, 1977).

Sparse biostratigraphic data confirm an Early Late Cambrian (Dresbachian) age for the upper part of the clastic sequence in the Mississippi Valley graben (wells 2, 4, 38, 39) (Grohskopf, 1955; Palmer, 1962; Missouri Geological Survey open file), but the age of the oldest part of the graben fill is unknown. The graben-fill clastic sequence is overlain by Upper Cambrian carbonate rocks which extend across the graben-boundary faults and rest, generally with a thin basal sandstone, on basement rocks of the upfaulted blocks (Woodward, 1961), indicating that the graben structures were filled with clastic sediment by Late Cambrian time. Thicknesses of the graben-fill sequences indicate more than a kilometer of pre-Late Cambrian displacement on some graben-boundary faults (Fig. 3), and some of the faults had later Paleozoic movement as well (Dever and others, 1977; Dever, 1986).

Cambrian-Ordovician Carbonate Rocks

A widespread Cambrian and Ordovician carbonate succession above the basal clastic unit records the maximum transgression during deposition of the Sauk sequence and indicates a persistent shallow-marine shelf that extended throughout the Black Warrior basin, across the Rome–Rough Creek–Mississippi Valley graben system, and farther north and west on the craton (Fig. 3; Plate 8C-2). The succession includes limestones and dolostones separated by partly diachronous facies boundaries; limestone is generally more common stratigraphically upward and toward the south (Fig. 3; Plate 8B). Parts of the succession include sandy carbonate rocks that contain well-rounded medium to coarse quartz sand grains and rare interbeds of quartz sandstone. Parts of the carbonate sequence are cherty. Locally, in the southern part of the Mississippi Valley graben, dark-colored fine-grained limestone suggests deeper water environments (Plate 8C-2).

Regionally, the Upper Cambrian and Lower Ordovician carbonate rocks are assigned to the Knox Group, and the Knox is unconformably overlain by Middle Ordovician limestones of the Chickamauga Group (Chickamauga Supergroup of Milici and Smith, 1969). The craton-wide pre-Middle Ordovician unconformity between the Sauk and Tippecanoe sequences (of Sloss, 1963) is not lithologically distinct in the Black Warrior basin, and the position of the boundary between Lower and Middle Ordovician strata is somewhat uncertain (Fig. 4). Biostratigraphic work, based mainly on ostracodes (from drill cuttings), indicates that Middle Ordovician faunas extend more than 680 m below the top of the carbonate succession (R. W. Harris, *in* Mississippi Geological Society, undated). A dominantly limestone succession (assigned to the Chickamauga) and an underlying thick dolostone (assigned to the Knox) are separated by sandy carbonate rocks which have been interpreted as marking the base of the Middle Ordovician, an interpretation that is consistent with the biostratigraphic correlations of Harris (Fig. 4) (Thomas, 1972a; J. R. Derby, *in* Ross and others, 1982). Alternatively, on the basis of regional lithostratigraphy, the top of the Knox Group (top of Lower Ordovician) has been placed at the top of a dolostone unit more than 700 m above the sandy carbonate rocks (Fig. 4) (Mellen, 1972; Kidd, 1975), and that correlation commonly has been used in the petroleum industry. More recent biostratigraphic work, using conodonts, indicates the Lower-Middle Ordovician boundary at a lithologically unmarked position within the carbonate succession between the stratigraphic levels previously identified as the top of the Knox Group (Fig. 4) (Amoco, unpublished data; J. E. Repetski and L. P. Alberstadt, unpublished data). These biostratigraphic data indicate more than 270 m of the Whiterockian Stage (lower Middle Ordovician). Probably little or no hiatus marks the Sauk-Tippecanoe boundary in the Black Warrior basin.

The Middle Ordovician in the Black Warrior basin is dominantly limestone, but it contains a dolostone unit that is progressively younger toward the south (Fig. 4; Plate 8B), suggesting a southward-prograding intertidal shoal (Thomas, 1972a). The time-transgressive base of the dolostone apparently crosses the projected level of the Lower-Middle Ordovician boundary as based on conodonts (Fig. 4); however, none of the biostratigraphic data are from the area of the older part of the dolostone facies. The carbonate succession includes thin beds of bentonite which apparently are similar in lithologic association to bentonites in the Middle Ordovician of the Appalachian foreland basin and the Appalachian fold-thrust belt; however, some bentonites in the Black Warrior basin are stratigraphically below the Lower-Middle Ordovician boundary as based on conodonts (Fig. 4).

The age of the top of the Ordovician limestone is not defined by biostratigraphic data, but Middle Ordovician fossils have been reported from near the top of the limestone in the subsurface in Mississippi (Fig. 4). In outcrop in the eastern part of the Black Warrior basin and on the Nashville dome, the upper part of the limestone succession includes Upper Ordovician rocks; but in some places, these are overlain by younger Ordovician clastic rocks (Wilson, 1949; Drahovzal and Neathery, 1971; Neathery and Drahovzal, 1985).

The Ordovician and older carbonate facies extends throughout the Black Warrior basin and Appalachian fold-thrust belt in eastern Mississippi and Alabama (Plate 8C-2), as well as the Arkoma basin where the section includes more sandy units. In contrast to the carbonate facies, in the Ouachita fold-thrust belt in Arkansas, the Upper Cambrian–Ordovician succession contains black graptolite-bearing shales, fine-grained limestones, bedded chert, and units of quartzose sandstone (Ham, 1959; Sterling and others, 1966; Viele, 1973; King, 1975; Thomas, 1976; Ethington and others, in press). The Ouachita succession is interpreted as an off-shelf deep-basin facies equivalent to the shallow-marine carbonate-shelf facies of the Black Warrior and Arkoma basins. Carbonate and quartz sand debris within the Ouachita succession suggests proximity to the edge of the shelf. Seismic reflection data indicate that autochthonous carbonate-shelf facies extend southward from the Arkoma basin beneath northward-translated allochthonous off-shelf facies in the Ouachita Mountains (Lillie and others, 1983). A similar relationship probably prevails in the subsurface along the western part of the Black Warrior basin of western Mississippi, where Ordovician rocks in the carbonate facies evidently extend southward beneath northward-translated frontal Ouachita thrust sheets. In central Mississippi, the thrust front cuts through the shelf edge; and farther east, Ordovician carbonate rocks are contained in Appalachian thrust sheets. Locally, in the Alabama Appalachian outcrops, the black, graptolite-bearing Middle Ordovician Athens Shale overlies the lower part of the carbonate sequence (Drahovzal and Neathery, 1971) and indicates a deep-water environment (Benson and Mink, 1983). The Athens Shale is restricted to the southeastern thrust sheets, and time-equivalent rocks belong to the carbonate-shelf facies in the northwestern Appalachian thrust sheets and adjacent Black Warrior basin.

Ordovician Clastic Rocks

In the Appalachian fold-thrust belt in eastern Alabama, the Middle and Upper Ordovician includes an eastward-thickening clastic wedge of red mudstone and sandstone (Greensport, Colvin Mountain, and Sequatchie formations) (Drahovzal and Neathery, 1971; Chowns and McKinney, 1980); maximum thickness of the wedge in the Appalachians in Alabama is approximately 130 m. The clastic wedge thins westward into the eastern part of the Black Warrior basin and is discontinuous farther west (Plate 8B, 8C-3). Facies relationships indicate that the clastic wedge prograded westward over the carbonate facies. The clastic sediments are interpreted to be the distal fringe of a clastic wedge derived from orogenic uplifts along the Appalachians northeast of Alabama, indicating the southernmost extent of effects of the Taconic orogeny in the foreland of the Appalachians (Thomas, 1977a). From the Alabama Appalachians westward along the Black Warrior and Arkoma foreland basins, no synorogenic clastic-wedge sediments derived from adjacent continental margins on the south or southwest are recognized within the Ordovician passive-margin shelf sequence.

In the central part of the Black Warrior basin, scattered deposits of sandstone, sandy limestone, and black shale less than 35 m thick overlie the Ordovician limestone (Fig. 4; Plate 8B, 8C-3) (Thomas, 1972a). The age of the clastic rocks is uncertain, but lithostratigraphic correlation suggests that the sandstone may constitute a westernmost discontinuous distal fringe of the Upper Ordovician Sequatchie Formation. The Sequatchie in southern Tennessee and northernmost Alabama includes channel-filling sandstones (Wilson, 1949; Neathery and Drahovzal, 1985). The sandstone in the western part of the Black Warrior basin overlies an eastward-pinching black shale (Thomas, 1972a). Although correlation is uncertain, the shale may be part of a widespread Upper Ordovician shale (Maquoketa, Cason, and Sylvan Shales) that extends from the central part of the craton around the Arkoma basin (Maher and Lantz, 1953; Ham, 1959; Thomas, 1972a).

Silurian

In the northern part of the Black Warrior basin, the Silurian includes a lower thin gray limestone; a middle unit of red, green, and gray limestones and red and green mudstones; and an upper green to gray limestone that is partly argillaceous to shaly (Jewell, 1969; Lundin and Newton, 1970; Kidd, 1975). These lithostratigraphic units are correlated with the Brassfield, Wayne, and Brownsport-Decatur formations, respectively, of the section exposed around the southwest side of the Nashville dome (Wilson, 1949). Biostratigraphic data document correlation to the Wenlockian and Ludlovian; however, no fossils have been obtained from the presumably older Brassfield-equivalent strata (Lundin and Newton, 1970; J. E. Repetski and L. P. Alberstadt, unpublished data).

Along the southeastern edge of the Black Warrior basin, the Silurian carbonate facies grades southeastward into clastic rocks of the hematite-bearing Red Mountain Formation, which in the Appalachian fold-thrust belt farther southeast is mainly sandstone and mudstone (Plate 8C-4). Most of the Red Mountain (Llandoverian and Wenlockian) is equivalent to the lower part of the shaly limestone succession to the northwest, but part of the Red Mountain is Pridolian (Berdan and others, 1986) and is as young as the Decatur (the youngest Silurian in the Black Warrior basin).

The Red Mountain clastic facies grades westward and northwestward from the Appalachian outcrops into a carbonate-dominated succession. The relatively thin red and green mudstones decrease southward from the Nashville dome (Plate 8C-4). In the southwestern part of the Black Warrior basin, the Silurian consists of gray limestone and dolostone, part of which is cherty (Fig. 4; Plate 8B, 8C-4) (Thomas, 1972a).

The maximum thickness of Silurian rocks is approximately 180 m in the south-central part of the Black Warrior basin, and the section thins to the east, north, and west. The Silurian is unconformably absent locally along the south limb of the Nashville dome and in the southeastern part of the Appalachian fold-thrust belt (Plate 8C-4). Both the top and base of the Silurian are marked by unconformities regionally; however, unconformable boundaries cannot be conclusively documented in the southwestern part of the Black Warrior basin (Plate 8B).

Devonian

Devonian chert as much as 350 m thick in the western part of the Black Warrior basin is in the stratigraphic position of the

Arkansas Novaculite of the Ouachita Mountains, the Penters Chert of the south limb of the Ozark dome, and cherty carbonate rocks around the southwest side of the Nashville dome. The chert unit thins northeastward across the Black Warrior basin and pinches out northeastward in the subsurface of western Alabama (Plate 8B, 8C-5) (Thomas, 1972a; Thomas and Drahovzal, 1973; Kidd, 1975). Toward the northeast and upward in the section, the chert grades into a cherty limestone in part. Along the southwest side of the Black Warrior basin, the chert thins southwestward (Fig. 4; Plate 8B, 8C-5) and locally (well 33) appears to be very thin or lacking entirely.

The chert contains Lower (Schoharie and Onondaga, upper Lower Devonian) and Middle Devonian shallow-marine fossils in Mississippi (Figs. 4, 5). Helderbergian (Lower Lower Devonian) ostracodes have been identified in drill samples from an interval between the Upper Devonian Chattanooga Shale and strata containing Silurian ostracodes in northwestern Alabama, near the northeastern limit of Devonian chert and cherty limestone (Lundin and Newton, 1970). The craton-wide unconformity between the Tippecanoe and Kaskaskia sequences is at the top of the Helderbergian (Sloss, 1963), but the extent of Helderbergian strata in the Black Warrior basin is poorly known. Possibly the Tippecanoe-Kaskaskia boundary is at or near the base of the Devonian chert throughout the basin, but little or no hiatus may mark the boundary in the southwestern part of the basin.

On the southeast in the Appalachian fold-thrust belt, the Lower and Middle Devonian Frog Mountain Formation consists of sandstone, mudstone, chert, and limestone. The formation is bounded above and below by unconformities and contains internal unconformities (Butts, 1926; Ferrill, 1984). A shallow-marine fauna documents an age range from Oriskany (middle Lower Devonian) to Hamilton (lower Middle Devonian) equivalents (Butts, 1926; Ferrill, 1984). The unconformity at the base of the Frog Mountain corresponds to the base of the Kaskaskia sequence. The Frog Mountain is irregularly distributed (Thomas and Drahovzal, 1973; Ferrill, 1984). No clastic equivalents of the Frog Mountain are known in the Black Warrior basin.

The Devonian chert in the Black Warrior basin is within a succession of shallow-marine deposits. In contrast, the Arkansas Novaculite in the Ouachita Mountains is generally interpreted as a deep-water chert (Viele, 1973; Viele and Thomas, in press). The Devonian Penters Chert on the south limb of the Ozark dome in northern Arkansas and Devonian chert and cherty carbonate units on the southwest limb of the Nashville dome in southern Tennessee reflect shallow-shelf environments. In Arkansas, an intermediate thin shaly facies separates the shallow-water Penters Chert on the north from the deep-water Arkansas Novaculite to the south (Thomas, 1977b); possibly the thinner chert at the southwest edge of the Black Warrior basin (Plate 8C-5) reflects a similar shelf-edge transition. Distribution of the Devonian chert in the Black Warrior basin suggests analogy with the shallow-marine Devonian Penters Chert and, therefore, a tectonic framework of deposition on a passive-margin shelf and shelf edge.

Upper Devonian–Lower Mississippian

In the eastern part of the Black Warrior basin in northern Alabama, the Upper Devonian Chattanooga Shale (black shale) and Lower Mississippian Maury Shale (green mudstone containing phosphate nodules) (Hass, 1956) have a combined thickness generally less than 15 m (Conant and Swanson, 1961). The shale units thin and pinch out to the west and south, and in Mississippi, only rare discontinuous black shale units mark the position of the Chattanooga (Fig. 5) (Mellen, 1955; Thomas, 1972a). In the eastern part of the Black Warrior basin, the Chattanooga Shale rests unconformably on Silurian, and locally Ordovician, rocks; toward the west, it oversteps the eastward-pinching Devonian chert (Plate 8B).

Mississippian Fort Payne–Tuscumbia

The youngest rocks in the carbonate-dominated part of the Paleozoic succession in the Black Warrior basin are cherty limestone and chert that are lithologically similar to and in the stratigraphic position of the Lower Mississippian Fort Payne Chert and Tuscumbia Limestone of Alabama outcrops and the Boone Formation of northern Arkansas outcrops. The Fort Payne Chert (Osagean) thickens northeastward and is overlain by

Figure 5. Generalized correlation cross section of lithostratigraphic subdivisions and lithofacies (black) and biostratigraphically controlled chronostratigraphic subdivisions (red) of Mississippian and Pennsylvanian rocks.

Biostratigraphic data: ***Well 49*** (location on Figure 2, Table 1) 1. Lower Pennsylvanian brachiopods; 2. Chesterian and Pennsylvanian brachiopods, bryozoans; 3. Chesterian brachiopods; 4. Chesterian (Lower Namurian) goniatites, pelecypods; 5. Upper Visean (possibly near Meramecian-Chesterian boundary) goniatites; 6. Chesterian brachiopods; 7. probable Chesterian brachiopods; 8. Meramecian or Chesterian brachiopods; 9. Osagean brachiopods, bryozoans, goniatites; 10. Lower Devonian (Schoharie and Onondaga) brachiopods, trilobites; 11. Upper Mississippian conodonts. ***Well 50*** (location on Figure 2, Table 1) 12. Chesterian and Pennsylvanian brachiopods, bryozoans, pelecypods; 13. Chesterian brachiopods; 14. Upper Devonian conodonts; 15. Lower Devonian (Schoharie and Onondaga) brachiopods. ***Well 51*** (location on Figure 2, Table 1) 16. Chesterian and Pennsylvanian brachiopods, bryozoans, corals. ***References to data sources:*** 1-9, 12, 13, 16—Gordon, 1953; 10, 15—Boucot, 1953; 11, 14—Hass, 1953. ***Correlations of Pennsylvanian strata:*** 17. palynomorphs, Upshaw, 1967; 18. megaflora, Lyons and others, 1985. ***Correlations of Mississippian outcrop sections:*** 19, 20—references cited in text.

Lithostratigraphic boundaries: X. diachronous base of Pottsville (base of lowest sandstone containing quartz pebbles). Y. diachronous base of Parkwood (base of lowest sandstone). Z. arbitrary lateral cut-off (Pride Mountain Formation below Hartselle Sandstone; equivalent beds included in Floyd Shale southwest of limit of Hartselle) (Mynot Sandstone Member and Tanyard Branch Member of Pride Mountain are laterally continuous with informally named Evans sand and Lewis sand, respectively).

Abbreviations: FM.—Formation; LS.—Limestone; SH.—Shale; SS.—Sandstone; MBR.—Member.

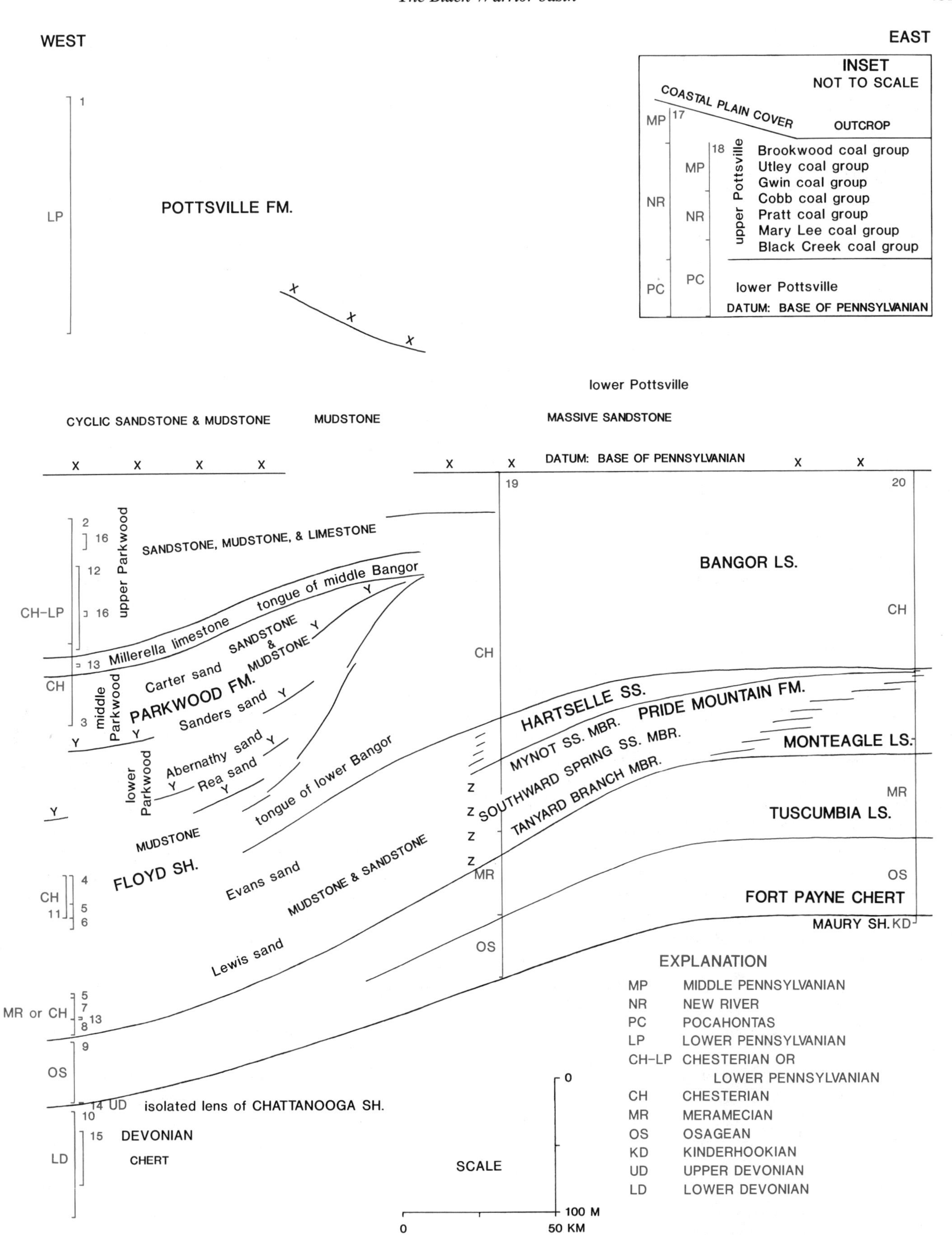

WEST
EAST
INSET
NOT TO SCALE
COASTAL PLAIN COVER
OUTCROP
Brookwood coal group
Utley coal group
Gwin coal group
Cobb coal group
Pratt coal group
Mary Lee coal group
Black Creek coal group
upper Pottsville
lower Pottsville
DATUM: BASE OF PENNSYLVANIAN
POTTSVILLE FM.
lower Pottsville
CYCLIC SANDSTONE & MUDSTONE
MUDSTONE
MASSIVE SANDSTONE
DATUM: BASE OF PENNSYLVANIAN
SANDSTONE, MUDSTONE, & LIMESTONE
upper Parkwood
tongue of middle Bangor
Millerella limestone
SANDSTONE & MUDSTONE
Carter sand
PARKWOOD FM.
middle Parkwood
Sanders sand
Abernathy sand
Rea sand
lower Parkwood
tongue of lower Bangor
MUDSTONE
FLOYD SH.
Evans sand
MUDSTONE & SANDSTONE
Lewis sand
BANGOR LS.
HARTSELLE SS.
PRIDE MOUNTAIN FM.
MYNOT SS. MBR.
SOUTHWARD SPRING SS. MBR.
TANYARD BRANCH MBR.
MONTEAGLE LS.
TUSCUMBIA LS.
FORT PAYNE CHERT
MAURY SH.
isolated lens of CHATTANOOGA SH.
DEVONIAN
CHERT
SCALE
0
100 M
0
50 KM
EXPLANATION
MP MIDDLE PENNSYLVANIAN
NR NEW RIVER
PC POCAHONTAS
LP LOWER PENNSYLVANIAN
CH–LP CHESTERIAN OR LOWER PENNSYLVANIAN
CH CHESTERIAN
MR MERAMECIAN
OS OSAGEAN
KD KINDERHOOKIAN
UD UPPER DEVONIAN
LD LOWER DEVONIAN

the northeastward-thickening Tuscumbia Limestone (Meramecian) (Fig. 5; Plate 8B) (Thomas, 1972b, 1974). The Tuscumbia differs from the Fort Payne in containing less chert and more coarse bioclastic limestone (Thomas, 1972b); the two formations aggregate a maximum thickness of approximately 130 m in the northeastern part of the Black Warrior basin. In the thinner sections on the southwest, the two formations are not lithologically distinct. West of the limit of Chattanooga Shale in the Black Warrior basin, the Fort Payne–Tuscumbia rests on the Devonian chert (Plate 8B). The Fort Payne–Tuscumbia also thins southeastward into the Appalachian fold-thrust belt.

The outline of the passive margin south of the Black Warrior basin appears identical for Cambrian to Lower Mississippian rocks. The Lower Mississippian limestones and cherts contain echinoderm-brachiopod faunas indicating shallow-shelf environments (Thomas, 1972b). The southwestward thinning and transition to generally finer grained limestones suggest a deeper shelf in that direction (Plate 8C-6).

Mississippian-Pennsylvanian Clastic Wedge: Regional Stratigraphy and Correlation

The Mississippian-Pennsylvanian succession above the Fort Payne–Tuscumbia in the Black Warrior basin constitutes a southwestward-thickening, northeastward-prograding clastic wedge of gray mudstone and sandstone (Plate 8B). The maximum preserved thickness of the clastic wedge in the structurally deepest part of the basin is more than 3200 m, and the entire succession reflects shallow-marine to deltaic deposition. The Mississippian-Pennsylvanian clastic wedge contrasts with the underlying Paleozoic passive-margin shelf facies in sediment types, sedimentation rates, sediment dispersal, and tectonic framework.

The Mississippian part of the clastic wedge (Floyd Shale, Pride Mountain Formation, Hartselle Sandstone, and Parkwood Formation) grades northeastward into a carbonate facies (Monteagle and Bangor Limestones), and the clastic and carbonate facies intertongue across a wide area in the eastern part of the Black Warrior basin (Fig. 5; Plate 8B) (Thomas, 1972a, 1972b). The facies boundary trends northwestward across the Black Warrior basin, approximately perpendicular to strike of Appalachian structures along the southeast side of the basin, and approximately parallel with the Ouachita structural front along the southwest side of the basin (Plate 8C). The clastic facies in the Pennsylvanian part of the wedge (Pottsville Formation) extends northeastward over the Mississippian carbonate facies (Fig. 5; Plate 8B).

Formal lithostratigraphic subdivision of the Mississippian-Pennsylvanian clastic wedge includes diachronous boundaries and arbitrary cut-offs because of the progradational history of the various components (Fig. 5) (see summary of evolution of nomenclature, Thomas, 1972b). In addition to the system of formal stratigraphic names, informal names for specific strata are used in the petroleum and coal industries (Fig. 5).

Biostratigraphic data are available locally from the subsurface in the northwestern part of the Black Warrior basin (Boucot, 1953; Gordon, 1953; Hass, 1953) and from outcrops around the eastern end of the basin (summarized by Butts, 1926; Drahovzal, 1967; Thomas, 1979) (Fig. 5). The basal beds of the clastic wedge are of late Meramecian (equivalent to Ste. Genevieve) age, but the Meramecian-Chesterian boundary is near the base of the clastic facies (Fig. 5). Most of the Floyd-Parkwood succession in the western part of the basin contains a Chesterian marine invertebrate fauna equivalent to that in the upper Monteagle and Bangor Limestones to the east (Fig. 5).

The upper part of the Parkwood Formation in the western part of the basin contains "an upper Mississippian (Chester) invertebrate fauna with Pennsylvanian elements" that is interpreted as possibly extending across the Mississippian-Pennsylvanian boundary (Fig. 5) (Gordon, 1953). In the same area, the lower Pottsville strata contain Lower Pennsylvanian brachiopods. In outcrops in the Appalachian fold-thrust belt, as well as in the northern part of the Black Warrior basin, the Parkwood in various places includes both Mississippian and Pennsylvanian marine invertebrates and plant fossils that range from the Mississippian-Pennsylvanian boundary into the Lower Pennsylvanian (Butts, 1926; summary by Thomas, 1979; Henry and others, 1981; Jennings and Thomas, 1987). The evident range of ages in the upper part of the Parkwood is consistent with the diachronous "base of Pottsville" as defined on the basis of the lowest sandstones containing quartz pebbles (Fig. 5) (Thomas, 1972a; Thomas and Hines, 1985).

The transitional assemblages in the biostratigraphic succession in the Black Warrior basin suggest no regional unconformity between Mississippian and Pennsylvanian strata. Local scour surfaces and lag gravels, common at numerous levels within the progradational clastic succession (Floyd-Parkwood-Pottsville), evidently are associated with local channels rather than a regional unconformity. In the eastern part of the Black Warrior basin, massive sandstones of the lower Pottsville overlie the Mississippian Bangor Limestone (Plate 8B); however, a thin interval of maroon and green mudstone, gray mudstone, and sandstone commonly separates the highest thick limestone from the base of the massive sandstone, suggesting rapid progradation of clastic sediment over the carbonate shelf (Thomas, 1972b). The base of the massive sandstone commonly rests on a scoured surface; whether the scour reflects local depositional systems (such as tidal channels in a barrier island) or a regional unconformity is unresolved. In the interior of the craton, the craton-wide unconformity between the Kaskaskia and Absaroka sequences separates Mississippian and Pennsylvanian rocks; however, toward the margins of the craton, the hiatus decreases and transitional units such as the Parkwood wedge into the succession (Sloss, 1963). The base of the Pottsville in the eastern part of the Black Warrior basin evidently coincides with the Kaskaskia-Absaroka boundary.

The entire Pottsville Formation exposed in the eastern part of the Black Warrior basin commonly is assigned to the Lower Pennsylvanian (Butts, 1926; Metzger, 1965; Upshaw, 1967), and

Middle Pennsylvanian (Atokan, Kanawha-equivalent) beds as indicated by palynology are restricted to the subsurface in the western part of the basin (Fig. 5). Alternative paleobotanical correlations place the contact between the Lower and Middle Pennsylvanian within the upper Pottsville in the eastern part of the basin (Fig. 5) (Lyons and others, 1985). Although not dated biostratigraphically, the youngest beds preserved in the basin below the Mesozoic strata of the Gulf Coastal Plain must be Middle Pennsylvanian or younger.

Mississippian-Pennsylvanian Clastic Wedge: Thickness, Depth of Burial, Sedimentation Rates, and Subsidence History of the Black Warrior Basin

The Mississippian part of the clastic wedge thickens southwestward to a maximum of nearly 500 m, and more than 2700 m of Pennsylvanian strata are preserved in the southwestern part of the basin. The youngest Pennsylvanian rocks are overlain unconformably by Cretaceous strata of the Gulf Coastal Plain, and some Pennsylvanian rocks must have been removed by pre-Mesozoic erosion. Maximum burial depth of the top of the Tuscumbia Limestone (the base of the clastic wedge) calculated from vitrinite reflectance, mudstone density, and sandstone porosity within the clastic wedge ranges from 5500 m in the southern part of the basin to 1850 m in the northern part (Hines, 1988). These values suggest erosion of more than 3500 m of cover from above the presently preserved strata in part of the basin; however, data for burial-depth calculations are not available for places where the greatest thickness is preserved.

The thickness (restored for compaction) of Mississippian strata deposited between late Meramecian and the end of Chesterian indicates a sedimentation rate of 2.8 to 3.1 cm/1000 yr (Hines, 1988). All of the sediment was deposited essentially at sea level, so the sedimentation rate reflects basin subsidence rate. Similar calculations for the Lower Pennsylvanian yield sedimentation and subsidence rates of 28.9 to 30.4 cm/1000 yr (Hines, 1988). Subsidence history and geometry of the Black Warrior basin are comparable to those calculated for foreland basins associated with thrust loading (Jordan, 1981; Speed and Sleep, 1982; Schedl and Wiltschko, 1984).

Mississippian-Pennsylvanian Clastic Wedge: Sediment Dispersal and Provenance

Paleogeographic reconstructions of various parts of the Mississippian-Pennsylvanian succession include northwest-trending barrier beaches facing seaward to the northeast; northeastward-prograding delta distributaries; and northwest-trending marine-reworked delta-front sandstones along the northeast sides of delta-distributary sandstones (Plate 8C) (for example, Ferm and Ehrlich, 1967; Thomas, 1972a, 1972b, 1974, 1979; Hobday, 1974; Horsey, 1981; Thomas and Mack, 1982). The paleogeographic reconstructions consistently incorporate dispersal of clastic sediment by northeastward-prograding delta systems from a source southwest of the Black Warrior basin. Southwestward thickening of the Mississippian-Pennsylvanian succession along with the persistence of shallow-marine to deltaic depositional systems indicates differential subsidence of the Black Warrior foreland basin and an approximately equal rate of sediment accumulation. The clastic facies prograded northeastward (in the direction of sediment transport) across the more rapidly subsiding part of the basin and onto a carbonate shelf. The history of basin subsidence and sediment dispersal conforms to models for evolution of foreland basins during tectonic loading at convergent plate margins (for example, Jordan, 1981; Speed and Sleep, 1982; Schedl and Wiltschko, 1984).

Sandstone petrography provides an independent test of the tectonic setting of the source of the Mississippian-Pennsylvanian clastic sediment (summarized here from Mack and others, 1981; Mack and others, 1983). Sandstones range from lithic arenite to quartz arenite, but variation in the proportion of quartz is directly related to the intensity of marine reworking within different depositional environments. The quartz population includes both monocrystalline and polycrystalline types. The polycrystalline quartz is mostly unstable, indicating significant input from a low-grade metasedimentary source (Young, 1976), but the relatively high proportion of monocrystalline quartz suggests mixing of recycled quartz from sedimentary source rocks (Mack and others, 1981).

The most abundant non-quartz detrital grains are fine-grained schist, phyllite, and slate, suggesting derivation from low-grade metapelitic rocks. Chert is the most common type of sedimentary rock fragment and is more abundant upward. Volcanic rock fragments, generally in mafic to intermediate compositional ranges, are present throughout the clastic wedge; however, the proportion of volcanic rock fragments to other types of rock fragments increases up section. Feldspar is a minor component. With two exceptions (chert and volcanic rock fragments), the non-quartz population is uniform throughout the vertical succession of the clastic wedge.

Composition of the sandstones indicates that the provenance of the clastic wedge included sedimentary, metamorphic, and volcanic rocks. Furthermore, these components were present together from the inception of sediment dispersal. Orogenic juxtaposition of an arc, subduction complex, and fold-thrust belt is suggested (Mack and others, 1983). The upward increase in volcanic rocks and chert may reflect progressive integration of a drainage system across an orogenic belt. Composition of the sandstones can best be interpreted to indicate that the provenance of the clastic wedge resulted from arc-continent collision associated with southward-directed subduction of the southern margin of North American crust (Mack and others, 1983). Such a provenance is consistent with the interpretation that the source of clastic sediment was southwest of the Black Warrior basin and is also consistent with tectonic models for foreland-basin subsidence.

An alternative interpretation, that sediment was dispersed from the craton into the Black Warrior basin by southeastward-prograding deltas (Swann, 1964; Welch, 1978; Cleaves and Broussard, 1980), implies delta progradation parallel with and along the clastic-carbonate facies boundary, requires basin subsidence at a passive margin or intracratonic location, and suggests a distant and/or cratonic provenance. Regional distribution of the clastic and carbonate facies, paleogeographic reconstructions, geometry and subsidence history of the basin, and sandstone petrography all support the interpretation that sediment was dispersed from an arc-continent collision orogen along the southwest side of the Black Warrior foreland basin.

Mississippian-Pennsylvanian Clastic Wedge: Lithofacies, Depositional Systems, and Paleogeography

Subdivisions. Lithofacies patterns, including succession of carbonate intertongues and distribution of sandstones, enable a five-fold vertical subdivision of the clastic wedge. The five subdivisions in ascending order are: Pride Mountain–Hartselle–lower Floyd, lower and middle Parkwood, upper Parkwood, lower Pottsville, and upper Pottsville (Fig. 5). Regional facies relations, distribution of lithosomes within the clastic wedge, sedimentary structures, and vertical sedimentary sequences provide the basis for interpretation of depositional systems and reconstruction of paleogeography within each of the subdivisions.

Pride Mountain–Hartselle–Lower Floyd. The lower part (upper Meramecian and lower Chesterian) of the clastic wedge is an extensive tongue of gray mudstone that contains four distinct units of quartzose sandstone (Fig. 5; Plate 8B). In the northeasternmost part of the Black Warrior basin, the clastic tongue grades northeastward into the Monteagle Limestone which is characterized by tidal oolite bars (Handford, 1978). The clastic tongue progrades over the limestone, and the Hartselle Sandstone at the top of the tongue extends farthest northeast. The Hartselle Sandstone is overlain by a transgressive tongue of the lower part of the Bangor Limestone that extends far to the southwest into the clastic wedge (Fig. 5; Plate 8B). The Pride Mountain–Hartselle–lower Floyd clastic tongue thickens southwestward across the eastern part of the Black Warrior basin, but it thins abruptly southward into the southern part of the basin (distinctly defined by southward thinning of the interval between the Tuscumbia and Bangor Limestones; Plate 8B) (Higginbotham, 1985). The lowest of the four sandstones extends throughout most of the basin (Plate 8C-7), but the upper three are excluded from the area on the south where the clastic tongue is thin (Plate 8C-8, 9). The thin mudstone in that area contains fossil leaf imprints and plant fragments (Holmes, 1981; Higginbotham, 1985).

Each of the four sandstones has a generally northwest-trending linear distribution in the eastern part of the basin. Each pinches out northeastward toward the carbonate facies and southeastward along trend into mudstone (Plate 8C-7, 8, 9). The original northwestward extent is unknown because of post-Mississippian erosion of the northern limb of the Black Warrior basin. The distribution of sandstone on the southwest is distinct for each of the four, reflecting differences in details of depositional systems.

The lower three sandstone units (Pride Mountain and equivalent lower Floyd sandstones) commonly include limestone interbeds, and locally grade into limestones that are variously argillaceous, bioclastic, oolitic, or sandy (Moser and Thomas, 1967; Thomas, 1972a, 1972b). Parts of the mudstone succession are calcareous and contain marine invertebrate fossils. The sandstones are characterized by ripple lamination, crossbedding, and bioturbation (Thomas, 1972b; Holmes, 1981; Higginbotham, 1985). Although a local basal scoured surface and lag gravels are associated with small channels, a gradational coarsening-upward basal sandstone is more common (Thomas, 1972b; Higginbotham, 1985). The sandstones in the northeastern part of the basin are interpreted as shallow-marine bars (Thomas, 1972b; Holmes, 1981); but local northeast-trending thick sandstones, in part characterized by a fining-upward succession indicated by geophysical well logs, suggest deltaic deposits in the western part of the basin for the lower and upper Pride Mountain sandstones (Plate 8C-7, 8) (Higginbotham, 1985). In the lower sandstone, deltaic facies along the southwestern side of the basin are bordered on the northeast by a wide area of shallow-marine bars (Plate 8C-7). In contrast, the deltaic facies of the upper sandstone is limited to the northwest part of the basin (north of the isopach thin of the entire tongue) and is nearly isolated from a narrow, northwest-trending band of shallow-marine bar sandstones (Plate 8C-8). The middle Pride Mountain sandstone is distributed similarly to the bar sandstone facies of the upper Pride Mountain sandstone, but no middle Pride Mountain deltaic facies has been recognized (Thomas, 1972b; Higginbotham, 1985).

The southwestern part of the Hartselle Sandstone includes horizontally laminated foreshore sandstones, crossbedded shoreface sandstones locally containing fossil logs and marine invertebrates, and local rooted mudstones containing leaf fossils (Thomas and Mack, 1982). On the northeast, the Hartselle consists of crossbedded and rippled marine-bar sandstone that, farther northeast, pinches out into the carbonate facies. Paleogeographic reconstruction for the Hartselle includes a northwest-trending barrier island facing seaward to the northeast (Plate 8C-9) (Thomas and Mack, 1982). On the southwest, the Hartselle Sandstone pinches out abruptly into mudstone, part of which contains plant fossils (Holmes, 1981) and thin coaly beds (Mellen and Moore, 1962) that suggest a coastal marsh in a back-barrier location. The Hartselle barrier-island sandstones were extensively reworked by marine transgression as indicated by widespread association of oscillation-rippled sandstone with barrier facies, and the Hartselle is overlain by the Bangor Limestone.

Lower and Middle Parkwood. The middle tongue of the Mississippian part of the clastic wedge progrades northeastward over the lower part of the Bangor Limestone, and is overlain by an extensive transgressive tongue of the middle Bangor (Fig. 5; Plate 8B). The clastic tongue includes four relatively persistent sandstones that pinch out northeastward within a mudstone succession, and the entire clastic tongue pinches out northeastward between the two tongues of Bangor Limestone. The northeastern limit of the clastic tongue is in the subsurface in the eastern part of the Black Warrior basin, but equivalent clastic rocks are exposed on the southeast in the Appalachian fold-thrust belt.

The Bangor Limestone persists across a relatively uniform shelf (East Warrior platform of Thomas, 1972b) in the eastern part of the Black Warrior basin (Plate 8B). A tongue of the lower Bangor extends far southwest and pinches out into the clastic facies. The base of the lower Bangor tongue approximates a time-stratigraphic horizon, suggesting rapid transgression. The lower Bangor tongue is a southwestward-thinning wedge, the top of which reflects gradual progradation of the lower and middle Parkwood clastic tongue (Fig. 5; Plate 8B). On the southwestern part of the platform, the Bangor consists mainly of bioclastic grainstones and thick massive oolitic grainstones. Southwestward in the direction of thinning, the limestone is finer grained and argillaceous in part, and the limestone grades southwestward into mudstone (Scott, 1978; Miesfeldt, 1985). The shape of the lower Bangor tongue has been interpreted to reflect a carbonate ramp (Scott, 1978); however, limestone in the toe of the ramp grades into prodelta mudstone (Plate 8B) (Miesfeldt, 1985). Calculations of uncompacted thickness and the association of depositional systems imply maximum water depth of less than 60 m (Miesfeldt, 1985).

The four lower and middle Parkwood sandstones generally thin and pinch out northeastward, but successively higher sandstones extend progressively farther northeast, documenting northeastward progradation of the clastic tongue (Plate 8B). Sandstone distribution defines delta distributary systems (Plate 8C-10, 11) (Thomas, 1972a; Welch, 1978; Shepard, 1979; Cleaves and Broussard, 1980; Nix, 1986). The sandstones are relatively thick on the southwest and west, but sandstone thickness varies abruptly. Geophysical well logs show a typical fining-upward sandstone above an abrupt contact with underlying mudstone. In contrast to sandstones on the southwest and west, the northeastern fringes of the lower three sandstones are marked locally by northwest-trending linear distribution (Plate 8C-10) and include coarsening-upward prograding sandstones interpreted as marine delta-front deposits. The uppermost sandstone is fringed by a wide area of marine-reworked sandstone that laps onto the carbonate facies (Plate 8C-11) (Miesfeldt, 1985; Nix, 1986).

The extensive limestone tongue (tongue of middle Bangor) above the lower and middle Parkwood clastic tongue indicates rapid transgression. Locally distributed sandstone lenses at the base of the limestone reflect marine reworking of the older deltaic deposits during transgression (Nix, 1986). The limestone tongue includes distinctive interbeds of maroon and green mudstone.

Upper Parkwood. The upper Parkwood clastic tongue interfingers with and grades northeastward into the upper part of the Bangor Limestone (Fig. 5; Plate 8B). Several thin, partly discontinuous limestone tongues extend widely to the southwest (Miesfeldt, 1985). Limestone and calcareous mudstone interbeds are more common in the lower part of the clastic tongue, whereas mudstone and sandstone extend farthest northeast in the upper part of the tongue, documenting northeastward progradation.

Sandstones in the northeastern part of the clastic tongue are irregularly distributed within the succession of mudstone and limestone. Some of the sandstones grade laterally into limestone. Thicker sandstones are restricted to the southwestern and western part of the basin. The thicker sandstones are interpreted as delta distributary sands, whereas the discontinuous sandstones on the northeast are interpreted as marine-bar sands (Plate 8C-12).

The upper part of the Bangor Limestone in the eastern part of the Black Warrior basin includes discontinuous beds of maroon and green mudstone (Thomas, 1972b). Separate from the northeastward-prograding clastic wedge in the southwestern part of the Black Warrior basin, a southwestward-prograding clastic wedge (Pennington Formation, Plate 8B, 8C-12) grades westward into the upper part of the Bangor Limestone at the eastern edge of the Black Warrior basin. The Pennington Formation is the lower part of the Pennington-Lee clastic wedge (of Thomas, 1977a) which is centered farther northeast in the Tennessee Appalachians. The Pennington and Parkwood grade from opposite directions into the Bangor Limestone in north-central Alabama; however, the two oppositely directed clastic wedges merge around the southeastern edge of the Bangor carbonate shelf in the Appalachian fold-thrust belt (Plate 8C-12) (Thomas, 1972b, 1979).

Lower Pottsville. The lower Pottsville consists of three distinct laterally equivalent facies: in outcrop and in the subsurface in the eastern part of the Black Warrior basin, massive sandstone; in the subsurface in the central part of the basin, a mudstone succession containing rare thin sandstones; and in the subsurface in the western part of the basin, a cyclic succession of fining-upward sandstones and mudstones (Plate 8B, 8C-13). The massive sandstone on the east is interpreted as a succession of stacked barrier-island sandstones (Hobday, 1974; Horsey, 1981; Engman, 1985). Discontinuous thin mudstone partings suggest six successive barrier complexes, in which internal variations define northwest-trending sandstone isolith lines (Engman, 1985). The barrier-sandstone facies grades southwestward into a mudstone-dominated facies interpreted to be back-barrier and lagoonal sediments (Plate 8B, 8C-13); the facies transition is associated with back-barrier coals (Thomas and Womack, 1983; Engman, 1985).

In the western part of the Black Warrior basin, the lower Pottsville is cyclic and includes several sandstones within a succession of mudstones. The sandstones generally fine upward and have gravel beds at the base (Thomas, 1972a). Most of the sandstone units have abrupt basal contacts with the underlying mudstone and gradational contacts with the overlying mudstone. The sandstones are interpreted as distributary deposits interbedded with interdistributary-bay mudstones that record episodic delta-lobe shifting and delta progradation. The deltaic sandstones pinch out eastward into the lagoonal mudstone succession in the central part of the basin (Plate 8C-13) (Engman, 1985).

In the western part of the basin, the cyclic sandstone-mudstone of the lower Pottsville is not lithologically distinct from the underlying sandstone-mudstone succession of the upper

TABLE 2. DATA SOURCES FOR PLATE 8

Wells listed in Table 1.

Data sources for other wells used in compiling Plate 8 include sample descriptions by author; geophysical well logs; published sample descriptions (Grohskopf, 1955; McGlamery, 1955; Milhous, 1959); unpublished industry reports; and state geological survey open files in Alabama, Arkansas, Kentucky, Mississippi, Missouri, and Tennessee.

References:

8A-2. Harrelson and Bicker, 1979; Neathery and Copeland, 1983; Denison, 1984.
8C-1. Thomas and Drahovzal, 1973; Denison, 1984; Howe, 1985.
8C-2. Butts, 1926; B. A. Ferrill, unpublished data.
8C-3. Wilson, 1949; Jewell, 1969; Thomas, 1972a; Kidd, 1975; Neathery and Drahovzal, 1985.
8C-4. Wilson, 1949; Jewell, 1969; Lundin and Newton, 1970; Thomas, 1972a; Kidd, 1975.
8C-5. Wilson, 1949; Lundin and Newton, 1970; Thomas, 1972a; Thomas and Drahovzal, 1973; Kidd, 1975; Ferrill, 1984.
8C-6. Thomas, 1972a, 1972b, 1974.
8C-7. Thomas, 1972b; Higginbotham, 1985.
8C-8. Thomas, 1972b; Higginbotham, 1985.
8C-9. Thomas, 1972b; Thomas and Mack, 1982.
8C-10. Thomas, 1972b; Nix, 1986.
8C-11. Thomas, 1972b; Osborne, 1985; Nix, 1986.
8C-12. Thomas, 1972b; Hines, 1988.
8C-13. Thomas, 1972a; Cleaves, 1981; Thomas and Womack, 1983; Engman, 1985.
8C-14. Thomas and Womack, 1983; Sestak, 1984; Thomas and Hines, 1985.
8C-15. Thomas and Womack, 1983; Sestak, 1984; Thomas and Hines, 1985.

Parkwood, except for the scattered limestone interbeds in the latter. The lithologically defined contact between Pottsville and Parkwood is diachronous (Fig. 5) (Thomas, 1972a).

Upper Pottsville. The productive coal beds in the eastern part of the Black Warrior basin are contained within a mudstone and sandstone succession in the upper part of the Pottsville Formation. The succession contains seven "coal groups," each of which contains several separate coal beds (Fig. 5) (Culbertson, 1964). The vertical clustering of coal beds within the "groups" is part of the expression of a cyclic succession that includes sandstones and mudstones. Parts of the mudstone succession between some of the coal groups include marine invertebrate fossils (Butts, 1926; Metzger, 1965).

The upper Pottsville is dominated by gray mudstone (Plate 8B). Sandstones in irregularly linear to elongate triangular distribution patterns trend northeastward (Plate 8C-14, 15) (Sestak, 1984). Many of the sandstones have abrupt basal contacts with the underlying mudstone and fine upward at the top, indicating distributary-channel sands (Sestak, 1984; Thomas and Hines, 1985). The extensive mudstones are interdistributary-bay deposits, and the bay-fill successions include coarsening-upward mudstone-siltstone units. Some thin sandstones within the mudstone succession are rippled and bioturbated, suggesting levee and crevasse-splay deposits (Thomas and Hines, 1985). The more extensive coal beds are within interdistributary-bay succession, although coal deposits ranging from backswamp to abandoned channel have been recognized. West- to northwest-trending, rippled and crossbedded to massive sandstones on the north and east are interpreted as parts of barrier systems, including tidal channels and tidal deltas, around the delta front (Plate 8C-14) (Sestak, 1984; Shadroui, 1986). Sandstone distribution indicates that the delta systems prograded northeastward throughout deposition of the upper Pottsville. In addition, the upper part of the upper Pottsville includes some northwestwardly prograding components, suggesting initiation of a sediment source on the southeast as a result of Appalachian orogenesis (Horsey, 1981, Sestak, 1984).

The stacked barrier sandstones of the lower Pottsville in the eastern part of the Black Warrior basin are overlapped by distributary, interdistributary, and marine-reworked delta-front deposits of the upper Pottsville. The vertical succession suggests limited transgression and regression of the shoreline location during deposition of the lower Pottsville, followed by resumption of progradation at the beginning of deposition of the upper Pottsville (Plate 8B, 8C-14, 15, 16). In the western part of the basin, no abrupt upward change in depositional style is evident within the cyclic sandstone-mudstone succession of the lower and upper Pottsville. The upper Pottsville exhibits an irregular upward increase in size and number of channel sandstones, as well as in numbers of gravel beds. These deposits suggest an upward transition from lower delta plain to upper delta plain, indicating continued progradation of the clastic wedge.

Mesozoic-Cenozoic Gulf Coastal Plain

Paleozoic strata in the western part of the Black Warrior basin, as well as those in the adjacent Appalachian-Ouachita fold-thrust belt, are truncated at an unconformity and overlain by a westward-thickening cover of Mesozoic and Cenozoic strata of the Gulf Coastal Plain (Fig. 2; Plate 8A-3). Both the unconformity and overlying beds dip southwestward toward the Mississippi

Embayment of the Gulf Coastal Plain. Stratigraphic distribution, as well as structure, indicates that the Mesozoic-Cenozoic strata were deposited in a tectonic framework different from that of the Black Warrior foreland basin.

The Mesozoic and Cenozoic rocks reflect deposition in coastal-plain to shallow-marine environments in a tectonic framework associated with the present Gulf of Mexico. The oldest rocks deposited in that framework are Upper Triassic graben-filling red beds; however, because of northward onlap, more extensive Upper Cretaceous strata unconformably overlie Paleozoic rocks of the Black Warrior basin.

TECTONIC EVOLUTION OF THE BLACK WARRIOR BASIN

The Black Warrior basin is underlain by Precambrian crystalline rocks of North American continental crust. The Paleozoic sedimentary succession includes a passive-margin carbonate facies in the Cambrian to Lower Mississippian and a synorogenic clastic wedge in the Upper Mississippian and Pennsylvanian. All of the Paleozoic rocks and structures are truncated and overstepped by Mesozoic and Cenozoic rocks of the Gulf Coastal Plain.

Late Precambrian to early Paleozoic rifting of the southern margin of North American continental crust is inferred from interpretations of opening of Iapetus Ocean. The rifted margin south of the Black Warrior basin is presently unknown in detail because no wells have penetrated rift-related rocks along the Ouachita margin. The trace of the Appalachian-Ouachita orogenic belt, as well as the distribution of Paleozoic sedimentary facies, conforms to an orthogonally zigzag outline of the Paleozoic eastern margin of North America that was framed by rift segments offset by transform faults (Thomas, 1977a). A northwest-trending segment of the margin south of the present Black Warrior basin is interpreted as a transform fault (Cebull and others, 1976; Thomas, 1976, 1977a), and intersections of the transform fault with rift segments outline the Alabama promontory of continental crust and the Ouachita embayment (Fig. 1).

Age of the rifted margin can be inferred from somewhat distant data points: the Blue Ridge in the Appalachians of Tennessee, North Carolina, and Georgia; the Southern Oklahoma fault system; and the Mississippi Valley graben (Fig. 1). Rifting associated with opening of Iapetus is indicated in the Blue Ridge by late Precambrian graben-fill sedimentary rocks (Ocoee Supergroup) and volcanic and sedimentary rocks (Mt. Rogers and Grandfather Mountain Formations) (King, 1970; Rankin, 1970, 1975). The late Precambrian syn-rift sedimentary and volcanic rocks overlie Grenville-age basement rocks, and are overlain at a post-rift unconformity by the Lower Cambrian Chilhowee Group, defining transition to a passive margin at ~570 Ma. Syn-rift igneous rocks associated with the Southern Oklahoma fault system range in age from approximately 570 to 525 Ma, and are overlain unconformably by the Upper Cambrian Reagan Sandstone (Ham and others, 1964; Hoffman and others, 1974; Gilbert, 1983; Coffman and others, 1986). The indicated age of rifting is inconsistent between the Blue Ridge and southern Oklahoma; however, these ages define possible limits for diachronous rifting along the continental margin adjacent to continental crust beneath the Black Warrior basin. Furthermore, the Cambrian graben-fill sedimentary sequence in the Rome–Rough Creek–Mississippi Valley graben system suggests crustal extension approximately synchronous with igneous activity along the Southern Oklahoma fault system and later than the post-rift unconformity in the Blue Ridge. Diachronous rifting may be a result of a spreading-center shift at the beginning of the Cambrian (Thomas, 1986b).

The position of the Black Warrior basin on the Alabama promontory of the southern margin of North American continental crust has significantly influenced the tectonic evolution of the basin. The lower Paleozoic passive-margin carbonate-bank facies extends throughout the Black Warrior basin and the Appalachian fold-thrust belt in Alabama. Presently allochthonous rocks within the Ouachita fold-thrust belt include a lower Paleozoic off-shelf deep-water facies that reflects deposition beyond the margin of continental crust within the Ouachita embayment. Presumably the edge of the passive-margin carbonate bank coincided approximately with the edge of continental crust. The passive-margin shelf setting prevailed from Cambrian into Mississippian from the Alabama promontory around the Ouachita embayment. In contrast, a convergent-margin orogenic belt northeast of the Alabama promontory supplied sediment to the Blount clastic wedge, the distal part of which prograded westward over the carbonate facies on the Alabama promontory during the Ordovician and Silurian (Plate 8C-3, 4), recording the most southerly expression of the Taconic orogeny in the Appalachian foreland (Thomas, 1977a).

Stratigraphy of the Cambrian to Mississippian passive-margin facies reflects the outline and evolution of the Alabama promontory, but does not indicate a "proto-Black Warrior basin." The transgressive Cambrian-Ordovician Sauk sequence indicates sea-level rise (Vail and others, 1977). The greater thickness of the basal Cambrian clastic sequence in the Alabama Appalachians indicates a greater subsidence rate adjacent to the rift-defined continental margin than that along the transform-defined continental margin southwest of the Alabama promontory or that farther toward the interior of the craton. An average subsidence rate (sediment accumulation rate) of 3.3 cm/1000 yr along the Appalachian margin during Cambrian and Early Ordovician contrasts with 1.8 cm/1000 yr along the Ouachita transform margin and 1.6 cm/1000 yr on the craton (Nashville dome). The apparent lack of hiatus between the Sauk and Tippecanoe sequences, as well as southwestward thickening of Middle Ordovician carbonate rocks (Plate 8B), adjacent to the southwestern transform margin contrasts with the greater hiatus and thinner succession farther cratonward and also in the Alabama Appalachians. Variations in both subsidence rates and sea level are suggested by the thin succession and numerous unconformities through the Upper Ordovician to Lower Mississippian part of the passive-margin stratigraphy. The average subsidence rate (<0.2 cm/1000 yr) is much less than that during the Cambrian and Ordovician. Contemporaneous off-shelf facies were deposited in the Ouachita embayment.

A change in the regional tectonic framework from passive to convergent active margin is indicated by the oldest sediments (late Meramecian) of the cratonward-prograding clastic wedge of Mississippian-Pennsylvanian age. Sediment dispersal patterns indicate sediment supply from a source southwest of the present Black Warrior basin. Composition of clastic-wedge sediments suggests an orogenic source terrane that resulted from an arc-continent collision. The Black Warrior basin is a peripheral foreland basin (in the terminology of Dickinson, 1974) that resulted from southward subduction of North American continental crust. Subsidence of the continental margin, presumably a result of tectonic loading, is reflected in greater subsidence rates in the southwestern part of the Black Warrior basin. In contrast, slower subsidence on the East Warrior platform in the eastern part of the basin is reflected in the Mississippian carbonate shelf. Sediment accumulation rate equaled subsidence rate as indicated by persistent shallow-marine to deltaic environments during deposition of the clastic wedge. Average subsidence rates in the southwestern part of the basin are 2.8 to 3.1 cm/1000 yr during the Mississippian and 28.9 to 30.4 cm/1000 yr during the Pennsylvanian (calculated for uncompacted thickness) (Hines, 1988). Greater thickness of synorogenic clastic-wedge rocks on the southwest, northwestward trend of isopach lines, and the ultimate southwestward-dipping homoclinal form of the basin all indicate subsidence under a thrust load advancing from the southwest. The geometry of the fill of the Black Warrior basin indicates a foreland basin related to the Ouachita fold-thrust belt rather than to the Appalachian fold-thrust belt; however, the southeastern edge of the foreland basin subsequently was deformed by Appalachian thrust faults. Widespread limestone tongues in the Mississippian may be a result of sea-level changes or of temporary interruption in supply of clastic sediment; successive episodes of progradation of the clastic wedge may reflect successive thrusting events. The down-to-southwest normal faults in the Black Warrior basin may have been initiated along with subsidence of the continental margin, but the time of fault movement has not yet been well documented. Possibly the basement faults originated during late Precambrian–early Paleozoic rifting and were reactivated during convergence and loading.

In contrast to the shallow-marine and deltaic environments of the Black Warrior foreland basin, the equivalent clastic sequence in the Ouachita fold-thrust belt consists of deep-water turbidites deposited within the Ouachita embayment. Similarities in stratigraphic succession, as well as composition of the rock-fragment populations, suggest that the turbidite facies in the Ouachitas was part of the same dispersal system within a regional clastic wedge as the deltaic facies in the Black Warrior basin (Graham and others, 1976; Thomas, 1976). Distribution of these large-scale depositional systems within one clastic wedge is consistent with a sediment source from an arc-continent collision orogen along the irregularly shaped southern margin of North America. Where the arc and subduction complex collided with North American continental crust, clastic sediments prograded directly onto the shelf in the Black Warrior foreland basin. In contrast, deep-water turbidites prograded into a remnant ocean basin between the arc and the southern margin of North American crust in the Ouachita embayment. Sediment dispersal patterns indicate progressive closure of the remnant ocean basin in the Ouachita embayment and westward progression of thrusting onto the shelf edge through time (Thomas, 1985; Houseknecht, 1986).

Orogenic activity culminated in large-scale cratonward thrusting in which thrust faults propagated cratonward into the older clastic-wedge sediments along the southwestern edge of the Black Warrior foreland basin. In the Appalachian fold-thrust belt along the southeastern side of the Black Warrior basin, the basal decollement is at the base of the shelf succession, and allochthonous rocks include the lower Paleozoic carbonate-shelf sequence and the eastern fringe of the shallow-marine to deltaic upper Paleozoic clastic wedge. In contrast, in the Ouachita fold-thrust belt, deep-water off-shelf facies of both the passive and convergent margins have been thrust over autochthonous passive-margin shelf facies. Between the Ouachita Mountains and the Appalachian structures, the frontal decollement of the fold-thrust belt crosses from the off-shelf facies to the base of the shelf facies.

The age of thrusting along the Appalachian-Ouachita orogen adjacent to the Black Warrior basin is constrained by the age of the youngest preserved rocks and by the onlapping Mesozoic rocks of the Gulf Coastal Plain. The youngest rocks known to be preserved in the Black Warrior basin are of Middle Pennsylvanian (Atokan) age. The youngest rocks involved in thrusting in the Appalachian fold-thrust belt in Alabama are no younger than Middle Pennsylvanian (Butts, 1926). Thrusting in the Ouachita Mountains post-dates Atokan deposition.

Following Appalachian-Ouachita orogenesis, opening of the Gulf of Mexico in the early Mesozoic is indicated by a northwest-trending system of faults that cut Paleozoic rocks; downthrown fault blocks are filled by Triassic-Jurassic clastic rocks, evaporites, and volcanic rocks. The rift-related faults pass south of the Black Warrior basin (Thomas and others, in press). As a result of post-rift onlap, Cretaceous strata unconformably overlie the Paleozoic rocks in the western part of the Black Warrior basin. Post-rift subsidence of the Gulf Coastal Plain toward the Gulf of Mexico is reflected in the present structural configuration of Coastal Plain strata above the Paleozoic structures of the Appalachian-Ouachita orogen and the Black Warrior foreland basin.

REFERENCES CITED

Benson, D. J., and Mink, R. M., 1983, Depositional history and petroleum potential of the Middle and Upper Ordovician of the Alabama Appalachians: Gulf Coast Association of Geological Societies Transactions, v. 33, p. 13–21.

Berdan, J. M., Boucot, A. J., and Ferrill, B. A., 1986, The first fossiliferous Pridolian beds from the southern Appalachians in northern Alabama, and the age of the uppermost Red Mountain Formation: Journal of Paleontology, v. 60, p. 180–185.

Bickford, M. E., and Mose, D. G., 1975, Geochronology of Precambrian rocks in the St. Francois Mountains, southeastern Missouri: Geological Society of America Special Paper 165, 48 p.

Boucot, A. J., 1953, Report on referred fossils [unpublished]: U.S. Geological Survey.

Buchanan, R. S., and Johnson, F. K., 1968, Bonanza gas field—A model for Arkoma basin growth faulting, *in* Cline, L. M., ed., Geology of the western Arkoma basin and Ouachita Mountains, Oklahoma: Oklahoma City Geological Society Guidebook, p. 75–85.

Butts, C., 1926, The Paleozoic rocks, *in* Geology of Alabama: Alabama Geological Survey Special Report 14, p. 41–230.

Cebull, S. E., Shurbet, D. H., Keller, G. R., and Russell, L. R., 1976, Possible role of transform faults in the development of apparent offsets in the Ouachita–southern Appalachian tectonic belt: Journal of Geology, v. 84, p. 107–114.

Chowns, T. M., and McKinney, F. K., 1980, Depositional facies in Middle-Upper Ordovician and Silurian rocks of Alabama and Georgia, *in* Frey, R. W., ed., Excursions in southeastern geology, v. 2: American Geological Institute, p. 323–348.

Cleaves, A. W., 1981, Resource evaluation of Lower Pennsylvanian (Pottsville) depositional systems of the western Warrior coal field, Alabama and Mississippi: Mississippi Mineral Resources Institute Technical Report No. 81-1, 125 p.

Cleaves, A. W., and Broussard, M. C., 1980, Chester and Pottsville depositional systems, outcrop and subsurface, in the Black Warrior basin of Mississippi and Alabama: Gulf Coast Association of Geological Societies Transactions, v. 30, p. 49–60.

Coffman, J. D., Gilbert, M. C., and McConnell, D. A., 1986, An interpretation of the crustal structure of the Southern Oklahoma aulacogen satisfying gravity data: Oklahoma Geological Survey Guidebook 23, p. 1–10.

Conant, L. C., and Swanson, V. E., 1961, Chattanooga Shale and related rocks of central Tennessee and nearby areas: U.S. Geological Survey Professional Paper 357, 91 p.

Copeland, C. W., and Raymond, D. E., 1985, Stratigraphic columns, Alabama, *in* Patchen, D. G., Avary, K. L., and Erwin, R. B., coordinators, Correlation of stratigraphic units in North America—Southern Appalachian region correlation chart: American Association of Petroleum Geologists Correlation Chart Series.

Culbertson, W. C., 1964, Geology and coal resources of the coal-bearing rocks of Alabama: U.S. Geological Survey Bulletin 1182-B, 79 p.

Denison, R. E., 1984, Basement rocks in northern Arkansas: Arkansas Geological Commission Miscellaneous Publication 18-B, p. 33–49.

Dever, G. R., Jr., Hoge, H. P., Hester, N. C., and Ettensohn, F. R., 1977, Stratigraphic evidence for late Paleozoic tectonism in northeastern Kentucky: Field Trip, Fifth Annual Meeting, Eastern Section, American Association of Petroleum Geologists, Lexington, Kentucky, Kentucky Geological Survey, 80 p.

Dever, G. R., Jr., 1986, Mississippian reactivation along the Irvine–Paint Creek fault system in the Rome trough, east-central Kentucky: Southeastern Geology, v. 27, p. 95–105.

Dickinson, W. R., 1974, Plate tectonics and sedimentation: Society of Economic Paleontologists and Mineralogists Special Publication 22, p. 1–27.

Drahovzal, J. A., 1967, The biostratigraphy of Mississippian rocks in the Tennessee Valley: Alabama Geological Society Guidebook, 5th Annual Field Trip, p. 10–24.

Drahovzal, J. A., and Neathery, T. L., 1971, Middle and Upper Ordovician stratigraphy of the Alabama Appalachians: Alabama Geological Society Guidebook, 9th Annual Field Trip, p. 1–62.

Engman, M. A., 1985, Depositional systems in the lower part of the Pottsville Formation, Black Warrior basin, Alabama [M.S. thesis]: Tuscaloosa, University of Alabama, 250 p.

Ethington, R. L., Finney, S. C., and Repetski, J. E., 1988, Biostratigraphy of the Paleozoic rocks of the Ouachita orogen, Arkansas, Oklahoma, west Texas, *in* Hatcher, R. D., Jr., Thomas, W. A., and Viele, G. W., eds., The Appalachian-Ouachita orogen in the United States: Boulder, Colorado, Geological Society of America, The Geology of North America, v. F-2 (in press).

Ferm, J. C., and Ehrlich, R., 1967, Petrology and stratigraphy of the Alabama coal fields, *in* Ferm, J. C., Ehrlich, R., and Neathery, T. L., A field guide to Carboniferous detrital rocks in northern Alabama (Geological Society of America 1967 Annual Meeting): Tuscaloosa, Alabama Geological Society, p. 11–15.

Ferrill, B. A., 1984, Frog Mountain Formation, southwestern Appalachian fold and thrust belt, Alabama [M.S. thesis]: Tuscaloosa, University of Alabama, 178 p.

Gilbert, M. C., 1983, Timing and chemistry of igneous events associated with the Southern Oklahoma aulacogen: Tectonophysics, v. 94, p. 439–455.

Gordon, M., Jr., 1953, Report on referred fossils [unpublished]: U.S. Geological Survey.

Graham, S. A., Ingersoll, R. V., and Dickinson, W. R., 1976, Common provenance for lithic grains in Carboniferous sandstones from Ouachita Mountains and Black Warrior basin: Journal of Sedimentary Petrology, v. 46, p. 620–632.

Grohskopf, J. G., 1955, Subsurface geology of the Mississippi Embayment of southeast Missouri: Missouri Geological Survey, v. 37, 133 p.

Haley, B. R., 1982, Geology and energy resources of the Arkoma basin, Oklahoma and Arkansas: University of Missouri-Rolla Journal, no. 3, p. 43–53.

Haley, B. R., and Hendricks, T. A., 1968, Geology of the Greenwood quadrangle, Arkansas-Oklahoma: Arkansas Geological Commission Information Circular 20-F, 15 p.

Ham, W. E., 1959, Correlation of pre-Stanley strata in the Arbuckle-Ouachita Mountain regions, *in* The geology of the Ouachita Mountains, a symposium: Dallas Geological Society and Ardmore Geological Society, p. 71–86.

Ham, W. E., Denison, R. E., and Merritt, C. A., 1964, Basement rocks and structural evolution of southern Oklahoma: Oklahoma Geological Survey Bulletin 95, 302 p.

Handford, C. R., 1978, Monteagle Limestone (Upper Mississippian)—Oolitic tidal-bar sedimentation in southern Cumberland Plateau: American Association of Petroleum Geologists Bulletin, v. 62, p. 644–656.

Harrelson, D. W., and Bicker, A. R., Jr., 1979, Petrography of some subsurface igneous rocks of Mississippi: Gulf Coast Association of Geological Societies Transactions, v. 29, p. 244–251.

Harris, L. D., 1975, Oil and gas data from the Lower Ordovician and Cambrian rocks of the Appalachian basin: U.S. Geological Survey Miscellaneous Investigations Series Map I-917 D.

Hass, W. H., 1953, Report on referred fossils [unpublished]: U.S. Geological Survey.

Hass, W. H., 1956, Age and correlation of the Chattanooga Shale and the Maury Formation: U.S. Geological Survey Professional Paper 286, 47 p.

Henry, T. W., Gillespie, W. H., Gordon, M., Jr., and Schweinfurth, S. P., 1981, Stratigraphic significance of plant and invertebrate fossils from Parkwood Formation, northern Alabama: Geological Society of America Abstracts with Programs, v. 13, p. 471–472.

Higginbotham, D. R., 1985, Regional stratigraphy, environments of deposition, and tectonic framework of Mississippian clastic rocks between the Tuscumbia and Bangor Limestones in the Black Warrior basin of Alabama and Mississippi [M.S. thesis]: Tuscaloosa, University of Alabama, 177 p.

Hines, R. A., Jr., 1988, Carboniferous evolution of the Black Warrior foreland basin, Alabama and Mississippi [Ph.D. thesis]: Tuscaloosa, University of

Alabama, 231 p.
Hobday, D. K., 1974, Beach- and barrier-island facies in the upper Carboniferous of northern Alabama: Geological Society of America Special Paper 148, p. 209–223.
Hoffman, P., Dewey, J. F., and Burke, K., 1974, Aulacogens and their genetic relation to geosynclines, with a Proterozoic example from Great Slave Lake, Canada: Society of Economic Paleontologists and Mineralogists Special Publication 19, p. 38–55.
Holmes, J. W., 1981, The depositional environment of the Mississippian Lewis sandstone in the Black Warrior basin of Alabama [M.S. thesis]: Tuscaloosa, University of Alabama, 172 p.
Horsey, C. A., 1981, Depositional environments of the Pennsylvanian Pottsville Formation in the Black Warrior basin of Alabama: Journal of Sedimentary Petrology, v. 51, p. 799–806.
Houseknecht, D. W., 1986, Evolution from passive margin to foreland basin: The Atoka Formation of the Arkoma basin, south-central U.S.A.: International Association of Sedimentologists Special Publication 8, p. 327–345.
Houseknecht, D. W., and Ethridge, F. G., 1978, Depositional history of the Lamotte Sandstone of southeastern Missouri: Journal of Sedimentary Petrology, v. 48, p. 575–586.
Howe, J. R., 1985, Tectonics, sedimentation, and hydrocarbon potential of the Reelfoot aulacogen [M.S. thesis]: Norman, University of Oklahoma, 109 p.
Jennings, J. R., and Thomas, W. A., 1987, Fossil plants from Mississippian-Pennsylvanian transition strata in the southern Appalachians: Southeastern Geology, v. 27, p. 207–217.
Jewell, J. W., 1969, An oil and gas evaluation of north Alabama: Geological Survey of Alabama Bulletin 93, 65 p.
Jordan, T. E., 1981, Thrust loads and foreland basin evolution, Cretaceous, western United States: American Association of Petroleum Geologists Bulletin, v. 65, p. 2506–2520.
Kane, M. F., Hildenbrand, T. G., and Hendricks, J. D., 1981, Model for the tectonic evolution of the Mississippi Embayment and its contemporary seismicity: Geology, v. 9, p. 563–568.
Kersting, J. J., 1982, The petrology and petrography of the Mt. Simon and pre-Mt. Simon sandstones: Evidence for possible Precambrian rifting in the central midcontinent [M.S. thesis]: Pittsburgh, Pennsylvania, University of Pittsburgh, 178 p.
Kidd, J. T., 1975, Pre-Mississippian subsurface stratigraphy of the Warrior basin in Alabama: Gulf Coast Association of Geological Societies Transactions, v. 25, p. 20–39.
Kidd, J. T., and Neathery, T. L., 1976, Correlation between Cambrian rocks of the southern Appalachian geosyncline and the interior low plateaus: Geology, v. 4, p. 767–769.
King, P. B., 1970, The Precambrian of the United States of America—Southeastern United States, *in* Rankama, K., ed., The geologic systems—The Precambrian, v. 4: New York, Interscience, p. 1–71.
King, P. B., 1975, The Ouachita and Appalachian orogenic belts, *in* Nairn, A.E.M., and Stehli, F. G., eds., The ocean basins and margins, v. 3: New York, Plenum, p. 201–241.
Koinm, D. N., and Dickey, P. A., 1967, Growth faulting in McAlester basin of Oklahoma: American Association of Petroleum Geologists Bulletin, v. 51, p. 710–718.
Kulander, B. R., and Dean, S. L., 1978, Gravity, magnetics, and structure of the Allegheny Plateau/western Valley and Ridge in West Virginia and adjacent states: West Virginia Geological and Economic Survey Report of Investigation RI-27, 91 p.
Lillie, R. J., Nelson, K. D., de Voogd, B., Brewer, J. A., Oliver, J. E., Brown, L. D., Kaufman, S., and Viele, G. W., 1983, Crustal structure of Ouachita Mountains, Arkansas: A model based on integration of COCORP reflection profiles and regional geophysical data: American Association of Petroleum Geologists Bulletin, v. 67, p. 907–931.
Lundin, R. F., and Newton, G. D., 1970, Ostracoda and the Silurian stratigraphy of northwestern Alabama: Geological Survey of Alabama Bulletin 95, 65 p.
Lyons, P. C., Meissner, C. R., Jr., Barwood, H. L., and Adinolfi, F. G., 1985, North American and European megafloral correlations with the upper part of the Pottsville Formation of the Warrior coal field, Alabama, U.S.A.: Compte Rendu, Tenth International Congress on Carboniferous Stratigraphy and Geology, v. 2, p. 203–245.
Mack, G. H., 1980, Stratigraphy and depositional environments of the Chilhowee Group (Cambrian) in Georgia and Alabama: American Journal of Science, v. 280, p. 497–517.
Mack, G. H., James, W. C., and Thomas, W. A., 1981, Orogenic provenance of Mississippian sandstones associated with southern Appalachian-Ouachita orogen: American Association of Petroleum Geologists Bulletin, v. 65, p. 1444–1456.
Mack, G. H., Thomas, W. A., and Horsey, C. A., 1983, Composition of Carboniferous sandstones and tectonic framework of southern Appalachian-Ouachita orogen: Journal of Sedimentary Petrology, v. 53, p. 931–946.
Maher, J. C., and Lantz, R. J., 1953, Correlation of pre-Atoka rocks in the Arkansas Valley, Arkansas: U.S. Geological Survey Oil and Gas Investigations Chart OC 51.
McGlamery, W., 1955, Subsurface stratigraphy of northwest Alabama: Alabama Geological Survey Bulletin 64, 503 p.
Mellen, F. F., 1947, Black Warrior basin, Alabama and Mississippi: American Association of Petroleum Geologists Bulletin, v. 31, p. 1801–1816.
Mellen, F. F., 1955, Subsurface outliers of Chattanooga Shale in northeastern Mississippi: American Association of Petroleum Geologists Bulletin, v. 39, p. 1864–1865.
Mellen, F. F., 1972, Regional Paleozoic stratigraphy in Mississippi between Ouachita and Appalachian Mountains: Discussion: American Association of Petroleum Geologists Bulletin, v. 56, p. 2457–2458.
Mellen, F. F., 1977, Cambrian System in Black Warrior basin: American Association of Petroleum Geologists Bulletin, v. 61, p. 1897–1900.
Mellen, F. F., and Moore, W. H., 1962, The Tula prospect, Lafayette County, Mississippi: Mississippi Geological Economic and Topographical Survey Bulletin 96, 29 p.
Metzger, W. J., 1965, Pennsylvanian stratigraphy of the Warrior basin, Alabama: Alabama Geological Survey Circular 30, 80 p.
Miesfeldt, M. A., 1985, Facies relationships between the Parkwood and Bangor formations in the Black Warrior basin [M.S. thesis]: Tuscaloosa, University of Alabama, 149 p.
Milhous, H. C., 1959, Well logs in Tennessee: Tennessee Division of Geology Bulletin 62, 606 p.
Milici, R. C., and Smith, J. W., 1969, Stratigraphy of the Chickamauga Supergroup in its type area: Tennessee Division of Geology Report of Investigations No. 24, 35 p.
Mississippi Geological Society, undated, Correlation section, Black Warrior basin, pre-Chester.
Missouri Geological Survey, 1979, Geologic map of Missouri, scale 1:500,000.
Moser, P. H., and Thomas, W. A., 1967, Pre-Hartselle Mississippian sandstone in Colbert County, Alabama: Alabama Geological Society Guidebook, 5th Annual Field Trip, p. 25–39.
Neathery, T. L., and Copeland, C. W., 1983, New information on the basement and lower Paleozoic stratigraphy of north Alabama: Alabama Geological Survey Open-File Report, 28 p.
Neathery, T. L., and Drahovzal, J. A., 1985, Lithostratigraphy of Upper Ordovician strata in Alabama: Geological Survey of Alabama Circular 124, 55 p.
Nix, M. A., 1986, Facies within the lower part of the Parkwood Formation in the Black Warrior basin of Mississippi and Alabama: Appalachian Basin Industrial Associates Program, v. 11, p. 93–107.
Osborne, W. E., 1985, Depositional environments and dispersal system of the Parkwood Formation (Carboniferous) on the northwest limb of the Cahaba syncline, Jefferson County, Alabama [M.S. thesis]: Tuscaloosa, University of Alabama, 160 p.
Palmer, A. R., 1962, *Glyptagnostus* and associated trilobites in the United States: U.S. Geological Survey Professional Paper 374-F, 49 p.
Palmer, A. R., 1971, The Cambrian of the Appalachian and eastern New England regions, eastern United States, *in* Holland, C. H., ed., Cambrian of the New World: New York, Interscience, p. 169–217.
Pfeil, R. W., and Read, J. F., 1980, Cambrian carbonate platform margin facies,

Shady Dolomite, southwestern Virginia, U.S.A.: Journal of Sedimentary Petrology, v. 50, p. 91–116.

Rankin, D. W., 1970, Stratigraphy and structure of Precambrian rocks in northwestern North Carolina, *in* Fisher, G. W., Pettijohn, F. J., Reed, J. C., Jr., and Weaver, K. N., eds., Studies of Appalachian geology: Central and southern: New York, Interscience, p. 227–245.

Rankin, D. W., 1975, The continental margin of eastern North America in the southern Appalachians: The opening and closing of the proto-Atlantic Ocean: American Journal of Science, v. 275-A, p. 298–336.

Riggs, K. A., 1976, Subsurface Precambrian granite, Lafayette County, Mississippi: American Association of Petroleum Geologists Bulletin, v. 60, p. 455–457.

Rodgers, J., 1968, The eastern edge of the North American continent during the Cambrian and Early Ordovician, *in* Zen, E., White, W. S., Hadley, J. B., and Thompson, J. B., Jr., eds., Studies of Appalachian geology: Northern and maritime: New York, Interscience, p. 141–149.

Ross, R. J., Jr., and 27 others, 1982, The Ordovician System in the United States: International Union of Geological Sciences Publication No. 12, 73 p.

Schedl, A., and Wiltschko, D. V., 1984, Sedimentological effects of a moving terrain: Journal of Geology, v. 92, p. 273–287.

Schwalb, H. R., 1982, Geologic-tectonic history of the area surrounding the northern end of the Mississippi Embayment: University of Missouri-Rolla Journal, no. 3, p. 31–42.

Scott, G. L., 1978, Deposition, facies patterns, and hydrocarbon potential of Bangor Limestone (Mississippian), northern Black Warrior basin, Alabama and Mississippi: Mississippi Geological Society Guidebook, 17th Field Trip, p. 34–54.

Sestak, H. M., 1984, Stratigraphy and depositional environments of part of the Pennsylvanian Pottsville Formation in the Black Warrior basin: Alabama and Mississippi [M.S. thesis]: Tuscaloosa, University of Alabama, 184 p.

Shadroui, J. M., 1986, Depositional environments of the Pennsylvanian Bremen Sandstone Member and associated strata, Pottsville Formation, north-central Alabama [M.S. thesis]: Tuscaloosa, University of Alabama, 172 p.

Shepard, B. K., 1979, Petrography and environments of deposition of the Carter sandstone (Mississippian) in the Black Warrior basin of Alabama and Mississippi [M.S. thesis]: Tuscaloosa, University of Alabama, 196 p.

Sloss, L. L., 1963, Sequences in the cratonic interior of North America: Geological Society of America Bulletin, v. 74, p. 93–114.

Soderberg, R. K., and Keller, G. R., 1981, Geophysical evidence for deep basin in western Kentucky: American Association of Petroleum Geologists Bulletin, v. 65, p. 226–234.

Speed, R. C., and Sleep, N. H., 1982, Antler orogeny and foreland basin: A model: Geological Society of America Bulletin, v. 93, p. 815–828.

Sterling, P. J., Stone, C. G., and Holbrook, D. F., 1966, General geology of eastern Ouachita Mountains, Arkansas: Kansas Geological Society Guidebook, 29th Field Conference, p. 177–194.

Swann, D. H., 1964, Late Mississippian rhythmic sediments of Mississippi Valley: American Association of Petroleum Geologists Bulletin, v. 48, p. 637–658.

Thomas, W. A., 1972a, Regional Paleozoic stratigraphy in Mississippi between Ouachita and Appalachian Mountains: American Association of Petroleum Geologists Bulletin, v. 56, p. 81–106.

Thomas, W. A., 1972b, Mississippian stratigraphy of Alabama: Alabama Geological Survey Monograph 12, 121 p.

Thomas, W. A., 1973, Southwestern Appalachian structural system beneath the Gulf Coastal Plain: American Journal of Science, v. 273-A, p. 372–390.

Thomas, W. A., 1974, Converging clastic wedges in the Mississippian of Alabama: Geological Society of America Special Paper 148, p. 187–207.

Thomas, W. A., 1976, Evolution of Ouachita-Appalachian continental margin: Journal of Geology, v. 84, p. 323–342.

Thomas, W. A., 1977a, Evolution of Appalachian-Ouachita salients and recesses from reentrants and promontories in the continental margin: American Journal of Science, v. 277, p. 1233–1278.

Thomas, W. A., 1977b, Structural and stratigraphic continuity of the Ouachita and Appalachian Mountains: Arkansas Geological Commission, Symposium on the geology of the Ouachita Mountains, v. 1, p. 9–24.

Thomas, W. A., 1979, Mississippian stratigraphy of Alabama: U.S. Geological Survey Professional Paper 1110-I, p. I1–I22.

Thomas, W. A., 1985, The Appalachian-Ouachita connection: Paleozoic orogenic belt at the southern margin of North America: Annual Review of Earth and Planetary Sciences, v. 13, p. 175–199.

Thomas, W. A., 1986a, A Paleozoic synsedimentary structure in the Appalachian fold-thrust belt in Alabama, *in* McDowell, R. C., and Glover, L., III, eds., The Lowry volume: Studies in Appalachian geology: Virginia Tech Department of Geological Sciences Memoir 3, p. 1–12.

Thomas, W. A., 1986b, Early Paleozoic southern margin of North America: Geological Society of America Abstracts with Programs, v. 18, p. 772.

Thomas, W. A., 1988, The Appalachian-Ouachita orogen beneath the Gulf Coastal Plain between the outcrops in the Appalachian and Ouachita Mountains, *in* Hatcher, R. D., Jr., Thomas, W. A., and Viele, G. W., eds., The Appalachian-Ouachita orogen in the United States: Boulder, Colorado, Geological Society of America, The Geology of North America, v. F-2 (in press).

Thomas, W. A., Chowns, T. M., Daniels, D. L., Neathery, T. L., Glover, L., III, and Gleason, R. J., 1988, Pre-Mesozoic paleogeologic map of Appalachian-Ouachita orogen beneath Atlantic and Gulf Coastal Plains, *in* Hatcher, R. D., Jr., Thomas, W. A., and Viele, G. W., eds., The Appalachian-Ouachita orogen in the United States: Boulder, Colorado, Geological Society of America, The Geology of North America, v. F-2, Plate 6 (in press).

Thomas, W. A., and Drahovzal, J. A., 1973, Regional Paleozoic stratigraphy of Alabama: Alabama Geological Society Guidebook, 11th Annual Field Trip, p. 66–91.

Thomas, W. A., and Hines, R. A., 1985, Primary geologic controls on coalbed methane content: University of Alabama School of Mines and Energy Development Research Report No. 72, 20 p.

Thomas, W. A., and Mack, G. H., 1982, Paleogeographic relationship of a Mississippian barrier-island and shelf-bar system (Hartselle Sandstone) in Alabama to the Appalachian-Ouachita orogenic belt: Geological Society of America Bulletin, v. 93, p. 6–19.

Thomas, W. A., and Womack, S. H., 1983, Coal stratigraphy of the deeper part of the Black Warrior basin in Alabama: Gulf Coast Association of Geological Societies Transactions, v. 33, p. 439–446.

Upshaw, C. F., 1967, Pennsylvanian palynology and age relationships in the Warrior basin, Alabama, *in* Ferm, J. C., Ehrlich, R., and Neathery, T. L., A field guide to Carboniferous detrital rocks in northern Alabama (Geological Society of America 1967 Annual Meeting): Tuscaloosa, Alabama Geological Society, p. 16–20.

Vail, P. R., Mitchum, R. M., Jr., and Thompson, S., III, 1977, Seismic stratigraphy and global changes of sea level, Part 4: Global cycles of relative changes of sea level: American Association of Petroleum Geologists Memoir 26, p. 83–97.

Van Schmus, W. R., Bickford, M. E., and Zietz, I., 1987, Early and middle Proterozoic provinces in the central United States, *in* Kröner, A., ed., Proterozoic lithospheric evolution: American Geophysical Union, Geodynamics Series, v. 17, p. 43–68.

Viele, G. W., 1973, Structure and tectonic history of the Ouachita Mountains, Arkansas, *in* DeJong, K. A., and Scholten, R., eds., Gravity and tectonics: New York, Wiley, p. 361–377.

Viele, G. W., and Thomas, W. A., 1988, Tectonic synthesis of the Ouachita orogenic belt, *in* Hatcher, R. D., Jr., Thomas, W. A., and Viele, G. W., eds., The Appalachian-Ouachita orogen in the United States: Boulder, Colorado, Geological Society of America, The Geology of North America, v. F-2 (in press).

Wasserburg, G. J., Wetherill, G. W., Silver, L. T., and Flawn, P. T., 1962, A study of the ages of the Precambrian of Texas: Journal of Geophysical Research, v. 67, p. 4021–4047.

Webb, E. J., 1980, Cambrian sedimentation and structural evolution of the Rome trough in Kentucky [Ph.D. thesis]: Cincinnati, Ohio, University of Cincinnati, 98 p.

Welch, S. W., 1959, Mississippian rocks of the northern part of the Black Warrior basin, Alabama and Mississippi: U.S. Geological Survey Oil and Gas Investigations Chart OC-62.

Welch, S. W., 1978, Deposition of the Carter-Sanders zone of the Black Warrior basin, Mississippi and Alabama: Mississippi Geological Society Guidebook, 17th Field Trip, p. 25–33.

Wilson, C. W., Jr., 1949, Pre-Chattanooga stratigraphy in central Tennessee: Tennessee Division of Geology Bulletin 56, 407 p.

Woodward, H. P., 1961, Preliminary subsurface study of southeastern Appalachian interior plateau: American Association of Petroleum Geologists Bulletin, v. 45, p. 1634–1655.

Young, S. W., 1976, Petrographic textures of detrital polycrystalline quartz as an aid to interpreting crystalline source rocks: Journal of Sedimentary Petrology, v. 46, p. 595–603.

MANUSCRIPT ACCEPTED BY THE SOCIETY FEBRUARY 9, 1988

ACKNOWLEDGMENTS

The research summarized here has been supported during the past 25 years by contributions of data from numerous organizations and individuals. Unpublished data have been contributed specifically for this chapter by Amoco Production Company, M. Gordon, Jr., J. E. Repetski, L. P. Alberstadt, and R. E. Denison. The detailed interpretations of stratigraphy and depositional systems of Mississippian-Pennsylvanian rocks are based in part upon the work of Graduate Research Assistants S. H. Womack, W. E. Osborne, B. A. Ferrill, S. P. Hertig, H. M. Sestak, R. A. Hines, D. R. Higginbotham, M. A. Miesfeldt, M. A. Engman, J. M. Shadroui, M. A. Nix, D. E. Leverett, J. L. Mintz, T. C. Block, J. B. Hoover, and J. J. Connors in the Stratigraphy Laboratory at the University of Alabama. The interpretation of basement structure beneath the Appalachian fold-thrust belt in Alabama (Figs. 1, 3; Plate 8A-2) incorporates work by B. A. Ferrill. Parts of this research have been supported by grants from the National Science Foundation (EAR-8109470; EAR-8218604), the Petroleum Research Fund of the American Chemical Society (17069-AC2), the Office of Surface Mining, the Appalachian Basin Industrial Associates, and the University of Alabama School of Mines and Energy Development. The compilation of Plate 8 was completed while I was a Visiting Scientist at the Institute for the Study of the Continents at Cornell University.

Printed in U.S.A.

The Geology of North America
Vol. D-2, Sedimentary Cover—North American Craton: U.S.
The Geological Society of America, 1988

Chapter 17

Conclusions

L. L. Sloss
Department of Geological Sciences, Northwestern University, Evanston, Illinois 60208

Early in the Decade of North American Geology, a planning document was assembled and published as Perspectives in Regional Geological Synthesis (Palmer, 1982). This collage of diverse contributions included an item devoted to the North American Craton in the United States that closed with several "unresolved questions" (Sloss, 1982a, p. 39):

> 1. What is the significance of episodic emergences of the craton responsible for the sequence-bounding unconformities? The natural corollary is the meaning of the several episodes of cratonic submergence. In both cases the magnitude of vertical movement far exceeds that which could be ascribed to any combination of eustatics and loading-unloading by sediment and water. Is there a demonstrable correlation with plate interactions?
> 2. Why is deformation of the craton sometimes dominated by flexure, at other times by fracture? When the fractural mode is operating, is the continental crust in extension or compression? That is, are the downdropped basins and block uplifts characteristic of Absaroka I deposition produced by thrusts of Wind River type or are they representative of distinctly different stress/strain patterns? In either case, what are the driving forces?
> 3. Is the tectonic mode prevailing on the craton during Sauk deposition as distinct from the younger Phanerozoic as is suggested here? If so, does this indicate that there are irreversible secular changes in cratonic behavior as a product of evolving crustal thickness, or rheologic change, or whatever?
> 4. Why do sedimentary basins and intervening arches remain fixed in positions for hundreds of millions of years? Is it demonstrable that basins are localized on or at the intersections of basement rifts and sutures?
> 5. What are the driving forces of basin subsidence? Again, eustatics plus surficial load cannot achieve the amplitudes recorded and, at distances from continental margins, thermal contraction or thermally induced phase changes are unsatisfactory hypotheses for many reasons.

The first three of these questions refer to the craton as a whole and can be addressed as an interrelated group partially resolved by new data, new concepts, and the evolving intellectual maturity of the community of tectonists. In the course of my own evolution toward intellectual maturity, I once produced a paper (Sloss, 1966) that demanded almost complete separation in chronology and genesis of tectonic conditions of cratonic interiors and tectonic events in "mobile belts" at cratonic margins. Mercifully, these pronouncements were quickly submerged as demonstrations of the validity of the global-tectonic model became manifest and as the enhanced resolution of the dating of events at continental margins clarified the kinship of plate-tectonic and craton-interior histories. In fact, our predecessors in tectonic theory had long discussed the linkage between orogenesis in "geosynclines" or "mobile belts" and marine transgression or regression on continental interiors. Such thinking goes back at least as far as the writings of Émile Haug (1890) and has been revived and reviewed by Johnson (1971, 1972).

Meanwhile, it has been suggested (e.g., Hays and Pitman, 1973) that the eustatic flooding of cratons, and concomitant localized subsidence induced by the load of water and sediments, is the product of acceleration of plate divergence leading to the reduction of ocean-basin volume by elevated mid-ocean ridge systems, although the possibility of alternative routes, such as mid-plate oceanic volcanism (e.g., Schlanger and others, 1981), may have achieved the same results. In any case, there is no longer significant support for the exclusion of cratons from the dynamics of plate tectonics. Dramatic examples of plate convergence, such as continent-continent collision at the Ouachita-Marathon margin during Absaroka I accumulation (Pennsylvanian–Early Permian) concurrent with the creation of mountainous blocks deep within the craton, demonstrate the interrelation between the tectonics of continental interiors and convergent margins. Similarly, there is a basis for belief that when more than one margin of a craton is in a state of divergence, as during the Late Proterozoic–Cambrian breakup of a supercontinent, extensional forces may penetrate far inboard, as indicated by such Sauk features as the Oklahoma and Mississippi Valley–Reelfoot aulacogenic rift systems and the related Rome Trough.

The argument for influence on the tectonics of the craton by compressional and extensional events at continental margins, as such influence is demonstrated by uplift of reverse- or wrench-fault–bounded blocks in one instance and by propagating rift systems in the other, is supported by well-constrained chronologies indicating synchronous behavior. Whether compression or extension has been the driving mechanism, it seems clear that in the circumstances discussed above the craton has responded to external forces (created by lateral motions of lithospheric plates).

The same cause/effect relationship can be called upon to

Sloss, L. L., 1988, Conclusions, *in* Sloss, L. L., ed., Sedimentary Cover—North American Craton; U.S.: Boulder, Colorado, Geological Society of America, The Geology of North America, v. D-2.

account for episodes of brittle, fracture-dominated behavior of the cratonic crust. One such episode in Pennsylvanian–Early Permian time coincides, as noted above, with convergence at the southern margin of the craton and may reflect repeated shocks administered by collision with an alien continent, as the Asian interior is wracked today by responses to the collision and subduction of the Indian subcontinent.

In contrast, episodes of broad undifferentiated uplift of cratons, such as characterized the times of erosion at the sub-Sauk (Late Proterozoic–Early and Middle Cambrian), sub-Tippecanoe (Early Ordovician), sub-Kaskaskia (Early Devonian), and sub-Zuni (Early and Middle Jurassic) unconformities, are less easily understood. The first and last of these (pre-Sauk and pre-Zuni) are times of continental breakup, but it is not intuitively obvious why the extensional concomitants of breakup should include uplift of cratonic interiors at great distances from divergent margins; although these margins may have been elevated by thermal effects related to ridge volcanism. Moreover, the cratonic interior of North America remained emergent until latest Cambrian in the Sauk case and until Late Cretaceous in the span of the Zuni sequence, requiring continuing and accelerating subsidence tens of millions of years after the supposed incidence of thermally driven uplift. The mid-Paleozoic emergent episodes (pre-Tippecanoe and pre-Kaskaskia) do not seem to accompany activity, either convergent or divergent, at any North American continental margins.

As was noted in the chapter on tectonic evolution of the craton (Sloss, Ch. 3, this volume), an earlier and feebly substantiated hypothesis (Sloss and Speed, 1974) proposed that cratonic uplift was a response to the accumulation of the products of radial heat flow (melt and partial melt) in the subcontinental asthenosphere below the insulating cover of tens of kilometers of continental crust. As a corollary, it was claimed that cratonic subsidence occurred when stored melt was drained from under cratons by volcanism at continental margins or by transfer to suboceanic asthenospheric circulation and eventual extraction at mid-ocean ridge systems. Thus, in this scenario, cratonic uplift would signify a slowdown in plate motion as melt was stored below a continent, and submergence would equate with accelerated plate motion fed by release of melt from continental asthenosphere.

If, as is demonstrable, widely separated cratons tend to exhibit a degree of synchrony in their episodes of uplift and downwarp, then some globally effective phenomenon would seem to be represented. At one time it was thought that sea-floor spreading rates at globally distributed ridge systems were subject to variation at time scales of tens of millions of years; such variation, if related to episodes of storage and extraction of melt below continents, would explain the apparent synchrony among cratons. At present, influenced by refinement of the dating of oceanic magnetic anomalies and other factors reviewed by Schlanger (1986), the level of support for long-term (10^7 m.y.) time-dependent variations in spreading rates has lessened, and some alternative driving mechanism for the episodic emergence and submergence of cratons is required. In the meantime, opinion derived from a wide variety of data and concepts (e.g., Courtillot and Besse, 1987; Gordon, 1987; Oxburgh and O'Nions, 1987) is developing a basis for consideration of episodic changes in the thermal budget of the earth as responses to secular variation in radial heat flow. If these thoughts prove tenable, students of cratons will be furnished with a basis in theory capable of accommodating a vast array of data without recourse to local or regional "solutions" to isolated problems.

Turning to the questions related to cratonic sedimentary basins and intervening arches and domes, investigators again encounter a wide diversity of poorly substantiated opinions; the most popular of these is supported, however frailly, by the elegance of the numerical modeling employed. Some time back, as an outgrowth of analysis of subsidence histories at passive margins clearly associated with cooling at the former positions of ridge crests (e.g., Sleep, 1971), and encouraged by fine tuning of data on cratonic basins (e.g., Haxby and others, 1976; Sleep and others, 1980), the thermal subsidence theory and its phase-change partners (e.g., Fowler and Nisbet, 1985) were extended to cratonic interiors. Several observations combine to make it difficult to accept thermal subsidence as a viable explanation of the depression of cratonic basins. Thermal subsidence above a localized (hot spot?) source follows upon thermal uplift, and there is no evidence that basins were the sites of uplift prior to basin genesis. Basin subsidence commonly extends over too long a time span to represent cooling, and three-dimensional data (as opposed to selected wells or cross sections) seldom fit the theoretical $t^{1/2}$ curve (e.g., Sloss, 1982b). Further, basin subsidence slows and accelerates episodically, in some instances exhibiting marked synchrony from basin to basin and craton to craton (Sloss, 1987). Basin architecture changes through time; that is, depocenters and axes of deposition shift about but commonly return to ancestral patterns after excursions marked by aberrant behavior. These are among the circumstances that serve to deny simple thermal contraction as a major cause of subsidence.

Rising sea levels and concomitant imposition of water plus sediment mass are unequivocally incapable of creating the magnitudes of cratonic submergence recorded in the Phanerozoic record (Bally, 1980), nor can these mechanisms account for the depths of subsidence of individual basins. The inadequacy of isostatic subsidence is particularly glaring where "starved" basins have been downwarped even though occupied by water alone (or, in basins evaporated to dryness, by no more than atmosphere). By the same token, cratonic uplift of arches and domes (and the Canadian Shield) cannot be explained by marine regression, exposure to erosion and the stripping of cover rocks, or by thinning of the emergent basement.

There is a class of cratonic sedimentary basins that has a clearly defined origin in precursor tectonic states involving thermal events. These are aulacogens, which begin as graben systems oriented at high angles to cratonic margins; volcanic and intrusive activity are common accompaniments of the initial faulting, giving evidence of elevated heat flow leading to subsequent flexure-dominated thermal subsidence. The manifest relationship between initial extensional rifting as a precedent to downwarp of aulacogens led to a somewhat uncritical acceptance of tensional rifting as the paramount explanation of the origin and location of other cratonic basins (e.g., Klein and Hsui, 1987). There are however, important differences between the positions of aulacogenic and craton interior basins in terms of tectonic geography and in their subsidence histories. Aulacogens such as those in southern Oklahoma (the precursor to the Anadarko Basin) and the Mississippi Valley–Reelfoot Graben System (ancestral to the Illinois Basin and the Mississippi Embayment) developed in latest Proterozoic to mid-Cambrian time, when the margins of the craton were in extensional modes. Through the Early Ordovician, both reentrants to the cratonic margin subsided rapidly at rates conforming to cooling curves and similar to those operating at coeval passive margins. In contrast, craton interior basins such as the Michigan and Williston Basins, rest on and are totally surrounded by regions of continental crust; these basins experienced only minimal subsidence over the Cambrian–Early Ordovician time span, postponing the acquisition of significant basinal geometry until mid-Ordovician time. It is assumed (e.g., Klein and Hsui, 1987) that initial subsidence of the Michigan Basin is a consequence of cooling along the arm of the Midcontinent Rift System that enters the Lower Peninsula of Michigan from the northwest. However, the rift system in question ceased activity hundreds of millions of years before the genesis of the Michigan Basin, and no Paleozoic basin exhibits any regard for the trend of the Midcontinent Rift in its course across the cratonic interior from Minnesota to Kansas; further, the axis of the rift in Michigan, a linear element at least 300 km long, is more than 50 km removed from the center of subsidence in the basin. Others (e.g., Ahern and Mrvicka, 1984) have applied similar reasoning to the Williston Basin, claiming the delay in subsidence is a consequence of the cooling of a deep-seated source.

Stretching and thinning of the continental crust (and upper mantle?) is a process appealed to by some theorists (e.g., McKenzie, 1978); such thinning under extension could be followed by thermal relaxation in the upwelled mantle, leading to subsidence. But there is little evidence to support a belief in episodes of cratonic extension corresponding to episodes of interior-basin subsidence. Indeed, as is pointed out in an earlier paragraph, times of basin subsidence commonly coincide with times of plate convergence at cratonic margins; this is not the setting for pervasive craton extension.

Consideration of the mechanics of the behavior of continental crust adjacent to the foreland basin created by sedimentational and tectonic loading at an active margin leads to recognition of a secondary area of subsidence (interior basin) separated from the margin by a relative upwarp (arch). Beaumont and others (1988) have exhaustively modeled the subsidence history of the Michigan and Illinois Basins and their easterly and southeasterly bounding arches in relationship to foreland-basin development along the Appalachian and Ouachita Orogens. The model results produce a reasonable mimic of the evolution of two basins and of the Findlay-Cincinnati-Nashville Arch trend. However, other basins (e.g., Williston, Hudson Bay) and other arches (e.g., Sweetgrass, Transcontinental) at great distances from Appalachian-Ouachita foreland basins share closely synchronous stages of tectonic evolution with the area modeled. Inasmuch as the same level of synchrony is exhibited by other cratons (e.g., Soares and others, 1978; Sloss, 1987) and without apparent yoking of subsidence and uplift to foreland-basin histories, it is difficult to accept this concept of control of localized cratonic interior elements by linkage to cratonic margins (except under the special circumstances of collision represented by late Paleozoic and early Cenozoic time).

It remains eminently reasonable to consider craton-interior basins to represent subsidence at preordained loci inherited from the ancient tectonic framework of continental basement. Gerhard and Anderson (this volume), for example, promote Precambrian wrench faulting for the establishment of the pattern of basins and arches repeatedly rejuvenated during Phanerozoic time. Some workers seek a still more ancient origin in discontinuities marking the sutures between lithospheric blocks assembled to form this and other cratons. These opinions remain difficult to confirm or deny in the current state of knowledge of the deep continental crust and upper mantle; in any case, although such thinking helps in understanding why basins are where they are and why they remain relatively fixed in position for major parts of Phanerozoic time, the big questions remain: why is basin subsidence episodic, and why do widely separated basins act in concert? Satisfactory answers to these persistent questions cannot be reached by contemplation and manipulation of data sets derived from individual basins or specific regions in isolation from information pertaining to other basins, other regions, and other cratons.

At the peril of appearing immodest, it is again necessary to recall the proposition (from Sloss and Speed, 1974) invoked previously in these concluding remarks. If cratons are subject to inflation by storage of heat and its concomitants below continents, then cooling or withdrawal of melt would result in deflation and the buckling of the elastic (or visco-elastic) crust in patterns predetermined by ancient elements of crustal architecture, whether these be interplate sutures, megashears, or whatever. Thus, basins and arches would appear, stagnate, and be reactivated by long-term (10^7 to 10^8 m.y.) pulses of radial heat flow. Different orientations of confinement or release at continental margins would find responses in the particular characteristics of the shapes, amplitudes, and positions of maximum subsidence of basins at specific times in cratonic history. Similarly, arches, representing a deficiency of subsidence relative to adjacent basins (the most common mode of arch tectonism) or exhibiting actual positive tendency, would be responsive in amplitude, position, and timing to revisions of the cratonic stress field during episodes of emergence and subsequent buckling accompanying deflation and downwarp.

To all of this must be added the influences of global sea levels and of long-wavelength (10^3 km) perturbations of the geoid (see papers in Mörner, 1980) as well as the paleopositions of cratons with respect to latitude and climatic belts. These are among the many factors contributing to the presence/absence, thickness, and facies of the sedimentary rocks on which so much emphasis is placed in reconstructing the history of a craton. It becomes increasingly probable that these many ingredients are intimately interrelated and that the sedimentologic, stratigraphic, structural, and geophysical data and observations on which we habitually base complex and commonly contradictory hypotheses are all second- or third-order consequences of dimly sensed processes deep in the interior of the heat engine we call Earth.

REFERENCES CITED

Ahern, J. L., and Mrvicka, S. R., 1984, A mechanical and thermal model for the evolution of the Williston basin: Tectonics, v. 5, p. 79–102.

Bally, A. W., 1980, Basins and subsidence; A summary, *in* Bally, A. W., Bender, P. L., McGetchin, T. R., and Walcott, R. I., eds., Dynamics of Plate Interiors: American Geophysical Union and Geological Society of America Geodynamics Series, v. 1, p. 5–20.

Beaumont, C., Quinlan, G., and Hamilton, J., 1988, Orogeny and stratigraphy; Numerical models of the Paleozoic of the Eastern Interior of North America: Tectonics, v. 9 (in press).

Courtillot, V., and Besse, J., 1987, Magnetic field reversals, polar wander, and core-mantle coupling: Science, v. 237, p. 1140–1147.

Fowler, C.M.R., and Nisbet, E. G., 1985, The subsidence of the Williston basin: Canadian Journal of Earth Sciences, v. 22, p. 408–415.

Gordon, R. G., 1987, Polar wandering and paleomagnetism: Annual Review of Earth and Planetary Sciences, v. 15, p. 567–593.

Haug, É., 1890, Les geosynclinaux at les aires continentales; Contribution a l'étude des transgressions at des regressions marines: Societé Géologique de France Bulletin, series 3, v. 28, p. 617–711.

Haxby, W. F., Turcotte, D. L., and Bird, J. M., 1976, Thermal and mechanical evolution of the Michigan basin: Tectonophysics, v. 36, p. 57–75.

Hays, J. D., and Pitman, W. C., III, 1973, Lithospheric plate motion, sea level changes, and climatic and ecological consequences: Nature, v. 246, p. 18–21.

Johnson, J. G., 1971, Timing and coordination of orogenic, epeirogenic, and eustatic events: Geological Society of America Bulletin, v. 82, p. 3263–3298.

——, 1972, Antler effect equals Haug effect: Geological Society of America Bulletin, v. 83, p. 2497–2498.

Klein, G. de V., and Hsui, A. T., 1987, Origin of cratonic basins: Geology, v. 15, p. 1094–1098.

McKenzie, D. P., 1978, Some remarks on the development of sedimentary basins: Earth and Planetary Sciences Letters, v. 40, p. 25–32.

Mörner, N.-A., 1980, Earth Rheology, Isostasy, and Eustasy: New York, Wiley and Sons, 599 p.

Oxburgh, E. R., and O'Nions, R. K., 1987, Helium loss, tectonics, and the terrestrial heat budget: Science, v. 237, p. 1583–1588.

Palmer, A. R., ed., 1982, Perspectives in Regional Geological Synthesis: Boulder, Colorado, Geological Society of America, The Decade of North American Geology, Special Publication 1, 176 p.

Schlanger, S. O., 1986, High frequency sea-level fluctuations in Cretaceous time, an emerging geophysical problem, *in* Mesozoic and Cenozoic Oceans: American Geophysical Union and Geological Society of America Geodynamics Series, v. 15, p. 61–74.

Schlanger, S. O., Jenkyns, H. C., and Premoli-Silva, I., 1981, Volcanism and vertical tectonics in the Pacific basin related to global Cretaceous transgressions: Earth and Planetary Science Letters, v. 52, p. 435–449.

Sleep, N., 1971, Thermal effects of the formation of Atlantic continental margins by continental breakup: Royal Astronomical Society, Geophysical Journal, v. 24, p. 325–350.

Sleep, N. H., Nunn, J. A., and Chou, L., 1980, Platform basins: Annual Review of Earth and Planetary Sciences, v. 8, p. 17–34.

Sloss, L. L., 1966, Orogeny and epeirogeny; The view from the craton: New York Academy of Sciences Transactions, series II, v. 28, p. 579–587.

——, 1982a, The Midcontinent Province; United States, *in* Palmer, A. R., ed., Perspectives in Regional Geological Synthesis: Boulder, Colorado, Geological Society of America, Decade of North American Geology Special Publication 1, p. 27–40.

——, 1982b, Subsidence of continental margins; The case for alternatives to thermal contraction: American Geophysical Union and Geological Society of America Geodynamics Series, v. 6, p. 197–200.

——, 1987, Williston in the family of cratonic basins, *in* Longman, M. W., ed., Williston Basin; Anatomy of a Cratonic Oil Province: Denver, Colorado, Rocky Mountain Association of Geologists, p. 1–8.

Sloss, L. L., and Speed, R. C., 1974, Relationship of cratonic and continent-margin tectonic episodes, *in* Dickinson, W. R., ed., Tectonics and sedimentation: Society of Economic Paleontologists and Mineralogists Special Publication 22, p. 38–55.

Soares, P. C., Landim, P.M.B.L., and Fulfaro, V. J., 1978, Tectonic cycles and sedimentary sequences in the Brazilian intracratonic basins: Geological Society of America Bulletin, v. 89, p. 181–191.

Manuscript Accepted by the Society January 27, 1988

Printed in U.S.A.

Index

[Italic page numbers indicate major references]

Typeset by WESType Publishing Services, Inc., Boulder, Colorado
Printed in U.S.A. by Malloy Lithographing, Inc., Ann Arbor, Michigan

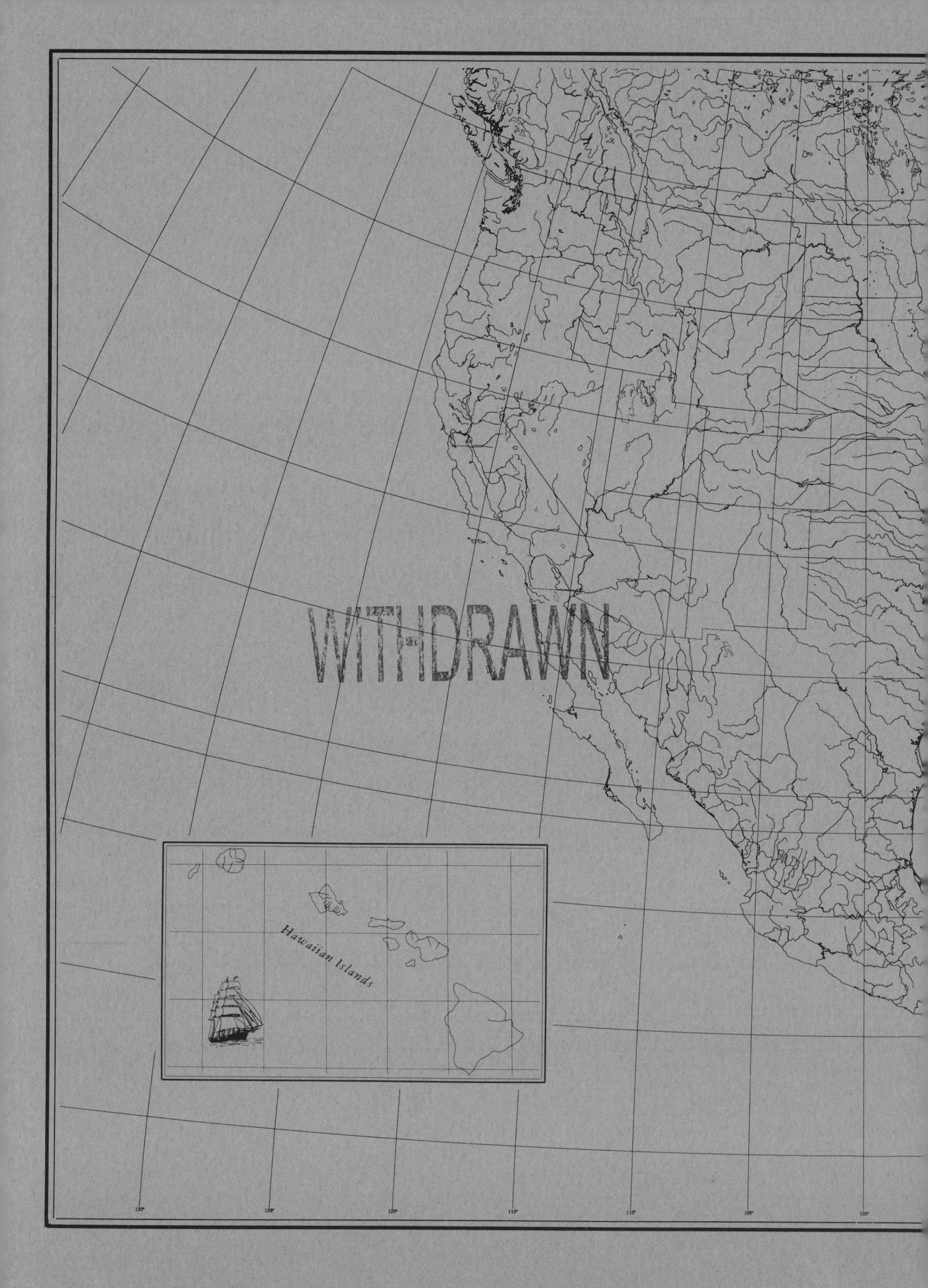
Hawaiian Islands
130°
125°
120°
115°
110°
105°
100°